The Chemistry and Technology of Petroleum

FOURTH EDITION

CHEMICAL INDUSTRIES

A Series of Reference Books and Textbooks

Founding Editor

HEINZ HEINEMANN
Berkeley, California

Consulting Editor

JAMES G. SPEIGHT
Laramie, Wyoming

1. *Fluid Catalytic Cracking with Zeolite Catalysts*, Paul B. Venuto and E. Thomas Habib, Jr.
2. *Ethylene: Keystone to the Petrochemical Industry*, Ludwig Kniel, Olaf Winter, and Karl Stork
3. *The Chemistry and Technology of Petroleum*, James G. Speight
4. *The Desulfurization of Heavy Oils and Residua*, James G. Speight
5. *Catalysis of Organic Reactions*, edited by William R. Moser
6. *Acetylene-Based Chemicals from Coal and Other Natural Resources*, Robert J. Tedeschi
7. *Chemically Resistant Masonry*, Walter Lee Sheppard, Jr.
8. *Compressors and Expanders: Selection and Application for the Process Industry*, Heinz P. Bloch, Joseph A. Cameron, Frank M. Danowski, Jr., Ralph James, Jr., Judson S. Swearingen, and Marilyn E. Weightman
9. *Metering Pumps: Selection and Application*, James P. Poynton
10. *Hydrocarbons from Methanol*, Clarence D. Chang
11. *Form Flotation: Theory and Applications*, Ann N. Clarke and David J. Wilson
12. *The Chemistry and Technology of Coal*, James G. Speight
13. *Pneumatic and Hydraulic Conveying of Solids*, O. A. Williams

The Chemistry and Technology of Petroleum

FOURTH EDITION

James G. Speight

CD&W Inc.
Laramie, Wyoming

CRC Press
Taylor & Francis Group
Boca Raton   London   New York

CRC Press is an imprint of the
Taylor & Francis Group, an informa business

CRC Press
Taylor & Francis Group
6000 Broken Sound Parkway NW, Suite 300
Boca Raton, FL 33487-2742

© 2007 by Taylor and Francis Group, LLC
CRC Press is an imprint of Taylor & Francis Group, an Informa business

International Standard Book Number-10: 0-8493-9067-2 (Hardcover)
International Standard Book Number-13: 978-0-8493-9067-8 (Hardcover)
Library of Congress Card Number 2006014100

Library of Congress Cataloging-in-Publication Data

Speight, J. G.
   The chemistry and technology of petroleum / James G. Speight. -- 4th ed.
      p. cm. -- (Chemical industries ; v 114)
   Includes bibliographical references and index.
   ISBN-13: 978-0-8493-9067-8 (alk. paper)
   ISBN-10: 0-8493-9067-2 (alk. paper)
   1. Petroleum. 2. Petroleum--Refining. I. Title. II. Series.

TP690.S74 2006
665.5--dc22   2006014100

Visit the Taylor & Francis Web site at
http://www.taylorandfrancis.com

and the CRC Press Web site at
http://www.crcpress.com

Preface to the Fourth Edition

The success of the first, second, and third editions of this text has been the primary reason for the decision to publish a fourth edition. In addition, the demand for petroleum products, particularly liquid fuels (gasoline and diesel fuel) and petrochemical feedstocks (such as aromatics and olefins), is increasing throughout the world. Traditional markets such as North America and Europe are experiencing a steady increase in demand whereas emerging Asian markets, such as India and China, are witnessing a rapid surge in demand for liquid fuels. This has resulted in a tendency for existing refineries to seek fresh refining approaches to optimize efficiency and throughput. In addition, the increasing use of heavier feedstock for refineries is forcing technology suppliers and licensors to revamp their refining technologies in an effort to cater to the growing customer base.

Further, the evolution in product specifications caused by various environmental regulations plays a major role in the development of petroleum refining technologies. In many countries, especially in the United States and Europe, gasoline and diesel fuel specifications have changed radically in the past half decade (since the publication of the third edition of this book) and will continue to do so in the future. Currently, reducing the sulfur levels of liquid fuels is the dominant objective of many refiners. This is pushing the technological limits of refineries to the maximum and the continuing issue is the elimination of sulfur in liquid fuels, as tighter product specifications emerge worldwide. These changing rules also have an impact on the market for heavy products such as fuel oil.

Refineries must, and indeed are eager to, adapt to changing circumstances and are amenable to trying new technologies that are radically different in character. Currently, refineries are also looking to exploit heavy (more viscous) crude oils and tar sand bitumen (sometimes referred to as extra heavy crude oil), provided they have the refinery technology capable of handling such feedstocks. Transforming the higher boiling constituents of these feedstock components into liquid fuels is becoming a necessity. It is no longer a simple issue of mixing the heavy feedstock with conventional petroleum to make up a blended refinery feedstock. Incompatibility issues arise that can, if not anticipated, close down a refinery or, at best, a major section of the refinery. Therefore, handling such feedstocks requires technological change, including more effective and innovative use of hydrogen within the refinery.

Heavier crude oil could also be contaminated with sulfur and metal containing molecules that must be removed to meet quality standards. A better understanding of how catalysts perform (both chemically and physically) with the feedstock is necessary to provide greater scope for process and catalyst improvements.

However, even though the nature of crude oil is changing, refineries are here to stay in the foreseeable future, since petroleum products satisfy wide-ranging energy requirements and demands that are not fully covered by alternate fossil fuel sources such as natural gas and coal. And the alternative (so-called renewable) energy technologies are not poised to supplement the demand for energy.

Therefore, it is the purpose of this book to provide the reader with a detailed overview of the chemistry and technology of petroleum as they evolve into the twenty-first century. With this in mind, many of the chapters that appeared in the third edition have been rewritten to include the latest developments in the refining industry. Updates on the evolving processes and new processes as well as various environmental regulations are presented. As part of this update, the chapters contain updates of the relevant processes that are used the industry evolves. The text still maintains its initial premise, to introduce the reader to the science of petroleum, beginning with its formation in the ground, eventually leading to the production of a wide variety of products and petrochemical intermediates. The text will also prove useful for those scientists and engineers already engaged in the petroleum industry as well as in the catalyst manufacturing industry who wish to gain a general overview or update of the science of petroleum.

Finally, as always, I am indebted to my colleagues in many different countries who have continued to engage me in lively discussions and who have offered many thought-provoking comments. Thanks are also due to those colleagues who have made constructive comments on the previous editions, which were of great assistance in writing this edition. For such discussions and commentary, I continue to be grateful.

Dr. James G. Speight
Laramie, Wyoming

Preface to the First Edition

For many years, petroleum has been regarded as the cheapest source of liquid fuels by many countries, especially the United States and Canada. However, with the recent energy crises and concern over future supplies of gaseous and liquid fuels in many parts of the world, particularly Western Europe and North America, we have seen a gradual acceptance by the petroleum industry and the general public of the inevitability that petroleum and natural gas will, at some time within the foreseeable future, be in very short supply.

As a result, petroleum technology is expanded to such an extent that wells that were previously regarded as nonproductive because of their inability to produce oil without considerable external stimulation are now reexamined with the object of, literally, recovering every last possible drop of petroleum.

Serious attempts are also underway to produce liquid fuels from unconventional sources, such as coal, oil shale, and oil sands (also variously referred to as tar sands or bituminous sands). Oil sands, in fact, have already been developed to such an extent that commercial production of a synthetic crude oil from the oils sands located in northeastern Alberta (Canada) has been underway for some ten years, with a second plant on-stream since 1978 and serious negotiations underway for other oil sands plants.

This expansion of liquid fuels technology has resulted in a vacuum in the labor output insofar as the universities have been unable to produce sufficient people with any form of training in petroleum technology and petroleum chemistry. However, it now appears that various universities, which have initiated research into the various aspects of petroleum science, are considering some kind of formal training in this area.

Thus it happened that during the winter of 1976–1977, the author organized a course entitled "An Introduction to the Chemistry of Petroleum" through the Faculty of Extension at the University of Alberta. In the early stages of preparation, it became apparent that, although several older books were available, there was no individual book that could serve as a teaching text for teachers and engineers as well as chemists. Therefore, this book is the outcome of the copious notes collected as a result of that course. The text introduces the reader to the science of petroleum, beginning with its formation in the ground, and eventually leads to analyses of the production of a wide variety of petrochemical intermediates as well as the more conventional fuel products. This book has also been written for those people already engaged in the petroleum industry (engineers and chemists) who wish to gain a general overview of the science of petroleum.

Although any text on petroleum must of necessity include some chemistry, attempts have been made, for the benefit of those readers without any formal college training in chemistry, to keep the chemical sections as simple as possible. In fact, there are, within the text, several pages of explanatory elementary organic chemistry for the benefit of such people.

At a time when the anglicized nations of the world are undergoing a transferal to the metric system of measurement, there are still those disciplines that are based on such scales as the Fahrenheit temperature scale as well as the foot measure instead of the meter. Accordingly, the text contains both the metric and nonmetric measures, but it should be noted that exact conversion is not often feasible, and thus conversion data are often taken to the nearest whole number. Indeed, conversions involving the two temperature scales—Fahrenheit and Celsius—are, at the high temperatures quoted in the text, often rounded off to the nearest 5°, especially when serious error would not arise from such a conversion.

For the sake of simplicity, illustrations contained in the text, especially in the chapter relating to petroleum refining, are line drawings, and no attempt has been made to draw to scale the various reactors, distillation towers, or other equipment.

The majority of the work on this text was carried out while the author was a staff member of the Alberta Research Council. Thus, the author wishes to acknowledge the assistance given by the many members of the Alberta Research Council. The author is particularly indebted to his colleagues J.F. Fryer and Dr. S.E. Moschopedis for their comments on the manuscript, as well as to P. Williams, M.A. Harris, and H. Radvanyi for typing the manuscript.

Dr. James G. Speight

Author

Dr. James G. Speight has a BSc and a PhD from the University of Manchester, England. He was employed by the Alberta Research Council (Edmonton, Alberta, Canada, 1967–1980), Exxon Research and Engineering Company (Linden and Annandale, New Jersey, 1980–1984), and by the Western Research Institute (Laramie, Wyoming) where he was chief scientific officer and executive vice president (1984–1990) and chief executive officer (1990–1998). He is currently a consultant–author–lecturer on energy and environmental issues with CD&W Inc. (Laramie, Wyoming, 1998–present).

Dr. Speight has more than 38 years of experience in areas associated with the properties and recovery of reservoir fluids as well as refining conventional petroleum, heavy oil, and tar sand bitumen. He has taught more than 60 courses and contributed to more than 400 publications, reports, and presentations.

Dr. Speight is the editor and founding editor of *Petroleum Science and Technology* (Taylor & Francis Publishers); the editor of *Energy Sources. Part A: Recovery, Utilization, and Environmental Effects* (Taylor & Francis Publishers); and the editor and founding editor of *Energy Sources. Part B: Economics, Planning, and Policy* (Taylor & Francis Publishers). He is also an adjunct professor of chemical and fuels engineering, University of Utah, an adjunct professor of chemistry and visiting professor, University of Trinidad and Tobago, and a visiting professor at the Technical University of Denmark (Lyngby, Denmark). Dr. Speight is the author–editor–compiler of more than 30 books and bibliographies related to fossil fuel processing and environmental issues.

Dr. Speight has received (1) a diploma of honor, National Petroleum Engineering Society for *Outstanding Contributions to the Petroleum Industry*, (2) a gold medal, Russian Academy of Sciences for *Outstanding Work in the Area of Petroleum Science* in 1996, (3) the Specialist Invitation Program Speakers Award, NEDO (New Energy Development Organization, Government of Japan) for *Contributions to Coal Research*, (4) Doctor of Sciences, Scientific Research Geological Exploration Institute (VNIGRI), St. Petersburg, Russia for *Exceptional Work in Petroleum Science*, (5) the Einstein Medal, Russian Academy of Sciences in recognition of *Outstanding Contributions and Service in the Field of Geologic Sciences*, and (6) a gold medal—Scientists without Frontiers, Russian Academy of Sciences in recognition of *Continuous Encouragement of Scientists to Work Together across International Borders*.

Table of Contents

Chapter 9

Chapter 25

Part I

History, Occurrence, and Recovery

1 History and Terminology

1.1 HISTORICAL PERSPECTIVES

Petroleum is perhaps the most important substance consumed in modern society. It provides not only raw materials for the ubiquitous plastics and other products, but also fuel for energy, industry, heating, and transportation. The word petroleum, derived from the Latin *petra* and *oleum*, means literally rock oil and refers to **hydrocarbons** that occur widely in the sedimentary rocks in the form of gases, liquids, semisolids, or solids.

From a chemical standpoint, petroleum is an extremely complex mixture of hydrocarbon compounds, usually with minor amounts of nitrogen-, oxygen-, and sulfur-containing compounds as well as trace amounts of metal-containing compounds (Chapter 7).

The fuels that are derived from petroleum supply more than half of the world's total supply of energy. Gasoline, kerosene, and diesel oil provide fuel for automobiles, tractors, trucks, aircraft, and ships. **Fuel oil** and **natural gas** are used to heat homes and commercial buildings, as well as to generate electricity. Petroleum products are the basic materials used for the manufacture of synthetic fibers for clothing and in plastics, paints, fertilizers, insecticides, soaps, and synthetic rubber. The uses of petroleum as a source of raw material in manufacturing are central to the functioning of modern industry.

Petroleum is a carbon-based resource. Therefore, the geochemical carbon cycle is also of interest to fossil fuel usage in terms of petroleum formation, use, and the buildup of atmospheric carbon dioxide (Chapter 28). Thus, more efficient use of petroleum is of paramount importance. Petroleum technology, in one form or another, is with us until suitable alternative forms of energy are readily available (Boyle, 1996; Ramage, 1997). Therefore, a thorough understanding of the benefits and limitations of petroleum recovery and processing is necessary and, hopefully, can be introduced within the pages of this book.

The history of any subject is the means by which the subject is studied in the hope that much can be learnt from the events of the past. In the current context, the occurrence and use of petroleum, petroleum derivatives (**naphtha**), **heavy oil**, and **bitumen** is not new. The use of petroleum and its derivatives was practiced in pre-Christian times and is known largely through historical use in many of the older civilizations (Henry, 1873; Abraham, 1945; Forbes, 1958a, 1958b; James and Thorpe, 1994). Thus, the use of petroleum and the development of related technology is not such a modern subject as we are inclined to believe. However, the petroleum industry is essentially a twentieth-century industry but to understand the evolution of the industry, it is essential to have a brief understanding of the first uses of petroleum.

The Tigris–Euphrates valley, in what is now Iraq, was inhabited as early as 4000 BC by the people known as the Sumerians who established one of the first great cultures of the civilized world. The Sumerians devised the cuneiform script, built the temple-towers known as ziggurats, an impressive law, literature, and mythology. As the culture developed, bitumen or **asphalt** was frequently used in construction and in ornamental works.

Although it is possible to differentiate between the words bitumen and asphalt in modern use, the occurrence of these words in older texts offers no such possibility. It is significant that the early use of bitumen was in the nature of cement for securing or joining together various objects, and it thus seems likely that the name itself was expressive of this application.

The word asphalt is derived from the Akkadian term *asphaltu* or *sphallo*, meaning to split. It was later adopted by the Homeric Greeks in the form of the adjective ασφαλήσες signifying firm, stable, secure, and the corresponding verb ασφαλίζωίσω meaning to make firm or stable, to secure. It is a significant fact that the first use of asphalt by the ancients was in the nature of cement for securing or joining together various objects, such as the bricks used for building and it thus seems likely that the name itself was expressive of this application. From the Greek, the word passed into late Latin (**asphaltum**, *aspaltum*), and thence into French (*asphalte*) and English (*aspaltoun*).

The origin of the word bitumen is more difficult to trace and subject to considerable speculation. The word was proposed to have originated in the Sanskrit language, where we find the words *jatu*, meaning **pitch**, and *jatukrit*, meaning pitch creating. From Sanskrit, the word *jatu* was incorporated into the Latin language as *gwitu* and is reputed to have eventually become *gwitumen* (pertaining to pitch). Another word, *pixtumen* (exuding or bubbling pitch) is also reputed to have been in the Latin language, although the construction of this Latin word form from which the word bitumen was reputedly derived, is certainly suspect. There is the suggestion that subsequent derivation of the word led to a shortened version (which eventually became the modern version) *bitûmen* thence passing via French into English. From the same root is derived the Anglo Saxon word *cwidu* (mastic, adhesive), the German work *kitt* (cement or mastic) and the equivalent word *kvada* which is found in the old Norse language as being descriptive of the material used to waterproof the long ships and other sea-going vessels. It is just as (perhaps even more than) likely that the word is derived from the Celtic *bethe* or *beithe* or *bedw* that was the birch tree that was used as a source of **resin (tar)**. The word appears in Middle English as *bithumen*. In summary, a variety of terms exist in ancient language from which, from their described use in texts, they can be proposed as having the meaning bitumen or asphalt (Table 1.1) (Abraham, 1945).

Using these ancient words as a guide, it is possible to trace the use of petroleum and its derivatives as described in ancient texts. And, preparing derivatives of petroleum was well within the area of expertise of the early scientists (perhaps *refiners* would be a better term) since alchemy (early chemistry) was known to consist of four subroutines: dissolving, melting, combining, and distilling (Cobb and Goldwhite, 1995).

Early references to petroleum and its derivatives occur in the Bible, although by the time the various books of the Bible were written, the use of petroleum and bitumen was established. Nevertheless, these writings do offer documented examples of the use of petroleum and related materials.

For example, in the *Epic of Gilgamesh* written more than 4500 years ago, a great flood causes the hero to build a boat that is caulked with bitumen and pitch (see for example, Kovacs, 1990). And, in a related story (it is not the intent here to discuss the similarities of the two stories) of Mesopotamia and just prior to the Flood, Noah is commanded to build an ark that also includes instructions for caulking the vessel with pitch (Genesis 6:14):

> Make thee an ark of gopher wood; rooms shalt thou make in the ark, and shalt pitch it within and without with pitch.

The occurrence of **slime** (bitumen) pits in the Valley of Siddim (Genesis 14:10), a valley at the southern end of the Dead Sea, is reported. There is also reference to the use of tar as a mortar when the Tower of Babel was under construction (Genesis 11:3):

> And they said one to another, Go to, let us make brick, and burn them thoroughly. And they had brick for stone, and slime had they for mortar.

TABLE 1.1
Linguistic Origins of Words Related to the Various Aspects
of Petroleum Technology

Language	Word	Possible Meaning
Sumerian	esir	petroleum
		bitumen
	esir-lah	hard/glossy asphalt
	esir-harsag	rock asphalt
	esir-é-a	mastic asphalt
	esir-ud-du-a	pitch
	kupru	slime, pitch
Sanskrit	jatu	bitumen
		pitch
	śilā-jatu	rock asphalt
	aśmajātam-jatu	rock asphalt
Assyrian/Akkadian	idd, ittû, it-tû-u	bitumen
		pitch
	amaru	bitumen
	sippatu	pitch
Hebrew	zephet	bitumen
	kopher or kofer	pitch
	hêmâr	pitch
Arabic and Turkish	seyali	bitumen
	zift or zipht	bitumen or pitch
	chemal	rock asphalt
	humar (houmar)	rock asphalt
	gasat (qasat)	rock asphalt
	ghir or gir	asphalt mastic
	kir or kafr	asphalt mastic or pitch
	mûmûia	bitumen
	neftgil	petroleum wax, mineral wax
Greek	maltha	soft asphalt
	asphaltos	bitumen
	pissasphaltos	rock asphalt
	pittasphaltos	rock asphalt
	pittolium	rock asphalt
	pissa or pitta	pitch
	ampelitis	mineral wax and asphaltites
Latin	maltha	soft asphalt
	bitumen liquidum	soft asphalt
	pix	pitch

In the Septuagint, or Greek version of the Bible, this work is translated as *asphaltos*, and in the Vulgate or Latin version, as bitumen. In the Bishop's Bible of 1568 and in subsequent translations into English, the word is given as slime. In the Douay translation of 1600, it is *bitume*, whereas in Luther's German version it appears as *thon*, the German word for clay.

Another example of the use of pitch (and slime) is given in the story of Moses (Exodus 2:3):

> And when she could not longer hide him, she took for him an ark of bulrushes, and daubed it with slime and with pitch, and put the child therein; and she laid it in the flags by the river's brink.

Perhaps the slime was a lower melting bitumen (bitumen mixed with solvent), whereas the pitch was a higher melting material; the one (slime) acting as a flux for the other (pitch). The lack of precise use of the words for bitumen and asphalt as well as for tar and pitch even now makes it unlikely that the true nature of the biblical tar, pitch, and slime will ever be known, but one can imagine their nature! In fact, even modern Latin dictionaries give the word bitumen as the Latin word for asphalt!

It is most probable that, in both these cases, the pitch and the slime were obtained from the seepage of **oil** to the surface, which was a fairly common occurrence in the area. And during biblical times, bitumen was exported from Canaan to various parts of the countries that surround the Mediterranean (Armstrong, 1997).

In terms of liquid products, there is an interesting reference (Deuteronomy 32:13) to bringing oil out of flinty rock. The exact nature of the oil is not described nor is the nature of the rock. The use of oil for lamps is also referenced (Matthew 23:3), but whether it was mineral oil (a petroleum derivative such as naphtha) or whether it was vegetable oil is not known.

Excavations conducted at Mohenjo-Daro, Harappa, and Nal in the Indus Valley indicated that an advanced form of civilization existed there. An asphalt mastic composed of a mixture of asphalt, clay, gypsum, and organic matter was found between two brick walls in a layer about 25 mm thick, probably a waterproofing material. Also unearthed was a bathing pool that contained a layer of mastic on the outside of its walls and beneath its floor.

In the Bronze Age, dwellings were constructed on piles in lakes close to the shore to better protect the inhabitants from the ravages of wild animals and attacks from marauders. Excavations have shown that the wooden piles were preserved from decay by coating with asphalt, and posts preserved in this manner have been found in Switzerland. There are also references to deposits of bitumen at Hit (the ancient town of Tuttul on the Euphrates River in Mesopotamia) and the bitumen from these deposits was transported to Babylon for use in construction (Herodotus, *The Histories*, Book I). There is also reference (Herodotus, *The Histories*, Book IV) to a Carthaginian story in which birds' feathers smeared with pitch are used to recover gold dust from the waters of a lake.

One of the earliest recorded uses of asphalt was by the pre-Babylonian inhabitants of the Euphrates Valley in southeastern Mesopotamia, present-day Iraq, formerly called Sumer and Akkad and, later, Babylonia. In this region there are various asphalt deposits, and uses of the material have become evident. For example, King Sargon Akkad (Agade) (ca. 2550 BC) was (for reasons that are lost in the annals of time) set adrift by his mother in a basket of bulrushes on the waters of the Euphrates, he was discovered by Akki the husbandman (the irrigator), whom he brought up to serve as gardener in the palace of Kish. Sargon eventually ascended the throne.

On the other hand, the bust of Manishtusu, King of Kish, an early Sumerian ruler (about 2270 BC), was found in the course of excavations at Susa in Persia, and the eyes, composed of white limestone, are held in their sockets with the aid of bitumen. Fragments of a ring composed of asphalt have been unearthed above the flood layer of the Euphrates at the site of the prehistoric city of Ur in southern Babylonia, ascribed to the Sumerians of about 3500 BC.

An ornament excavated from the grave of a Sumerian king at Ur consists of a statue of a ram with the head and legs carved out of wood over which gold foil was cemented with asphalt. The back and flanks of the ram are coated with asphalt in which the hair is embedded. Another art of decoration consisted of beating thin strips of gold or copper, which were then fastened to a core of asphalt mastic. An alternative method was to fill a cast metal object with a core of asphalt mastic, and such specimens have been unearthed at Lagash and Nineveh. Excavations at Tell-Asmar, 50 miles northeast of Baghdad, revealed the use of asphalt by the Sumerians for building purposes.

Mortar composed of asphalt has also been found in excavations at Ur, Uruk, and Lagash, and excavations at Khafaje have uncovered floors composed of a layer of asphalt that has been identified as asphalt, mineral filler (loam, limestone, and marl), and vegetable fibers (straw). Excavations at the city of Kish (Persia) in the palace of King Ur-Nina showed that the foundations consist of bricks cemented together with an asphalt mortar. Similarly, in the ancient city of Nippur (about 60 miles south of Baghdad), excavations show Sumerian structures composed of natural stones joined together with asphalt mortar. Excavation has uncovered an ancient Sumerian temple in which the floors are composed of burnt bricks embedded in asphalt mastic that still shows impressions of the reeds with which it must originally have been mixed.

The Epic of Gilgamesh (written before 2500 BC) and transcribed on to clay tablets during the time of Ashurbanipal, King of Assyria (668 to 626 BC), makes reference to the use of asphalt for building purposes. In the eleventh tablet, Ut-Napishtim relates the well-known story of the Babylonian flood, stating that he smeared

. . . the inside of a boat with six sar of kupru and the outside with three sar . . .

Kupru may have meant that the pitch or bitumen was mixed with other materials (perhaps even a solvent such as distillate from petroleum) to give it the appearance of slime as mentioned in the Bible. In terms of measurement, *sar* is a word of mixed origin and appears to mean an interwoven or wickerwork basket. Thus, an approximate translation is that the inside of the boat was smeared (coated, caulked) with six baskets full of pitch and the outside of the boat was smeared (coated, caulked) with three baskets full of pitch. There are also indications from these texts that that asphalt mastic was sold by volume (by the *gur*). On the other hand, bitumen was sold by weight (by the mina or shekel).

Use of asphalt by the Babylonians (1500 to 538 BC) is also documented. The Babylonians were well versed in the art of building, and each monarch commemorated his reign and perpetuated his name by the construction of buildings or other monuments. For example, the use of bitumen mastic as a sealant for water pipes, water cisterns, and in outflow pipes leading from flush toilets in cities such as Babylon, Nineveh, Calah, and Ur has been observed and the bitumen lines are still evident (Speight, 1978).

Bitumen was used as mortar from very early times, and sand, gravel, or clay were employed in preparing these mastics. Asphalt-coated tree trunks were often used to reinforce wall corners and joints, for instance in the temple tower of Ninmach in Babylon. In vaults or arches, a mastic-loam composite was used as mortar for the bricks, and the keystone was usually dipped in asphalt before it was set in place. The use of **bituminous** mortar was introduced into the city of Babylon by King Hammurabi, but the use of bituminous mortar was abandoned toward the end of Nebuchadnezzar's reign in favor of lime mortar to which varying amounts of asphalt were added. The Assyrians recommended the use of asphalt for medicinal purposes, as well as for building purposes, and perhaps there is some merit in the fact that the Assyrian moral code recommended that asphalt, in the molten state, be poured onto the heads of delinquents. Pliny, the Roman author, also notes that bitumen could be used to stop bleeding, heal wounds, drive away snakes, treat cataracts as well as a wide variety of other diseases, and straighten out eyelashes which inconvenience the eyes. One can appreciate the use of bitumen to stop bleeding but its use to cure other ailments is questionable and one has to consider what other agents were used concurrently with bitumen.

The Egyptians were the first to adopt the practice of embalming their dead rulers and wrapping the bodies in cloth. Before 1000 BC, asphalt was rarely used in mummification, except to coat the cloth wrappings and thereby protect the body from the elements. After the viscera was removed, the cavities were filled with a mixture of resins and spices, the corpse immersed in a bath of potash or soda, dried, and finally wrapped. From 500 to about 40 BC, asphalt was generally used both to fill the corpse cavities, as well as to coat the cloth wrappings. The word *mûmûia* first made its appearance in Arabian and Byzantine literature about 1000 AD, signifying bitumen. In fact, it is believed it was through the spread of the Islamic Empire that Arabic science and the use of bitumen were brought to western Europe.

In Persian, the term bitumen is believed to have acquired the meaning equivalent to **paraffin wax** that might be symptomatic of the nature of some of the **crude oils** in the area. Alternatively, it is also possible that the destructive distillation of bitumen to produce pitch produced paraffins that crystallized from the mixture over time. In Syriac, the term alluded to substances used for mummification. In Egypt, resins were used extensively for purposes of embalming up to the Ptolemaic period, when asphalts gradually came into use.

The product *mûmûia* was used in prescriptions, as early as the 12th century, by the Arabian physician Al Magor, for the treatment of contusions and wounds. Its production soon became a special industry in Alexandria. The scientist Al-Kazwînî alluded to the healing properties of *mûmûia*, and Ibn Al-Baitâr gives an account of its source and composition. Engelbert Kämpfer (1651 to 1716 AD) in his treatise *Amoenitates Exoticae* gives a detailed account of the gathering of *mûmûia,* the different grades and types, and its curative properties in medicine. As the supply of mummies was of course limited, other expedients came into vogue. The corpses of slaves or criminals were filled with asphalt, swathed, and artificially aged in the sun. This practice continued until the French physician, Guy de la Fontaine, exposed the deception in 1564 AD.

Many other references to bitumen occur throughout the Greek and Roman empires and from then to the Middle Ages early scientists (alchemists) frequently alluded to the use of bitumen. In later times, both Christopher Columbus and Sir Walter Raleigh (depending on the country of origin of the biographer) have been credited with the discovery of the asphalt deposit on the island of Trinidad and apparently used the material to caulk their ships.

The use of petroleum has also been documented in China: as early as 600 BC (Owen, 1975), petroleum was encountered when drilling for salt and mention of petroleum as an impurity in the salt is also noted in documents of the third century AD. It is presumed that the petroleum that contaminated the salt might be similar to that found in Pennsylvania and was, therefore, a more conventional type rather than the heavier type.

There was also an interest in the thermal product of petroleum (nafta; naphtha) when it was discovered that this material could be used as an illuminant and as a supplement to asphalt incendiaries in warfare. For example, there are records of the use of mixtures of pitch and/or naphtha with sulfur as a weapon of war during the Battle of Palatea, Greece, in the year 429 BC (Forbes, 1959). There are references to the use of a liquid material, **naft** (presumably the volatile fraction of petroleum which we now call naphtha and which is used as a solvent or as a precursor to gasoline), as an incendiary material during various battles of the pre-Christian era (James and Thorpe, 1994). This is the so-called Greek fire, a precursor and chemical cousin to napalm. Greek fire is also recorded as being used in the period 674 to 678 AD, when the city of Constantinople was saved by the use of the fire against an Arab fleet (Davies, 1996). In 717 to 718 AD, Greek fire was again used to save the city of Constantinople from attack by another Arab fleet; again with deadly effect (Dahmus, 1995). After this time, the Byzantine navy of 300 triremes frequently used Greek fire against all comers (Davies, 1996).

This probably represents the first documented use of the volatile derivatives of petroleum that led to a continued interest in petroleum. Greek fire was a viscous liquid that ignited on contact with water and was sprayed from a pump-like device on to the enemy. One can imagine the early users of the fire attempting to ignite the liquid before hurling it toward the enemy.

However, the hazards that can be imagined from such tactics could become very real, and perhaps often fatal, to the users of the Greek fire if any spillage occurred before ejecting the fire toward the enemy. The later technology for the use of Greek fire probably incorporated heat-generating chemicals such as quicklime (CaO) (Cobb and Goldwhite, 1995), which was suspended in the liquid and which, when coming into contact with water (to produce [Ca(OH)$_2$], released heat that was sufficient to cause the liquid to ignite. One assumes that the users of the fire were extremely cautious during periods of rain or, if at sea, during periods of turbulent weather.

As an aside, the use of powdered lime in warfare is also documented. The English used it against the French on August 24, 1217 with disastrous effects for the French. As was usual for that time, there was a difference of opinion between the English and the French that resulted in their respective ships meeting at the east end of the English Channel. Before any other form of engagement could occur, the lime was thrown from the English ships and carried by the wind to the French ships where it made contact with the eyes of the French sailors. The burning sensation in the eyes was too much for the French sailors and the English prevailed with the capture of much booty (Powicke, 1962).

The combustion properties of bitumen (and its fractions) were known in Biblical times. There is the reference to these properties (Isaiah 34:9) when it is stated that:

> And the stream thereof shall be turned into pitch, and the dust thereof into brimstone, and the land thereof shall become burning pitch.
> It shall not be quenched night nor day; the smoke thereof shall go up forever: from generation to generation it shall lie waste; none shall pass through it for ever and for ever.

One might surmise that the effects of the burning bitumen and sulfur (brimstone) were long lasting and quite devastating.

Approximately 2000 years ago, Arabian scientists developed methods for the distillation of petroleum, which were introduced into Europe by way of Spain. This represents another documented use of the volatile derivatives of petroleum which led to a continued interest in petroleum and its derivatives as medicinal materials and materials for warfare, in addition to the usual construction materials.

The Baku region of northern Persia was also reported (by Marco Polo in 1271 to 1273) as having an established commercial petroleum industry. It is believed that the prime interest was in the kerosene fraction that was then known for its use as an illuminant. By inference, it has to be concluded that the distillation and perhaps the thermal decomposition of petroleum were established technologies. If not, Polo's diaries might well have contained a description of the stills or the reactors.

In addition, bitumen was investigated in Europe during the Middle Ages (Bauer, 1546, 1556), and the separation and properties of bituminous products were thoroughly described. Other investigations continued, leading to a good understanding of the sources and use of this material even before the birth of the modern petroleum industry (Forbes, 1958a, 1958b).

There are also records of the use of petroleum spirit, probably a higher boiling fraction than naphtha that closely resembled modern-day liquid paraffin, for medicinal purposes. In fact, the so-called liquid paraffin continued to be prescribed up to modern times. The naphtha of that time was obtained from shallow wells or by the destructive distillation of asphalt.

Parenthetically, the destructive distillation operation may be likened to modern coking operations (Chapter 17) in which the overall objective is to convert the feedstock into distillates for use as fuels. This particular interest in petroleum and its derivatives continued with an increasing interest in nafta (naphtha), because of its use as an illuminant and as a supplement to asphaltic incendiaries for use in warfare.

To continue such references is beyond the scope of this book, although they do give a flavor of the developing interest in petroleum. However, it is sufficient to note that there are many other references to the occurrence and use of bitumen or petroleum derivatives up to the beginning of the modern petroleum industry (Cook and Despard, 1927; Mallowan and Rose, 1935; Nellensteyn and Brand, 1936; Mallowan, 1954; Marschner et al., 1978).

In summary, the use of petroleum and related materials has been observed for almost 6000 years. During this time, the use of petroleum has progressed from the relatively simple use of asphalt from Mesopotamian seepage sites to the present-day refining operations that yield a wide variety of products (Chapter 28) and petrochemicals (Chapter 30).

1.2 MODERN PERSPECTIVES

The modern petroleum industry began in the later years of the 1850s with the discovery, in 1857, and subsequent commercialization of petroleum in Pennsylvania in 1859 (Bell, 1945). The modern refining era can be said to have commenced in 1862 with the first appearance of petroleum distillation (Table 1.2). The story of the discovery of the character of petroleum is somewhat circuitous but worthy of mention, in the historical sense (Burke, 1996).

At a time when the carbonation of water was investigated, Joseph Priestley became involved in attempting to produce such a liquid since it was to be used as a cure for scurvy on Captain Cook's second expedition in 1771. Priestley decided to make a contribution to the success of the expedition and set himself to invent a drink that would cure scurvy. During his

TABLE 1.2
Process Development Since the Commencement of the Modern Refining Era

Year	Process Name	Purpose	By-products
1862	Atmospheric distillation	Produce kerosene	Naphtha, cracked residuum
1870	Vacuum distillation	Lubricants	Asphalt, residua
1913	Thermal cracking	Increase gasoline yield	Residua, fuel oil
1916	Sweetening	Reduce sulfur	Sulfur
1930	Thermal reforming	Improve octane number	Residua
1932	Hydrogenation	Remove sulfur	Sulfur
1932	Coking	Produce gasoline	Coke
1933	Solvent extraction	Improve lubricant viscosity index	Aromatics
1935	Solvent dewaxing	Improve pour point	Wax
1935	Catalytic polymerization	Improve octane number	Petrochemical feedstocks
1937	Catalytic cracking	Higher octane gasoline	Petrochemical feedstocks
1939	Visbreaking	Reduce viscosity	Increased distillate yield
1940	Alkylation	Increase octane number	High-octane aviation fuel
1940	Isomerization	Produce alkylation feedstock	Naphtha
1942	Fluid catalytic cracking	Increase gasoline yield	Petrochemical feedstocks
1950	Deasphalting	Increase cracker feedstock	Asphalt
1952	Catalytic reforming	Convert low-quality naphtha	Aromatics
1954	Hydrodesulfurization	Remove sulfur	Sulfur
1956	Inhibitor sweetening	Remove mercaptans	Disulfides and sulfur
1957	Catalytic isomerization	Convert to high octane products	Alkylation feedstocks
1960	Hydrocracking	Improve quality and reduce sulfur	Alkylation feedstocks
1974	Catalytic dewaxing	Improve pour point	Wax
1975	Resid hydrocracking	Increase gasoline yield	Cracked residua

experiments at a brewery near his home in Leeds, he had discovered the properties of the carbon dioxide (he called it fixed air) given off by the fermenting beer vats. One of these properties was that when water was placed in a flat dish for a time above the vats, it acquired a pleasant, acidulous taste that reminded Priestley of seltzer mineral waters.

Experiments convinced him that the medicinal qualities of seltzer might be due to the air dissolved in it. Pouring water from one glass to another for 3 min in the fixed air above a beer vat achieved the same effect. By 1772, he had devised a pumping apparatus that would impregnate water with fixed air, and the system was set up on board Cook's ships *Resolution* and *Adventure* in time for the voyage. It was a great success. Meanwhile, Priestley's politics continued to dog him. His support for the French Revolution was seen as particularly traitorous, and in 1794, a mob burned down his house and laboratory. So Priestley took ship for Pennsylvania, where he settled in Northumberland, honored by his American hosts as a major scientific figure. Then one night, while dining at Yale, he met a young professor of chemistry. The result of their meeting would change the life of twentieth-century America.

It may have been because the young man at dinner that night, Benjamin Silliman, was a hypochondriac (rather than the fact that he was a chemist) that subsequent events took the course they did. Silliman imagined he suffered from lethargy, vertigo, nervous disorders and whatever else he could think of. In common with other invalids, he regularly visited health spas like Saratoga Springs, New York (at his mother's expense), and he knew that such places were only for the rich. So his meeting with Priestley moved him to decide to make the mineral-water cure available to the common people (also at his mother's expense).

In 1809, he set up in business with an apothecary named Darling, assembled apparatus to impregnate 50 bottles of water a day and opened two soda-water fountains in New York City, one at the Tontine Coffee House and one at the City Hotel. The decor was hugely expensive (a lot of gilt), and they only sold 70 glasses on opening day. But Darling was optimistic. A friend of Priestley visited and declared that drinking the waters would prevent yellow fever. In spite of Silliman's hopes that the business would make him rich, by the end of the summer the endeavor was a disastrous flop. It would be many more decades before the soda fountain became a cultural icon in America!

Silliman cast around for some other way to make money. Two years earlier, he had analyzed the contents of a meteor that had fallen on Weston, Connecticut, and this research had enhanced his scientific reputation. So he decided to offer his services (as a geologist) to mining companies. His degree had been in law: he was as qualified for geology as he was to be a professor of chemistry at Yale. The geology venture prospered, and by 1820 Silliman was in great demand for field trips, on which he took his son, Benjamin, Jr. When he retired in 1853, his son took up where he had left off, as professor of General and Applied Chemistry at Yale (this time, with a degree in the subject). After writing a number of chemistry books and being elected to the National Academy of Sciences, Benjamin, Jr., took up lucrative consulting posts, as his father had done, with the Boston City Water Company and various mining enterprises.

In 1855, one of these asked him to research and report on some mineral samples from the new Pennsylvania Rock Oil Company. After several months of work Benjamin, Jr., announced that about 50% of the black tar-like substance could be distilled into first-rate burning oils (which would eventually be called kerosene and paraffin) and that an additional 40% of what was left could be distilled for other purposes, such as lubrication and gaslight. On the basis of this single report, a company was launched to finance the drilling of the Drake Well at Oil Creek, Pennsylvania, and in 1857 it became the first well to produce petroleum. It would be another 50 years before Silliman's reference to other fractions available from the oil through extra distillation would provide gasoline for the combustion engine of the first automobile. Silliman's report changed the world because it made possible an entirely new form of transportation and helped turn the United States into an industrial superpower. But back to the future.

After completion of the first well (by Edwin Drake), the surrounding areas were immediately leased and extensive drilling took place. Crude oil output in the United States increased from approximately 2000 barrels (1 barrel, bbl = 42 US gallons = 35 Imperial gallons = 5.61 foot^3 = 158.8 L) in 1859 to nearly 3,000,000 bbl in 1863 and approximately 10,000,000 bbl in 1874. In 1861 the first cargo of oil, contained in wooden barrels, was sent across the Atlantic to London, and by the 1870s, refineries, tank cars, and pipelines had become characteristic features of the industry, mostly through the leadership of Standard Oil that was founded by John D. Rockefeller (Johnson, 1997). Throughout the remainder of the nineteenth century, the United States and Russia were the two areas in which the most striking developments took place.

At the outbreak of World War I in 1914, the two major producers were the United States and Russia, but supplies of oil were also being obtained from Indonesia, Rumania, and Mexico. During the 1920s and 1930s, attention was also focused on other areas for oil production, such as the Middle East, and Indonesia. At this time, European and African countries were not considered major oil-producing areas. In the post-1945 era, Middle Eastern countries continued to rise in importance because of new discoveries of vast reserves. The United States, although continuing to be the biggest producer, was also the major consumer and thus was not a major exporter of oil. At this time, oil companies began to roam much farther in the search for oil, and significant discoveries in Europe, Africa, and Canada thus resulted.

However, what is more pertinent to the industry is that throughout the millennia in which petroleum has been known and used, it is only in the last decade or so that some attempts have been made to standardize the nomenclature and terminology. But confusion may still exist. Therefore, it is the purpose of this chapter to provide some semblance of order into the disordered state that exists in the segment of petroleum technology that is known as *terminology*.

1.3 DEFINITIONS AND TERMINOLOGY

Terminology is the means by which various subjects are named so that reference can be made in conversations and in writings and so that the meaning is passed on. Definitions are the means by which scientists and engineers communicate the nature of a material to each other and to the world, through either the spoken or the written word. Thus, the definition of a material can be extremely important and have a profound influence on how the technical community and the public perceive that material.

The definition of petroleum has been varied, unsystematic, diverse, and often archaic. Further, the terminology of petroleum is a product of many years of growth. Thus, the long established use of an expression, however inadequate it may be, is altered with difficulty, and a new term, however precise, is at best adopted only slowly.

It is essential that the definitions and the terminology of petroleum science and technology be given prime consideration because of the need for a thorough understanding of petroleum and the associated technologies. This will aid in a better understanding of petroleum, its constituents, and its various fractions. Of the many forms of terminology that have been used not all have survived, but the more commonly used are illustrated here. Particularly troublesome, and more confusing, are those terms that are applied to the more viscous materials, for example the use of the terms bitumen and asphalt. This part of the text attempts to alleviate much of the confusion that exists, but it must be remembered that the terminology of petroleum is still open to personal choice and historical usage.

Petroleum is a mixture of gaseous, liquid, and solid hydrocarbon compounds that occur in sedimentary rock deposits throughout the world and also contains small quantities of nitrogen-, oxygen-, and sulfur-containing compounds as well as trace amounts of metallic constituents (Bestougeff, 1967; Colombo, 1967; Thornton, 1977; Speight, 1990).

Petroleum is a naturally occurring mixture of hydrocarbons, generally in a liquid state, which may also include compounds of sulfur, nitrogen, oxygen, metals, and other elements (ASTM, 2005b). Petroleum has also been defined (ITAA, 1936) as

1. Any naturally occurring hydrocarbon, whether in a liquid, gaseous, or solid state
2. Any naturally occurring mixture of hydrocarbons, whether in a liquid, gaseous, or solid state
3. Any naturally occurring mixture of one or more hydrocarbons, whether in a liquid, gaseous, or solid state and one or more of the following, that is to say, hydrogen sulfide, helium, and carbon dioxide

The definition also includes any petroleum as defined by paragraph (1), (2), or (3) that has been returned to a natural **reservoir**.

In the crude state petroleum has minimal value, but when refined it provides high-value liquid fuels, solvents, lubricants, and many other products (Purdy, 1957). The fuels derived from petroleum contribute approximately one-third to one-half of the total world energy supply and are used not only for transportation fuels (i.e., gasoline, diesel fuel, and aviation fuel, among others) but also to heat buildings. Petroleum products have a wide variety of uses that vary from gaseous and liquid fuels to near-solid machinery lubricants. In addition, the residue of many refinery processes, asphalt—a once-maligned by-product—is now a premium value product for highway surfaces, roofing materials, and miscellaneous waterproofing uses.

Crude petroleum is a mixture of compounds boiling at different temperatures that can be separated into a variety of different generic fractions by distillation (Chapter 16). And the terminology of these fractions has been bound by utility and often bears little relationship to composition.

The molecular boundaries of petroleum cover a wide range of boiling points and carbon numbers of hydrocarbon compounds and other compounds containing nitrogen, oxygen, and sulfur, as well as metallic (porphyrinic) constituents. However, the actual boundaries of such a petroleum map can only be arbitrarily defined in terms of boiling point and carbon number (Chapter 9). In fact, petroleum is so diverse that materials from different sources exhibit different boundary limits, and for this reason alone it is not surprising that petroleum has been difficult to map in a precise manner.

Since there is a wide variation in the properties of crude petroleum (Table 1.3), the proportions in which the different constituents occur vary with origin (Gruse and Stevens, 1960; Koots and Speight, 1975). Thus, some crude oils have higher proportions of the lower boiling components and others (such as heavy oil and bitumen) have higher proportions of higher boiling components (asphaltic components and **residuum**).

For the purposes of terminology, it is preferable to subdivide petroleum and related materials into three major classes (Table 1.4):

1. Materials that are of natural origin
2. Materials that are manufactured
3. Materials that are integral fractions derived from natural or manufactured products

TABLE 1.3
Typical Variations in the Properties of Petroleum

Petroleum	Specific Gravity	API Gravity	Residuum >1000°F (% v/v)
U.S. Domestic			
California	0.858	33.4	23.0
Oklahoma	0.816	41.9	20.0
Pennsylvania	0.800	45.4	2.0
Texas	0.827	39.6	15.0
Texas	0.864	32.3	27.9
Other Countries			
Bahrain	0.861	32.8	26.4
Iran	0.836	37.8	20.8
Iraq	0.844	36.2	23.8
Kuwait	0.860	33.0	31.9
Saudi Arabia	0.840	37.0	27.5
Venezuela	0.950	17.4	33.6

1.4 NATIVE MATERIALS

1.4.1 PETROLEUM

Petroleum and the equivalent term *crude oil* cover a wide assortment of materials consisting of mixtures of hydrocarbons and other compounds containing variable amounts of sulfur, nitrogen, and oxygen, which may vary widely in volatility, specific gravity, and viscosity. Metal-containing constituents, notably those compounds that contain vanadium and nickel, usually occur in the more viscous crude oils in amounts up to several thousand parts per million and can have serious consequences during processing of these feedstocks (Gruse and Stevens, 1960; Speight, 1984). Because petroleum is a mixture of widely varying constituents

TABLE 1.4
Subdivision of Petroleum and Similar Materials into Various Subgroups

Natural Materials	Derived Materials	Manufactured Materials
Natural gas	Saturates	Synthetic crude oil
Petroleum	Aromatics	Distillates
Heavy oil	Resins	Lubricating oils
Bitumen[a]	Asphaltenes	Wax
Asphaltite	Carbenes[b]	Residuum
Asphaltoid	Carboids[b]	Asphalt
Ozocerite (natural wax)		Coke
Kerogen		Tar
Coal		Pitch

[a]Bitumen from tar sand deposits.
[b]Usually thermal products from petroleum processing.

and proportions, its physical properties also vary widely (Chapter 9) and the color from colorless to black.

Petroleum occurs underground, at various pressures depending on the depth. Because of the pressure, it contains considerable natural gas in solution. Petroleum underground is much more fluid than it is on the surface and is generally mobile under reservoir conditions because the elevated temperatures (the geothermal gradient) in subterranean formations decrease the viscosity. Although the geothermal gradient varies from place to place, it is generally of the order of 25°C to 30°C/km (15°F/1000 ft or 120°C/1000 ft, i.e., 0.015°C per foot of depth or 0.012°C per foot of depth).

Petroleum is derived from aquatic plants and animals that lived and died hundreds of millions of years ago. Their remains mixed with mud and sand in layered deposits that, over the millennia, were geologically transformed into sedimentary rock. Gradually the organic matter decomposed and eventually formed petroleum (or a related precursor), which migrated from the original source beds to more porous and permeable rocks, such as **sandstone** and siltstone, where it finally became entrapped. Such entrapped accumulations of petroleum are called reservoirs. A series of reservoirs within a common rock structure or a series of reservoirs in separate but neighboring formations is commonly referred to as an oil field. A group of fields is often found in a single geologic environment known as a sedimentary basin or province.

The major components of petroleum (Chapter 7) are hydrocarbons, compounds of hydrogen and carbon that display great variation in their molecular structure. The simplest hydrocarbons are a large group of chain-shaped molecules known as the paraffins. This broad series extends from methane, which forms natural gas, through liquids that are refined into gasoline, to crystalline waxes. A series of ring-shaped hydrocarbons, known as the naphthenes, range from volatile liquids such as naphtha to high molecular weight substances isolated as the **asphaltene** fraction. Another group of ring-shaped hydrocarbons is known as the aromatics; the chief compound in this series is benzene, a popular raw material for making petrochemicals.

Nonhydrocarbon constituents of petroleum include organic derivatives of nitrogen, oxygen, sulfur, and the metals nickel and vanadium. Most of these impurities are removed during refining.

Geologic techniques (Chapter 5) can determine only the existence of rock formations that are favorable for oil deposits, not whether oil is actually there. Drilling is the only sure way to ascertain the presence of oil. With modern rotary equipment, wells can be drilled to depths of more than 30,000 ft (9000 m). Once oil is found, it may be recovered (brought to the surface) by the pressure created by natural gas or water within the reservoir. Crude oil can also be brought to the surface by injecting water or steam into the reservoir to raise the pressure artificially, or by injecting such substances as carbon dioxide, polymers, and solvents to reduce crude oil viscosity. Thermal recovery methods are frequently used to enhance the production of heavy crude oils, whose extraction is impeded by viscous resistance to flow at reservoir temperatures.

Crude oil is transported to refineries by pipelines, which can often carry more than 500,000 barrels per day, or by ocean-going tankers. The basic refinery process is distillation (Chapter 16), which separates the crude oil into fractions of differing volatility. After the distillation, other physical methods are employed to separate the mixtures, including absorption, adsorption, solvent extraction, and crystallization. After physical separation into such constituents as light and heavy naphtha, kerosene, and light and heavy gas oils, selected petroleum fractions may be subjected to conversion processes, such as thermal cracking (i.e., coking; Chapter 17) and catalytic cracking (Chapter 18). In the most general terms, cracking breaks the large molecules of heavier gas oils into the smaller molecules that form the lighter, more valuable naphtha fractions.

Reforming (Chapter 23) changes the structure of straight-chain paraffin molecules into branched-chain *iso*-paraffins and ring-shaped aromatics. The process is widely used to raise the octane number of gasoline (Chapter 26) obtained by distillation of paraffinic crude oils.

1.4.2 HEAVY OIL

There are also other types of petroleum that are different from conventional petroleum in that they are much more difficult to recover from the subsurface reservoir. These materials have a much higher viscosity (and lower API gravity) than conventional petroleum, and primary recovery of these petroleum types usually requires thermal stimulation of the reservoir (Chapter 5 and Chapter 6).

When petroleum occurs in a reservoir that allows the crude material to be recovered by pumping operations as a free-flowing dark to light-colored liquid, it is often referred to as conventional petroleum.

Heavy oils are more difficult to recover from the subsurface reservoir than light oils. The definition of heavy oils is usually based on the API gravity or viscosity, and the definition is quite arbitrary although there have been attempts to rationalize the definition based on viscosity, API gravity, and density.

For many years, petroleum and heavy oil were very generally defined in terms of physical properties. For example, heavy oils were considered to be those crude oils that had gravity somewhat less than 20° API, generally falling into the API gravity range 10° to 15°. For example, Cold Lake heavy crude oil has an API gravity equal to 12° and extra heavy oils, such as tar sand bitumen, usually have an API gravity in the range 5° to 10° (Athabasca bitumen = 8° API). Residua would vary depending on the temperature at which distillation was terminated but usually vacuum residua are in the range 2° to 8° API (Speight, 2000; Speight and Ozum, 2002; and references cited therein).

Heavy oils have a much higher viscosity (and lower API gravity) than conventional petroleum, and primary recovery of these petroleum types usually requires thermal stimulation of the reservoir. The generic term heavy oil is often applied to a crude oil that has less than 20° API and usually, but not always, a sulfur content higher than 2% by weight (Speight, 2000). Further, in contrast to conventional crude oils, heavy oils are darker in color and may even be black.

The term *heavy oil* has also been arbitrarily used to describe both the heavy oils that require thermal stimulation of recovery from the reservoir and the bitumen in **bituminous sand** (*q.v.*, **tar sand**) formations from which the heavy bituminous material is recovered by mining operation.

Extra heavy oils are materials that occur in the solid or near-solid state and are generally incapable of free flow under reservoir conditions (q.v., bitumen).

1.4.3 BITUMEN

The term bitumen (also, on occasion, referred to as **native asphalt,** and extra heavy oil) includes a wide variety of reddish brown to black materials of semisolid, viscous to brittle character that can exist in nature with no mineral impurity or with mineral matter contents that exceed 50% by weight. Bitumen is frequently found filling the pores and crevices of sandstone, limestone, or argillaceous sediments, in which case the organic and associated mineral matrix is known as **rock asphalt** (Abraham, 1945; Hoiberg, 1964).

Bitumen is a naturally occurring material that is found in deposits where the permeability is low and passage of fluids through the deposit can only be achieved by prior application of fracturing techniques. Tar sand bitumen is a high-boiling material with little, if any, material boiling below 350°C (660°F) and the boiling range approximates the boiling range of an atmospheric residuum.

Tar sands have been defined in the United States (FE-76-4) as

> ... the several rock types that contain an extremely viscous hydrocarbon which is not recoverable in its natural state by conventional oil well production methods including currently used enhanced recovery techniques. The hydrocarbon-bearing rocks are variously known as bitumen-rocks oil, impregnated rocks, oil sands, and rock asphalt.

The recovery of the bitumen depends to a large degree on the composition and construction of the sands. Generally, the bitumen found in tar sand deposits is an extremely viscous material that is immobile under reservoir conditions and cannot be recovered through a well by the application of secondary or enhanced recovery techniques.

The expression tar sand is commonly used in the petroleum industry to describe sandstone reservoirs that are impregnated with a heavy, viscous black crude oil that cannot be retrieved through a well by conventional production techniques (FE-76-4, given earlier). However, the term *tar sand* is actually a misnomer; more correctly, the name tar is usually applied to the heavy product remaining after the destructive distillation of coal or other organic matter (Speight, 1994).

The bitumen in tar sand formations requires a high degree of thermal stimulation for recovery to the extent that some thermal decomposition may have to be induced. Current recovery operations of bitumen in tar sand formations involve use of a mining technique (Chapter 6).

It is incorrect to refer to native bituminous materials as tar or pitch. Although the word tar is descriptive of the black, heavy bituminous material, it is best to avoid its use with respect to natural materials and to restrict the meaning to the volatile or near-volatile products produced in the destructive distillation of such organic substances as coal (Speight, 1994). In the simplest sense, pitch is the distillation residue of the various types of tar.

Thus, alternative names, such as bituminous *sand* or **oil sand**, are gradually finding usage, the former name (bituminous sands) is more technically correct. The term *oil sand* is also used in the same way as the term *tar sand*, and these terms are used interchangeably throughout this text.

However, to define bitumen, heavy oil, and conventional petroleum, the use of a single physical parameter such as viscosity is not sufficient. Physical properties such as API gravity, elemental analysis, and composition fall short of giving an adequate definition. It is the properties of the bulk deposit and, most of all, the necessary recovery methods that form the basis of the definition of these materials. Only then is it possible to classify petroleum, heavy oil, and tar sand bitumen (Chapter 2).

1.4.4 WAX

Naturally occurring wax, often referred to as **mineral wax**, occurs as a yellow to dark brown, solid substance that is composed largely of paraffins (Wollrab and Streibl, 1969). Fusion points vary from 60°C (140°F) to as high as 95°C (203°F). They are usually found associated with considerable mineral matter, as a filling in veins and fissures or as an interstitial material in porous rocks. The similarity in character of these native products is substantiated by the fact that, with minor exceptions where local names have prevailed, the original term **ozokerite** (*ozocerite*) has served without notable ambiguity for mineral wax deposits (Gruse and Stevens, 1960).

Ozokerite (ozocerite), from the Greek meaning odoriferous wax, is a naturally occurring hydrocarbon material composed chiefly of solid paraffins and cycloparaffins (i.e., hydrocarbons)

(Wollrab and Streibl, 1969). Ozocerite usually occurs as stringers and veins that fill rock fractures in tectonically disturbed areas. It is predominantly paraffinic material (containing up to 90% nonaromatic hydrocarbons) with a high content (40% to 50%) of normal or slightly branched paraffins as well as cyclic paraffin derivatives. Ozocerite contains approximately 85% carbon, 14% hydrogen, and 0.3% each of sulfur and nitrogen and is, therefore, predominantly a mixture of pure hydrocarbons; any nonhydrocarbon constituents are in the minority.

Ozocerite is soluble in solvents that are commonly employed for the dissolution of petroleum derivatives, e.g., toluene, benzene, carbon disulfide, chloroform, and ethyl ether. In the present context, note that the term *migrabitumen* signifies secondary bitumen (secondary macerals) generated from fossil organic material during diagenesis and catagenesis (Chapter 3). These materials are usually amorphous solids and can be classified into several subgroups (Chapter 3).

1.4.5 ASPHALTITE

Asphaltites are a variety of naturally occurring, dark brown to black, solid, nonvolatile bituminous substances that are differentiated from bitumen primarily by their high content of material insoluble in the common organic solvents and high yields of thermal coke (Yurum and Ekinci, 1995). The resultant high temperature of fusion (approximate range 115°C to 330°C, 240°F to 625°F) is characteristic. The names applied to the two rather distinct types included in this group are now accepted and used for the most part without ambiguity.

Gilsonite was originally known as uintaite from its discovery in the Uinta Basin of western Colorado and eastern Utah. It is characterized by a bright luster and a carbon residue in the range 10% to 20% by weight. The mineral occurs in nearly vertical veins varying from about an inch to many feet in width and is relatively free of occluded inorganic matter. Samples taken from different veins and across the larger veins may vary somewhat in softening point, solubility characteristics, sulfur content, and so on, but the variation is not great. It is evident in all instances that it is essentially the same material, and it is therefore appropriate to apply a single name to this mineral. However, caution should be exercised in using the same term without qualification for similar materials until it can be shown that they are equivalent to gilsonite.

The second recognized type in this category is **grahamite**, which is very much like gilsonite in external characteristics but is distinguished from the latter by its black streak, relatively high fixed carbon value (35% to 55%), and high temperature of fusion, which is accompanied by a characteristic intumescence. The undifferentiated term *grahamite* must be used with caution; similarities in the characteristics of samples from different areas do not necessarily imply any chemical or genetic relationship.

A third but rather broad category of asphaltite includes a group of bituminous materials known as glance pitch, which physically resemble gilsonite but have some of the properties of grahamite. They have been referred to as intermediates between the two; although the possibility exists that they are basically different from gilsonite and may represent something between bitumen and grahamite.

1.4.6 ASPHALTOID

Asphaltoids are a further group of brown to black, solid bituminous materials of which the members are differentiated from the asphaltites by their infusibility and low solubility in carbon disulfide. These substances have also been designated **asphaltic pyrobitumen**, as they decompose on heating into bitumen-like materials. However, the term **pyrobitumen** does not convey the impression intended; thus the members of this class are referred to as asphaltoids since they closely resemble the asphaltites.

Pyrobitumen is a naturally occurring solid organic substance that is distinguishable from bitumen (*q.v.*) by being infusible and insoluble. When heated, pyrobitumen generates, or transforms into, bitumen-like liquid and gaseous hydrocarbon compounds. Pyrobitumen may be either asphaltic or nonasphaltic. The asphaltic pyrobitumen, derived from petroleum, is relatively hard, and has a specific gravity below 1.25. They do not melt when heated but swell and decompose (intumesce).

There is much confusion regarding the classification of asphaltoids, although four types are recognized: elaterite, wurtzilite, **albertite**, and imposonite—in the order of increasing density and fixed carbon content. In fact, it is doubtful that the asphaltoid group can ever be clearly differentiated from the asphaltites. It is even more doubtful that the present subdivisions will ever have any real meaning, nor is it clear that the materials have any necessary genetic connection. Again, caution should be exercised in the use of the names, and due care should be applied to the qualification of the particular sample.

1.4.7 BITUMINOUS ROCK AND BITUMINOUS SAND

Bituminous rock and bituminous sand (see also bitumen, page 16) are those formations in which the bituminous material is found as a filling in veins and fissures in fractured rocks or impregnating relatively shallow sand, sandstone, and limestone strata. The deposits contain as much as 20% bituminous material, and if the organic material in the rock matrix is bitumen, it is usual (although chemically incorrect) to refer to the deposit as rock asphalt to distinguish it from bitumen that is relatively mineral free. A standard test (ASTM, 2005a) is available for determining the bitumen content of various mixtures with inorganic materials, although the use of word bitumen as applied in this test might be questioned and it might be more appropriate to use the term *organic residues* to include tar and pitch.

If the material is of the asphaltite-type or asphaltoid-type, the corresponding terms, rock asphaltite or rock asphaltoid, should be used. Bituminous rocks generally have a coarse, porous structure, with the bituminous material in the voids. A much more common situation is that in which the organic material is present as an inherent part of the rock composition insofar as it is a diagenetic residue of the organic material detritus that was deposited with the sediment. The organic components of such rocks are usually refractory and are only slightly affected by most organic solvents.

A special class of bituminous rocks that has achieved some importance is the so-called **oil shale**. These are argillaceous, laminated sediments of generally high organic content that can be thermally decomposed to yield appreciable amounts of oil, commonly referred to as shale oil. Oil shale does not yield shale oil without the application of high temperatures and the ensuing thermal decomposition that is necessary to decompose the organic material (**kerogen**) in the shale.

Sapropel is an unconsolidated sedimentary deposit, rich in bituminous substances. It is distinguished from peat in being rich in fatty and waxy substances and poor in cellulosic material. When consolidated into rock, sapropel becomes oil shale, bituminous shale, or boghead coal. The principal components are certain types of algae that are rich in fats and waxes. Minor constituents are mineral grains and decomposed fragments of spores, fungi, and bacteria. The organic materials accumulate in water under reducing conditions.

1.4.8 KEROGEN

Kerogen is the complex carbonaceous (organic) material that occurs in sedimentary rocks and shales. It is for the most part insoluble in the common organic solvents. When kerogen occurs in shale, the entire material is often referred to as oil shale. This, like the term *oil sand*, is a misnomer insofar as the shale does not contain oil; oil sand (like the more correct term *bituminous sand* implies) contains a viscous nonvolatile material that can be isolated without thermal decomposition. A **synthetic crude oil** is produced from oil shale by the application of heat so that the kerogen is thermally decomposed (cracked) to produce the lower molecular weight products. Kerogen is also reputed to be a precursor of petroleum (Chapter 4).

For comparison with tar sand, oil shale is any fine-grained sedimentary rock containing solid organic matter (q.v., kerogen) that yields oil when heated (Scouten, 1990). Oil shales vary in their mineral composition. For example, clay minerals predominate in true shales, whereas other minerals (e.g., dolomite and calcite) occur in appreciable but subordinate amounts in the carbonates. In all shale types, layers of the constituent mineral alternate with layers of kerogen.

1.4.9 NATURAL GAS

The generic term *natural gas* applies to gases commonly associated with petroliferous (petroleum-producing, petroleum-containing) geologic formations. Natural gas generally contains high proportions of methane (a single carbon hydrocarbon compound, CH_4) and some of the higher molecular weight higher paraffins (C_nH_{2n+2}) generally containing up to six carbon atoms may also be present in small quantities (Table 1.5). The hydrocarbon constituents of natural gas are combustible, but nonflammable nonhydrocarbon components such as carbon dioxide, nitrogen, and helium are often present in the minority and are regarded as contaminants.

In addition to the natural gas found in petroleum reservoirs, there are also those reservoirs in which natural gas may be the sole occupant. The principal constituent of natural gas is methane, but other hydrocarbons, such as ethane, propane, and butane, may also be present. Carbon dioxide is also a common constituent of natural gas. Trace amounts of rare gases, such as helium, may also occur, and certain natural gas reservoirs are a source of these rare gases. Just as petroleum can vary in composition, so can natural gas. Differences in natural gas composition occur between different reservoirs, and two wells in the same field may also yield gaseous products that are different in composition (Speight, 1990).

Natural gas has been known for many centuries, but its initial use was probably more for religious purposes rather than as a fuel. For example, gas wells were an important aspect of religious life in ancient Persia because of the importance of fire in their religion. In classical

TABLE 1.5
Constituents of Natural Gas

Name	Formula	Vol.%
Methane	CH_4	>85
Ethane	C_2H_6	3–8
Propane	C_3H_8	1–5
Butane	C_4H_{10}	1–2
Pentane[a]	C_5H_{12}	1–5
Carbon dioxide	CO_2	1–2
Hydrogen sulfide	H_2S	1–2
Nitrogen	N_2	1–5
Helium	He	<0.5

[a]Pentane and higher molecular weight hydrocarbons, including benzene and toluene.

times these wells were often flared and must have been, to say the least, awe inspiring (Lockhart, 1939; Forbes, 1964).

There is also the possibility that the voices of the gods recorded by the ancients were actually natural gas forcing its way through fissures in the earth's surface (Scheil and Gauthier, 1909; Schroder, 1920). Gas wells were also known in Europe in the Middle Ages, and oil was reportedly ejected from the wells, as in the phenomena observed at the site near the town of Mineo, Sicily. From other such documentation, it can be surmised that the combustible material, or the source of the noises in the earth, was actually natural gas.

Just as petroleum was used in antiquity, natural gas was also known in antiquity. However, the use of petroleum has been relatively well documented because of its use in warfare and as mastic for walls and roads (Henry, 1873; Abraham, 1945; Forbes, 1958a,b; James and Thorpe, 1994). The use of natural gas in antiquity is somewhat less well documented, although historical records indicate that the use of natural gas (for other than religious purposes) dates back to about 250 AD when it was used as a fuel in China. The gas was obtained from shallow wells and was distributed through a piping system constructed from hollow bamboo stems. There is other fragmentary evidence for the use of natural gas in certain old texts, but the use is usually inferred since the gas is not named specifically. However, it is known that natural gas was used on a small scale for heating and lighting in northern Italy during the early seventeenth century. From this it might be conjectured that natural gas found some use from the seventeenth century to the present day, recognizing that gas from coal would be a strong competitor.

Natural gas was first discovered in the United States in Fredonia, New York, in 1821. In the years following this discovery, natural gas usage was restricted to its environs since the technology for storage and transportation (bamboo pipes notwithstanding) was not well developed and, at that time, natural gas had little or no commercial value. In fact, in the 1930s when petroleum refining was commencing an expansion in technology that is still continuing, natural gas was not considered a major fuel source and was only produced as an unwanted by-product of crude oil production.

The principal gaseous fuel source at that time (i.e., the 1930s) was the gas produced by the surface gasification of coal. In fact, each town of any size had a plant for the gasification of coal (hence the use of the term *town gas*). Most of the natural gas produced at the petroleum fields was vented to the air or burned in a flare stack; only a small amount of the natural gas from the petroleum fields was pipelined to industrial areas for commercial use. It was only in the years after World War II that natural gas became a popular fuel commodity, leading to the recognition that it has at the present time.

There are several general definitions that have been applied to natural gas. Thus, **lean gas** is gas in which methane is the major constituent. **Wet gas** contains considerable amounts of the higher molecular weight hydrocarbons. Sour gas contains hydrogen sulfide whereas sweet gas contains very little, if any, hydrogen sulfide. Residue gas is natural gas from which the higher molecular weight hydrocarbons have been extracted and **casing head gas** is derived from petroleum, but is separated at the separation facility at the well-head

To further define the terms *dry* and *wet* in quantitative measures, the term *dry natural gas* indicates that there is less than 0.1 gallon (1 US gallon $= 264.2$ m^3) of gasoline vapor (higher molecular weight paraffins) per 1000 ft^3 (1 ft$^3 = 0.028$ m^3). The term *wet natural gas* indicates that there are such paraffins present in the gas, in fact more than 0.1 gal/1000 ft^3.

Associated or dissolved natural gas occurs either as free gas or as gas in solution in the petroleum. Gas that occurs as a solution in the petroleum is dissolved gas whereas the gas that exists in contact with the petroleum (**gas cap**) is associated gas.

Other components such as carbon dioxide (CO_2), hydrogen sulfide (H_2S), mercaptans (thiols; R-SH), as well as trace amounts of other constituents may also be present. Thus, there is no single composition of components which might be termed *typical* natural gas. Methane and ethane constitute the bulk of the combustible components; carbon dioxide (CO_2) and nitrogen (N_2) are the major noncombustible (inert) components.

1.5 MANUFACTURED MATERIALS

1.5.1 WAX

The term paraffin wax is restricted to the colorless, translucent, highly crystalline material obtained from the light lubricating fractions of paraffinic crude oils (wax distillates). The commercial products melt in the approximate range of 50°C to 65°C (120°F to 150°F). Dewaxing of heavier fractions leads to semisolid material, generally known as **petrolatum**, and solvent de-oiling of the petroleum or of heavy, waxy residua results in dark-colored waxes of a sticky, plastic to hard nature. The waxes are composed of fine crystals and contain, in addition to *n*-paraffins, appreciable amounts of *iso*-paraffins and cyclic hydrocarbon compounds substituted with long-chain alkyl groups. The melting points of the commercial grades are in the 70°C to 90°C (160°F to 195°F) range (Brooks et al., 1938).

Highly paraffinic waxes are also produced from peat, lignite, or shale oil residua, and paraffin wax known as **ceresin**, which are quite similar to the waxes from petroleum, may also be prepared from ozokerite.

1.5.2 RESIDUUM (RESIDUA)

A residuum (pl. residua, also shortened to resid, pl. resids) is the residue obtained from petroleum after nondestructive distillation has removed all the volatile materials. The temperature of the distillation is usually maintained below 350°C (660°F) since the rate of thermal decomposition of petroleum constituents is minimal below this temperature, but the rate of thermal decomposition of petroleum constituents is substantial above 350°C (660°F).

Residua are black, viscous materials and are obtained by distillation of a crude oil under atmospheric pressure (atmospheric residuum) or under reduced pressure (vacuum residuum) (Chapter 16). They may be liquid at room temperature (generally atmospheric residua) or almost solid (generally vacuum residua) depending on the nature of the crude oil (Chapter 16).

When a residuum is obtained from a crude oil and thermal decomposition has commenced, it is more usual to refer to this product as pitch (see also page 23). The differences between parent petroleum and the residua are due to the relative amounts of various constituents present, which are removed or remain by virtue of their relative volatility.

The chemical composition of a residuum from an asphaltic crude oil is complex. Physical methods of fractionation usually indicate high proportions of asphaltenes and resins, even in amounts up to 50% (or higher) of the residuum. In addition, the presence of ash-forming metallic constituents, including such organometallic compounds as those of vanadium and nickel, is also a distinguishing feature of residua and the heavier oils. Further, the deeper the cut into the crude oil, the greater is the concentration of sulfur and metals in the residuum and the greater the deterioration in physical properties (Chapter 16).

1.5.3 ASPHALT

Asphalt (Chapter 26) is manufactured from petroleum and is a black or brown material that has a consistency varying from a viscous liquid to a glassy solid. To a point, asphalt can resemble bitumen, hence the tendency to refer to bitumen (incorrectly) as native asphalt. It is recommended that there be differentiation between asphalt (manufactured) and bitumen (naturally occurring) other than by use of the qualifying terms *petroleum* and *native,* since the origins of the materials may be reflected in the resulting physicochemical properties of the two types of materials. It is also necessary to distinguish between the asphalt which originates from petroleum by refining and the product in which the source of the asphalt is a material other than petroleum, e.g., wurtzilite asphalt (Bland and Davidson, 1967). In the absence of a qualifying word, it is assumed that the term asphalt refers to the product manufactured from petroleum.

When the asphalt is produced simply by the distillation of an asphaltic crude oil, the product can be referred to as **residual asphalt** or straight run asphalt. If the asphalt is prepared by solvent extraction of residua or by light hydrocarbon (propane) precipitation, or if blown or otherwise treated, the term should be modified accordingly to qualify the product (e.g., **propane asphalt, blown asphalt**).

Asphalt softens when heated and is elastic under certain conditions. The mechanical properties of asphalt are of particular significance when it is used as a binder or adhesive. The principal application of asphalt is in road surfacing, which may be done in a variety of ways. Light oil dust layer treatments may be built up by repetition to form a hard surface, or a granular aggregate may be added to an asphalt coat, or earth materials from the road surface itself may be mixed with the asphalt.

Other important applications of asphalt include canal and reservoir lining, dam facings, and sea works. The asphalt so used may be a thin, sprayed membrane, covered with earth for protection against weathering and mechanical damage, or thicker surfaces, often including riprap (crushed rock). Asphalt is also used for roofs, coatings, floor tiles, soundproofing, waterproofing, and other building-construction elements and in a number of industrial products, such as batteries. For certain applications an asphaltic emulsion is prepared, in which fine globules of asphalt are suspended in water.

1.5.4 Tar and Pitch

Tar is a product of the destructive distillation of many bituminous or other organic materials and is a brown to black, oily, viscous liquid to semisolid material.

Tar is most commonly produced from bituminous coal and is generally understood to refer to the product from coal, although it is advisable to specify **coal tar** if there is the possibility of ambiguity. The most important factor in determining the yield and character of the coal tar is the carbonizing temperature. Three general temperature ranges are recognized, and the products have acquired the designations: low-temperature tar (approximately 450°C to 700°C; 540°F to 1290°F); mid-temperature tar (approximately 700°C to 900°C; 1290°F to 1650°F); and high-temperature tar (approximately 900°C to 1200°C; 1650°F to 2190°F). Tar released during the early stages of the decomposition of the organic material is called primary tar since it represents a product that has been recovered without the secondary alteration that results from prolonged residence of the vapor in the heated zone.

Treatment of the distillate (boiling up to 250°C, 480°F) of the tar with caustic soda causes separation of a fraction known as tar acids; acid treatment of the distillate produces a variety of organic nitrogen compounds known as tar bases. The residue left following removal of the heavy oil, or distillate, is pitch, a black, hard, and highly ductile material.

In the chemical-process industries, pitch is the black or dark brown residue obtained by distilling coal tar, wood tar, fats, fatty acids, or fatty oils.

Coal tar pitch is a soft to hard and brittle substance containing chiefly aromatic resinous compounds along with aromatic and other hydrocarbons and their derivatives; it is used chiefly as road tar, in waterproofing roofs and other structures, and to make electrodes. Wood tar pitch is a bright, lustrous substance containing resin acids; it is used chiefly in the manufacture of plastics and insulating materials and in caulking seams. Pitch derived from fats, fatty acids, or fatty oils by distillation is usually a soft substance containing polymers and decomposition products; it is used chiefly in varnishes and paints and in floor coverings.

PitchLake is the name that has been applied to large surface deposits of bitumen. Guanoco Lake in Venezuela covers more than 1100 acres (445 ha) and contains an estimated 35,000,000 bbl of bitumen. It was used as a commercial source of asphalt from 1891 to 1935. Smaller deposits occur commonly where tertiary marine sediments outcrop on the surface; an example is the tar pits at Rancho La Brea in Los Angeles (*brea* and tar have been used synonymously with bitumen). Although most pitch lakes are fossils of formerly active seeps, some, such as Pitch Lake on the island of Trinidad, continue to be supplied with fresh crude oil seeping from a subterranean source. Pitch Lake covers 115 acres and contains an estimated 40,000,000 bbl of bitumen.

1.5.5   COKE

Coke is the solid carbonaceous material produced from petroleum during thermal processing. It is often characterized by its high carbon content (95% + by weight) and a honeycomb type of appearance. The color varies from gray to black, and the material is insoluble in organic solvents.

1.5.6   SYNTHETIC CRUDE OIL

Coal, oil shale, bitumen, and heavy oil can be upgraded through thermal decomposition by a variety of processes to produce a marketable and transportable product. These products vary in nature, but the principal product is a hydrocarbon that resembles conventional crude oil: hence the use of the terms *synthetic crude oil* and *syncrude*. The synthetic crude oil, although it may be produced from one of the less conventional conversion processes, can actually be refined by the usual refinery system.

By comparison, the unrefined synthetic crude oil from bitumen will generally resemble petroleum more closely than either the synthetic crude oil from coal or the synthetic crude oil from oil shale. Unrefined synthetic crude oil from coal can be identified by a high content of phenolic compounds, whereas the unrefined synthetic crude oil from oil shale will contain high proportions of nitrogen-containing compounds.

1.6   DERIVED MATERIALS

1.6.1   ASPHALTENES, CARBENES, AND CARBOIDS

All the petroleum-related materials mentioned in the preceding sections are capable of being separated into several fractions (Chapter 8) that are sufficiently distinct in character to warrant the application of individual names (Pfeiffer, 1950).

Treatment of petroleum, residua, heavy oil, or bitumen with a low-boiling liquid hydrocarbon results in the separation of brown to black powdery materials known as asphaltenes (Figure 1.1). The reagents for effecting this separation are *n*-pentane and *n*-heptane, although other low molecular weight liquid hydrocarbons have been used (van Nes and van Westen, 1951; Mitchell and Speight, 1973).

Asphaltenes separated from petroleum, residua, heavy oil, and bitumen dissolve readily in benzene, carbon disulfide, chloroform, or other chlorinated hydrocarbon solvents. However, in the ease of the higher molecular weight native materials or petroleum residua that have been heated intensively or for prolonged periods, the *n*-pentane-insoluble or *n*-heptane-insoluble fraction may not dissolve completely in the aforementioned solvents. Definition

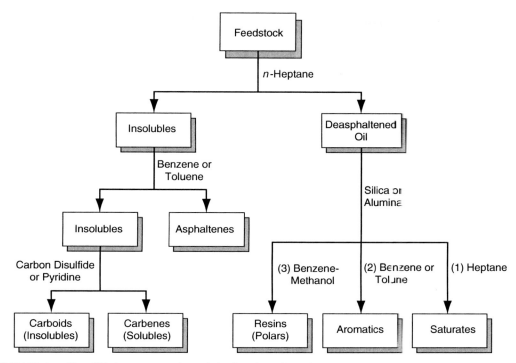

FIGURE 1.1 Simplified representation of the fractionation of petroleum.

of the asphaltene fraction has therefore been restricted to that of the *n*-pentane- or *n*-heptane-insoluble material that dissolves in such solvents as benzene.

The benzene- or toluene-insoluble material is collectively referred to as **carbenes** and **carboids**, and the fraction soluble in carbon disulfide (or pyridine), but insoluble in benzene is defined as carbenes (and not the long-barreled gun used to shoot bison). By difference, carboids are insoluble in carbon disulfide (or pyridine). However, because of the different solvents that might be used in place of benzene, it is advisable to define the carbenes (or carboids) by prefixing them with the name of the solvent used for the separation.

1.6.2 RESINS AND OILS

The portion of petroleum that is soluble in, for example, pentane or heptane is often referred to as **maltenes** (malthenes). This fraction can be further subdivided by percolation through any surface-active material, such as Fuller's earth or alumina to yield an oil fraction. A more strongly adsorbed, red to brown semisolid material known as resins remains on the adsorbent until desorbed by a solvent such as pyridine or a benzene/methanol mixed solvent (Figure 1.1). The oil fraction can be further subdivided into an aromatics fraction and a **saturates** fraction. Several other methods have been proposed for separating the resin fraction (Chapter 8); for example, a common procedure in the refining industry (Chapter 19) that can also be used in laboratory practice involves precipitation by liquid propane.

The resin fraction and the oil (maltenes) fraction have also been referred to collectively as **petrolenes**, thereby adding further confusion to this system of nomenclature. However, it has been accepted by many workers in petroleum chemistry that the term *petrolenes* may be applied to that part of the *n*-pentane-soluble (or *n*-heptane soluble) material that is low boiling (<300°C, <570°F, 760 mm) and can be distilled without thermal decomposition. Consequently, the term *maltenes* is now arbitrarily assigned to the pentane-soluble portion of petroleum that is relatively high boiling (>300°C, 760 mm).

Different feedstocks have different amounts of the asphaltene, resin, and oil fractions (Chapter 8), which can also lead to different yields of thermal coke as produced in the Conradson carbon residue or Ramsbottom carbon residue tests (Chapter 9). Such differences have effects on the methods chosen for refining the different feedstocks.

The fraction precipitated by propane may also contain acid material, often referred to as asphaltic acids or asphaltogenic acids. These acids can probably be regarded simply as cyclic and noncyclic organic acids of high molecular weight. These acids usually appear in the resin fraction, but if they have been removed or are absent, the resins are said to be neutral. Removal of the resins from maltenes leaves the oils fraction.

1.7 OIL PRICES

At this stage of the introduction to crude oil science and technology and in keeping with the historical aspects that have been discussed earlier, a word (or two) about crude oil pricing and the historical perspectives is warranted.

Currently, oil is the primary energy source in the world (Figure 1.2) (BP, 2002). For a century, the world has depended on low-cost oil to stimulate and maintain economic growth (Yergin, 1991). However, sustaining the rate of economic growth is open to question because the volume of oil that can ultimately be recovered is subject to much speculation because of the uncertainties of reserve estimation (Chapter 3) and this, in turn, affects the price of oil.

At this stage, while dealing with petroleum reserves and resources, it is appropriate to deal with a related topic and that is crude oil prices. However, it is not the intent here to move into predictions of the future. There is nothing difficult about making predictions. But accurate predictions of future events are always difficult! It is difficult to be correct! Everyone can justify with amazingly accurate 20–20 hindsight why their predictions were incorrect. But they are never incorrect. These erstwhile mediums will use statistics to show that, after several rounds of mathematical manipulation, their predictions were very close. Even though the outcome bears no relationship to what really happened. It is easy to make a statement that oil prices will continue to rise (after all, the pessimist is never disappointed), but the predictability lies in determining when and by how much.

Therefore, it is the purpose of this section to forgo any predictions. It is, however, the purpose of this section, to present a brief history of oil prices from which the readers can make their own predictions.

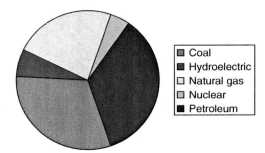

FIGURE 1.2 Distribution of world energy resources. (From World Energy Outlook 2005, International Energy Agency.)

1.7.1 PRICING STRATEGIES

References to the oil price are usually either references to the spot price of either West Texas (light) crude oil traded on the New York Mercantile Exchange (NYMEX) for delivery in Cushing, Oklahoma; or the price of Brent crude oil traded on the International Petroleum Exchange (IPE) for delivery at Sullom Voe.

Briefly, the price of a barrel of oil is highly dependent on both its grade (which is determined by factors such as its specific gravity or API and its sulfur content) and location. The vast majority of oil will not be traded on an exchange but on an over-the-counter basis, typically with reference to a marker crude oil grade that is typically quoted via the pricing agency. IPE claims that 65% of traded oil is priced off their Brent benchmarks. The Energy Information Administration (EIA) uses the Imported Refiner Acquisition Cost, the weighted average cost of all oil imported into the United States as the world oil price.

1.7.2 OIL PRICE HISTORY

Crude oil prices have seen wide price swings over the past decade whether it is due to apparent shortage or oversupply. At the time of writing, prices are close to $80 barrel which makes the so-called average price of approximately $21 per barrel (a number per barrel placed on the table by some economists) seems meaningless. Just as the determination of the so-called average structure is meaningless (Chapter 10 and Chapter 11), so is an average price per barrel of oil. Even when adjusted for inflation to current dollars, an average price per barrel bears little relationship to reality, especially when reality is a much higher price per barrel. The analogy often used is that of a scientist or engineer standing with their left foot in a pail of boiling water and their right foot in a pail of ice water and declaring that they are comfortable because is at average temperature!

The very long-term view of petroleum pricing can be considered in the same way. Average prices, even when adjusted for inflation, do not help the consumer who has to bear the brunt of the price increases.

Historically (or, some might say, hysterically), crude oil prices varied from $2.50 to $3.00 from 1948 through the end of the 1960s. The price rose from $2.50 in 1948 to about $3.00 in 1957. From 1958 to 1970 prices were stable at about $3.00 per barrel. Organization of Petroleum Exporting Countries (OPEC) was formed in 1960 with five founding members Iran, Iraq, Kuwait, Saudi Arabia, and Venezuela. By the end of 1971, six other nations (namely Qatar, Indonesia, Libya, United Arab Emirates, Algeria, and Nigeria) had swelled the membership ranks of OPEC.

Throughout this period, the petroleum exporting countries found increasing demand for their crude oil. In 1972, the price of crude oil was about $3.00 per barrel and by the end of 1974 the price of oil had quadrupled to over $12.00. The Yom Kippur War started with an attack on Israel by Syria and Egypt on October 5, 1973. Many countries in the western world showed strong support for Israel and, as a result, several of the Middle Eastern oil exporting nations imposed an embargo on those countries by decreasing oil.

From 1974 to 1978, the price of crude oil was relatively flat ranging from $12.21 per barrel to $13.55 per barrel. When adjusted for inflation the price over that period of time exhibited a moderate decline. Then events in Iran and Iraq (the overthrow of the Shah of Iran and the Iran–Iraq war) led to another round of crude oil price increases and crude oil prices rose to $35 per barrel in 1981.

The higher prices resulted in, among other actions usually involving energy conservation, increased exploration and production in the non-OPEC world. In mid-1985, oil prices were linked to the spot market for crude and by early 1986 (with increased production by some

OPEC members), crude oil prices moved downward to $8 to $10 per barrel. The price of crude oil rose again in 1990 with the Iraqi invasion of Kuwait and the ensuing Gulf War, but following the war, crude oil prices entered a steady decline. The price cycle then turned up and from 1990 to 1997 world oil consumption increased by more than six million barrels per day. The price increases came to a rapid end when, due to the downward trend in several Asian economies, higher OPEC production went downward.

A low point was reached in January 1999 after increased oil production from Iraq coincided with the Asian financial crisis, which reduced demand. The prices then rapidly increased, more than doubling by September 2000, then fell until the end of 2001 before steadily increasing, reaching $40 to $50 per barrel by September 2004. In October 2004, the price of crude oil exceeded $53 per barrel and for December delivery exceeded $55 per barrel. Crude oil prices surged to a record high above $60 a barrel in June 2005, sustaining a rally built on strong demand for gasoline and diesel and on concerns about refiners' ability to keep up. This trend continued into early August 2005 as crude oil prices surged past $65 per barrel.

1.7.3 FUTURE OF OIL

The Hubbert theory assumes that oil reserves will not be replenished (i.e., that abiogenic replenishment is negligible) and predicts that future world oil production must inevitably reach a peak and then decline as these reserves are exhausted. Controversy surrounds the theory since as predictions for the time of the global peak is dependent on past production and discovery data used in the calculation.

For the United States, the prediction turned out to be correct and, after U.S. oil production peaked in 1971, it lost its excess production capacity, OPEC was able to manipulate oil prices. Since then oil production in several other countries has also peaked. However, for a variety of reasons, it is difficult to predict the oil peak in any given region. Based on available production data, proponents have previously (and incorrectly) predicted the peak for the world to be in 1989, 1995, or in the 1995 to 2000 period. Other predictions have chosen 2007 and beyond for the peak of oil production.

1.7.4 EPILOG

In summary, the petroleum industry is indeed at the verge of a major decision period with the onset of processing high volumes of heavy crude oil and residua. Several technology breakthroughs have made this possible, but many technical challenges remain.

But more important, several trends that should have been established in the wake of decreasing crude prices have never been put into practice. For example, and most important, the failure of politicians to recognize the need for a measure of energy independence through the development of alternate resources as well the development of technologies that would assist in maximizing oil recovery.

Some would argue that the periods of oil price decline were the impetus to start development of better technology and expertise.

REFERENCES

Abraham, H. (1945). *Asphalts and Allied Substances*. Van Nostrand, New York.
Armstrong, K. (1997). *Jerusalem: One City, Three Faiths*. Balantine Books, New York. p. 3.
ASTM (2005a). ASTM D4. Test method for bitumen content. *Annual Book of Standards*. Volume 04.04. American Society for Testing and Materials, Philadelphia, PA.

ASTM (2005b). ASTM D4175. Standard terminology relating to petroleum, petroleum products, and lubricants. *Annual Book of Standards*. Volume 05.03. American Society for Testing and Materials, Philadelphia, PA.

Bauer, G. (Georgius Agricola) (1546). Book IV, *De Natura Fossilium*, Basel, Switzerland.

Bauer, G. (Georgius Agricola) (1556). Book XII, *De Re Metallica*, Basel, Switzerland.

Bell, H.S. (1945). *American Petroleum Refining*, Van Nostrand, New York.

Bestougeff, M. (1967). In *Fundamental Aspects of Petroleum Geochemistry*. Nagy B. and Colombo U. eds. Elsevier, Amsterdam. Chapter 3.

Bland, W.F. and Davidson, R.L. (1967). *Petroleum Processing Handbook*. McGraw-Hill, New York.

Boyle, G. ed. (1996). *Renewable Energy: Power for a Sustainable Future*. Oxford University Press, Oxford, England.

BP (2002). BP Statistical Review of World Energy. British Petroleum Company, London, June.

Brooks, B.T., Dunstan, A.E., Nash, A.W., and Tizard, H. (1938). *The Science Petroleum*. Oxford University Press, Oxford, England.

Burke, J. (1996). *The Pinball Effect*. Little, Brown and Co., New York. p. 128 et seq.

Cobb, C. and Goldwhite, H. (1995). *Creations of Fire: Chemistry's Lively History from Alchemy to the Atomic Age*. Plenum Press, New York.

Colombo, U. (1967). *In Fundamental Aspects of Petroleum Geochemistry*. Nagy B. and Colombo U. eds. Elsevier, Amsterdam. Chapter 8.

Cook, A.B. and Despard, C. (1927). *J. Inst. Petroleum Technol*. 13:124.

Dahmus, J. (1995). *A History of the Middle Ages*. Barnes and Noble, New York. pp. 135 and 136; previously published with the same title by Doubleday Book Co., 1958.

Davies, N. (1996). *Europe: A History*. Oxford University Press. Oxford, England. pp. 245 and 250.

Forbes, R.J. (1958a). *A History of Technology*, Oxford University Press, Oxford, England.

Forbes, R.J. (1958b). *Studies in Early Petroleum Chemistry*. E.J. Brill, Leiden, The Netherlands.

Forbes, R.J. (1959). *More Studies in Early Petroleum Chemistry*. E.J. Brill, Leiden, The Netherlands.

Forbes, R.J. (1964). *Studies in Ancient Technology*. E.J. Brill, Leiden, The Netherlands.

Gruse, W.A. and Stevens, D.R. (1960). *The Chemical Technology of Petroleum*. McGraw-Hill, New York.

Henry, J.T. (1873). *The Early and Later History of Petroleum*. Volumes I and II. APRP Co., Philadelphia, PA.

Hoiberg, A.J. (1964). *Bituminous Materials: Asphalts, Tars, and Pitches*. John Wiley and Sons, New York.

ITAA (1936). Income Tax Assessment Act. Government of the Commonwealth of Australia.

James, P. and Thorpe, N. (1994). *Ancient Inventions*. New York: Ballantine Books.

Johnson, P. (1997). A History of the American People Harper Collins Publishers Inc., New York. p. 602.

Koots, J.A. and Speight, J.G. (1975). *Fuel* 54:179.

Kovacs, M.G. (1990). *The Epic of Gilgamesh*. Stanford University Press, Stanford. Tablet XI.

Lockhart, L. (1939). *J. Inst. Petroleum* 25:1.

Mallowan, M.E.L. (1954). *Iraq* 16(2):115.

Mallowan, M.E.L. and Rose, J.C. (1935). *Iraq* 2(1):1.

Marschner, R.F., Duffy, L.J., and Wright, H. (1978). *Paleorient* 4:97.

Mitchell, D.L. and Speight, J.G. (1973). *Fuel* 52:149.

Nellensteyn, F.J. and Brand, J. (1936). *Chem. Weekbl.* 261.

Owen, E.W. (1975). In *A History of Exploration for Petroleum*. Memoir No. 6 American Association of Petroleum Geologists, Tulsa, OK. p. 2.

Pfeiffer, J.H. (1950). *The Properties of Asphaltic Bitumen*. Elsevier, Amsterdam.

Powicke, F.M. (1962). *The Thirteenth Century: 1216–1307*. 2nd edn. Oxford University Press, Oxford, England. p. 13.

Purdy, G.A. (1957). *Petroleum: Prehistoric to Petrochemicals*. Copp Clark Publishing, Toronto.

Ramage, J. (1997). *Energy: A Guidebook*. Oxford University Press, Oxford. England.

Scheil, V. and Gauthier, A. (1909). Annales de Tukulti Ninip II, Paris.

Schroder, O. (1920). Keilschrifttexte aus Assur Vershiedenen, xiv. Leipzig.

Scouten, C.S. (1990). In *Fuel Science and Technology Handbook*. Marcel Dekker, New York.

Speight, J.G. (1978). Personal observations at archeological digs at the cities of Babylon, Calah, Nineveh, and Ur.

Speight, J.G. (1984). *Characterization of Heavy Crude Oils and Petroleum Residues*. Kaliaguine S. and Mahay A. eds. Elsevier, Amsterdam. p. 515.

Speight, J.G. (1990). *Fuel Science and Technology Handbook*. Marcel Dekker, New York.

Speight, J.G. (1994). *The Chemistry and Technology of Coal*. 2nd edn. Marcel Dekker, New York.

Speight, J.G. (2000). *The Desulfurization of Heavy Oils and Residua*. 2nd edn. Marcel Dekker, New York.

Speight, J.G. and Ozum, B. (2002). *Petroleum Refining Processes*. Marcel Dekker, New York.

Thornton, D.P. Jr (1977). In *Energy Technology Handbook*. Considine D.M., Ed. McGraw-Hill, New York. pp. 3–12.

van Nes, K. and van Westen, H.A. (1951). *Aspects of the Constitution of Mineral Oils*. Elsevier, Amsterdam.

Wollrab, V. and Streibl, M. (1969). In *Organic Geochemistry*. Eginton G., and Murphy M.T.J., eds. Springer-Verlag, New York. p. 576.

Yergin, D. (1991). *The Prize: The Epic Quest for Oil, Money, and Power*. Simon & Schuster, New York.

Yurum, Y. and Ekinci, E. (1995). In *Composition, Geochemistry and Conversion of Oil Shales*. Snape C., ed. Kluwer Academic Publishers, Dordrecht, The Netherlands. p. 329.

2 Classification

2.1 INTRODUCTION

By definition (Chapter 1), **petroleum** (also called **crude oil**) is a mixture of gaseous, liquid, and solid hydrocarbon compounds. Petroleum occurs in sedimentary rock deposits throughout the world and also contains small quantities of nitrogen-, oxygen-, and sulfur-containing compounds as well as trace amounts of metallic constituents (Bestougeff, 1967; Colombo, 1967; Hobson and Pohl, 1973; Thornton, 1977; Considine and Considine, 1984). Thus, the classification of petroleum as a hydrocarbon mixture should follow from this definition. But some clarification is required.

As already pointed out (in Chapter 1), the definition of petroleum-associated materials has been varied, unsystematic, diverse, and often archaic and it is only recently that some attempt has been made to define these materials in a meaningful manner. However, attempts to classify petroleum have also evolved and it is the purpose of this chapter to review these methods and present them for further consideration.

Petroleum itself is defined (Chapter 1) as a naturally occurring mixture of hydrocarbons, generally in a liquid state, which may also include compounds of sulfur, nitrogen, oxygen, metals, and other elements (ASTM, 2005). In more specific terms, petroleum has also been defined (ITAA, 1936) as

1. Any naturally occurring hydrocarbon, whether in a liquid, gaseous, or solid state
2. Any naturally occurring mixture of hydrocarbons, whether in a liquid, gaseous or solid state
3. Any naturally occurring mixture of one or more hydrocarbons, whether in a liquid, gaseous, or solid state and one or more of the following, that is to say, hydrogen sulfide, helium, and carbon dioxide

The definition includes any petroleum as defined by paragraph (1), (2), or (3) that has been returned to a natural reservoir.

One of the disadvantages in attempting to classify petroleum, compared to the classification of coal, is that the elemental (ultimate) composition of petroleum (Chapter 6) is not reported to the same extent as it is for coal (Lowry, 1945, 1963; Meyers, 1981; Hessley, 1990; Speight, 1994). The proportions of the elements in petroleum vary only slightly over fairly narrow limits in spite of the wide variation in physical properties from the lighter, more mobile (conventional) crude oil to bitumen at the other extreme. Thus, as a result of the wide variety of organic constituents and physical conditions that can play a role in the formation of petroleum, it is not surprising that there exists a wide variation in composition and properties resulting in a variety of attempts to classify petroleum.

It is the purpose of this chapter to review these attempts at petroleum classification and examine their evolution in terms of modern usage and to place **heavy oil** and tar sand bitumen in the correct perspective.

Like coal, petroleum exhibits a wide range of physical properties, but these properties are less well defined than they are in coal classification systems. The general classification of petroleum into conventional petroleum, heavy oil, and extra heavy oil involves not only an inspection of several properties but also some acknowledgment of the method of recovery (Chapter 5). However, there is a correlation that can be made between various physical properties. Whereas properties such as viscosity, density, boiling point, and color of petroleum may vary widely, the ultimate or elemental analysis varies, as already noted, over a narrow range for a large number of petroleum samples (Chapter 7 and Chapter 8).

In terms of the elemental composition of petroleum, the carbon content is relatively constant, whereas the hydrogen and heteroatom contents are responsible for the major differences between petroleum. The nitrogen, oxygen, and sulfur can be present in only trace amounts in some petroleum, which as a result consists primarily of hydrocarbons. On the other hand, crude oil containing 9.5% heteroatom constituents may contain essentially no true hydrocarbon constituents insofar as the constituents contain at least one or more nitrogen, oxygen, or sulfur atoms within the molecular structures.

The molecular species in petroleum vary from simple hydrocarbon species (in the majority) to very complex organic molecules containing atoms of carbon, hydrogen, nitrogen, oxygen, and sulfur (in the minority) as well as trace amounts of metals such as vanadium, nickel, iron, and copper (Chapter 7). The complexity of petroleum is further illustrated by the number of potential isomers, i.e., molecules having the same atomic formula, which can exist for a given number of paraffin carbon atoms, which increases rapidly as molecular weight increases:

Carbon Atoms per Hydrocarbon	Number of Isomers
4	2
8	18
12	355
18	60,523

This same increase in the number of isomers with molecular weight also applies to the other molecular types present. Since the molecular weight of the molecules found in petroleum can vary from that of methane (CH_4; molecular weight $= 16$) to several thousand (Chapter 7), it is clear that the heavier nonvolatile fractions can contain virtually unlimited numbers of molecules. However, in reality the number of molecules in any specified fraction is limited by the nature of the precursors of petroleum, their chemical structures, and the physical conditions that are prevalent during the maturation (conversion of the precursors) processes.

The foregoing comments lead to the conclusion that a detailed analysis may apply to a specific petroleum sample. However, the generic name petroleum describes more accurately what could be in a sample of petroleum rather than what is actually in it.

The original methods of classification arose because of commercial interest in petroleum type and were a means of providing refinery operators with a rough guide to processing conditions. It is therefore not surprising that systems based on a superficial inspection of a physical property, such as specific gravity or API (Baumé) gravity (Chapter 9), are easily applied, and are actually used to a large extent in expressing the quality of crude oils. Such a system is approximately indicative of the general character of a crude oil as long as materials of one general type are under consideration. For example, among crude oils from a particular area, an oil of 40° API (specific gravity = 0.825) is usually more valuable than one of 20° API (specific gravity = 0.934), because it contains more light fractions (e.g., gasoline) and fewer heavy, undesirable asphaltic constituents.

2.2 CLASSIFICATION SYSTEMS

2.2.1 CLASSIFICATION AS A HYDROCARBON RESOURCE

Petroleum is referred to generically as a fossil energy resource and is further classified as a **hydrocarbon resource** and, for illustrative (or comparative) purposes in this text, coal and oil shale kerogen have also been included in this classification. However, the inclusion of coal and oil shale under the broad classification of hydrocarbon resources has required (incorrectly) that the term *hydrocarbon* be expanded to include the macromolecular nonhydrocarbon heteroatomic species that constitute coal and oil shale kerogen. Use of the term *organic sediments* would be more correct (Figure 2.1). The inclusion of coal and oil shale kerogen in the category hydrocarbon resources is due to the fact that these two natural resources (coal and oil shale kerogen) will produce hydrocarbons on high-temperature processing. Therefore, if coal and oil shale kerogen are to be included in the term *hydrocarbon resources*, it is more appropriate that they be classed as hydrocarbon-producing resources under the general classification of organic sediments (Figure 2.2).

Thus, fossil energy resources divide into two classes: (1) naturally occurring hydrocarbons (petroleum, natural gas, and natural waxes), and (2) hydrocarbon sources (oil shale and coal) which may be made to generate hydrocarbons by the application of conversion processes. Both classes may aptly be described as organic sediments.

Petroleum contains high proportions of individual hydrocarbons (Bestougeff, 1967). The mineral waxes, such as **ozocerite**, can also be shown on this scheme (Figure 2.2), but because of their character (solid), fall at the lower end of the scale. This should not be construed to mean that the mineral wax is the same heteroatomic material as coal and kerogen but is a result of the physical state.

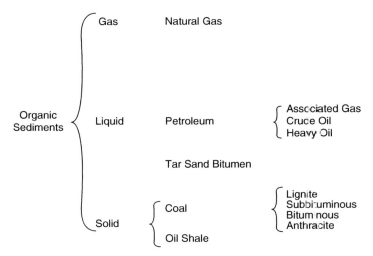

FIGURE 2.1 Subdivision of the earth's organic sediments.

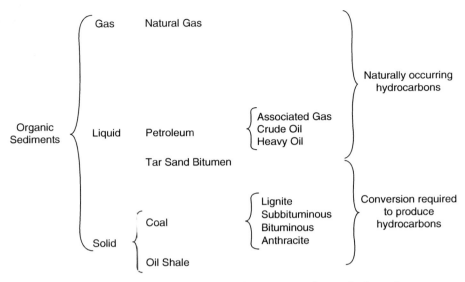

FIGURE 2.2 Classification of the earth's organic sediments according to hydrocarbon occurrence and production.

In summary, the classification of petroleum and natural gas as naturally occurring mixtures of hydrocarbons occurs by virtue of the fact that they can be separated into their original hydrocarbon constituents that have not been altered by any applied process. The hydrocarbon constituents, separated from petroleum and natural gas, are the hydrocarbon constituents that existed in the reservoir. Naturally occurring hydrocarbons are major contributors to the composition of petroleum and natural gas. Coal and kerogen do not enjoy this means of separation and methods of thermal decomposition must be applied before hydrocarbons are produced. And these hydrocarbon products, generated by the thermal process, are not naturally occurring hydrocarbons.

2.2.2 CLASSIFICATION BY CHEMICAL COMPOSITION

Composition refers to the specific mixture of chemical compounds that constitute petroleum. The composition of these materials is related to the nature and mix of the organic material that generated the hydrocarbons. Composition is also subject to the influence of natural processes such as migration (movement of oil from source rock to reservoir rock), biodegradation (alteration by the action of microbes), and water washing (effect of contact with water flowing in the subsurface) upon that composition. Thus, petroleum is the result of the metamorphosis of natural products as a result of chemical and physical changes imparted by the prevailing conditions at a particular locale.

The composition of petroleum obtained from the well is variable and depends not only on the original composition of the oil **in situ** but also on the manner of production and the stage reached in the life of the well or reservoir. In general terms, petroleum (conventional crude oil) ranges from a brownish green to black liquid having a specific gravity (at 60°F, 15.6°C) that varies from about 0.75 to 1.00 (57° to 10° API), with the specific gravity of most crude oils falling in the range 0.80 to 0.95 (45° to 17° API). The boiling range of petroleum varies from about 20°C (68°F) to above 350°C (660°F), above which active decomposition ensues when distillation is attempted. Petroleum can contain from 0% to 35% or more of gasoline, as well as varying proportions of kerosene hydrocarbons and higher boiling constituents up to the viscous and nonvolatile compounds present in lubricant oil and in asphalt.

Thus, petroleum varies in composition from one oil field to another, from one well to another in the same field, and even from one level to another in the same well. This variation can be in both molecular weight and the types of molecules present in petroleum. Petroleum may well be described as a mixture of organic molecules drawn from a wide distribution of molecular types that lie within a wide distribution of molecular weight

Petroleum is often described as a hydrocarbon resource. However, by definition, a hydrocarbon contains carbon and hydrogen only (Morrison and Boyd, 1973; Fessenden and Fessenden, 1990). On the other hand, if an organic compound contains nitrogen, or sulfur, or oxygen, or metals, it is a heteroatomic compound and not a hydrocarbon. Organic compounds containing heteroelements (elements such as nitrogen, oxygen, and sulfur), in addition to carbon and hydrogen, are defined in terms of the locations of these heteroelements within the molecule. In fact, it is to a large extent, the heteroatomic function that determines the chemical and physical reactivity of the heteroatomic compounds (Roberts and Cesario, 1964; Morrison and Boyd, 1973; Fessenden and Fessenden, 1990). In addition, the chemical and physical reactivity of the heteroatomic compounds is quite different from the chemical and physical reactivity of the hydrocarbons.

Naturally occurring hydrocarbons are those molecules, which contain carbon and hydrogen only and which occur in nature. The term *naturally occurring hydrocarbons* does not apply to those hydrocarbons that are produced by applied processes such as the thermal decomposition of coal to produce liquids and gases or the retorting (thermal decomposition) of oil shale kerogen to produce shale oil.

Petroleum is a naturally occurring hydrocarbon insofar as it contains compounds that are composed of carbon and hydrogen only (Hobson and Pohl, 1973; Considine and Considine, 1984), which do not contain any heteroatoms (nitrogen, oxygen, and sulfur as well as compounds containing metallic constituents, particularly vanadium, nickel, iron, and copper). The hydrocarbons found in petroleum are classified into the following types:

1. *Paraffins*, i.e., saturated hydrocarbons with straight or branched chains, but without any ring structure
2. *Cycloparaffins* (**naphthenes**), i.e., saturated hydrocarbons containing one or more rings, each of which may have one or more paraffin side-chains (more correctly known as **alicyclic hydrocarbons**)
3. *Aromatics*, i.e., hydrocarbons containing one or more aromatic nuclei such as benzene, naphthalene, and phenanthrene ring systems that may be linked up with (substituted) naphthalene rings or paraffin side-chains

On this basis, petroleum may have some value in the crude state but, when refined, provides fuel gas, petrochemical gas (methane, ethane, propane, and butane), transportation fuel (gasoline, diesel fuel, aviation fuel) solvents, lubricants, asphalt, and many other products.

In addition to the hydrocarbon constituents, petroleum does contain heteroatomic (nonhydrocarbon) species but they are in the minority compared with the macromolecular heteroatom-containing kerogen (Costantinides and Arich, 1967; Tissot and Welte, 1978).

A widely used classification of petroleum distinguishes between crude oils either on a paraffin base or on an asphalt base and arose because paraffin wax separates from some crude oils on cooling, but other oils show no separation of paraffin wax on cooling. The terms *paraffin base* and *asphalt base* were introduced and have remained in common use (van Nes and van Westen, 1951).

The presence of paraffin wax is usually reflected in the paraffin nature of the constituent fractions, and a high asphaltic content corresponds with the so-called naphthene properties of the fractions. As a result, the misconception that paraffin-base crude oils consist mainly of paraffins and asphalt-base crude oils mainly of cyclic (or naphthene) hydrocarbons has arisen. In addition to paraffin- and asphalt-base oils, a mixed base had to be introduced for those oils that leave a mixture of bitumen and paraffin wax as a residue by nondestructive distillation.

In practice, a distinction is often made between light and heavy crude oils (indicating the proportion of low-boiling material present), which, in combination with the preceding distinction (paraffin, asphaltic, and so on), doubles the number of possible classes.

2.2.3 Correlation Index

An early attempt to give the classification system a quantitative basis suggested that a variety of crude should be called asphaltic if the distillation residue contained less than 2% wax. A division according to the chemical composition of the 250°C to 300°C (480°C to 570°F) fraction has also been used (Table 2.1), but the difficulty that arises in using such a classification is that in the fractions boiling above 200°C (390°F) the molecules can no longer be placed in one group, because most of them are of a typically mixed nature. Purely naphthene or aromatic molecules occur very seldom; cyclic compounds generally contain paraffin side-chains and often even aromatic and naphthene rings side by side. More direct chemical information is often desirable and can be supplied by means of the correlation index (CI).

The CI, developed by the U.S. Bureau of Mines, is based on the plot of specific gravity vs. the reciprocal of the boiling point in degrees Kelvin (°K = °C + 273). For pure hydrocarbons, the line described by the constants of the individual members of the normal paraffin series is given a value of CI = 0 and a parallel line passing through the point for the values of benzene is given as CI = 100. Thus,

$$CI = 473.7d - 456.8 + 48,640/K,$$

where K for a petroleum fraction is the average boiling point determined by the standard Bureau of Mines distillation method and d is the specific gravity (Gruse and Stevens, 1960).

Values for the index between 0 and 15 indicate a predominance of paraffin hydrocarbons in the fraction. A value from 15 to 50 indicates predominance of either naphthenes or of mixtures of paraffins, naphthenes, and aromatics. An index value more than 50 indicates a predominance of aromatic species.

TABLE 2.1
Classification by Chemical Composition

Composition of 250°C–300°C (480°F–570°F) Fraction					
Paraffin (%)	Naphthene (%)	Aromatic (%)	Wax (%)	Asphalt (%)	Crude Oil Classification
>46, <61	>22, <32	>12, <25	<10	<6	Paraffin
>42, <45	>38, <39	>16, <20	<6	<6	Paraffin–naphthene
>15, <26	>61, <76	>8, <13	0	<6	Naphthene
>27, <35	>36, <47	>26, <33	<1	<10	Paraffin–naphthene–aromatic
<8	>57, <78	>20, <25	<0.5	<20	Aromatic

It is also possible to describe a crude oil by an expression of its chemical composition on the basis of the correlation index figures for its middle portions. A diagram has been constructed using this principle. The horizontal axis represents a progression from paraffin oils to aromatic oils, with 0 and 100 representing the extremes from *n*-paraffins to benzene. The position of a crude on this scale is determined by the average correlation index for its fractions boiling between 200°C (390°F) at atmospheric pressure and 275°C (525°F) at 40 mm, and the paraffin, or cyclic, nature of the fractions, and thereby the nature of the bulk of the crude, are expressed directly. The height of the vertical bar above the horizontal indicates the wax content of the heavy gas oil and light lubricating fractions of that crude, and hence the crystalline wax content of the crude can be assessed. The length of each bar below the horizontal is a measure of the carbon residue (Conradson) of the nonvolatile residuum and thus serves to indicate the asphalt content of each residuum and, thus, the content of each variety of crude. By combining these chemical indications with a qualitative expression of the results achieved by conventional refining, reasonable assessment can be obtained of petroleum in relation to both composition and application.

2.2.4 DENSITY

Density (specific gravity) has been, since the early years of the industry, the principal, and often the only specification of crude oil products and was taken as an index of the proportion of gasoline and, particularly, kerosene present. As long as only one kind of petroleum was in use, the relations were approximately true, but as crude oils having other properties were discovered and came into use, the significance of density measurements disappeared. Nevertheless, crude oils of particular types are still rated by gravity, as are gasoline and naphtha within certain limits of other properties. The use of density values has been advocated for quantitative applications using a scheme based on the American Petroleum Institute (API) gravity of the 250°C to 275°F (480°F to 525°F, 1760 mm) and the 275°C to 300°C (525°F to 570°F, 40 mm) distillation fractions (Table 2.2). Indeed, investigation of crude oils from worldwide sources showed that 85% fell into one of the three classes: paraffin, intermediate, or naphthene base.

It has also been proposed to classify heavy oils according to characterization gravity. This is defined as the arithmetic average of the instantaneous specific gravity of the distillates

---

TABLE 2.2
Classification According to API Gravity

Fraction				
250°C–270°C (480°F–520°F)		275°C–300°C (525°F–570°F)		
API Gravity	Type	API Gravity	Type	Classification
>40.0	Paraffin	>30.0	Paraffin	Paraffin
>40.0	Paraffin	20.1–29.9	Intermediate	Paraffin–intermediate
33.1–39.9	Intermediate	>30.0	Paraffin	Intermediate–paraffin
33.1–39.9	Intermediate	20.1–29.9	Intermediate	Intermediate
33.1–39.9	Intermediate	<20.0	Naphthene	Intermediate–naphthene
<33.0	Naphthene	20.1–29.9	Intermediate	Naphthene–intermediate
<33.0	Naphthene	<20.0	Naphthene	Naphthene
>44.0	Paraffin	<20.0	Naphthene	Paraffin–naphthene
33.0	Naphthene	>30.0	Paraffin	Naphthene–paraffin

boiling at 177°C (350°F), 232°C (450°F), and 288°C (550°F) vapor line temperature at 25 mm pressure in a true boiling point distillation.

In addition, a method of petroleum classification based on other properties, as well as the density, of selected fractions has been developed. The method consists of a preliminary examination of the aromatic content of the fraction boiling up to 145°C (295°F), as well as that of the asphaltene content, followed by a more detailed examination of the chemical composition of the naphtha (b.p. <200°C <390°F). For this examination, a graph is used that is a composite of curves expressing the relation between percentage distillate from the naphtha, the aniline point, refractive index, specific gravity, and the boiling point. The aniline point after acid extraction is included to estimate the paraffin–naphthene ratio.

2.2.5   API Gravity

Conventional crude oil and heavy oil have also been defined very generally in terms of physical properties. For example, heavy oils were considered those petroleum-type materials that had gravity somewhat less than 20° API, with the heavy oils falling into the API gravity range of 10° to 15° (e.g., Cold Lake crude oil = 12° API) and bitumen falling into the 5° to 10° API range (e.g., Athabasca bitumen = 8° API). Residua vary depending on the temperature at which distillation is terminated. Atmospheric residua are usually in the 10° to 15° API range of, and vacuum residua are in the range of 2° to 8° API (Speight, 1986, 2000).

There have been several recent noteworthy attempts to classify crude oil (Bestougeff et al., 1984; Danyluk et al., 1984; Gibson, 1984; Khayan, 1984) using one or more of the general physical properties of crude oils. One method uses divisions by API gravity, which is already accepted by most workers, it also uses viscosity data (Khayan, 1984). This method is essentially a more formal attempt in which specific numbers are applied without recognition of the implications of these numbers.

In the author's experience, the assignment of specific numbers to the classification of petroleum is fraught with difficulty. For example, heavy oil was considered those petroleum-type materials that had gravity somewhat less than 20° API, and generally fell into the API gravity range of 10° to 15° with tar sand bitumen falling into the 5° to 10° API range (Speight, 2000). Using such lines of demarcation does not circumvent the question that must arise when one considers a material with an API gravity equal to 9.9 and one material with an API gravity equal to 10.1. Nor does the line of demarcation make allowance for the limitations of the accuracy of the analytical method. Cleary the use of one physical parameter be it API gravity or any other physical property for that matter, is inadequate to the task of classifying conventional petroleum, heavy oil, and tar sand bitumen.

2.2.6   Viscosity

At the same time, and in concert with the use of API gravity, the line of demarcation between petroleum and heavy oil vis-à-vis tar sand bitumen has been drawn at 10,000 centipoises. Briefly, materials having viscosity less than 10,000 centipoises (cp) are conventional petroleum and heavy oil, whereas tar sand bitumen has a viscosity greater than 10,000 cp. Use of such a scale requires a fine line of demarcation between the various crude oils, heavy oils, and bitumen to the point where it would be confusing to differentiate between a material having a viscosity of 9950 cp and one having a viscosity of 10,050 cp. Further, the inaccuracies (i.e., the limits of experimental error) of the method of measuring viscosity also increase the potential for misclassification.

In the author's experience, the viscosity of tar sand bitumen is usually in excess of 50,000 cp and higher than 100,000 cp. But even using a higher line of demarcation does not circumvent the use of one physical property and the difference between a material having viscosity equal to 49,900 and 50,100 cp (or 99,900 and 100,100 cp). Cleary, the use of one physical parameter be it API gravity or viscosity is inadequate to the task of classifying conventional petroleum, heavy oil, and tar sand bitumen.

2.2.7 CARBON DISTRIBUTION

A method for the classification of crude oils can only be efficient, first, if it indicates the distribution of components according to volatility, and second, if it indicates the character-istic properties of the various distillate fractions. The distribution according to volatility has been considered the main property of petroleum, and any fractionating column with a sufficient number of theoretical plates may be used for recording a curve in which the boiling point of each fraction is plotted against the percentage by weight.

However, for the characterization of the various fractions of petroleum, the use of the $n.d.M$ method (n = refractive index, d = density, M = molecular weight; see Chapter 9) is suggested. This method enables determination of the carbon distribution and thus indicates the percentage of carbon in aromatic structure ($\%C_A$), the percentage of carbon in naphthene structure ($\%C_N$), and the percentage of carbon in paraffin structure ($\%C_P$). The yields over the various boiling ranges can also be estimated; for example, in the lubricating oil fractions the percentage of carbon in paraffin structure can be divided into two parts, giving the percentage of carbon in normal paraffins ($\%C_{nP}$) and the percentage of carbon in paraffin side-chains. The percentage of normal paraffins present in lubricating oil fractions can be calculated from the percentage of normal paraffin carbon ($\%C_{nP}$) by multiplication by a factor that depends on the hydrogen content of the fractions.

It is also possible to extrapolate the carbon distribution to the gasoline range on the one hand and to the residue on the other hand. A high value of $\%C_A$ at 500°C (930°F) boiling point usually indicates a high content of asphaltenes in the residue, whereas a high value of $\%C_{nP}$ at 500°C (930°F) boiling point usually indicates a waxy residue.

2.2.8 VISCOSITY–GRAVITY CONSTANT

This parameter, along with the Universal Oil Products characterization factor, has been used to some extent as a means of classifying crude oils. Both parameters are usually employed to give an indication of the paraffin character of the crude oil, and both have been used, if a subtle differentiation can be made, as a means of petroleum characterization rather than for petroleum classification.

Nevertheless, the viscosity–gravity constant (VGC) was one of the early indices proposed to characterize (or classify) oil types:

$$VGC = 10d - 1.0752 \log{(\nu - 38)}/(10 - \log{(\nu - 38)})$$

In this equation, d is the specific gravity 60°/60°F and ν is the Saybolt viscosity at 39°C (100°F). For heavy oil, where the low-temperature viscosity is difficult to measure, an alternative formula has been proposed in which the 99°C (210°F) Saybolt viscosity is used, viz:

$$VGC = d - 0.24 - 0.022 \log{(\nu - 35.5)}/0.755$$

The two do not agree well for low-viscosity oils. However, the viscosity–gravity constant is of particular value in indicating a predominantly paraffin or cyclic composition. The lower the index number, the more paraffin the feedstock; for example, naphthene lubricating oil distillates have VGC equal to approximately 0.876 and the raffinate obtained by solvent extraction of lubricating oil distillate has VGC equal to approximately 0.840.

2.2.9 UOP CHARACTERIZATION FACTOR

This factor is perhaps one of the more widely used derived characterization or classification factors and is defined by the formula

$$K = (T_B)^{1/3}/d$$

where T_B is the average boiling point in degrees Rankine (°F + 460) and d is the specific gravity 60°/60°F. This factor has been shown to be additive on a weight basis. It was originally devised to show the thermal cracking characteristics of heavy oils; thus, highly paraffin oils have K in the range 12.5 to 13.0 and cyclic (naphthene) oils have K in the range 10.5 to 12.5.

Finally, all the classification systems mentioned here are based on the assumption that petroleum can be more or less characterized by the properties of one or of a few fractions. However, the properties of certain fractions of a crude oil are definitely not always reflected in those of other fractions of the same oil. Indeed, some crude oils have a different chemical character in low-boiling and high-boiling fractions, and any method of classification in which the properties of a certain fraction are extrapolated to the whole crude oil must be applied with caution, as serious errors can arise.

2.2.10 RECOVERY METHOD

The generic term *heavy oil* is often applied to petroleum that has an API gravity of less than 20° and those materials having an API gravity less than 10° have been referred to as bitumen. Following this convenient generalization, there has also been an attempt to classify petroleum, heavy oil, and tar sand bitumen using a modified API gravity or viscosity scale (Figure 2.3).

In order to classify petroleum, heavy oil, and bitumen the use of a single parameter such as viscosity is not enough. Other properties such as API gravity, elemental analysis, composition, and the properties of the fluid in the reservoir as well as the method of recovery need to be acknowledged.

The United States Congress has defined tar sands as the several rock types that contain an extremely viscous hydrocarbon which is not recoverable in its natural state by conventional oil well production methods including currently used enhanced recovery techniques (US Congress, 1976). Further, a detailed definition of the bitumen contained in tar sand deposits is also necessary.

As defined earlier (Chapter 1), bitumen is a naturally occurring material that is found in deposits where the permeability is low and passage of fluids through the deposit can only be achieved by prior application of fracturing techniques. Tar sand bitumen is a high-boiling material with little, if any, material boiling below 350°C (660°F) and the boiling range

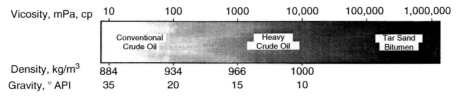

FIGURE 2.3 Classification of petroleum, heavy oil, and bitumen by API gravity and viscosity.

approximate the boiling range of an atmospheric residuum. It might also be added that the deposits in which bitumen is found require prior fracturing techniques to overcome the low permeability and allow the passage of fluids.

2.2.11 POUR POINT

Tar sand bitumen is a naturally occurring material that is immobile in the deposit and cannot be recovered by the application of enhanced oil recovery technologies, including steam-based technologies. On the other hand, heavy oil is mobile in the reservoir and can be recovered by the application of enhanced oil recovery technologies, including steam-based technologies.

Since the most significant property of tar sand bitumen is its immobility under the conditions of temperatures and pressure in the deposit, the interrelated properties of API gravity (ASTM D287) and viscosity (ASTM D445) may present an indication of the mobility of oil or immobility of bitumen, but in reality these properties only offer subjective descriptions of the oil in the reservoir. The most pertinent and objective representation of this oil or bitumen mobility is the **pour point** (ASTM D97).

By definition, the pour point is the lowest temperature at which oil will move, pour, or flow when it is chilled without disturbance under definite conditions (ASTM D97). In fact, the pour point of oil when used in conjunction with the reservoir temperature gives a better indication of the condition of the oil in the reservoir than the viscosity. Thus, the pour point and reservoir temperature present a more accurate assessment of the condition of the oil in the reservoir, being an indicator of the mobility of the oil in the reservoir. Indeed, when used in conjunction with reservoir temperature, the pour point gives an indication of the liquidity of the heavy oil or bitumen and, therefore, the ability of the heavy oil or bitumen to flow under reservoir conditions. In summary, the pour point is an important consideration because, for efficient production, additional energy must be supplied to the reservoir by a thermal process to increase the reservoir temperature beyond the pour point.

For example, Athabasca bitumen with a pour point in the range 50°C to 100°C (122°F to 212°F) and a deposit temperature of 4°C to 10°C (39°F to 50°F) is a solid or near solid in the deposit and will exhibit little or no mobility under deposit conditions. Pour points of 35°C to 60°C (95°F to 140°F) have been recorded for the bitumen in Utah with formation temperatures of the order of 10°C (50°F). This indicates that the bitumen is solid within the deposit and therefore immobile. The injection of steam to raise and maintain the reservoir temperature above the pour point of the bitumen and to enhance bitumen mobility is difficult, in some cases almost impossible. Conversely, when the reservoir temperature exceeds the pour point, the oil is fluid in the reservoir and therefore mobile. The injection of steam to raise and maintain the reservoir temperature above the pour point of the bitumen and to enhance bitumen mobility is possible and oil recovery can be achieved.

A method that uses the pour point of the oil and the reservoir temperature adds a specific qualification to the term *extremely viscous* as it occurs in the definition of tar sand. In fact, when used in conjunction with the recovery method, pour point offers more general applicability to the conditions of the oil in the reservoir or the bitumen in the deposit and comparison of the two temperatures (pour point and reservoir temperatures) shows promise and warrants further consideration.

2.3 MISCELLANEOUS SYSTEMS

Predicting the behavior of crude oil during production and refining operations requires a detailed knowledge of the structure of crude oil (Chapter 11). The classical approach of correlating properties with structural types has proved to be of some value. However, determining the structural types that will dominate the properties of crude oil is often difficult. Thus, other classification systems have evolved.

For example, crude oil can be assessed or defined by its relative distillation characteristics (Table 2.3) (Chapter 8 and Chapter 16), but the properties and character of the residuum from one specific petroleum depend on the temperature at which the distillation of the volatile constituents is terminated. Nevertheless, these materials are very difficult to classify since the constituents of heavy oils and residua fall into a boiling range in which very little is known about the individual model compounds.

It is even more confusing when an increase in boiling point is directly equated to an increase in the number of rings. Such a system often fails to acknowledge that an increase in boiling point of any ring compound can also (and more likely in the case of petroleum constituents) be equated to an increase in the number or size of the substituents on the ring (Speight, 2005).

Another classification method (Burg et al., 1997) involves the use of chromatographic data. This method evaluates the individual interactions that can take place at the molecular level to determine the strength of these interactions. The result is the classification of crude oil on the basis of polarity and the relation of polarity to behavior. Petroleum wax has also been classified using techniques such as gas chromatography, Fourier transform infrared spectroscopy, proton magnetic resonance, urea adduction, and solid liquid chromatography (Mohamed and Zaky, 2005). The multitechnique approach used for wax reflects the potential problems that can arise for classifying petroleum, especially when the complexity of petroleum vis-à-vis wax is considered. The multitechnique approach also indicates the lack of authenticity when no physical parameter is used to classify petroleum.

There have also been several attempts to classify kerogen in addition to the general type classification (Chapter 4) that already exists. However, an interesting set of criteria derived from pyrolysis–gas chromatography (Horsfield, 1984) has shown that hydrocarbon

TABLE 2.3
Properties of Tia Juana Crude Oil and Its 650°F, 950°F, and 1050°F Residua

		Residua		
	Whole Crude	**650°F⁺**	**950°F⁺**	**1050°F⁺**
Yield, vol.%	100.0	48.9	23.8	17.9
Sulfur, wt.%	1.08	1.78	2.35	2.59
Nitrogen, wt.%		0.33	0.52	0.60
API gravity	31.6	17.3	9.9	7.1
Carbon residue, wt.%				
Conradson		9.3	17.2	21.6
Metals				
Vanadium, ppm		185		450
Nickel, ppm		25		64
Viscosity				
Kinematic				
At 100°F	10.2	890		
At 210°F		35	1010	7959
Furol				
At 122°F		172		
At 210°F			484	3760
Pour point, °F	−5	45	95	120

generation can be quantified. It is also interesting to note from this work that hydrogen deficiency in the original kerogen is not directly equated to a low oil-producing propensity. This, and other work, will not only assist in more accurate descriptions (classification) of kerogen but also offers the potential for a more detailed description of organic facies).

In summary, any attempt to classify petroleum, heavy oil, and bitumen on the basis of a single property is no longer sufficient to define the nature and properties of petroleum and petroleum-related materials (Yen, 1992), perhaps being an exercise in futility. Such an attempt would be analogous to classifying the human race on the basis of the physical characteristics of a sumo wrestler, thereby omitting many other anthropological characteristics.

2.4 RESERVOIR CLASSIFICATION

A conventional petroleum reservoir is a dome or vault of impermeable rock, formed by the folding or faulting of the rock layers or by the rise of salt domes, with permeable rocks beneath it. In those permeable rocks there may be a layer of natural gas on top, with petroleum below. Beneath the oil is usually a layer of rock soaked with water or brine. However, not all reservoirs have the above characteristics and there are also unconventional or continuous reservoirs in which the oil is not trapped as described earlier. Therefore, reservoir characterization is an essential part of petroleum technology and offers an understanding or indication of the applicable method of recovery.

2.4.1 IDENTIFICATION AND QUANTIFICATION

Petroleum reservoir characterization is the process of identifying and quantifying those properties of a given petroleum reservoir which affect the distribution and migration of fluids within that reservoir. These aspects are controlled by the geological history of the reservoir. Further, the ultimate goal of a hydrocarbon reservoir characterization study is the development of a reasonable physical description of a given reservoir. This physical description can then be used as a basis for simulation studies, which, in turn, are used to assess the effectiveness of various recovery strategies. An accurate physical description of the reservoir will often lead to the maximum production of hydrocarbons from the reservoir.

The type of reservoir from which it originates and the method of recovery appear to be the most appropriate method for classifying petroleum, heavy oil, and tar sand bitumen. Thus, a general classification scheme has been developed for petroleum reservoirs contained in U.S. Department of Energy's tertiary-oil-recovery information system (Toris). This resulted in the classification and description of 2,300 light-oil reservoirs (greater than 20° API), collectively containing 308 billion bbl of **original oil-in-place** (OOIP) (Ray et al., 1991).

The reservoirs fell within 173 combinations defined by their lithology, depositional environment, structural deformation, and diagenetic overprint. These groups were then collapsed into a smaller number of classes while maintaining meaningful descriptions of the processes that produce reservoir heterogeneity at the interwell scale. Using a basic classification system relying on lithology and depositional environment, with subclasses reflecting postdeposition processes, would suffice. As a result, 22 geologic classes, 16 clastic, and 6 carbonate, with structural subclasses for clastic and diagenetic subclasses for carbonates, were defined.

Sixteen clastic classes were derived from 28 clastic depositional systems described in the classifier. These classes contain silici-clastic rocks deposited in the paleoenvironmental setting indicated by their names. Although some are fairly uniform environments, such as those of the eolian class, others are complex, as in the various deltaic environments.

For most reservoirs, relatively refined description of the depositional processes was possible (e.g., fluvial-dominate deltas), but for some, the unavailability of data or the complexity of the depositional processes required broader, undifferentiated classes (e.g., fluvial, strand plain, delta).

Heterogeneity due to the postdeposition structural and diagenetic history of a reservoir can have an overriding influence on the flow of oil and other fluids. These descriptive reservoir modifiers provided the basis for defining subclasses.

In clastic reservoirs, variations in types of structural controls on heterogeneity relative to diagenesis suggested that clastic reservoirs could be classified by combining depositional environment with structural, rather that diagenetic, elements. Compaction and cementation was indicated as the principal diagenetic event in 89% of clastic reservoirs analyzed. Structural modifiers include fracturing, faulting, and folding, all of which can greatly affect reservoir heterogeneity. The term *structured* was adopted to describe interwell heterogeneity that result from these structural overprints on the reservoir. In combination, one-third of the clastic reservoirs have some sort of structural overprint. The resulting structured and unstructured subclassifications were useful in describing this lithology.

Six carbonate classes were derived from 20 individually described carbonate depositional systems defined in the classifier. These carbonate reservoirs classified were deposited in marine or near-marine settings. The class names are descriptive of the location or conditions under which deposition occurred. However, the shallow shelf/restricted carbonate class contains reservoir rock deposited in the near-shore subtidal as well as the shallow shelf environment. In carbonate reservoirs, diagenetic factors have significant effects on heterogeneity. Therefore, diagenetic factors were used as the basis of subclasses. Structural features of the described carbonate reservoirs were not found to vary substantially (88% of the carbonate reservoirs are described as unstructured). Therefore, structural features were not used to define carbonate subclasses.

The occurrence and variability of the diagenetic descriptors, however, justified the differentiation of carbonated reservoirs into three carbonate diagenetic subclasses. These subclasses are dolomitization, massive dissolution, and the other. The subclass other combines compaction/cementation, grain enhancement, and silicification. Five of the six carbonate classes are divided into three subclasses. The sixth, slope-basin, contains only reservoirs described by the other diagenetic processes, the single subclass in the class.

The grouping of the reservoirs into classes created a smaller number of research targets, yet the distinctness of the reservoir is preserved. The results of the classification effort present a focus for specific studies.

Reservoirs within these classes are expected to manifest distinct types of reservoir heterogeneity as a consequence of their similar lithology and depositional histories. The creation of subclasses assists in the analysis of the impact of postdeposition events on reservoir heterogeneity.

2.4.2 FUTURE

Future energy resources will be found in what are currently considered to be unconventional reservoirs, especially low-permeability reservoirs in shale, siltstone, fine-grained sand, and carbonates. However, there is inadequate geologic data to evaluate the contribution that such reservoirs will make to petroleum reserves in the future.

These unconventional resources are probably very large, but their characteristics and distribution are not yet well understood. Further research on modeling the geometry and distribution of porosity and permeability, as well as determining the chemical and physical sensitivities of hydrocarbon reservoirs is necessary. An important focus is the low-permeability (tight) gas reservoirs, heavy oil reservoirs, and tar sand deposits. In addition, understanding deep formations and reservoirs (>15,000 ft; >4572 m) is essential if deep-lying

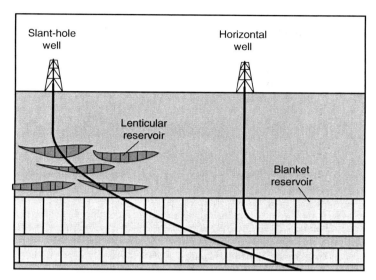

FIGURE 2.4 Representation of deviated well drilling.

resources are to be recovered. In fact, detailed estimates of the amount of petroleum and natural oil and gas recoverable under varying scenarios of economics and technology are needed to guide long-range energy policies. The resources that are currently termed unconventional resources will play a critical role in the energy base for the remainder of this century.

One aspect of resource recovery from difficult formations (reservoirs) that is having some success is deviated well drilling (Figure 2.4). Deviated well drilling involves drilling horizontal and slant-hole wells to better intersect vertical fractures in tight formations (http://energy.usgs.gov/factsheets/Petroleum/drilling.html). Natural vertical fractures are important factors in the economic production of gas from these rocks because the permeability of the natural fractures is almost always much higher than the nonfractured rock. However, most of the gas resources reside in the rock pores and move out of the rock to the wellbore via fractures.

REFERENCES

ASTM 2005. Standard ASTM D4175. Standard terminology relating to petroleum, petroleum products, and lubricants. *Annual Book of Standards*, Volume 05.03. American Society for Testing and Materials, Philadelphia, PA.

Bestougeff, M. 1967. In *Fundamental Aspects of Petroleum Geochemistry* Nagy B. and Colombo, U. eds. Elsevier, Amsterdam. Chapter 3.

Bestougeff, M.A., Burollet, P.F., and Byramjee, R.J. 1984. *The Future of Heavy Crude and Tar Sands*. Meyer R.F., Wynn J.C., and Olson J.C., eds. McGraw-Hill, New York. p. 12.

Burg, P., Selves, J.L., and Colin, J.P. 1997. *Fuel* 76: 85.

Colombo, U. 1967. In *Fundamental Aspects of Petroleum Geochemistry*. Nagy B. and Colombo U., eds. Elsevier, Amsterdam. Chapter 8.

Considine, D.M. and Considine, G.D. 1984. *Encyclopedia of Chemistry*, 4th edn. Van Nostrand Reinhold Co., New York.

Costantinides, G. and Arich, G. 1967. In *Fundamental Aspects of Petroleum Geochemistry*. Nagy B. and Colombo U., eds. Elsevier, Amsterdam. Chapter 4.

Danyluk, M.D., Galbraidt, B.E., and Omana, R.A. 1984. *The Future of Heavy Crude and Tar Sands*. Meyer R.F., Wynn J.C., and Olson J.C., eds. McGraw-Hill, New York. p. 3.

Fessenden, R.J. and Fessenden, J.S. 1990. *Organic Chemistry*, 4th ed. Brooks/Cole Publishing Co., Pacific Grove, CA.

Gibson, B.J. 1984. *The Future of Heavy Crude and Tar Sands*. (Meyer R.F., Wynn J.C., and Olson J.C., eds.). McGraw-Hill, New York. p. 17.

Gruse, W.A. and Stevens, D.R. 1960. *The Chemical Technology of Petroleum*. McGraw-Hill, New York.

Hessley, R.A. 1990. In *Fuel Science and Technology Handbook*. Speight J.G., ed. Marcel Dekker, New York.

Hobson, G.D. and Pohl, W. 1973. *Modern Petroleum Technology*, 4th edn. Applied Science Publishers, Barking, England.

Horsfield, B. 1984. *Adv. Geochem.* 1: 247.

ITAA 1936. *Income Tax Assessment Act*. Government of the Commonwealth of Australia.

Khayan, M. 1984. *The Future of Heavy Crude and Tar Sands*. Meyer R.F., Wynn J.C., and Olson J.C., eds. McGraw-Hill, New York. p. 7.

Lowry, H.H. ed. 1945. *Chemistry of Coal Utilization*, Volumes I and II. Lowry H.H., ed. John Wiley & Sons, New York.

Lowry, H.H. ed. 1963. *Chemistry of Coal Utilization*, Supplementary Volume. Lowry H.H., ed. John Wiley & Sons, New York.

Meyers, R.A. 1981. In *Coal Handbook* Meyers R.A. ed. Marcel Dekker, New York. p. 7 et seq.

Mohamed, N.H. and Zaky, M.T. 2005. *Petro. Sci. Technol.* 23: 483–493.

Morrison, R.T. and Boyd, R.N. 1973. *Organic Chemistry*, 3rd edn. Allyn and Bacon, Boston, MA. p. 40.

Ray, R.M., Brashear, J.P., and Biglarbigi, K. 1991. *Oil Gas J.* September: 89.

Speight, J.G. 1986. *Ann. Rev. Energy* 11: 253.

Speight, J.G. 1994. *The Chemistry and Technology of Coal*, 2nd ed. Marcel Dekker, New York.

Speight, J.G. 2000. *The Desulfurization of Heavy Oils and Residua*, 2nd ed. Marcel Dekker, New York.

Speight, J.G. ed. 2005. *Lange's Handbook of Chemistry*, 16th edn. McGraw-Hill, New York.

Thornton, D.P., Jr. 1977. In *Energy Technology Handbook* Considine D.M., ed. McGraw-Hill, New York. pp. 3–12.

Tissot, B.P. and Welte, D.H. 1978. *Petroleum Formation and Occurrence*. Springer-Verlag, New York.

ongress U. 1976. *Public Law FEA-76-4*. United States Congress, Washington, DC.

van Nes, K. and van Westen, H.A. 1951. *Aspects of the Constitution of Mineral Oils*. Elsevier, Amsterdam.

Yen, T.F. 1992. In *The Second International Conference on the Future of Heavy Crude and Tar Sands*. Meyer R.F., Wynn J.C., and Olson J.C., eds. McGraw-Hill, New York. p. 412.

3 Origin and Occurrence

3.1 INTRODUCTION

Petroleum is by far the most commonly used source of energy, especially as the source of liquid fuels (Figure 3.1), and its use is projected to continue at least at the current levels for at least two decades (Figure 3.2). However, there is petroleum and there is petroleum and all are not equal. In recent years, the average quality of crude oil has worsened. This is reflected in a progressive decrease in API gravity (i.e., increase in density) and a rise in sulfur content (Swain, 1991, 1993, 1997, 2000). However, it is now believed that there has been a recent tendency for the quality of crude oil feedstock to stabilize. Be that as it may, the nature of crude oil refining has changed considerably.

The declining **reserves** of light crude oil have resulted in an increasing need to develop options to upgrade the abundant supply of known **heavy oil** reserves. In addition, there is considerable focus and renewed efforts on adapting recovery techniques to the production of heavy oil and tar sand bitumen (Chapter 6).

Fossil fuels are those fuels, namely coal, petroleum (including heavy oil and **bitumen**), **natural gas**, and oil shale produced by the decay of plant remains over geological time (Speight, 1990). They are carbon-based and represent a vast source of energy. Resources such as heavy oil and bitumen in tar sand formations are also discussed in this text. These represent an unrealized potential, with liquid fuels from petroleum forming only a fraction of those that could ultimately be produced from heavy oil and tar sand bitumen.

In fact, at the present time, most of the energy consumed by humans is produced from fossil fuels (petroleum, ca. 38% to 40%; coal, ca. 31% to 35%; natural gas, ca. 20% to 25%) with the remainder of the energy requirements to be met by nuclear and hydroelectric sources. The nuclear power industry (Zebroski and Levenson, 1976; Rahn, 1987) is truly an industry where the future is uncertain or, at least, at the crossroads. As a result, fossil fuels (in varying amounts depending on the source of information) are projected to be the major sources of energy for the next 50 years.

Petroleum is scattered throughout the earth's crust, which is divided into natural groups or strata, categorized in order of their antiquity (Table 3.1). These divisions are recognized by the distinctive systems of organic debris (as well as fossils, minerals, and other characteristics) that form a chronological time chart that indicates the relative ages of the earth's strata. It is generally acknowledged that carbonaceous materials such as petroleum occur in all these geological strata from the pre-Cambrian to the recent, and the origin of petroleum within these formations is a question that remains open to conjecture and the basis for much research.

Indeed, petroleum technology, in one form or another, is with us until suitable alternative forms of energy are readily available. Therefore, a thorough understanding of the benefits and limitations of petroleum recovery and processing is necessary and will be introduced within the pages of this book.

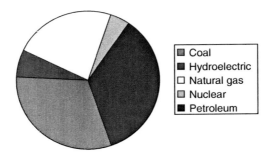

FIGURE 3.1 Distribution of world energy resources.

3.2 ORIGIN

There are two theories on the origin of carbon fuels: the abiogenic theory and the biogenic theory (Kenney et al., 2001, 2002). The two theories have been intensely debated since the 1860s, shortly after the discovery of widespread occurrence of petroleum. It is not the intent of this section to sway the readers in their views of the origin of petroleum and natural gas. The intent is to place before the reader both points of view from which the reader can do further research and decide.

3.2.1 ABIOGENIC ORIGIN

There have been several attempts at formulating theories that describe the detail of the origin of petroleum, of which the early postulates started with inorganic substances as source material. For example, in 1866, Berthelot considered acetylene the basic material and crude oil constituents were produced from the acetylene. Initially, inorganic carbides were formed by the action of alkali metals on carbonates after which acetylene was produced by the reaction of the carbides with water.

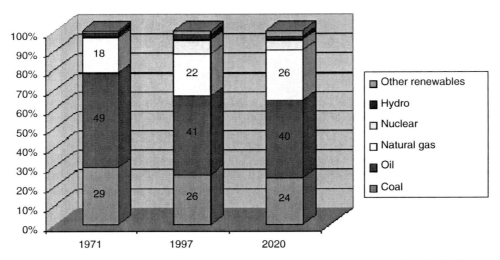

FIGURE 3.2 Trends and the projected trend for the use of fossil fuel resources and other fuel resources until 2020.

TABLE 3.1
Geological Timescale

Era	Period	Epoch	Approximate Duration (Millions of Years)	Approximate No. of Years Ago (Millions of Years)
Cenozoic	Quaternary	Holocene	10,000 years ago to the present	
		Pleistocene	2	0.01
	Tertiary	Pliocene	11	2
		Miocene	12	13
		Oligocene	11	25
		Eocene	22	36
		Paleocene	71	58
Mesozoic	Cretaceous		71	65
	Jurassic		54	136
	Triassic		35	190
Paleozoic	Permian		55	225
	Carboniferous		65	280
	Devonian		60	345
	Silurian		20	405
	Ordovician		75	425
	Cambrian		100	500
Precambrian			3380	600

Source: World Energy Outlook 2005, *International Energy Agency*, Washington DC.

$$CaCO_3 + \text{alkali metal} \rightarrow CaC_2 \text{ (calcium carbide)}$$

$$CaC_2 + H_2O \rightarrow HC\equiv CH \text{ (acetylene)} \rightarrow \text{petroleum}$$

Mendelejeff, who proposed that the action of dilute acids or hot water on mixed iron and manganese carbides produces a mixture of hydrocarbons from which petroleum evolved, described another theory in which acetylene is considered to be the basic material:

$$Fe_3C + H_2O + H^+ \rightarrow \text{hydrocarbons} \rightarrow \text{petroleum}$$

$$Mn_3C + H_2O + H^+ \rightarrow \text{hydrocarbons} \rightarrow \text{petroleum}$$

There are also several recent theories related to the formation of petroleum from nonbiogenic sources in the earth (Gold and Soter, 1980, 1982, 1986; Gold, 1984, 1985; Osborne, 1986; Szatmari, 1989).

Thus, the idea of abiogenic petroleum origin proposes that large amounts of carbon exist naturally in the planet, some in the form of hydrocarbons. **Hydrocarbons** are less dense than aqueous pore fluids, and migrate upward through deep fracture networks. Thermophilic, rock-dwelling microbial life-forms are in part responsible for the biomarkers found in petroleum. However, their role in the formation, alteration, or contamination of the various hydrocarbon deposits is not yet understood. Thermodynamic calculations and experimental studies confirm that *n*-alkanes (common petroleum components) do not spontaneously evolve from methane at pressures typically found in sedimentary basins, and so the theory of an abiogenic origin of hydrocarbons suggests deep generation (below 200 km).

However, it is now generally accepted, but not conclusively proven, that petroleum formation predominantly arises from the decay of organic matter in the earth. It is therefore from this scientific aspect that petroleum formation is referenced in this text. Nevertheless, alternative theories should not be dismissed until it can be conclusively established that petroleum formation is due to one particular aspect of geochemistry.

From the chemical point of view the inorganic theories are interesting because of their historical importance, but these theories have not received much attention. Geological and chemical methods have demonstrated the optical activity of petroleum constituents, the presence of thermally labile organic compounds, and the almost exclusive occurrence of oil in sedimentary rocks. Other theories have attempted to correlate the occurrence of coal strata and oil in the earth's crust with coal being the precursor to crude oil or both fossil fuels are formed from the same precursors at the same time but are the result of divergent paths. Chemical and geological investigations do not support this concept and the idea is considered obsolete.

3.2.2 BIOGENIC ORIGIN

Petroleum is a naturally occurring hydrocarbon mixture but hydrocarbons that are synthesized by living organisms usually account for less than 20% of the petroleum (Hunt, 1996). The remainder of the hydrocarbons in petroleum is produced by a variety of processes that convert other organic material to hydrocarbons as part of the maturation processes generally referred to as **diagenesis**, **catagenesis**, and **metagenesis**. These three processes are a combination of bacteriological action and low-temperature reactions that convert the source material into petroleum. During these processes, migration of the liquid products from the source sediment to the reservoir rock may also occur.

Most geologists view crude oil and natural gas as the products of compression and heating of ancient vegetation over geological time scales. According to this theory, it is formed from the decayed remains of prehistoric marine animals and terrestrial plants. Over many centuries this organic matter, mixed with mud, is buried under thick sedimentary layers of material. The resulting high levels of heat and pressure cause the remains to metamorphose, first into a waxy material known as kerogen, and then into liquid and gaseous hydrocarbons in a process known as catagenesis. These then migrate through adjacent rock layers until they become trapped underground in porous rocks called reservoirs, forming an oil field, from which the liquid can be extracted by drilling and pumping.

These reactions are thought to be very temperature sensitive: reactions that produce recognizable oil commence at about 130°C (266°F), and those that continue the breakdown of oil to natural gas commence at about 180°C (356°F). The range of 130°C to 150°C (266°F to 302°F) is generally considered the oil window. Though this corresponds to different depths for different locations around the world, a typical depth for an oil window might be 13,125 to 16,400 ft (4000 to 5000 m). Three conditions must be present for oil reservoirs to form: a rich source rock, a migration conduit, and a trap (seal) that forms the reservoir.

The reactions that produce oil and natural gas are often modeled as first-order breakdown reactions, where kerogen breaks down to oil and natural gas by a large set of parallel reactions, and oil eventually breaks down to natural gas by another set of reactions.

During the past 600 million years, incompletely decayed plant and animal remains have become buried under thick layers of rock. It is believed that petroleum consists of the remains of these organisms but it is the small microscopic plankton organism remains that are largely responsible for the relatively high organic carbon content of fine-grained sediments like shale that are believed to be the principle source rocks for petroleum.

Chemically, it is generally proposed that petroleum is formed through the progressive chemical change of materials provided by microscopic aquatic organisms that were incorporated over eons in marine or near-marine sedimentary rocks. In fact, the details of petroleum genesis (diagenesis, catagenesis, and metagenesis) have long been a topic of interest. However, the details of this transformation and the mechanism by which petroleum is expelled from the source sediment and accumulates in the reservoir rock are still uncertain.

Transformation of some of this sedimentary material to petroleum probably began soon after deposition, with bacteria playing a role in the initial stages and clay particles serving as catalysts. Heat within the strata may have provided energy for the reaction, temperatures increasing more or less directly with depth. Some evidence indicates that most petroleum has formed at temperatures not exceeding about 100°C to 120°C (210°F to 250°F), with the generation of petroleum hydrocarbons beginning as low as 65°C (150°F).

The formation of petroleum hydrocarbons by long-term thermal reactions was advocated at an early date. In 1888, Engler demonstrated that pressure distillation of fats yields, an oil product with high olefin content. This highly unsaturated material (**protopetroleum**) was believed to have been formed from the fatty components of the organic debris in sediments by mild thermal cracking and **polymerization**. Paraffin hydrocarbons were assumed to be formed by decarboxylation of fatty acids and the olefins underwent isomerization to cyclic hydrocarbons.

Early theories of petroleum formation such as this are no longer valid, but there appears to be no reason that low-temperature decarboxylation, deamination, **cyclization**, hydrogenation, isomerization, or other types of reactions might not proceed under the conditions known to exist in nature. Indeed, at relatively low temperature (200°C to 250°C, 390°F to 480°F) polyene structures (i.e., –C=C–C=C–C=C– etc.) can be cyclized to produce aromatic hydrocarbons. For example, m-xylene has been found in the pyrolysis products of bixin and capsanthin, and other carotenoids have yielded not only m-xylene but also toluene and 2,6-dimethylnaphthalene.

Following from this, in 1911 Engler was the first author to postulate that an organic substance other than coal was the source material of petroleum. He invoked the concept of three separate development stages. In the first stage, animal and vegetable deposits accumulate at the bottom of inland seas (lagoon conditions) and are then decomposed by bacteria; the carbohydrates and the bulk of the protein are converted into water-soluble material or gases and thus removed from the site. The fats, waxes, and other fat-soluble and stable materials (rosins, cholesterol, and others) remain.

In the second stage, high temperatures and pressures cause carbon dioxide to evolve from compounds containing a carboxyl group, and water is produced from the hydroxy acids and alcohols to leave a **bituminous** residue. Continued application of the heat and pressure causes light cracking, producing a liquid product with high olefin content (protopetroleum). Engler also produced experimental evidence which showed that distillation of fats under pressure brought about the formation of a petroleum type of material, and he assumed that time and high pressure offset the fact that the temperature in oil source rocks is lower than that used experimentally.

In the third stage, the unsaturated components of the protopetroleum are polymerized under the influence of contact catalysts and thus the polyolefins are converted into paraffins and/or cycloparaffins (naphthenes). Aromatics were presumed to be formed either directly during cracking, by cyclization through condensation reactions, or even during the decomposition of protein. It was also proposed that grahamite and gilsonite (see Chapter 1) were formed from petroleum by polymerization and oxidation reactions. The essential elements of this theory have survived. However, the main objection to a theory of this type is that the high-temperature sequence and the composition of the end product obtained in these experiments (paraffins and unsaturated hydrocarbons) differ essentially from that of petroleum, which consists chiefly of paraffins, cycloparaffins (naphthenes), and aromatics.

Thus, it is generally believed that the generation of petroleum is associated with the deposition of organic detritus. Detritus deposition occurs during the development of, fine-grained sedimentary rocks that occur in marine, near-marine, or even nonmarine environments (Vassoyevitch et al., 1969; Hood et al., 1975; Tissot and Welte, 1978; Brooks and Welte, 1984; Bjoroy et al., 1987). Petroleum is believed to be the product arising from the decay of plant and animal debris that was incorporated into sediments at the time of deposition. However, the details of this transformation and the mechanism by which petroleum is expelled from the source sediment and accumulates in the reservoir rock are still uncertain but progress has been made (Nagy and Colombo, 1967; Hobson and Tiratsoo, 1975; Hunt, 1996).

Nevertheless, the composition of petroleum is greatly influenced not only by the nature of the precursors that eventually form petroleum but also by the relative amounts of these precursors (that are dependent on the local flora and fauna) that occur in the source material. Hence, it is not surprising that petroleum composition can vary with the location and age of the field in addition to any variations that occur with the depth of the individual well. Two adjacent wells are more than likely to produce petroleum with very different characteristics.

3.2.2.1 Deposition of Organic Matter

Only a minute fraction of the carbon cycling through the biosphere becomes incorporated into sediments and of this only a small part is preserved to become fossil fuels. The actual mechanism by which this is achieved in aqueous environments involves photosynthesis by (1) phytoplankton, in cooler waters, and (2) phytobacteria, in warmer waters. The photosynthetic process is ultimately responsible for producing nearly all the organic matter in the oceans (Sorokin, 1971; Menzel, 1974).

The relative importance of different life forms to primary productivity in the aquasphere (water systems) through geological time has been assessed (Tappan and Loeblich, 1970), but it is difficult to determine any precise quantitative relationship between the fossil record and organic productivity. Many productive species have no hard body parts that are easily preserved, and hence their productivity may not be reflected in the surviving fossils. The fossil record indicates differences between production in marine and lacustrine environments but it is difficult to deny the possibility of petroleum formation in nonmarine (lacustrine) situations.

Photosynthetic activity is controlled by light, temperature, and the availability of nutrients. Light does not seem to be the major controlling factor, except in deep water, polar regions, or turbid coastal regions (Menzel, 1974). The availability of carbon dioxide, which is essential for photosynthesis, is not limiting in the eutrophic zone, the top 200 m, where productivity is highest.

In fact, nutrient availability (often nitrogen as ammonium, NH_4^+, or as nitrate, NO_3^-) in marine environments or as phosphate (PO_3^{2-}) in lacustrine environments seems to be the most common limit on photosynthetic productivity (Strickland, 1965). Productivity in lakes follows the same general pattern as in oceans, but deposition in the confines of a lake is much more subject to local climatic variations and periodic changes in input (e.g., volcanic ash) than is marine deposition.

3.2.2.2 Establishment of Source Beds

Flowing water can erode particles from rocks and transport them to sites where the current is less strong. The water also contains suspended organic matter, which will settle with the minerals and hence serve as the oil-generating substance. There are arguments in favor of marine plankton as the prominent source material, but local conditions may favor the accumulation of other organisms, such as marine algae, the remains of larger marine animals, or even material from terrestrial sources.

Both the nature and the quantity of the organic and inorganic settling matter may vary. As a consequence, it must be expected that the characteristics of the crude oils found in different reservoirs will vary. The nature of crude oil is related not only to the nature of the source material but also to the relative amounts of the different constituents in the source material. In addition, the temperature, pressure, and any other physical conditions prevailing in the area will also affect the character of the petroleum.

For example, high-wax crude oils are commonly associated with source material containing high proportions of lipids of terrestrial higher plants and of microbial organisms. On the other hand, high-sulfur crude oils are frequently related to carbonate-type source rocks. In addition, alteration of the petroleum in the reservoir may be equally responsible for variations in composition. In short, crude oil alteration will very likely make the original character of the oil difficult to assess, thereby also masking the true nature of the precursors.

The debris that settles at the bottom of the sea is attacked by bottom-dwelling (benthic) organisms that feed both on the sediment and on any water-soluble material. The benthic bacteria are largely responsible for the decomposition of organic compounds and the synthesis of new organic compounds through enzymatic transformations. The remaining material is partly transformed by bacteria and buried under the steadily increasing cover of sediments.

There are, indeed, indications that the oxygen content of the organic matter decreases during these bacterial conversions and the chemical composition of the source material becomes more oriented toward a petroleum type of material. However, the number of hydrocarbon constituents is not as great as in the ultimate petroleum, and the molecular weight of the components are somewhat higher because it is presumed that the organic matter must be present in the solid or semisolid state to be retained in the soft mud.

$$CH_3(CH_2)_nCH_2CO_2H \rightarrow CH_3(CH_2)_nCH_3 SO_4^{2-} \rightarrow S^{2-} \rightarrow H_2S$$

The minimum percentage of organic matter required for a sediment to serve as source bed is uncertain, but it is assumed, of course, that sediments with as high an organic content as possible are preferred. The ultimate percentage of organic matter in the mud at the bottom of the sea depends not only on the amount of settling mineral matter and the number of dead organisms in the water but also on the type of bacterial decomposition. Whether or not the surroundings are anaerobic (i.e., lacking oxygen) also affects the amount of organic material in the bottom mud. Aerobic bacteria do their destructive work much more rapidly and thoroughly than their anaerobic counterparts, and hence in an anaerobic environment the organic content of the sediment is far higher than under aerobic conditions. It is not surprising, therefore, that the so-called marine sapropel, originating as a black mud with a high percentage of organic matter in an anaerobic environment, is considered to be the preferred type of source bed. However, it is by no means certain that sediments with relatively low organic matter content ($\leq 10\%$) cannot serve as source beds.

As the pressure in the sediment increases, the water content diminishes from 70% to 80% w/w to less than 10% w/w, depending on the depth and the type of sediment. At this time, there is also the onset of anaerobic bacterial decomposition, which usually continues for a considerable length of time, during which biochemical transformations occur (Eglinton and Murphy, 1969). It has been estimated (Moore, 1969) that 60% to 70% w/w of the sedimentary organic carbon is typically liberated as carbon dioxide during this process; the majority of the remaining organic carbon is converted into new products, and the end result is a very complex mixture.

One of the major issues that relate to petroleum formation is the identification of the source beds because there is considerable uncertainty about the organic content of the source beds and the relation of this material to petroleum. There is some doubt about the events occurring in the source bed after the formation of the protopetroleum and the events immediately preceding and immediately after the onset of migration of the material to the reservoir rock.

The role of the evolving protopetroleum in the source bed must also be questioned. If the system evolves as a complete entity until the petroleum is virtually formed, any action that causes the removal of the protopetroleum may cause the onset of a different type of chemistry in the bed. The chemical alteration of the young petroleum (the protopetroleum) during secondary migration and after entrapment in the reservoir rock adds further complexity to the issue of source bed identification and the behavior of the precursors in the source bed.

Kerogen is a known thermal source of hydrocarbons but evidence for the production of hydrocarbons from kerogen under conditions that might approximate those in source beds is not necessarily forthcoming. If the source beds are fairly shallow, i.e., <20,000 ft deep in the earth, temperatures of 200°C (390°F) or less are likely to be operative. Production of hydrocarbons from kerogen at elevated temperatures (>300°C, >570°F) acknowledges that these higher temperatures will increase the rate of reaction, but fails to acknowledge that the higher temperatures has a great likelihood of changing the chemistry of the reaction.

3.2.2.3 Nature of the Source Material

The organic compounds that constitute petroleum are generated from finely disseminated organic matter in source beds; thus the first appearance of petroleum is also in a dispersed form. Processes of primary and especially secondary migration, which finally lead to the formation of petroleum accumulations, must therefore encompass mechanisms that concentrate dispersed petroleum compounds.

Hydrocarbons have contributed to the formation of crude oil to a limited extent as they occur in living plants in concentrations up to several tenths percent on a dry weight basis (Henderson et al., 1968; Atkinson and Zuckerman, 1981; Hunt, 1996). Paraffinic and naphthenic hydrocarbon components undoubtedly survive after burial to become ultimately incorporated in any crude oil formed. The polyenes, which comprise the terpene hydrocarbons and the carotenoid pigments (Tappan and Loeblich, 1970), also survive for a time. It has been demonstrated (Erdman, 1965) that thermal alteration of a polyterpene (β-carotene) and an amino acid (phenylalanine) can be natural precursors to the alkylaromatic (i.e., alkylbenzenes and alkylnaphthalenes) constituents of petroleum.

Fatty acids appear to be a ready source of petroleum constituents because they are converted to hydrocarbons by the simple elimination of carbon dioxide (Warren and Storrer, 1895):

$$RCO_2H \rightarrow RH + CO_2$$

In fact, fatty acids have been the object of many experiments, and thermal decarboxylation is a proven mechanism for hydrocarbon production (Bogomolov et al., 1960, 1961, 1963; Shimoyama and Johns, 1971, 1972; Almon, 1974). Degradation of other naturally occurring oxygen-containing species, such as alcohols (Sakikana, 1951), esters, ketones (Frost, 1940; Demorest et al., 1951), and aldehydes (Levi and Nicholls, 1958), has also been examined as a means of hydrocarbon generation.

The natural fatty acids are largely of the straight-chain type; many are unsaturated and exhibit wide variations in the arrangement of the double bonds. However, the fats of the simplest and most primitive organisms are usually composed of very complex mixtures of fatty acids, in contrast to the higher plants and animals for which the component acids are few in number.

Fats are mixtures of various glycerides and are made up not only of the symmetrical compounds, that is, glycerides, of which the three fatty acid radicals are the same but are also of mixed compounds that contain two or three different acyl radicals in the molecule. Fats from aquatic organisms are largely of the unsaturated type and, on a weight basis, constitute an important part of both freshwater and marine organisms; it is presumed that they represent a significant contribution to the total organic fraction of sediments.

Carbohydrates form an important part of all plants and also contribute to the total organic content of the sediments, particularly in the near-shore area. Carbohydrates are quite varied in character, ranging from the simple sugars to polysaccharides, such as glycogen, starch, and cellulose.

The simpler members of the carbohydrates are water soluble and hence form a highly acceptable substrate for bacteria. Thus, part of the total carbohydrate is destroyed by oxidative organisms, another part is consumed by animal forms and converted into tissue, and finally, some of the carbohydrate is trapped in sediments. The fraction that escapes immediate destruction is probably largely composed of the water-soluble and biochemically more resistant compound types, but again, very little is known of the fate of these materials after burial.

Proteins are another important and highly complex component of all living matter. They are polymeric substances composed of one or more of some 25 amino acids linked to one another through the carboxyl carbon and the nitrogen, and chemically, the proteins are very reactive. In the native state, proteins are hydrolyzed by either alkaline or acid media to yield water-soluble products; as for the carbohydrates, it is likely that they are consumed in the aquatic environment either by oxidative destruction or by consumption as food by animal life.

Native protein is rare outside the living cell owing to the ease with which it loses structural organization (denaturation). In many instances solubility is greatly diminished after denaturation, and incorporation in the sediment is distinctly favored. Studies on the long-term thermal stability of amino acids indicate that these substances should survive under the anticipated low-temperature earth conditions for long periods of time. Indeed, amino acids in recent sediments, as well as in shales, are estimated to be some 30 million years old.

Lignin is a mixture of complex, high molecular weight, amorphous substances and forms the cell wall structure of plants, particularly those of woody type (Sarkanen and Ludwig, 1971; Philp et al., 1983). Despite the extensive work on the lignin of the higher land plants, its structure has not been defined except that it is probably built up of phenylpropane units containing methoxyl and hydroxyl groups. In sediments it is frequently found as a complex with protein or carbohydrate, but so little is known about the lignin of marine plants that the ligneous substances in sediments cannot definitely be typed as to source.

Sterols, which are based on a saturated, condensed ring structure are widely distributed in animals and plants and are also believed to be source material for petroleum although very little is known of their natural degradation reactions. Additional classes of compounds that also make up the source material are lipids and terpenoids and, possibly, the humic substances, specifically humic acids and, to a lesser extent, fulvic acids (Gaffney et al., 1996).

Briefly, humic acids are those organic compounds found in the environment that cannot be classified as any other chemical class of compounds (e.g., polysaccharides, proteins, etc.). They are traditionally defined according to solubility. Fulvic acids are those organic materials that are insoluble in water at all pH values. Humic acids are those materials that are insoluble at acidic pH values (pH < 2) but are soluble at higher pH values. Humin is the fraction of natural organic materials that is insoluble in water at all pH values. These definitions reflect the traditional methods for separating the different fractions from the original mixture.

The humic content of soils varies from 0% to almost 10% w/w. In surface waters, the content of humic material, expressed as dissolved organic carbon varies from 0.1 to 50 ppm in dark-water swamps. In ocean waters, the dissolved organic carbon varies from 0.5 to 1.2 ppm at the surface, and the dissolved organic carbon samples from deep groundwater vary from 0.1 to 10 ppm (1). In addition, about 10% of the dissolved organic carbon in surface waters is found in suspended matter, either as organic or organically coated inorganic particulate matter (Gaffney et al., 1996).

Humic materials have a wide range of molecular weights and sizes, ranging from a few hundred to as much as several hundred thousand atomic mass units. In general, fulvic acids are of lower molecular weight than humic acids, and soil-derived materials are larger than aquatic materials (Stevenson, 1982). Humic materials vary in composition depending on their source, location, and method of extraction; however, their similarities are more pronounced than their differences. The range of the elemental composition of humic materials is relatively narrow, being approximately 40% to 60% carbon, 30% to 50% oxygen, 4% to 5% hydrogen, 1% to 4% nitrogen, 1% to 2% sulfur, and 0% to 0.3% phosphorus (MacCarthy and Suffet, 1989). Humic acids contain more hydrogen, carbon, nitrogen, and sulfur and less oxygen than fulvic acids. Studies on humins have shown that they are similar to humic acids except that they are strongly bound to metals and clays, rendering them insoluble (Schnitzer and Kahn, 1972).

It is generally believed, and some evidence does exist, that humic materials consist of a skeleton of alkyl/aromatic units cross-linked mainly by oxygen and nitrogen groups with the major functional groups being carboxylic acid, phenolic and alcoholic hydroxyls, ketone, and quinone groups (Schulten, 1996). The structures of fulvic acids are somewhat more aliphatic and less aromatic than humic acids; and fulvic acids are richer in carboxylic acid, phenolic acid, and ketonic groups (Schulten, 1996). This is responsible for their higher solubility in water at all pH values. Humic acids, being more highly aromatic, become insoluble when the carboxylic groups are protonated at low pH values. This structure allows the humic materials to function as surfactants, with the ability to bind both hydrophobic and hydrophilic materials. This function in combination with their colloidal properties makes humic and fulvic materials effective agents in transporting both organic and inorganic contaminants in the environment.

Porphyrins, which are complex derivatives of the basic material, porphine (Figure 3.3), occur as part of the nitrogen-containing constituents of crude oil. These compounds are recognized as the degradation products of the chlorophyll (photosynthetic pigments of plants and some bacteria) and of the hemes (haems) and hematins (haematins), the respiratory pigments of both plants and animals. Most, if not all, of the porphyrin material in crude oils is completed with a metal, of which vanadium is the most important, followed by nickel; iron and copper may also be present. The vanadium content of crude oil ranges from a few parts per million for paraffinic petroleum to several hundred for heavy oil and even higher for tar sand bitumen (Gruse and Stevens, 1960; Tissot and Welte, 1978).

In general, the organic detritus of sediments varies with the environment just as the aquatic organisms vary in their chemical makeup. This is very important in understanding the differences in petroleum types: gross differences and structural variations of petroleum constituents are now believed to arise because of required variations in the source materials and also variations in the conditions (geophysical) under which the petroleum was formed.

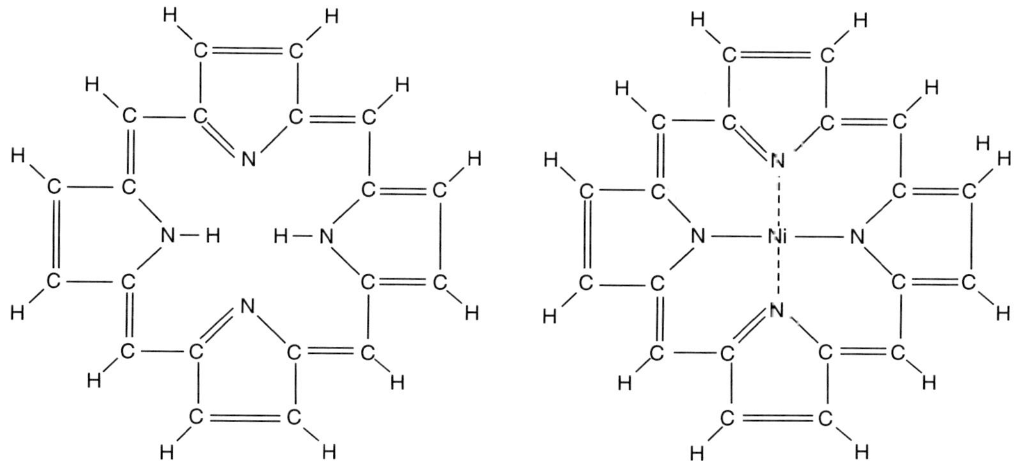

FIGURE 3.3 Porphine and the nickel-containing derivative.

3.2.2.4   Transformation of Organic Matter into Petroleum

The least well-understood phase of petroleum genesis is the chemistry of the transformation of the organic matter into petroleum (Snowdon and Powell, 1982). The discovery of prolific bacterial growths in recent sediments suggests that biological activity is a means of generating crude oil. Bacteria are reputed to attack carbohydrates and proteins readily, but their effect on certain lipids, proteins, and lignin, for example, is relatively unknown (Brooks and Welte, 1984). There are also theories that thermal effects have played a major role in the generation of petroleum (Califet and Oudin, 1966; Barker and Wang, 1988). But the role of the thermal effects is not identified and such theories remain to be proven.

The overall result of this activity is the production of such compounds as carbon dioxide (CO_2), hydrogen sulfide (H_2S), ammonia (NH_3), acids ($R \cdot CO_2H$), alcohols ($R \cdot OH$), and amines (e.g., primary amines, $R \cdot NH_2$). An important inorganic bacterial reaction in recent sediments is the reduction of sulfate ion to sulfide:

$$-SO_4^{2+} \rightarrow > S^{2-}$$

Except for the production of methane, bacteria tend to leave organic compounds bearing carboxyl ($-CO_2H$), hydroxyl ($-OH$), amine ($-NH_2$), and sulfhydryl ($-SH$) groups rather than effect the complete removal of oxygen, nitrogen, or sulfur atoms to produce hydrocarbons. From the observation that tyrosine could be fermented to yield either phenol or *p*-cresol, it was assumed that phenylalanine might be made to yield benzene or toluene, but using the same cultures, no trace of either hydrocarbon could be detected.

The abundant production of methane (CH_4) originally suggested that the higher paraffins may also be produced in this manner. However, analysis of the products of a wide variety of microbial fermentation processes failed to show anything more than trace amounts of ethane (C_2H_6) or higher molecular weight paraffins (C_nH_{2n+2}, where $n > 3$).

Hydrocarbons are synthesized by bacteria as a part of their cell substance. Extracts of autotrophic bacteria cultures (which use molecular hydrogen as the sole source of energy and carbon dioxide or carbonate as the sole source of carbon) were demonstrated to contain at least 25% paraffins and cycloparaffins (naphthenes). It appears likely that this is, at least, one of the sources of some of the heavier hydrocarbons in recent sediments. The possibility that biocatalysts or enzymes are involved in the formation of hydrocarbons cannot be discounted at the present time. A wide variety of such organic catalysts is produced by bacteria; it is possible that, under the prevailing low-temperature anaerobic conditions, these substances may function for some time after the organisms that produced them have ceased to exist.

Ring formation (cyclization) and polymerization reactions have been noted to occur in the same temperature range for several unsaturated fatty acids of marine organisms, and it is conceivable that such polyene-type reactions may well contribute substantially to the naphthene-aromatic fraction of crude oil. Chemical reactions of this type are particularly attractive because the clays and other fine-grained materials, which are always present in the sediment, may act as catalysts for hydrocarbon synthesis as well as for the degradation of the organic source material to hydrocarbons. In fact, it has been suggested that any of several naturally occurring minerals might be active in promoting polymerization, and the idea has been developed further to include hydro-polymerization, isomerization, and cyclization to explain the variety of hydrocarbon types present in petroleum.

The most serious objection to be raised against catalytic investigations is that the majority of laboratory studies have been carried out in the absence of water, whereas water is always present in sediments. The adsorption of water on the mineral components may substantially reduce any catalytic activity and also seriously alter their catalytic nature. However, some components of crude oil, as well as source material types, can be adsorbed by clays from an aqueous medium. It has been shown that a wide variety of substances will enter the clay lattice to form highly stable complexes. For example, protein in combination with clay is substantially more resistant to bacterial destruction than protein alone; possibly, in the adsorbed condition, time is available for slow chemical reaction that might otherwise not occur.

It is appropriate at this point to introduce the concept of organic **facies**, which are stratigraphic rock units differentiated from adjacent or associated units by appearance or by characteristics that usually reflect the origin (Bates and Jackson, 1980; Demaison et al., 1984).

Organic facies I (strongly oil prone) are typical of the strongly anoxic environments of stratified lacustrine and marine locales. The strata are laminated, and the absence of bioturbation (mixing) indicates that benthic organisms were absent from the depositional environment. Organic facies II (oil prone) are also typical of anoxic environments but may include moderately oxic environments that had high rates of deposition. The carbon contents of the type II facies are generally lower than those of type I facies and are often in the range of 1%–10% total organic carbon. On the other hand, organic facies III (gas prone) are typical of the mildly oxic conditions in coal swamps or shallow marine environments.

Any planktonic or algal material deposited in such an environment is usually degraded quickly by benthic organisms. Preserved material of aquatic origin is usually thoroughly bioturbated. Finally, organic facies IV (nonsource) are typical of aquatic environments in which organic matter spends a long time in the oxic zone where aerobic bacteria are active. Such environments occur even at great ocean depths with circulation, as well as in shallow, circulating seas with a high energy input. Low organic contents in the rocks reflect efficient degradation even where the initial organic productivity was high.

3.2.2.5 Accumulation in Reservoir Sediments

At an early date it was recognized that hydrocarbon accumulations require the existence of a reservoir rock to store the fluids and a cap to prevent their escape. In addition to these readily appreciated requirements, there must also be a source rock in which the oil and gas are formed.

Reservoir rock must possess fluid-holding capacity (**porosity**) and also fluid-transmitting capacity (**permeability**); a variety of different types of openings in rocks are responsible for these properties in reservoir rocks. The most common are the pores between the grains of which the rocks are made or the cavities inside fossils, openings formed by solution or fractures, and joints that have been created in various ways. The relative proportion of the different kinds of openings varies with the rock type, but pores usually account for the bulk of the storage space. The effective porosity for oil storage results from continuously connected openings. These openings alone provide the property of permeability, and although a rock must be porous to be permeable, there is no simple quantitative relationship between the two.

Reservoir rocks tend to show far greater variations in permeability than in porosity, and in addition, these two properties, as measured on core samples from reservoir rocks, are not always identical with the values indicated for the rock in bulk underground. The differences arise from the nonrepresentative nature of cores, especially when there are wide variations in the sizes of the openings in the rocks and irregularities in their distribution. Porosity is generally in the range of 5% to 30%; whereas permeability is commonly between 0.005 darcy and several darcys as measured on small samples. It should be noted that pores may be, at best, only a millimeter or so in width, whereas fossil and solution cavities may sometimes be 30 to 50 times wider. Many joints and fractures are probably only a millimeter across, although they may extend for considerable distances.

The processes by which petroleum migrates into pools are not well defined. However, it is known that crude oil and gas occur in pools or fields in which the oil or gas occupies pore space in the rock. Some pools are large, extending laterally over many square kilometers, with their vertical extent ranging from several meters to as much as several hundred meters. The crude oil or gas occupying the pore space in the pool has displaced the water that was initially present in the pores. The presence of a pool of oil or gas therefore implies that the oil or gas has migrated into the pool.

Petroleum accumulations are generally found in relatively coarse-grained porous and permeable rocks that contain little or no insoluble organic matter. It is highly unlikely that petroleum originated from organic matter of which now no trace remains. Rather, it appears that petroleum constituents are generated through geochemical action on organic detritus that is usually found in fine-grained sedimentary rocks. In addition, it may be anticipated that some of the organic material would also change sufficiently from the original organic material to remain in the sediment through, and beyond, the oil-generating stage. Thus, it is generally assumed that the place of origin of oil and gas is not identical to the locations where it is found. The oil had to migrate to present reservoirs from the place of origin.

When sediments such as mud and clay are deposited, they contain water. As the layers of sediment accumulate, the increasing load of material causes compaction of the sediment and part of the water is expelled from the pores. Migration probably begins at this stage and may involve a substantial horizontal component. Initially, the oil hydrocarbons may be present in the water either suspended as tiny globules or dissolved, hydrocarbons being slightly soluble in water.

Petroleum constituents may migrate through one or more formations that have permeability and porosity similar to those of the ultimate reservoir rock. It is the occurrence of an impermeable, or a very low permeability, barrier that causes the formation of oil and gas accumulations. Since practically all pores in the subsurface are water saturated, the movement of petroleum constituents within the network of capillaries and pores must take place in the presence of the aqueous pore fluid. Such movement may be due to active water flow or may occur independently of the aqueous phase, either by displacement or by diffusion. There may be a single phase (oil and gas dissolved in water) or a multiphase (separate water and

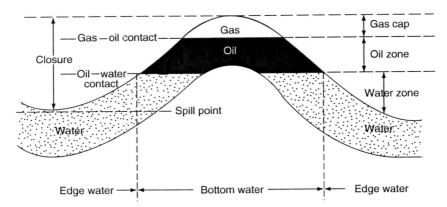

FIGURE 3.4 Typical anticlinal petroleum trap.

hydrocarbon phases) fluid system. The loss of hydrocarbons from a trap is referred to as dismigration.

The specific gravity of gas and petroleum, the latter generally between specific gravity (at $60°F$, $15.6°C$) that varies from about 0.75 to 1.00 ($57°$ to $10°$ API), with the specific gravity of most crude oils falling in the range of 0.80 to 0.95 ($45°$ to $17°$ API), are considerably lower than those of saline pore waters (specific gravity: 1.0 to 1.2). Thus, petroleum accumulations are usually found in structural highs where reservoir rocks of suitable porosity and permeability are covered by a dense, relatively impermeable cap rock, such as an evaporite or shale. A reservoir rock sealed by a cap rock in the position of a geological high, such as an *anticline*, is known as a structural petroleum trap (Figure 3.4). Other types of **traps**, such as sand lenses, reefs, and pinch-outs of more permeable and porous rock units, are also known and occur in various fields. In all these situations, the changes in permeability and porosity determine the location of oil or a gas accumulation.

The predominant theory assumes that as the sedimentary layers superimposed in the source bed became thicker, the pressure increased and the compression of the source bed caused liquid organic matter to migrate to sediments with a higher permeability, which as a rule is sand or porous limestone. Thus, several mechanisms have been postulated for the migration of petroleum from the source rock to the reservoir rock (Table 3.2), and there is differentiation between the **primary migration** mechanism and **secondary migration** mechanism.

TABLE 3.2
General Mechanisms for Petroleum Migration

Geological Event	Migration Effect
Basin development	Downward fluid flow
Mature basin	Sediments move downward
Hydrocarbon generation	Thermal effects on source sediment
Hydrocarbons dissolve	Pore fluid become saturated
Geothermal gradient changes	Isotherms depressed
Pore fluids cool	Hydrocarbons form separate phase
Hydrocarbons separate	Move to top of carrier fluid (water)
Updip migration	Buoyancy effect
Intermittent faulting	Hydrocarbon migration to traps

Primary migration is the movement of hydrocarbons (oil and natural gas) from mature, organic-rich source rocks to a point where the oil and gas can collect as droplets or as a continuous phase of liquid hydrocarbon. Secondary migration is often attributed to various aspects of buoyancy and hydrodynamics (Schowalter, 1979; Barker, 1980) and is the movement of the hydrocarbons as a single, continuous fluid phase through water-saturated rocks, fractures, or faults followed by accumulation of the oil and gas in sediments (traps) from which further migration is prevented.

However, it must be emphasized that the various mechanisms that have been proposed for petroleum migration do not compete with each other but are in fact complementary to each other. The prevailing conditions in the subterranean strata may dictate that a particular migration mechanism is favored and this is therefore the dominant mechanism, with alternative migration mechanisms playing lesser roles. In another given situation, the role of the dominant migration mechanism is reversed.

It is believed that, during migration, petroleum does not move through the bulk of the nonsource rock and shale bodies, but through faults and fractures that may be in the form of a channel network that permits leakage of oil and gas from one zone to another. During this migration, the composition of the oil may be changed through physical causes such as filtration and adsorption, in a manner analogous to the chromatographic separation of petroleum (Chapter 8). Reactions with minerals such as elemental sulfur or even with sulfur-containing minerals (e.g., sulfates) may also occur.

Many examples of the effect of adsorption on crude oils exist (Chapter 8). The passage of whole, or fractions of, crude oils through adsorbents is well documented, and the changes in composition are authenticated. In terms of interaction with minerals, one such example of alteration during migration exists in northern Iraq, close to the city of Mosul (Speight, 1978). A heavy oil (Qayarah) is produced (by steam stimulation) from shallow formations that are located to the north of a sulfur bed. The sulfur is extracted commercially from the bed but must have an extraneous bituminous material removed as part of the purification procedure. The Qayarah crude oil has in excess of 8% w/w sulfur, of which one-quarter (i.e., approximately 2% w/w) is free sulfur. The current conjecture is that the oil (or oil precursor) has migrated from the Kirkuk area (to the south) through the sulfur bed, thereby contaminating the sulfur, which in turn causes chemical reactions to occur that produce the heavier constituents as well as picking up free sulfur during the migration.

Any porous and permeable stratum will suffice as the reservoir, and a very common reservoir rock is a porous or fractured limestone (especially of the reef, bioherm type); several such reservoir structures are found throughout the earth. Most reservoir rocks are sedimentary rocks, almost always the coarser grained of the sedimentary rocks: sand, sandstone, limestone, and dolomite. A less common reservoir is a fractured shale or even igneous or metamorphic rock. It is only rarely that shales act as reservoir rocks, and again fractures and other relatively wide openings are believed to confer the required reservoir properties on an otherwise unsuitable rock. It may be that petroleum found in at least some reef structures is indigenous because of the large concentration of organisms in reefs, and there is often no other obvious source of the oil.

Source beds do not generally coincide with reservoir rocks and the belief is that petroleum, before it comes to rest in a trap, migrates large distances. Various examples are known of vertical migration, and migration upward for some kilometers is considered possible. Nevertheless, the converse may also be true, and the in situ theory (juxtaposition of source bed and reservoir rock) advocates that petroleum migrates very little, if at all and that even the amount of vertical migration is negligible.

Once the oil has been transferred to the reservoir rock, it is free to move under any force that may be applied. Gravitational forces are presumed to be dominant, thereby causing the oil, gas, and water to segregate according to their relative densities in the upper parts of the porous stratum. Favorable locations where the oil can accumulate may be anywhere along the path of fluid travel, and even though the porous zones adjacent to a compacting mud are commonly sand (sands form important reservoir rocks), almost any porous and permeable stratum will suffice.

Once the oil has accumulated in the reservoir rock, gravitational forces are presumed to be dominant, thereby causing the oil, gas, and water to segregate according to their relative densities in the upper parts of the reservoir (Landes, 1959). If the pores in the reservoir rock are of uniform size and evenly distributed, there are transition zones, from the pores occupied entirely by water to pores occupied mainly by oil to those pores occupied mainly by gas. The thickness of the water–oil transition zone depends on the densities and interfacial tension of the oil and water as well as on the size of the pores. Similarly, there is some water in the pores in the upper gas zone (the **gas cap**), which has at its base a transition zone from pores occupied largely by gas to pores filled mainly by oil.

The cap rock and basement rock, which have a far lower permeability than the reservoir rock and are impermeable to oil and gas, act as a seal to prevent the escape of oil and gas from the reservoir rock. Typical cap and basement rocks are clay and shale, that is, strata in which the pores are much finer than those of reservoir rocks. Other rocks, such as marl and dense limestone, can also serve as cap and basement rocks provided that pores if any are very small. There are cases in which evaporites (salt, anhydrite, and gypsum) act as effective sealants. The cap rock has a far lower permeability than the reservoir rock, but it is equally true that cap rocks have very high capillary pressures and reservoir rocks have much lower capillary pressures. The capillary pressure is the pressure required to cause a fluid to displace from the openings in a rock by another fluid with which it is not miscible. Capillary pressure is dependent on the size of the openings, the interfacial tension between the two fluids, and the contact angle for the system.

The distribution of the fluids in a reservoir rock is dependent on the densities of the fluids as well as on the properties of the rock. If the pores are of uniform size and evenly distributed, there is

1. Upper zone where the pores are filled mainly by gas (the gas cap)
2. Middle zone in which the pores are occupied principally by oil with gas in solution
3. Lower zone with its pores filled by water

Such accumulations usually occur at folds in the earth's strata (anticlines), which may be several miles in length. A certain amount of water (approximately 10% to 30%) occurs along with the oil in the middle zone. There is a transition zone from the pores occupied entirely by water to pores occupied mainly by oil in the reservoir rock, and the thickness of this zone depends on the densities and interfacial tension of the oil and water as well as on the sizes of the pores. Similarly, there is some water in the pores in the upper gas zone that has at its base a transition zone from pores occupied largely by gas to pores filled mainly by oil.

The water found in the oil and gas zones is known generally as interstitial water. It usually occurs as collars around grain contacts, as a filling of pores with unusually small throats connecting with adjacent pores, or, to a much smaller extent, as wetting films on the surface of the mineral grains when the rock is preferentially wet by water. The water may occur as wetting films, or collars, around the sand grains as well as in some completely filled pores. The three-dimensional network allows continuity to exist for the hydrocarbons by means of connections on every side of the sand grains. The so-called gas-oil and oil-water contacts are generally horizontal but have been known to exist as a very gentle incline. On occasion, part of an accumulation of the oil or gas has its lower boundary marked, not by the water-bearing zone of the reservoir rock but by an adjacent sealing rock that has characteristics similar to those of the cap rock. When the pressure and temperature conditions are suitable in relation to the proportions and the nature of the gas and oil, there may be no gas cap but only oil, with dissolved gas overlying the water.

Oil and gas cannot be retained as an accumulation unless there is a trap, and this requires that the boundary between the cap rock or other sealing agent and the reservoir rock generally be convex upward, but the exact form of the boundary varies widely. The simplest forms are the flat-lying convex lens, the anticline, and the dome, each of which has a convex upper surface (Figure 3.4). Many oil and gas accumulations are trapped in anticlines or domes, structures that are generally more easily detected than some other types of traps. Some traps are formed by the reservoir rock being cut off at its upper end by a fault that places sealing rock against the fractured end. Alternatively, the upper end may have been eroded away during a period of unconformity, resulting in the subsequent covering of the eroded top of the reservoir rock by a sealing rock. There are also examples in which the reservoir rock wedges out at its upper end as an original depositional feature due to lateral variation in deposition or abuts against an old land surface (stratigraphic trap). Traps associated with salt intrusions are of various kinds; limestone reefs can also serve as reservoir rocks and give rise to overlying traps of anticlinal formed as a result of differential compaction. Examples are also known in which the reservoir rock extends to the surface of the earth but oil and gas are sealed in it by a clogging of the pores by bitumen or by natural cements. Many reservoirs display more than one of the factors contributing to the entrapment of hydrocarbons.

The distinction between a structural trap and a stratigraphic trap is often blurred. For example, an anticlinal trap may be related to an underlying buried limestone reef. Beds of sandstone may wedge out against an anticline because of depositional variations or intermittent erosion intervals. Salt domes, formed by flow of salt at substantial depths, have also created numerous traps that are both a structural trap and a stratigraphic trap.

3.2.2.6   In Situ Transformation of Petroleum

Petroleum, as we know it, is a very complex mixture of organic compounds and is thermodynamically unstable under a variety of geological conditions Therefore, petroleum is also susceptible to alteration after it has collected in a reservoir or in sediments (Evans et al., 1971). Therefore, it is important, at this point, to address the issue of the alteration of petroleum once it has accumulated in the reservoir or as it accumulates in sediments on its journey to the reservoir.

Alteration of reservoir petroleum is accepted for most of the world oil accumulations. Alteration may be related to the relative instability of petroleum or to the traps which may be susceptible to incursion of chemical agents, such as oxygen. Physical effects, such as those caused when the level of burial of the trap changes as a result of further subsidence or erosion, may also play a role. Examples of chemical alteration are thermal maturation and microbial degradation of the reservoir oil. Examples of physical alteration of petroleum are the preferential loss of low-boiling constituents by diffusion or the addition of new constituents to the oil-in-place by migration of these constituents from a source outside the reservoir.

It is difficult to draw a precise distinction between chemical and physical processes since it is often more than likely that the two processes are often interrelated and may even occur simultaneously (Evans et al., 1971). It is therefore pertinent at this point to present a brief acknowledgment of the various means by which petroleum can be altered in a reservoir.

3.2.2.6.1   Thermal Alteration
The alteration of reservoir petroleum occurs over geological time; to what degree (no pun intended) and to what extent this involves thermal forces remains speculative.

Although the geothermal gradient varies from place to place, it is generally on the order of 25°C to 30°C/km (15°F/1000 ft or 120°C/1000 ft, i.e., 0.015°C per ft of depth or 0.012°C per ft of depth), i.e., approximately one degree for every 100 ft below the surface. Thus, with increasing depth of the reservoir, there is a tendency for crude oil to become lighter insofar as it contains increasing amounts of low-molecular weight hydrocarbons and decreasing amounts of the higher molecular weight constituents.

The pyrolysis of hydrocarbon waxes yields oily products, and it has been postulated that the formation of a waxy protopetroleum occurs. Prolonged exposure to moderate temperatures and high pressures yields olefins, which then rearrange or polymerize to produce a wide variety of hydrocarbons containing branched chains and rings. However, it has been estimated that a temperature in excess of 200°C (390°F) would be required to bring about any significant amount of conversion for such a process within a geologically acceptable period.

Current estimates of the increase in temperature with depth (0.015°C, 0.012°F per ft of depth) indicate the depths required to attain such a temperature, assuming, of course, that the temperature gradient remained approximately the same during the oil-forming period. Indeed, current concepts of the principal phase of oil formation favor a low-temperature process, whether the process is bacterial or chemical.

However, much of the laboratory work has been carried out at temperatures of the order of 300°C (570°F). The geothermal gradient varies from place to place, but is generally in the range of 25°C/km to 30°C/km (15°F/1000 ft or 12°C/1000 ft, i.e., 0.015°C per ft of depth or 0.012°C per ft of depth). A temperature of the order of 240°C represents a source rock at about 20,000 ft. Then it is assumed that laboratory simulations of petroleum formation using temperatures of the order of 300°C (570°F) provide a quicker response than geologic time at lower temperatures; after all who can plan laboratory experiments that may have to last several million years.

It may be interpreted that all petroleum generation involving thermal alteration of the sedimentary matter starts at extreme depths; decarboxylation of acidic functions will commence at lower temperatures and, therefore, closer to the surface. Even if the petroleum-forming reactions can proceed at 200°C (390°F), using the current thermal gradient, depths in excess of 20,000 ft are required to attain such a temperature.

The thermal alteration of kerogen (Chapter 4) has also been the subject of much experimentation with the result that the formation of petroleum from kerogen is postulated to involve the following path:

$$kerogen \rightarrow bitumen \rightarrow gas + oil + residue$$

The term *bitumen* used in this equation is not to be confused with the bitumen that occurs in tar sand deposits (Chapter 1).

Acceptance of such a scheme ignores the many reactions that can occur concurrently (Miknis et al; 1987; Burnham, 1995; Hunt, 1996) and assumes that kerogen, because it is a source of hydrocarbons through pyrolytic chemistry, is an intermediate in the formation of petroleum. There are other options available (Chapter 4).

It is true that an increase in temperature will increase the rate of a chemical reaction. In most cases, an increase in a reaction temperature of 10°C (18°F) will double the reaction rate. During all of these assumptions, there does not appear to be any recognition of the fact that not only does the reaction rate increase but the reaction chemistry may also change. In fact, the three phases of the maturation sequences (i.e., diagenesis, catagenesis, and metagenesis) invoke the concept of different chemical reactions occurring within the three temperature ranges (Hunt, 1996). Therefore, if temperatures in excess of 200°C (390°F) are considered unlikely, even extreme (and they must be questioned), invoking the occurrence of low temperature pathways becomes self-evident.

Nevertheless, a temperature gradient does exist and it is recognized that the effect of increasing the temperature of a reservoir is to gradually increase the content of compounds containing fewer than 15 carbon atoms at the expense of the heavier (C_{15+}) liquids, which themselves become increasingly paraffin in nature. The overall result of this thermal maturation process is the production of lighter oil, generally with a lower sulfur content but with a higher paraffin content.

The occurrence in many crude oils of sulfur-containing compounds, often in substantial quantities, requires a source of sulfur other than the minor amounts present in the source material. The most reasonable mechanism for the formation of these sulfur-containing bodies appears to be a thermal reaction between elemental sulfur, and possibly also hydrogen sulfide, and the other organic components of the sediments, including the hydrocarbons. Presumably these reactions may continue even after the oil has accumulated in the reservoir, if elemental sulfur or hydrogen sulfide is still present, with resulting slow alteration in the character of the petroleum (see also page 58).

The thermal maturation process is essentially in a series of disproportionation reactions, for example, there are hydrogen transfer reactions, and in the more polar constituents of the oil, decarboxylation, dehydration, and desulfurization reactions produce carbon dioxide, water, and hydrogen sulfide. There is also a simultaneous production of gases, such as methane and other light hydrocarbons (Evans et al., 1971). Thus, the presence of the more thermodynamic pyrobitumen types of materials plus low molecular weight hydrocarbon gases and the absence of intermediate oil constituents are usually an indication of disproportionation reactions during thermal alteration. In fact, the thermal alteration process is essentially a disproportionation reaction in which there is hydrogen transfer from, presumably, the more aromatic polar species to the more aliphatic species.

3.2.2.6.2 Deasphalting

Another relatively common method of petroleum alteration involves **deasphalting** in the reservoir. Deasphalting (Chapter 8) is the precipitation of asphaltenes from crude oils by admixture with or dissolution in the oil of large amounts of light hydrocarbon or gaseous hydrocarbons that vary from methane to the various heptane isomers. Other hydrocarbons, having up to 16 carbon atoms (hexadecane, $C_{16}H_{34}$), are also known to precipitate an asphaltene fraction or an asphaltic fraction (asphaltenes plus resins) from crude oils (Chapter 8) but are not known to have substantial effects in the reservoir. Such deasphalting can occur as a natural process among the heavier crude oils whenever considerable amounts of the lower molecular weight hydrocarbons are generated in substantial quantities because of thermal alteration of the oil or as a result of gas incursion from secondary migration.

The overall effects of gas deasphalting are often difficult to distinguish from those of thermal maturation since both processes usually occur concomitantly and the net change in composition is that the oils become lighter (Evans et al., 1971). The yield of asphaltenes is related to the amount of the liquid hydrocarbon used for the deasphalting (Mitchell and Speight, 1973; Speight et al., 1984), the amount of asphaltenes precipitated in a reservoir can also be correlated with the amount of gas dissolved in the oils. The solution **gas–oil ratio** (GOR) for these oils is a measure of the amount of gas injected into the reservoir. In addition, the API gravity of the oil is an approximate measure of asphaltene content insofar as the API gravity decreases with increasing asphaltene content.

There has also been the suggestion that gravity segregation occurs in connection with reservoir deasphalting. For example, in reservoirs that have a high vertical profile, assuming a small temperature differential between the top and the bottom of the reservoir, the crude oil may be progressively heavier with increasing depth. The material at the higher points of such reservoirs could conceivably contain much more gas in solution than oil in the lower parts of the reservoir. Thus more asphaltenes are precipitated at the top of the reservoir than at the bottom of the reservoir (Evans et al., 1971). That gravity segregation occurs is not in doubt since there is evidence that the bitumen in the lower regions of **oil sand** formations contains higher proportion of asphaltenes than the bitumen at the top of the oil sand formation. Whether this is due to some past occurrence during the maturation of the oil, similar to the mechanism already suggested, or to some other means is not clear at this time.

3.2.2.6.3 Biodegradation and Water Washing

The microbial alteration of crude oil (biodegradation) and alteration due to water washing, that is, the removal of water-soluble compounds, are common methods of petroleum alteration. Both processes are frequently observed in combination since they are both due to the action of moving subsurface water. The biodegradation of crude oil is a selective utilization of certain types of hydrocarbons by microorganisms (Evans et al., 1971; Bailey et al., 1973a,b; Deroo et al., 1974; Connan et al., 1975).

The processes and effects of water washing are less well described than the biodegradation effects since there is a distinct lack of specific examples of the water washing process. Further, water washing normally should have less severe effects on the composition of a crude oil, and although the two processes may occur independently, they are apparently in most cases parallel to each other. Water washing results in the removal of more soluble hydrocarbons from crude oils. In general, light hydrocarbons are more easily dissolved and selectively removed from petroleum than are the higher molecular weight hydrocarbons.

Biodegradation and water washing can be expected wherever reservoirs are close to the surface and where they are accessible to surface-derived waters. An example of such crude oil alteration is based on crude oil samples from the western Canada sedimentary basin (Deroo et al., 1974). Systematic changes have been observed, from what are believed to be normal, unaltered oils in the deeper reservoirs, to heavier oils in the more shallow reservoirs, and to drastically altered oils very close to the surface, that is, the sandstone reservoirs impregnated with heavy oils, such as the Athabasca oil sands. There are also instances of an increase in sulfur content of the crude oil along with increasing bacterial degradation. It is not yet clear whether sulfur is added to the crude oils by the action of bacteria or whether the sulfur compounds are not attacked by the bacteria and thus remain after the bacteria have preferentially eliminated the hydrocarbon moieties.

In summary, the ultimate composition of petroleum may be influenced strongly by alteration subsequent to the process of accumulation. The three major alteration processes are thermal maturation, deasphalting, and degradation associated with the action of surface-derived formation waters (Bailey et al., 1974).

3.2.3 Differences between the Abiogenic Theory and the Biogenic Theory

There are several differences between the biogenic and abiogenic theories and these are listed below

1. Raw material
   • Biogenic: remnants of buried plant and animal life.
   • Abiogenic: deep carbon deposits from when the planet formed or subducted material.
2. Events before conversion
   • Biogenic: Large quantities of plant and animal life were buried. Sediments accumulating over the material slowly compressed it and covered it. At a depth of several hundred meters, catagenesis converted it to bitumen and kerogen.

- Abiogenic: At depths of hundreds of kilometers, carbon deposits are a mixture of hydrocarbon molecules that leak upward through the crust. Much of the material becomes methane.

3. Conversion to petroleum and methane
   - Biogenic: Catagenesis occurs as the depth of burial increases and the heat and pressure breaks down kerogen to form petroleum. Significant advances in the understanding of chemical processes and organic reactions and improved knowledge about the effects of heating and pressure during burial and diagenesis of organic sediments support biogenic processes.
   - Abiogenic: When the material passes through temperatures at which extremophile microbes can survive some of it will be consumed and converted to heavier hydrocarbons.

4. Evidence supporting abiogenic theory

   Cold planetary formation: In the late 19th century, it was believed that the Earth was extremely hot, possibly completely molten, during its formation. One reason for this was that a cooling, shrinking, planet was necessary to explain geologic changes such as mountain formation. A hot planet would have caused methane and other hydrocarbons to be out-gassed and oxidized into carbon dioxide and water, thus there would be no carbon remaining under the surface. Planetary science now recognizes that formation was a relatively cool process until radioactive materials accumulate together deep in the planet.

   Known hydrocarbon sources: Carbonaceous chondrite meteorites contain carbon and hydrocarbons. Heated under pressure, this material would release hydrocarbon fluids in addition to creating solid carbon deposits. Further, at least ten bodies in our solar system are known to contain at least traces of hydrocarbons. In 2004, the Cassini spacecraft confirmed methane clouds and hydrocarbons on Titan, a moon of Saturn.

   Unusual deposits: Hydrocarbon deposits have been found in places which are poorly explained by biogenic theory. Some oil fields are being refilled from deep sources, although this does not rule out a deep biogenic source rock. The White Tiger field in Vietnam and many wells in Russia, in which oil and natural gas are being produced from granite basement rock. As this rock is believed to have no oil-producing sediments under it, the biogenic theory requires the oil to have leaked in from source rock dozens of kilometers away.

   Deep microbes: Microbial life has been discovered 4.2 km deep in Alaska and 5.2 km deep in Sweden.

5. Evidence supporting biogenic theory

   It was once argued that the abiogenic theory does not explain the detection of various biomarkers in petroleum. Microbial consumption does not yet explain some trace chemicals found in deposits. Materials that suggest certain biological processes include tetracyclic diterpane, sterane, hopane, and oleanane. Although microorganisms exist deep underground and some metabolize carbon, some of these biomarkers are only known so far to be created in surface plants. This shows that some petroleum deposits may have been in contact with ancient plant residues, though it does not show that either is the origin of the other.

3.2.4 Relationship of Petroleum Composition and Properties

The geochemistry of petroleum is an extremely complex subject. It is not the intention here to examine in any detail the individual constituents of crude oil that have been reported elsewhere (Evans et al., 1971; Deroo et al., 1974; Ho et al., 1974; Orr, 1977; Hunt, 1996;

Brooks, 1981). It is intended to indicate the general trends that occur when a wide range of crude oils is examined.

As has already been mentioned, the passage of whole, or fractions of, crude oils through adsorbents is well documented, and the changes in composition are authenticated. In terms of interaction with minerals, one such example of alteration during migration exists in northern Iraq, close to the city of Mosul (Speight, 1978). Qayarah heavy oil is produced (by steam stimulation) from shallow formations that are located to the north of a sulfur bed. The sulfur is extracted commercially from the bed but must have an extraneous bituminous material removed as part of the purification procedure. Qayarah heavy oil has in excess of 8% w/w sulfur, of which one-quarter (i.e., approximately 2% w/w) is free sulfur. The current conjecture is that the oil (or oil precursor) has migrated from the Kirkuk area (to the south) through the sulfur bed, thereby contaminating the sulfur, which in turn causes chemical reactions to occur that produce the heavier constituents as well as picking up free sulfur during the migration.

Of all the properties, specific gravity, or American Petroleum Institute gravity (Chapter 9), is the variable usually observed. The changes may simply reflect compositional differences, such as the gasoline content or asphalt content, but analysis may also show significant differences in sulfur content or even in the proportions of the various hydrocarbon types.

Elemental sulfur is a common component of sediments and, if present in the reservoir rock, it dissolves in the crude oil and reacts slowly with it to produce various sulfur compounds or hydrogen sulfide, which may react further with certain components of the oil. These reactions are probably much the same as those that occur in the source bed and are presumed to be largely responsible for the sulfur content of a petroleum; these reactions are accompanied by a darkening of the oil and a significant rise in specific gravity and viscosity.

The variation in the character of crude oils with depth of burial is also of interest. An increase in the lighter constituents and a decrease in the density and in the amount of heavier constituents with increasing depth for sands of the same age has been noted. In sands of the same depth, an increase in the lighter constituents and a decrease in density with age have also been noted. It has been suggested that the precursors of such oils are heavy naphthenic types, which under the influences of pressure and temperature are slowly transformed into lighter, more paraffinic oils. However, there do appear to be cases in which many oils are less paraffinic and of greater density than the overlying, younger crude oils; differences in the source materials or in regional geology or later chemical alterations may be cited as possible controlling factors. There may also be a correlation between the composition and the nature of the sediments. There is also the possibility that young, heavy oils are converted into lighter crude oils over the course of geological time by mild thermal cracking, but if such changes occur, catalysis may be necessary and thermal alteration is possible when the oil has been exposed to significant temperatures.

Inspection of the data from a suite of Alberta crude oils indicates that a general relationship exists between the API gravity and depth of the reservoir, with the API gravity increasing with the depth of the reservoir. The decrease in API gravity with depth is in keeping with the concept that crude oil maturation involves formation of the lighter constituents. A decrease in the sulfur content of crude oil is also observed with depth. This is not surprising since the sulfur content of crude oil is related to the proportion of the higher molecular weight (asphaltene and resin) fractions in crude oil.

However, it is worthy of note that asphaltene solubility is susceptible to minor changes in the nature of the solvent and also to minor amounts of insoluble organic material and other gases in the solvent (Girdler, 1965; Mitchell and Speight, 1973; Speight and Moschopedis, 1981). This is obviously very important in reservoirs where a substantial gas cap exists above the oil; the constituents of the gas may not only dissolve in the oil but may also cause asphaltic material (asphaltenes and resins) to precipitate within the reservoir, thereby altering the character of the oil.

Thus, there is the distinct possibility, if not certainty, that the original character of petroleum can be substantially altered by events following the migration process. Each of the suggested mechanisms for this alteration (i.e., thermal alteration, deasphalting, and biogenic alteration) may play a role in the alteration of a particular crude oil. As for the choice of a migration mechanism, the choice of the prevalent in situ alteration process depends on the prevalent conditions to which the petroleum is subject. A particular mechanism for alteration may be the prime operative, or all may contribute to determining the ultimate character of the petroleum. The determining factor is site specificity.

The processes involved in petroleum maturation are extremely complex but several trends are notable in the maturation process:

1. Mature crude oils are low in asphaltic fraction (asphaltenes plus resins) and, consequently, lower in nitrogen, oxygen, and sulfur.
2. Asphaltene constituents from mature crude oils are more aromatic (lower atomic H/C ratio) than asphaltenes from immature crude oils.
3. A concentration effect (the extent of which depends on the reactions involved in the maturation process) causes the majority of the heteroelements (nitrogen, oxygen, and sulfur) to be located in the higher molecular weight fractions.

In summary, the petroleum formation process is generally accepted as the transformation of complex organic substances dispersed in source sediments. Like most typical organic reactions, the rate of petroleum generation increases with increasing temperature and, consequently, with increasing burial depths of the sediments. But the limits of burial and the temperatures to which the petroleum precursors have been subjected are not well defined, except by assumptions leading to laboratory experiments that are subject to question.

Whatever the conditions prevalent during maturation, nonhydrocarbon organic constituents also experience compositional changes during burial diagenesis. The highly complex nonhydrocarbon components, such as asphaltene constituents, are believed to be the products of condensation and disproportionation reactions, which proceed simultaneously with those reactions responsible for the formation of light hydrocarbons. In contrast to the hydrocarbon diagenesis pattern, asphaltenes become more complex with increasing maturity. This is confirmed by the observed increase in aromaticity.

In addition to maturation by thermal means, petroleum also undergoes nonthermal transformations after their accumulation in reservoirs. These changes are usually restricted to shallow oil deposits located near surface outcrops or near faults, which provide access to surface waters containing dissolved oxygen.

3.3   OCCURRENCE

The reservoir rocks that yield crude oil range in age from pre-Cambrian to Recent geologic time, but rocks deposited during the Tertiary, Cretaceous, Permian, Pennsylvanian, Mississippian, Devonian, and Ordovician periods are particularly productive. In contrast, rocks of Jurassic, Triassic, Silurian, and Cambrian age are less productive and rocks of pre-Cambrian age yield petroleum only under exceptional circumstances.

Most of the crude oil currently recovered is produced from underground reservoirs. However, surface seepage of crude oil and natural gas are common in many regions. In fact, it is the surface seepage of oil that led to the first use of the high boiling material (bitumen) in the Fertile Crescent (Chapter 1). It may also be stated that the presence of active seeps in an area is evidence that oil and gas are still migrating.

The majority of crude oil reserves identified to date are located in a relatively small number of very large fields, known as giants. In fact, approximately three hundred of the largest oil fields contain almost 75% of the available crude oil. Although most of the world's nations produce at least minor amounts of oil, the primary concentrations are in Saudi Arabia, Russia, the United States (chiefly Texas, California, Louisiana, Alaska, Oklahoma, and Kansas), Iran, China, Norway, Mexico, Venezuela, Iraq, Great Britain, the United Arab Emirates, Nigeria, and Kuwait. The largest known reserves are in the Middle East.

3.3.1 RESERVES

The definitions that are used to describe petroleum reserves are often misunderstood because they are not adequately defined at the time of use. Therefore, as a means of alleviating this problem, it is pertinent at this point to consider the definitions used to describe the amount of petroleum that remains in subterranean reservoirs.

Petroleum is a resource; in particular, petroleum is a fossil fuel resource (Chapter 1). A resource is the entire commodity that exists in the sediments and strata, whereas the reserves represent that fraction of a commodity that can be recovered economically. However, the use of the term *reserves* as being descriptive of the resource is subject to much speculation. In fact, it is subject to word variations. For example, reserves are classed as proved, unproved, probable, possible, and undiscovered.

Proven reserves are those reserves of petroleum that are actually found by drilling operations and are recoverable by means of current technology. They have a high degree of accuracy and are frequently updated as the recovery operation proceeds. They may be updated by means of reservoir characteristics, such as production data, pressure transient analysis, and reservoir modeling.

Probable reserves are those reserves of petroleum that are nearly certain but about which a slight doubt exists. **Possible reserves** are those reserves of petroleum with an even greater degree of uncertainty about recovery but about which there is some information. An additional term **potential reserves** is also used on occasion; these reserves are based on geological information about the types of sediments where such resources are likely to occur and they are considered to represent an educated guess. Then, there are the so-called undiscovered reserves, which are little more than figments of the imagination. The terms *undiscovered reserves* or *undiscovered resources* should be used with caution, especially when applied as a means of estimating reserves of petroleum reserves. The data are very speculative and are regarded by many energy scientists as having little value other than unbridled optimism.

The term *inferred reserves* is also commonly used in addition to, or in place of, potential reserves. Inferred reserves are regarded as of a higher degree of accuracy than potential reserves, and the term is applied to those reserves that are estimated using an improved understanding of reservoir frameworks. The term also usually includes those reserves that can be recovered by further development of recovery technologies.

The differences between the data obtained from these various estimates can be considerable, but it must be remembered that any data about the reserves of petroleum (and, for that matter, about any other fuel or mineral resource) will always be open to questions about the degree of certainty. Thus, in reality, and in spite of the use of self-righteous word-smithing, proven reserves may be a very small part of the total hypothetical or speculative amounts of a resource.

At some time in the future, certain resources may become reserves. Such a reclassification can arise as a result of improvements in recovery techniques which may either make the resource accessible or bring about a lowering of the recovery costs and render winning of the resource an economical proposition. In addition, other uses may also be found for a commodity, and the increased demand may result in an increase in price. Alternatively, a large deposit may become exhausted and unable to produce any more of the resource, thus forcing production to focus on a resource that is lower grade but has a higher recovery cost.

It is very rare that petroleum (the exception being tar sand deposits, from which most of the volatile material has disappeared over time) does not occur without an accompanying cover of gas (Figure 3.4). It is therefore important, when describing reserves of petroleum, to also acknowledge the occurrence, properties, and character of the gaseous material, more commonly known as natural gas.

More recently, the Society for Petroleum Engineers has developed a resource classification system (Figure 3.5) that moves away from systems in which all quantities of petroleum that are estimated to be initially-in-place are used. Some users consider only the estimated recoverable portion to constitute a resource. In these definitions, the quantities estimated to be initially-in-place are (1) total petroleum-initially-in-place, (2) discovered petroleum-initially-in-place, and (3) undiscovered petroleum-initially-in-place. The recoverable portions of petroleum are defined separately as (1) reserves, (2) contingent resources, and (3) prospective resources. In any case and whatever the definition, reserves are a subset of resources and are those quantities of petroleum that are discovered (i.e., in known accumulations), recoverable, commercial and remaining.

The total petroleum-initially-in-place is that quantity of petroleum that is estimated to exist originally in naturally occurring accumulations. The total petroleum-initially-in-place is, therefore, that quantity of petroleum that is estimated, on a given date, to be contained in known accumulations, plus those quantities already produced therefrom, plus those estimated quantities in accumulations yet to be discovered. The total petroleum-initially-in-place may be subdivided into discovered petroleum-initially-in-place and undiscovered petroleum-initially-in-place, with discovered petroleum-initially-in-place being limited to known accumulations.

It is recognized that the quantity of petroleum-initially-in-place may constitute potentially recoverable resources since the estimation of the proportion that may be recoverable can be subject to significant uncertainty and will change with variations in commercial circumstances, technological developments and data availability. A portion of those quantities classified as unrecoverable may become recoverable resources in the future as commercial circumstances change, technological developments occur, or additional data are acquired.

Discovered petroleum-initially-in-place is that quantity of petroleum that is estimated, on a given date, to be contained in known accumulations, plus those quantities already produced therefrom. Discovered petroleum-initially-in-place may be subdivided into commercial and subcommercial categories, with the estimated potentially recoverable portion being classified as reserves and contingent resources respectively (as defined below).

Reserves are those quantities of petroleum which are anticipated to be commercially recovered from known accumulations from a given date forward (http://www.spe.org/spe/jsp/basic/0,,1104_1718,00.html).

Estimated recoverable quantities from known accumulations that do not fulfill the requirement of commerciality should be classified as contingent resources (as defined later). The definition of commerciality for an accumulation will vary according to local conditions and circumstances and is left to the discretion of the country or company concerned. However, reserves must still be categorized according to specific criteria and, therefore, **proved reserves** will be limited to those quantities that are commercial under current economic conditions,

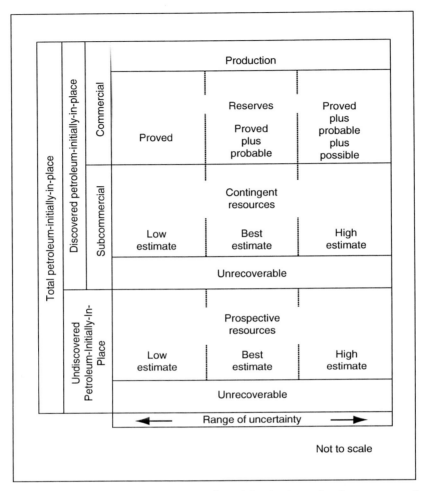

FIGURE 3.5 Representation of Resource Estimation. (The horizontal axis represents the range of uncertainty in the estimated potentially recoverable volume for an accumulation, whereas the vertical axis represents the level of status/maturity of the accumulation. The vertical axis can be further subdivided to classify accumulations on the basis of the commercial decisions required to move an accumulation towards production.) (From United States Geological Survey, Washington DC.)

whereas probable and possible reserves may be based on future economic conditions. In general, quantities should not be classified as reserves unless there is an expectation that the accumulation will be developed and placed on production within a reasonable timeframe.

In certain circumstances, reserves may be assigned even though development may not occur for some time. An example of this would be where fields are dedicated to a long-term supply contract and will only be developed as and when they are required to satisfy that contract.

Contingent resources are those quantities of petroleum that are estimated, on a given date, to be potentially recoverable from known accumulations, but which are not currently considered as commercially recoverable. Some ambiguity may exist between the definitions of contingent resources and unproved reserves. This is a reflection of variations in current industry practice, but if the degree of commitment is not such that the accumulation is expected to be developed and placed on production within a reasonable timeframe, the estimated recoverable volumes for the accumulation may be classified as contingent resources. Contingent resources may include, for example, accumulations for which there is currently no viable market, or where commercial recovery is dependent on the development of new technology, or where evaluation of the accumulation is still at an early stage.

Undiscovered petroleum-initially-in-place is that quantity of petroleum that is estimated, on a given date, to be contained in accumulations yet to be discovered. The estimated potentially recoverable portion of undiscovered petroleum-initially-in-place is classified as prospective resources, which are those quantities of petroleum that are estimated, on a given date, to be potentially recoverable from undiscovered accumulations.

Estimated ultimate recovery (EUR) is the quantity of petroleum which is estimated, on a given date, to be potentially recoverable from an accumulation, plus those quantities already produced therefrom. Estimated ultimate recovery is not a resource category but it is a term that may be applied to an individual accumulation of any status/maturity (discovered or undiscovered).

Petroleum quantities classified as reserves, contingent resources or prospective resources should not be aggregated with each other without due consideration of the significant differences in the criteria associated with their classification. In particular, there may be a significant risk that accumulations containing Contingent Resources or Prospective Resources will not achieve commercial production.

The range of uncertainty (Figure 3.5) reflects a reasonable range of estimated potentially recoverable volumes for an individual accumulation. Any estimation of resource quantities for an accumulation is subject to both technical and commercial uncertainties, and should, in general, be quoted as a range. In the case of reserves, and where appropriate, this range of uncertainty can be reflected in estimates for proved reserves (1P), proved plus probable reserves (2P) and proved plus probable plus possible reserves (3P) scenarios. For other resource categories, the terms *low estimate*, *best estimate*, and *high estimate* are recommended.

The term *best estimate* is used as a generic expression for the estimate considered the closest to the quantity that will actually be recovered from the accumulation between the date of the estimate and the time of abandonment. If probabilistic methods are used, this term would generally be a measure of central tendency of the uncertainty distribution. The terms *low estimate* and *high estimate* should provide a reasonable assessment of the range of uncertainty in the best estimate.

For undiscovered accumulations (prospective resources) the range will, in general, be substantially greater than the ranges for discovered accumulations. In all cases, however, the actual range will be dependent on the amount and quality of data (both technical and commercial) that is available for that accumulation. As more data become available for a specific accumulation (e.g., additional wells, reservoir performance data) the range of uncertainty in the estimated ultimate recovery for that accumulation should be reduced.

The low estimate, best estimate, and high estimate of potentially recoverable volumes should reflect some comparability with the reserves categories of proved reserves, proved plus probable reserves, and proved plus probable plus possible reserves, respectively. Although there may be a significant risk that subcommercial or undiscovered accumulations will not achieve commercial production, it is useful to consider the range of potentially recoverable volumes independently of such a risk.

3.3.2 CONVENTIONAL PETROLEUM

At the present time, several countries are recognized as producers of petroleum and have available reserves. These available reserves have been defined (Campbell, 1997), but not quite in the manner outlined earlier. For example, on a worldwide basis the produced conventional crude oil is estimated to be approximately 784 billion (784×10^9) bbl with approximately 836 billion bbl remaining as reserves. It is also estimated that there are 180 billion bbl which remain to be discovered with approximately 1 trillion (1×10^{12}) bbl yet-to-be-produced. The annual depletion rate is estimated to be 2.6%.

The United States is an oil-based culture (Williamson and Daum, 1959; Williamson et al., 1963), one of the largest importers of petroleum and, as the imports of crude oil into the United States continue to rise, it is interesting, perhaps frightening, that projections made in 1990 (in the Second Edition of this book) are remarkably close to the current import scenarios (*Oil & Gas Journal*, 2005). The United States now imports approximately 65% of its daily crude oil (and crude oil product) requirements. As recent events have shown, there seems to be little direction in terms of stability of supply or any measure of self-sufficiency in liquid fuels precursors, other than resorting to military action. This is particularly important for the United States refineries since a disruption in supply could cause major shortfalls in feedstock availability.

In addition, the crude oils available today to the refinery are quite different in composition and properties to those available some 50 years ago (Swain, 1991, 1993, 1997, 2000). The current crude oils are somewhat heavier, as they have higher proportions of nonvolatile (asphaltic) constituents. In fact, by the standards of yesteryear, many of the crude oils currently in use would have been classified as heavy feedstock, bearing in mind that they may not approach the definitions used today for heavy crude oil (Chapter 1). Changes in feedstock character, such as this tendency to heavier materials, require adjustments to refinery operations to handle these heavier crude oils to reduce the amount of coke formed during processing and to balance the overall product slate.

3.3.3 NATURAL GAS

Natural gas is the gaseous mixture associated with petroleum reservoirs and is predominantly methane, but does contain other combustible hydrocarbon compounds as well as nonhydrocarbon compounds (Speight, 1993). In fact, associated natural gas is believed to be the most economical form of ethane (Farry, 1998). Natural gas has no distinct odor and the main use is for fuel, but it can also be used to make chemicals and **liquefied petroleum gas** (LPG).

The gas occurs in the porous rock of the earth's crust either alone or with accumulations of petroleum (Figure 3.4). In the latter case, the gas forms the gas cap, which is the mass of gas trapped between the liquid petroleum and the impervious cap rock of the petroleum reservoir. When the pressure in the reservoir is sufficiently high, the natural gas may be dissolved in the petroleum and is released upon penetration of the reservoir as a result of drilling operations.

There are several general definitions that have been applied to natural gas. Thus, **lean gas** is gas in which methane is the major constituent. **Wet gas** contains considerable amounts of the higher molecular weight hydrocarbons. Sour gas contains hydrogen sulfide, whereas sweet gas contains very little, if any, hydrogen sulfide. Residue gas is natural gas from which the higher molecular weight hydrocarbons have been extracted and **casing head gas** is derived from petroleum but is separated at the separation facility at the wellhead.

To further define the terms *dry* and *wet* in quantitative measures, the term *dry* natural gas indicates that there is less than 0.1 gallon (1 (US) gallon = 264.2 m^3) of gasoline vapor (higher molecular weight paraffins) per 1000 ft^3 (1 ft^3 = 0.028 m^3). The term *wet* natural gas indicates that there are such paraffins present in the gas, in fact more than 0.1 gal/1000 ft^3.

Associated or dissolved natural gas occurs either as free gas or as gas in solution in the petroleum. Gas that occurs as a solution in the petroleum is dissolved gas, whereas the gas that exists in contact with the petroleum (gas cap) is associated gas.

Other components such as carbon dioxide (CO_2), hydrogen sulfide (H_2S), mercaptans (thiols; R–SH), as well as trace amounts of other constituents, may also be present. Thus, there is no single composition of components that might be termed *typical natural gas*. Methane and ethane constitute the bulk of the combustible components; carbon dioxide (CO_2) and nitrogen (N_2) are the major noncombustible (inert) components.

Some natural gas wells also produce helium, which can occur in commercial quantities; nitrogen and carbon dioxide are also found in some natural gases. Gas is usually separated at as high a pressure as possible, reducing compression costs when the gas is to be used for gas lift or delivered to a pipeline. After gas removal, lighter hydrocarbons and hydrogen sulfide are removed as necessary to obtain a crude oil of suitable vapor pressure for transport yet retaining most of the natural gasoline constituents.

In addition to composition and thermal content (Btu/scft, Btu/ft^3), natural gas can also be characterized on the basis of the mode of the natural gas found in reservoirs where there is no or at best only minimal amounts of crude oil.

Thus there is nonassociated natural gas, which is found in reservoirs in which there is no, or at best only minimal amounts of, crude oil. Nonassociated gas is usually richer in methane but is markedly leaner in terms of the higher molecular weight hydrocarbons and condensate.

Conversely, there is also associated natural gas (dissolved natural gas) which occurs either as free gas or as gas in solution in the crude oil. Gas that occurs as a solution with the crude petroleum is dissolved gas, whereas the gas that exists in contact with the crude petroleum (gas cap) is associated gas. Associated gas is usually leaner in methane than the nonassociated gas but is richer in the higher molecular weight constituents.

Another product is gas condensate, which contains relatively high amounts of the higher molecular weight liquid hydrocarbons. These hydrocarbons may occur in the gas phase in the reservoir.

The most preferred type of natural gas is the nonassociated gas. Such gas can be produced at high pressure, whereas associated, or dissolved, gas must be separated from petroleum at lower separator pressures, which usually involves increased expenditure for compression. Thus, it is not surprising that such gas (under conditions that are not economically favorable) is often flared or vented.

The nonhydrocarbon constituents of natural gas can be classified as two types of materials:

1. Diluents, such as nitrogen, carbon dioxide, and water vapors
2. Contaminants, such as hydrogen sulfide or other sulfur compounds

The diluents are noncombustible gases that reduce the heating value of the gas and are on occasion used as fillers, when it is necessary to reduce the heat content of the gas. On the other hand, the contaminants are detrimental to production and transportation equipment in addition to being obnoxious pollutants. Thus, the primary reason for gas processing is to remove the unwanted constituents of natural gas.

The major diluents or contaminants of natural gas are

1. Acid gas, which is predominantly hydrogen sulfide although carbon dioxide does occur to a lesser extent
2. Water, which includes all entrained free water or water in condensed forms
3. Liquids in the gas, such as higher boiling hydrocarbons as well as pump lubricating oil, scrubber oil, and, on occasion, methanol
4. Any solid matter that may be present, such as fine silica (sand) and scaling from the pipe

As with petroleum, natural gas from different wells varies widely in composition and analyses (Speight, 1993) and the proportion of nonhydrocarbon constituents can vary over a very wide range. Thus, a particular natural gas field could require production, processing, and handling protocols different from those used for gas from another field.

Liquefied petroleum gas is composed of propane (C_3H_8), butanes (C_4H_{10}), or mixtures thereof, small amounts of ethane (C_2H_6) and pentane (C_5H_{12}) may also be present as impurities. On the other hand, natural gasoline (like refinery gasoline) consists mostly of pentane (C_5H_{12}) and higher molecular weight hydrocarbons.

The term *natural gasoline* has also on occasion in the gas industry been applied to mixtures of LPG, pentanes, and higher molecular weight hydrocarbons. Caution should be taken not to confuse natural gasoline with the term *straight-run gasoline* (often also incorrectly referred to as natural gasoline), which is the gasoline distilled unchanged from petroleum.

The proven reserves of natural gas are in excess of 3600 trillion cubic feet (1 Tcf $= 1 \times 10^{12}$). Approximately 300 Tcf exist in the United States and Canada (BP Statistical Review of World Energy, 2002; BP Review of World Gas, 2002). It should also be remembered that the total gas resource base (like any fossil fuel or mineral resource base) is dictated by economics. Therefore, when resource data are quoted, some attention must be given to the cost of recovering those resources. Most important, the economics must also include a cost factor that reflects the willingness to secure total, or a specific degree of, energy independence.

3.3.4 HEAVY OIL

When petroleum occurs in a reservoir that allows the crude material to be recovered by pumping operations as a free-flowing dark- to light-colored liquid, it is often referred to as conventional petroleum. Heavy oils are the other types of petroleum that are different from conventional petroleum insofar as they are much more difficult to recover from the subsurface reservoir. These materials have a much higher viscosity (and lower API gravity) than conventional petroleum, and primary recovery of these petroleum types usually requires thermal stimulation of the reservoir.

The definition of heavy oil is usually based on the API gravity or viscosity, and the definition is quite arbitrary although there have been attempts to rationalize the definition based upon viscosity, API gravity, and density (Chapter 2).

Thus, the generic term *heavy oil* is often applied to petroleum that has an API gravity of less than 20° and usually, but not always, sulfur content higher than 2% by weight. Further, in contrast to conventional crude oils, heavy oils are darker in color and may even be black. The term *heavy oil* has also been arbitrarily used to describe both the heavy oils that require thermal stimulation of recovery from the reservoir and the bitumen in bituminous sand (tar sand) formations from which the heavy bituminous material is recovered by a mining operation.

3.3.5 BITUMEN (EXTRA HEAVY OIL)

In addition to conventional petroleum and heavy crude oil, there remains an even more viscous material that offers some relief to the potential shortfalls in supply; such material is the bitumen (extra heavy oil) found in tar sand (**oil sand**) deposits (Chapter 1). It is therefore worth noting here the occurrence and potential supply of these materials. However, many of these reserves are only available with some difficulty and optional refinery scenarios will be necessary for conversion of these materials to liquid products (Chapter 23 and Chapter 24), because of the substantial differences in character between conventional petroleum and tar sand bitumen (Table 3.3).

Tar sands, also variously called oil sands or bituminous sands (Chapter 1), are loose-to-consolidated sandstone or a porous carbonate rock, impregnated with bitumen, a high-boiling asphaltic material with an extremely high viscosity that is immobile under reservoir conditions.

As a historical note, it is evident that the Cree Indians, native to the Athabasca region of northeastern Alberta, have known about the tar sands for well over 200 years. The documents by an itinerant New England fur trader note that the Indians used a sticky substance oozing from the river banks to waterproof their canoes.

The first scientific interest in tar sands was taken by the Canadian Government in 1890. In 1897 to 1898, the sands were first drilled at Pelican Rapids on the Athabasca River, 129 km southwest of Fort McMurray. Up to 1960, the development was mainly small commercial enterprises. Between 1957 and 1967, three extensive pilot-plant operations were conducted in the Athabasca region, each leading to a proposal for a commercial venture. Shell Canada, Ltd. tested in sift recovery of bitumen; a group headed by Cities Service Athabasca, Ltd. operated a pilot plant at Mildred Lake based on a hot-water extraction process; and Great Canadian Oil Sands, Ltd. operated a pilot plant at the present commercial plant site near Mildred Lake. In addition, Mobil Oil Company has conducted tests on in sift recovery, and Petrofina established a pilot facility across the Athabasca River from the site of the Great Canadian Oil Sands (GCOS, now Suncor) plant. The operations of the latter two pilot plants remain proprietary.

TABLE 3.3
Comparison of Tar Sand Bitumen (Athabasca) and Crude Oil Properties

Property	Bitumen (Athabasca)	Crude Oil
Specific gravity	1.03	0.85–0.90
Viscosity (cp)		
38°C/100°F	750,000	<200
100°C/212°F	11,300	
Pour point (°F)	>50	ca. −20
Elemental analysis (wt.%)		
Carbon	83.0	86.0
Hydrogen	10.6	13.5
Nitrogen	0.5	0.2
Oxygen	0.9	<0.5
Sulfur	4.9	<2.0
Ash	0.8	0.0
Nickel (ppm)	250	<10.0
Vanadium (ppm)	100	<10.0
Fractional composition (wt.%)		
Asphaltenes (pentane)	17.0	<10.0
Resins	34.0	<20.0
Aromatics	34.0	>30.0
Saturates	15.0	>30.0
Carbon residue (wt.%)		
Conradson	14.0	<10.0

Although the origin of petroleum has been discussed elsewhere, it is appropriate to deal here with the origin of bitumen. The mechanisms of bitumen formation differ somewhat from those of petroleum formation.

There are two schools of thought regarding the origin of the bitumen. One is that the bitumen was formed locally and has neither migrated a great distance, nor been subjected to large overburden pressures. Since under these conditions the bitumen cannot have been subjected to thermal degradation, it is geologically young and therefore dense and viscous. The opposing theory assumes a remote origin for the bitumen, both geographically and in geological time. The bitumen precursor is assumed to have migrated into the sand deposit, which may have been originally filled with water. After the bitumen precursor migrated, the overburden pressures were relieved, and the light portions of the crude evaporated, leaving behind a dense viscous residue. A more recent theory also assumes a remote origin for the bitumen but postulates that bacteria carried into the reservoirs in oxygenated, meteoric waters destroyed the light hydrocarbons.

The remote-origin theory would explain the water layer surrounding sand grains in the Athabasca deposit. However, because the metals and porphyrin contents of bitumen are similar to those of some conventional Alberta crude oils of the Lower Cretaceous age and because Athabasca bitumen has a relatively low coking temperature, the bitumen may be of the Lower Cretaceous age. This is the age of the McMurray formation, which is geologically young. This evidence supports the theory that the bitumen was formed in situ and is a precursor, rather than a residue of some other material or oil precursor. The issue remains unresolved.

On an international note, the bitumen in tar sand deposits represents a potentially large supply of energy. However, many of the reserves are available only with some difficulty and optional refinery scenarios will be necessary for conversion of these materials to liquid products because of the substantial differences in character between conventional petroleum and tar sand bitumen (Table 3.3).

Because of the diversity of available information and the continuing attempts to delineate the various world tar sand deposits, it is virtually impossible to present accurate numbers that reflect the extent of the reserves in terms of the barrel unit. Indeed, investigations into the extent of many of the world's deposits are continuing at such a rate that the numbers vary from one year to the next. Accordingly, the data quoted here must be recognized as approximate, with the potential of being quite different at the time of publication.

Throughout this text, frequent reference is made to tar sand bitumen, but because commercial operations have been in place for over 30 years (Spragins, 1978) it is not surprising that more is known about the Alberta (Canada) tar sand reserves than any other reserves in the world. Therefore, when discussion is made of tar sand deposits, reference is made to the relevant deposit, but when the information is not available, the Alberta material is used for the purposes of the discussion.

Tar sand deposits are widely distributed throughout the world (Phizackerley and Scott, 1967; Demaison, 1977; Meyer and Dietzman, 1981; Speight, 1990, 1997). The various deposits have been described as belonging to two types:

1. Materials found in stratigraphic traps
2. Deposits located in structural traps

There are, inevitably, gradations, and combinations of these two types of deposits and a broad pattern of deposit entrapment are believed to exist. In general terms, the entrapment characteristics for the very large tar sand deposits all involve a combination of stratigraphic and structural traps, and there are no very large ($>4 \times 10^9$ bbl) oil sand accumulations either in purely structural or in purely stratigraphic traps.

The potential reserves of bitumen that occur in tar sand deposits have been variously estimated on a world basis as being in excess of 3 trillion (3×10^{12}) barrels and the reserves that have been estimated for the United States has been estimated to be in excess of 52 million (52×10^6) barrels. That commercialization has taken place in Canada does not mean that commercialization is imminent for other tar sands deposits. There are considerable differences between the Canadian and the U.S. deposits that could preclude across-the-board application of the Canadian principles to the U.S. sands (Speight, 1990).

Various definitions have been applied to energy reserves (Chapter 1), but the crux of the matter is the amount of a resource that is recoverable using current technology. Although tar sands are not a principal energy reserve, they certainly are significant with regard to projected energy consumption over the next several generations.

Thus, in spite of the high estimations of the reserves of bitumen, the two conditions of vital concern for the economic development of tar-sand deposits are the concentration of the resource, or the percent bitumen saturation, and its accessibility, usually measured by the overburden thickness. Recovery methods are based either on mining combined with some further processing or operation on the oil sands in situ. The mining methods are applicable to shallow deposits, characterized by an overburden ratio (i.e., overburden depth to thickness of tar-sand deposit). For example, indications are that for the Athabasca deposit, no more than 15% of the in-place deposit is available within current concepts of the economics and technology of open-pit mining; this 10% portion may be considered as the proven reserves of bitumen in the deposit.

REFERENCES

Almon, W.R. 1974. Petroleum-forming reactions: clay catalyzed fatty acid decarboxylation, Ph.D. Thesis. University of Missouri–Columbia.

Atkinson, G. and Zuckerman, J.J. 1981. *Origin and Chemistry of Petroleum*. Pergamon Press, New York.

Bailey, N.J.L., Krouse, H.H., Evans, C.R., and Rogers, M.A. 1973b. *Bull. Am. Assoc. Petrol. Geol.* 57: 1276.

Bailey, N.J.L., Evans, C.R., and Milner, C.W.D. 1974. *Bull. Am. Assoc. Petrol. Geol.* 58: 2284.

Barker, C. 1980. *Problems of Petroleum Migration*. Roberts, W.H. and Cordell, R.J. eds. Studies in petroleum geology, Bulletin No. 10. American Association of Petroleum Geologists, Tulsa, OK.

Barker, C. and Wang, L. 1988. *J. Anal. Appl. Pyrol.* 13: 9.

Bates, R.L. and Jackson, J.L. 1980. *Glossary of Geology*. American Geological Institute, Falls Church.

Bjoroy, M., Hall, P.B., Loberg, R., McDermott, J.A., and Mills, N. 1987. *Adv. Org. Geochem.* 13: 221.

Bogomolov, A.I., Panina, K.I., and Khotinseva, L.I. 1960. Publication No. 155. *VNIGRI*, St. Petersburg, Russia. *Geokh. Sbornik.* 6: 163.

Bogomolov, A.I., Panina, K.I., and Khotinseva, L.I. 1961. Publication No. 174. *VNIGRI*, St. Petersburg, Russia. *Geokh. Sbornik.* 7: 17.

Bogomolov, A.I., Panina, K.I., and Khotinseva, L.I. 1963. Publication No. 212. *VNIGRI*, St. Petersburg, Russia. *Geokh. Sbornik.* 8: 66.

BP Review of World Gas 2002. British Petroleum Company, London. September.

BP Statistical Review of World Energy 2002. British Petroleum Company, London. June.

Brooks, J. 1981. *Organic Maturation Studies and Fossil Fuel Exploration*. Academic Press, London.

Brooks, J. and Welte, D.H. 1984. *Advances in Petroleum Geochemistry*, Volume I. Academic Press, New York.

Burnham, A.K. 1995. *Composition, Geochemistry, and Conversion of Oil Shales*. Snape, C. ed. Kluwer Academic Publishers, Dordrecht, The Netherlands. p. 211.

Califet, Y. and Oudin, J.L. 1966. *Advances in Organic Geochemistry*. Hobson, G.D. and Speers, G.C. eds. Pergamon Press, New York.

Campbell, C.J. 1997. *Oil Gas J.* 95(52): 33.

Connan, J., Le Tran, K., and van der Weide, B. 1975. *Proc. Ninth World Petrol. Congr.* 2: 171.

Demaison, G.J. 1977. *The Oil Sands of Canada–Venezuela*. Redford, D.A. and Winestock, A.G. eds. Special Volume No. 17. Canadian Institute of Mining and Metallurgy, Montreal. p. 9.

Demaison, G., Holck, A.J., Jones, R.W., and Moore, G.T. 1984. *Proc. Eleventh World Petrol. Congr.* 2: 17.

Demorest, M., Mooberry, D., and Danforth, D. 1951. *Ind. Eng. Chem.* 43: 2569.

Deroo, G., Tissot, B., McCrossan, R.G., and Der, F. 1974. *Memoir No. 3*. Canadian Society of Petroleum Geologists. pp. 148 and 184.

Eglinton, G. and Murphy, M.J.T. 1969. *Organic Geochemistry*. Springer-Verlag, New York.

Erdman, J.G. 1965. *Memoir No. 4*. American Association of Petroleum Geologists. p. 20.

Evans, C.R., Rogers, M.A., and Bailey, N.J.L. 1971. *Chem. Geol.* 8: 147.

Farry, M. 1998. *Oil Gas J.* 96(23): 115.

Frost, A.V. 1940. *Zur. Fiz. Khim.* 4(9): 56.

Gaffney, J.S., Marley, N.A., and Clark, S.B. 1996. *Humic and Fulvic Acids*. Gaffney, J.S. Marley, N.A. and Clark, S.B. eds. Symposium Series No. 651. American Chemical Society, Washington, DC. Chapter 1.

Girdler, R.A. 1965. *Proc. Assoc. Asphalt Paving Technol.* 34: 45.

Gold, T. 1984. *Sci. Am.* 251(5): 6.

Gold, T. 1985. *Annu. Rev. Energy* 10: 53.

Gold, T. and Soter, S. 1980. *Sci. Am.* 242(6): 154.

Gold, T. and Soter, S. 1982. *Energy Explorat. Exploit.* 1(1): 89.

Gold, T. and Soter, S. 1986. *Chem. Eng. News* 64(16): 1.

Gruse, W.A. and Stevens, D.R. 1960. *Chemical Technology of Petroleum*. McGraw-Hill, New York.

Henderson, W., Eglinton, G., Simmonds, P., and Lovelock, J.E. 1968. *Nature (London)* 219: 1012.

Ho, T.Y., Rogers, M.A., Drushel, H.V., and Koons, C.B. 1974. *Bull. Am. Assoc. Petrol. Geol.* 58: 2338.

Hobson, G.D. and Tiratsoo, E.N. 1975. *Introduction to Petroleum Geology*. Scientific Press, Beaconsfield, England.

Hood, A., Gutjahr, C.C.M., and Heacock, R.L. 1975. *Bull. Am. Assoc. Petrol. Geol.* 59: 986.

Hunt, J.M. 1996. *Petroleum Geochemistry and Geology*, 2nd Edn. Freeman, W.H. and Co., New York.

Kenney, J., Shnyukov, A., Krayushkin, V., Karpov, I., Kutcherov, V., and Plotnikova, I. 2001. *Energia* 22(3): 26–34.

Kenney, J., Kutcherov, V., Bendeliani, N., and Alekseev, V. 2002. *Proc. Natl Acad. Sci. USA* 99: 10976–10981.

Landes, K.K. 1959. *Petroleum Geology*. John Wiley & Sons Inc., New York.

Levi, L. and Nicholls, R.V. 1958. *Ind. Eng. Chem.* 50: 1005.

MacCarthy, P. and Suffet, I.H. eds. 1989. *Aquatic Humic Substances: Influence on Fate and Treatment of Pollutants*. Advances in chemistry series No. 219. American Chemical Society, Washington, DC.

Menzel, D.W. 1974. *The Sea, Marine Chemistry*. Goldberg, D. ed. John Wiley & Sons Inc., New York. 5: 659.

Meyer, R.F. and Dietzman, W.D. 1981. *The Future of Heavy Crude and Tar Sands*. Meyer, R.F. and Steele, C.T. eds. McGraw-Hill, New York. p. 16.

Miknis, F.P., Turner, T.F., Berdan, G.L., and Conn, P.J. 1987. *Energy Fuels* 1: 477.

Mitchell, D.L. and Speight, J.G. 1973. *Fuel* 52: 149.

Moore, L.R. 1969. *Organic Geochemistry*. Eglinton G. and Murphy, M.T.J. eds. Springer-Verlag, Berlin, Germany. p. 265.

Nagy, B. and Colombo, V. 1967. *Fundamental Aspects of Petroleum Geochemistry*. Elsevier, Amsterdam.

Oil & Gas Journal 2005. Statistics: API imports of crude oil and products. 103(48).

Orr, W.L. 1977. Preprints. *Div. Fuel Chem. Am. Chem. Soc.* 22(3): 86.

Osborne, D. 1986. *Atlantic Monthly* (February): 39.

Philp, R.P., Gilbert, T.D., and Friedrich, J. 1983. *Chemical and Geochemical Aspects of Fossil Energy Extraction*. Yen T.F. Kawahara F.K., and Hertzberg, R. eds. Ann Arbor Science Publishers Inc., Michigan.

Phizackerley, P.H. and Scott, L.O. 1967. *Proc. Seventh World Petrol. Congr.* 3: 551.

Rahn, F.J. 1987. *McGraw-Hill Encyclopedia of Science and Technology*, Volume 12. Parker, S.P. ed. McGraw-Hill, New York. p. 171.

Sakikana, N. 1951. *J. Chem. Soc.* 72: 280.

Sarkanen, K.V. and Ludwig, C.H. 1971. *Lignins: Occurrence, Structure, Formation, and Reactions*. John Wiley & Sons Inc., New York.

Schnitzer, M. and Kahn, S.U. 1972. *Humic Substances in the Environment*. Marcel Dekker, New York.

Schowalter, T.T. 1979. *Bull. Am. Assoc. Petrol. Geol.* 63: 723.

Schulten, H.R. 1996. *Humic and Fulvic Acids*. Gaffney J.S. Marley N.A., and Clark, S.B., eds. Symposium Series No. 651. American Chemical Society, Washington. DC.

Shimoyama, A. and Johns, W.D. 1971. *Nature (London)* 232: 140.

Shimoyama, A. and Johns, W.D. 1972. *Geochim. Cosmochim. Acta* 36: 87.

Snowdon, L.R. and Powell, T.G. 1982. *Bull. Am. Assoc. Petrol. Geol.* 66: 775.

Sorokin, J.I. 1971. *Rev. Geosci. Hydrobiol.* 56: 1.

Speight, J.G. 1978. Personal observations.

Speight, J.G. ed. 1990. *Fuel Science and Technology Handbook*, Marcel Dekker. New York.

Speight, J.G. 1993. *Gas Processing: Environmental Aspects and Methods*. Butterworth-Heinemann, Oxford, UK.

Speight, J.G. 1997. *Kirk-Othmer Encyclopedia of Chemical Technology*, 4th edn, Volume 23. p. 717.

Speight, J.G. and Moschopedis, S.E. 1981. *The Future of Heavy Crude and Tar Sands*. McGraw-Hill, New York. p. 603.

Speight, J.G., Long, R.B., and Trowbridge, T.R. 1984. *Fuel* 63: 616.

Spragins, F.K. 1978. *Development in Petroleum Science, No. 7. Bitumens, Asphalts and Tar Sands*. Yen, T.F. and Chilingarian, G.V. eds. Elsevier, New York. p. 92.

Stevenson, F.J. 1982. *Humus Chemistry: Genesis, Composition, and Reactions*. John Wiley & Sons Inc., New York.

Strickland, I.D.H. 1965. *Chem. Oceanogr.* 1(12): 478.

Swain, E.J. 1991. *Oil Gas J.* 89(36): 59.

Swain, E.J. 1993. *Oil Gas J.* 91(9): 62.

Swain, E.J. 1997. *Oil Gas J.* 95(45): 79.

Swain, E.J. 2000. *Oil Gas J.* (March): 13.

Szatmari, P. 1989. *Bull. Am. Assoc. Petrol. Geol.* 73(8): 989.

Tappan, H. and Loeblich, A.R., Jr. 1970. *Symposium on the Palynology of the Late Cretaceous and Early Tertiary*. Kosanke, R.M. and Cross, S.T. eds. Geological Society of America. p. 247.

Tissot, B.P. and Welte, D.H. 1978. *Petroleum Formation and Occurrence*. Springer-Verlag, New York.

Vassoyevitch, N.B., Korchagina, Y.I., Lopatin, N.V., and Chernyshev, V.V. 1969. *Int. Geol. Rev.* 12: 1276.

Warren, J. and Storrer, P. 1895. *Proc. Am. Assoc. Arts Sci.* 2(9): 177.

Williamson, H.F. and Daum, A.R. 1959. *The American Petroleum Industry: Volume I, The Age of Illumination*. Northwestern University Press, Evanston, IL.

Williamson, H.F., Andreano, R.L., Daum, A.R., and Klose, G.C. 1963. *The American Petroleum Industry: Volume II, The Age of Energy*. Northwestern University Press, Evanston, IL.

Zebroski, E. and Levenson, M. 1976. *Annu. Rev. Energy* 1: 101.

4 Kerogen

4.1 INTRODUCTION

Precise details regarding the twin problems of origin, and migration and accumulation of petroleum have yet to be fully answered. Recent advances in analytical chemistry and geochemistry have advanced knowledge and understanding, but issues remain to be resolved. Therefore, in any text dealing with the science of petroleum, there must, of necessity, be a section dealing with **kerogen**. This will help the reader to understand kerogen and its place as a naturally occurring organic material. It is not the intent of this chapter to replace the excellent texts already available on the nature of kerogen (Tissot and Welte, 1978; Durand, 1980; Pelet and Durand, 1984; Vandenbroucke, 2003). It is hoped that this chapter will assist the reader to understand the nature of kerogen as well as understand the role that kerogen might play in petroleum science. It is anticipated that this chapter will also serve as an introduction to the more complex constituents of petroleum and aid in an understanding of the nature of the high molecular weight polar constituents (i.e., the asphaltene constituents and the resin constituents) of petroleum (see Chapter 8, Chapter 9, and Chapter 11).

Petroleum scientists have always searched for new trustful technologies to provide valuable evidence regarding the origin of hydrocarbons found in sedimentary basins. Present-day analysis of petroleum systems, when performed integrated with direct geochemistry, provides support to the theory that crude oil and natural gas accumulations have a biological origin. The biogenic precursor to oil and gas is believed to be kerogen.

Kerogen is the naturally occurring, solid, insoluble organic matter that occurs in source rocks and can yield oil upon heating. Typical organic constituents of kerogen are algae and woody plant material. Kerogen has a high molecular weight relative to **bitumen**, or soluble organic matter. Bitumen (not to be confused with tar sand bitumen, see Chapter 1) forms from kerogen during petroleum generation. Kerogen is described as Type I, consisting of mainly algal and amorphous (but presumably algal) kerogen and highly likely to generate oil; Type II, mixed terrestrial and marine source material that can generate waxy oil; and Type III, woody terrestrial source material that typically generates gas. A fourth type, Type IV kerogen is also now recognized (please see later).

The term *kerogen* is used throughout this text to mean the carbonaceous material that occurs in sedimentary rocks, carbonaceous shale, and oil shale. This carbonaceous material is, for the most part, insoluble in common organic solvents. A soluble fraction, bitumen, coexists with the kerogen. The bitumen is often also referred to as extra heavy oil (see Chapter 1).

4.2 PROPERTIES

Kerogen is a solid, waxy, organic substance that forms when pressure and heat from the Earth act on the remains of plants and animals. Kerogen converts to various liquid and gaseous hydrocarbons at a depth of seven or more kilometers and a temperature between $50°C$ and

100°C (122°F and 212°F) (USGS, 1995). The name kerogen is also generally used for organic matter in sedimentary rocks that is insoluble in common organic and inorganic solvents. Kerogen also yields oil when the shale containing kerogen is heated to temperatures sufficient to cause destructive distillation.

The formation of kerogen is open to speculation. It can only be assumed that the similar types of plant debris that went into the formation of petroleum also played a role in the formation of kerogen. Of the main chemical families constituting organic matter from living organisms (Chapter 3), it has been postulated that lipids and lignin are generally more resistant to decomposition and may be considered the chemical species more likely to produce kerogen (Tissot et al., 1978; Erdman, 1981). Indeed, lipids (with the exception of the ester links), contain a high proportion of carbon–carbon bonds, which require considerable energy for cleavage. Lignin, consisting of polymers and copolymers of phenylpropenyl alcohols (Chapter 3), decomposes only very slowly under attack by specific organisms (lignolytic fungi in particular). It is assumed that the other chemical species that make up the original organic detritus decompose at a much quicker rate to produce the constituents that eventually become part of the petroleum.

Once incorporated into sediments, the organic matter is buried under increasing depths as deposition of the mineral matter continues (sedimentation). Within the sediment, the physicochemical and biological environment is then gradually modified by the following events: (1) compaction; (2) decrease in water content; (3) cessation of bacterial activity; (4) transformation of the mineral phase; and (5) to some extent, but largely unknown, an increase in temperature. Under these conditions, the skeletal structures of the lignin and lipids could be preserved to a significant degree. Should this be the case, there is the distinct possibility that oil and kerogen are produced from the organic material by simultaneous or closely consecutive processes (Figure 4.1). There is also a theory that lignin derivatives do not usually form oil but are more likely to produce coal (Speight, 1994). However, it must be remembered that these statements are theories, and proof, other than that obtained by laboratory experiments, is difficult to obtain.

As already noted, kerogen has an implied role in the formation of petroleum and the term *kerogen* has also been used generally to indicate the material that is a precursor to petroleum. However, caution is advised in choosing the correct definition since there is a distinct possibility that kerogen, far from being a precursor to petroleum, is one of the byproducts of the petroleum generation and maturation processes and may not be a direct precursor to petroleum.

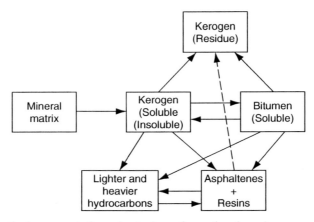

FIGURE 4.1 Hypothetical representation of petroleum formation from kerogen.

The role played by kerogen in the petroleum maturation process is not fully understood, although it is believed to be considerable (Tissot and Welte, 1978; Durand, 1980; Pelet and Durand, 1984; Hunt, 1996). What obviously needs to be addressed more fully in terms of kerogen participation in petroleum generation is the potential to produce petroleum constituents from kerogen by low-temperature processes rather than by processes that involve the use of temperatures in excess of 250°C (>480°F).

Petroleum precursors and petroleum itself is indeed subject to elevated temperatures in the subterranean formations due to the geothermal gradient. Although the geothermal gradient varies from place to place, it is generally of the order of 25°C to 30°C/km (15°F/1000 ft or 120°C/1000 ft i.e., 0.015°C per foot of depth or 0.012°C per foot of depth). This leaves serious question about whether or not the material has been subjected to temperatures in excess of 250°C (>480°F).

Such experimental work is interesting insofar as it shows similar molecular moieties in kerogen and petroleum (thereby confirming similar origins for kerogen and petroleum). However, the absence of geological time in the laboratory is not a reason to increase the temperature and it must be remembered that application of high temperatures (>250°C, <480°F) to a reaction not only increases the rate of reaction (thereby making up for the lack of geological time), but can also change the nature and the chemistry of a reaction. In such a case, the geochemistry is altered. Further, the introduction of a pseudoactivation energy in which the activations energy of the kerogen conversion reactions are reduced leave much to be desired, because of the assumption required to develop this pseudoactivation energy equation(s). In fact, not only will the oil window (the oil-producing phase) vary from kerogen-type to kerogen-type; it is not valid to use a fixed set of kinetic parameters within each of these groups (Whelan and Farrington, 1992). Therefore, a thorough investigation is needed to determine the chance of such high temperatures being present during the main phase, or even various phases, of petroleum generation.

By far the major part of the organic material present in most rocks (or shales) occurs as kerogen, an insoluble organic solid of variable composition, which is usually finely dispersed throughout the mineral matrix. Typical kerogen-containing shales are nonporous, impermeable strata containing approximately 5% to 20% w/w organic material and the remainder (80% to 95% w/w) as the mineral matrix. The organic material is often further defined as insoluble material (kerogen) and material (bitumen) extractable from the mineral matrix by organic solvents. The term *bitumen*, as used here, is a term of convenience and should not be confused with naturally occurring bitumen, also referred to as extra heavy oil (Chapter 1). It would be more correct to add a qualifier so that the name of the extractable material becomes kerogen–bitumen.

In order to explain the role of kerogen in the formation of petroleum, the biogenic theory of petroleum formation is espoused as opposed to the abiogenic theory of petroleum formation (Chapter 3). In simple terms, petroleum has formed throughout much of the Earth's history; in fact, oil is being formed in some parts of the Earth even today. Almost all oil and gas comes from tiny decayed plants, algae, and bacteria. At certain times in the Earth's history, conditions for oil formation have been particularly favorable. For example, in the Jurassic period (one of the oil-forming periods about 150 million years ago), the seas and swampy areas were rich in microscopic plants and animals.

When these died, they slowly sank to the bottom forming thick layers of organic material. This, in turn, became covered in layers of mud that trapped the organic material. The layers of mud prevented air from reaching the organic material. Without air, the organic material did not decay in the same way as organic material decays in the presence of air such as, for example, in a compost heap. As the layers of mud grew in thickness, pressure increased and the temperature also increased. The increasing temperature, pressure and anaerobic bacteria (microorganisms that can live without oxygen) started acting on the organic material. As this happened, the material was changed (matured) into petroleum and natural gas.

The Van Krevelen diagram (a plot of the atomic hydrogen–carbon ratio vs. the atomic oxygen–carbon ratio) (Figure 4.2), derived from the elemental analysis of kerogen and coal, is a very practical means of studying kerogen composition. The position of kerogen in the H/C–O/C diagram is related to the total quantity of hydrocarbons, which in turn is a function of the relative amounts of aromatic hydrocarbon structures. The data for kerogen analysis in the H/C–O/C diagram can be considered to describe the evolutionary path for kerogen from different precursors. Oil and gas are believed to have formed during this evolutionary path. Analysis of the minor elements, sulfur and nitrogen, is much more difficult to simulate and may require a more detailed framework.

The general characteristics of the four types of kerogen are:

Type I Kerogen

- Alginite
- Hydrogen/carbon atomic ratio >1.25
- Oxygen/carbon atomic ratio <0.15
- Tendency to readily produce liquid hydrocarbons
- Derived principally from lacustrine algae
- Has few cyclic or aromatic structures
- Formed mainly from proteins and lipids

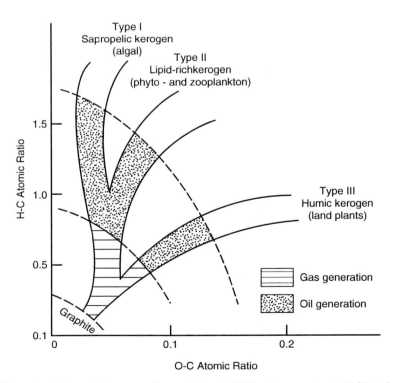

FIGURE 4.2 Hypothetical evolutionary pathways (atomic H/C ration vs. atomic O/C ratio) of kerogen.

Type II Kerogen

- Hydrogen/carbon atomic ratio <1.25
- Oxygen/carbon atomic ratio 0.03 to 0.18
- Tends to produce a mix of gas and oil
- Several types: exinite, cutinite, resinite, and liptinite
- Exinite is formed from pollen and spores
- Cutinite is formed from terrestrial plant cuticle
- Resinite is terrestrial plant resins, animal decomposition resins
- Liptinite is formed from terrestrial plant lipids and marine algae

Type III Kerogen

- Hydrogen/carbon atomic ratio <1.0
- Oxygen/carbon atomic ratio 0.03 to 0.3
- Material is thick, resembling wood or coal
- Tends to produce coal and gas
- Has very low hydrogen because of the extensive ring and aromatic systems

Formed from terrestrial plant matter that is lacking in lipids or waxy matter; forms from cellulose, the carbohydrate polymer that forms the rigid structure of terrestrial plants, lignin, another carbohydrate polymer (polysaccharide) that binds the strings of cellulose together, and terpenes and phenolic compounds in the plant

Type IV Kerogen (residue)

- Hydrogen/carbon atomic ratio <0.5
- Contains mostly decomposed organic matter in the form of polycyclic aromatic hydrocarbons
- Little or no potential to produce hydrocarbons

4.3 COMPOSITION

In very general terms, the hydrogen content of kerogen falls between that of petroleum and that of coal, but this varies considerably with the source, so that a range of values is found. This has been suggested as reflecting an overlap between terrestrial and aquatic origin. In fact, high lipid content, consistent with the occurrence of aquatic plants in the source material, appears to be diminished in kerogen by lignin of terrestrial origin (Whitehead, 1982; Scouten, 1990; and references cited therein). In fact, kerogen is best represented as a macromolecule that contains considerable amounts of carbon and hydrogen. Further, it is the macromolecular and heteroatom nature of kerogen with up to 400 heteroatoms (nitrogen plus oxygen plus sulfur) for every 1000 carbon atoms occurring as an integral part of the macromolecule that classifies kerogen as a naturally occurring heteroatomic material.

4.4 CLASSIFICATION

In addition to being classified as a naturally occurring heteroatomic material, kerogen can be subclassified into three different types (I, II, and III). These types of kerogen originate because of the different kinds of debris deposited in the sediment and also because of the conditions that prevail in that sediment over geological time. As it is initially deposited in a recent sediment, each type of debris may have a characteristic range of composition that can depend on local conditions, such as the types of flora and fauna that contribute to the debris. As the sediment is buried deeper or hotter and for a longer time, the organic material in the sediment undergoes maturation to give oil, gas, or a mixture of the two (Tissot and Welte, 1978; Hunt, 1996).

Type I kerogen is rich in lipid-derived aliphatic chains and has a relatively low content of polynuclear aromatic systems and of heteroatomic systems. The initial atomic H/C ratio is high (1.5 or more), and the atomic O/C ratio is generally low (0.1 or less). This type of kerogen is generally of lacustrine origin. Organic sources for the type I kerogen include the lipid-rich products of algal blooms and the finely divided and extensively reworked lipid-rich biomass deposited in stable stratified lakes.

Type II kerogen is characteristic of the marine oil shales. The organic matter in this type of kerogen is usually derived from a mixture of zooplankton, phytoplankton, and bacterial remains that were deposited in a reducing environment. Atomic H/C ratios are generally lower than for type I kerogen, but the O/C atomic ratios are generally higher for type II kerogen than for type I kerogen. Organic sulfur levels are also generally higher in the type II kerogen. The oil-generating potential of type II kerogen is generally lower than those of the type I kerogen (i.e., less of the organic material is liberated as oil upon heating a type II kerogen at the same level of maturation).

Type III kerogen is characteristic of coals and coaly shales. Easily identified fossilized plants and plant fragments are common, indicating that this type of kerogen is derived from woody terrestrial material. These materials have relatively low atomic H/C ratios (usually <1.0) and relatively high atomic O/C ratios (>0.2). Aromatic and heteroaromatic contents are high, and ether units (especially of the diaryl ethers) are important, as might be anticipated for a lignin-derived material. Oil-generating potential is low, but gas-generating potential is high.

At the beginning of the type I path, the kerogen types have a strongly aliphatic nature; at the beginning of the type III path, the kerogen consists largely of aromatic structures that carry oxygen functions. At the beginning of the type II path, and in general for intermediate paths between type I and type III, elemental analysis supplies little information about the chemical structure.

4.5   ISOLATION

The first step in any study of the behavior and structure of kerogen has generally been the isolation of a kerogen concentrate (Forsman and Hunt, 1958; Forsman, 1963; Robinson, 1969; Saxby, 1976). A variety of methods can be employed to isolate fractions of organic material without altering the structure of the native kerogen. There are also those methods intended for degradation of the organic material in a controlled manner. The terminology of the material isolated by such methods is often based on the method employed. Therefore, an understanding of these methods assists in understanding the terminology. For example, a particular method may result in the generation of hydrocarbon products as well as more complex products that are heteroatomic or high in molecular weight.

Physical methods to produce an organic-rich kerogen concentrate are of interest because exposure of the kerogen to the strong acid or base is avoided, thereby lessening the chance of chemical alteration. Such methods generally involve the potential for contamination of the kerogen with materials used to effect the separation. However, in many cases the potential impact of such contaminants can be limited by using only one or a small number of known and easily identified chemical species. Among the more important physical methods for kerogen concentration are sink–float, oil agglomeration, and froth flotation methods (Luts, 1928; Hubbard et al., 1952; Vadovic, 1983).

By far, the most common technique for kerogen isolation involves acid demineralization of the shale to produce the kerogen concentrate. To dissolve the mineral matrix, a series of successive treatments with a hydrochloric acid–hydrofluoric acid mixture (at approximately 65°C, 150°F) is employed (Durand and Nicase, 1980). On the other hand, demineralization with a reduced chance of organic alteration has been achieved by carrying out the treatment at a lower temperature (20°C, 70°F) and for shorter times (Scouten et al., 1987). The use of base to dissolve silicates has also been investigated followed by an acid treatment to dissolve carbonates (McCollum, 1987).

The isolation of kerogen from mineral matrices also depends on the extent of the interactions between the kerogen and the various minerals. From the results of model compound–model mineral tests, interactions between acid clay minerals and nitrogen-containing organic compounds have been identified as much stronger than other likely candidates for kerogen–mineral interactions (Siskin et al., 1987a,b). The importance of this finding led to the use of differential wetting, a phenomenon typically associated with physical separation methods, as critical to the success of kerogen separation. Thus, efficient kerogen recovery can be achieved by adding an organic solvent that wets and swells the kerogen, thereby diminishing some of the nitrogen–mineral interactions and aiding the physical sink–float separation. Thus, both chemical and physical aspects are important for production under mild conditions of a kerogen concentrate with, presumably, minimal structural alterations.

The laboratory sink–float methods offer mild conditions to minimize chemical alteration and can produce a kerogen concentrate with low ash content. There are disadvantages to this technique, however, which include (1) rejection of organic compounds (leading to low recovery of the kerogen) and (2) the possibility of kerogen fractionation along with mineral rejection. The oil agglomeration method (Quass, 1939) relies on selective wetting of kerogen particles by an oily paste material such as hexadecane (see also Himus and Basak, 1949; Smith and Higby, 1960; Robinson, 1969).

4.6 METHODS FOR PROBING KEROGEN STRUCTURE

4.6.1 ULTIMATE (ELEMENTAL) ANALYSIS

Although not strictly a method for probing the structure of kerogen, elemental analysis offers valuable information about the atomic constituents of kerogen. The elemental analysis of kerogen is a method for characterizing the origin and evolution of sedimentary organic matter. Elemental analysis also establishes a framework within which other physicochemical methods can be used more effectively.

The Van Krevelen diagram (a plot of the atomic hydrogen–carbon ratio vs. the atomic oxygen–carbon ratio) (Figure 4.2), derived from the elemental analysis of kerogen and coal, is a very practical means of studying kerogen composition. The position of kerogen in the H/C–O/C diagram is related to the total quantity of hydrocarbons, which in turn is a function of the relative amounts of aromatic hydrocarbon structures. The data for kerogen analysis in the H/C–O/C diagram can be considered to describe the evolutionary path for kerogen from different precursors. Oil and gas are believed to be formed during this evolutionary path. Analysis of the minor elements, sulfur and nitrogen, is much more difficult to simulate and may require a more detailed framework.

Inferences regarding kerogen structure are drawn from the results of bitumen (kerogen extract) analyses. The conclusions are generally based on the premise that the bitumen is analogous to original organic matter, that is, the bitumen represents units of the precursor that did not bound into the insoluble three-dimensional macromolecular network of the kerogen. It is also assumed that the bitumen is representative of units of the kerogen structure that have been cleaved more or less intact, with little or no structural alteration, from the kerogen by thermal treatment.

Many different compound types have been identified (by extraction procedures) as part of the kerogen matrix, but their mode of inclusion in kerogen is open to speculation. For example, the compounds isolated from kerogen include paraffins (Cummins and Robinson, 1964; Anderson et al., 1969), steranes (Anderson et al, 1969), cycloalkanes (Anders and Robinson, 1971), aromatics (Anders et al., 1973), and polar compounds (Anders et al., 1975). Kerogen generally contains only a small amount of bitumen (<15% w/w of the total organic matter), but the bitumen usually has a higher hydrogen content than the corresponding kerogen. This corresponds to a lower proportion of aromatics, as well as nitrogen-, oxygen-, and sulfur-containing compounds. This is an obvious limitation to due usefulness of structural inferences drawn from the bitumen composition.

The volatile oil generally represents a much larger fraction of the original organic material (usually 50% or more of the available organic carbon). The methods employed for analysis of the product oils are similar to those used for petroleum (Uden et al., 1978; Fenton et al., 1981; Holmes and Thompson, 1981; Williams and Douglas, 1981; Regtop et al., 1982). As a consequence of their thermal treatment, the volatile oils are usually richer in aromatics and olefins than the starting kerogen, but are relatively deficient in nitrogen- and sulfur-containing compounds. As for petroleum coking (Speight, 1970, 1987), the nitrogen- and sulfur-containing species are concentrated in the nonvolatile char. The volatile oils are a greater reflection of the thermal treatment used to produce the oil and, consequently, are not too reliable in terms of an accurate picture of kerogen structure. Thus, although many inferences about kerogen structure have been drawn from bitumen and oil analyses, their limitations must be recognized.

4.6.2 FUNCTIONAL GROUP ANALYSIS

Attempts to characterize the oxygen functional groups in kerogen have focused on acid demineralization (successive treatments with hydrochloric acid and hydrofluoric acid) to prepare a kerogen concentrate. The concentrate has then been treated by wet chemical methods to determine the distribution of oxygen functional groups (Fester and Robinson, 1964, 1966; Robinson and Dineen, 1967).

4.6.3 OXIDATION

Oxidative degradation, one of the primary methods of structural determination used in natural product chemistry, has also been employed to examine kerogen structure (Vitorovic, 1980). Alkaline permanganate and chromic acid have been the two most widely used oxidants, although ozone, periodate, nitric acid, perchloric acid, air or oxygen, hydrogen peroxide, and electrochemical oxidation (among many other reagents) have also been used.

Alkaline permanganate oxidation of kerogen has been carried out in two very different ways. Older work (Bone et al., 1930) generally involved the use of the carbon balance method developed for studies of coal. The products of this exhaustive oxidation are carbon dioxide, oxalic acid ($HO_2C–CO_2H$, from aromatic rings), nonvolatile benzene polycarboxylic acids, and unoxidized organic carbon. However, because aliphatic material is oxidized mainly to carbon dioxide, this method is not well suited to probe the structure of kerogen that is highly aliphatic. This led to the development of stepwise procedures to give products that retain more structural information about the starting kerogen.

The careful development of the stepwise alkaline permanganate method represented attempts to minimize unwanted secondary oxidation of the first-formed product by adding the oxidant (KMnO$_4$ in 1% aqueous KOH) in small portions. In some cases, the acids obtained from stepwise oxidation proved to be of such high molecular weight that they precipitated upon acidification, were insoluble in ether, and were difficult to characterize. In these cases, the precipitated acids were subjected to further stepwise oxidation to produce the desired ether-soluble acids of lower molecular weight.

In general terms, alkaline permanganate oxidizes alkylbenzenes, alkylthiophenes, and alkylpyridines (but not alkylfurans) to the corresponding carboxylic acids. This is not true when the aromatic ring bears an electron-donating group (e.g., −OH, −OR, or −NH$_2$). In such cases, degradation of the aromatic portion is usually rapid. Condensed aromatics are also attacked and benzene polycarboxylic acids are produced. In addition, caution is advised since benzene itself also reacts (slowly) in hot alkaline permanganate solutions. Olefins are rapidly converted into the corresponding glycol, which are then cleared to carboxylic acids.

$$RCH = CHR' \rightarrow RCH(OH)CH(OH)R' \rightarrow RCO_2H + RCO_2H$$

Cyclic olefins yield dicarboxylic acids. Enolizable ketones are also cleaved, presumably via the enol.

Tertiary and benzylic carbon–hydrogen groups are reacted to afford tertiary alcohols. In simple alkyl systems, the presence of an alcohol group markedly accelerates the rate of this reaction. Primary and secondary alcohols are oxidized to the corresponding acids and ketones, and alkaline permanganate oxidation degrades the porphyrin nucleus, giving pyrrole-2,4-dicarboxylic acid derivatives under mild conditions. The porphyrinic side chains −CH$_3$, −CH$_2$CH$_3$, CH$_2$CH$_2$CO$_2$H, −COCH$_3$, and −CH(OH)CH$_3$ persist in the degradation products, but the −CH=CH$_2$ and −CHO side chains are both oxidized to –CO$_2$H.

The structural information obtained from the oxidation of kerogen by chromic acid (and other chromium-containing oxidants) is usually similar to that obtained with alkaline permanganate (Wiberg, 1965; Lee, 1980; Vitorovic, 1980). However, for any particular technique the recovery of organic carbon in the oxidation products may be as low as 10% of the original carbon. Consequently, the alkaline permanganate procedure is often considered superior for elucidating kerogen structure.

Nitric acid has also been used for the oxidation of kerogen (Robinson et al., 1963). However, it must be recognized that nitric acid reacts in different ways depending on the temperature, time, and concentration. Thus, in addition to the anticipated oxidation reactions, aromatic structures in the kerogen or even in the products are nitrated. However, nitric acid has been successfully used for investigating aliphatic structural units, and the data are often complementary to other structural studies.

Oxidants such as ozone (Rogers, 1973), air (Robinson et al., 1965), oxygen (Robinson et al., 1963), and hydrogen peroxide (Kinney and Leonard, 1961) have also been used for the oxidative degradation of kerogen. As in other cases, it is strongly recommended that structural data not be compiled in an absolute manner on the basis of one oxidant. The data from the various methods should be employed in a complementary manner so that an overall model can be compiled that explains the behavior of the kerogen under different reaction conditions.

4.6.4 THERMAL METHODS

Under favorable conditions, oxidation methods can give reasonably high recoveries of organic material, but structural alteration does occur and some features are obliterated. A particularly attractive approach to minimize the formation of intractable residues, and the obliteration of important structural features, has been to heat the kerogen at moderate temperatures for prolonged periods and then to extract the degraded kerogen (Hubbard and Robinson, 1950; Bock et al., 1984a,b). Thus, it may be possible to achieve complete conversion of the kerogen and recover >90% w/w as lower molecular weight (liquid) products. However, product alteration is also a feature of thermal reactions, but there is the distinct possibility that the thermal fragments that were allowed to escape from the reaction zone (by virtue of their volatility) would preserve some of the original (skeletal) features of the kerogen. Overall, the mild heat–soak–extraction method is a complementary method to the oxidation studies for providing lower molecular weight products for structural studies.

The use of micropyrolysis coupled with gas chromatography–mass spectrometry (GC–MS) can also provide valuable information about the structural units in kerogen (Schmidt-Collerus and Prien, 1974). In addition to the on-line micropyrolysis–GC–MS studies, larger samples of kerogen have been pyrolyzed to obtain products that were fractionated by chromatography (ion exchange, complexation with ferric chloride ($FeCl_3$), and silica gel) into compound classes, and then by gel permeation chromatography into fractions of increasing molecular weight. These samples can then be investigated by conventional mass spectrometric techniques as well as by other spectroscopic studies.

4.6.5 Acid-Catalyzed Hydrogenolysis

The use of hydrogenolysis in the presence of stannous chloride ($SnCl_2$) to degrade kerogen (Hubbard and Fester, 1958) affords good yields of liquid products for characterization purposes. Under these conditions, the majority of the heteroatoms are generally removed: nitrogen as ammonia, oxygen as carbon dioxide or as water, and sulfur as hydrogen sulfide. From these data, it was concluded that the nitrogen, oxygen, and sulfur functionalities comprise internuclear links (as opposed to their existence in ring systems) in the kerogen structure. These conclusions may appear to be at variance with other data, but this may be a result of the severe degradation conditions employed in the experiments, and examination under milder conditions is warranted.

4.7 STRUCTURAL MODELS

The need to gather a very large mass of information about kerogen structure into a compact form, useful for guiding research and development, has led to models for kerogen structure. These models are not intended to depict the molecular structure of kerogen. At least not in the sense that the double helix describes the structure of deoxyribonucleic acid (DNA), or even in the sense that synthetic polymers are described in terms of monomers joined to form chains that have well-defined structures. The kerogen models represent attempts, based on the available data, to depict a collection of skeletal fragments and functional groups as a three-dimensional network in the most reasonable manner possible.

However, it must be recognized that no single analytical technique provides sufficient information to construct a precise model of the macromolecular structure of kerogen. Thus, most workers now use a multidimensional (multiple-technique) approach, but as is quite often the case, the techniques used by different workers have different strengths and may emphasize different features of kerogen structure. Sample-to-sample variation and the use of different isolation techniques also complicate the issue.

Several models have been proposed for the structure of kerogen in which a multidimensional approach has been employed. Such approaches are extremely valuable since they bring together the results of several analytical methods. Indeed, the success of such approaches in the deduction of structural types in kerogen is also paralleled by the use of a similar approach to the deduction of the structural types that occur in the asphaltene fraction (Chapter 11). Such structural models are of interest, because they give an overall picture of the perception of the kerogen structure and the models may even allow predictions of properties and behavior. If the model does not match such behavior and properties, the model must be reworked since it then becomes of little, if any, value.

A model derived for kerogen was based on the chromic acid oxidative degradation already discussed (Simoneit and Burlingame, 1974 and references cited therein). Additional information was provided by studies of the bitumen, again primarily by mass spectrometry, and was incorporated into the structural model that included regions of undefined structure containing trapped organic compounds of unknown nature and bearing side chains linked to the main structure by nonhydrolyzable C–C and hydrolyzable ester linkages. Ester linkage was also believed to be present and the model also includes an alicyclic (naphthene) ring. To understand this model, it is important to recall that the oxidation products (acids and ketones), upon which most of this structure is based, represent only a fraction of total organic carbon.

Another model for kerogen was developed on the basis of the data derived from the stepwise alkaline permanganate oxidation of kerogen, which produced high yields of carboxylic acids (Djurcic et al., 1971). Based on the oxidation results, a cross-linked macromolecular network structure was proposed. The most striking feature of this model is the predominance of straight-chain groups in the backbone of the network. The network bears both linear and branched side chains; branching points are indicated in the model by open circles. This model accommodates many important experimental observations, including reversible swelling and gel-like rubbery behavior in the swollen state, but does not satisfactorily account for the aromatic carbons observed by carbon 13 magnetic resonance spectroscopy or the nitrogen and sulfur contents determined by elemental analysis.

Another kerogen model (Schmidt-Collerus and Prien, 1974) was assembled from the subunits identified by micropyrolysis–mass spectrometry studies. Key features of this model include formulation as a three-dimensional macromolecular network and a very uniform hydrocarbon portion comprised mostly of small alicyclic and partially hydrogenated aromatic subunits with few heterocyclic rings. Long-chain alkylene and isoprenoid units and ethers serve as interconnecting bridges in this structure. Entrapped species (bitumen) include long-chain alkanes and both n-alkyl and branched carboxylic acids. This model provides a useful view of the types and role of hydrocarbon units but de-emphasizes heteroatom functional groups and rings, presumably because groups containing these elements would not be detected efficiently by the micropyrolysis technique.

The structure of kerogen has also been probed by a wide variety of techniques, including stepwise alkaline permanganate and dichromate-acetic acid oxidation, electrochemical oxidation and reduction (in nonaqueous ethylenediamine–lithium chloride), and x-ray diffraction techniques (Barakat and Yen, 1988 and references cited therein). It was concluded for that particular sample of kerogen that

1. Aromaticity was low but isolated carbon–carbon double bonds were possible.
2. Structure largely comprised three-to-four-ring naphthenes.
3. Oxygen was present mostly as esters and as ethers.
4. Kerogen structure comprises a three-dimensional network and ethers serve as cross-links in this network (see the data of Fester and Robinson, 1966).
5. Additional linkages provided by disulfides, nitrogen heterocyclic groups, unsaturated isoprenoid chains, hydrogen bonding, and charge-transfer interactions.

Using these components as building blocks, a multipolymer network was envisaged (Yen, 1974, 1976). It was also pointed out that the extractable bitumen molecules could resid, more or less freely depending on their size, within the network.

To further account for the observed variations in the products obtained from the individual steps of stepwise permanganate oxidation, it was suggested that a core plus shell arrangement existed for the individual kerogen particles. The core was visualized as a cross-linked region containing most of the alkyl and alkylene chains and the bulk of the kerogen as naphthenic ring structures. On the other hand, the shell is more tightly cross-linked and contains most of the heteroatom functional groups and heterocyclic rings. This is interesting from the geochemical viewpoint, since the outer shell of this model is that part of the kerogen in contact with the mineral matrix; heteroatom functions tend to interact more strongly with minerals than do the hydrocarbon chains. Organic–mineral interactions in the resulting composite would then be ideally situated to hinder physical separation of minerals from kerogen. This picture is consistent with the data of other workers (Siskin et al., 1987a,b) on chemically assisted oil shale enrichment.

Another hypothetical model for kerogen is based on the results of a multidimensional approach to probing kerogen structure, which also includes a detailed analysis of the functional groups in the kerogen (Scouten et al., 1987; Siskin et al., 1995). A comparison with other kerogen models serves to illustrate some of the key features of this model for kerogen. Aliphatic material is the most obvious feature of this model, and the aliphatic moieties are longer and more linear than those in other kerogen models. In addition, the aliphatic moieties are present, both as alkylene bridges and as alkyl side chains, and secondary structure due to paraffin–paraffin interactions is important in the kerogen. Naphthenic and partially hydrogenated aromatic rings also make an important contribution to the aliphatic moieties. The average ring system is only slightly larger than that in other kerogen models, but the size distribution is appreciably broader; a significant number of the larger four- to five-ring systems are present.

Yet another approach to deriving models for kerogen structure involves a more generalized procedure in which models representative of the three types of kerogen and of the asphaltenes from the corresponding oils as a function of maturity were developed (Tissot and Espitalié, 1975; Behar and Vandenbroucke, 1986). Emphasis in this work was placed on elucidating the chemistry of maturation for the three kerogen types and representing kerogen at the beginning of diagenesis (excluding the early stages of diagenesis, which is probably dominated by microbial action) (Behar and Vandenbroucke, 1986). These models provide an interesting view of the structural relationships between the three types of kerogen (Vandenbroucke, 2003). It is of interest to note the similarities in the models of asphaltenes from kerogen (especially in terms of the size of the polynuclear aromatic systems) with current postulates of the structures of petroleum asphaltenes (Chapter 11).

4.8   KEROGEN MATURATION

The theory that the petroleum precursors form a mix that is often referred to as **protopetroleum** (also referred to as primordial precursor soup or petroleum porridge) is an acceptable generalization. The continuation of the concept to the postulation of kerogen as a petroleum precursor has received considerable attention. There is little doubt that kerogen is a descendent of organic detritus (Cane, 1976), but the precise role played by kerogen in the generation of petroleum is still open to speculation. Kerogen, when heated under a variety of conditions, produces oil. It does not follow, however, that similar natural processes were responsible for the production of oil that is recovered from various reservoirs today.

Some of the temperatures employed in modern investigations are usually much higher than those anticipated on the basis of the geothermal gradient, that is, an increase in temperature of 0.012°C (0.015°C) for every foot of depth, and that might have occurred during the maturation process (Landes, 1966). The argument that increasing the temperature merely increases the rate of reaction is certainly open to criticism. Increased temperatures are known not only to increase reaction rates, but also to alter the chemistry of the reactions. In summary, there may be several inconsistencies in the theory that kerogen is the direct precursor to petroleum.

After the relatively rapid alterations that take place shortly after the initial deposition of organic matter in sediment, the surviving organic matter undergoes additional changes. Kerogen is presumably produced from part of the organic detritus and oil from the other part. Kerogen maturation involves the loss of hydrogen and oxygen; hydrogen is lost primarily as methane and other light hydrocarbon gases, water, and hydrogen gas; oxygen is lost primarily as water and carbon oxides (Brooks, 1981; Brooks and Welte, 1984). Maturation increases with increased exposure of the organic matter to time, temperature, and pressure. The catalytic effects of minerals in the sediment can accelerate and affect this process, as can the presence of water (Durand, 1980; Burnham, 1995; Siskin and Katritzky, 1995).

Kerogen is transformed into a variety of lower molecular weight compounds when subjected to higher temperature and pressure. An ideal situation to study this transformation is in a sedimentary sequence in which all geological and geochemical parameters, except burial, remain constant. The microscopic, chemical, and physical methods of the progressive transformation of kerogen are a function of maximum burial depth (Karweil, 1956; Louis and Tissot, 1967; Tissot et al., 1971, 1974; Durand et al., 1972; Espitalité et al., 1973).

Transformations occurring over geological periods are in essence not accessible to human experiment, as we shall never be able to account for the millions of years involved in these natural processes. Therefore in the laboratory it has been found necessary, perhaps erroneously, to increase temperatures to accelerate the reactions and compensate for time. Because a valid basis for comparison has not yet been established, it is not possible to determine the extent to which the laboratory processes are representative of natural transformations.

Although the degradation of kerogen and the associated weight loss are continuous and progressive, three successive stages may be distinguished as occurring in the laboratory heating studies of kerogen (type II):

First, at approximately 350°C (660°F), there is a weight loss due mostly to the production of water and carbon dioxide. Elemental analysis and infrared spectroscopy show a loss of oxygen associated with a diminution of the carbonyl (C=O band). The elemental composition of the treated sample is compared with that of natural samples buried to depths of 1000–1500 m (3300–5000 ft). This third statement should not be taken as conclusive proof of the viability of the high-temperature maturation theory. After all, it is possible to burn toast to a product having the elemental composition of a low-rank coal. And the action of concentrated sulfuric acid on sugar also gives a product with the composition of a low-rank coal. But there is no claim that coal originated from the immersion of the plant precursors in a lake of concentrated sulfuric acid. This might offer a new experience to the concept of **acid rain** (Chapter 31).

From about 350°C to 500°C (660°F to 930°F), kerogen encounters its major degradation stage. The products are mostly hydrocarbons (particularly aliphatic hydrocarbons), and the elemental composition changes rapidly. The atomic H/C ratio decreases to approximately 0.5 and dealkylation of aromatic systems occurs. The carbonyl (C=O) band progressively disappears: acids are removed first and then esters, which are more stable, whereas ketones are progressively eliminated. This step of evolution has been correlated with the catagenesis occurring in sedimentary basins. But again, and for reasons given in the preceding paragraph, the viability of the comparison is suspect.

Finally, at temperatures above 500°C (930°F), major structural rearrangements occur. Aromatic systems are reputed to gather to form clusters or aggregates within the range of 100–200 Å. Such transformations are compared with the data from the examination of samples from deep natural samples. This process has been compared with the natural interval known as **metagenesis**. And, the comment at the end of the preceding paragraph still stands.

It is necessary to recognize, however, that the quantitative aspects of kerogen evolution vary from one type to the other as a result of differences in the original composition of kerogen and that these generalizations may not reflect the true nature of the chemistry that occurs during the maturation process. These conclusions are based on gross changes to analyzable chemical entities. The detailed chemistry is more difficult to analyze.

The total amount of products generated upon heating, as measured by the weight loss, is greatest for type I kerogen and least for type III kerogen. The proportion of hydrocarbon products is higher for type I kerogen and type II kerogen and lower for type III kerogen. In addition, the light to medium hydrocarbon ratio (gas–oil ratio) is low for type I kerogen, moderate for type II kerogen, and high for type III kerogen. In contrast to the hydrocarbon products, the proportion of carbon dioxide generated is higher for type III kerogen, and the transformation of type III kerogen is progressive and extends over a wide temperature range. The transformation interval is much narrower and the temperature of maximum weight loss is higher for type I kerogen.

Thus, the main feature in laboratory-simulated kerogen evolution is the emergence of a carbon order that progressively extends over wider areas and becomes stronger with increasing temperature. Elimination of the steric hindrances to ordering results in the formation of a wide range of compounds, including medium- to low-molecular-weight hydrocarbons, carbon dioxide, water, and hydrogen sulfide.

Interesting as these observations may be, natural modifications observed in sedimentary organic matter are brought about at relatively low temperatures with compensation by the long geological periods involved. In addition, the occurrence of catalytic effects should also be invoked to explain the low-temperature reaction processes that can occur over geological time. Thus, there are important differences between pyrolysis and natural evolution.

The concept of the low-temperature maturation of kerogen has been developed further (Quigley et al., 1987) and has resulted in the classification of kerogen into reactive and nonreactive types. Thus kerogen is postulated as composed of labile, refractory, and inert moieties that can decompose into gas and oil depending on the prevailing conditions. It is assumed that the general categories of materials or organisms that are shown to produce the oil or the kerogen are consistent with those chemicals assumed to be precursors to petroleum (Chapter 3). In fact, there has been the suggestion that asphaltenes contain the same chemical moieties as kerogen (Behar et al., 1984; Orr, 1986; Pelet et al., 1986). This can also be interpreted as a divergence in the maturation path in which kerogen takes one route and the protopetroleum another.

What also makes this hypothesis interesting, and perhaps more realistic than others, is the inclusion of a provision for the generation of petroleum without the intermediate formation of kerogen. I believe this is more in keeping with petroleum formation than any theory that requires kerogen as a necessary intermediate in the formation of petroleum.

REFERENCES

Anders, D.E. and Robinson, W.E. 1971. *Geochim. Cosmochim. Acta* 35: 661.
Anders, D.E., Doolittle, F.C., and Robinson, W.E. 1973. *Geochim. Cosmochim. Acta* 37: 1213.
Anders, D.E., Doolittle, F.G., and Robinson, W.E. 1975. *Geochim. Cosmochim. Acta* 39: 1423.

Anderson, Y.C., Gardner, P.M., Whitehead, E.V., Anders, D.E., and Robinson, W.E. 1969. *Geochim. Cosmochim. Acta* 33: 1304.

Barakat, A.O. and Yen, T.F. 1988. *Energ. Fuels* 2: 105.

Behar, F. and Vandenbroucke, M. 1986. *Rev. Inst. Fran. Petrol.* 41: 173.

Behar, F., Pelet, R., and Roucache, J. 1984. *Org. Geochem.* 6: 595.

Bock, J., McCall, P.P., Robbins, M.L., and Siskin, M. 1984a. United States Patent 4,458,757.

Bock, J., McCall, P.P., Robbins, M.L., and Siskin, M. 1984b. United States Patent 4,461,696.

Bone, W.A., Horton, L., and Ward, S.G. 1930. *Proc. Roy. Soc.* (Lond.) 127A: 480.

Brooks, J. 1981. *Organic Maturation Studies and Fossil Fuel Exploration*. Academic Press, London.

Brooks, J. and Welte D. 1984. *Advances in Petroleum Geochemistry*. Academic Press, London.

Burnham, A.K. 1995. In *Composition, Geochemistry and Conversion of Oil Shales*. Snape, C. ed., Kluwer Academic, Dordrecht. p. 211.

Cane, R.F. 1976. In *Oil Shale*. Yen, T.F. and Chilingarian, G.V. eds, Elsevier, Amsterdam.

Cummins, J.J. and Robinson, W.E. 1964. *J. Chem. Eng. Data* 9: 304.

Djurcic, M.V., Murphy, R.C., Vitorovic, D., and Biemann, K. 1971. *Geochim. Cosmochim. Acta* 35: 1201.

Durand, B. 1980. *Kerogen: Insoluble Organic Matter from Sedimentary Rocks*. Editions Technip, Paris.

Durand, B., Espitalité, J., Nicase, G., and Combaz, A. 1972. *Rev. Inst. Fran. Petrol.* 27: 865.

Erdman, J.G. 1981. In *Origin and Chemistry of Petroleum*. Atkinson, G. and Zuckerman, J.J. eds., Pergamon Press, New York.

Espitalité, J., Durand, B., Roussel, J.C., and Souron, C. 1973. *Rev. Inst. Fran. Petrol.* 28: 37.

Fenton, M.D., Henning, H., and Ryden, R.L. 1981. In *Oil Shale, Tar Sands and Related Materials*. Stauffer, H.C. ed., American Chemical Society, Washington, D.C. p. 315.

Fester, J.I. and Robinson, W.E. 1964. *Anal. Chem.* 36: 1392.

Fester, J.I. and Robinson, W.E. 1966. *Coal Science*. Advances in chemistry series No. 55. American Chemical Society, Washington, D.C. p. 22.

Forsman, J.P. 1963. In *Organic Geochemistry*. Breger, I.A. ed., Pergamon Press, Oxford, UK. p. 148.

Forsman, J.P. and Hunt, J.M. 1958. In *Habitat of Oil*. Weeks, L.G. ed., American Association of Petroleum Geologists. Tulsa, OK. p. 747.

Himus, G. and Basak, G.C. 1949. *Fuel* 28: 57.

Holmes, S.A. and Thompson, L.F. 1981. *Proceedings. 14th Oil Shale Symposium*. Gary, J.H. ed., Colorado School of Mines Press, Golden, CO. p. 235.

Hubbard, A.S. and Fester, J.I. 1958. *Ind. Eng. Chem.* 3: 147.

Hubbard, A.B. and Robinson, W.E. 1950. Report of Investigations No. 4744. United States Bureau of Mines, Washington, D.C.

Hubbard, A.B., Smith, H.N., Heady, H.H., and Robinson, W.E. 1952. Report of Investigations No. 5725. United States Bureau of Mines, Washington. D.C.

Hunt, J.M. 1996. *Petroleum Geochemistry and Geology*. 2nd edn. W.H. Freeman, San Francisco.

Karweil, J. 1956. *Z. Deutsch. Geol. Gesell.* 107: 132.

Kinney, C.R. and Leonard, J.T. 1961. *J. Chem. Eng. Data* 6: 474.

Landes, K.K. 1966. *Oil Gas J.* 64(18): 172.

Lee, D.G. 1980. *The Oxidation of Organic Compounds by Permanganate Ion and Hexavalent Chromium*. Open Count Publishing, La Salle, IL.

Louis, M. and Tissot, B. 1967. *Proc. 7th World Petrol. Cong* 2: 47

Luts, K. 1928. *Brennstoff-Chem.* 9: 217.

McCollum, J.D. 1987. *Preprints. Div. Petrol. Chem. Am. Chem. Soc.* 32(1): 74.

Orr, W.L. 1986. *Org. Geochem.* 10: 499

Pelet, R., and Durand, B. 1984. In *Magnetic Resonance: Introduction, Advanced Topics, and Applications to Fossil Energy*. Perakis, L. and Fraissard, J. P. eds., D. Reidel, Norwell, MA.

Pelet, R., Behar, F., and Monin, J.C. 1986. *Org. Geochem.* 10: 481.

Quass, F.W. 1939. *J. Inst. Petrol.* 25: 81.

Quigley, T.M., Mackenzie, A.S., and Gray, J.R. 1987. In *Migration of Hydrocarbons in Sedimentary Basins*. Doligez, B. ed., Editions Technip, Paris.

Regtop, R.A., Crisp, P.T., and Ellis, J. 1982. *Fuel* 61: 185.

Robinson, W.E. 1969. In *Organic Geochemistry*. Eglinton, G. and Murphy, M.T.J. eds., Springer-Verlag, Berlin. p. 181.

Robinson, W.E. and Dineen, G U. 1967. *Proceedings. 7th World Petroleum Congress*. Elsevier, Amsterdam. p. 669.

Robinson, W.E., Lawlor, D.L., Cummins, J.J., and Fester, J.I. 1963. Report of Investigations No. 6166. United States Bureau of Mines, Washington, D.C.

Robinson, W.E., Cummins, J.J., and Dineen, G.U. 1965. *Geochim. Cosmochim. Acta* 29: 249.

Rogers, M.P. 1973. *Bibliography of Oil Shale and Shale Oil*. Bureau of Mines Publications. Laramie Energy Research Center, United States Bureau of Mines, Laramie, WY.

Saxby, J.D. 1976. In *Oil Shale*. Yen, T.F. and Chilingarian, G.V. eds., Elsevier, Amsterdam. p. 103.

Schmidt-Collerus, J.J. and Prien, C.H. 1974. Preprints. *Div. Fuel Chem. Am. Chem. Soc.* 19(2): 100.

Scouten, C.G., Siskin, M., Rose, K.D., Aczel, T., Colgrove, S.G., and Pabst, R.E. 1987. *Proceedings. 4th Australian Workshop on Oil Shale*. Brisbane, Australia. p. 94.

Simoneit, B.R.T. and Burlingame, A.L. 1974. *In Advances in Organic Geochemistry* 1973. Tissot, B. and Bienner, F. eds., Editions Technip, Paris. p. 191.

Siskin, M. and Katritzky, A.R. 1995. In *Composition, Geochemistry and Conversion of Oil Shales*. Snape, C. ed., Kluwer Academic, Dordrecht. p. 313.

Siskin, M., Brons, G., and Payack, J.F. 1987a. Preprints. *Div. Petrol. Chem. Am. Chem. Soc.* 32(1): 75.

Siskin, M., Brons, G., and Payack, J.F. 1987b. *Energ. Fuels* 1: 100.

Siskin, M., Scouten, C.G., Rose, K.D., Aczel, T., Colgrove, S.G., and Pabst, R.E., Jr. 1995. In *Composition, Geochemistry and Conversion of Oil Shales*. Kluwer Academic, Dordrecht. p. 143

Smith, J.W. and Higby, L.W. 1960. *Anal. Chem.* 32: 17.

Speight J.G. 1970. Fuel 49: 134.

Speight, J.G. 1987. Preprints. *Div. Petrol. Chem. Am. Chem. Soc.* 32(2): 413

Speight, J.G. 1994. *The Chemistry and Technology of Coal*. 2nd edn. Marcel Dekker, New York.

Tissot, B. and Espitalité, J. 1975. *Rev. Inst. Fran. Petrol.* 30: 743.

Tissot, B. and Welte, D.H. 1978. *Petroleum Formation and Occurrence*. Springer-Verlag, New York.

Tissot, B., Califet-Debyser, Y., Deroo, G., and Oudin, N.L. 1971. *AAPG Bull.* 55: 2177.

Tissot, B., Durand, B., Espitalité, J., and Combaz, A. 1974. *AAPG Bull.* 58: 499.

Tissot, B., Deroo, G., and Hood, A. 1978. *Geochim. Cosmochim. Acta* 42: 1469.

Uden, P.C., Siggia, S., and Jensen, H.B. 1978. *Analytical Chemistry of Liquid Fuel Sources*. Advances in Chemistry Series No. 170. American Chemical Society, Washington, D.C.

USGS 1995. United States Geological Survey. *Dictionary of Mining and Mineral-Related Terms*. 2nd edn. Bureau of Mines & American Geological Institute. Mines Bureau Special Publication SP 96-1.

Vadovic, C.J. 1983. In *Geochemistry and Chemistry of Oil Shales*. Miknis, F.P. and McKay, J.F eds., Symposium Series No. 230. American Chemical Society, Washington, D.C. p. 385.

Vandenbroucke, M. 2003. Oil and gas science and technology, *Rev. Inst. Fran. Petrol. 58: 243.*

Vitorovic, D. 1980. In *Kerogen*. Durand, B. ed., Editions Tecnip, Paris. p. 301.

Whelan, J.K. and Farrington, J.W. 1992. In *Organic Matter: Productivity, Accumulation, and Preservation in Recent and Ancient Sediments*. Columbia University Press: New York.

Whitehead, E.V. 1982. In *Petroanalysis'81. Advances in Analytical Chemistry in the Petroleum Industry 1975–1982*. Crump, G.B. ed., John Wiley & Sons, London, UK.

Wiberg, K.B. 1965. In *Oxidation in Organic Chemistry*. Wiberg, K.B. ed., Academic Press, New York. p. 69.

Williams, P.F.V. and Douglas, A.G. 1981. In *Organic Maturation Studies and Petroleum Exploration*. Brooks, J. ed., Academic Press, London, UK. p. 255.

Yen, T.F. 1974. Preprints. *Div. Fuel Chem. Am. Chem. Soc.* 19(2): 109.

Yen, T.F. 1976. In *Oil Shale*. Yen, T.F and Chilingarian, G.V. eds., Elsevier, New York. p. 129.

5 Exploration, Recovery, and Transportation

5.1 INTRODUCTION

Generally, the first stage in the extraction of crude oil is to drill a well into the underground reservoir. Often many wells (multilateral wells) are drilled into the same reservoir, to ensure that the extraction rate is economically viable. In addition, some wells (secondary wells) may be used to pump water, steam, acids or various gas mixtures into the reservoir to raise or maintain the reservoir pressure, and so maintain an economic extraction rate.

If the underground pressure in the oil reservoir is sufficient, the oil will be forced to the surface under this pressure (primary recovery). **Natural gas** (associated natural gas) is often present, which also supplies needed underground pressure (primary recovery). In this situation, it is sufficient to place an arrangement of valves (the Christmas tree) on the well head to connect the well to a pipeline network for storage and processing.

Over the lifetime of the well the pressure will fall, and at some point there will be insufficient underground pressure to force the oil to the surface. Secondary oil recovery uses various techniques to aid in recovering oil from depleted or low-pressure reservoirs. Sometimes pumps, such as beam pumps (horsehead pumps) and electrical submersible pumps are used to bring the oil to the surface. Other secondary recovery techniques increase the reservoir's pressure by water injection, natural gas reinjection and gas lift, which injects air, carbon dioxide, or some other gas into the reservoir.

Enhanced oil recovery (**EOR**, tertiary oil recovery) relies on methods that reduce the viscosity of the oil to increase. Tertiary recovery is started when secondary oil recovery techniques are no longer enough to sustain production. For example, thermally enhanced oil recovery methods are those in which the oil is heated, making it easier to extract; usually steam is used for heating the oil.

Conventional primary and secondary recovery processes are ultimately expected to produce about one-third of the original oil discovered, although recoveries from individual reservoirs can range from less than 5% to as high as 80% of the original oil in place. This broad range of recovery efficiency is a result of variations in the properties of the specific rock and fluids involved from reservoir to reservoir (Table 5.1), as well as the kind and level of energy that drives the oil to producing wells, where it is captured.

Conventional oil production methods may be unsuccessful because the management of the reservoir is poor or because reservoir heterogeneity prevents the recovery of crude oil in an economical manner. Reservoir heterogeneity, such as fractures and faults, can cause reservoirs to drain inefficiently by conventional methods. Also, highly cemented or shale zones can produce barriers to the flow of fluids in reservoirs and lead to high residual oil saturation. Reservoirs containing crude oils with low API gravity often cannot be produced efficiently without application of enhanced oil recovery methods because of the high viscosity of the

TABLE 5.1
Effect of Reservoir Heterogeneity on Petroleum Recovery

Reservoir Heterogeneity Type	Reservoir Continuity	Sweep Efficiency		ROS in Swept Zones	Rock–Fluid Interaction
		Horizontal	Vertical		
Sealing fault	×	×			
Semisealing fault	(×)	×	×		
Nonsealing fault	(×)	×	×		
Boundaries genetic units	×	×	×		
Permeability zonation within genetic units		(×)	×	(×)	
Baffles within genetic units		(×)	×	(×)	
Lamination, cross-bedding		(×)	(×)	×	
Microscopic heterogeneity				×	(×)
Textural types				×	×
Mineralogy					×
Fracturing					
Tight		(×)		×	
Open		×	×	×	

× = strong effect; (×) = moderate effect; ROS = residual oil saturation.

crude oil. In some cases, the reservoir pressure is depleted prematurely by poor reservoir management practices to create reservoirs with low energy and high oil saturation.

As might be expected, the type of exploration technique employed depends upon the nature of the site. In other words, and as for many environmental operations, the recovery techniques applied to a specific site are dictated by the nature of the site and are, in fact, site specific. For example, in areas where little is known about the subsurface, preliminary reconnaissance techniques are necessary to identify potential reservoir systems that warrant further investigation. Techniques for reconnaissance that have been employed to make inferences about the subsurface structure include satellite and high-altitude imagery and magnetic and gravity surveys.

Once an area has been selected for further investigation, more detailed methods (such as the seismic reflection method) are brought into play. Drilling is the final stage of the exploratory program and is in fact the only method by which a petroleum reservoir can be conclusively identified. However, in keeping with the concept of site specificity, drilling may be the only option in some areas for commencement of the exploration program. The risk involved in the drilling operation depends upon previous knowledge of the site subsurface. Thus, there is the need to relate the character of the exploratory wells at a given site to the characteristics of the reservoir.

5.2 EXPLORATION

Exploration for petroleum originated in the latter part of the nineteenth century when geologists began to map land features to search out favorable places to drill for oil (Landes, 1959; Hobson and Tiratsoo, 1975). Of particular interest to geologists were outcrops that provided evidence of alternating layers of porous and impermeable rock. The porous rock (typically a sandstone, limestone, or dolomite) provides the reservoir for petroleum; the impermeable rock (typically clay or shale) acts as a trap and prevents migration of the petroleum from the reservoir.

By the early twentieth century, most of the areas where surface structural characteristics offered the promise of oil had been investigated and the era of subsurface exploration for oil began in the early 1920s (Forbes, 1958). New geological and geophysical techniques were developed for areas where the strata were not sufficiently exposed to permit surface mapping of the subsurface characteristics. In the 1960s, the development of geophysics provided methods for exploring below the surface of the earth.

The principles used are basically magnetism (magnetometer), gravity (gravimeter), and sound waves (seismograph). These techniques are based on the physical properties of materials that can be utilized for measurements and include those that are responsive to the methods of applied geophysics. Further, the methods can be subdivided into those that focus on gravitational properties, magnetic properties, seismic properties, electrical properties, electromagnetic properties, and radioactive properties. These geophysical methods can be subdivided into the following two groups: (1) those without depth control and (2) those with depth control.

In the first group, the measurements incorporate spontaneous effects from both local and distant sources over which the observer has no control. For example, gravity measurements are affected by the variation in the radius of the earth with latitude. They are also affected by the elevation of the site relative to sea level, the thickness of the earth's crust, and the configuration and density of the underlying rocks, as well as by any abnormal mass variation that might be associated with a mineral deposit. In the last stages of assessment, the interpretation always depends upon the geological knowledge of the interpreter.

In the second group of measurements (those with depth control), seismic or electric energy is introduced into the ground and variations in transmissibility with distance are observed and interpreted in terms of geological quantities. Thus, depths to geological horizons with marked differences in transmissibility can be computed on a quantitative basis and the physical nature of these horizons deduced. The accuracy, ease of interpretation, and applicability of all methods falling into this group are not the same, and there are natural and economic conditions under which the measurements of the first group are preferable for exploration studies despite their inherent limitations.

However, it must be recognized that geophysical exploration techniques cannot be applied indiscriminately. Knowledge of the geological parameters likely to be associated with the mineral or subsurface condition being studied is essential, both in choosing the method to be applied and in interpreting the results obtained. Further, not all the techniques described here may be suitable for petroleum exploration. Nevertheless, the techniques that are described here are included as it is valuable to know their nature and how they might be applied to subsurface exploration.

It should also be noted that such terms as *geophysical borehole logging* can imply the use of one or more of the geophysical exploration techniques. This procedure involves drilling a well and using instruments to log or make measurements at various levels in the hole by such means as gravity (**density**), electrical resistivity, or radioactivity. In addition, formation samples (cores) are taken for physical and chemical tests.

5.2.1 GRAVITY METHODS

Gravity methods are based on the measurement of physical quantities related to the gravitational field, which in turn are affected by differences in the density and the disposition of underlying geological bodies. In oil and gas exploration, in which no direct density control is associated with the material being sought, exploration is based on the mapping of geological structures to determine situations that might localize the material being sought. In such cases, the significant density values are salt 2.1 to 2.2, igneous rocks 2.5 to 3.0, and sedimentary rocks 1.6 to 2.8. The last value increases with depth owing to consolidation and geological age, and as a result, structural deformation associated with faults and folding can be detected. Compaction of sediments over edges or knolls on the underlying crystalline rock surface also leads to a local increase in mass, as does the development of calcareous cap rock over the heads of intrusive salt columns.

Thus, the gravimeter detects differences in gravity and gives an indication of the location and density of underground rock formations. Differences from the normal can be caused by geological and other influences, and such differences provide an indication of subsurface structural formations. In the early days of gravity prospecting, both the torsion balance and the pendulum apparatus were extensively employed, but these have been supplanted by spring balance systems (gravimeters). The latter can be read in a matter of minutes, in contrast to the several hours required in obtaining readings with the earlier instruments.

There is a variety of gravimeters, but those in common use consist essentially of a weighted boom that pivots about a hinge point. The boom is linked to a spring system so that the unit is essentially unstable and hence very sensitive to slight variations in gravitational attraction. Deflections of the boom from a central (zero) position are measured by observing the change in the tension in the spring system required to bring the boom back to that position. Readings are taken from a graduated dial on the head of the instrument that is attached to the spring system through a screw. There must be an accurate calibration of the screw, reading dial, and spring response for the readings to have gravitational significance.

Gravimeters can also be employed for use in shallow water. Thus, use of watertight housings with automatic leveling and electronic reading devices allows gravimeter surveys to be carried out in aqueous environments. Other gravimeters have been developed for use in submarines and on gyro-stabilized platforms on surface ships as well as in aircraft.

5.2.2 Magnetic Methods

Magnetic methods are based upon measuring the magnetic effects produced by varying concentrations of ferromagnetic minerals, such as magnetite. Instruments used for magnetic prospecting vary from the simple mining compass used in the seventeenth century to sensitive airborne magnetic units permitting intensity variations to be measured with an accuracy greater than 1/10,000 part of the earth's field.

The magnetometer is a specially designed magnetic compass and detects minute differences in the magnetic properties of rock formations, thus helping to find structures that might contain oil, such as the layers of sedimentary rock that may lie on top of the much denser igneous, or basement, rock. The data give clues to places that might conceal anticlines or other oil-favorable structures. Of even more value is the determination of the approximate total thickness of the sedimentary rock, which can save unwarranted expenditure later or more costly geophysics or even the drilling of a well when the sediment may not contain sufficient oil to warrant further investigation. Most magnetometer surveys used now are performed by the use of aircraft, which permits large-scale surveys to be made rapidly and surveys over regions that may be otherwise inaccessible.

One of the most widely used magnetic instruments is the Schmidt vertical magnetometer. It consists of a pair of blade magnets balanced horizontally on a quartz knife edge. The balance is oriented at right angles to the magnetic meridian. The deflection from the horizontal is observed, giving the variation in magnetic vertical intensity with gravity. The torsion fiber magnetometer is also a vertical component instrument but has an operating range greater than the Schmidt instrument. It also has an advantage in that it is easier and quicker to read. The instrument values are referred to a base and corrected for temperature and diurnal variation and for the normal geographic variation of the earth's magnetic field. The nuclear precession magnetometer is another continuous recording magnetic instrument that measures the earth's total magnetic field by observing the free precession (progressive movement) frequency of the protons in a sample of water.

The interpretation of magnetic measurements is subject to the same fundamental drawbacks as noted for gravity measurements. The drawbacks are as follows:

1. Contrast in physical properties of the formations
2. Depth of origin and integrated contributions from many sources
3. Changes in strength and direction of the earth's field with location
4. Canceling effect related to proximity of opposite induced poses at the boundaries of finite geological bodies

However, the method has proved valuable in exploration for magnetic mineral deposits, in the determination of geological structural trends, and in estimating the probable depth of the crystalline rock floor beneath sedimentary rock areas.

5.2.3 SEISMIC METHODS

Seismic methods are based on determinations of the time interval that elapses between the initiation of a sound wave from detonation of a dynamite charge or other artificial shock and the arrival of the vibration impulses at a series of seismic detectors (geophones). The arrivals are amplified and recorded along with time marks (0.01 sec intervals) to give the seismogram. The method depends upon whether (1) the velocity within each of the layers penetrated at depth is greater than that in the layers above; (2) the layers are bounded by plane surfaces; and (3) the material within each layer is essentially homogeneous.

The seismograph measures the shock waves from explosions initiated by triggering small controlled charges of explosives at the bottom of shallow holes in the ground. The formation depth is determined by the time elapsed between the explosion and detection of the reflected wave at the surface.

The depths and media reached by seismic waves depend on the distance between shot point and receiving points. The first impulses or breaks in a seismogram are caused by waves that have traveled quickly between the shot point and any receiving point. At short distances this is usually also the shortest path, but beyond a certain distance it is quicker for a refracted pulse to travel via a longer path involving underlying layers with a higher velocity. From a plot of travel time as a function of surface distance, data are obtained for determining both the velocity of the material and number of layers present. From the distances at which changes in velocity are indicated, the depth of each layer can be computed.

In general, the deeper, older formations as a result of higher compression have a higher density and also a higher seismic velocity than the overlying material. Observed differences in velocity not only define the direction of slope of the rock surfaces but also provide information for computing the degree of slope present. For what might be termed *normal conditions* (increase in velocity with depth), the error determined in depths is usually less than 10% with this method.

Seismic geophysical work is also carried out on the water, greatly aiding the search for oil on the continental shelves and other areas covered by water. A marine seismic project moves continually, with detectors towed behind the boat at a constant speed and a fairly constant depth. Explosive charges are detonated at a position and time determined by the speed of the boat, so that a continuous survey of the reflecting horizons can be obtained.

5.2.4 ELECTRICAL METHODS

Electrical prospecting methods depend upon differences in electrical conductivity between the geological bodies under study and the surrounding rocks. In general, metallic minerals, particularly the sulfides, range in resistivity from 1.0 to several Ω-cm, whereas consolidated sediments of low water content average about 10^4 S-cm, igneous rocks range from 10^4 to 10^6 Ω-cm, and saturated unconsolidated sediments from 10^2 to 10^4 Ω-cm. The resistivity of the last depends largely on the amount and electrolytic nature (salinity) of the included water.

On the other hand, the self-potential method makes use of the fact that most metallic sulfide minerals are easily oxidized by downward-percolating groundwater. As a result of this surface oxidation, the elements of a simple chemical battery are established and an electrical current flows down through the ore body and back to the surface through the surrounding water-saturated ground, which acts as the electrolyte. It is possible to locate these localized electrical fields and, hence, ore bodies by mapping points of equal electrical potential at the surface using nonpolarizing electrodes and a sensitive ammeter, or a milli-ammeter. Alternatively, measuring the potential differences between successive profile stakes forming a grid over an area using a potentiometer can also be employed.

A special application of electrical methods is in the study of subsurface stratigraphy by measuring the potential differences between the surface and an electrode lowered in a borehole and also by measuring variations in electrical resistivity with depth (electrical logging). This method produces a measure of porosity and permeability, as the data are affected markedly by the ability of the drilling fluid to penetrate the formation. The resistivity measurements define the position of formation boundaries and the lithological character of the sediments.

Three resistivity logs are usually taken: (1) one having a shallow penetration to define the location of the formation boundaries and two others having (2) intermediate and (3) deep penetration. These last two logs are used to determine the extent to which the drilling fluid has penetrated into the formations and the true resistivity of the formation present. The various measurements taken in conjunction provide a valuable tool not only for studying conditions in a given well, but also for carrying out correlation studies between wells and thus defining geological structure and horizontal changes in lithology.

5.2.5 ELECTROMAGNETIC METHODS

Electromagnetic methods are based upon the concept that an alternating magnetic field causes an electrical current to flow in conducting material. Measurements are carried out by connecting a source of alternating current to a coil of wire, which acts as a source for a magnetic field similar to that which will be produced by a short magnet located on the axis of the coil. A receiving system consisting of a second coil connected to a voltmeter is mounted, so that there is free rotation about a horizontal axis.

The receiving coil should be mounted so that rotation is on an axis perpendicular to that of the induced magnetic field. In this case, the induced voltage (in the absence of a conductor) will vary from zero (when the coil plane is parallel to the plane of the applied field) to a maximum (when the coil plane is perpendicular to the plane of the applied field). However, if a conductor is present, the induced current in the conductor sets up a secondary magnetic field that distorts the primary field and gives a value that is not horizontal except directly over the conductor. By using an inclinometer to record the angle of the moving search coil when in the null position, the location of a conductor can be determined as the crossover (inflection) point on a profile across the body.

Another variation of this method is to have both the receiver and the transmitting coils in the horizontal plane. In this arrangement, the voltage developed over nonconducting ground is a function of the construction of the coils that are usually moved across the ground with a constant separation. The presence of a conductor is indicated by changes in the voltage values from the normal values for this configuration.

5.2.6   RADIOACTIVE METHODS

In the disintegration of radioactive minerals three spontaneous emissions take place, the election of an electron (β-ray), a helium nucleus (α-ray), and short-wavelength electromagnetic radiation (γ-rays). The instruments used in radioactive exploration are the Geiger counter and the scintillometer. In addition to prospecting for radioactive minerals, the radioactive method is extensively applied in borehole studies of subsurface stratigraphy as might be deemed necessary when prospecting for oil. Different sedimentary rocks are naturally characterized by different concentrations of radioactive materials. Shale and volcanic ash give the highest γ-ray count and limestone, the lowest γ-ray count.

5.2.7   BOREHOLE LOGGING

Another valuable exploration method is geophysical borehole logging, which involves drilling a well and the use of instruments to log or make measurements at various levels in the hole by such means as electrical resistivity, radioactivity, acoustics, or density. In addition, formation samples (cores) are taken for physical and chemical tests.

The use of electrical logging is based on the fact that the resistivity of a rock layer is a function of its fluid content. Oil-filled sand has very high resistivity. The method consists of passing a current between an electrode at the surface and one that is lowered into the hole, the latter being uncased and filled with drilling mud. Any change in the resistivity conditions around the moving electrode affects the flow of current and voltage distribution around it. Voltage fluctuations can be measured by a pair of separate electrodes used in conjunction with the moving electrode.

The natural radioactive properties of many constituents of rock have made it possible to develop and use nuclear radiation detectors (radioactive logging) in the borehole or even in holes that have already been cased. Two commonly used methods are γ-ray and neutron logging. In the first case, the natural radiation from the rock is used. In the second, a neutron source is employed to excite the release of radiation from the rock. The neutron source is usually a mixture of beryllium and radium, but it can be a miniature Van der Graaff particle accelerator. The neutron method is a means for determining the relative porosity or rock formations; the γ-ray log helps define shale.

The acoustic logging method is quite similar to surface seismic work. Instead of explosives, an electrically operated acoustic pulse generator is used. In one instrument, the generator is separated from the receiver by an acoustic insulator. The design permits automatic selection and recording of the travel times of the onsets of pulses that travel through the rock wall of the hole as the instrument moves down or up. Signals are recorded continuously at the surface, being transmitted through a cable on which the instrument is suspended. The velocity log provided by the instrument helps to define beds and evaluate formation porosity.

Density can now be logged with a new technique that uses radioactivity (density logging). The instrument consists of a radioactive cobalt source of γ-rays and a Geiger counter as a detector, which is shielded from the source. The rock formation is bombarded with the γ-rays, some of which are scattered back from the formation and enter the detector. The degree to which the original radiation is adsorbed is a function of the density of the rock.

Test well sampling is another important method used in the search for oil (core sampling). Well data obtained from the examination of formation samples taken from various depths in the borehole are of considerable value in deciding further exploratory work. These samples can be cores, which have been taken from the hole by a special coring device or drill cuttings screened from the circulating drilling mud. The major purpose of sample examination is to identify the various strata in the borehole and compare their positions with the standard stratigraphic sequence of all the sedimentary rocks occurring in the specific basin where the hole has been drilled.

5.3 DRILLING OPERATIONS

Generally, the first stage in the extraction of crude oil is to drill a well into the underground reservoir. Often many wells (called multilateral wells) are drilled into the same reservoir, to ensure that the extraction rate is economically viable. Also, some wells (secondary wells) may be used to pump water, steam, acids, or various gas mixtures into the reservoir to raise or maintain the reservoir pressure, and hence maintain an economic extraction rate.

Drilling for oil is a complex operation and has evolved considerably over the past 100 years. The older cable tool method, used extensively until 1900, involves raising and dropping a heavy bit and drill stem attached by cable to a cantilever arm at the surface. It pulverizes the rock and earth, gradually forming a hole. The cable tool system is generally preferred only for penetrating hard rock at shallow depths and when oil reservoirs are expected at shallow depths. The weight of the column is usually enough to attain penetration but can be augmented by a hydraulic pressure cylinder at the surface.

5.3.1 PREPARING TO DRILL

Once the site has been selected, it must be surveyed to determine its boundaries, and environmental impact studies may need to be performed. Lease agreements, titles, and right-of way accesses for the land must be obtained and evaluated legally. For offshore sites, legal jurisdiction must be determined. Once the legal issues have been settled, the crew goes about preparing the land; preparation is essential and involves the following steps:

1. L and is cleared and leveled, and access roads may be built.
2. Because water is used in drilling, there must be a source of water nearby. If there is no natural source, a water well is necessary.
3. Reserve pit, which is used to dispose of rock cuttings and drilling mud during the drilling process and which is lined with plastic to protect the environment, is created. If the site is an ecologically sensitive area, such as a marsh or wilderness, then the cuttings and mud must be disposed offsite, it may have to be trucked away instead of being placed in a pit.

Once the land has been prepared, several holes must be dug to make way for the rig and the main hole. A rectangular pit (a cellar) is dug around the location of the actual drilling hole. The cellar provides a workspace around the hole for the workers and drilling accessories. The crew then begins drilling the main hole, often with a small drill truck rather than the main rig. The first part of the hole is larger and shallower than the main portion, and is lined with a large-diameter conductor pipe. Additional holes are dug off to the side to temporarily store equipment, after which the rig equipment can be brought in and set up.

5.3.2 Drilling Rig

Depending upon the remoteness of the drill site and its access, equipment may be transported to the site by truck, helicopter, or barge. Some rigs are built on ships or barges for work on inland water where there is no foundation to support a rig (as in marshes or lakes). Once the equipment is at the site, the rig is set up (Figure 5.1). The anatomy of a drilling rig, although simple in a schematic representation is, in reality, quite complex and consists of the following systems:

1. Power system:
   a. Large diesel engines to provide the main source of power
   b. Electrical generators powered by diesel engines to provide electrical power
2. Mechanical system that is driven by electric motors:
   a. Hoisting system that is used for lifting heavy loads and consists of a mechanical winch (draw works) with a large steel cable spool, a block-and-tackle pulley, and a receiving storage reel for the cable
   b. Turntable that is part of the drilling apparatus
3. Rotating equipment (used for rotary drilling):
   a. Swivel, i.e., a large handle, which holds the weight of the drill string; allows the string to rotate and makes a pressure-tight seal on the hole

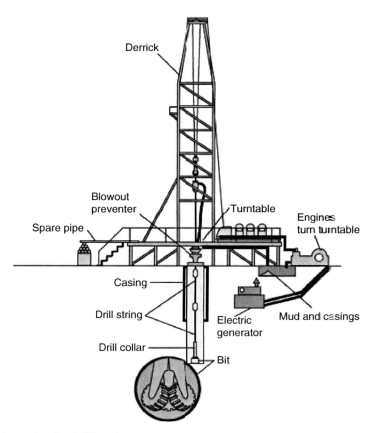

FIGURE 5.1 Schematic of a drilling rig.

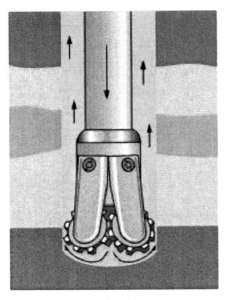

FIGURE 5.2 Schematic of mud circulating around a drill bit.

    b. Kelly, i.e., a four- or six-sided pipe that transfers rotary motion to the turntable and drill string

    c. Turntable or rotary table drives the rotating motion using power from electric motors

    d. Drill string that consists of drill pipe (connected sections of about 30 ft/10 m) and drill collars (larger diameter, heavier pipe that fits around the drill pipe and places weight on the drill bit)

    e. Drill bit(s) at the end of the drill that actually cuts up the rock and comes in many shapes and materials (tungsten carbide steel, diamond) that are specialized for various drilling tasks and rock formations

4. Casing, which is a large-diameter concrete pipe that lines the drill hole, prevents the hole from collapsing, and allows drilling mud to circulate (Figure 5.2)

5. Circulation system—pumps drilling mud (mixture of water, clay, weighting material and chemicals, used to lift rock cuttings from the drill bit to the surface) under pressure through the kelly, rotary table, drill pipes and drill collars (Figure 5.3):

    a. Pump—sucks mud from the mud pits and pumps it to the drilling apparatus

    b. Pipes and hoses—connect pump to drilling apparatus

    c. Mud-return line—returns mud from hole

    d. Shale shaker—shaker/sieve that separates rock cuttings from the mud

    e. Shale slide—conveys cuttings to the reserve pit

    f. Reserve pit—collects rock cuttings separated from the mud

    g. Mud pits—where drilling mud is mixed and recycled

    h. Mud-mixing hopper—where new mud is mixed and then sent to the mud pits

6. Derrick—support structure that holds the drilling apparatus; tall enough to allow new sections of drill pipe to be added to the drilling apparatus as drilling progresses

7. Blowout preventer—high-pressure valves (located under the land rig or on the sea floor) that seal the high-pressure drill lines and relieve pressure, when necessary, to prevent a blowout (uncontrolled gush of gas or oil to the surface, often associated with fire)

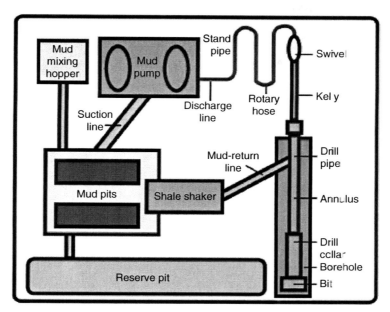

FIGURE 5.3 A mud circulatory system.

5.3.3 DRILLING RIG COMPONENTS

Although there are many variations in design, all modern rotary drilling rigs have essentially the same components (Figure 5.1). The hoisting, or draw works, raises and lowers the drill pipe and casing, which can weigh as much as 200 tons (200,000 kg). The height of the derrick depends on the number of joints of drill pipe to be withdrawn as a unit before being unscrewed. The rotary table located in the middle of the rig floor rotates the drill column. The table also gaps the drill stem when the hoist is disconnected and pipe sections are inserted or removed. It imparts rotary motion to the drill stem through the kelly attached to the upper end of the column. The kelly fits into a shaped hole in the center of the rotary table. The couplings between pipe sections are called tool joints.

The drilling bit is connected to drill collars at the bottom of the stem. These are thick steel cylinders, 20 to 29 ft (6 to 9 m) long; as many as 10 may be screwed together. They concentrate weight at the bottom of the column and exert tension on the more flexible pipe above, reducing the tendency of the hole to go off-line and the drill pipe to fracture. Drill bits have many designs and the variations include the number of blades, type of metal, and shape of cutting or brading components. To alleviate the problem of dull wearout or jamming, a mud-circulating system is one of the most important parts of a rig (Ranney, 1979). This system maintains the drilling mud in proper condition, free of rock cuttings or other abrasive materials that might cause problems with the drilling operation, as well as retains the proper physical and chemical characteristics of the mud.

5.3.4 DRILLING

Drilling an oil well is tedious and is often accompanied by difficulties. Some problems result from formation penetration and the occurrence of high-pressure gas, fissures, or unexpected high pressures in permeable rock. Others result from metallurgical or mechanical failures in the bit, the drill stem, the draw works, or the mud system. Many tools and techniques have been developed to solve most problems; probably the best known is the fish used to recover broken bits. Another technique involves intentional deviation of the borehole to avoid difficult formations, to go around an unrecoverable fish, or sometimes to restore the direction of the hole after an accidental deviation.

Once the rig is set up, drilling operations commence. First, from the starter hole, a surface hole is drilled to a preset depth, which is somewhere above the location of the oil trap. There are five basic steps to drilling the surface hole:

1. Place the drill bit, collar, and drill pipe in the hole
2. Attach the kelly and turntable and begin drilling
3. As drilling progresses, circulate mud through the pipe and out of the bit to float the rock cuttings out of the hole
4. As the hole increases in depth, add new sections (joints) of drill pipes
5. Remove (trip out) the drill pipe, collar and bit when the preset depth (anywhere from a few hundred to a couple-thousand feet) is reached

When the preset depth is reached, the casing pipe sections are run into the hole and cemented to prevent the hole from collapsing. The casing pipe has spacers around the outside to keep it centered in the hole. The cement is pumped down the casing pipe using a bottom plug, a cement slurry, a top plug, and drilling mud. The pressure from the drilling mud causes the cement slurry to move through the casing and fill the space between the outside of the casing and the hole. Finally, the cement is allowed to harden and is then tested for such properties as hardness, alignment, and a proper seal.

Drilling continues in stages and when the rock cuttings from the mud reveal the oil sand from the reservoir rock, the final depth may have been reached. At this point, the drilling apparatus is removed from the hole and tests are preformed to confirm that the final depth has been reached. These tests include the following:

1. Well logging: lowering electrical and gas sensors into the hole to take measurements of the rock formations there
2. Drill-stem testing: lowering a device into the hole to measure the pressures, which will reveal whether reservoir rock has been reached
3. Core sampling: taking samples of rock to look for characteristics of reservoir rock

5.4   WELL COMPLETION

Once the final depth has been reached, the well is completed to allow oil to flow into the casing in a controlled manner. First, a perforating gun is lowered into the well to the production depth. The gun has explosive charges to create holes in the casing through which oil can flow. After the casing has been perforated, a small-diameter pipe (tubing) is run into the hole as a conduit for oil and gas to flow up the well and a packer is run down the outside of the tubing. When the packer is set at the production level, it is expanded to form a seal around the outside of the tubing. Finally, a multivalve structure (the Christmas tree; Figure 5.4) is installed at the top of the tubing and cemented to the top of the casing. The Christmas tree allows them to control the flow of oil from the well.

Tight formations are occasionally encountered and it becomes necessary to encourage flow. Several methods are used, one of which involves setting off small explosions to fracture the rock. If the formation is mainly limestone, hydrochloric acid is sent down the hole to dissolve channels in the rock. The acid is inhibited to protect the steel casing. In sandstone, the preferred method is hydraulic fracturing.

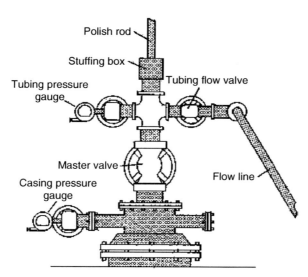

FIGURE 5.4 The Christmas tree—a collection of control valves at the wellhead.

A fluid with a viscosity high enough to hold coarse sand in suspension is pumped at very high pressure into the formation, fracturing the rock. The grains of sand remain, helping to hold the cracks open.

5.5 RECOVERY

Recovery, as applied in the petroleum industry, is the production of oil from a reservoir. There are several methods by which this can be achieved, which range from recovery because of reservoir energy (i.e., the oil flows from the well hole without assistance) to enhanced recovery methods in which considerable energy must be added to the reservoir to produce the oil. However, the effect of the method on the oil and on the reservoir (Table 5.2) must be considered before application.

TABLE 5.2
Recovery Process Parameters and Their Potential Adverse Effects Leading to Sludge and Sediment Formation

Property	Comments
Carbon dioxide injection	Lowers pH; can change oil composition leading to phase separation of sludge or sediment and blocking of channels.
Miscible flooding	Hydrocarbon-rich gases lower the solubility parameter and solvent power of the oil and cause separation of asphaltene material.
Organic chemicals	Can lower the solubility parameter and solvent power of the oil and cause separation of asphaltene material; blocking of channels.
Acidizing	Interaction of crude oil constituents upsetting molecular balance and deposition of sludge or sediment; blocking of channels.
Pressure decrease	Can change composition of oil medium leading to phase separation of asphaltene material as sludge or sediment; blocking of channels.
Temperature decrease	Can change composition of oil medium leading to phase separation of asphaltene material as sludge or sediment; blocking of channels.

This chapter, for the most part, deals with those recovery methods that are applied to the recovery of conventional crude oil and, in some cases, to recovery of heavy oil. Methods that are being proposed for the recovery of bitumen from tar deposits are presented elsewhere (Chapter 6).

Thus, once the well is completed, the flow of oil into the well commences. For limestone reservoir rock, acid is pumped down the well and out the perforations. The acid dissolves channels in the limestone that lead oil into the well. For sandstone reservoir rock, a specially blended fluid containing proppants (sand, walnut shells, aluminum pellets) is pumped down the well and out the perforations. The pressure from this fluid makes small fractures in the sandstone that allow oil to flow into the well, while the proppants hold these fractures open. Once the oil starts flowing, the oil rig is removed from the site and production equipment is set up to extract the oil from the well.

A well is always carefully controlled in its flush stage of production to prevent the potentially dangerous and wasteful gusher. This is actually a dangerous condition, and is (hopefully) prevented by the blowout preventer and the pressure of the drilling mud. In most wells, acidizing or fracturing the well starts the oil flow.

As already noted (Chapter 3), crude oil accumulates over geological time in porous underground rock formations called reservoirs that are at varying depths in the earth's crust, and in many cases elaborate, expensive equipment is required to get it from there. The oil is usually found trapped in a layer of porous sandstone, which lies just beneath a dome-shaped or folded layer of some nonporous rock such as limestone. In other formations the oil is trapped at a fault or break in the layers of the crust.

Generally, crude oil reservoirs exist with an overlying **gas cap**, in communication with aquifers, or both. The oil resides together with water and free gas in very small holes (pore spaces) and fractures. The size, shape, and degree of interconnection of the pores vary considerably from place to place in an individual reservoir. Below the oil layer, the sandstone is usually saturated with salt water. The oil is released from this formation by drilling a well and puncturing the limestone layer on either side of the limestone dome or fold. If the peak of the formation is tapped, only the gas is obtained. If the penetration is made too far from the center, only salt water is obtained. Oil wells may be either on land or under water. In North America, many wells are offshore in the shallow parts of the oceans. The crude oil or unrefined oil is typically collected from individual wells by small pipelines.

The oil in such formation is usually under such great pressure that it flows naturally, and sometimes with great force, from the well. However, in some cases, this pressure later diminishes so that the oil must be pumped from the well. Natural gas or water is sometimes pumped into the well to replace the oil that is withdrawn. This is called repressurizing the oil well.

Thus, the anatomy of a reservoir is complex and is site specific, microscopically and macroscopically. Because of the various types of accumulations and the existence of wide ranges of both rock and fluid properties, reservoirs respond differently and must be treated individually.

Conventional crude oil is a brownish green to black liquid of specific gravity in a range from about 0.810 to 0.985, with a boiling range from about 20°C (68°F) to above 350°C (660°F), above which active decomposition ensues when distillation is attempted. The oils contain from 0% to 35% or more of gasoline, as well as varying proportions of kerosene hydrocarbons and higher boiling constituents up to the viscous and nonvolatile compounds present in lubricant oil and in asphalt. The composition of the crude oil obtained from the well is variable and depends not only on the original composition of the oil **in situ**, but also on the manner of production and the stage reached in the life of the well or reservoir.

For a newly opened formation and under ideal conditions, the proportions of gas may be so high that the oil is, in fact, a solution of liquid in gas that leaves the reservoir rock so efficiently that a core sample will not show any obvious oil content. A general rough indication of this situation is a high ratio of gas to oil produced. This ratio may be zero for fields in which the rock pressure has been dissipated. The oil must be pumped out to as much as 50,000 ft.[3] or more of gas per barrel of oil in the so-called condensate reservoirs, in which a very light crude oil (0.80 specific gravity or lighter) exists as vapor at high pressure and elevated temperature.

New methods to drill for oil are continually being sought, including directional or horizontal drilling techniques, to reach oil under ecologically sensitive areas, and using lasers to drill oil wells.

Directional drilling is also used to reach formations and targets not directly below the penetration point or drilling from shore to locations under water (Figure 5.5). A controlled deviation may also be used from a selected depth in an existing hole to attain economy in drilling costs. Various types of tools are used in directional drilling along with instruments to help orient their position and measure the degree and direction of deviation; two such tools are the whipstock and the knuckle joint. The whipstock is a gradually tapered wedge with a chisel-shaped base that prevents rotation after it has been forced into the bottom of an open hole. As the bit moves down, it is deflected by the taper by about 5° from the alignment of the existing hole.

Approximately one-third of the world's crude oil is produced from offshore fields, usually from steel drilling platforms set on the ocean floor. In shallow, calm waters, these may be little more than a wellhead and workspace. The larger ocean rigs, however, include not only the well equipment but also processing equipment and extensive crew quarters. Recent developments in ocean drilling include the use of floating tension leg platforms that are tied to the sea floor by giant cables and drill ships, which can hold a steady position above a sea floor well using constant, computer-controlled adjustments. Subsea satellite platforms, where all the necessary equipment is located on the ocean bed at the well site, have been used for small fields located in producing areas. In Arctic areas, islands are built from dredged gravel and sand to provide platforms capable of resisting drifting ice fields.

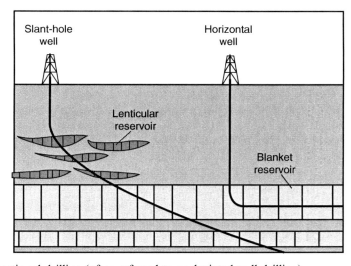

FIGURE 5.5 Directional drilling (often referred to as deviated well drilling).

Drilling does not end when production commences and continues after a field enters production. Extension wells must be drilled to define the boundaries of the crude oil pool. In-field wells are necessary to increase recovery rates, and service wells are used to reopen wells that have become clogged. Additionally, wells are often drilled at the same location but to different depths, to test other geological structures for the presence of crude oil.

Recovery of oil when a well is first opened is usually by natural flow forced by the pressure of the gas or fluids that are contained within the deposit. There are several means that serve to drive the petroleum fluids from the formation, through the well, and to the surface, and these methods are classified as either natural or applied flow.

5.5.1 Primary Recovery (Natural Methods)

If the underground pressure in the oil reservoir is sufficient, then the oil will be forced to the surface under this pressure. Gaseous fuels or natural gas are usually present, which also supply needed underground pressure. In this situation, it is sufficient to place a complex arrangement of valves (the Christmas tree) at the well head to connect the well to a pipeline network for storage and processing. This is called **primary oil recovery**.

Thus, primary oil production (primary oil recovery) is the first method of producing oil from a well and depends upon natural reservoir energy to drive the oil through the complex pore network to producing wells. If the pressure on the fluid in the reservoir (reservoir energy) is great enough, the oil flows into the well and up to the surface. Such driving energy may be derived from liquid expansion and evolution of dissolved gases from the oil as reservoir pressure is lowered during production, expansion of free gas, or a gas cap, influx of natural water, gravity, or combination of these effects.

Crude oil moves out of the reservoir into the well by one or more of three processes. These processes are: dissolved gas drive, gas cap drive, and water drive. Early recognition of the type of drive involved is essential to the efficient development of an oil field.

In dissolved gas drive, the propulsive force is the gas in solution in the oil, which tends to come out of solution because of the pressure released at the point of penetration of a well. Dissolved gas drive is the least efficient type of natural drive as it is difficult to control the gas–oil ratio; the bottom-hole pressure drops rapidly, and the total eventual recovery of petroleum from the reservoir may be less than 20%.

If gas overlies the oil beneath the top of the trap, it is compressed and can be utilized (gas cap drive) to drive the oil into wells situated at the bottom of the oil-bearing zone (Figure 5.6). By producing oil only from below the gas cap, it is possible to maintain a high gas–oil ratio in the reservoir until almost the very end of the life of the pool. If, however, the oil deposit is not systematically developed, so that bypassing of the gas occurs, an undue proportion of oil is left behind. The usual recovery of petroleum from a reservoir in a gas cap field is 40% to 50%.

Usually the gas in a gas cap (associated natural gas) contains methane and other hydrocarbons that may be separated out by compressing the gas. A well-known example is natural gasoline that was formerly referred to as **casinghead gasoline** or natural gas gasoline. However, at high pressures, such as those existing in the deeper fields, the density of the gas increases and the density of the oil decreases until they form a single phase in the reservoir. These are the so-called retrograde condensate pools, because a decrease (instead of an increase) in pressure brings about condensation of the liquid hydrocarbons. When this reservoir fluid is brought to the surface and the condensate is removed, a large volume of residual gas remains. The modern practice is to cycle this gas by compressing it and inject it back into the reservoir, thus maintaining adequate pressure within the gas cap, and condensation in the reservoir is prevented. Such condensation prevents recovery of the oil, for the low percentage of liquid saturation in the reservoir precludes effective flow.

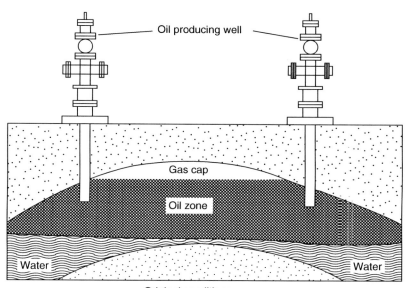

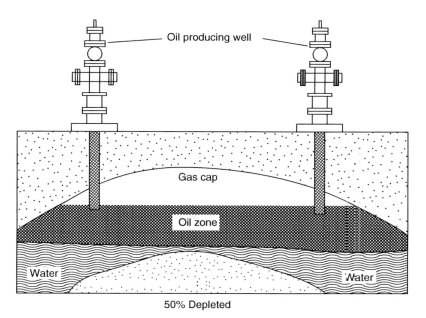

FIGURE 5.6 The gas cap drive mechanism.

The most efficient propulsive force in driving oil into a well is natural water drive, in which the pressure of the water forces the lighter recoverable oil out of the reservoir into the producing wells. In anticlinal accumulations, the structurally lowest wells around the flanks of the dome are the first to come into water. Then the oil–water contact plane moves upward until only the wells at the top of the anticline are still producing oil; eventually, these also must be abandoned as the water displaces the oil (Figure 5.7).

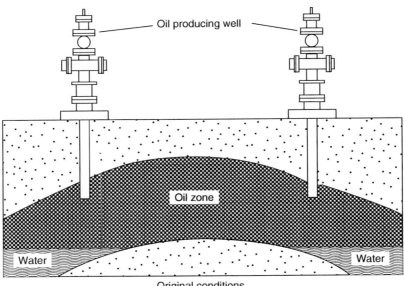

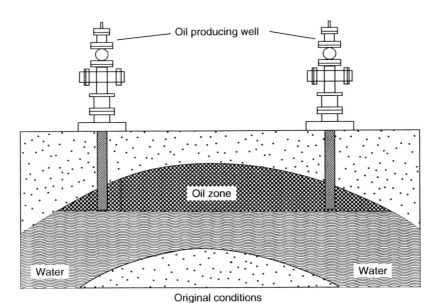

FIGURE 5.7 The water drive mechanism.

In a water drive field, it is essential that the removal rate be adjusted so that the water moves up evenly as space is made available for it by the removal of the hydrocarbons. An appreciable decline in bottom-hole pressure is necessary to provide the pressure gradient required to cause water influx. The pressure differential needed depends on the reservoir permeability; the greater the permeability, the lesser the difference in pressure necessary. The recovery of petroleum from the reservoir in properly operated water drive pools may run as high as 80%. The force behind the water drive may be hydrostatic pressure, the expansion of the reservoir water, or a combination of both. Water drive is also used in certain submarine fields.

Gravity drive is an important factor when oil columns of several thousands of feet exist, as they do in some North American fields. Further, the last bit of recoverable oil is produced in many pools by **gravity drainage** of the reservoir. Another source of energy during the early stages of withdrawal from a reservoir containing under-saturated oil is the expansion of that oil as the pressure reduction brings the oil to bubble point (the pressure and temperature at which the gas starts to come out of solution).

For primary recovery operations, no pumping equipment is required. If the reservoir energy is not sufficient to force the oil to the surface, then the well must be pumped. In either case, nothing is added to the reservoir to increase or maintain the reservoir energy or to sweep the oil toward the well. The rate of production from a flowing well tends to decline as the natural reservoir energy is expended. When a flowing well is no longer producing at an efficient rate, a pump is installed.

The recovery efficiency for primary production is generally low when liquid expansion and solution gas evolution are the driving mechanisms. Much higher recoveries are associated with reservoirs with water and gas cap drives and with reservoirs in which gravity effectively promotes drainage of the oil from the rock pores. The overall recovery efficiency is related to how the reservoir is delineated by production wells. Thus, for maximum recovery by primary recovery, it is often preferable to sink several wells into a reservoir, thereby bringing about recovery by a combination of the methods outlined here.

5.5.2   SECONDARY RECOVERY

Over the lifetime of the well the pressure will fall, and at some point there will be insufficient underground pressure to force the oil to the surface. If economical, and it often is, the remaining oil in the well is extracted using secondary oil recovery methods. It is at this point that secondary recovery methods must be applied.

Secondary oil recovery methods use various techniques to aid in recovering oil from depleted or low-pressure reservoirs. Sometimes pumps on the surface or submerged (electrical submersible pumps (ESPs)), are used to bring the oil to the surface. Other secondary recovery techniques increase the reservoir's pressure by water injection and gas injection, which injects air or some other gas into the reservoir.

Together, primary recovery and secondary recovery allow 25% to 35% of the reservoir's oil to be recovered.

Primary (or conventional) recovery can leave as much as 70% of the petroleum in the reservoir. Such effects as microscopic trapping and bypassing are the more obvious reasons for the low recovery. There are two main objectives in secondary crude oil production. One is to supplement the depleted reservoir energy pressure, and the second objective is to sweep the crude oil from the injection well toward and into the production well. In fact, secondary oil recovery involves the introduction of energy into a reservoir to produce more oil. For example, the addition of materials to reduce the interfacial tension of the oil results in higher recovery of oil.

The most common follow-up, or secondary recovery, operations usually involve the application of pumping operations or of injection of materials into a well to encourage movement and recovery of the remaining petroleum. The pump (generally known as the horsehead pump or the sucker-rod pump) (Figure 5.8) provides mechanical lift to the fluids in the reservoir. The most commonly recognized oil-well pump is the reciprocating or plunger pumping equipment (also called a sucker-rod pump), which is easily recognized by the horsehead beam pumping jacks. A pump barrel is lowered into the well on a string of 6 in. (inner

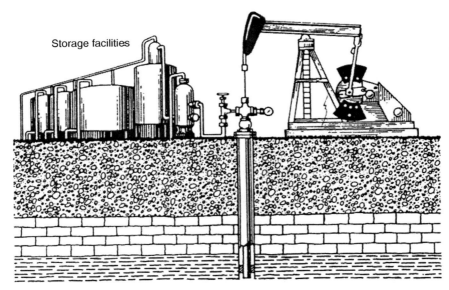

FIGURE 5.8 The horsehead pump. (From United States Geological Survey, Washington DC.)

diameter) steel rods known as sucker rods. The up-and-down movement of the sucker rods forces the oil up the tubing to the surface. A walking beam powered by a nearby engine may supply this vertical movement, or it may be brought about through the use of a pump jack, which is connected to a central power source by means of pull rods. Electrically powered centrifugal pumps and submersible pumps (both pump and motor are in the well at the bottom of the tubing) have proven their production capabilities in numerous applications.

There are also secondary oil recovery operations that involve the injection of water or gas into the reservoir. When water is used the process is called a **waterflood**; with gas, a gasflood. Separate wells are usually used for injection and production. The injected fluids maintain reservoir pressure or repressure the reservoir after primary depletion and displace a portion of the remaining crude oil to production wells. In fact, the first method recommended for improving the recovery of oil was probably the reinjection of natural gas, and there are indications that gas injection was utilized for this purpose before 1900 (Craft and Hawkins, 1959; Frick, 1962). These early practices were implemented to increase the immediate productivity and are therefore classified as pressure maintenance projects. Recent gas injection techniques have been devised to increase the ultimate recovery, thus qualifying as secondary recovery projects.

In secondary recovery, the injected fluid must dislodge the oil and propel it toward the production wells. Reservoir energy must also be increased to displace the oil. Using techniques such as gas and water injection does not change the state of oil. Similarly, there is no change in the state of the oil during miscible fluid displacement technologies. The analogy that might be used is that of a swimmer (in water) in which there is no change to the natural state of the human body.

Thus, the success of secondary recovery processes depends on the mechanism by which the injected fluid displaces the oil (**displacement efficiency**) and on the volume of the reservoir that the injected fluid enters (conformance or **sweep efficiency**). In most proposed secondary projects, water does both these things more effectively than gas. It must be decided whether the use of gas offers any economic advantages because of availability and relative ease of injection. In reservoirs with high permeability and high vertical span, the injection of gas may result in high recovery factors as a result of gravity segregation, as described in a later section. However, if the reservoir lacks either adequate vertical permeability or the possibility for gravity segregation, a frontal drive similar to that used for water injection can be used (dispersed gas injection). Thus, dispersed gas injection is anticipated to be more effective in reservoirs that are relatively thin and have little dip. Injection into the top of the formation (or into the gas cap) is more successful in reservoirs with higher vertical permeability (200 md or more) and enough vertical relief to allow the gas cap to displace the oil downward.

Vaporization is another recovery mechanism used to inject gas into oil reservoirs. A portion of the oil affected by the dry injection gas is vaporized into the oil and transported to the production wells in the vapor phase. In some instances, this mechanism has been responsible for a substantial amount of the secondary oil produced.

During the withdrawal of fluids from a well, it is usual practice to maintain pressure in the reservoir at or near the original levels by pumping either gas or water into the reservoir as the hydrocarbons are withdrawn. This practice has the advantage of retarding the decline in the production of individual wells and considerably increasing the ultimate yield. It also may bring about the conservation of gas that otherwise would be wasted, and the disposal of brines that otherwise might pollute surface and near-surface potable waters.

In older fields, it was not the usual practice to maintain reservoir pressure, and it is now necessary to obtain petroleum from these fields by means of secondary recovery projects. Thus, several methods have been developed to obtain oil from reservoirs where previous economic policies dictated that ordinary production systems were no longer viable.

Considerable experimentation has been carried on in the use of different types of input gas. Examples are wet casinghead gas, enriched gas, **liquefied petroleum gas** (LPG), such as butane and propane, high-pressure gas, and even nitrogen. High-pressure gas not only pushes oil through the reservoir, but may also produce a hydrocarbon exchange so that the concentration of liquid petroleum gases in the oil is increased.

Water injection is still predominantly a secondary recovery process (waterflood). Probably, the principal reason for this is that reservoir formation water is ordinarily not available in volume during the early years of an oil field and pressure maintenance water from outside the field may be too expensive. When a young field produces considerable water, it may be injected back into the reservoir primarily for the purpose of nuisance abatement, but reservoir pressure maintenance is a valuable by-product.

Nevertheless, some passages in the formation are larger than others, and the water tends to flow freely through these, by passing smaller passages where the oil remains. A partial solution to this problem is possible by miscible fluid flooding. Liquid butane and propane are pumped into the ground under considerable pressure, dissolving the oil and carrying it out of the smaller passages; additional pressure is obtained by using natural gas.

5.5.3 ENHANCED OIL RECOVERY

Enhanced oil recovery (tertiary oil recovery) (please see Chapter 6) is the incremental ultimate oil that can be recovered from a petroleum reservoir over oil that can be obtained by primary and secondary recovery methods.

The viscosity (or the API gravity) of petroleum (Chapter 9) is an important factor that must be taken into account when heavy oil is recovered from a reservoir. In fact, certain reservoir types, such as those with very viscous crude oils and some low-permeability carbonate (limestone, dolomite, or chert) reservoirs, respond poorly to conventional secondary recovery techniques.

In these reservoirs, it is desirable to initiate enhanced oil recovery (EOR) operations as early as possible. This may mean considerably abbreviating conventional secondary recovery operations or bypassing them altogether. Thermal floods using steam and controlled in situ combustion methods are also used. Thermal methods of recovery reduce the viscosity of the crude oil by heat so that it flows more easily into the production well. Thus, tertiary techniques are usually variations of secondary methods with a goal of improving the sweeping action of the invading fluid.

Thus, enhanced oil recovery methods are designed to reduce the viscosity of the crude oil (i.e., to reduce the pour point of the crude oil relative to the temperature of the reservoir), thereby increasing oil production. Enhanced oil recovery methods are applied or started when secondary oil recovery techniques are no longer enough to sustain production.

Thermally enhanced oil recovery methods are tertiary recovery techniques that heat the oil and make it easier to extract. Steam injection is the most common form of this process and is used extensively to increase oil production. In situ combustion is another form of thermally enhanced oil recovery, but instead of using steam to reduce the crude oil viscosity, some of the oil is burned to heat the surrounding oil. Detergents are also used to decrease oil viscosity.

A significant amount of laboratory research and field testing has been devoted to developing enhanced oil recovery methods as well as defining the requirements for a successful recovery and the limitations of the various methods (Figure 5.9). The intent of enhanced oil recovery is to increase the effectiveness of oil removal from the pores of the rock (displacement efficiency) and to increase the volume of rock contacted by injected fluids (sweep efficiency) (Schumacher, 1980).

To understand the phenomenon of enhanced oil recovery, it is helpful to understand the condition in the reservoir after other recovery operations have been exhausted. The oil remaining after conventional recovery operations is retained in the pore space of reservoir rock at a lower concentration than originally existed. In portions of the reservoir that have been contacted or swept by the injection fluid, the residual oil remains as droplets (or ganglia) trapped in either individual pores or clusters of pores. It may also remain as films partly

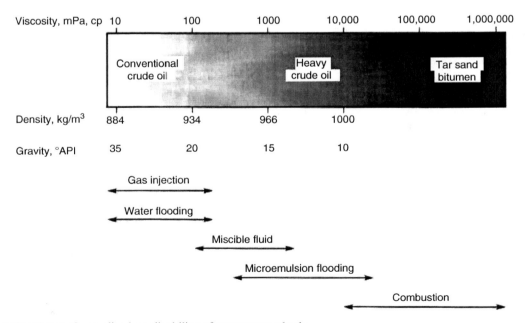

FIGURE 5.9 Generalized applicability of recovery methods.

coating the pore walls. Entrapment of this residual oil is predominantly due to capillary and surface forces and to pore geometry.

In the pores of those volumes of reservoir rock that were not well swept by displacing fluids, the oil continues to exist at higher concentrations and may exist as a continuous phase. This macroscopic bypassing of the oil occurs because of reservoir heterogeneity, the placement of wells, and the effects of viscous, gravity, and capillary forces, which act simultaneously in the reservoir. The resultant effect depends upon conditions at individual locations.

The higher the mobility of the displacing fluid relative to that of the oil (i.e., the higher the **mobility ratio**), the greater the propensity for the displacing fluid to bypass oil. Because of fluid density differences, gravity forces cause vertical segregation of the fluids in the reservoir so that water tends to under-run, and gas tends to over-ride, the oil-containing rock. These mechanisms can be controlled or utilized only to a limited extent in primary and secondary recovery operations.

For example, during waterflood, the capillary forces that cause the displacement of oil by water can also result in the trapping of residual oil. Particularly important is the faster movement of water through the smaller pore because (1) its smaller diameter increases the capillary force and (2) the oil volume displaced by water is far less. Water moving more rapidly through the small pore will reach a common outlet before all the oil is displaced from the upper large pore. A net capillary force is then exerted on the downstream end of the large pore, which can be likened to closing a back door. Further displacement of oil ceases, trapping oil between the two interfaces. Thus, most of the emphasis in developing chemically enhanced methods has been toward recovering such residual oil as might remain after waterflood.

Enhanced oil recovery processes use thermal, chemical, or fluid phase behavior effects to reduce or eliminate the capillary forces that trap oil within pores, to thin the oil or otherwise improve its mobility or to alter the mobility of the displacing fluids. In some cases, the effects of gravity forces, which ordinarily cause vertical segregation of fluids of different densities, can be minimized or even used to advantage. The various processes differ considerably in complexity, the physical mechanisms responsible for oil recovery, and the amount of experience that has been derived from field application. The degree to which enhanced oil recovery methods are applicable in the future will depend on development of improved process technology. It will also depend on improved understanding of fluid chemistry, phase behavior, and physical properties; and on the accuracy of geology and reservoir engineering in characterizing the physical nature of individual reservoirs (Borchardt and Yen, 1989).

Chemical methods include polymer flooding, surfactant (**micellar** or polymer and **microemulsion**) flooding, and alkaline flood processes.

Polymer flooding (**polymer augmented waterflooding**) is waterflooding in which organic polymers are injected with the water to improve horizontal and vertical sweep efficiency. The process is conceptually simple and inexpensive, and its commercial use is increasing despite relatively small potential incremental oil production. Surfactant flooding is complex and requires detailed laboratory testing to support field project design. As demonstrated by field tests, it has excellent potential for improving the recovery of low-viscosity to moderate-viscosity oil. Surfactant flooding is expensive and has been used in few large-scale projects. Alkaline flooding has been used only in reservoirs containing specific types of high-acid-number crude oils.

The terms *microemulsion* and *micellar solution* are used to describe concentrated, surfactant-stabilized dispersions of water and hydrocarbons that are used to enhance oil recovery. At concentrations above a certain critical value, the surfactant molecules in solution form aggregates called micelles. These micelles are capable of solubilizing fluids in their cores and are called swollen micelles. Spherical micelles have size ranges from 10^{-6} to 10^{-4} mm. The micellar solution or microemulsion is homogeneous, transparent or translucent, and stable to phase separation. The term *soluble oil* is often used to describe an oil external system having little or no dispersed water.

Although used to describe the process, neither microemulsion nor micellar solution accurately describes all compositions that are used in *emulsion flooding*. In fact, many systems used do not have an identifiable external or continuous phase and these terms do not always apply.

Microemulsion flooding (micellar/emulsion flooding) refers to a fluid injection process in which a stable solution of oil, water, and one or more surfactants along with electrolytes of salts is injected into the formation and is displaced by a mobility buffer solution (Reed and Healy, 1977; Dreher and Gogarty, 1979). Injecting water in turn displaces the mobility buffer. Depending on the reservoir environment, a preflood may or may not be used. The microemulsion is the key to the process. Oil and water are displaced ahead of the microemulsion **slug**, and a stabilized oil and water bank develops. The displacement mechanism is the same under secondary and tertiary recovery conditions. In the secondary case, water is the primary produced fluid until the oil bank reaches the well.

In microemulsion flooding, two approaches have developed to enhance oil recovery (Gogarty, 1976). In the first process, a relatively low-concentration surfactant microemulsion is injected at large pore volumes of 15% to 60% to reduce the interfacial tension between the water and oil, thereby increasing oil recovery. In the second process, a relatively small pore volume, from 3% to 20% of a high-concentration surfactant microemulsion is injected. With the high concentration of surfactant in the microemulsion, the micelles solubilize the oil and water in the displacing microemulsion. Consequently, the high-concentration system may initially displace the oil in a miscible-like manner. However, as the high-concentration slug moves through the reservoir, it is diluted by the formation fluids and the process ultimately or gradually reverts to a low-concentration flood. However, this initial displacement forms an oil bank, which is very important in establishing displacement efficiency. Low-concentration systems typically contain 2% to 4% surfactant, whereas high-concentration systems contain 8% w/w to 12% w/w.

Mobility control is important for the success of the process. The mobility of the microemulsion can be matched to that of the stabilized water–oil bank by controlling the microemulsion viscosity. The mobility buffer following the microemulsion slug prevents rapid slug deterioration from the rear and thus minimizes the slug size required for efficient oil displacement. Water external emulsions and aqueous solutions of high-molecular-weight polymers have been used as mobility buffers.

Microemulsion flooding can be applied over a wide range of reservoir conditions. Generally, wherever a waterflood has been successful, microemulsion flooding may also be applicable. In cases where waterflooding was a failure because of poor mobility relationships, microemulsion flooding might be technically successful because of the required mobility control. Of course, if waterflooding was a failure because of certain reservoir conditions, such as fracturing or very high permeability streaks, microemulsion flooding will most likely also fail.

In microemulsion flooding, the slug must be designed for specific reservoir conditions of temperature, resident water salinity, and crude oil type. If the temperature is very high, a fluid-handling problem may result in the field because of the increased vapor pressure of the hydrocarbon in microemulsion.

In analyzing the applicability of microemulsion–polymer flooding to a given reservoir, the need for a thorough understanding of the reservoir and fluid characteristics cannot be overemphasized. As mentioned, such characteristics as the nature of the oil and water content, relative permeability, mobility ratios, formation fractures, and variations in permeability, porosity, formation continuity, and rock mineralogy can have a dramatic effect on the success or failure of the process.

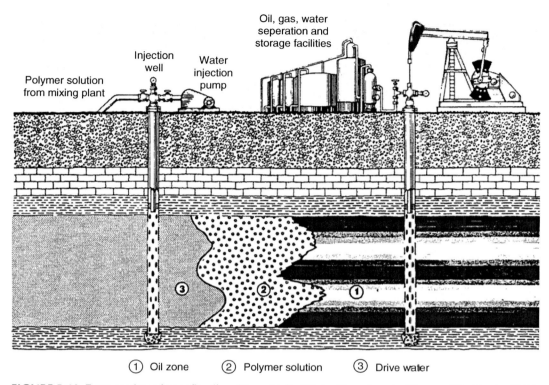

FIGURE 5.10 Recovery by polymer flooding. (From United States Department of Energy, Washington DC.)

Conventional waterflooding can often be improved by the addition of polymers (polymer flooding) to injection water (Figure 5.10) to improve the mobility ratio between the injected and in-place fluids. The polymer solution affects the relative flow rates of oil and water and sweeps a larger fraction of the reservoir than water alone, thus contacting more of the oil and moving it to production wells. Polymers currently in use are produced both synthetically (polyacrylamides) and biologically (polysaccharides). The polymers may also be cross-linked in situ to form highly viscous fluids that will divert the subsequently injected water into different reservoir strata.

Polymer flooding has its greatest utility in heterogeneous reservoirs and those that contain moderately viscous oils. Oil reservoirs with adverse waterflood mobility ratios have a potential for increased oil recovery through better horizontal sweep efficiency. Heterogeneous reservoirs may respond favorably as a result of improved vertical sweep efficiency. Because the microscopic displacement efficiency is not affected, the increase in recovery over water-flood will likely be modest and limited to the extent that sweep efficiency is improved, but the incremental cost is also moderate. Currently, polymer flooding is being used in a significant number of commercial field projects. The process may be used to recover oils of higher viscosity than those for which a surfactant flood might be considered.

Polymer solutions must be stable for a prolonged period at reservoir conditions. Mechanical, chemical, thermal, and microbial effects can degrade polymers. However, degradation can be minimized or even prevented by using specific equipment or methods (Table 5.3).

Stability problems may occur as a result of oxygen contamination of the polymer solutions. Such contamination can lower the screen factor of polyacrylamide solutions by as much as 30%. In field operations, the loss of mobility reduction due to oxygen may be more serious, as

TABLE 5.3
Polymer Degradation in Service

| Type of Degradation | Susceptibility | | | Cause | Remarks |
	Polyacrylamide	Xanthan Gum			
Chemical	High	Moderate		The cations Na^+, Ca^{2+}, Mg^{2+}	Divalent ions are more detrimental
Chemical	High	High		Transition metal ions	Aggravated by high temperatures and high pH
Chemical	High	High		Oxygen and oxidizing agents	Aggravated by high temperatures
Chemical	High	High		Hydrolysis by acidic or basic chemicals	Aggravated under aerobic conditions or high temperatures
Thermal	High ($>250°F$)	High ($>160°F$)		High temperature	Aggravated under aerobic conditions or high pH
Microbial	Moderate	High		Yeasts, bacteria, fungi	Aggravated by warm temperatures or under aerobic conditions
Mechanical shear	High	Low		Intense shear stress and high flux such as that occurring with flow through valves, orifices, and low-permeability formations	Aggravated by di- and trivalent cations

control of the reservoir fluid composition can be difficult. Sodium hydrosulfite in low concentrations is an effective oxygen collector for polyacrylamide solutions. However, sodium hydrosulfite tends to catalyze polymer deterioration when free oxygen and decomposed polymers are present. Therefore, the proper use of sodium hydrosulfite is imperative to avoid severe polymer degradation. In addition, caution is necessary to prevent oxygen from reentering the system once sodium hydrosulfite has been added to the makeup water.

Surfactant flooding (Figure 5.11) is a multiple-slug process involving the addition of surface-active chemicals to water. These chemicals reduce the capillary forces that trap the oil in the pores of the rock. The surfactant slug displaces the major part of the oil from the reservoir volume contacted, forming a flowing oil–water bank that is propagated ahead of the surfactant slug. The principal factors that influence the surfactant slug design are interfacial properties, slug mobility in relation to the mobility of the oil–water bank, the persistence of acceptable slug properties and slug integrity in the reservoir, and cost.

A slug of water containing polymer in solution follows the surfactant slug. The polymer solution is injected to preserve the integrity of the more costly surfactant slug and to improve the sweep efficiency. Both these goals are achieved by adjusting the polymer solution viscosity in relation to the viscosity of the surfactant slug to obtain a favorable mobility ratio. The polymer solution is then followed by injection of drive water, which continues until the project is completed.

Each reservoir has unique fluid and rock properties, and specific chemical systems must be designed for each individual application. The chemicals used, their concentrations in the

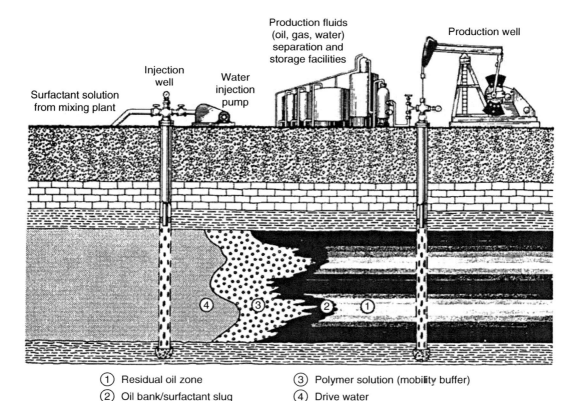

① Residual oil zone   ③ Polymer solution (mobility buffer)
② Oil bank/surfactant slug   ④ Drive water

FIGURE 5.11 Recovery by surfactant flooding. (From United States Department of Energy, Washington DC.)

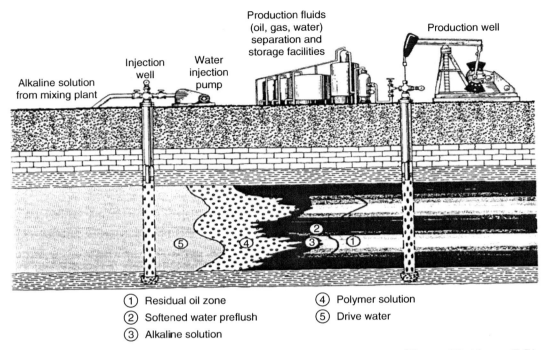

FIGURE 5.12 Recovery by alkaline flooding. (From United States Department of Energy, Washington DC.)

slugs, and the slug sizes depend upon the specific properties of the fluids and the rocks involved and upon economic considerations.

Alkaline flooding (Figure 5.12) adds inorganic alkaline chemicals, such as sodium hydroxide, sodium carbonate, or sodium orthosilicates to the water to enhance oil recovery by one or more of the following mechanisms: interfacial tension reduction, spontaneous emulsification, or wettability alteration (Morrow, 1996). These mechanisms rely on the in situ formation of surfactants during the neutralization of petroleum acids in the crude oil by the alkaline chemicals in the displacing fluids.

Although emulsification in alkaline flooding processes decreases injection fluid mobility to a certain degree, emulsification alone may not provide adequate sweep efficiency. Sometimes, polymer is included as an ancillary mobility control chemical in an alkaline waterflood to augment any mobility ratio improvements due to alkaline-generated emulsions.

Other variations on this theme include the use of steam and the means of reducing interfacial tension by the use of various solvents (Ali, 1974; Ali and Abad, 1975). The solvent approach has had some success when applied to bitumen recovery from mined tar sand but application to nonmined material losses of solvent and dissolved bitumen is always an issue. However, this approach should not be rejected out of hand since a novel concept may arise that guarantees minimal (acceptable) losses of bitumen and solvent.

Miscible fluid displacement (miscible displacement) is an oil displacement process in which an alcohol, a refined hydrocarbon, a condensed petroleum gas, carbon dioxide, liquefied natural gas, or even exhaust gas is injected into an oil reservoir, at pressure levels such that the injected gas or alcohol and reservoir oil are miscible; the process may include the concurrent, alternating, or subsequent injection of water.

The procedures for miscible displacement are the same in each case and involve the injection of a slug of solvent that is miscible with the reservoir oil, followed by injection of either a liquid or a gas to sweep up any remaining solvent. It must be recognized that the miscible slug of solvent becomes enriched with oil as it passes through the reservoir and its composition changes, thereby reducing the effective scavenging action. However, changes in the composition of the fluid can also lead to wax deposition (Weingarten and Euchner, 1986; Majeed et al., 1990; Pedersen et al., 1991; Erickson et al., 1993; Pan and Firoozabadi, 1996; Calange et al., 1997) as well as deposition of asphaltene constituents (Leontaritis et al., 1987 and references cited therein; Leontaritis, 1989; Chung, 1992; Nghiem et al., 1993; Nor-Aziam and Adewumi, 1993; Kamath et al., 1994; Deo et al., 1995; Rassamdana et al., 1999). Therefore, caution is advised.

Microscopic observations of the leading edge of the miscible phase have shown that the displacement takes place at the boundary between the oil and the displacing phase (Koch and Slobod, 1957). The small amount of oil that is bypassed is entrained and dissolved in the rest of the slug of miscible fluids; mixing and diffusion occur to permit complete recovery of the remaining oil. If a second miscible fluid is used to displace the first, another zone of displacement and mixing follows. The distance between the leading edge of the miscible slug and the bulk of pure solvent increases with the distance traveled, as mixing and reservoir heterogeneity cause the solvent to be dispersed.

Other parameters affecting the miscible displacement process are reservoir length, injection rate, porosity, and permeability of reservoir matrix, size and mobility ratio of miscible phases, gravitational effects, and chemical reactions. Miscible floods using carbon dioxide, nitrogen, or hydrocarbons as miscible solvents have the greatest potential for enhanced recovery of low-viscosity oils. Commercial hydrocarbon-miscible floods have been operated since the 1950s, but **carbon dioxide-miscible flooding** on a large scale is relatively recent and is expected to make the most significant contribution to miscible enhanced recovery in the future.

Carbon dioxide is capable of displacing many crude oils, thus permitting recovery of most of the oil from the reservoir rock that is contacted (carbon dioxide-miscible flooding) (Figure 5.13). The carbon dioxide is not initially miscible with the oil. However, as the carbon dioxide

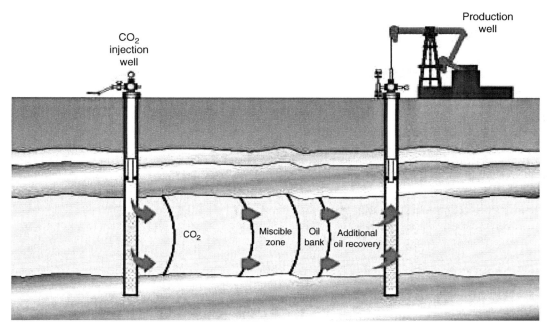

FIGURE 5.13 Recovery by carbon dioxide-miscible flooding. (From United States Department of Energy, Washington DC.)

contacts the in situ crude oil, it extracts some of the hydrocarbon constituents of the crude oil into the carbon dioxide and carbon dioxide is also dissolved in the oil. Miscibility is achieved at the displacement front when no interfaces exist between the hydrocarbon-enriched carbon dioxide mixture and the carbon dioxide-enriched oil. Thus, by a dynamic (multiple-contact) process involving interphase mass transfer, miscible displacement overcomes the capillary forces that otherwise trap oil in pores of the rock.

The reservoir operating pressure must be kept at a level high enough to develop and maintain a mixture of carbon dioxide and extracted hydrocarbons such that, at reservoir temperature, it will be miscible with the crude oil. Impurities in the carbon dioxide stream, such as nitrogen or methane, increase the pressure required for miscibility. Mixing due to reservoir heterogeneity and diffusion tends to locally alter and destroy the miscible composition, which must then be regenerated by additional extraction of hydrocarbons. In field applications, both miscible and near-miscible displacements may actually proceed simultaneously in different parts of the reservoir.

The volume of carbon dioxide injected is specifically chosen for each application and usually ranges from 20% to 40% of the reservoir pore volume. In the later stages of the injection program, carbon dioxide may be driven through the reservoir by water or a lower-cost inert gas. To achieve higher sweep efficiency, water and carbon dioxide are often injected in alternate cycles.

In some applications, particularly in carbonate (limestone, dolomite, and chert) reservoirs where it is likely to be used most frequently, carbon dioxide may prematurely break through to producing wells. When this occurs, remedial action using mechanical controls in injection and production wells may be taken to reduce carbon dioxide production. However, substantial carbon dioxide production is considered normal. Generally, this produced carbon dioxide is reinjected, often after processing to recover valuable light hydrocarbons.

For some reservoirs, miscibility between the carbon dioxide and the oil cannot be achieved and is dependent upon the oil properties (Figure 5.14). However, carbon dioxide can still be used to recover additional oil. The carbon dioxide swells crude oils, thus increasing the volume of pore space occupied by the oil and reducing the quantity of oil trapped in the pores. It also reduces the oil viscosity. Both effects improve the mobility of the oil. Carbon dioxide-immiscible flooding has been demonstrated in both pilot and commercial projects, but overall it is expected to make a relatively small contribution to enhanced oil recovery.

The solution gas–oil ratio for carbonated crude oil should be measured in the normal way and plotted as gas–oil ratio in volume per volume vs. pressure. The greater the solubility of carbon dioxide in the oil, the larger is the increase in the solution gas–oil ratio. In fact, the increase in the gas–oil ratio usually parallels the increase in the oil formation volume factor due to swelling. It should be noted that the gas in any gas–oil ratio experiment is not carbon dioxide, but contains hydrocarbons that have vaporized from the liquid phase. Consequently, whether the gas–oil ratio is measured in a pressure–volume–temperature cell or from a slim tube experiment, compositional analysis must be carried out to obtain the composition of the gas as well as that of the equilibrium liquid phase. If actual measured values are not available, the correlation developed for crude oil containing dissolved gases can be used, but this gives only approximate values at best. As the density of pure gases is a function of pressure and temperature, for crude oil saturated with gases, the density in the mixing zone must be specified as a function of pressure and mixing zone composition.

Hydrocarbon gases and condensates have been used for over 100 commercial and pilot miscible floods. Depending upon the composition of the injected stream and the reservoir crude oil, the mechanism for achieving miscibility with reservoir oil can be similar to that obtained with carbon dioxide (dynamic or multiple-contact miscibility), or the miscible solvent and in situ oil may be miscible initially (first-contact miscibility). Except in special circumstances, these light hydrocarbons are generally too valuable to be used commercially.

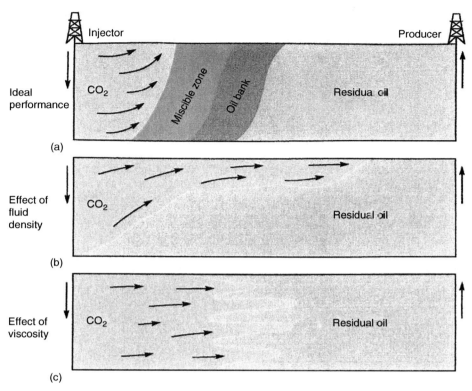

FIGURE 5.14 Effect of oil properties on recovery by carbon dioxide-miscible recovery. (From United States Department of Energy, Washington DC.)

Nitrogen and flue gases have also been used for commercial miscible floods. Minimum miscibility pressures for these gases are usually higher than for carbon dioxide, but in high-pressure, high-temperature reservoirs where miscibility can be achieved, these gases may be a cost-effective alternative to carbon dioxide.

Thermal methods for oil recovery have been found to be most useful when the oil in the reservoir has a high viscosity. For example, heavy oil is usually highly viscous (hence the use of the adjective heavy), with a viscosity ranging from approximately 100 cp to several million centipoises at reservoir conditions (Figure 5.15). In addition, oil viscosity is also a function of temperature and API gravity (Speight, 2000 and references cited therein). Thus, for heavy crude oil samples with API gravity ranging from 4° to 21° API (1.04 to 0.928 kg/m³):

$$\log \log (\mu \sigma + \alpha) = A - B \log (T + 460)$$

In this equation, $\mu \sigma$ is oil viscosity in centipoises, T is temperature in °F, A and B are constants, and α is an empirical factor used to achieve a straight-line correlation at low viscosity. This equation is usually used to correlate kinematic viscosity in centistokes, in which case, an α of 0.6 to 0.8 is suggested (dynamic viscosity in centipoises equals kinematic viscosity in centistokes times density in g/ml).

An alternative equation for correlating viscosity data is:

$$\mu = a e^{b/T^*}$$

where a and b are constants, and T^* is the absolute temperature.

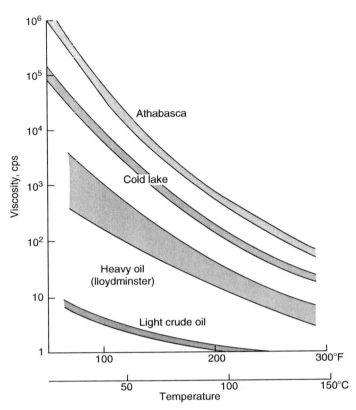

FIGURE 5.15 Variation of petroleum, heavy oil, and tar sand bitumen viscosity with temperature. (From United States Department of Energy, Washington DC.)

Thermally-enhanced oil recovery processes add heat to the reservoir to reduce oil viscosity or to vaporize the oil. In both instances, the oil is made more mobile so that it can be more effectively driven to producing wells. In addition to adding heat, these processes provide a driving force (pressure) to move oil to producing wells.

Thermal recovery methods include **cyclic steam injection**, steam flooding, and in situ combustion. The steam processes are the most advanced of all enhanced oil recovery methods in terms of field experience and thus have the least uncertainty in estimating performance, provided a good reservoir description is available. Steam processes are most often applied in reservoirs containing viscous oils and tars, usually in place of rather than following secondary or primary methods. Commercial application of steam processes has been underway since the early 1960s. In situ combustion has been field tested under a wide variety of reservoir conditions, but few projects have proved economical and advanced to commercial scale.

Steam drive injection (steam injection) has been commercially applied since the early 1960s. The process occurs in two steps: (1) steam stimulation of production wells, i.e., direct steam stimulation (Figure 5.16), and (2) steam drive by steam injection to increase production from other wells (indirect steam stimulation).

When there is some natural reservoir energy, steam stimulation normally precedes steam drive. In steam stimulation, heat is applied to the reservoir by the injection of high-quality steam into the produce well. This cyclic process, also called **huff and puff** or **steam soak**, uses

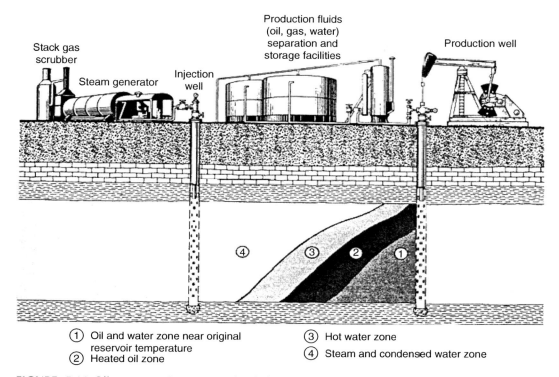

FIGURE 5.16 Oil recovery by steam stimulation. (From United States Department of Energy, Washington DC.)

the same well for both injection and production. The period of steam injection is followed by production of reduced viscosity oil and condensed steam (water). One mechanism that aids production of the oil is the flashing of hot water (originally condensed from steam injected under high pressure) back to steam as pressure is lowered when a well is put back on production.

When natural reservoir drive energy is depleted and productivity declines, most cyclic steam injection projects are converted to steam drives. In some projects, producing wells are periodically steam stimulated to maintain high production rates. Normally, stream drive projects are developed on relatively close well spacing to achieve thermal communication between adjacent injection and production wells. To date, steam methods have been applied almost exclusively in relatively thick reservoirs containing viscous crude oil.

Cyclic steam injection is the alternating injection of steam and production of oil with condensed steam from the same well or wells. Thus, steam generated at surface is injected in a well and the same well is subsequently put back on production.

A cyclic steam injection process includes three stages. The first stage is injection, during which a measured amount of steam is introduced into the reservoir. The second stage (the soak period) requires that the well be shut in for a period of time (usually several days) to allow uniform heat distribution to reduce the viscosity of the oil (alternatively, to raise the reservoir temperature above the pour point of the oil). Finally, during the third stage, the now-mobile oil is produced through the same well. The cycle is repeated until the flow of oil diminishes to a point of no return.

Cyclic steam injection is used extensively in heavy-oil reservoirs, tar sand deposits, and in some cases to improve injectivity before steam flooding or in situ combustion operations. Thus cyclic steam injection is also called steam soak or the huff 'n' puff method.

In practice, steam is injected into the formation at greater than fracturing pressure (150–1600 psi for Athabasca sands), followed by a soak period after which production is commenced (Winestock, 1974; Daily Oil Bull., 1975; Mungen and Nichols, 1975; Burger, 1978). The technique has also been applied to the California tar sand deposits (Bott, 1967) and in some heavy-oil reservoirs north of the Orinoco deposits (Ballard et al., 1976; Franco, 1976). The steam flooding technique has been applied, with some degree of success, to the Utah tar sands (Watts et al., 1982) and has been proposed for the San Miguel (Texas) tar sands (Hertzberg et al., 1983).

In situ combustion is normally applied to reservoirs containing low-gravity oil, but has been tested over perhaps the widest spectrum of conditions of any enhanced oil recovery process. In the process, heat is generated within the reservoir by injecting air and the burning part of the crude oil. This reduces the oil viscosity and partially vaporizes the oil in place, and the oil is driven out of the reservoir by a combination of steam, hot water, and gas drive. Forward combustion involves movement of the hot front in the same direction as the injected air. Reverse combustion involves movement of the hot front opposite to the direction of the injected air.

The relatively small portion of the oil that remains after these displacement mechanisms have acted becomes the fuel for the in situ combustion process. Production is obtained from wells offsetting the injection locations. In some applications, the efficiency of the total in situ combustion operation can be improved by alternating water and air injection. The injected water tends to improve the utilization of heat by transferring heat from the rock behind the combustion zone to the rock immediately ahead of the combustion zone.

The performance of in situ combustion is predominantly determined by the following four factors:

1. Quantity of oil that initially resides in the rock to be burned
2. Quantity of air required to burn the portion of the oil that fuels the process
3. Distance to which vigorous combustion can be sustained against heat losses
4. Mobility of the air or combustion product gases

In many field projects, the high gas mobility has limited recovery through its adverse effect on the sweep efficiency of the burning front. Because of the density contrast between air and reservoir liquids, the burning front tends to over-ride the reservoir liquids. To date, combustion has been most effective for the recovery of viscous oils in moderately thick reservoirs in which reservoir dip and continuity provide effective gravity drainage or operational factors permit close well spacing.

Using combustion to stimulate oil production is regarded as attractive for deep reservoirs (Finken and Meldau, 1972; Terwilliger, 1975) and, in contrast to steam injection, usually involves no loss of heat. The duration of the combustion may be short (<30 days) or more prolonged (approximately 90 days), depending upon requirements. In addition, backflow of the oil through the hot zone must be prevented or coking occurs.

Both forward and reverse combustion methods have been used with some degree of success when applied to tar sand deposits. The forward combustion process has been applied to the Orinoco deposits (Terwilliger et al., 1975) and in the Kentucky sands (Terwilliger, 1975). The reverse combustion process has been applied to the Orinoco deposit (Burger, 1978) and the Athabasca deposit (Mungen and Nichols, 1975). In tests such as these, it is essential to control the airflow (Wilson et al., 1963) and to mitigate the potential for spontaneous ignition (Wilson et al., 1963; Dietz and Weijdema, 1968; Burger, 1978). There has also been some success in the application of the reverse combustion technique to the Missouri tar sands (Trantham and Marx, 1966). A modified combustion approach has been applied to the

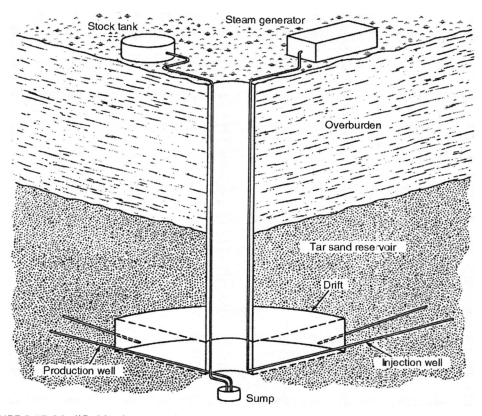

FIGURE 5.17 Modified in situ extraction process.

Athabasca deposit (Mungen and Nichols, 1975). The technique involved a heat-up phase and a production (or blowdown phase) followed by a displacement phase using a fireflood–waterflood (COFCAW) process.

In modified in situ extraction processes (Figure 5.17), combinations of in situ and mining techniques are used to access the reservoir. A portion of the reservoir rock must be removed to enable application of the in situ extraction technology. The most common method is to enter the reservoir through a large-diameter vertical shaft, excavate horizontal drifts from the bottom of the shaft, and drill injection and production wells horizontally from the drifts. Thermal extraction processes are then applied through the wells. When the horizontal wells are drilled at or near the base of the tar sand reservoir, the injected heat rises from the injection wells through the reservoir, and drainage of produced fluids to the production wells is assisted by gravity.

5.6 PRODUCTS AND PRODUCT QUALITY

Fluids produced from a well are seldom pure crude oil: in fact, a variety of materials may be produced by oil wells in addition to liquid and gaseous hydrocarbons. The natural gas itself may contain as impurities one or more nonhydrocarbon substances. The most abundant of these impurities is hydrogen sulfide, which imparts a noticeable odor to the gas. A small amount of this compound is considered advantageous as it gives an indication of leaks and where they occur. A larger amount, however, makes the gas obnoxious and difficult to market. Such gas is referred to as sour gas (Chapter 1) and much of it is used in the manufacture of carbon black. A few natural gases contain helium, and this element does in fact occur in commercial quantities in certain gas fields; nitrogen and carbon dioxide are also found in some natural gases. Gas is usually separated at as high a pressure as possible, reducing compression costs when the gas is to be used for gaslift or delivered to a pipeline. Lighter hydrocarbons and hydrogen sulfide are removed as necessary to obtain a crude oil of suitable vapor pressure for transport, yet retaining most of the natural gasoline constituents.

By far, the most abundant extraneous material is water. Many wells, especially during their declining years, produce vast quantities of salt water, and disposing of it is both a serious and an expensive problem. Further, the brine may be corrosive, which necessitates frequent replacement of casing, pipe, and valves, or it may be saturated so that the salts tend to precipitate upon reaching the surface. In either case, the water produced with the oil is a source of continuing trouble. Finally, if the reservoir rock is an incoherent sand or poorly cemented sandstone, large quantities of sand are produced along with the oil and gas. On its way to the surface, the sand has been known to scour its way completely through pipes and fittings.

It must also be remembered that in any field where primary production is followed by a secondary or enhanced production method, there will be noticeable differences in properties between the fluids produced (Thomas et al., 1983). The differences in elemental composition may not reflect these differences to any great extent (Zou et al., 1989), but more significant differences will be evident from an inspection of the physical properties. One issue that arises from the physical property data is that such oils may be outside the range of acceptability for refining techniques other than thermal options. In addition, overloading of thermal process units will increase as the proportion of the heavy oil in the refinery feedstock increases. Obviously there is a need for more and more refineries to accept larger proportions of heavy crude oils as the refinery feedstock and have the capability to process such materials.

In summary, the technologies applied to oil recovery involve different concepts, some of which can cause changes to the oil during production.

Technologies such as alkaline flooding, microemulsion (micellar/emulsion) flooding, polymer augmented waterflooding, and carbon dioxide-miscible/immiscible flooding do not require or cause any change to the oil. The steaming technologies may cause some steam distillation that can augment the process when the steam distilled material moves with the steam front and acts as a solvent for oil ahead of the steam front (Pratts, 1986). Again, there is no change to the oil, although there may be favorable compositional changes to the oil insofar as lighter fractions are recovered and heavier materials remain in the reservoir (Richardson et al., 1992).

The technology in which changes do occur involves combustion of the oil in situ. The concept of any combustion technology requires that the oil be partially combusted and that thermal decomposition occur to other parts of the oil. This is sufficient to cause irreversible chemical and physical changes to the oil to the extent that the product is markedly different to the oil in place. Recognition of this phenomenon is essential before combustion technologies are applied to oil recovery.

Although improvement in properties may not appear to be too drastic, it usually is sufficient to have major advantages for refinery operators. Any incremental increase in the units of hydrogen/carbon ratio can save amounts of costly hydrogen during upgrading. The same principles are also operative for reductions in the nitrogen, sulfur, and oxygen contents. This latter occurrence also improves catalyst life and activity as well as reduces the metals content.

In short, in situ recovery processes (although less efficient in terms of bitumen recovery relative to mining operations) may have the added benefit of leaving some of the more obnoxious constituents (from the processing objective) in the ground.

5.7 TRANSPORTATION

Most oil fields are at a considerable distance from the refineries that convert crude oil into usable products, and therefore the oil must be transported in pipelines and tankers (Figure 5.18). However, most crude oil needs some form of treatment near the reservoir before it can be carried considerable distances through the pipelines or in the tankers. Railroad cars and motor vehicles are also used to a large extent for the transportation of petroleum products.

Fluids produced from a well are seldom pure crude oil. In fact, the oil often contains quantities of gas, saltwater, or even sand. Separation must be achieved before transportation. Separation and cleaning usually take place at a central facility that collects the oil produced from several wells. Gas can be separated conveniently at the wellhead. When the pressure of the gas in the crude oil as it comes out at the surface is not too great, a simple flow tank fitted with baffles can be used to separate the gas from the oil at atmospheric pressure. If a considerable amount of gas is present, particularly if the crude oil is under considerable pressure, a series of flow tanks is necessary. The natural gas itself may contain as impurities one or more nonhydrocarbon substances. The most abundant of these impurities is hydrogen sulfide, which imparts a noticeable odor to the gas. A small amount of this compound is considered advantageous as it gives an indication of leaks and where they occur, as mentioned earlier (Bell, 1945).

Another step that needs to be taken in the preparation of crude oil for transportation is the removal of excessive quantities of water. Crude oil at the wellhead usually contains emulsified water in proportions that may reach amounts approaching 80% to 90%. It is generally required that crude oil to be transported by pipeline contain substantially less water than may appear in the crude at the wellhead. In fact, water contents from 0.5% to 2.0% have been specified as the maximum tolerable amount in a crude oil to be moved by pipeline. It is therefore necessary to remove the excess water from the crude oil before transportation.

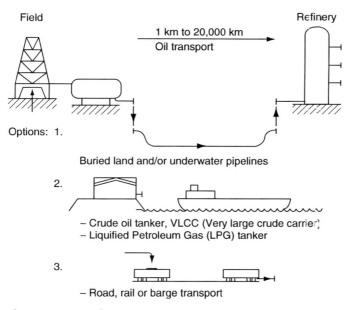

FIGURE 5.18 Petroleum transportation.

In an emulsion, the globules of one phase are usually surrounded by a thin film of an emulsifying agent that prevents them from congregating into large droplets. In the case of an oil–water emulsion, the emulsifying agent may be part of the heavier (asphaltic), more polar constituents. The film may be broken mechanically, electrically, or by the use of demulsifying agents and the proportion of water in the oil reduced to the specified amounts, thereby rendering the crude oil suitable for transportation.

The transportation of crude oil may be further simplified by blending crude oils from several wells, thereby homogenizing the feedstock to the refinery. It is usual practice, however, to blend crude oils of similar characteristics, although fluctuations in the properties of the individual crude oils may cause significant variations in the properties of the blend over a period of time. However, the technique of blending several crude oils before transportation, or even after transportation but before refining, may eliminate the frequent need to change the processing conditions that would perhaps be required to process each of the crude oils individually.

The arrival of large quantities of petroleum oil at import and refining centers has brought about the need for storage facilities. The usual form of crude oil storage is the collection of large cylindrical steel storage tanks (tank farm) that are a familiar sight at most refineries and shipping terminals. The tanks vary in size, but some are capable of holding up to 950,000 barrels of oil. Crude oil may also be stored in such geological features as salt domes. The domes have been previously leached or hollowed out into huge underground caves, such as those used by the U.S. Strategic Petroleum Reserve in Louisiana and Texas. Other underground storage facilities include disused coal mines and artificial caverns. Natural gas is, on occasion, stored in old reservoirs from which the gas has been recovered. The gas is pumped under pressure into the reservoir at times of low gas demand, so that it can be retrieved later to meet peak demand.

Large-scale transportation of crude oil, refined petroleum products, and natural gas is usually accomplished by pipelines and tankers, whereas smaller-scale distribution, especially of petroleum products, is carried out by barges, trucks, and rail tank cars. In fact, the transportation from the source of the crude oil to the market is as old as the industry itself. Even the bitumen used in Babylon and other cities of the Fertile Crescent had to be transported from the seepage at Hit to the place where it would be used.

In more modern times, the transportation of crude oil from fields to refineries and of products to market centers was at one time essentially dependent upon rail transportation. By the early 1970s, the use of railroad tank cars had diminished to the point at which only a little over 1% of the total petroleum tonnage was hauled by the railroads. Pipeline mileage increased to become the major means of transportation.

There were two principal technological trends during that period, which increased pipeline use: (1) more widespread use was made of large-diameter pipe; and (2) more efficient diesel pumps became available for installation at stations along the line, replacing steam-driven models. In addition, there were other developments such as (1) the introduction of welded rather than screwed-in couplings; (2) the use of high carbon steel to replace lap-welded pipe; and (3) the replacement of diesel equipment by electrically powered pumps. At present, the pipeline used for petroleum transportation may be up to 48 in. in diameter and may cover many thousands of miles.

Pipelines may be used to transport different types of crude oil (batch transportation). When the different batches must be kept separated to prevent mixing, slugs of kerosene, water, or occasionally inflatable rubber balls are used to separate the batches. However, there is also the possibility that the batches can be transported through the pipelines without such separators. The properties of each batch may be such that mixing, other than to form a narrow interface, is prevented. It is frequently necessary to pass cylindrical steel cleaners (pigs) through the pipelines, between pumping stations, to maintain the pipeline clear of deposits.

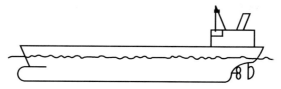

Type/Cargo	Length	Beam	Draft	dwt[a]	Speed
1 VLCC/crude	332 (m)	58 (m)	22 (m)	298,000	15.5 knot
2 Multi product/crude heated tanks	244 (m)	42 (m)	14.6 (m)	105,000	14 knot
3 Clean/dirty product	183 (m)	32 (m)	12.2 (m)	47,000	15.6 knot
4 LPG carrier	209 (m)	31.4 (m)	12.5 (m)	47,000	—
5 LNG spherical tanks	272 (m)	47.2 (m)	11.4 (m)	67,000	18.5 knot
6 LNG 'oblong' tanks	287 (m)	41.8 (m)	11.3 (m)	71,470	19.2 knot

[a]dwt (deadweight tonnes) is the cargo capacity of the vessel.

FIGURE 5.19 Definition of tankers for petroleum transportation.

Tank trucks are used for both lock and intermediate hauling from manufacturing and distance hauling from manufacturing and terminal points to individual domestic, commercial, and industrial consumers that maintain storage tanks on their premises Because of costs, most bulk deliveries by truck fall within a radius of 300 miles.

Seagoing tankers (Figure 5.19), on the other hand, can be sent to any destination where a port can accommodate them and can be shifted to different routes according to need.

The seagoing tanker fleets that are owned, or used, by the world's oil companies are also responsible for the movement of a considerable portion of the world's crude oil. In fact, seagoing tankers form one of the most characteristic features associated with the transportation of petroleum. Many of these ships are of such a size that there are few ports that can handle them. Instead, these large ships (very large crude carriers (VLCCs), and ultra large crude carriers (ULCCs)) spend their time sailing the seas between different points, filling up and off-loading without ever entering port. Special loading jetties, artificial islands, or large buoys moored far offshore have been developed to load or off-load these tankers.

In general, the larger the tanker the lower its unit cost of transportation. As a result, the size of tankers during the 1960s and 1970s grew steadily. During the 1930s and 1940s, the average size of tankers was about 12,000 dwt (dwt = the number of tons of cargo, stores, and fuel that a ship can usually carry). By the 1950s a 33,000 dwt tanker was considered standard size. At present, the VLCC and ULCC classes involve tankers of over 200,000 and 300,000 dwt. A few tankers in the 500,000 dwt range also exist, and some have even exceeded this size.

The cargo space in the tankers is usually divided into two, three, or four rows of cargo tanks by longitudinal bulkheads. These are further divided into individual tanks (from 25 to 40 ft long in large vessels) by transverse bulkheads. Access to the tanks is through oil-tight hatches on the deck, cargo being loaded or discharged by means of the ship's own pumps, which may have a capacity in excess of 4000 ton/h. There has lately been some discussion about the wisdom of building such tankers with double hulls. In theory, the single-hull tanker has a better chance of staying afloat, but will spill some of the cargo into the sea.

Over the past two decades, it has also become evident that crude oil tankers are usually shorter lived than most other cargo ships. Crude oil cargoes can deposit corrosive sludge on the bottom of the hold, and gasoline cargoes can also have a corrosive effect on the steel of the tanks. As a result, a tanker may last only 12 years instead of the 20 year life of a cargo ship, although protective coatings have been developed that help to withstand the corrosive effects of petroleum products.

Transportation is a major aspect of oil sands exploitation (Demaison, 1977). There are four major aspects of liquid fuels production from an oil sand resource: (1) ore recovery; (2) bitumen separation; (3) bitumen conversion to synthetic crude oil; and (4) refining the synthetic crude oil to usable liquid fuels. Currently, the two commercial plants carry out the first three stages on-site, and in this respect the plants are completely self-contained. The synthetic crude oil is then shipped by pipeline to a more conventional refinery site (e.g., Edmonton) for fuel upgrading to liquid fuels. There are, however, constraints on the character of liquids that may be shipped by pipeline. The synthetic crude oil conveniently meets the specifications for pipeline shipment, but should an alternative means of bitumen upgrading be established, that is **visbreaking** (viscosity breaking) (Chapter 17), pipeline specifications must be met.

However, it may be deemed desirable to ship the whole bitumen feed by pipeline, in which case dilution with naphtha is necessary. The naphtha may actually be produced at the recovery site by construction of a nominal conversion operation. However, it is to be anticipated that a feed with a viscosity in excess of 15,000 cSt at 35°C (100°F) will require in excess of 0.5 bbl naphtha per barrel of bitumen. At higher temperatures, the amount of naphtha required is reduced, but light ends and even water may have to be removed to prevent undue pressure buildup in the pipeline if shipping temperatures of about 95°C (200°F) are considered.

One other aspect of transportation is the shipment of bitumen separated from tar sand (or the whole oil sand or even bitumen-enriched oil sand, produced by, say, a less efficient once-through hot water separation) in trucks or trains. Currently, economic constraints related to the amount of material that would have to be moved to enable even a nominal conversion or upgrading operation to run continuously (hazards of weather and mechanical constraints notwithstanding) have caused these types of operation to be downgraded in priority.

Finally, it is also possible for bitumen to be emulsified and shipped (by pipeline) as an emulsion. This particular idea has received some attention, especially in regard to bitumen recovery by aqueous flooding methods. The idea is to produce the bitumen from the formation as an oil-in-water emulsion at the remote site, followed by shipping of the emulsion to an oil recovery and upgrading site.

Natural gas is also transported by seagoing vessels. The gas is either transported under pressure at ambient temperatures (e.g., propane and butanes) or at atmospheric pressure, but with the cargo under refrigeration (e.g., liquefied petroleum gas). For safety reasons, petroleum tankers are constructed with several independent tanks so that rupture of one tank will not necessarily drain the whole ship, unless it is a severe bow-to-stern (or stern-to-bow) rupture. Similarly, gas tankers also contain several separate tanks.

Natural gas presents different transportation requirement problems. Before World War II, its use was limited by the difficulty in transporting it over long distances. The gas found in oil fields was frequently burned off; and unassociated (dry) gas was usually abandoned. After the war, new steel alloys permitted the laying of large-diameter pipes for gas transport in the United States. The discovery of the Groningen field in the Netherlands in the early 1960s and the exploitation of huge deposits in Soviet Siberia in the 1970s and 1980s led to a similar expansion of pipelines and natural gas use in Europe.

Natural gas is much more expensive to ship than crude oil because of its lower density. Most natural gas moves by pipeline, but in the late 1960s, tanker shipment of cryogenically liquefied natural gas (LNG) began, particularly from the producing nations in the Pacific to

Japan. Special alloys are required to prevent the tanks from becoming brittle at the low temperatures (−161°C, −258°F) required to keep the gas liquid.

Thus, the means by which natural gas is transported depends upon several factors:

1. Physical characteristics of the gas to be transported, whether in the gaseous or the liquid phase
2. Distance over which the gas will be moved
3. Such features as the geological and geographic characteristics of the terrain, including land and sea operations
4. Complexity of the distribution systems
5. Environmental regulations that are relevant to the mode of transportation

In the last case, such factors as the possibility of pipeline rupture as well as the effect of the pipeline itself on the ecosystems need to be addressed. In general, and aside from any economic factors, it is possible to construct and put in place a system capable of transporting natural gas in the gaseous or liquid phase that allows system flexibility.

There are many such pipeline systems throughout the world and the United States (Considine, 1977). However, natural gas pipeline companies must meet environmental and legal standards. Economic standards are also a necessity: it would be extremely foolhardy (and economic suicide) if a company were to construct several pipelines when one such system would suffice. Construction of a pipeline system involves not only environmental and legal considerations, but also compliance with the regulations of the local, state, or federal authorities.

In general, many of the pipeline systems available use pipe material up to 48 in. in diameter (although lately, the larger diameter pipe has become more favorable) and sections of pipe may be up to 40 ft long. Protective coatings are usually applied to the pipe to prevent corrosion of the pipe from outside influences.

The gas pressure in long-distance pipelines may vary up to 5000 psi, but pressures up to 1500 psi are more usual. To complete the pipeline, it is necessary to install a variety of valves and regulators that can be opened or closed to adjust the flow of gas. The system must also be capable of shut down of any section in which an unexpected rupture may be caused by natural events (such as weather) or even by unnatural events (such as sabotage). Most of the valves or regulators in the pipeline system can now be operated by remote control, so that in the event of a rupture the system can be closed down. This is especially valuable where it may take a repair crew considerable time to reach the site of the breakdown.

The trend in recent years has been to expand the pipeline system into marine environments where the pipeline is actually under a body of water. This has arisen mainly because of the tendency for petroleum and natural gas companies to expand their exploration programs to the sea. Lines are now laid in marine locations where depths exceed 500 ft and cover distances of several hundred miles to the shore. Excellent examples of such operations include the drilling operations in the Texas gulf and in the North Sea.

One early concern with the laying of pipelines under a body of water arose because of the buoyancy of the pipe and the subsequent need to place the pipe in a permanent position on the floor of the lake bed or sea bed. In such instances, the negative buoyancy of the pipe can be overcome by the use of a weighted coating (e.g., concrete) on the pipe. Other factors, such as laying the pipe without too much stress (which would otherwise induce a delayed rupture), as well as the anchoring and positioning of the pipe on the seabed, are major issues that need to be addressed.

REFERENCES

Ali, S.M.F. 1974. *Oil Sands Fuel of the Future*. Hills L.V. ed., Canadian Society of Petroleum Geologists, Calgary, Alberta. p. 199.

Ali, S.M.F. and Abad, B. 1975. *Proceedings*. Twenty-Sixth Annual Meeting Petroleum Society. Canadian Institute of Mining, Banff, Alberta, Canada.

Ballard, J.R., Lanfranchi, E.E., and Vanags, P.A. 1976. *Proceedings*. Twenty-Seventh Annual Meeting Petroleum Society. Canadian Institute of Mining, Calgary, Alberta, Canada, June.

Bell, H.S. 1945. *American Petroleum Refining*. Van Nostrand, New York.

Borchardt, J.K. and Yen, T.F. 1989. *Oil Field Chemistry*. Symposium Series No. 396. American Chemical Society, Washington, DC.

Bott, R.C. 1967. *J. Petrol. Technol.* 19: 585.

Burger, J. 1978. *Developments in Petroleum Science, No. 7, Bitumens, Asphalts and Tar Sands*. Chilingarian G.V. and Yen T.F. eds., Elsevier, New York. p. 191.

Calange, S., Ruffier-Meray, V., and Behar, E. 1997. Onset of Crystallization Temperature and Deposit Amount for Waxy Crudes: Experimental Determinations and Thermodynamic Modeling. Paper No. SPE 37239. *Proceedings*. Annual Technical Conference and Exhibition, Houston, TX, February 18 to 21.

Chung, T-H. 1992. Thermodynamic Modeling for Organic Solids Precipitation. Paper No. SPE 24851. *Proceedings*. 67th Annual Technical Conference, Washington, DC, October 4 to 7.

Considine, D.M. 1977. *Energy Technology Handbook*. McGraw-Hill, New York.

Craft, B.C. and Hawkins, M.F. 1959. *Applied Petroleum Reservoir Engineering*. Prentice-Hall, Englewood Cliffs, New Jersey.

Daily Oil Bull. 1975. Calgary, Alberta, Canada, August 14.

Demaison, G.J. 1977. *The Oil Sands of Canada–Venezuela*. Redford D.A. and Winestock A.G. eds., Special Volume No. 17. Canadian Institute of Mining and Metallurgy. p. 9.

Deo, M.D., Miharia, A., and Kumar, R. 1995. Solids Precipitation in Reservoirs Due to Non-Isothermal Injections. Paper No. 28967. SPE, San Antonio, Texas, February 14 to 17.

Dietz, D.N. and Weijdema, L. 1968. *Producers Monthly* 32(5): 10.

Dreher, K.D. and Gogarty, W.B. 1979. *J. Rheol.* 23(2): 209.

Erickson, D.D., Nielsen, V.G., and Brown, T.S. 1993. Thermodynamic Measurement and Prediction of Paraffin Precipitation in Crude Oil. Paper No. SPE 26604. *Proceedings*. 68th Annual Technical Conference and Exhibition, Houston, Texas, October 3 to 6.

Finken, R.E., and Meldau, R.F. 1972. *Oil Gas J.* 70(29): 108.

Forbes, R.I. 1958. *A History of Technology*. Oxford University Press, Oxford, England.

Franco, A. 1976. *Oil Gas J.* 74(14): 132.

Frick, T.C. 1962. *Petroleum Production Handbook*, Volume II. McGraw-Hill, New York.

Gogarty, W.B. 1976. *J. Petrol. Technol.* 93.

Hertzberg, R., Hojabri, F., and Ellefson, L. 1983. Preprint No. 35e. Summer National Meeting. American Institute of Chemical Engineers, Denver, CO, August 28 to 31.

Hobson, G.D. and Tiratsoo, E.N. 1975. *Introduction to Petroleum Geology*. Scientific Press, Beaconsfield, England.

Kamath, V.A., Kakade, M.G., and Sharma, G.D. 1994. An Improved Molecular Thermodynamic Model for Asphaltene Equilibria. *Asphaltene Particles in Fossil Fuel Exploration, Recovery, Refining, and Production Processes*. Sharma M.K. and Yen T.F. eds., Plenum Press, New York.

Koch, H.A. and Slobod, R.L. 1957. *Trans. Am. Inst. Mech. Eng.* 210: 40.

Landes, K.K. 1959. *Petroleum Geology*. John Wiley & Sons Inc., New York.

Leontaritis, K.J. 1989. Asphaltene Deposition: A Comprehensive Description of the Problem, Manifestations, and Modeling Approaches. Paper No. SPE 18892. Proceedings. Symposium on Production Operations, Oklahoma City, OK, March 13 and 14.

Leontaritis, K.J., Mansoori, G.A., and Mansoori, U. 1987. Asphaltene Flocculation during Oil Production and Processing: A Thermodynamic Colloidal Model. Paper No. SPE 16258. *Proceedings*. International Symposium on Oilfield Chemistry, San Antonio, Texas, February 4 to 6.

Majeed, A., Bringedal, B., and Overå, S. 1990. *Oil Gas J.* (June 18): 63–69.

Morrow, N.R. ed., 1996. *Proceedings*. 3rd International Symposium on Evaluation of Reservoir Wettability and Its Effect on Oil Recovery. University of Wyoming, Laramie, WY.

Mungen, R. and Nichols, J.H. 1975. *Proc. 9th World Petrol. Congr.* 5: 29.

Nghiem, L.X., Hassam, M.S., Nutakki, R., and George, A.E.D. 1993. Efficient Modeling of Asphaltene Precipitation. Paper No. SPE 26642. *Proceedings*. 68th Annual Technical Conference. Houston, TX, October 4 to 7.

Nor-Aziam, N. and Adewumi, M.A. 1993. Development of Asphaltene Phase Equilibria Predictive Model. Paper No. SPE 26905. *Proceedings*. Eastern Regional Conference and Exhibition, Pittsburgh, PA, November 2 to 4.

Pan, H. and Firoozabadi, A. 1996. Pressure and Composition Effects on Wax Precipitation: Experimental Data and Model Results. Paper No. SPE 36740. *Proceedings*. Annual Technical Conference and Exhibition, Denver, CO, October 6 to 9.

Pedersen, K.S., Skovborg, P., and Rønningsen, H.P. 1991. *Energy Fuels* 5: 924–932.

Pratts, M. 1986. *Thermal Recovery*, Volume 7. Society of Petroleum Engineers, New York.

Ranney, M.W. 1979. *Crude Oil Drilling Fluids*. Noyes Data Corp., Park Ridge, NJ.

Rassamdana, H., Farhani, M., Dabir, B., Mozaffarian, M., and Sahimi, M. 1999. *Energy Fuels* 13: 176–187.

Reed, R.L. and Healy, R.N. 1977. *Improved Oil Recovery by Surfactant and Polymer Flooding*. Shah D.O. and Schechter R.S., eds., Academic Press, New York.

Richardson, W.C., Fontaine, M.F., and Haynes, S. 1992. Paper No. SPE 24033. Western Regional Meeting, Bakersfield, CA, March 30–April 1.

Schumacher, M.M. 1980. *Enhanced Recovery of Residual and Heavy Oils*. Noyes Data Corp., Park Ridge, NJ.

Speight, J.G. 2000. *Desulfurization of Heavy Oils and Residua*, 2nd Edn. Marcel Dekker Inc., New York.

Terwilliger, P.L. 1975. Paper 5568. *Proceedings*. 50th Annual Fall Meeting. Society of Petroleum Engineers, American Institute of Mechanical Engineers.

Terwilliger, P.L., Clay, R.R., Wilson, L.A., and Gonzalez-Gerth, E. 1975. *J. Petrol. Technol.* 27: 9.

Thomas, K.P., Barbour, R.V., Branthaver, J.F., and Dorrence, S.M. 1983. *Fuel* 62: 438.

Trantham, J.S. and Marx, J.W. 1966. *J. Petrol. Technol.* 18: 109.

Watts, K.C., Hutchinson, H.L., Johnson, L.A., Barbour, R.V., and Thomas, K.P. 1982. *Proceedings*. 54th Annual Fall Meeting, Society of Petroleum Engineers, American Institute of Mechanical Engineers, New Orleans, September 26 to 29.

Weingarten, J.S. and Euchner, J.A. 1986. Methods for Predicting Wax Precipitation and Deposition. SPE Paper No. 15654. *Proceedings*. 61st Annual Technical Conference and Exhibition, New Orleans, LA. October 5 to 8.

Wilson, L.A., Reed, R.L., Reed, D.W., Clay, R.R., and Harrison, N.H. 1963. *Soc. Petrol. Eng. J.* 3: 127.

Winestock, A.G. 1974. *Oil Sands Fuel of the Future*. Hills L.V. ed, Canadian Society of Petroleum Geologists, Calgary, Alberta, Canada. p. 190.

Zou, J., Gray, M.R., and Thiel, J. 1989. *AOSTRA J. Res.* 5: 75.

6 Recovery of Heavy Oil and Tar Sand Bitumen

6.1 INTRODUCTION

Heavy oil and bitumen (the primary component of **tar sand**) are often defined (loosely and incorrectly) in terms of API gravity. A more appropriate definition of bitumen, which sets it apart from heavy oil and conventional petroleum, is based on the definition offered by the U.S. government as the extremely viscous hydrocarbon which is not recoverable in its natural state by conventional oil well production methods including currently used enhanced recovery techniques (Chapter 1).

By inference, conventional petroleum and heavy oil (recoverable by conventional oil well production methods including currently used enhanced recovery techniques) are different to tar sand bitumen. Be that as it may, at some stage of production, conventional petroleum (in the later stages of recovery) and heavy oil (in the earlier stages of recovery) may require the application of **enhanced oil recovery** methods for recovery.

Initially, conventional crude oil is produced from the oil-bearing formations by drilling wells into the formation and recovery of the oil by any one of several possible methods (Figure 6.1). The oil is driven from the formation up through the wells (production wells) by energy stored in the formation, such as the pressure of water, dissolved natural gas (primary recovery). If this natural energy of the formation is expended, then energy must be injected into the formation in order to stimulate production through addition of energy to the formation through added water and gas (**secondary recovery**), followed by other more energy intensive methods of recovery (enhanced recovery) (Figure 6.1) (Chapter 5) (Chakma et al., 1991; Islam et al., 1994). Further, crude oil recovery depends upon several factors that, in turn, are site specific (Figure 6.2) and a variety of selection criteria are involved (Figure 6.3, Figure 6.4, and Figure 6.5).

In reservoirs that contain heavy oil, it is often desirable to initiate enhanced oil recovery (EOR) operations as early as possible. In tar sand deposits that contain bitumen, it is often desirable to initiate EOR operations as early as possible. This may mean considerably abbreviating conventional secondary recovery operations or bypassing them altogether. Thermal floods using steam and controlled **in situ combustion** methods are also used. Thermal methods of recovery reduce the viscosity of the crude oil by heat so that it flows more easily into the production well (Pratts, 1986). Thus, advanced techniques are usually variations of secondary methods with a goal of improving the sweeping action of the invading fluid.

The technologies applied to oil recovery involve different concepts, some of which can cause changes to the oil during production. Technologies such as alkaline flooding, micro-emulsion (micellar/emulsion) flooding, polymer-augmented water flooding, and carbon dioxide miscible/immiscible flooding do not require or cause any change to the oil. The steaming technologies may cause some steam distillation that can augment the process when

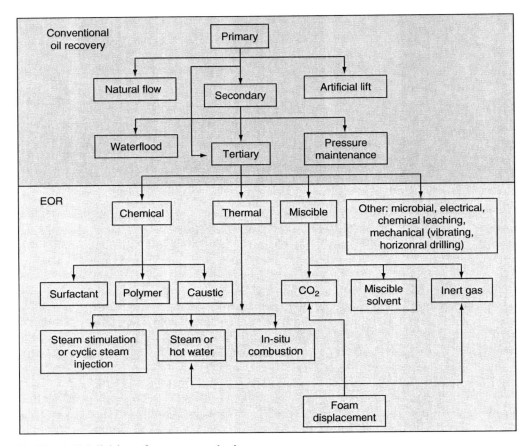

FIGURE 6.1 Subdivision of recovery methods.

the steam-distilled material moves with the steam front and acts as a solvent for oil ahead of the steam front. Again, there is no change to the oil although there may be favorable compositional changes to the oil insofar as lighter fractions are recovered and heavier materials remain in the reservoir.

The technology in which changes do occur involves combustion of the oil in situ. The concept of any combustion technology requires that the oil be partially combusted and that thermal decomposition occur to other parts of the oil. This is sufficient to cause irreversible

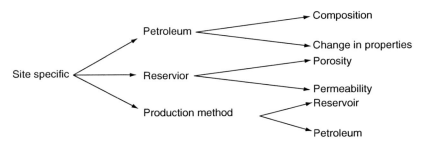

FIGURE 6.2 Recovery is site specific and depends upon several variable factors.

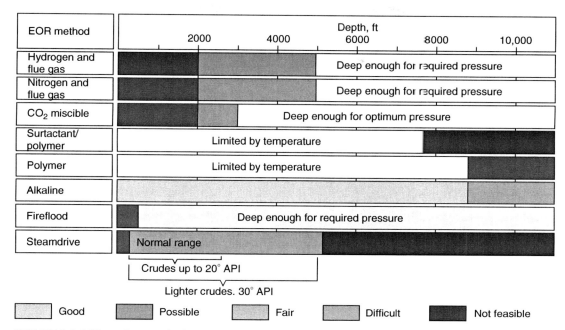

FIGURE 6.3 Effect of reservoir depth on recovery.

chemical and physical changes to the oil, to the extent that the product is markedly different from the oil in place indicating **upgrading** of the bitumen during the process. Recognition of this phenomenon is essential before combustion technologies are applied to oil recovery.

Further, in any field where primary production is followed by a secondary or enhanced recovery method, there is the potential for noticeable differences in properties between the

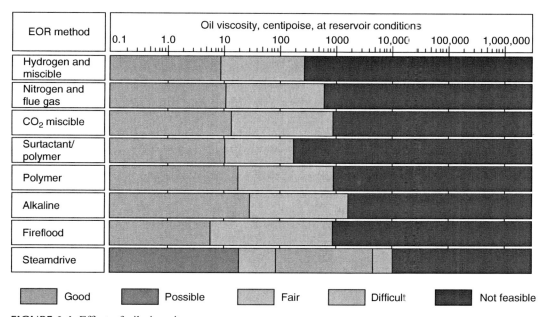

FIGURE 6.4 Effect of oil viscosity on recovery.

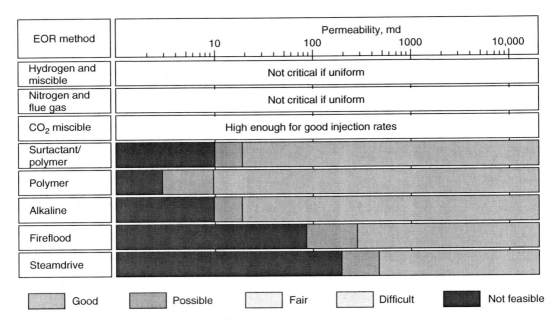

FIGURE 6.5 Effect of permeability on oil recovery.

fluids produced. Significant differences may render the product outside of the range of acceptability for the usual refining options and force a higher demand for thermal process (i.e., coking) units. Thus, overloading thermal process units will increase as the proportion of the heavy oil in the refinery feedstock increases.

Enhanced oil recovery (Chapter 5) is the incremental ultimate oil that can be economically recovered from a petroleum reservoir over oil that can be economically recovered by conventional primary and secondary methods. The intent of EOR is to increase the effectiveness of oil removal from pores of the rock (displacement efficiency) and to increase the volume of rock contacted by injected fluids (sweep efficiency).

EOR processes (Chapter 5) use thermal, chemical, or fluid phase behavior effects to reduce or eliminate the capillary forces that trap oil within pores, to thin the oil or otherwise improve its mobility, or to alter the mobility of the displacing fluids. In some cases, the effects of gravity forces, which ordinarily cause vertical segregation of fluids of different densities, can be minimized or even used to advantage. The various processes differ considerably in complexity, the physical mechanisms responsible for oil recovery, and the amount of experience that has been derived from field application. The degree to which the EOR methods are applicable in the future will depend on development of improved process technology. It will also depend on improved understanding of fluid chemistry, phase behavior, and physical properties; and on the accuracy of geology and reservoir engineering in characterizing the physical nature of individual reservoirs.

Variations of the EOR theme include the use of steam and solvents as the means of reducing interfacial tension. The solvent approach has had some success when applied to bitumen recovery from mined tar sand, but when applied to nonmined material, phenomenal losses of solvent and bitumen are always a major obstacle. This approach should not be rejected out of hand as a novel concept may arise that guarantees minimal (acceptable) losses of bitumen and solvent. In fact, **miscible fluid displacement (miscible displacement)** is a process in which an alcohol, a refined hydrocarbon, a condensed petroleum gas, carbon dioxide, liquefied natural gas, or even exhaust gas is injected into an oil reservoir, at pressure levels

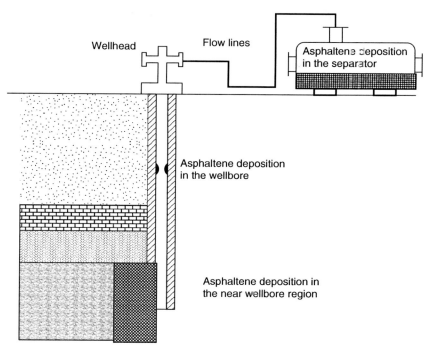

FIGURE 6.6 Positions where asphaltenes (and wax) deposition can occur.

such that the injected gas or alcohol and reservoir oil are miscible; the process may include the concurrent, alternating, or subsequent injection of water.

The procedures for miscible displacement are the same in each case and involve the injection of a **slug** of solvent that is miscible with the reservoir oil followed by injection of either a liquid or a gas to sweep up any remaining solvent. As the miscible slug of solvent becomes enriched with oil as it passes through the reservoir, the composition changes, thereby reducing the effective scavenging action. However, changes in the composition of the fluid can also lead to wax deposition as well as deposition of asphaltene constituents (Figure 6.6). Therefore, caution is advised.

Microscopic observations of the leading edge of the miscible phase have shown that the displacement takes place at the boundary between the oil and the displacing phase. The small amount of oil that is bypassed is entrained and dissolved in the rest of the slug of miscible fluids; mixing and diffusion occur to permit complete recovery of the remaining oil. If a second miscible fluid is used to displace the first, another zone of displacement and mixing follows. The distance between the leading edge of the miscible slug and the bulk of pure solvent increases with the distance traveled, as mixing and reservoir heterogeneity cause the solvent to be dispersed.

Other parameters affecting the miscible displacement process are reservoir length, injection rate, porosity, and permeability of reservoir matrix, size and mobility ratio of miscible phases, gravitational effects, and chemical reactions.

Thermal recovery methods (Figure 6.7) (Chapter 5) have found most use when heavy oil or bitumen has an extremely high viscosity under reservoir conditions. For example, most heavy oils are highly viscous, with a viscosity ranging from a thousand centipoises to a million centipoises or more at reservoir conditions. In addition, oil viscosity is also a function of temperature and API gravity (Chapter 5) (Speight, 2000 and references cited therein).

Thermal Recovery

This is accomplished either by hot fluid injection (hot water or steam) or in situ combustion (burning a part of the crude oil in place). Variations of these methods improve production of crudes by heating them, thereby improving their mobility and ease of recovery by fluid injection.

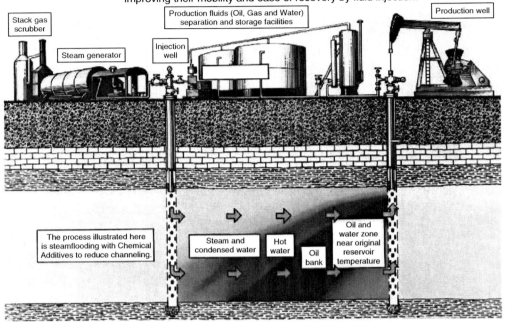

FIGURE 6.7 Oil recovery by thermal methods. (From United States Department of Energy, Washington DC.)

Thermal enhanced oil recovery processes (i.e., cyclic steam injection, steam flooding, and in situ combustion) add heat to the reservoir to reduce oil viscosity "and" or "or" to vaporize the oil. In both instances, the oil is made more mobile so that it can be more effectively driven to producing wells. In addition to adding heat, these processes provide a driving force (pressure) to move oil to producing wells.

In situ combustion may make a comeback with a new concept. THAI (toe-to-heel air injection) (Figure 6.8) is based on the geometry of horizontal wells that may solve the problems

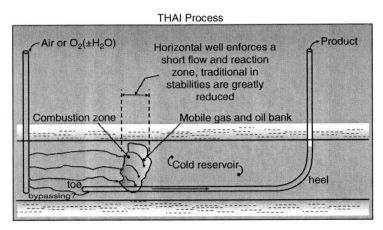

FIGURE 6.8 The THAI process.

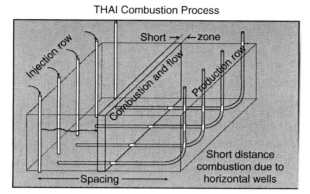

FIGURE 6.9 Configuration of the THAI process.

that have plagued conventional in situ combustion. The well geometry (Figure 6.9) enforces a short flow path so that any instability issues associated with conventional combustion are reduced or even eliminated.

In situ conversion, or underground refining, is a promising new technology to tap the extensive reservoirs of heavy oil and deposits of bitumen. The new technology (Gregoli and Rimmer, 2000; Gregoli et al., 2000) features the injection of high-temperature, high-quality steam and hot hydrogen into a formation containing heavy hydrocarbons to initiate conversion of the heavy hydrocarbons into lighter hydrocarbons. In effect, the heavy hydrocarbons undergo partial underground refining that converts them into a synthetic crude oil (or **syncrude**). The heavier portion of the syncrude is treated to provide the fuel and hydrogen required by the process, and the lighter portion is marketed as conventional crude oil.

Thus, below ground, superheated steam and hot hydrogen are injected into a heavy oil or bitumen formation, which simultaneously produces the heavy oil or bitumen and converts it in situ (i.e., within the formation) into syncrude. Above ground, the heavier fraction of the syncrude is separated and treated on-site to produce the fuel and hydrogen required by the process, while the lighter fraction is sent to a conventional refinery to be made into petroleum products (Gregoli and Rimmer, 2000; Gregoli et al., 2000).

The potential advantages of an in situ process for bitumen and heavy oil include (1) leaving the carbon forming precursors in the ground, (2) leaving the heavy metals in the ground, (3) reducing sand handling, and (4) bringing a partially upgraded product to the surface. The extent of the upgrading can, hopefully, be adjusted by adjusting the exposure of the bitumen or heavy oil to the underground thermal effects.

In the modified in situ extraction processes, combinations of in situ and mining techniques are used to access the reservoir. A portion of the reservoir rock must be removed to enable application of the in situ extraction technology. The most common method is to enter the reservoir through a large-diameter vertical shaft, excavate horizontal drifts from the bottom of the shaft, and drill injection and production wells horizontally from the drifts. Thermal extraction processes are then applied through the wells. When the horizontal wells are drilled at or near the base of the tar sand reservoir, the injected heat rises from the injection wells through the reservoir, and drainage of produced fluids to the production wells is assisted by gravity.

Generally, as opposed to heavy oil recovery, bitumen recovery requires a higher degree of thermal stimulation because bitumen, in its immobile state, is extremely difficult to move to a production well. Extreme processes are required, usually in the form of a degree of thermal conversion that produces free-flowing product oil that will flow to the well and reduce the resistance of the bitumen to flow.

Bitumen recovery processes can be conveniently divided into two categories: (1) mining methods also called oil mining and (2) nonmining methods.

In the former type of process, the tar sand must first be removed from the application by a mining technique and then transported to a bitumen recovery center. In the latter type of process, usually (but not always correctly) termed in situ, bitumen (or a portion of the bitumen in place) is recovered from the formation by a suitable thermal method, leaving the formation somewhat less disturbed than when the mining method is employed.

6.2 OIL MINING

The alternative to in situ processing is to mine the tar sands, transport them to a processing plant, extract the bitumen value, and dispose of the waste sand. Such a procedure is often referred to as oil mining. This is the term applied to the surface or subsurface excavation of petroleum-bearing formations for subsequent removal of the heavy oil or bitumen by washing, flotation, or retorting treatments. Oil mining also includes recovery of heavy oil by drainage from reservoir beds to mine shafts or other openings driven into the rock, or by drainage from the reservoir rock into mine openings driven outside the tar sand but connected with it by bore holes or mine wells.

Oil mining is not new. Mining of petroleum and bitumen has occurred in the Sinai Peninsula, the Euphrates valley, and in Persia prior to 5000 BC. In addition, subsurface oil mining was used in the Pechelbronn oil field in Alsace, France, as early as 1735. This early mining involved the sinking of shafts to the reservoir rock, only 100 to 200 ft (30 to 60 m) below the surface and the excavation of the tar sand in short drifts driven from the shafts. These tar sands were hoisted to the surface and washed with boiling water to release the bitumen. The drifts were extended as far as natural ventilation permitted. When these limits were reached, the pillars were removed and the openings filled with waste. This type of mining continued at Pechelbronn until 1866, when it was found that oil could be recovered from deeper, and more prolific, sands by letting it drain in place through mine openings with no removal of sand to the surface for treatment. Nevertheless, mining for petroleum is a new challenge facing the petroleum industry.

Oil mining methods should be applied in reservoirs that have significant residual oil saturation and have reservoir or fluid properties that make production by conventional methods inefficient or impossible. The high well density in improved oil mining usually compensates for the inefficient production caused by reservoir heterogeneity. However, close well spacing can also magnify the deleterious effects of reservoir heterogeneity. If a high-permeability streak exists with a lateral extent that is less than the inter-well spacing of conventional wells but is comparable to that of improved oil mining, the channeling is more unfavorable for the improved oil mining method.

6.2.1 TAR SAND MINING

The bitumen occurring in oil sand deposits poses a major recovery problem. The material is notoriously immobile at formation temperatures and must therefore require some stimulation (usually by thermal means) in order to ensure recovery. Alternately, proposals have been noted, which advocate bitumen recovery by solvent flooding or by the use of emulsifiers. There is no doubt that with time, one or more of these functions may come to fruition, but for the present the two commercial operations rely on the mining technique.

The oil mining method of recovery has received considerable attention since it was chosen as the technique of preference for the only two commercial bitumen recovery plants in operation in North America. In situ processes have been tested many times in the United States, Canada, and other parts of the world and are ready for commercialization. There are also conceptual schemes that are a combination of both mining (above-ground recovery) and in situ (nonmining recovery) methods.

Engineering a successful oil mining project must address a number of items because there must be sufficient recoverable resources, the project must be conducted safely, and the project should be engineered to maximize recovery within economic limits. The use of a reliable screening technique is necessary to locate viable candidates. Once the candidate is defined, this should be followed by an exhaustive literature search covering local geology, drilling, production, completion, and secondary and tertiary recovery operations.

The reservoir properties, which can affect the efficiency of heavy oil or bitumen production by mining technology, can be grouped into three classes:

1. *Primary properties*, i.e., those properties that have an influence on the fluid flow and fluid storage properties and include rock and fluid properties, such as porosity, permeability, wettability, crude oil viscosity, and pour point
2. *Secondary properties*, i.e., those properties that significantly influence the primary properties, including pore size distribution, clay type, and content
3. *Tertiary properties*, i.e., those other properties that mainly influence oil production operation (fracture breakdown pressure, hardness, and thermal properties) and mining operations (e.g., temperature, subsidence potential, and fault distribution)

There are also important rock mechanical parameters of the formation in which a tunnel is to be mined and from where all oil mining operations will be conducted. These properties are mostly related to the mining aspects of the operations, and not all are of equal importance in their influence on mining technology. Their relative importance also depends on the individual reservoir.

Many of the candidate reservoirs for application of improved oil mining are those with high oil saturation resulting from the adverse effects of reservoir heterogeneity. Faulting, fracturing, and barriers to fluid flow are features that cause production of shallow reservoirs by conventional methods to be inefficient. Production of heterogeneous reservoirs by underground oil production methods requires consideration of the manner in which fractures alter the flow of fluids.

In a highly fractured formation with low matrix permeability, the fluid conductivity of the fracture system may be many times that of the matrix rock. In a highly fractured reservoir with low matrix permeability and reasonably high porosity, the fracture system provides the highest permeability to the flow of oil, but the matrix rock contains greater volume of the oil in place. The rate of the flow of oil from the matrix rock into the fracture system, the extent and continuity of the fracture system, and the degree to which the production wells effectively intersect the fracture system determine the production rate. Special consideration must be given to these factors in predicting production rates in fractured reservoirs. Under favorable circumstances, higher production rates may be achieved in fractured reservoirs by improving mining methods than in less heterogeneous reservoirs. Other reservoirs that are good candidates for oil mining are those that are shallow, have high oil saturation, have a nearby formation that is competent enough to support the mine, and cannot be efficiently produced by conventional methods.

Surface mining is the mining method that is currently being used by Suncor Energy and Syncrude Canada Limited to recover oil sand from the ground. Surface mining can be used in mineable oil sand areas that lie under 75 m (250 ft) or less of overburden material. Only 7% of the Athabasca Oil Sands deposit can be mined using the surface mining technique, as the other 93% of the deposit has more than 75 m of overburden. This other 93% will have to be mined using different mining techniques.

The first step in surface mining is the removal of muskeg and overburden. Muskeg is a water-soaked area of decaying plant material that is 1 to 3 m thick and lies on top of the overburden material. Before the muskeg can be removed, it must be drained of its water content. The process can take up to 3 years to complete. Once the muskeg has been drained and removed, the overburden must also be removed. Overburden is a layer of clay, sand, and silt that lies directly above the oil sands deposit. Overburden is used to build dams and dykes around the mine and will eventually be used for land reclamation projects. When all of the overburden is removed, the oil sand is exposed and is ready to be mined.

There are two methods of mining currently in use in the Athabasca Oil Sands. Suncor Energy uses only the truck and shovel method of mining, whereas Syncrude uses the truck and shovel method of mining, as well as draglines and bucket-wheel reclaimers. These enormous draglines and bucket-wheels are being phased out and soon will be completely replaced with large trucks and shovels. The shovel scoops up the oil sand and dumps it into a heavy hauler truck. The heavy hauler truck takes the oil sand to a conveyor belt that transports the oil sand from the mine to the extraction plant. Presently, there are extensive conveyor belt systems that transport the mined oil sand from the recovery site to the extraction plant. With the development of new technologies, these conveyors are being phased out and replaced with hydro-transport technology. Hydrotransport is a combination of ore transport and preliminary extraction. After the bituminous sands have been recovered using the truck and shovel method, they are mixed with water and caustic soda to form a slurry and this is pumped along a pipeline to the extraction plant. The extraction process thus begins with the mixing of the water, and agitation is needed to initiate bitumen separation from the sand and clay.

Mine spoils need to be disposed of in a manner that assures physical stabilization. This means appropriate slope stability for the pile against not only gravity, but also earthquake forces. Since return of the spoils to the mine excavations is seldom economical, the spoil pile must be designed as a permanent structure whose outline blends into the landscape. Straight, even lines in the pile must be avoided.

Even though estimates of the recoverable oil from the Athabasca deposits are only of the order of 27 × 109 bbl of synthetic crude oil (representing <10% of the total in-place material), this is, for the Canadian scenario, approximately six times the estimated volume of recoverable conventional crude oil. In addition, the comparative infancy of the development of the alternative options almost ensured the adoption of the mining option for the first two (and even later) commercial ventures.

Underground mining options have also been proposed, but for the moment have not been developed because of the fear of collapse of the formation onto any operation or equipment. This particular option should not, however, be rejected out of hand, because a novel aspect or the requirements of the developer (which remove the accompanying dangers) may make such an option acceptable. Currently, bitumen is recovered commercially from tar and deposits by a mining technique. This produces tar sand that is sent to the processing plant for separation of the bitumen from the sand prior to upgrading.

6.2.2   HOT-WATER PROCESS

Tar sand, as mined commercially in Canada, contains an average of 10% to 12% bitumen, 83% to 85% mineral matter, and 4% to 6% water. A film of water coats most of the mineral matter, and this property permits extraction by the hot-water process.

The hot-water process is, to date, the only successful commercial process to be applied to bitumen recovery from mined tar sands in North America (Clark, 1944; Carrigy, 1963a, b;

Fear and Innes, 1967; Speight and Moschopedis, 1978). Many process options have been tested with varying degrees of success and one of these options may even supersede the hot-water process.

The process utilizes the linear and the nonlinear variation of bitumen density and water density, respectively, with temperature, so that the bitumen that is heavier than water at room temperature becomes lighter than water at 80°C (180°F). Surface-active materials in the tar sand also contribute to the process. The essentials of the hot-water process involve conditioning, separation, and scavenging.

Thus, in the hot-water extraction process, the tar sand feed is introduced into a conditioning drum. In this step the oil sand is heated, mixed with water, and agglomeration of the oil particles begins. The conditioning is carried out in a slowly rotating drum that contains a steam-sparging system for temperature control as well as mixing devices to assist in lump size reduction and a size ejector at the outlet end. The oil sand lumps are reduced in size by ablation and mixing action. The conditioned pulp has the following characteristics: (1) solids 60% to 85%; and (2) pH 7.5 to 8.5.

In the conditioning step, also referred to as mixing or pulping, tar sand feed is heated and mixed with water to form a pulp of 60% to 85% by weight solids at 80°C to 90°C (175°F to 196°F). First, the lumps of tar sand as mined are reduced in size by ablation, i.e., successive layers of lump are warmed and sloughed off revealing cooler layers. The conditioned pulp is screened through a double-layer vibrating screen. Water is then added to the screened material (to achieve more beneficial pumping conditions) and the pulp enters the separation cell through a central feed well and distributor. The bulk of the sand settles in the cell and is removed from the bottom as tailing, but the majority of the bitumen floats to the surface and is removed as froth. A middlings stream (mostly of water with suspended fines and some bitumen) is withdrawn from approximately midway up the side of the cell wall. Part of the middlings is recycled to dilute the conditioning-drum effluent for pumping. Clays do not settle readily and generally accumulate in the middlings layer. High concentrations of clays increase the viscosity and can prevent normal operation in the separation cell. Thus, it is necessary to withdraw a drag stream to act as a purge; this is usually done at high-clay concentrations but may not be as essential with a low-clay oil sand charge.

Under certain operating conditions, it may be necessary to withdraw a middling stream to the scavenger cells (air flotation cells to recover bitumen from the drag stream). The froth from the scavenger unit usually has a high mineral and water content that can be removed by gravity settling in froth settlers, after which the froth is combined with the froth from the main separation cell from the centrifuge plant for dewatering and demineralizing. Before the centrifuging operation, the froth is de-aerated and naphtha added to lower the viscosity for a more efficient water and mineral removal operation.

The separation cell acts like two settlers, one on top of the other. In the lower settler the sand settles down, whereas in the upper settler the bitumen floats. The bulk of the sand in the feed is removed from the bottom of the separation cell as tailings. A large portion of the feed bitumen floats to the surface of the separation cell and is removed as froth. A middlings stream consists mostly of water with some suspended fine minerals and bitumen particles. A portion of the middlings may be returned for mixing with the conditioning-drum effluent in order to dilute the separation-cell feed for pumping. The remainder of the middlings is called the dragstream, which is withdrawn from the separation cell to be rejected after processing in the scavenger cells. Tar sand feed contains a certain portion of fine minerals that, if allowed to build up in concentration in the middlings, increases the viscosity and eventually disrupts settling in the separation cell. The dragstream is required as a purge in order to control the fines concentration in the middlings. The amounts of water that can enter with the feed and leave with the separation-cell tailings and froth are relatively fixed. Thus, the size of the dragstream determines the makeup water requirement for the separation cell. The separation cell is an open vessel with straight sides and a cone bottom. Mechanical rakes on the bottom move the sand toward the center for discharge. Wiper arms rotating on the surface push the froth to the outside of the separation cell where it overflows into launders for collection.

The combined froth from the separation cell and scavenging operation contains an average of about 10% by weight mineral material and up to 40% by weight water. Dewatering and demineralizing are accomplished in two stages of centrifuging; in the first stage, the coarser mineral material is removed but much of the water remains. The feed then passes through a filter to remove any additional large-size mineral matter that would plug up the nozzles of the second stage centrifuges.

The third step in the hot-water process is scavenging. Depending on the dragstream size and composition, enough bitumen may leave the process in the dragstream to make another recovery step economical. Froth flotation with air is usually employed. The scavenger froth is combined with the separation-cell froth to be further treated and upgraded to synthetic crude oil. Tailings from the scavenger cell join the separation-cell tailings stream and go to waste. Conventional froth-flotation cells are suitable for this step.

Froth from the hot-water process may be mixed with a hydrocarbon diluent, e.g., coker naphtha, and centrifuged. The Suncor process employs a two-stage centrifuging operation and each stage consists of multiple centrifuges of conventional design installed in parallel. The bitumen product contains 1% to 2% by weight mineral (dry bitumen basis) and 5% to 15% by weight water (wet diluted basis). Syncrude also utilizes a centrifuge system with naphtha diluent.

One of the major problems that comes from the hot-water process is the disposal and control of the tailings. The fact is that each ton of oil sand in place has a volume of about 16 cft, which will generate about 22 cft of tailings giving a volume gain of the order of 40%. If the mine produces about 200,000 tons of oil sand per day, the volume expansion represents a considerable solids disposal problem. Tailings from the process consist of about 49% to 50% by weight of sand, 1% by weight of bitumen, and about 50% by weight of water. The average particle size of the sand is about 200 am and it is a suitable material for dike building. Accordingly, Suncor used this material to build the sand dike, but for fine sand, the sand must be well compacted.

Environmental regulations in Canada or the United States will not allow the discharge of tailings streams into (1) the river; (2) on to the surface; or (3) on to any area where groundwater domains or the river may be contaminated. The tailings streams is essentially high in clays and contains some bitumen, hence the current need for tailings ponds, where some settling of the clay occurs. In addition, an approach to acceptable reclamation of the tailings ponds will have to be accommodated at the time of site abandonment.

The structure of the dike may be stabilized on the upstream side by beaching. This gives a shallow slope, but consumes sand during the season when it is impossible to build the dike. In remote areas such as the Fort McMurray (Alberta) site, the dike can only be built in above-freezing weather because (1) frozen water in the pores of the dike will create an unstable layer and (2) the vapor emanating from the water creates a fog, which can create a work hazard. The slope of the tailings dike is about 2.5:1 depending on the amount of fines in the material. It may be possible to build 2:1 slopes with coarser material, but steeper slopes must be stabilized quickly by benching. After discharge from the hot-water separation system, it is preferable that attempts be made to separate the sand, sludge, and water; hence, the tailings pond. The sand is used to build dikes and the runoff that contains the silt, clay, and water collects in the pond. Silt and some clay settle out to form sludge and some of the water is recycled to the plant.

In summary, the hot-water separation process involves extremely complicated surface chemistry with interfaces among various combinations of solids (including both silica sand and aluminosilicate clays), water, bitumen, and air. The control of pH seems to be critical with the preferred range being 8.0 to 8.5, which is achievable by use of any of the monovalent bases. Polyvalent cations must be excluded because they tend to flocculate the clays and thus raise the viscosity of the middlings in the separation cell.

6.2.3 OTHER PROCESSES

It is conceivable that the problems related to bitumen mining and bitumen recovery for the tar sand may be alleviated somewhat by the development of process options that require considerably less water in the sand or bitumen separation step. Such an option would allow a more gradual removal of the tailings ponds.

The proposed cold-water process for bitumen separation from mined tar sand has also been recommended (Miller and Misra, 1982). The process uses a combination of cold water and solvent and the first step usually involves disintegration of the tar sand charge that is mixed with water, diluent, and reagents. The diluent may be a petroleum distillate fraction such as kerosene and is added in approximately a 1:1 weight ratio to the bitumen in the feed. The pH is maintained at 9 to 9.5 by the addition of wetting agents and approximately 0.77 kg of soda ash per ton of tar sand. The effluent is mixed with more water, and in a raked classifier, the sand is settled from the bulk of the remaining mixture. The water and oil overflow the classifier and are passed to thickeners where the oil is concentrated. Clay in the tar sand feed has a distinct effect on the process; it forms emulsions that are hard to break and are wasted with the underflow from the thickeners.

The sand-reduction process is a cold-water process without solvent. In the first step, the tar sand feedstock is mixed with water at approximately 20°C (68°F) in a screw conveyor in a ratio of 0.75 to 3 ton per ton of tar sand (the lower range is preferred). The mixed pulp from the screw conveyor is discharged into a rotary-drum screen, which is submerged in a water-filled settling vessel. The bitumen forms agglomerates that are retained by an 840 μm (20-mesh) screen. These agglomerates settle and are withdrawn as oil products. The sand readily passes through the 840 μm (20 mesh) screen and is withdrawn as waste stream. The process is called sand reduction because its objective is the removal of sand from the tar sand to provide a feed suitable for a fluid coking process; about 80% of sand is removed. Nominal composition of the oil product is 58% by weight (bitumen), 27% by weight mineral matter, and 15% by weight water.

The spherical agglomeration process resembles the sand-reduction process. Water is added to tar sands and the mixture is ball-milled. The bitumen forms dense agglomerates of 75% to 87% by weight bitumen, 12% to 25% by weight sand, and 1% to 5% by weight water.

An oleophilic sieve process (Kruyer, 1982, 1983) offers the potential for reducing tailings pond size because of a reduction in the water requirements. The process is based on the concept that when a mixture of an oil phase and an aqueous phase is passed through a sieve made from oleophilic materials, the aqueous phase and any hydrophilic solids pass through the sieve, but the oil adheres to the sieve surface on contact. The sieve is in the form of a moving conveyor, the oil is captured in a recovery zone, and recovery efficiency is high.

An anhydrous **solvent extraction process** for bitumen recovery has been attempted and usually involves the use of a low-boiling hydrocarbon. The process generally involves up to four steps. In the mixer step, fresh tar sand is mixed with recycle solvent that contains some bitumen and small amounts of water and mineral. Solvent-to-bitumen weight ratio is adjusted to approximately 0.5. The drain step consists of a three-stage counter-current wash. Settling and draining time is approximately 30 min for each stage. After each extraction step, a bed of sand is formed and the extract is drained through the bed until the interstitial pore volume of the bed is emptied. From time to time, the bed is plugged with fine mineral or emulsion. In these cases, the drainage rate is essentially zero and the particular extraction stage is ineffective. The last two steps of the process are devoted to solvent recovery. Stripping of the solvent from the bitumen is straightforward. The solvent recovery from the solids holds the key to the economic success of an anhydrous process.

Another above-ground method of separating bitumen from mined tar sand involves direct heating of the tar sand without previous separation of the bitumen (Gishler, 1949). Thus, the bitumen is not recovered as such but is an upgraded overhead product. In the process, the sand is crushed and introduced into a vessel, where it is contacted with either hot (spent) sand or with hot product gases that furnish part of the heat required for cracking and volatilization. The volatile products are passed out of the vessel and are separated into gases and (condensed) liquids. The coke that is formed as a result of the thermal decomposition of the bitumen remains on the sand, which is then transferred to a vessel for coke removal by burning in air. The hot flue gases can be used either to heat incoming tar sand or as refinery fuel. As expected, processes of this type yield an upgraded product but require various arrangements of pneumatic and mechanical equipment for solids movement around the refinery.

One approach toward recovering a significant portion of heavy oil from a reservoir is to use a combination of petroleum recovery and mining technologies. Through the use of mining technology, access is developed below the petroleum reservoir within or beneath a permeability barrier. Underground development consists of providing room for subsurface drilling and petroleum production operations as well as a life-supporting atmosphere and safe working conditions. Wells are drilled at inclinations from the horizontal to the vertical upward into the reservoir. Wells produce as a result of a combination of pressure depletion and gravity drainage.

In improved mining, directional (horizontal or slant) wells are drilled into the reservoir from a mine in an underlying formation to drain oil by pressured depletion and gravity drainage. In the process of gravity drainage extraction of liquid crude oil, the wells are completed so that only the forces acting within the reservoir are used. The forces acting on the reservoir are left intact, perhaps maintained or increased. A large number of closely spaced wells can be drilled into a reservoir from an underlying tunnel more economically than the same number of wells from the surface. In addition, only one pumping system is required in underground drainage, whereas at the surface each well must have a pumping system. The objective of using a large number of wells is to produce each well slowly so that the gas or oil and water or oil interfaces move toward each other efficiently. By maintaining the reservoir pressures because of forces acting on the reservoir, it is then assured that the oil production is provided by the internal forces due to gravity (the buoyancy effect) and capillary effects.

The recovery efficiency is improved by applying an enhanced oil recovery method. The production drain holes should be surveyed while drilling, and the drilling should be accomplished using mud-operated drills with a bent sub for control of direction. The drain holes must be controlled so that a network of uniformly spaced holes conforming to the data output from the computer modeling can be drilled in the production zone. The drill string should be equipped with check valves to minimize the backflow of mud during the installation of additional drill pipe. The return mud line should be equipped with a blow line to safely vent to the surface any formation gas encountered. All drilling should be accomplished working through a blowout preventer or diverter. Drill cuttings should be contained in a closed system and not allowed to encumber the mine atmosphere.

Large vertical shafts sunk from the surface are generally the means through which underground openings can be excavated. These shafts are one means of access to offer an outlet for removal of excavated rock, provide sufficient opening for equipment, provide ventilation, and allow the removal of oil and gas products during later production. These requirements plus geological conditions and oil reservoir dimensions determine the shaft size. It is expected that an access shaft will range from 8 to 20 ft in diameter.

Completing wells from a level of drifts beneath the reservoir is the most economical application of gravity drainage. In this case, only one pumping system is required rather than installing a pump in each well as is necessary in wells drilled from the surface. When wells are drilled from beneath the producing formation in an oil–water system, the developer has two options for completion. The casing may be set totally through the formation and the region opposite the oil saturation perforated to permit production. If desired, the easing may pass only through the water-saturated zone and then a jet slotting process can be used to wash out or drill slots in the oil-saturated zone to increase the productivity of the individual well. Other shaft configurations include those for mining in weak rock and shafts for pumping drainage.

Excavation of shafts and tunnels during the mining process results in waste rock that must be stored or disposed of at the ground surface. The waste rock cannot be dumped in a heap near the mine shaft as an eyesore for future generations and a source of air and water pollution for untold years. Air, water, and esthetics must be conserved, not only for the present but also against reasonable future contingencies, either natural or man-made.

6.3   NONMINING METHODS

The gravity of tar sand bitumen is usually less than 10° API depending upon the deposit, and viscosity is very high. Whereas conventional crude oils may have a viscosity of several poise (at 40°C, 105°F), the tar sand bitumen has a viscosity of the order of 50,000 to 1,000,000 cp or more at formation temperatures (approximately 0°C to 10°C, 32°F to 50°F, depending upon the season). This offers a formidable (but not insurmountable) obstacle to bitumen recovery.

The successful recovery technique that is applied to one deposit or resource is not necessarily the technique that will guarantee success for another deposit. There are sufficient differences between the tar sand deposits of the United States and the Canadian deposits that general applicability is not guaranteed. Hence, caution is advised when applying the knowledge gained from one resource to the issues of another resource Although the principles may at first sight appear to be the same, the technology must be adaptable.

In principle, the nonmining recovery of bitumen from tar sand deposits is an enhanced recovery technique and requires the injection of a fluid into the formation through an injection wall. This leads to the in situ displacement of the bitumen from the recovery and bitumen production at the surface through an egress (production well). There are, however, several serious constraints that are particularly important and relate to bulk properties of the tar sand and the bitumen. In fact, both must be considered in toto in the context of bitumen recovery by nonmining techniques. For example, such processes need a relatively thick layer of overburden to contain the driver substance within the formation between injection and production wells.

One of the major deficiencies in applying mining techniques to bitumen recovery from tar sand deposits is (next to the immediate capital costs) the associated environmental problems.

Moreover, in most of the known deposits, the vast majority of the bitumen lies in formations in which the overburden or pay zone ratio is too high. Therefore, it is not surprising that over the last two decades a considerable number of pilot plants have been applied to the recovery of bitumen by nonmining techniques from tar sand deposits where the local terrain and character of the tar sand may not always favor a mining option.

In principle, the nonmining recovery of bitumen from tar sand deposits requires the injection of a fluid into the formation through an injection well, the in situ displacement of the bitumen from the reservoir, and bitumen production at the surface through an egress (production well). There are, of course, variants around this theme, but the underlying principle remains the same.

There are, however, several serious constraints that are particularly important and relate to bulk properties of the tar sand and the bitumen. In fact, both must be considered in the context of bitumen recovery by nonmining techniques. For example, the Canadian deposits are unconsolidated sands with a porosity ranging up to about 45%, whereas other deposits may range from predominantly low-porosity, low-permeability consolidated sand to, in a few instances, unconsolidated sands. In addition, the bitumen properties are not conducive to fluid flow under deposit conditions. Nevertheless, where the general nature of the deposits prohibits the application of a mining technique, a nonmining method may be the only feasible bitumen recovery option.

Another general constraint to bitumen recovery by nonmining methods is the relatively low injectivity of tar sand formations. Thus, it is usually necessary to inject displacement or recovery finds at a pressure such that fracturing (parting) is achieved. Such a technique therefore changes the reservoir profile and introduces a series of channels through which fluids can flow from the injection well to the production well. On the other hand, the technique may be disadvantageous insofar as the fracture occurs along the path of least resistance, giving undesirable (i.e., inefficient) flow characteristics within the reservoir between the injection and production wells, leaving a large part of the reservoir relatively untouched by the displacement of recovery fluids.

In principle, the nonmining recovery of bitumen from tar sand deposits is an enhanced oil recovery technique and requires the injection of a fluid into the formation through an injection well. This leads to the in situ displacement of the bitumen from the reservoir and bitumen production at the surface through an egress (production) well. There are, however, several serious constraints that are particularly important and relate to the bulk properties of the tar sand and the bitumen. In fact, both must be considered in toto in the context of bitumen recovery by nonmining techniques.

In **steam stimulation**, heat and drive energy are supplied in the form of steam injected through wells into the tar sand formation. In most instances, the injection pressure must exceed the formation fracture pressure in order to force the steam into the tar sands and into contact with the oil. When sufficient heating has been achieved, the injection wells are closed for a soak period of variable length and then allowed to produce, first applying the pressure created by the injection and then using pumps as the wells cool and production declines.

Steam can also be injected into one or more wells with production coming from other wells (steam drive). This technique is very effective in heavy oil formations, but has found little success during application to tar sand deposits because of the difficulty in connecting injection and production wells. However, once the flow path has been heated, the steam pressure is cycled, alternately moving steam up into the oil zone, then allowing oil to drain down into the heated flow channel to be swept to the production wells.

If the viscous bitumen in a tar sand formation can be made mobile by admixture of either a hydrocarbon diluent or an emulsifying fluid, a relatively low temperature secondary recovery process is possible (emulsion steam drive). If the formation is impermeable, communication problems exist between injection and production wells. However, it is possible to apply a solution or dilution process along a narrow fracture plane between injection and production wells.

6.3.1 STEAM-BASED PROCESSES

The steam processes are the most advanced of all enhanced oil recovery methods in terms of field experience and thus have the least uncertainty in estimating performance, provided that a good reservoir description is available. Steam processes are most often applied in reservoirs containing viscous oils and tars, usually in place of rather than following secondary or primary methods. Commercial application of steam processes has been underway since the early 1960s. In situ combustion has been field tested under a wide variety of reservoir conditions, but few projects have proven economical and advanced to commercial scale.

Steam drive injection (steam injection) has been commercially applied since the early 1960s. The process occurs in two steps: (1) steam stimulation of production wells, i.e., direct steam stimulation, and (2) steam drive by steam injection to increase production from other wells (i.e., indirect steam stimulation).

When there is some natural reservoir energy, steam stimulation normally precedes steam drive. In steam stimulation, heat is applied to the reservoir by the injection of high-quality steam into the produce well. This cyclic process, also called **huff and puff** or steam soak, uses the same well for both injection and production (Figure 6.10). The period of steam injection is followed by production of reduced viscosity oil and condensed steam (water). One mechanism that aids production of the oil is the flashing of hot water (originally condensed from steam injected under high pressure) back to steam, as pressure is lowered when a well is put back on production.

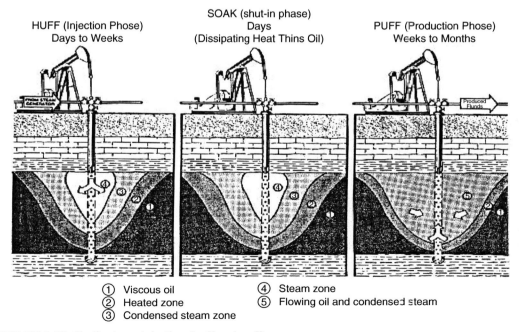

① Viscous oil ④ Steam zone
② Heated zone ⑤ Flowing oil and condensed steam
③ Condensed steam zone

FIGURE 6.10 Cyclic steam injection (huff and puff).

Cyclic steam injection is the alternating injection of steam and production of oil with condensed steam from the same well or wells. This process is predominantly a vertical well process, with each well alternately injecting steam and producing heavy oil and steam condensate. In practice, steam is injected into the formation at greater than fracturing pressure followed by a soak period after which production is commenced. The heat injected warms the heavy oil and lowers its viscosity. A heated zone is created through which the warmed heavy oil can flow back into the well. This is a well-developed process; a major limitation is that less than 30% (usually less than 20%) of the initial oil in place can be recovered.

Steam drive involves the injection of steam through an injection well into a reservoir and the production of the mobilized bitumen and steam condensate from a production well. Steam drive is usually a logical follow-up to cyclic steam injection. Steam drive requires sufficient effective permeability (with the immobile bitumen in place) to allow injection of the steam at rates sufficient to raise the reservoir temperature to mobilize the bitumen.

Two expected problems inherent in the steam drive process are steam override and reservoir plugging. Any in situ thermal process tends to override (migrate to the top of the effected interval) because of differential density of the hot and cold fluids. These problems can be partially mitigated by rapid injection of steam at the bottom or below the target interval through a high-permeability water zone or fracture. Each of these options will raise the temperature of the entire reservoir by conduction, and, to a lesser degree, by convection. The bitumen will be at least partially mobilized and the effectiveness of the following injection of steam into the target interval will be enhanced.

For a successful steam drive project, the porosity of the reservoir rock should be at least 20%; the permeability should be at least 100 MD; and the bitumen saturation should be at least 40%. The reservoir oil content should be at least 800 bbl per acre-foot. The depth of the reservoir should be less than 3000 ft and the thickness should be at least 30 ft; other preferential parameters have also been noted on the basis of success with several heavy oil reservoirs.

Other variations on this theme include the use of steam and the means of reducing interfacial tension by the use of various solvents. The solvent extraction approach has had some success when applied to bitumen recovery from mined tar sand but when applied to unmined material, losses of solvent and bitumen are always a major obstacle. This approach should not be rejected out of hand, since a novel concept may arise, which guarantees minimal (acceptable) losses of bitumen and solvent.

6.3.2 COMBUSTION PROCESSES

In situ combustion is normally applied to reservoirs containing low-gravity oil but has been tested over perhaps the widest spectrum of conditions of any enhanced oil recovery process. In the process, heat is generated within the reservoir by injecting air and burning part of the crude oil. This reduces the oil viscosity and partially vaporizes the oil in place, and the oil is driven out of the reservoir by a combination of steam, hot water, and gas drive. Forward combustion involves movement of the hot front in the same direction as the injected air; reverse combustion involves movement of the hot front opposite to the direction of the injected air.

During in situ combustion or fire flooding, energy is generated in the formation by igniting bitumen in the formation and sustaining it in a state of combustion or partial combustion. The high temperatures generated decrease the viscosity of the oil and make it more mobile. Some cracking of the bitumen also occurs and an upgraded product rather than bitumen itself is the fluid recovered from the production wells.

The relatively small portion of the oil that remains after the displacement mechanisms have acted becomes the fuel for the in situ combustion process. Production is obtained from wells offsetting the injection locations. In some applications, the efficiency of the total in situ combustion operation can be improved by alternating water and air injection. The injected water tends to improve the utilization of heat by transferring heat from the rock behind the combustion zone to the rock immediately ahead of the combustion zone.

The performance of in situ combustion is predominantly determined by four factors:

1. Quantity of oil that initially resides in the rock to be burned
2. Quantity of air required to burn the portion of the oil that fuels the process
3. Distance to which vigorous combustion can be sustained against heat losses and
4. Mobility of the air or combustion product gases

In many field projects, the high gas mobility has limited recovery through its adverse effect on the sweep efficiency of the burning front. Because of the density contrast between air and reservoir liquids, the burning front tends to override the reservoir liquids. To date, combustion has been most effective for the recovery of viscous oils in moderately thick reservoirs in which reservoir dip and continuity provide effective gravity drainage or operational factors permit close well spacing.

The use of combustion to stimulate oil production is regarded as attractive for deep reservoirs. In contrast to steam injection, it usually involves no loss of heat. The duration of the combustion may be less than 30 days or much as 90 days depending on requirements. In addition, backflow of the oil through the hot zone must be prevented or coking will occur.

Forward combustion involves movement of the hot front in the same direction as the injected air whereas reverse combustion involves movement of the hot front opposite to the direction of the injected air. In forward combustion, the hydrocarbon products released from the zone of combustion move into a relatively cold portion of the formation. Thus, there is a definite upper limit of the viscosity of the liquids that can be recovered by a forward-combustion process. On the other hand, since the air passes through the hot formation before reaching the combustion zone, burning is complete; the formation is left completely cleaned of hydrocarbons.

Forward combustion is particularly applicable to reservoirs containing mobile heavy oil with a high effective permeability. Even though lower effective reservoir permeability is required for air injection compared with steam injection, the reservoir ahead of the combustion front is subject to plugging as the vaporized fluids cool and condense. Consequently, a relatively high permeability (400 to 1000 MD) and relatively low bitumen saturation (45% to 65% of pore volume) are most favorable for this process. The combustion process yields a partially upgraded product because the temperature gradient ahead of the combustion front mobilizes the lighter hydrocarbon components that move toward the cooler portion of the reservoir and mix with unheated bitumen. This mixture is eventually produced through a production well. The heavier components (e.g., coke) are left on the sand grains and are consumed as fuel for the combustion. Under certain operating conditions, a significant cost saving is attained by injecting oxygen or oxygen-enriched air rather than atmospheric air, because of reduced compression costs and a lower produced gas or oil ratio.

Reverse combustion is particularly applicable to reservoirs with lower effective permeability (in contrast with forward combustion). It is more effective because lower permeability would cause the reservoir to be plugged by the mobilized fluids ahead of a forward combustion front. In the reverse combustion process, the vaporized and mobilized fluids move through the heated portion of the reservoir behind the combustion front. The reverse combustion partially cracks the bitumen, consumes a portion of the bitumen as fuel, and deposits residual coke on the sand grains. In the process, part of the bitumen will be consumed as fuel and part will be deposited on the sand grains as coke leaving 40% to 60% recoverable. This coke deposition serves as a cementing material, reducing movement and production of sand.

A modified combustion approach has been applied to the Athabasca deposit. The technique involved a heat-up phase, production (or blow-down phase), followed by a displacement phase using a combined fire flood and water flood (COFCAW process). In this manner, over a total 18 month period (heat-up: 8 month; blow-down: 4 month; displacement: 6 month), 29,000 bbl of upgraded oil was produced from an estimated 90,000 bbl of oil in place.

The addition of water or steam to an in situ combustion process can result in a significant increase in the overall efficiency of that process. Two major benefits may be derived. Heat transfer in the reservoir is improved because the steam and condensate have greater heat-carrying capacity than combustion gases and gaseous hydrocarbons. Sweep efficiency may also be improved because of the more favorable mobility ratio of steam to bitumen compared with gas to bitumen.

Modes of application include injection of alternate slugs of air (oxygen) and water or co-injection of air (oxygen) and steam. Again, the combination of air (oxygen) injection and steam or water injection increases injectivity costs that may be justified by increased bitumen recovery.

Process efficiency is affected by reservoir heterogeneity that will reduce horizontal sweep. The underburden and overburden must provide effective seals to avoid loss of injected air and produced bitumen. Process efficiency is enhanced by the presence of some interstitial water saturation. The water is vaporized by the combustion and enhances the heat transfer by convection. The combustion processes are subject to override because of differences in the densities of injected and reservoir fluids. Production wells should be monitored for, and equipped to cool, excessively high temperatures ($>1095°C$, $>2000°F$) that may damage down-hole production tools and tubulars.

Applying a preheating phase before the bitumen recovery phase may significantly enhance the steam or combustion extraction processes. Preheating can be particularly beneficial if the saturation of highly viscous bitumen is sufficiently great to lower the effective permeability to the point of production being precluded by reservoir plugging. Preheating partially mobilizes the bitumen by raising its temperature and lowering its viscosity. The result is a lower required pressure to inject steam or air and move the bitumen.

Preheating may be accomplished by several methods. Conducting a reverse combustion phase in a zone of relatively high effective permeability and low bitumen saturation is one method. Steam or hot gases may be rapidly injected into a high-permeability zone in the lower portion of the reservoir. In the fracture-assisted steam technology (**FAST**) process, steam is injected rapidly into an induced horizontal fracture near the bottom of the reservoir to preheat the reservoir. This process has been applied successfully in three pilot projects in southwest Texas. Shell has accomplished the same preheating goal by injecting steam into a high-permeability bottom-water zone in the Peace River (Alberta) field. Electrical heating of the reservoir by radio-frequency waves may also be an effective method.

Using combustion to stimulate oil production is regarded as attractive for deep reservoirs and, in contrast to steam injection, usually involves no loss of heat. The duration of the combustion may be less than 30 days) or approximately 90 days, depending upon require-ments. In addition, backflow of the oil through the hot zone must be prevented or coking occurs.

Using combustion to stimulate bitumen production is regarded as attractive for deep reservoirs and, in contrast to steam injection, usually involves no loss of heat. The duration of the combustion may be short (days) depending upon requirements. In addition, backflow of the oil through the hot zone must be prevented or coking will occur. A variation of the combustion process involves use of a heat-up phase, then a blow-down (production) phase, followed by a displacement phase using a fire–water flood (a combination of forward combustion and waterflood, **COFCAW**).

Finally, by all definitions, the quality of the bitumen from tar sand deposits is poor as a refinery feedstock. As in any field in which primary recovery operations are followed by secondary or enhanced recovery operations and there is a change in product quality, such is also the case for tar sand recovery operations. Thus, product oils recovered by the thermal stimulation of tar sand deposits show some improvement in properties over those of the bitumen in place.

Although this improvement in properties may not appear to be too drastic, it usually is sufficient to have major advantages for refinery operators. Any incremental increase in the units of hydrogen to carbon ratio can save amounts of costly hydrogen during upgrading. The same principles are also operative for reductions in the nitrogen, sulfur, and oxygen contents. This latter occurrence also improves catalyst life and activity as well as reduces the metal content. In short, in situ recovery processes (although less efficient in terms of bitumen recovery relative to mining operations) may have the added benefit of leaving some of the more obnoxious constituents (from the processing objective) in the ground.

6.3.3 OTHER PROCESSES

Many innovative concepts in heavy oil production have been developed and the major new technologies that have positively affected the heavy oil industry in the last 10 years are:

1. Horizontal well technology for shallow applications (1000 m), often combined with gravity drainage approaches
2. Cold production using long horizontal wells that are often combined with multilaterals with aggregate lengths exceeding several kilometers
3. Gravity-driven processes, particularly steam-assisted gravity drainage (SAGD), vapor-assisted petroleum extraction (VAPEX), and inert gas injection (IGI), all using horizontal wells to establish stable, gravity-assisted oil recovery
4. Cold heavy oil production with sand, or CHOPS technology, where sand production is encouraged and managed as a means of enhancing well productivity
5. Pressure pulse flow enhancement technology, both as a reservoir-wide method and as a work-over method
6. Improvements in upgrading viscous, sulfur-rich, heavy crude oil feedstocks

The oil industry pioneered drilling shallow (500 to 3500 ft deep) horizontal wells at cost-per-meter values that are now only 1.2 to 1.3 times those of vertical wells. These wells, in the shallowest cases (150 to 1500 ft deep) are often drilled using masts inclined to reduce curvature build rates required to turn the corner from vertical to horizontal. In "reduced footprint" developments, where there are a number of producing wells and injecting wells on the same pad (savings in roads, services . . .), many horizontal wells (2 to 6) may be drilled from a small area, no larger than a hectare.

Coiled tubing drilling and work-over have been introduced and perfected in the last decade, further reducing costs of horizontal well drilling. Good seismic control and cuttings analysis allow precise steering in thin zones ≤20′) to place the well in the optimum position in the reservoir. In the production phase, the long drainage length of a single well, as much as 1200 m in many cases, allows much more effective production, also giving higher production percentages of original oil in place (OOIP) when used in gravity drainage technologies.

Long horizontal wells with several multi lateral branches have been used widely in the development of heavy oils of Venezuela, where production rates as high as 2000 to 2500 bbl/day in some wells has been achieved through the use of aggregate horizontal lengths as large as 10,000 m in oil of 1200 to 5000 cp viscosity. Unfortunately, this technology can only achieve 8% to 15% recovery, and only from the best high permeability zones. Thus, for the more efficient development of these resources, other technologies will necessarily be implemented in the future.

Horizontal wells using cold primary production in heavy oils have been widely used in Canada since the late 1980s, but the successes (usually well published) have been substantially offset by the failures (rarely published). Further, low recovery factors (seldom close to 10%), early water breakthrough (usually impossible to plug when it happens), short well life (30% to 40% decline per year), and other factors such as expensive work-over if sand plugging takes place, have combined to make this a technology that has little attraction in Canada.

IGI is a technology for conventional oils in reservoirs where good vertical permeability exists, or where it can be created through propped hydraulic fracturing. It is generally viewed as a top-down process with nitrogen or methane injection through vertical wells at the top of the reservoirs, creating a gas–oil interface that is slowly displaced toward long horizontal production wells (Figure 6.11). As with all gravity drainage processes, it is essential to balance the injection and production volumes precisely so that the system does not become pressure driven, but remains in the gravity-dominated flow regime (Mearos et al., 1990).

High recovery ratios are achieved because, in the absence of elevated pressure gradients, a thin oil film is maintained between the gas and water phases. The film is maintained because the sum of the oil–water and gas–oil surface tensions is always less than the water–gas surface tension. Thus, the thin film configuration is thermodynamically stable, and allows the oil to drain to values far lower than the residual oil saturation value.

The interfaces in IGI are gravity-stabilized because of the difference in phase densities, so that at slow drainage rates the interfaces remain approximately horizontal without viscous fingering. In one configuration, the horizontal wells are produced under a back-pressure equal to the pressure in the underlying water phase, so no water coning can occur. During production, if the water cut increases, the production rate is reduced so that the interface becomes stable. Alternatively, if gas is injected too quickly, gas coning can develop, and if this

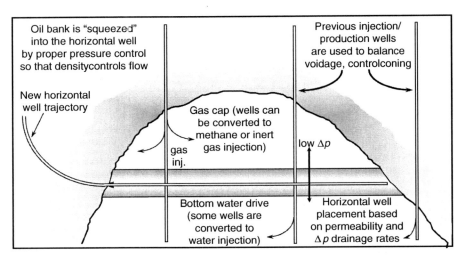

FIGURE 6.11 Inert gas injection.

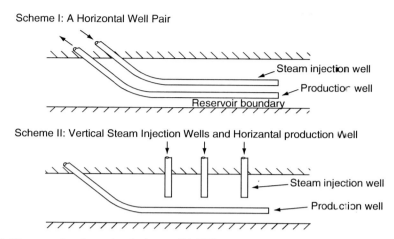

Scheme I: A Horizontal Well Pair

Steam injection well

Production well

Reservoir boundary

Scheme II: Vertical Steam Injection Wells and Horizantal production Well

Steam injection well

Production well

FIGURE 6.12 Steam assisted gravity drainage (SAGD).

is observed, the gas injection rate must be reduced to sustain stability. The process is continued until the oil zone is "pinched down" to the horizontal well, achieving the high recovery ratios possible with gravity drainage methods. These principles are fundamental to all gravity-dominated processes and failure to adhere to them will drive the system into conditions of instability (such as coning and fingering).

In reservoirs with excellent vertical permeability, the bottom-water zone can also be injected with water to cause the oil–water interface to rise slowly toward the production well. First implemented in Canada, IGI is used extensively in carbonate pinnacle reefs with excellent vertical permeability, and recovery ratios exceeding 80% are systematically achieved. As with all gravity drainage processes, it is generally necessary to place the horizontal wells as low in the structure as possible.

SAGD (Figure 6.12) was developed first in Canada for reservoirs where the immobile bitumen occurs (Dusseault et al., 1998). This process uses paired horizontal wells. Low-pressure steam, continuously injected through the upper well, creates a steam chamber along the walls of which the heated bitumen flows and is produced in the lower well.

Several variations of this process have been developed. One variation uses a single horizontal well, with steam injection through a central pipe and production along the annulus. Another variation involves steam injection through existing vertical wells and production through an underlying horizontal well. The key benefits of the SAGD process are an improved steam–oil ratio and high ultimate recovery (on the order of 60% to 70%). The outstanding technical issues relate to low initial oil rate, artificial lifting of bitumen to the surface, horizontal well operation, and the extrapolation of the process to reservoirs having low permeability, low pressure, or bottom water.

In the process, a pair of horizontal wells, separated vertically by about 15 to 20 ft are drilled at the bottom of a thick unconsolidated sandstone reservoir. Steam, perhaps along with a mixture of hydrocarbons that dissolve into the oil and help reduce its viscosity, is injected into the upper well. The heat reduces the oil viscosity to values as low as 1 to 10 cp (depending on temperature and initial conditions) and develops a "steam chamber" that grows vertically and laterally (Figure 6.13). The steam and gases rise because of their low density, and the oil and condensed water are removed through the lower well. The gases produced during SAGD tend to be methane with some carbon dioxide and traces of hydrogen sulfide.

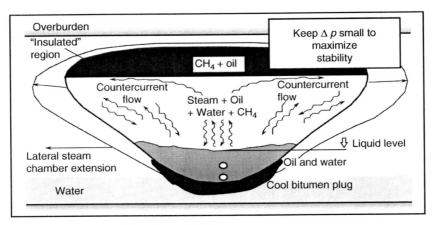

FIGURE 6.13 The chemistry and physics of SAGD.

To a small degree, the noncondensable gases tend to remain high in the structure, filling the void space, and even acting as a partial "insulating blanket" that helps to reduce vertical heat losses as the chamber grows laterally. At the pore scale, and at larger scales as well, flow is through countercurrent, gravity-driven flow, and a thin and continuous oil film is sustained, giving high recoveries estimated to be as large as 70% to 80% in suitable reservoirs.

Operating the production and injection wells at approximately the same pressure as the reservoir eliminates viscous fingering and coning processes, and also suppresses water influx or oil loss through permeable streaks. This keeps the steam chamber interface relatively sharp, and reduces heat losses considerably. Injection pressures are much lower than the fracture gradient, which means that the chances of breaking into a thief zone, an instability problem that plagues all high-pressure steam injection processes, such as cyclic steam soak, are essentially zero.

Thus, the SAGD process, as all gravity-driven processes, is extremely stable because the process zone grows only by gravity segregation, and there are no pressure-driven instabilities such as channeling, coning, and fracturing. It is vital in the SAGD process to maintain a volume balance, replacing each unit volume withdrawn with a unit volume injected, to maintain the processes in the gravity-dominated domain. If bottom-water influx develops, this indicates that the pressure in the water is larger than the pressure in the steam chamber, and steps must be taken to balance the pressures. Because it is not possible to reduce the pressure in the water zone, the pressure in the steam chamber and production well region must be increased. This can be achieved by increasing the operating pressure of the steam chamber through the injection rate of steam or through reduction of the production rate from the lower well. After some time, the pressures will become more balanced and the water influx ceases. Thereafter, maintaining a volume balance carefully is necessary.

Clearly, a low-pressure gradient between the bottom water and the production well must be sustained. If pressure starts to build up in the steam chamber zone, then loss of hot water can take place as well. In such cases, the steam chamber pressure must be reduced and perhaps also the production rate increased slightly to balance the pressures. In all these cases, the system tends to return to a stable configuration because of the density differences between the phases.

SAGD seems to be relatively insensitive to shale streaks and similar horizontal barriers, even up to several meters thick (3 to 6 ft), that otherwise would restrict vertical flow rates. This occurs because as the rock is heated, differential thermal expansion causes the shale to be placed under a tensile stress, and vertical fractures are created, which serve as conduits for steam (up) and liquids (down). As high temperatures hit the shale, the kinetic energy in the water increases, and adsorbed water on clay particles is liberated. Thus, instead of expanding thermally, dehydration (loss of water) occurs and this leads to volumetric shrinkage of the shale barriers. As the shale shrink, the lateral stress (fracture gradient) drops until the pore pressure exceeds the lateral stress, which causes vertical fractures to open. Thus, the combined processes of gravity segregation and shale thermal fracturing make SAGD so efficient that recovery ratios of 60% to 70% are probably achievable even in cases where there are many thin shale streaks, although there are limits on the thickness of shale bed that can be traversed in a reasonable time.

Heat losses and deceleration of lateral growth mean that there is an economic limit to the lateral growth of the steam chamber. This limit is thought to be a chamber width of four times (4×) the vertical zone thickness. For thinner zones, horizontal well pairs would therefore have to be placed close together, increasing costs as well as providing lower total resources per well pair. In summary, the zone thickness limit (net pay thickness) must be defined for all reservoirs.

The cost of heat is a major economic constraint on all thermal processes. Currently, steam is generated with natural gas, and when the cost of natural gas rises, operating costs rise considerably. Thermally, SAGD is about twice as efficient as cyclic steam stimulation, with steam–oil ratios that are now approaching two (instead of four for cyclic steam soak), for similar cases. Combined with the high recovery ratios possible, SAGD will likely displace pressure-driven thermal process in all cases where the reservoir is reasonably thick.

Finally, because of the lower pressures associated with SAGD, in comparison to high-pressure processes such as cyclic steam soak and steam drive, greater wellbore stability should be another asset, reducing substantially the number of sheared wells that are common in cyclic steam soak projects.

Cold heavy oil production with sand (CHOPS) is now widely used as a production approach in unconsolidated sandstones.

The process results in the development of high-permeability channels (wormholes) in the adjacent low cohesive strength sands, facilitating the flow of oil foam that is caused by solution gas drive. The key benefits of the process are improved reservoir access, order-of-magnitude higher oil production rates (as compared to primary recovery) and lower production costs. The outstanding technical issues involve sand handling problems, field development strategies, wormhole plugging for water shut-off, low ultimate recovery, and sand disposal. Originally, cold production mechanisms were thought to apply only to vertical wells with high-capacity pumps. It is now believed that these mechanisms may also apply to horizontal wells and lighter (heavy) oils.

Thus, instead of blocking sand ingress by screens or gravel packs, sand is encouraged to enter the wellbore by aggressive perforation and swabbing strategies. Vertical or slightly inclined wells (vertical to 45°) are operated with rotary progressive cavity pumps (rather than reciprocating pumps) and old fields are converting to higher-capacity progressive cavity pumps, giving production boosts to old wells. Productivity increases over conventional production and a CHOPS process can ensure that as much as 12% to perhaps as much as 25% of the original oil in place can be recovered, rather than the 0% to 5% typical of primary production without sand in such cases. Finally, because massive sand production creates a large disturbed zone, the reservoir may be positively affected for later implementation of thermal processes.

The CHOPS process increases productivity for the following reasons: (1) if the sand can move or is unconsolidated, the basic permeability to fluids is enhanced, (2) as more sand is produced, a growing zone of greater permeability is generated, similar to a large-radius well that gives better production, (3) gas coming out of solution in heavy oil does not generate a continuous gas phase; rather, bubbles flow with the fluid and do not coalesce, but expand down gradient, generating an "internal" gas drive, referred to as "foamy flow." This also helps to locally destabilize the sand, sustaining the process, (4) continuous sand production means that asphaltene or fines plugging of the near-wellbore environment potentially do not occur, so there is no possibility of an effect to impair productivity, and (5) as sand is removed, the overburden weight acts to shear and destabilize the sand, helping to drive sand and oil toward the wellbore.

Typically, a well placed on-CHOPS production will initially produce a high percentage of sand, greater than 20% by volume of liquids. However, this generally drops after some weeks or months. The huge volumes of sand are disposed of by slurry fracture injection or salt cavern placement or by sand placement in a landfill in an environmentally acceptable manner.

Pressure pulsing technologies (PPT) involving a radically new aspect of porous media mechanics was discovered and developed into a production enhancement method in the period 1997 to 2003. Pressure pulse flow enhancement technology (PPT) is based on the discovery that large amplitude pressure pulses that are dominated by low-frequency wave energy generate enhanced flow rates in porous media. For example, in preliminary experiments in heavy oil reservoirs in Alberta, PPT has reduced the rate of depletion, increased the oil recovery ratio, and prolonged the life of wells. Also, it has been found that very large amplitude pressure pulses applied for 5 to 30 h to a blocked producing well can reestablish economic production in a CHOPS well for many months, even years.

The mechanism by which PPT works is to generate a porosity dilation wave (a fluid displacement wave similar to a tsunami); this generates pore-scale dilation and contraction so that oil and water flow into and out of pores, leading to periodic fluid accelerations in the pore throats. As the porosity dilation wave moves through the porous medium at a velocity of about 50 to 100 ft/s (40 to 80 m/s), the small expansion and contraction of the pores with the passage of each packet of wave energy helps unblock pore throats, increase the velocity of liquid flow, overcome part of the effects of capillary blockage, and reduce some of the negative effects of instability due to viscous fingering, coning, and permeability streak channeling.

Although a very new concept dating only since 1999 in small-scale field experiments, PPT promises to be a major adjunct to a number of oil production processes, particularly all pressure-driven processes, where it will both accelerate flow rates as well as increase oil recovery factors. It is also used now in environmental applications to help purge shallow aquifers of nonmiscible phases such as oil, with about six successful case histories to date.

VAPEX is a new process in which the physics of the process are essentially the same as for SAGD and the configuration of wells is generally similar. The process involves the injection of vaporized solvents such as ethane or propane to create a vapor-chamber through which the oil flows due to gravity drainage (Butler and Mokrys, 1991; Butler and Mokrys, 1995a, b; Butler and Jiang, 2000). The process can be applied in paired horizontal wells, single horizontal wells, or a combination of vertical and horizontal wells. The key benefits are significantly lower energy costs, potential for in situ upgrading, and application to thin reservoirs, with bottom water or reactive mineralogy.

Because of the slow diffusion of gases and liquids into viscous oils, this approach, used alone, perhaps will be suited only for less viscous oils although preliminary tests indicate that there are micro-mechanisms that act so that the VAPEX dilution process is not diffusion rate limited and the process may be suitable for the highly viscous tar sand bitumen (Yang and Gu, 2005a, b).

Nevertheless, VAPEX can undoubtedly be used in conjunction with SAGD methods. As with SAGD and IGI, a key factor is the generation of a three-phase system with a continuous gas phase so that as much of the oil as possible can be contacted by the gaseous phases, generating the thin oil film drainage mechanism. As with IGI, vertical permeability barriers are a problem, and must be overcome through hydraulic fracturing to create vertical permeable channels, or undercut by the lateral growth of the chamber beyond the lateral extent of the limited barrier, or "baffle."

Hybrid approaches that involve the simultaneous use of several technologies are evolving and will see greater applications in the future. Some of the evolving options that will be tried at the field scale in the next decade are listed here:

1. Mixture of steam and miscible and noncondensable hydrocarbons is being field tested as a hybrid SAGD–VAPEX approach, with apparent good success and reduction of steam–oil ratios.
2. Single horizontal laterally offset wells can be operated as moderate pressure cyclic steam stimulation wells in combination with SAGD pairs to widen the steam chamber and reduce steam–oil ratios by about 20%.
3. Simultaneous CHOPS and SAGD, with CHOPS used in offset wells until steam breakthrough occurs. Then the CHOPS wells are converted to slow gas and hot-water (or steam) injection wells to control the process. The high-permeability zones generated by CHOPS should accelerate the SAGD recovery process.
4. Incorporating PPT along with CHOPS has already been field tested with economic success, and PPT has potential applications in other hybrid approaches.
5. PPT may aid in partially stabilizing waterflood through reducing the viscous fingering and coning intensity.

In addition to hybrid approaches, the new production technologies, along with older, pressure-driven technologies, will be used in successive phases to extract more oil from reservoirs, even from reservoirs that have been abandoned after primary exploitation. Old reservoirs can be redeveloped with horizontal wells, even linking up the wells to bypassed oil because of the physics of oil film spreading between water and gas phases. These staged approaches hold the promise of significantly increasing recoverable reserves worldwide, not just in heavy oil cases.

Microbial enhanced oil recovery (**MEOR**) processes (Table 6.1) involve the use of reservoir microorganisms or specially selected natural bacteria to produce specific metabolic events that lead to enhanced oil recovery.

In microbial enhanced oil recovery processes, microbial technology is exploited in oil reservoirs to improve recovery (Clark et al., 1981; Stosur, 1991; Banat, 1995). From a microbiologist's perspective, microbial enhanced oil recovery processes are somewhat akin to in situ bioremediation processes. Injected nutrients, together with indigenous or added microbes, promote in situ microbial growth or generation of products that mobilize additional oil and move it to producing wells through reservoir repressurization, interfacial tension, or oil viscosity reduction and selective plugging of the most permeable zones (Bryant et al., 1989; Bryant and Lindsey, 1996). Alternatively, the oil-mobilizing microbial products may be produced by fermentation and injected into the reservoir.

This technology requires consideration of the physicochemical properties of the reservoir in terms of salinity, pH, temperature, pressure, and nutrient availability (Khire and Khan, 1994a, b). Only bacteria are considered promising candidates for microbial enhanced oil recovery. Molds, yeasts, algae, and protozoa are not suitable due to their size or inability to grow under the conditions present in reservoirs. Many petroleum reservoirs have high

TABLE 6.1
Types of Microbial Processes for Oil Recovery

Process	Production Problem	Types of Activity or Product Needed
Well bore cleanup (improve oil drainage into well bore)	Paraffin and scale deposits	Emulsifiers, biosurfactants, solvents, acids, hydrocarbon degradation
Well stimulation (stimulate release of oil entrapped by capillaries and brine)	Formation damage, pore damage	Gas, acids, solvents, biosurfactants
	High water production	Biomass and polymer production
Enhanced waterflood (reduce permeability variation and block water channels)	Poor displacement efficiency	Biosurfactants, solvents, polymers
	Poor sweep efficiency	Biomass and polymer production
	Scouring	Nitrate reduction

concentrations of sodium chloride (Jenneman, 1989) and require the use of bacteria that can tolerate these conditions (Shennan and Levi, 1987). Bacteria producing biosurfactants and polymers can grow at sodium concentrations up to 8% and selectively plug sandstone to create a biowall to recover additional oil (Raiders et al., 1989).

Organisms that participate in oil recovery produce a variety of fermentation products, e.g., carbon dioxide, methane, hydrogen, biosurfactants, and polysaccharides from crude oil, pure hydrocarbons, and a variety of nonhydrocarbon substrates. Organic acids produced through fermentation readily dissolve carbonates and can greatly enhance permeability in limestone reservoirs, and attempts have been made to promote anaerobic production.

The microbial enhanced oil recovery process may modify the immediate reservoir environment in a number of ways that could also damage the production hardware or the formation itself (http://mmbr.asm.org/cgi/content/full/67/4/503#R280#R280). Certain sulfate reducers can produce H_2S, which can corrode pipeline and other components of the recovery equipment. Thus, despite numerous microbial enhanced oil recovery tests, considerable uncertainty remains regarding process performance. Ensuring success requires an ability to manipulate environmental conditions to promote growth or product formation by the participating microorganisms. Exerting such control over the microbial system in the subsurface is itself a serious challenge. In addition, conditions vary from reservoir to reservoir, which calls for reservoir-specific customization of the microbial enhanced oil recovery process, and this alone has the potential to undermine microbial process economic viability.

Microbial enhanced oil recovery differs from chemical enhanced oil recovery in the method by which the enhancing products are introduced into the reservoir (Figure 6.14 and Figure 6.15). Thus, in oil recovery by the cyclic microbial method, a solution of nutrients and microorganisms is introduced into the reservoir during injection. The injection well is then shut for an incubation period allowing the microorganisms to produce carbon dioxide gas and surfactants that assist in mobilization of the oil. The well is then opened and oil and oil products resulting from the treatment are produced. The process is repeated as often as oil can be produced from the well. Oil recovery by microbial flooding also involves the use of microorganisms but in this case, the reservoir is usually conditioned by a water flush after which a solution of microorganisms and nutrients is injected into the formation. As this solution is pushed through the reservoir by water drive, gases and surfactants are formed, the oil is mobilized and pumped through the well. However, microbes produce the necessary chemical reactions in situ (Table 6.2) whereas surface

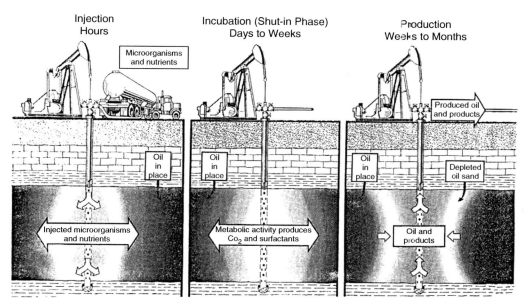

FIGURE 6.14 Cyclic microbial oil recovery. (From United States Department of Energy, Washington DC.)

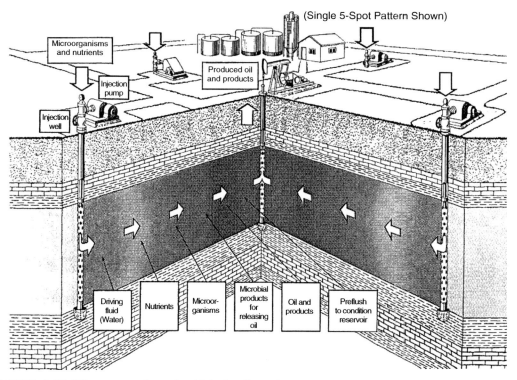

FIGURE 6.15 Oil recovery by microbial flooding. (From United States Department of Energy, Washington DC.)

TABLE 6.2
Microbial Products and Their Contribution to Enhanced Oil Recovery

Microbial	Effect
Acids	• Modification of reservoir rock
	• Improvement of porosity and permeability
	• Reaction with calcareous carbon dioxide production
Biomass	• Selective or nonselective plugging
	• Emulsification through adherence to hydrocarbons
	• Modification of solid surfaces, e.g., wetting
	• Degradation and alteration of oil
	• Reduction of oil viscosity and oil pour point
	• Desulfurization of oil
Gases (CO_2, CH_4, H_2)	• Reservoir repressurization
	• Oil swelling
	• Viscosity reduction
	• Increase of permeability due to solubilization of carbonate rocks by CO_2
Solvents	• Oil dissolution
Surface active agent	• Lowering interfacial tension
Polymers	• Mobility control
	• Selective or nonselective plugging

injected chemicals may tend to follow areas of higher permeability, resulting in decreased sweep efficiency, there is need for caution and astute observation of the effects of the microorganisms on the reservoir chemistry.

The mechanism by which microbial EOR processes work can be quite complex and may involve multiple biochemical processes. In selective plugging approaches, microbial cell mass or biopolymers plug high-permeability zones and lead to a redirection of the water flood. In other processes, biosurfactants are produced in situ, which leads to increased mobilization of residual oil. In still other processes, microbial production of carbon dioxide and organic solvents reduces the oil viscosity as the primary mechanism for EOR.

In a microbial EOR process, conditions for microbial metabolism are supported via injection of nutrients. In some processes, this involves injecting a fermentable carbohydrate into the reservoir. Some reservoirs also require inorganic nutrients as substrates for cellular growth or for serving as alternative electron acceptors in place of oxygen or carbohydrates.

The stimulation of oil production by in situ bacterial fermentation is thought to proceed by one or a combination of the following mechanisms:

1. Improvement of the relative mobility of oil to water by biosurfactants and biopolymers.
2. Partial repressurization of the reservoir by methane and carbon dioxide.
3. Reduction of oil viscosity through the dissolution of organic solvents in the oil phase.
4. Increase of reservoir permeability and widening of the fissures and channels through the etching of carbonaceous rocks in limestone reservoirs by organic acids produced by anaerobic bacteria.
5. Cleaning the wellbore region through the acids and gas from in situ fermentation in which the gas pushes oil from dead space and dislodges the debris that plugs the pores. The average pore size is increased and, as a result, the capillary pressure near the wellbore is made more favorable for the flow of oil.

6. Selective plugging of highly permeable zones by injecting slime-forming bacteria followed by sucrose solution, which initiates the production of extra cellular slimes and aerial sweep efficiency is improved.

The target for EOR processes is the quantity of unrecoverable oil in known reservoirs and bitumen and in known deposits. One of the major attributes of microbial EOR technologies is its low cost but there must be the recognition that microbial EOR is a single process. Further, reports on the deleterious activities of microorganisms in the oil field contribute to the skepticism of employing technologies using microorganisms. It is also clear that scientific knowledge of the fundamentals of microbiology must be coupled with an understanding of the geological and engineering aspects of oil production in order to develop microbial enhanced oil recovery technology.

Finally, recent developments in upgrading of heavy oil and bitumen (Chapter 23) indicate that the near future could see a reduction of the differential cost of upgrading heavy oil. These processes are based on a better understanding of the issues of asphaltene solubility effects at high temperatures, incorporation of a catalyst that is chemically precipitated internally during the upgrading, and improving hydrogen addition or carbon rejection.

REFERENCES

Banat, I.M. 1995. *Biores. Technol.* 51: 1–12. DOI: 10.1016/0960-8524(94)00101-6.

Bryant, R.S. and Lindsey, R.P. 1996. World-wide applications of microbial technology for improving oil recovery. In *Proceedings of the SPE Symposium on Improved Oil Recovery*. Society of Petroleum Engineers, Richardson, TX. pp. 27–134.

Bryant, R.S., Donaldson, E.C., Yen, T.F., and Chilingarian, G.V. 1989. Microbial enhanced oil recovery. In *Enhanced Oil Recovery II: Processes and Operations*. Donaldson, E.C. Chilingarian, G.V. and Yen, T.F. eds. Elsevier, Amsterdam. pp. 423–450.

Butler, R.M. and Jiang, Q. 2000. *J. Canad. Petrol. Technol.* 39: 48–56.

Butler, R.M. and Mokrys, I.J. 1991. *J. Canad. Petrol. Technol.* 30(1): 97–106.

Butler, R.M. and Mokrys, I.J. 1995a. Process and apparatus for the recovery of hydrocarbons from a hydrocarbon deposit. U.S. Patent 5,407,009. April 18.

Butler, R.M. and Mokrys, I.J. 1995b. Process and apparatus for the recovery of hydrocarbons from a hydrocarbon deposit. U.S. Patent 5,607,016. March 4.

Carrigy, M.A. 1963a. *Bulletin No. 14*. Alberta Research Council, Edmonton, Alberta, Canada.

Carrigy, M.A. 1963b. *The Oil Sands of Alberta*. Information Series No. 45. Alberta Research Council, Edmonton, Alberta, Canada.

Chakma, A., Islam, M.R., and Berruti, F. eds. 1991. *Enhanced Oil Recovery*. AIChE Symposium Series No. 280, Volume 87, American Institute of Chemical Engineers, New York. 147p.

Clark, K.A. 1944. *Trans. Can. Inst. Min. Met.* 47: 257.

Clark, J.B., Munnecke, D.M., and Jenneman, G.E. 1981. *Dev. Ind. Microbiol* 5: 695–701.

Dusseault, M.B., Geilikman M.B., and Spanos T.J.T. 1998. *J. Petrol. Technol.* 50(9): 92–94.

Fear, J.V.D. and Innes, E.D. 1967. *Proc. 7th World Petrol. Congr.* 3: 549.

Gishler, P.E. 1949. *Can. J. Res.* 27: 104.

Gregoli, A.A. and Rimmer, D.P. 2000. Production of synthetic crude oil from heavy hydrocarbons recovered by in situ hydrovisbreaking. U.S. Patent 6,016,868. January 25.

Gregoli, A.A., Rimmer, D.P., and Graue, D.J. 2000. Upgrading and recovery of heavy crude oils and natural bitumen by in situ hydrovisbreaking. U.S. Patent 6,016,867. January 25.

Islam, M.R., Chakma, A., and Jha, K.N. 1994. *Petrol. Sci. Eng.* 11: 213–226.

Jenneman, G.E. 1989. *Dev. Petrol. Sci.* 22: 37–74.

Khire, J.M. and Khan, M.I. 1994a. *Enzyme Microb. Technol.* 16: 170–172.

Khire, J.M. and Khan, M.I. 1994b. *Enzyme Microb. Technol.* 16: 258–259.

Kruyer, J. 1982. *Proceedings*. Second International Conference on Heavy Crude and Tar Sands. Caracas, Venezuela, February 7–17.

Kruyer, J. 1983. *Preprint No. 3d*. Summer National Meeting. American Institute of Chemical Engineers. Denver, CO, August 28–31.

Miller, J.C. and Misra, M. 1982. *Fuel Proc. Technol*. 6: 27.

Pratts, M. 1986. *Thermal Recovery*, Volume 7. Society of Petroleum Engineers, New York.

Raiders, R.A., Knapp, R.M., and McInerney, M.J. 1989. *J. Ind. Microbiol*. 4: 215–230.

Shennan, J.L. and Levi, J.D. 1987. In situ microbial enhanced oil recovery. *Biosurfactants and Biotechnology*. Kosaric, N. Cairns, W.L. and Gray, N.C.C. eds. Marcel Dekker, New York. p. 163–180.

Speight, J.G. 2000. *The Desulfurization of Heavy Oils and Residua*. 2nd Ed. Marcel Dekker, New York.

Speight, J.G. and Moschopedis, S.E. 1978. *Fuel Proc. Technol*. 1: 261.

Stosur, G.J. 1991. *Crit. Rep. Appl. Chem*. 33: 341–373.

Yang, C. and Gu, Y. 2005a. A novel experimental technique for studying solvent mass transfer and oil swelling effect in a vapor extraction (VAPEX) process. *Paper No. 2005–099. Proceedings*. 56th Annual Technical Meeting. The Canadian International Petroleum Conference, Calgary, June 7–9.

Yang, C. and Gu, Y. 2005b. Effects of solvent–heavy oil interfacial tension on gravity drainage in the VAPEX process. *Paper No. SPE 97906*. Society of Petroleum Engineers International Thermal Operations and Heavy Oil Symposium, Calgary, Alberta, Canada, November 1–3.

Part II

Composition and Properties

from samples for which the origin is carefully identified (Wallace et al., 1988). It is to be hoped that this program continues, as it will provide a valuable database for tar sand and bitumen characterization.

Like conventional petroleum, of the data that are available, the elemental composition of oil sand bitumen is generally constant and, like the data for petroleum, falls into a narrow range (Speight, 1990 and references cited therein):

Carbon: 83.4 $\pm$ 0.5%
Hydrogen: 10.4 $\pm$ 0.2%
Nitrogen: 0.4 $\pm$ 0.2%
Oxygen: 1.0 $\pm$ 0.2%
Sulfur: 5.0 $\pm$ 0.5%
Metals (Ni and V): >1000 ppm

The major exception to these narrow limits is the oxygen content of bitumen, which can vary from as little as 0.2% to as high as 4.5%. This is not surprising, since when oxygen is estimated by difference the analysis is subject to the accumulation of all of the errors in the other elemental data. In addition, bitumen is susceptible to aerial oxygen and the oxygen content is very dependent upon the sample history. In addition, the ultimate composition of the Alberta bitumen does not appear to be influenced by the proportion of bitumen in the oil sand or by the particle size of the oil sand minerals.

Bitumen from U.S. tar sand has an ultimate composition similar to that of the Athabasca bitumen (Speight, 1990 and references cited therein). As already noted earlier, when the many localized or regional variations in maturation conditions are assessed it is perhaps surprising that the ultimate compositions are so similar.

Several generalities can be noted from the ultimate composition; these can only give indications of how the material might behave during processing (Chapter 8).

The viscosity of bitumen is related to its hydrogen-to-carbon atomic ratio and hence the required supplementary heat energy for thermal extraction processes. It also affects the bitumen's distillation curve or thermodynamic characteristics, its gravity, and its pour point. Atomic hydrogen-to-carbon ratios as low as 1.3 have been observed for tar sand bitumen, although an atomic hydrogen-to-carbon ratio of 1.5 is more typical. The higher the hydrogen–carbon ratio of bitumen, the higher is its value as refinery feedstock because of the lower hydrogen requirements. Elements related to the hydrogen–carbon ratio are distillation curve, bitumen gravity, pour point, and bitumen viscosity.

The occurrence of sulfur in bitumen as organic or elemental sulfur or in produced gas as compounds of oxygen and hydrogen is an expensive nuisance. It must be removed from the bitumen at some point in the upgrading and refining process. Sulfur contents of some tar sand bitumen can exceed 10% w/w. Elements related to sulfur content are hydrogen content, hydrogen–carbon ratio, nitrogen content, distillation curve, and viscosity.

The nitrogen content of tar sand bitumen can be as high as 1.3% by weight and nitrogen-containing constituents complicate the refining process by poisoning the catalysts employed in the refining process. Elements related to nitrogen content are sulfur content, hydrogen content, hydrogen–carbon ratio, bitumen viscosity, distillation profile, and viscosity.

7.3 CHEMICAL COMPONENTS

Petroleum contains an extreme range of organic functionality and molecular size. In fact, the variety is so great that it is unlikely that a complete compound-by-compound description for even a single crude oil would be possible. As already noted, the composition of petroleum can vary with the location and age of the field in addition to any variations that occur with the depth of the individual well. Two adjacent wells are more than likely to produce petroleum with very different characteristics.

In very general terms (and as observed from elemental analyses), petroleum, heavy oil, bitumen, and residua are a complex composition of: (1) hydrocarbons; (2) nitrogen compounds; (3) oxygen compounds; (4) sulfur compounds; and (5) metallic constituents. However, this general definition is not adequate to describe the composition of petroleum as it relates to the behavior of these feedstocks. Indeed the consideration of hydrogen-to-carbon atomic ratio, sulfur content, and API gravity are no longer adequate to the task of determining refining behavior.

Further, the molecular composition of petroleum can be described in terms of three classes of compounds: saturates, aromatics, and compounds bearing heteroatoms (sulfur, oxygen, or nitrogen). Within each class, there are several families of related compounds: (1) saturated constituents include normal alkanes, branched alkanes, and cycloalkanes (paraffins, iso-paraffins, and naphthenes, in petroleum terms), (2) alkene constituents (olefins) are rare to the extent of being considered an oddity, (3) monoaromatic constituents range from benzene to multiple fused ring analogs (naphthalene, phenanthrene, etc.), (4) thiol (mercaptan) constituents contain sulfur as do thioethers and thiophene forms, (5) nitrogen-containing and oxygen-containing constituents are more likely to be found in polar forms (pyridines, pyrroles, phenols, carboxylic acids, amides, etc.) than in nonpolar forms (such as ethers). The distribution and characteristics of these molecular species account for the rich variety of crude oils.

Feedstock behavior during refining is better addressed through consideration of the molecular makeup of the feedstock (perhaps, by analogy, just as genetic makeup dictates human behavior). The occurrence of amphoteric species (i.e., compounds having a mixed acid or base nature) is not addressed nor is the phenomenon of molecular size or the occurrence of specific functional types, which can play a major role in the interactions between the constituents of a feedstock. All of these items are important in determining feedstock behavior during refining operations.

An understanding of the chemical types (or composition) of any feedstock can lead to an understanding of the chemical aspects of processing the feedstock. Processability is not only a matter of knowing the elemental composition of a feedstock; it is also a matter of understanding the bulk properties as they relate to the chemical or physical composition of the material. For example, it is difficult to understand, a priori, the process chemistry of various feedstocks from their elemental composition alone. From such data, it might be surmised that the major difference between a heavy crude oil and a more conventional material is the H/C atomic ratio alone. This property indicates that a heavy crude oil (having a lower H/C atomic ratio and being more aromatic in character) would require more hydrogen for upgrading to liquid fuels. This is, indeed, true but much more information is necessary to understand the processability of the feedstock.

The hydrocarbon content of petroleum may be as high as 97% by weight (e.g., in the lighter paraffinic crude oils) or as low as 50% by weight or less as illustrated by the heavy asphaltic crude oils. Nevertheless, crude oils with as little as 50% hydrocarbon components are still assumed to retain most of the essential characteristics of the hydrocarbons. It is, nevertheless, the nonhydrocarbon (sulfur, oxygen, nitrogen, and metal) constituents that play a large part in determining the processability of the crude oil (Rossini et al., 1953; Brooks et al., 1954; Lochte and Littmann, 1955; Schwartz and Brasseaux, 1958; Brandenburg and Latham, 1968; Rall et al., 1972). But there is more to the composition of petroleum than the hydrocarbon content. The inclusion of organic compounds of sulfur, nitrogen, and oxygen serves only to present crude oils as even more complex mixtures, and the appearance of appreciable amounts of these nonhydrocarbon compounds causes some concern in the refining of crude oils. Even though the concentration of nonhydrocarbon constituents (i.e., those organic compounds containing one or more sulfur, oxygen, or nitrogen atoms) in certain fractions may be quite small, they tend to concentrate in the higher boiling fractions of petroleum. Indeed, their influence on the processability of the petroleum is important, irrespective of their molecular size and the fraction in which they occur.

The presence of traces of nonhydrocarbon compounds can impart objectionable characteristics to finished products, leading to discoloration or lack of stability during storage. On the other hand, catalyst poisoning and corrosion are the most noticeable effects during refining sequences when these compounds are present. It is therefore not surprising that considerable attention must be given to the nonhydrocarbon constituents of petroleum as the trend in the refining industry, of late, has been to process more heavy crude oil as well as residua that contain substantial proportions of these nonhydrocarbon materials.

7.3.1 HYDROCARBON CONSTITUENTS

The isolation of pure compounds from petroleum is an exceedingly difficult task, and the overwhelming complexity of the hydrocarbon constituents of the higher molecular weight fractions as well as the presence of compounds of sulfur, oxygen, and nitrogen, are the main causes for the difficulties encountered. It is difficult on the basis of the data obtained from synthesized hydrocarbons to determine the identity or even the similarity of the synthetic hydrocarbons to those that constitute many of the higher boiling fractions of petroleum. Nevertheless, it has been well established that the hydrocarbon components of petroleum are composed of paraffinic, naphthenic, and aromatic groups (Table 7.1). Olefin groups are not

TABLE 7.1
Hydrocarbon and Heteroatom Types in Petroleum

Class	Compound Types
Saturated hydrocarbons	*n*-Paraffins *Iso*-paraffins and other branched paraffins Cycloparaffins (naphthenes) Condensed cycloparaffins (including steranes, hopanes) Alkyl side chains on ring systems
Unsaturated hydrocarbons	Olefins not indigenous to petroleum; present in products of thermal reactions
Aromatic hydrocarbons	Benzene systems Condensed aromatic systems Condensed aromatic-cycloalkyl systems Alkyl side chains on ring systems
Saturated heteroatomic systems	Alkyl sulfides Cycloalkyl sulfides Alkyl side chains on ring systems
Aromatic heteroatomic systems	Furans (single-ring and multi-ring systems) Thiophenes (single-ring and multi-ring systems) Pyrroles (single-ring and multi-ring systems) Pyridines (single-ring and multi-ring systems) Mixed heteroatomic systems Amphoteric (acid–base) systems Alkyl side chains on ring systems

usually found in crude oils, and acetylenic hydrocarbons are very rare indeed. It is therefore convenient to divide the hydrocarbon components of petroleum into the following three classes:

1. *Paraffins*, which are saturated hydrocarbons with straight or branched chains, but without any ring structure
2. *Naphthenes*, which are saturated hydrocarbons containing one or more rings, each of which may have one or more paraffinic side chains (more correctly known as **alicyclic hydrocarbons**)
3. *Aromatics*, which are hydrocarbons containing one or more aromatic nuclei, such as benzene, naphthalene, and phenanthrene ring systems, which may be linked up with (substituted) naphthene rings or paraffinic side chains

7.3.1.1 Paraffin Hydrocarbons

The proportion of paraffins in crude oil varies with the type of crude, but within any one crude oil, the proportion of paraffinic hydrocarbons usually decreases with increasing molecular weight and there is a concomitant increase in aromaticity and the relative proportion of heteroatoms (nitrogen, oxygen, and sulfur) (Figure 7.1).

The relationship between the various hydrocarbon constituents of crude oils is one of hydrogen addition or hydrogen loss (Figure 7.2). This is an extremely important aspect of petroleum composition and there is no reason to deny the occurrence of these inter-

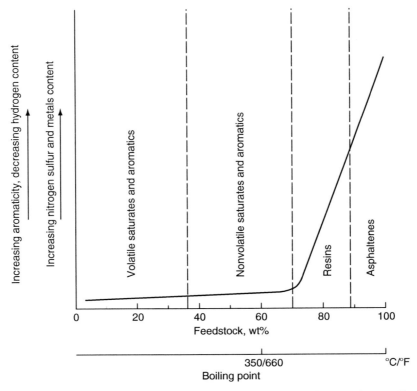

FIGURE 7.1 Variation of compound types in petroleum with boiling point. For heavy oil and tar sand bitumen the line from zero to 70% is steeper.

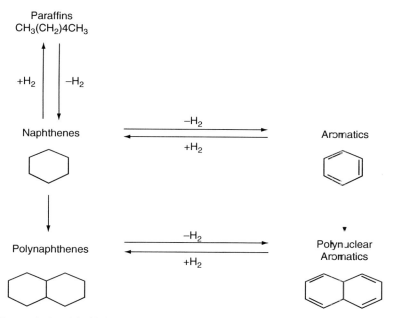

FIGURE 7.2 Interrelationship if the hydrocarbon types in petroleum.

conversion schemes during the formation, maturation, and in situ alteration of petroleum. Indeed, a scheme of this type lends even more credence to the complexity of petroleum within the hydrocarbon series alone and also supports current contentions that the high molecular weight constituents (resin constituents and asphaltene constituents) are structurally related to the lower-boiling constituents rather than proposals that invoke the existence of highly condensed polynuclear aromatic systems.

The abundance of the different members of the same homologous series varies considerably in absolute and relative values. However, in any particular crude oil or crude oil fraction, there may be a small number of constituents forming the greater part of the fraction, and these have been referred to as the predominant constituents (Rossini and Mair, 1959; Bestougeff, 1961). This generality may also apply to other constituents and is dependent upon the nature of the source material as well as the relative amounts of the individual source materials prevailing during maturation conditions (Chapter 2).

Normal paraffin hydrocarbons (*n*-paraffins, straight-chain paraffins) occur in varying proportions in most crude oils. In fact, paraffinic petroleum may contain up to 20% to 50% by weight *n*-paraffins in the gas oil fraction. However, naphthenic or asphaltic crude oils sometimes contain only very small amounts of normal paraffins.

Considerable quantities of *iso*-paraffins have been noted to be present in the straight-run gasoline fraction of petroleum. The 2- and 3-methyl derivatives are the most abundant, and the 4-methyl derivative is present in small amounts, if at all, and it is generally accepted that the slightly branched paraffins predominate over the highly branched materials. It seems that the *iso*-paraffins occur throughout the boiling range of petroleum fractions. The proportion tends to decrease with increasing boiling point; it appears that if the *iso*-paraffins are present in lubricating oils, their amount is too small to have any significant influence on the physical properties of the lubricating oils.

As the molecular weight (or boiling point) of the petroleum fraction increases, there is a concomitant decrease in the amount of free paraffins in the fraction. In certain types of crude oil, there may be no paraffins at all in the vacuum gas oil fraction. For example, in the paraffinic crude oils, free paraffins will separate as a part of the asphaltene fraction, but in the naphthenic crude oils, free paraffins are not expected in the gas oil and asphaltene fractions.

The vestiges of paraffins in the asphaltenes fractions occur as alkyl side chains on aromatic and naphthenic systems. And, these alkyl chains can contain 40 or more carbon atoms (Chapter 9 and Chapter 10).

7.3.1.2 Cycloparaffin Hydrocarbons (Naphthenes)

Although only a small number of representatives have been isolated so far, cyclohexane derivatives, cyclopentane derivatives, and decahydronaphthalene (decalin) derivatives are largely represented in oil fractions. Petroleum also contains polycyclic naphthenes, such as terpenes, and such molecules (often designated bridge-ring hydrocarbons) occur even in the heavy gasoline fractions (boiling point 150°C to 200°C, 300°F to 390°F). Naphthene rings may be built up of a varying number of carbon atoms, and among the synthesized hydrocarbons there are individual constituents with rings of the three-, four-, five-, six-, seven-, and eight-carbon atoms. It is now generally believed that crude oil fractions contain chiefly five- and six-carbon rings. Only naphthenes with five- and six-membered rings have been isolated from the lower boiling fractions. Thermodynamic studies show that naphthene rings with five and six carbon atoms are the most stable. The naphthenic acids contain chiefly cycle pentane as well as cyclohexane rings.

Cycloparaffins (naphthenes) are represented in all fractions in which the constituent molecules contain more than five carbon atoms. Several series of cycloparaffins, usually containing five- or six-membered rings or their combinations, occur as polycyclic structures. The content of cycloparaffins in petroleum varies up to 60% of the total hydrocarbons. However, the cycloparaffin content of different boiling range fractions of a crude oil may not vary considerably and generally remains within rather close limits. Nevertheless, the structure of these constituents may change within the same crude oil, as a function of the molecular weight or boiling range of the individual fractions as well as from one crude oil to another.

The principal structural variation of naphthenes is the number of rings present in the molecule. The mono- and bicyclic naphthenes are generally the major types of cycloparaffins in the lower boiling fractions of petroleum, with boiling point or molecular weight increased by the presence of alkyl chains. The higher boiling point fractions, such as the lubricating oils, may contain two to six rings per molecule.

As the molecular weight (or boiling point) of the petroleum fraction increases, there is a concomitant increase in the amount of cycloparaffinic (naphthenic) species in the fraction. In the asphaltic (naphthenic) crude oils, the gas oil fraction can contain considerable amounts of naphthenic ring systems that increase even more in consideration of the molecular types in the asphaltenes. However, as the molecular weight of the fraction increases, the occurrence of condensed naphthene ring systems and alkyl-substituted naphthene ring systems increases.

There is also the premise that the naphthene ring systems carry alkyl chains that are generally shorter than the alkyl substituents carried by aromatic systems. There are indications from spectroscopic studies that the short chains (methyl and ethyl) appear to be characteristic substituents of the aromatic portion of the molecule, whereas a limited number (one or two) of longer chains may be attached to the cycloparaffin rings. The total number of chains, which is in general four to six, as well as their length, increases according to the molecular weight of the naphthenoaromatic compounds.

In the asphaltene constituent, free condensed naphthenic ring systems may occur but general observations favor the occurrence of combined aromatic–naphthenic systems that are variously substituted by alkyl systems. There is also general evidence, that the aromatic systems are responsible for the polarity of the asphaltenes constituents. The heteroatoms are favored to occur on or within the aromatic (pseudo-aromatic) systems (Chapter 10 and Chapter 11).

7.3.1.3 Aromatic Hydrocarbons

The concept of the occurrence of identifiable aromatic systems in nature is a reality and the occurrence of monocyclic and polycyclic aromatic systems in natural product chemicals is well documented (Sakarnen and Ludwig, 1971; Durand, 1980; Weiss and Edwards, 1980). However, one source of aromatic systems that is often ignored is petroleum (Eglinton and Murphy, 1969; Tissot and Welte, 1978; Speight, 1980; Brooks and Welte, 1984). Therefore, attempts to identify such systems in the nonvolatile constituents of petroleum should be an integral part of the repertoire of the petroleum chemist as well as the domain of the natural product chemist.

Crude oil is a mixture of compounds, and aromatic compounds are common to all petroleum and it is the difference in extent that becomes evident upon examination of a series of petroleum. By far the majority of these aromatics contain paraffinic chains, naphthene rings, and aromatic rings side by side.

There is a general increase in the proportion of aromatic hydrocarbons with increasing molecular weight. However, aromatic hydrocarbons without the accompanying naphthene rings or alkyl-substituted derivatives seem to be present in appreciable amounts only in the lower petroleum fractions. Thus, the limitation of instrumentation notwithstanding, it is not surprising that spectrographic identification of such compounds has been concerned with these low-boiling aromatics.

All known aromatics are present in gasoline fractions but the benzene content is usually low compared to the benzene homologs, such as toluene and the xylene isomer. In addition to the 1- and 2-methylnaphthalenes, other simple alkylnaphthalenes have also been isolated from crude oil. Aromatics without naphthene rings appear to be relatively rare in the heavier fractions of petroleum (e.g., lubricating oils). In the higher molecular weight fractions, the rings are usually condensed together. Thus, components with two aromatic rings are presumed to be naphthalene derivatives and those with three aromatic rings may be phenanthrene derivatives. Currently, and because of the consideration of the natural product origins of petroleum, phenanthrene derivatives are favored over anthracene derivatives.

In summation, all hydrocarbon compounds that have aromatic rings, in addition to the presence of alkyl chains and naphthenic rings within the same molecule, are classified as aromatic compounds. Many separation procedures that have been applied to petroleum (Chapter 9) result in the isolation of a compound as an aromatic even if there is only one such ring (i.e., six carbon atoms) that is substituted by many more than six nonaromatic carbon atoms.

It should also be emphasized that in the higher boiling point petroleum fractions, many polycyclic structures occur in naphthenoaromatic systems. The naphthenoaromatic hydrocarbons, together with the naphthenic hydrocarbon series, form the major content of higher boiling point petroleum fractions. Usually, the different naphthenoaromatic components are classified according to the number of aromatic rings in their molecules. The first to be distinguished is the series with an equal number of aromatic and naphthenic rings. The first members of the bicyclic series C_9 to C_{11} are the simplest, such as the 1-methyl-, 2-methyl, and 4-methylindanes and 2-methyl- and 7-methyltetralin. Tetralin and methyl-, dimethyl-, methyl ethyl-, and tetramethyltetralin have been found in several crude oils, particularly in the heavier, naphthenic, crude oils.

Of special interest in the present context are the aromatic systems that occur in the nonvolatile asphaltene fraction (Speight, 1994b). These polycyclic aromatic systems are complex molecules that fall into a molecular weight and boiling range where very little is known about model compounds (Chapter 10). There has not been much success in determining the nature of such systems in the higher boiling constituents of petroleum, i.e., the residua or nonvolatile constituents. In fact, it has been generally assumed that as the boiling point of a petroleum fraction increases, so does the number of condensed rings in a polycyclic aromatic system. To an extent, this is true, but the simplicities of such assumptions cause an omission of other important structural constituents of the petroleum matrix, the alkyl substituents, the heteroatoms, and any polycyclic systems that are linked by alkyl chains or by heteroatoms.

The active principle is that petroleum is a continuum (Corbett and Petrossi, 1978; Long, 1979, 1981) (Chapter 10 and Chapter 11) and has natural product origins (Chapter 3). As such, it might be anticipated that there is a continuum of aromatic systems throughout petroleum that might differ from volatile to nonvolatile fractions but which, in fact, are based on natural product systems. It might also be argued that substitution patterns of the aromatic nucleus that are identified in the volatile fractions, or in any natural product counterparts, also apply to the nonvolatile fractions.

The application of thermal techniques to study the nature of the volatile thermal fragments from petroleum asphaltenes has produced some interesting data relating to the nature of the aromatic systems in crude oil (Speight, 1971; Ritchie et al., 1979a, b; Schucker and Keweshan, 1980; Gallegos, 1981; Raul, 1982). These thermal techniques have produced strong evidence for the presence of small (one- to four-ring) aromatic systems (Speight and Pancirov, 1984; Speight, 1987). There was a preponderance of single-ring (cycloparaffin and alkylbenzene) species as well as the domination of saturated material over aromatic material.

Further studies using pyrolysis or gas chromatography or mass spectrometry (py/gc/ms) (Speight and Pancirov, 1984) showed that different fractions of an asphaltene would produce the same type of polycyclic aromatic systems in the volatile matter but the distribution was not constant (Chapter 10). It was also possible to compute the hydrocarbon distribution, from which a noteworthy point here is preponderance of single-ring (cycloparaffin and alkylbenzene) species, as well as the domination of saturated material over aromatic material. The emphasis on low molecular weight material in the volatile products is to be anticipated on the basis that more complex systems remain as nonvolatile material and, in fact, are converted to coke.

One other point worthy of note is that the py/gc/ms program does not accommodate nitrogen and oxygen species, whether or not they are associated with aromatic systems. This matter is resolved, in part, not only by the concentration of nitrogen and oxygen in the nonvolatile material (coke), but also by the overall low proportions of these heteroatoms originally present in the asphaltenes (Speight, 1971; Speight and Pancirov, 1984). The major drawback to the use of the py/gc/ms technique to study the aromatic systems in asphaltenes is the amount of material that remains as a nonvolatile residue.

Of all of the methods applied to determining the types of aromatic systems in petroleum asphaltenes, one with considerable potential, but given the least attention, is ultraviolet spectroscopy (Friedel and Orchin, 1951; Braude and Nachod, 1955; Rao, 1961; Jaffe and Orchin, 1966; Lee et al., 1981; Bjorseth, 1983).

Typically, the ultraviolet spectrum of an asphaltene shows two major regions with very little fine structure. Interpretation of such a spectrum can only be made in general terms (Friedel, 1959; Boyd and Montgomery, 1963; Posadov et al., 1977). However, the technique can add valuable information about the degree of condensation of polycyclic aromatic ring systems through the auspices of high performance liquid chromatography (HPLC) (Lee et al., 1981; Bjorseth, 1983; Monin and Pelet, 1983; Felix et al., 1985; Killops and Readman, 1985; Speight, 1986). Indeed, when this approach is taken, the technique not only confirms the complex nature of the asphaltene fraction but also allows further detailed identifications to be made of the individual functional constituents of asphaltenes. An asphaltene, or asphaltene fraction, produces a multicomponent chromatogram (Chapter 10) but subfractions produce a less complex and much narrower chromatograph that may even approximate a single peak, which may prove much easier to monitor by the ultraviolet detector.

These data provide strong indications of the ring-size distribution of the polycyclic aromatic systems in petroleum asphaltenes. For example, from an examination of various functional subfractions, it was shown that amphoteric species and basic nitrogen species contain polycyclic aromatic systems having two-to-six rings per system. On the other hand, acid subfractions (phenolic or carboxylic functions) and neutral polar subfractions (amides or imino functions) contain few if any polycyclic aromatic systems having more than three rings per system (Chapter 10).

In all cases, the evidence favored the preponderance of the smaller (one-to-four) ring systems (Speight, 1986). But, perhaps what is more important about these investigations is that the data show that asphaltenes are a complex mixture of compound types, which confirms fractionation studies and cannot be represented by any particular formula that is construed to be average. Therefore, the concept of a large polycyclic aromatic ring system as the central feature of asphaltene molecules must be abandoned.

In summary, the premise is that petroleum is a natural product and that the aromatic systems are based on identifiable structural systems that are derived from natural product precursors.

7.3.1.4   Unsaturated Hydrocarbons

The presence of olefins ($R \cdot CH = CH \cdot R^1$) in petroleum has been under dispute for many years, because there are investigators who claim that olefins are actually present; in fact, these claims usually refer to distilled fractions, and it is very difficult to entirely avoid cracking during the distillation process. Nevertheless, evidence for the presence of considerable proportions of olefins in Pennsylvanian crude oils has been obtained; spectroscopic and chemical methods showed that the crude oils, as well as all distillate fractions, contained up to 3% w/w olefins. Hence, although the opinion that petroleum does not contain olefins requires some revision, it is perhaps reasonable to assume that the Pennsylvanian crude oils may hold an exceptional position and that olefins are present in crude oil in only a few special cases. The presence of dienes ($R \cdot CH = CH - CH = R'$) and acetylenes ($RC \equiv CR'$) is considered to be extremely unlikely.

In summary, a variety of hydrocarbon compounds occur throughout petroleum. Although the amount of any particular hydrocarbon varies from one crude oil to another, the family from which that hydrocarbon arises is well represented.

7.3.2   Nonhydrocarbon Constituents

The previous sections present some indication of the types and nomenclature of the organic hydrocarbons that occur in various crude oils. Thus, it is not surprising that petroleum, which contains only hydrocarbons, is in fact, an extremely complex mixture. The phenomenal increase in the number of possible isomers for the higher hydrocarbons makes it very difficult, if not impossible in most cases, to isolate individual members of any one series having more than, say, 12 carbon atoms.

Inclusion of organic compounds of nitrogen, oxygen, and sulfur serves only to present crude oil as an even more complex mixture than was originally conceived. Nevertheless, considerable progress has been made in the isolation or identification of the lower molecular weight hydrocarbons, as well as accurate estimations of the overall proportions of the hydrocarbon types present in petroleum. Indeed, it has been established that, as the boiling point of the petroleum fraction increases, not only the number of the constituents, but the molecular complexity of the constituents also increases (Figure 7.1) (Speight, 2000 and references cited therein).

Crude oils contain appreciable amounts of organic nonhydrocarbon constituents, mainly sulfur-, nitrogen-, and oxygen-containing compounds and, in smaller amounts, organometallic compounds in solution and inorganic salts in colloidal suspension. These constituents appear throughout the entire boiling range of the crude oil, but tend to concentrate mainly in the heavier fractions and in the nonvolatile residues.

Although their concentration in certain fractions may be quite small, their influence is important. For example, the decomposition of inorganic salts suspended in the crude can cause serious breakdowns in refinery operations; the thermal decomposition of deposited inorganic chlorides with evolution of free hydrochloric acid can give rise to serious corrosion problems in the distillation equipment. The presence of organic acid components, such as mercaptans and acids, can also promote metallic corrosion. In catalytic operations, passivation or poisoning of the catalyst can be caused by deposition of traces of metals (vanadium and nickel) or by chemisorption of nitrogen-containing compounds on the catalyst, thus necessitating the frequent regeneration of the catalyst or its expensive replacement.

The presence of traces of nonhydrocarbons may impart objectionable characteristics in finished products, such as discoloration, lack of stability on storage, or a reduction in the effectiveness of organic lead antiknock additives. It is thus obvious that a more extensive knowledge of these compounds and of their characteristics could result in improved refining methods and even in finished products of better quality. The nonhydrocarbon compounds, particularly the porphyrins and related compounds, are also of fundamental interest in the elucidation of the origin and nature of crude oils. Further, a knowledge of their surface-active characteristics is of help in understanding problems related to the migration of oil from the source rocks to the actual reservoirs.

7.3.2.1 Sulfur Compounds

Sulfur compounds are among the most important heteroatomic constituents of petroleum, and although there are many varieties of sulfur compounds (Table 7.2), the prevailing conditions during the formation, maturation, and even in situ alteration may dictate that only preferred types exist in any particular crude oil. Nevertheless, sulfur compounds of one type or another are present in all crude oils (Thompson et al., 1976). In general, the higher the density of the crude oil, the lower the API gravity of the crude oil and the higher the sulfur content; the total sulfur in the crude oil can vary from 0.04% w/w for light crude oil to about 5.0% for heavy crude oil and tar sand bitumen. However, the sulfur content of crude oils produced from broad geographic regions varies with time, depending on the composition of newly discovered fields, particularly those in different geological environments.

The presence of sulfur compounds in finished petroleum products often produces harmful effects. For example, in gasoline, sulfur compounds are believed to promote corrosion of engine parts, especially under winter conditions, when water containing sulfur dioxide from the combustion may accumulate in the crankcase. In addition, mercaptans in hydrocarbon solution cause the corrosion of copper and brass in the presence of air and also affect lead susceptibility and color stability. Free sulfur is also corrosive, as are sulfides, disulfides, and thiophenes, which are detrimental to the octane number response to tetraethyl lead. However, gasoline with a sulfur content between 0.2% and 0.5% has been used without obvious harmful effect. In diesel fuels, sulfur compounds increase wear and can contribute to the formation of engine deposits. Although a high sulfur content can sometimes be tolerated in industrial fuel

TABLE 7.2
Nomenclature and Types of Organic Sulfur Compounds

RSH   Thiols (Mercaptans)

RSR′   Sulfides

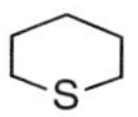

Cyclic Sulfides

RSSR′   Disulfides

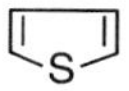

Thiophene

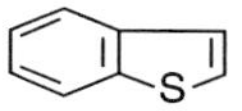

Benzothiophene

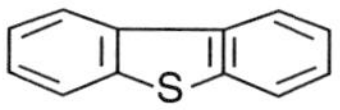

Dibenzothiophene

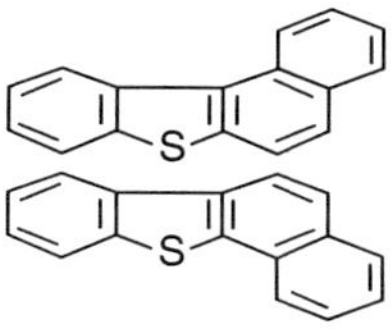

Naphtobenzothiophene

oils, the situation for lubricating oils is that a high content of sulfur compounds in lubricating oils seems to lower resistance to oxidation and increases the deposition of solids.

The distribution of sulfur in the various fractions of crude oils has been studied many times, beginning in 1891. Although it is generally true that the proportion of sulfur increases with the boiling point during distillation (Speight, 2000 and references cited therein), the middle fractions may actually contain more sulfur than higher boiling fractions as a result of decomposition of the higher molecular weight compounds during the distillation. High sulfur content is generally considered harmful in most petroleum products, and the removal of sulfur compounds or their conversion to less deleterious types is an important part of refinery practice. The distribution of the various types of sulfur compounds varies markedly among crude oils of diverse origin, but fortunately some of the sulfur compounds in petroleum undergo thermal reactions at relatively low temperatures. If elemental sulfur is present in the oil, a reaction, with the evolution of hydrogen sulfide, begins at about 150°C (300°F) and is very rapid at 220°C (430°F), but organically bound sulfur compounds do not yield hydrogen sulfide until higher temperatures are reached. Hydrogen sulfide is, however, a common constituent of many crude oils, and some crude oils with >1% sulfur are often accompanied by a gas having substantial properties of hydrogen sulfide.

Various thiophene derivatives have also been isolated from a variety of crude oils; benzothiophene derivatives are usually present in the higher boiling petroleum fractions. On the other hand, disulfides are not regarded as true constituents of crude oil but are generally formed by oxidation of thiols during processing:

$$2R-SH + [O] \rightarrow R-S-S-R + H_2O$$

Although sulfur is the most important (certainly the most abundant) heteroatom (i.e., non-hydrocarbon) present in petroleum with respect to the current context, other nonhydrocarbon atoms can exert a substantial influence, not only on the nature and properties of the products but also on the nature and efficiency of the process. Such atoms are nitrogen, oxygen, and metals, and because of their influence on the process, some discussion of each is warranted here.

7.3.2.2 Oxygen Compounds

Oxygen in organic compounds can occur in a variety of forms ($R–OH$, $Ar–OH$, $R–O–R'$, $R–CO_2H$, $AR–CO_2H$, $R–CO_2R$, $Ar–CO_2R$, $R_2C=O$ as well as the cyclic furan derivatives, where R and R′ are alkyl groups and Ar is an aromatic group) in nature, so it is not surprising that the more common oxygen-containing compounds occur in petroleum. The total oxygen content of crude oil is usually less than 2% w/w, although larger amounts have been reported, but when the oxygen content is phenomenally high it may be that the oil has suffered prolonged exposure to the atmosphere either during or after production. However, the oxygen content of petroleum increases with the boiling point of the fractions examined; in fact, the nonvolatile residua may have oxygen contents up to 8% w/w. Although these high molecular weight compounds contain most of the oxygen in petroleum, little is known concerning their structure, but those of lower molecular weight have been investigated with considerably more success and have been shown to contain carboxylic acids and phenols.

The presence of acid substances in petroleum first appears to have been reported in 1874, and it was established 9 years later that these substances contained carboxyl groups and were carboxylic acids. These were termed *naphthenic acids*. Although alicyclic (naphthenic) acids appear to be the more prevalent, it is now well known that aliphatic acids are also present. In addition to the carboxylic acids, alkaline extracts from petroleum contain phenols.

It has generally been concluded that the carboxylic acids in petroleum with fewer than eight carbon atoms per molecule are almost entirely aliphatic in nature; monocyclic acids begin at C_6 and predominate above C_{14}. This indicates that the structures of the carboxylic acids correspond with those of the hydrocarbons with which they are associated in the crude oil. In the range in which paraffins are the prevailing type of hydrocarbon, the aliphatic acids may be expected to predominate. Similarly, in the ranges in which monocycloparaffins and dicycloparaffins prevail, one may expect to find principally monocyclic and dicyclic acids, respectively.

In addition to the carboxylic acids and phenolic compounds, the presence of ketones, esters, ethers, and anhydrides has been claimed for a variety of crude oils. However, the precise identification of these compounds is difficult because most of them occur in the higher molecular weight nonvolatile residua. They are claimed to be products of the air blowing of the residua, and their existence in virgin petroleum may yet need to be substantiated.

Although comparisons are frequently made between the sulfur and nitrogen contents and such physical properties as the API gravity, it is not the same with the oxygen contents of crude oils. It is possible to postulate, and show, that such relationships exist. However, the ease with which some of the crude oil constituents can react with oxygen (aerial or dissolved) to incorporate oxygen functions into their molecular structure often renders the exercise somewhat futile, if meaningful deductions are to be made.

7.3.2.3 Nitrogen Compounds

Nitrogen in petroleum may be classified arbitrarily as basic and nonbasic. The basic nitrogen compounds (Table 7.3), which are composed mainly of pyridine homologs and occur throughout the boiling ranges, have a decided tendency to exist in the higher boiling fractions

TABLE 7.3
Nomenclature and Types of Organic Nitrogen Compounds

Nonbasic

Pyrrole	C_4H_5N	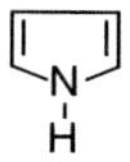
Indole	C_8H_7N	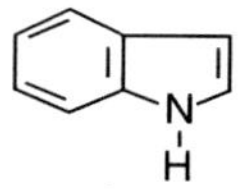
Carbazole	$C_{12}H_9N$	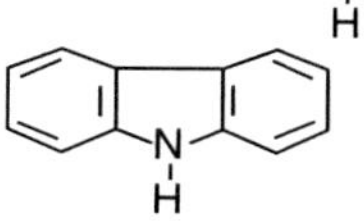
Benzo(a)carbazole	$C_{16}H_{11}N$	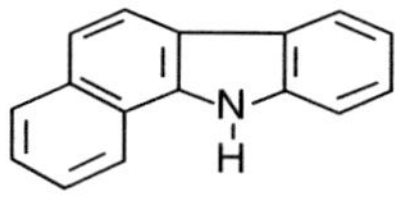 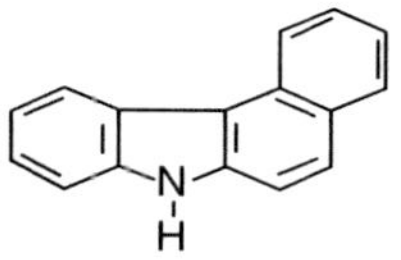

Basic

Pyridine	C_5H_5N	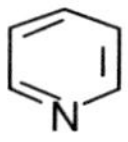
Quinoline	C_9H_7N	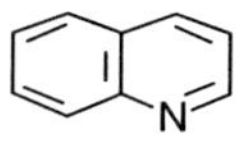
Indoline	C_8H_9N	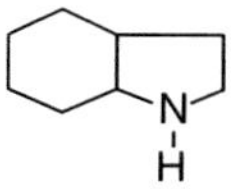
Benzo(f)quinoline	$C_{13}H_9N$	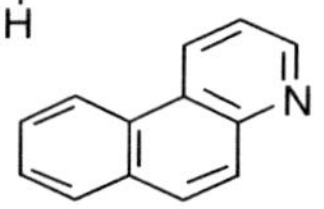

and residua. The nonbasic nitrogen compounds, which are usually of the pyrrole, indole, and carbazole types, also occur in the higher boiling fractions and residua.

In general, the nitrogen content of crude oil is low and generally falls within the range 0.1% to 0.9%, although early work indicates that some crude oil may contain up to 2% nitrogen. However, crude oils with no detectable nitrogen or even trace amounts are not uncommon, but in general the more asphaltic the oil, the higher its nitrogen content. Insofar as an approximate correlation exists between the sulfur content and API gravity of crude oils (Speight, 2000), there also exists a correlation between nitrogen content and the API gravity of crude oil. It also follows that there is an approximate correlation between the nitrogen content and the carbon residue: the higher the nitrogen content, the higher the carbon residue. The presence of nitrogen in petroleum is of much greater significance in refinery operations than might be expected from the small amounts present. Nitrogen compounds can be responsible for the poisoning of cracking catalysts, and they also contribute to gum formation in such products as domestic fuel oil. The trend in recent years toward cutting deeper into the crude to obtain stocks for catalytic cracking has accentuated the harmful effects of the nitrogen compounds, which are concentrated largely in the higher boiling portions.

Basic nitrogen compounds with a relatively low molecular weight can be extracted with dilute mineral acids; equally strong bases of higher molecular weight remain unextracted because of unfavorable partitioning between the oil and aqueous phases. A method has been developed in which the nitrogen compounds are classified as basic or nonbasic, depending on whether they can be titrated with perchloric acid in a 50:50 solution of glacial acetic acid and benzene. Application of this method has shown that the ratio of basic to total nitrogen is approximately constant (0 to 30 $\pm$ 0.05), irrespective of the source of the crude. Indeed, the ratio of basic to total nitrogen was found to be approximately constant throughout the entire range of distillate and residual fractions. Nitrogen compounds extractable with dilute mineral acids from petroleum distillates were found to consist of alkyl pyridine derivatives, alkyl quinoline derivatives, and alkyl isoquinoline derivatives carrying alkyl substituents, as well as pyridine derivatives in which the substituent was a cyclopentyl or cyclohexyl group. The compounds that cannot be extracted with dilute mineral acids contain the greater part of the nitrogen in petroleum and are generally of the carbazole, indole, and pyrrole types.

7.3.2.4 Metallic Constituents

Metallic constituents are found in every crude oil and the concentrations have to be reduced to convert the oil to transportation fuel. Metals affect many upgrading processes and cause particular problems because they poison catalysts used for sulfur and nitrogen removal as well as other processes such as catalytic cracking.

The trace metals Ni and V are generally orders of magnitude higher than other metals in petroleum, except when contaminated with coproduced brine salts (Na, Mg, Ca, and Cl) or corrosion products gathered in transportation (Fe).

The occurrence of metallic constituents in crude oil is of considerably greater interest to the petroleum industry than might be expected from the very small amounts present. Even minute amounts of iron, copper, and particularly nickel and vanadium in the charging stocks for catalytic cracking affect the activity of the catalyst and result in increased gas and coke formation and reduced yields of gasoline. In high-temperature power generators, such as oil-fired gas turbines, the presence of metallic constituents, particularly vanadium in the fuel, may lead to ash deposits on the turbine rotors, thus reducing clearances and disturbing their balance. More particularly, damage by corrosion may be very severe. The ash resulting from the combustion of fuels containing sodium and especially vanadium reacts with refractory furnace linings to lower their fusion points and so cause their deterioration.

Thus, the ash residue left after burning of a crude oil is due to the presence of these metallic constituents, part of which occur as inorganic water-soluble salts (mainly chlorides and sulfates of sodium, potassium, magnesium, and calcium) in the water phase of crude oil emulsions. These are removed in the desalting operations, either by evaporation of the water and subsequent water washing or by breaking the emulsion, thereby causing the original mineral content of the crude to be substantially reduced. Other metals are present in the form of oil-soluble organometallic compounds as complexes, metallic soaps, or in the form of colloidal suspensions, and the total ash from desalted crude oils is of the order of 0.1 to 100 mg/L. Metals are generally found only in the nonvolatile portion of crude oil (Altgelt and Boduszynski, 1994; Reynolds, 1998).

Two groups of elements appear in significant concentrations in the original crude oil associated with well-defined types of compounds. Zinc, titanium, calcium, and magnesium appear in the form of organometallic soaps with surface-active properties adsorbed in the water or oil interfaces and act as emulsion stabilizers. However, vanadium, copper, nickel, and part of the iron found in crude oils seem to be in a different class and are present as oil-soluble compounds. These metals are capable of complexing with pyrrole pigment compounds derived from chlorophyll and hemoglobin and are almost certain to have been present in plant and animal source materials. It is easy to surmise that the metals in question are present in such form, ending in the ash content. Evidence for the presence of several other metals in oil-soluble form has been produced, and thus zinc, titanium, calcium, and magnesium compounds have been identified in addition to vanadium, nickel, iron, and copper. Examination of the analyses of a number of crude oil samples for iron, nickel, vanadium, and copper indicates a relatively high vanadium content, which usually exceeds that of nickel, although the reverse can also occur.

Distillation concentrates the metallic constituents in the residues (Reynolds, 1998), although some can appear in the higher boiling distillates, but the latter may be due in part to entrainment. Nevertheless, there is evidence that a portion of the metallic constituents may occur in the distillates by volatilization of the organometallic compounds present in the petroleum. In fact, as the percentage of overhead obtained by vacuum distillation of a reduced crude is increased, the amount of metallic constituents in the overhead oil is also increased. The majority of the vanadium, nickel, iron, and copper in residual stocks may be precipitated along with the asphaltenes by hydrocarbon solvents. Thus, removal of the asphaltenes with *n*-pentane reduces the vanadium content of the oil by up to 95% with substantial reductions in the amounts of iron and nickel.

7.3.2.5 Porphyrins

Porphyrins are a naturally occurring chemical species that exist in petroleum and usually occur in the nonbasic portion of the nitrogen-containing concentrate (Bonnett, 1978; Reynolds, 1998). They are not usually considered among the usual nitrogen-containing constituents of petroleum, nor are they considered a metallo-containing organic material that also occurs in some crude oils. As a result of these early investigations, there arose the concept of porphyrins as biomarkers that could establish a link between compounds found in the geosphere and their corresponding biological precursors (Treibs, 1934; Glebovskaya and Volkenshtein, 1948).

Porphyrins are derivatives of porphine [that consists of four pyrrole molecules joined by methine (–CH=) bridges] (Figure 7.3). The methine bridges establish conjugated linkages between the component pyrrole nuclei, forming a more extended resonance system. Although the resulting structure retains much of the inherent character of the pyrrole components, the larger conjugated system gives increased aromatic character to the porphine molecule (Falk, 1964; Smith, 1975). Further, the imine functions (–NH–) in the porphine system allow metals such as nickel to be included into the molecule through chelation (Figure 7.4).

A large number of different porphyrin compounds exist in nature or have been synthesized. Most of these compounds have substituents other than hydrogen on many of the ring carbons. The nature of the substituents on porphyrin rings determines the classification of a specific porphyrin compound into one of various types according to one common system of nomenclature (Bonnett, 1978). Porphyrins also have well-known trivial names or acronyms that are often in more common usage than the formal system of nomenclature.

When one or two double bonds of a porphyrin are hydrogenated, a chlorin or a phlorin is the result. Chlorins are components of chlorophylls and possess an isocyclic ring formed by two methylene groups bridging a pyrrolic carbon to a methine carbon. Geological porphyrins that contain this structural feature are assumed to be derived from chlorophylls.

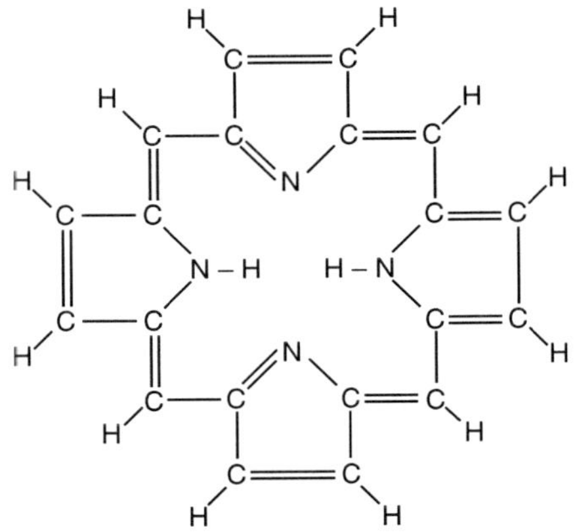

FIGURE 7.3 Porphine—the basic structure of porphyrins.

Etioporphyrins are also commonly found in geological materials and have no substituents (other than hydrogen) on the methine carbons. Benzoporphyrins and tetrahydrobenzoporphyrins also have been identified in geological materials. These compounds have either a benzene ring or a hydrogenated benzene ring fused onto a pyrrole unit.

Almost all crude oil, heavy oil, and bitumen contain detectable amounts of vanadyl and nickel porphyrins. More mature, lighter crude oils usually contain only small amounts of these compounds. Heavy oils may contain large amounts of vanadyl and nickel porphyrins. Vanadium concentrations of over 1000 ppm are known for some crude oil, and a substantial amount of the vanadium in these crude oils is chelated with porphyrins. In high-sulfur crude oil of marine origin, vanadyl porphyrins are more abundant than nickel porphyrins. Low-sulfur crude oils of lacustrine origin usually contain more nickel porphyrins than vanadyl porphyrins.

Of all the metals in the periodic table, only vanadium and nickel have been proven definitely to exist as chelates in significant amounts in a large number of crude oils and tar sand bitumen. The existence of iron porphyrins in some crude oil has been claimed (Franceskin et al., 1986). Geochemical reasons for the absence of substantial quantities of porphyrins

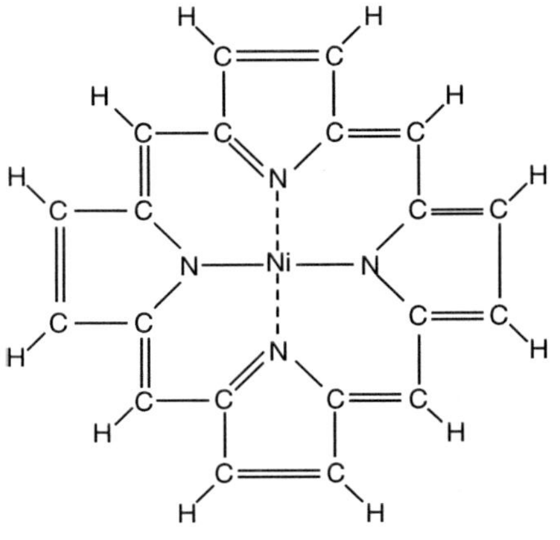

FIGURE 7.4 Nickel chelate of porphine.

chelated with metals other than nickel and vanadium in most crude oils and tar sand bitumen have been advanced (Hodgson et al., 1967; Baker, 1969; Baker and Palmer, 1978; Baker and Louda, 1986; Filby and Van Berkel, 1987; Quirke, 1987).

If the vanadium and nickel contents of crude oils are measured and compared with porphyrin concentrations, it is usually found that not all the metal content can be accounted for as porphyrins (Dunning et al., 1960; Reynolds, 1998). In some crude oils, as little as 10% w/w of total metals appears to be chelated with porphyrins. Only rarely can all measured nickel and vanadium in a crude oil be accounted for as porphyrinic (Erdman and Harju, 1963). Currently, some investigators believe that part of the vanadium and nickel in crude oils is chelated with ligands that are not porphyrins. These metal chelates are referred to as nonporphyrin metal chelates or complexes (Crouch et al., 1983; Fish et al., 1984; Reynolds et al., 1987).

On the other hand, the issue of nonporphyrin metal in crude oils has been questioned (Goulon et al., 1984). There is the possibility that in such systems as the heavy crude oils, in which intermolecular associations are important, measurement of the porphyrin concentrations are unreliable and there is a tendency to undercount the actual values. However, for the purposes of this chapter it is assumed that nonporphyrin chelates exist in fossil fuels, but the relative amount is as yet unknown.

Finally, during the fractionation of petroleum (Chapter 9), the metallic constituents (metalloporphyrins and nonporphyrin metal chelates) are concentrated in the asphaltene fraction. The deasphaltened oils (petrolenes and maltenes) (Chapter 1) contain smaller concentrations of porphyrins than the parent materials and usually very small concentrations of nonporphyrin metals.

7.4 CHEMICAL COMPOSITION BY DISTILLATION

Although distillation is presented in more detail elsewhere (Chapter 9 and Chapter 16), it is appropriate to mention distillation here insofar as it is a method by which the constituents of petroleum can be separated and identified.

Distillation is a means of separating chemical compounds (usually liquids) through differences in their vapor pressures (Chapter 8). In the mixture, the components evaporate, such that the vapor has a composition determined by the chemical properties of the mixture. Distillation of a given component is possible, if the vapor has a higher proportion of the given component than the mixture. This is caused by the given component having a higher vapor pressure and a lower boiling point than the other components.

By the nature of the process, it is theoretically impossible to completely separate and purify the individual components of petroleum when the possible number of isomers are considered for the individual carbon numbers that occur within the paraffin family (Table 7.4). When other types of compounds are included, such as the aromatic derivatives and heteroatom derivatives, even though the maturation process might limit the possible number of isomeric permutations (Tissot and Welte, 1978), the potential number of compounds in petroleum is still (in a sense) astronomical.

However, petroleum can be separated into a variety of fractions on the basis of the boiling points of the petroleum constituents. Such fractions are primarily identified by their respective boiling ranges and, to a lesser extent, by chemical composition. However, it is often obvious that as the boiling ranges increase, the nature of the constituents remains closely similar and it is the carbon number of the substituents that caused the increase in boiling point. It is through the recognition of such phenomena that molecular design of the higher boiling constituents can be achieved (Figure 7.5). Invoking the existence of structurally

TABLE 7.4
Boiling Point of the *n*-Isomers of the Various Paraffins and the Number
of Possible Isomers for That Particular Carbon Number

Number of Carbon Atoms	*n*-Isomer Boiling Point (°C)	(°F)	Number of Isomers
5	36	97	3
10	174	345	75
15	271	519	4,347
20	344	651	366,319
25	402	755	36,797,588
30	450	841	4,111,846,763
40	525	977	62,491,178,805,831

different constituents in the nonvolatile fractions from those identifiable constituents in the lower boiling fractions is unnecessary and (considering the nature of the precursors and maturation paths, Chapter 3) is unnecessary and irrational (Speight, 1994b). For example, the predominant types of condensed aromatic systems in petroleum are derivatives of phenanthrene and it is to be anticipated that the higher peri-condensed homologs (Figure 7.6) shall be present in resin constituents and asphaltene constituents rather than the derivatives of a kata-condensed polynuclear aromatic system (Figure 7.7).

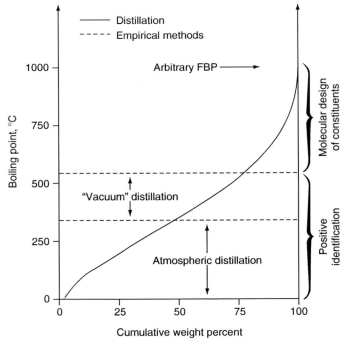

FIGURE 7.5 Separation of petroleum constituents by distillation.

Chrysene

Picene

1,2,5,6-Dibenzanthracene

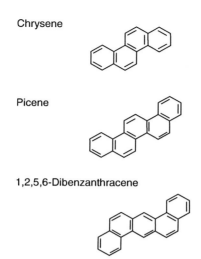

FIGURE 7.6 Examples of peri-condensed aromatic systems.

7.4.1 GASES AND NAPHTHA

Methane is the main hydrocarbon component of petroleum gases with lesser amounts of ethane, propane, butane, isobutane, and some C_4^+ hydrocarbons. Other gases, such as hydrogen, carbon dioxide, hydrogen sulfide, and carbonyl sulfide, are also present.

Saturated constituents with lesser amounts of mono- and di-aromatics dominate the naphtha fraction. Whereas naphtha covers the boiling range of gasoline, most of the raw petroleum naphtha molecules have low octane number. However, most raw naphtha is processed further and combined with other process naphtha and additives to formulate commercial gasoline.

Within the saturated constituents in petroleum gases and naphtha, every possible paraffin from methane (CH_4) to n-decane (n-$C_{10}H_{22}$), is present. Depending upon the source, one of these low boiling paraffins may be the most abundant compound in a crude oil reaching several percent. The iso-paraffins begin at C4 with isobutane as the only isomer of n-butane. The number of isomers grows rapidly with carbon number and there may be increased difficulty in dealing with multiple isomers during analysis.

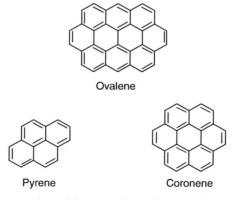

Ovalene

Pyrene   Coronene

FIGURE 7.7 Examples of kata-condensed (cata-condensed) aromatic systems.

In addition to aliphatic molecules, the saturated constituents consist of cycloalkanes (naphthenes) with predominantly five- or six-carbon rings. Methyl derivatives of cyclopentane and cyclohexane, which are commonly found at higher levels than the parent unsubstituted structures may be present (Tissot and Welte, 1978). Fused ring dicycloalkanes such as *cis*-decahydronaphthalene (*cis*-decalin) and *trans*-decahydronaphthalene (*trans*-decalin) and hexahydroindan are also common, but bicylic naphthenes separated by a single bond, such as cyclohexyl cyclohexane, are not.

The numerous aromatic constituents in petroleum naphtha begin with benzene, but its C_1 to C_3 alkylated derivatives are also present (Tissot and Welte, 1978). Each of the alkyl benzene homologs through the 20 isomeric C4 alkyl benzenes have been isolated from crude oil along with various C_5-derivatives (Mair, 1964). Benzene derivatives having fused cycloparaffin rings (naphtheno-aromatics), such as indane and tetralin, have been isolated along with a number of their methyl derivatives. Naphthalene is included in this fraction, while the 1- and 2-methyl naphthalenes and higher homologs of fused two-ring aromatics appear in the mid-distillate fraction.

Sulfur-containing compounds are the only heteroatom compounds to be found in this fraction (Mair, 1964; Rall et al., 1972). Usually, the total amount of the sulfur in this fraction accounts for less than 1% of the total sulfur in the crude oil. In naphtha from high-sulfur (sour) petroleum, 50% to 70% of the sulfur may be in the form of mercaptans (thiols). Over 40 individual thiols have been identified, including all the isomeric C_1 to ~6 compounds plus some C_7 and C_8 isomers plus thiophenol (Rall et al., 1972). In naphtha from low-sulfur (sweet) crude oil, the sulfur is distributed between sulfides (thioethers) and thiophenes. In these cases, the sulfides may be in the form of both linear (alkyl sulfides) and five- or six-ring cyclic (thiacyclane) structures. The sulfur structure distribution tends to follow the distribution of hydrocarbons; i.e., naphthenic oils with a high cyclocyloalkane content tend to have a high thiacyclane content. Typical alkyl thiophene derivatives in naphtha have multiple short side chains or exist as naphthenothiophenes (Rall et al., 1972). Methyl and ethyl disulfides have been confirmed to be present in some crude oils in analyses that minimized their possible formation by oxidative coupling of thiols (Aksenov and Kanayanov, 1980; Freidlina and Skorova, 1980).

7.4.2 MIDDLE DISTILLATES

Saturated species are the major component in the mid-distillate fraction of petroleum but aromatics, which now include simple compounds with up to three aromatic rings, and heterocyclic compounds are present and represent a larger portion of the total. Kerosene, jet fuel, and diesel fuel are all derived from raw middle distillate, which can also be obtained from cracked and hydroprocessed refinery streams.

Within the saturated constituents, the concentration of *n*-paraffins decreases regularly from C_{11} to C_{20}. Two isoprenoid species (pristane = 2,6,10,14-tetramethylpentadecane and phytane = 2,6,10,14-tetramethylhexadecane) are generally present in crude oils in sufficient concentration to be seen as irregular peaks alongside the *n*-C_{17} and *n*-C_{18} peaks in a gas chromatogram. These isoprene derivatives, believed to arise as fragments of ancient precursors, have relevance as simple biomarkers, to the genesis of petroleum. The distribution of pristane and phytane, relative to their neighboring *n*-C_{17} and *n*-C_{18} peaks, has been used to aid the identification of crude oils and to detect the onset of biodegradation. The ratio of pristane to phytane has also been used for the assessment of the oxidation and reduction environment in which ancient organisms were converted into petroleum (Hunt, 1979).

Mono- and di-cycloparaffins with five or six carbons per ring constitute the bulk of the naphthenes in the middle distillate boiling range, decreasing in concentration as the carbon number increases (Tissot and Welte, 1978) and the alkylated naphthenes may have a single long side chain as well as one or more methyl or ethyl groups (Hood et al., 1959). Similarly substituted three-ring naphthenes have been detected by gas chromatography and adamantane has been found in crude oil (Lanka and Hala, 1958; Hunt, 1979; Lee et al., 1981; Richardson and Miller, 1982).

The most abundant aromatics in the mid-distillate boiling fractions are di- and tri-methyl naphthalenes. Other one- and two-ring aromatics are undoubtedly present in small quantities as either naphtheno or alkyl homologs in the C_{11}–C_{20} range. In addition to these homologs of alkylbenzenes, tetralin, and naphthalenes, the mid-distillate contains some fluorene derivatives and phenanthrene derivative (Lee et al., 1990). The phenanthrene structure appears to be favored over that of the anthracene structure (Tissot and Welte, 1978) and this appears to continue through the higher boiling fractions of petroleum (Speight, 1994b).

The five-membered heterocyclic constituents in the mid-distillate range are primarily the thiacyclane derivatives, benzothiophene derivatives, and dibenzothiophene derivatives with lesser amounts dialkyl-, diaryl-, and aryl-alkyl sulfides (Aksenov and Kanayanov, 1980; Freidlina and Skorova, 1980). Alkylthiophenes are also present. As with the naphtha fractions, these sulfur species account for a minimal fraction of the total sulfur in the crude.

Although only trace amounts (usually ppm levels) of nitrogen are found in the middle distillate fractions, both neutral and basic nitrogen compounds have been isolated and identified in fractions boiling below 343°C (650°F) (Hirsch et al., 1974). Pyrrole derivatives and indole derivatives account for about two-thirds of the nitrogen, whereas the remainder is found in the basic alkylated pyridine and alkylated quinoline compounds.

The saturate constituents contribute less to the vacuum gas oil (VGO) than the aromatic constituents, but more than the polars that are now present at percentage rather than trace levels. Vacuum gas oil is occasionally used as a heating oil, but most commonly it is processed by catalytic cracking to produce naphtha or extraction to yield lubricating oil.

Within the vacuum gas oil, saturates, distribution of paraffins, *iso*-paraffins and naphthenes is highly dependent upon the petroleum source. Generally, the naphthene constituents account for approximately two-thirds or 60% of the saturate constituents, but the overall range of variation is from <20% to >80%. In most samples, the *n*-paraffins from C_{20}–C_{44} are still present in sufficient quantity to be detected as distinct peaks in gas chromatographic analysis. Some (but not all) crude oils show a preference for odd-numbers alkanes. Both the distribution and the selectivity toward odd-numbered hydrocarbons are considered to reflect differences in the petrogenesis of the crude oil.

The bulk of the saturated constituents in vacuum gas oil consist of *iso*-paraffins and especially naphthene species although isoprenoid compounds, such as squalane (C_{30}) and lycopane (C_{40}), have been detected. Analytical techniques show that the naphthenes contain from one to more than six fused rings accompanied by alkyl substitution. For mono- and di-aromatics, the alkyl substitution typically involves several methyl and ethyl substituents. Hopanes and steranes have also been identified and are also used as internal markers for estimating biodegradation of crude oils during bioremediation processes (Prince et al., 1994).

The aromatics in vacuum gas oil may contain one to six fused aromatic rings that may bear additional naphthene rings and alkyl substituents in keeping with their boiling range. Mono- and di-aromatics account for about 50% of the aromatics in petroleum vacuum gas oil samples. Analytical data show the presence of up to four fused naphthenic rings on some aromatic compounds. This is consistent with the suggestion that these species originate from the aromatization of steroids. Although present at lower concentration, alkyl benzenes and naphthalenes show one long side chain and multiple short side chains.

The fused ring aromatic compounds (having three or more rings) in petroleum include phenanthrene, chrysene, and picene as well as fluoranthene, pyrene, benzo(a)pyrene, and benzo(ghi)perylene.

The most abundant reported individual phenanthrene compounds appear to be the 3-derivatives. In addition, phenanthrene derivatives outnumber anthracene derivatives by as much as 100:1. In addition, chrysene derivatives are favored over pyrene derivative.

Heterocyclic constituents are significant contributors to the vacuum gas oil fraction. In terms of sulfur compounds, thiophene and thiacyclane sulfur predominate over sulfide sulfur. Some molecules even contain more than one sulfur atom. The benzothiophenes and dibenzothiophenes are the prevalent thiophene forms of sulfur.

In the vacuum gas oil range, the nitrogen-containing compounds include higher molecular weight pyridines, quinolines, benzoquinoline derivatives, amides, indoles, carbazole, and molecules with two nitrogen atoms (diaza compounds) whereas three and four aromatic rings are especially prevalent (Green et al., 1989). Typically, about one-third of the compounds are basic, i.e., pyridine and its benzologs whereas the reminder are present as neutral species (amides and carbazoles). Although benzo- and dibenzo-quinolines found in petroleum are rich in sterically hindered structures, hindered and unhindered structures have been found to be present at equivalent concentrations in source rocks. This has been rationalized as "geochromatography" in which the less polar (hindered) structures moved more readily to the reservoir (Yamamoro, 1993).

Oxygen levels in the vacuum gas oil parallel the nitrogen content. Thus, the most commonly identified oxygen compounds are the carboxylic acids and phenols, collectively called naphthenic acids (Seifert and Teeter, 1970).

7.4.3 Vacuum Residua (1050°F$^+$)

This fraction, the vacuum bottoms is the most complex of petroleum. Vacuum residua contain the majority of the heteroatoms originally in the petroleum and molecular weight of the constituents range, as near as can be determined subject to method of dependence, up to several thousand. The fraction is so complex that the characterization of individual species is virtually impossible, no matter what claims have been made or will be made. Separation of vacuum residua by group type becomes difficult and confused because of the multi-substitution of aromatic and naphthenic species, as well as by the presence of multiple functionalities in single molecules.

Classically, n-pentane or n-heptane precipitation is used as the initial step for the characterization of vacuum residuum. The insoluble fraction, the pentane- or heptane-asphaltenes, may be as much as 50% by weight of a vacuum residuum. The pentane- or heptane-soluble portion (maltenes) of the residuum is then fractionated chromatographically into several solubility or adsorption classes for characterization. However, in spite of claims to the contrary, the method is not a separation by chemical type. Kit is a separation by solubility and adsorption.

The separation of the asphaltene constituents does, however, provide a simple way to remove some of the highest molecular weight and most polar components, but the asphaltene fraction is so complex that compositional detail based on average parameters is of questionable value.

The use of ion exchange chromatography (McKay et al., 1976; Green et al., 1989) has offered some indication of chemical types within the complex high molecular weight fractions.

For the 565°C$^+$ (l050°F$^+$) fractions of petroleum, the levels of nitrogen and oxygen may begin to approach the concentration of sulfur. These elements consistently concentrate in the most polar fractions to the extent that every molecule contains more than one heteroatom. At this point, structural identification is somewhat fruitless and characterization techniques are used to confirm the presence of the functionalities found in lower boiling fractions such as, for example, acids, phenols, nonbasic (carbazole-type) nitrogen, and basic (quinoline-type) nitrogen.

Several models have been proposed based on the observed functionalities, apparent molecular weight, and elemental analysis of the fraction, but whether or not these models offer insights into the nature and behavior of the asphaltene constituents remains open to speculation and question (Speight, 1994b).

The nickel and vanadium that are concentrated into the vacuum residuum appear to occur in two forms: (1) porphyrins and (2) nonporphyrins (Reynolds, 1998). Because the metallo-porphyrins can provide insights into petroleum maturation processes, they have been studied extensively and several families of related structures have been identified. On the other hand, the nonporphyrin metals remain not clearly identified although some studies suggest that some of the metals in these compounds still exist in a tetrapyrrole (porphyrin-type) environment (Fish and Komlenic, 1984; Pearson and Green, 1993).

It is more than likely that, in a specific residuum molecule, the heteroatoms are arranged in different functionalities, making an incredibly complex molecule. Considering how many different combinations are possible, the chances of determining every structure in a residuum are very low. Because of this seemingly insurmountable task, it may be better to determine ways of utilizing the residuum rather attempting to determine (at best questionable) molecular structures.

REFERENCES

Aksenov, V.S. and Kanayanov, V.F. 1980. *Regularities in Composition and Structures of Native Sulfur Compounds from Petroleum.* Proceedings. 9th International Symposium on Organic Sulfur Chemistry, Riga, USSR, June 9–14.

Altgelt, K.H. and Boduszynski, M.M. 1994. *Compositional Analysis of Heavy Petroleum Fractions.* Marcel Dekker, New York.

Baker, E.W. 1969. *Organic Geochemistry.* Eglinton, G. and Murphy, M.T.J. eds. Springer-Verlag, New York.

Baker, E.W. and Palmer, S.E. 1978. *The Porphyrins. Volume I: Structure and Synthesis.* Part Dolphin, A.D. ed. Academic Press, New York.

Baker, E.W. and Louda, J.W. 1986. *Biological Markers in the Sedimentary Record.* Johns, R.B. ed. Elsevier, Amsterdam.

Bestougeff, M.A. 1961. *J. Etud. Methodes Sep. Immed. Chromatogr.* Comptes Rendus (Paris). p. 55.

Bjorseth, A. 1983. *Handbook of Polycyclic Aromatic Hydrocarbons.* Marcel Dekker, New York.

Bonnett, R. 1978. In *The Porphyrins. Volume I: Structure and Synthesis.* Part Dolphin, A.D. ed. Academic Press, New York.

Boyd, M.L. and Montgomery, D.S. 1963. *J. Inst. Petrol.* 49: 345.

Brandenburg, C.R. and Latham, D.R. 1968. *J. Chem. Eng. Data* 13: 391.

Braude, E.A. and Nachod, F.C. 1955. *Determination of Organic Structures by Physical Methods.* Academic Press, New York.

Brooks, B.T., Kurtz, S.S., Jr., Boord, C.E., and Schmerling, L. 1954. *The Chemistry of Petroleum Hydrocarbons.* Reinhold, New York.

Brooks, J. and Welte, D.H. 1984. *Advances in Petroleum Geochemistry.* Academic Press, New York.

Bunger, J.W., Thomas, K.P., and Dorrence, S.M. 1979. *Fuel* 58: 183.

Camp, F.W. 1976. *The Tar Sands of Alberta.* Cameron Engineers, Denver, CO.

Charbonnier, R.P., Draper, R.G., Harper, W.H., and Yates, A. 1969. Analyses and Characteristics of Oil Samples from Alberta. *Information Circular IC 232.* Department of Energy Mines and Resources, Mines Branch, Ottawa, Canada.

Corbett, L.W. and Petrossi, U. 1978. *Ind. Eng. Chem. Prod. Res. Dev.* 17: 342.

Crouch, F.W., Sommer, C.S., Galobardes, J.F., Kraus, S., Schmauch, E.M., Galobardes, M., Fatmi, A., Pearsall, K., and Rogers, L.B. 1983. *Sep. Sci. Technol.* 18: 603.

Draper, R.G., Kowalchuk, E., and Noel, G. 1977. Analyses and Characteristics of Crude Oil Samples Performed Between 1969 and 1976. *Report ERP/ERL 77-59 (TR).* Energy, Mines, and Resources, Ottawa, Canada.

Dunning, H.N., Moore, J.W., Bieber, H., and Williams, R.B. 1960. *J. Chem. Eng. Data*. 5: 547.

Durand, B. ed. 1980. *Kerogen: Insoluble Organic Matter from Sedimentary Rocks*. Editions Technip, Paris, France.

Eglinton, G. and Murphy, B. 1969. *Organic Geochemistry: Methods and Results*. Springer-Verlag, New York.

Erdman, J.G. and Harju, P.H. 1963. *J. Chem. Eng. Data* 8: 252.

Falk, J.E. 1964. *Porphyrins and Metalloporphyrins*. Elsevier, New York.

Felix, G., Bertrand, C., and Van Gastel, F. 1985. *Chromatographia* 20: 155.

Filby, R.H. and Van Berkel, G.J. 1987. *Metal Complexes in Fossil Fuels*. Filby, R.H. and Branthaver, J.F. eds. Symposium Series No. 344. American Chemical Society, Washington, DC. p. 2.

Fish, R.H. and Komlenic, J. 1984. *Anal. Chem.* 56: 510.

Fish, R.H., Konlenic, J.J., and Wines, B.K. 1984. *Anal. Chem.* 56: 2452.

Franceskin, P.J., Gonzalez-Jiminez, M.G., DaRosa, F., Adams, O., and Katan, L. 1986. *Hyperfine Interact*. 28: 825.

Freidlina, I.K. and Skorova, A.E. eds. 1980. *Organic Sulfur Chemistry*. Pergamon Press, New York.

Friedel, R.A. 1959. *J. Chem. Phys.* 31: 280.

Friedel, R.A. and Orchin, M. 1951. *Ultraviolet Spectra of Aromatic Compounds*. John Wiley & Sons Inc., New York.

Gallegos, E.J.J. 1981. *Chromatogr. Sci.* 19: 177.

Glebovskaya, F.A. and Volkenshtein, M.V. 1948. *J. Gen. Chem. USSR*. 18: 1440.

Goulon, J., Retournard, A., Frient, P., Goulon-Ginet, C., Berthe, C.K., Muller, J.R., Poncet, J.L., Guilard, R., Escalier, J.C., and Neff, B. 1984. *J. Chem. Soc. Dalton Trans*. 1095.

Green, J.A., Green, J.B., Grigsby, R.D., Pearson, C.D., Reynolds, J.W., Sbay, I.Y., Sturm, O.P., Jr., Thomson, J.S., Vogh, J.W., Vrana, R.P., Yu, S.K.Y., Diem, B.H., Grizzle, P.L., Hirsch, D.E., Hornung, K.W., Tang, S.Y., Carbongnani, L., Hazos, M., and Sanchez, V. 1989. *Analysis of Heavy Oils: Method Development and Application to Cerro Negro Heavy Petroleum*. NIPER-452 (DE90000200), Volumes I and II. Research Institute, National Institute for Petroleum and Energy Research (NIPER), Bartlesville, OK.

Gruse, W.A. and Stevens, D.R. 1960. *The Chemical Technology of Petroleum*. McGraw-Hill, New York.

Hirsch, D.E., Cooley, J.E., Coleman, H.J., and Thompson, C.J. 1974. *Qualitative Characterization of Aromatic Concentrates of Crude Oils from GPC Analysis*. Report 7974. Bureau of Mines, U.S. Department of the Interior, Washington, DC.

Hodgson, G.W., Baker, B.L., and Peake, E. 1967. *Fundamental Aspects of Petroleum Geochemistry*. Nagy, B. and Columbo, U. eds. Elsevier, Amsterdam. Chapter 5.

Hood, R.I., Clerc, R.J., and O'Neal, M.I. 1959. *J. Inst. Petrol*. 45: 168.

Jaffe, H.H. and Orchin, M. 1966. *Theory and Applications of Ultraviolet Spectroscopy*. John Wiley & Sons Inc., New York.

Killops, S.D. and Readman, J.W. 1985. *Org. Geochem*. 8: 247.

Lanka, S.Y. and Hala, S.I. 1958. *Erdol Kohle* 11: 698.

Lee, M.L., Novotny, M.S., and Bartle, K.D. 1981. *Analytical Chemistry of Polycyclic Aromatic Compounds*. Academic Press, New York.

Lee, S.W., Coulombe, S., and Glavinccvski, B. 1990. *Energy Fuels* 4: 20.

Lochte, H.L. and Littmann, E.R. 1955. *The Petroleum Acids and Bases*. Chemical Publishing Co., New York.

Long, R.B. 1979. Preprints. *Div. Petrol. Chem. Am. Chem. Soc*. 24(4): 891.

Long, R.B. 1981. *The Chemistry of Asphaltenes*. Bunger, J.W. and Li, N. eds. Advances in Chemistry Series No. 195. American Chemical Society, Washington, DC.

Mair, B.J. 1964. *Oil Gas J*. (September. 14): 130.

McKay, J.F., Weber, J.H., and Latham, D.R. 1976. *Anal. Chem.* 48: 891.

Meyer, R.F. and Steele, C.F, ed. 1981. *The Future of Heavy Crude and Tar Sands*. McGraw-Hill, New York.

Monin, J.C. and Pelet, R. 1983. *Advances in Organic Geochemistry*. Bjoroey, M. ed. John Wiley & Sons Inc., New York.

Pearson, C.D. and Green, J.B. 1993. *Energy Fuels* 7: 338.

Posadov, I.A., Pokonova, J.V., Khusidman, M.B., Gitlin, I.G., and Proskuryakov, V.A. 1977. *Zhur. Priklad. Khim.* 50: 594.

Prince, R.O., Elmendoff, D.L., Lute, B.R., Hsu, C.S., Hath, CUE., Sunnis. B.P., Decherd, G.H.I., Douglas, DO'S., and Butler, EEL. 1994. *Environ. Sci. Technol.* 28: 142.

Quirke, J.M.E. 1987. *Metal Complexes in Fossil Fuels.* Filby, R.H. and Branthaver, J.F. eds. Symposium Series No. 344. American Chemical Society, Washington, DC. p. 74.

Rall, H.T., Thompson, C.J., Coleman. H.J., and Hopkins, R.L. 1972. Bulletin 659. Bureau of Mines, U.S. Department of the Interior, Washington, DC.

Rao, C.N.R. 1961. *Ultraviolet and Visible Spectroscopy: Chemical Applications.* Butterworths, London, England.

Raul, P.R. 1982. *Trends Anal. Chem.* 1: 237.

Reynolds, J.G. 1998. *Petroleum Chemistry and Refining.* Speight, J.G. ed. Taylor & Francis Publishers, Washington, DC. Chapter 3.

Reynolds, J.G., Biggs, W.E., and Bezman, S.A. 1987. *Metal Complexes in Fossil Fuels.* Filby, R.H. and Branthaver, J.F. eds. Symposium Series No. 344. American Chemical Society, Washington, DC. p. 205.

Richardson, J.S. and Miller, O. 1982. *Anal. Chem.* 54: 765.

Ritchie, R.G.S., Roche, R.S., and Steedman, W. 1979a. *Fuel* 58: 523.

Ritchie, R.G.S., Roche, R.S., and Steedman, W. 1979b. *Chem. Ind.* p. 25.

Rossini, F.D. and Mair, B.J. 1959. *Proc. Fifth World Petrol. Congr.* 5: 223.

Rossini, F.D., Mair, B.J., and Streif, A.J. 1953. *Hydrocarbons from Petroleum.* Reinhold, New York.

Sakarnen, K.V. and Ludwig, C.H. 1971. *Lignins: Occurrence, Formation, Structure and Reactions.* John Wiley & Sons Inc., New York.

Schwartz, R.D. and Brasseaux, D.J. 1958. *Anal. Chem.* 30: 1999.

Schucker, R.C. and Keweshan, C.F. 1980. Preprints. *Div. Fuel Chem. Am. Chem. Soc.* 25: 155.

Seifert, W.K. and Teeter, R.M. 1970. *Anal. Chem.* 42: 750.

Smith, K.M. 1975. *Porphyrins and Metalloporphyrins.* Elsevier, New York.

Speight, J.G. 1971. *Fuel* 49: 134.

Speight, J.G. 1981. Paper presented at *Div. Geochem. Am. Chem. Soc.* New York meeting.

Speight, J.G. 1986. Preprints. *Div. Petrol. Chem. Am. Chem. Soc.* 31(4): 818.

Speight, J.G. 1987. Preprints. *Div. Petrol. Chem. Am. Chem. Soc.* 32(2): 413.

Speight, J. G. 1990. *Fuel Science and Technology Handbook.* Marcel Dekker, New York.

Speight, J.G. 1994a. *The Chemistry and Technology of Coal,* 2nd Edition. Marcel Dekker, New York.

Speight, J.G. 1994b. *Asphaltenes and Asphalts. I: Developments in Petroleum Science, 40.* Yen, T.F. and Chilingarian, G.V. eds. Elsevier, Amsterdam. Chapter 2.

Speight, J.G. 2000. *Desulfurization of Heavy Oils and Residua.* Marcel Dekker, New York.

Speight, J.G. and Pancirov, R.J. 1984. *Liq. Fuels Technol.* 2: 287.

Thompson, C.J., Ward, C.C., and Ball, J.S. 1976. *Characteristics of World's Crude oils and Results of API Research Project 60.* Report BERC/RI-76/8. Bartlesville Energy Technology Center, Bartlesville, OK.

Tissot, B.P. and Welte, D.H. 1978. *Petroleum Formation and Occurrence.* Springer-Verlag, New York.

Treibs, A. 1934. *Analen.* 509: 103.

Wallace, D., Starr, J., Thomas, K.P., and Dorrence. S.M. 1988. *Characterization of Oil Sands Resources.* Alberta Oil Sands Technology and Research Authority, Edmonton, Alberta, Canada.

Weiss, V. and Edwards, J.M. 1980. *The Biosynthesis of Aromatic Compounds.* John Wiley & Sons, Inc., New York.

Yamamoro, M. 1993. *Adv. Org. Geochem.* 19: 389.

8 Fractional Composition

8.1 INTRODUCTION

Refining **petroleum** involves subjecting the feedstock to a series of physical and chemical processes (Chapter 14) as a result of which a variety of products are generated. In some of the processes, e.g., distillation, the constituents of the feedstock are isolated unchanged, whereas in other processes, e.g., cracking, considerable changes are brought about to the constituents.

Recognition that refinery behavior is related to the composition of the feedstock has led to a multiplicity of attempts to establish petroleum and its fractions as composition of matter. As a result, various analytical techniques have been developed for the identification and quantification of every molecule in the lower boiling fractions of petroleum. It is now generally recognized that the name petroleum does not describe a composition of matter but rather a mixture of various organic compounds which includes a wide range of molecular weights and molecular types that exist in balance with each other (Speight, 1994; Long and Speight, 1998). There must also be some questions of the advisability (perhaps futility is a better word) of attempting to describe every molecule in petroleum. The true focus should be to what ends these molecules can be used.

Thus, investigations of the character of petroleum need to be focused on the influence of its character on refining operations and the nature of the products that will be produced. Further, one means by which the character of petroleum has been studied is through its fractional composition. However, the fractional composition of petroleum varies markedly with the method of isolation or separation, thereby leading to potential complications (especially in the case of the heavier feedstocks) in the choice of suitable processing schemes for these feedstocks. Crude oil can be fractionated into three or four general fractions: (1) **asphaltene** fraction, (2) **resin** fraction, (3) aromatics fraction, and (4) **saturates** fraction (Figure 8.1). Thus, it is possible to compare interlaboratory investigations and hence to apply the concept of predictability to refining sequences and potential products.

Investigations of the character of petroleum through fractionation studies has been practised for more than 170 years (Boussingault, 1837), although modern fractionation techniques are essentially a twentieth-century approach to examining petroleum composition (Rostler, 1965; Altgelt and Gouw, 1979). In fact, the fractionation of petroleum has evolved to such an extent that it is now possible to determine with a high degree of accuracy, the types of compounds present in crude oil.

The fractionation methods available to the petroleum industry allow a reasonably effective degree of separation of hydrocarbon mixtures. However, the problems are separating the petroleum constituents without alteration in their molecular structure and obtaining these constituents in a substantially pure state. Thus, the general procedure is to employ techniques that segregate the constituents according to molecular size and molecular type.

It is more generally true, however, that the success of any attempted fractionation procedure involves not only the application of one particular technique, but also the

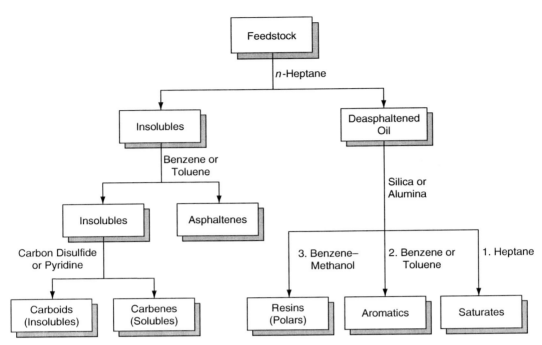

FIGURE 8.1 General fractionation scheme for petroleum.

utilization of several integrated techniques, especially those techniques involving the use of chemical and physical properties to differentiate among the various constituents. For example, the standard processes of physical fractionation used in the petroleum industry are those of distillation and solvent treatment, as well as adsorption by surface-active materials. Chemical procedures depend on specific reactions, such as the interaction of olefins with sulfuric acid or the various classes of adduct formation. Chemical fractionation is often, but not always, successful because of the complex nature of crude oil. This may result in unprovoked chemical reactions that have an adverse affect on the fractionation and the resulting data. Indeed, caution is advised when using methods that involve chemical separation of the constituents.

The order in which the several fractionation methods are used is determined not only by the nature or composition of the crude oil, but also by the effectiveness of a particular process and its compatibility with the other separation procedures to be employed. Although there are wide variations in the nature of crude oil (Chapter 1 and Chapter 2), there have been many attempts to devise standard methods of petroleum fractionation. However, the various laboratories are inclined to adhere firmly to, and promote, their own particular methods. Recognition that no one particular method may satisfy all the requirements of petroleum fractionation is the first step in any fractionation study. This is due, in the main part, to the complexity of petroleum not only from the distribution of the hydrocarbon species (Chapter 7), but also from the distribution of the heteroatom (nitrogen, oxygen, and sulfur) species.

It is the purpose of this chapter to present an overview of those methods that have been applied to the separation of petroleum. This leads not only to an understanding of the separation of petroleum, but also to an understanding of the character of petroleum. Both factors bear a very strong relationship to the processability of petroleum (van Nes and van Westen, 1951; Traxler, 1961; Speight, 1980; Fisher, 1987).

8.2 DISTILLATION

Distillation is a means of separating chemical compounds (usually liquids) through differences in their vapor pressures. The theory of distillation has occupied several large texts and is discussed only briefly here (Halvorsen and Skogestad, 2000; Petyluk, 2004; Lei et al., 2005).

Simple distillation is effective only when separating a volatile liquid from a nonvolatile substance or when separating two liquids that differ in boiling point by 50°C or more. If the liquids comprising the mixture that is being distilled have boiling points that are closer than 50° to one another, the distillate collected will be richer in the more volatile compound, but not to the degree necessary for complete separation of the individual compounds. Thus, in the mixture, the components evaporate and the vapor has a composition determined by the chemical properties of the mixture. Distillation of a given component is possible, if the vapor has a higher proportion of the given component than the mixture. This is caused by the given component having a higher vapor pressure and a lower boiling point than the other components.

The minimum in distillation is flash distillation, where either the temperature is rapidly increased or pressure reduced, and vapor and liquid fractions are thus obtained, which may be processed as such. The device used in distillation is referred to as a still and consists of a minimum of a reboiler (pot) in which the source material is heated, a condenser in which the heated vapor is cooled back to the liquid state, and a receiver in which the concentrated or purified liquid is collected.

The equipment may affect separation by one of two main methods. First, the vapors given off by the heated mixture may consist of two liquids with significantly different boiling points. Thus, the vapor that is given off is a vast majority of one or the other liquid, which after condensation and collection effects the separation. Second, fractional distillation may be necessary and is more effective at separating liquids that have similar boiling points. This method relies on a gradient of temperatures existing in the condenser stage of the equipment. Often in this technique, a vertical condenser, or column, is used and by removal of the distillation products that are liquid at different heights up the column, it is possible to separate liquids that have different boiling points. The greater the distance over which the temperature gradient in the condenser is applied, the easier and more complete the separation.

The basic idea behind fractional distillation is the same as simple distillation, only the process is repeated many times. If simple distillation was performed on a mixture of liquids with similar volatilities, the resulting distillate would be more concentrated in the more volatile compound than the original mixture, but it would still contain a significant amount of the higher boiling compound. If the distillate of this simple distillation was distilled again, the resulting distillate would again be even more concentrated in the lower boiling compound, but still a portion of the distillate would be the higher boiling compound. If this process is repeated several times, a fairly pure distillate will eventually result. This, however, would take a very long time. In fractional distillation, the vapors formed from the boiling mixture rise into the fractionating column where they condense on the column's packing. This condensation is tantamount to a single run of simple distillation; the condensate is more concentrated in the lower boiling compound than the mixture in the distillation flask. As vapors continue to rise through the column, the liquid that has condensed will vaporize. Each time this occurs, the resulting vapors are more and more concentrated in the more volatile substances. The length of the fractionating column and the material it is packed with, impact the number of times the vapors will condense before passing into the condenser; the number of times the column will support this is referred to as the number of theoretical plates of the column.

Since the procedures of simple distillation are so similar to those involved in fractional distillation, the apparatus that are used in the procedures are also very similar. The only difference between the equipment used in fractional distillation and that used in simple distillation is that with fractional distillation, a packed fractionating column is attached to the top of the distillation flask and beneath the condenser. This provides the surface area on which rising vapors condense, and subsequently vaporize.

Thus, the fractionating column supplies a temperature gradient over which distillation occurs. In an ideal situation, the temperature in the distillation flask would be equal to the boiling point of the mixture of liquids and the temperature at the top of the fractionating column would be equal to the boiling point of the lower boiling compound; all of the lower boiling compound would be distilled away before any of the higher boiling compound. In reality, fractions of the distillate must be collected because as the distillation proceeds, the concentration of the higher boiling compound in the distillate being collected steadily increases. Fractions of the distillate, which are collected over a small temperature range, will be essentially purified; several fractions should be collected as the temperature changes and these portions of the distillate should be distilled again to amplify the purification that has already occurred.

Distillation is a common method for the fractionation of petroleum that is used in the laboratory as well as in refineries. The technique of distillation has been practised for many centuries, and the stills that have been employed have taken many forms (Speight and Ozum, 2002; Chapter 13). It was recognized in the early days of the refining industry, that the desirable product (**kerosene** as a **lamp oil**) could be separated by distillation. Thus, it is not surprising that distillation became the process of choice for petroleum refining and the process has evolved from simple distillation units to the complex multiplate still used in the refining industry (Chapter 16).

A modern refinery uses units that provide continuous distillation in which the feedstock is sent to the still all the time and products are drawn out at the same time. The idea in continuous distillation is that the amount going into the still and the amount leaving the still should always equal each other at any given point of time. Older refineries (and even some modern process units) employ batch distillation in which the amount of feedstock entering the still and the amount of products leaving the still is not usually the same all the time and the still is refilled from time to time. Thus, the distiller fills the still pot at the start, then heats it, as time goes by the vapors are condensed to yield the products. When the designated quantities of products have been collected, the distiller stops the still and empties it out, ready for a new batch.

Separation by distillation takes place according to volatility, not necessarily according to molecular weight. Comparison of the boiling points of 2-hydroxypyridine (280°C/760 mm Hg, 535°F/760 mm Hg) and 4-hydroxypyridine (257°C/10 mm Hg, 495°F/10 mm Hg) illustrates that molecular structure (and the consequences of this structure) influences boiling point. In more general terms, and without any attendant bonding influences, the boiling point usually increases with molecular weight in each homologous series. However, the differences between boiling points in the different homologous series are quite substantial. For example, the two-ring **naphthene** decahydronaphthalene (decalin) with 10 carbon atoms (molecular weight 138) boils at 195°C (383°F), but the 10-carbon-atom normal paraffin decane (molecular weight 142) boils at 174°C (345°F). Thus, if a liquid is contained in a closed space, it emits vapor until a certain pressure, of the vapor, is reached that is related to the temperature of the system; the vapor is then saturated. The vapor pressure of a liquid substance in contact with its own liquid is constant and is independent of the amount of liquid and of vapor present in the system. The vapor pressure is usually expressed in terms of the height of a mercury column (in millimeters or inches) that produces an equivalent pressure.

The vapor pressure of a liquid increases with increasing temperature, and when the vapor pressure is equal to the total pressure exerted on the surface of the liquid, the liquid boils. Thus, the boiling point of a liquid may be defined as the temperature at which the vapor pressure of the liquid is equal to the external pressure exerted on the liquid surface. This external pressure may be exerted by atmospheric air, by other gases, by vapor and air, and so on. The boiling point at a pressure of 760 mm air is usually referred to as the normal boiling point.

The boiling point of a pure liquid has a definite and constant value at a constant pressure, but the boiling point of an impure liquid depends to some extent on the nature of the impurities. If the impurities are nonvolatile, the liquid boils at a constant temperature and the impurities remain behind when the liquid has been distilled. However, if the impurities are volatile, the boiling point rises gradually as the liquid distills or may remain constant at a particular stage of the distillation, because of the formation of a constant boiling point mixture of two or more components.

The common feature of all distillation processes is the tendency for the concentration of the more volatile component in the vapor phase to be greater than the concentration in the liquid phase when the two phases have been in contact. In simple distillation, the enrichment of the more volatile component is achieved by partially vaporizing a liquid mixture, either by raising the temperature or by reducing the pressure, and allowing the two phases to separate.

The study of most distillation problems involves an estimate of the distribution of the components between the liquid and vapor phases. The distribution of any component of the mixture of solution between liquid and vapor phases is defined by the equilibrium ratio (for ideal solutions):

$$K_A = Y_A/X_A = P_A$$

Y_A is mole fraction of component A in the vapor phase, X_A is mole fraction of component A in the liquid phase, P_A is vapor pressure of pure component A at the system temperature, and P is the total pressure. Thus, the distribution of a component between vapor and solution may be expressed as a function of temperature and pressure. If an ideal solution contains components of different vapor pressures at a specific temperature, the vapor phase in equilibrium with the liquid phase at this temperature is relatively richer in the more volatile components. The liquid phase is relatively richer in the less volatile components, the components with lower vapor pressures, and a partial separation of components may be achieved. The vapor phase that separates is said to distill at this temperature, and a solution of continuous boiling points, such as petroleum, can be separated by multiple stages of distillation into fractions; each fraction has a relatively narrow boiling range but may, in actual fact, contain many constituents.

Among the prevalent distillation methods, there are two major categories: (1) column distillation and (2) short-path distillation (Chapter 9).

Column distillation (batch mode) is performed at high pressures, at atmospheric pressure, and at reduced pressures. High pressures are used mainly in large-scale refinery distillations with low-boiling distillates and result in higher distillation temperatures.

In short-path distillation (also called molecular distillation) (continuous mode), a very high vacuum is applied and the sample passes rapidly as a very thin film over a heated surface. The lighter molecules evaporate and are condensed on a cooled surface that is located within one inch (2 to 3 cm) of the condensing surface. The vacuum must be high enough to ensure that the mean-free-path length of a distillate molecule is shorter than the distance between the heated and the cooled surfaces.

The bulk of the distillation procedures performed in the laboratory are carried out in packed columns. The efficiency of a distillation column is measured in terms of its number of theoretical trays or plates and the higher this number, the higher its efficiency. Generally, the number of theoretical plates is proportional to the column length. For most column packings in laboratory distillations, the height of a theoretical plate is roughly equal to the diameter of the column.

A theoretical plate in distillation is a hypothetical section of a column which produces the same difference in composition of the ascending distillate as exists at equilibrium between a liquid mixture and its vapor. It acts as an ideal bubble-cap tray. The packing provides a large surface area for the descending reflux. As the rising vapor comes in contact with the descending liquid, some of its higher-boiling component transfers to the reflux, and some of the lower-boiling component transfers from the liquid to the vapor.

Therefore, the vapor arriving at the top of a theoretical plate section contains a lower amount of high-boiling material than it had when it came in at the bottom, and its lighter component is correspondingly enriched. On the other hand, the reflux leaving the theoretical plate at the bottom contains a higher amount of high-boiling material than when it came in at the top. Now we compare the composition of the liquid phase at the top with that of the liquid phase at the bottom. If the section indeed has the length of a theoretical plate, then the composition of the liquid phase at its top is equal to the (theoretical) vapor composition in equilibrium with the liquid at the bottom.

8.2.1 ATMOSPHERIC PRESSURE

Distillation is a common method for the fractionation of petroleum that is used in the laboratory as well as in refineries. The technique of distillation has been practised for many centuries and it was recognized in the early days of the refining industry, that the desirable product (kerosene as a lamp oil) could be separated by distillation. Thus, it is not surprising that distillation became the process of choice for petroleum refining and the process has evolved from the simple distillation units to the complex multiplate still used in the refining industry (Chapter 13).

Distillation has found wide applicability in petroleum science and technology, but it is generally recognized that the fractions separated by distillation are only rarely, if at all, suitable for designation as a petroleum product. Each usually requires some degree of refining, which of course varies with the impurities in the fraction and the desired properties of the finished product (Chapter 20). Nevertheless, distillation is the most important fractionating process for the separation of petroleum hydrocarbons; it is an essential part of any refinery operation (Bland and Davidson, 1967).

However, insofar as petroleum is a mixture of several thousand (or even more) individual chemical compounds, there is little commercial emphasis on the isolation of the individual components. The aim of the distillation of petroleum is predominantly an assessment of the nature and volatility of the material through separation into several fractions of substantially broad boiling ranges.

The initial fractionation of crude oil essentially involves distillation of the material into various fractions, as illustrated by the distillation procedure used for the boiling range specifications of petroleum (Chapter 8). The fractions into which crude oil is commonly separated (Chapter 16) do vary depending on the nature and composition of the crude oil; indeed, as the terminology indicates, there is considerable overlap between the various fractions.

The kerosene (stove oil) and light gas oil fractions are often referred to as **middle distillates** and usually represent the last fractions to be separated by distillation at atmospheric pressure. This leaves the fractions from the heavy gas oil and higher boiling material that are collectively called **reduced crude**.

An atmospheric distillation unit (atmospheric pipe still) is a distillation unit that contains a large number of theoretical plates, and the side streams are taken off at different heights up the distillation tower. The petroleum is partially or totally vaporized in a furnace and then fed to the bottom of the atmospheric still. The distillation tower separates the petroleum by boiling point, i.e., molecular weight, with the lower molecular weight more volatile constituents concentrating at the top of the column and the higher molecular weight less volatile constituents concentrating at the bottom of the column. Naphtha is the chemical precursor of gasoline and boils over the same range, while the atmospheric residuum has an initial boiling point on the order of 350°C (660°F). The atmospheric residuum can be distilled further in a vacuum distillation unit.

8.2.2 REDUCED PRESSURES

Separation of the reduced crude into the constituent fractions requires that the next-stage distillation be carried out under reduced pressure. The higher boiling constituents undergo thermal decomposition at temperatures above 350°C (660°F). This will result in molecular fragmentation leading to volatile products (that were not indigenous to the crude oil) and to coke.

To avoid these thermal reactions, it is necessary to reduce the pressure at which distillation is performed, and since the vapor pressure and temperature are related, the lowering of pressure is accompanied by a corresponding decrease in the boiling points of the individual constituents. For example, a specific compound boiling near 350°C (660°F) at 1 atm (760 mm Hg) may boil over 100°C (180°F) lower (approximately 250°C, 480°F) at 25 to 30 mm Hg, and the danger of thermal decomposition, as with the other thermal interactions, is markedly reduced, if not eliminated.

The vacuum distillation unit (vacuum pipe still) is a distillation unit that contains a smaller number of theoretical plates than the atmospheric distillation unit and, again, the side streams are taken off at different heights up the distillation tower. The atmospheric residuum fractionated to produce overhead fractions such as vacuum gas oils and lubricating oil distillates, and the **bottoms** is a vacuum residuum with an initial boiling point of about 565°C (1050°F), calculated to atmospheric pressure.

The amounts of the various fractions obtained by distillation depend on the nature of the crude oil, which is reflected in the boiling point profile (Figure 8.2). Clearly, petroleum character is very important to the refiner in meeting the product demands of customers as well as in determining what further processing the distillation products may require. For example, the distillates from one crude oil (e.g., Louisiana crude oil) are suitable feedstocks for lubricating oils and specialty products. On the other hand, the distillates from another crude oil (e.g., Bachaquero crude oil) are more suitable for use as fuel oil. Thus, the boiling range of petroleum along with its quality is so important to a refiner in meeting the product slate that most refineries must be able to meet any changes in feedstock quality by adapting the refining options.

In the laboratory (but not always practical in the refinery), lower pressures can be attained by use of an empty column or a spinning band column. With these, the pressure drop is small enough that low pressures can be maintained in the reboiler. For instance, with a distillate pressure of 0.1 mm Hg, one may have a pressure of 0.5 to 1 mm Hg in the reboiler that, theoretically, allows the collection of distillates with an atmospheric equivalent temperature cut point of as high as 560°C (1050°F). The actual observed boiling points during this distillation are, of course, much lower. Despite the low pressure, the reboiler may have to be heated as high as 370°C (700°F) for such high-boiling distillates.

From the actual boiling points obtained at reduced pressure, the so-called atmospheric equivalent temperatures (AET) are calculated at which the material would boil under

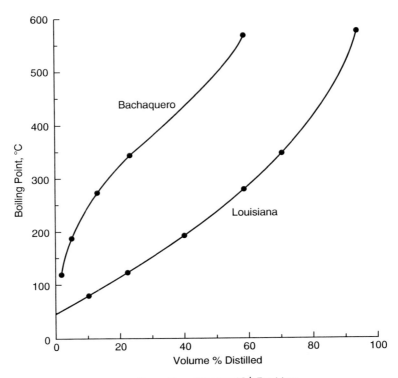

FIGURE 8.2 Distillation profiles and residua properties for Louisiana and Bachaquero crude oils.

Properties of the 565°C$^+$ Residua:

	Louisiana	Bachaquero
Gravity, API	13.1	2.8
Sulfur, wt%	0.9	3.71
Nitrogen, wt%	0.4	0.6
Conradson Carbon, wt%	15.8	27.5
Nickel, ppm	20	100
Manadium, ppm	8	900
Pour Point, °C		55

atmospheric pressure if it was stable and would not decompose. This provides a common basis for the categorization and direct comparison of petroleum components across the entire volatility range, accessible by atmospheric and vacuum distillation.

Within the distillates, the physical and chemical properties are known to change only gradually with the boiling point, allowing interpolations and even extrapolations with reasonable certainty. Whenever the range of distillates is expanded, first by vacuum distillation, then by short-path distillation, the new distillate portions of the previously nondistillable residua follow the same patterns as the previous distillates. The relationship of carbon-number or molecular weight, or of sulfur and nitrogen concentrations, could be extended smoothly into the new regions. Such continuity is no surprise, but highly likely, considering the nature of the petroleum precursors and the maturation chemistry. Thus, the data can be extrapolated across the dividing line between distillates and residua, even into those ranges that are still beyond exact measurements.

As an alternative to empty column or spinning band distillations or to cut even deeper into the stock, high-vacuum short-path distillation can be used. Other names for this technique are molecular distillation or wiped-film distillation, although the latter is often performed with insufficient vacuum and belongs then to a different category.

The most important feature of the short-path still is a very high vacuum of at least 10−3 mm Hg that ensures that the mean-free-path length of the molecules in the gas phase is approximately 2–3 cm, which is the distance between the evaporator and the condenser surfaces. These conditions afford much lower distillation temperatures than possible in regular open columns at the same surface temperature. The sample is spread into a very thin film on the evaporating surface for quick evaporation and a short residence time. The combination of high vacuum, short distance, and short residence time allow very deep distillation without decomposition. Modern versions of these short-path stills can fractionate oils up to atmospheric equivalent boiling points of 700°C (1300°F) with fairly high throughput rates. The small DISTACT laboratory short-path still, for example, has a rate of about 100–800 mL/h with a residence time of less than a minute. Large production plants can operate with throughputs of upto 300 L/m^2/h.

8.2.3 AZEOTROPIC AND EXTRACTIVE DISTILLATION

Chemical bonding between the components of the mixture creates properties unique to the mixture. If the system forms azeotropes, as in a benzene and cyclohexane system, a different problem arises—the azeotropic composition limit the separation, and for a better separation this azeotrope must be bypassed in some way. At the azeotropic point, the mixture contains the given component in the same proportion as the vapor, so that evaporation does not change the purity, and distillation does not affect separation. For example, ethyl alcohol and water form an azeotrope (azeotropic mixture) at 78.2°C.

If the separation of individual components from petroleum itself or from petroleum products is required, there are means by which this can be accomplished. For example, when a constant-boiling mixture of hydrocarbons contains components whose vapor pressure is affected differently by the addition of, say, a nonhydrocarbon compound, distillation of the hydrocarbon mixture in the presence of a nonhydrocarbon additive may facilitate separation of the hydrocarbon components.

In general, the nonhydrocarbon additive is a polar organic compound and should also have the ability to form a binary minimum constant-boiling (or azeotropic) mixture with each of the hydrocarbons. Thus, it is often possible to separate compounds that have very close boiling points by means of azeotropic distillation.

However, when the added compound is relatively nonvolatile, it exists almost entirely in the liquid phase, and the process is actually **extractive distillation**.

Therefore, extractive distillation is distillation in the presence of a miscible, high boiling, relatively nonvolatile component, the solvent, which forms no azeotropes with the other components in the mixture. It is widely used in the chemical and petrochemical industries for separating azeotropic (close boiling) and other constituents in a mixture.

In extractive distillation, the solvent is specially chosen to interact differently with the components of the original mixture, thereby altering their relative volatility. Because these interactions occur predominantly in the liquid phase, the solvent is continuously added near the top of the extractive distillation column so that an appreciable amount is present in the liquid phase on all of the trays below. The mixture to be separated is added through second feed point further down the column. In the extractive column, the component having the greater volatility, not necessarily the component having the lowest boiling point, is taken overhead as a relatively pure distillate. The other component leaves with the solvent via the column bottoms. The solvent is separated from the remaining components in a second distillation column and then recycled back to the first column.

The separation of petroleum by distillation into fractions results in a concentration effect in which the heteroatom constituents (metals included) occur for the most part in the residua (Long and Speight, 1990; Reynolds, 1998). Sulfur, because of its ubiquitous molecular nature is often an exception to this generalization and occurs in most distillation fractions.

Finally, a comment about distillation and its use in the fractionation of petroleum: the issues that arise when distillation is employed relate to the composition of the nonvolatile residue. For the purpose of illustration, if it is assumed that atmospheric residua (boiling range $>345°C$, $>650°F$) have a $>C_{20}$ cutoff and vacuum residua (boiling range $>565°C$, $1050°F$) have a $>C_{35}$ cutoff, distillation can leave as much as 60% w/w of the original oil unfractionated. It is because of this limitation that other methods of fractionation have been sought.

In summary, and as already noted, the distribution of a component between vapor and solution can be expressed as a function of temperature and pressure. Thus, at a specific temperature, the vapor phase in equilibrium with the liquid phase at this temperature is relatively richer in the more volatile components. The liquid phase is relatively richer in the less volatile components, the components with lower vapor pressures, and separation of components may be achieved.

8.3 SOLVENT TREATMENT

The use of solvents invokes the concept of the solubility (or insolubility) of a solute in the chosen solvent. The solubility of a solute is the maximum quantity of solute that can dissolve in a certain quantity of solvent or quantity of solution at a specified temperature. The main factors that have an effect on solubility are: (1) the nature of the solute and solvent, (2) temperature, and (3) pressure.

The rate of solution is a measure of how fast a substance dissolves. Some of the factors determining the rate of solution are: (1) size of the particles, (2) stirring, (3) the amount of solute already dissolved, and (4) temperature.

In order for a solvent to dissolve a solute, the particles of the solvent must be able to separate the particles of the solute and occupy the intervening spaces. Polar solvent molecules can effectively separate the molecules of other polar substances. This happens when the positive end of a solvent molecule approaches the negative end of a solute molecule. A force of attraction then exists between the two molecules. The solute molecule is pulled into solution when the force overcomes the attractive force between the solute molecule and its neighboring solute molecule. Ethyl alcohol and water are examples of polar substances that readily dissolve in each other. Polar solvents can generally dissolve solutes that are ionic.

Fractionation of petroleum by distillation is an excellent means by which the volatile constituents can be isolated and studied. However, the nonvolatile residuum, which may actually constitute from 1% to 60% of the petroleum, cannot be fractionated by distillation without the possibility of thermal decomposition, and as a result, alternative methods of fractionation have been developed.

The distillation process separates light (lower molecular weight) and heavy (higher molecular weight) constituents by virtue of their volatility and involves the participation of a vapor phase and a liquid phase. These are, however, physical processes that involve the use of two liquid phases, usually a solvent phase and an oil phase.

Solvent methods have also been applied to petroleum fractionation on the basis of molecular weight. The major molecular weight separation process used in the laboratory as well as in the refinery is solvent precipitation. Solvent precipitation occurs in a refinery in a deasphalting unit (Chapter 19) and is essentially an extension of the procedure for separation

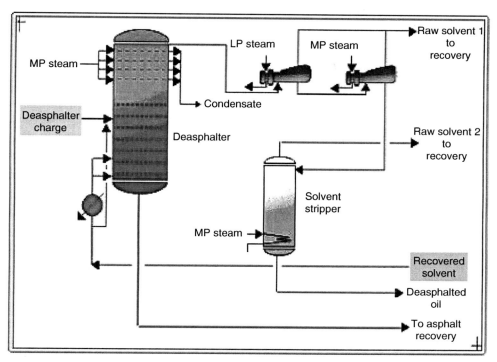

FIGURE 8.3 A deasphalting unit. (From OSHA Technical Manual, Section IV, Chapter 2, Petroleum Refining Processes.)

by molecular weight, although some separation by polarity might also be operative. The deasphalting process is usually applied to the higher molecular weight fractions of petroleum such as atmospheric and vacuum residua for the production of asphalt or demetallized deasphalted oil.

In the deasphalting process (Figure 8.3), which uses liquid propane or liquid butane and mixture thereof as a solvent, high molecular weight fractions (usually residua) are separated to produce heavy lubricating oil, catalytic cracking feedstock, and asphalt. Feedstock and liquid propane are pumped to an extraction tower at precisely controlled mixtures, temperatures (65°C to 95°C, 150°F to 250°F), and pressures of 350 to 600 psi. Separation occurs in a rotating disc contactor, based on differences in solubility, after which the products are steam stripped to recover the propane for recycle. Deasphalting also removes some sulfur and nitrogen compounds, metals, and the carbon residue-forming constituent from the feedstock.

These fractionation techniques can also be applied to cracked residua, asphalt, bitumen, and even to virgin petroleum, but in the last case, the possibility of losses of the lower boiling constituents is apparent; hence, the recommended procedure for virgin petroleum is, first, distillation followed by fractionation of the residua.

The separation of petroleum fractions by the use of solvents has, to some extent, already been considered, but it is necessary at this point to briefly review the basis of solvent extraction for a better understanding of the application of this method to the fractionation of petroleum and petroleum products.

The simplest application of solvent extraction consists of mixing petroleum with another liquid, which results in the formation of two phases. This causes distribution of the petroleum constituents over the two phases; the dissolved portion is referred to as the extract, and the nondissolved part of the petroleum is referred to as the raffinate.

The ratio of the concentration of any particular component in the two phases is known as the distribution coefficient K:

$$K = C_1/C_2$$

C_1 and C_2 are the concentrations in the various phases. The distribution coefficient is usually constant and may vary only slightly, if at all, with the concentration of the other components. In fact, the distribution coefficients may differ for the various components of the mixture to such an extent that the ratio of the concentrations of the various components in the solvent phase differs from that in the original petroleum; this is the basis for solvent extraction procedures.

It is generally molecular type, not molecular size, which is responsible for the solubility of species in various solvents. Thus, solvent extraction separates petroleum fractions according to type, although within any particular series there is a separation according to molecular size. Lower molecular weight hydrocarbons of a series (the light fraction) may well be separated from their higher molecular weight homologues (the heavy fraction) by solvent extraction procedures.

In general, it is advisable that selective extraction be employed with fairly narrow boiling range fractions. However, the separation achieved after one treatment with the solvent is rarely complete and several repetitions of the treatment are required. Such repetitious treatments are normally carried out by movement of the liquids counter-currently through the extraction equipment (countercurrent extraction), which affords better yields of the extractable materials.

The list of compounds that have been suggested as selective solvent for the fractionation of petroleum is extensive, but before any extraction process is attempted, it is necessary to consider the following criteria:

1. Differences in the solubility of the petroleum constituents in the solvent should be substantial.
2. Solvent should be significantly less or more dense than the petroleum (product) to be separated to allow easier countercurrent flow of the two phases.
3. Separation of the solvent from the extracted material should be relatively easy.

It may also be advantageous to consider other properties, such as viscosity, surface tension, and the like, as well as the optimal temperature for the extraction process. Thus, aromatics can be separated from naphthene and paraffinic hydrocarbons by the use of selective solvents. Further, aromatics with differing numbers of aromatic rings that may exist in various narrow boiling fractions can also be effectively separated by solvent treatment.

8.3.1 ASPHALTENE SEPARATION

8.3.1.1 Influence of Solvent Type

The systematic separation of petroleum by treatment with solvents has been practised for several decades (Girdler, 1965 and references cited therein). If chosen carefully, solvents effect a separation between the constituents of conventional petroleum, heavy oil, residua, and tar sand bitumen according to differences in molecular weight and aromatic character. The nature and the quantity of the components separated depend on the conditions of the experiment, namely, the degree of dilution temperature and the nature of the solvent.

On the basis of the solubility in a variety of solvents, it has become possible to distinguish among the various constituents of petroleum, heavy oil, and tar sand bitumen (Figure 8.1).

Highly paraffin crude oil may contain only small portions of asphaltenes. Crude oil generally does not contain **carboids** and **carbenes** that are, for the purposes of this text, considered to be the products of thermal processes. Hence, residua from cracking distillation or from cracking processes may contain 2% w/w by weight or more of carboids and carbenes.

Thus, the separation of crude oil into two fractions, asphaltenes and maltenes, is conveniently brought about by means of low-molecular-weight paraffinic hydrocarbons, which were recognized to have selective solvency for hydrocarbons, and simple relatively low molecular weight hydrocarbon derivatives. The more complex, higher molecular weight compounds are precipitated particularly well by addition of 40 volumes of n-pentane or n-heptane in the methods generally preferred at present (Speight et al., 1984; Speight, 1994), although hexane is used on occasion (Yan et al., 1997). It is no doubt a separation of the chemical components with the most complex structures from the mixture, and this fraction, which should correctly be called n-pentane asphaltenes or n-heptane asphaltenes is qualitatively and quantitatively reproducible (Figure 8.1).

If the **precipitation** method (deasphalting) involves the use of a solvent and a residuum and is essentially a leaching of the heavy oil from the insoluble residue, this process may be referred to as extraction. However, under the prevailing conditions now in laboratory use, the term *precipitation* is perhaps more correct and descriptive of the method.

Variation in the solvent type also causes significant changes in asphaltene yield (Figure 8.4). The solvent power of the solvents (i.e., the ability of the solvent to dissolve asphaltenes) increases in the order

2-Methylparaffin (iso-paraffin) < n-paraffin < terminal olefin

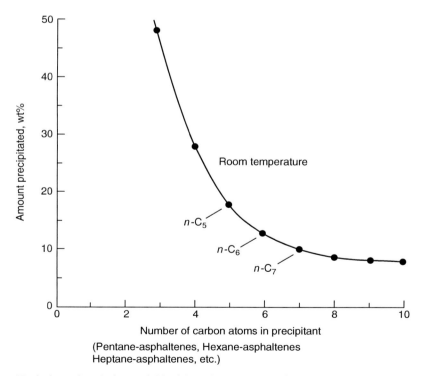

FIGURE 8.4 Variation of asphaltene yield with carbon number of the liquid hydrocarbon.

Cycloparaffins (naphthenes) have a remarkable effect on asphaltene yield and give results totally unrelated to those from any other nonaromatic solvent (Mitchell and Speight, 1973). For example, when cyclopentane, cyclohexane, or their methyl derivatives are employed as precipitating media, only about 1% of the material remains insoluble.

To explain those differences it was necessary to consider the solvent power of the precipitating liquid, which can be related to molecular properties (Hildebrand et al., 1970). Thus the solvent power of nonpolar solvents has been expressed as a **solubility parameter**, δ, and equated to the internal pressure of the solvent, that is, the ratio between the surface tension and the cubic root of the molar volume V:

$$\delta_1 = \gamma^3 \sqrt{V}$$

Alternatively, the solubility parameter of nonpolar solvents can be related to the energy of vaporization ΔR^v and the molar volume,

$$\delta_2 = (\Delta E^v/V)^{1/2}$$

or

$$\delta_2 = (\Delta H^v - RT/V)^{1/2}$$

ΔH^v is the heat of vaporization, R is the gas constant, and T is the absolute temperature.

Consideration of this approach shows that there is indeed a realtionship between the solubility parameters for a variety of solvents and the amount of precipitate (Mitchell and Speight, 1973). The introduction of a polar group (heteroatom function) into the molecule of the solvent has significant effects on the quantity of precipitate. For example, treatment of a residuum with a variety of ethers or treatment of asphaltenes with a variety of solvents illustrates this point (Speight, 1979). In the latter instance, it was not possible to obtain data from the addition of the solvent to the whole feedstock *per se*, since the majority of the nonhydrocarbon materials were not miscible with the feedstock. It is nevertheless interesting that, as with the hydrocarbons, the amount of precipitate, or asphaltene solubility, can be related to the solubility parameter.

The solubility parameter allows an explanation of certain apparent anomalies, for example, the insolubility of asphaltenes in pentane and the near complete solubility of the materials in cyclopentane. Moreover, the solvent power of various solvents is in agreement with the derivation of the solubility parameter; for any one series of solvents the relationship between amount of precipitate (or asphaltene solubility) and the solubility parameter δ is quite regular.

In any method used to isolate asphaltenes as a separate fraction, standardization of the technique is essential. For many years, the method of asphaltene separation was not standardized, and even now it remains subject to the preferences of the standards organizations of different countries. The use of both *n*-pentane and *n*-heptane has been widely advocated, and although *n*-heptane is becoming the deasphalting liquid of choice, this is by no means a hard-and-fast rule. And it must be recognized that large volumes of solvent may be required to effect a reproducible separation, similar to amounts required for consistent asphaltene separation. It is also preferable that the solvents be of sufficiently low boiling point that complete removal of the solvent from the fraction can be effected and, most important, the solvent must not react with the feedstock. Hence, there has been a preference for hydrocarbon liquids. Although several standard methods are available, they are not unanimous in the particular hydrocarbon liquid or in ratio of hydrocarbon liquid to feedstock.

Method	Deasphalting Liquid	Volume (mL/g)
ASTM D893	*n*-pentane	10
ASTM D2006	*n*-pentane	50
ASTM D2007	*n*-pentane	10
IP 143	*n*-heptane	30
ASTM D3279	*n*-heptane	100
ASTM D4124	*n*-heptane	100

However, it must be recognized that some of these methods were developed for use with feedstocks other than heavy oil and tar sand bitumen. Therefore, adjustments in the methods may be necessary to ensure efficient separation.

In general petroleum research, *n*-pentane and *n*-heptane are the solvents of choice in the laboratory (other solvents can be used) (Speight, 1979) and cause the separation of asphaltenes as brown-to-black powdery materials. In the refinery, supercritical low molecular weight hydrocarbons (e.g., liquid propane, liquid butane, or mixtures of both) are the solvents of choice and the product is a semisolid (tacky) to solid asphalt. The amount of asphalt that settles out of the paraffin or residuum mixture depends on the size of the paraffin, the temperature, and the paraffin-to-feedstock ratio (Figure 8.4) (Girdler, 1965; Corbett and Petrossi, 1978; Speight et al., 1984).

Insofar as industrial solvents are very rarely one compound, it was also of interest to note that the physical characteristics of two different solvent types, in this case benzene and *n*-pentane, are additive on a mole-fraction basis (Mitchell and Speight, 1973) and also explain the variation of solubility with temperature. The data also show the effects of blending a solvent with the bitumen itself and allowing the resulting solvent-heavy oil blend to control the degree of bitumen solubility. Varying proportions of the hydrocarbon alter the physical characteristics of the oil to such an extent that the amount of precipitate (asphaltenes) can be varied accordingly within a certain range.

8.3.1.2   Influence of the Degree of Dilution

At constant temperature, the quantity of precipitate first increases with the increasing ratio of solvent to feedstock and then reaches a maximum (Figure 8.5). In fact, there are indications that when the proportion of solvent in the mix is <35%, little or no asphaltenes are precipitated (Mitchell and Speight, 1973).

8.3.1.3   Influence of Temperature

When pentane and the lower molecular weight hydrocarbon solvents are used in great excess, the quantity of precipitate and the composition of the precipitate changes with increasing temperature (Mitchell and Speight, 1973; Andersen, 1994).

One particular example is the separation of asphaltenes from using *n*-pentane. At ambient temperatures (21°C, 70°F), the yield of asphaltenes is 17% w/w but at 35°C (95°F), 22.5% by weight, asphaltenes are produced using the same feedstock–pentane ratio. This latter precipitate is in fact asphaltenes plus resins; similar effects have been noted with other hydrocarbon solvents at temperatures up to 70°C (160°F). These results are self-explanatory when it is realized that the heat of vaporization ΔH^v and the surface tension γ, from which the solubility parameters are derived, both decrease with increasing temperature.

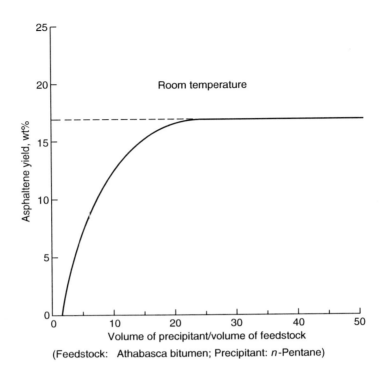

(Feedstock: Athabasca bitumen; Precipitant: n-Pentane)

FIGURE 8.5 Variation of asphaltene yield with amount of liquid hydrocarbon added.

8.3.1.4 Influence of Contact Time

Contact time between the hydrocarbon and the feedstock (especially feedstocks such as heavy oil, residua, and tar sand bitumen) also plays an important role in asphaltene separation (Speight et al., 1984). Yields of the asphaltenes reach a maximum after approximately 8 h, which may be ascribed to the time required for the asphaltene particles to agglomerate into particles of a filterable size as well as the diffusion-controlled nature of the process, since the heavier feedstocks also need time for the hydrocarbon to penetrate their mass.

8.3.2 FRACTIONATION

After removal of the asphaltene fraction, further fractionation of petroleum is also possible by variation of the hydrocarbon solvent. For example, liquefied gases, such as propane and butane, precipitate as much as 50% by weight of the residuum or bitumen. The precipitate is a black, tacky, semisolid material, in contrast to the pentane-precipitated asphaltenes, which are usually brown, amorphous solids. Treatment of the propane precipitate with pentane then yields the insoluble brown, amorphous asphaltenes and soluble, near-black, semisolid resins, which are, as near as can be determined, equivalent to the resins isolated by adsorption techniques (page 223 et seq.).

There are also claims that solvent treatment at low temperatures ($-4°C$ to $-20°C$, $-4°F$ to $25°F$) brings about fractionation of the maltenes. The hydrocarbon solvents, pentane and hexane have been claimed adequate for this purpose but may not be successful with maltenes from bitumen or from residua. The author has had considerable success using acetone at $-4°C$ ($25°F$) for the fractionation of maltenes from material other than cracked residua.

Other miscellaneous fractionation procedures involving the use of solvents are available. One method includes a procedure using polar solvents for separation into five fractions. These fractions are:

1. Asphaltene constituents, insoluble in hexane
2. Hard resin constituents, insoluble in 80:20 iso-butyl alcohol–cyclohexane mixture
3. Wax constituents, insoluble in a 1:2 mixture of acetone and methylene chloride at 0°C (−18°F)
4. Soft resin constituents, insoluble in iso-butyl alcohol
5. Oil constituents, the balance of the sample that remains soluble in iso-butyl alcohol

The only drawback to this particular scheme appears to be the use of hexane as the precipitating medium. Use of this solvent does not completely precipitate the asphaltenes in a sample and perhaps the fraction designated hard resins may contain considerable portions of asphaltenes.

A fractionation procedure was also devised using n-butanol and acetone as the solvents and consists of (1) separation of an asphaltic fraction by n-butanol, and (2) separation of the butanol-soluble portion into a paraffinic fraction and a cyclics fraction by chilling an acetone solution of the two, but it has the unfortunate result that all three fractions obtained may contain asphaltenes. It is difficult to compare the method with any other or even establish any correlation with previous experience. Even precipitation of the asphaltenes with n-pentane and subtracting the asphaltenes from the asphaltic, arriving at a fourth fraction (asphaltic resins), did not appear to correct the faults of the method since the asphaltic fraction does not contain all the asphaltenes of the specimen.

Another method of fractionation consists of stepwise separation into the following fractions:

1. *Asphaltene constituents*, precipitated by n-pentane
2. *Resin constituents*, precipitated with propane and subdivided by fractionation with aniline into soft resins and hard resins
3. *Wax constituents*, precipitated with methyl iso-butyl ketone
4. *Oil constituents*, remaining fraction separated with acetone into paraffinic oils and naphthenic oils

A strong feature of this method is the subdivision of the resins fraction by solubility in aniline and the subdivision of the paraffinic fraction into the three components: wax, paraffinic oils, and naphthenic oils.

It is unfortunate that, with the exception of asphaltene precipitation, no standard method exists for the fractionation of crude oil, residua, bitumen, by solvent treatment. The procedures described here each have their own individual merits, but there has not been any serious effort to apply these methods to a wide variety of carbonaceous liquids to assess their general applicability. Fractionation by means of solvents alone (perhaps with the exception of propane, which requires pressure equipment) would be convenient indeed, provided that facile separation of the solvent and the products could be achieved at a later stage.

Another all-solvent procedure involves the use of acetone, which discharges a resin fraction, and dimethylformamide, which then discharges a saturates fraction.

The use of all-solvent methods for the separation of petroleum allows the fractionation of feedstocks to be achieved without loss of material (on the adsorbent) and produces fractions of varying polarity (Speight, 1989). The obvious benefit of the all-solvent techniques is the complete recovery of material, thereby allowing a more quantitative and qualitative examination of the feedstocks.

The disadvantages of an all-solvent separation technique are that, first, in some instances, low temperatures (e.g., 0°C to −10°C and the like) are advocated as a means of effecting oil fractionation with solvents (Rostler, 1965; Speight, 1979). Such requirements may cause inconvenience in a typical laboratory operation by requiring a permanently cool temperature during the separation. Second, it must be recognized that large volumes of solvent may be required to effect a reproducible separation in the same manner as the amounts required for consistent asphaltene separation (ASTM D2006, ASTM D2007, ASTM D4124, ASTM D893; IP 143). Finally, it is also essential that the solvent is of sufficiently low boiling point, so that complete removal of the solvent from the product fraction can be effected. Although not specifically included in the three main disadvantages of the all-solvent approach, it should also be recognized that the solvent must not react with the feedstock constituents. In addition, caution is still required to ensure that there is no interaction between the solvent and the solute.

8.4 ADSORPTION

Adsorption is the bonding of molecules or particles to a surface. On the other hand, *absorption* is the filling of pores in a solid. The bonding to the surface is usually (but not always) weak and reversible. Compounds that contain functional groups are, very often, strongly adsorbed on activated carbon.

The most common industrial adsorbents are activated **clay**, carbon, silica gel, and alumina, because they present enormous surface areas per unit weight. Clay is a naturally occurring mineral. Roasting organic material to decompose it to granules of carbon produces activated carbon. Coconut shell, wood, and bone are common sources of activated carbon. Silica gel is a matrix of hydrated silicon dioxide. Alumina is mined or precipitated aluminum oxide and hydroxide.

Temperature effects on adsorption are profound, and measurements are usually at a constant temperature. Graphs of the data are called isotherms. Most steps using adsorbents have little variation in temperature.

8.4.1 CHEMICAL FACTORS

As already stated (Chapter 7), petroleum is a complex mixture of paraffin, naphthene, and aromatic hydrocarbons as well as nitrogen-, oxygen-, and sulfur-containing compounds and traces of a variety of metal-containing compounds, and the amounts of nonhydrocarbon compounds increase with molecular weight. By definition, the saturate fraction consists (or should consist) of paraffins and cycloparaffins (naphthenes). The single-ring naphthenes, or cycloparaffins, present in petroleum are primarily alkyl-substituted cyclopentane and cyclohexane. The alkyl groups are usually quite short, with methyl, ethyl, and isopropyl groups, the predominant substituents. As the molecular weight of the naphthenes increases, the naphthene fraction contains more condensed rings with six-membered rings predominating. However, five-membered rings are still present in the complex higher molecular weight molecules.

The aromatics fraction consists of those compounds containing an aromatic ring and vary from monoaromatics (containing one benzene ring in a molecule) to diaromatics (substituted naphthalene) to triaromatics (substituted phenanthrene). Higher condensed ring systems (tetraaromatics, pentaaromatics) are also known but are somewhat less prevalent than the lower ring systems, and each aromatic type will have increasing amounts of condensed ring naphthene attached to the aromatic ring as molecular weight is increased.

However, depending on the adsorbent employed for the separation, a compound having an aromatic ring (i.e., six aromatic carbon atoms) carrying side chains consisting *in toto* of more than six carbon atoms (i.e., more than six nonaromatic carbon atoms) will appear in the aromatic fraction.

The typical nitrogen compounds found in petroleum are generally divided into two groups, basic and nonbasic (Chapter 6), each of which has alkyl chains and other ring systems. The basic nitrogen compounds cause difficulty with the many acid-catalyzed processes used in petroleum refining. For example, in catalytic cracking, the basic nitrogen adsorbs on the catalytic acid sites and reduces the cracking activity of the catalyst. Further, in reactions catalyzed by liquid acids, the presence of basic nitrogen compounds in the feed increases acid consumption and, thus, the cost of the process. Typical of such processes are alkylation, isomerization, and olefin absorption.

Another type of nitrogen compound is the **porphyrins** (Chapter 6), which consist of four pyrrole rings connected together with methylene bridges at the carbons next to the nitrogen atoms. They are normally found in trace quantities in the high molecular weight fractions of petroleum as metal complexes.

The oxygen compounds found in petroleum fractions are often products of exposure to air. However, some naturally occurring oxygenated compounds do exist in petroleum; these are typically phenols, naphthenic acids, and esters (Chapter 6). They increase in quantity and complexity as molecular weight increases, just as sulfur and nitrogen compounds do.

A variety of sulfur compounds occur in petroleum (Chapter 6) and include mercaptans (**–SH), sulfides (–S–), and disulfides (–S–S–). They are much less thermally stable than the thiophene derivatives and often lose hydrogen sulfide on heating. They can also react thermally to form more stable sulfur compounds. In the chromatographic separation, these sulfur compounds are found in the polar aromatics fraction, even though they may not all be aromatic. Thiophenes tend to exhibit aromatic behavior and are collected with the aromatics when adsorption separation is used on petroleum fractions.

8.4.2   FRACTIONATION METHODS

8.4.2.1   General Methods

Separation of petroleum, heavy oil, tar sand bitumen, and residua by adsorption chromatography essentially commences with the preparation of a porous bed of finely divided solid, the adsorbent (Hoiberg, 1964). The adsorbent is usually contained in an open tube (column chromatography); the sample is introduced at one end of the adsorbent bed and induced to flow through the bed by means of a suitable solvent. As the sample moves through the bed, the various components are held (adsorbed) to a greater or lesser extent depending on the chemical nature of the component. Thus, those molecules that are strongly adsorbed spend considerable time on the adsorbent surface rather than in the moving (solvent) phase, but components that are slightly adsorbed move through the bed comparatively rapidly.

Numerous factors randomly affect the process of migration through a bed, and in fact, the total distance traveled in a given time by different molecules of the same material is not constant. Nevertheless, the suitable choice of a bed and a moving (solvent) phase allows adequate separation of even multicomponent mixtures to be achieved. Thus, the overall adsorption chromatographic process may be viewed as migration of different compounds along the bed, which varies with compound structure as well as a range of migrations by different molecules of the same compound.

The fractionation of petroleum and residua by adsorption on such materials as Fuller's earth, animal charcoal, and various types of clay dates back to the beginning of the twentieth century (Hoiberg, 1964). These materials effect an arbitrary separation of the material into a number of fractions that have variously been described as oil, resins, hard resins, and soft resins, to mention only the more commonly used terms.

It is essential that, before application of the adsorption technique to the petroleum, the asphaltenes first be completely removed, for example, by any of the methods outlined in the previous section. The prior removal of the asphaltenes is essential insofar as they are usually difficult to remove from the earth or clay and may actually be irreversibly adsorbed on the adsorbent.

Nevertheless, careful monitoring of the experimental procedures and the nature of the adsorbent have been responsible for the successes achieved with this particular technique. Early procedures consisted of warming solutions of the petroleum fraction with the adsorbent and subsequent filtration. This procedure has continued to the present day, and separation by adsorption is used commercially in plant operations in the form of clay treatment of crude oil fractions and products (Chapter 19).

In the laboratory very little use is made of the technique of warming a solution of the sample with the adsorbent. Rather, a chromatographic technique is employed in which the sample is washed through a column of the adsorbent using various solvents.

Early investigations involved filtration of crude oils through a column of Fuller's earth and it was observed that the gasoline components appeared in the initial part of the filtrate. Subsequent investigations showed that if light oil is drawn upward by a pump through a column of earth, the light aliphatic hydrocarbons accumulate in the top section. The aromatic constituents did not rise as high and the nitrogen- and sulfur-containing compounds were largely retained on the adsorbent.

A later chromatographic method involved separation into four principal fractions— asphaltenes, asphaltic resins, dark oils, water white oils, and a fifth fraction constituting the balance of the specimen. The method consists of precipitation of asphaltenes with 40 volumes of *n*-pentane (or *n*-heptane) and elution of the *n*-pentane (or *n*-heptane) soluble fraction from a chromatographic column of Fuller's earth. The elution technique gave a series of fractions:

1. *Oils* eluted with *n*-pentane
2. *Dark oils* eluted with methylene chloride
3. *Resins* with methyl ethyl ketone
4. *Hard resins* that were desorbed with an acetone–chloroform mixture

The proportions of each fraction are subject to the ratio of Fuller's earth to *n*-pentane soluble materials. For example, a change in the ratio from 10:1 to 25:1 causes a decrease in the percentage of water white oils by a factor of 3:2, and the percentage of dark oils and asphaltic resins increases by about the same factor. The method, like all chromatographic procedures proposed for the fractionation of crude oils, is dominated by equilibrium conditions, and does not give fractions that are different components but only blends of the same components in different proportions.

Other methods of fractionation by the use of adsorbents include separation of the maltene fraction by elution with *n*-heptane from silica gel into two fractions named aromatics and nonaromatics and is, in fact, a separation into the two broad groups called resins and oils in other methods. The silica gel method may also be modified to produce three fractions: (1) nonaromatics eluted with *n*-heptane, (2) aromatics eluted with benzene, and (3) compounds that contain oxygen as well as sulfur and nitrogen, eluted with pyridine. Prior separation of the asphaltenes renders the procedure especially suitable and convenient for use with heavy oil and bitumen. Other modifications include successive elution with *n*-pentane, benzene, carbon tetrachloride, and ethanol.

Alumina has also been used as an adsorbent and involves (1) precipitation of asphaltenes with normal pentane, (2) elution of oils from alumina with pentane, and (3) elution of resins from alumina with a methanol–benzene mixture. In fact, the choice of the adsorbent appears to be arbitrary, as does the choice of the various solvents or solvent blend. The use of ill-defined adsorbents, such as earths or clays is a disadvantage, in that certain components of the petroleum may undergo changes (for e.g., polymerization) caused by the catalytic nature of the adsorbent and can no longer be extracted quantitatively. Further, extraction of the adsorbed components may require the use of solvents of comparatively high solvent power, such as chloroform or pyridine, which may be difficult to remove from the product fractions.

It is also advisable, once a procedure using an earth or clay has been established, that the same type of adsorbent be employed for future fractionation since the ratio of the product fractions varies from adsorbent to adsorbent. It is also very necessary that the procedure be used with caution and that the method not only be reproducible but quantitative recoveries be guaranteed; reproducibility with only, say, 85% of the material recoverable is not a criterion of success.

There are two procedures that have received considerable attention over the years and these are: (1) the United States Bureau of Mines–American Petroleum Institute (USBM–API) method (Figure 8.6) and (2) the saturates–aromatics–resins–asphaltenes (SARA) method. This latter method is often also called the saturates–aromatics–polars–asphaltenes (SAPA) method. These two methods are used for representing the standard methods of petroleum fractionation. Other methods are also noted, especially when the method has added further meaningful knowledge to compositional studies.

The USBM–API method employs ion-exchange chromatography and coordination chromatography with adsorption chromatography to separate heavy oils and residua into seven broad fractions; acids, bases, neutral nitrogen compounds, saturates, and mono-, di-, and polyaromatic compounds. The acid and base fractions are isolated by ion-exchange chromatography, the neutral nitrogen compounds by complexation chromatography using ferric chloride, and the saturates and aromatics by adsorption chromatography on activated alumina (Jewell et al., 1972a) or on a combined alumina–silica column (Hirsch et al., 1972; Jewell et al., 1972b).

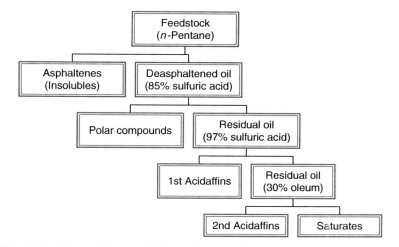

FIGURE 8.6 The United States Bureau of Mines–American Petroleum Institute (USBM–API) method.

The feedstock sample can be separated into chemically significant fractions that are suitable for analysis according to the compound type (Jewell et al., 1972b; McKay et al., 1975). Although originally conceived for the separation of distillates, this method has been successfully applied for determining the composition of heavy oil and bitumen (Cummins et al., 1975; McKay et al., 1976; Bunger, 1977). Originally, the method required distillation of a feedstock into narrow boiling point fractions, but whole feedstocks, such as the bitumen from Utah or Athabasca tar sands (Selucky et al., 1977), have been separated into classes of compounds without previous distillation or removal of asphaltenes. The latter finding is supported by the separation of asphaltenes using ion-exchange materials (McKay et al., 1977; Francisco and Speight, 1984).

The SARA method (Jewell et al., 1974) is essentially an extension of the API method that allows more rapid separations by placing the two ion-exchange resins and the $FeCl_3$–clay–anion-exchange resin packing into a single column. The adsorption chromatography of the nonpolar part of the same is still performed in a separation operation. Since the asphaltene content of petroleum (and synthetic fuel) feedstocks is often an important aspect of processability, an important feature of the SARA method is that the asphaltenes are separated as a group. Perhaps more important is that the method is reproducible and applicable to a large variety of the most difficult feedstocks, such as residue tar sand bitumen, shale oil, and coal liquids.

Both the USBM–API and SARA methods require some caution if the asphaltenes are first isolated as a separate fraction. For example, the asphaltene yield varies with the hydrocarbon used for the separation and with other factors (Girdler, 1965; Mitchell and Speight, 1973; Speight et al., 1984). An inconsistent separation technique can give rise to problems resulting from residual asphaltenes in the deasphaltened oil undergoing irreversible adsorption on the solid adsorbent.

The USBM–API and SARA methods are widely used separation schemes for studying the composition of heavy petroleum fractions and other fossil fuels, but several other schemes have also been used successfully and have found common usage in investigations of feedstock composition. For example, a simple alternative to the SARA sequence is the chromatographic preseparation of a deasphaltened sample on deactivated silica or alumina with pentane (or hexane) into saturated materials followed by elution with benzene for aromatic materials and with benzene–methanol for polar materials (resins). This allows further chromatography into narrower (more similar) fractions without mutual interference on the adsorbent.

The selection of any separation procedure depends primarily on the information desired about the feedstock. For example, separation into multiple fractions to examine the minute details of feedstock composition requires a complex sequence of steps. An example of such a separation scheme involves the fractionation of Athabasca tar sand bitumen into four gross fractions and subfractionation of these four fractions (Boyd and Montgomery, 1963). This allowed the investigators to study the distribution of the functional types within the bitumen.

Other investigators (Oudin, 1970) have reported that a combination chromatography using alumina and silica gel is suitable for deasphaltened oils. A more complex scheme, also involving the use of silica, resulted in the successful separation of hexane-deasphalted crude oil. There is also a report (Al-Kashab and Neumann, 1976) of the direct fractionation of hydrocarbon and hetero-compounds from deasphaltened residua on a dual alumina–silica column with subsequent treatment of the polar fraction with cation- and anion-exchange resins into basic, acid, and neutral materials. The method also includes chromatography of the asphaltenes, but only with highly polar asphaltene samples are the basic and acid compounds first removed with ion-exchange resins. The remainder of the feedstock is separated into saturates, aromatics, and hetero compounds using alumina–silica adsorption. Separation of the saturates into *n*-alkanes and iso-alkanes plus cycloalkanes is achieved by the use of urea ($H_2N\cdot C=O\cdot NH_2$) and thiourea ($H_2N\cdot C=S\cdot NH_2$).

One of the problems of such a fractionation scheme is the initial separation of the feedstocks into two ill-defined fractions (colloids and dispersant) without first removing the asphaltenes. As already noted, asphaltenes are specifically defined by the method of separation. They are less well defined using such liquids as ethyl acetate in place of the more often used hydrocarbons, such as pentane and heptane (Speight, 1979). The use of ethyl acetate undoubtedly leads to asphaltene material in the dispersoids and nonasphaltene material in the colloids. Application of a more standard deasphalting technique would undoubtedly improve this method and provide an excellent insight into feedstock composition.

One of the common issues related to the use of any adsorption-based fractionation scheme is the nature of the adsorbent. In the early reports of petroleum fractionation (Pfeiffer, 1950) clays often appeared as an adsorbent to effect the separation of the feedstock into various constituent fractions. However, clay (Fuller's earth, attapulgus clay, and the like) is often difficult to define with any degree of precision from one batch to another. Variations in the nature and properties of the clay can, and will, cause differences not only in the yields of composite fractions, but also in the distribution of the compound types in those fractions. In addition, irreversible adsorption of the more polar constituent to the clay can be a serious problem when further investigations of the constituent fractions are planned.

One option for resolving this problem has been the use of more standard adsorbents, such as alumina and silica. These materials are easier to define and are often accompanied by guarantees of composition and type by various manufacturers. They also tend to irreversibly adsorb less of the feedstock than a clay. Once the nature of the adsorbent is guaranteed, reproducibility becomes a reality. Without reproducibility, the analytic method does not have credibility.

8.4.2.2 ASTM Methods

There are three ASTM methods that provide for the separation of a feedstock into four or five constituent fractions and it is interesting to note that as the methods have evolved there has been a change from the use of pentane (ASTM D2006 and ASTM D2007) (Figure 8.7 and Figure 8.8) to heptane (ASTM D4124) (Figure 8.9) to separate asphaltenes. This is,

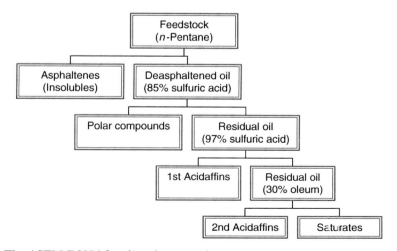

FIGURE 8.7 The ASTM D2006 fractionation procedure.

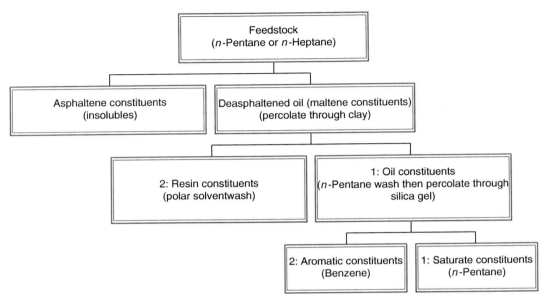

FIGURE 8.8 The ASTM D2007 fractionation procedure.

in fact, in keeping with the production of a more consistent fraction that represents these higher molecular weight, more complex constituents of petroleum (Girdler, 1965; Speight et al., 1984).

Two of the methods (ASTM D2007 and D4124) use adsorbents to fractionate the deasphaltened oil, but the third method (ASTM D2006) advocates the use of various grades of sulfuric acid to separate the material into compound types. Caution is advised in the application of this method since the method does not work well with all feedstocks. For example, when the sulfuric acid method (ASTM D2006) is applied to the separation of heavy feedstocks, complex emulsions can be produced.

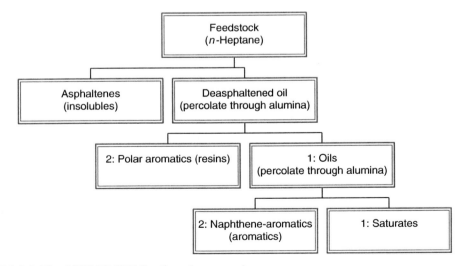

FIGURE 8.9 The ASTM D4124 fractionation procedure.

Obviously, there are precautions that must be taken when attempting to separate heavy feedstocks or polar feedstocks into constituent fractions. The disadvantages in using ill-defined adsorbents are that adsorbent performance differs with the same feed and, in certain instances, may even cause chemical and physical modification of the feed constituents. The use of a chemical reactant like sulfuric acid should only be advocated with caution since feeds react differently and may even cause irreversible chemical changes or emulsion formation. These advantages may be of little consequence when it is not, for various reasons, the intention to recover the various product fractions in toto or in the original state, but in terms of the compositional evaluation of different feedstocks, the disadvantages are very real.

In summary, the terminology used for the identification of the various methods might differ. However, in general terms, group-type analysis of petroleum is often identified by the acronyms for the names: PONA (paraffins, olefins, naphthenes, and aromatics), PIONA (paraffins, iso-paraffins, olefins, naphthenes, and aromatics), PNA (paraffins, naphthenes, and aromatics), PINA (paraffins, iso-paraffins, naphthenes, and aromatics), or SARA (saturates, aromatics, resins, and asphaltenes). However, it must be recognized that the fractions produced by the use of different adsorbents will differ in content and will also be different from fractions produced by solvent separation techniques.

The variety of fractions isolated by these methods and the potential for the differences in composition of the fractions makes it even more essential that the method is described accurately and that it reproducible not only in any one laboratory, but also between various laboratories.

8.5   CHEMICAL METHODS

The most common methods that have been used for the separation procedure, other than the use of ferric chloride on an adsorbent (page 226), involve the use of sulfuric acid and urea adduction. Although to be truthful, the urea adduction method is in reality a physical method, but insofar as it provides a separation that is specific to certain constituents of petroleum it is included here. There are many other methods that have been used for the chemical separation of petroleum that are now included as refinery processes. Acid treatment (particularly the sulfuric acid method) and the urea addition method are still widely used in the laboratory and hence the inclusion of these methods here.

8.5.1   Acid Treatment

The method of chemical separation commonly applied is treatment with sulfuric acid. Marcusson and Eickmann made an early reference to the use of sulfuric acid in 1908 and used a procedure (Figure 8.10) to precipitate asphaltene constituents from asphaltic materials by treatment of the sample with low-boiling naphtha, followed by fractionation of the naphtha-soluble material with concentrated sulfuric acid. The precipitate produced by the sulfuric acid treatment was actually material that had been converted to an asphaltene type of product by interaction of the asphaltic constituents with the sulfuric acid. It is nevertheless possible that some of the acid-precipitated material originated as asphaltenes that were incompletely precipitated by the naphtha. The constitution of the naphtha was unknown; most likely it was not pure *n*-pentane and it may even have contained hexane(s) or higher paraffins. The addition of only 20 volumes of solvent to heavy feedstocks is not a sufficient amount to completely precipitate asphaltene material.

However, the method has served as a demonstration of the type of separation that can be obtained by means of sulfuric acid. A later refinement of this principle by Rostler and

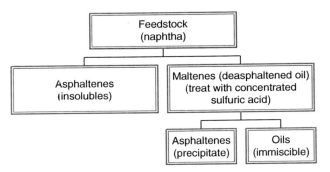

FIGURE 8.10 The Marcusson–Eickmann fractionation procedure.

Sternberg in 1962 led to the development of a technique that proposes a resolution of crude oils, crude oil residua, and asphaltic or bituminous materials into five broad fractions (ASTM D2006) (Figure 8.11). These fractions are distinctly different in chemical reactivity as measured by response of the fractions to cold sulfuric acid of increasing strength (sulfur trioxide concentration). The fractionation involves, like most other methods, initial separation of the sample into asphaltenes and the fraction referred to as maltenes. The fractionation of maltenes into resins and oils is often too vague for identification purposes, and the nature, or composition, of these two fractions may differ from one oil to another.

The chemical precipitation method is claimed to provide the needed subdivision of the resins and the oils to yield chemically related fractions. The names given to the individual fractions are descriptive of the steps used in the procedure. For example, the name paraffins is used for the saturated nonreactive fraction, whereas the next two groups in ascending reactivity are second acidaffins and first acidaffins. The term *second acidaffins* denotes the group of hydrocarbons having affinity for strong acid as represented by fuming sulfuric acid. The term *first acidaffins* denotes that the constituents have an affinity for ordinary concentrated sulfuric acid. The next fraction, nitrogen bases, separated by sulfuric acid of 85%

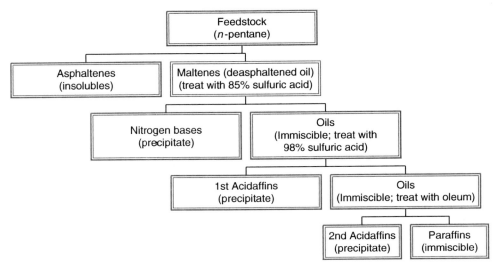

FIGURE 8.11 The Rostler–Sternberg fractionation procedure.

concentration, includes the most reactive components and may contain, among other components, substantially all the nitrogen-containing compounds.

The fractionation accomplished by the sulfuric acid method is presumed to be the subdivision of crude oils or asphaltic materials into five groups of components by virtue of their chemical makeup. The method was accepted by the American Society for Testing and Materials as a standard method of test for *Characteristic Groups in Rubber Extender and Processing Oils by the Precipitation Method* (ASTM D2006). However, there is still some doubt regarding the applicability of the method to a wide variety of petroleum, petroleum residua, and other bituminous materials.

In the author's experience, application of the method to conventional crude oil, heavy oil, and bitumen has led, in some instances, to emulsion formation and hence to difficulties in the separation procedure. It must also be remembered that the use of strong grades of sulfuric acid can, and presumably does, lead to sulfonation of the constituents in many instances, and regeneration of these constituents to their natural state may be difficult, if not impossible. Thus, it is not feasible to compare the sulfuric acid method to the relatively simple precipitation of the basic constituents of petroleum (as their hydrochlorides) by passage of dry hydrogen chloride gas through a solution of the material in a solvent, such as carbon disulfide. It therefore appears that a number of refinements of the sulfuric acid method are desirable since it is apparent that not all carbonaceous liquids react in the same manner with sulfuric acid.

8.5.2 MOLECULAR COMPLEX FORMATION

The formation of crystalline molecular complexes between urea ($H_2N \cdot C = O \cdot NH_2$) or thiourea ($H_2N \cdot C = S \cdot NH_2$) and hydrocarbons has been known since the 1940s, and it is not surprising that the technique has received considerable attention with respect to its use in petroleum chemistry.

The previously described techniques of adsorption and distillation differentiate molecules by class and volatility (or size within any one homologous series), respectively, whereas adduct formation separates on the basis of molecular shape and, to a lesser extent, by size and class. When combined with the older fractionation methods, adduct formation can often be useful for solving separation problems, provided the limitations of the method are realized. Urea and thiourea adduction are not completely selective, as was first supposed, and there is an overlapping of structural types that adduct, especially among the higher molecular weight hydrocarbons.

Adduct formation may be achieved merely by bringing together the adduct-former and reagent under a wide range of reaction conditions. Preferably, the reactant is an inert hydrocarbon solvent, a reagent solvent, referred to as the activator, present in varying amounts to increase the rate of reaction, and the crystalline product that precipitates can be conveniently separated by filtration. Further, the adducts may be decomposed easily and the adducting material recovered by one of several procedures.

8.5.2.1 Urea Adduction

For a hydrocarbon to form an adduct with urea, it is mandatory that the compound contain a long, unbranched chain, usually a chain of at least six carbon atoms, under the conditions most often employed, such as ambient temperature and atmospheric pressure. As side chains or ring structures are added to the molecule, the chain must be lengthened, if adduction is to occur. Generally, the larger the size or number of the substituents or the farther the substituent is from the terminal carbon atom, the longer must be the unbranched portion of the chain. Olefinic unsaturation usually has little effect on adduct formation, but the stability of the adducts of *n*-olefins are somewhat less than those of the corresponding *n*-paraffins. Thus, *n*-pentane does not form an adduct under normal conditions. However, an adduct can be obtained at low temperature with some pressure, and *n*-hexane is the first member of the *n*-alkane series that forms an adduct at room temperature and atmospheric pressure; under similar conditions, 1-octene is the lowest olefin reported to adduct. There is usually no upper limit to chain length other than that imposed by the solubility of the hydrocarbon; however, as the reaction temperature is increased to meet the requirements of hydrocarbon solubility and preferred reaction rate, the stability of the urea lattice in the adduct decreases.

In any given homologous series, the ease of adduct formation and adduct stability increases with increasing straight-chain length. Single methyl side chains, such as those present in compounds such as 3-methylheptane [$CH_3CH_2CH(CH_3)CH_2CH_2CH_2CH_3$] interfere with adduct formation, presumably because the dimensions of the molecules at the methyl group are too great to allow the hydrocarbons to fit properly into the urea adduct channel. However, in the presence of an adducting *n*-paraffin, such as *n*-decane, the methyl-substituted paraffins are assisted into the adduct (induction). The 3-methyleicosane [CH_3CH_2 $CH(CH_3)$ $CH_2(CH_2)_{15}CH_3$] presumably because of its sufficiently long unbranched chain, forms adducts readily.

When the carbon chain is sufficiently long, ring structures do not prevent adduct formation. For example, 1-phenyloctadecane [$C_6H_5CH_2(CH_2)_{16}CH_3$] forms an adduct with urea but 1-phenyloctane [$C_6H_5CH_2(CH_2)_6CH_3$] does not participate in adduct formation.

The stability of the adducts formed from the phenyl- or cyclohexyl-paraffins are usually considerably less than those of the corresponding straight-chain hydrocarbons with the same number of carbon atoms.

Many classes of organic compounds form adducts, and some examples of these are ketones, acids, esters, halides, mercaptans, and ethers; both saturated and unsaturated structures adduct, provided the chain is sufficiently long. The carbonyl oxygen in ketones, acids, and esters does not appear to interfere with adduct formation and, to some extent, even aids the process, since shorter carbon chains are adducted in oxygenated compounds. Thus acetone, with a straight chain of three carbon atoms, *n*-butyric acid with four carbon atoms, and their higher homologues form complexes with urea. Among the alkyl halides, 1-bromohexane, but not 2-bromohexane, forms an adduct; 2-bromodecane can be adducted, indicating that an unsubstituted chain of eight carbon atoms is sufficient to overcome substituent effects.

8.5.2.2 Thiourea Adduction

The relationship among structure, adductability, and adduct stability is not as well defined for the thiourea-adductible compounds as for the urea-reactive hydrocarbons. The stability of any thiourea adduct is quite low, even at 0°C, and corresponds approximately to the stability of the urea completes of the lower *n*-paraffins. The high stability of the urea completes of the higher *n*-paraffins has no parallel in thiourea completes. Among the higher hydrocarbons containing a ring or branching and a long chain, there is a lesser tendency for the thiourea adducts to form than in the lower molecular weight homologs, as the alkyl chain apparently reduces the stability. The higher *n*-paraffins adduct fairly readily and become more stable with increasing molecular weight, but the stability of these adducts is still of the same low order of magnitude as the lower molecular weight iso-paraffins and naphthenes.

8.5.2.3 Adduct Composition

Analysis of the completes for their mob ratio of reagent to reactant may be made by several methods. The urea (or thiourea) content of the crystalline complex may be established by nitrogen determination, or the amount of organic component may be determined from the carbon content or by the weight loss upon dissociation of the adduct. Also, measurement of concentration changes in the reaction liquid is applicable to certain mixtures.

For urea adducts of normal hydrocarbons, the mol ratio of urea to hydrocarbon in the adduct can be represented by the equation:

$$m = 0.65n + 1.5,$$

where m is the mol ratio of urea to hydrocarbon and n is the number of carbon atoms in the hydrocarbon. For all practical purposes, the use of the ratio of 3.3 g of urea per gram of normal hydrocarbon is more convenient. The greater variety of hydrocarbon structures adducting with thiourea than with urea adversely affects development of an equation relating the mol ratio of thiourea to reactant. For a given number of carbon atoms, the more compact molecules are generally associated with less thiourea in the adduct than are the less compact molecules. The weight ratio decreases from 2.7 to 2.8 for aliphatic compounds and monocyclic naphthenes, to 2.5 for dicyclic naphthenes, to 2.2 to 2.3 for condensed ring systems; the weight ratio for unsaturated compounds is generally very similar to that of the corresponding saturated structure.

8.5.2.4 Adduct Structure

X-ray diffraction patterns have been obtained for many different urea complexes, and it appears that the hydrocarbon in the adduct does not make any contribution to the structure. In the adduct, the urea molecules are held together by spirals of hydrogen bonds between an oxygen atom and a nitrogen atom of adjacent molecules; these adjacent molecules are turned 120°C with respect to one another. A distance of 3.7 Å separates adjacent nonhydrogen-bonded molecules along the axis of the spiral, and the edge length of the unit cell is 4.8 Å. In this molecular arrangement there exists a hexagonal channel, or canal, into which the reactant molecule must fit, which has been calculated as 4.9 Å in diameter. The diameter of an n-paraffin chain is of the order of 3.8 × 4.2 Å, thus allowing the n-paraffins to fit into the channel.

The crystal structure of urea in the adduct is entirely different from that of the pure reagent. There are six urea molecules in the hexagonal unit cell of the adduct, whereas pure urea is a tetragonal, close-packed crystal with no canal or available free space in which other molecules could be enclosed. Thus, the change in crystal structure from a tetragonal to a hexagonal system occurs at some time during the adduction process, and it is considered likely that the urea molecules grow in a spiral around the hydrocarbon. X-ray studies also indicate that many of the thiourea adducts have crystalline structures similar to one another and quite analogous to those of the urea adducts. Although pure thiourea crystals have an orthorhombic structure, the unit cell of the thiourea complexes is usually rhombohedral. A few cases are known in which the unit cell of the complex is orthorhombic, and with the thiourea adducts different crystalline forms can apparently be precipitated under different reaction conditions. The channel concept also applies to the thiourea adducts as it does to those of urea. However, the larger size of the sulfur atom in thiourea in relation to the oxygen in urea results in a channel with a larger cross section. The unit cell constants of the thiourea adducts apparently vary with the nature of the adducted molecule, and hence the channel cross-sectional dimensions also vary and are presumed to be of the order of 5.8 to 6.8 Å.

It is generally accepted that, in the urea adducts, the urea molecules are connected into spirals by hydrogen bonds between the oxygen and the amino groups of adjacent urea molecules, resulting in a channel into which the adducting compound can fit. The size of the channel limits the molecules that may adduct to those having cross-sectional dimensions equal to, or less than, those of the channel in the urea adduct, but in certain cases, distortion of the urea lattice may occur and slightly larger molecules can adduct. The structure of the thiourea adducts is similar to that of the urea adducts. The larger size of the sulfur atom results in a channel of somewhat larger cross-sectional dimensions, hence allowing larger molecules to adduct with thiourea than are able to do so with urea. In general, urea forms adducts with organic compounds containing a long unbranched chain, such as the *n*-paraffins, whereas thiourea forms complexes with compounds that contain a moderate amount of branching of cyclization. However, especially among the higher molecular weight hydrocarbons, urea adducts can be formed from *n*-paraffins.

8.5.2.5 Adduct Properties

The amount of dissociation that occurs when an adduct is contacted with solvents is a function of the solvent type. These types may be classified as:

1. Hydrocarbon solvents (benzene)
2. Urea (or thiourea) solvents (water)
3. Hydrocarbon and urea (or thiourea) solvents (methanol or benzene–methanol mixtures)

As expected, the last two types cause the greatest dissociation. Solvents in which urea or thiourea are relatively insoluble, but in which the hydrocarbon is readily dissolved, exert relatively little influence. With any solvent, increasing temperature increases the amount of dissociation.

The most practical method of decomposition of the adducts is by solution, generally with hot water. The adducted hydrocarbons form an immiscible layer on top of the aqueous urea solution and may be readily separated. Volatile reactants may be recovered by heating the adduct, either dry or with steam, and collecting the liberated hydrocarbon as it is released. Actually, a crude fractional dissociation may be accomplished in this manner or by partial solvent extraction. The least stable adduct formers are released first and may be collected. Separations of compounds from mixtures are carried out, in effect, on the basis of their stability. The most desirable results are generally obtained if all the possible material is first precipitated as the adduct; the recovered adducted material is then fractionally readducted. For such fractionation, the reprecipitation may be performed by:

1. Using a quantity of reagent insufficient to adduct all the material first adducted
2. At a different temperature using a more optimum reagent to reactant ratio
3. By applying both techniques at the same time

In a system containing essentially only one homologous series, such as the *n*-paraffins in paraffin waxes, the use of a fractional adduction technique furnishes a separation on the basis of molecular weight.

In a narrow molecular weight distillation fraction containing several adductible hydrocarbon types, separation can be accomplished as a result of the difference in stability of the adducts of the different hydrocarbon homologous series. With urea the *n*-paraffins react first, followed by the slightly branched iso-paraffins and by the more highly branched iso-paraffins and the cyclic structures last. However, working with a wide molecular weight fraction, such as petroleum waxes, the stability of the urea adducts of lower *n*-paraffins present are of the same order of magnitude as those of the adducts of higher iso-paraffins and monocyclic compounds; a mixture of types is of different molecular weights. Hence. fractional distillation of the hydrocarbon samples, either before or after adduction, allows better analysis of the various fractions.

Fractionation may also be effected by selective decomposition of the adduct or by selective replacement. In the former method, the adduct is extracted using a solvent of relatively poor dissociating power whereby the least stable adducts tend to dissociate first. In selective replacement, the hydrocarbons of the less stable adducts are gradually displaced by slightly more stable adductors, and these in turn are displaced by more stable adduct formers until the most stable *n*-paraffins are employed.

The relative instabilities of the thiourea adducts renders the method less useful than the urea method. In addition, the less selective nature of thiourea adduction and the fact that the large differences in stability within a homologous series with the urea adducts do not apply to the thiourea adducts.

8.6   USE OF THE DATA

In the simplest sense, petroleum can be considered composites of four major operational fractions. However, it must never be forgotten that the nomenclature of these fractions lies within the historical development of petroleum science and that the fraction names are operational and are related more to the general characteristics than to the identification of specific compound types. Nevertheless, once a convenient fractionation technique has been established, it is possible to compare a variety of different feedstocks, varying from a conventional petroleum to a propane asphalt (Corbett and Petrossi, 1978).

It is noteworthy here that, throughout the history of studies related to petroleum composition, there has been considerable attention paid to the asphaltene fraction and the resin fraction. This is due in no small part to the tendency of the asphaltenes to be responsible for high yields of thermal coke and also for shortened catalyst lifetimes in refinery operations (Speight, 2000). In fact, it is the unknown character of the asphaltenes that has also been responsible for drawing the attention of investigators for the last five decades (Speight, 1984, 1994). Residua contain the majority of all of the potential coke-forming constituents and catalyst poisons that were originally in the crude oil because the distillation process is a concentration process. Most of the coke formers and catalyst poisons are nonvolatile and, hence, the asphaltene fraction contains most of the coke-forming constituents and catalyst poisons that were originally present in a heavy oil or in a residuum.

One of the early findings of composition studies was that the behavior and properties of any material are dictated by composition (Boduszynski, 1989; Speight, 2000). Although the early studies were primarily focused on the composition and behavior of asphalt, the techniques developed for those investigations have provided an excellent means of studying heavy feedstocks (Tissot, 1984). Later studies have focused not only on the composition of petroleum and its major operational fractions but on further fractionation, which allows different feedstocks to be compared on a relative basis to provide a very simple but convenient feedstock map.

Such a map does not give any indication of the complex interrelationships of the various fractions (Koots and Speight, 1975), although predictions of feedstock behavior is possible using such data. It is necessary to take the composition studies one step further using subfractionation of the major fractions to obtain a more representative indication of petroleum composition.

Thus, by careful selection of an appropriate technique, it is possible to obtain an overview of petroleum composition that can be used for behavioral predictions. By taking the approach one step further and by assiduous collection of various subfractions, it becomes possible to develop the petroleum map and add an extra dimension to compositional studies (Long and Speight, 1989). Petroleum and heavy feedstocks then appear more as a continuum than as four specific fractions.

Such a concept has also been applied to the asphaltene fraction of petroleum (Long, 1981) in which asphaltenes are considered a complex state of matter based on molecular weight and polarity (Long, 1981). The advantage of such a concept is that it can be used to explain differences in asphaltene yield with different hydrocarbons (pentane and heptane) and also differences in the character of asphaltenes from petroleum and from coal liquids.

Further, petroleum can be viewed as consisting of two continuous distributions, one of molecular weight and the other of molecular type. Using data from molecular weight studies and elemental analyses, the number of nitrogen and sulfur atoms in the aromatic and polar aromatic fractions were also exhibited. These data showed that not only can every molecule in the resins and asphaltenes have more than one sulfur or nitrogen, but also some molecules probably contain both sulfur and nitrogen. As the molecular weight of the aromatic fraction decreases, the sulfur and nitrogen contents of the fractions also decrease. In contrast to the sulfur containing molecules, which appear in both the naphthene aromatics and the polar aromatic fractions, the oxygen compounds present in the heavy fractions of petroleum are normally found in the polar aromatics fraction.

Other work (Long and Speight, 1989) involved the development of a different type of compositional map using the molecular weight distribution and the molecular type distribution as coordinates. The separation involved the use of an adsorbent such as clay, and the fractions were characterized by solubility parameter as a measure of the polarity of the molecular types. The molecular weight distribution can be determined by gel permeation chromatography. Using these two distributions, a map of composition can be prepared using molecular weight and solubility parameter as the coordinates for plotting the two distributions. Such a composition map can provide insights into many separation and conversion processes used in petroleum refining.

The molecular type was characterized by the polarity of the molecules, as measured by the increasing the adsorption strength on an adsorbent. At the time of the original concept, it was unclear how to characterize the continuum in molecular type or polarity. For this reason, the molecular type coordinate of their first maps was the yield of the molecular types ranked in order of increasing polarity. However, this type of map can be somewhat misleading because the areas are not related to the amounts of material in a given type. The horizontal distance on the plot is a measure of the yield and there is not a continuous variation in polarity for the horizontal coordinate. It was suggested that the solubility parameter of the different fractions could be used to characterize both polarity and adsorption strength.

In order to attempt to remove some of these potential ambiguities, more recent developments of this concept have focused on the solubility parameter. The simplest map that can be derived using the solubility parameter is produced with the solubility parameters of the solvents used in solvent separation procedures, and equating these parameters to the various fractions. However, the solubility parameter boundaries determined by the values for the eluting solvents that remove the fractions from the adsorbent offer a further step in the evolution of petroleum maps (Long and Speight, 1989).

Measuring the overall solubility parameter of a petroleum fraction is a time-consuming chore. Therefore, it is desirable to have a simpler, less time-consuming measurement that can be made on petroleum fractions that will correlate with the solubility parameter and thus give an alternative continuum in polarity. In fact, the hydrogen-to-carbon atomic ratio and other properties of petroleum fractions that can be correlated with the solubility parameter (Speight, 1994) also provide correlation for the behavior of crude oils.

REFERENCES

Al-Kashab, K. and Neumann, H.J. 1976. *Strassen Tieflau* 30(3): 44.

Altgelt, K.H. and Gouw, T.H. eds., 1979. *Chromatography in Petroleum Analysis*. Marcel Dekker, New York.

Andersen, S.I. 1994. *Fuel Sci. Technol. Int.* 12: 51.

Bland, W.F. and Davidson, R.L. 1967. *Petroleum Processing Handbook*. McGraw-Hill, New York.

Boduszynski, M.M. 1989. Preprints. *Div. Petrol. Chem. Am. Chem. Soc.* 34(2): 329.

Boussingault, M. 1837. *Ann. Chem. Phys.* 64: 141.

Boyd, M.L. and Montgomery, D.S. 1963. *J. Inst. Pet.* 49: 345.

Bunger, J.W. 1977. Preprints, *Div. Petrol. Chem. Am. Chem. Soc.* 22(2): 716.

Corbett, L.W. and Petrossi, U. 1978. *Ind. Eng. Chem. Prod. Res. Dev.* 17: 342.

Cummins, J.J., Poulson, R.E., and Robinson, W.E. 1975. Preprints. *Div. Fuel Chem. Am. Chem. Soc.* 20(2): 154.

Fisher, I.P. 1987. *Fuel* 66: 1192.

Francisco, M.A. and Speight, J.G. 1984. Preprints. *Div. Fuel Chem. Am. Chem. Soc.* 29(1): 36.

Girdler, R.B. 1965. *Proc. Assoc. Asphalt Paving Technol.* 34: 45.

Halvorsen, I.J. and Skogestad, S. 2000. Distillation Theory. *Encyclopedia of Separation Science*. Wilson Ian D. (Editor-in-Chief). Academic Press Inc., New York. pp. 1117–1134.

Hildebrand, J.H., Prausnitz, J.M., and Scott, R.L. 1970. *Regular Solutions*. Van Nostrand-Reinhold, New York.

Hirsch, D.E., Hopkins, R.L., Coleman, H.J., Cotton, F.O., and Thompson, D.J. 1972. *Anal. Chem.* 44: 915.

Hoiberg, A.J. 1964. *Bituminous Materials: Asphalts, Tars and Pitches*. Interscience, New York.

Jewell, D.M., Weber, J.H., Bunger, J.W., Plancher, H., and Latham, D.R. 1972a. *Anal. Chem.* 44: 1391.

Jewell, D.M., Ruberto, R.G., and Davis, B.E. 1972b. *Anal. Chem.* 44: 2318.

Jewell, D.M., Albaugh, E.W., Davis, B.E., and Ruberto, R.G. 1974. *Ind. Eng. Chem. Fund.* 13: 278.

Koots, J.A. and Speight, J.G. 1975. *Fuel* 54: 179.

Lei, Z., Chen. B and Ding, Z. 2005. *Special Distillation Processes*. Elsevier, Amsterdam.

Long, R.B. 1981. *The Chemistry of Asphaltenes*. Bunger, J.W. and Li, N. eds., Advances in Chemistry Series No. 195. American Chemical Society, Washington, DC. p. 17.

Long, R.B. and Speight, J.G. 1989. *Rev. Inst. Fran. Petrol.* 44: 205.

Long, R.B. and Speight, J.G. 1990. *Rev. Inst. Fran. Petrol.* 45: 553.

Long, R.B. and Speight, J.G. 1998. *Petroleum Chemistry and Refining*. Taylor & Francis Publishers, Washington, DC. Chapter 1.

McKay, J.F., Weber, J.H., and Latham, D.R. 1975. *Fuel* 54: 50.

McKay, J.F., Weber, J.H., and Latham, D.R. 1976. *Anal. Chem.* 48: 891.

McKay, J.F., Amend, P.J., Cogswell, T.E., Harnsberger, P.M., Erickson, R.B., and Latham, D.R. 1977. Preprints. *Div. Petrol. Chem. Am. Chem. Soc.* 22(2): 708.

Mitchell, D.L. and Speight, J.G. 1973. *Fuel* 52: 149.

Oudin, J.L. 1970. *Rev. Inst. Fran. Petrol.* 25: 470.

Petyluk, F.B. 2004. *Distillation Theory and its Application to Optimal Design of Separation Units*. Cambridge University Press, London.

Pfeiffer, J.P. ed., 1950. *The Properties of Asphaltic Bitumen*. Elsevier, Amsterdam.

Reynolds, J.G. 1998. *Petroleum Chemistry and Refining*. Speight, J.G. ed., Taylor & Francis Publishers, Washington, DC. Chapter 3.

Rostler, F.S. 1965. In *Bituminous Materials: Asphalts, Tars, and Pitches*, Volume II, Part I. Hoiberg, A.J. ed., Interscience, New York. p. 151.

Selucky, M.L., Chu, Y., Ruo, T.C.S., and Strausz, O.P. 1977. *Fuel* 56: 369

Speight, J.G. 1979. Information Series No. 84. Alberta Research Council, Edmonton, Alberta, Canada.

Speight, J.G. 1980. *The Chemistry and Technology of Petroleum*. Marcel Dekker, New York.

Speight, J.G. 1984. *Characterization of Heavy Crude Oils and Petroleum Residues*. Tissot, B.P. ed., Editions Technip, Paris. p. 32.

Speight, J.G. 1989. *Neftekhimiya* 29: 610.

Speight, J.G. 1994. *Asphaltenes and Asphalts: Developments in Petroleum Science*, 40. Yen, T.F. and Chilingarian, G.V. eds. Elsevier, Amsterdam. Chapter 2.

Speight, J.G. 2000. *The Desulfurization of Heavy Oils and Residua*, 2nd Edn. Marcel Dekker, New York.

Speight, J.G. and Ozum, B. 2002. *Petroleum Refining Processes*. Marcel Dekker, New York.

Speight, J.G., Long, R.B., and Trowbridge, T.D. 1984. *Fuel* 63: 616.

Tissot, B.P. ed., 1984. *Characterization of Heavy Crude Oils and Petroleum Residues*. Editions Technip, Paris.

Traxler, R.N. 1961. *Asphalt: Its Composition, Properties and Uses*. Reinhold, New York.

van Nes, K. and van Westen, H.A. 1951. *Aspects of the Constitution of Mineral Oils*. Elsevier, Amsterdam.

Yan, J., Plancher, H., and Morrow, N.R. 1997. *Paper No. SPE 37232*. SPE International Symposium on Oilfield Chemistry. Houston, TX.

9 Petroleum Analysis

9.1 INTRODUCTION

Petroleum exhibits a wide range of physical properties and several relationships can be made between various physical properties (Speight, 2001). Whereas properties such as **viscosity, density, boiling point**, and color of petroleum may vary widely, the ultimate or elemental analysis varies, as already noted, over a narrow range for a large number of petroleum samples. The carbon content is relatively constant, while the hydrogen and heteroatom contents are responsible for the major differences between petroleum samples. Coupled with the changes brought about to the feedstock constituents by refinery operations, it is not surprising that petroleum characterization is a monumental task.

Petroleum refinery processes can be conveniently divided into three different types (Chapter 14 and Chapter 15):

1. Separation: division of the feedstock into various streams (or fractions) depending on the nature of the crude material
2. Conversion: that is, the production of saleable materials from the feedstock by skeletal alteration, or even by alteration of the chemical type of the feedstock constituents
3. Finishing: purification of the various product streams by a variety of processes that remove impurities from the product

In some case, a fourth category can be added and includes processes such as the **reforming** (molecular rearrangement) processes. For the purposes of this text, reforming processes are included in the finishing processes because that is precisely what they are: processes designed to finish various refinery streams and render them ready for sale as defined products.

The separation and finishing processes may involve distillation or treatment with a *wash* solution. The conversion processes are usually regarded as those processes that change the number of carbon atoms per molecule (thermal decomposition), alter the molecular hydrogen–carbon ratio (**aromatization, hydrogenation**), or even change the molecular structure of the material without affecting the number of carbon atoms per molecule (**isomerization**) (Chapter 15).

Although it is possible to classify refinery operations using the three general terms just outlined, the behavior of various feedstocks in these refinery operations is not simple. The atomic ratios from ultimate analysis give an indication of the nature of a feedstock and the generic hydrogen requirements to satisfy the refining chemistry (Chapter 15), but it is not possible to predict with any degree of certainty how the feedstock will behave during refining. Any deductions made from such data are pure speculation and are open to much doubt.

The chemical composition of a feedstock is a much truer indicator of refining behavior. Whether the composition is represented in terms of compound types or in terms of generic compound classes, it can enable the refiner to determine the nature of the reactions. Hence, chemical composition can play a large part in determining the nature of the products that arise from the refining operations. It can also play a role in determining the means by which a particular feedstock should be processed (Nelson, 1958; Ali et al., 1985; Wallace et al., 1988).

As indicated elsewhere (Chapter 7, Chapter 8, and Chapter 11), petroleum is an exceedingly complex and structured mixture consisting predominantly of hydrocarbons and containing sulfur, nitrogen, oxygen, and metals as minor constituents. Although sulfur has been reported in elemental form in some crude oils, most of the minor constituents occur in combination with carbon and hydrogen.

The physical and chemical characteristics of crude oils and the yields and properties of products or fractions prepared from them vary considerably and are dependent on the concentration of the various types of hydrocarbons and minor constituents present. Some types of petroleum have economic advantages as sources of fuels and lubricants with highly restrictive characteristics because they require less specialized processing than that needed for production of the same products from many types of crude oil. Others may contain unusually low concentrations of components that are desirable fuel or lubricant constituents, and the production of these products from such crude oils may not be economically feasible.

Evaluation of petroleum for use as a feedstock usually involves an examination of one or more of the physical properties of the material. By this means, a set of basic characteristics can be obtained that can be correlated with utility. To satisfy specific needs with regard to the type of petroleum to be processed, as well as to the nature of the product, various standards organizations, such as the American Society for Testing and Materials in North America and the Institute of Petroleum in Britain, have devoted considerable time and effort to the correlation and standardization of methods for the inspection and evaluation of petroleum and petroleum products. A complete discussion of the large number of routine tests available for petroleum fills an entire book. However, it seems appropriate that in any discussion of the physical properties of petroleum and petroleum products reference be made to the corresponding test, and accordingly, the various test numbers have been included in the text.

9.2   PETROLEUM ASSAY

An efficient assay is derived from a series of test data that give an accurate description of petroleum quality and allow an indication of its behavior during refining. The first step is, of course, to assure adequate (correct) sampling by use of the prescribed protocols (ASTM D4057).

Thus, analyses are performed to determine whether each batch of crude oil received at the refinery is suitable for refining purposes. The tests are also applied to determine if there has been any contamination during wellhead recovery, storage, or transportation that may increase the processing difficulty (cost). The information required is generally crude oil dependent or specific to a particular refinery and is also a function of refinery operations and desired product slate. To obtain the necessary information, two different analytical schemes are commonly used and these are: (1) an inspection assay and (2) a comprehensive assay (Table 9.1).

Inspection assays usually involve determination of several key bulk properties of petroleum (e.g., API gravity, sulfur content, pour point, and distillation range) as a means of determining if major changes in characteristics have occurred since the last comprehensive assay was performed.

For example, a more detailed inspection assay might consist of the following tests: API gravity (or density or relative density), sulfur content, pour point, viscosity, salt content, water and sediment content, trace metals (or organic halides). The results from these tests

TABLE 9.1
Recommended Inspection Data Required for Petroleum and Heavy Feedstocks (Including Residua)

Petroleum	Heavy Feedstocks
Density, specific gravity	Density, specific gravity
API gravity	API gravity
Carbon, wt.%	Carbon, wt.%
Hydrogen, wt.%	Hydrogen, wt.%
Nitrogen, wt.%	Nitrogen, wt.%
Sulfur, wt.%	Sulfur, wt.%
	Nickel, ppm
	Vanadium, ppm
	Iron, ppm
Pour point	Pour point
Wax content	
Wax appearance temperature	
Viscosity (various temperatures)	Viscosity (various temperatures)
Carbon residue of residuum	Carbon residue
	Ash, wt.%
Distillation profile:	Fractional composition:
All fractions plus vacuum residue	Asphaltenes, wt.%
	Resins, wt.%
	Aromatics, wt.%
	Saturates, wt.%

with the archived data from a comprehensive assay provide an estimate of any changes that have occurred in the crude oil that may be critical to refinery operations. Inspection assays are routinely performed on all crude oils received at a refinery.

On the other hand, the comprehensive (or full) assay is more complex (as well as time-consuming and costly) and is usually performed only when a new field comes on stream, or when the inspection assay indicates that significant changes in the composition of the crude oil have occurred. Except for these circumstances, a comprehensive assay of a particular crude oil stream may not (unfortunately) be updated for several years. A full petroleum assay may involve at least determinations of (1) carbon residue yield, (2) density (**specific gravity**), (3) sulfur content, (4) distillation profile (volatility), (5) metallic constituents, (6) viscosity, and (7) pour point, as well as any tests designated necessary to understand the properties and behavior of the crude oil under examination.

The inspection assay tests discussed above are not exhaustive, but are the ones most commonly used and provide data on the impurities present as well as a general idea of the products that may be recoverable. Other properties that are determined on an as needed basis include, but are not limited to, the following: (1) vapor pressure (Reid method) (ASTM D323, IP 69, IP 402), (2) total acid number—(ASTM D664, IP 177), and **the aniline point** (or mixed aniline point) (ASTM D611, IP 2).

The **Reid vapor pressure** test method (ASTM D323, IP 69) measures the vapor pressure of volatile petroleum. The Reid vapor pressure differs from the true vapor pressure of the sample due to some small sample vaporization and the presence of water vapor and air in the confined space. The acid number is the quantity of base, expressed in milligrams of potassium hydroxide per gram of sample that is required to titrate a sample in this solvent to a green or green-brown end point, using p-naphtholbenzein indicator solution. The strong acid number is the quantity of base, expressed as milligrams of potassium hydroxide per gram of sample, required to titrate a sample in the solvent from its initial meter reading to a meter reading corresponding to a freshly prepared nonaqueous acidic buffer solution or a well defined inflection point as specified in the test method (ASTM D664, IP 177). The aniline point (or mixed aniline point) (ASTM D611, IP 2) has been used for the characterization of crude oil, although it is more applicable to pure hydrocarbons and in their mixtures and is used to estimate the aromatic content of mixtures. Aromatic mixtures exhibit the lowest aniline points and paraffin mixtures have the highest aniline points. Cycloparaffins and olefins exhibit values between these two extremes. In any hydrocarbon homologous series, the aniline point increases with increasing molecular weight.

Using the data derived from the test assay, it is possible to assess petroleum quality and to acquire a degree of predictability of performance during refining. However, knowledge of the basic concepts of refining will help the analyst understand the production and, to a large extent, the anticipated properties of the product, which in turn is related to storage, sampling, and handling the products.

Therefore, the judicious choice of a crude oil to produce any given product is just as important as the selection of the product for any given purpose. Thus, initial inspection of petroleum (Table 9.2) will provide information relative to the most logical means of refining or correlation of various properties to structural types present and hence attempted classification of the petroleum (Chapter 2). Indeed, careful evaluation of petroleum from physical property data is a major part of the initial study of any petroleum destined as a refinery feedstock. Proper interpretation of the data resulting from the inspection of crude oil requires an understanding of their significance. In the following section, an indication of the physical properties that may be applied to petroleum, or even petroleum product, evaluation will be presented.

TABLE 9.2
Variation of Density and API Gravity of Various Residua with Temperature and Pressure

Source	Temperature			14.21	2,843	5,685	8,528	11,371	14,214
Residuum	°C	°F	Pressure: psi	14.21	2,843	5,685	8,528	11,371	14,214
			Pressure: atm.	0.97	193	387	580	774	967
			Pressure: MPa	0.098	19.6	39.2	58.8	78.4	98.0
California	25	77	Density, g/cm^3	1.014	1.023	1.031	1.038	1.045	1.051
			API gravity	8.0	6.8	5.7	4.8	3.9	3.3
	45	113	Density, g/cm^3	1.002	1.011	1.020	1.028	1.035	1.041
			API gravity	9.7	8.5	7.2	6.1	5.2	4.4
	65	149	Density, g/cm^3	0.990	1.000	1.009	1.017	1.025	1.032
			API gravity	11.4	10.0	8.7	7.6	6.6	5.6
Venezuela	25	77	Density, g/cm^3	1.024	1.032	1.040	1.048	1.054	1.061
			API gravity	6.7	5.6	4.6	3.5	2.7	1.9
	45	113	Density, g/cm^3	1.012	1.020	1.029	1.037	1.044	1.051
			API gravity	8.3	7.2	6.0	5.0	4.0	3.1
	65	149	Density, g/cm^3	1.000	1.009	1.018	1.027	1.034	1.041
			API gravity	10.0	8.7	7.5	6.3	5.3	4.4

9.3 PHYSICAL PROPERTIES

9.3.1 ELEMENTAL (ULTIMATE) ANALYSIS

The analysis of petroleum for the percentages of carbon, hydrogen, nitrogen, oxygen, and sulfur is perhaps the first method used to examine the general nature, and perform an evaluation, of a feedstock. The atomic ratios of the various elements to carbon (i.e., H/C, N/C, O/C, and S/C) are frequently used for indications of the overall character of the feedstock. It is also of value to determine the amounts of trace elements, such as vanadium, nickel and other metals, in a feedstock since these materials can have serious deleterious effects on catalyst performance during refining by catalytic processes.

However, it has become apparent, with the introduction of the heavier feedstocks into refinery operations, that these ratios are not the only requirement for predicting feedstock character before refining. The use of more complex feedstocks (in terms of chemical composition) has added a new dimension to refining operations. Thus, although atomic ratios, as determined by elemental analyses, may be used on a comparative basis between feedstocks, now there is no guarantee that a particular feedstock will behave as predicted from these data. Product slates cannot be predicted accurately, if at all, from these ratios.

The ultimate analysis (elemental composition) of petroleum is not reported to the same extent as for coal (Speight, 1994). Nevertheless, there are ASTM procedures for the ultimate analysis of petroleum and petroleum products but many such methods may have been designed for other materials.

For example, carbon content can be determined by the method designated for coal and coke (ASTM D3178) or by the method designated for municipal solid waste (ASTM E777). There are also methods designated for:

1. *Hydrogen content* (ASTM D1018, ASTM D3178, ASTM D3343, ASTM D3701, and ASTM E777)
2. *Nitrogen content* (ASTM D3179, ASTM D3228, ASTM D3431, ASTM E148, ASTM E258, and ASTM E778)
3. *Oxygen content* (ASTM E385)
4. *Sulfur content* (ASTM D124, ASTM D1266, ASTM D1552, ASTM D1757, ASTM D2662, ASTM D3177, ASTM D4045 and ASTM D4294)

Of the data that are available, the proportions of the elements in petroleum vary only slightly over narrow limits:

Carbon	83.0 to 87.0%
Hydrogen	10.0 to 14.0%
Nitrogen	0.1 to 2.0%
Oxygen	0.05 to 1.5%
Sulfur	0.05 to 6.0%
Metals (Ni and V)	<1000 ppm

And yet, there is a wide variation in physical properties from the lighter more mobile crude oil at one extreme to the heavier asphaltic crude oils at the other extreme. The majority of the more aromatic species and the heteroatoms occur in the higher boiling fractions of feedstocks. The heavier feedstocks are relatively rich in these higher boiling fractions (Chapter 8).

Of the ultimate analytical data, more has been made of the sulfur content than any other property. For example, the sulfur content (ASTM D124, D1552, and D4294) and the API gravity represent the two properties that have, in the past, had the greatest influence on determining the value of petroleum as a feedstock.

The sulfur content varies from about 0.1 wt.% to about 3 wt.% for the more conventional crude oils to as much as 5% to 6% for heavy oil and bitumen. Residua, depending on the sulfur content of the crude oil feedstock, may be of the same order or even have a higher sulfur content. Indeed, the very nature of the distillation process by which residua are produced, that is, removal of distillate without thermal decomposition, dictates that the majority of the sulfur, which is located predominantly in the higher molecular weight fractions, be concentrated in the residuum.

9.3.2 DENSITY AND SPECIFIC GRAVITY

The density and specific gravity of crude oil (ASTM D70, ASTM D71, ASTM D287, ASTM D941, ASTM D1217, ASTM D1298, ASTM D1480, ASTM D1481, ASTM D1555, ASTM D1657, ASTM D4052, IP 235, IP 160, IP 249, IP 365) are two properties that have found wide use in the industry for preliminary assessment of the character and quality of crude oil.

Density is the mass of a unit volume of material at a specified temperature and has the dimensions of grams per cubic centimeter (a close approximation to grams per milliliter). Specific gravity is the ratio of the mass of a volume of the substance to the mass of the same volume of water and is dependent on two temperatures, those at which the masses of the sample and the water are measured. When the water temperature is 4°C (39°F), the specific gravity is equal to the density in the centimeter–gram–second (cgs) system, since the volume of 1 g of water at that temperature is, by definition, 1 mL. Thus, the density of water, for example, varies with temperature, and its specific gravity at equal temperatures is always unity. The standard temperatures for a specific gravity in the petroleum industry in North America are 60/60°F (15.6/15.6°C).

In the early years of the petroleum industry, density was the principal specification for petroleum and refinery products; it was used to give an estimation of the gasoline and, more particularly, the kerosene present in the crude oil. However, the derived relationships between the density of petroleum and its fractional composition were valid only if they were applied to a certain type of petroleum and lost some of their significance when applied to different types of petroleum. Nevertheless, density is still used to give a rough estimation of the nature of petroleum and petroleum products. Although density and specific gravity are used extensively, the (American Petroleum Institute) API gravity is the preferred property. This property was derived from the Baumé scale:

$$\text{Degrees Baume} = (140/\text{sp gr}60/60°F) - 130$$

However, a considerable number of hydrometers calibrated according to the Baumé scale were found at an early period to be in error by a consistent amount, and this led to the adoption of the equation:

$$\text{Degrees API} = (141.5/\text{sp gr}60/60°F) - 131.5$$

The specific gravity of petroleum usually ranges from about 0.8 (45.3° API) for the lighter crude oils to over 1.0 (less than 10° API) for heavy crude oil and bitumen.

Specific gravity is influenced by the chemical composition of petroleum, but quantitative correlation is difficult to establish. Nevertheless, it is generally recognized that increased amounts of aromatic compounds result in an increase in density, whereas an increase in saturated compounds results in a decrease in density. Indeed, it is also possible to recognize certain preferred trends between the density of petroleum and one or another of the physical properties. For example, an approximate correlation exists between the density (API gravity) and sulfur content, **Conradson carbon residue**, viscosity, and nitrogen content (Speight, 2000).

Density, specific gravity, or API gravity may be measured by means of a hydrometer (ASTM D287, ASTM D1298, ASTM D1657, IP 160), a pycnometer (ASTM D70, ASTM D941, ASTM D1217, ASTM D1480, and ASTM D1481), by the displacement method (ASTM D712), or by means of a digital density meter (ASTM D4052, IP 365) and a digital density analyzer (ASTM D5002).

The pycnometer method (ASTM D70, ASTM D941, ASTM D1217, ASTM D1480, ASTM D1481) for determining density is reliable, precise and requires relatively small test samples. However, because of the time required, other methods such using the hydrometer (ASTM D1298), the density meter (ASTM D4052), and the digital density analyzer (ASTM D5002) are often preferred. However, surface tension effects can affect the displacement method and the density meter method loses some of its advantage when measuring the density of heavy oil and bitumen.

The pycnometer method (ASTM D70, ASTM D941, ASTM D1217, ASTM D1480, and ASTM D1481) is routinely used to measure the density of samples being charged to a distillation flask, where volume charge is needed, but the volume is not conveniently measured. The volume may be found weighing the sample, and determining the sample density. It is also used in routine measurements of material properties. It is worthy of note that even a small amount of solids in the sample will influence its measured density. For example, one per cent by weight solids in the sample can raise the density by 0.007 g/cm^3.

The densimeter method (ASTM D4052) uses an instrument that measures the total mass of a tube by determining its natural frequency of vibration. This frequency is a function of the dimensions and the elastic properties of the tube, and the weight of the tube and contents. Calibration with water and air provides data for the determination of the instrument constraints which allow conversion of the natural frequency of vibration to sample density.

The variation of density with temperature (Table 9.3), effectively the coefficient of expansion, is a property of great technical importance, since most petroleum products are sold by volume and specific gravity is usually determined at the prevailing temperature (21°C, 70°F) rather than at the standard temperature (60°F, 15.6°C). The tables of gravity corrections (ASTM D1555) are based on an assumption that the coefficient of expansion of all petroleum products is a function (at fixed temperatures) of density alone. Recent work has focused on the calculation and predictability of density using new mathematical relationships (Gomez, 1989, 1992).

However, not all of these methods are suitable for measuring the density or specific gravity of heavy oil and bitumen, although some methods lend themselves to adaptation.

The API gravity of a feedstock (ASTM D287) is calculated directly from the specific gravity. The specific gravity of bitumen shows a fairly wide range of variation. The largest degree of variation is usually due to local conditions that affect material close to the faces, or exposures, occurring in surface oil sand beds. There are also variations in the specific gravity of the bitumen found in beds that have not been exposed to weathering or other external factors. The range of specific gravity usually varies over the range of the order of 0.995 to 1.04.

A very important property of the Athabasca bitumen (which also accounts for the success of the hot water separation process) is the variation in density (specific gravity) of the bitumen with temperature. Over the temperature range 30°C to 130°C (85°F to 265°F) the bitumen is lighter than water. Flotation of the bitumen (with aeration) on the water is facilitated, hence the logic of the hot water separation process (Berkowitz and Speight, 1975; Spragins, 1978; Speight, 2005) (Chapter 6).

TABLE 9.3
Surface Tension of Various Hydrocarbons

Hydrocarbon	Surface Tension			
	20°C (68°F)	38°C (100°F)	93°C (200°F)	
n-Pentane	16.0	14.0	8.0	dyn/cm
	16.0	14.0	8.0	mN/m
n-Hexane	18.4	16.5	10.9	dyn/cm
	18.4	16.5	10.9	mN/m
n-Heptane	20.3	18.6	13.1	dyn/cm
	20.3	18.6	13.1	mN/m
n-Octane	21.8	20.2	14.9	dyn/cm
	21.8	20.2	14.9	mN/m
Cyclopentane	22.4			dyn/cm
	22.4			mN/m
Cyclohexane	25.0			dyn/cm
	25.0			mN/m
Tetralin	35.2			dyn/cm
	35.2			mN/m
Decalin	29.9			dyn/cm
	29.9			mN/m
Benzene	28.8			dyn/cm
	28.8			mN/m
Toluene	28.5			dyn/cm
	28.5			mN/m
Ethylbenzene	29.0			dyn/cm
	29.0			mN/m
n-Butyl benzene	29.2			dyn/cm
	29.2			mN/m

9.3.3 VISCOSITY

Viscosity is the force in dynes required to move a plane of 1 cm^2 area at a distance of 1 cm from another plane of 1 cm^2 area through a distance of 1 cm in 1 sec. In the cgs system, the unit of viscosity is the poise or centipoise (0.01 P). Two other terms in common use are *kinematic viscosity* and *fluidity*. The kinematic viscosity is the viscosity in centipoises divided by the specific gravity, and the unit is the stoke (cm^2/sec), although centistokes (0.01 cSt) is in more common usage; fluidity is simply the reciprocal of viscosity. The viscosity (ASTM D445, D88, D2161, D341, and D2270) of crude oils varies markedly over a very wide range. Values vary from less than 10 cP at room temperature to many thousands of centipoises at the same temperature.

In the early days of the petroleum industry, viscosity was regarded as the body of petroleum, a significant number for lubricants or for any liquid pumped or handled in quantity. The changes in viscosity with temperature, pressure, and rate of shear are pertinent not only in lubrication but also for such engineering concepts as heat transfer. The viscosity and relative viscosity of different phases, such as gas, liquid oil, and water, are determining influences in producing the flow of reservoir fluids through porous oil-bearing formations. The rate and amount of oil production from a reservoir are often governed by these properties.

Many types of instruments have been proposed for the determination of viscosity. The simplest and most widely used are capillary types (ASTM D445), and the viscosity is derived from the equation:

$$\mu = \pi r^r P/8nl,$$

where r is the tube radius, l the tube length, P the pressure difference between the ends of a capillary, n the coefficient of viscosity, and μ the quantity discharged in unit time. Not only are such capillary instruments the most simple, but when designed in accordance with known principles and used with known necessary correction factors, they are probably the most accurate viscometers available. It is usually more convenient, however, to use relative measurements, and for this purpose the instrument is calibrated with an appropriate standard liquid of known viscosity.

Batch flow times are generally used; in other words, the time required for a fixed amount of sample to flow from a reservoir through a capillary is the datum actually observed. Any features of technique that contribute to longer flow times are usually desirable. Some of the principal capillary viscometers in use are those of Cannon-Fenske, Ubbelohde, Fitzsimmons, and Zeitfuchs.

The Saybolt universal viscosity (SUS) (ASTM D88) is the time in seconds required for the flow of 60 mL of petroleum from a container, at constant temperature, through a calibrated orifice. The Saybolt furol viscosity (SFS) (ASTM D88) is determined in a similar manner, except that a larger orifice is employed.

As a result of the various methods for viscosity determination, it is not surprising that much effort has been spent on interconversion of the several scales, especially converting Saybolt to kinematic viscosity (ASTM D2161),

$$\text{Kinematic viscosity} = a \times \text{Saybolt s} + b/\text{Saybolt s},$$

where a and b are constants.

The Saybolt universal viscosity equivalent to a given kinematic viscosity varies slightly with the temperature at which the determination is made, because the temperature of the calibrated receiving flask used in the Saybolt method is not the same as that of the oil. Conversion factors are used to convert kinematic viscosity from 2 to 70 cSt at 38°C (100°F) and 99°C (210°F) to equivalent Saybolt universal viscosity in seconds. Appropriate multipliers are listed to convert kinematic viscosity over 70 cSt. For a kinematic viscosity determined at any other temperature, the equivalent Saybolt universal value is calculated by use of the Saybolt equivalent at 38°C (100°F) and a multiplier that varies with the temperature:

$$\text{Saybolt s at } 100°\text{F}(38°\text{C}) = \text{cSt} \times 4.635$$

$$\text{Saybolt s at } 210°\text{F}(99°\text{C}) = \text{cSt} \times 4.667$$

Various studies have also been made on the effect of temperature on viscosity since the viscosity of petroleum, or a petroleum product, decreases as the temperature increases. The rate of change appears to depend primarily on the nature or composition of the petroleum, but other factors, such as volatility, may also have a minor effect. The effect of temperature on viscosity is generally represented by the equation:

$$\log\log(n + c) = A + B\log T,$$

where n is absolute viscosity, T is temperature, and A and B are constants. This equation has been sufficient for most purposes and has come into very general use. The constants A and B vary widely with different oils, but c remains fixed at 0.6 for all oils with a viscosity over 1.5 cSt; it increases only slightly at lower viscosity (0.75 at 0.5 cSt). The viscosity–temperature characteristics of any oil, so plotted, thus create a straight line, and the parameters A and B are equivalent to the intercept and slope of the line. To express the viscosity and viscosity–temperature characteristics of an oil, the slope and the viscosity at one temperature must be known; the usual practice is to select 38°C (100°F) and 99°C (210°F) as the observation temperatures.

Suitable conversion tables are available (ASTM D341), and each table or chart is constructed in such a way that for any given petroleum or petroleum product the viscosity–temperature points result in a straight line over the applicable temperature range. Thus, only two viscosity measurements need be made at temperatures far enough apart to determine a line on the appropriate chart from which the approximate viscosity at any other temperature can be read. The charts can be applicable only to measurements made in the temperature range in which a given petroleum oil is a Newtonian liquid. The oil may cease to be a simple liquid near the cloud point because of the formation of wax particles or, near the boiling point, because of vaporization. However, the charts do not give accurate results when either the cloud point or boiling point is approached but they are useful over the Newtonian range for estimating the temperature at which oil attains a desired viscosity.

Since the viscosity–temperature coefficient of lubricating oil is an important expression of its suitability, a convenient number to express this property is very useful, and hence, a **viscosity index** (ASTM D2270) was derived. It is established that naphthenic oils have higher viscosity–temperature coefficients than the paraffinic oils at equal viscosity and temperatures. The Dean and Davis scale was based on the assignment of a zero value to a typical naphthenic crude oil and that of 100 to a typical paraffinic crude oil; intermediate oils were rated by the formula:

$$\text{Viscosity index} = (L - U)/(L - H \times 100),$$

where L and H are the viscosities of the zero and 100 index reference oils, both having the same viscosity at 99°C (210°F), and U is that of the unknown, all at 38°C (100°F). Originally, the viscosity index was calculated from Saybolt viscosity data, but subsequently, figures were provided for kinematic viscosity.

The viscosity of petroleum fractions increases on the application of pressure, and this increase may be very large. The pressure coefficient of viscosity correlates with the temperature coefficient, even when oils of widely different types are compared. A plot of the logarithm of the kinematic viscosity against pressure for several oils has given reasonably linear results up to about 20,000 psi, and the slopes of the isotherms are such that extrapolated values for a given oil intersect. At higher pressures, the viscosity decreases with increasing temperature, as at atmospheric pressure; in fact, viscosity changes of small magnitude are usually proportional to density changes, whether these are caused by pressure or by temperature.

The classification of lubricating oil by viscosity is a matter of some importance. A useful system is that of the Society of Automotive Engineers (SAE). Each oil class carries an index designation. For those classes designated by letter and number, maximum viscosity and minimum viscosity are specified at -18°C (0°F); those designated by number only are specified in viscosity at 99°C (210°F). Viscosity is also used in specifying several grades of fuel oils and in setting the requirement for kerosene and insulating oil.

Because of the importance of viscosity in determining the transport properties of petroleum, recent work has focused on the development of an empirical equation for predicting the dynamic viscosity of low molecular weight and high molecular weight hydrocarbon vapors at atmospheric pressure (Gomez, 1995). The equation uses molar mass and specific temperature as the input parameters and offers a means of estimation of the viscosity of a wide range of petroleum fractions. Other work has focused on the prediction of the viscosity of blends of lubricating oils as a means of accurately predicting the viscosity of the blend from the viscosity of the base oil components (Al-Besharah et al., 1989)

9.3.4 SURFACE AND INTERFACIAL TENSION

Surface tension is a measure of the force acting at a boundary between two phases. If the boundary is between a liquid and a solid or between a liquid and a gas (air) the attractive forces are referred to as surface tension, but the attractive forces between two immiscible liquids are referred to as interfacial tension.

Temperature and molecular weight have a significant effect on surface tension (Table 9.4 and Table 9.5). For example, in the normal hydrocarbon series, a rise in temperature leads to a decrease in the surface tension, but an increase in molecular weight increases the surface tension. A similar trend, that is, an increase in molecular weight causing an increase in surface tension, also occurs in the acrylic series and, to a lesser extent, in the alkylbenzene series.

The surface tension of petroleum and petroleum products has been studied for many years. The narrow range of values (approximately 24–38 dyn/cm) for such widely diverse materials as gasoline (26 dyn/cm), kerosene (30 dyn/cm), and the lubricating fractions (34 dyn/cm) has rendered the surface tension of little value for any attempted characterization. However, it is generally acknowledged that nonhydrocarbon materials dissolved in an oil reduce the surface tension: polar compounds, such as soaps and fatty acids, are particularly active. The effect is marked at low concentrations up to a critical value beyond which further additions cause little change; the critical value corresponds closely with that required for a monomolecular layer on the exposed surface, where it is adsorbed and accounts for the lowering. Recent work has focused on the predictability of surface tension using mathematical relationships (Gomez, 1987):

$$\text{Dynamic surface tension} = 681.3/K(1 - T/13.488^{1.7654} \times sg^{2.1250})^{1.2056}$$

TABLE 9.4
Effect of Temperature on the Surface Tension of Athabasca Bitumen

Temperature		Surface Tension	
°C	°F	dyn/cm	mN/m
21.1	70.0	35.3	35.3
23.3	74.0	34.7	34.7
30.0	86.0	30.1	30.1
43.3	110.0	27.3	27.3
51.7	125.0	28.0	28.0
65.6	150.0	25.4	25.4
73.9	165.0	22.5	22.5
82.2	180.0	21.0	21.0
87.8	190.0	18.9	18.9
95.6	204.0	20.0	20.0
104.0	219.0	19.2	19.2
123.9	255.0	18.2	18.2

TABLE 9.5
Distillation Profile of Conventional Crude Oil (Leduc, Woodbend, Upper Devonian, Alberta, Canada) and Selected Properties of the Fractions

	Boiling Range			Cumulative	Specific	API	Sulfur	Carbon Residue
	°C	°F	wt.%	wt.%	Gravity	Gravity	wt.%	(Conradson)
Whole crude oil				100.0	0.828	39.4	0.4	1.5
fraction*								
1	0–50	0–122	2.6	2.6	0.650	86.2		
2	50–75	122–167	3.0	5.6	0.674	78.4		
3	75–100	167–212	5.2	10.8	0.716	66.1		
4	100–125	212–257	6.6	17.4	0.744	58.7		
5	125–150	257–302	6.3	23.7	0.763	54.0		
6	150–175	302–347	5.5	29.2	0.783	49.2		
7	175–200	347–392	5.3	34.5	0.797	46.0		
8	200–225	392–437	5.0	39.5	0.812	42.8		
9	225–250	437–482	4.7	44.2	0.823	40.4		
10	250–275	482–527	6.6	50.8	0.837	37.6		
11	<200	<392	5.4	56.2	0.852	34.6		
12	200–225	392–437	4.9	61.1	0.861	32.8		
13	225–250	437–482	5.2	66.3	0.875	30.2		
14	250–275	482–527	2.8	69.1	0.883	28.8		
15	275–300	527–572	6.7	75.4	0.892	27.0		
Residuum	>300	>572	22.6	98.4	0.929	20.8		6.6
Distillation loss			1.6					

*Distillation at 765 mm Hg then at 40 mm Hg for fractions 11–15.

K is the Watson characterization factor (page 40), sg is the specific gravity (page 244), and T is the temperature in K.

A high proportion of the complex phenomena shown by emulsions and foams can be traced to these induced surface tension effects. Dissolved gases, even hydrocarbon gases, lower the surface tension of oils, but the effects are less dramatic and the changes probably result from dilution. The matter is presumably of some importance in petroleum production engineering in which the viscosity and surface tension of the reservoir fluid may govern the amount of oil recovered under certain conditions.

On the other hand, although petroleum products show little variation in surface tension, within a narrow range the interfacial tension of petroleum, especially of petroleum products, against aqueous solutions provides valuable information (ASTM D971). Thus, the interfacial tension of petroleum is subject to the same constraints as surface tension, that is, differences in composition, molecular weight, and so on. When oil–water systems are involved, the pH of the aqueous phase influences the tension at the interface; the change is small for highly refined oils, but increasing pH causes a rapid decrease for poorly refined, contaminated, or slightly oxidized oils.

A change in interfacial tension between oil and alkaline water has been proposed as an index for following the refining or deterioration of certain products, such as turbine and insulating oils. When surface or interfacial tensions are lowered by the presence of solutes, which tend to concentrate on the surface, some time is required to obtain the final concentration and hence the final value of the tension. In such systems, dynamic and static tension must be distinguished; the first concerns the freshly exposed surface with nearly the same composition as the body of the liquid; it usually has a value only slightly less than that of the pure solvent. The static tension is that existing after equilibrium concentrations have been reached at the surface.

The interfacial tension between oil and distilled water provides an indication of compounds in the oil that have an affinity for water. The measurement of interfacial tension has received special attention because of its possible use in predicting when an oil in constant use will reach the limit of its serviceability. This interest is based on the fact that oxidation decreases the interfacial tension of the oil. Further, the interfacial tension of turbine oil against water is lowered by the presence of oxidation products, impurities from the air or rust particles, and certain antirust compounds intentionally blended in the oil. Thus, a depletion of the antirust additive may cause an increase in interfacial tension, whereas the formation of oxidation products or contamination with dust and rust lowers the interfacial tension.

In following the performance of oil in service, a decrease in interfacial tension indicates oxidation, if it is known that antirust additives and contamination with dust and rust are absent. In the absence of contamination and oxidation products, an increase in interfacial tension indicates a depletion trend in the antirust additive. Very minor changes over appreciable periods of time signify satisfactory operating conditions. The addition of makeup oil to a system introduces further complications in following the effects of service on the interfacial tension of a particular charge of oil.

9.3.5 METALS CONTENT

Heteroatoms (nitrogen, oxygen, sulfur, and metals) are found in every crude oil and the concentrations have to be reduced to convert the oil to transportation fuel. The reason is that if nitrogen and sulfur are present in the final fuel during combustion, nitrogen oxides (NO_x) and sulfur oxides (SO_x) form, respectively. In addition, metals affect many upgrading processes adversely, poisoning catalysts in refining and causing deposits in combustion.

Heteroatoms do affect every aspect of refining. Sulfur is usually the most concentrated and is fairly easy to remove; many commercial catalysts are available that routinely remove 90% of the sulfur. Nitrogen is more difficult to remove than sulfur, and there are fewer catalysts that are specific to nitrogen. Metals cause particular problems because they poison catalysts used for sulfur and nitrogen removal as well as other processes such as catalytic cracking.

A variety of tests (ASTM D1026, D1262, D1318, D1368, D1548, D1549, D2547 D2599, D2788, D3340, D3341, and D3605) have been designated for the determination of metals on petroleum products. At the time of writing, the specific test for the determination of metals in whole feeds has not been designated. However, this task can be accomplished by combustion of the sample so that only inorganic ash remains. The ash can then be digested with an acid and the solution examined for metal species by atomic absorption (AA) spectroscopy or by inductively coupled argon plasma (ICP) spectrometry.

Heavy oils and residua contain relatively high proportions of metals either in the form of salts or as organometallic constituents (such as the metallo-porphyrins), which are extremely difficult to remove from the feedstock. Indeed, the nature of the process by which residua are produced virtually dictates that all the metals in the original crude oil are concentrated in the residuum (Speight, 2000). Those metallic constituents that may actually volatilize under the distillation conditions and appear in the higher boiling distillates are the exceptions here. The deleterious effect of metallic constituents on the catalyst is known, and serious attempts have been made to develop catalysts that can tolerate a high concentration of metals without serious loss of catalyst activity or catalyst life.

9.4 THERMAL PROPERTIES

9.4.1 VOLATILITY

The volatility of a liquid or liquefied gas may be defined as its tendency to vaporize, that is, to change from the liquid to the vapor or gaseous state. Because one of the three essentials for combustion in a flame is that the fuel be in the gaseous state, volatility is a primary characteristic of liquid fuels.

The vaporizing tendencies of petroleum and petroleum products are the basis for the general characterization of liquid petroleum fuels, such as liquefied petroleum gas, natural gasoline, motor and aviation gasoline, naphtha, kerosene, gas oil, diesel fuel, and fuel oil (ASTM D2715). A test (ASTM D6) also exists for determining the loss of material when crude oil and asphaltic compounds are heated. Another test (ASTM D20) is a method for the distillation of road tars that might also be applied for estimating the volatility of high molecular weight residues.

For some purposes, it is necessary to have information on the initial stage of vaporization. To supply this need, flash and fire, vapor pressure, and evaporation methods are available. The data from the early stages of the several distillation methods are also useful. For other uses, it is important to know the tendency of a product to partially vaporize or to completely vaporize, and in some cases, to know if small quantities of high-boiling components are present. For such purposes, chief reliance is placed on the distillation methods.

The **flash point** of petroleum or a petroleum product is the temperature to which the product must be heated under specified conditions to give off sufficient vapor to form a mixture with air that can be ignited momentarily by a specified flame (ASTM D56, D92, and D93). The **fire point** is the temperature to which the product must be heated under the prescribed conditions of the method to burn continuously when the mixture of vapor and air is ignited by a specified flame (ASTM D92).

From the viewpoint of safety, information about the flash point is of most significance at or slightly above the maximum temperatures (30°C to 60°C, 86°F to 140°F) that may be encountered in storage, transportation, and use of liquid petroleum products, in either closed or open containers. In this temperature range, the relative fire and explosion hazard can be estimated from the flash point. For products with flash point below 40°C (104°F), special precautions are necessary for safe handling. Flash points above 60°C (140°F) gradually lose their safety significance until they become indirect measures of some other quality.

The flash point of a petroleum product is also used to detect contamination. A substantially lower flash point than expected for a product is a reliable indicator that a product has become contaminated with a more volatile product, such as gasoline. The flash point is also an aid in establishing the identity of a particular petroleum product.

A further aspect of volatility that receives considerable attention is the vapor pressure of petroleum and its constituent fractions. The vapor pressure is the force exerted on the walls of a closed container by the vaporized portion of a liquid. Conversely, it is the force that must be exerted on the liquid to prevent it from vaporizing further (ASTM D323). The vapor pressure increases with temperature for any given gasoline, liquefied petroleum gas, or other product. The temperature at which the vapor pressure of a liquid, either a pure compound of a mixture of many compounds, equals 1 atm (14.7 psi, absolute) is designated as the boiling point of the liquid.

In each homologous series of hydrocarbons, the boiling points increase with molecular weight, and structure also has a marked influence since it is a general rule that branched paraffin isomers have lower boiling points than the corresponding *n*-alkane. In any given series, steric effects notwithstanding, there is an increase in boiling point with an increase in carbon number of the alkyl side chain. This particularly applies to alkyl aromatic compounds where alkyl-substituted aromatic compounds can have higher boiling points than polycondensed aromatic systems. And this fact is very meaningful when attempts are made to develop hypothetical structures for asphaltene constituents (Chapter 11).

The boiling points of petroleum fractions are rarely, if ever, distinct temperatures; it is, in fact, more correct to refer to the boiling ranges of the various fractions. To determine these ranges, the petroleum is tested in various methods of distillation, either at atmospheric pressure or at reduced pressure. In general, the limiting molecular weight range for distillation at atmospheric pressure without thermal degradation is 200 to 250, whereas the limiting molecular weight range for conventional vacuum distillation is 500 to 600.

As an early part of characterization studies, a correlation was observed between the quality of petroleum products and their hydrogen content since gasoline, kerosene, diesel fuel, and lubricating oil are made up of hydrocarbon constituents containing high proportions of hydrogen. Thus, it is not surprising that tests to determine the volatility of petroleum and petroleum products were among the first to be defined. Indeed, volatility is one of the major tests for petroleum products and it is inevitable that all products will, at some stage of their history, be tested for volatility characteristics.

Distillation involves the general procedure of vaporizing the petroleum liquid in a suitable flask either at atmospheric pressure (ASTM D86, D216, D285, D447, and D2892) or at reduced pressure (ASTM D1160), and the data are reported in terms of one or more of the following seven items:

1. *Initial boiling point* is the thermometer reading in the neck of the distillation flask when the first drop of distillate leaves the tip of the condenser tube. This reading is materially affected by a number of test conditions, namely, room temperature, rate of heating, and condenser temperature.

2. *Distillation temperatures* are usually observed when the level of the distillate reaches each 10% mark on the graduated receiver, with the temperatures for the 5% and 95% marks often included. Conversely, the volume of the distillate in the receiver, that is, the percentage recovered, is often observed at specified thermometer readings.

3. *End-point* or *maximum temperature* is the highest thermometer reading observed during distillation. In most cases, it is reached when the entire sample has been vaporized. If a liquid residue remains in the flask after the maximum permissible adjustments are made in heating rate, this is recorded as indicative of the presence of very high boiling compounds.

4. *Dry point* is the thermometer reading at the instant the flask becomes dry and is for special purposes, such as for solvents and for relatively pure hydrocarbons. For these purposes, dry point is considered more indicative of the final boiling point than end point or maximum temperature.

5. *Recovery* is the total volume of distillate recovered in the graduated receiver and *residue* is the liquid material, mostly recondensed vapors, left in the flask after it has been allowed to cool at the end of distillation. The residue is measured by transferring it to an appropriate small graduated cylinder. Low or abnormally high residues indicate the absence or presence, respectively, of high-boiling components.

6. *Total recovery* is the sum of the liquid recovery and residue; **distillation loss** is determined by subtracting the total recovery from 100%. It is, of course, the measure of the portion of the vaporized sample that does not condense under the conditions of the test. Like the initial boiling point, distillation loss is affected materially by a number of test conditions, namely, condenser temperature, sampling and receiving temperatures, barometric pressure, heating rate in the early part of the distillation, and others.

Provisions are made for correcting high distillation losses for the effect of low barometric pressure because of the practice of including distillation loss as one of the items
in some specifications for motor gasoline.

7. *Percentage evaporated* is, the percentage recovered at a specific thermometer reading or
   other distillation temperatures, or the converse. The amounts that have been evaporated are usually obtained by plotting observed thermometer readings against the
   corresponding observed recoveries plus, in each case, the distillation loss. The initial
   boiling point is plotted with the distillation loss as the percentage evaporated. Distillation
   data are considerably reproducible, particularly for the more volatile products.

One of the main properties of petroleum that serves to indicate the comparative ease with
which the material can be refined is the volatility (Chapter 14). Investigation of the volatility
of petroleum is usually carried out under standard conditions, thereby allowing comparisons
to be made between data obtained from various laboratories. Thus, nondestructive distillation
data (U.S. Bureau of Mines method) show that, not surprisingly, bitumen is a higher boiling
material than the more conventional crude oils (Table 9.6 and Table 9.7). There is usually little,
or no, gasoline (naphtha) fraction in bitumen and the majority of the distillate falls in the gas
oil-lubrication distillate range (>260°C, >500°F). In excess of 50% of each bitumen is
nondistillable under the conditions of the test, and the yield of the nonvolatile material
corresponds very closely to the asphaltic (asphaltenes plus resins) content of each feedstock.

Detailed fractionation of the sample might be of secondary importance. Thus, it must be
recognized that the general shape of a one-plate distillation curve is often adequate for

TABLE 9.6
**Distillation Profile of Bitumen (Athabasca, McMurray Formation, Upper Cretaceous,
Alberta Canada) and Selected Properties of the Fractions**

Feedstock	Boiling Range °C	Boiling Range °F	wt.%	Cumulative wt.%	Specific Gravity	API Gravity	Sulfur wt.%	Carbon Residue (Conradson)
Whole bitumen				100.0	1.030	5.9	5.8	19.6
fraction*								
1	0–50	0–122	0.0	0.0				
2	50–75	122–167	0.0	0.0				
3	75–100	167–212	0.0	0.0				
4	100–125	212–257	0.0	0.0				
5	125–150	257–302	0.9	0.9				
6	150–175	302–347	0.8	1.7	0.809	43.4		
7	175–200	347–392	1.1	2.8	0.823	40.4		
8	200–225	392–437	1.1	3.9	0.848	35.4		
9	225–250	437–482	4.1	8.0	0.866	31.8		
10	250–275	482–527	11.9	19.9	0.867	31.7		
11	<200	<392	1.6	21.5	0.878	29.7		
12	200–225	392–437	3.2	24.7	0.929	20.8		
13	225–250	437–482	6.1	30.8	0.947	17.9		
14	250–275	482–527	6.4	37.2	0.958	16.2		
15	275–300	527–572	10.6	47.8	0.972	14.1		
Residuum	>300	>572	49.5	97.3				39.6

*Distillation at 762 mm Hg and then at 40 mm Hg for fractions 11–15.

TABLE 9.7
Heat of Combustion of Canadian Heavy Oil and Bitumen

Heavy Oil or Bitumen	Heat of Combustion		
	Btu/lb	cal/g	kJ/kg
Athabasca, Mildred Lake	18,030	10,025	41,940
Carbonate, Grosmont	17,570–17,650	9,765–9,810	40,865–41,050
Cold Lake, Clearwater	17,975–18,300	9,990–10,170	41,810–42,530
Lloydminster	17,975–18,285	9,990–10,165	41,810–42,530
Peace River	17,750–18,020	9,880–10,020	41,350–42,530
Wabasca	17,875–18,400	9,935–10,230	41,580–42,800

making engineering calculations, correlating with other physical properties, and predicting the product slate (Nelson, 1958).

There is also another method that is increasing in popularity for application to a variety of feedstocks and that is the method commonly known as simulated distillation (ASTM D2887). The method has been well researched in terms of method development and application (Hickerson, 1975; Green, 1976; Stuckey, 1978; Vercier and Mouton, 1979; Thomas et al., 1983; Romanowski and Thomas, 1985; Schwartz et al., 1987; Thomas et al., 1987; Trestianu et al., 1985). The benefits of the technique include good comparisons with other ASTM distillation data as well as the application to higher boiling fractions of petroleum. In fact, data output include the provision of the corresponding Engler profile (ASTM D86) as well as the prediction of other properties such as vapor pressure and flash point (DeBruine and Ellison, 1973). When it is necessary to monitor product properties, as is often the case during refining operations, such data provide a valuable aid to process control and on-line product testing.

For a more detailed distillation analysis of feedstocks and products, a low-resolution, temperature-programmed gas chromatographic analysis has been developed to simulate time-consuming true boiling point distillation. The method relies on the general observation that hydrocarbons are eluted from a nonpolar adsorbent in the order of their boiling points. The regularity of the elution order of the hydrocarbon components allows the retention times to be equated to distillation temperatures (Green et al., 1964) and the term *simulated distillation by gas chromatography* (or simdis) is used throughout the industry to refer to this technique.

Simulated distillation by gas chromatography is often applied in the petroleum industry to obtain true boiling point data for distillates and crude oils (Butler, 1979). Two standardized methods (ASTM D2887 and ASTM D3710) are available for the boiling point determination of petroleum fractions and gasoline, respectively. The ASTM D2887 method utilizes nonpolar, packed gas chromatographic columns in conjunction with flame ionization detection. The upper limit of the boiling range covered by this method is approximately 540°C (1000°F) atmospheric equivalent boiling point. Recent efforts in which high temperature gas chromatography were used have focused on extending the scope of the ASTM D2887 method for higher boiling petroleum materials to 800°C (1470°F) atmospheric equivalent boiling point.

9.4.2 LIQUEFACTION AND SOLIDIFICATION

Petroleum and the majority of petroleum products are liquids at ambient temperature, and problems that may arise from solidification during normal use are not common. Nevertheless, the melting point is a test (ASTM D87 and D127) that is widely used by suppliers of wax and by wax consumers; it is particularly applied to the highly paraffinic or crystalline waxes. Quantitative prediction of the melting point of pure hydrocarbons is difficult, but the melting point tends to increase qualitatively with the molecular weight and with symmetry of the molecule.

Unsubstituted and symmetrically substituted compounds (e.g., benzene, cyclohexane, p-xylene, and naphthalene) melt at higher temperatures relative to the paraffin compounds of similar molecular weight: the unsymmetrical isomers generally melt at lower temperatures than the aliphatic hydrocarbons of the same molecular weight.

Unsaturation affects the melting point principally by its alteration of symmetry; thus ethane (−172°C, −278°F) and ethylene (−169.5°C, −273°F) differ only slightly, but the melting points of cyclohexane (6.2°C, 21°F) and cyclohexane (−104°C, −155°F) contrast strongly. All types of highly unsymmetrical hydrocarbons are difficult to crystallize; asymmetrically branched aliphatic hydrocarbons as low as octane and most substituted cyclic hydrocarbons comprise the greater part of the lubricating fractions of petroleum, crystallize slowly, if at all, and on cooling merely take the form of glass-like solids.

Although the melting points of petroleum and petroleum products are of limited usefulness, except to estimate the purity or perhaps the composition of waxes, the reverse process, solidification, has received attention in petroleum chemistry. In fact, solidification of petroleum and petroleum products has been differentiated into four categories, namely, freezing point, congealing point, cloud point, and pour point.

Petroleum becomes more or less a plastic solid when cooled to sufficiently low temperatures. This is due to the congealing of the various hydrocarbons that constitute the oil. The cloud point of a petroleum oil is the temperature at which paraffin wax or other solidifiable compounds present in the oil appear as a haze when the oil is chilled under definitely prescribed conditions (ASTM D2500 and ASTM D3117). As cooling is continued, all petroleum oils become more and more viscous and flow becomes slower and slower. The pour point of a petroleum oil is the lowest temperature at which the oil pours or flows under definitely prescribed conditions, when it is chilled without disturbance at a standard rate (ASTM D97).

The solidification characteristics of a petroleum product depend on its grade or kind. For grease, the temperature of interest is that at which fluidity occurs, commonly known as the dropping point. The dropping point of grease is the temperature at which the grease passes from a plastic solid to a liquid state and begins to flow under the conditions of the test (ASTM D566 and ASTM D2265). For another type of plastic solid, including petrolatum and microcrystalline wax, both melting point and congealing point are of interest.

The melting point of wax is the temperature at which the wax becomes sufficiently fluid to drop from the thermometer; the congealing point is the temperature at which melted petrolatum ceases to flow when allowed to cool under definitely prescribed conditions (ASTM D938).

For paraffin wax, the solidification temperature is of interest. For such purposes, the melting point is the temperature at which the melted paraffin wax begins to solidify, as shown by the minimum rate of temperature change, when cooled under prescribed conditions. For pure or essentially pure hydrocarbons, the solidification temperature is the freezing point, the temperature at which a hydrocarbon passes from a liquid to a solid state (ASTM D1015 and ASTM D1016).

The relationship of cloud point, pour point, melting point, and freezing point to one another varies widely from one petroleum product to another. Hence, their significance for different types of product also varies. In general, cloud, melting, and freezing points are of more limited value and each is of narrower range of application than the pour point.

The cloud point of petroleum or a petroleum product is the temperature at which paraffin wax or other solidifiable compounds present in the oil appear as a haze when the sample is chilled under definitely prescribed conditions (ASTM D2500, ASTM D3117, IP 219, IP 444, IP 445, IP 446).

To determine the cloud point and the pour point (ASTM D97, ASTM D5327, ASTM D5853, ASTM D5949, ASTM D5950, ASTM D5985, IP 15, IP 219, IP 441), the oil is contained in a glass test tube fitted with a thermometer and immersed in one of three baths containing coolants. The sample is dehydrated and filtered at a temperature 25°C (45°F) higher than the anticipated cloud point. It is then placed in a test tube and cooled progressively in coolants held at −1°C to +2°C (30°F to 35°F), −18°C to −20°C (−4°F to 0°F), and −32°C to −35°C (−26°F to −31°F), respectively. The sample is inspected for cloudiness at temperature intervals of 1°C (2°F). If conditions or oil properties are such that reduced temperatures are required to determine the pour point, alternate tests are available that accommodate the various types of samples. Related to the cloud point, the wax appearance temperature or wax appearance point is also determined (ASTM D3117, IP 389).

The pour point of petroleum or a petroleum product is determined using this same technique (ASTM D97, IP 15) and it is the lowest temperature at which the oil pours or flows. It is actually 2°C (3°F) above the temperature at which the oil ceases to flow under these definitely prescribed conditions, when it is chilled without disturbance at a standard rate. To determine the pour point, the sample is first heated to 46°C (115°F) and cooled in air to 32°C (90°F) before the tube is immersed in the same series of coolants as used for the determination of the cloud point. The sample is inspected at temperature intervals of 2°C (3°F) by withdrawal and holding horizontal for 5 sec until no flow is observed during this time interval.

Cloud and pour points are useful for predicting the temperature at which the observed viscosity of oil deviates from the true (Newtonian) viscosity in the low-temperature range. They are also useful for identification of oils or when planning the storage of oil supplies, as low temperatures may cause handling difficulties with some oils.

The pour point of a crude oil was originally applied to crude oil that had a high wax content. More recently, the pour point, like the viscosity, is determined principally for use in pumping and pipeline design calculations. Difficulty occurs in these determinations with waxy crude oils that begin to exhibit irregular flow behavior when wax begins to separate. These crude oils possess viscosity relationships that are difficult to predict in a pipeline operation. In addition, some waxy crude oils are sensitive to heat treatment, which can also affect their viscosity characteristics. This complex behavior limits the value of viscosity and pour point tests on waxy crude oils. At the present time, long crude oil pipelines and the increasing production of waxy crude oils make an assessment of the pumpability of a wax-containing crude oil through a given system a matter of some difficulty, which that can often only be resolved after field trials. Consequently, considerable work is in progress to develop a suitable laboratory pumpability test (such as that described in IP 230) that gives an estimate of minimum handling temperature and minimum line or storage temperature.

9.4.3 CARBON RESIDUE

Petroleum products are mixtures of many compounds that differ widely in their physical and chemical properties. Some of them may be vaporized in the absence of air at atmospheric pressure without leaving an appreciable residue. Other nonvolatile compounds leave a carbonaceous residue when destructively distilled under such conditions. This residue is known as carbon residue when determined in accordance with prescribed procedure.

The carbon residue is a property that can be correlated with several other properties of petroleum (Speight, 2000); hence it also presents indications of the volatility of the crude oil and the coke-forming (or gasoline-producing) propensity. However, tests for carbon residue are sometimes used to evaluate the carbonaceous depositing characteristics of fuels used in certain types of oil-burning equipment and internal combustion engines.

The mechanical design and operating conditions of such equipment have such a profound influence on carbon deposition during service that comparison of carbon residues between oils should be considered as giving only a rough approximation of relative deposit-forming tendencies. A more precise relationship between carbon residue and hydrogen content, H/C atomic ratio, nitrogen content, and sulfur content has been shown to exist (see also Nelson, 1974). These data can provide more precise information about the anticipated behavior of a variety of feedstocks in thermal processes (Roberts, 1989).

Because of the extremely small values of carbon residue obtained by the Conradson and Ramsbottom methods when applied to the lighter distillate fuel oils, it is customary to distill such products to 10% residual oil and determine the carbon residue thereof. Such values may be used directly in comparing fuel oils, as long as it is kept in mind that the values are that for a residuum oil and are not to be compared with the carbon residue of the whole feedstock.

There are two older methods for determining the carbon residue of a petroleum or petroleum product: the Conradson method (ASTM D189) and the Ramsbottom method (ASTM D524). Both are applicable to the relatively nonvolatile portion of petroleum and petroleum products, which partially decompose when distilled at a pressure of 1 atmosphere. However, crude oil that contains ash-forming constituents will have an erroneously high carbon residue by either method, unless the ash is first removed from the oil; the degree of error is proportional to the amount of ash.

A third method, involving micropyrolysis of the sample, is also available as a standard test method (ASTM D4530). The method requires smaller sample amounts and was originally developed as a thermogravimetric method. The carbon residue produced by this method is often referred to as the **microcarbon residue** (MCR). Agreements between the data from the three methods are good, making it possible to interrelate all of the data from carbon residue tests (Long and Speight, 1989).

Even though the three methods have their relative merits, there is a tendency to advocate use of the more expedient microcarbon method to the exclusion of the Conradson and Ramsbottom methods, because of the lesser amounts required in the microcarbon method, which is somewhat less precise in practical technique.

9.4.4 ANILINE POINT

The aniline point of a liquid was originally defined as the consolute or critical solution temperature of the two liquids, that is, the minimum temperature at which they are miscible in all proportions. The term is now most generally applied to the temperature at which exactly equal parts of the two are miscible. This value is more conveniently measured than the original value and is only a few tenths of a degree lower for most substances.

Although it is an arbitrary index (ASTM D611), the aniline point is of considerable value in the characterization of petroleum products. For oils of a given type, it increases slightly with molecular weight; for those of given molecular weight it increases rapidly with increasing paraffinic character. As a consequence, it was one of the first properties proposed for the group analysis of petroleum products with respect to aromatic and naphthene content. It is used, alternately, even in one of the more recent methods. The simplicity of the determination makes it attractive for the rough estimation of aromatic content when that value is important for functional requirements, as in the case of the solvent power of naphtha and the combustion characteristics of gasoline and diesel fuel.

9.4.5 Specific Heat

Specific heat is defined as the quantity of heat required to raise a unit mass of material through one degree of temperature (ASTM D2766).

Specific heats are extremely important engineering quantities in refinery practice, because they are used in all calculations on heating and cooling petroleum products. Many measurements have been made on various hydrocarbon materials, but the data for most purposes may be summarized by the general equation:

$$C = 1/d \ (0.388 + 0.00045t)$$

C is the specific heat at $t°F$ of an oil whose specific gravity 60/60°F is d; thus, specific heat increases with temperature and decreases with specific gravity.

9.4.6 Latent Heat

There are two properties that represent phase transformations: the latent heat of fusion and the latent heat of vaporization. The latent heat of fusion, defined as the quantity of heat necessary to change a unit weight of solid to a liquid without any temperature change, has received only intermittent attention but, nevertheless, some general rules have been formulated. For hydrocarbons, latent heats of fusion commence at approximately 15 cal/g for methane, rising to 40 cal/g for octane, then gradually approaching a limiting value of 55 cal/g. Branched paraffins usually have a lower latent heat of fusion than the normal isomers; paraffin wax has a latent heat of fusion in the range 50 to 60 cal/g.

The latent heat of vaporization, defined as the amount of heat required to vaporize a unit weight of a liquid at its atmospheric boiling point, is perhaps the most important property of the two and has received considerably more attention because of its connection with equipment design. The latent heat of vaporization at the atmospheric boiling point generally increases with increasing molecular weight and, for the normal paraffins, generally decreases with increasing temperature and pressure.

9.4.7 Enthalpy or Heat Content

Enthalpy is the heat energy necessary to bring a system from a reference state to a given state. Enthalpy is a function only of the end states and is the integral of the specific heats with respect to temperature between the limit states, plus any latent heats of transition that occur within the interval. The usual reference temperature is 0°C (32°F). Enthalpy data are easily obtained from specific heat data by graphic integration, or, if the empirical equation given for specific heat is sufficiently accurate, from the equation:

$$H = 1/d(0.388 + 0.000225t^2 - 12.65)$$

Generally, only differences in enthalpy are required in engineering design, that is, the quantity of heat necessary to heat (or cool) a unit amount of material from one temperature to another. Such calculations are very simple since the quantities are arithmetically additive, and the enthalpy for such a change of state is merely the difference between the enthalpies of the end states.

9.4.8 Thermal Conductivity

The thermal conductivity K of hydrocarbons (in cgs units) is given by the equation:

$$K = 0.28/d(l - 0.00054) \times 10^{-3}$$

where d is the specific gravity. The value for solid paraffin wax is about 0.00056, nearly independent of temperature and wax type; the oil equation holds satisfactorily for waxes above the melting point.

9.4.9 PRESSURE–VOLUME–TEMPERATURE RELATIONSHIPS

Hydrocarbon vapors, like other gases, follow the ideal gas law ($PV = RT$) only at relatively low pressures and high temperatures, that is, far from the critical state. Several more empirical equations have been proposed to represent the gas laws more accurately, such as the well-known van der Waals equation, but they are either inconvenient for calculation or require the experimental determination of several constants. A more useful device is to use the simple gas law and to induce a correction, termed the *compressibility factor*, μ, so that the equation takes the form:

$$PV = \mu RT$$

For hydrocarbons, the compressibility factor is very nearly a function only of the reduced variables of state, that is, a function of the pressure and temperature divided by the respective critical values. The compressibility factor method functions excellently for pure compounds but may become ambiguous for mixtures, because the critical constants have a slightly different significance. However, the use of pseudocritical temperature and pressure values are generally lower than the true values, permitting the compressibility factor to be employed in such cases.

9.4.10 HEAT OF COMBUSTION

The gross heats of combustion of crude oil and its products are given with fair accuracy by the equation:

$$Q = 12,400 - 2100d^2$$

where d is the 60/60°F specific gravity. Deviation is generally less than 1%, although many highly aromatic crude oils show considerably higher values; the ranges for crude oil is 10,000 to 11,600 cal/g and the heat of combustion of heavy oil and tar sand bitumen is considerably higher.

For gasoline, the heat of combustion is 11,000 to 11,500 cal/g and for kerosene (and diesel fuel) it falls in the range 10,500 to 11,200 cal/g. Finally, the heat of combustion for fuel oil is of the order of 9500 to 11,200 cal/g. Heats of combustion of petroleum gases may be calculated from the analysis and data of the pure compounds. Experimental values for gaseous fuels may be obtained by measurement in a water flow calorimeter, and heats of combustion of liquids are usually measured in a bomb calorimeter.

For thermodynamic calculation of equilibria useful in hydrocarbon research, combustion data of extreme accuracy are required, because the heats of formation of water and carbon dioxide are large in comparison with those in the hydrocarbons. Great accuracy is also required of the specific heat data for the calculation of free energy or entropy. Much care must be exercised in selecting values from the literature for these purposes, since many of those available were determined before the development of modern calorimetric techniques.

9.4.11 CRITICAL PROPERTIES

The temperature, pressure, and volume at the critical state are of considerable interest in petroleum physics, particularly in connection with modern high-pressure, high-temperature refinery operations and in correlating pressure–temperature–volume relationships for other states. Critical data are known for most of the lower molecular weight pure hydrocarbons, and standard methods are generally used for such determinations.

The critical point of a pure compound is the equilibrium state in which its gaseous and liquid phases are indistinguishable and coexistent; they have the same intensive properties. However, localized variations in these phase properties may be evident experimentally. The definition of the critical point of a mixture is the same. However, mixtures generally have a maximum temperature or pressure at other than the true critical point; maximum here denotes the greatest value at which two phases can coexist in equilibrium.

9.5 ELECTRICAL PROPERTIES

9.5.1 CONDUCTIVITY

From the fragmentary evidence available, the electrical conductivity of hydrocarbons is quite small. For example, the normal hydrocarbons (from hexane up) have an electrical conductivity smaller than 10^{-16} Ω/cm; benzene itself has an electrical conductivity of 4.4×10^{-17} Ω/cm, and cyclohexane has an electrical conductivity of 7×10^{-18} Ω/cm. It is generally recognized that hydrocarbons do not usually have an electrical conductivity larger than 10^{-18} Ω/cm. Thus, it is not surprising that the electrical conductivity of hydrocarbon oils is also exceedingly small (ASTM D3114), of the order of 10^{-19} to 10^{-12} Ω/cm.

Available data indicate that the observed conductivity is frequently more dependent on the method of measurement and the presence of trace impurities than on the chemical type of the oil. Conduction through oils is not ohmic; that is, the current is not proportional to field strength: in some regions it is observed to increase exponentially with the latter. Time effects are also observed, the current being at first relatively large and decreasing to a smaller steady value. This is partly because of electrode polarization and partly because of ions removed from the solution. Most oils increase in conductivity with rising temperatures.

9.5.2 DIELECTRIC CONSTANT

The dielectric constant, ε, of a substance may be defined as the ratio of the capacity of a condenser with the material between the condenser plates C to that with the condenser empty and under vacuum C_0:

$$\varepsilon = C/C_0$$

The dielectric constant of petroleum and petroleum products may be used to indicate the presence of various constituents, such as asphaltenes, resins, or oxidized materials. Further, the dielectric constant of petroleum products that are used in equipment, such as condensers, may actually affect the electrical properties and performance of that equipment (ASTM D877).

The dielectric constant of hydrocarbons, and hence most crude oils and their products, is usually low and decreases with an increase in temperature. It is also noteworthy that for hydrocarbons, hydrocarbons fractions, and products, the dielectric constant is approximately equal to the square of the refractive index. Polar materials have dielectric constants greater than the square of the refractive index.

9.5.3 DIELECTRIC STRENGTH

The dielectric strength, or breakdown voltage (ASTM D877), is the greatest potential gradient or potential that an insulator can withstand without permitting an electric discharge.

The property is, in the case of oils as well as other dielectric materials, somewhat dependent on the method of measurement, that is, on the length of path through which the breakdown occurs, the composition, shape, and condition of the electrode surfaces, and the duration of the applied potential difference.

The standard test used in North America is applied to oils of petroleum origin for use in cables, transformers, oil circuit breakers, and similar apparatus. Oils of high purity and cleanliness show nearly the same value under standard conditions, generally ranging from 30 to 35 kV. For alkanes, dielectric strength has been shown to increase linearly with liquid density, and the value for a mineral oil fits the data well. For *n*-heptane a correlation was found between the dielectric strength and the density changes with temperature. There are many reasons that the dielectric strength of an insulator may fail. The most important appears to be the presence of some type of impurity, produced by corrosion, oxidation, thermal or electrical cracking, or gaseous discharge; invasion by water is a common trouble.

9.5.4 DIELECTRIC LOSS AND POWER FACTOR

A condenser insulated with an ideal dielectric shows no dissipation of energy when an alternating potential is applied. The charging current, technically termed the *circulating current*, lags exactly 90° in phase angle of the applied potential, and the energy stored in the condenser during each half-cycle is completely recovered in the next. No real dielectric material exhibits this ideal behavior; that is, some energy is dissipated under alternating stress and appears as heat. Such a lack of efficiency is broadly termed *dielectric loss*.

Ordinary conduction comprises one component of dielectric loss. Here the capacitance-held charge is partly lost by short circuit through the medium. Other effects in the presence of an alternating field occur, and a dielectric of zero conductivity may still exhibit losses. Suspended droplets of another phase undergo spheroidal oscillation by electrostatic induction effects and dissipate energy as heat as a consequence of the viscosity of the medium. Polar molecules oscillate as electrets and dissipate energy on collision with others. All such losses are of practical importance when insulation is used in connection with alternating-current equipment.

The measure of the dielectric loss is the power factor. This is defined as the factor k in the relation:

$$k = W/EI$$

W is the power in watts dissipated by a circuit portion under voltage *E* and passing current *I*.

From AC theory, the power factory is recognized as the cosine of the phase angle between the voltage and current where a pure sine wave form exists for both; it increases with a use in temperature. When an insulating material serves as the dielectric of a condenser, the power factor is an intrinsic property of the dielectric. For practical electrical equipment, low-power factors for the insulation are of course always desirable; petroleum oils are generally excellent in this respect, having values of the order of 0.0005, comparable with fused quartz and polystyrene resins. The power factor of pure hydrocarbons is extremely small. Traces of polar impurities, however, cause a striking increase. All electrical oils, therefore, are drastically refined and handled with care to avoid contamination; insoluble oxidation products are particularly undesirable.

9.5.5 STATIC ELECTRIFICATION

Dielectric liquids, particularly light naphtha, may acquire high static charges on flowing through or being sprayed from metal pipes. The effect seems to be associated with colloidally dispersed contaminants, such as oxidation products, which can be removed by drastic filtration or adsorption. Since a considerable fire hazard is involved, a variety of methods have been studied for minimizing the danger.

For large-scale storage, avoidance of surface agitation and the use of floating metal roofs on tanks are beneficial. High humidity in the surrounding atmosphere is helpful in lowering the static charge, and radioactive materials have been used to try to induce discharge to ground. A variety of additives have been found that increase the conductivity of petroleum liquids, thus lowering the degree of electrification; chromium salts of alkylated salicylic acids and other salts of alkylated sulfosuccinic acids are employed in low concentrations, say 0.005%.

9.6 OPTICAL PROPERTIES

9.6.1 REFRACTIVE INDEX

The refractive index is the ratio of the velocity of light in a vacuum to the velocity of light in the substance. The measurement of the refractive index is very simple (ASTM D1218), requires small quantities of material, and, consequently, has found wide use in the characterization of hydrocarbons and petroleum samples.

For closely separated fractions of similar molecular weight, the values increase in the order: paraffin, naphthene, and aromatic. For polycyclic naphthenes and polycyclic aromatics, the refractive index is usually higher than that of the corresponding monocyclic compounds. For a series of hydrocarbons of essentially the same type, the refractive index increases with molecular weight, especially in the paraffin series. Thus, the refractive index can be used to provide valuable information about the composition of hydrocarbon (petroleum) mixtures; as with density, low values indicate paraffinic materials and higher values indicate the presence of aromatic compounds. However, the combination of refractive index and density may be used to provide even more definite information about the nature of a hydrocarbon mixture and, hence, the use of the refractivity intercept ($n - d/2$; ASTM D2159).

The refractive and specific dispersion, as well as the molecular and specific refraction, have all been advocated for use in the characterization of petroleum and petroleum products.

The refractive dispersion of a substance is defined as the difference between its refractive indices at two specified wavelengths of light. Two lines, commonly used to calculate dispersions are, the C (6563 Å, red) and F (4861 Å, blue) lines of the hydrogen spectrum. The specific dispersion is the refractive dispersion divided by the density at the same temperature:

$$\text{Specific dispersion} = n_F - n_C/d$$

This equation is of particular significance in petroleum chemistry because all the saturated hydrocarbons, naphthene and paraffin, have nearly the same value irrespective of molecular weight, whereas aromatics are much higher and unsaturated aliphatic hydrocarbons are intermediate.

Specific refraction is the term applied to the quantity defined by the expression:

$$n - 1/(n^2 + 2)d = C,$$

where n is the refractive index, d is the density, and C is a constant independent of temperature.

Molecular refraction is the specific refraction multiplied by molecular weight; its particular usefulness lies in the fact that it is very nearly additive for the components of a molecule; that is, numerical values can be assigned to atoms and structural features, such as double bonds and rings. The value for any pure compound is then approximately the sum of such component constants for the molecule.

9.6.2 Optical Activity

The occurrence of optical activity in petroleum is universal and is a general phenomenon not restricted to a particular type of crude oil, such as the paraffinic or naphthenic crude oils. Petroleum is usually dextrorotatory, that is, the plane of polarized light is rotated to the right, but there are known *laevorotatory* crude oils, that is, the plane of polarized light is rotated to the left, and some crude oils have been reported to be optically inactive.

Examination of the individual fractions of optically active crude oils shows that the rotatory power increases with molecular weight (or boiling point) to pronounced maxima and then decreases again. The rotatory power appears to be concentrated in certain fractions, the maximum lying at a molecular weight of about 350 to 400; this maximum is about the same for all crude oils. The occurrence of optically active compounds in unaltered natural petroleum has been a strong argument in favor of a rather low temperature origin of petroleum from organic raw materials (Chapter 2).

A magnetic field causes all liquids to exhibit optical rotation, usually in the same direction as that of the magnetizing current; this phenomenon is known as the Faraday effect, θ and it may be expressed by the relation:

$$\theta = pth$$

θ is the total angle of rotation, t is the thickness of substance through which the light passes, and h is the magnetic field; the constant p is an intrinsic property of the substance, usually termed the *Verdet constant* (minutes of arc/cm per G); there have been some attempts to use the Verdet constant in studying the constitution of hydrocarbons by physical property correlation.

9.7 SPECTROSCOPIC METHODS

Spectroscopic studies have played an important role in the evaluation of petroleum and of petroleum products for the last three decades and many of the methods are now used as standard methods of analysis for refinery feedstocks and products. Application of these methods to feedstocks and products is a natural consequence for the refiner.

The methods include the use of mass spectrometry to determine the (1) hydrocarbon types in middle distillates (ASTM D2425); (2) hydrocarbon types of gas oil saturate fractions (ASTM D2786); and (3) hydrocarbon types in low-olefin gasoline (ASTM D2789); (d) aromatic types of gas oil aromatic fractions (ASTM D3239). **Nuclear magnetic resonance spectroscopy** has been developed as a standard method for the determination of hydrogen types in aviation turbine fuels (ASTM D3701). X-ray fluorescence spectrometry has been applied to the determination of lead in gasoline (ASTM D2599), as well as to the determination of sulfur in various petroleum products (ASTM D2622 and ASTM D4294).

Infrared spectroscopy is used for the determination of benzene in motor and aviation gasoline (ASTM D4053), while ultraviolet spectroscopy is employed for the evaluation of mineral oils (ASTM D2269) and for determining the naphthalene content of aviation turbine fuels (ASTM D1840).

Other techniques include the use of flame emission spectroscopy for determining trace metals in gas turbine fuels (ASTM D3605) and the use of absorption spectrophotometry for the determination of the alkyl nitrate content of diesel fuel (ASTM D4046). Atomic absorption has been employed as a means of measuring the lead content of gasoline (ASTM D3237) and also for the manganese content of gasoline (ASTM D3831), as well as for determining barium, calcium, magnesium, and zinc contents of lubricating oils (ASTM D4628). Flame photometry has been employed as a means of measuring the lithium or sodium content of lubricating greases (ASTM D3340) and the sodium content of residual fuel oil (ASTM D1318).

Nowhere is the contribution of spectroscopic studies more emphatic than in application to the delineation of structural types in the heavier feedstocks. This has been necessary because of the unknown nature of these feedstocks by refiners. One particular example is the n.d.M. method (ASTM D3238) which is designed for the carbon distribution and structural group analysis of petroleum oils. Later investigators have taken structural group analysis several steps further than the n.d.M. method.

It is also appropriate at this point to give a brief description of other methods that are used for the identification of the constituents of petroleum (Yen, 1984).

It is not intended to convey here that any one of these methods can be used for identification purposes. However, although these methods may fall short of complete acceptability as methods for the characterization of individual constituents of feedstocks, they can be used as methods by which an overall evaluation of the feedstock may be obtained in terms of molecular types.

9.7.1 INFRARED SPECTROSCOPY

Conventional infrared spectroscopy yields information about the functional features of various petroleum constituents. For example, infrared spectroscopy will aid in the identification of N–H and O–H functions, the nature of polymethylene chains, the C–H out-of-place bending frequencies, and the nature of any polynuclear aromatic systems (Yen, 1973).

With the recent progress of Fourier transform infrared (FTIR) spectroscopy, quantitative estimates of the various functional groups can also be made. This is particularly important for application to the higher molecular weight solid constituents of petroleum (i.e., the asphaltene fraction). It is also possible to derive structural parameters from infrared spectroscopic data and these are: (1) saturated hydrogen to saturated carbon ratio; (2) paraffinic character; (3) naphthenic character; (4) methyl group content; and (5) paraffin chain length.

In conjunction with proton magnetic resonance (see next section), structural parameters such as the fraction of paraffinic methyl groups to aromatic methyl groups can be obtained.

9.7.2 NUCLEAR MAGNETIC RESONANCE

Nuclear magnetic resonance has frequently been employed for general studies and for the structural studies of petroleum constituents (Bouquet and Bailleul, 1982; Hasan et al., 1989). In fact, proton magnetic resonance (PMR) studies (along with infrared spectroscopic studies) were, perhaps, the first studies of the modern era that allowed structural inferences to be made about the polynuclear aromatic systems that occur in the high molecular weight constituents of petroleum.

In general, the proton (hydrogen) types in petroleum fractions can be subdivided into three types (Brown and Ladner, 1960) or into five types (Yen and Erdman, 1962).

The Brown and Ladner approach classifies the hydrogen types into (1) aromatic ring hydrogen; (2) aliphatic hydrogen adjacent to an aromatic ring; and (3) aliphatic hydrogen remote from an aromatic ring. The Yen and Erdman approach subdivides the hydrogen distribution into (1) aromatic hydrogen; (2) substituted hydrogen next to an aromatic ring; (3) naphthenic hydrogen; (4) methylene hydrogen; and (5) terminal methyl hydrogen remote from an aromatic ring. Other ratios are also derived from which a series of structural parameters can be calculated.

However, it must be remembered that the structural details of the carbon backbone obtained from proton spectra are derived by inference, but it must be recognized that protons at peripheral positions can be obscured by intermolecular interactions. This, of course, can cause errors in the ratios that can have a substantial influence on the outcome of the calculations (Ebert et al., 1987; Ebert, 1990).

It is in this regard that carbon-13 magnetic resonance (CMR) can play a useful role. Since carbon magnetic resonance deals with analyzing the carbon distribution types, the obvious structural parameter to be determined is the aromaticity, f_a. A direct determination from the various carbon type environments is one of the better methods for the determination of aromaticity (Snape et al., 1979). Thus, through a combination of proton and carbon magnetic resonance techniques, refinements can be made on the structural parameters and for the solid state high-resolution CMR technique, additional structural parameters can be obtained (Weinberg et al., 1981).

9.7.3 MASS SPECTROMETRY

Mass spectrometry can play a key role in the identification of the constituents of feedstocks and products (Aczel, 1989). The principal advantages of mass spectrometric methods are: (1) high reproducibility of quantitative analyses; (2) the potential for obtaining detailed data on the individual components and carbon number homologues in complex mixtures; and (3) a minimal sample size is required for analysis. The ability of mass spectrometry to identify individual components in complex mixtures is unmatched by any modern analytical technique. Perhaps, the exception is gas chromatography.

However, there are disadvantages arising from the use of mass spectrometry and these are: (1) the limitation of the method to organic materials that are volatile and stable at temperatures up to 300°C (570°F); and (2) the difficulty of separating isomers for absolute identification. The sample is usually destroyed, but this is seldom a disadvantage.

Nevertheless, in spite of these limitations, mass spectrometry does furnish useful information about the composition of feedstocks and products, even if this information is not as exhaustive as might be required. There are structural similarities that might hinder identification of individual components. Consequently, identification by type or by homologue will be more meaningful, since similar structural types may be presumed to behave similarly in processing situations. Knowledge of the individual isomeric distribution may add only a little to the understanding of the relationships between composition and processing parameters.

Mass spectrometry should be used discriminately where a maximum amount of information can be expected. The heavier nonvolatile feedstocks are for practical purposes, beyond the useful range of routine mass spectrometry. At the elevated temperatures necessary to encourage volatility, thermal decomposition will occur in the inlet and any subsequent analysis would be biased to the low molecular weight end and to the lower molecular products produced by thermal decomposition.

9.8 CHROMATOGRAPHIC METHODS

9.8.1 GAS CHROMATOGRAPHY

Gas–liquid chromatography (GLC) is a method for separating the volatile components of various mixtures. It is, in fact, a highly efficient fractionating technique, and it is ideally suited to the quantitative analysis of mixtures when the possible components are known and the interest lies only in determining the amounts of each present. In this type of application, gas chromatography has taken over much of the work previously done by the other techniques; it is now the preferred technique for the analysis of hydrocarbon gases, and gas chromatographic in-line monitors are having increasing application in refinery plant control.

Thus, it is not surprising that gas chromatography has been used extensively for individual component identification, as well as percentage composition, in the gaseous boiling ranges (ASTM D2163, ASTM D2426, ASTM D2504, ASTM D2505, ASTM D2593, ASTM D2597, ASTM D2712, ASTM D4424, ASTM D4864, ASTM D5303, ASTM D6159, IP 264, IP 318, IP 337, IP 345), in the gasoline boiling range (e.g., ASTM D2426, ASTM D2427, ASTM D3525, ASTM D3606, ASTM D3710, ASTM D4420, ASTM D4815, ASTM D5134, ASTM D5441, ASTM D5443, ASTM D5501, ASTM D5580, ASTM D5599, ASTM D5623, ASTM D5845, ASTM D5986, IP 425), in higher boiling ranges such as diesel fuel (ASTM D3524), aviation gasoline (ASTM D3606), engine or motor oil (ASTM D5480), and wax (ASTM D5442), as well as for the boiling range distribution of petroleum fractions (ASTM D2887, ASTM D5307), light hydrocarbons in stabilized crude oil (IP 344), or the purity of solvents using capillary gas chromatography (ASTM D2268). There are also recommendations for calibrating and checking gas chromatographic analyzers (IP 353).

The evolution of gas–liquid chromatography has been a major factor in the successful identification of petroleum constituents. It is, however, almost impossible to apply this technique to the higher boiling petroleum constituents because of the comparatively low volatility. It is this comparative lack of volatility in the higher molecular weight, asphaltic constituents of petroleum that brought about another type of identification procedure, namely, carbon-type analysis.

The technique has proved to be an exceptional and versatile instrumental tool for analyzing compounds that are of low molecular weight and that can be volatilized without decomposition. However, these constraints limit the principal applicability in petroleum science to feedstock identification when the composition is known to be in the low to medium boiling range. The use of this technique for direct component analysis in the heavy fractions of petroleum is subject to many limitations (Speight, 2001).

For example, the number of possible components of a certain molecular weight range increases markedly with increasing molecular weight. Further, there is a corresponding sharp decrease in physical property differences between isometric structures as the molecular weight increases. Thus, it is very difficult, and on occasion almost impossible, to separate and identify single components in the heavier fractions of petroleum by gas chromatography. Indeed, the molecular weights of the constituents dictate that long residence times are necessary. This is inevitably accompanied by the requirement of increased column temperature, which decreases the residence time on the column but, at the same time, increases the possibility of thermal decomposition.

The instrumentation for gas–liquid chromatography is fairly straightforward and involves passing a carrier gas through a controller to the column (packed with an adsorbent) at the opening of which is a sample injector. The carrier gas then elutes the components of the mixture through the column to the detector, at the end of which may be another gas flow monitor. Any gas, such as helium, argon, nitrogen, or hydrogen that is easily distinguishable from the components in the mixture may be used as the carrier gas.

Column dimensions vary, but for analytic purposes a packed column may be 6 ft. (2 m) long by 3 in. (6 mm) in diameter. It is also necessary to use a dissolving liquid as part of the column substance. This remains stationary on the adsorbent and affects partition of the components of the mixture. The solid support is usually a porous material that allows passage of the gas. For example, kieselguhr (diatomaceous earth), which can absorb up to 40% by weight of a liquid without appearing to be overly moist, is commonly used. The supporting material should not adsorb any of the components of the mixture and must therefore be inert.

Individual components of mixtures are usually identified by their respective retention times, that is, the time required for the component to traverse through the column under the specified conditions. Although tables for retention time data are available, it is more common in practice to determine the retention times of the pure compounds. The retention time of any component is itself a function of the many variables of column operation, such as the flow rate of the carrier gas and column temperature, and exact duplication of other operator's conditions may be difficult, if not impossible.

The sample size used in gas chromatography may vary upward from a microliter, and there is no theoretical upper limit to the size of the sample that may be handled if the equipment is built to accommodate it. The technique can be used for the analysis of mixtures of volatile, vaporizable compounds boiling at any temperature between absolute zero ($-273°C$, $459°F$) and $450°C$ ($840°F$). Identification of any substance that can be heated sufficiently without decomposing to give a vapor pressure of a few millimeters mercury is also possible.

The use of gas–liquid chromatography for direct component analysis in the higher boiling fractions of petroleum, such as residua, is beset by many problems, not the least of which is the low volatility (Chapter 9) and high adsorption tendencies (Chapter 8 and Chapter 9) of many of the high molecular weight constituents of petroleum. The number of possible components in any given molecular weight range increases markedly with the molecular weight (Speight, 2002), and there is a significant drop in the differences in physical properties among similar structural entities. This limits the ability of gas–liquid chromatography, and unless the sample has been fractionated by other techniques to reduce the complexity, complete component analysis is difficult, if not impossible.

The mass spectrometer identifies chemical compounds principally in terms of molecular type and molecular weight, and for many problems, therefore, it becomes necessary to use additional means of identification; the integrated gas–liquid chromatography infrared system is a very valuable complement to the mass spectrometer technique.

Considerable attention has also been given to trapping devices to collect gas chromatographic fractions for examination by one or more of the spectroscopic techniques. At the same time, developments in preparative gas–liquid chromatography have contributed even more to the compositional studies of petroleum and its products. With column size of 4 to 6 in. of diameter and capable of dealing with sample sizes of 200 mL or more, there is every possibility that gas–liquid chromatography will replace distillation in such areas as standard crude oil assay work.

Gas–liquid chromatography also provides a simple and convenient method for determining n-paraffin distribution throughout the petroleum distillate range. In this method, the n-paraffins are first separated by activated chemical destruction of the sieve with hydrofluoric acid, and the identity of the individual paraffins is determined chromatographically. This allows n-paraffin distribution throughout the boiling range $170°C$ to $500°C$ ($340°F$ to $930°F$) to be determined.

Gas chromatographic process analyzers have become very important in petroleum refineries. In some refineries, more samples are analyzed automatically by process chromatographs than are analyzed with laboratory instruments. These chromatographs are usually fully automatic. In some cases, after an analysis, the instrument even makes automatic adjustments to the refinery unit. The chromatographs usually determine from 1 to 10 components, and the analyses are repeated at short intervals (15 to 20 min) over 24 h.

A more recent, very important development in gas chromatography is its combination with a mass spectrometer as the detector. The technique in which gas chromatography is combined with spectrometry (GC/MS) has proved to be a powerful tool for identifying many compounds at very low levels in a wide range of boiling matrix. By the combination of the two techniques in one instrument, the onerous trapping of fractions from the gas chromatographic column is avoided and higher sensitivities can be attained. In passing through the gas chromatographic column, the sample is separated more or less according to its boiling point.

In view of the molecular characterizing nature of spectrometric techniques, it is not surprising that considerable attention has been given to the combined use of gas–liquid chromatography and these techniques. In recent years, the use of the **mass spectrometer** to monitor continuously the effluent of a chromatographic column has been reported, and considerable progress has been made in the development of rapid scan infrared spectrometers for this purpose. The mass spectrometer, however, has the advantage that the quantity of material required for the production of a spectrum is considerably less than that necessary to produce an infrared spectrum.

Although insufficient component resolution is observed in most cases, the eluting compounds at any time are usually closely related to each other in boiling point or molecular weight or both and are free from interfering lower and higher molecular weight species. Because of the reduced complexity of the gas chromatographic fractions, mass spectrometric scans carried out at regular intervals yield simpler spectra from which compound classes can be more easily determined.

Pyrolysis gas chromatography can be used for information on the gross composition of heavy petroleum fractions. In this technique, the sample under investigation is pyrolyzed and the products are introduced into a gas chromatography system for analysis. There has also been extensive use of pyrolysis gas chromatography by geochemists to correlate crude oil with source rock and to derive geochemical characterization parameters from oil-bearing strata.

In the technique of inverse gas–liquid chromatography, the sample under study is used as the stationary phase and a number of volatile test compounds are chromatographed on this column. The interaction coefficient determined for these compounds is a measure of certain qualities of the liquid phase. The coefficient is therefore indicative of the chemical interaction of the solute with the stationary phase. The technique has been used largely for studies of asphalt.

9.8.2 SIMULATED DISTILLATION

Gas–liquid chromatography has also been found useful for the preparation of simulated distillation curves. By integrating increments of the total area of the chromatogram and relating these to the boiling points of the components within each increment, which are calculated from the known boiling points of the easily recognizable n-paraffins, simulated boiling point data are produced.

Distillation is the most widely used separation process in the petroleum industry (Chapter 14). In fact, knowledge of the boiling range of crude feedstocks and finished products has been an essential part of the determination of feedstock quality since the start of the refining industry. The technique has been used for control of plant and refinery processes as well as for predicting product slates. Thus, it is not surprising that routine laboratory scale distillation tests have been widely used for determining the boiling ranges of crude feedstocks and a whole slate of refinery products (Chapter 9).

There are some limitations to the routine distillation tests. For example, although heavy crude oils contain volatile constituents, it is not always advisable to use distillation for identification of these volatile constituents. Thermal decomposition of the constituents of petroleum is known to occur at approximately 350°C (660°F). However, thermal decomposition of the constituents of the heavier, but immature, crude oil has been known to commence at temperatures as low as 200°C (390°F). Thus, thermal alteration of the constituents and erroneous identification of the decomposition products as natural constituents is always a possibility.

On the other hand, the limitations to the use of distillation as an identification technique may be economic, and detailed fractionation of the sample may also be of secondary importance. There have been attempts to combat these limitations, but it must be recognized that the general shape of a one-plate distillation curve is often adequate for making engineering calculations, correlating with other physical properties, and predicting the product slate (Nelson, 1958).

However, a low-resolution, temperature-programmed gas chromatographic analysis has been developed to simulate the time-consuming true boiling point distillation (ASTM D2887). The method relies on the general observation that hydrocarbons are eluted from a nonpolar adsorbent in the order of their boiling points. The regularity of the elution order of the hydrocarbon components allows the retention times to be equated to distillation temperatures (Green et al., 1964), and the term *simulated distillation by gas chromatography* (or simdis) is used throughout the industry to refer to this technique. The method has been well researched in terms of method development and application (Hickerson, 1975; Green, 1976; Stuckey, 1978; Vercier and Mouton, 1979; Thomas et al., 1983; Romanowski and Thomas, 1985; MacAllister and DeRuiter, 1985; Schwartz et al., 1987; Thomas et al., 1987). The benefits of the technique include good comparisons with other ASTM distillation data as well as application to higher boiling fractions of petroleum (Speight, 2002).

The full development of simulated distillation as a routine procedure has been made possible by the massive expansion in gas chromatographic instrumentation (such as the introduction of automatic temperature programming) since the 1960s. In fact, a fully automated simulated distillation system, under computer control, can operate continuously to provide finished reports in a choice of formats that agree well with true boiling point data. For example, data output includes the provision of the corresponding Engler profile (ASTM D86) as well as the prediction of other properties, such as vapor pressure and flash point (DeBruine and Ellison, 1973).

Simulated distillation by gas chromatography is applied in the petrochemical industry to obtain true boiling point distributions of distillates and crude oils (Butler, 1979). Two standardized methods, ASTM D2887 and D3710, are available for the boiling point determination of petroleum fractions and gasoline, respectively. The ASTM D2887 method utilizes nonpolar, packed gas chromatographic columns in conjunction with flame ionization detection. The upper limit of the boiling range covered by this method is to approximately 540°C (1000°F) atmospheric equivalent boiling point. Recent efforts in which high-temperature gas chromatography was used, have focused on extending the scope of the ASTM D2887 methods for higher boiling petroleum materials to 800°C (1470°F) atmospheric equivalent boiling point.

9.8.3 ADSORPTION CHROMATOGRAPHY

Adsorption chromatography has helped to characterize the group composition of crude oils and hydrocarbon products since the beginning of this century.

The type and relative amount of certain hydrocarbon classes in the matrix can have a profound effect on the quality and performance of the hydrocarbon product and two standard test methods have been used predominantly over the years (ASTM D2007, ASTM D4124). The fluorescent indicator adsorption (FIA) method (ASTM D1319) has served for over 30 years as the official method of the petroleum industry for measuring the paraffinic, olefinic, and aromatic content of gasoline and jet fuel. The technique consists of displacing a sample under *iso*-propanol pressure through a column packed with silica gel in the presence of fluorescent indicators specific to each hydrocarbon family. Despite its widespread use, fluorescent indicator adsorption has numerous limitations (Suatoni and Garber, 1975; Miller et al., 1983; Norris and Rawdon, 1984).

The segregation of individual components from a mixture can be achieved by application of adsorption chromatography in which the adsorbent is either packed in an open tube (column chromatography) or shaped in the form of a sheet (thin-layer chromatography, TLC). A suitable solvent is used to elute from the bed of the adsorbent. Chromatographic separations are usually performed for the purpose of determining the composition of a sample. Even with such complex samples as petroleum, much information about the chemical structure of a fraction can be gained from the separation data (Chapter 8).

In the present context, the challenge is the nature of the heteroatomic species in the heavier feedstocks. It is these constituents that are largely responsible for coke formation and catalyst deactivation during refining operations. Therefore, it is these constituents that are the focus of much of the study. An ideal integrated separation scheme for the analysis of the heteroatomic constituents should therefore meet several criteria:

1. The various compound types should be concentrated into a reasonable number of discrete fractions, and each fraction should contain specific types of the heteroatomic compounds. It is also necessary that most of the hetero-compounds be separated from the hydrocarbons and sulfur compounds that may constitute the bulk of the sample.
2. Perhaps most important, the separation should be reproducible insofar as the yields of the various fractions and the distribution of the compound types among the fractions should be constant within the limits of experimental error.
3. The separation scheme should be applicable to high-boiling distillates and heavy feedstocks such as residua since **heteroatomic compounds** often predominate in these feed-stocks.
4. The separation procedures should be relatively simple to perform and free of complexity.
5. Finally, the overall separation procedure should yield quantitative or, at worst, near quantitative recovery of the various heteroatomic species present in the feedstock. There should be no significant loss of these species to the adsorbent or, perhaps more important, any chemical alteration of these compounds. Should chemical alteration occur, it will give misleading data that have could serious effects on refining predictions or on geochemical observations.

Group type analysis by means of chromatography has been applied to a wide variety of petroleum types and products (Chapter 10). These types of analysis are often abbreviated by the names PONA (paraffins, olefins, naphthenes, and aromatics), PIONA (paraffins, iso-paraffins, olefins, naphthenes, and aromatics), PNA (paraffins, naphthenes, and aromatics), PINA (paraffins, iso-paraffins, naphthenes, and aromatics), or SARA (saturates, aromatics, resins, and asphaltenes).

The USBM–API (U.S. Bureau of Mines–American Petroleum Institute) method allows fractionation of petroleum samples into acids, bases, neutral nitrogen compounds, saturates, and mono-, di-, and polyaromatic compounds (Chapter 7). Multidimensional techniques, that is, the combination of two or more chromatographic techniques, can be very useful to gain further information about the individual components of chemical groups. Compounds can be isolated and identified from complex matrices and detailed fingerprinting of petroleum constituents is feasible (Altgelt and Gouw, 1979).

9.8.4 Gel Permeation Chromatography

There are two additional techniques that have evolved from the more recent development of chromatographic methods.

The first technique, gel filtration chromatography (GFC), has been successfully employed for application to aqueous systems by biochemists for more than three decades. This technique was developed using soft, cross-linked dextran beads. The second technique, gel permeation chromatography (GPC), employs semi-rigid, cross-linked polystyrene beads. In either technique, the packing particles swell in the chromatographic solvent, forming a porous gel structure.

The distinction between the methods is based on the degree of swelling of the packing; the dextran swells to a much greater extent than the polystyrene. Subsequent developments of rigid porous packings of glass, silica, and silica gel have led to their use and classification as packings for gel permeation chromatography.

Gel permeation chromatography, also called size exclusion chromatography (SEC), in its simplest representation consists of employing a column or columns packed with gels of varying pore sizes in a liquid chromatograph Carbognani, 1997). Under conditions of constant flow, the solutes are injected onto the top of the column, whereupon they appear at the detector in order of decreasing molecular weight. The separation is based on the fact that the larger solute molecules cannot be accommodated within the pore systems of the gel beads and thus, are eluted first. On the other hand, the smaller solute molecules have increasing volume within the beads, depending upon their relative size, and require more time to elute.

Thus it is possible, with careful flow control, calibration, injection, and detection (usually by refractive index or UV absorption), to obtain an accurate chromatographic representation of the molecular weight distribution of the solute (Carbognani, 1997). This must of course assume that there is no chemical or physical interaction between the solute and the gel that negates the concept of solute size and pore size. For example, highly polar, small molecules that could associate in solution and are difficult to dissociate could conceivably appear in the incorrect molecular weight range.

In theory, gel permeation chromatography is an attractive technique for the determination of the number of average molecular weight (M_n) distribution of petroleum fractions. However, it is imperative to recognize that petroleum contains constituents of widely differing polarity, including nonpolar paraffins and naphthenes (alicyclic compounds), moderately polar aromatics (mononuclear and condensed), and polar nitrogen, oxygen, and sulfur species. Each particular compound type interacts with the gel surface to a different degree. The strength of the interaction increases with increasing polarity of the constituents and with decreasing polarity of the solvent. It must therefore be anticipated that the ideal linear relationship of log M_n against elution volume V_e that may be operative for nonpolar hydrocarbon species cannot be expected to remain in operation. It must also be recognized that the lack of realistic standards of known number average molecular weight distribution and of chemical nature similar to that of the constituents of petroleum for calibration purposes may also be an issue. However, gel permeation chromatography has been employed in the study of petroleum constituents, especially the heavier constituents, and has yielded valuable data (Oelert, 1969; Baltus and Anderson, 1984; Hausler and Carlson, 1985; Reynolds and Biggs, 1988; Speight, 2001 and references cited therein).

The adoption of gel permeation chromatography represents a novel approach to the identification of the constituents since the method is not limited by the vapor pressure of the constituents. However, the situation is different with heavy petroleum samples. These are not homologous mixtures differing only in molecular weight. In any particular crude oil, a large variety of molecular species, varying from paraffinic molecules to the polynuclear aromatic ring systems, may not follow the assumed physical relationships that the method dictates from use with polymers (Altgelt, 1968, 1970).

Gel permeation chromatography is the separation method that comes closest to differentiating by molecular weight only, and is almost unaffected by chemical composition (hence the alternate name size exclusion chromatography). The method is that it separates by molecular size and has been used to measure molecular weights (Altgelt and Guow, 1979), although there is some question about the value of the data when the method is applied to asphaltenes (Speight et al., 1985 and references cited therein).

Size exclusion chromatography is usually practised with refractive index detection and yields a mass profile (concentration vs. time or elution volume) that can be converted to a mass vs. molecular weight plot by means of a calibration curve. The combination of size exclusion chromatography with element specific detection has widened this concept to provide the distribution of hetero-compounds in the sample as a function of elution volume and molecular weight.

The use of size exclusion chromatography with reverse phase high-performance liquid chromatography (HPLC) with a graphite furnace atomic absorption (GFAA) detector has been described for measuring the distribution of vanadium and nickel in high molecular weight petroleum fractions, including the asphaltene fraction. Using variants of this technique, inductively coupled and direct current plasma atomic emission spectroscopy (ICP and DCP), the method was extended and improved the former size exclusion chromatography–graphite furnace atomic absorption (SEC–GFAA) method, allowing the separation to be continuously monitored.

The combination of gel permeation chromatography with another separation technique also allows the fractionation of a sample separately by molecular weight and by chemical structure. This is particularly advantageous for the characterization of the heavier fractions of petroleum materials because there are limitations to the use of other methods (Altgelt, 1968, 1970). Thus, it is possible to obtain a matrix of fractions differing in molecular weight and in chemical structure. It is also considered advisable to first fractionate a feedstock by gel permeation chromatography to avoid overlap of the functionality that might occur in different molecular weight species in the separation by other chromatographic methods.

The combination of gel permeation chromatography with another separation technique also allows the fractionation of a sample separately by molecular weight and by chemical structure. This is particularly advantageous for the characterization of the heavier fractions of petroleum materials because there are limitations to the use of other methods (Altgelt, 1968, 1970). Thus, it is possible to obtain a matrix of fractions differing in molecular weight and in chemical structure. It is also considered advisable to first fractionate a feedstock by gel permeation chromatography to avoid overlap of the functionality that might occur in different molecular weight species in the separation by other chromatographic methods.

In short, the gel permeation chromatographic technique concentrates all of a specific functional type into one fraction, recognizing that there will be a wide range of molecular weight species in that fraction. This is especially true when the chromatographic feedstock is a whole feed rather than a distillate fraction.

9.8.5 Ion-Exchange Chromatography

Ion-exchange chromatography is widely used in the analyses of petroleum fractions for the isolation and preliminary separation of acid and basic components (Speight, 2001 and references cited therein). This technique has the advantage of greatly improving the quality of a complex operation, but it can be a very time consuming separation.

Ion-exchange resins are prepared from aluminum silicates, synthetic resins, and polysaccharides. The most widely used resins have a skeletal structure of polystyrene cross-linked with varying amounts of divinylbenzene. They have a loose gel structure of cross-linked polymer chains through which the sample ions must diffuse to reach most of the exchange sites. Since ion-exchange resins are usually prepared as beads that are several hundred micrometers in diameter, most of the exchange sites are located at points quite distant from the surface. Because of the polyelectrolyte nature of these organic resins, they can absorb large amounts of water or solvents and swell to volumes considerably larger than the dried gel. The size of the species that can diffuse through the particle is determined by the intermolecular spacing between the polymeric chains of the three-dimensional polyelectrolyte resin.

Cation-exchange chromatography is now used primarily to isolate the nitrogen constituents in a petroleum fraction (Drushel and Sommers, 1966; McKay et al., 1974). The relative importance of these compounds in petroleum has arisen because of their deleterious effects in many petroleum refining processes. They reduce the activity of cracking and hydrocracking catalysts and contribute to gum formation, color, odor, and poor storage properties of the fuel. However, not all basic compounds isolated by cation-exchange chromatography are nitrogen compounds. Anion-exchange chromatography is used to isolate the acid components (such as carboxylic acids and phenols) from petroleum fractions.

9.8.6 HIGH-PERFORMANCE LIQUID CHROMATOGRAPHY

HPLC, particularly in the normal phase mode, has found great utility in separating different hydrocarbon group types and identifying specific constituent types (Colin and Vion, 1983; Miller et al., 1983; Chartier et al., 1986). However, a severe shortcoming of most high-performance liquid chromatographic approaches to a hydrocarbon group type of analysis is the difficulty in obtaining accurate response factors applicable to different distillate products. Unfortunately, accuracy can be compromised when these response factors are used to analyze hydrotreated and hydrocracked materials having the same boiling range. In fact, significant changes in the hydrocarbon distribution within a certain group type cause the analytic results to be misleading for such samples because of the variation in response with carbon number exhibited by most routinely used HPLC detectors (Drushel, 1983).

Of particular interest is the application of the HPLC technique to the identification of the molecular types in nonvolatile feedstocks such as residua., The molecular species in the asphaltene fraction have been of particular interest (Chmielowiec et al., 1980; Alfredson, 1981; Bollet et al., 1981; Colin and Vion, 1983; George and Beshai, 1983; Felix et al., 1985; Coulombe and Sawatzky, 1986), leading to identification of the size of polynuclear aromatic systems in the asphaltene constituents.

Several recent high-performance liquid chromatographic separation schemes are particularly interesting since they also incorporate detectors not usually associated with conventional hydrocarbon group types of analyses (Matsushita et al., 1981; Miller et al., 1983; Norris and Rawdon, 1984; Rawdon, 1984; Lundanes and Greibokk, 1985; Schwartz and Brownlee, 1986). The ideal detector for a truly versatile and accurate hydrocarbon group type of analysis is one that is sensitive to hydrocarbons but demonstrates a response independent of carbon number. More recent work (Hayes and Anderson, 1987) has demonstrated the merits of the dielectric constant detector as an integral part of a hydrocarbon group analyzer system.

In general, the amount of information that can be derived from any chromatographic separation, however effective, depends on the detectors (Hayes and Anderson, 1987). As the field of application for high-performance liquid chromatography has increased, the limitations of commercially available conventional detectors, such as ultraviolet or visible absorption (UV/VIS) and refractive index (RI) have become increasingly restrictive to the growth of the technique. This has led to search for detectors capable of producing even more information. The so-called hyphenated techniques are the outcome of this search.

The general advantages of high-performance liquid chromatography method are: (1) each sample is analyzed as received; (2) the boiling range of the sample is generally immaterial; (3)

the total time per analysis is usually of the order of minutes; and (4) the method can be adapted for on-stream analysis.

9.8.7 SUPERCRITICAL FLUID CHROMATOGRAPHY

A supercritical fluid is defined as a substance above its critical temperature that has properties not usually found at ambient temperatures and pressures. Use of a fluid under supercritical conditions conveys upon the fluid extraction capabilities that allows the opportunity to improve recovery of a solute.

In supercritical fluid chromatography, the mobile phase is a substance maintained at a temperature a few degrees above its critical point. The physical properties of this substance are intermediate to those of a liquid and of a gas at ambient conditions. Hence, it is preferable to designate this condition as the supercritical phase.

In a chromatographic column, the supercritical fluid usually has a density about one-third to one-fourth of that of the corresponding liquid when used as the mobile phase; the diffusivity is about 1/100 that of a gas and about 200 times that of the liquid. The viscosity is of the same order of magnitude as that of the gas. Thus, for chromatographic purposes, such a fluid has more desirable transport properties than a liquid. In addition, the high density of the fluid results in a 1000-fold better solvency than that of a gas. This is especially valuable for analyzing high-molecular-weight compounds.

A primary advantage of chromatography using supercritical mobile phases results from the mass transfer characteristics of the solute. The increased diffusion coefficients of supercritical fluids compared with liquids can lead to greater speed in separations or greater resolution in complex mixture analyses. Another advantage of supercritical fluids compared with gases is that they can dissolve thermally labile and nonvolatile solutes and, upon expansion (decompression) of this solution, introduce the solute into the vapor phase for detection. Although supercritical fluids are sometimes considered to have superior solvating power, they usually do not provide any advantages in solvating power over liquids, given a similar temperature constraint. In fact, many unique capabilities of supercritical fluids can be attributed to the poor solvent properties obtained at lower fluid densities. This dissolution phenomenon is increased by the variability of the solvent power of the fluid with density, as the pressure or temperature changes.

The solvent properties that are most relevant for supercritical fluid chromatography are the critical temperature, polarity, and any specific solute–solvent intermolecular interactions (such as hydrogen bonding) that can enhance solubility and selectivity in a separation. Nonpolar or low-polarity solvents with moderate critical temperatures (e.g., nitrous oxide, carbon dioxide, ethane, propane, pentane, xenon, sulfur hexafluoride, and various Freons) have been well explored for use in supercritical fluid chromatography. Carbon dioxide has been the fluid of choice in many supercritical fluid chromatography applications because of its low critical temperature (31°C, 88°F), nontoxic nature, and lack of interference with most detection methods (Lundanes et al., 1986).

9.9 MOLECULAR WEIGHT

Even though refining produces, in general, lower molecular weight species than those originally in the feedstock, there is still the need to determine the molecular weight of the original constituents as well as the molecular weights of the products as a means of understanding the process. For those original constituents and products, e.g., resins and asphaltenes, that have little or no volatility, vapor pressure osmometry (VPO) has been proven to be of considerable value.

A particularly appropriate method involves the use of different solvents (at least two), and the data are then extrapolated to infinite dilution (Schwager et al., 1979). There has also been the use of different temperatures for a particular solvent after which the data are extrapolated to room temperature (Speight et al., 1985; Speight, 1987). In this manner, different solvents are employed and the molecular weight of a petroleum fraction (particularly the asphaltenes) can be determined, for which it can be assumed that there is little or no influence from any intermolecular forces. In summary, the molecular weight may be as close to the real value as possible.

In fact, it is strongly recommended that to negate concentration effects and temperature effects, the molecular weight determination be carried out at three different concentrations at three different temperatures. The data for each temperature are then extrapolated to zero concentration and the zero concentration data at each temperature are then extrapolated to room temperature (Speight, 1987).

9.10   USE OF THE DATA

The data derived from the evaluation techniques described here can be employed to give the refiner an indication of the means by which the crude feedstock should be processed as well as for the prediction of product properties (Dolbear et al., 1987; Adler and Hall, 1988; Wallace and Carrigy, 1988; Al-Besharah et al., 1989). Other properties (Table 9.1) may also be required for further feedstock evaluation, or, more likely, for comparison between feedstocks even though they may not play any role in dictating which refinery operations are necessary.

Nevertheless, it must be emphasized that to proceed from the raw evaluation data to full-scale production is not the preferred step; further evaluation of the processability of the feedstock is usually necessary through the use of a pilot-scale operation. To take the evaluation of a feedstock one step further, it may then be possible to develop correlations between the data obtained from the actual plant operations (as well as the pilot plant data) with one or more of the physical properties determined as part of the initial feedstock evaluation.

However, it is essential that when such data are derived, the parameters employed should be carefully specified. For example, the data presented in the tables were derived on the basis of straight-run residua having API gravity less than 18°. The gas oil end point was of the order of 470°C to 495°C (875°F to 925°F), the gasoline end point was 205°C (400°F), and the pressure in the coke drum was standardized at 35 to 45 psi. Obviously, there are benefits to the derivation of such specific data, but the numerical values, although representing only an approximation, may vary substantially when applied to different feedstocks (Speight, 1987).

Evaluation of petroleum from known physical properties may also be achieved by use of the refractivity intercept. Thus, if refractive indices of hydrocarbons are plotted against the respective densities, straight lines of constant slope are obtained, one for each homologous series; the intercepts of these lines with the ordinate of the plot are characteristic, and the refractivity intercept is derived from the formula:

$$\text{Refractivity intercept} = n - d/2$$

The intercept cannot differentiate accurately among all series, which restricts the number of different types of compounds that can be recognized in a sample. The technique has been applied to nonaromatic olefin-free materials in the gasoline range by assuming additivity of the constant on a volume basis.

Following from this, an equation has been devised that is applicable to straight-run lubricating distillates if the material contains between 25% and 75% of the carbon present in naphthenic rings:

$$\text{Refractivity intercept} = 1.0502 - 0.00020\%C_N$$

Although not specifically addressed in this chapter, the fractionation of petroleum (Chapter 8) also plays a role, along with the physical testing methods, of evaluating petroleum as a refinery feedstock.

For example, by careful selection of an appropriate technique it is possible to obtain a detailed overview of feedstock or product composition that can be used for process predictions. Using the adsorbent separation as an example it becomes possible to develop one or more petroleum maps and determine how a crude oil might behave under specified process conditions.

This concept has been developed to the point where various physical parameters act as the ordinates and abscissa. However, it must be recognized that such *maps* do not give any indication of the complex interactions that occur between, for example, such fractions as the asphaltenes and resins (Koots and Speight, 1975; Speight, 1994), but it does allow predictions of feedstock behavior. It must also be recognized that such a representation varies for different feedstocks.

REFERENCES

Aczel, T. 1989. Preprints. *Div. Petrol. Chem. Am. Chem. Soc.* 34(2): 318.

Adler, S.B. and Hall, K.R. 1988. *Hydrocarbon Processing.* 71(11): 71.

Al-Besharah, J.M., Mumford, C.J., Akashah, S.A., and Salman, O. 1989. *Fuel* 68: 809.

Ali, M.F., Hasan, M., Bukhari, A., and Saleem, M. 1985. *Oil Gas J.* 83(32): 71.

Alfredson, T.V. 1981. *J. Chromatogr.* 218: 715.

Altgelt, K.H. 1968. Preprints. *Div. Petrol. Chem. Am. Chem. Soc.* 13(3): 37.

Altgelt, K.H. 1970. *Bitumen, Teere, Asphalte, Peche.* 21: 475.

Altgelt, K.H. and Gouw, T.H. 1979. *Chromatography in Petroleum Analysis.* Marcel Dekker, New York.

ASTM 2005. *Annual Book of Standards.* American Society for Testing and Materials, West Conshohocken, PA.

Baltus, R.E. and Anderson, J.L. 1984. *Fuel* 63: 530.

Berkowitz, N. and Speight, J.G. 1975. *Fuel* 54: 138.

Bollet, C., Escalier, J.C., Souteyrand, C., Caude, M., and Rosset, R. 1981. *J. Chromatogr.* 206: 289.

Bouquet, M. and Bailleul, A. 1982. *Petroananlysis '81. Advances in Analytical Chemistry in the Petroleum Industry 1975–1982.* Crump, G.B. ed. John Wiley & Sons, Chichester, UK.

Brown, J.K. and Ladner, W.R. 1960. *Fuel* 39: 87.

Butler, R.D. 1979. *Chromatography in Petroleum Analysis.* Altgelt, K.H. and Gouw, T.H. eds., Marcel Dekker, New York.

Carbognani, L. 1997. *J. Chromatogr. A.* 788: 63–73.

Chartier, P., Gareil, P., Caude, M., Rosset, R., Neff, B., Bourgognon, H.F., and Husson, J.F. 1986. *J. Chromatogr.* 357: 381.

Chmielowiec, J., Beshai, J.E., and George, A.E. 1980. *Fuel* 59: 838.

Colin, J.M. and Vion, G. 1983. *J. Chromatogr.* 280: 152.

Coulombe, S. and Sawatzky, H. 1986. *Fuel* 65: 552.

DeBruine, W. and Ellison, R.J. 1973. *J. Petrol. Inst.* 59: 146.

Dolbear, G.E., Tang, A., and Moorehead, E.L. 1987. *Metal Complexes in Fossil Fuels.* Filby, R.H. and Branthaver, J.F. eds. Symposium Series No. 344. American Chemical Society, Washington, DC. p. 220.

Drushel, H.V. 1983. *J. Chromatogr. Sci.* 21: 375.

Drushel, H.V. and Sommers, A.L. 1966. *Anal. Chem.* 38: 19.

Ebert, L.B. 1990. *Fuel Sci. Technol. Intl.* 8: 563.

Ebert, L.B., Mills, D.R., and Scanlon, J.C. 1987. Preprints. *Div. Petrol. Chem. Am. Chem. Soc.* 32(2): 419.

Felix, G., Bertrand, C., and Van Gastel, F. 1985. *Chromatographia* 20(3): 155.

George, A.E. and Beshai, J.E. 1983. *Fuel* 62: 345.

Gomez, J.V. 1987. *Oil Gas J.* (December 7): 68.

Gomez, J.V. 1989. *Oil Gas J.* (March 27): p. 66.

Gomez, J.V. 1992. *Oil Gas J.* (July 13): 49.

Gomez, J.V. 1995. *Oil Gas J.* (February 6): 60.

Green, L.E. 1976. *Hydrocarbon Process.* 55(5): 205.

Green, L.E., Schmauch, L.J., and Worman, J.C. 1964. *Anal. Chem.* 36: 1512.

Hasan, M., Ali, M.F., and Arab, M. 1989. *Fuel* 68: 801.

Hausler, D.W. and Carlson, R.S. 1985. Preprints. *Div. Petrol. Chem. Am. Chem. Soc.* 30(1): 28.

Hayes, P.C. and Anderson, S.D. 1987. *J. Chromatogr.* 387: 333.

Hickerson, J.F. 1975. *Special Publication No. STP 577.* American Society for Testing and Materials, Philadelphia. p. 71.

IP 2005. Standard No. 143. *Standard Methods for Analysis and Testing of Petroleum and Related Products 1997.* Institute of Petroleum, London, UK.

Koots, J.A. and Speight, J.G. 1975. *Fuel* 54: 179.

Long, R.B. and Speight, J.G. 1989. *Rev. Inst. Franç. Petrol.* 44: 205.

Lundanes, E. and Greibokk, T. 1985. *J. Chromatogr.* 349: 439.

Lundanes, E. Iversen, B., and Greibokk, T. 1986. *J. Chromatogr.* 366: 391.

MacAllister, D.J. and DeRuiter, R.A. 1985. *Paper SPE 14335.* 60th Annual Technical Conference. Society of Petroleum Engineers, Las Vegas. September 22–25.

Matsushita, S., Tada, Y., and Ikushige. 1981. *J. Chromatogr.* 208: 429.

McKay, J.F., Cogswell, T.E., and Latham, D.R. 1974. Preprints. *Div. Petrol. Chem. Am. Chem. Soc.* 19(1): 25.

Miller, R.L., Ettre, L.S., and Johansen, N.G. 1983. *J. Chromatogr.* 259: 393.

Nelson, W.L. 1958. *Petroleum Refinery Engineering.* McGraw-Hill, New York.

Nelson, W.L. 1974. *Oil Gas J.* 72(6): 72.

Norris, T.A. and Rawdon, M.G. 1984. *Anal. Chem.* 56: 1767.

Oelert, H.H. 1969. *Erdoel Kohle* 22: 536.

Rawdon, M. 1984. *Anal. Chem.* 56: 831.

Reynolds, J.G. and Biggs, W.R. 1988. *Fuel Sci. Technol. Intl.* 6: 329.

Roberts, I. 1989. Preprints. *Div. Petrol. Chem. Am. Chem. Soc.* 34(2): 251.

Romanowski, L.J. and Thomas, K.P. 1985. Report No. DOE/FE/60177–2326. United States Department of Energy, Washington, DC.

Schwager, I., Lee, W.C., and Yen, T.F. 1979. *Anal. Chem.* 51: 1803.

Schwartz, H.E. and Brownlee, R.G. 1986. *J. Chromatogr.* 353: 77.

Schwartz, H.E., Brownlee, R.G., Boduszynski, M.M., and Su, F. 1987. *Anal. Chem.* 59: 1393.

Snape, C.E., Ladner, W.R., and Bartle, K.D. 1979. *Anal. Chem.* 51: 2189.

Speight, J.G. 1987. Preprints, *Am. Chem. Soc. Div. Petrol. Chem.* 32(2): 413.

Speight, J.G. 1994. *Asphaltenes and Asphalts, I. Developments in Petroleum Science, 40.* Yen, T.F. and Chilingarian, G.V. eds, Elsevier, Amsterdam. Chapter 2.

Speight, J.G. 2000. *The Desulfurization of Heavy Oils and Residua*, 2nd Edn. Marcel Dekker Inc., New York.

Speight, J.G. 2001. *Handbook of Petroleum Analysis.* John Wiley & Sons Inc., Hoboken, NJ.

Speight, J.G. 2002. *Handbook of Petroleum Product Analysis.* John Wiley & Sons Inc., Hoboken, NJ.

Speight, J.G. 2005. *Natural Bitumen (Tar Sands) and Heavy Oil.* In Coal, Oil Shale, Natural Bitumen, Heavy Oil and Peat, from *Encyclopedia of Life Support Systems (EOLSS),* Developed under the Auspices of the UNESCO, EOLSS Publishers, Oxford, UK, [http://www.eolss.net].

Speight, J.G., Wernick, D.L., Gould, K.A., Overfield, R.E., Rao, B.M.L., and Savage, D.W. 1985. *Rev. Inst. Franç. Petrol.* 40: 27.

Spragins, F.K. 1978. *Development in Petroleum Science, No. 7, Bitumens, Asphalts and Tar Sands.* Yen, T.F. and Chilingarian, G.V. eds., Elsevier, New York. p. 92.

Stuckey, C.L. 1978. *J. Chromatogr. Sci.* 16: 482.

Suatoni, J.C. and Garber, H.R. 1975. *J. Chromatogr. Sci.* 13: 367.

Thomas, K.P., Barbour, R.V., Branthaver, J.F., and Dorrence, S.M. 1983. *Fuel* 62: 438.

Thomas, K.P., Harnsberger, P.M., and Guffey, F.D. 1987. Report No. DOE/MC/11076–2451. United States Department of Energy, Washington, DC.

Trestianu, S., Zilioli, G., Sironi, A., Saravalle, C., Munari, F., Galli, M., Gaspar, G., Colin, J.M., and Jovelin, J.L. 1985. *J. High Resol. Chromatogr. Chromatogr. Commn.* 8: 771.

Vercier, P. and Mouton, M. 1979. *Oil Gas J.* 77(38): 121.

Wallace, D. and Carrigy, M.A. 1988. *The Third UNITAR/UNDP International Conference on Heavy Crude and Tar Sands.* Meyer, R.F. ed., Alberta Oil Sands Technology and Research Authority, Edmonton, Alberta, Canada.

Wallace, D., Starr, J., Thomas, K.P., and Dorrence. S.M. 1988. *Characterization of Oil Sand Resources.* Alberta Oil Sands Technology and Research Authority, Edmonton, Alberta, Canada.

Weinberg, V.L., Yen, T.F., Gerstein, B.C., and Murphy, P.D. 1981. Preprints. *Div. Petrol. Chem. Am. Chem. Soc.* 26(4): 816.

Yen, T.F. 1973. *Fuel* 52: 93.

Yen, T.F. 1984. *The Future of Heavy Crude and Tar Sands.* Meyer, R.F. Wynn. J.C. and Olson, J.C. eds. McGraw-Hill, New York.

Yen, T.F. and Erdman, J.G. 1962. Preprints. *Div. Petrol. Chem. Am. Chem. Soc.* 7(3): 99.

10 Structural Group Analysis

10.1 INTRODUCTION

Petroleum processing requires knowledge of feedstock properties (Chapter 8 and Chapter 9), because petroleum varies markedly in properties and composition according to the source. Knowledge of petroleum properties is required for optimization of existing processes as well as for the development and design of new processes.

In the early days of petroleum processing, there was a lesser need to understand the character and behavior of petroleum in the detail that is currently required. Refining involved distillation of the valuable kerosene fraction that was required as an illuminant. After the commercialization of the internal combustion engine, the desired product became gasoline and it was also obtained by distillation. Even when crude oil that contained little natural gasoline was used, cracking (i.e., thermal decomposition with simultaneous removal of distillate) became the modus operandi.

However, with the startling demands on the petroleum industry during, and after World War II, and the emergence of the age of petrochemicals and plastics, the petroleum industry needed to produce materials not even considered as products in the decade before the war. Thus, petroleum refining took on the role of technological innovator as new and better processes were invented and advances in the use of materials for reactors were developed. In addition, it became a necessity to find out more about petroleum, so that the refiner might be able to enjoy the luxury of predictability and plan a product slate that was based on market demand. An often difficult task when the character of the crude oil was unknown. The idea that petroleum refining should be a *hit-and-miss* affair was not acceptable.

Valuable information can be obtained from the **true boiling point** (TBP) curve, which is a function of percent weight distilled and temperature, that is, a boiling point distribution (Figure 10.1). However, there are boiling point limitations on this function that are well below the final boiling point (FBP) of a crude oil. In addition to the boiling point distribution, it is possible to measure bulk physical properties, such as specific gravity and viscosity that have assisted in the establishment of certain empirical relationships for petroleum processing from the TBP curve. Many of these relationships include assumptions that are based on experience with a range of feedstocks. However, movement of the refining industry to feedstocks that contain higher proportions of coke-forming materials emphasizes the need for more definitive data that would enable more realistic predictions to be made of crude oil behavior in refinery operations.

The history of analysis of the constituents in petroleum started over 140 years ago and the rapid advances in analytic techniques have allowed the identification of large numbers of petroleum constituents. At this time, the major chemical types of compounds that exist in crude oil have been identified and many members of the various homologous series have been separated or conclusively identified by various techniques (Table 10.1) (Chapter 7) (see Rossini et al., 1953 for historical details).

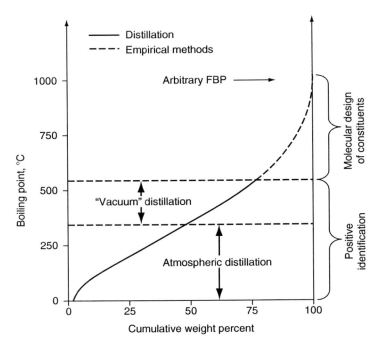

FIGURE 10.1 Differentiation of the identification of petroleum constituents according to boiling point.

TABLE 10.1
Various Compound Classes and Compound Types in Petroleum

Class	Compound Types
Saturated hydrocarbons	*n*-Paraffins *iso*-Paraffins and other branched paraffins Cycloparaffins (naphthenes) Condensed cycloparaffins (including steranes, hopanes) Alkyl side chains on ring systems
Unsaturated hydrocarbons	Olefins not indigenous to petroleum; present in products of thermal reactions
Aromatic hydrocarbons	Benzene systems Condensed aromatic systems Condensed aromatic-cycloalkyl systems Alkyl side chains on ring systems
Saturated heteroatomic systems	Alkyl sulfides Cycloalkyl sulfides Alkyl side chains on ring systems
Aromatic heteroatomic systems	Furans (single-ring and multi-ring systems) Thiophenes (single-ring and multi-ring systems) Pyrroles (single-ring and multi-ring systems) Pyridines (single-ring and multi-ring systems) Mixed heteroatomic systems Amphoteric (acid–base) systems Alkyl side chains on ring systems

It is the purpose of this chapter to review the methods that can be used for the structural group analysis of petroleum fractions. Methods are also included that are usually employed to emphasize a particular set of molecular parameters that are derived from physical properties or from particular chemical methods. Each method should be recognized for the particular capabilities that it offers, and used in conjunction with one or more of the other methods described in this chapter. In addition, the structural data is usually obtained from a combination of two or more methods (Speight, 2001 and references cited therein; Speight, 2002 and references cited therein; Nasritdinov, 2003). This is preferred, as a multidimensional approach yields more valuable information about a particular fraction than the data from one method alone.

Because of the nature of the subject matter of this chapter, there is overlap with the subject matter of other chapters. This is inevitable, and the readers are referred to other relevant chapters to place the subject matter of this chapter in the proper perspective.

10.2   METHODS FOR STRUCTURAL GROUP ANALYSIS

The precursors to petroleum are as diverse as the plant chemicals themselves, with the added recognition that there may have been some evolution of these chemicals over geological time as plants, and their constituents, evolved to modern-day counterparts. This is not to suggest that petroleum is a collection of plant chemicals and should be considered as a composite of carbohydrates, proteins, fatty acids, and any other chemical that can be derived from the pages of a natural product text book. Petroleum (especially the higher molecular weight nonvolatile constituents) is an extremely complex material, wherein the original precursors and the constituents of the ensuing **protopetroleum** (Chapter 3) have undergone considerable change through chemical and physical interactions with their environment.

Further, local and regional variations in the nature of the original plant precursors and in the maturation conditions offer a means of variation of petroleum from field to field and can be used to explain the variations in petroleum types and constituents. What this means is that a derivative of a chemical entity that is predominant in petroleum from one field may be an entity of somewhat lesser importance in another crude oil, even in crude oils of one particular type.

As a result, as composition studies in petroleum advance to the higher molecular-weight ranges, it is impractical to consider the determination of the individual compounds and very difficult, if not impossible, to segregate and identify fractions of the required simplicity.

However, a problem occasionally encountered in petroleum technology is that of establishing the structure of a single compound. Correlation between spectra and structure are useful for this purpose, and although is may never be possible to write the structural formula from spectral data alone, in combination with other forms of data, a choice between several alternate structures is often possible.

For this reason, investigators must be content with information on molecular types of hydrocarbons and on structural groups and it is perhaps in this field that spectroscopy has its most valuable application.

Although gas–liquid chromatography and other techniques have been applied successfully to the identification of a considerable number of petroleum constituents, they are, however, mainly limited to the so-called front end, that is, the volatile portion, of petroleum. Since the majority of crude oils contain significant proportions of a nonvolatile residuum, it is not possible to identify individual components in this part of the petroleum by a technique that requires some degree of volatility of the constituents. It is because of this that other methods of identification have been pursued.

However, despite many attempts, the separation of higher boiling fractions and residua into individual components appears to be an altogether hopeless enterprise. Even the preparation and identification of uniform fractions containing exclusively molecules of the same size and type are extremely complicated and success is not guaranteed. Pursuit of the composition of both virgin petroleum fractions and refinery products in terms of paraffinic, naphthenic, and aromatic contents has given rise to structural group analysis.

Structural group analysis is, in fact, the determination of the statistical distribution of these structural elements in the oil fraction, irrespective of the way in which the elements are combined in molecules. Thus, structural group analysis occupies a position midway between ultimate analysis, in which atoms are the components, and molecular analysis, in which molecules are the components. A method for structural group analysis seems to be complete only if structural elements are chosen in such a way that the sum of all equals 100% (or unity), for instance by considering the distribution of carbon in aromatic, naphthenic, and paraffinic locations in petroleum, its fractions, and products.

Structural group analysis has been widely applied to the analysis of petroleum and its fractions. Comparative data have been collected pertaining to the character of crude oils and variations in composition of fractions within the same crude, as well as the composition of intermediate and finished products. These data are often helpful in identifying products of unknown origin or in giving valuable indications concerning their manufacture. Much benefit has also been derived from structural group analysis in following physical separation processes, such as solvent extraction, solvent dewaxing, and percolation, and in elucidating overall reaction schemes in conversion processes, such as cracking.

When more thorough knowledge of hydrocarbon types is required than can be derived from the average proportion of structural elements of petroleum, structural group analysis may also render good service. However, statistical analysis often requires assumptions that may be without true scientific foundation. The failure to recognize the distinction between matters of statistical significance and practical importance leads to interpretation of the data on the basis of statistical significance alone and without consideration of proper experimental design of the test for the detection of systemic errors of a magnitude that has serious consequences for the end result.

One of the basic ideas underlying structural group analysis is the conception of a high molecular weight petroleum constituent as complex in structure and hence complex in properties. High molecular weight petroleum constituents may contain (and very frequently do contain) aromatic as well as naphthenic rings and paraffinic side chains, which is reflected in their physical and chemical properties. For instance, the viscosity and viscosity index (Chapter 8) of such hydrocarbons will to a large extent be determined by the relative quantity of each structural element; the same is true for the behavior of these hydrocarbons in physical separation processes, such as solvent extraction and adsorptive percolation. Since these complex molecules cannot be described as aromatics, naphthenes, or paraffins, it is logical that their composition should be expressed in terms of structural fragments.

Structural group analysis may be seen as an analytical method, giving information somewhere between that obtained by elemental analysis on the one hand and by analysis for individual hydrocarbons (molecular type analysis) on the other. Elemental analysis, as such, is unattractive since it gives only limited information about the constitution of petroleum because of the remarkable constancy of the elemental composition (Chapter 6). Thus, instead of molecules or atoms, certain structural types are considered components of the oil.

Methods for structural group analysis usually involve the determination of physical constants of the sample. However, since there is no simple relation between physical properties and chemical composition, a reliable correlation can only be obtained by studying properties of a great variety of oil fractions or pure compounds according to exact methods, laborious though that may be. The data collected statistically in this way may form the basis for chemical analysis by physical constants. The better the representation and the greater the number of these basic data, the more reliable will be the resulting method for structural group analysis. However, it should be remembered that structural group analysis is not the ultimate answer when applied to the heavier petroleum fractions or residua. Nevertheless, its importance lies in the straight correlation existing between the information derived from such analysis and physical properties, but it is often sufficient to get an overall description of the material in terms of its average structural group composition.

High-boiling petroleum fractions, petroleum products, and residua can be analyzed in terms of groups of hydrocarbons, and four classes are generally recognized:

1. Aromatic, if it contains at least one aromatic ring
2. Olefinic, if it contains at least one olefinic bond
3. Naphthenic, if it contains at least one naphthenic ring
4. Paraffinic, if it contains neither an aromatic nor a naphthenic ring nor an olefinic double bond

Aromatic hydrocarbons are also subdivided according to *aromatic type*, a term that describes compounds having the same number and grouping of aromatic rings. If an aromatic hydrocarbon contains two aromatic rings, three types may be distinguished:

1. Rings may be condensed, that is, fused together, to form the naphthalene nucleus
2. Rings may be joined by an inter-ring bond as in the biphenyl nucleus
3. Rings are separated by one or more nonaromatic carbon atoms, such as diphenyl-methane

This nomenclature is open to extension to polyaromatics with more than two aromatic rings. Alkyl chains and naphthene rings are generally not considered in discussions of aromatic type.

If a hydrocarbon contains structural groups of various types, it can be placed in more than one class. In such cases, in consequence of the preceding definitions, the total of aromatics + olefins + naphthenes + paraffins in a petroleum fraction may be considerably over 100%. To avoid this complication, other classes of hydrocarbons may be defined, such as naphthe-noaromatics. Owing to the rapidly increasing number and complexity of compounds in the higher boiling ranges, that part of the crude oil that might be broadly designated as the lubricant portion is in practice hardly suitable for this type of analysis.

There are two ways of reporting the results of a structural group analysis. One method is to determine the number of rings or other structural groups relative to a hypothetical average molecule of the sample, that is, a molecule containing the structural groups in the proportions found by structural group analysis. Such figures referring to the average number of aromatic rings (R_A), naphthene rings (R_N) and the total number of rings ($R_T = R_A + R_N$) is designated ring content. The other method is to determine the number of carbon atoms in aromatic ($\%C_A$), naphthenic ($\%C_N$), and paraffinic structures ($\%C_P$), all expressed per 100 carbon atoms in the sample. Such figures may be designated carbon distribution because the analysis gives the distribution of carbon over various structures, such as aromatic, naphthenic, and paraffinic structures. If the mean molecular weight of the sample is known and if an assumption is made about the type of rings present, the ring content can be recalculated as carbon distribution; the converse is also applicable.

Identification of the constituents of petroleum by molecular type may proceed in a variety of ways but generally can be classified into three methods: (1) spectroscopic techniques,

(2) chemical techniques, and (3) physical property methods. Various structural parameters are derived from a particular property by a sequence of mathematical manipulations. It is difficult to completely separate these three methods of structural elucidation and there must, by virtue of need and relationship be some overlap. Thus, although this review is more concerned with the use of spectroscopic methods applied to the issues of petroleum structure, there will also be reference to the other two related methods.

The end results of these methods are, at best, indications of the structural types present in the material. These indications are, in turn, dependent on the assumptions used to develop the method. However, there is the unfortunate tendency of researchers to then attempt to interrelate these structural types into a so-called average structure.

It is always the case that when mathematical manipulations are employed to derive average structures, the structures will only be as reliable as the assumptions used for the mathematical procedure. Then the literal interpretation which leads to the conclusion, and the insistence, that such structures exist in coal only leads to more confusion. Representation is one matter, adherent belief is another. Indeed, the complexity of the nonvolatile constituents of petroleum makes the construction of average structures extremely futile and, perhaps, misleading.

In fact, the heterogeneous chemical structures of the wide range of plant chemicals that formed the starting material for coal promise, but do not guarantee, an almost unlimited range of chemical structures within the various types of coal. Thus, it is perhaps best to consider the nonvolatile constituents of petroleum as a variety of chemical entities that dictate reactivity under specific conditions.

Nevertheless, indications of the methods that allow petroleum, specifically nonvolatile constituents of petroleum, to be defined in terms of structural entities are presented here.

10.2.1 Physical Property Methods

There have been many proposals for the structural group analysis of petroleum and petroleum products, which are usually based on inspection of the elemental analyses and physical properties of the material. The overall result has been the acceptance of several of these methods on the basis of their convenience and relative simplicity, and these methods are described here.

10.2.1.1 Direct Method

This method involves direct determination of the required physical properties of the petroleum sample or petroleum product and, unlike other methods, does not usually require prior separation of the oil into aromatic and saturated fractions.

Thus, by means of elemental analysis (Chapter 6) and molecular weight determination (Chapter 8) before and after hydrogenation of the sample, the percentage carbon in aromatic structure ($\%C_A$) and the average number of rings R_T can be estimated. If an oil is hydrogenated so that only aromatic rings are converted into naphthene rings, each aromatic carbon atom takes up one hydrogen atom. Therefore,

$$\%C_A = 1191(H'M' - HM)/(100 - H)M$$

H is the percentage hydrogen of the oil fraction, M is the average molecular weight, and H' and M' are the corresponding data for the hydrogenated product. In several cases, the molecular weights can be omitted, since the difference between M and M' may be negligible. The hydrogen content of the hydrogenated product, which is presumed to contain naphthenic and paraffinic carbon only, is an exact measure of the number of rings. As compared with paraffins, each ring closure involves a reduction by two hydrogen atoms. If R_T is the number of rings of the hypothetical molecule,

$$R_T = 1 = (0.08326 - 0.005793H')M'$$

To derive a complete set of carbon distribution and ring content figures from $\%C_A$ and R_T, an assumption must be made about the type of rings present. It has been assumed that all rings are six membered and, in the case of the polycyclic materials, are kata-condensed; this type of condensation only occurs in such a way that the rings have two carbon atoms in common, as, for example, naphthalene, anthracene, or tetracene. However, the occurrence of systems based on anthracene and tetracene is now considered less likely than that of occurrence of a system based on phenanthrene and chrysene in crude oil. The natural product precursors of petroleum (Chapter 3) are considered more likely to produce the linear condensed aromatic systems.

However, for C_R, that is, the average number of ring carbon atoms per molecule, assuming that only kata-condensed six-membered rings are present,

$$C_R = 4R_T + 2R_{TS}$$

R_{TS} is the number of substantial rings ($R_{TS} = R_T$ if <1 and $R_{TS} = 1$ if $R_T > 1$). In a kata-condensed system, it is possible to differentiate between the rings, insofar as one ring (a substantial ring) has six carbon atoms and the additional rings contribute only four carbon atoms each.

Thus, the percentage carbon in ring structures is given by the expression

$$\%CP = 240{,}200(2R_T + R_{TS})/M(100 - H),$$

and the carbon distribution is

$$\%C_P = 100 - \%C_R$$

$$\%C_N = \%C_R - \%C_A,$$

where C_N is naphthene carbon and C_A is aromatic carbon. Further, since

$$\%C_A = 240{,}200(2R_A + R_{AS})/M(100 - H),$$

where R_{AS} is the number of substantial aromatic rings ($R_{AS} = R_A$ if $R_A < 1$ and $R_{AS} = 1$ if $R_A > 1$), then for $R_A < 1$,

$$R_A = \%C_A M(100 - H)/720{,}600,$$

and for $R_A > 1$,

$$R_A = [\%C_A M(100 - H)/480{,}400] - 0.5$$

The naphthene ring content is deduced from the relationship

$$R_N = R_T - R_A$$

It is perhaps unfortunate that the direct method is based on the assumption that heteroelements (e.g., nitrogen, oxygen, and sulfur) are not present in the petroleum sample. However, if these elements are present, corrections can be applied, but this detracts from the basis of the method, since rings containing heteroelements may be cleaved during the hydrogenation process. Nevertheless, the method provides a means whereby $\%C_A$ and R_T can be determined and has been considered one of the more reliable methods for structural group analysis.

10.2.1.2 Waterman Ring Analysis

This method requires the determination of the refractive index, density, molecular weight, and aniline point (see Chapter 8). In essence, the elemental analysis used in the direct method is replaced by the specific refraction.

Thus, since molecular refraction r_M is additive for a hydrocarbon C_xH_y,

$$r_M = Mr_D{}^{20}$$

$$= xr_C = yr_H,$$

where r_C and r_H represent the atomic refraction of carbon and hydrogen, and M is the molecular weight. Since

$$y = MH/100.8$$

$$x = M(100 - H)/1201.0,$$

H is the percentage of hydrogen, then

$$\%H = (100.8 \times 1201.0r_D^{20} - 100.8r_C)/(1201r_H - 100.8r_C)$$

However, correlation of hydrogen content and specific refraction for a large series of petroleum fractions leads to:

$$\%H = 110.48r_D^{20} - 22.078$$

The total number of rings R_T is given by

$$R_T = 1 + (0.08326 - 0.005793H')M'$$

Then,

$$R_T = 1 + (0.2122 - 0.6401r_{20}^D)M$$

Since

$$\%C_R = 240.200(2R_T + R_{TS})/M(100 - H)$$

Therefore,

$$\%C_R = 240,200(2R_T + 2R_{TS})/M(77.922 - 100.48r_D^{20})$$

$$R_{TS} = R_T, \text{ if } R_T < 1$$

And,

$$R_{TS} = 1 \text{ if } R_T > 1$$

Thus, a relation between specific refraction, molecular weight, and percentage carbon in naphthene rings for saturated hydrocarbons or atomic- and olefin-free oil fractions can be defined (Brooks et al., 1954, p. 451). Similarly, a graphic relation between aniline point, specific refraction, and molecular weight of saturated oil fractions has also been defined (Brooks et al., 1954, p. 453).

The most important sources of error of the Waterman ring analysis are the specific refraction, the molecular weight, and the aniline point. And, while the accuracy of the Waterman ring analysis is questionable, the method has found wide acceptance.

10.2.1.3 Density Method

This method is based on the use of the density d, specific refraction r_{LL} (Chapter 8), and the molecular weight M (Chapter 8). This method is based on correlation of oil composition, determined by the direct method, with the physical properties of a large number of petroleum fractions obtained by various means. The procedure used for this method is analogous to that employed for the Waterman ring analysis, but the main difference is that the density is used, instead of the aniline point.

Thus, the density corresponding to the observed molecular weight M and specific refraction r_{LL} is determined (Brooks et al., 1954, p. 453); the difference Δd between this value and the observed value is noted. Thus,

$$\%C_A = 420\Delta d/(1 + 3.2)d$$

and C_R is obtained graphically; thence,

$$\%C_P = 100 - \%C_R$$

$$\%C_N = \%C_R - \%C_A$$

Compared with the Waterman ring analysis, the experimental procedure is simplified, the applicability is more general, and there is better agreement with data obtained by the direct method. Disadvantages, which are also inherent in the Waterman ring analysis, are that the molecular weight range is limited to 200–500 and extrapolation of the data to the higher range is uncertain; indeed, a high aromatic content causes the specific refraction to fall outside the correlation. The accuracy of the density method is in general fairly good, although according to the basic data used, the application should be limited to fractions with $\%C_A < 1.5\%C_N$ or with $R_A < 0.5R_T$.

10.2.1.4 *n.d.M.* Method

Linear relationships between the composition of petroleum fractions (as determined by the direct method) and the refractive index (Chapter 8), density (Chapter 8), and molecular weight (Chapter 8) led to the development of a method based on these three physical characteristics of petroleum fractions.

Thus, as the name ($n.d.M$ method) implies, the refractive index, n, density, d, and molecular weight M, of the sample are determined; if the sulfur content of the sample is expected to be greater than 0.206, this value should also be determined. Substitution of the data into the formulae allows an estimation of the carbon distribution and ring content.

The $n.d.M$ method is especially intended for petroleum fractions boiling above the gasoline range and for similar products after extraction, hydrogenation, or other treatment. The validity of the method is good for petroleum fractions with up to $\%C_R = 75\%$ (aromatic + naphthenic), provided that $\%C_A$ (as found by the $n.d.M$ method) is not higher than 1.5 times $\%C_N$. The method is also valid for samples containing up to four rings per molecule, if not more than half of these are aromatic. Application to oils with a relatively high naphthene ring content is presumed not to cause serious errors. Although the presence of foreign elements influences its accuracy unfavorably, the method is applicable to samples containing sulfur up to 2%, oxygen up to 0.5%, and nitrogen up to 0.5%.

10.2.1.5 Dispersion–Refraction Method

This method is analogous to the direct method, but the hydrogenation and the examination of the hydrogenated product are replaced by the determination of specific dispersion (Chapter 8), bromine number (ASTM D1159; IP 129; IP 130), and specific refraction (Lorentz–Lorenz) (Chapter 8) of the original fraction (Van Nes and van Westen, 1951, p. 344). The analytic results obtained by the dispersion–refraction method are similar to those derived from the direct method, except olefin double bonds, is also measured. The method is based on data derived from pure hydrocarbons having fewer than 18 carbon atoms per molecule and requires the following determinations: carbon and hydrogen content, refractive index, density, dispersion, molecular weight, and bromine number.

Calculation of the structural parameters is fairly involved (Van Nes and van Westen, 1951, p. 355), and the accuracy of the method does not seem to be very high. However, it should be remembered that the method can be used on materials derived from cracking processes as determination of the olefin content is provided for, by inclusion of the bromine number.

10.2.1.6 Density–Temperature Coefficient Method

For the analysis of paraffin–naphthene mixtures with a density below 0.861, it is assumed that in the density–density temperature coefficient diagram, the portion of the intercept (at constant density) between the paraffin and naphthene line is divided by the sample point into parts proportional to paraffin and naphthene content. In this way, the following equation for mixtures of paraffins and naphthenes in the regions below 0.861 density was derived:

$$\text{Wt.\% rings} = [190.0d - 217.9 - 10^5 dd/dt]/(0.593d - 0.249),$$

where d is the density at 20°C (68°F) (Chapter 8) and dd/dt is the change in density per degree change in temperature.

For mixtures having a density above 0.861, a compromise was reached between the condensed and noncondensed naphthenes, assuming an equal distribution of these two types. The sample point was assumed to divide the line between the limiting paraffin point and a point on the naphthene ring line above 0.861 into parts proportional to paraffin and naphthene content, and for this region the following formula was derived:

$$\text{Wt.\% rings} = [102.8d - 142.8 - 10^5 \text{dd/dt}]/0.262$$

Simplification of the experimental procedure could be achieved by deriving the density coefficient from the molecular weight, which in turn can be estimated from other physical properties, such as the density and mid-boiling point, or viscosity at $38\,^\circ$C ($100\,^\circ$F) and $99\,^\circ$C ($210\,^\circ$F). However, since:

$$-10^5 \text{dd/dt} = 53.5 + 3360/M$$

if $d < 0.861$,

$$\text{Wt.\% rings} = [190.0d - 164.4 + 3360/M](0.593d - 0.249)$$

and if $d > 0.861$,

$$\text{Wt.\% rings} = (102.8d - 89.3 + 3360M)/0.262$$

For the analysis of aromatic mixtures containing no naphthene rings, a procedure is followed that is quite analogous to the ring-chain analysis for naphthenes. The line between the limiting paraffin point and a point of the aromatic ring line in the density–density temperature coefficient diagram is divided by the sample point. It is assumed that the parts are proportional to aromatic ring and paraffinic chain content. The relationship between wt. aromatic ring, density coefficient, and density is only given graphically graphs are used for condensed ring aromatics, for noncondensed ring aromatics, and for mixtures having an equal distribution of these two types (Van Nes and van Westen, 1951, pp. 357–364).

The applicability of the method for analyzing aromatic fractions is limited to fractions having no naphthene rings along with the aromatic rings and since it is generally accepted that aromatic molecules in the higher boiling straight-run petroleum fractions nearly all contain naphthene rings, the method derived for alkylaromatics is probably not applicable to gas oil and lubricating oil factions.

10.2.1.7 Molecular Weight–Refractive Index Method

The samples to be analyzed by the molecular weight–refractive index method must be separated beforehand into aromatic and paraffin–naphthene fractions by, for example, silica gel adsorption chromatography (Chapter 9). As the name indicates, only the molecular weight M (Chapter 8) and the refractive index n_D^{20} (Chapter 8) are the required parameters, and by using data derived from use of pure hydrocarbons, the following relationships appear to hold:

Aromatic oils, noncondensed rings:

$$\%C_A = [1060[(n_D^{20} - 1.4750)M + 8.79]^{0.85}]/M - 2 + 1.01[n_D^{20} - 1.4750)M + 8.79]^{0.85}$$

$$R_A = 0.126[(n_D^{20} - 1.4750)M + 8.79]^{0.85}$$

Aromatic oils, kata-condensed rings:

$$\%C_A = [2150[(n_D^{20} - 1.4750)M + 8.79]^{0.56}]/M - 3 + 2.3[(n_D^{20} - 1.4750)M + 8.79]^{0.56}$$

$$R_A = 0.165[(n_D^{20} - 1.4750)M + 8.79]^{0.74}$$

Mixtures of condensed and noncondensed rings:

$$\%C_A = [1650[n_D^{20} - 1.4750) + 8.79]^{0.67}]/M - 2 + 1.65[(n_D^{20} - 1.4750)M + 8.79)^{0.67}$$

$$R_A = 0.151[(n_D^{20} - 1.4750)M + 8.79]^{0.78}$$

Nonaromatic oils, condensed and noncondensed rings:

$$\%C_R = [2920[(n_D^{20} - 1.4750)M + 8.79)^{0.73}]/M - 2 + 0.57[(n_D^{20} - 1.4750)M + 8.79)^{0.86}$$

$$R_T = 0.284[(n_D^{20} - 1.4750)M + 8.79]^{0.86}$$

Application of the $M - n_D{}^{20}$ method to saturated portions of oil fractions gives results, which in general are in good agreement with those obtained by other methods.

10.2.1.8 Miscellaneous Methods

A method has also been proposed for determining the number of aromatic rings R_A, and naphthenic rings R_N, in aromatic fractions of petroleum and involves measurement of the molecular weight M (Chapter 8), density d (Chapter 8), and refractive dispersion s (Chapter 8). From the measured data, the functions

$$F(s, M) = (s - 98)(M + 12)10^{-3},$$

and

$$F(d, M) = (d - 0.854)(M + 12)$$

are calculated, and R_A and R_N are estimated graphically (Brooks et al., 1954, p. 465).

The method provides a valuable contribution to the existing methods for structural group analysis. The procedure is simple and rapid, and the method is claimed to hold for aromatic concentrates from both nondestructive distillation and cracking of petroleum fractions; average deviations of 0.1 of an aromatic ring and about 0.2 of a naphthene ring are usual. If there are more than three aromatic rings per molecule, the results are uncertain because of a lack of basic data and knowledge concerning the types of petroleum hydrocarbons with four or more aromatic rings. Saturated hydrocarbons, olefins, some noncondensed polycyclic aromatics, and nonhydrocarbons are reputed to introduce serious errors into the analysis.

Other methods include the derivation of a linear relation between percentage carbon in aromatic structure $\%C_A$ (Chapter 8), refractive index n_D^{20} (Chapter 8), density (Chapter 8), and aniline point (AP) (Chapter 8):

$$\%C_A = 1039.4n_D^{20} - 470.4d^{20} - 0.315AP - 1094.3$$

This formula holds good only if $\%C_A < 30$. When the calculated value for $\%C_A$ exceeds 30, a corrected value must be found by using the formula

$$\%C_A(\text{corr.}) = 0.5\%C_A(\text{calc.}) + 15$$

Alternatively, it has been suggested that the molecular weight determination of the *n.d.M* method be replaced by kinematic viscosity measurements (Chapter 8) leading to a *n.d.V* method.

An equation has been devised that is applicable to lubricating distillates that have not been subjected to thermal cracking, temperatures $<350°C$ ($<660°F$). If the naphthenic carbon is of the order of 25%–75% of the total carbon, a relationship exists between the refractivity intercept (Chapter 8) and the number of carbons in naphthenic locations (C_N):

$$\text{Refractivity intercept} = 1.0502 - 0.00020 \times \%C_N$$

Finally, it has also been proposed that for substances such as **asphaltenes**, which contain condensed ring systems, the following relationships be applied:

$$R = 0.11(9.9C - 3.1H - 3.70 + 1.5N + 14S - M/d)$$

where R is the number of rings, C, H, O, N, and S are the numbers of carbon, hydrogen, oxygen, nitrogen, and sulfur atoms, M is the molecular weight, and d is the density.

However, if it is not possible to determine the molecular weight, the relationship has been modified to

$$C/R = 9.2/(9.9 + 3.1H/C + 3.70/C + 1.5N/C + 14S/C - 1200/\%C,d),$$

C/R is the ring condensation index, H/C, O/C, N/C, and S/C are the various atomic ratios calculated from elemental analyses, $\%C$ is the percentage of carbon obtained by elemental analysis, and d is the density.

The method assumes that atoms other than carbon and hydrogen are not major components, olefins are absent, the number of rings is greater than one, and the rings are the fused naphthalene type rather than biphenyl type.

10.2.2 SPECTROSCOPIC METHODS

A decrease in the volatility of the petroleum constituents in residua precludes the use of gas–liquid chromatography for identification purposes, and it is at this stage that methods of structural analysis involving physical property determinations become more predominant.

With the startling advances in spectroscopic techniques after World War II, chemists and technologists have been able to gain a better understanding of the chemical nature of petroleum and, therefore, of the chemical changes which occur during refining. Consequently, new fields of research have been opened involving the structural determinations of fossil fuels, and their derived products, by a variety of physical, chemical, and spectroscopic techniques. Indeed, a large number of the individual constituents of petroleum have been identified; this has been an aid not only to the refiner but also to the chemist and geochemist whose interests lie in the ways and means by which petroleum was formed from its precursors. Scientifically interesting as this may be, the real message lies in understanding the formation of petroleum and the types of formations where more reserves might be discovered.

The application of spectroscopic methods to structural analysis is a fairly recent innovation but has received considerable attention, nevertheless. As composition studies in petroleum advance to the higher molecular weight ranges, the physical nature of the material often requires that the structural analysis be performed by a more convenient and suitable technique. For example, gas–liquid chromatography may well suffice for the identification of the volatile constituents of petroleum, as well as for volatile petroleum products.

However, a point will certainly be reached at which even the determination of the various physical properties, outlined in the previous section, may become tedious and impractical, if not impossible. Such a point is usually reached when investigators turn their attention to the asphaltic constituents of petroleum; the asphaltenes, the resins, and the higher boiling or even nonvolatile portion of the oily constituents.

It is perhaps for this reason that spectroscopic methods of structural analysis have come into fairly common usage. This is not to say that distinct divisions should be made and a particular method for structural analysis be assigned to a particular fraction. It is in fact preferred that as many methods of structural analysis as possible be applied to any particular fraction to gain an overall picture of the structural entities present. However, it is expected that the chemical and physical nature of the various fractions will most certainly preclude the use of one or more of the methods proposed for structural analysis.

It is also appropriate at this point to give a brief description of other methods that are used for the identification of the constituents of petroleum (Yen, 1984). It is not intended to convey here that any one of these methods can be used for identification purposes. However, although these methods may fall short of acceptability as methods for the identification of individual constituents, they are recognized as methods by which an overall evaluation of the feedstock may be obtained in terms of molecular types. It is appropriate to describe these methods briefly at this point, to place them in their context relative to the other methods already described.

10.2.2.1 Infrared Spectroscopy

Infrared absorption spectroscopy is an excellent method for providing detailed information about the chemical constitution of organic materials. Infrared absorption spectroscopy is one of the most widely used techniques for examining the chemical constitution of organic materials. Whether the material being examined is a single compound or a mixture, infrared absorption spectroscopy offers valuable information about the hydrocarbon skeleton and about the functional groups in petroleum (Wen et al., 1978).

Indeed, studies of the infrared spectra of a variety of compounds have demonstrated systematic correlation with structure. Thus, it is not surprising that attempts have been made to estimate the structural entities present in petroleum and other carbonaceous materials by means of infrared spectroscopy.

When properly prepared and scanned with suitable instruments, the infrared spectra of petroleum fractions show a number of well-defined bands (Table 10.2) (Rao, 1961). For the purposes of structural group analysis, however, the bands at 1380, 1465, 2880, and $2920\,\mathrm{cm}^{-1}$ are usually employed. This allows an estimation of the number of methyl and methylene groups present in the samples.

Thus, initially, the absorbtivity, in grams per liter per centimeter, is calculated as

$$a_b - A_b/cl$$

where a_b is the absorbtivity $(\mathrm{g/L \times cm^{-1}})$ at the wavelength b, A_b is the absorbance of the solution at wavelength b, c is the concentration of the solution (g/L), and l is the cell thickness in centimeters.

Following from this, the absorbtivity per gram carbon in the standard group absorbing is derived in the following manner:

$$a_c = Ma_b/12.01n$$

where a_c is the absorbtivity $(\mathrm{g/L\,cm^{-1}})$ per gram carbon of the type absorbing, M is the gram molecular weight, n is the number of carbon atoms of the type absorbing per molecule, 12.01 is the gram atomic weight of carbon, and a_b is the absorbtivity derived previously. By use of

TABLE 10.2
Infrared Absorption Frequencies

Bond	Compound Type	Frequency Range, cm^{-1} (Wave Number)
O—H	Monomeric alcohols and phenols	3640–3160(s, br) stretch
O—H	Hydrogen-bonded alcohols and phenols	3600–3200(b) stretch
N—H	Amines, amides	3560–3320(m) stretch; nonhydrogen bonded
N—H	Amines	3400–3100(m) stretch; hydrogen bonded
C—H	Alkynes	3333–3267(s) stretch
C—H	Aromatic rings	3100–3000(m) stretch
C—H	Alkenes	3080–3020(m) stretch
O—H	Carboxylic acids	3000–2500(b) stretch
C—H	Alkanes	2960–2850(s) stretch
C≡N	Nitriles	2260–2220(v) stretch
C≡C	Alkynes	2260–2100(w, sh) stretch
C—H	Phenyl ring substitution	2000–1600(w)
C=O	Aldehydes, ketones, carboxylic acids, esters	1760–1670(s) stretch
C=C	Alkenes	1680–1640(m, w) stretch
NO$_2$	Nitro-compounds	1660–1500(s) asymmetrical stretch
N—H	Amines	1650–1580(m) bend
C=C	Aromatic rings	1600–1500(w) stretch
C—H	Alkanes	1470–1350(v) scissoring and bending
NO$_2$	Nitro-compounds	1390–1260(s) symmetrical stretch
C—H	CH$_3$ deformation	1380(m or w, can be a doublet); *iso*-propyl, *t*-butyl
C—N	Amines	1340–1020(m) stretch
C—O	Alcohols, ethers, carboxylic acids, esters	1260–1000(s) stretch
C—H	Alkenes	1000–675(s) bend
C—H	Phenyl ring substitution	870–675(s) bend
C—H	Alkynes	700–610(b) bend
C—Cl	Chloro-compounds	600–800(m, w)
C—Br	Bromo-compounds	500–600(m, w)
C—I	Iodo-compounds	500(m, w)

v, variable; m, medium; s, strong; b, broad; w, weak; wave number = 1/wavelength.

the a_c term for specific types of carbon (1.50 at 1465 cm^{-1} for methyl; 3.60–4.00 at 2880 cm^{-1} for methylene), the number of carbon atoms of that type present in the sample may be calculated.

Another infrared spectroscopic procedure that has been employed to gain insight into the carbon distribution in petroleum residua and related materials involves determination of the molecular extinction coefficient E for the band under investigation.

$$E = (1/cl) \log(I_0/I),$$

where c is the molarity of the solution, l is the path length in centimeters, and I_0/I is determined directly from the infrared spectrum and is actually the ratio of the intensity of the incident light I_0 to the intensity of the transmitted light I.

The fraction of hydrogen atoms h_a, attached to aromatic carbons can be derived in the following manner:

$$h_a = H_A/(H_A + H_S)$$

H_A is the number of aromatic hydrogen atoms and H_S, is the number of saturated hydrogen atoms.

$$h_a = [E_{3030}/(E_{3030} + E_{2920})]^k,$$

E_{3030} and E_{2920} are the extinction coefficients of the 3030 and 2920 cm^{-1} absorption bands, respectively. Studies with standard alkylaromatic compounds and polymers have indicated that there is good agreement between the theoretical and observed hydrogen distribution, when k is unity.

The proportion of carbon atoms in methyl groups to the total number of carbon atoms can be determined in a similar manner. Thus,

$$C_{Me}/C_T = H/C \times C_{1380}/E_{3030} + E_{2920},$$

C_{Me} is methyl carbon, C_T is total carbon, and H/C is the atomic hydrogen–carbon ratio determined from elemental analysis, E_{1380} is the extinction coefficient of the band at 1380 cm^{-1} and E_{3030} and E_{2920} are the extinction coefficients of the bands at 3030 and 2920 cm^{-1}, respectively.

Infrared spectroscopy has also been used extensively in the elucidation of the structural entities present in asphaltenes and resins (Erdman, 1965; Witherspoon and Winniford, 1967; Monin and Pelet, 1984; Tripathi et al., 1984; Hasan et al., 1988).

For example, the bands at about 3030 cm^{-1} have been assigned to aromatic carbon–hydrogen stretching and have demonstrated that this band may shift to values as high as 3052 cm^{-1} with decreasing ring number. In addition to the carbonyl functions, the infrared spectra of asphaltenes show a strong carbon–hydrogen band at 2940 cm^{-1}, a peak characteristic of CH$_2$ groups at 1470 cm^{-1}, and peaks for methyl groups at 1380 cm^{-1}. The group of peaks at about 760 cm^{-1}, 814 cm^{-1}, and 870 cm^{-1} are believed to be due to substituted aromatic ring structures. The aromatic ring band at 1600 cm^{-1} is preeminent, while the broad and strong absorption between 1330 and 1110 cm^{-1} has been assigned to oxygen-containing species (Stewart, 1957). Peaks at 1300 and 1135 cm^{-1} have been attributed to sulfur–oxygen bonds, but a characteristic peak at 1040 cm^{-1} has not been adequately diagnosed. The band at 725 cm^{-1}, usually attributed to methylene chains having four or more units, is often present, but varies in intensity depending upon the petroleum source. The band at 740 cm^{-1} has been suggested to arise from alkylbenzenes, the band at approximately 1600 cm^{-1} has been assigned to aromatic carbon–carbon bonds (Nagy and Gagnon, 1961; Millson, 1967).

In addition, the aromaticity (1600 cm^{-1} band) of the asphaltenes is significantly greater than that of the maltene fraction but, from infrared spectroscopy) increases only slightly with molecular size (Altgelt, 1965). This is in general agreement with the data from elemental analysis and from carbon-13 resonance spectroscopy, where the aromaticity approaches an asymptotic limit for decane-precipitated asphaltenes (Speight, 1994) although the distribution of aromaticity appears to be dependent upon the functional types within an asphaltene (Gould and Long, 1986).

The first use of infrared spectroscopy in a quantitative approach to the structural analysis of petroleum occurred in 1956 (Brandes, 1956), in which the distributions of aromatic, naphthenic, and paraffinic carbon were calculated for a series of gas oils. Following from this, the 1610 and 720 cm^{-1} infrared absorption bands have also been used to calculate the carbon distribution (Fischer and Schram, 1959).

The conclusion was that asphaltenes consist of condensed aromatic ring systems substituted by paraffinic chains, with naphthenic rings also present, but only to a minor extent. Similar results were also reported for other bituminous and asphaltic materials (Chelton and

Traxler, 1959; Romberg et al., 1959). However, it was shown that use of the 1600 cm^{-1} band for estimation of weight per cent aromatic carbon has, in some instances, led to values in excess of 100% and any discrepancy may arise from interference by carbonyl groups attached to the aromatic system.

The application of applied infrared spectroscopic techniques to the elucidation of the structures of petroleum resins and asphaltenes, illustrated the presence of many structural groups and their relative positions to one another (Yen and Erdman, 1962a). These studies also were suggestive of a different type of ring condensation for petroleum asphaltenes to that found in coal and it was concluded that petroleum asphaltenes were primarily peri-condensed, like phenanthrene, while the coals were kata-condensed, like anthracene.

Further development of the method for structural group analysis, using a large number of hydrocarbons, allowed three chemical and two physical properties to be expressed in terms of five structural groups (Montgomery and Boyd, 1959; Boyd and Montgomery, 1963). The chemical properties included the carbon and hydrogen content as well as the number of aromatic carbons present per molecule, and the physical properties required for the analysis are the molar volume and molar refraction. However, it was concluded that the 1600 cm^{-1} absorption band was not capable of producing reliable estimates of aromatic carbon in the heavier feedstocks because the fractions are of such complex composition. The results did indicate that the infrared data were self-consistent for the estimation of methylene and methyl groups in the bituminous fractions.

Infrared spectroscopy has also been used to identify various functional groups including hydroxyl (3450 cm^{-1}) and carbonyl (1695–1725 cm^{-1}) (McKay et al., 1978). In combination with nitrogen titration data (Jacobson and Gray, 1987) better representations of the structural groups can be made (Petersen, 1967).

Conventional **infrared spectroscopy** also yields information about the functional features of various petroleum constituents. For example, infrared spectroscopy aids in the identification of N—H and O—H functions, the nature of polymethylene chains, the C—H out-of-place bending frequencies, and the nature of any polynuclear aromatic systems (Yen, 1973). With the recent progress of Fourier transform infrared (FTIR) spectroscopy, quantitative estimates of the various functional groups can also be made. This is particularly important for application to the higher molecular weight solid constituents of petroleum (i.e., the asphaltene fraction).

It is also possible to derive structural parameters from infrared spectroscopic data and these are H_S/C_S saturated hydrogen to saturated carbon ratios C_P/C, paraffinic character; C_N/C, naphthenic character; %Me, methyl content; and C_{MP}/C_{SU}, chain length. In conjunction with proton magnetic resonance (see next section), further structural parameters, such as the fraction of methyl groups (C_{SMe}/C_{Me}), can be obtained.

10.2.2.2 Nuclear Magnetic Resonance Spectroscopy

Nuclear magnetic resonance spectroscopy has also found considerable use in the structural analysis of petroleum and petroleum products. The technique essentially identifies and counts hydrogen atoms according to their chemical and physical environments (for example, see Speight, 1970).

Nuclear magnetic resonance spectroscopy has proved to be of great value in petroleum research because it identifies and counts hydrogen atoms according to their chemical and physical environments (Wen et al., 1978; Dereppe and Moreaux, 1985; Ramaswamy et al., 1986; Gray et al, 1989; Majid et al., 1989). Moreover, proton magnetic resonance allows rapid and nondestructive determination of the total hydrogen content and distribution of hydrogen among the chemical functional groups present in the sample (Cookson et al., 1986). The technique is also useful as a means of estimating molecular weight (Leon, 1987) and for studying processing effects (Giavarini and Vecchi, 1987).

Early attempts to apply proton magnetic resonance spectroscopy to the study of petroleum showed that hydrogen could be differentiated according to types (Williams, 1959). As a result, structural parameters were estimated for the aromatic systems (Gardner et al., 1959). In addition, it was also demonstrated that the information derived by use of proton magnetic resonance was similar to that obtained using other spectroscopic methods (Ramsey et al., 1967).

Thus, the nuclear magnetic resonance spectrum of a petroleum fraction can be divided into five major regions, each identifiable with a hydrogen type (for example, see Speight, 1970). From the areas under each region, it is possible to calculate the fractional distribution of the hydrogen atoms among aromatic hydrogen, H_A, and paraffinic methyl hydrogen, H_{Me}. To calculate the average carbon distribution from the hydrogen distribution, it is assumed that H_N, H_{My}, and H_{Me} are derived from the presence of their respective methylene ($-CH_2-$) groups. The principal departure from this assumption occurs in the fused-ring naphthenic contributions to H_{My}, and H_{Me}.

For example, the nuclear magnetic resonance spectrum for the decalin isomers show that when the bridge hydrogen atoms are on the same side, like in, the *cis* configuration, the entire naphthenic resonance concentrates essentially in a single peak at $J = 8.59$, within the range assigned to H_N. However, when the hydrogen atoms are on opposite sides of the molecule, the *trans* configuration, approximately one-third of the hydrogen signal is located at higher J values, concentrating mainly in the range assigned to H_{Me}. Thus, H_N may be low by up to 20%.

It is also necessary to assume that each carbon atom attached to an aromatic ring carries two hydrogen atoms. The principal deviation here may be in the lower boiling fractions, that is, the nonaliphatic material, where substituents on aromatic nuclei may be predominantly methyl groups. An increase in the molecular weight of the fraction is usually accompanied by, among other things, an increase in the size of the alkyl chains attached to an aromatic ring. Hence, the assumption that each benzylic carbon carries two hydrogen atoms may be essentially correct for the higher boiling fractions that occur in the heavy feedstocks.

Nuclear magnetic resonance has frequently been employed for general studies and for the structural studies of petroleum constituents (Bouquet and Bailleul, 1982; Hasan et al., 1989). In fact, proton magnetic resonance (PMR) studies (along with infrared spectroscopic studies) were perhaps the first studies of the modem era that allowed structural inferences to be made about the polynuclear aromatic systems that occur in the high-molecular-weight constituents of petroleum.

In general, the proton (hydrogen) types in petroleum fractions can be subdivided into three types (Brown and Ladner, 1960) or into five types (Yen and Erdman, 1962b). The first approach classifies the hydrogen types into H_{ar}^*, aromatic ring hydrogen; $H^{*''}$, aliphatic hydrogen adjacent to aromatic rings; and H_o^*, aliphatic hydrogen not adjacent to aromatic rings. The second approach subdivides the hydrogen distribution into aromatic hydrogen H_A ($H_A/H = H_{ar}^* = h_a)''$-substituted hydrogen next to aromatics ($H''/H = H^{*''}$), H'', naphthenic hydrogen H_N, methylene hydrogen H_R, and terminal methyl hydrogen not close to aromatic system H_{Sme} ($H_o^* = H_N/H + H_R/H + H_{Sme}/H$).

Following the calculation of the fractional distribution of the protons, the carbon distribution in atoms per molecule may be obtained from the following relationships:

$$C_S = H_T[(H_a/2)/(H_N/2) + (H_{My}/2) + (H_{Me}/3)]$$

$$C_{SA} = H_T \times H_a/2$$

$$C_A = C_T - C_S$$

$$C_P = H_T[H_A + (H_a/2)]$$

$$C_I = C_A - C_P$$

$$C_{My} = H_T \times H_{My}/2$$

$$C_{Me} = H_T \times H_{me}/3$$

$$C_N = C_S - (C_{SA} + C_R)$$

$$R_A = (C_I + 2)/2$$

C_S: total saturated carbon atoms per molecule
C_{SA}:   total saturated carbon atoms to an aromatic ring
C_A: total aromatic carbons per molecule
C_T: total carbon atoms per molecule from analysis and molecular weight
C_P: peripheral carbon in a condensed aromatic sheet or the total number of hydrogen atoms present if the sheet is completely unsubstituted
C_I: internal carbon in a condensed aromatic sheet
C_{My}:   total paraffinic methylene carbon atoms per molecule in locations other than α to an aromatic ring
C_{Me}:   total paraffinic methyl carbon atoms per molecule in locations other than α to an aromatic ring
C_N: total naphthenic carbon atoms per molecule
H_T: total hydrogen atoms per molecule from analysis and molecular weight
R_A: aromatic rings per molecule

It is also possible to estimate certain structural parameters for the aromatics within fractions. Thus, the ratio C_{SA}/C_P is the average number of carbon atoms directly attached to an aromatic sheet divided by the total peripheral carbon atoms. Expressed in another manner, the ratio C_{SA}/C_P is the degree of substitution of the aromatic sheet. The C_S/C_{SA} ratio is the total number of saturated carbon atoms divided by the number of carbon atoms attached to the edge of the aromatic sheet. Alternatively, the C_S/C_{SA} ratio represents the average number of carbon atoms attached to a position of the edge of an aromatic sheet. The C_P/C_A ratio is the peripheral carbon atoms per aromatic sheet divided by the total aromatic carbon atoms for that sheet. Alternatively, this ratio is an estimate of the shape of the aromatic sheets; and $(C_S-C_{Me})/C_{Me}$ is the ratio of methylene carbon to methyl carbon, that is, an estimate of the degree of branching in the saturated moieties of the molecule; it can also be written $C_S/C_{Me}-1$.

The C_P/C_A parameter, which indicates the shape of the aromatic centers of the molecules, is perhaps a little misleading, and literal interpretation can lead to misconceptions about the structures present in, say, residua. It is perhaps more pertinent if this and the other parameters are interpreted on a comparative basis.

On the basis of data derived from PMR investigations, a variety of hypothetical structures have been, and continue to be, proposed for asphaltenes isolated from a variety of crude oils (Winniford and Bersohn, 1962; Yen and Erdman, 1962c; Speight, 1970; Ali et al., 1990).

The general indications were that large proportions (up to 50%) of the hydrogen in natural asphaltenes occurred in methylene groups; aliphatic methyl hydrogen atoms were second in abundance, followed by benzylic hydrogen atoms, and, lastly, the aromatic hydrogen atoms. The relative lack of aromatic and benzylic hydrogen in petroleum asphaltenes appeared to support the occurrence of condensed polynuclear aromatic ring systems. These results also appeared to be in agreement with the kind of structures suggested by Yen and co-workers from their x-ray diffraction studies of asphaltenes (Yen et al., 1961c). Hence, it was advocated that asphaltenes are composed of polynuclear aromatic ring systems having 10–20 fused rings bearing aliphatic and naphthenic side chains.

Other workers (Ferris et al., 1966; Speight, 1970) investigated asphaltenes and resins and concluded, by virtue of data derived from PMR spectroscopy (substantiated by electron spin resonance and x-ray diffraction) that the resins contained from 6–16 condensed aromatic rings in a single sheet; the latter size appeared to be the maximum condensed structure in asphalt. In the asphaltene molecules, the condensed sheet was repeated one or more times and was linked through saturated chains or rings. The solid-state asphaltenes appeared to form stacks of condensed aromatic sheets which may include individual sheets from several molecules. In benzene, or other effective solvents, the low-molecular-weight stacks appeared to disintegrate to form true solutions, but the higher, less soluble fractions, appeared to remain associated.

Application of PMR to the analysis of aromatic fractions from crude petroleum oils and quantitative determinations of the methyl, methylene, and methine carbons in the fractions led to the suggestion that uncertainties existed due to band overlaps occurring in the spectra (Williams and Chamberlain, 1963).

In summary, PMR was one of the first spectroscopic techniques that allowed the derivation of total structural types for asphaltenes. However, later work has thrown some doubt on the validity of these conclusions; not so much because of the efforts and conclusions of the investigators, but more because of the deficiencies of the technique and the potential masking of some of the hydrogen atoms. This has the potential to skew the structural parameters, thereby producing data that are not truly consistent with the molecular types in the asphaltene fraction.

However, it must be recognized that the structural parameters derived by magnetic resonance may present a deceptively simple picture and may be of extremely limited significance for a complex mixture such as the asphaltene fraction.

With the evolution of magnetic resonance and the development of the carbon-13 procedure, there has been further detail added to the structure of petroleum fractions (Severin and Glintzer, 1984; Snape et al., 1984).

The application of carbon-13 magnetic resonance to the issue of structural types in petroleum has allowed the more precise identification of carbon atoms in various locations and, hence, a more reasonable calculation of the structural parameters. But what is more important, carbon-13 magnetic resonance can be used to show the differences in aromatic carbon content of asphaltenes from different crude oils.

In this context, it would be of interest to fractionate asphaltenes (from different crude oils) by a standard method, such as ion exchange chromatography (e.g., see McKay et al., 1978; Francisco and Speight, 1984). From the data, it would be possible to determine the distribution of the amphoteric constituents, basic constituents, acidic constituents, and less neutral polar species in each asphaltene and whether or not the difference in aromatic carbon content was due to variations in the relative proportions of the subfractions.

Carbon-type analysis does not allow the identification of individual constituents, but presents an overall picture of the carbon framework and (if the term can be used without too literal a translation) actually presents an *average* molecular structure of these higher boiling materials.

Regardless of the method for separating the hydrogen types, the most derived structural parameters take the form of ratios of the various hydrogen types. The H_I/C_A (H_{aru}/C_{ar}) ratio indicates extent of condensation of the unsubstituted aromatic system and the $C_{su}H_I(\sigma)$ ratio gives the degree of substitution. The $C_s/C_{su}(n)$ ratio represents the average chain length, the H_N/C_N ratio is the degree of condensation of naphthenic clusters and f_a factor is the fraction of carbon that is aromatic. C_A is the total number of aromatic carbon atoms in the sample, R_S is the number of substituted aromatic ring carbons and R_A is the number of aromatic rings.

The structural details of the carbon backbone obtained from proton spectra are derived by inference, but it must be recognized that protons at peripheral positions can be obscured by intermolecular interactions. This, of course, can cause errors in the ratios that can have a substantial influence on the outcome of the calculations (Ebert et al., 1987). It is in this regard that carbon-13 magnetic resonance (CMR) can play a useful role.

Since CMR deals with analyzing the carbon distribution types, the obvious structural parameter to be determined is the aromaticity f_a. A direct determination from the various carbon environments is one of the better methods for the determination of aromaticity. Certain regions are often designated for certain types of carbon atoms (Snape et al., 1979): heterocyclic + quinone, 155–160; pyrrole-like, 130–150; mono- to polyaromatics, 120–130; εC, 295; δC, 29; γC, 32; βC, 23); and αC, 14.

Through a combination of proton and carbon magnetic resonance techniques, refinements can be made on the structural parameters. However, with the solid-state high-resolution carbon magnetic resonance technique, a number of additional structural parameters can be obtained. For example, quaternary aromatic carbons, tertiary aromatic carbons, ratio of primary to quaternary aliphatic carbons, and the ratio of secondary to tertiary aliphatic can be estimated (Weinberg et al., 1981).

10.2.2.3 Mass Spectrometry

Mass spectrometry plays a key role in the identification of the constituents of petroleum and of petroleum products (Aczel, 1979, 1989). Mass spectrometric methods are particularly well suited to the analyses of complex mixtures of relatively simple hydrocarbons that comprise the bulk of petroleum and its products.

The principal advantages of mass spectrometric methods are (1) high reproducibility of quantitative analyses; (2) the potential for obtaining detailed data on the individual components and carbon number homologues in complex mixtures; and (3) a minimal sample size is required for analysis. The ability of mass spectrometry to identify individual components in complex mixtures is unmatched by any modern analytic technique, perhaps the exception is gas chromatography.

In general, most petroleum samples are very complex mixtures containing literally hundreds and even thousands of individual components. These components tend to form several homologous series of the same general formulae and extend from near gaseous materials to nonvolatile components. Thus, there are disadvantages arising from the use of mass spectrometry: (1) the limitation of the method to organic materials that are volatile and stable at temperatures up to 300°C (570°F); and (2) the difficulty of separating isomers for absolute identification. The sample is usually destroyed, but this is seldom a disadvantage.

Despite these limitations, mass spectrometry furnishes useful and adequate information on the composition of petroleum streams, even if this information is not as exhaustive as might be required. Structural similarities hindering identification of individual components make, in effect, identification by type or by homologue only more meaningful, as these similar structures behave similarly in most processing situations and knowledge of the individual isomeric distribution would add little to our understanding of the relationships between composition and processing parameters.

The availability of high-resolution instrumentation extends the scope of mass spectrometry for both qualitative and quantitative applications. The main advantage of high-resolution techniques is the separation of fragment and molecular ions of the same nominal molecular weight but different elemental composition. In addition, most high-resolution instruments are capable of furnishing precise mass measurements, so that the formulae can be calculated from these measurements.

Mass spectrometry should be used cautiously when a maximum amount of information can be expected. However, heavy oils and residua are for practical purposes beyond the useful range of routine mass spectrometry, but many of the discrete fractions that can now be obtained could be studied qualitatively by high-resolution mass spectral techniques. Many of the higher molecular weight constituents in heavy oils and residual are nonvolatile. At elevated temperatures, decomposition could occur in the inlet and any subsequent analysis would be biased to the low-molecular-weight end and to the lower molecular products produced by thermal decomposition.

Mass spectrometry adds further knowledge to the structural analysis of petroleum by aiding the researcher in the calculation of ring distribution per molecule ions in a mixture. However, there are questions about the application of the technique to the nonvolatile fractions of petroleum. The lack of volatility, which is almost self-explanatory, does not allow accurate identification of the total asphaltene.

Nevertheless, the petroleum industry has long been the leader in the use of mass spectroscopy to solve its analytical problems. As soon as instruments became capable of producing repeatable spectra in the early 1940s, hydrocarbon mixtures were analyzed by mass spectrometry with speed and precision not approached by other techniques available at that time. As heat inlet systems evolved, analysis of gases and low-boiling liquids was followed by type analyses of higher-boiling petroleum fractions, and advances in mass-spectroscopic analysis of petroleum and its derivatives paralleled improved instrumentation.

Early work (Lumpkin, 1956) involved development of a procedure for type analysis of the saturate portion of petroleum materials of high molecular weight and employed characteristic fragment masses of paraffins, noncondensed naphthenes, and condensed naphthenes containing 2–6 condensed rings per molecule to delineate these types. This was followed (Lumpkin, 1964) by an analysis of a trinuclear aromatic portion of a coker gas oil. Six condensed aromatic hydrocarbon types, five sulfur, two oxygen, and one nitrogen compound were identified and a quantitative analysis based on low-voltage sensitivity showed that derivatives of alkyl dibenzothiophene, phenanthrene, and fluorene were the major types present.

Other workers (Clerk and O'Neal, 1961) also investigated high-molecular-weight petroleum fractions using a technique developed for examining high-resolution mass spectra of high-molecular-weight solids, and concluded that asphaltic materials consisted principally of condensed ring structures, both aromatic and cyclic, to which one long chain was attached. If two-ring systems were found in a molecule, they were presumably located at opposite ends of a long chain; substitution of the rings was believed to be limited to methyl and ethyl groups. It was postulated that molecules of these types were very probably the principal components of the so-called resin fraction in petroleum and asphalt.

Mass spectrometry has also been employed to investigate the aromatic structural unit of asphaltenes resins (Dickie and Yen, 1967, 1968a,b).

An analysis of the nitrogen and oxygen compounds from crude oil using high-resolution mass spectrometry as the principal analytical tool showed a variety of compound-types to be present. These included derivatives of indole, carbazole, benzcarbazole, pyridine, quinoline, pyridone, quinolone, dibenzofuran, naphthobenzofuran, dinaphthofuran, phenol, carboxylic acids, aliphatic ketones, and esters (Snyder et al., 1968; Snyder, 1969).

Further studies using **pyrolysis**/gas chromatography/mass spectrometry (py/gc/ms) (Speight and Pancirov, 1984) showed that different subfractions of an asphaltene fraction would produce the same type of polycyclic aromatic systems in the volatile matter, but the distribution was not constant. It was also possible to compute the hydrocarbon distribution. From which a noteworthy point is, preponderance of single-ring (cycloparaffin and alkylbenzene) species as well as the domination of saturated material over aromatic material. The emphasis on low molecular weight material in the volatile products is to be anticipated on the basis that more complex systems remain as nonvolatile material and, in fact, are converted to coke.

One other point worthy of note is that the py/gc/ms program does not accommodate nitrogen and oxygen species, whether or not they are associated with aromatic systems. This matter is resolved, in part, not only by the concentration of nitrogen and oxygen in the nonvolatile material (coke), but also by the overall low proportions of these heteroatoms originally present in the asphaltenes. The major drawback to the use of the py/gc/ms technique to the study of the aromatic systems in asphaltenes is the amount of material that remains as a nonvolatile residue.

The most widely accepted theory for the origin of petroleum is that it is derived from a biological source material and is supported by the presence of **porphyrins** in petroleum (Chapter 6). Porphyrins are assumed to be products from the degradation of chlorophyll. High-resolution mass spectra of porphyrins extracted from several asphaltenes have been examined (Baker, 1966) and shown to contain 7–18 methylene groups per molecule, while absorption spectroscopy gave no indication of oxygen functional groups. Further work (Baker et al., 1967) involving an examination of asphaltenes from a variety of crude oils confirmed the presence of cycloalkyl- and alkylporphyrins and the visible spectra were indicative of either incomplete beta-substitution or bridge substitution.

10.2.2.4 Electron Spin Resonance

Spectrophotometry in the electron spin resonance range (ESR) is useful to supply supplementary information to that obtained by other methods. Parameters that can be obtained by such methods include I, color intensity; g, Lande g-factor; N_g, spin numbers; T_1, spin-lattice relaxation time; P_C, population of aromatic carbon per spin; J, spin-excitation energy; and N_d, doublet-spin concentration.

The intensely dark color of the resin and asphaltene fractions of petroleum is believed to be closely associated with the organic free radicals, which in turn are usually associated with the aromatic structure (Erdman, 1962). There are also indications that the porphyrins in asphaltenes can be treated as single sheets of alkyl-substituted polynuclear aromatic hydrocarbons, which can be fused to multiple naphthenic rings. Association may also play an important part in the micellar structure of the porphyrins, remembering that separated porphyrins will behave differently from the porphyrins in petroleum.

The free-radical signal can be used as a measure of the asphaltene content of petroleum hydrocarbons. Asphaltenes exhibit a signal equivalent to $2-4 \times 10^{18}$ free radicals per gram, or about one per 100 asphaltene molecules, assuming a molecular weight of about 2000 for the asphaltenes (Flinn et al., 1961; Wen et al., 1978; Champagne et al., 1985). The postulate is that the free radicals contribute to the tendency toward micelle formation in the asphaltenes and their results indicate that the free-radical sites were essentially confined to the asphaltene fraction and the resins contain only 2% of the total free radicals present.

There has also been some focus on the metal species in asphaltenes (Tynan and Yen, 1969; Shepherd and Graham, 1986). The data indicate that there are, in addition to other metal species, two different types of vanadium, one bound and the other free, occurring naturally in petroleum, both as tetravalent vanadium. It was suggested that the observed vanadium occurring in an asphaltene fraction is associated with the asphaltene molecule itself, most probably the aromatic portion.

10.2.2.5   Ultraviolet Spectroscopy

Of all of the methods applied to determining the types of aromatic systems in petroleum asphaltenes, one with considerable potential is ultraviolet spectroscopy (Friedel and Orchin, 1951; Braude and Nachod, 1955; Rao, 1961; Jaffe and Orchin, 1966; Brown and Searl, 1979; Lee et al., 1981; Bjorseth, 1983).

Typically, the ultraviolet spectrum of the asphaltene fraction shows two major regions with very little fine structure. Interpretation of such a spectrum can only be made in general terms (Friedel, 1959; Boyd and Montgomery, 1963; Posadov et al., 1977). More recent studies have shown that the types of chromophores remain constant throughout an asphaltene, but the number increases with the molecular weight of the fraction (Yokota et al., 1986).

This technique can add valuable information about the degree of condensation of polycyclic aromatic ring systems when used in conjunction with high performance liquid chromatography (HPLC) (Suatoni and Swab, 1976; Dark and McGough, 1978; Lee et al., 1981; Bjorseth, 1983; Monin and Pelet, 1983; Felix et al., 1985; Killops and Readman, 1985; Yokota et al., 1986; Verhasselt, 1992; Akhlaq, 1993; Ali and Nofal, 1994). Indeed, when this approach is taken, the technique not only confirms the complex nature of the asphaltene fraction, but also allows further detailed identifications to be made of the individual functional constituents of asphaltenes.

The asphaltene fraction is a complex mix of a variety of constituents that is reflected in a multi-component chromatogram (Figure 10.2) that differs for each asphaltene sample and is in keeping with the fractionation of asphaltene constituents by functionality (Figure 10.3). In addition, asphaltene subfractions produce a less complex and much narrower chromatograph that may even approximate a single peak which may prove much easier to monitor by the ultraviolet detector. These data provide strong indications of the ring-size distribution of the polycyclic aromatic systems in petroleum asphaltenes. For example, from an examination of various functional subfractions, it becomes evident that amphoteric species and basic nitrogen species contain polycyclic aromatic systems having two-to-six rings per system. On the other hand, acid subfractions (containing phenolic and/or carboxylic functions) and neutral polar subfractions (containing amide and/or imino functions) contain few, if any, polycyclic aromatic systems having more than three rings per system. In short, when the need is to consider the detailed chemistry and reactivity of the asphaltene fraction-constituents, such data are not supportive of the concept of average structures or average molecular weight as determined by a variety of methods.

In all cases, the evidence favored the preponderance of the smaller (one-to-four) ring systems (Speight, 1986). But, perhaps what is more important about these investigations is that the data show that asphaltenes are a complex mixture of compound types which confirms

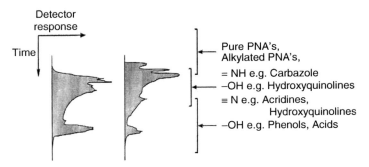

FIGURE 10.2 HPLC profiles of two different asphaltene fractions.

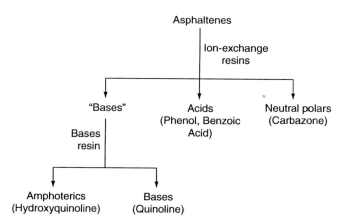

FIGURE 10.3 Fractionation of asphaltene fractions in accordance with functionality.

fractionation studies and, further, the constituents of the asphaltene cannot be represented by any particular formula that is construed to be average. Therefore, the concept of a large polycyclic aromatic ring system as the central feature of asphaltene molecules must be abandoned.

The technique is also useful on a before refining and after refining basis. It allows molecular changes in, for example, the asphaltene fraction to be tracked and provides information relative to the chemistry of coke formation.

10.2.2.6 X-Ray Diffraction

X-ray diffraction techniques also provide information about the structural qualities of organic materials, but the data provide information about the spatial periodicity within a substance. However, there is one necessary assumption in the application of this method and it requires that, within a given fraction, the larger portion of the molecules contain within and among themselves certain repeated structural features, such as sheets of condensed aromatic rings.

The x-ray diffraction patterns of petroleum fractions are diffuse compared to those of the well-ordered carbons, and with the exception of the x-ray diffraction patterns of the asphaltenes, it is usually very difficult to extract information from the diffraction patterns. However, it must be remembered that the x-ray diffraction technique is, by necessity, used for solid materials, such as asphaltenes.

Nevertheless, after plotting the diffraction patterns onto a straight baseline, it is possible to make estimates of the aromaticity. Thus, the fraction of aromatic carbon f_a can be determined using the relationship

$$f_A = C_{\text{aromatic}}/C_{\text{total}}$$

$$= A_{002}/(A_{002} + A_\gamma)$$

where A represents the areas under the respective peaks.

Early investigations (Nellensteyn, 1938) indicated the presence of a band corresponding to 3.5 C in the x-ray diffraction pattern of an asphaltene fraction and compared the similarity of this feature to that of amorphous carbons, while evidence of crystallinity (molecular order) in asphaltenes was also indicated (Williford, 1943). Later work (Labout, 1950) showed petroleum asphaltenes to have an x-ray pattern, characteristic of amorphous substances.

Use of the well-established Scherrer equation,

$$L = kl/(B\cos\theta)$$

L is either the crystallite height L_c or the crystallite diameter L_a, B is the half-peak width, is one-half of the diffraction angle (2θ), k is a constant (for L_c, $k = 1.00$; for L_a, $k = 1.84$), and λ is the wavelength of the incident beam. This, of course, assumes that asphaltenes have repeating aromatic units with a tendency to stack in layers, as in the case of carbon and graphite. Interlayer distances $c/2$ are calculated from the position of the maximum of the (002) band. The data derived using this equation refer mainly to the structures found in the asphaltene fraction of petroleum. The data indicate that layer diameters L_a are of the order of 6–15 D, interlayer distances $c/2$ fall into the range 3.5–3.8 Å, and stack heights L_c are about 10–16 Å, which suggests that there are three to five aromatic lamellae per stack.

Other investigations (Young, 1954; Padovani et al., 1959; Bestougeff, 1967; Pollack and Yen, 1969) indicated that the asphaltenes might consist of condensed aromatic sheets carrying naphthenic and paraffinic substituents and could be explained on the basis of polynuclear aromatic sheets of 9–15X diameter, spaced about 3.6X units apart. A structure consisting of about four sheets could be obtained before the natural heterogeneity of asphaltenes prevented further accumulation. However, accumulation of other nuclei was still possible because the asphaltene molecules were considered to contain two or more polynuclear aromatic sheets sigma-bonded by aliphatic carbon chains. Chains orientation of aliphatic groups, spaced about 5.7X apart in the manner of a saturated carbon chain or loose net of naphthenic rings, was also indicated.

For materials with limited order, such as pyrolytic carbons, the *x-ray diffraction* (XRD) technique can be used to determine the repeating sequences of the molecular arrangement. For a dark-colored, solid material in particular, the vectors of the d spacings between 0.5 and 10 Å can be easily derived from a reduced x-ray spectrum. The crystallite parameters often used as structural parameters are (Yen et al., 1961b; Yen and Erdman, 1963) L_a, layer diameter; L_c, cluster diameter; d_m, interlayer distance; d_i, interchain distance; and M_e, effective layer number. One of the reputed advantages of the XRD technique is that it can also be used for a measurement of the aromaticity f_a of the sample. The calculation is based on the relative intensity of the (γ) band and the (002) band of the pattern.

X-ray diffraction used at very small angles, small angle scattering (SAS), has also been used for the characterization of heavy fractions of petroleum (Pollack and Yen, 1970). The range of small angle scattering is in the molecular dimension from 20 to 2000 Å. Some of the useful parameters are f, structure number; O_r, surface parameter; l, heterogeneity length; l_c, coherence length; and r_M, radius of gyration. There has been serious criticism of the use of the method for determining absolute structural parameters of petroleum materials (Ebert et al., 1984; Ebert, 1990).

Thus, it is not surprising that XRD studies have been used in attempts to elucidate the structure of the micelle (Wen et al., 1978). Molecular weight data (Chapter 8 and Chapter 10) show that asphaltenes associate, even in dilute solution and under the conditions often prescribed for determination of molecular weight. This has raised the issue of asphaltene–asphaltene interactions as well as the means by which asphaltenes exist in petroleum.

Small angle x-ray scattering (SANS) has also been used as a means of determining particle size and asphaltene molecular weight (Espinat et al., 1984; Herzog et al., 1988; Storm et al., 1993). And, it is interesting to note that the SANS technique gives good agreement with other methods for the molecular weight of the fractions from gel permeation chromatography for the lower molecular weight fractions, but was unusable for the higher molecular weight fractions.

What is perhaps more important from the findings was that the particle size of the asphaltenes was found to increase as a result of deasphalting. This raises the issue of the changes caused to the asphaltene constituents during deasphalting (perhaps analogous to denaturing the protein during the boiling of an egg). Further, it questions the validity of the literal interpretations that have been made of the physical character of asphaltenes that have been separated from petroleum. Therefore, deductions about the nature of the micelle, made on the basis of the behavior of separated asphaltenes, are certainly open to question and re-investigation.

An x-ray technique (Diamond, 1957, 1958, 1959) can be used to estimate the size groups of aromatic lamellae (sheets) that constitute the molecular structure of synthetic carbons, coals, asphaltenes, and the like. After correction for polarization and absorption, 31 evenly spaced intensity values between $S = 0.66$ and 0.96 (where $S = 2 \sin \theta / \lambda$) are treated with Diamond's matrix H_5. Corrections for Compton scattering due to hetero (e.g., noncarbon) atoms in the layers are also applied, and a histogram is derived. Each histogram is made up of seven parts showing, from left to right, size groups consisting of nonaromatic material A and aromatic lamellae of varying diameters, 5.8, 8.4, 10.0, 15.0, 20.0, and 20.0 Å.

10.2.3 HETEROATOM SYSTEMS

Although the major focus of spectroscopic techniques appears to have been delineation of the hydrocarbon skeleton of the nonvolatile constituents of petroleum, efforts have also been made to determine the nature and location of the nitrogen, oxygen, and sulfur atoms in the aromatic systems (McKay et al., 1978).

PMR studies do not take into account the heteroatoms which had to be accommodated by convenient incorporation into the hydrocarbon systems. In fact, placement of the hetero-atoms into the structures derived by PMR would require significant modifications of the molecular models to, perhaps, much larger structures. Indeed, this may be cited as a major disadvantage of many of the models that have been derived to date.

It is, therefore, necessary to make reference to the spectroscopic methods that have been used to determine the nature of the heteroatom species. Some mention has already been made in the previous sections. This section will confirm what has already been noted. But, it is necessary to note that even though nonspectroscopic methods may have been the prime method, spectroscopy usually played a role in confirming the nature of the heteroatom.

10.2.3.1 Nitrogen

Nitrogen species occur readily in crude oils and it must be anticipated that similar species will occur within the asphaltenes. Studies on the disposition of nitrogen in petroleum asphaltenes indicated the existence of nitrogen as various heterocyclic types (Helm et al., 1957; Ball et al., 1959; Clerc and O'Neal, 1961; Nicksic and Jeffries-Harris, 1968; Moschopedis and Speight, 1976b, 1979; Schmitter et al., 1983; Jacobson and Gray, 1987). Much of the nitrogen is believed to be in aromatic locations (Mitra-Kirtley et al., 1993). There are also reports where the organic nitrogen in petroleum asphaltenes has been defined in terms of basic and nonbasic types (Nicksic and Jeffries-Harris, 1968) and there is evidence for the occurrence of carbazole nitrogen in asphaltenes (Clerc and O'Neal, 1961; Moschopedis and Speight, 1979).

More recent assumptions include nitrogen in asphaltenes on the basis of not more than one nitrogen atom per molecule (Guiochon, 1982). Such an assumption may be difficult to justify on a natural product basis where molecules with two or more nitrogen atoms are known but it would certainly serve as a reasonable guide for molecular studies where generalizations are often necessary.

Recent studies (Schmitter et al., 1984) have also brought to light the occurrence of four-ring aromatic nitrogen species in petroleum. These findings are of particular interest since they correspond to the ring systems that have been tentatively identified by application of HPLC to the basic nitrogen fraction of asphaltenes.

Application of XANES spectroscopy to the determination of nitrogen species in asphaltenes confirmed the absence of saturated amines, with pyrrolic nitrogen and pyridinic nitrogen being the two major types of nitrogen (Mullins, 1995). In fact, the technique showed the predominance of pyrrolic nitrogen in the samples examined.

10.2.3.2 Oxygen

Of the heteroatom-containing species in petroleum, there are more data pertaining to the locations of the oxygen atoms than to the sulfur and nitrogen atoms. However, the majority of the data relates to oxygen functions in blown (oxidized) asphalt and residua that may be of little relevance to the oxygen functions in the native materials.

Oxygen has been identified in carboxylic, phenolic and ketonic locations (Nicksic and Jeffries-Harris, 1968; Moschopedis and Speight, 1976a,b; Ritchie et al., 1979a), but is not usually regarded as being located primarily in heteroaromatic ring systems. In the context of polyhydroxy aromatic nuclei existing in Athabasca asphaltenes, it is of interest to note that pyrolysis at 800°C (1470°F) results in the formation of resorcinols (Ritchie et al., 1979a) implying that such polyhydroxy aromatic rings systems may indeed exist in the asphaltenes.

A very important aspect of the presence of oxygen functional groups is the identification of these functions and definition of their role in determining the physical properties of the system. The physical properties can also influence the structure of the micelle (Moschopedis and Speight, 1976a; Branthaver et al., 1993; Bukka et al., 1993).

10.2.3.3 Sulfur

Sulfur occurs in petroleum as benzothiophenes, dibenzothiophenes and naphthenebenzothiophenes and many of these sulfur types occur in the asphaltene fraction (Clerc and O'Neal, 1961; Nicksic and Jeffries-Harris, 1968; Speight and Pancirov, 1984; Ngassoum et al., 1986; Rose and Francisco, 1988). More highly condensed thiophene-types may also occur in the asphaltene fraction, but are precluded from identification by low volatility. Other forms of sulfur that occur in asphaltenes include the alkyl–alkyl sulfides, alkyl–aryl sulfides and aryl–aryl sulfides (Yen, 1974; Waldo et al., 1992).

Application of FTIR spectroscopy and sulfur-33 magnetic resonance spectroscopy has provided valuable data about the distribution of the sulfur types in petroleum, particularly the partition of sulfur into aromatic types and aliphatic types (Novelli et al., 1984). It is particularly interesting to note that sulfur exists in a variety of locations that may be dependent upon the natural product precursors and the mechanism of maturation.

On a natural product basis, sulfur (as thiophene sulfur) is more difficult to identify. The presence of sulfur, other than the sulfur-containing amino acids cysteine and methionine, is still speculative, although there are means by which inorganic sulfur can be incorporated into petroleum as organic sulfur (Hobson and Pohl, 1973; Orr, 1977; Orr and Sinninghe Damste, 1990).

An interesting aspect of sulfur identification has arisen through the application of x-ray absorption spectroscopy to answer the question about sulfur forms in asphaltenes (George et al., 1990; Kelemen et al., 1990). The data gave a clear demonstration of the existence of nonvolatile sulfide and thiophene sulfur in the asphaltenes.

Application of XANES spectroscopy to the determination of sulfur species in asphaltenes confirmed the presence of thiophene sulfur and sulfide sulfur (Mullins, 1995). In fact, the technique showed the predominance of thiophene sulfur and, to a lesser extent, sulfide sulfur in the samples examined.

10.2.3.4　Metals

Metals are not found in any fractions that boil below 540°C (1000°F) (Reynolds, 1997). Typically, the asphaltenes have the highest concentration of vanadium and other metals. Most often, in characterizing metal complexes in crude oils, an asphaltene precipitation is performed that concentrates the vanadium (as well as nickel and iron). The nickel and nitrogen distributions are very similar to the vanadium distributions, in general, with the asphaltene fraction having the highest concentrations. The metals are found bound in two types, **metallo-petroporphyrins** and metallo-nonporphyrins. The former have the distinctive ultraviolet–visible Soret band due to the highly conjugated porphyrin ring system, and the latter is comprised of whatever metal concentrations are not accountable by the ultraviolet–visible Soret band.

The different types of metals led to some controversy because only metallo-petroporphyrins have been unequivocally identified. This had led to many studies over the years which indicated that the metallo-nonporphyrin compounds are vanadium bound with four nitrogen atoms in the first coordination sphere as porphyrins or with other heteroatoms (combinations of nitrogen, sulfur, and oxygen), not in the porphyrin environment.

Metals (i.e., nickel and vanadium) are much more difficult to integrate into the asphaltene system. The nickel and vanadium occur as porphyrins (Baker, 1969; Yen, 1975; Malhotra and Buckmaster, 1985; Shepherd and Graham, 1986) but whether or not these are an integral part of the asphaltene structure is not known. Some of the porphyrins can be isolated as a separate stream from petroleum (Branthaver and Dorrence, 1978; Reynolds et al., 1989; Branthaver, 1990).

10.3　MISCELLANEOUS METHODS

There also are a number of chemical methods that can be used to obtain structural information about the constituents of petroleum. For example, derivation of the functional groups using such techniques as silylation of hydroxyls, esterification of acid functions, and quarternarization of amines have been employed successfully. Examination of the products from oxidative degradation or hydrogenolysis as well as the products of alkylation, dealkylation, or scission of carbon–nitrogen, carbon–oxygen, or carbon–sulfur bonds are useful methods. Catalytic hydrogenation or dehydrogenation in naphthenic–aromatic systems via naphthenic intermediates is also of some value.

Questions about the nature of the metal species in petroleum have been ongoing for the past two decades. Metals are important because of the manner in which they can interact with catalysts causing deactivation of the catalyst and also appearing to promote coke formation.

Inductively coupled plasma emission spectroscopy has been used for the determination of trace elements in ash residues from various crude oils (Saban et al., 1984). The data were applied to determining the origin and occurrence of crude oils as an aid in oil exploration. Nevertheless, the technique does offer valuable information about the occurrence of trace elements in petroleum.

Inductively coupled plasma emission spectroscopy has also been employed, in conjunction with graphite furnace atomic absorption spectroscopy to investigate the metals types in petroleum (Reynolds et al., 1984). The data showed that a considerable percentage (50–80%) of the metals in a suite of crude oils existed in the nonporphyrin form. These nonporphyrin metal species were relatively low molecular weight (<400) species that were liberated when the tertiary structure of the asphalt material was denatured by extraction.

Application of x-ray absorption spectroscopy (EXAFS/XANES) and ultraviolet–visible spectrometric analysis to the issue of the chemical environment of vanadium in various asphaltenes (Goulon et al., 1984) indicated a contradiction in terms of the amount of metals in porphyrins environments and that the conclusions were method dependent.

The results of small angle neutron scattering (SANS) studies (Overfield et al., 1989) have indicated a true molecular weight below 6000 and indications that this decreases with increasing temperature. Such conclusions can benefit the processing of asphaltene-containing crude oil (such as heavy oil) since they offer some insight into the structure of the micelle and the behavior of asphaltenes under processing conditions.

REFERENCES

Aczel, T. 1979. *Practical Mass Spectrometry*. Middleditch, B.S. ed. Plenum Press, New York.

Aczel, T. 1989. Preprints. *Div. Petrol. Chem. Am. Chem. Soc.* 34(2): 318.

Akhlaq, M.S. 1993. *J. Chromatogr.* 644: 253–258.

Ali, L.H., Al-Ghannam, K.A., and Al-Rawi, J.M. 1990. *Fuel* 69: 519.

Ali, M.A. and Nofal, W.A. 1994. *Fuel Sci. Technol. Int.* 12: 21–33.

Altgelt, K.H. 1965. *J. Appl. Polym. Sci.* 9: 3389.

Baker, E.W. 1966. *J. Am. Chem. Soc.* 88: 2311.

Baker, E.W. 1969. *Organic Geochemistry*. G. Eglinton and Murphy, M.T.J. eds. Springer-Verlag, New York.

Baker, E.W., Yen, T.F., Dickie, J.P., Rhodes, R.E., and Clark, L.F. 1967. *J. Am. Chem. Soc.* 89: 3631.

Ball, J.S., Latham, D.R., and Helm, R.V. 1959. *J. Chem. Eng. Data* 4: 167.

Bestougeff, M.A. 1967. *Bull. Soc. Chim. Fran.* p. 4773.

Bjorseth, A. 1983. *Handbook of Polycyclic Aromatic Hydrocarbons*. Marcel Dekker, New York.

Bouquet, M. and Bailleul, A. 1982. *Petroanalysis '81. Advances in Analytical Chemistry in the Petroleum Industry 1975–1982*. Crump, G.B. ed., John Wiley & Sons, Chichester, UK.

Boyd, M.L. and Montgomery, D.S. 1963. *J. Inst. Petrol.* 49: 345.

Brandes, G.V. 1956. *Brennstoff-Chem.* 37: 263.

Branthaver, J.F. 1990. *Fuel Science and Technology Handbook*. Speight, J.G. ed., Marcel Dekker, New York.

Branthaver, J.F. and Dorrence, S.M. 1978. *Analytical Chemistry of Liquid Fuel Sources: Tar Sands, Oil Shale, Coal, and Petroleum*. Uden, P.C. Siggia, S. and Jensen, H.B. eds. Advances in Chemistry Series No. 170. American Chemical Society, Washington, DC. Chapter 10.

Branthaver, J.F., Petersen, J.C., Robertson, R.E., Duvall, J.J., Kim, S.S., Harnsberger, P.M., Mill, T., Ensley, E.K., Barbour, F.A., and Schabron, J.F. 1993. *Binder Characterization and Evaluation*. strategic highway research program. summary report national research council, washington, DC. Volume 2. Chapter 5.

Braude, E.A. and Nachod, F.C. 1955. *Determination of Organic Structures by Physical Methods*. Academic Press, New York.

Brooks, B.T., Boord, C.E., Kurtz, S.S., Jr., and Schmerling, L, eds. 1954. *The Chemistry of Petroleum Hydrocarbons*, Volume 1. Reinhold, New York.

Brown, J.K. and Ladner, W.R. 1960. *Fuel* 39: 87.

Brown, R.A. and Searl, T.D. 1979. *Chromatography in Petroleum Analysis*. Altgelt, K.A. and Gouw,T.H. eds. Marcel Dekker, New York. Chapter 15.

Bukka, K., Miller, J.D., Hanson, F.V., Misra, M., and Oblad, A.G. 1993. *Fuel* 73: 257.

Champagne, P.J., Manolakis, E., and Ternan, M. 1985. *Fuel* 64: 423.

Chelton, H.M. and Traxler, R.N. 1959. *Proceedings. 5th World Petroleum Congress*. Section V. Paper 19.

Clerc, R.J. and O'Neal, M.J. 1961. *Anal. Chem.* 33: 380.

Cookson, D.J., Lloyd, C.P., and Smith, B.E. 1986. *Fuel* 65: 1247.

Dark, W.A. and McGough, R.R. 1978. *J. Chromatogr. Sci.* 16: 610.

Dereppe, J.M. and Moreaux, C. 1985. *Fuel* 64: 1174.

Diamond, R. 1957. *Acta Crystallogr.* 10: 359.

Diamond, R. 1958. *Acta Crystallogr.* 11:129.

Diamond, R. 1959. *Phil. Trans. R. Soc.* 252A: 193.

Dickie, J.P. and Yen, T.F. 1967. Preprints. *Div. Petrol. Chem. Am. Chem. Soc.* 12: B-117.

Dickie, J.P. and Yen, T.F. 1968a. Preprints. *Div. Petrol. Chem. Am. Chem. Soc* 13: F-140.

Dickie, J.P. and Yen, T.F. 1968b. *J. Inst. Petrol.* 54: 50.

Ebert, L.B. 1990. *Fuel Sci. Technol. Int.* 8: 563.

Ebert, L.B., Scanlon, J.C., and Mills, D.R. 1984. *Liq. Fuels Technol.* 2: 257.

Ebert, L.B., Mills, D.R., and Scanlon, J.C. 1987. Preprints. *Div. Petrol. Chem. Am. Chem. Soc.* 32(2): 419.

Erdman, J.G. 1962. *Proc. Am. Petrol. Inst.* 42: 33.

Erdman, J.G. 1965. *Hydrocarbon Analysis*. Special Publication No. STP 389. American Society for Testing and Materials, Philadelphia, PA. p. 259.

Espinat, D., Tchoubar, D., Boulet, R., and Freund, E. 1984. *Characterization of Heavy Crude Oils and Petroleum Residues*. Tissot, B. ed., Editions Technip, Paris. p. 147.

Felix. G., Bertrand, C., and Van Gastel, F. 1985. *Chromatographia* 20(3): 155.

Ferris, S.W., Black, E.P., and Clelland, J.B. 1966. Preprints. *Div. Petrol. Chem. Am. Chem. Soc.* 11: B-130.

Fischer, K.A. and Schram, A. 1959. *Erdoel Kohle* 12: 368.

Flinn, R.A., Beuther, H., and Schmid, B.K. 1961. *Petrol. Refiner* 40: 139.

Francisco, M.A. and Speight, J.G. 1984. Preprints. *Div. Fuel Chem. Am. Chem. Soc.* 29(1): 36.

Friedel, R.A. 1959. *J. Chem. Phys.* 3l: 280.

Friedel, R.A. and Orchin, M. 1951. *Ultraviolet Spectra of Aromatic Compounds*. John Wiley & Sons Inc., New York.

Gardner, R.A., Hardman, H.F., Jones, A.J., and Williams, R.B. 1959. *J. Chem. Eng. Data* 4: 155.

George, G.N., Gorbaty, M.L., and Kelemen, S.R. 1990. *Geochemistry of Sulfur in Fossil Fuels*. Orr, W.L. and White, C.M. eds., Symposium Series No. 429. American Chemical Society, Washington, DC. Chapter 12.

Giavarini, C. and Vecchi, C. 1987. *Fuel* 66: 868.

Gould, K.A. and Long, R.B. 1986. *Fuel* 65: 572.

Goulon, J., Esselin, C., Friant, P., Berthe, C., Muller, J-F., Poncet, J-L., Guilard, R., Escalier, J-C., and Neff, B. 1984. *Characterization of Heavy Crude Oils and Petroleum Residues*. Tissot, B. ed., Editions Technip, Paris. p. 3.

Gray, M.R., Choi, J.H.K., and Egiebor, N.O. 1989. *Fuel Sci. Technol. Int.* 7: 599.

Guiochon, G. 1982. *Petroanalysis'81*. Crump, G.C. ed. John Wiley & Sons Inc., Chichester, UK.

Hasan, M., Siddiqui, M.N., and Arab, M. 1988. *Fuel* 67: 1131.

Hasan, M., Ali, M.F., and Arab, M. 1989. *Fuel* 68: 801.

Helm, R.V., Latham, D.R., Ferris, C.R., and Ball, J.S. 1957. *Chem. Eng. Data Ser.* 2: 95.

Herzog, P., Tchoubar, D., and Espinat, D. 1988. *Fuel* 67: 245.

Hobson, G.D. and Pohl, W., eds., 1973. *Modern Petroleum. Technology*. Applied Science Publishers Ltd. London.

Jacobson, J.M. and Gray, M.R. 1987. *Fuel* 66: 749.

Jaffe, H.H. and Orchin, M. 1966. *Theory and Applications of Ultraviolet Spectroscopy*. John Wiley & Sons Inc., New York.

Kelemen, S.R., George, G.N., and Gorbaty, M.L. 1990. *Fuel* 69: 939.

Killops, S.D. and Readman, J.W. 1985. *Org. Geochem.* 8: 247.

Labout, J.W.A. 1950. *The Properties of Asphaltic Bitumen* Pfeiffer, J.P. ed, Elsevier, New York. p. 24.

Lee, M.L., Novotny, M.S., and Bartle, K.D. 1981. *Analytical Chemistry of Polycyclic Aromatic Compounds*. Academic Press Inc., New York.

Leon, V. 1987. *Fuel* 66: 1445.

Lumpkin, H.E. 1956. *Anal. Chem.* 28: 1946.

Lumpkin, H.E. 1964. *Anal. Chem.* 36: 2399.

Majid, A., Bornais, J., and Hutchison, R.A. 1989. *Fuel Sci. Technol. Int.* 7: 507.

Malhotra, V.M. and Buckmaster, H.A. 1985. *Fuel* 64: 335.

McKay, J.F., Amend, P.J., Cogswell, T.E., Harnsberger, P.M., Erickson, R.B., and Latham, D.R. 1978. *Analytical Chemistry of Liquid Fuel Sources: Tar Sands, Oil Shale, Coal, and Petroleum.* Uden, P.C. Siggia, S. and Jensen, H.B. eds., Advances in Chemistry Series No. 170. American Chemical Society, Washington, DC. Chapter 9.

Millson, M.F. 1967. Report No. FD 67/84. Department of Energy Mines Resources, Mines Branch, Ottawa, Ontario, Canada.

Mitra-Kirtley, S., Mullins, O.C., van Elp, J., and Cramer, S.P. 1993. *Fuel* 72: 133.

Monin, J.C. and Pelet, R. 1983. *Advances in Organic Geochemistry.* Bjoroey, M. ed., John Wiley & Sons Inc., New York.

Monin, J.C. and Pelet, R.O. 1984. *Characterization of Heavy Crude Oils and Petroleum Residues.* Tissot, B. ed., Editions Technip, Paris. p. 104.

Montgomery, D.S. and Boyd, M.L. 1959. *Anal. Chem.* 31: 1290.

Moschopedis, S.E. and Speight, J.G. 1976a. *Fuel* 55: 187.

Moschopedis, S.E. and Speight, J.G. 1976b. *Fuel* 55: 334.

Moschopedis, S.E. and Speight, J.G. 1979. Preprints. *Div. Petrol. Chem. Am. Chem. Soc.* 24(4): 1007.

Mullins, O.C. 1995. *Asphaltenes: Fundamentals and Applications.* Sheu, E.Y. and Mullins, O.C. eds., Plenum Press, New York. Chapter II.

Nagy, B. and Gagnon, G.C. 1961. *Geochim. Cosmochim. Acta* 23: 155.

Nasritdinov, S. 2003. *Chemistry and Technology of Fuels and Oils.* 39(5): 292–298.

Nellensteyn, F.J. 1938. *The Science of Petroleum.* Dunstan, A.E. Nash, A.W. Brooks, B.T. and Tizard, H.T. eds., Oxford University Press, Oxford, UK. Volume IV. p. 2760.

Ngassoum, M.B., Faure, R., Ruiz, J.M., Lena, L., Vincent, E.J., and Neff, B. 1986. *Fuel* 65: 142.

Nicksic, S.W. and Jeffries-Harris, M.J. 1968. *J. Inst. Pet. London* 54: 107.

Novelli, J-M., Ngassoum, M., Faure, R., Ruiz, J-M., Lena, L., Vincent E-J., and Escalier, J-C. 1984. *Characterization of Heavy Crude Oils and Petroleum Residues.* Tissot, B. ed., Editions Technip, Paris. p. 356.

Orr, W.L. 1977. Preprints. *Div. Fuel Chem. Am. Chem. Soc.* 22(3): 86.

Orr, W.L. and Sinninghe Damste, J.S. 1990. *Geochemistry of Sulfur in Fossil Fuels.* Orr, W.L. and White, C.M. eds., Symposium Series No. 429. American Chemical Society, Washington, DC. Chapter 1.

Overfield, R.E., Sheu, E.Y., Sinha, S.K., and Liang, K.S. 1989. *Fuel Sci. Technol. Int.* 7: 611.

Padovani, C., Berti, V., and Prinetti, A. 1959. *Proceedings. 5th World Petroleum Congress.* Session V. Paper 21.

Petersen, J.C. 1967. *Fuel* 44: 295.

Pollack, S.S. and Yen, T.F. 1969. Preprints. *Div. Petrol. Chem. Am. Chem. Soc.* 14: B-118.

Pollack, S.S. and Yen, T.F. 1970. *Anal. Chem.* 42: 623.

Posadov, I.A., Pokonova, J.V., Khusidman, M.B., Gitlin, I.G., and Proskuryakov, V.A. 1977. *Zhur. Priklad. Khim.* 50: 594.

Ramaswamy, V., Sarowha, S.L.S., and Singh, I.D. 1986. *Fuel* 65: 1281.

Ramsey, J.W., McDonald, F.R., and Petersen, J.C. 1967. *Ind. Eng. Chem. Prod. Res. Dev.* 6: 231.

Rao, C.N.R. 1961. *Ultraviolet and Visible Spectroscopy: Chemical Applications.* Butterworths, London.

Reynolds, J.G. 1997. *Petroleum Chemistry and Refining.* Speight, J.G. ed., Taylor & Francis, Washington, DC. Chapter 3.

Reynolds, J.G., Biggs, W.R., Fetzer, J.C., Gallegos, E.J., Fish, R.H., Komlenic, J.J., and Wines, B.K. 1984. *Characterization of Heavy Crude Oils and Petroleum Residues.* Tissot, B. ed. Editions Technip, Paris. p. 153.

Reynolds, J.G., Jones, E.L., Bennett, J.A., and Biggs, W.R. 1989. *Fuel Sci. Technol. Int.* 7: 625.

Ritchie, R.G.S., Roche, R.S., and Steedman, W. 1979a. *Fuel* 58: 523.

Romberg, J.M., Mesmith, S.D., and Traxler, R.N. 1959. *J. Chem. Eng. Data* 4: 159.

Rose, K.D. and Francisco, M.A. 1988. *J. Am. Chem. Soc.* 110: 637.

Rossini, F.D., Mair, B.J., and Streiff, A.J. 1953. *Hydrocarbons from Petroleum*. Reinhold Publishing Corp., New York.

Saban, M., Vitorovic, O., and Vitorovic, D. 1984. *Characterization of Heavy Crude Oils and Petroleum Residues*. Tissot, B. ed., Editions Technip, Paris. p. 122.

Schmitter, J.M., Ignatiadis, I., and Arpino, P.J. 1983. *Geochim. Cosmochim. Acta* 47: 1975.

Schmitter, J.M., Garrigues, P., Ignatiadis, I., De Vazelhes, R., Perin, F., Ewald, M., and Arpino, P. 1984. *Org. Geochem.* 6: 579.

Severin, D. and Glintzer, O. 1984. *Characterization of Heavy Crude Oils and Petroleum Residues*. Tissot, B. ed., Editions Technip, Paris. p. 19.

Shepherd, R.A. and Graham, W.R.M. 1986. *Fuel* 65: 1612.

Snape, C.E., Ladner, W.R., and Bartle, K.D. 1979. *Anal. Chem.* 51: 2189.

Snape, C.E., Ladner, W.R., Bartle, K.D., and Taylor, N. 1984. *Characterization of Heavy Crude Oils and Petroleum Residues*. Tissot, B. ed., Editions Technip, Paris. p. 315.

Snyder, L.R. 1968. *Principles of Adsorption Chromatography*. Marcel Dekker, New York.

Snyder, L.R. 1969. *Anal. Chem.* 41: 1084.

Speight, J.G. 1970. *Fuel* 49: 76.

Speight, J.G. 1986. Preprints. *Div. Petrol. Chem. Am. Chem. Soc.* 31(4): 818.

Speight, J.G. 1994. *Asphaltenes and Asphalts. Developments in Petroleum Science, 40*. Yen, T.F. and Chilingarian, G.V. eds., Elsevier, Amsterdam. Chapter 2.

Speight, J.G. 2001. *Handbook of Petroleum Analysis*. John Wiley & Sons Inc., Hoboken, NJ.

Speight, J.G. 2002. *Handbook of Petroleum Product Analysis*. John Wiley & Sons Inc., Hoboken, NJ.

Speight, J.G. and Pancirov, R.J. 1984. *Liq. Fuels Technol.* 2: 287.

Stewart, J.E. 1957. *J. Res. Nat. Bur. Std.* 58: 265.

Storm, D.A., Sheu, E.Y., and DeTar, M.M. 1993. *Fuel* 72: 977.

Suatoni, J.C. and Swab, R.E. 1976. *J. Chromatogr. Sci.* 14 535.

Tripathi, G.K., Jassal, J.K., Mathur, V.N., Dawar, J.C., Katiyar, A.K., Sharma, A.K., and Chandra, K. 1984. *Characterization of Heavy Crude Oils and Petroleum Residues*. B. Tissot (ed.), Editions Technip, Paris. p. 128.

Tynan, E.C. and Yen, T.F. 1969. *Fuel* 48: 191.

Verhasselt, A.F. 1992. *Fuel Sci. Technol. Int.* 10: 581.

Waldo, G.S., Mullins, O.C., Penner-Hahn, J.E., and Cramer, S.P. 1992. *Fuel* 71: 53.

Weinberg, V.L., Yen, T.F., Gerstein, B.C., and Murphy, P.D. 1981. Preprints. *Div. Petrol. Chem. Am. Chem. Soc.* 26(4): 816.

Wen, C.S., Chilingarian, G.V., and Yen, T.F. 1978. *Bitumens, Asphalts and Tar Sands*. Chilingarian, G.V. and Yen, T.F. eds, Elsevier, Amsterdam. Chapter 7.

Williams, R.B. 1959. *Spectrochim. Acta.* 14: 24.

Williams, R.B. and Chamberlain, N.F. 1963. *Proceedings. 6th World Petroleum Congress, Frankfurt, Germany*. Section V. Paper 17.

Williford, C. 1943. *Texas Agr. Exp. Sta. Bull.* 73: 7.

Winniford, R.S. and Bersohn, M. 1962. Preprints, *Am. Chem. Soc. Div. Fuel Chem.* (September): 21.

Witherspoon, P.A. and Winniford, R.S. 1967. *Fundamental Aspects of Petroleum Geochemistry*. B. Nagy and U. Colombo eds., Elsevier, Princeton, NJ. p. 261.

Yen, T.F. 1973. *Fuel* 52: 93.

Yen, T.F. 1974. *Energy Sources* 1: 447.

Yen, T.F. 1975. *The Role of Trace Metals in Petroleum*. Ann Arbor Science Publishers Inc., Ann Arbor, MI.

Yen, T.F. 1984. *The Future of Heavy Crude and Tar Sands*. Meyer, R.F. Wynn, J.C. and Olson, J.C. eds., McGraw-Hill, New York.

Yen, T.F. and Erdman, J.G. 1962a. Preprints. *Div. Petrol. Chem. Am. Chem. Soc.* 7: 5.

Yen, T.F. and Erdman, J.G. 1962b. Preprints. *Div. Petrol. Chem.. Am. Chem. Soc.* 7(3): 99.

Yen, T.F. and Erdman, J.G. 1962c. Preprints, *Am Chem. Soc. Div. Petrol. Chem.* 7: 99.

Yen. T.F. and Erdman, J.G. 1963. *Encyclopedia of X Rays and Gamma Rays*. Reinhold, New York.

Yen, T.F., Erdman, J.G., and Pollack, S.S. 1961b. *Anal. Chem.* 33: 1587.
Yen, T.F., Erdman, J.G., and Pollack, S.S. 1961c. *Anal. Chem.* 33: 1587.
Yokota, T., Scriven, F., Montgomery, D.S., and Strausz, O.P. 1986. *Fuel* 65: 1142.
Young, R.S. 1954. *Bull. Am. Assoc. Petrol. Geol.* 38: 2017.

11 Asphaltene Constituents

11.1 INTRODUCTION

As the boiling points of petroleum fractions increase, the complexity of the constituents in these fractions also increases, and the differences between the main classes of hydrocarbons become less pronounced. The high-boiling fractions of petroleum (and this may commence with the **saturates** and the aromatics fractions, depending on the boiling profile of the crude oil) may be considered as having an extremely complicated composition. In fact, any attempted isolation of individual compounds is a frustrating and an impossible task.

Nevertheless, the properties of any material are dictated by its chemical composition, and once the chemical composition of a material has been established, it becomes possible to correlate composition and properties. Petroleum is no exception to this generality; it is recognized that the constituents of petroleum contribute to its physical state as well as to its behavior during processing. It is therefore desirable to investigate the general composition of crude oil and it is not surprising that the separation of the constituents of crude oil by virtue of their physical properties has received considerable attention.

It is generally recognized that crude oils are composed of four major fractions (saturates, aromatics, **resins**, and **asphaltenes**) that differ from one another sufficiently in solubility and adsorptive character and that the separation can be achieved by the application of relevant methods (Chapter 8). Indeed, although these four fractions are chemically complex, the methods of separation have undergone several modifications to such an extent that the evolution of the separation techniques is a study in itself (Chapter 8). And because the fractions are in a balanced realtionship in crude oil, the chemical and physical character of these constituents needs reference in this chapter.

It will become immediately apparent from the contents of this chapter that the majority of the work carried out on the **asphalt** constituents of petroleum concentrated on the asphaltene fraction. Parallel work on the resin and oil fractions is extremely sketchy and has not been considered worthwhile, insofar as it is generally believed that the constituents of the asphaltene fraction have markedly adverse effects on the processability of crude oils. Further, it is the asphaltenes that also play a role in the rheological properties of asphalt (Speight, 1995, 1999).

For the purposes of this chapter, the resin constituents are regarded as those materials soluble in n-pentane or n-heptane (i.e., whichever hydrocarbon is used for the separation of asphaltenes), but insoluble in liquid propane. In addition, resin constituents are also those materials soluble in n-pentane or n-heptane but that cannot be extracted from the earth by n-pentane or n-heptane. The **oils** fraction is, therefore, that fraction of crude oil soluble in n-pentane or in n-heptane that is extractable from the earth by n-pentane or by n-heptane. The oils fraction is composed of saturates and aromatics fractions (Chapter 1).

Nevertheless, attempts have been made to collate the results of work carried out on the resin and oil fractions, as it is now evident that petroleum is a delicately balanced physical system

(Chapter 12). Each fraction depends on the others for complete mobility and maintaining the stability of the petroleum.

Questions may also be raised regarding the usefulness of attempting to determine the exact structure of each complex material, such as asphaltenes, resins, and nonvolatile oils. Perhaps the answers lie not only in determining the locations of the heteroatoms (nitrogen, oxygen, and sulfur). It is the hope that this allows modifications to the various processes resulting in easy elimination of these atoms. It will also allow evaluation of the structural types present in the hydrocarbon portions of these molecules in the hope of deriving as much benefit as possible during the processing of these complex materials.

Whether these reasons are sufficiently valid remains to be seen, but considerable work has been done and is even now in progress, to evaluate the influence of these high molecular weight materials during petroleum recovery and during petroleum refining (Speight, 1994; Yan et al., 1997).

11.2 SEPARATION

The asphalt constituents of petroleum usually appear as a dark brown to black, sticky, semisolid fraction that is the result of various refining processes, but what constitutes the asphalt fraction of petroleum is a matter of conjecture. The residuum from vacuum distillation is most frequently referred to as the asphalt portion of, as is the material precipitated by liquid propane–liquid butane mixtures during the **deasphalting** stage of petroleum refining. In both cases, the residue usually consists of asphaltene constituents, resin constituents, and possibly a heavy gas oil fraction, which may be lubricating stock material and cannot be distilled. Hence, for the purposes of this chapter, the asphalt constituents of petroleum are considered asphaltene constituents, resin constituents, and the portion of the oil fraction that cannot be distilled without the onset of thermal decomposition.

Separation of the asphalt portion of petroleum in asphaltenes, resins, and oils can be achieved by a variety of methods (Chapter 8). Further discussion about these techniques is not warranted here. However, it should be noted that, unless otherwise stated, the asphalt materials described in this chapter are those materials separated by nondestructive distillation or by solvent–adsorbent treatment. Materials separated by chemical methods or by methods that could involve substantial structural changes within the molecular framework of the constituents are not considered here.

The asphaltene fraction is, by definition, a solubility class (Chapter 1 and Chapter 8). The fraction is precipitated from feedstocks by the addition of 40 volumes of the liquid hydrocarbon (Table 11.1) (Chapter 1 and Chapter 8) (Girdler, 1965; Mitchell and Speight, 1973; Speight et al., 1982, 1984; Andersen and Birdi, 1990; Speight, 1994; Cimino et al., 1995).

TABLE 11.1
Standard Methods for Asphaltene Precipitation

Method	Precipitant	Volume Precipitant Per Gram of Sample
ASTM D893	n-pentane	10 mL
ASTM D2006	n-pentane	50 mL
ASTM D2007	n-pentane	10 mL
ASTM D3279	n-heptane	100 mL
ASTM D4124	n-heptane	100 mL
IP 143	n-heptane	30 mL

Source: ASTM Annual Book of Standards, 1980–2005.

The asphaltene fraction is a dark brown to black friable solid that has no definite melting point and usually foams and swells on heating to leave a carbonaceous residue. They are obtained from petroleum by the addition of a nonpolar solvent (such as a hydrocarbon) with a surface tension lower than that of 25 dyne cm^{-1} at 25°C (77°F), such as low-boiling petroleum naphtha, petroleum ether, n-pentane, iso-pentane, n-heptane, liquefied petroleum gases and the like (Chapter 8) (Mitchell and Speight, 1973) and, in fact, propane is used commercially in processing petroleum residua for asphalt production (Chapter 28). On the other hand, the asphaltene constituents are soluble in liquids with a surface tension above 25 dyne cm^{-1}, such as pyridine, carbon disulfide, carbon tetrachloride, and benzene.

Thus, the nomenclature of petroleum fractions is an operational aid (Chapter 1) and is not usually based on chemical or structural features. However, the asphaltene fraction can be defined in terms of functional group composition (Figure 11.1) (Francisco and Speight, 1984), but it cannot be defined by the separation method and the solvent employed. In fact, there is no one parameter that is operational in the separation of asphaltenes; the relevant parameters for asphaltene separation are physical and chemical in nature and include:

1. Polarity (the presence of functional groups derived from the presence of heteroatoms in the asphaltenes) (Long, 1979, 1981; Speight, 1994)
2. Aromaticity (the presence of polynuclear aromatic systems in the asphaltenes) (Girdler, 1965; Mitchell and Speight, 1973)
3. Molecular weight (molecular size) (Long, 1979, 1981)
4. Three-dimensional structure (the micelle) of the asphaltene constituents as they exist in relationship with the other constituents of crude oil (Speight, 1992, 1994);
5. Solvent power of the precipitating or extracting liquid used for the separation (Chapter 8) (Girdler, 1965; Mitchell and Speight, 1973)
6. Time required to allow the precipitating or extracting liquid to penetrate the micelle, which is dependent upon the ability of the hydrocarbon liquid to penetrate the micelle, indicating that the process is diffusion-controlled (Speight et al., 1984)
7. Ratio of the precipitating or extracting liquid to crude oil that dictates the yield and character of the asphaltene product (Girdler, 1965; Mitchell and Speight, 1973; Speight et al., 1984)
8. Temperature, which may reduce the induction period that is a requirement of diffusion-controlled processes (Mitchell and Speight, 1973)

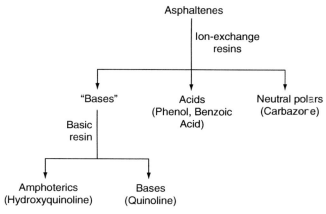

FIGURE 11.1 Separation of asphaltene constituents based on polarity.

9. Pressure, as employed in several refinery processes (Chapter 14) as a means of maintaining the low-boiling liquid hydrocarbon in the liquid phase

Other parameters may be defined as subsets of those enumerated earlier. It is also worthy of note that, in order to remove entrained resin material, precipitation of the asphaltenes from benzene or toluene is often necessary (Speight et al., 1984; Ali et al., 1985). But, none of these parameters applied to the separation of petroleum can be related to the separation of distinct chemical types.

Despite this, there are still reports of asphaltenes being isolated from crude oil by much lower proportions of the precipitating medium, which can lead to errors not only in the determination of the amount of asphaltenes in the crude oil but also in the determination of the compound type. For example, when insufficient proportions of the precipitating medium are used, resins may also appear within the asphaltene fraction by adsorption onto the asphaltenes from the supernatant liquid and can be released by reprecipitation (Speight et al., 1984; Ali et al., 1985). These resins would be isolated at a later stage of the separation procedure by adsorption chromatography if standardized parameters were employed for the separation of the asphaltenes (Speight et al., 1984; ASTM 2005). Questionable isolation techniques throw serious doubt on any conclusions drawn from subsequent work on the isolated material (Andersen and Speight, 2001).

In fact, it is now accepted that to ensure stable asphaltene yields, it is necessary to employ the following parameters:

1. Excess of the liquid hydrocarbon (>30 mL hydrocarbon per gram feedstock)
2. Use of *n*-pentane or *n*-heptane as the liquid hydrocarbon; volatility constraints and consistency of the of the asphaltene fraction favor the use of *n*-heptane
3. Prolonged contact time (8 to 10 h) is preferable (Chapter 8)
4. Precipitation sequence to remove any constituents that separate with the asphaltene constituents

The precipitation sequence involves dissolution of the asphaltenes in benzene or toluene (10 mL per g asphaltene), followed by the addition of the hydrocarbon (50 mL precipitant per mL toluene or benzene) to the solution. This sequence should be repeated three times to remove adsorbed lower molecular weight resin material and to provide consistency of the asphaltene fraction.

Briefly, the precipitation of asphaltenes can be ascribed to changes in crude oil composition caused by the addition of lower-boiling components that disturbs the complex equilibrium, keeping the asphaltenes in solution or in a peptized state. In the case of addition of low-boiling liquid hydrocarbons to the feedstock, the hydrocarbon causes a change in the solubility parameter of the oil medium that, in turn, changes the tolerance of the medium for the complex micelle structure. As this occurs, the lower molecular weight and less polar constituents of the micelle are extracted into the liquid, leaving the asphaltene constituents without any surrounding (dispersing) sheath. Separation then ensues. If the process is thermal (i.e., in excess of the thermal decomposition temperature), a different sequence of events ensues leading to the formation of an insoluble phase and ultimately to coke (Chapter 15).

The focal point of the asphaltene separation method is the isolation of the asphaltene constituents as a discrete fraction. However, a question that often arises is related to whether or not the asphaltenes self-associate in the crude oil or if the asphaltene constituents interact directly with the other crude oil constituents. In addition, there is also a question related to precipitation of the asphaltene constituents or extraction of the nonasphaltene constituents from the micelle that is responsible for the dispersion of the asphaltenes in the oil.

The term *resin* generally implies material that has been eluted from various solid adsorbents, whereas the term **maltenes** (or **petrolenes**) (Chapter 1) indicates a mixture of the resin constituents and oil constituents obtained in the filtrates from the asphaltene precipitation. Thus, after the asphaltenes are precipitated, adsorbents are added to the *n*-pentane solutions of the resins and oils, by which process the resins are adsorbed and subsequently recovered by use of a more polar solvent and the oils remain in solution (Chapter 8).

Resin constituents are soluble in the liquids that precipitate asphaltene constituents and are usually soluble in most organic liquids, except in the lower alcohols and acetone, but they are precipitated by liquid propane and liquid butanes. The resin constituents often coprecipitate with the asphaltenes in controlled propane deasphalting procedures, and the product, called **propane asphalt**, contains appreciable amounts of adsorbed resins and has the properties of a low-melting point asphalt. The resins are dark, semisolid or solid, very adhesive materials of high molecular weight. Their composition can vary depending on the kind of precipitating liquid and on the temperature of the liquid system. They become quite fluid on heating but often show pronounced brittleness when cold.

The oils fraction (comprising the saturates fraction and the aromatics fraction) comprises the lowest molecular weight fraction of petroleum and is the dispersion medium for the peptized asphaltenes. Although the oils fraction is sometimes colored, fractions obtained by chromatography are colorless and are similar to white medicinal oils or lubricating oils of high purity. The oils fraction may be quite viscous because of the presence of paraffin waxes that vary over a wide range for crude oil and other feedstocks. The oils fraction is miscible with low molecular weight liquid hydrocarbons (such as *n*-pentane and *n*-heptane), as well as most organic solvents, and may be separated into various hydrocarbon subfractions by suitable solvent or adsorbent methods.

11.3   COMPOSITION

Asphaltene fractions isolated from different sources are remarkably constant in terms of ultimate composition, although there can be variations in terms of heteroatom content because of local and regional variations in the plant precursors and mineralogical composition of the nearby geological formations. Even though the nature of the source material and subtle regional variations in maturation conditions serve to differentiate one crude oil (and hence one asphaltene) from another, it might be suggested that the maturation process tends to be a molecular equalizer over the time periods involved. Thus, differences between the asphaltene constituents reflect differences in the relative amounts of the functional molecular types present and the way in which the structural types are combined in the various asphaltenes. This is reflected, in part, by changes in the functional group composition of asphaltene fractions from different crude oils.

The elemental composition of asphaltenes isolated by use of excess (greater than 40) volumes of *n*-pentane as the precipitating medium show that the amounts of carbon and hydrogen usually vary over only a narrow range (Speight, 1994). These values correspond to a hydrogen-to-carbon atomic ratio of 1.15 $\pm$ 0.5%, although values outside this range are sometimes found. The near constancy of the H/C ratio is surprising when the number of possible molecular permutations involving the heteroelements is considered. In fact, this property, more than any other, is the cause for the general belief that unaltered asphaltene constituents from crude oil have a definite composition. Further, it is still believed that asphaltenes are precipitated from petroleum by hydrocarbon solvents because of this composition, not only because of solubility properties.

In contrast to the carbon and hydrogen contents of asphaltenes, notable variations occur in the proportions of the heteroelements, in particular in the proportions of oxygen and sulfur.

Oxygen contents vary from 0.3% to 4.9% and sulfur contents vary from 0.3% to 10.3%. On the other hand, the nitrogen content of the asphaltenes has a somewhat lesser degree of variation (0.6% to 3.3% at the extremes). However, exposing asphaltenes to atmosphere oxygen can substantially alter the oxygen content and exposing a crude oil to elemental sulfur or even to sulfur-containing minerals can result in excessive sulfur uptake. Perhaps oxygen and sulfur contents vary more markedly than do nitrogen contents because of these conditions.

The use of *n*-heptane as the precipitating medium yields a product that is substantially different from the *n*-pentane-insoluble material. For example, the hydrogen-to-carbon atomic ratio of the *n*-heptane precipitate is lower than that of the *n*-pentane precipitate (Speight, 1994, 2001). This indicates a higher degree of aromaticity in the *n*-heptane precipitate. Nitrogen-to-carbon, oxygen-to-carbon, and sulfur-to-carbon ratios are usually higher in the *n*-heptane precipitate, indicating higher proportions of the heteroelements in this material (Speight and Moschopedis, 1981a).

Some interesting compositional relationships can also be derived from the data. These data confirm the consistency of asphaltene composition but show that when composition varies, this is usually due to heteroatom content. It is also possible to show from the data how the asphaltenes compare with the ranges of kerogen maturation (Figure 11.2). The natural asphaltene constituents will (with few exceptions) occur on this diagram in an area bounded by an H/C ration ratio of 1.00 to 1.50 and an O/C atomic ratio of 0.00 to 0.05. However, any conclusions drawn from this relationship must be treated with caution because there is some question about the precise role of kerogen in the maturation process (Chapter 4).

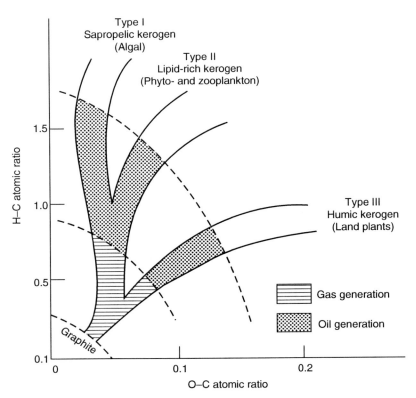

FIGURE 11.2 The kerogen maturation diagram as illustrated using H/C and O/C atomic ratios.

For example, it has not been conclusively demonstrated that kerogen is first formed, then acts as the precursor to oil. It generates oil by pyrolysis, but this is another dimension. It is more likely that kerogen is formed simultaneously with oil but from the more polar constituents of the **protopetroleum** (Chapter 3). Not all the polar constituents react to form kerogen and some of the polar constituents may be carried by the crude oil into the reservoir. The highly polar material remains in the shale bed to form kerogen. At this time, there is also the potential for interaction of the liquid with minerals in the strata. Inclusion of nonhydro-carbon material is always a possibility. Any heteroatom material in the maturating liquid could then form asphaltenes during the maturation process. Thus, the origin of the asphaltene is more likely to occur within the oil rather than as a breakdown of kerogen during the maturation process.

However, it has been reported that the pentane asphaltenes from a series of crude oils have similar atomic hydrogen-to-carbon ratios as the source kerogen, but have significantly lower atomic oxygen-to-carbon ratios and slightly lower atomic sulfur-to-carbon ratios (Orr, 1985). However, the atomic nitrogen-to-carbon ratios are almost the same for the asphaltenes and the source kerogen. In addition, with increasing maturity, the atomic hydrogen-to-carbon ratio in asphaltenes decreases much like the maturation changes in kerogen. Whether or not this is direct evidence for the production of asphaltenes from kerogen remains to be proven. It is considered likely to be an indicator of the similarities of the starting materials for the asphaltenes and kerogen as well as an indicator of similar maturation conditions.

One aspect of asphaltene characterization that has provided strong evidence for the complexity of petroleum asphaltenes arises from composition studies using fractionation techniques (Bestougeff and Mouton, 1977; Speight, 1986a). Indeed, the asphaltene fraction can be subdivided by a variety of techniques (Bestougeff and Darmois, 1947, 1948; Bestougeff and Mouton, 1977; Francisco and Speight, 1984). Of specific interest is the observation that when asphaltenes are fractionated on the basis of aromaticity and polarity, it appears that the more aromatic species contain higher amounts of nitrogen (Speight, 1984a), suggesting that the nitrogen species are located predominantly in aromatic systems (Speight, 1994).

The solvent-based fractionation of the asphaltene component of petroleum showed that it is possible to obtain asphaltene fractions characterized by different degrees of aromaticity or heteroatom content by using benzene or pentane (toluene or pentane) or benzene or methanol (toluene or methanol) mixtures in variable ratios (Speight, 1979; Andersen, 1997). The use of mixtures of a polar and a nonpolar solvent in order to fractionate an asphaltene sample will tend to direct the fractionation by introducing polar forces and hydrogen bonding, as well as dispersion forces, as factors determining which components of the asphaltene sample are soluble in the mixture.

In contrast, the exclusive use of saturated hydrocarbons whose solvent power derives entirely from the dispersion forces should, in principle, permit the fractionation to be carried out with variation of only a single component of the solubility parameter. It is known that the quantity of asphaltenes precipitated by linear paraffins decreases with increasing carbon number until a limiting value is reached above n-heptane (n-C_7H_{16}) to n-nonane (n-C_9H_{20}) (Girdler, 1965; Mitchell and Speight, 1973).

The cycloparaffins have better solvating power toward the asphaltenes and often fail to induce asphaltene separation from crude oil. It is possible to fractionate asphaltenes using mixtures of solvents, such as cyclohexane and n-alkanes, and any differences in the precipitation threshold underscores the point that the solubility properties of the asphaltene fractions of different crude oils can vary markedly (Acevedo et al., 1995).

It appears, therefore, that the tendency of asphaltenes to precipitate from a given crude oil because of a variation in the solubility parameter of the precipitating solvent mixture is more closely related to the aromaticity and polarity of the asphaltenes than to their dimensions.

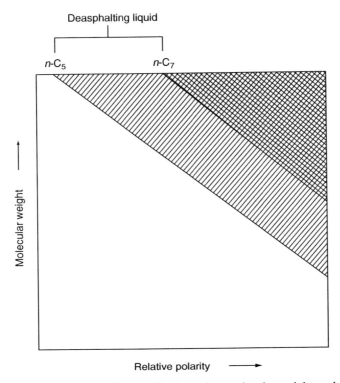

FIGURE 11.3 Representation of the asphaltene faction using molecular weight and polarity.

However, it is possible to conclude (Acevedo et al., 1995) that molecular weight, aromaticity, and polarity combine to determine asphaltene solubility in hydrocarbon media.

The fractionation of asphaltenes into a variety of functional (and polar) types (Figure 11.1) (Francisco and Speight, 1984) has confirmed the complexity of the asphaltene fraction. Other studies (Schmitter et al., 1984) have brought to light the occurrence of four-ring aromatic nitrogen species in petroleum. These findings are of particular interest as they correspond to the ring systems that have been tentatively identified by application of the HPLC technique to the basic nitrogen fraction of asphaltenes (Speight, 1986b, 1994).

The molecular weight or polarity concept in which asphaltene fractions are defined on the basis of molecular weight and polarity (Figure 11.3) can be used to indicate the relative nature of carbenes and carboids (thermal decomposition products of asphaltenes), which have lower molecular weight and are more polar than the asphaltene constituents from which they are formed (Speight, 1994). Further, the slope of the line as illustrated is purely arbitrary and will vary with the source and composition of different fractions (Figure 11.4). However, it

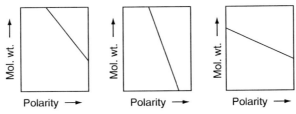

FIGURE 11.4 Representation of the molecular weight/polarity diagram for different asphaltene fractions.

should not be construed that this two-dimensional diagram is truly representative of the nature of asphaltenes and a three-dimensional diagram involving functionality or molecular weight might be more appropriate for showing the chemical and physical differences between the various asphaltenes, carbenes, and carboids. Nevertheless, the concept does emphasize the lower molecular weight and increased polarity of the various asphaltene constituents.

Unfortunately, in many studies, too little emphasis has been placed on determining the nature and location of the nitrogen, oxygen, and sulfur atoms in the asphaltene structure.

Studies on the disposition of nitrogen in petroleum asphaltenes indicated the existence of nitrogen as various heterocyclic types (Clerc and O'Neal, 1961; Nicksic and Jeffries-Harris, 1968; Moschopedis and Speight, 1976b, 1979; Jacobson and Gray, 1987; Mullins, 1995). The more conventional (from the organic chemist's viewpoint) nitrogen species i.e., primary, secondary, and tertiary aromatic amines have not been established as present in petroleum asphaltenes. There are also reports in which the organic nitrogen in petroleum asphaltenes has been defined in terms of basic and nonbasic types (Nicksic and Jeffries-Harris, 1968). Spectroscopic investigations (Moschopedis and Speight, 1979) suggest that carbazoles occur in asphaltenes; this supports earlier mass spectroscopic evidence (Clerc and O'Neal, 1961) for the occurrence of carbazole nitrogen in asphaltenes. The application of x-ray absorption near-edge structure (XANES) spectroscopy to the study of asphaltenes has led to the conclusion that a large portion of the nitrogen present is aromatic, in pyrrolic rather than pyridinic form (Mitra-Kirtley et al., 1993).

Thermal studies have shown that only 1% of the nitrogen is lost during the thermal treatment, but substantially more sulfur (23%) and almost all the oxygen (81%) is lost as a result of this treatment. The tendency for nitrogen and sulfur to remain in the nonvolatile residue produced during thermal decomposition, as opposed to the relatively facile elimination of oxygen, supports the concept that nitrogen and sulfur have stability because of their location in ring systems.

Oxygen has been identified in carboxylic, phenol, and ketone functions (Nicksic and Jeffries-Harris, 1968; Petersen et al., 1974; Moschopedis and Speight, 1976a,b; Ritchie et al., 1979; Speight and Moschopedis, 1981b; Rose and Francisco, 1987) locations but is not usually regarded as located primarily in heteroaromatic ring systems.

Some evidence for the location of oxygen within the asphaltene fraction has been obtained by infrared spectroscopy. Examination of dilute solutions of the asphaltenes in carbon tetrachloride show that at low concentration (0.01% wt/wt) of asphaltenes, a band occurs at 3585 cm^{-1}, which is within the range anticipated for free nonhydrogen-bonded phenolic hydroxyl groups. In keeping with the concept of hydrogen bonding, this band becomes barely perceptible, and the appearance of the broad absorption in the range 3200 to 3450 cm^{-1} becomes evident at concentrations above 1% by weight.

Other evidence for the presence and nature of oxygen functions in asphaltenes has been derived from infrared spectroscopic examination of the products after interaction of the asphaltenes with acetic anhydride. Thus, when asphaltenes are heated with acetic anhydride in the presence of pyridine, the infrared spectrum of the product exhibits prominent absorptions at 1680, 1730, and 1760 cm^{-1}. These changes in the infrared spectrum of the asphaltene fraction as a result of treatment with refluxing acetic anhydride suggest acetylation of free and hydrogen-bonded phenolic hydroxyl groups present in the asphaltene constituents (Moschopedis and Speight, 1976a,b).

The absorption bands at 1680 and 1760 cm^{-1} are attributed to a nonhydrogen-bonded carbonyl of a ketone, for example, a diaryl ketone or a quinone, and to the carbonyl of a phenolic acetate, respectively. If the 1680 cm^{-1} band is in fact the result of the presence of nonhydrogen-bonded ketones or quinones, the appearance of this band in the infrared spectra of the products could conceivably arise by acetylation of a nearby hydroxyl function. This hydroxyl function may have served as a hydrogen-bonding partner to the ketone (or quinone), thereby releasing this function and causing a shift from about 1600 cm^{-1} to higher frequencies.

The 1730 cm^{-1} band is the third prominent feature in the spectrum of the acetylated products. It is also ascribed to phenolic acetates and, as with the 1760 cm^{-1} band, falls within the range 1725 to 1760 cm^{-1} assigned to esters of polyfunctional phenols. This suggests that a considerable portion of the hydroxyl groups present in the asphaltenes may occur as collections of two or more hydroxyl functions on the same aromatic ring. Alternate sites include adjacent peripheral sites on a condensed ring system or sites adjacent to a carbonyl function in a condensed ring system. In the context of polyhydroxyaromatic nuclei existing in asphaltenes, it is of interest to note that pyrolysis at 800°C (1470°F) results in the formation of resorcinols (Ritchie et al., 1979), implying that such functions may indeed exist in the asphaltenes.

Sulfur occurs as benzothiophenes, dibenzothiophenes, and naphthene benzothiophenes (Clerc and O'Neal, 1961; Nicksic and Jeffries-Harris, 1968; Drushel, 1970; Yen, 1974; Speight and Pancirov, 1984; Rose and Francisco, 1988; Keleman et al., 1990); more highly condensed thiophene types may also exist but are precluded from identification by low volatility. Other forms of sulfur that occur in asphaltenes include the alkyl–alkyl sulfides, alkyl–aryl sulfides, and aryl–aryl sulfides (Yen, 1974).

Metals (i.e., nickel and vanadium) are much more difficult to integrate into the asphaltene system. Nickel and vanadium occur as porphyrins (Baker, 1969; Yen, 1975), but whether these are an integral part of the asphaltene structure is not known. Some of the porphyrins can be isolated as a separate stream from petroleum (Branthaver, 1990; Reynolds, 1997).

In summary, asphaltenes contain similar functionalities but with some variation in degree (Speight, 1981a). For example, asphaltenes from the more paraffinic crude oils appeared (through this cursory examination) to contain less quinone and amide systems than those from the heavy feedstocks.

The occurrence of small polynuclear aromatic systems in nature is not completely unknown. For example, they have been identified in unpolluted sediments (Tissot and Welte, 1978), and the sterane ring system, which can aromatize to polynuclear aromatic systems, is well known. In addition, polyhydroxyaromatic compounds and polyhydroxyquinone compounds are also common in nature (Weiss and Edwards, 1980).

Amide and carbazole types have also been recognized in natural product systems (Weiss and Edwards, 1980). Sulfur (as thiophene sulfur) is more difficult to identify, and its presence in biological systems, other than the sulfur-containing amino acids (cysteine, cystine, and methionine) is still speculative, although there are means by which inorganic sulfur can be incorporated into petroleum (Hobson and Pohl, 1973; Orr, 1977). It is also worthy of note here that asphaltenes have been postulated to contain charged species within their structure (Nicksic and Jeffries-Harris, 1968). This, of course, is analogous to the occurrence of zwitterion (dipolar ion) structures in proteins and peptides (Lehninger, 1970).

Thus it is apparent that asphaltene functionality can be defined in terms of several commonly occurring natural product types that are more in keeping with the recognized maturation pathways than any of the structures heretofore conceived.

11.4 MOLECULAR WEIGHT

Perhaps the most confusing aspect of the character of the asphaltene fraction is measurement of the molecular weight. The asphaltene faction is a complex composition of functional types and molecular weight data represent an average number. Further, there are those who will advocate (with some justification) in one breath that that asphaltene fraction contains 20,000 constituents. In the next breath, these same advocates will announce that they know the correct and precise molecular weight of the fraction. Such pronouncement border on the ludicrous and nothing can be further from the truth.

Determining the molecular weights of asphaltenes is a problem because they have a low solubility in the liquids often used for determination. Also, adsorbed resins lead to discrepancies in molecular weight determination, and precipitated asphaltenes should be reprecipitated several times before the determination. Thus, careful precipitation and careful choice of the determination method are both very important for obtaining meaningful results (Speight et al., 1985).

The molecular weights of asphaltene fractions span a wide range from a few hundred to several million leading to speculation about self-association (Sakhanov and Vassiliev, 1927; Mack, 1932; Katz, 1934; Lerer, 1934; Hillman and Barnett, 1937; Pfeiffer and Saal, 1940; Grader, 1942; Kirby, 1943; Labout, 1950; Ray et al., 1957; Winniford, 1963; Wales and van der Waarden, 1964; Altgelt, 1968; Markhasin et al., 1969; Reerink, 1973; Koots and Speight, 1975; Speight et al., 1985).

The tendency of the asphaltene constituents to form aggregates in hydrocarbon solution is one of their most characteristic features and complicates the determination of asphaltene molecular weight (Winniford, 1963; Speight and Moschopedis, 1977). The average molecular weights measured by means of vapor pressure osmometry (VPO) or size exclusion chromatography (SEC) are significantly influenced by the conditions of the analysis (temperature, asphaltene concentration, solvent polarity) (Altgelt, 1968; Moschopedis et al., 1976; Speight and Moschopedis, 1980; Speight, 1981b; Speight et al., 1985; Acevedo et al., 1992). For this reason, molecular weights are reported in the literature as relative values only and these values may be quite different from the molecular weights of unassociated molecules.

The influence of the asphaltene concentration on the values of molecular weight measured is significant and it is important to use as high a dilution as possible in order to measure a molecular weight that at least approaches that of the unassociated asphaltene molecules, remembering that the asphaltene fraction is a complex mixture of different species and that any molecular weight of unassociated species will be an average molecular weight of different chemical species.

The existence of asphaltenes aggregates in hydrocarbon solvents has been demonstrated by means of small-angle neutron scattering (SANS) studies. The physical dimensions and shape of the aggregates are functions of the solvent used and the temperature of the investigation (Ravey et al., 1988; Overfield et al., 1989; Thiyagarajan et al., 1995). In addition, surface tension measurements have been used to study the self-association of asphaltenes in pyridine and nitrobenzene (Sheu et al., 1992). A discontinuous transition in the surface tension as a function of asphaltene concentration was interpreted as the critical asphaltene concentration above which self-association occurs.

The study of asphaltene molecular weights by vapor pressure osmometry shows that the molecular weights of various asphaltenes are dependent not only on the nature of the solvent but also on the solution temperature at which the determinations were performed (Moschopedis et al., 1976). However, data from later work involving molecular weight determinations by the cryoscopic method (Speight and Moschopedis, 1977) indicate that the molecular nature of asphaltenes is not conducive to the determination of absolute molecular weights by any one method.

For any one method, the observed molecular weights suggest that asphaltenes form molecular aggregates, even in dilute solution, and this association is influenced by solvent polarity, asphaltene concentration, and the temperature at which the determination is made. The precise mechanism of the association has not been conclusively established, but hydrogen bonding and the formation of charge-transfer complexes have been cited as responsible for intermolecular association (Yen, 1974). In fact, intermolecular hydrogen bonding could be involved in asphaltene association and may have a significant effect on observed molecular weights (Moschopedis and Speight, 1976c; Speight et al., 1985).

It is also interesting to note that use of a solvent of low dielectric constant (benzene) does not cause any variation in the molecular weight when the concentration of the asphaltenes is varied over the range 2% to 7% by weight. However, use of a solvent of higher dielectric constant (pyridine) caused significant variation in the observed molecular weights over this particular range. Molecular weight determinations at lower (<2%) and higher (>8%) concentrations are subject to instrument sensitivity and solubility interference, respectively. Extrapolation of the pyridine data to infinite dilution, when asphaltene–asphaltene interaction may be assumed negligible, affords molecular weights of the same order (about 1800) as those recorded using nitrobenzene as the solvent. On the other hand, extrapolation of the pyridine data to higher concentrations suggests that molecular weights of the order of those recorded in a solvent of lower dielectric constant (e.g., benzene) may be obtained.

The higher molecular weights recorded when solvents of low polarity are employed are undoubtedly the result of intermolecular association between the asphaltene nuclei, but solvents of high dielectric constant are able to bring about dissociation of these asphaltene agglomerations to what, in fact, appear to be single asphaltene particles. Further, this observation precludes asphaltene structures that invoke the concept of a polymer molecule to account for the high molecular weights observed in nonpolar solvents.

Determination of asphaltene molecular weight is subject to the presence of occluded resin constituents (Ali et al., 1985) and removal of the resin constituents gives rise to higher observed molecular weights of the purified asphaltenes. In addition, and as noted previously for the whole asphaltenes (Moschopedis et al., 1976), the molecular weights of the purified asphaltenes also varied with the solvent used for the determination; that is, solvents of high dielectric constant decrease the observed molecular weights. Further, extraction of freshly precipitated asphaltene fractions using a Soxhlet extractor and different solvents, followed by molecular weight determinations of the insoluble material show a decrease in the asphaltene molecular weight with the dielectric constant of the solvent.

Thus, the tendency of asphaltenes to undergo association or dissociation depending upon the nature of the solvent also appears true for the series of higher molecular weight fractions. However, it should be noted here that although the results with asphaltenes available from several crude oils (Moschopedis et al., 1976) suggest that molecular weight varies with the dielectric constant of the solvent, there may be other factors that may in part also be responsible for this phenomenon. The final phenomenon that influences the molecular weight of the asphaltene is the relative polarity of the solvent used in the precipitation technique.

It is of interest to speculate at this point about the means by which the resin material is retained by the asphaltenes. For example, asphaltenes may be analogous to coal insofar as it appears that asphaltenes have adsorption characteristics and may even exhibit a distinct physical structure or may even participate in clathrate systems. If this is the case, swelling by a solvent such as pyridine would undoubtedly free resin material from the asphaltenes. On the other hand, if the resin material is retained purely by a surface adsorption phenomenon, it may be expected to be removed from the asphaltenes by the repetitive precipitation. Both effects could play a part in the retention of resin (i.e., precipitant-soluble) material by the asphaltenes.

Obviously, many other facets of asphaltene precipitation need to be investigated, but it is obvious from these data that it is extremely difficult to obtain clean (resin-free) asphaltenes without a multiple precipitation technique. The amount of resin may have some influence on the asphaltene yield (and, therefore, on an estimation of crude oil composition). However, the resin occluded within (or adsorbed onto) the asphaltene during the separation procedure has a considerable effect on the observed molecular weight and, presumably, on the degree of association of the asphaltenes. Indeed, the speculative concept that asphaltenes release the final vestiges of this resin only upon swelling by a solvent such as pyridine is also worthy of consideration.

Attention should be focused upon the delineation of various structural types in the asphaltenes. As petroleum is actually a continuum of structural types, there is little to be inferred from this study other than the influence on degree of association or on the degree of polymerization (for those concepts based on the concept of a polymer structure).

If, however, the structural types vary from the resins to the asphaltenes, it may be necessary to reassess the concepts of molecular types that have been proposed heretofore. In addition, the concept that the resins may be entrapped within the asphaltene matrix (and may even be analogous to a clathrate compound) warrants further investigation.

It is worthy of mention at this point, as the theory of the physical structure of petroleum is based upon the results of the method, that the specific viscosity of asphaltene solutions has also been employed to determine asphaltene molecular weights.

The method essentially requires a determination of the relative viscosity, n_r, of known solute–solvent systems, when the constant k is derived by the relationship

$$\log n_r = ckM$$

where $c =$ concentration (g per 100 mL) and $M =$ molecular weight. Thus, subsequent application of the relationship to solute–solvent systems involving asphaltenes and maltenes allowed Mack, in 1932, to derive molecular weights of the order of 5000 at 0°C (32°F) to 1800 at 120°C (250°F). Thus, it is not surprising that the postulate of the existence of asphaltene clusters within the maltenes evolved and has remained to this time.

Application of this procedure to systems involving Athabasca asphaltenes allows derivation of values for the constant k, but transposition of k from one solute–solvent system to another gives inconsistent results. For example, the constants derived from naphthalene–benzene or biphenyl–benzene systems give unexpectedly low molecular weights when applied to asphaltene–benzene systems. Indeed application of these constants to petrolene-solute systems gives molecular weights of naphthalene and biphenyl in error by some several hundred per cent.

Application of the constant k for the biphenyl–benzene system to the asphaltene–petrolene system gives molecule weights within an acceptable range. However, the validity is suspect because of the unsatisfactory results obtained when the constants from one solute–solvent system are transposed to a different solute–solvent system in which the molecular weight of the solute is known. This is emphasized to an even greater extent by application of the constant derived from the asphaltene–benzene system to the asphaltene–petrolene system, which yields molecular weights of the order of 17,000 to 60,000.

In view of these inconsistencies, the viscosity method of determining asphaltene molecular weights must be regarded as suspect when the constants k, as derived for one solute–solvent system, is applied to a completely different solute–solvent system. Indeed, Mack's original work involved transposition of data from earlier work (Kendall and Monroe, 1917) and must be regarded as questionable, as these authors consider their own results to be indecisive because data from different solvent systems containing the same solute are in conflict. Even though preliminary application of some of the viscosity data yields asphaltene molecular weights within an acceptable range, the remaining data indicate several inconsistencies in the method, and recognition of this causes some revision of previous theories of the physical structure of crude oil.

In summary, asphaltene molecular weights are variable (Yen, 1974; Speight et al., 1985; Speight, 1994). There is a tendency to associate even in dilute solution in nonpolar solvents. However, data obtained using highly polar solvents indicate that the molecular weights, in solvents that prevent association, usually fall in the range 2000 ± 500.

11.5   REACTIONS

The relative reactivity of petroleum constituents can be assessed on the basis of bond energies and the asphaltene constituents are no exception to this generalization. Although the use of bond energy data is a method for predicting the reactivity or the stability of specific bonds under designed conditions, it must be remembered that the reactivity of a particular bond is also subject to its environment. Thus, it is not only the reactivity of the constituents of petroleum that are important in processing behavior, it is also the stereochemistry of the constituents as they relate to one another that is also of some importance (Chapter 12). It must be appreciated that the stereochemistry of organic compounds is often a major factor in determining reactivity and properties.

Thus, it is not surprising that the chemical reactions of the constituents of the asphaltene fraction have received some attention and provide valuable information about the potential chemical conversion that can be performed with a view to the chemical functions within the asphaltenes. However, these reactions also provide valuable information about the structural types in these constituents and also what effects changes in the hydrocarbon structure or functionality will have on the properties of the asphaltene constituents. This information is valuable when perturbations to the petroleum system are given consideration vis-à-vis the potential effects that can cause deposition of sediment after the delicate balance of the system has been changed or even destroyed (Chapter 12).

The desirability of removing the asphaltene fraction from a crude oil has been advocated many times, the principal reasons being the production of a cracking stock, low in metal impurities and heteroatom compounds, and a low carbon residue. As a result of the structural studies just discussed, it is evident that petroleum asphaltenes are agglomerations of compounds of a particular type. Thus, it is not surprising that asphaltenes undergo a wide range of interactions of a chemical and physical nature based not only on their condensed aromatic structure but also on the attending alkyl and naphthenic moieties (Speight, 1984b).

Asphaltenes can be thermally decomposed under conditions similar to those employed for visbreaking (viscosity breaking; about 470°C, 880°F) to afford, on the one hand, light oils that contain higher (to $>C_{40}$) paraffins and, on the other hand, coke:

Asphaltene fraction $\rightarrow H_2$, CO, CO_2, H_2S, SO_2, $H_2O + CH_2 = CH_2$, CH_4, CH_3CH_3, CH_3
$\times (CH_2)_n CH_3$

The thermal decomposition of asphaltenes provides an excellent example of inconsistencies in the derivation of structural types from spectroscopic materials (i.e., magnetic resonance) in which alkyl side chains are deduced to contain approximately four carbon atoms (Speight, 1970a, 1971). Asphaltene pyrolysis (350°C to 800°C, 660°F to 1470°F) produces substantial amounts of alkanes (having up to 40 carbon atoms in the molecule) in the distillate that can only be presumed to reflect the presence of such chains in the original asphaltene. Transalkylation studies (Farcasiu et al., 1983) also provide evidence for longer alkyl chains. Obviously, recognition of the inconsistencies of the spectroscopic method with respect to the paraffinic moieties must lead to the recognition of similar inconsistencies when considering the aromatic nucleus.

The application of thermal techniques to study the nature of the volatile thermal fragments from petroleum asphaltenes has produced some interesting data relating to the polynuclear aromatic systems (Speight, 1971; Ritchie et al., 1979; Schucker and Keweshan, 1980; Gallegos, 1981;). These thermal techniques have produced strong evidence for the presence of small (one-ring to four-ring) polynuclear aromatic systems (Speight and Pancirov, 1984) and now, application of the technique to the various functional fractions has confirmed the general but unequal distribution of these systems throughout asphaltenes.

Each asphaltene fraction produced the same type of polynuclear aromatic systems (i.e., alkyl derivatives of benzene, naphthalene, phenanthrene, chrysene, benzothiophene, and dibenzothiophene) in the volatile matter but the distribution was not constant. It was also possible to compute the hydrocarbon distribution; a noteworthy point here is the overall preponderance of single-ring (cycloparaffin and alkylbenzene) species, as well as the domination of saturated material over aromatic material. The preponderance of the low-molecular-weight material in the volatile products is anticipated on the basis that more complex systems remain as nonvolatile material and, in fact, are converted to coke. One other point worthy of note is that the pyrolysis–gas chromatography–mass spectrometry (py/gc/ms) program does not accommodate nitrogen and oxygen species. This matter is resolved, in part, by the concentration of nitrogen and oxygen in the nonvolatile material (coke) and the overall low proportions of these heteroatoms originally present in the asphaltenes.

A major drawback to the application of py/gc/ms technique to the study of the polynuclear aromatic systems in asphaltene constituents is the amount of material that remains as a nonvolatile residue. Aside from speculation about the polynuclear aromatic systems in the residue, it should be noted that the majority of the nitrogen (>90%), oxygen (>50%), and sulfur (>60%) in the natural asphaltene remains in the coke (Speight, 1971; Speight and Pancirov, 1984).

Paraffins are not the only hydrocarbon products of the thermal reactions of asphaltenes. The reaction paths are extremely complex; spectroscopic investigations indicate an overall dealkylation of the aromatics to methyl (predominantly) or ethyl (minority) groups. This is in keeping with a mass spectroscopic examination of asphaltene fractions (by direct introduction into the ionization chamber), which indicates a progressive increase with increasing temperature (50°C to 350°C, 120°F to 660°F) of ions attributable to low- molecular- weight hydrocarbons. Higher temperatures (500°C, 932°F) promote the formation of benzene and naphthalene nuclei as the predominant aromatics in the light oil, but unfortunately an increase in coke production is noted.

In conclusion, thermal decomposition of asphaltenes affords a light oil having a similar composition to that from the heavy oil and a hydrocarbon gas composed of the lower paraffins, which, after the removal of the by-products (water, ammonia, and hydrogen sulfide) has good burning properties. The formation of these paraffins can be ascribed to the generation of hydrogen within the system that occurs during the pyrolysis of condensed aromatic structures.

Oxidation of asphaltenes with common oxidizing agents, such as acid and alkaline peroxide, acid dichromate, and alkaline permanganate, is a slow process. The occurrence of a broad band centered at $3420\ cm^{-1}$ and a band at $1710\ cm^{-1}$ in the infrared spectra of the products indicates the formation of phenolic and carboxyl groups during the oxidation. Elemental analyses of the products indicate that there are two predominant oxidation routes, notably

1. The oxidation of naphthene moieties to aromatics as well as the oxidation of active methylene groups to ketones
2. Severe oxidation of naphthene and aromatic functions resulting in degradation of these systems to carboxylic acid functions

Oxidation of asphaltenes in solution, by air, and in either the presence or absence of a metal salt, is also possible (Moschopedis and Speight, 1978). There is some oxygen uptake, as can be seen from the increased O/C atomic ratios, but the most obvious effect is the increase in the amount of *n*-heptane-insoluble material. And analysis of the data shows that it is the higher heteroatom (more polar constituents) of the asphaltenes that are more susceptible to oxidation, leaving the suggestion that the polarity of the constituents may be determined by the incorporation of the heteroatoms into ring systems.

Air-blowing of asphaltenes at various temperatures brings about significant oxygen uptake. This is accompanied by a marked decrease in the molecular weight (vapor pressure osmometry, benzene solution) of the product. This indicates that intermolecular hydrogen bonding of oxygen functionality may play a part in the observed high molecular weights and physical structure of petroleum (Moschopedis and Speight, 1978; see also Taft et al., 1996).

Asphaltenes may also be hydrogenated to produce resins and oils at elevated temperatures (>250°C). Chemical hydrogenation under much milder conditions, for example with lithium–ethylenediamine or sodium—liquid ammonia, also produces lower molecular weight species together with marked reductions in the sulfur and oxygen contents.

It may appear at first sight that sulfur and oxygen exist as linkages among hydrocarbon segments of asphaltene molecules. Although this may be true, in part, it is also very likely, in view of what has been discussed previously, that the lower molecular weights reflect changes in molecular association brought about by the elimination of oxygen and sulfur.

Aromatics undergo condensation with formaldehyde to afford a variety of products. This process can be extended to the introduction of various functions into the asphaltene molecules, such as sulfomethylation, that is, introduction of the $-CH_2SO_3H$ group. This latter process, however, usually proceeds more readily if functional groups are present within the asphaltene molecule.

Thus, oxidation of asphaltenes produces the necessary functional groups, and subsequently sulfomethylation can be conveniently achieved. Sulfomethylation of the oxidized asphaltenes occurence can be confirmed from three sources:

1. Overall increases in the sulfur contents of the products relative to those of the starting material
2. The appearance of a new infrared absorption band at 1030 cm^{-1} attributable to the presence of sulfonic acid group(s) in the molecule(s)
3. The water solubility of the products, a characteristic of this type of material. These sulfomethylated oxidized asphaltenes even remain in solution after parent oxidized asphaltenes can be precipitated from alkaline solution by acidification to pH 6.5

The facile sulfomethylation reaction indicates the presence in the starting materials of reactive sites ortho or para to a phenolic hydroxyl group. The related reaction, sulfonation, is also a feasible process for oxidized asphaltenes. The ease with which this reaction proceeds suggests the presence of quinoid structures in the oxidized materials. Alternatively, active methylene groups in the starting materials facilitate sulfonation as such groups have been known to remain intact after prolonged oxidation.

Halogenation of asphaltenes occurs readily to afford the corresponding halo-derivatives; the physical properties of the halogenated materials are markedly different from those of the parent asphaltenes. For example, the unreacted asphaltenes are dark brown, amorphous, and readily soluble in benzene, nitrobenzene, and carbon tetrachloride, but the products are black, shiny, and only sparingly soluble, if at all, in these solvents.

There are also several features that distinguish the individual halogen reactions from one another. For example, during chlorination of asphaltenes there is a cessation of chlorine uptake by the asphaltenes after 4 h. Analytical data indicate that more than 37% of the total chlorine in the final product is introduced during the first 0.5 h, reaching the maximum after 4 h. Further, the H/C ratio of 1.22 in the parent asphaltenes [(H + Cl)/C ratio in the chlorinated materials] remains constant during the first 2 h of chlorination, by which time chlorination is 88% complete (Moschopedis and Speight, 1971). This is interpreted as substitution of hydrogen atoms by chlorine in the alkyl moieties of the asphaltenes; the condensed aromatic sheets remain unaltered as substitution of aryl hydrogen appears to occur readily only in the presence of a suitable catalyst, such as $FeCl_3$, or at elevated temperatures. It is only after more or less complete reaction of the alkyl chains that addition to the aromatic rings occurs, as evidenced by the increased atomic (H + Cl)/C ratios in the final stages of chlorination.

Bromine uptake by the asphaltenes is also complete in a comparatively short time (<8 h). However, in contrast to the chlorinated products, the atomic (H + halogen)/C ratio remains fairly constant (1.23 and 1.21 in the bromoasphaltenes or 1.22 in the unreacted asphaltenes) over prolonged periods (up to 24 h) of the bromination.

Iodination of asphaltenes is different insofar as a considerable portion of the iodine, recorded initially as iodine uptake, can be removed by extraction with ether or with ethanol, whereas very little weight loss is recorded after prolonged exposure of the material to a high vacuum. The net result is the formation of a product with an atomic (H + I)/C ratio of 1.24 after an 8 h reaction; a more prolonged reaction period affords a product with a (H + I)/C ratio of 1.17. This latter may be the result of iodination of the alkyl or naphthenic moieties of the asphaltenes with subsequent elimination of hydrogen iodide. Alternatively, dehydrogenation of naphthene rings to aromatic systems or coupling of aromatic nuclei would also account for lower (H + I)/C ratios. In fact, this latter phenomenon could account, in part, for the insolubility of the products in solvents that are normally excellent for dissolving the unchanged asphaltenes. However, it will be appreciated that these aforementioned reactions are only a few of many possible reactions that can occur, and undoubtedly halogenation of the asphaltenes is much more complex than would appear from the product data.

Halogenation of the asphaltenes can also be achieved by use of sulfonyl chloride, iodine monochloride, and N-bromusuccinimide, or, indirectly, via the Gomberg reaction.

Attempts to introduce hydrophilic groups into asphaltenes by **hydrolysis** of the halo-derivatives with either aqueous sodium hydroxide or with aqueous sodium sulfite are not feasible. The products from the reaction of halogenated Athabasca asphaltenes with these aqueous solutions are insoluble in strong aqueous alkali. Partial reaction does occur and is evident from the decreased atomic (H + Cl)/C ratios and the increased oxygen-to-carbon atomic ratios of the products relative to those of the untreated halo-asphaltenes (Moschopedis and Speight, 1971). However, it was not possible to establish conclusively the presence of sulfonic acid group(s) in the product from the sodium sulfite reaction by assignment of infrared absorption bands to this particular group.

It is also conceivable that the decreased atomic (H + Cl)/C ratios may be due, in part, to elimination of the elements of hydrogen chloride during the reactions, not solely, as intended, to substitution of chlorine by −OH and by −SO_3H. Olefin formation by elimination of hydrogen halide from alkyl or aralkyl halides in strongly basic media is well known, and the elimination of hydrogen chloride from β-phenylethyl chloride in the presence of aqueous sodium sulfite to yield styrene has also been recorded. On the other hand, α-phenylpropyl and α-phenylbutylbromides react normally with aqueous sodium sulfite to afford the corresponding sulfonic acids.

Interactions of asphaltenes with metal chlorides yield products containing organically bound chlorine, but the analytic data indicate that dehydrogenation processes occur simultaneously. There is, of course, no clear way that the extent of the dehydrogenation can be estimated, but it is presumed to be a dehydrogenation condensation rather than olefin formation.

Inter- or intramolecular condensations are the preferred reaction route, insofar as the solubility and apparent complexity of the products vary markedly from those of the starting materials, and these differences cannot be attributed wholly to the incorporation of chlorine atoms into the constituents of the asphaltenes or heavy oil. Indeed, the data accumulated are indicative of a condensation dehydrogenation or, in part, loss of alkyl substituents, for example, lower molecular weight hydrocarbons, during the reactions. As an illustration of the former, the cokes produced during the thermal cracking (450°C, 840°F) of the asphaltenes or heavy oil have H/C ratios in the range of 0.59 to 0.77. The majority of the insoluble materials produced in the asphaltene–metal chloride reactions have only slightly higher (H + Cl)/C ratios (0.88 to 1.10).

The facile reactions of asphaltenes with metal chlorides have also been applied as a possible means of petroleum deasphalting. The results indicate that, without exception, the asphaltenes react more rapidly than the heavy oil to produce insoluble organic material. It is also worthy of note here that, in the case of the heavy oil, the soluble organic material contains only traces (<1%, usually of the order of 0.5%) of organically bound chlorine. This is in contrast to the results described when higher ratios of metal chloride to heavy oil are employed and the reaction is presumed to go to completion over 24 h.

Thus, application of the process to a heavy feedstock shows that maximum yields of asphaltene materials are removed from the feedstock–solvent mix at low concentrations of metal chloride. The resulting deasphalted heavy oil contains only traces (<0.5% wt/wt) of organically bound chlorine and little or no (<0.1% wt/wt) mineral matter.

It is therefore apparent that asphaltene removal from a crude oil is greatly facilitated by addition of reactive metal salts, which remove the necessity of handling large volumes of solvents and the like as in the conventional deasphalting process. Although the precipitated asphaltene materials contain substantial amounts (10% to 20% and even up to 30% wt/wt) of organically bound chlorine, this is nevertheless not a disadvantage, as other work has shown that these materials lend themselves to a variety of uses, especially thermal decomposition to good grade coke.

Reactions of asphaltenes with sulfur have also received some attention and have yielded interesting results. For example, treatment of the asphaltenes with oxygen or with sulfur at 150 to 250°C (300 to 480°F) yields a condensed aromatic product [H/C = 0.97; H/C (asphaltenes) = 1.20] containing very little additional sulfur. The predominant reaction here appears to be condensation between the aromatic and aliphatic moieties of the asphaltene constituents caused by elemental sulfur, which are in turn converted to hydrogen sulfide. Condensation appears to proceed in preference to molecular degradation, and treatment of the condensed products at 200°C to 300°C (390°F to 570°F) for 1 to 5 h again affords good grade cokes (H/C = 0.54 to 0.56). In all instances, the final products contain only very low amounts of elements other than carbon and hydrogen (ΣNOS <5% wt/wt), a desirable property of good grade coke.

Attempts to phosphorylate asphaltenes with phosphoric acid, phosphorous trichloride, or phosphorous oxychloride have been partially successful insofar as it is possible to introduce up to 3% wt/wt phosphorus into the asphaltenes. However, application of these same reactions to oxidized asphaltenes increased the uptake of phosphorus quite markedly, since there is up to 10% wt/wt in the product. Subsequent reaction of the phosphorus-containing products is necessary to counteract the acidity of the phosphorus functions.

Other reactions of asphaltenes have also been performed, but the emphasis has mainly been on the formation of more condensed materials to produce good grade cokes. For example, thermal treatment of the halogenated derivatives affords aromatic cokes [H/C = 0.58; coke from the thermal conversion of asphaltenes at ~460°C (860°F) has H/C = 0.77] containing less than 1% wt/wt halogen. Other investigations also show that treatment of the halo-asphaltenes with suitable metal catalysts—copper at 200°C to 300°C (390°F to 570°F) for 1 to 5 h or sodium at 80°C to 100°C (175°F to 212°F) for 1 to 5 h yields aromatic (H/C = 0.55 to 0.86, respectively) coke-like materials having 0.5% to 3% wt/wt halogen. Residual halogen may finally be removed by treatment at 300°C (570°F) for 5 h.

Other chemical modifications include metallation of the asphaltenes or halo-asphaltenes using metal or metallo-organics, followed by carboxylation to yield acid functions in the product. Interaction of the asphaltenes with m-dinitrobenzene affords a product that, when treated with hydroxylamine (NH$_2$OH) or an amine (RNH$_2$), yields nitrogen-enriched products.

Reaction of asphaltenes with maleic anhydride and subsequent hydrolysis also yields a product bearing carboxylic acid functions. Asphaltenes react with diazo compounds and the procedure may be used to introduce functional groups into the asphaltene molecule to modify the properties of these materials.

11.6   SOLUBILITY PARAMETER

The solubility parameter (Table 11.2) is a measure of the ability of the compound to act as a solvent or nonsolvent and as well as a measure of the intermolecular and intramolecular forces between the solvent molecule and between the solvent and solute molecules.

TABLE 11.2
Hansen Solubility Parameters for Selected Molecules

Compound	dD	dP	dH
Carbon disulfide	20.5	0.0	0.6
CHCl$_3$	17.8	3.1	5.7
Toluene	18.0	1.4	2.0
Cyclopentane	16.4	0.0	1.8
THF	16.8	5.7	8.0
Cyclohexane	16.8	0.0	0.2
Carbon tetrachloride	17.8	0.0	0.6
o-Xylene	17.8	1.0	3.1
trans-Decalin	18.0	0.0	0.6
p-Diethylbenzene	18.0	0.0	0.6
Mesitylene	18.0	0.0	0.6
Butylbenzoate	18.3	2.9	5.5
Benzene	18.4	0.0	2.0
p-Divinylbenzene	18.6	1.0	7.0
Styrene	18.6	1.0	4.1
Succinaldehyde	18.6	1.0	4.1
cis-Decalin	18.8	0.0	0.0
Thiophene	18.9	2.4	7.8
1,4-Dioxane	19.0	1.8	7.4
Naphthalene	19.2	2.0	5.9
Tetrahydronaphthalene	19.6	2.0	2.9
Diphenyl ether	19.6	3.2	5.8
1-Methylnaphthalene	20.6	0.8	4.7
Biphenyl	21.4	1.0	2.0

Any change in the physical and chemical properties of crude oil can be a key factor in the stability of the system. For example, as the degree of conversion in a **thermal process** increases, the solubility power of the medium toward the heavy and polar molecules decreases due to the formation of saturated products (Speight, 1994). This is reflected in a relative change in the solubility parameters of the dispersed and solvent phases leading to a phase separation (Speight, 1992, 1994; Wiehe, 1993, 1994).

The most prevalent thermodynamic approach to describing asphaltene solubility has been the application of the solubility parameter or the concept of cohesive energy density. The application of solubility parameter data to correlate asphaltene precipitation, and, hence, crude oil–solvent interaction, has been used on prior occasions (Labout, 1950; Mitchell and Speight, 1973; Wiehe, 1995). The solubility parameters of asphaltene constituents can be estimated from the properties of the solvent used for separation (Yen, 1984; Long and Speight, 1989) or be measured by the titration method (Andersen and Speight, 1992) or even from the atomic hydrogen–carbon ratio (Figure 11.5).

The solubility parameter difference that results in a phase separation of two materials, such as asphaltenes in a solvent, can be estimated using the Scatchard–Hildebrand equation:

$$\ln a_a = \ln x_a + M_a / \left\{ RT\rho_a \left[\Phi_s^2 (\delta_s - \delta_a)^2 \right] \right\}$$

where a_a is the activity of the solute a, x_a is the mole fraction solubility of a, M_a is the molecular weight of a, ρ_a is the density of a, Φ_s is the volume fraction of solvent, and $(\delta_s - \delta_a)$ is the difference between the solubility parameters of the solute a and the solvent s. Assuming that the activity of the asphaltenes a_a is 1 (solid asphaltenes in equilibrium with dissolved asphaltenes) and the volume fraction of an excess of solvent is essentially 1, the equation can be rearranged into a form that can be used to gain insight into the solubility of asphaltenes:

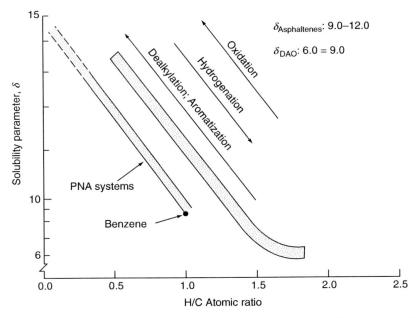

FIGURE 11.5 Estimation of the solubility parameters of asphaltene constituents from the atomic hydrogen–carbon ratio (compared to polynuclear aromatic systems) and changes caused by chemical–thermal effects.

$$\ln x_a = -M_a/RT_a[(\delta_s - \delta_a)^2]$$

Assuming a density for asphaltenes of 1.28 g/cc and a molecular weight of 1000 g/mol, the solubility of asphaltenes as a function of the differences between solubility parameters of the asphaltenes and precipitating solvent can be calculated. Thus, the solubility of asphaltenes can be shown to decrease as the difference between solubility parameters increases, with the limit of solubility attained at a difference of about 3. Thus, if the asphaltenes are part of a polarity and molecular weight continuum in a crude oil, their precipitation is not as straightforward as it would be for a single species with a particular solubility parameter.

Additional information can be gleaned by calculating the solubility parameter difference at several molecular weights ranging from 50 to 10,000 g/mole at which the solubility of asphaltenic or other material is a mole fraction of 0.001 (0.1% or 1000 ppm). As a result, a phase diagram that is a function of molecular weight and solubility parameter difference can be defined. Both polarity and molecular weight of asphaltenes in a solvent define the solubility boundaries and explain conceptually how asphaltenes are precipitated from the mixture in crude oils that can be considered a type of continuum of molecular weights and polarities. Further, as the molecular weight of a particular solute decreases, there is an increased tolerance of polarity difference between solute and solvent under miscible conditions.

Incompatibility phenomena can be explained by the use of the solubility parameter for asphaltenes and other petroleum fractions (Speight, 1992, 1994). As an extension of this concept, there is sufficient data to draw an approximate correlation between hydrogen-to-carbon atomic ratio and the solubility parameter, δ, for hydrocarbon solvents and petroleum constituents (Speight, 1994). In general, hydrocarbon liquids can dissolve poly-nuclear aromatic hydrocarbons where there is, usually, less than a 3-point difference between the lower solubility parameter of the solvent and the solubility parameter of the hydrocarbon.

By this means, the solubility parameter of asphaltenes can be estimated to be a range of values that is also in keeping with the asphaltenes, composed of a mixture of different compound types with the accompanying variation in polarity. Removal of alkyl side chains from the asphaltenes will decrease the H/C ratio and increase the solubility parameter thereby bringing about a concurrent decrease of the asphaltene product in the hydrocarbon solvent. In fact, on the molecular weight polarity diagram for asphaltenes, carbenes and carboids can be shown as lower molecular weight, highly polar entities in keeping with molecular fragmentation models.

11.7   STRUCTURAL ASPECTS

The molecular nature of the nonvolatile fractions of petroleum has been the subject of numerous investigations (Speight, 1972, 1994; Yen, 1972), but determining the actual structures of the constituents of the asphaltene fraction has proved to be difficult. It is the complexity of the asphaltene fraction that has hindered the formulation of the individual molecular structures. Nevertheless, various investigations have brought to light some significant facts about asphaltene structure. There are indications that asphaltenes consist of condensed aromatic nuclei that carry alkyl and alicyclic systems with heteroelements (i.e., nitrogen, oxygen, and sulfur) scattered throughout in various, including heterocyclic, locations.

Other basic generalizations have also been noted: with increasing molecular weight of the asphaltene fraction, both aromaticity and the proportion of heteroelements increase (Yen, 1970, 1971; Koots and Speight, 1975). In addition, the proportion of asphaltenes in petroleum varies with source, depth of burial, the specific (or API) gravity of the crude oil, and the sulfur content of the crude oil as well as nonasphaltene sulfur (Koots and Speight, 1975). However, many facets of asphaltene structure remain unknown, and it is the purpose of this chapter to bring together the pertinent information on asphaltene structure as well as the part played by asphaltenes in the physical structure of petroleum.

In fact, it is this overall lack of structural detail that indicates that current, and future, chemical research should focus on several items. These are:

1. Skeletal structure
2. Size of the polynuclear aromatic systems (Chapter 11)
3. Internuclear bonds that need to be thermally or catalytically cleaved to produce more identifiable lower molecular weight fragments (Chapter 7 and Chapter 15)
4. Heteroatom types and subsequent ease or difficulty of removal (Chapter 7)
5. Metallic constituents and their occurrence in porphyrinic or nonporphyrinic locations (Chapter 7)
6. Physicochemical relationships that exist between the asphaltenes and the other constituents of the oil (Chapter 12)

Despite numerous investigations, determination of the actual molecular structures that exist in petroleum has proved difficult. It is no doubt the great complexity of these materials, which are actually isolated by physical phenomena rather than by virtue of their chemical structures, that has hindered the formulation of individual molecular structures. Nevertheless, the various investigations have brought to light some significant facts about the structures that exist in petroleum.

There are indications that the asphaltenes consist of condensed aromatic nuclei that carry alkyl and cylcoalkyl (naphthenic) substituents; heteroatoms (i.e., sulfur, nitrogen, and oxygen) are scattered throughout the hydrocarbon systems. In addition, it appears that with increasing molecular weight, both aromaticity and the proportion of the heteroatoms increase. There is also fragmentary evidence that, with increasing molecular weight, the heteroatoms become incorporated into the more stable heterocyclic systems. This is especially true for sulfur, in which dibenzothiophene systems are believed to predominate in the higher molecular weight asphalts; it is presumed that similar arrangements occur for the nitrogen atoms; oxygen, may also occur in cyclic systems or in various functional groups.

If the residuum is produced by a process in which thermal decomposition has occurred, there may be traces of thiols (mercaptans, R–SH) in the mix that would, along with any hydrogen sulfide generated during the process, impart an objectionable odor to the residuum. The occurrence of thiols in heavy oils that have not been subjected to thermal treatment is also possible, but is generally not considered likely.

The structural nature of the asphaltenes has been open to question for some time. Early postulates invoked the concept of simple aromatic structures, whereas other postulates consider the possibility of polymers of aromatic and naphthenic ring systems (see, for e.g. Hillman and Barnett, 1937). A considerable amount of structural work has been carried out since that time on asphaltenes from various crude oils. In fact, attempts to derive information about the carbon skeleton of the asphaltenes have mainly been centered on the use of infrared and nuclear magnetic resonance spectroscopic techniques (Chapter 10).

The data obtained by these methods supported the occurrence of condensed polynuclear aromatic ring systems bearing alkyl side chains. The condensed aromatic ring systems appear to vary from smaller systems composed of about 6 rings to the more ponderous, 15- to 20-ring systems. The high molecular weight of asphaltenes is accounted for by suggesting that the unit is repeated one or more times and is linked through saturated chains or rings.

However, it is somewhat difficult to visualize the existence of structures of this type as part of the asphaltene molecule, irrespective of their means of derivation. Indeed, all the methods employed for structural analysis involve, at some stage or another, assumptions that, although based upon molecular types identified in the more volatile fractions of petroleum, must be of questionable validity, even though they may appear to serve the purpose and produce acceptable data.

Of all the methods applied to determining the types of polynuclear aromatic systems in petroleum asphaltenes (Speight and Long, 1981; Tissot, 1984), one with considerable potential, but given the least attention, is ultraviolet spectroscopy (Bjorseth, 1983). Typically, the ultraviolet spectrum of a petroleum asphaltene shows two major regions with very little fine structure, and interpretation of such a spectrum can only be made in general terms. What is often not realized is that the technique can add valuable information about the degree of condensation of polynuclear aromatic ring systems through the auspices of high-performance liquid chromatography (Lee et al., 1981; Bjorseth, 1983; Speight, 1986b).

For the whole asphaltene, before fractionation into functional types, the high-performance liquid chromatography–ultraviolet spectroscopy chromatogram confirms the complexity of the asphaltene fraction (Figure 11.6). In addition, the technique offers the attractive option of investigating the nature of asphaltenes on a before processing and after processing basis. This allows researchers to make more detailed studies of asphaltene chemistry during processing.

For more detailed information on ring size distribution, fractionation of the asphaltene is advocated. A high-performance liquid chromatographic investigation of a mixture of standard polynuclear aromatic systems with the UV detector at fixed wavelengths from 240 to 360 nm confirmed the applicability of the technique to determine the presence of ring size distribution using the different selectivity of the polynuclear aromatic systems (Speight, 1986b). The one-ring and two-ring polynuclear aromatic systems were more prominent in the chromatogram at 240 mm but were not at all evident in the chromatogram at wavelengths above 300 nm. The converse was true for the three-ring to six-ring systems. This was confirmed by examination of a standard solution of polynuclear aromatic (one-ring to seven-ring) systems, which confirmed that one-ring to three-ring systems gave prominent ultraviolet detector signals at <300 nm but gave no signals at >300 nm. On the other hand, compounds with four- to seven-ring systems still gave signals at >300 nm, but gave no signals at 365 nm.

These data provide strong indications of the ring size distribution of the polynuclear aromatic systems in petroleum asphaltenes. For example, amphoteric species and basic nitrogen species contain polynuclear aromatic systems having two to six rings per system. On the other hand, the acid subfractions (phenolic or carboxylic functions) and the neutral

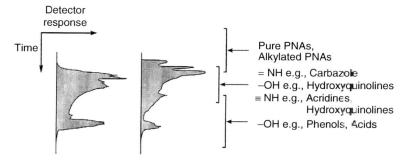

FIGURE 11.6 The complexity of the asphaltene fraction as shown by HPLC; the figure also refutes the concept of using an average structure to determine the behavior and properties of asphaltene constituents.

polar subfractions (amides and imino functions) contain few if any polynuclear aromatic systems with more than three rings per system. In no case was there any strong or conclusive evidence for polynuclear aromatic ring systems containing more than six condensed rings. In all cases, the evidence favored the preponderance of the smaller one- to four-ring systems (Speight, 1986b; Speight, 1994).

It must not be forgotten that the method is subject to the limitation of the sensitivity of polynuclear aromatic systems and the fact that some of the asphaltene material (<2% wt/wt) was irreversibly adsorbed on the adsorbent. In this latter case, the missing material is polar material and any deviation from the ring size distribution just outlined is not believed to be significant enough to influence the general conclusions about the nature of the polynuclear aromatic systems in petroleum asphaltenes.

These observations require some readjustment to the previous postulates of the asphaltene structure. Previously conceived hypotheses in which the polynuclear aromatic system is large (>10 rings) are considered unlikely, despite their frequent and recent occurrence in the literature. The thermolysis products are indicative of a much more open structure than has previously been recognized. Indeed, the failure to detect (during the chromatographic examination) any strong evidence for the existence of large multiring polynuclear aromatic systems in the asphaltene fractions is evidence against structures invoking this concept.

The manner in which these moieties occur within the asphaltene fraction must, for the present, remain largely speculative. Above all, it must be remembered that asphaltenes are a solubility class and, as such, may be an accumulation of thousands of (similar or dissimilar) structural entities. Hence, caution is advised against combining a range of identified products into one (albeit hypothetical) structure. For example, it would be presumptuous (if not ludicrous) to suggest that, on the basis of the present findings, petroleum asphaltenes are a polymeric analog of alkylbenzenes, naphthalenes, phenanthrenes, benzothiophenes, or dibenzothiophenes.

Attempts have also been made to describe the total structures of asphaltenes in accordance with magnetic resonance data and the results of other spectroscopic and analytical techniques (Speight, 1994 and references cited therein). It is difficult, however, to visualize these postulated structures as part of the asphaltene molecule. Indeed, such structures are not consistent with the proposed maturation of petroleum. It is considered more likely that the structures of the constituents of asphaltene fractions will be more closely related to the structural moieties found in natural products. Such complex structures as derived in the early studies may have several shortcomings that do not explain asphaltene behavior.

Nevertheless, these studies served as pioneering efforts by which later studies were stimulated. The fact is that all methods employed for structural analysis involve, at some stage or another, assumptions that, although based on data concerning the more volatile fractions of petroleum, are of questionable validity when applied to asphaltenes.

The concept of asphaltene constituents as a sterane-sulfur polymer or even a regular hydrocarbon polymer has arisen because of the nature of the products obtained by reaction of an asphaltene fraction with potassium naphthalide (Ignasiak et al., 1977). It was erroneously assumed that this particular organometallic reagent, one of several known to participate in rapid, complex reactions with organic substrates, cleaved only carbon–sulfur–carbon bonds, but not carbon–carbon bonds. However, potassium naphthalide cleaves carbon–carbon bonds in various diphenylmethanes, and the cleavage of carbon–carbon bonds in 1,2-diarylethanes has been documented.

In each case, the isolation of well-defined organic reaction products confirms the nature of the reaction. Further, the possibility of transmetallation from the arylnaphthalide to the aromatic centers of the asphaltenes complicates the situation and undoubtedly leads to more complex reactions and to reaction products of questionable composition. Formulating the structure of the unknown reactant (asphaltenes) under such conditions is extremely difficult. Indeed, it is evident that reacting asphaltenes with any particularly active reagents (e.g., the alkali aryls) leads to complex reactions, and it may be difficult, if not impossible, to predict accurately the course of these reactions.

In fact, the reaction of potassium naphthalide with tetrahydrofuran under conditions identical to those reported in which asphaltenes were also present, produces a light brown amorphous powder (Speight and Moschopedis, 1980). This product could erroneously be identified as a major degradation product of the asphaltene constituents had any asphaltene been present. Errors of this type have only added confusion to the already complex area of asphaltene structure.

There have also been attempts to substantiate the concept of a sulfur-containing polymer by virtue of thermal decomposition in the presence of tetralin (Ignasiak and Strausz, 1978). However, aliphatic carbon–carbon bonds (such as those in 1,2-diphenylethane) cleave under similar conditions, and reactions involving alteration of the hydrocarbon structure also occur. Indeed, an investigation of the nature of thermal dissociation of tetralin indicates that the reaction is quite complex, and the presence of an added material (such as coal or asphaltene) renders any attempt to rationalize the reaction in simple terms, of extremely dubious value.

Asphaltenes have also been subjected to x-ray analyses to gain an insight into their macromolecular structure (Yen et al., 1961). The method is reputed to yield information about the dimension of the unit cell, such as inter-lamellar distance ($c/2$), layer diameter (L_a), height of unit cell (L_c), and number of lamellae (Nc) contributing to the micelle.

A major constraint in the acceptance of proposed structures for asphaltenes is the validity of the proposed structures in terms of the complex maturation process (Tissot and Welte, 1978). During this maturation process, original molecules may remain unchanged or may be modified as they become part of petroleum. In addition, the character of the heteroatoms (nitrogen, oxygen, and sulfur) casts serious doubts upon the authenticity of the hypothetical large polynuclear aromatic systems that have been postulated as a major part of the asphaltene structure.

There is now adequate evidence to show that asphaltenes contain smaller polynuclear aromatic systems than has previously been, or even continues to be acknowledged. In addition, spectroscopic studies (before and after derivative formation) have indicated the presence of carbazole nitrogen and amide nitrogen. Nonaqueous potentiometric titration has confirmed the presence of these nitrogen types and has also provided evidence for the presence of pyridine nitrogen and indole nitrogen in asphaltenes. Oxygen fractions vary from the phenol oxygen and polyhydroxyphenols to quinone and amide oxygen.

These observations require some readjustments to the previous postulates of the asphaltene structure. Thus, previously conceived structures in which the sole aromatic system is a large polynuclear alkylated system are considered unlikely in view of these data. In fact, the current evidence is indicative of a much more open (rather than highly condensed) structure. From these investigations, a key structural feature of petroleum asphaltenes is the occurrence of small polynuclear aromatic systems, benzothiophenes, dibenzothiophenes, and n-paraffins within the asphaltene molecule.

The manner in which these moieties appear within the asphaltene molecule must for the present remain largely speculative, but above all, it must be remembered that asphaltenes are a solubility class and, as such, may be an accumulation of (literally) thousands of structural entities. Hence, caution is advised against combining a range of identified products into one (albeit hypothetical) structure.

Highly condensed polynuclear aromatic structures, which have been used to explain the high yields of thermal coke from asphaltenes, are not consistent with the natural product origins of petroleum asphaltenes. Such entities require conditions of temperature and pressure for their formation that are more extreme than those currently recognized as responsible for petroleum maturation. There is evidence that can be used in support of the formation of a nonvolatile residue (coke) being formed from small condensed (≤ 6) ring systems rather than requiring a large condensed aromatic ring system. Heat-resistant polymers containing aromatic and heterocyclic units, such as polyquinolines and polyquinoxalines, have a strong tendency to form large condensed systems during pyrolysis and finally, carbonize. Formulas that include the occurrence of smaller polynuclear aromatic hydrocarbon systems in asphaltenes are consistent with such behavior when it is recognized that the majority of the nitrogen, oxygen, and sulfur species accumulate in the nonvolatile residue.

Thermal studies using model compounds confirm that volatility of the fragments is a major influence in carbon residue formation and a pendant-core model for the high molecular weight constituents of petroleum has been proposed (Wiehe, 1994). In such a model, the scission of alkyl side chains occurs, thereby leaving a polar core of reduced volatility that commences to produce a carbon residue (Speight, 1970b, 1994; Wiehe, 1994).

The danger in all of these studies is in the attempts to link together fragmented molecules produced by thermal processes into a structure that is believed to be real. Functionality and polynuclear aromatic ring systems are real, but combination of these parameters into one or even several structures can be misleading. Molecular models can be an excellent aid to understanding process chemistry and physics. As long as the model allows predictability of process behavior, use of the model is warranted. However, caution is advised. Because of the complexity of the asphaltene fraction, there is no one method that will guarantee the development of a suitable molecular model. The effort must be multidimensional, taking into account all the properties and characteristics of the material. Extreme caution is advised when average structures are used, as averages can be meaningless and, worse case, misleading. Indeed, even when a multidimensional approach is employed for the derivation of a model there is no guarantee that the model is a true reflection of the molecular types in the asphaltene fraction. It must also be recognized that any model developed may not be the only, or final, answer to the problem. As techniques improve and knowledge evolves, so must the model.

The constituents of the asphaltene fraction are macromolecules insofar as the molecular weights are generally accepted to be at least several hundred and more, often approaching 2000. In addition, asphaltenes exist as complex mixtures of components that are separated from petroleum by treatment with a low-boiling liquid hydrocarbon (Mitchell and Speight, 1973; Speight et al., 1984).

A number of molecular models have been proposed for asphaltenes and are described in detail elsewhere (Speight, 1994), and repetition is not warranted here. It is sufficient to state that any molecular models derived for asphaltenes must be in keeping with behavioral characteristics. Efforts were made to describe the asphaltene fraction in terms of a single, representative asphaltene molecule or molecules (Speight, 1994) incorporating, in the correct proportions, all of the chemical and elemental constituents known to be present in a given asphaltenic matrix.

Although this approach gives an idea of the structural complexity of the compounds making up the asphaltene component, it obscures the highly differentiated chemical nature of the molecules in this petroleum fraction (Long, 1979, 1981). The idealized molecular structures defined by this method, however, are of some utility for inferring the chemistry and physical-chemical properties of the asphaltenes (Speight, 1992, 1994).

Briefly, there are essentially three types of models:

1. Those that help to visualize the three-dimensional architecture and stereochemistry but are not to scale

2. Framework-type models that indicate correct bond distances and bond angles and can be used to measure distances between nonbonded atoms in molecules but do not show the atoms as such
3. Space-filling models that provide a fairly realistic three-dimensional representation of what the molecule actually looks like

Attempts to model asphaltenes have usually employed pictographic representations of the first kind of models. Minor efforts have been made to represent asphaltenes using models of types 2 and 3.

A serious shortcoming, common to all molecular models, is that they have fixed bond angles and that rotation about single bonds is excessively facile, especially in the first two types of models. In contrast, in the real situation, there are relatively easily definable bond angles and substantial barriers to rotation about single bonds. Moreover, if it is desired to measure intramolecular and intermolecular bond distances, the model must first be fixed in its actual conformation. For molecules having a number of single bonds, this may be quite inconvenient (even though mechanical devices to stop bond rotation are available) and it may be difficult to set the actual torsion angles with any kind of precision.

As a result of all these difficulties, modeling of the macromolecules in the asphaltene fraction has proved to be speculative, although some success has been achieved in attempting to understand the nature of the constituents and their behavior in thermal processes (Speight, 1994). Success notwithstanding, the chemical dynamics are still speculative because of the speculative nature of the model.

The properties of macromolecules depend on both the chemical structure and physical structure. The term **primary structure** describes the chemical sequence of atoms in a macromolecule. The ordering of the atoms in space relative to each other is referred to as the **secondary structure**, and the three-dimensional structure of a molecule is called the **tertiary structure**.

By analogy, the primary structure of asphaltenes is the two-dimensional structure derived by a variety of analytical techniques and is often presented on paper as an average structure. The secondary and tertiary structure, and perhaps an often ignored but extremely important aspect of asphaltene chemistry and physics, is the micelle structure, which represents the means by which asphaltenes exist in crude oil.

A model that may be more in keeping with the amphoteric constituents of the asphaltenes would, on the basis of the preceding concept, have smaller polynuclear aromatic systems. Such a model would not only have to be compatible with the other constituents of petroleum but also be able to represent the thermal chemistry and other process operations. It must also be recognized that such a model can have the large size dimensions that have been proposed for asphaltenes. It can also be a molecular chameleon insofar as it can vary in these dimensions and shape depending upon the freedom of rotation about one or more of the bonds and whether or not it has been isolated from the crude oil. It is not possible to compose a model that represents all the properties of asphaltenes. The molecular size, as deduced from molecular weight measurements, prohibits this. It will be necessary to compose models that are representative of the various constituent fractions of an asphaltene. Then, perhaps, we will truly have representative models.

REFERENCES

Acevedo, S., Escobar, G., Gutierrez, L.B., and D'Aquino, J. 1992. *Fuel* 71: 1077.
Acevedo, S., Ranaudo, M.A., Escobar, G., Gutierrez, L.B., and Gutierrez, X. 1995. *Asphaltenes: Fundamentals and Applications*. Sheu, E.Y. and Mullins, O.C. eds. Plenum Press, New York. Chapter IV.

Ali, L.H., Al-Ghannan, A., and Speight, J.G. 1985. *Iraqi J. Sci.* 26: 41.

Altgelt, K.H. 1968. Preprints. *Div. Petrol. Chem. Am. Chem. Soc.* 13(3): 37.

Andersen, S.I. 1997. *Petrol. Sci. Technol.* 15: 185.

Andersen, S.I. and Birdi, K.S. 1990. *Fuel Sci. Technol. Int.* 8: 593.

Andersen, S.I. and Speight, J.G. 1992. Preprints. *Div. Fuel Chem. Am. Chem. Soc.* 37(3): 1335.

Andersen, S.I. and Speight, J.G. 2001. *Petrol. Sci. Technol.* 15: 185.

ASTM 2005. *Annual Book of Standards*. American Society for Testing and Materials, Philadelphia, PA.

Baker, E.W. 1969. *Organic Geochemistry*. Eglinton, G. and Murphy, M.T.J. eds. Springer-Verlag, New York.

Bestougeff, M.A. and Darmois, R. 1947. *Comptes Rend.* 224: 1365.

Bestougeff, M.A. and Darmois, R. 1948. *Comptes Rend.* 227: 129.

Bestougeff, M.A. and Mouton, Y. 1977. *Bull. Liason Lab. Pont. Chaus.* Special Volume. p. 79.

Bjorseth, A. 1983. *Handbook of Polycyclic Aromatic Hydrocarbons*. Marcel Dekker, New York.

Branthaver, J.F. 1990. *Fuel Science and Technology Handbook*. J.G. Speight (ed.). Marcel Dekker, New York.

Cimino, R., Correrra, S., Del Bianco, A., and Lockhart, T.P. 1995. *Asphaltenes: Fundamentals and Applications*. Sheu, E.Y. and Mullins, O.C. eds. Plenum Press, New York. Chapter III.

Clerc, R.J. and O'Neal, M.J. 1961. *Anal. Chem.* 33: 380.

Drushel, H.V. 1970. Preprints *Div. Petrol. Chem. Am. Chem. Soc.* 15: C13.

Farcasiu, M., Forbus, T.R., and LaPierre, R.B. 1983. Preprints. *Div. Petrol. Chem. Am. Chem. Soc.* 28: 279.

Francisco, M.A. and Speight, J.G. 1984. Preprints. *Div. Fuel. Chem. Am. Chem. Soc.* 29(1): 36.

Gallegos, E.J.J. 1981. *Chromatogr. Sci.* 19: 177.

Girdler, R.B. 1965. *Proc. Assoc. Asphalt Paving Technol.* 34: 45.

Grader, R. 1942. *Oel Kohle* 38: 867.

Hillman, E. and Barnett, B. 1937. *Proc. 4th Ann. Meet. ASTM* 37(2): 558.

Hobson, G.D. and Pohl, W. 1973. *Modern Petroleum Technology*. Applied Science Publishers, London.

Ignasiak, T. and Strausz, O.P. 1978. *Fuel* 57: 617.

Ignasiak, T., Kemp-Jones, A.V., and Strausz, O.P. 1977. *J. Org. Chem.* 42: 312.

Jacobson, J.M. and Gray, M.R. 1987. *Fuel* 66: 749.

Katz, M. 1934. *Can. J. Res.* 10: 435.

Keleman, S.R., George, G.N., and Gorbaty, M.L. 1990. *Fuel* 69: 939.

Kendall, J. and Monroe, K.P. 1917. *J. Am. Chem. Soc.* 39: 1787.

Kirby, W. 1943. *Chem. Ind.* 62: 58.

Koots, J.A. and Speight, J.G. 1975. *Fuel* 54: 179.

Labout, J.W.A. 1950. *Properties of Asphaltic Bitumen*. Pfeiffer, J.P. ed. Elsevier, Amsterdam.

Lee, M.L., Novotny, M.S., and Bartle, K.D 1981. Analytical *Chemistry of Polycyclic Aromatic Compounds*. Academic Press Inc., New York.

Lehninger, A.L. 1970. *Biochemistry*. Worth Publishers, New York.

Lerer, M. 1934. *Ann. Comb. Liq.* 9: 511.

Long, R.B. 1979. Preprints. *Div. Petrol. Chem. Am. Chem. Soc.* 24(4): 891.

Long, R.B. 1981. *The Chemistry of Asphaltenes*. Bunger, J.W. and Li, N., eds. Advances in Chemistry Series No. 195. American Chemical Society, Washington, DC.

Long, R.B. and Speight, J.G. 1989. *Rev. Inst. Franç. Petrol.* 44: 205.

Mack, C. 1932. *Phys. Chem.* 36: 2901.

Markhasin, I.L., Svirskaya, O.D., and Strads, L.N. 1969. *Kolloid Zeit.* 31: 299.

Mitchell, D.L. and Speight, J.G. 1973. *Fuel* 52: 149.

Mitra-Kirtley, S., Mullins, O.C., van Elp, J., George, S.J., Chen, J., and Cramer, S.P. 1993. *J. Am. Chem. Soc.* 115: 252.

Moschopedis, S.E. and Speight, J.G. 1971. *Fuel* 50: 60.

Moschopedis, S.E. and Speight, J.G. 1976a. *Fuel* 55: 187.

Moschopedis, S.E. and Speight, J.G. 1976b. *Fuel* 55: 334.

Moschopedis, S.E. and Speight, J.G. 1976c. *Proc. Assoc. Asphalt Paving Technol.* 45: 78.

Moschopedis, S.E. and Speight, J.G. 1978. *Fuel* 57: 235.

Moschopedis, S.E. and Speight, J.G. 1979. Preprints. *Div. Petrol. Chem. Am. Chem. Soc.* 24(4): 1007.

Moschopedis, S.E., Fryer, J.F., and Speight, J.G. 1976. *Fuel* 55: 227.

Mullins, O.C. 1995. *Asphaltenes: Fundamentals and Applications.* Sheu, E.Y. and Mullins, O.C. eds. Plenum Press, New York. Chapter II.

Nicksic, S.W. and Jeffries-Harris, M.J. 1968. *J. Inst. Petrol.* 54: 107.

Orr, W.L. 1977. Preprints. *Div. Fuel Chem. Am. Chem. Soc.* 22(3): 86.

Orr, W.L. 1985. *Adv. Org. Geochem.* 10: 499.

Overfield, R.E., Sheu, E.Y., Sinha, S.K., and Liang, K.S. 1989. *Fuel Sci. Technol. Int.* 7: 611.

Petersen, J.C., Barbour, F.A., and Dorrence, S.M. 1974. *Proc. Assoc. Asphalt Paving Technol.* 43:162.

Pfeiffer, J.P. and Saal, R.N. 1940. *Phys. Chem.* 44: 139.

Ravey, J.C., Decouret, G., and Espinat, D. 1988. *Fuel* 67: 1560.

Ray, B.R., Witherspoon, P.A., and Grim, R.E. 1957. *Phys. Chem.* 61: 1296.

Reerink, H. 1973. *Ind. Eng. Chem. Prod. Res. Dev.* 12: 82.

Reynolds, J.G. 1997. *Petroleum Chemistry and Refining.* Taylor & Francis Publishers, Washington, DC. Chapter 3.

Ritchie, R.G.S., Roche, R.S., and Steedman, W. 1979. *Fuel* 58: 523.

Rose, K.D. and Francisco, M.A. 1987. *Energy Fuels* 1: 233.

Rose, K.D. and Francisco, M.A. 1988. *J. Am. Chem. Soc.* 110: 637.

Sakhanov, A. and Vassiliev, N. 1927. *Petrol. Zeit.* 23: 1618.

Schmitter, J.M., Garrigues, P., Ignatiadis, I., De Vazelhes, R., Perin, F., Ewald, M., and Arpino, P. 1984. *Org. Geochem.* 6: 579.

Schucker, R.C. and Keweshan, C.F. 1980. Preprints. *Div. Fuel Chem. Am. Chem. Soc.* 25: 155.

Sheu, E.Y., DeTar, M.M., Storm, D.A., and DeCanio, S.J. 1992. *Fuel* 71: 299.

Speight, J.G. 1970a. *Fuel* 49: 76.

Speight, J.G. 1970b. *Fuel* 49: 134.

Speight, J.G. 1971. *Fuel* 50: 102.

Speight, J.G. 1972. *Appl. Spectrosc. Rev.* 5: 211.

Speight, J.G. 1979. *Studies on Bitumen Fractionation.* Information Series No. 84. Alberta Research Council, Edmonton, Alberta, Canada.

Speight, J.G. 1981a. Paper presented at the *Division of Geochemistry.* American Chemical Society New York, August 23–28.

Speight, J.G. 1981b. Preprints *Div. Petrol. Chem. Am. Chem. Soc.* 26: 825.

Speight, J.G. 1984a. *Characterization of Heavy Crude Oils and Petroleum Residues.* Tissot, B. ed. Editions Technip, Paris.

Speight, J.G. 1984b. *Neftekhimiya* 24: 147.

Speight, J.G. 1986a. *Ann. Rev. Energy* 11: 253.

Speight, J.G. 1986b. Preprints. *Div. Petrol. Chem. Am. Chem. Soc.* 31(4): 818.

Speight, J.G. 1992. *Proceedings.* 4th International Conference on the Stability and Handling of Liquid Fuels. U.S. Department of Energy (DOE/CONF-911102). p. 169.

Speight, J.G. 1994. *Asphaltenes and Asphalts. I. Developments in Petroleum Science, 40.* Yen, T.F. and Chilingarian, G.V. eds. Elsevier, Amsterdam. Chapter 2.

Speight, J.G. 1995. Asphalt. *Encyclopedia of Energy Technology and the Environment.* p. 321.

Speight, J.G. 1999. Asphalt. *Kirk–Othmer Encyclopaedia of Chemical Technology,* 5th Edition. p. 192.

Speight, J.G. 2001. *The Desulfurization of Heavy Oils and Residua,* 2nd Edition. Marcel Dekker, New York.

Speight, J.G. and Long, R.B. 1981. *Atomic and Nuclear Methods in Fossil Energy Research.* Filby, R.H. ed. Plenum Press, New York. p. 295.

Speight, J.G. and Moschopedis, S.E. 1977. *Fuel* 56: 344.

Speight, J.G. and Moschopedis, S.E. 1980. *Fuel* 59: 440.

Speight, J.G. and Moschopedis, S.E. 1981a. *Chemistry of Asphaltenes.* Bunger, J.W. and Li, N.C. eds. Advances in Chemistry Series No. 195. American Chemical Society, Washington, DC.

Speight, J.G. and Moschopedis, S.E. 1981b. Preprints. *Div. Petrol. Chem. Am. Chem. Soc.* 26(4): 907.

Speight, J.G. and Pancirov, R.J. 1984. *Liq. Fuels Technol.* 2: 287.

Speight, J.G., Long, R.B., and Trowbridge, T.D. 1982. Preprints. *Div. Fuel Chem. Am. Chem. Soc.* 27(3/4): 268.

Speight, J.G., Long, R.B., and Trowbridge, T.D. 1984. *Fuel* 63: 616.

Speight, J.G., Wernick, D.L., Gould, K.A., Overfield, R.E., Rao, B.M.L., and Savage, D.W. 1985. *Rev. Inst. Franç. Petrol.* 40: 51.

Taft, R.W., Berthelot, M., Laurence, C., and Leo, A.J. 1996. *Chemtech* (July): 20.

Thiyagarajan, P., Hunt, J.E., Winansa, R.E., Anderson, K.B., and Miller, J.T. 1995. *Energy Fuels* 9: 629.

Tissot, B. 1984. *Characterization of Heavy Crude Oils and Petroleum Residues*. Editions Technip, Paris.

Tissot, B.P. and Welte, D.H. 1978. *Petroleum Formation and Occurrence*. Springer-Verlag, New York.

Wales, M. and van der Waarden, M. 1964. Preprints. *Div. Petrol. Chem. Am. Chem. Soc.* 9(2): B-21.

Weiss, V. and Edwards, J.M. 1980. *The Biosynthesis of Aromatic Compounds*. John Wiley & Sons Inc., New York.

Wiehe, I.A. 1993. *Ind. Eng. Chem. Res.* 32: 2447.

Wiehe, I.A. 1994. *Energy Fuels* 8: 536.

Wiehe, I.A. 1995. *In. Eng. Chem. Res.* 34: 661.

Winniford, R.S. 1963. *Inst. Petrol, Rev.* 49: 215.

Yan, J., Plancher, H., and Morrow, N.R. 1997. *Paper No. SPE 37232*. SPE International Symposium on Oilfield Chemistry. Houston, TX.

Yen, T.F. 1970. *Fuel* 49: 134.

Yen, T.F. 1971. Preprints. *Div. Fuel Chem. Am. Chem. Soc.* 15(1): 57.

Yen, T.F. 1972. Preprints. *Div. Petroleum Chem. Am. Chem. Soc.* 17(4): F102.

Yen, T.F. 1974. *Energy Sources* 1: 447.

Yen, T.F. 1975. *The Role of Trace Metals in Petroleum*. Ann Arbor Science Publishers Inc., Ann Arbor, MI.

Yen, T.F. 1984. *The Future of Heavy Crude Oil and Tar Sands*. Meyer, R.F., Wynn, J.C., Olson, J.C. eds. McGraw-Hill, New York. p. 412.

Yen, T.F., Erdman, J.C., and Pollack, S.S. 1961. *Anal. Chem.* 33: 1587.

12 Structure of Petroleum

12.1 INTRODUCTION

The structure of petroleum has been a topic of some interest for at least five decades, ever since the first theory of petroleum structure was published (Pfeiffer and Saal, 1940). And, the structure of petroleum has become a much more important issue with the increased use of the heavier crude oils. Studies of petroleum structure present indications of the interactions and interrelationships of the constituents. In fact, the implications of petroleum structure as it relates to recovery operations and to refinery operations has become of importance.

However, in order to discuss the chemical and physical structure of petroleum it is necessary to give consideration to the chemical and physical nature of the constituents of petroleum. The chemical nature of the various constituents of petroleum has been addressed elsewhere, with some reference to the physical nature of petroleum (Chapter 6 and Chapter 7) and it is not the intent to repeat those discourses here. It is, however, necessary to understand the interrelationships of the various constituents. Further, it is essential to note that the concept of petroleum is not only accepted as a continuum among the lower molecular weight ranges, but also that the continuum is complete and continues into the higher molecular weight ranges. It is for this reason, that the chemical and physical nature of the resin and gas oil fractions should also be given consideration.

The concept that petroleum is a continuum of structural types that may be further defined in terms of polarity and molecular weight is not new and has been generally accepted. The initial focus in this concept was on the hydrocarbon types that occurred in the volatile **saturates** and aromatics fractions, leaving the much less volatile polar aromatic fraction (Figure 12.1), leaving the nature of the nonvolatile polar aromatics fraction subject to structural group analysis (Chapter 10) and, therefore, open to speculation about the true nature of these constituents (Chapter 11). Further work on nonvolatile gas oils has shown that the continuum is ubiquitous, at least sufficient to include such fractions. By inference, and considering the processes by which petroleum is formed, it is to be anticipated that a continuum is the operative factor in petroleum structure.

The concept that asphaltene fraction is located at the high molecular weight end of this continuum has also been postulated and is now becoming accepted (Speight, 1994). Similar reasoning is appropriate to the constituents of the resin fraction. In addition, there is evidence that the continuum of hydrocarbon types, or more correctly hydrocarbon fragments, continues from the gas oil fractions into the resins and asphaltenes.

Data from pyrolysis investigations and from high performance liquid chrometography-ultraviolet (HPLC–UV) investigations show that the condensed aromatic hydrocarbon fragments in the resins and asphaltenes match those in the gas oil fraction. It now appears that the asphaltene constituents are not different types (i.e., polynuclear aromatic sheets containing twelve or more condensed rings) to the remainder of the oil. The asphaltene constituents are compatible with the structural types, with the lower molecular weight fractions and also

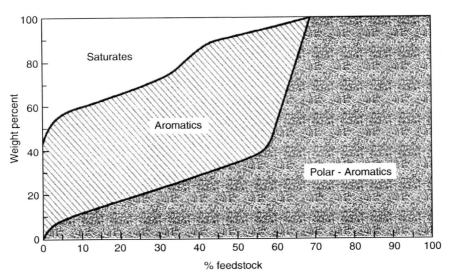

FIGURE 12.1 Representation of the constituents of petroleum; volatility decreases from left to right.

compatible with natural product origins of petroleum (Speight, 1986, 1994). However, in addition to the increasing molecular weight of the resin and asphaltenes constituents, it is the frequency of occurrence of the heteroatom functions that increases. This is the main difference between the constituents of the resin and asphaltenes fractions and the remainder of the constituents in petroleum.

Thus, there is no reason to refute the existence of the continuum from the lower molecular weight species to the highest molecular weight, and highest polarity species. Indeed, there is sufficient evidence to extol and embrace the concept of a complete continuum of structural types throughout the whole range of petroleum constituents, be they hydrocarbon or heteroatom in nature.

The issues relating to the composition and structural types that occur in the gas oil and resin fractions has not been previously addressed, although considerable detail has been given about the constituents of the asphaltene fraction (Chapter 10). Thus, it is appropriate at this point that some recognition be given to the molecular types found in the gas oil and resin fractions.

12.2 MOLECULAR SPECIES IN PETROLEUM

12.2.1 VOLATILE FRACTIONS

Petroleum is not a uniform material insofar as the proportions of the various chemical constituents can vary, not only with the location and age of the oil field but also with the depth of the individual well. However, on a molecular basis, petroleum is a complex mixture of hydrocarbons plus organic compounds of sulfur, oxygen, and nitrogen, as well as compounds containing metallic constituents, particularly vanadium, nickel, iron, and copper. The hydrocarbon content may be as high as 97%, for example in the lighter paraffinic crude oils, or as low as 50% or less as illustrated by the heavier asphaltic crude oils. Nevertheless, crude oils with as little as 50% hydrocarbon components are still assumed to retain most of the essential characteristics of the hydrocarbons.

12.2.2 RESIN CONSTITUENTS

The separation of the resin constituents from the **deasphaltened** oil is a different problem, and it is in this part of the separation scheme that opinions regarding isolation of the resin constituents from deasphaltened oil most frequently differ.

It must always be remembered that the definition of the nonvolatile constituents of petroleum (i.e., the asphaltene constituents, the resin constituents, and, to some extent, part of the oils fraction insofar as nonvolatile oils occur in residua and other heavy feedstocks) is an operational aid. It is difficult to base such separations on chemical or structural features (Girdler, 1965; Speight et al., 1982, 1984; Speight, 1994). This is particularly true for the asphaltene fraction, for which the separation procedure not only dictates the yield but can also dictate the quality of the fraction insofar as the yield and character of this is dependent upon the hydrocarbon used for the separation (Girdler, 1965; Mitchell and Speight, 1973). The technique employed also dictates whether the asphaltene contains coprecipitated resins (Ali et al., 1985; Speight, 1994). This is based on the general definition that asphaltenes are insoluble in *n*-pentane (or in *n*-heptane) but resins are soluble in *n*-pentane (or in *n*-heptane).

12.2.2.1 Composition

Elemental constitutions of a suite of petroleum resins (Koots and Speight, 1975), isolated by the same procedure and therefore comparable, show that the proportions of carbon and hydrogen, like those of the asphaltene constituents, vary over a narrow range: $85 \pm 3\%$ carbon and $10.5 \pm 1\%$ hydrogen. The proportions of nitrogen ($0.5 \pm 0.15\%$) and oxygen ($1.0 \pm 0.2\%$) also appear to vary over a narrow range, but the amount of sulfur (0.4% to 5.1%) varies over a much wider range.

There are notable increases in the H/C ratios of the resins, relative to those of the asphaltenes (Chapter 11) (Speight, 1994). Presumably, this indicates that aromatization is less advanced in the resins than in the asphaltenes. It may actually be that the asphaltenes are maturation products of the resins, and if this is the case, it indicates that one of the maturation processes involves aromatization of the nonaromatic portion of the resin. There is also a tendency to increase proportions of nitrogen, oxygen, and sulfur in the resins relative to the asphaltenes. There are, however, exceptions to this trend, notably in the proportions of sulfur present in the asphaltenes and resins.

12.2.2.2 Resins (Structure)

Initial postulates of resin structure invoked the concept of long paraffinic chains with naphthenic rings interspersed throughout. Other structures used the idea of condensed aromatic and naphthenic ring systems and allowed the interspersion of heteroatoms throughout the molecule.

Structural investigations of resin fractions have not been carried out to the same extent as they have been on the asphaltene fractions. Nevertheless, the data available from, for the most part, the molecular magnetic resonance method indicate that the resin molecules are of lower molecular weight than the asphaltenes from the same crude oil.

Infrared spectroscopic investigation of the resins also indicates the presence of hydrogen-bonded hydroxyl groups. In addition, at high concentrations (10%), a band is evident at 3490 cm^{-1} that is assigned to N–H functions in pyrroles or indoles. Acetylation of the resins, combined with detailed infrared spectroscopic examination, has also established the presence of ester functions and acid functions as well as carbonyl (ketone or quinone) functions. As with the asphaltenes, the presence of ether or sulfur–oxygen functions cannot be discounted on the basis of the available evidence.

There is also evidence that the structural aspects of the constituents of the resin fraction may differ very little from those of the corresponding asphaltene fraction, the main difference being the proportion of aromatic carbon within each fraction (Koots and Speight, 1975; Selucky et al., 1981). It has been postulated that resins and asphaltenes are small fragments of kerogen (Tissot, 1984) or at least have the same origins as the kerogen (Chapter 4) and, therefore, a relationship might be anticipated.

Identification of the constituents of complex mixtures (such as petroleum residua) by molecular type may be achieved by a variety of methods (Chapter 9). These methods can be generally classified into three groups:

1. Spectroscopic techniques
2. Physical property methods
3. Chemical techniques

Using these methods, various structural parameters are derived from a particular property by a sequence of mathematical manipulations. It is difficult to completely separate these three methods of structural elucidation and there must, by virtue of need and relationship, be some overlap.

The end results of the application of these methods are indications of the structural types present in the material. However, there is the unfortunate tendency of researchers to then attempt to interrelate these structural types into a so-called average structure and there is caution advised in the use of average structures (Chapter 10).

Much work has been carried out on the application of the spectroscopic and physical property methods to the deduction of the structural types in the high boiling fractions of petroleum. Application of chemical reactivity to the elucidation and description of the structural types in petroleum has received less attention. And yet, such reactivity cannot be ignored. It is for this reason that the reactivity of the resin and nonvolatile oils fractions is considered here.

The chemical reactions of the resins have received much less attention than the reactions of the asphaltenes, and usually only fragmentary reports are available. Like the chemical reactions of the asphaltene constituents, the chemical reactions of the resin constituents give indications of the structural types within these constituents. The reactions also indicate the effects that change in the hydrocarbon structure or functionality will have on the properties of the resin constituents.

Although much work has been reported on the thermal decomposition of asphaltenes, a lesser amount of work is reported for the resins. The data that are available indicate that, like the asphaltenes, the resins can be decomposed thermally to produce, on the one hand, a hydrocarbon-type of distillate and, on the other hand, coke-like material. The prevalent conditions determine the relative proportions of the two products.

One area of resin chemistry that has received attention is the interaction with oxygen. Thus, the **oxidation** of resins in benzene solution with air in the presence or absence of various metal salts proceeds readily to yield asphaltene products (Moschopedis and Speight, 1978b). Substantial uptake of oxygen occurs, and from the atomic heteroatom–carbon ratios in the starting material and products, it appears that preferential reaction of the more polar entities occurs. Resins also undergo condensation with formaldehyde, which is especially rapid after introduction of oxygen functions by oxidation. Resins can also be sulfonated to yield water-soluble or oil-soluble sulfonates. Resins react with nitric acid to yield complex mixtures of oxidation and nitration products. Reactions with sulfur cause dehydrogenation as well as the formation of complex sulfides. The overall result is the production of higher molecular weight material with low hydrogen content.

Resins also react with diazo compounds, which not only produce asphaltene-type products but also can be used to introduce various functional groups into the molecule. Resins react with acetic anhydride to afford a variety of acetylated products.

12.2.2.3 Molecular Weight

The molecular weights of resin fractions are substantially lower than those of asphaltenes and do not usually vary, except for the limits of experimental error, with the nature of the solvent or the temperature of the determination. It may therefore be presumed that the molecular weights of resins, as determined by various methods, are in fact true molecular weights and that forces that result in intermolecular association contribute very little, if anything, to their magnitude.

12.2.3 NONVOLATILE OILS

12.2.3.1 Composition

Elemental compositions of the oil fractions are markedly different from those of the asphaltenes and resins. The hydrogen-to-carbon ratio of the oils fraction is much higher than those of the asphaltenes and of the resins. This indicates the presence of more paraffinic (or naphthenic) constituents in the oil fraction. The heteroatoms, with the exception of sulfur, are noticeably absent or are present at the most in very small amounts.

12.2.3.2 Structure

Structurally, the oils may be composed of a single naphthene ring containing one or more side chains of varying lengths. The greater the number of these aliphatic side chains, the nearer is the molecule to a true paraffin hydrocarbon, and the longer the side chains, the higher is the solidification temperature of the hydrocarbons. In addition to the naphthene rings, alkylated aromatic hydrocarbons are also believed to be present. However, it is generally believed that this picture of oil structure is oversimplified and that the higher boiling, or nonvolatile oil fractions contain more complex structures of the type illustrated for the resins, in which paraffinic chains and saturated ring systems coexist in a variety of ways. It is also considered likely that the saturated ring structures can be replaced by aromatic ring structures in the resin formulas.

Of the three fractions that constitute the asphaltic portion of petroleum, the oils have received least attention. Nevertheless, some aspects of their chemistry have been studied and are outlined here.

This fraction is thermally stable at ordinary temperatures and is quite resistant to attack by many chemicals. At elevated temperatures, fission of side chains takes place. In the presence of oxygen (or air), carbon dioxide, water, and products containing carbonyl $>C=O$ and hydroxyl $-O-H$ groups are found. If the temperature is sufficiently high, the products containing oxygen functions are unstable and dehydrogenation is the overall result.

Just as the asphaltenes and resins react with oxygen at low temperatures ($<100°C$, $<212°F$), so do the oils. Thus, oxidation in benzene solution by air, in either the presence or absence of metal salts, proceeds smoothly to yield asphaltene and resin products (Moschopedis and Speight, 1978b). In general, majority of the oil recovered after oxidation has lower atomic N/C and S/C ratios than the untreated oil; this indicates that it is the more polar components of the oil fraction that are susceptible to oxidation. It is also evident that the presence of catalytic amounts of metal salts causes acceleration in the oxidation rate of the oil fraction.

Halogens react with the oils relatively slowly, especially if there is no olefinic ($>C=C<$) or aromatic unsaturation; essentially substitution reactions occur. At elevated temperatures, dehydrohalogenation occurs, giving rise to resin and asphaltene products.

Dilute acids have little, if any, effect on the oils; concentrated nitric acid brings about oxidation; concentrated sulfuric acid or oleum ($H_2SO_4 + SO_3$) produces predominantly oil-soluble sulfonates.

12.2.3.3 Molecular Weight

Molecular weights of oil fractions usually vary over the range 250 to 600, and it is only rarely, if ever, that molecular weights in excess of 600 are recorded.

The nature of the solvent has no effect on the molecular weight of the oils; this is also true of the temperature at which the determination is carried out. It is therefore presumed that recorded molecular weights of oils are true molecular weights and are not the result of molecular association caused by intermolecular forces.

12.3 CHEMICAL AND PHYSICAL STRUCTURE OF PETROLEUM

Thus, chemical models for asphaltenes that might be more in keeping with maturation pathways and with behavioral characteristics might well be selected from the two extremes of the molecular weight–polarity diagram. Such models would have smaller polynuclear aromatic systems and also span the range of functional types as well as molecular sizes (Speight, 1994). It must also be recognized that such models can have the large size dimensions that have been proposed for asphaltene constituents but will also be molecular chameleons insofar as they can vary in dimensions depending upon the angle of rotation about an axis and the freedom of rotation about one, or more, of the bonds. The ability of the asphaltene molecules to adapt to a physical environment is very important in recovery operations. Adsorption of the asphaltene on to the reservoir rock in parallel to the plane of the rock surface (with two or more points of contact between the asphaltene and the rock) or at a right-angle to the rock surface (Speight, 1994) with the remainder of the molecule suspended in the oil phase can make a difference to the behavioural characteristics of the oil–rock system.

An early hypothesis of the physical structure of petroleum (Pfeiffer and Saal, 1940) suggested that asphaltenes are the centers of micelles formed by adsorption, or even by absorption of part of the maltene fraction, i.e., resin material, on to the surfaces or into the interiors of the asphaltene particles. Thus, most of those substances with greater molecular weight and with the most pronounced aromatic nature are situated closest to the nucleus and are surrounded by lighter constituents of lesser aromatic nature. The transition of the intermicellar (dispersed or oil) phase is gradual and almost continuous.

And, there is no reason to reject this hypothesis. It has survived the test of time and the current discussion focuses on the actual mechanics of micelle structure.

It is also known that, in the absence of the resins, the asphaltenes are incompatible with the remainder of the crude oil (Swanson, 1942; Witherspoon and Munir, 1960; Koots and Speight, 1975). This indicates that the resin fraction is, under normal operating conditions, essential for the dispersability of the asphaltene constituents and represents one of several factors that need to be taken into account when attempting to resolve the issues of **incompatibility** and **sediment** formation (Mushrush and Speight, 1995).

Continued attention to this aspect of asphaltene chemistry has led to the postulate that asphaltene–asphaltene clusters form the center of the micelle, but other options such as a single asphaltene–resin cluster at the center of the micelle has also been proposed.

The relationships of petroleum constituents are an essential factor in determining the structure of petroleum systems. There is a multitude of structural types in petroleum and much effort has been given to the identification of these constituents (Chapter 9 and Chapter 10). However, it is not only the structures of the constituents that determine the structure of petroleum but also the means by which these constituents relate to one another. Remembering that the asphaltene constituents are not soluble or dispersible in the deresined deasphaltened oil, this involves the transition from one phase to another to present a homogenous mixture.

For example, the reactivity of molecular species can be assessed on the basis of bond energies (Chapter 12), but the reactivity of a particular bond is also subject to its environment (Chapter 10) and can change the order of reactivity that has been assessed on the basis of relative bond energy. In the present context, the relationships of petroleum constituents to one another are also subject to the environment. And a major determinant of this relationship is the stereochemistry of the constituents as they relate to one another. It must be appreciated that the stereochemistry of organic compounds is often a major factor in determining reactivity and properties (Eliel and Wilen, 1994).

The evidence available in the literature appears to indicate that the hydrocarbon structures and some features, such as the various condensed ring systems in different crude oils, are similar (from the asphaltenes and resins to the constituents of the oil fraction). A wide variety of source materials are involved in petroleum genesis (Chapter 3). Thus, on a molecular scale, there are substantial structural differences between the constituents of different crude oils as dictated by regional variations in the composition of the first-formed **protopetroleum** and the maturation conditions.

The structural differences in different crude oils are reflected in the difficulty with which resins from one crude oil peptize asphaltenes from a different crude oil and the instability of the blend (Koots and Speight, 1975). For reference, in asphaltene science as in colloid science, the terms *peptized*, *dispersed*, and *solubilized* are often used interchangeably to describe the means by which asphaltenes exist within petroleum.

Some researchers, in addition to deriving broad generalities from the use of a variety of analytical techniques, are attempting to assign specific molecular configurations to the asphaltene constituents. Perhaps it is of little value to petroleum technology and certainly beyond the scope of the available methods to derive such formulas.

However, a variety of structures exist in the asphaltene fraction (in which there is a decided hydrogen deficiency). But the close relationships of the various hydrocarbon series comprising the asphaltene constituents, resin constituents, and oil constituents give rise to overlapping of fractions into neighboring series, both in molecular weight and in the atomic hydrogen-to-carbon atomic ratio.

The asphaltene constituents have been proposed to be the final (excluding **carbenes** and **carboids**, those organic fractions of petroleum that are insoluble in toluene or benzene) maturation products from the original protopetroleum. Oxidation experiments certainly indicate that the reaction sequence is as follows (Moschopedis and Speight, 1976c):

$$Oils \rightarrow Resins \rightarrow Asphaltenes$$

The oxygenated water passing through petroleum contains sediments that can have some effect on the above relationship, but the effect of oxidation during the maturation processes is unknown and, at best, speculative.

Other than high-temperature experiments that are subject to wonder and criticism (Chapter 4), there is no hard evidence that petroleum formed directly from kerogen. It may also be considered just as probable that petroleum formed (Chapter 3) simultaneously with kerogen (Chapter 4), at which time the petroleum moved away and carried with the other kerogen relative, the dispersible asphaltenes.

A relatively high degree of aromaticity is generally prevalent in the asphaltenes and the resins. In some resins, however, the hydrocarbons show an increase in aliphatic material (more side chains, for example) until, with considerable saturation, the oils, which contain numerous alkyl chains of varying length, are reached. The degree of aromaticity is important when the resins are desorbed. A high aromaticity of the maltenes (i.e., that part of petroleum remaining after the asphaltenes have been removed, often referred to as the deasphalted or deasphaltened oil) indicates good solvency for the asphaltenes. In fact, the solvent power of the maltenes is one of the most important factors in determining the physicochemical behavior of the petroleum colloid system (Speight, 1994; Sheu and Storm, 1995).

The means by which asphaltenes associate in solution and in crude oil have been the subject of many investigations. Much of the work related to asphaltene association has arisen from molecular weight measurements from which deductions have been made about the asphaltene micelle, but there is still considerable conjecture about the true nature of the micelle. Specifically, there is some debate about whether the micelle in petroleum is composed of homogeneous material, insofar as it is composed only of asphaltene molecules, or is composed of asphaltene and resin molecules (Pfeiffer and Saal, 1940; Dickie and Yen, 1967; Koots and Speight, 1975; Speight, 1994).

The pronounced tendency of asphaltenes to form aggregates in hydrocarbon solution is one of their most characteristic traits. Thus, although a number of experimental methods indicate that isolated asphaltene monomers have an average molecular weight on the order of 1500 to 2500 (vapor pressure osmometry in polar solvents), numerous methods show that asphaltenes associate spontaneously in most hydrocarbon media, resulting in observed molecular weights of 10,000 or more. Thus, it is not surprising that this tendency for asphaltenes to aggregate in hydrocarbon solvents has been applied to the existence of asphaltenes in crude oil. The debate is about whether this line of thinking is correct.

However, dilute solutions (0.01% to 0.5% by weight) of an asphaltene fraction in a nonpolar organic solvent (e.g., carbon tetrachloride, CCl_4) exhibit the free hydroxyl absorption band (about 3585 cm^{-1}) in the infrared. At higher concentrations (>1.0% wt/wt), this band becomes less distinguishable, with concurrent onset of the hydrogen-bonded hydroxyl absorption (3200–3450 cm^{-1}). On addition of a dilute solution (0.1% to 1% wt/wt) of the corresponding resins to the asphaltene solutions, the free hydroxyl absorption is reduced markedly or disappears, which indicates the occurrence of preferential intermolecular hydrogen bonding between the asphaltenes and resins.

Further, molecular weight data show that apparent molecular weights of asphaltenes determined in benzene or carbon tetrachloride are reduced from the range 4800 ± 200 to the range 1800 ± 100 when phenol is added to the solutions before molecular weight determination. The reduced value (1800 ± 100) is near that calculated on the basis of the single theoretical asphaltene structure and is significantly higher than the calculated molecular weight (about 900) of the phenol–asphaltene mixture. Addition of measured quantities of phenol to suspensions of asphaltenes in either a petroleum naphtha fraction or in the oil fraction from the bitumen causes solubilization of part of the asphaltenes.

The means by which asphaltenes and resins interact to exist in petroleum remains the subject of speculation, but hydrogen bonding (Moschopedis and Speight 1976a; Acevedo et al., 1985) and the formation of charge-transfer complexes (Yen, 1974) have been cited as the causative mechanisms. Indeed, the influence of hydrogen bonding interactions in determining molecular structure is established (Taft et al., 1996) and to ignore the potential for hydrogen bonding interactions here would be a gross injustice. Finally, there is also evidence that asphaltenes participate in charge-transfer complexes (Penzes and Speight, 1974; Speight and Penzes, 1978), but the exact chemical or physical manner in which they would form in petroleum is still open to discussion. Therefore, it might be more reasonable to consider that the asphaltene constituents exits as aggregations only when separated and flocculated and that in the crude oil the micelle does not consist of aggregates of asphaltene molecules (Pelet et al., 1985; Speight, 1994).

The original concept of the asphaltene–resin micelle invoked the concept of asphaltene–asphaltene association to form a graphite-like stack (Chapter 11) (Speight, 1994 and references cited therein), which acted as the micelle core that, in turn, was stabilized by the resins. Indeed, application of various physical methods to the study of asphaltenes aggregation has shown that the physical dimensions and shape of the aggregates are functions of the solvent used and the temperature of the investigation (Ravey et al., 1988; Overfield et al. 1989; Thiyagarajan et al., 1995). In addition, surface tension measurements have been used to study the self-association of asphaltenes in pyridine and nitrobenzene (Sheu et al., 1992). A discontinuous transition in the surface tension as a function of asphaltene concentration was interpreted as the critical asphaltene concentration, above which self-association occurs.

When resin constituents and asphaltene constituents are present together in solution, hydrogen bonding studies show resin–asphaltene interactions are preferred over asphaltene–asphaltene interactions (Moschopedis and Speight, 1976a). If the same intermolecular forces can, with any degree of justification, be projected to petroleum, asphaltene constituents in petroleum are single entities, are peptized and effectively dispersed, by resin constituents. However, whatever the means by which the individual molecular species are included in the micelle, the structure is recognized as being complex (Bardon et al., 1996).

Therefore, petroleum is a complex system with each fraction dependent upon other systems for complete mobility and solubility (Koots and Speight, 1975). It is presumed that the resins associate with the asphaltenes in the manner of an electron donor–acceptor. In addition, there could well be several points of structural similarity between the asphaltenes and resins that would have an adverse effect on the ability of the resins to associate with asphaltenes from a different crude oil (Koots and Speight, 1975; Moschopedis and Speight, 1976a).

It therefore appears that, when resins and asphaltenes are present together, hydrogen bonding may be one of the mechanisms by which resin–asphaltene interactions are achieved, and resin–asphaltene interactions are preferred over asphaltene–asphaltene interactions. It has also been reported that hydrogenation and concurrent desulfurization cause a decrease in asphaltene molecular weights, and the reduction in molecular weight is caused by a decrease in hydrogen bonding interactions as a direct result of loss of the heteroelement.

If the composition and properties of the precipitated asphaltenes reflect those of the micelles in solution, the latter should be considered as mixed micelles. Asphaltene constituents vary in character (Chapter 10). Resins are also complex mixtures of constituents. Therefore, homogeneous micelles cannot be predicted. This is a reasonable assumption in view of the large quantities of soluble resins found in the precipitated solid. It should be noted that if mole fractions are used as a composition measure, the one for resins in the precipitates would be greater than that for asphaltenes.

Considering the micelles to be composed of asphaltene species that are peptized by resins and asphaltenes is a more reasonable approach. An important corollary of petroleum composition is that the mole fraction of resins is always larger than that of asphaltenes and hence the micelles are expected to be richer in resins. The micelle center would be formed from polar asphaltene molecules. These molecules would be surrounded by other more soluble asphaltene molecules, which would be placed between the center and the periphery. The inclusion of other asphaltene species is not seen as a graphite-type stack, as has been proposed previously. It is more likely an association of convenience that facilitates the association of the central asphaltene with resin species. In many cases, more than one asphaltene molecule per micelle is considered unlikely.

However, in such a model, the substances with higher molecular weights and with the most pronounced aromatic nature are situated closest to the nucleus and are surrounded by lighter constituents of less aromatic nature. The transition of the intermicellar (dispersed or oil) phase is gradual and almost continuous. As asphaltenes are incompatible with the oil fraction

(Swanson, 1942; Koots and Speight, 1975), asphaltene dispersion is attributable mainly to the resins (polar aromatics).

Empirical observations indicate that the resins play an important role in stabilizing asphaltenes in crude oil (Koots and Speight, 1975 and references cited therein). Under unfavorable solvent conditions, the asphaltene species are prone to further aggregation into clusters that are unstable and precipitate from the crude oil (Mitchell and Speight, 1973; Koots and Speight, 1975; Yen, 1990).

Asphaltene constituents ↔ Micelle ↔ Micelle clusters/Flocs

This model requires that the asphaltene micelles are composed of an insoluble molecular core that associates with the resins, thereby providing steric stabilization against flocculation and precipitation. Thus, the phase separation of asphaltenes upon the addition of nonpolar solvents to the crude oil can be rationalized in terms of reduction of the **solubility parameter** or polarity of the hydrocarbon medium (Speight, 1992, 1994; Wiehe 1992, 1993). Such a change results in solubilization of the resins and leads to dissociation of the resin–asphaltene complexes with the result that there is a destabilization of the unattended asphaltene constituents, followed by flocculation or precipitation or phase separation.

The vast differences in molecular structure among individual asphaltene molecules prevent these agglomerates from forming a discotic phase, which requires the monomers to have identical molecular structure. Instead, the asphaltenes randomly stack to form globular micelles as proposed in this model. It was also proposed that the elementary micelle units can further cluster into larger particles similar to flocs.

Acceptance of these models and the concept of a micelle is easier to understand if the constituents of the resin and asphaltene fractions are viewed as two adjacent portions of a single broad compositional continuum that contains the polar, aromatic, and higher molecular weight components of petroleum. This allows the further acceptance of structural similarities in the asphaltenes and resins that facilitate formation of the micelles (Speight, 1992, 1994).

Thus, the means by which asphaltenes exist in crude oil is assumed to involve asphaltene–resin interactions leading to the formation of micelles (Koots and Speight, 1975), but there are still many unknowns related to the existence of asphaltenes in crude oil.

On this basis, the stability of petroleum can be represented by a three-phase system in which the asphaltene constituents, the aromatic constituents (including the resin constituents), and the saturate constituents are delicately balanced (Speight, 1992, 1994). Various factors, such as oxidation, can have an adverse effect on the system, leading to instability or incompatibility as a result of changing the polarity, and bonding arrangements, of the species in crude oil.

An alternative explanation for the stabilizing influence of the resin constituents and for the stability of asphaltene constituents in hydrocarbon media in general, is that individual asphaltene species and micelles are in thermodynamic equilibrium (Cimino et al., 1995). Thus, the degree of association is determined by the relative energies of solvation of the monomers and the micelles and the entropy and enthalpy changes associated with the self-association of the asphaltene monomers.

According to this model, the stabilization of the asphaltene micelles by the resin fraction derives from the contribution of the latter to the solvent power (polarity and aromaticity) of the medium. Destabilization of the asphaltenes, leading to phase separation, occurs when the solvating power of the medium toward the asphaltene monomers and micelles is reduced to the point at which they are no longer fully soluble.

The size of the asphaltene agglomerate varies as a function of the temperature, the asphaltene concentration, and the identity of the solvent. This establishes that the self-association of the asphaltenes is reversible and that the molecular size in the solution state results from a true thermodynamic equilibrium between primary particles and reversible aggregates, the equilibrium constant depending on some property of the solvent. But again, the correlation of the behavior of asphaltenes in hydrocarbon solvents to their behavior in crude oil is still extremely speculative.

Perhaps the study of asphaltene constituents that have been removed from petroleum is analogous to studying a hard-boiled egg. The system (i.e., the egg protein and the asphaltene phase relationships) has been denatured by the application of a physical technique.

Perhaps the most appropriate thought when studying asphaltene constituents vis-à-vis the structure of petroleum is to recall two theories that have some standing in the scientific community: (1) the Heisenberg uncertainty principle, viz: once a system is studied it changes by virtue of being studied, and (2) chaos theory, viz: no matter how well you plan, the unexpected is inevitable. Therefore, it must be recognized that separation of the asphaltene constituents from petroleum undoubtedly changes these molecular entities in terms of their intermolecular relationship with the other constituents of petroleum and even with their intramolecular relationships between other asphaltene constituents.

Thus, when we are told that the asphaltene fraction has 20,000 or more constituents and that a researcher has determined the precise molecular weight and the precise behavior of the fraction, a modicum of skepticism must arise.

12.4 STABILITY OR INSTABILITY OF THE CRUDE OIL SYSTEM

Petroleum can be considered to be a delicately balanced system insofar as the different fractions that contain hydrocarbons (saturates and aromatics) as well as heteroatom constituents (Figure 12.1). The heteroatom constituents tend to concentrate in the higher molecular weight fractions (the asphaltenes and resins). The nitrogen, oxygen, and sulfur species that are in near-neutral molecular locales will also occur in the saturate fraction and in the aromatics fraction, remembering that the nomenclature is not necessarily precise and that the composition of each fraction is a function of the separation process.

Instability usually occurs (in heavy oil) with the separation or precipitation of asphaltene constituents. This may be caused by well stimulation, such as acidizing, which involves a drastic shift in local chemical equilibria, pH, and liberation of carbon dioxide. It may also increase the concentration of some ions, such as iron, which will promote the formation of asphaltenic sediment. Similarly, the separation of wax from paraffinic crude oil will also occur under certain conditions (some are defined, some are not) and cause problems during recovery by blocking the channels in the reservoir rock. However, the focus of this section is on the separation of the asphaltene constituents during recovery and refining.

As a recap (please see Chapter 10), the asphaltene fraction is a complex mixture of species of varying molecular weight and polarity (Chapter 11). Thus, carbenes and carboids are lower molecular weight, highly polar species that are predominantly products of thermal processes and might not occur in the typical recovery process. However, the application of thermal techniques, such as fire flooding, to petroleum recovery can produce such species and they will either deposit on the reservoir rock or appear as suspended solids in the oil. In fact, it is recognized that it is the polar species in the crude oil that govern the oil–rock interactions in the reservoir (Bruning, 1991) from which many sediments can arise.

With the exception of the carbenes and the carboids, the fractions are compatible provided there are no significant disturbances or changes made to the system. Such changes are:

1. Alteration of the natural abundance of the different fractions
2. Chemical or physical alteration of the constituents as might occur during recovery, especially changes that might be brought on by thermal processes
3. Alteration of the polar group distribution as might occur during oxidation by exposure to aerial oxygen during the recovery process.

In the reservoir, asphaltene incompatibility can cause blockages of the pores and channels through which the oil must move during recovery operations (Islam, 1994; Park et al., 1994; Leontaritis, 1996).

All of these incidents cause disturbances to the petroleum system. However, when such disturbances occur, it is the higher molecular weight constituents that are most seriously affected, eventually leading to incompatibility (precipitation, sediment formation, **sludge** formation) depending upon the circumstances. Thus, the dispersability of the higher molecular weight constituents becomes an issue that needs attention. And one of the ways by which this issue can be understood is to be aware of the chemical and physical character of the higher molecular weight constituents. By such means, the issue of dispersability, and the attending issue of incompatibility can be understood and even predicted.

The asphaltene fraction is chemically complex, but it can be conveniently represented on the basis of molecular weight and polarity (Chapter 11) (Long, 1979, 1981). For different crude oils, the slope of the line representing the distribution of molecular weight and the variation in polarity will vary (Speight, 1994). Asphaltenes cannot be crystallized in the usual sense of the word. However, asphaltenes can be subfractionated by the use of a variety of techniques. The fractions vary in molecular weight and also in terms of the functional group types and functional group content (Francisco and Speight, 1984). These data lend support to, and reaffirm, the concept that asphaltenes are complex mixtures of molecular sizes and various functional types (Chapter 10).

In addition, any variation of the major parameters (precipitant, precipitant or oil ratio, time, and temperature) can cause substantial variations in the nature and amount of the separated asphaltenic material. It can vary from a dark brown amorphous solid to a black tacky deposit, either of which under the prevalent conditions could be termed an *asphaltene* (Chapter 10).

Petroleum is a balanced system insofar as the components of the system exist in harmonious balance with each other. Changes to the system can result in the formation of an insoluble (solid) phase and therefore the effects that cause the separation of this immiscible phase need to be addressed.

Separate, or insoluble, phases are produced when external effects perturb the system. Such effects might arise during recovery operations, during refining operations, or during storage.

The presence of asphaltenes is known to have a significant effect on the processability of crude oil. Indeed, the presence of asphaltenes is particularly felt when circumstances permit their separation (precipitation, phase separation) from the oil medium leading to the deposition of solids during recovery operations. Phase separation will cause coke formation during processing as well as failure of an asphalt pavement by loss of physical structure of the asphalt-aggregate system (Speight, 1992).

Thus, there is a need to understand the structure and stability of crude oil and one way in which this can be achieved is by an investigation of asphaltene precipitation or flocculation, by titration, using solvent or nonsolvent mixtures. This may give an indication of both asphaltene and maltene (deasphaltened oil) properties (Heithaus, 1962; Bichard, 1969; Fuhr et al., 1991).

At the point of incipient precipitation, i.e., the point at which separation of asphaltenes from a crude oil becomes apparent, the precipitated material is, presumably, a conglomeration of species based on molecular size and polarity of the type that constitute the asphaltenes (Long, 1979, 1981). This phenomenon has, however, not been addressed in any detail and is certainly worthy of investigation in order to increase the understanding of crude oil (asphaltene or maltene) relationships.

Thus, in order to investigate the mechanism of asphaltene precipitation, the change in precipitate with changes in the composition of nonsolvent or solvent have been examined (Andersen and Speight, 1992). Immiscibility is known to occur at solubility differences between solvent and solute of about 3.5 to 4 $MPa^{1/2}$. Hence, the solubility parameter of the precipitated material can be estimated to fall within the range 20.2 to 20.7 $MPa^{1/2}$, which is in accordance with previous work (Hirschberg et al., 1984). Using the data further, it is predictable that there will be no precipitation of material above a flocculation threshold of 0.33 ($\delta_{eff} = 16.2$ $Mpa^{1/2}$ for the total oil).

As a note, and to avoid unnecessary confusion, there is another scale by which the solubility parameter can be measured. The numbers are precisely half of those used above. Either scale is adequate to the task but must be defined to avoid any misinterpretation of the data.

The material precipitated from n-heptane to toluene insoluble constituents may have a δ-distribution between 19 $Mpa^{1/2}$ and 22 $Mpa^{1/2}$. This observation is in disagreement with the data from the titration experiments that give an upper limit of 20.7 $MPa^{1/2}$, and hence were not able to predict the presence of toluene insoluble material. One titration experiment using 2,4-dimethylpyridine showed a flocculation threshold of 0.42 (18.3 $MPa^{1/2}$) indicating an upper limit of 22.3 $Mpa^{1/2}$.

This is to be anticipated as, on the basis of molecular weight data, asphaltene constituents associate less in pyridine (Moschopedis et al., 1976; Speight et al., 1985). Hence, the interactions leading to dispersion of the least soluble part are not present, and the compounds behave as single molecular entities. The latter is possible as the asphaltene concentration of this sample is below the critical micelle concentration (4.32 g/L) at the beginning of the titration (Andersen and Birdi, 1991a,b). Hence, if solubility parameters are determined from titration experiments, the data are relative to the conditions such as solvent system and equilibrium time.

According to the group contribution approach for calculation of the solubility parameter (δ), an increase in molecular weight does not strictly imply an increase in the magnitude of the solubility, as long chain paraffin systems have low δ values (Barton, 1983). Hence, the precipitation may be governed by molecular weight differences between the solvent and the solute.

High-performance liquid chromatography coupled with selective elution chromatography has shown that a high concentration of toluene in the precipitant leads to shorter peak retention time, and changes take place on the low molecular weight side of the eluted profile. The largest change in profiles was seen when going from pure heptane to a mixture containing 10% toluene. The n-heptane asphaltenes had a close to bimodal profile whereas for 10% (or higher amounts) of toluene, the profile is a single tailing peak, indicating that the incremental material, soluble in toluene or heptane but insoluble in heptane, is composed of lower molecular weight species. This suggests that asphaltenes may be composed of different molecular weight types as suggested from fractionation studies (Bestougeff and Darmois, 1947, 1948; Speight, 1971, 1975).

Hydrogen bonding is involved, to some extent, in asphaltene association and dispersion (Moschopedis and Speight, 1976a). Indeed, an examination of the Fourier transform infrared spectra, particularly in the range 3100 to 3700 cm^{-1}, showed that the main change in the nature of the precipitates from that obtained with pure heptane is the occurrence of less hydrogen bonded structures.

In the spectrum of the precipitate from the 40% toluene mixture, distinct, relatively large, bands for free hydroxyl and carboxylic hydroxyl groups are found. In the presence of the coprecipitated resin-type material, these bands are extremely weak, indicating that dispersion of any toluene insoluble material may take place through association with the coprecipitate. A similar concept has been proposed for resin–asphaltene interaction (Koots and Speight, 1975). Coprecipitated material was in all samples of lower molecular weight. Discrepancies between methods (equilibrium precipitation and flocculation titration) employed to determine the incipient precipitation of asphaltenes have been found and may be related to slow kinetics of precipitation (Speight et al., 1982, 1984).

Solubility parameters of asphaltenes can be measured by the titration method but relative to the experimental parameters such as time, solvent and solute (asphaltene or oil) concentration. The recovery of a toluene insoluble fraction from a toluene soluble fraction shows the complexity of the asphaltene aggregation during precipitation. Evidence of rearrangements of molecular interactions is suggested.

The self-association of asphaltenes has led to the concept of the critical micelle concentration, due to the polarity of the asphaltenes. Indeed, polarity is believed to play a role in the behavior of crude oils and also offers a means of predictability of crude oil behavior (Bruning, 1991). In addition, combined with solubility parameter studies, knowledge of the critical micelle concentration may also be an aid to a more precise definition of crude oil behavior during refining operations. It may also be of particular assistance in developing methods of coke mitigation as well as understanding the behavior of asphalt (Speight, 1992).

The critical micelle concentrations (CMC) of asphaltene-type materials have been determined more recently, using a variety of different methods such as surface tension (Rogacheva et al., 1980; Menon and Wasan, 1986) and calorimetric titration (Andersen and Birdi, 1991a,b). The data for the values of the critical micelle concentration for the asphaltenes found by these methods were in the range of 1 to 5 g/L in toluene.

A very recent publication (Sheu et al., 1992) confirmed that the self-association of asphaltenes occurred in organic solvents using surface tension measurements. The data were also used to show that a critical concentration exists in organic solvents above which asphaltene self-association will occur, in a manner similar to surfactant systems. The reported magnitude of the critical micelle concentration and effects of differences in solvent are in agreement with the previous works.

However, it is considered doubtful that a well-defined critical micelle concentration can be determined for the resin (or pentane soluble) fraction of petroleum, especially from surface tension–concentration data. In fact, inflection points noted in the surface tension concentration curves (Sheu et al., 1992) may be subject to question because of the method of analysis. A more correct analysis of the data can be obtained based on the Gibbs excess equation:

$$G_a = (C_a/RT) \times (dr/dC)$$

G_a is the Gibbs surface excess, C_a is the concentration of compound a, and r the surface tension of the solution.

According to this equation, the determination of a critical micelle concentration determination would be better served by an examination of surface tension, r, versus ln C_a (Chattoraj and Birdi, 1984). Thus, for the resins, the Gibbs analysis shows a different perspective and precludes the concept of resin micelles, thereby confirming earlier molecular weight investigations whereby the data showed little or no association of the molecular species in resin (nonasphaltene) species. This different perspective would have little, if any, effect on the data from asphaltene solutions as the surface tension, θ, value is of a constant magnitude above the critical micelle concentration.

Reconsideration of these recent data (Sheu et al. 1992; Andersen and Speight, 1993) using the Gibbs equation (given earlier) failed to provide a distinct inflection point. Moreover, the surface tension–concentration relationships given (Sheu et al., 1992) for the resins may be obtained for various solutes that do not form micelles, i.e., alcohol–water solutions, where the solute has a positive Gibbs surface excess (Shaw, 1980).

Hence, it can be concluded that the nonasphaltene material is not conducive to the initiation of micelle formation, but it may have a positive surface excess due to polarity. Indeed, there is already evidence (Andersen and Birdi, 1991a,b) that the nonasphaltene soluble fraction exerts no significant influence on the critical micelle concentration, but the data do not preclude the participation of the resins in the micelle. They are not proactive in the formation of the micelle. Hence, the resins might be regarded as inert in terms of micellization.

Monolayer compression studies have revealed that the pentane soluble material does not generally exhibit surface activity (or at best low activity) at the air–water interface. On the other hand, pentane- and heptane-asphaltenes show a relatively strong surface activity at the air–water interface, indicating that the nonasphaltene fraction lacks the structural features necessary in micellization (Speight and Moschopedis, 1980; Leblanc and Thyrion, 1989).

Thus, the occurrence of a micelle and a critical micelle concentration in the deasphaltened oil is certainly suspect and such a concept might even be denied. Another of the issues with the present data of Sheu et al., (1992) relates to the particular fractions chosen for their study.

Petroleum is a continuum and each component fraction depends upon the other for the stability of the whole oil. The omission of a critical fraction in these referenced studies (Sheu et al., 1992) may have serious consequences on the data and the magnitude of the conclusions.

For example, questions related to the role played by this fraction in the structure and stability of the micelle are not answered. Further, there is also some doubt about the viability of the data as they relate to petroleum macrostructure that, in turn, is very real in terms of the structure and the behavior of petroleum in refining operations, especially as the critical micelle concentration relates to coke formation.

It therefore appears that another means of data analysis would be preferable in studies of critical micelle concentration. The fractions chosen for such a study must be chosen carefully, and based on the precept that petroleum is a continuum, and conclusions derived from separate fractions must be subject to question.

It is the general consensus that it is the polar, i.e., heteroatom, constituents of petroleum that are responsible for the formation of suspended organic solids during a variety of recovery processes (Islam, 1994; Park et al., 1994). However, an area that remains largely undefined, insofar as the chemistry and physics are still speculative, is the phenomenon of the incompatibility of the crude oil constituents, as might occur during these operations. The formation of a suspended solid phase during recovery (as well as during refining operations) is related to the chemical and physical structure of petroleum; the latter is greatly influenced by the former.

By way of a brief series of definition, the failure of petroleum fractions to mix and the separation of a separate phase is usually referred to as incompatibility. When incompatibility occurs, the constituents (usually the asphaltenes and the resins) that form a separate phase are variously referred to as precipitate, sediment, or sludge depending upon the nature of the material and the causes of the separation.

Asphaltenes and asphaltene-related materials are known to deposit as sediments during recovery operations in the vicinity of production wells during miscible floods, after acid stimulation, or during pressure changes (Burke et al., 1990; Islam, 1994; Park et al., 1994). Many reservoirs produce without any such problems until the oil stability is perturbed during later stages of oil production.

The parameters that govern sediment formation and deposition of asphaltenic materials from petroleum are related to the composition of the crude oil (Chapter 13). It must also be recognized that the material that deposits from the crude oil as a separate phase is more aromatic and richer in heteroatom compounds than the original crude oil. In fact, in some cases, especially when oxidation has occurred, the deposited material is more aromatic and richer in heteroatoms than the asphaltenes.

In terms of the crude oil parameters that influence sediment formation, there has been considerable focus on the asphaltene content as well as the chemistry and physics of the asphaltenes relationship to the remainder of the oil. For example, although the asphaltene content of petroleum oils varies over a wide range (Koots and Speight, 1975), asphaltene content is not the single determining influence on sediment formation. It has been noted that asphaltene/resin ratios are usually below unity in oils that are stable and higher than unity in oils that exhibit ready precipitation of asphaltenic material (Sachanen, 1945).

Even though it is generally understood that asphaltene content of a crude oil increases with decreasing API gravity, asphaltene precipitation has been reported in light oils as well. Such an effect (gas deasphalting) arises from the increased solubility of hydrocarbon gases in the petroleum as reservoir pressure increases during maturation. Incompatibility will also occur when asphaltenes interact with reservoir rock, especially acidic functions of rocks, through the functional groups (e.g., the basic nitrogen species) just as they interact with adsorbents. And, there is the possibility for interaction of the asphaltene with the reservoir rock through the agency of a single functional group in which the remainder of the asphaltene molecule remains in the liquid phase (vertical association relative to the rock surface). On the other hand, the asphaltene constituents can react with the rock at several points of contact (horizontal association relative to the rock surface), thereby enhancing the bonding to the rock and, in some cases, effecting recovery operations to an even greater extent. Both modes of reaction can entrap other species (such as resins and aromatics) within the space between the rock and the asphaltene.

Incompatibility might also play a detrimental role during recovery operations when it occurs as a result of aerial oxidation. In general, it is the more polar species that oxidize first with or without the presence of catalysts leaving an oil that is relatively free of heteroatom species (Moschopedis and Speight, 1978a). Thus, oxidation is a means for the production of highly polar species in crude oil leading to the deposition of polar sediments and is analogous to the deasphalting procedure. Thus, after incorporation of oxygen to an oil-dependent limit, significant changes occur to asphaltenes and resins. These changes are not so much due to oxidative degradation but to the incorporation of oxygen functions that interfere with the natural order of intramolecular and intermolecular structuring, leading to the separation of asphaltenic material (Moschopedis and Speight, 1976c, 1977).

Thus, with alterations in these parameters, the formation of sediment can occur although the nature of the sediment will vary and be dependent upon the process of formation. There are also several process-related destabilizing forces that can cause precipitation of asphaltenic material and each involves disturbance to the equilibrium that exists within petroleum.

For example, carbon dioxide causes the destabilization of the petroleum equilibrium by lowering pH, by changing oil composition, and by creating turbulence. Usually, asphaltene precipitation increases as the volume of carbon dioxide available to the crude oil increases during the later stages of carbon dioxide injection or stimulation. The most noticeable primary locations of asphaltene deposition are the wellbore and the pump regions. In addition, flooding of a rich gas (miscible flooding) destabilizes the asphaltene–crude oil mixture by lowering the solvent power of the solution. The hydrocarbon gases used in such applications effectively cause deasphalting (gas deasphalting, solvent deasphalting) of the crude oil. The negative effect of rich gas is at a maximum near the bubble point; this effect is alleviated after the bubble point is reached. Similarly, organic chemicals such as isopropyl alcohol, methyl alcohol, acetone, and even some glycol, alcohol, or surfactant based solvents, which do not have an aromatic component, may selectively precipitate asphaltenes and resins.

A decrease in the pressure is another important factor that influences the onset of solids deposition from petroleum. In fact, the effect of pressure is particularly noticeable just above the bubble point for crude oils that are rich in light ends. Depending on the location of the pressure decrease, deposition may occur in different parts of the reservoir as well as in the wellbore and in the production stream. Further, a decrease in pressure is usually accompanied by a decrease in temperature that can also cause physicochemical instability leading to the separation of asphaltenic material from the oil. Pressure change alone can also invoke similar asphaltene precipitation.

The asphaltenic material in crude oil is electrically charged through the existence of zwitterions or polarization within the molecular species (Preckshot et al., 1943; Katz and Beu, 1945; Penzes and Speight, 1974; Speight and Penzes, 1978; Fotland and Anfindsen; 1996). Therefore, any process (such as the flow through reservoir channels or through a pipe) that can induce a potential across, or within, the oil will also result in the electro-deposition of asphaltenic material through the disturbance of the stabilizing electrical forces. In addition, neutralization of the molecular charge will also result in the formation of sediment.

The mechanism of asphaltene precipitation is very complex (Speight et al., 1982, 1984; Andersen and Speight, 1993; Buckley, 1996) and controversy as to the nature of asphaltene solutions persists. However, petroleum consists of a mixture of saturates, aromatics, resins, and asphaltenes (Chapter 8) and it is necessary to consider each of the constituents of this system as a continuous or discrete mixture interacting with the other. The omission of any one fraction could lead to errors in the outcome (Andersen and Speight, 1993).

There are two general approaches to the consideration of the nature of asphaltenes in oil (Hirschberg et al., 1984; Mansoori et al, 1987; Park et al., 1994; Islam, 1994). The first approach considers asphaltenes to be dissolved in oil in a true liquid state. In this case, asphaltene precipitation is considered to depend on thermodynamic conditions of temperature, pressure, and composition. This particular approach recognizes asphaltene precipitation as a thermodynamically reversible process. The second approach considers asphaltenes to be solid particles that are suspended colloidally in the crude oil and are stabilized by resin molecules; consequently, the deposition process is considered to be irreversible. If the material precipitated is a reacted derivative of the asphaltenes (i.e., a sediment), this may be true, but if the precipitated material is unreacted asphaltene, then this assumption may need a correction. One question of major interest during recovery operations is the timing of sediment formation or deposition and the amount of the organic material that will separate from the oil under specific conditions.

Two different models have been proposed to explain the behavior of petroleum and the potential for solids deposition during recovery operations (Kawanaka et al., 1989; Islam, 1994; Park et al., 1994). The continuous thermodynamic model utilizes the theory of heterogeneous polymer solutions and is utilized for the predictions of the onset point and amount of organic deposits from petroleum crude. A *steric colloidal model* that is capable of predicting the beginning of organic deposition has also been developed. A combination of these two models results in a fractal aggregation model. These efforts have generally been adequate to predict the asphaltene–oil interaction problems (phase behavior or flocculation), wherever it may occur during oil production and processing.

In the continuous thermodynamic model, the degree of dispersion of the high molecular weight organic constituents in petroleum depends upon the chemical composition of the petroleum. Precipitation of the high molecular weight material can be explained by a change in the molecular equilibria that exist in petroleum through a change in the balance of oil composition. Moreover, the precipitation process is considered to be reversible. Indeed, the reconstitution of petroleum after fractionation has been demonstrated (Koots and Speight, 1975) and lends support to this model.

The ratio of polar to nonpolar molecules and the ratio of high to low molecular weight molecules in a complex mixture such as petroleum are the two factors primarily responsible for maintaining mutual solubility. The stability of the system is altered by the addition of miscible solvents causing the high molecular weight and the polar molecules to separate from the mixture either in the form of another liquid phase, or as a solid precipitate. Hydrogen bonding and the sulfur-containing or nitrogen-containing segments of the separated molecules could start to aggregate and, as a result, produce a solid phase that separates from the oil.

In the steric colloidal model, the high molecular weight materials in petroleum are considered to be solid particles of different sizes suspended colloidally in the oil and stabilized by other petroleum constituents (i.e., resins) adsorbed on their surface.

The original hypothesis of the physical structure of petroleum (Pfeiffer and Saal, 1940) in which resins played a role in the stabilization of the asphaltenes and later confirmation of the role of the resins in petroleum (Swanson, 1942; Witherspoon and Munir, 1960; Koots and Speight, 1975) is supportive of this model.

In the fractal aggregation model, it is assumed that π–π interactions are the principal means by which asphaltenes associate. This assumption may not be completely valid because of the evidence that favors hydrogen bonding between the molecular species and the observation that asphaltene–resin interactions may predominate over asphaltene–asphaltene interactions in petroleum. The concept that asphaltene–asphaltene interactions may be the predominant interactions is true for solutions of asphaltenes. This is reflected in the molecular weight data (Speight et al., 1985). However, there is no guarantee that these interactions are predominant in petroleum (especially with evidence that indicates the high potential for other interactions) (Speight, 1994).

Inasmuch as the high molecular weight constituents of petroleum have a wide range of polarity and molecular weight distribution, such compounds have been considered to act as heterogeneous polydisperse polymers. However, knowing with a new understanding of the nature of asphaltenes (Speight, 1994), there may be some difficulties with this assumption. Nevertheless, in order to predict the phase behavior of these constituents, it has also been assumed that the properties of these constituents depend on their molecular weights. The model also assumes that the asphaltenes are partly dissolved and partly in the colloidal state, thereby accounting for both the solubility and colloidal effect of high molecular weight organic constituents in the lower molecular weight constituents. The proposed models can provide the tool for making satisfactory prediction of the phase behavior of the deposition of high molecular weight materials.

Because the issue of the deposition of asphaltenic materials problem is complex, it is necessary to attempt an understanding of the deposition mechanism before an accurate and representative model can be formulated. Utilization of kinetic theory of fractal aggregation has enabled the development of the fractal aggregation model. This model allows researchers to describe several situations, such as phase behavior of heavy organic deposition, the mechanism of the association of the high molecular weight constituents, the geometrical aspects of aggregates, the size distribution of the sediments, and the solubility of the high molecular weight constituents in the solution under the influence of a miscible solvent.

Another potential model involves use of the solubility parameter of the asphaltenes and the surrounding medium. It is known that the solubility of asphaltenes varies with the solubility parameter of the surrounding liquid medium (Mitchell and Speight, 1973) and calculation or estimation of the solubility parameters of various liquids is known. From these data, it is possible to estimate the point at which asphaltenic material is precipitated when the composition of the oil is changed by the addition of a hydrocarbon liquid (Mitchell and Speight, 1973). In fact, the solubility parameter has been used successfully to determine the character of heavy oils and investigate the separation of asphaltenic material (Wiehe, 1996). The solubility parameter concept also recognized the gradation of polarity of the asphaltenes as selective precipitation occurs during the addition of a nonsolvent when the most polar constituents are precipitated first (Andersen and Speight, 1992).

The models that apply the solubility parameter concept calculate the interaction through an assumption of the total crude oil as formed by asphaltenes and deasphaltened oil; hence, the system is regarded as a two-component system. The changes in phase equilibrium are caused by the changes in the solubility parameter of either of the two pseudo-components, which may happen either by dissolution of gas or alkane in the deasphaltened oil phase or by changes in temperatures. The amount of precipitated asphaltenes is calculated as the differences in asphaltenes present in the oil and the solubility of asphaltenes at the saturation point. In the different models, the change in the composition of the deasphaltened oil is taken as significant for the phase equilibrium, and various methods, i.e., cubic equations of state, are applied to determine the properties of this fraction.

Further development of this concept (Speight, 1994) has led to the graphical representation of the solubility parameters of polynuclear aromatic systems and estimation of the solubility parameter of the asphaltenes based on hydrogen/carbon ratios (Figure 12.2). Further development of this knowledge can allow the progress of asphaltene deposition to be followed when changes on the solubility parameters of the asphaltene constituents or the solubility parameter of the liquid medium occur (Figure 12.3) (Speight, 1994) and the region of sediment formation (or the region of instability and incompatibility) is estimated using a simplified phase diagram.

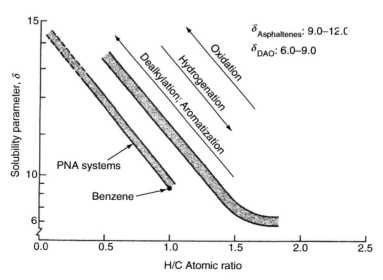

FIGURE 12.2 Estimation of the solubility parameters of asphaltene constituents from the atomic hydrogen–carbon ratio (compared to polynuclear aromatic systems) and changes caused by chemical–thermal effects.

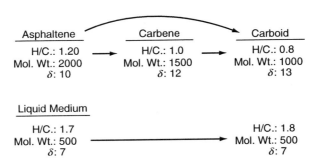

FIGURE 12.3 Representation of changes that occur during the thermal reaction of petroleum.

In general, these models are, to a degree, applicable to the prediction of heavy organic deposition (asphaltene, paraffin wax, as resin) from petroleum due to changes in pressure, temperature, and composition. However, the use of assumptions that do not reflect, or recognize, what might be the actual chemical and physical structure of petroleum can lead to errors in the data. Further modeling must involve recognition of the more modern concepts of the structure of petroleum as well as application of the models to the predictability of the location and amount of the deposition of the sediments inside the producing wells and oil-transport pipelines.

Petroleum is a complex system that depends upon the relationship of the constituent fractions to each other and the relationships are dictated by molecular interactions. Thus, recovery chemistry can only be generalized because of the intricate and complex nature of the molecular species that make up the crude oil leading to difficulties in analyzing not only the recovered material but also the original oil in place. Moreover, the incompatibility of crude oils with each other is a continuing issue and the occurrence of suspended organic solids during recovery (especially thermal) reduces the efficiency of a variety of processes.

Asphaltene precipitation or the mere presence of asphaltenes may invoke many implications in recovering asphaltic crude oils. Asphaltene precipitation may occur under various thermally or nonthermally enhanced oil recovery schemes or even primary production conditions.

These models are applicable to the prediction of sediment (i.e., asphaltene, resin, wax) formation and deposition from petroleum due to changes in pressure, temperature, and composition. Further modeling must involve an understanding of the chemistry of these materials and reflect the more modern approach to the physicochemical structure of petroleum. Only then will it be possible to more correctly predict the onset of precipitation as well as the location and amount of the sediment deposition in the producing wells and in oil-transport pipelines.

12.5   EFFECTS ON RECOVERY AND REFINING

The separation of solid sediments from reservoir fluids during oil production is an annoying and frustrating occurrence that can result in the plugging of the formation, well bores, and production facilities and it is necessary to take remedial actions (Park et al., 1994). Although the separation of asphaltenic sediments during refining operations can now be understood and there is some degree of predictability of phase separation (the separation of a sediment or deposition of asphaltenic material) (Magaril and Akensova, 1968; Magaril and Ramazaeva,

1969; Magaril et al., 1970; Wiehe, 1992; Speight, 1994), the deposition of asphaltenic material during recovery operations is less well understood.

It is the general consensus that it is the polar (heteroatom-containing) constituents (two fractions that are designated as asphaltenes and the resins) of petroleum that are responsible for the formation of suspended organic solids during a variety of recovery processes (Islam, 1994; Park et al., 1994). However, recovery chemistry and physics are areas that remain largely undefined insofar as the chemistry and physics are still speculative. Thus, precise chemical and physical models are difficult to describe and the factors leading to the incompatibility of petroleum constituents, as might occur during such operations, remain unknown.

As a first statement of what is known, petroleum is a delicately balanced system insofar as the different fractions are compatible provided there are no significant disturbances or changes made to the system. The changes that can occur, which will upset the balance of the petroleum system are: (1) alteration of the natural abundances of the different fractions, as might occur when gases are dissolved in the crude oil under reservoir pressure and when these dissolved gases are released at the time when a reservoir is first penetrated; (2) chemical alteration of the constituents as might occur during recovery processes, especially changes that might be brought on by thermal recovery processes; and (3) alteration of the distribution of the polar functional groups as might occur during oxidation or the elimination of polar functions during recovery operations, as might occur when exposure of petroleum to air occurs.

When such disturbances occur, it is the higher molecular weight constituents that are most seriously affected. This can lead to incompatibility, which is variously referred to as precipitation, sludge formation, sediment formation, and deposition of asphaltenic material, depending on the circumstances. When incompatibility occurs, it is the higher molecular weight polar constituents (the asphaltenes and, in some instances, the resins) that form a separate phase as a result of the loss of dispersability. In the reservoir, asphaltene incompatibility can cause blockages of the pores and channels through which the oil must move during recovery operations.

Thus, the dispersability of the higher molecular weight constituents becomes an issue that needs attention. And one of the ways by which this issue can be understood is to be aware of the chemical and physical character of the polar constituents of crude oil. By such means, the issue of dispersability, fluid–fluid interactions, oil–rock interactions, and the attending issue of incompatibility can be understood and perhaps predicted.

12.5.1 Effects on Recovery Operations

There are several mechanisms by which asphaltene constituents can be discharged from crude oil as sediment and cause problems during recovery operations. However, because so little is known about these events that can occur in a reservoir, it is difficult to foresee such events. In fact, there is no general rule that can be applied to the predictability of the separation of asphaltenic sediments. Each crude oil will behave differently under reservoir conditions and, therefore, predictability of behavior is difficult, at best speculative. Any adverse effects on the system can increase the potential for asphaltene incompatibility in the reservoir.

When it occurs, the deposition of asphaltenic material (sediment) can occur during crude oil production and can lead to serious problems. Deposition can occur in the reservoir and in the well tubing, thence being carried downstream through the lines into the separators and other downstream equipment. Apart from the loss of production, the cost of removing asphaltenic sediments can be very expensive and significantly alter the economics of a project.

What is perhaps more important is that any one of a number of operating conditions can result in the formation of asphaltenic sediments.

Thus, when a reservoir at the end of a primary depletion period is brought back to operation above the bubble point pressure, as may happen during a water-flooding operation, part of the gas cap formed during blowdown will be redissolved in the oil. In some cases, especially when the gas contains substantial amounts of propane and butane, phase separation can occur. This is analogous to the propane and butane deasphalting of a residuum (atmospheric or vacuum) that is employed in the production of asphalt during refinery operations. It is also analogous to the gas deasphalting that can occur during maturation of petroleum in a reservoir (before the reservoir is penetrated and goes into production) when it is believed that dissolution of hydrocarbon gases in the oil can deposit asphaltenic material on the reservoir rock (Evans et al., 1971).

Following this, there has also been the suggestion that gravity segregation occurs in connection with gas deasphalting in a reservoir. For example, in reservoirs that have a high vertical profile, assuming a temperature differential between the top and bottom of the reservoir, the petroleum may be increasingly heavier with depth. The material at the higher points of such reservoirs can contain more gas in solution than the petroleum in the lower points of the reservoir. Thus, more asphaltenic material is precipitated at the top of the reservoir than at the bottom of the reservoir.

Some enhanced recovery processes can also prompt the separation of asphaltene sediment. For example, carbon dioxide flooding and miscible hydrocarbon flooding have been known to produce sediments during operations (Islam, 1994). In carbon dioxide flooding, one or more banks are injected into the reservoir followed by water as the driving fluid. The major operative mechanism in carbon dioxide flooding appears to be swelling of the crude oil, miscibility, viscosity reduction, and most important, reduced interfacial tension. The deposition of asphaltenic material appears to be a result of displacement of the asphaltenes material by the carbon dioxide that competes with the asphaltenes for solubility in the crude oil. This is almost the reverse of the situation that occurs when salt is added to carbonated aqueous solutions and tends to displace the carbon dioxide from the aqueous system.

A similar effect can be observed in miscible hydrocarbon flooding during which a rich gas (i.e., gas with a high content of ethane, propane, butane, etc.) is injected into the reservoir (Islam, 1994). Again, this can approximate the deasphalting process that is used to manufacture asphalt and the injection of the low-boiling hydrocarbon will discharge asphaltenic material from the crude oil.

The adsorption of asphaltenic material during the separation of petroleum into its component fractions has become a standard operation for petroleum analysis. It must therefore be presumed that a similar effect can be operative under certain reservoir conditions, analogous to those prevalent when carbon dioxide and miscible hydrocarbon flooding technologies are applied to crude oil recovery. Thus, interactions of the polar constituents with the reservoir rock must also be included in any acknowledgment of the deposition of asphaltenic material. In addition, the deposition of asphaltenic material on to the reservoir rock can alter the properties of the rock (Islam, 1994) to the extent that water-wet rock can become oil-wet rock, which ultimately can have an adverse effect on enhanced oil recovery processes.

Any asphaltenic material adsorbed from the crude oil by the reservoir rock may not be readily desorbed from the rock (depending upon the nature of the rock) and may even be desorbed from the rock as insoluble material. If this asphaltenic material becomes mobile as sediment in the oil, it can move to the pore channels to the recovery well, thereby blocking the channels, making recovery difficult if not impossible.

Incompatibility is also possible when asphaltenes interact with reservoir rock, especially acidic functions of rocks, through the functional groups (e.g., the basic nitrogen species) just as they interact with adsorbents. And, there is the possibility for interaction of the asphaltene with the reservoir rock through the agency of a single functional group in which the remainder of the asphaltene molecule remains in the liquid phase. On the other hand, there is also the possibility that the asphaltene constituents can react with the rock at several points of contact thereby enhancing the bonding to the rock and, in some cases, effecting recovery operations to an even greater extent.

Incompatibility might also play a detrimental role during recovery operations is as a result of oxidation. In general, it is the more polar species that oxidize first. And, after incorporation of oxygen to a limit, significant changes can occur in asphaltene molecular weight (Moschopedis and Speight, 1976b; Moschopedis and Speight, 1978a). In fact, studies of the oxidation of the constituent fractions of petroleum have also shown that it is the most polar fractions that are more reactive; the less polar constituents are much slower to oxidize (Moschopedis and Speight, 1978a). And, the analysis of the products indicates that oxygen is incorporated as polar functions. In addition, the analysis of sediment material produced during recovery operations is consistent with this concept.

It must be emphasized here that the resin constituents incorporate oxygen to produce (on the basis of the separation method by which each fraction is defined) asphaltenes. The molecular weights of the products are higher than those of the starting materials. This can arise through two phenomena: (1) it is the high molecular weight polar species that are oxidized principally, and (2) the high molecular weight can arise from some association between the product species.

Another phenomenon that has been observed is that in the initial stages of the mild oxidation (without molecular degradation) the molecular weight of the asphaltenes (VPO/benzene) is reduced from an unoxidized molecular weight of 4920 to a molecular weight in the product asphaltenes of 1645. This change in molecular weight is due to oxidative degradation but far the incorporation of oxygen functions within the asphaltenes that interfere with the intermolecular association between individual asphaltene molecules, which is known to occur in nonpolar solvents such as benzene. In fact, it is quite reasonable to suggest that the molecular weight of 1645 might be close to a representative molecular weight of an individual hypothetical asphaltene molecule.

Nevertheless, should a similar phenomenon (i.e., nondegradation during oxidation) occur during recovery operations, the result could well be a phase separation of asphaltene material. This in turn could lead to loss of recovery efficiency of the crude oil because of the phase separation of the asphaltene constituents leading to blockage of the channel systems of the rock.

There are also other implications of the chemical reactions of the asphaltenic material produced by exposure of crude oil constituents to air. It may not be possible to model such phenomena unless the chemistry of the oxidation is known. Modeling of the deposition of asphaltenic materials has been studied with some degree of detail (Islam, 1994; Park et al., 1994). If the deposition of asphaltenic material is purely a physical effect, the models should allow some degree of predictability. However, the participation of chemical phenomena may make any model based on physical effects difficult to apply with a high degree of precision.

Those recovery processes that involve thermal systems are particularly susceptible to the deposition of asphaltenic material.

The thermolysis of asphaltenes produces a variety of products not the least of which are paraffinic species as well as highly polar aromatic species. The paraffinic products render the oil medium less polar (more paraffinic) and the increased polarity of the asphaltenic core material can only lead to deposition of aromatic polar species in a manner analogous to the deposition of sediment during, for example, visbreaking. Again there is the tendency for such deposits to hinder production. Although the temperatures in thermal recovery processes (with the exception of fire flooding processes) rarely reach those (>350°C; >660°F), such temperatures are possible. But another effect can also occur.

Steam flooding of a reservoir can lead to steam distillation (in the reservoir) of the more volatile constituents of the crude oil. In many cases, it is quite possible that the lower molecular weight aromatic species might be more volatile in steam that many of the paraffinic constituents of the oil. The overall effect of such a phenomenon is an increase in the paraffinic character of the oil and, hence, precipitation of asphaltenic material. During steam flooding, the precipitated asphaltenic material can be altered or unaltered material.

It is also necessary in any such discussion of the deposition of asphaltenic material to consider the electrical nature of asphaltenes.

Other studies (Preckshot et al., 1943; Katz and Beu, 1945; Penzes and Speight, 1974) have shown that asphaltenes in solution carry an electric charge. One postulate of asphaltene structure invoked the concept of zwitterions to explain the possibility of charged species within the asphaltenes. If the presence of such species is true, there is also the possibility that the induction of charges during the flow of crude oil through pipelines will, under some circumstances, lead to the deposition of asphaltenic material. In other circumstances, such as the use of well stimulation fluids that contain acidic species, there is also the potential to create "salts" of the asphaltenic species (i.e., asphaltenes $H^+ X^-$ or resins $H^+ X^-$). Whether this be through the agency of the charges species within the asphaltenic molecules or through the agency of the basic (pyridine-type) nitrogen that is know on to exist in the ring systems of the asphaltenic constituents is unknown. But it will be sufficient to change the natural order of the constituents within the crude oil leading to the deposition of incompatible (insoluble) material.

In summary, there are several effects that can occur during recovery processes that will lead to the deposition of asphaltenic material. When this happens, problems such as interruption or even cessation of the recovery operations can occur. In fact, when the increased popularity of asphaltene-containing crude oil is taken into consideration, it is surprising that problems of the deposition of asphaltenic material are not more widespread, assuming that all such occurrences are reported.

12.5.2 EFFECTS ON REFINING OPERATIONS

When petroleum is heated to temperatures in excess of 350°C (660°F), the rate of thermal decomposition of the constituents increases significantly. The higher the temperature, the shorter the time to achieve a given conversion, and the severity of the process conditions is a combination of the residence time of the crude oil constituents in the reactor and the temperature needed to achieve a given conversion.

Thus, the challenges facing process chemistry and physics are determining (1) the means by which petroleum constituents thermally decompose, (2) the nature of the products of thermal decomposition, (3) the subsequent decomposition of the primary thermal products, (4) the interaction of the products with each other, (5) the interaction of the products with the original constituents, and (6) the influence of the products on the composition of the liquids.

Polynuclear aromatic systems that are denuded of the attendant hydrocarbon moieties are somewhat lesser soluble in the surrounding hydrocarbon medium than their parent systems (Bjorseth, 1983; Dias, 1987, 1988). This is due to the solubilizing effect of the alkyl moieties and the enrichment of the liquid medium in paraffinic constituents. Again, there is an analogy with the deasphalting process, except that the paraffinic material is a product of the thermal decomposition of the asphaltene molecules and is formed **in situ** rather than being added separately.

Instability or incompatibility, resulting in the separation of solids during refining, can occur during a variety of processes, either by intent (such as in the deasphalting process) or inadvertently when the separation is detrimental to the process. Thus, separation of solids occurs whenever the solvent characteristics of the liquid phase are no longer adequate to maintain polar and high molecular weight material in solution. Examples of such occurrences are: (1) asphaltene separation when the paraffin content or character of the liquid medium increases, (2) sludge or sediment formation in a reactor, which occurs when the solvent characteristics of the liquid medium change so that asphaltic or wax materials separate, and (3) coke formation, which occurs at high temperatures and commences when the solvent power of the liquid phase is not sufficient to maintain the coke precursors in solution.

Thus, there is a need to define adequately the nature of a refinery feedstock before use as well as the nature of the products in order to anticipate the onset of instability or incompatibility. The occurrence of such phenomena during thermal processes will be discussed in this chapter as it relates to the thermal chemistry of crude oil constituents.

The formation of solid sediments or coke during thermal processes is a major limitation on processing. Further, the presence of different types of solids shows that solubility controls the formation of solids. And the tendency for solid formation changes in response to the relative amounts of the light ends, middle distillates, and residues and to changes in chemical composition during the process. In fact, the prime mover in the formation of incompatible products during the processing of feedstocks containing asphaltenes is the nature of the primary thermal decomposition products, particularly those designated as carbenes and carboids (Speight, 1992, 1994).

One of the postulates of coke formation involves the production of coke by a sequence of polymerization and condensation steps from the lightest (lower molecular weight) to the heaviest (higher molecular weight) fractions. In addition, there is an induction period during which no coke is formed, and many kinetic models that contain sequences of direct chemical reactions to coke fail to predict the induction period. This induction period has been observed experimentally by many previous investigators (Levinter et al., 1966, 1967; Magaril and Aksenova, 1968, 1970; Valyavin et al., 1979) and makes visbreaking and the Eureka processes possible. The postulation that coke formation is triggered by the phase separation of asphaltenes (Magaril et al., 1971) led to the use of linear variations of the concentration of each fraction with reaction time, resulting in the assumption of zero-order kinetics rather than first-order kinetics.

During the coke induction period, the reactant asphaltenes form only lower molecular weight products and, as long as the asphaltenes remain dissolved, the heptane-soluble materials can provide sufficient abstractable hydrogen to terminate asphaltene-free radicals, making asphaltene radical–asphaltene radical recombination infrequent. In support of this assumption, it is known (Langer et al., 1961) that partially hydrogenated refinery process streams provide abstractable hydrogen and, as a result, inhibit coke formation during residuum thermal conversion. Thus, the heptane-soluble fraction of a residuum, which contains naturally occurring, partially hydrogenated aromatics, can provide abstractable hydrogen during thermal reactions.

The asphaltene concentration varies little in the coke induction period, but decreases once coke begins to form. Observing this, it might be concluded that asphaltenes are unreactive, but it is the high reactivity of the asphaltenes down to the asphaltene core that offsets the generation of asphaltene cores from the heptane-soluble materials to keep the overall asphaltene concentration nearly constant.

Thus, the available data suggest that coke formation is a complex process involving both chemical reactions and thermodynamic behavior. Reactions that contribute to this process are cracking of side chains from aromatic groups, dehydrogenation of naphthenes to form aromatics, condensation of aliphatic structures to form aromatics, condensation of aromatics to form higher fused-ring aromatics, and dimerization or oligomerization reactions. Loss of side chains always accompanies thermal cracking, and dehydrogenation and condensation reactions are favored by hydrogen-deficient conditions.

There is a need to investigate fully the degree of association of asphaltenes in thermal systems and to determine whether it has any influence on the coke-forming tendencies of the asphaltenes.

REFERENCES

Acevedo, S., Mends, B., Rajas, A., Lairs, I., and Rives, H. 1985. *Fuel* 64: 1741.

Ali, L.H., Al-Ghannan, A., and Speight, J.G. 1985. *Iraqi J. Sci.* 26: 41.

Andersen, S.I. and Birdi, K.S. 1991a. *Proceedings. International Symposium on the Chemistry of Bitumens.* Western Research Institute, Laramie, WY. p. 236.

Andersen, S.I. and Birdi, K.S. 1991b. *J. Coll. Interface Sci.* 161.

Andersen, S.I. and Speight, J.G. 1992. Preprints. *Div. Fuel. Chem. Am. Chem. Soc.* 37(3): 1335.

Andersen, S.I. and Speight, J.G. 1993. *Fuel.* 72: 1343.

Bardon, C., Barre, L., Espinat, D., Guille, V., Li, M.H., Lambard, J., Ravey, J.C., Rosenberg, E., and Zemb, T. 1996. *Fuel Sci. Technol. Int.* 14: 203.

Barton, A.F.M. 1983. *Handbook of Solubility Parameters and Other Cohesion Parameters.* CRC Press, Boca Raton, FL.

Bestougeff, M.A. and Darmois, R. 1947. *Comptes Rend.* 224: 1365.

Bestougeff, M.A. and Darmois, R. 1948. *Comptes Rend.* 227: 129.

Bichard, J.A. 1969. *Proceedings.* Annual Meeting. Canadian Society for Chemical Engineering. Edmonton, Alberta, Canada.

Bjorseth, A. 1983. *Handbook of Polycyclic Aromatic Hydrocarbons.* Marcel Dekker Inc., New York.

Bruning, I.M.R.A. 1991. *Oil Gas J.* 89(3): 38.

Buckley, J. 1996. *Fuel Sci. Technol. Int.* 14: 55.

Burke, N.E., Hobbs, R.E., and Kashou, S.F. 1990. *J. Petrol. Technol.* 42: 1440.

Chattoraj, D.K. and Birdi, K.S. 1984. *Adsorption & the Gibbs Surface Excess.* Plenum Press, New York.

Cimino, R., Correrra, S., Del Bianco, A., and Lockhart, T.P. 1995. In *Asphaltenes: Fundamentals and Applications.* Sheu, E.Y. and Mullins, O.C. eds. Plenum Press, New York. Chapter III.

Dias, J.R. 1987. *Handbook of Polycyclic Hydrocarbons. Part A. Benzenoid Hydrocarbons.* Elsevier, New York.

Dias, J.R. 1988. *Handbook of Polycyclic Hydrocarbons. Part B. Polycyclic Isomers and Analogs of Benzenoid Hydrocarbons.* Elsevier, New York.

Dickie, J.P. and Yen, T.F. 1967. *Anal. Chem.* 39: 1847.

Eliel, E. and Wilen, S. 1994. *Stereochemistry of Organic Compounds.* John Wiley & Sons Inc., New York.

Evans, C.R., Rogers, M.A., and Bailey, N.J.L. 1971. *Chem. Geol.* 8: 147.

Fotland, P. and Alfindsen, H. 1996. *Fuel Sci. Technol. Int.* 14: 101.

Francisco, M.A. and Speight. J.G. 1984. Preprints. *Div. Fuel. Chem. Am. Chem.. Soc.* 29(1): 36.

Fuhr, B.J., Cathrea, C., Coates, L., Kalra, A., and Majeed, A.I. 1991. *Fuel* 70: 1293.

Girdler, R.B. 1965. *Proc. Assoc. Asphalt Paving Technol.* 34: 45.

Heithaus, J.J. 1962. *J. Inst. Pet.* 48: 45.

Hirschberg, A., de Jong, L.N.J., Schnipper, B.A., and Meijjer, J.G. Soc. 1984. *Pet. Eng. J.* p. 283.

Islam, M.R. 1994. *Asphaltenes and Asphalts. Developments in Petroleum Science, 40.* Yen, T.F. and Chilingarian, G.V. eds. Elsevier, Amsterdam. Chapter 11.

Katz, D.L. and Beu, K.E. 1945. *Ind. Eng. Chem.* 43: 1165.

Kawanaka, S., Leontaritis, K.J., Park, S.J., and Mansoori, G.A. 1989. *Oil Field Chemistry: Enhanced Recovery and Production Simulation.* Borchardt, J.K. and Yen, T.F. eds. Symposium Series No. 396. American Chemical Society, Washington, DC. Chapter 24.

Koots, J.A. and Speight, J.G. 1975. *Fuel* 54: 179.

Langer, A.W., Stewart, J., Thompson, C.E., White, H.T., and Hill, R.M. 1961. *Ind. Eng. Chem.* 53: 27.

Leblanc, R.M. and Thyrion, F.C. 1989. *Fuel* 68:260.

Leontaritis, K.J. 1996. *Fuel Sci. Technol. Int.* 14: 13.

Levinter, M.E., Medvedeva, M.I., Panchenkov, G.M., Aseev, Y.G., Nedoshivin, Y.N., Finkelshtein, G.B., and Galiakbarov, M.F. 1966. *Khim. Tekhnol. Topl. Masel.* 9: 31.

Levinter, M.E., Medvedeva M.I., Panchenkov, G.M., Agapov, G.I., Galiakbarov, M.F., and Galikeev, R.K. 1967. *Khim. Tekhol. Topl. Masel.* 4: 20.

Long, R.B. 1979. Preprints. *Div. Petrol. Chem. Am. Chem. Soc.* 24(4): 891.

Long, R.B. 1981. *The Chemistry of Asphaltenes.* Bunger, J.W. and Li, N.C. eds. Advances in Chemistry Series No. 195. American Chemical Society, Washington, DC.

Magaril, R.Z. and Akensova, E.I. 1968. *Int. Chem. Eng.* 8: 727.

Magaril, R.Z. and Aksenova, E.I. 1970. *Khim. Tekhnol. Topl. Masel.* No. 7: 22.

Magaril, R.Z. and Ramazaeva, L.F. 1969. *Izv. Vyssh. Ucheb. Zaved. Neft Gaz* 12(1): 61.

Magaril, R.Z., Ramazaeva, L.F., and Aksenova, E.I. 1970. *Khim. Tekhnol. Topl. Masel.* 15(3): 15.

Magaril, R.Z., Ramazaeva, L.F., and Aksenova, E.I. 1971. *Int. Chem. Eng.* 11: 250.

Mansoori, G.A., Jiang, T.S., and Kawanaka, S. 1987. *Arabian J. Sci. Eng.* 13(1): 17.

Menon, V.B. and Wasan, D.T. 1986. *Colloid Surf.* 19:89.

Mitchell, D.L. and Speight, J.G. 1973. *Fuel* 52: 149.

Moschopedis, S.E. and Speight. J.G. 1976a. *Fuel* 55: 187.

Moschopedis, S.E. and Speight, J.G. 1976b. *Fuel* 55: 334.

Moschopedis, S.E. and Speight, J.G. 1976c. *Proc. Assoc. Asphalt Paving Technol.* 45: 78.

Moschopedis, S.E. and Speight, J.G. 1977. *J. Mater. Sci.* 12: 990.

Moschopedis, S.E. and Speight, J.G. 1978a. *Fuel* 57: 25.

Moschopedis, S.E. and Speight, J.G. 1978b. *Fuel* 57: 237.

Moschopedis, S.E., Fryer, J.F., and Speight, J.G. 1976. *Fuel* 55: 227.

Mushrush, G.W. and Speight, J.G. 1995. *Petroleum Products: Instability and Incompatibility.* Taylor & Francis, Washington, DC.

Overfield, R.E., Sheu, E.Y., Sinha, S.K., and Liang, K.S. 1989. *Fuel Sci. Technol. Int.* 7: 611.

Park, S.J., Escobedo, J., and Mansoori, G.A. 1994. *Asphaltenes and Asphalts. Developments in Petroleum Science, 40.* Yen, T.F. and Chilingarian, G.V. eds. Elsevier, Amsterdam. Chapter 8.

Pelet, R., Behar, F., and Monin, J.C. 1985. *Adv. Org. Geochem.* 10: 481.

Penzes, S. and Speight, J.G. 1974. *Fuel* 53: 192.

Pfeiffer, J.P. and Saal, R.N. 1940. *Phys. Chem.* 44: 139.

Preckshot, G.W., Delisle, N.G., Cottrell, C.E., and Katz, D.L. 1943. *AIME Trans.* 151: 188.

Ravey, J.C., Decouret, G., and Espinat, D. 1988. *Fuel* 67: 1560.

Rogacheva, O.V., Rimaev, R.N., Gubaidullin, V.Z., and Khakimov, D.K. 1980. *Colloid J. USSR.* 490.

Sachanen, A.N. 1945. *The Chemical Constituents of Petroleum.* Reinhold, New York.

Selucky, M.L., Kim, S.S., Skinner, F., and Strausz, O.P. 1981. *The Chemistry of Asphaltenes.* Bunger, J.W., and Li, N.C. editors Advances in Chemistry Series No. 195. American Chemical Society, Washington. DC.

Shaw, D.J. 1980. *Introduction to Colloid and Surface Chemistry.* 3rd Edition. Butterworths, London.

Sheu, E.Y., DeTar, M. M., Storm, D. A., and DeCanio, S. J. 1992. *Fuel* 71: 299.

Sheu, E.Y. and Storm, D.A. 1995. *Asphaltenes: Fundamentals and Applications.* Sheu, E.Y. and Mullins, O.C., eds. Plenum Press, New York. Chapter I.

Speight, J.G. 1971. *Fuel* 50: 102.

Speight, J.G. 1975. *Proc. Natl. Sci. Found. Symp. Fund. Org. Chem. Coal.* Knoxville, TN, p. 125.

Speight, J.G. 1986. Preprints. *Div. Petrol. Chem. Am. Chem. Soc.* 31(4): 818.

Speight, J.G. 1992. *Proceedings.* 4th International Conference on the Stability and Handling of Liquid Fuels. U.S. Department of Energy (DOE/CONF-911102). p. 169.

Speight, J.G. 1994. *Asphaltenes and Asphalts. I. Developments in Petroleum Science, 40.* Yen, T.F. and Chilingarian, G.V. eds. Elsevier, Amsterdam. Chapter 2.

Speight, J.G. and Moschopedis, S.E. 1980. *Fuel* 59: 440.

Speight, J.G. and Penzes, S. 1978. *Chem. Ind.* 729.

Speight, J.G., Long, R.B., and Trowbridge, T.D. 1982. Preprints. *Div. Fuel Chem. Am. Chem. Soc.* 27(3/4): 268.

Speight, J.G., Long, R.B., and Trowbridge, T.D. 1984. *Fuel* 63: 616.

Speight, J.G., Wernick, D.L., Gould, K.A., Overfield, R.E., Rao, B.M.L., and Savage, D.W. 1985. *Rev. Inst. Franç. Petrol.* 40: 51.

Swanson, J.M. 1942. *J. Phys. Chem.* 46: 141.

Taft, R.W., Berthelot, M., Laurence, C., and Leo, A.J. 1996. *Chemtech.* (July): 20.

Thiyagarajan, P., Hunt, J.E., Winansa, R.E., Anderson, K.B., and Miller, J.T. 1995. *Energy Fuels* 9: 629.

Tissot, B. 1984. *Characterization of Heavy Crude Oils and Petroleum Residues.* Editions Technip, Paris, France.

Valyavin, G.G., Fryazinov, V.V., Gimaev, R.H., Syunyaev, Z.I., Vyatkin, Y.L., and Mulyukov, S.F. 1979. *Khim. Tekhol. Topl. Masel.* 8: 8.

Wiehe, I.A. 1992. *Ind. Eng. Chem. Res.* 31: 530.

Wiehe, I.A. 1993. *Ind. Eng. Chem. Res.* 32: 2447.

Wiehe, I.A. 1996. *Fuel Sci. Technol. Int.* 14: 289.

Witherspoon, P.A. and Munir, Z.A. 1960. *Producers Monthly* 24(August): 20.

Yen, T.F. 1974. *Energy Sources* 1: 447.

Yen, T.F. 1990. Preprints *Div. Petrol. Chem. Am. Chem. Soc.* 35: 314.

13 Instability and Incompatibility

13.1 INTRODUCTION

The study of the properties of petroleum would not be complete without some attention to the phenomena of **instability** and **incompatibility**. Both result in the formation of degradation products or other undesirable changes in the original properties of petroleum products. And it is the analytical methods that provide the data that point to the reason for problems in the refinery or for the failure of products to meet specifications and to perform as desired.

As already stated (Chapter 1), petroleum products have been in use for over 5000 years (Abraham, 1945; Forbes, 1958a, b, 1959; James and Thorpe, 1994). Certain derivatives of petroleum (the nonvolatile residua and **asphalt**) could be used for caulking and as an adhesive for jewelry or for construction purposes as well as the use of asphalt for medicinal purposes. Other derivatives (the volatile **naft** or naphtha) were used as incendiary materials or as illuminants.

Current refineries (Chapter 14) are a complex series of manufacturing plants that can be subdivided into (1) separation processes, (2) conversion processes, and (3) finishing processes. Incompatibility can occur in many of these processes and the product can also exhibit incompatibility and instability. Further, there are a myriad of other products that have evolved through the short life of the petroleum industry. And the complexities of product composition have matched the evolution of the products (Hoffman, 1992). In fact, it is the complexity of product composition that has served the industry well and, at the same time, had an adverse effect on product use.

Product complexity and the means by which the product is evaluated (ASTM, 2005) have made the industry unique among industries. But product complexity has also brought to the fore, issues such as instability and incompatibility. Product complexity becomes even more disadvantageous when various fractions from different types of crude oil are blended or are allowed to remain under conditions of storage (prior to use) and a distinct phase separates from the bulk product. The adverse implications of this for refining the fractions to saleable products increase (Batts and Fathoni, 1991; Por, 1992; Mushrush and Speight, 1995).

Therefore, it is appropriate here to define some of the terms that are used in the liquid fuels field, so that their use later in the text will be more apparent and will also alleviate some potential for misunderstanding. The general scientific areas of instability and incompatibility are complex and have been considered to be nothing better than a black art, because not all the reactions that contribute to instability and incompatibility have been defined (Wallace, 1964). Nevertheless, recent studies over the past three decades have made valuable contributions to our understanding of instability and incompatibility in fuels. But, for the most part, gaps remain in our knowledge of the chemistry and physics of instability and incompatibility.

Briefly, the term *incompatibility* refers to the formation of a precipitate (or **sediment**) or separate phase when two liquids are mixed. The term *instability* is often used in reference to the formation of color, sediment, or gum in the liquid over a period of time and is usually due

TABLE 13.1
Examples of Properties Related to Instability in Petroleum and Petroleum Products

Property	Comments
Asphaltene constituents	Influence oil–rock interactions
	Separates from oil when gases are dissolved
	Thermal alteration can cause **phase separation**
Heteroatom constituents	Provide polarity to oil
	Preferential reaction with oxygen
	Preferential thermal alteration
Aromatic constituents	May be incompatible with paraffinic medium
	Phase separation of paraffinic constituents
Nonasphaltene constituents	Thermal alteration causes changes in polarity
	Phase separation of polar species

to chemical reactions, such as **oxidation**, and is chemical rather than physical. This term may be used to contrast the formation of a precipitate in the near term (almost immediately).

The phenomenon of instability is often referred to as incompatibility, and more commonly known as **sludge** formation, and sediment formation, or deposit formation. In petroleum and its products, instability often manifests itself in various ways (Table 13.1) (Stavinoha and Henry, 1981; Hardy and Wechter, 1990a; Ruzicka and Nordenson, 1990; Power and Mathys, 1992). Hence, there are different ways of defining each of these terms, but the terms are often used interchangeably.

Gum formation (ASTM D525, IP 40) alludes to the formation of soluble organic material, whereas sediment is the insoluble organic material. **Storage stability** (or **storage instability**) (ASTM D381, ASTM D4625, IP 131, IP 378) is a term used to describe the ability of the liquid to remain in storage over extended periods of time without appreciable deterioration as measured by gum formation and the formation sediment. **Thermal stability** is also defined as the ability of the liquid to withstand relatively high temperatures for short periods of time without the formation of sediment (i.e., carbonaceous deposits or coke) (Brinkman and White, 1981). Thermal oxidative stability is the ability of the liquid to withstand relatively high temperatures for short periods of time in the presence of oxidation and without the formation of sediment or deterioration of properties (ASTM D3241) and there is standard equipment for various oxidation tests (ASTM D4871). Stability is also defined as the ability of the liquid to withstand long periods at temperatures up to 100°C (212°F) without degradation. Determination of the reaction threshold temperature for various liquid and solid materials might be beneficial (ASTM D2883).

Existent-gum is the name given to the nonvolatile residue present in the fuel as received for test (ASTM D381, IP 131). In this test, the sample is evaporated from a beaker maintained at a temperature of 160°C to 166°C (320°F to 331°F) with the aid of a similarly heated jet of air. This material is distinguished from the potential gum that is obtained by aging the sample at an elevated temperature.

Thus, potential gum is determined by the accelerated gum test (ASTM D873, IP 138) that is used as a safeguard of storage stability and can be used to predict the potential for gum formation during prolonged storage. In this test, the fuel is heated for 16 h with oxygen under pressure in a bomb at 100°C (212°F). After this time, both the gum content and the solids precipitate are measured. A similar test, using an accelerated oxidation procedure is also in use for determining the oxidative stability of diesel fuel (ASTM D2274), steam turbine oil (ASTM D2272), distillate fuel oil (ASTM D2274) and lubricating grease (ASTM D942).

Dry sludge is defined as the material separated from petroleum and petroleum products by filtration and which is insoluble in heptane. Existent dry sludge is the dry sludge in the original sample as received and is distinguished from the accelerated dry sludge obtained after aging the sample by chemical addition or heat. The existent dry sludge is distinguished from the potential dry sludge that is obtained by aging the sample at an elevated temperature.

The existent dry sludge is operationally defined as the material separated from the bulk of a crude oil or crude oil product by filtration and which is insoluble in heptane. The test is used as an indicator of process operability and as a measure of potential downstream fouling.

An analogous test, the thin film oven test (TFOT) (ASTM D1754), and an ageing test (IP 390) are used to indicate the rate of change of various physical properties such as penetration (ASTM D5), viscosity (ASTM D2170), and ductility (ASTM D113), after a film of asphalt or bitumen has been heated in an oven for 5 h at 163°C (325°F) on a rotating plate. A similar test is available for the stability of engine oil by thin film oxygen uptake (TFOUT) (ASTM D4742).

This test establishes the effects of heat and air based on changes incurred in the earlier physical properties measured before and after the oven test. The allowed rate of changes in the relevant bitumen properties, after the exposure of the tested sample to the oven test, are specified in the relevant specifications (ASTM D3381).

Attractive as they may be, any tests that involve accelerated oxidation of the sample must be used with caution and consideration of the chemistry. Depending on the constituents of the sample, it is quite possible that the higher temperature and extreme conditions (oxygen under pressure) may not be truly representative of the deterioration of the sample under storage conditions. The higher temperature and the oxygen under pressure might change the chemistry of the system and produce products that would not be produced under ambient storage conditions. An assessment of the composition of the fuel before storage and application of the test will assist in this determination.

In general, fuel instability and fuel incompatibility can be related to the heteroatom-containing compounds (i.e., nitrogen-, oxygen-, and sulfur-containing compounds) that are present. The degree of unsaturation of the fuel (i.e., the level of olefinic species) also plays a role in determining instability or incompatibility. And, recent investigations have also implicated catalytic levels of various oxidized intermediates and acids as especially deleterious for middle distillate fuels.

Fuel incompatibility can have many meanings. The most obvious example of incompatibility (nonmiscible) is the inability of a hydrocarbon fuel and water to mix. In the present context, incompatibility usually refers to the presence of various polar functions (i.e., heteroatom function groups containing nitrogen, or oxygen, or sulfur, and even various combinations of the heteroatoms) in the crude oil.

Instability reactions are usually defined in terms of the formation of filterable and nonfilterable sludge (sediments, deposits, and gums), an increased peroxide level, and the formation of color bodies. Color bodies in and of themselves do not predict instability. However, the reactions that initiate color body formation can be closely linked to heteroatom-containing (i.e., nitrogen-, oxygen-, and sulfur-containing) functional group chemistry.

Thus, petroleum constituents are incompatible when sludge, semi-solid, or solid particles (for convenience here, these are termed *secondary products* to distinguish them from the actual petroleum product) are formed during and after blending. This phenomenon usually occurs prior to use. If the secondary products are marginally soluble in the blended petroleum product, use might detract from solubility of the secondary products and they will appear as sludge or sediment that can be separated by filtration or by extraction (ASTM D4310, IP

53, IP 375). When the secondary products are truly insoluble, they separate and settle out as a semisolid or solid phase floating in the fuel or are deposited on the walls and floors of containers. In addition, secondary products usually increase the viscosity of the petroleum product. Standing at low temperatures will also cause a viscosity change in certain fuels and lubricants (ASTM D2532). Usually the viscosity change might be due to separation of paraffins as might occur when diesel fuel, and similar, engines are allowed to cool and stand unused overnight in low temperature climates.

In the current context, the meaning of the term *incompatibility* is found when it is applied to petroleum recovery and petroleum refining. The application occurs when a product is formed in the formation or well pipe (recovery) or in a reactor (refining) and the product is incompatible with (immiscible with or insoluble in) the original petroleum or its products. Such an example is the formation and deposition of wax and other solids during recovery or the formation of coke precursors and even of coke during many thermal and catalytic operations. Coke formation is considered to be an initial phase separation of an insoluble, solid, coke precursor prior to coke formation proper. In the case of crude oils, sediments and deposits are closely related to sludge, at least as far as compositions are concerned. The major difference appears to be in the character of the material.

There is also the suggestion (often, but not always, real) in that the sediments and deposits originate from the inorganic constituents of petroleum. They may be formed from the inherent components of the crude oil (i.e., the metalloporphyrin constituents) or from the ingestion of contaminants by the crude oil during the initial processing operations. For example, crude oil is known to pick up iron and other metal contaminants from contact with pipelines and pumps.

Sediments can also be formed from organic materials but the usual inference is that these materials are usually formed from inorganic materials. The inorganic materials can be salt, sand, rust, and other contaminants that are insoluble in the crude oil and which settle to the bottom of the storage vessel. For example, gum typically forms by way of a hydroperoxide intermediate that induces polymerization of olefins. The intermediates are usually soluble in the liquid medium. However, gum that has undergone extensive oxidation reactions tends to be higher in molecular weight and much less soluble. In fact, the high molecular weight sediments that form in fuels are usually the direct result of autoxidation reactions. Active oxygen species involved include both molecular oxygen and hydroperoxides. These reactions proceed by a free radical mechanism and the solids produced tend to have increased incorporation of heteroatom and are thus also more polar, so increasingly less soluble in the fuel.

The most significant and undesirable instability change in fuel liquids is the formation of solids, termed *filterable sediment*. Filterable sediments can plug nozzles, filters, coke heat exchanger surfaces and otherwise degrade engine performance. These solids are the result of free radical autoxidation reactions. Although slight thermal degradation occurs in nonoxidizing atmospheres, the presence of oxygen or active oxygen species, such as hydroperoxides, will greatly accelerate oxidative degradation, as well as significantly lower the temperature at which undesirable products are formed. Solid deposits that form as the result of short-term high-temperature reactions share many similar chemical characteristics with filterable sediment that form in storage.

The soluble sludge or sediment precursors that form during processing or use may have a molecular mass in the several hundred range. For this soluble precursor to reach a molecular weight sufficient to precipitate (or to phase-separate), one of two additional reactions must occur. Either the molecular weight must increase drastically as a result of condensation reactions leading to the higher molecular weight species. Or the polarity of the precursor must increase (without necessarily increasing the molecular weight) by incorporation of additional oxygen, sulfur, or nitrogen functional groups. Additionally, the polarity may increase because of the removal of nonpolar hydrocarbon moieties from the polar core, as occurs during cracking reactions. In all three cases, insoluble material will form and separate from the liquid medium.

Additives are chemical compounds intended to improve some specific properties of fuels or other petroleum products. Different additives, even when added for identical purposes, may be incompatible with each other, for example, react and form new compounds. Consequently, a blend of two or more fuels, containing different additives, may form a system in which the additives react with each other and so deprive the blend of their beneficial effect.

The chemistry and physics of incompatibility can, to some extent, be elucidated (Wallace, 1969; Hardy and Wechter, 1990b; Por, 1992; Power and Mathys, 1992; Mushrush and Speight, 1995), but many unknowns remain. In addition to the chemical aspects, there are also aspects such as the attractive force differences, such as

1. Specific interactions between like or unlike molecules (e.g., hydrogen bonding and electron donor–acceptor phenomena)
2. Field interactions such as dispersion forces and dipole–dipole interactions
3. Any effects imposed on the system by the size and shape of the interacting molecular species

Such interactions are not always easy to define and, thus, the measurement of instability and incompatibility has involved visual observations, solubility tests, hot filtration sediment (HFS), and gum formation. However, such methods are often considered to be after-the-fact methods insofar as they did not offer much in the way of predictability. In refinery processes (Chapter 14), predictability is not just a luxury, it is a necessity. The same principle must be applied to the measurement of instability and incompatibility. Therefore, methods are continually sought to aid in achieving this goal.

It is the purpose of this chapter to document some of the more prominent methods used for determining instability and incompatibility. No preference is shown, and none is given to any individual methods. It is the choice of the individual experimentalist to choose the method on the basis of the type of fuel, the immediate needs, and the projected utilization of the data. As elsewhere, it the use of the data that often detracts from an otherwise sound method.

In addition to the gravimetric methods, there have also been many attempts to use crude oil and product characteristics and their relation to the sludge and deposit formation tendencies. In some cases, a modicum of predictability is the outcome but, in many cases, the data appear as preferred ranges and are subject to personal interpretation. Therefore, caution is advised.

13.2 INSTABILITY AND INCOMPATIBILITY IN PETROLEUM

The instability or incompatibility of crude oil and of crude oil products is manifested in the formation of sludge, sediment, and general darkening in color of the liquid (ASTM D1500, IP 17).

Sludge (or sediment) formation takes one of the following forms: (1) material dissolved in the liquid; (2) precipitated material; and (3) material emulsified in the liquid. Under favorable conditions, sludge or sediment will dissolve in the crude oil or product with the potential of increasing the viscosity. Sludge or sediment, which is not soluble in the crude oil (ASTM D96, ASTM D473, ASTM D1796, ASTM D2273, ASTM D4007, ASTM D4807, ASTM D4870), may either settle at the bottom of the storage tanks or remain in the crude oil as an emulsion. In most of the cases, the smaller part of the sludge or sediment will settle satisfactorily, the larger part will stay in the crude oil as emulsions. In any case there is a need of breaking the emulsion, whether it is a water-in-oil emulsion or whether it is the sludge itself, which has to be separated into the oily phase and the aqueous phase. The oily phase can be then processed with the crude oil and the aqueous phase can be drained out of the system.

Phase separation can be accomplished by either the use of suitable surface active agents allowing for sufficient settling time, or by use of a high voltage electric field for breaking such emulsions after admixing water at a rate of about 5% and at a temperature of about 100°C (212°F)

Emulsion breaking, whether the emulsion is due to crude oil–sludge emulsions, crude oil–water emulsions or breaking the sludge themselves into organic (oily) and inorganic components are of a major importance from operational as well as commercial aspects. With some heavy fuel oil products and heavy crude oils, phase separation difficulties often arise (Ruzicka and Nordenson, 1990: Por, 1992; Mushrush and Speight, 1995). Also, some crude oil emulsions may be stabilized by naturally occurring substances in the crude oil. Many of these polar particles accumulate at the oil–water interface, with the polar groups directed toward the water and the hydrocarbon groups toward the oil. A stable interfacial skin may be so formed; particles of clay or similar impurities, as well as wax crystals present in the oil may be embedded in this skin and make the emulsion very difficult to break (see Schramm, 1992 and references cited therein).

Chemical and electrical methods for sludge removal and for water removal, often combined with chemical additives, have to be used for breaking such emulsions. Each emulsion has its own structure and characteristics: water-in-oil emulsions, where the oil is the major component or oil-in-water emulsions, where the water is the major component. The chemical and physical nature of the components of the emulsion plays a major role in their susceptibility to the various surface-active agents used for breaking them.

Therefore, appropriate emulsion breaking agents have to be chosen very carefully, usually with the help of previous laboratory evaluations. Water or oil soluble demulsifiers, the latter is often nonionic surface-active alkylene oxide adducts, are used for this purpose. But, as had been said in the foregoing, the most suitable demulsifier has to be chosen for each case from a large number of such substances in the market, by a previous laboratory evaluation.

13.3 FACTORS INFLUENCING INSTABILITY AND INCOMPATIBILITY

A number of experimental methods are available for estimation of the factors that influence instability or incompatibility. These factors have been explored and attempts have been made to estimate the character of the fuel or product with varied results.

13.3.1 ELEMENTAL ANALYSIS

The ultimate analysis (elemental composition) of petroleum and its products is not reported to the same extent as for coal (Speight, 1994a and references cited therein). Nevertheless, there are ASTM procedures (ASTM, 2005) for the ultimate analysis of petroleum and petroleum products (Chapter 9), but many such methods may have been designed for other materials.

Of the data that are available, the proportions of the elements in petroleum vary only slightly over narrow limits: carbon 83.0% to 87.0%, hydrogen 10.0% to 14.0%, nitrogen 0.10% to 2.0%, oxygen 0.05% to 1.5%, sulfur 0.05% to 6.0% (Chapter 9). And yet, there is a wide variation in physical properties from the lighter more mobile crude oils at one extreme to the extra heavy crude oil at the other extreme (Chapter 1). In terms of the instability and incompatibility of petroleum and petroleum products, the heteroatom content appears to represent the greatest influence. In fact, it is not only the sulfur and nitrogen content of crude oil that are important parameters in respect of the processing methods which have to be used to produce fuels of specification sulfur concentrations, but also the type of sulfur and nitrogen species in the oil. There could well be a relation between nitrogen and sulfur content and crude oil (or product) stability; higher nitrogen and sulfur crude oils are suspect of higher sludge forming tendencies.

13.3.2 Density and Specific Gravity

In the earlier years of the petroleum industry, density and specific gravity (with the API gravity) were the principal specifications for feedstocks and refinery products (Chapter 8). They were used to give an estimate of the most desirable product, i.e., kerosene, in crude oil. At the present time, a series of standard tests exists for determining density and specific gravity (Chapter 9) (Speight, 2001, 2002).

There is no direct relation between the density and specific gravity of crude oils to their sludge forming tendencies, but crude oil having a higher density (thus, a lower API gravity) is generally more susceptible to sludge formation, presumably because of the higher content of the polar or asphaltic constituents.

13.3.3 Volatility

Petroleum can be subdivided by distillation into a variety of fractions of different **boiling ranges** or **cut points** (Chapter 9 and Chapter 16). In fact distillation was, and still is, the method for feedstock evaluation for various refinery options. Indeed, volatility is one of the major tests for petroleum products and it is inevitable that the majority of all products will, at some stage of their history, be tested for volatility characteristics (Chapter 8).

As an early part of characterization studies, a correlation was observed between the quality of petroleum fractions and their heteroatom content (Figure 13.1) since gasoline, kerosene, diesel fuel, and lubricating oil are made up of hydrocarbon constituents containing high

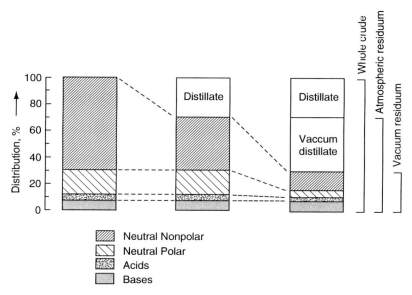

FIGURE 13.1 Schematic representation of the distribution of polar (heteroatomic) groups after distillation.

proportions of hydrogen. Thus, it is not surprising that tests to determine the volatility of petroleum and petroleum products were among the first to be defined.

The very nature of the distillation process by which residua are produced (Chapter 16), i.e., removal of distillate without thermal decomposition, dictates that the majority of the heteroatoms which are predominantly in the higher molecular weight fractions, will be concentrated in the higher boiling products and the residuum (Chapter 13; see also Asphalt, Chapter 21) (Speight, 2000). Thus, the inherent nature of the crude oil and the means by which it is refined can seriously influence the stability and incompatibility of the products.

Heavier crude oils, yielding higher amounts of residua, tend to form more sludge during storage than light crude oils.

13.3.4 Viscosity

The viscosity of a feedstock varies with the origin and type of the crude oil and also with the nature of the chemical constituents, particularly the polar functions, where intermolecular interactions can occur. For example, there is a gradation of viscosity between conventional crude oil, heavy oil, and bitumen (Chapter 8).

Viscosity is a measure of fluidity properties and consistencies at given temperatures (Chapter 9). Heavier crude oils, i.e., crude oils of lower API gravity, have usually higher viscosity. Increases of viscosity during storage indicate either an evaporation of volatile components or formation of degradation products dissolving in the crude oil.

13.3.5 Asphaltene Content

Asphaltenes are dark brown to black friable solids that have no definite melting point and usually intumesce on heating with decomposition to leave a carbonaceous residue. They are obtained from petroleum by the addition of a nonpolar solvent (such as a liquid hydrocarbon). Liquids used for this purpose are *n*-pentane and *n*-heptane (Table 13.2) (Chapter 7 and Chapter 10) (Speight, 1994b). Usually, the asphaltene fraction is removed by filtration through paper, but more recently, a membrane method has come into use (ASTM D4055). Liquid propane is used commercially in processing petroleum residues for asphaltenes and resins (Chapter 19). Asphaltenes are soluble in liquids such as benzene, toluene pyridine, carbon disulfide, and carbon tetrachloride.

The asphaltene fraction of feedstocks (Figure 13.2) is particularly important because as the proportion of this fraction increases, there is a concomitant increase in thermal coke yield and

TABLE 13.2
Standard Methods for Asphaltene Precipitation

Method	Precipitant	Volume Precipitant Per Gram of Sample
ASTM D893	*n*-Pentane	10 mL
ASTM D2006	*n*-Pentane	50 mL
ASTM D2007	*n*-Pentane	10 mL
ASTM D3279	*n*-Heptane	100 mL
ASTM D4124	*n*-Heptane	100 mL
IP 143	*n*-Heptane	30 mL

Source: ASTM Annual Book of Standards, 1980–2005.

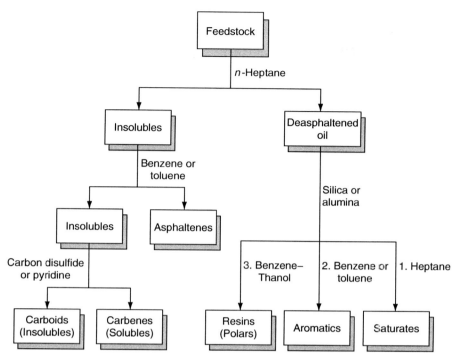

FIGURE 13.2 General fractionation scheme and nomenclature of petroleum fraction; carbenes and carboids are thermally-generated product fractions.

an increase in hydrogen demand as well as catalyst deactivation. The constituents of the asphaltenes form coke quite readily, which is of particular interest in terms of the compatibility or incompatibility of the coke precursors (Speight, 1994b).

The effect of asphaltenes and the micelle structure, the state of dispersion also merit some attention. The degree of dispersion of asphaltenes is higher in the more naphthenic or aromatic crude oils, because of the higher solvency of naphthenes and aromatics over paraffinic constituents. This phenomenon also acts in favor of the dissolution of any sludge that may form, thereby tending to decrease sludge deposition. However, an increase in crude oil often accompanies sludge dissolution.

Higher the asphaltene content, greater the tendency of the crude oil to form sludge, especially when blended with other noncompatible stocks.

13.3.6 POUR POINT

The **pour point** (Chapter 9) defines the cold properties of crude oils and petroleum and petroleum products, i.e., the minimal temperature at which they still retain their fluidity (ASTM D97). Therefore, pour point also indicates the characteristics of crude oils: higher the pour point, more paraffinic is the oil and vice versa. Higher pour point crude oils are waxy, and therefore, they tend to form wax-like materials that enhance sludge formation.

To determine the pour point (ASTM D97, ASTM D5327, ASTM D5853, ASTM D5949, ASTM D5950, ASTM D5985, IP 15, IP 219, IP 441), the sample is contained in a glass test tube fitted with a thermometer and immersed in one of three baths containing coolants. The sample is dehydrated and filtered at a temperature 25°C (45°F) higher than the anticipated cloud point. It is then placed in a test tube and cooled progressively in coolants held at −1°C to +2°C (30°F to

35°F), −18°C to −20°C (−4°F to 0°F) and −32°C to −35°C (−26°F to −31°F), respectively. The sample is inspected for cloudiness at temperature intervals of 1°C (2°F). If conditions or oil properties are such that reduced temperatures are required to determine the pour point, alternate tests are available that accommodate the various types of samples.

13.3.7 ACIDITY

The acidity of petroleum or petroleum products is usually measured in terms of the **acid number** which is defined as the number of milli-equivalents per gram of alkali required to neutralize the acidity of the petroleum sample (ASTM D664, ASTM D974, ASTM D3242).

Acidity due to the presence of inorganic constituents is not expected to be present in crude oils, but organic acidity might be found. Acidic character is composed of contributions from strong organic acids and other organic acids. Values more than 0.15 mg potassium hydroxide per gram are considered to be significantly high. Crude oils of higher acidities may exhibit a tendency of instability.

The acid imparting agents in crude oils are naphthenic acids and hydrosulfides (thiols, mercaptans, R–SH). These are sometimes present in the crude oil, originally in small and varying concentrations. Normally, the total acidity of crude oils is in the range of 0.1 to 0.5 mg potassium hydroxide per gram, although higher values are not exceptional.

Free hydrogen sulfide is often present in crude oils, a concentration of up to 10 ppm is acceptable in spite of its toxic nature. However, higher hydrogen sulfide concentrations are sometimes present, 20 ppm posing serious safety hazards. Additional amounts of hydrogen sulfide can form during the crude oil processing, when hydrogen reacts with some organic sulfur compounds converting them to hydrogen sulfide. In this case it is referred to as potential hydrogen sulfide, contrary to free hydrogen sulfide.

Acidity can also form by bacterial action insofar as some species of aerobic bacteria can produce organic acids from organic nutrients. On the other hand, anaerobic sulfate-reducing bacteria can generate hydrogen sulfide, which, in turn, can be converted to sulfuric acid (by bacterial action).

13.3.8 METALS (ASH) CONTENT

The majority of crude oils contain metallic constituents that are often determined as combustion ash (ASTM D482). This is particularly so for the heavier feedstocks. These constituents, of which nickel and vanadium are the principal metals, are very influential in regard to feedstock behavior in processing operations.

The metal (inorganic) constituents of petroleum or a liquid fuel arise from either those present in the crude oil originally or those picked up by the crude oil during storage and handling. The former are mostly metallic substances like vanadium, nickel, sodium, iron, silica etc.; the latter may be contaminants such as sand, dust, and corrosion products.

Incompatibility, leading to deposition of the metals (in any form) on to the catalyst leads to catalyst deactivation, whether it is by physical blockage of the pores or destruction of reactive sites. In the present context, the metals must first be removed if erroneously high carbon residue data are to be avoided. Alternatively, they can be estimated as ash by complete burning of the coke after carbon residue determination.

Metals content above 200 ppm are considered to be significant, but the variations are very large. The higher the ash content, the higher is the tendency of the crude oil to form sludge or sediment.

13.3.9 Water Content, Salt Content, and Bottom Sediment and Water (BS&W)

Water content (ASTM D4006, ASTM D4007, ASTM D4377, ASTM D4928), salt content (ASTM D3230), and bottom sediment and water (ASTM D96, ASTM D1796, ASTM D4007) indicate the concentrations of aqueous contaminants, present in the crude either originally or picked up by the crude during handling and storage. Water and salt content of crude oils produced at the field can be very high, forming sometimes its major part. The salty water is usually separated at the field, usually by settling and draining, surface-active agents electrical emulsion breakers (desalters) are sometimes employed. The water and salt contents of crude oil supplied to the buyers is a function of the production field. Water content below 0.5%, salt content up to 20 pounds per 1000 barrels, and bottom sediment and water up to 0.5% are considered to be satisfactory.

Although the centrifuge methods are still employed (ASTM D96, ASTM D1796, ASTM D2709 and ASTM D4007), many laboratories prefer the Dean and Stark adaptor (Figure 13.3) (ASTM D95). The apparatus consists of a round-bottom flask of capacity 50 mL connected to a Liebig condenser by a receiving tube of capacity 25 mL, graduated in 0.1 mL. A weighed amount, corresponding to approximately 100 mL of oil, is placed in the flask with 25 mL of dry toluene. The flask is heated gently until the 25 mL of toluene has distilled into the graduated tube. The water distilled with the toluene, separates to the bottom of the tube where the volume is recorded as mL, or the weight as mg, or percent.

To determine the sediment in petroleum or in a petroleum product, the method involves solvent extraction using a Soxhlet extractor (Figure 13.4).

The Karl Fischer titration method (ASTM D1744), and the colorimetric Karl Fischer titration method (ASTM D4298) still find wide application in many laboratories for the determination of water in liquid fuels, specifically the water content of aviation fuels.

Higher the bottom sediment and water content, higher the sludge and deposit formation rates that can be expected in the stored crude oil.

13.4 METHODS FOR DETERMINING INSTABILITY AND INCOMPATIBILITY

Stability and instability and compatibility and incompatibility can be estimated using several tests. These tests are

1. Compatibility spot tests (ASTM D2781, ASTM D4740)
2. Thermal stability test data (ASTM D873, ASTM D3241)
3. Existent and potential sludge formation (hot filtration test) (ASTM D4870)
4. Asphaltene content (ASTM D3729)
5. Rate of viscosity increase (ASTM D445) of stored residual fuel oil samples at various temperatures with and without exposure to air
6. Color (ASTM D1500)

In addition to the test for the stability and compatibility of residual fuels (ASTM D4740), the other most frequently used stability test is the existent and accelerated dry sludge content determination of residual fuels (the hot filtration test) (Figure 13.5). Hot filtration test results of up to 0.2% w/w are considered to be satisfactory, results above 0.4% w/w indicate poor stability, but differing values might be required, depending on the intended use of the product.

One test, or property, that is somewhat abstract in its application, but which is becoming more meaningful and popular is the **solubility parameter** (Chapter 10 and Chapter 11). The solubility parameter allows estimations to be made of the ability of liquids to become miscible

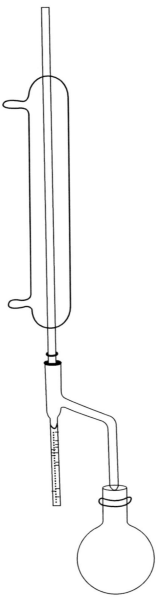

FIGURE 13.3 Dean and Stark distillation. (From Speight J.G., 2001. *Handbook of Petroleum Analysis.* John Wiley & Sons Inc., Hoboken, NJ. With permission.)

on the basis of miscibility of model compound types, where the solubility parameter can be measured or calculated.

Although the solubility parameter is often difficult to define when complex mixtures are involved, there has been some progress. For example, petroleum fractions have been assigned a similar solubility parameter to that of the solvent used in the separation. However, there is also the concept (Speight, 1992) that the solubility parameter of petroleum fractions may be defined somewhat differently insofar as they can be estimated from data such as the atomic hydrogen-to-carbon ratios. Whichever method is the best estimate may be immaterial,

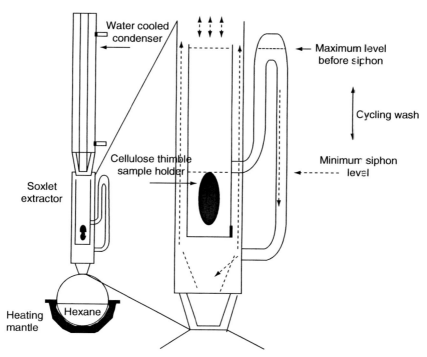

FIGURE 13.4 Determination of sediment by Soxhlet extraction. (From Speight J.G., 2001. *Handbook of Petroleum Analysis*. John Wiley & Sons Inc., Hoboken, NJ. With permission.)

as long as the data are used to the most appropriate benefit and allow some measure of predictability.

Bottle tests constitute the predominant test method and the test conditions have varied in volume, type of glass or metal, vented and unvented containers, type of bottle closure. Other procedures have involved stirred reactor vessels under both air pressure or under oxygen pressure, and small volumes of fuel employing a cover slip for solid deposition (ASTM D4625). All these procedures are gravimetric in nature.

There are several accelerated fuel stability tests that can be represented as a time–temperature matrix (Coordinating Research Council, 1979; Goetzinger et al., 1983; Hazlett, 1992). A graphical representation shows that the majority of the stability tests depicted fall close to the solid line, which represents a doubling of test time for each 10°C (18°F) change in temperature. The line extrapolates to approximately 1 year of storage under ambient conditions. Temperatures at 100°C (212°F) or higher, present special chemical problems.

The heteroatom content of the deposits formed in stability studies varied as the source or type of dopant and fuel liquid source varied. This would be an anticipated result. The color changes of both the fuel and the deposits formed are more difficult to interpret.

It is also worthy of note that **fractionation** of petroleum and its products may also give some indication of instability. There are many schemes by which petroleum and related materials might be fractionated (Chapter 7). It is not the intent to repeat the details of these in this chapter. However, a brief overview is necessary, since fractional composition can play a role in stability and incompatibility phenomena.

Petroleum can be fractionated into four broad fractions by a variety of techniques (Chapter 7), although the most common procedure involves precipitation of the asphaltene fraction and the use of adsorbents to fractionate the deasphalted oil. The fractions are named

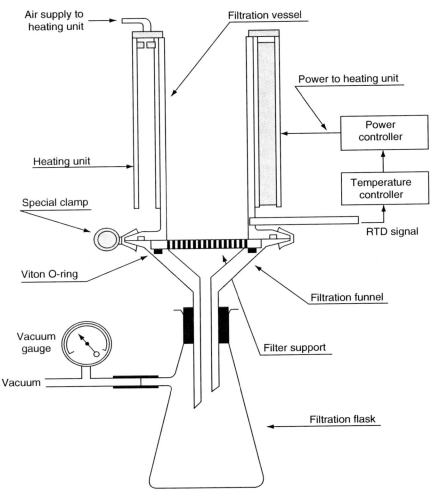

FIGURE 13.5 Schematic representation of the hot filtration apparatus. (From Speight J.G., 2001. *Handbook of Petroleum Analysis*. John Wiley & Sons Inc., Hoboken, NJ. With permission.)

for convenience and the assumption that fractionation occurs by specific compound type is not quite true.

In general terms, studies of the composition of the incompatible materials often involves determination of the distribution of the organic functional groups by selective fractionation that is analogous to the deasphalting procedure and subsequent fractionation of the maltenes (Chapter 7):

1. Heptane soluble materials: often called maltenes or petrolenes in petroleum work
2. Heptane insoluble material, benzene (or toluene) soluble material, often referred to as asphaltenes
3. Benzene (or toluene) insoluble material: referred to as carbenes and carboids when the fraction is a thermal product
4. Pyridine soluble material (carbenes) and pyridine insoluble (carboids)

Carbon disulfide and tetrahydrofuran have been used in place of pyridine. The former (carbon disulfide), although having an obnoxious odor and therefore not much different from pyridine, is easier to remove because of the higher volatility. The latter (tetrahydrofuran) is not as well established in petroleum science as it is in the coal–liquid related research. Thus, it is more than likely that the petroleum researcher will use carbon disulfide or, pyridine, or some suitable alternate solvent. It may also be necessary to substitute cyclohexane as an additional step for treatment of the heptane insoluble materials prior to treatment with benzene (or toluene). The use of quinoline has been suggested in place of pyridine but this solvent presents issues associated with its high boiling point.

Whichever solvent separation scheme is employed, there should be ample description of the procedure, so that the work can be repeated not only in the same laboratory by a different researcher but also in different laboratories by various researchers. Thus, for any particular feedstock and solvent separation scheme, the work should be repeatable and reproducible within the limits of experimental error.

Fractionation procedures allow a before-and-after inspection of any feedstock or product and can give an indication of the means by which refining or use changes the composition of the feedstock. In addition, fractionation also allows studies to be made of the interrelations between the various fractions. For example, the most interesting phenomenon (in the present context) to evolve from the fractionation studies is the relationship between the asphaltenes and the resins.

In petroleum, the asphaltenes and resins have strong interactions to the extent that the asphaltenes are immiscible (insoluble or incompatible) with the remaining constituents in the absence of the resins (Koots and Speight, 1975). And there appears to be points of structural similarity (for a crude oil) between the asphaltene and the resin constituents, thereby setting the stage for a more-than-is-generally-appreciated complex relationship, but confirming the hypothesis that petroleum is a continuum of chemical species, including the asphaltenes (Chapter 11).

This sets the stage for the incompatibility of the asphaltenes in any operation in which the asphaltene or resin constituents are physically or chemically altered. Disturbance of the asphaltene–resin relationships can be the stimulation by which, for example, some or all of the asphaltene constituents form a separate insoluble phase leading to such phenomena as coke formation (in thermal processes) or asphalt instability during use.

There is a series of characterization indices that also present indications of whether or not a petroleum product is stable or unstable. For example, the **characterization factor** indicates the chemical character of the crude oil and has been used to indicate whether a crude oil was paraffinic in nature or whether it was a naphthenic or aromatic crude oil.

The characterization factor (sometimes referred to as the **Watson characterization factor**) (Chapter 8 and Chapter 9) is a relationship between boiling point and specific gravity:

$$K = T_b^{1/3}/d$$

T_b is the cubic average boiling point, degrees Rankine (°F + 460) and d is the specific gravity at 15.6°C (60°F).

The characterization factor was originally devised to illustrate the characteristics of various feedstocks. Highly paraffinic oils have $K = 12.5$–13.0, whereas naphthenic oils have $K = 10.5$–12.5. In addition, if the characterization factor is above 12, the liquid fuel or product might, because of its paraffinic nature, be expected to form waxy deposits during storage.

The **viscosity–gravity constant** (**vgc**) was one of the early indices proposed to classify petroleum on the basis of composition. It is particularly valuable for indicating a predominantly paraffinic or naphthenic composition. The constant is based on the differences between the density and specific gravity for the various hydrocarbon species:

$$\text{vgc} = [10d - 1.0752 \log(v - 380)]/[10 - \log(v - 38)]$$

where d is the specific gravity and v is the Saybolt viscosity at 38°C (100°F). For viscous crude oils (and viscous products) where the viscosity is difficult to measure at low temperature, the viscosity at 991°C (2101°F) can be used:

$$\text{vgc} = [d - 0.24 - 0.022 \log(v - 35.5)]/0.755$$

In both cases, the lower index number is indicative of a more paraffinic sample. For example, a paraffinic sample may have a vgc on the order of 0.840, whereas the corresponding naphthenic sample may have an index on the order of 0.876.

The obvious disadvantage is the closeness of the indices, almost analogous to comparing crude oil character by specific gravity only where most crude oils fall into the range $d = 0.800–1.000$. The API gravity expanded this scale from 5–60, thereby adding more meaning to the use of specific gravity data.

In a similar manner, the correlation index which is based on a plot of the specific gravity (d) versus the reciprocal of the boiling point (K) in °K (°K = degrees Kelvin = °C + 273) for pure hydrocarbons adds another dimension to the numbers:

$$\text{Correlation Index (CI)} = 473.7d - 456.8 + 48640/K$$

In the case of a petroleum fraction, K is the average boiling point determined by the standard distillation method.

The line described by the constants of the individual members of the normal paraffin series is given a value of CI = 0 and a parallel line passing through the point for benzene is given a value of CI = 100 (Chapter 8 and Chapter 9). Value between 0 and 15 indicate a predominance of paraffinic hydrocarbons in the sample and values from 15–20 indicate predominance either of naphthenes or of mixtures of paraffins or naphthenes or aromatics; an index value above 50 indicates a predominance of aromatics in the fraction.

13.5   EFFECT OF ASPHALTENE CONSTITUENTS

The focus of incompatibility studies has usually been on the whole crude oil and specifically on the asphaltene fraction. As already stated (Chapter 12), the asphaltene constituents are the highest molecular weight and most polar fractions found in crude oil. The characteristics of the asphaltene constituents and the amount in crude oil depend to a greater or lesser extent on the crude oil source. During refining of crude oil, asphaltene constituents will end up in a high percentage in the residual fuels as the light fractions (gasoline, jet fuel etc.) are removed from the oil through cracking and visbreaking.

The problems with the asphaltene constituents has increased due to the need to extract even the heaviest crude oils as well as the trend to extract large amounts of light fractions out of crude oil by, among other methods, cracking and visbreaking. Some example of problems due to flocculation and sedimentation of asphaltene constituents and reacted asphaltene constituents are

- *Oil recovery*
  Wellbore plugging and pipeline deposition.
- *Visbreaking and cracking processes*
  Degraded asphaltene constituents are more aromatic (loss of aliphatic chains) and less soluble.
- *Emulsion formation*
  If there is a high degree of water contamination, emulsions may be formed. Asphaltene constituents are responsible for the undesired stabilization of emulsions, since asphaltene constituents are highly polar and surface active.
- *Preheating*
  Preheating of fuel oil prior to combustion encourages reaction and precipitation of the reacted asphaltene constituents leading to coking.
- *Combustion of fuel oil*
  A high content (>6% by weight) of asphaltene constituents in the fuel oil results in ignition delay and poor combustion, further leading to boiler fouling, poor heat transfer, stack solid emission, and corrosion problems.
- *Blending of fuel oils as well as blending of fuel oils with heavy crude oils*
  The change of medium during mixing may cause destabilization of asphaltene (and reacted asphaltene) constituents.
- *Storage problems of fuel oils and visbroken residua*

Sedimentation and plugging can occur due to oxidation of the asphaltene constituents. The increased polarity may cause asphaltene aggregation. Further, the sludge formation can be accelerated if there is bacterial infestation in the fuel.

Thus, the deposition of asphaltene constituents is the consequence of instability of the crude oil or the crude oil product (e.g., fuel oil). The asphaltene constituents are stabilized by resin constituents (Chapter 12) and maintained in the crude oil due to this stabilization. Asphaltene dispersants are substitutes for the natural resin constituents and are able to keep the asphaltene constituents dispersed to prevent flocculation or aggregation and phase separation. Dispersants will also clean up sludge in the fuel system and they have the ability to adhere to the surface of materials that are insoluble in the oil and convert them into stable colloidal suspensions.

At the other end of the molecular weight scale and refining are the heteroatoms (particularly nitrogen, sulfur, and trace metals) (Chapter 6) that are present in petroleum and might also be expected to be present in liquid fuels and other products from petroleum. And, indeed, this is often the case, although there may have been some skeletal changes induced by the refining processes. Oxygen is much more difficult to define in petroleum and liquid fuels. However, it must be stressed that instability and incompatibility is not directly related to the total nitrogen, oxygen, or sulfur content. The formation of color or sludge or sediment is a result of several factors. Perhaps the main factor is the location and nature of the heteroatom which, in turn, determines reactivity (Por, 1992; Mushrush and Speight, 1995).

The instability and incompatibility of petroleum products is a precursor to either the formation of degradation products or the occurrence of undesirable changes in the properties of the fuel. Individually, the components of a product may be stable and in compliance with specifications, but their blend may exhibit poor stability properties, making them unfit for use.

Incompatibility in the distillate products is important to consumers as well as the producer. Distillate products that are made from the refining process, based on straight-run distillation, show very little incompatibility problems. However, at present and in the future, problems for refiners will continue to increase as the quality of the available crude decreases worldwide. This decrease coupled with the inevitable future use of fuel liquids from both coal and shale sources will exacerbate the present problems for the producers.

Incompatibility in petroleum products can be linked to the presence of several different deleterious heteroatomic compound classes (Mushrush and Speight, 1995). The incompatibility observed in a fuel is dependent on the blending stocks employed in its production. The light cycle oils, produced during catalytic process, contain the unstable heteroatomic species that are responsible for the observed deterioration in fuel. The solution is to use straight-run distillate product, but a deficiency in supply and an option is to use chemical additives that overcome the incompatibilities of the variant chemical composition of the light cycle oil.

In general, the reaction sequence for sediment formation can be envisaged as it is dependent on the most reactive of the various heteroatomic species that are present in fuels (Pedley et al., 1987). The worst-case scenario would consist of a high olefin fuel with both high indole concentration and a catalytic trace of sulfonic acid species. This reaction matrix would lead to rapid degradation. However, just as there is no one specific distillate product, there is also no one mechanism of degradation. In fact, the mechanism and the functional groups involved will give a general but not specific mode of incompatibility (Taylor and Wallace, 1967; Taylor, 1969; Cooney et al., 1985; Hiley and Pedley, 1987; Mushrush et al., 1990, 1991). The key reaction in many incompatibility processes is the generation of the hydroperoxide species from dissolved oxygen. Once the hydroperoxide concentration starts to increase, macromolecular incompatibility precursors form in the fuel. Acid or base catalyzed condensation reactions then rapidly increase both the polarity, incorporation of heteroatoms, and the molecular weight.

When various feedstocks are blended at the refinery, incompatibility can be explained by the onset of acid–base catalyzed condensation reactions of the various organo-nitrogen compounds in the individual blending stocks. These are usually very rapid reactions with practically no observed induction time period (Hardy and Wechter, 1990b).

When the product is transferred to a storage tank or some other holding tank incompatibility can occur by the free-radical hydroperoxide induced polymerization of active olefins. This is a relatively slow reaction, because the observed increase in hydroperoxide concentration is dependent on the dissolved oxygen content (Taylor, 1976; Mayo and Lan, 1987; Mushrush et al., 1994).

The third incompatibility mechanism involves degradation when the product is stored for prolonged periods, as might occur during stockpiling of fuel for military use (Brinkman et al., 1980; Stavinoha and Westbrook, 1980; Brinkman and Bowden, 1982; Goetzinger et al., 1983; Cooney et al., 1985; Hazlett and Hall, 1985;). This incompatibility process involves (1) the buildup of hydroperoxide moieties after the gum reactions; (2) a free-radical reaction with the various organo-sulfur compounds present that can be oxidized to sulfonic acids; and (3) reactions such as condensations between organo-sulfur and nitrogen compounds and esterification reactions.

Incompatibility during refining can occur in a variety of processes, either by intent (such as in the deasphalting process) or inadvertently, when the separation is detrimental to the process. Thus, separation of solids occurs whenever the solvent characteristics of the liquid phase are no longer adequate to maintain polar or high molecular weight material in solution.

Examples of such occurrences are (1) asphaltene separation which occurs when the paraffinic nature of the liquid medium increases (Chapter 7 and Chapter 10), (2) wax separation which occurs when there is a drop in temperature or the aromaticity of the liquid medium increases (Chapter 7 and Chapter 19), (3) sludge or sediment formation in a reactor which occurs when the solvent characteristics of the liquid medium change so that asphaltic or wax materials separate (Chapter 7, Chapter 10, and Chapter 19), coke formation which occurs at high temperatures and commences when the solvent power of the liquid phase is not sufficient to maintain the coke precursors in solution (Chapter 7, Chapter 10, and Chapter 12), and sludge or sediment formation in fuel products which occurs because of the interplay of several chemical and physical factors (Mushrush and Speight, 1995).

REFERENCES

Abraham, H. 1945. *Asphalt and Allied Substances*. 5th edn. Van Nostrand Inc., New York. Volume I. p. 1.

ASTM. 2005. *Annual Book of Standards*. American Society for Testing and Materials, West Conshohocken, PA.

Batts, B.D. and Fathoni, A.Z. 1991. *Energy Fuels* 5: 2.

Brinkman, D.W., Bowden, J.N., and Giles, H.H. 1980. *Crude Oil and Finished Fuel Storage Stability: An Annotated Review*. NTIS No. DOE/BETC/RI-79/13.

Brinkman, D.W. and Bowden, J.N. 1982. *Fuel* 61: 1141.

Brinkman, D.W. and White, E.W. 1981. *Distillate Fuel Stability and Cleanliness*. Stavinoha, L.L. and Henry, C.P. eds. Special Technical Publication No. 751. American Society for Testing and Materials, Philadelphia. p. 84.

Cooney, J.V., Beal, E.J., and Hazlett, R.N. 1985. *Ind. Eng. Chem. Prod. Res. Dev*. 24: 294.

Coordinating Research Council, 1979. *CRC Literature Survey on Thermal Oxidation Stability of Jet Fuel*. Report No. 509. Coordinating Research Council Inc., Atlanta, GA.

Forbes, R.J. 1958a. *A History of Technology*. Oxford University Press, Oxford, UK. Volume V. p. 102.

Forbes, R.J. 1958b. *Studies in Early Petroleum Chemistry*. Brill, E.J. Leiden, The Netherlands.

Forbes, R.J. 1959. *More Studies in Early Petroleum Chemistry*. Brill, E.J. Leiden, The Netherlands.

Goetzinger, J.W., Thompson, C.J., and Brinkman, D.W. 1983. *A Review of Storage Stability Characteristics of Hydrocarbon Fuels*. U.S. Department of Energy, Report No DOE/BETC/IC-83-3.

Hardy, D.R. and Wechter, M.A. 1990a. *Energy Fuels* 4: 270.

Hardy, D.R. and Wechter, M.A. 1990b. *Fuel* 69: 720.

Hazlett, R.N. 1992. *Thermal Oxidation Stability of Aviation Turbine Fuels*. Monograph No. 1. American Society for Testing and Materials, Philadelphia. PA.

Hazlett, R.N. and Hall, J.M. 1985. *Chemistry of Engine Combustion Deposits*, Ebert, L.B. ed. Plenum Press, New York.

Hiley, R.W. and Pedley, J.F. 1987. *Fuel* 67: 1124.

Hoffman, H.L. 1992. *Petroleum Processing Handbook*. McKetta, J.J. ed. Marcel Dekker Inc., New York. p. 2.

James, P. and Thorpe, N. 1994. *Ancient Inventions*. Ballantine Books, New York.

Koots, J.A. and Speight, J.G. 1975. *Fuel* 54: 179.

Mayo, F.R. and Lan, B.Y. 1987. *Ind. Eng. Chem. Prod. Res.* 26: 215.

Mushrush, G.W., Beal, E.J., Hazlett, R.N., and Hardy, D.R. 1990. *Energy Fuels* 5: 258.

Mushrush, G.W., Hazlett, R.N., Pellenbarg, R.E., and Hardy, D.R. 1991. *Energy Fuels* 5: 258.

Mushrush, G.W. and Hardy, D.R. 1994. Preprints. *Div. Fuel Chem. Am. Chem. Soc.* 39: 904

Mushrush, G.W. and Speight, J.G. 1995. *Petroleum Products: Instability and Incompatibility*. Taylor & Francis Publishers, Washington, DC.

Pedley, J.F., Hiley, R.W., and Hancock, R.A. 1987. *Fuel* 66: 1645.

Por, N. 1992. *Stability Properties of Petroleum Products*. Israel Institute of Petroleum and Energy, Tel Aviv, Israel.

Power, A.J. and Mathys, G.I. 1992. *Fuel* 71: 903.

Ruzicka, D.J. and Nordenson, S. 1990. *Fuel* 69: 710.

Schramm, L.L. ed. 1992. *Emulsions: Fundamentals and Applications in the Petroleum Industry*. Advances in Chemistry Series No. 231. American Chemical Society, Washington, DC.

Speight, J.G. 1992. *Proceedings,* 4th International Conference on the Stability and Handling of Liquid Fuels. U.S. Department of Energy, Washington, DC. Volume 1. p. 169.

Speight, J.G. 1994a. *The Chemistry and Technology of Coal*. 2nd edn. Marcel Dekker Inc., New York.

Speight, J.G. 1994b. *Asphaltenes and Asphalts*. I. *Developments in Petroleum Science, 40*. Yen, T.F. and Chilingarian, G.V. eds. Elsevier, Amsterdam. Chapter 2.

Speight, J.G. 2000. *The Desulfurization of Heavy Oils and Residua*. 2nd edn. Marcel Dekker Inc., New York.

Speight, J.G. 2001. *Handbook of Petroleum Analysis*. John Wiley & Sons Inc., New York.

Speight, J.G. 2002. *Handbook of Petroleum Product Analysis*. John Wiley & Sons Inc., New York.

Stavinoha, L.L. and Henry, C.P. eds. 1981. *Distillate Fuel Stability and Cleanliness*. Special Technical Publication No. 751. American Society for Testing and Materials, Philadelphia.

Stavinoha, L.L. and Westbrook, S.R. 1980. *Accelerated Stability Test Techniques for Diesel Fuels*. U.S. Department of Energy, Report No. DOE/BC/10043-12.

Taylor, W.F. 1969. *Ind. Eng. Chem. Prod. Re. Dev.* 8: 375.

Taylor, W.F. 1976. *Ind. Eng. Chem. Prod. Re. Dev.* 15: 64.

Taylor, W.F. and Wallace, T.J. 1967. *Ind. Chem. Prod. Res. Dev.* 6: 258.

Wallace, T.J. 1964. *Advances in Petroleum Chemistry and Refining*. McKetta J.J. Jr. ed. Interscience, New York. p. 353.

Wallace, T.J. 1969. *Advances in Petroleum Chemistry and Refining*. McKetta J.J. Jr. ed. Volume IX. Chapter 8. p. 353.

Part III

Refining

14 Introduction to Refining Processes

14.1 INTRODUCTION

In a very general sense, petroleum refining can be traced back over 5000 years to the times when asphalt materials and oils were isolated from areas where natural seepage occurred (Abraham, 1945; Forbes, 1958; Hoiberg, 1960). Any treatment of the asphalt (such as hardening in the air prior to use) or of the oil (such as allowing for more volatile components to escape prior to use in lamps) may be considered to be refining under the general definition of refining. However, petroleum refining as we know it is a very recent science and many innovations evolved during the twentieth century.

Briefly, petroleum refining is the separation of petroleum into fractions and the subsequent treating of these fractions to yield marketable products (Kobe and McKetta, 1958; Nelson, 1958; Gruse and Stevens, 1960; Bland and Davidson, 1967; Hobson and Pohl, 1973). In fact, a refinery is essentially a group of manufacturing plants which vary in number with the variety of products produced (Figure 14.1). Refinery processes must be selected and products manufactured to give a balanced operation in which petroleum is converted into a variety of products in amounts that are in accord with the demand for each. For example, the manufacture of products from the lower-boiling portion of petroleum automatically produces a certain amount of higher-boiling components. If the latter cannot be sold as, say, heavy fuel oil, these products will accumulate until refinery storage facilities are full. To prevent the occurrence of such a situation, the refinery must be flexible and be able to change operations as needed. This usually means more processes: thermal processes to change an excess of heavy fuel oil into more gasoline with coke as the residual product, or a vacuum distillation process to separate the heavy oil into lubricating oil stocks and asphalt.

As the basic elements of crude oil, hydrogen and carbon form the main input into a refinery, combining into thousands of individual constituents, and the economic recovery of these constituents varies with the individual petroleum according to its particular individual qualities, and the processing facilities of a particular refinery. In general, crude oil, once refined, yields three basic groupings of products that are produced when it is broken down into cuts or fractions (Table 14.1). The gas and gasoline cuts form the lower-boiling products and are usually more valuable than the higher-boiling fractions and provide gas (liquefied petroleum gas), naphtha, aviation fuel, motor fuel and feedstocks, for the petrochemical industry. **Naphtha**, a precursor to gasoline and solvents, is extracted from both the light and middle range of distillate cuts and is also used as a feedstock for the petrochemical industry. The middle distillates refer to products from the middle boiling range of petroleum and include **kerosene**, diesel fuel, distillate fuel oil, and light gas oil. Waxy distillate and lower-boiling lubricating oils are sometimes included in the middle distillates. The remainder of the crude oil includes the higher-boiling lubricating oils, gas oil, and residuum (the nonvolatile

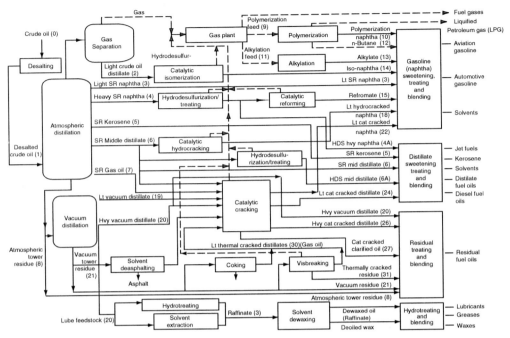

FIGURE 14.1 Schematic overview of a refinery. (From OSHA Technical Manual, Section IV, Chapter 2, Petroleum Refining Processes.)

fraction of the crude oil). The residuum can also produce heavy lubricating oils and waxes, but is more often sued for asphalt production. The complexity of petroleum is emphasized insofar as the actual proportions of light, medium and heavy fractions vary significantly from one crude oil to another.

TABLE 14.1
Crude Petroleum Is a Mixture of Compounds That Can be Separated into Different Generic Boiling Fractions

Fraction	Boiling Range[a]	
	°C	°F
Light naphtha	−1–150	30–300
Gasoline	−1–180	30–355
Heavy naphtha	150–205	300–400
Kerosene	205–260	400–500
Light gas oil	260–315	400–600
Heavy gas oil	315–425	600–800
Lubricating oil	>400	>750
Vacuum gas oil	425–600	800–1100
Residuum	>510	>950

[a]For convenience, boiling ranges are converted to the nearest 5°.

The refining industry has been the subject of the four major forces that affect most industries and which have hastened the development of new petroleum refining processes: (1) the demand for products such as gasoline, diesel, fuel oil, and jet fuel, (2) feedstock supply, specifically the changing quality of crude oil and geopolitics between different countries and the emergence of alternate feed supplies such as bitumen from tar sand, natural gas, and coal, (3) environmental regulations that include more stringent regulations in relation to sulfur in gasoline and diesel, and (4) technology development such as new catalysts and processes.

In the early days of the twentieth century, refining processes were developed to extract kerosene for lamps. Any other products were considered unusable and were usually discarded. Thus, first refining processes were developed to purify, stabilize, and improve the quality of kerosene. However, the invention of the internal combustion engine led (at about the time of World War I) to a demand for gasoline, for use in increasing quantities as a motor fuel for cars and trucks. This demand on the lower-boiling products increased, particularly when the market for aviation fuel developed. Thereafter, refining methods had to be constantly adapted and improved to meet the quality requirements and needs of car and aircraft engines.

Since then, the general trend throughout refining has been to produce more products from each barrel of petroleum and to process those products in different ways to meet the product specifications for use in modern engines. Overall, the demand for gasoline has rapidly expanded and demand has also developed for gas oils and fuels for domestic central heating, and fuel oil for power generation, as well as for light distillates and other inputs, derived from crude oil, for the petrochemical industries.

As the need for the lower-boiling products developed, petroleum yielding the desired quantities of the lower-boiling products became less available and refineries had to introduce conversion processes to produce greater quantities of lighter products from the higher-boiling fractions. The means by which a refinery operates in terms of producing the relevant products, depends not only on the nature of the petroleum feedstock but also on its configuration (i.e., the number of types of processes that are employed to produce the desired product slate) and the refinery configuration is, therefore, influenced by the specific demands of a market. Therefore, refineries need to be constantly adapted and upgraded to remain viable and responsive to ever-changing patterns of crude supply and product market demands. As a result, refineries have been introducing increasingly complex and expensive processes to gain higher yields of lower-boiling products from the higher-boiling fractions and residua.

To convert crude oil into desired products in an economically feasible and environmentally acceptable manner, refinery process for crude oil are generally divided into three categories: (1) separation processes, of which distillation is the prime example, (2) conversion processes, of which **coking** and **catalytic cracking** are prime examples, and (3) finishing processes, of which hydrotreating to remove sulfur is a prime example.

The simplest refinery configuration is the topping refinery, which is designed to prepare feedstocks for petrochemical manufacture or for the production of industrial fuels in remote oil-production areas. The topping refinery consists of tankage, a distillation unit, recovery facilities for gases and light hydrocarbons, and the necessary utility systems (steam, power, and water-treatment plants). Topping refineries produce large quantities of unfinished oils and are highly dependent on local markets, but the addition of hydrotreating and reforming units to this basic configuration results in a more flexible hydroskimming refinery, which can also produce desulfurized distillate fuels and high-octane gasoline. These refineries may produce up to half of their output as residual fuel oil, and they face increasing market loss as the demand for low-sulfur (even no-sulfur) high-sulfur fuel oil increases.

The most versatile refinery configuration today is known as the conversion refinery. A conversion refinery incorporates all the basic units found in both the topping and hydroskimming refineries, but it also features gas oil conversion plants such as catalytic cracking and hydrocracking units, olefin conversion plants such as alkylation or polymerization units, and, frequently, coking units for sharply reducing or eliminating the production of residual fuels. Modern conversion refineries may produce two-thirds of their output as unleaded gasoline, with the balance distributed between liquefied petroleum gas, jet fuel, diesel fuel, and a small quantity of coke. Many such refineries also incorporate solvent extraction processes for manufacturing lubricants and petrochemical units with which to recover propylene, benzene, toluene, and xylenes for further processing into polymers.

Finally, the yields and quality of refined petroleum products produced by any given oil refinery depends on the mixture of crude oil used as feedstock and the configuration of the refinery facilities. Light or sweet crude oil is generally more expensive and has inherent great yields of higher value low-boiling products such naphtha, gasoline, jet fuel, kerosene, and diesel fuel. Heavy sour crude oil is generally less expensive and produces greater yields of lower value higher-boiling products that must be converted into lower-boiling products.

The configuration of refineries may vary from refinery to refinery. Some refineries may be more oriented toward the production of gasoline (large reforming and catalytic cracking), whereas the configuration of other refineries may be more oriented toward the production of middle distillates such as jet fuel and gas oil. This chapter presents an introduction to petroleum refining in order for the reader to place each process in the correct context of the refinery.

14.2   DEWATERING AND DESALTING

Before the separation of petroleum into its various constituents can proceed, there is the need to clean the petroleum. This is often referred to as desalting and dewatering, in which the goal is to remove water and the constituents of the brine that accompany the crude oil from the reservoir to the wellhead during recovery operations.

Petroleum is recovered from the reservoir mixed with a variety of substances: gases, water, and dirt (minerals). Thus, refining actually commences with the production of fluids from the well or reservoir and is followed by pretreatment operations that are applied to the crude oil, either at the refinery or prior to transportation. Pipeline operators, for instance, are insistent on the quality of the fluids put into the pipelines; therefore, any crude oil to be shipped by pipeline or, for that matter, by any other form of transportation must meet rigid specifications with regard to water and salt content. In some instances, sulfur content, nitrogen content, and viscosity may also be specified.

Field separation, which occurs at a field site near the recovery operation, is the first attempt to remove the gases, water, and dirt that accompany crude oil coming from the ground. The separator may be no more than a large vessel that gives a quieting zone for gravity separation into three phases: gases, crude oil, and water containing entrained dirt.

Desalting is a water-washing operation performed at the production field and at the refinery site for additional crude oil cleanup (Figure 14.2). If the petroleum from the separators contains water and dirt, water washing can remove much of the water-soluble minerals and entrained solids. If these crude oil contaminants are not removed, they can cause operating problems during refinery processing, such as equipment plugging and corrosion as well as catalyst deactivation.

The usual practice is to blend crude oils of similar characteristics, although fluctuations in the properties of the individual crude oils may cause significant variations in the properties of

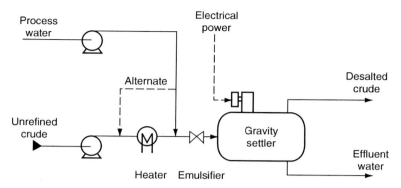

FIGURE 14.2 An electrostatic desalting unit. (From OSHA Technical Manual, Section IV, Chapter 2, Petroleum Refining Processes.)

the blend over a period of time. Blending several crude oils prior to refining can eliminate the frequent need to change the processing conditions that may be required to process each of the crude oils individually.

However, simplification of the refining procedure is not always the end result. Incompatibility of different crude oils, which can occur if, for example, a paraffinic crude oil is blended with heavy asphaltic oil, can cause sediment formation in the unrefined feedstock or in the products, thereby complicating the refinery process.

14.3 EARLY PROCESSES

Distillation was the first method by which petroleum was refined. In the early stages of refinery development, when illuminating and lubricating oils were the main products, distillation was the major and often only refinery process. At that time, gasoline was a minor, but more often unwanted, product. As the demand for gasoline increased, conversion processes were developed, because distillation could no longer supply the necessary quantities of this volatile product.

The original distillation method involved a batch operation in which the still was a cast-iron vessel mounted on brickwork over a fire and the volatile materials were passed through a pipe or gooseneck which led from the top of the still to a condenser. The latter was a coil of pipe (worm) immersed in a tank of running water. Heating a batch of crude petroleum caused the more volatile, lower-boiling components to vaporize and then condense in the worm to produce naphtha. As the distillation progressed, the higher-boiling components became vaporized and were condensed to produce kerosene, the major petroleum product of the time. When all of the possible kerosene had been obtained, the material remaining in the still was discarded. The still was then refilled with petroleum and the operation repeated. The capacity of the stills at that time was usually several barrels (bbl) of petroleum (1 bbl = 42 U.S. gallons = 34.97 Imperial gallons = 158.9 L of petroleum). It often required 3 days or even more to run (distill) a batch of crude oil.

The simple distillation, as practised in the 1860s and 1870s, was notoriously inefficient. The kerosene was, more often that not, contaminated by naphtha, which distilled during the early stages, or by heavy oil, which distilled from the residue during the final stages of the process. The naphtha generally rendered the kerosene so flammable, explosions accompanied that ignition. On the other hand, the presence of higher-boiling constituents adversely affected the excellent burning properties of the kerosene and created a great deal of smoke. This condition could be corrected by redistilling (**rerunning**) the kerosene, during which process the more volatile fraction (front-end) was recovered as additional naphtha, whereas the kerosene residue (tail) remaining in the still was discarded.

The 1880s saw the introduction of the continuous distillation of petroleum. The method employed a number of stills coupled together in a row (**battery**) and each still was heated separately and was hotter than the preceding one. The stills were arranged so that oil flowed by gravity from the first to the last. Crude petroleum in the first still was heated so that a light naphtha fraction distilled from it before the crude petroleum flowed into the second still, where a higher temperature caused the distillation of a heavier naphtha fraction. The residue then flowed to the third still where an even higher temperature caused kerosene to distill. The oil thus progressed through the battery to the last still, where destructive distillation (thermal decomposition; cracking) was carried out to produce more kerosene. The residue from the last still was removed continuously for processing into lubricating oils or for use as fuel oil.

In the early 1900s, a method of partial (or selective) condensation was developed to allow a more exact separation of petroleum fractions. A partial condenser (van Dyke tower) was inserted between the still and the conventional water-cooled condenser. The lower section of the tower was packed with stones and insulated with brick, so that the heavier less volatile material entering the tower condensed and drained back into the still. Noncondensed material passed into another section, where more of the less volatile material was condensed on air-cooled tubes and the condensate was withdrawn as a petroleum fraction. The volatile (**overhead**) material from the air-cooled section entered a second tower that also contained air-cooled tubes and often produced a second fraction. The volatile material remaining at this stage was then condensed in a water-cooled condenser to yield a third fraction. The van Dyke tower is essentially one of the first stages in a series of improvements that ultimately led to the distillation units found in modern refineries, which separate petroleum fractions by fractional distillation.

Petroleum refineries were originally designed and operated to run within a narrow range of crude oil feedstock and to produce a relatively fixed slate of petroleum products. Since the 1970s, refiners had to increase their flexibility in order to adapt to a more volatile environment. Several possible paths may be used by refiners to increase their flexibility within existing refineries. Examples of these paths are change in the severity of operating rules of some process units by varying the range of inputs used, thus achieving a slight change in output. Alternatively, refiners can install new processes and this alternate scenario offers the greatest flexibility, but is limited by the constraint of strict complementarily of the new units with the rest of the existing plant and involves a higher risk than the previous ones. It is not surprising that many refiners decide to modify existing processes.

Whatever the choice, refinery practice continues to evolve and (as will be seen in the relevant chapter) new processes are installed in line with the older modified process. The purpose of this chapter is to present to the reader a general overview of refining that, when taken into the context of the following chapters will show some of the differences that occur in refineries.

14.4 DISTILLATION

In the early stages of refinery development, when illuminating and lubricating oils were the main products, distillation was the major, and often only, refinery process. At that time, gasoline was a minor product, but as the demand for gasoline increased, conversion processes were developed, because distillation could no longer supply the necessary quantities.

It is possible to obtain products ranging from gaseous materials taken off at the top of the distillation column to a nonvolatile residue or reduced crude (**bottoms**), with correspondingly lighter materials at intermediate points. The reduced crude may then be processed by vacuum,

TABLE 14.2
Comparison of Visbreaking with Delayed Coking and Fluid Coking

Visbreaking
Purpose: to reduce viscosity of fuel oil to acceptable levels (Conversion is not a prime purpose)
Mild (470°C to 495°C; 880°F to 920°F) heating at pressures of 50 to 200 psi
    Reactions quenched before going to completion
    Low conversion (10%) to products boiling less than 220°C (430°F)
Heated coil or drum (soaker)

Delayed Coking
Purpose: to produce maximum yields of distillate products
Moderate (480°C to 515°C; 900°F to 960°F) heating at pressures of 90 psi
    Reactions allowed to proceed to completion
    Complete conversion of the feedstock
Soak drums (845°F to 900°F) used in pairs (one on stream and one off stream being decoked)
Coked until drum solid
Coke removed hydraulically from off-stream drum
Coke yield: 20%–40% by weight (dependent upon feedstock)
Yield of distillate boiling below 220°C (430°F): ca. 30% (but feedstock dependent)

Fluid Coking
Purpose: to produce maximum yields of distillate products
Severe (480°C to 565°C; 900°F to 1050°F) heating at pressures of 10 psi
    Reactions allowed to proceed to completion
    Complete conversion of the feedstock
Oil contacts refractory coke
Bed fluidized with steam; heat dissipated throughout the fluid bed
Higher yields of light ends ($<C_5$) than delayed coking
Less coke make than delayed coking (for one particular feedstock)

or steam, distillation in order to separate the high-boiling lubricating oil fractions without the danger of decomposition, which occurs at high (>350°C, >660°F) temperatures. Atmospheric distillation may be terminated with a lower-boiling fraction (cut), if it is felt that vacuum or steam distillation will yield a better-quality product, or if the process appears to be economically more favorable. Not all crude oils yield the same distillation products (Table 14.2), and the nature of the crude oil dictates the processes that may be required for refining.

14.4.1 HISTORICAL DEVELOPMENT

Distillation was the first method by which petroleum was refined. The original technique involved a batch operation in which the still was a cast-iron vessel mounted on brickwork over a fire and the volatile materials were passed through a pipe or gooseneck which led from the top of the still to a condenser. The latter was a coil of pipe (worm) immersed in a tank of running water.

Heating a batch of crude petroleum caused the more volatile, lower-boiling components to vaporize and then condense in the worm to form naphtha. As the distillation progressed, the higher-boiling components became vaporized and were condensed to produce kerosene: the major petroleum product of the time. When all of the possible kerosene had been obtained, the material remaining in the still was discarded. The still was then refilled with petroleum and the operation repeated.

The capacity of the stills at that time was usually several barrels of petroleum and it often required 3 or more days to distill (*run*) a batch of crude oil. The simple distillation as practised in the 1860s and 1870s was notoriously inefficient. The kerosene was more often that not contaminated by naphtha, which distilled during the early stages, or by heavy oil, which distilled from the residue during the final stages of the process. The naphtha generally rendered the kerosene so flammable, explosions accompanied that ignition. On the other hand, the presence of heavier oil adversely affected the excellent burning properties of the kerosene and created a great deal of smoke. This condition could be corrected by redistilling (rerunning) the kerosene, during which process the more volatile fraction (front-end) was recovered as additional naphtha, while the kerosene residue (tail) remaining in the still was discarded.

The 1880s saw the introduction of the continuous distillation of petroleum. The method employed a number of stills coupled together in a row and each still was heated separately and was hotter than the preceding one. The stills were arranged so that oil flowed by gravity from the first to the last. Crude petroleum in the first still was heated, so that a light naphtha fraction distilled from it before the crude petroleum flowed into the second still, where a higher temperature caused the distillation of a heavier naphtha fraction. The residue then flowed to the third still where an even higher temperature caused kerosene to distill. The oil thus progressed through the battery to the last still, where destructive distillation (thermal decomposition; cracking) was carried out to produce more kerosene. The residue from the last still was removed continuously for processing into lubricating oils or for use as fuel oil.

In the early 1900s, a method of partial (or selective) condensation was developed to allow a more exact separation of petroleum fractions. A partial condenser was inserted between the still and the conventional water-cooled condenser. The lower section of the tower was packed with stones and insulated with brick so that the heavier less volatile material entering the tower condensed and drained back into the still. Noncondensed material passed into another section where more of the less volatile material was condensed on air-cooled tubes and the condensate was withdrawn as a petroleum fraction. The noncondensable (overhead) material from the air-cooled section entered a second tower that also contained air-cooled tubes and often produced a second fraction. The volatile material remaining at this stage was then condensed in a water-cooled condenser to yield a third fraction. The van Dyke tower is essentially one of the first stages in a series of improvements which ultimately led to the distillation units found in modern refineries, which separate petroleum fractions by fractional distillation.

14.4.2 MODERN PROCESSES

14.4.2.1 Atmospheric Distillation

The present-day petroleum distillation unit is, like the battery of the 1800s, a collection of distillation units but, in contrast to the early battery units, a tower is used in the modern-day refinery (Figure 14.3) and brings about a fairly efficient degree of fractionation (separation).

The feed to a distillation tower is heated by flow-through pipes arranged within a large furnace. The heating unit is known as a pipe still heater or pipe still furnace, and the heating unit and the fractional distillation tower make up the essential parts of a distillation unit or pipe still. The pipe still furnace heats the feed to a predetermined temperature—usually a temperature at which a predetermined portion of the feed will change into vapor. The vapor is held under pressure in the pipe in the furnace until it discharges as a foaming stream into the fractional distillation tower. Here, the unvaporized or liquid portion of the feed descends to the bottom of the tower to be pumped away as a bottom nonvolatile product, while the vapors pass up the tower to be fractionated into gas oils, kerosene, and naphtha.

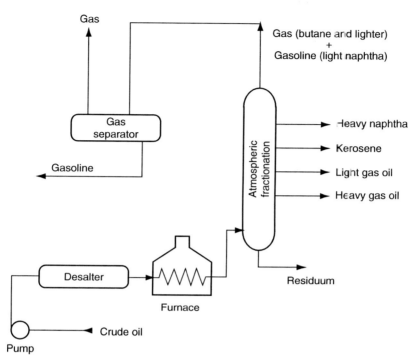

FIGURE 14.3 An atmospheric distillation unit. (From OSHA Technical Manual. Section IV, Chapter 2, Petroleum Refining Processes.)

Pipe still furnaces vary greatly and, in contrast to the early units where capacity was usually 200 to 500 bbl per day, can accommodate 25,000 bbl, or more, of crude petroleum per day. The walls and ceiling are insulated with firebrick and the interior of the furnace is partially divided into two sections: a smaller convection section where the oil first enters the furnace and a larger section (fitted with heaters) where the oil reaches its highest temperature.

Another twentieth century innovation in distillation is the use of heat exchangers which are also used to preheat the feed to the furnace. These exchangers are bundles of tubes arranged within a shell, so that a feedstock passes through the tubes in the opposite direction from a heated feedstock passing through the shell. By this means, cold crude oil is passed through a series of heat exchangers where hot products from the distillation tower are cooled, before entering the furnace and as a heated feedstock. This results in a saving of heater fuel and is a major factor in the economical operation of modern distillation units.

All of the primary fractions from a distillation unit are equilibrium mixtures and contain some proportion of the lighter constituents characteristic of a lower-boiling fraction. The primary fractions are stripped of these constituents (stabilized) before storage or further processing.

14.4.2.2 Vacuum Distillation

Vacuum distillation as applied to the petroleum refining industry is truly a technique of the twentieth century and has since wide use in petroleum refining. Vacuum distillation evolved because of the need to separate the less volatile products, such as lubricating oils, from the petroleum without subjecting these high-boiling products to cracking conditions. The boiling point of the heaviest cut obtainable at atmospheric pressure is limited by the temperature (ca. 350°C; ca. 660°F) at which the residue starts to decompose (crack). When the feedstock is required for the manufacture of lubricating oils, further fractionation without cracking is desirable and this can be achieved by distillation under vacuum conditions.

Operating conditions for vacuum distillation (Figure 14.4) are usually 50 to 100 mm of mercury (atmospheric pressure = 760 mm of mercury). In order to minimize large fluctuations in pressure in the vacuum tower, the units are necessarily of a larger diameter than the atmospheric units. Some vacuum distillation units have diameters of the order of 45 ft (14 m). By this means, a heavy gas oil may be obtained as an overhead product at temperatures of about 150°C (300°F), and lubricating oil cuts may be obtained at temperatures of 250°C to 350°C (480°F to 660°F), feed and residue temperatures are kept below the temperature of 350°C (660°F), above which cracking will occur. The partial pressure of the hydrocarbons is effectively reduced still further by the injection of steam. The steam added to the column, principally for the stripping of asphalt in the base of the column, is superheated in the convection section of the heater.

The fractions obtained by vacuum distillation of the reduced crude (atmospheric residuum) from an atmospheric distillation unit depend on whether or not the unit is designed to produce lubricating or vacuum gas oils. In the former case, the fractions include (1) heavy gas oil, which is an overhead product and is used as catalytic cracking stock or, after suitable treatment, a light lubricating oil, (2) lubricating oil (usually three fractions—light, intermediate, and heavy), which is obtained as a side-stream product, and (3) asphalt (or residuum), which is the bottom product and may be used directly as, or to produce, asphalt and which may also be blended with gas oils to produce a heavy fuel oil.

In the early refineries, distillation was the prime means by which products were separated from crude petroleum. As the technologies for refining evolved into the twentieth century, refineries became much more complex (Figure 14.1), but distillation remained the prime means by which petroleum is refined. Indeed, the distillation section of a modern refinery (Figure 14.3 and Figure 14.4, see also Figure 14.1) is the most flexible section in the refinery, since conditions can be adjusted to process a wide range of refinery feedstocks from the

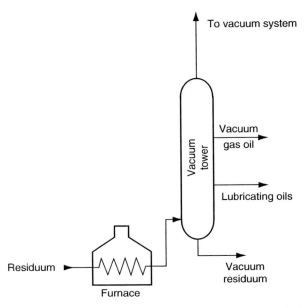

FIGURE 14.4 A vacuum distillation unit. (From OSHA Technical Manual, Section IV, Chapter 2, Petroleum Refining Processes.)

lighter crude oils to the heavier more viscous crude oils. However, the maximum permissible temperature (in the vaporizing furnace or heater) to which the feedstock can be subjected is 350°C (660°F). Thermal decomposition occurs above this temperature which, if it occurs within a distillation unit, can lead to coke deposition in the heater pipes or in the tower itself with the resulting failure of the unit.

The contained use of atmospheric and vacuum distillation has been a major part of refinery operations during this century and no doubt will continue to be employed throughout the remainder of the century as the primary refining operation.

14.4.2.3 Azeotropic and Extractive Distillation

As the twentieth century evolved, distillation techniques in refineries became more sophisticated to handle a wider variety of crude oils, to produce marketable products or feedstocks for other refinery units. However, it became apparent that the distillation units in the refineries were incapable of producing specific product fractions. In order to accommodate this type of product demand, refineries have, in the latter half of the twentieth century, incorporated azeotropic distillation and extractive distillation in their operations.

All compounds have definite boiling temperatures, but a mixture of chemically dissimilar compounds will sometimes cause one or both of the components to boil at a temperature other than that expected. A mixture that boils at a temperature lower than the boiling point of any of the components is an azeotropic mixture. When it is desired to separate close-boiling components, the addition of a nonindigenous component will form an azeotropic mixture with one of the components of the mixture, thereby lowering the boiling point by the formation of an azeotrope and facilitate separation by distillation.

The separation of these components of similar volatility may become economic if an entrainer can be found that effectively changes the relative volatility. It is also desirable that the entrainer be reasonably cheap, stable, nontoxic, and readily recoverable from the components. In practice, it is probably this last-named criterion that limits severely the application of extractive and azeotropic distillation. The majority of successful processes are those in which the entrainer and one of the components separate into two liquid phases on cooling if direct recovery by distillation is not feasible. A further restriction in the selection of an azeotropic entrainer is that the boiling point of the entrainer be in the range of 10°C to 40°C (18°F to 72°F) below that of the components.

14.5 THERMAL METHODS

14.5.1 HISTORICAL DEVELOPMENT

Cracking was used commercially in the production of oils from coal and shales before the petroleum industry began, and the discovery that the heavier products could be decomposed to lighter oils was used to increase the production of kerosene and was called cracking distillation.

The precise origins of cracking distillation are unknown. It is rumored that, in 1861, a stillman had to leave his charge for a longer time than he intended (the reason is not known) during which time the still overheated. When he returned, he noticed that the distillate in the collector was much more volatile than anticipated at that particular stage of the distillation. Further investigation lead to the development of cracking distillation (i.e., thermal degradation with the simultaneous production of distillate).

Cracking distillation (thermal decomposition with simultaneous removal of distillate) was recognized as a means of producing the valuable lighter product (kerosene) from heavier nonvolatile materials. In the early days of the process (1870 to 1900), the technique was very simple—a batch of crude oil was heated until most of the kerosene had been distilled from it and the overhead material had become dark in color. At this point, distillation was discontinued and the heavy oils were held in the hot zone, during which time some of the high molecular weight components were decomposed to produce lower molecular weight products. After a suitable time, distillation was continued to yield light oil (kerosene) instead of the heavy oil that would otherwise have been produced.

The yields of kerosene products were usually markedly increased by means of cracking distillation, but the technique was not suitable for gasoline production. As the need for gasoline arose in the early 1900s, the necessity of prolonging the cracking process became apparent and a process known as pressure cracking evolved.

Pressure cracking was a batch operation in which, as an example, gas oil (200 bbl) was heated to about 425°C (800°F) in stills that had been reinforced to operate at pressures as high as 95 psi (6.4 atm). The gas oil was held under maximum pressure for 24 h, while fires maintained the temperature. Distillation was then started and during the next 48 h produced a lighter distillate (100 bbl) which contained the gasoline components. This distillate was treated with sulfuric acid to remove unstable gum-forming components and then redistilled to produce a cracked gasoline (boiling range)

The large-scale production of cracked gasoline was first developed by Burton in 1912. The process employed batch distillation in horizontal shell stills and operated at about 400°C (ca. 750°F) and 75 to 95 psi. It was the first successful method of converting heavier oils into gasoline. Nevertheless, heating a bulk volume of oil was soon considered cumbersome, and during the years 1914 to 1922, a number of successful continuous cracking processes were developed. By these processes, gas oil was continuously pumped through a unit that heated the gas oil to the required temperature, held it for a time under pressure, and then discharged the cracked material into distillation equipment where it was separated into gases, gasoline, gas oil, and tar.

The tube-and-tank cracking process is not only typical of the early (post-1900) cracking units, but is also one of the first units on record in which the concept of reactors (soakers) being on-stream or off-stream is realized. Such a concept departs from the true batch concept, and it allowed a greater degree of continuity. In fact, the tube-and-tank cracking unit may be looked upon as a forerunner of the delayed coking operation.

In the tube-and-tank process, a feedstock (at that time, a gas oil) was preheated by exchange with the hot products from the unit pumped into the cracking coil, which consisted of several hundred feet of very strong pipe that lined the inner walls of a furnace, where oil or gas burners raised the temperature of the gas oil to 425°C (800°F). The hot gas oil passed from the cracking coil into a large reaction chamber (soaker), where the gas oil was held under temperature and pressure conditions long enough for the cracking reactions to be completed. The cracking reactions formed coke which, in the course of several days, filled the soaker. The gas oil stream was then switched to a second soaker, and the first soaker was cleaned out by drilling operations similar to those used in drilling an oil well.

The cracked material (other than coke) left the on-stream soaker to enter an evaporator (tar separator) maintained under a much lower pressure than the soaker where, because of the lower pressure, all of the cracked material, except the tar, became vaporized. The vapor left the top of the separator where it was distilled into separate fractions—gases, gasoline, and gas oil. The tar that was deposited in the separator was pumped out for use as asphalt or as a heavy fuel oil.

Early in the development of tube-and-tank **thermal cracking**, it was found that adequate yields of gasoline could not be obtained by a passage of the stock through the heating coil once; attempts to increase the conversion in one pass brought about undesirable high yields of gas and coke. It was better to crack to a limited extent, remove the products, and recycle the rest of the oil (or a distilled fraction free of tar) for repeated partial conversion. The high-boiling constituents once exposed to cracking were so changed in composition as to be more refractory than the original feedstock.

With the onset of the development of the automobile, the most important part of any refinery became the gasoline-manufacturing facilities. Among the processes that have evolved for gasoline production are thermal cracking, catalytic cracking, thermal reforming, catalytic reforming, polymerization, alkylation, coking, and distillation of fractions directly from crude petroleum.

When kerosene was the major product, gasoline was the portion of crude petroleum too volatile to be included in kerosene. The refiners of the 1890s and early 1900s had no use for it and often dumped an accumulation of gasoline into the creek or river that was usually nearby. As the demand for gasoline increased with the onset of World War I and the ensuing 1920s, more crude oil had to be distilled not only to meet the demand for gasoline but also to reduce the overproduction of the heavier petroleum fractions, including kerosene.

The problem of how to produce more gasoline from less crude oil was solved in 1913 by the incorporation of cracking units into refinery operations in which fractions heavier than gasoline were converted into gasoline by thermal decomposition. The early (pre-1940) processes employed for gasoline manufacture were processes in which the major variables involved were feedstock type, time, temperature, and pressure, which need to be considered to achieve the cracking of the feedstock to lighter products with minimal coke formation.

As refining technology evolved throughout the 20th century, the feedstocks for cracking processes became the residuum or heavy distillate from a distillation unit. In addition, the residual oils produced as the end-products of distillation processes and even some of the heavier virgin oils, often contain substantial amounts of asphaltic materials, which preclude use of the residuum as fuel oils or lubricating stocks. However, subjecting these residua directly to thermal processes has become economically advantageous, since, on the one hand, the end result is the production of lower-boiling saleable materials; on the other hand, the asphaltic materials in the residua are regarded as the unwanted coke-forming constituents.

As the thermal processes evolved and catalysts were employed with more frequency, poisoning of the catalyst with a concurrent reduction in the lifetime of the catalyst became a major issue for refiners. To avoid catalyst poisoning, it became essential that as much of the nitrogen and metals (such as vanadium and nickel) as possible should be removed from the feedstock. The majority of the heteroatoms (nitrogen, oxygen, and sulfur) and the metals are contained in, or associated with, the asphaltic fraction (residuum). It became necessary that this fraction be removed from cracking feedstocks.

With this as the goal a number of thermal processes, such as tar separation (flash distillation), vacuum flashing, visbreaking, and coking, came into wide usage by refiners and were directed at upgrading feedstocks by removal of the asphaltic fraction. The method of deasphalting with liquid hydrocarbon, gases such as propane, butane, or *iso*-butane, became a widely used refinery operation in the 1950s and was very effective for the preparation of residua for cracking feedstocks. In this process, the desirable oil in the feedstock is dissolved in the liquid hydrocarbon and asphaltic materials remain insoluble.

Operating conditions in the deasphalting tower depend on the boiling range of the feedstock and the required properties of the product. Generally, extraction temperatures can range from 55°C to 120°C (130°F to 250°F), with a pressure of 400 to 600 psi. Hydrocarbon to oil ratios of 6:1 to 10:1 by volume are normally used.

14.5.2 MODERN PROCESSES

14.5.2.1 Thermal Cracking

One of the earliest conversion processes used in the petroleum industry is the thermal decomposition of higher-boiling materials into lower-boiling products. This process is known as thermal cracking and the exact origins of the process are unknown. The process was developed in the early 1900s to produce gasoline from the "unwanted" higher-boiling products of the distillation process. However, it was soon learned that the thermal cracking process also produced a wide slate of products varying from highly volatile gases to non-volatile coke.

The heavier oils produced by cracking are light and heavy gas oils as well as a residual oil which could also be used as heavy fuel oil. Gas oils from catalytic cracking were suitable for domestic and industrial fuel oils or as diesel fuels when blended with straight-run gas oils. The gas oils produced by cracking were also a further important source of gasoline. In a once-through cracking operation, all of the cracked material is separated into products and may be used as such. However, the gas oils produced by cracking (cracked gas oils) are more resistant to cracking (more refractory) than gas oils produced by distillation (straight-run gas oils), but could still be cracked to produce more gasoline. This was achieved using a later innovation (post-1940), involving a recycle operation, in which the cracked gas oil was combined with fresh feed for another trip through the cracking unit. The extent to which recycling was carried out affected the yield of gasoline from the process.

The majority of the thermal cracking processes use temperatures of 455°C to 540°C (850°F to 1005°F) and pressures of 100 to 1000 psi; the Dubbs process may be taken as a typical application of an early thermal cracking operation. The feedstock (reduced crude) is pre-heated by direct exchange with the cracking products in the fractionating columns. Cracked gasoline and heating oil are removed from the upper section of the column. Light and heavy distillate fractions are removed from the lower section and are pumped to separate heaters. Higher temperatures are used to crack the more refractory light distillate fraction. The streams from the heaters are combined and sent to a soaking chamber where additional time is provided to complete the cracking reactions. The cracked products are then separated in a low-pressure flash chamber, where a heavy fuel oil is removed as bottoms. The remaining cracked products are sent to the fractionating columns.

Mild cracking conditions, with a low conversion per cycle, favor a high yield of gasoline components, with low gas and coke production, but the gasoline quality is not high, whereas more severe conditions give increased gas and coke production and reduced gasoline yield (but of higher quality). With limited conversion per cycle, the heavier residues must be recycled, but these recycled oils become increasingly refractory upon repeated cracking, and if they are not required as a fuel oil stock they may be coked to increase gasoline yield or refined by means of a hydrogen process.

The thermal cracking of higher-boiling petroleum fractions to produce gasoline is now virtually obsolete. The antiknock requirements of modern automobile engines together with the different nature of crude oils (compared with those of 50 or more years ago) have reduced the ability of the thermal cracking process to produce gasoline on an economic basis. Very few new units have been installed since the 1960s and some refineries may still operate the older cracking units.

14.5.2.2 Visbreaking

Visbreaking (viscosity breaking) is essentially a process of the post-1940 era and was initially introduced as a mild thermal cracking operation that could be used to reduce the viscosity of

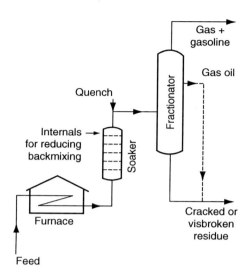

FIGURE 14.5 A soaker visbreaker. (From OSHA Technical Manual, Section IV, Chapter 2, Petroleum Refining Processes.)

residua to allow the products to meet fuel oil specifications. Alternatively, the visbroken residua could be blended with lighter product oils to produce fuel oils of acceptable viscosity. By reducing the viscosity of the residuum, visbreaking reduces the amount of light heating oil that is required for blending to meet the fuel oil specifications. In addition to the major product, fuel oil, material in the gas oil and gasoline boiling range are produced. The gas oil may be used as additional feed for catalytic cracking units, or as heating oil.

In a typical visbreaking operation (Figure 14.5), a crude oil residuum is passed through a furnace where it is heated to a temperature of 480°C (895°F), under an outlet pressure of about 100 psi. The heating coils in the furnace are arranged to provide a soaking section of low heat density, where the charge remains until the visbreaking reactions are completed and the cracked products are then passed into a flash-distillation chamber. The overhead material from this chamber is then fractionated to produce a low-quality gasoline as an overhead product and light gas oil as bottom. The liquid products from the flash chamber are cooled with a gas oil flux and then sent to a vacuum fractionator. This yields a heavy gas oil distillate and a residual tar of reduced viscosity.

14.5.2.3 Coking

Coking is a thermal process for the continuous conversion of heavy, low-grade oils into lighter products. Unlike visbreaking, coking involves complete thermal conversion of the feedstock into volatile products and coke (Table 14.2). The feedstock is typically a residuum and the products are gases, naphtha, fuel oil, gas oil, and coke. The gas oil may be the major product of a coking operation and serves primarily as a feedstock for catalytic cracking units. The coke obtained is usually used as fuel but specialty uses, such as electrode manufacture, production of chemicals and metallurgical coke are also possible and increases the value of the coke. For these uses, the coke may require treatment to remove sulfur and metal impurities.

After a gap of several years, the recovery of heavy oils, through secondary recovery techniques from oil sand formations caused a renewal of interest in these feedstocks in the 1960s and, thereafter, for coking operations. Further, the increasing attention paid to

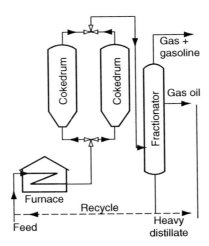

FIGURE 14.6 A delayed coker. (From OSHA Technical Manual, Section IV, Chapter 2, Petroleum Refining Processes.)

reducing atmospheric pollution has also served to direct some attention to coking, since the process not only concentrates pollutants such as feedstock sulfur in the coke, but also can usually yield volatile products that can be conveniently desulfurized.

Investigations of technologies that result in the production of coke are almost as old as the refining industry itself, but the development of the modern coking processes can be traced to the 1930s with many units being added to refineries in the 1940 to 1970 era. Coking processes generally use longer reaction times than the older thermal cracking processes and, in fact, may be considered to be descendents of the thermal cracking processes.

Delayed coking is a semicontinuous process (Figure 14.6) in which the heated charge is transferred to large soaking (or coking) drums, which provide the long residence time needed to allow the cracking reactions to proceed to completion. The feed to these units is normally an atmospheric residuum, although cracked residua are also used.

The feedstock is introduced into the product fractionator where it is heated and lighter fractions are removed as side streams. The fractionator bottoms, including a recycle stream of heavy product, are then heated in a furnace whose outlet temperature varies from 480°C to 515°C (895°F to 960°F). The heated feedstock enters one of a pair of coking drums where the cracking reactions continue. The cracked products leave as overheads, and coke deposits form on the inner surface of the drum. To give continuous operation, two drums are used; while one is on stream, the other is being cleaned. The temperature in the coke drum ranges from 415°C to 450°C (780°F to 840°F) with pressures from 15 to 90 psi.

Overhead products go to the fractionator, where naphtha and heating oil fractions are recovered. The nonvolatile material is combined with preheated fresh feed and returned to the reactor. The coke drum is usually on stream for about 24 h before becoming filled with porous coke, after which the coke is removed hydraulically. Normally, 24 h is required to complete the cleaning operation and to prepare the coke drum for subsequent use on stream.

Fluid coking is a continuous process (Figure 14.7) which uses the fluidized-solids technique to convert atmospheric and vacuum residua to more valuable products. The residuum is coked by being sprayed into a fluidized bed of hot, fine coke particles, which permits the coking reactions to be conducted at higher temperatures and shorter contact times than can be employed in delayed coking. Moreover, these conditions result in decreased yields of coke; greater quantities of more valuable liquid product are recovered in the fluid coking process.

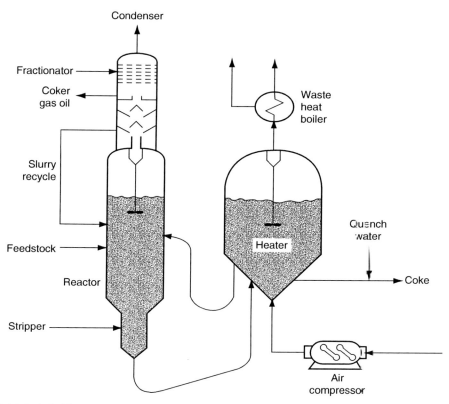

FIGURE 14.7 A fluid coker.

Fluid coking uses two vessels, a reactor and a burner; coke particles are circulated between these to transfer heat (generated by burning a portion of the coke) to the reactor. The reactor holds a bed of fluidized coke particles, and steam is introduced at the bottom of the reactor to fluidize the bed.

Flexicoking (Figure 14.8) is also a continuous process that is a direct descendent of fluid coking. The unit uses the same configuration as the fluid coker, but has a gasification section in which excess coke can be gasified to produce refinery fuel gas. The flexicoking process was designed during the late 1960s and the 1970s as a means by which excess coke-make could be reduced in view of the gradual incursion of the heavier feedstocks in refinery operations. Such feedstocks are notorious for producing high yields of coke (>15% by weight) in thermal and catalytic operations.

14.6 CATALYTIC METHODS

14.6.1 Historical Development

There are many processes in a refinery that employ a catalyst to improve process efficiency (Table 14.3). The original incentive arose from the need to increase gasoline supplies in the 1930s and 1940s. Since cracking could virtually double the volume of gasoline from a barrel of crude oil, cracking was justifiable on this basis alone.

In the 1930s, thermal cracking units produced approximately 50% of the total gasoline. The octane number of this gasoline was about 70 compared with 60 for straight-run (distilled)

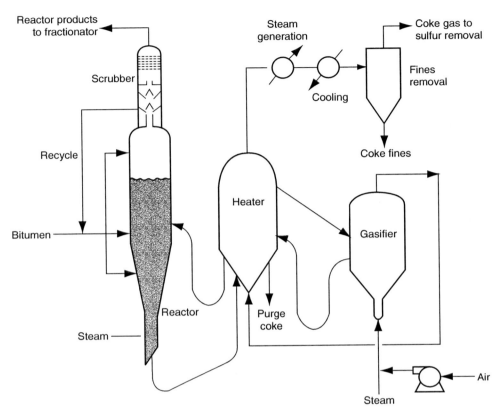

FIGURE 14.8 Flexicoking process.

TABLE 14.3
Summary of Catalytic Cracking Processes

Conditions
Solid acidic catalyst (silica–alumina, zeolite, etc.)
Temperature: 480°C to 540°C (900°F to 1000°F (solid/vapor contact)
Pressure: 10 to 20 psi
Provisions needed for continuous catalyst replacement with heavier feedstocks (residua)
Catalyst may be regenerated or replaced

Feedstocks
Gas oils and residua
Residua pretreated to remove salts (metals)
Residua pretreated to remove high molecular weight (asphaltic constituents)

Products
Lower molecular weight than feedstock
Some gases (feedstock and process parameters dependent)
Iso-paraffins in product
Coke deposited on catalyst

Variations
Fixed bed
Moving bed
Fluidized bed

gasoline. The thermal reforming and polymerization processes that were developed during the 1930s could be expected to further increase the octane number of gasoline to some extent, but an additional innovation was needed to increase the octane number of gasoline to enhance the development of more powerful automobile engines.

In 1936, a new cracking process opened the way to higher-octane gasoline—this process was catalytic cracking. This process is basically the same as thermal cracking, but differs by the use of a catalyst, which is not (in theory) consumed in the process, and directs the course of the cracking reactions to produce more of the desired higher-octane hydrocarbon products.

Catalytic cracking has a number of advantages over thermal cracking: (a) the gasoline produced has a higher octane number; (b) the catalytically cracked gasoline consists largely of *iso*-paraffins and aromatics, which have high octane numbers and greater chemical stability than monoolefins and diolefins which are present in much greater quantities in thermallycracked gasoline. Substantial quantities of olefinic gases suitable for polymer gasoline manufacture and smaller quantities of methane, ethane, and ethylene are produced by catalytic cracking. Sulfur compounds are changed in such a way that the sulfur content of catalytically cracked gasoline is lower than in thermally cracked gasoline. Catalytic cracking produces less heavy residual or tar and more of the useful gas oils than does thermal cracking. The process has considerable flexibility, permitting the manufacture of both motor and aviation gasoline and a variation in the gas oil yield to meet changes in the fuel oil market.

The last 40 years have seen substantial advances in the development of catalytic processes. This has involved not only rapid advances in the chemistry and physics of the catalysts themselves but also major engineering advances in reactor design. For example, the evolution of the design of the catalyst beds from fixed beds to moving beds to fluidized beds. Catalyst chemistry and physics and bed design have allowed major improvements in process efficiency and product yields.

14.6.2 MODERN PROCESSES

Catalytic cracking is another innovation that truly belongs to the twentieth century and is regarded as the modern method for converting high-boiling petroleum fractions, such as gas oil, into gasoline and other low-boiling fractions. Thus, catalytic cracking in the usual commercial process involves contacting a gas oil faction with an active catalyst under suitable conditions of temperature, pressure, and residence time, so that a substantial part (>50%) of the gas oil is converted into gasoline and lower-boiling products, usually in a single-pass operation.

However, during the cracking reaction, carbonaceous material is deposited on the catalyst, which markedly reduces its activity, and removal of the deposit is very necessary. This is usually accomplished by burning the catalyst in the presence of air, until catalyst activity is reestablished.

The several processes currently employed in catalytic cracking differ mainly in the method of catalyst handling, although there is overlap with regard to catalyst type and the nature of the products.

The catalyst, which may be an activated natural or synthetic material, is employed in bead, pellet, or microspherical form and can be used as a fixed bed, moving bed, or fluid bed. The fixed-bed process was the first process to be used commercially and uses a static bed of catalyst in several reactors, which allows a continuous flow of feedstock to be maintained. Thus, the cycle of operations consists of (1) flow of feedstock through the catalyst bed, (2) discontinuance of feedstock flow and removal of coke from the catalyst by burning, and (3) insertion of the reactor on stream. The moving-bed process uses a reaction vessel (in which cracking takes place) and a kiln (in which the spent catalyst is regenerated) and catalyst movement between the vessels is provided by various means.

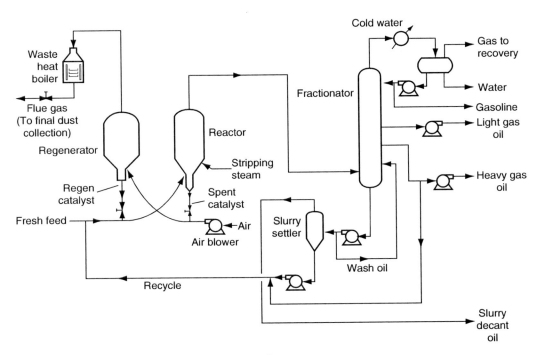

FIGURE 14.9 A fluid catalytic cracking (FCC) unit.

The fluid-bed process (Figure 14.9) differs from the fixed-bed and moving-bed processes, insofar as the powdered catalyst is circulated essentially as a fluid with the feedstock. The several fluid catalytic cracking processes in use differ primarily in mechanical design. Side-by-side reactor–regenerator construction along with unitary vessel construction (the reactor either above or below the regenerator) are the two main mechanical variations.

14.6.3 CATALYSTS

Natural clays have long been known to exert a catalytic influence on the cracking of oils, but it was not until about 1936 that the process using silica–alumina catalysts was developed sufficiently for commercial use. Since then, catalytic cracking has progressively supplanted thermal cracking as the most advantageous means of converting distillate oils into gasoline. The main reason for the wide adoption of catalytic cracking is the fact that a better yield of higher-octane gasoline can be obtained than by any known thermal operation. At the same time, the gas produced consists mostly of propane and butane with less methane and ethane. The production of heavy oils and tars, higher in molecular weight than the charge material, is also minimized, and both the gasoline and the uncracked "cycle oil" are more saturated than the products of thermal cracking.

The major innovations of the twentieth century lie not only in reactor configuration and efficiency, but also in catalyst development. There is probably no oil company in the United States that does not have some research and development activity related to catalyst development. Much of the work is proprietary and, therefore, can only be addressed here in generalities.

The cracking of crude oil fractions occurs over many types of catalytic materials, but high yields of desirable products are obtained with hydrated aluminum silicates. These may be either activated (acid-treated) natural clays of the bentonite type of synthesized silica–alumina or silica–magnesia preparations. Their activity to yield essentially the same products may be enhanced to some extent by the incorporation of small amounts of other materials such as the oxides of zirconium, boron (which has a tendency to volatilize away on use), and thorium. Natural and synthetic catalysts can be used as pellets or beads and also in the form of powder; in either case, replacements are necessary because of attrition and gradual loss of efficiency. It is essential that they be stable to withstand the physical impact of loading and thermal shocks, and that they withstand the action of carbon dioxide, air, nitrogen compounds, and steam. They should also be resistant to sulfur and nitrogen compounds and synthetic catalysts, or certain selected clays, appear to be better in this regard than average untreated natural catalysts.

The catalysts are porous and highly adsorptive and their performance is affected markedly by the method of preparation. Two chemically identical catalysts having pores of different size and distribution may have different activity, selectivity, temperature coefficients of reaction rates, and responses to poisons. The intrinsic chemistry and catalytic action of a surface may be independent of pore size, but small pores produce different effects because of the manner in which hydrocarbon vapors are transported into and out of the pore systems.

14.7 HYDROPROCESSES

14.7.1 HISTORICAL DEVELOPMENT

The use of hydrogen in thermal processes is perhaps the single most significant advance in refining technology during the twentieth century. The process uses the principle that the presence of hydrogen during a thermal reaction of a petroleum feedstock will terminate many of the coke-forming reactions and enhance the yields of the lower-boiling components such as gasoline, kerosene and jet fuel (Table 14.4).

Hydrogenation processes for the conversion of petroleum fractions and petroleum products may be classified as destructive and nondestructive. Destructive hydrogenation (hydrogenolysis or hydrocracking) is characterized by the conversion of the higher molecular weight constituents in a feedstock to lower-boiling products. Such treatment requires severe processing conditions and the use of high hydrogen pressures to minimize polymerization and condensation reactions that lead to coke formation.

Nondestructive or simple hydrogenation is generally used for the purpose of improving product quality without appreciable alteration of the boiling range. Mild processing conditions are employed so that only the more unstable materials are attacked. Nitrogen, sulfur, and oxygen compounds undergo reaction with the hydrogen to remove ammonia, hydrogen sulfide, and water, respectively. Unstable compounds which might lead to the formation of gums, or insoluble materials, are converted to more stable compounds.

14.7.2 MODERN PROCESSES

Hydrotreating (Figure 14.10) is carried out by charging the feed to the reactor, together with hydrogen in the presence of catalysts such as tungsten–nickel sulfide, cobalt–molybdenum–alumina, nickel oxide–silica–alumina, and platinum–alumina. Most processes employ cobalt–molybdena catalysts which generally contain about 10% of molybdenum oxide and less than 1% of cobalt oxide supported on alumina. The temperatures employed are in the range of 260°C to 345°C (500°F to 655°F), while the hydrogen pressures are about 500 to 1000 psi.

TABLE 14.4
Summary of Hydrocracking Process Operations

Conditions
Solid acid catalyst (silica–alumina with rare earth metals, various other options)
Temperature: 260°C to 450°C (500°F to 845°F (solid/liquid contact)
Pressure: 1000 to 6000 psi hydrogen
Frequent catalysts renewal for heavier feedstocks
Gas oil: catalyst life up to three years
Heavy oil or tar sand bitumen: catalyst life less than one year

Feedstocks
Refractory (aromatic) streams
Coker oils,
Cycle oils
Gas oils
Residua (as a full hydrocracking or hydrotreating option)
In some cases, asphaltic constituents (S, N, and metals) removed by deasphalting

Products
Lower molecular weight paraffins
Some methane, ethane, propane, and butane
Hydrocarbon distillates (full range depending on the feedstock)
Residual tar (recycle)
Contaminants (asphaltic constituents) deposited on the catalyst as coke or metals

Variations
Fixed bed (suitable for liquid feedstocks)
Ebullating bed (suitable for heavy feedstocks)

The reaction generally takes place in the vapor phase but, depending on the application, may be a mixed-phase reaction. Generally, it is more economical to hydrotreat high-sulfur feedstocks prior to catalytic cracking than to hydrotreat the products from catalytic cracking.

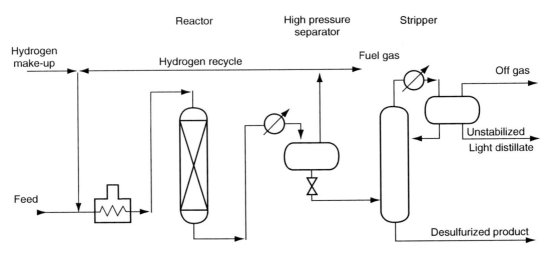

FIGURE 14.10 A distillate hydrotreater for hydrodesulfurization. (From OSHA Technical Manual, Section IV, Chapter 2, Petroleum Refining Processes.)

The advantages are that (1) sulfur is removed from the catalytic cracking feedstock, and corrosion is reduced in the cracking unit, (2) carbon formation during cracking is reduced so that higher conversions result, and (3) the cracking quality of the gas oil fraction is improved.

Hydrocracking is similar to catalytic cracking, with hydrogenation superimposed and with the reactions taking place either simultaneously or sequentially. Hydrocracking was initially used to upgrade low-value distillate feedstocks, such as cycle oils (high aromatic products, from a catalytic cracker, which are usually not recycled to extinction for economic reasons), thermal and coker gas oils, and heavy-cracked and straight-run naphtha. These feedstocks are difficult to process by either catalytic cracking or reforming, since they are characterized usually by a high polycyclic aromatic content or by high concentrations of the two principal catalyst poisons: sulfur and nitrogen compounds.

The older hydrogenolysis type of hydrocracking practised in Europe during, and after, World War II, used tungsten or molybdenum sulfides as catalysts and required high reaction temperatures and operating pressures, sometimes in excess of about 3000 psi (203 atm) for continuous operation. The modern hydrocracking processes (e.g., Figure 14.11) were initially developed for converting refractory feedstocks (such as gas oils) to gasoline and jet fuel, but process and catalyst improvements and modifications have made it possible to yield products from gases and naphtha to furnace oils and catalytic cracking feedstocks.

A comparison of hydrocracking with hydrotreating is useful in assessing the part played by these two processes in refinery operations. Hydrotreating of distillates may be defined simply as the removal of nitrogen-, sulfur-, and oxygen-containing compounds by selective hydrogenation. The hydrotreating catalysts are usually cobalt and molybdenum or nickel and

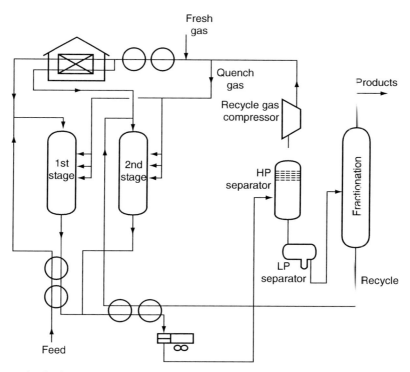

FIGURE 14.11 A single-stage or two-stage (optional) hydrocracking unit. (From OSHA Technical Manual, Section IV, Chapter 2, Petroleum Refining Processes.)

molybdenum (in the sulfide) form impregnated on an alumina base. The hydrotreated operating conditions are such that appreciable hydrogenation of aromatics will not occur at −1000 to 2000 psi hydrogen and about 370°C (700°F). The desulfurization reactions are usually accompanied by small amounts of hydrogenation and hydrocracking.

Hydrocracking is an extremely versatile process which can be used in many different ways such as conversion of the high-boiling aromatic streams which are produced by catalytic cracking or by coking processes. To take full advantage of hydrocracking, the process must be integrated in the refinery with other process units.

The commercial processes for treating, or finishing, petroleum fractions with hydrogen all operate in essentially the same manner. The feedstock is heated and passed with hydrogen gas through a tower or reactor filled with catalyst pellets. The reactor is maintained at a temperature of 260°C to 425°C (500°F to 800°F) at pressures from 100 to 1000 psi, depending on the particular process, the nature of the feedstock and the degree of hydrogenation required. After leaving the reactor, excess hydrogen is separated from the treated product and recycled through the reactor after removal of hydrogen sulfide. The liquid product is passed into a stripping tower where steam removes dissolved hydrogen and hydrogen sulfide and, after cooling, the product is taken to product storage or, in the case of feedstock preparation, pumped to the next processing unit.

14.7.2.1 Hydrofining

Hydrofining is a process that first went on-stream in the 1950s and is one example of the many hydroprocesses available. It can be applied to lubricating oils, naphtha, and gas oils. The feedstock is heated in a furnace and passed with hydrogen through a reactor containing a suitable metal oxide catalyst, such as cobalt and molybdenum oxides on alumina. Reactor operating conditions range from 205°C to 425°C (400°F to 800°F) and from 50 to 800 psi, and depend on the kind of feedstock and the degree of treating required. Higher-boiling feedstocks, high sulfur content, and maximum sulfur removal require higher temperatures and pressures.

After passing through the reactor, the treated oil is cooled and separated from the excess hydrogen which is recycled through the reactor. The treated oil is pumped to a stripper tower where hydrogen sulfide, formed by the hydrogenation reaction, is removed by steam, vacuum, or flue gas, and the finished product leaves the bottom of the stripper tower. The catalyst is not usually regenerated; it is replaced after about one year's use.

14.8 REFORMING

14.8.1 Historical Development

When the demand for higher-octane gasoline developed during the early 1930s, attention was directed to ways and means of improving the octane number of fractions within the boiling range of gasoline. Straight-run (distilled) gasoline frequently had very low octane numbers, and any process that would improve the octane numbers would aid in meeting the demand for higher octane number gasoline. Such a process (called thermal reforming) was developed and used widely, but to a much lesser extent than thermal cracking. Thermal reforming was a natural development from older thermal cracking processes; cracking converts heavier oils into gasoline, whereas reforming converts (reforms) gasoline into higher-octane gasoline. The equipment for thermal reforming is essentially the same as for thermal cracking, but higher temperatures are used.

14.8.2 MODERN PROCESSES

14.8.2.1 Thermal Reforming

In carrying out thermal reforming, a feedstock such as 205°C (400°F) end-point naphtha or a straight-run gasoline is heated to 510°C to 595°C (950°F to 1100°F) in a furnace, much the same as a cracking furnace, with pressures from 400 to 1000 psi (27 to 68 atm). As the heated naphtha leaves the furnace, it is cooled or quenched by the addition of cold naphtha. The material then enters a fractional distillation tower where any heavy products are separated. The remainder of the reformed material leaves the top of the tower to be separated into gases and reformate. The higher octane number of the reformate is due primarily to the cracking of longer-chain paraffins into higher-octane olefins.

The products of thermal reforming are gases, gasoline, and residual oil or tar, the latter being formed in very small amounts (about 1%). The amount and quality of the gasoline, known as reformate, is very dependent on the temperature. A general rule is: the higher the reforming temperature, the higher the octane number, but the lower the yield of reformate.

Thermal reforming is less effective and less economical than catalytic processes and has been largely supplanted. As it used to be practised, a single-pass operation was employed at temperatures in the range of 540°C to 760°C (1000°F to 1140°F) and pressures of about 500 to 1000 psi (34 to 68 atm). The degree of octane number improvement depended on the extent of conversion, but was not directly proportional to the extent of crack per pass. However at very high conversions, the production of coke and gas became prohibitively high. The gases produced were generally olefinic and the process required either a separate gas polymerization operation or one in which C3 to C4 gases were added back to the reforming system.

More recent modifications of the thermal reforming process due to the inclusion of hydrocarbon gases with the feedstock are known as gas reversion and polyforming. Thus, olefinic gases produced by cracking and reforming can be converted into liquids boiling in the gasoline range by heating them under high pressure. Since the resulting liquids (polymers) have high octane numbers, they increase the overall quantity and quality of gasoline produced in a refinery.

14.8.2.2 Catalytic Reforming

The catalytic reforming process was commercially nonexistent in the United States before 1940. The process is really a process of the 1950s and showed phenomenal growth in the 1953 to 1959 time period.

Like thermal reforming, catalytic reforming converts low-octane gasoline into high-octane gasoline (reformate). When thermal reforming could produce reformate with research octane numbers of 65 to 80 depending on the yield, catalytic reforming produces reformate with octane numbers on the order of 90 to 95. Catalytic reforming is conducted in the presence of hydrogen over hydrogenation–dehydrogenation catalysts, which may be supported on alumina or silica–alumina. Depending on the catalyst, a definite sequence of reactions takes place, involving structural changes in the feedstock. This more modern concept actually rendered thermal reforming somewhat obsolescent.

The commercial processes available for use can be broadly classified as the moving-bed, fluid-bed and fixed-bed types. The fluid-bed and moving-bed processes used mixed nonprecious metal oxide catalysts in units equipped with separate regeneration facilities. Fixed-bed processes use predominantly platinum-containing catalysts in units equipped for cycle, occasional, or no regeneration.

Catalytic reformer feeds are saturated (i.e., not olefinic) materials; in the majority of cases that feed may be a straight-run naphtha, but other byproduct low-octane naphtha (e.g., coker naphtha) can be processed after treatment to remove olefins and other contaminants. Hydro-cracker naphtha that contains substantial quantities of naphthenes is also a suitable feed.

Dehydrogenation is a main chemical reaction in catalytic reforming, and hydrogen gas is consequently produced in large quantities. The hydrogen is recycled through the reactors where the reforming takes place to provide the atmosphere necessary for the chemical reactions and also prevents the carbon from being deposited on the catalyst, thus extending its operating life. An excess of hydrogen above whatever is consumed in the process is produced, and, as a result, catalytic reforming processes are unique in that they are the only petroleum refinery processes to produce hydrogen as a byproduct.

Catalytic reforming is usually carried out by feeding a naphtha (after pretreating with hydrogen if necessary) and hydrogen mixture to a furnace, where the mixture is heated to the desired temperature, 450°C to 520°C (840°F to 965°F), and then passed through fixed-bed catalytic reactors at hydrogen pressures of 100 to 1000 psi (7 to 68 atm) (Figure 14.12). Normally, pairs of reactors are used in series with heaters which are located between adjoining reactors in order to compensate for the endothermic reactions taking place. Sometimes as many as four or five reactors are kept on stream in series, whereas one or more is regenerated.

The on-stream cycle of any one reactor may vary from several hours to many days, depending on the feedstock and reaction conditions.

14.8.2.3 Catalysts

The composition of a reforming catalyst is dictated by the composition of the feedstock and the desired reformate. The catalysts used are principally molybdena–alumina, chromia–alumina, or platinum on a silica–alumina or alumina base. The nonplatinum catalysts are widely used in

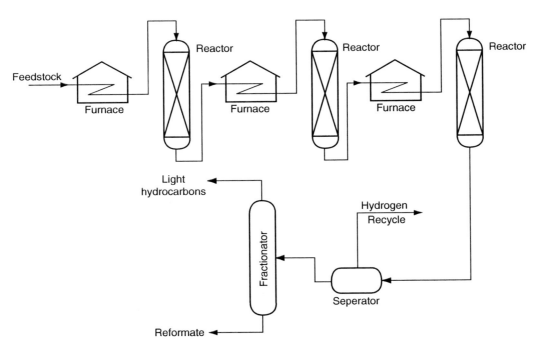

FIGURE 14.12 Catalytic reforming. (From OSHA Technical Manual, Section IV, Chapter 2, Petroleum Refining Processes.)

regenerative process for feeds containing, for example, sulfur, which poisons platinum catalysts, although pretreatment processes (e.g., hydrodesulfurization) may permit platinum catalysts to be employed.

The purpose of platinum on the catalyst is to promote dehydrogenation and hydrogenation reactions, i.e., the production of aromatics, participation in hydrocracking, and rapid hydrogenation of carbon-forming precursors. For the catalyst to have an activity for isomerization of both paraffins and naphthenes—the initial cracking step of hydrocracking—and to participate in paraffin dehydrocyclization, it must have an acid activity. The balance between these two activities is most important in a reforming catalyst. In fact, in the production of aromatics from cyclic saturated materials (naphthenes), it is important that hydrocracking be minimized to avoid loss of the desired product and, thus, the catalytic activity must be moderated relative to the case of gasoline production from a paraffinic feed, where dehydrocyclization and hydrocracking play an important part.

14.9 ISOMERIZATION

Catalytic reforming processes provide high-octane constituents in the heavier gasoline fraction but the normal paraffin components of the lighter gasoline fraction, especially butanes, pentanes and hexanes, have poor octane ratings. The conversion of these normal paraffins to their isomers (isomerization) yields gasoline components of high octane rating in this lower-boiling range. Conversion is obtained in the presence of a catalyst (aluminum chloride activated with hydrochloric acid), and it is essential to inhibit side reactions such as cracking and olefin formation.

14.9.1 HISTORICAL DEVELOPMENT

Isomerization, another "child of the twentieth century," found initial commercial applications during World War II for making high-octane aviation gasoline components and additional feed for alkylation units. The lowered alkylate demands in the post World War II period led to the majority of the butane isomerization units being shut down. In recent years, the greater demand for high-octane motor fuel has resulted in new butane isomerization units being installed.

The earliest process of note was the production of *iso*-butane, which is required as an alkylation feed. The isomerization may take place in the vapor phase, with the activated catalyst supported on a solid phase, or in the liquid phase with a dissolved catalyst. In the process, pure butane or a mixture of isomeric butanes (Figure 14.13), is mixed with hydrogen (to inhibit olefin formation) and passed to the reactor, at 110°C to 170°C (230°F to 340°F) and 200 to 300 psi (14 to 20 atm). The product is cooled, the hydrogen separated and the cracked gases are then removed in a stabilizer column. The stabilizer bottom product is passed to a superfractionator where the normal butane is separated from the *iso*-butane.

14.9.2 MODERN PROCESSES

Present isomerization applications in petroleum refining are used with the objective of providing additional feedstock for alkylation units or high-octane fractions for gasoline blending (Table 14.5). Straight-chain paraffins (*n*-butane, *n*-pentane, *n*-hexane) are converted to respective *iso*-compounds by continuous catalytic (aluminum chloride, noble metals) processes. Natural gasoline or light straight-run gasoline can provide feed by first fractionating as a preparatory step. High volumetric yields (>95%) and 40% to 60% conversion per pass are characteristic of the isomerization reaction.

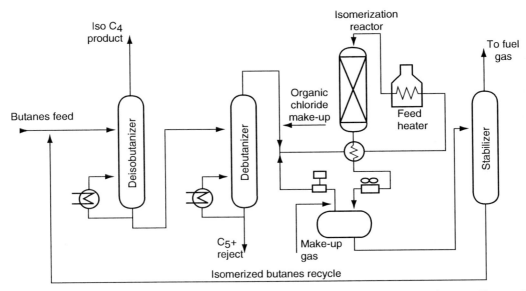

FIGURE 14.13 A butane isomerization unit. (From OSHA Technical Manual, Section IV, Chapter 2, Petroleum Refining Processes.)

14.9.3 CATALYSTS

During World War II, aluminum chloride was the catalyst used to isomerize butane, pentane, and hexane. Since then, supported metal catalysts have been developed for use in high-temperature processes which operate in the range of 370°C to 480°C (700°F to 900°F) and 300 to 750 psi of (20 to 51 atm), while aluminum chloride and hydrogen chloride are universally used for the low-temperature processes.

TABLE 14.5
Component Streams for Gasoline

		Boiling Range	
Stream	**Producing Process**	**°C**	**°F**
Paraffinic			
Butane	Distillation	0	32
	Conversion		
Iso-pentane	Distillation	27	81
	Conversion		
	Isomerization		
Alkylate	Alkylation	40–150	105–300
Isomerate	Isomerization	40–70	105–160
Naphtha	Distillation	30–100	85–212
Hydrocrackate	Hydrocracking	40–200	105–390
Olefinic			
Catalytic naphtha	Catalytic cracking	40–200	105–390
Cracked naphtha	Steam cracking	40–200	105–390
Polymer	Polymerization	60–200	140–390
Aromatic			
Catalytic reformate	Catalytic reforming	40–200	105–390

Nonregenerable aluminum chloride catalyst is employed with various carriers in a fixed-bed or liquid contactor. Platinum or other metal catalyst processes utilized fixed-bed operation and can be regenerable or nonregenerable. The reaction conditions vary widely depending on the particular process and feedstock, 40°C to 480°C) (100°F to 900°F) and 150 to 1000 psi (10 to 68 atm).

14.10 ALKYLATION PROCESSES

The combination of olefins with paraffins to form higher *iso*-paraffins is termed *alkylation*. Since olefins are reactive (unstable) and are responsible for exhaust pollutants, their conversion to high-octane *iso*-paraffins is desirable when possible. In refinery practice, only *iso*-butane is alkylated, by reaction with *iso*-butene or normal butene and *iso*-octane is the product. Although alkylation is possible without catalysts, commercial processes use aluminum chloride, sulfuric acid, or hydrogen fluoride as catalysts, when the reactions can take place at low temperatures, minimizing undesirable side reactions, such as polymerization of olefins.

Alkylate is composed of a mixture of *iso*-paraffins which have octane numbers that vary with the olefins from which they were made. Butylenes produce the highest octane numbers, propylene the lowest and pentylenes the intermediate values. All alkylates, however, have high octane numbers (>87) which makes them particularly valuable.

14.10.1 HISTORICAL DEVELOPMENT

Alkylation is another twentieth century refinery innovation, and developments in petroleum processing in the late 1930s and during World War II were directed toward production of high-octane liquids for aviation gasoline. The sulfuric acid process was introduced in 1938, and hydrogen fluoride alkylation was introduced in 1942. Rapid commercialization took place during the war to supply military needs, but many of these plants were shut down at the end of the war.

In the mid 1950s, aviation-gasoline demand started to decline, but motor-gasoline quality requirements rose sharply. Wherever practical, refiners shifted the use of alkylate to premium motor fuel. To aid in the improvement of the economics of the alkylation process and also the sensitivity of the premium gasoline pool, additional olefins were gradually added to alkylation feed. New plants were built to alkylate propylene and the butylenes (butanes) produced in the refinery rather than the butane–butylene stream formerly used.

14.10.2 MODERN PROCESSES

The alkylation reaction as now practised in petroleum refining is the union, through the agency of a catalyst, of an olefin (ethylene, propylene, butylene, and amylene) with *iso*-butane to yield high-octane branched-chain hydrocarbons in the gasoline boiling range. Olefin feedstock is derived from the gas produced in a catalytic cracker, while *iso*-butane is recovered by refinery gases or produced by catalytic butane isomerization.

To accomplish this, either ethylene or propylene is combined with *iso*-butane at 50°C to 280°C (125°F to 450°F) and 300 to 1000 psi (20 to 68 atm) in the presence of metal halide catalysts such as aluminum chloride. Conditions are less stringent in catalytic alkylation; olefins (propylene, butylenes or pentylenes) are combined with *iso*-butane in the presence of an acid catalyst (sulfuric acid or hydrofluoric acid) at low temperatures and pressures (1°C to 40°C, 30°F to 105°F and 14.8 to 150 psi; 1 to 10 atm) (Figure 14.14).

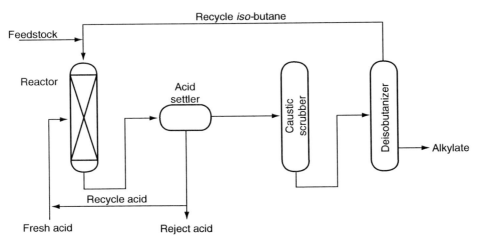

FIGURE 14.14 An alkylation unit (sulfuric acid catalyst). (From OSHA Technical Manual, Section IV, Chapter 2, Petroleum Refining Processes.)

14.10.3 CATALYSTS

Sulfuric acid, hydrogen fluoride, and aluminum chloride are the general catalysts used commercially. Sulfuric acid is used with propylene and higher-boiling feeds, but not with ethylene, because it reacts to form ethyl hydrogen sulfate. The acid is pumped through the reactor and forms an air emulsion with reactants, and the emulsion is maintained at 50% acid. The rate of deactivation varies with the feed and *iso*-butane charge rate. Butene feeds cause less acid consumption than the propylene feeds.

Aluminum chloride is not widely used as an alkylation catalyst, but when employed, hydrogen chloride is used as a promoter and water is injected to activate the catalyst as an aluminum chloride or hydrocarbon complex. Hydrogen fluoride is used for alkylation of higher-boiling olefins and the advantage of hydrogen fluoride is that it is more readily separated and recovered from the resulting product.

14.11 POLYMERIZATION PROCESSES

14.11.1 HISTORICAL DEVELOPMENT

In the petroleum industry, polymerization is the process by which olefin gases are converted to liquid products which may be suitable for gasoline (polymer gasoline) or other liquid fuels. The feedstock usually consists of propylene and butylenes from cracking processes or may even be selective olefins for dimer, trimer, or tetramer production.

Polymerization is a process that can claim to be the earliest process to employ catalysts on a commercial scale. Catalytic polymerization came into use in the 1930s and was one of the first catalytic processes to be used in the petroleum industry.

14.11.2 MODERN PROCESSES

Polymerization may be accomplished thermally or in the presence of a catalyst at lower temperatures. Thermal polymerization is regarded as not being as effective as catalytic polymerization, but has the advantage that it can be used to "polymerize" saturated materials that cannot be induced to react by catalysts. The process consists of vapor-phase cracking of, for example, propane and butane followed by prolonged periods at high temperature (510°C to 595°C, 950°F to 1100°F) for the reactions to proceed to near completion.

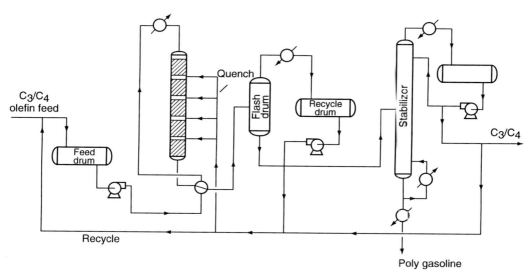

FIGURE 14.15 A polymerization unit.

Olefins can also be conveniently polymerized by means of an acid catalyst (Figure 14.15). Thus, the treated, olefin-rich feed stream is contacted with a catalyst (sulfuric acid, copper pyrophosphate, phosphoric acid) at 150°C to 220°C (300°F to 425°F) and 150 to 1200 psi (10 to 81 atm), depending on feedstock and product requirement.

14.11.3 CATALYSTS

Phosphates are the principal catalysts used in polymerization units; the commercially used catalysts are liquid phosphoric acid, phosphoric acid on kieselguhr, copper pyrophosphate pellets, and phosphoric acid film on quartz. The latter is the least active, but the most used and easiest one to regenerate simply by washing and recoating; the serious disadvantage is that tar must occasionally be burnt off the support. The process using liquid phosphoric acid catalyst is far more responsible to attempts to raise production by increasing temperature than the other processes.

14.12 SOLVENT PROCESS

14.12.1 DEASPHALTING

Solvent deasphalting processes are a major part of refinery operations (Bland and Davidson, 1967; Hobson and Pohl, 1973; Gary and Handwerk, 2001; Speight and Ozum, 2002) and are not often appreciated for the tasks for which they are used. In the solvent deasphalting processes, an alkane is injected into the feedstock to disrupt the dispersion of components and causes the polar constituents to precipitate. Propane (or sometimes propane and butane mixtures) is extensively used for deasphalting and produces a deasphalted oil (DAO) and propane deasphalter asphalt (PDA or PD tar) (Dunning and Moore, 1957). Propane has unique solvent properties; at lower temperatures (38°C to 60°C; 100°F to 140°F), paraffins are very soluble in propane and at higher temperatures (about 93°C; 200°F) all hydrocarbons are almost insoluble in propane.

A solvent deasphalting unit (Figure 14.16) processes the residuum from the vacuum distillation unit and produces deasphalted oil (DAO), used as feedstock for a fluid catalytic cracking unit, and the asphaltic residue (deasphalter tar, deasphalter bottoms) which, as a residual fraction, can only be used to produce asphalt or as a blend stock or visbreaker

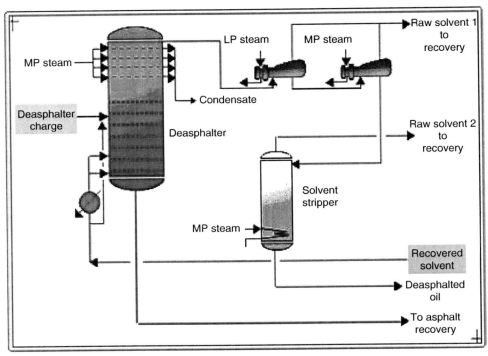

FIGURE 14.16 A deasphalting unit. (From OSHA Technical Manual, Section IV, Chapter 2, Petroleum Refining Processes.)

feedstock for low-grade fuel oil. Solvent deasphalting processes have not realized their maximum potential. With on-going improvements in energy efficiency, such processes would display their effects in a combination with other processes. Solvent deasphalting allows removal of sulfur and nitrogen compounds, as well as metallic constituents, by balancing yield with the desired feedstock properties (Ditman, 1973).

14.12.2 DEWAXING

Paraffinic crude oils often contain microcrystalline or paraffin waxes. The crude oil may be treated with a solvent such as methyl-ethyl-ketone (MEK) to remove this wax before it is processed. This is not a common practice, however and **solvent dewaxing** processes are designed to remove wax from lubricating oils to give the product good fluidity characteristics at low temperatures (e.g., low pour points) rather than from the whole crude oil. The mechanism of solvent dewaxing involves either the separation of wax as a solid that crystallizes from the oil solution at low temperature or the separation of wax as a liquid that is extracted at temperatures above the melting point of the wax through preferential selectivity of the solvent. However, the former mechanism is the usual basis for commercial dewaxing processes.

In the 1930s, two types of stocks, naphthenic and paraffinic, were used to make motor oils. Both types were solvent extracted to improve their quality, but in the high-temperature conditions encountered in service, the naphthenic type could not stand up as well as the paraffinic type. Nevertheless, the naphthenic type was the preferred oil, particularly in cold weather, because of its fluidity at low temperatures. Previous to 1938, the highest quality lubricating oils were of the naphthenic type and were phenol treated to pour points of $-40°C$ to $-7°C$ ($-40°F$ to $20°F$), depending on the viscosity of the oil. Paraffinic oils were also

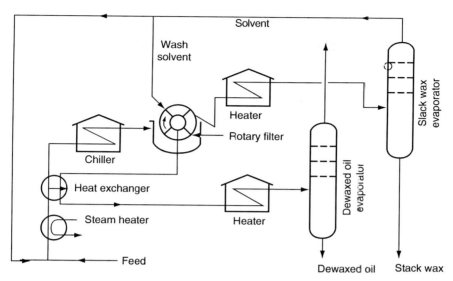

FIGURE 14.17 A solvent dewaxing unit. (From OSHA Technical Manual, Section IV, Chapter 2, Petroleum Refining Processes.)

available and could be phenol treated to higher-quality oil, but their wax content was so high that the oils were solid at room temperature.

The lowest viscosity paraffinic oils were dewaxed by the cold press method to produce oils with a pour point of 2°C (35°F). The light paraffin distillate oils contained a paraffin wax that crystallized into large crystals when chilled and could thus readily be separated from the oil by the cold press filtration method. The more viscous paraffinic oils (intermediate and heavy paraffin distillates) contained amorphous or microcrystalline waxes, which formed small crystals that plugged the filter cloths in the cold press and prevented filtration. Because the wax could not be removed from intermediate and heavy paraffin distillates, the high-quality, high-viscosity lubricating oils in them could not be used except as cracking stock.

Methods were therefore developed to dewax these high-viscosity paraffinic oils. The methods were essentially alike in that the waxy oil was dissolved in a solvent that would keep the oil in solution; the wax separated as crystals when the temperature was lowered. The processes differed chiefly in the use of the solvent. Commercially used solvents were naphtha, propane, sulfur dioxide, acetone–benzene, trichloroethylene, ethylene dichloride–benzene (Barisol), methyl ethyl ketone–benzene (**benzol**), methyl-*n*-butyl ketone, and methyl-*n*-propyl ketone.

The process as now practised involves mixing the feedstock with one to four times its volume of the ketone (Figure 14.17) (Scholten, 1992). The mixture is then heated until the oil is in solution and the solution is chilled at a slow, controlled rate in double-pipe, scraped-surface exchangers. Cold solvent, such as filtrate from the filters, passes through the two-inch annular space between the inner and outer pipes and chills the waxy oil solution flowing through the inner 6-in. pipe.

14.13 REFINING HEAVY FEEDSTOCKS

Petroleum refining is now in a significant transition period as the industry moves into the twenty-first century. Although the demand for petroleum and petroleum products has shown a sharp growth in recent years (Chapter 3), this might be the last century for petroleum refining, as we know it. The demand for transportation fuels and fuel oil is forecast to continue to show a steady growth in the future. The simplest means to cover the demand growth in low-boiling products is to increase the imports of light crude oils and low-boiling petroleum products, but these steps may be limited in the future.

Over the past three decades, crude oils available to refineries have generally decreased in API gravity. There is, nevertheless, a major focus in refineries on the ways in which heavy feedstocks might be converted into low-boiling high-value products (Khan and Patmore, 1997). Simultaneously, the changing crude oil properties are reflected in changes such as an increase in asphaltene constituents, an increase in sulfur, metal, and nitrogen contents. Pretreatment processes for removing such constituents or at least negating their effect in thermal process will also play an important role.

Heavy oil, tar sand bitumen, and residua are generally characterised by low API gravity (high density) and high viscosity, high initial boiling point, high carbon residue, high nitrogen content, high sulfur content, and high metals content (Chapter 8). In addition to these properties, the heavy feedstocks also have an increased molecular weight and reduced hydrogen content (Figure 14.18). However, in order to adequately define heavy oil and tar sand bitumen (Chapter 1), reference must also be made to the method of recovery (Chapter 5).

The limitations of processing these heavy feedstocks depend to a large extent on the amount of higher molecular weight constituents (i.e., asphaltene constituents) that contain the majority of the heteroatom constituents (Figure 14.19) (Chapter 11). These constituents are responsible for high yields of thermal and catalytic coke (Chapter 9 and Chapter 15). The majority of the metal constituents in crude oils are associated with the asphaltene constituents. Part of these metals forms organometallic complexes. The rest are found in organic or inorganic salts that are soluble in water or in crude. In recent years, attempts have been made to isolate and to study the vanadium present in petroleum porphyrins, mainly in asphaltene fractions.

When catalytic processes are employed, complex molecules (such as those that may be found in the original asphaltene fraction or those formed during the process) are not sufficiently mobile (or are too strongly adsorbed by the catalyst) to be saturated by hydro-

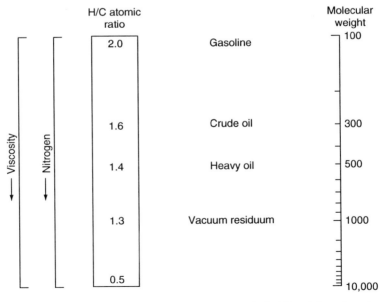

FIGURE 14.18 Relative hydrogen content (through the atomic H/C ratio) and molecular weight of refinery feedstocks.

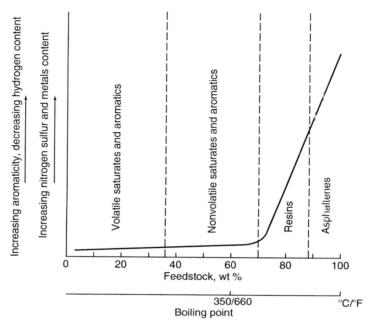

FIGURE 14.19 Relative distribution of heteroatoms in the various fractions.

genation. The chemistry of the thermal reactions of some of these constituents (Chapter 15) dictates that certain reactions, once initiated, cannot be reversed and proceed to completion. Coke is the eventual product. These deposits deactivate the catalyst sites and eventually interfere with the hydroprocess.

However, the essential step required of refineries is the upgrading of heavy feedstocks, particularly residua (McKetta, 1992; Dickenson et al., 1997). In fact, the increasing supply of heavy crude oils is a matter of serious concern for the petroleum industry. In order to satisfy the changing pattern of product demand, significant investments in refining conversion processes will be necessary to profitably utilize these heavy crude oils. The most efficient and economical solution to this problem will depend to a large extent on individual country and company situations. However, the most promising technologies will likely involve the conversion of vacuum bottom residual oils, asphalt from deasphalting processes, and super-heavy crude oils into useful low-boiling and middle distillate products.

Upgrading heavy oil upgrading and residua began with the introduction of desulfurization processes (Speight, 1984, 2000). In the early days, the goal was desulfurization but, in later years, the processes were adapted to a 10% to 30% partial conversion operation, as intended to achieve desulfurization and obtain low-boiling fractions simultaneously, by increasing severity in operating conditions. Refinery evolution has seen the introduction of a variety of residuum cracking processes based on thermal cracking (Table 14.6) (Chapter 17), catalytic cracking (Chapter 18), and **hydroconversion** (Chapter 20 and Chapter 21). Those processes are different from one another in cracking method, cracked product patterns and product properties, and will be employed in refineries according to their respective features. Thus, refining heavy feedstocks has become a major issue in modern refinery practice and several process configurations have evolved to accommodate the heavy feedstocks (RAROP, 1991; Shih and Oballa, 1991; Khan and Patmore, 1997).

TABLE 14.6
Recent Process Concepts and Configurations for Refining Heavy Feedstocks

Process	Comments
Solvent Processes	
DEMEX process	Less selective solvent than propane
MDS process	Solvent deasphalting and desulfurization for feedstock to catalytic cracker
ROSE process	Deasphaltening
Thermal processes	
ASCOT process	Combination of delayed coking and deep solvent deasphalting
CHERRY-P process	Feedstock slurred with coal
ET-II process	Feedstock mixed with high-boiling recycle oil
Catalytic cracking processes	
ART process	Efficient catalyst-feedstock contact
HOC process	Residuum first desulfurized
HOT process	Steam-iron reaction to produce hydrogen in the cracker

Technologies for upgrading heavy crude oils such as heavy oil, bitumen, and residua can be broadly divided into **carbon rejection** and **hydrogen addition** processes. Carbon rejection redistributes hydrogen among the various components, resulting in fractions with increased H/C atomic ratios and fractions with lower H/C atomic ratios. On the other hand, hydrogen addition processes involve reaction of heavy crude oils with an external source of hydrogen and result in an overall increase in H/C ratio. Within these broad ranges, all the more common upgrading technologies can be subdivided as follows:

1. *Carbon rejection*
   Visbreaking, coking, and fluid catalytic cracking
2. *Hydrogen addition*
   Hydrovisbreaking and catalytic hydrocracking
3. *Separation processes*
   Distillation and deasphalting

Thus, the options for refiners processing heavy high sulfur will be a combination of upgrading schemes and byproduct utilization. Residue upgrading options include: (1) deep cut vacuum distillation, (2) solvent deasphalting, (3) residue hydroprocessing, and residue catalytic cracking, in addition to options that focus on the well-established visbreaking and coking technologies. These process options for upgrading heavy oils and residua will be described in more detail in the respective chapters, since a detailed description of every process would be repetitive at this point.

For the present, using a schematic refinery operation (Figure 14.1), new processes for the conversion of residua and heavy oils will probably be used in concert with visbreaking, with some degree of hydroprocessing as a primary conversion step. Other processes may replace or augment the deasphalting units in many refineries. An exception, which may become the rule, is the upgrading of bitumen from tar sands (Speight, 2005). The bitumen is subjected to either delayed coking or fluid coking as the primary upgrading step (Figure 14.20) with some prior distillation or topping. After primary upgrading, the product streams are hydrotreated and combined to form a synthetic crude oil that is shipped to a conventional refinery for further

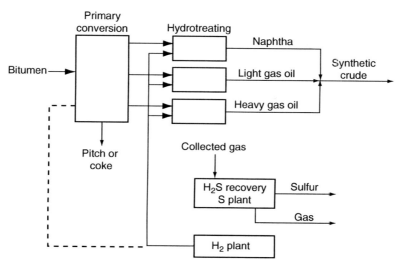

FIGURE 14.20 Processing sequence for tar sand bitumen.

processing. Conceivably, a heavy feedstock could be upgraded in the same manner and, depending on the upgrading facility, upgraded further for sales.

Finally, there is not one single heavy oil upgrading solution that will fit all refineries. Market conditions, existing refinery configuration, and available crude prices, all can have a significant effect on the final configuration. A proper evaluation, however, is not a simple undertaking for an existing refinery. The evaluation starts with an accurate understanding of the market for the various products along with corresponding product values at various levels of supply. The next step is to select a set of crude oils that adequately cover the range of crude oils that may be expected to be processed. It is also important to consider new unit capital costs as well as incremental capital costs for revamp opportunities along with the incremental utility, support and infrastructure costs. The costs, although estimated at the start, can be better assessed once the options have been defined leading to the development of the optimal configuration for refining the incoming feedstocks.

14.14 PETROLEUM PRODUCTS

Petroleum products (Chapter 26), in contrast to petrochemicals (Chapter 27), are those bulk fractions that are derived from petroleum and have commercial value as a bulk product. In the strictest sense, petrochemicals are also petroleum products, but they are individual chemicals that are used as the basic building blocks of the chemical industry.

The use of petroleum and its products was established in pre-Christian times and is known largely through documentation by many of the older civilizations (Chapter 1) and, thus, use of petroleum and the development of related technology is not such a modern subject as we are inclined to believe. However, there have been many changes in emphasis on product demand since petroleum first came into use some five to six millennia ago (Chapter 1). It is these changes in product demand that have been largely responsible for the evolution of the industry, from the asphalt used in ancient times to the gasoline and other liquid fuels of today.

Petroleum is an extremely complex mixture of hydrocarbon compounds, usually with minor amounts of nitrogen-containing, oxygen-containing, and sulfur-containing compounds as well as trace amounts of metal-containing compounds (Chapter 6). In addition, the properties of petroleum vary widely (Chapter 1 and Chapter 8). Thus, petroleum is not used in its raw state. A variety of processing steps is required to convert petroleum from its raw state to products that have well-defined properties (Table 26.3).

The constant demand for products, such as liquid fuels, is the main driving force behind the petroleum industry. Other products, such as lubricating oils, waxes, and asphalt, have also added to the popularity of petroleum as a national resource. Indeed, fuel products that are derived from petroleum supply more than half of the world's total supply of energy. Gasoline, kerosene, and diesel oil provide fuel for automobiles, tractors, trucks, aircraft, and ships. Fuel oil and natural gas are used to heat homes and commercial buildings, as well as to generate electricity. Petroleum products are the basic materials used for the manufacture of synthetic fibers for clothing and in plastics, paints, fertilizers, insecticides, soaps, and synthetic rubber. The uses of petroleum as a source of raw material in manufacturing are central to the functioning of modern industry.

Product complexity has made the industry unique among industries. Indeed, current analytical techniques that are accepted as standard methods for, as an example, the aromatics content of fuels (ASTM D1319, ASTM D2425, ASTM D2549, ASTM D2786, ASTM D2789), as well as proton and carbon nuclear magnetic resonance methods, yield different information. Each method will yield the "% aromatics" in the sample, but the data must be evaluated within the context of the method.

Product complexity becomes even more meaningful when various fractions from different types of crude oil, as well as fractions from synthetic crude oil, are blended with the corresponding petroleum stock. The implications for refining the fractions to saleable products increase. However, for the main part, the petroleum industry was inspired by the development of the automobile and the continued demand for gasoline and other fuels. Such a demand has been accompanied by the demand for other products: diesel fuel for engines, lubricants for engine and machinery parts, fuel oil to provide power for the industrial complex, and asphalt for roadways.

Unlike processes, products are more difficult to place on an individual evolutionary scale. Processes changed and evolved to accommodate the demand for, say, higher-octane fuels, longer-lasting asphalt, or lower sulfur coke. In this section, a general overview of some petroleum products is presented to show the raison d'être of the industry. Another consideration that must be acknowledged is the change in character and composition of the original petroleum feedstock (Chapter 3 and Chapter 7). In the early days of the petroleum industry, several products were obtained by distillation and could be used without any further treatment. Nowadays, the different character and composition of the petroleum dictates that any liquids obtained by distillation must go through one or more of the several available product improvement processes (Chapter 18). Such changes in feedstock character and composition have caused the refining industry to evolve in a direction, such that changes in the petroleum can be accommodated.

It must also be recognized that adequate storage facilities for the gases, liquids, and solids that are produced during the refining operations are also an essential part of a refinery. Without such facilities, refineries would be incapable of operating efficiently.

The customary processing of petroleum does not usually involve the separation and handling of pure hydrocarbons. Indeed, petroleum-derived products are always mixtures: occasionally simple, but more often very complex. Thus, for the purposes of this chapter, such materials as the gross fractions of petroleum (e.g., gasoline, naphtha, kerosene, and the like) which are usually obtained by distillation and refining are classed as petroleum products; asphalt and other solid products (e.g., wax) are also included in this division.

14.15 PETROCHEMICALS

The petrochemical industry began in the 1920s, as suitable byproducts became available through improvements in the refining processes. It developed parallel with the oil industry and has rapidly expanded since the 1940s, with the oil refining industry providing plentiful cheap raw materials.

A petrochemical is any chemical (as distinct from fuels and petroleum products) manufactured from petroleum (and natural gas) and used for a variety of commercial purposes. The definition, however, has been broadened to include the whole range of aliphatic, aromatic, and naphthenic organic chemicals, as well as carbon black and such inorganic materials as sulfur and ammonia. Petroleum and natural gas are made up of hydrocarbon molecules, which are composed of one or more carbon atoms, to which hydrogen atoms are attached. Currently, through a variety of intermediates (Table 14.7) oil and gas are the main sources of the raw materials (Table 14.8) because they are the least expensive, most readily available, and can be processed most easily into the primary petrochemicals. Primary petrochemicals include: olefins (ethylene, propylene and butadiene) aromatics (benzene, toluene, and the isomers of xylene); and methanol. Thus, petrochemical feedstocks can be classified into three general groups: olefins, aromatics, and methanol; a fourth group includes inorganic compounds and synthesis gas (mixtures of carbon monoxide and hydrogen). In many instances, a specific chemical included among the petrochemicals may also be obtained from other sources, such as coal, coke, or vegetable products. For example, materials such as benzene

TABLE 14.7
Hydrocarbon Intermediates Used in the Petrochemical Industry

Carbon Number	Hydrocarbon Type		
	Saturated	Unsaturated	Aromatic
1	Methane		
2	Ethane	Ethylene	
		Acetylene	
3	Propane	Propylene	
4	Butanes	*n*-Butenes	
		Iso-butene	
		Butadiene	
5	Pentanes	*Iso*-pentenes	
		(*Iso*-amylenes)	
		Iso-prene	
6	Hexanes	Methylpentenes	Benzene
	Cyclohexane		
7		Mixed heptenes	Toluene
8		Di-*iso*-butylene	Xylenes
			Ethylbenzene
			Styrene
9			Cumene
12		Propylene tetramer	
		Tri-*iso*-butylene	
18			Dodecylbenzene
6–18		*n*-Olefins	
11–18	*n*-Paraffins		

TABLE 14.8
Sources of Petrochemical Intermediates

Hydrocarbon	Source
Methane	Natural gas
Ethane	Natural gas
Ethylene	Cracking processes
Propane	Natural gas, catalytic reforming, cracking processes
Propylene	Cracking processes
Butane	Natural gas, reforming and cracking processes
Butene(s)	Cracking processes
Cyclohexane	Distillation
Benzene	Catalytic reforming
Toluene	Catalytic reforming
Xylene(s)	Catalytic reforming
Ethylbenzene	Catalytic reforming
Alkylbenzenes	Alkylation
>C_9	Polymerization

and naphthalene can be made from either petroleum or coal, whereas ethyl alcohol may be of petrochemical or vegetable origin.

As stated earlier above, some of the chemicals and compounds produced in a refinery are destined for further processing and as raw material feedstocks for the fast growing petrochemical industry. Such nonfuel uses of crude oil products are sometimes referred to as its nonenergy uses. Petroleum products and natural gas provide two of the basic starting points for this industry; methane from natural gas, and naphtha and refinery gases.

Petrochemical intermediates are generally produced by chemical conversion of primary petrochemicals to form more complicated derivative products. Petrochemical derivative products can be made in a variety of ways: directly from primary petrochemicals; through intermediate products which still contain only carbon and hydrogen; and, through intermediates which incorporate chlorine, nitrogen or oxygen in the finished derivative. In some cases, they are finished products; in others, more steps are needed to arrive at the desired composition.

Of all the processes used, one of the most important is polymerization. It is used in the production of plastics, fibers and synthetic rubber, the main finished petrochemical derivatives. Some typical petrochemical intermediates are: vinyl acetate for paint, paper and textile coatings, vinyl chloride for polyvinyl chloride (PVC), resin manufacture, ethylene glycol for polyester textile fibers, styrene which is important in rubber and plastic manufacturing. The end products number in the thousands, some going on as inputs into the chemical industry for further processing. The more common products made from petrochemicals include adhesives, plastics, soaps, detergents, solvents, paints, drugs, fertilizer, pesticides, insecticides, explosives, synthetic fibers, synthetic rubber, and flooring and insulating materials.

REFERENCES

Abraham, H. 1945. *Asphalts and Allied Substances,* Volume I. Van Nostrand, New York.
Bland, W.F. and Davidson, R.L. 1967. *Petroleum Processing Handbook*. McGraw-Hill, New York.
Dickenson, R.L., Biasca, F.E., Schulman, B.L., and Johnson, H.E. 1997. *Hydrocarbon Processing* 76(2): 57.

Ditman, J.G. 1973. *Hydrocarbon Processing* 52(5): 110.

Dunning, H.N. and Moore, J.W. 1957. *Petroleum Refiner* 36(5): 247–250.

Forbes, R.J. 1958. *A History of Technology*, V. Oxford University Press, Oxford, UK.

Gary, J.H. and Handwerk, G.E. 2001. *Petroleum Refining: Technology and Economics,* 4th ed. Marcel Dekker Inc., New York.

Gruse, W.A. and Stevens, D.R. 1960. *Chemical Technology of Petroleum*. McGraw-Hill, New York.

Hobson, G.D. and Pohl, W. 1973. *Modern Petroleum Technology*. Applied Science Publishers, Barking, UK.

Hoiberg, A.J. 1960. *Bituminous Materials: Asphalts, Tars and Pitches*, I & II. Interscience, New York.

Khan, M.R. and Patmore, D.J. 1997. *Petroleum Chemistry and Refining*. J.G. Speight, ed. Taylor & Francis, Washington, DC. Chapter 6.

Kobe, K.A. and McKetta, J.J. 1958. *Advances in Petroleum Chemistry and Refining*. Interscience, New York.

McKetta, J.J. ed. 1992. *Petroleum Processing Handbook*. Marcel Dekker Inc., New York.

Nelson, W.L. 1958. *Petroleum Refinery Engineering*. McGraw-Hill, New York.

RAROP. 1991. *RAROP Heavy Oil Processing Handbook*. Research Association for Residual Oil Processing. Noguchi (Chairman), T. Ministry of Trade and International Industry (MITI), Tokyo, Japan.

Scholten, G.G. 1992. *Petroleum Processing Handbook*. J.J. McKetta, ed. Marcel Dekker Inc., New York. p. 565.

Shih, S.S. and Oballa, M.C. eds. 1991. *Tar Sand Upgrading Technology*. Symposium Series No. 282. American Institute for Chemical Engineers, New York.

Speight, J.G. 1984. *Catalysis on the Energy Scene*. Kaliaguine ,S. and Mahay. A. eds. Elsevier, Amsterdam.

Speight, J.G. 2000. *The Desulfurization of Heavy Oils and Residua,* 2nd edn. Marcel Dekker Inc., New York.

Speight, J.G. 2005. Natural Bitumen (Tar Sands) and Heavy Oil. *Coal, Oil Shale, Natural Bitumen, Heavy Oil and Peat*. Encyclopedia of Life Support Systems (EOLSS), Developed under the Auspices of the UNESCO, EOLSS Publishers, Oxford, UK, [http://www.eolss.net].

Speight, J.G. and Ozum, B. 2002. *Petroleum Refining Processes*. Marcel Dekker Inc., New York.

15 Refining Chemistry

15.1 INTRODUCTION

Crude oil is rarely used in its raw form but must instead be processed into its various products, generally as a means of forming products with hydrogen content different from that of the original feedstock. Thus, the chemistry of the refining process is concerned primarily with the production, not only of better products but also of saleable materials.

Crude oil contains many thousands of different compounds that vary in molecular weight from methane (CH_4, 16) to those with a molecular weight of more than 2000 (Boduszynski 1987, Speight, 1994). This broad range in molecular weights results in boiling points that range from $-160°C$ ($-288°F$) to temperatures of the order of nearly $1100°C$ ($2000°F$). Many of the constituents of crude oil are paraffins. Remembering that the word paraffin was derived from the Latin *parum affinis* meaning little affinity or little reactivity, it must have come as a great surprise that hydrocarbons, paraffins included, can undergo a diversity of reactions (Smith, 1994; Laszlo, 1995).

The major refinery products are liquefied petroleum gas (LPG), gasoline, jet fuel, solvents, kerosene, middle distillates (known as **gas oil** outside the United States), residual fuel oil, and asphalt. In the United States, with its high demand for gasoline, refineries typically upgrade their products much more than in other areas of the world, where the heavy end products, like residual fuel oil, are used in industry and power generation.

Understanding refining chemistry not only allows an explanation of the means by which these products can be formed from crude oil, but also offers a chance of predictability. This is very necessary when the different types of crude oil accepted by refineries are considered. And the major processes by which these products are produced from crude oil constituents involve thermal decomposition.

There are various theories relating to the thermal decomposition of organic molecules and this area of petroleum technology has been the subject of study for several decades (Hurd, 1929; Fabuss et al., 1964; Fitzer et al., 1971). The relative reactivity of petroleum constituents can be assessed on the basis of bond energies, but the thermal stability of an organic molecule is dependent upon the bond strength of the weakest bond. And even though the use of bond energy data is a method for predicting the reactivity or the stability of specific bonds under designed conditions, the reactivity of a particular bond is also subject to its environment. Thus, it is not only the reactivity of the constituents of petroleum that are important in processing behavior, it is also the stereochemistry of the constituents as they relate to one another that is also of some importance (Chapter 11). It must be appreciated that the stereochemistry of organic compounds is often a major factor in determining reactivity and properties (Eliel and Wilen, 1994).

In the present context, it is necessary to recognize that (*parum affinis* or not), most hydrocarbons decompose thermally at temperatures above about $650°F$ ($340°C$), so the high boiling points of many petroleum constituents cannot be measured directly and must be estimated from other measurements. And in the present context, it is as well that hydro-carbons decompose at elevated temperatures. Thereby lies the route to many modern prod-ucts. For example, in a petroleum refinery, the highest value products are transportation fuels:

1. Gasoline (b.p. <220°C, <425°F)
2. Jet fuel (b.p. 175°C to 290°C, 350°F to 550°F)
3. Diesel (175°C to 370°C, 350°F to 700°F)

These fuels are produced by thermal decomposition of a variety of hydrocarbons, high molecular weight paraffins included. Less than one third of a typical crude oil distills in these ranges and thus, the goal of refining chemistry might be stated simply as the methods by which crude oil is converted to these fuels. It must be recognized that refining involves a wide variety of chemical reactions, but the production of liquid fuels is the focus of a refinery.

Refining processes involve the use of various thermal and catalytic processes to convert molecules in the heavier fractions to smaller molecules in fractions distilling at these lower temperatures (Jones, 1995). This efficiency translates into a strong economic advantage, leading to widespread use of conversion processes in refineries today. However, in order to understand the principles of **catalytic cracking**, understanding the principles of adsorption and reaction on solid surfaces is valuable (Samorjai, 1994; Masel, 1995)

A refinery is a complex network of integrated unit processes for the purpose of producing a variety of products from crude oil. Refined products establish the order in which the individual refining units will be introduced, and the choice from among several types of units and the size of these units are dependent upon economic factors. The trade-off among product types, quantity, and quality influences the choice of one kind of processing option over another.

Each refinery has its own range of preferred crude oil feedstock from which a desired distribution of products is obtained. Nevertheless, refinery processes can be divided into three major types:

1. *Separation*: division of crude oil into various streams (or fractions) depending on the nature of the crude material
2. *Conversion*: production of saleable materials from crude oil, usually by skeletal alter-ation, or even by alteration of the chemical type of the crude oil constituents
3. *Finishing*: purification of various product streams by a variety of processes that essen-tially remove impurities from the product; for convenience, processes that accomplish molecular alteration, such as **reforming**, are also included in this category

The separation and finishing processes may involve distillation or even treatment with a wash solution, either to remove impurities or, in the case of distillation, to produce a material boiling over a narrower range and the chemistry of these processes is quite simple (Chapter 13, Chapter 18, and Chapter 19). The inclusion of reforming processes in this category is purely for descriptive purposes rather than being representative of the chemistry involved. Reforming processes produce streams that allow the product to be finished, as the term applies to product behavior and utility.

Conversion processes are, in essence, processes that change the number of carbon atoms per molecule, alter the molecular hydrogen-to-carbon ratio, or change the molecular structure of the material without affecting the number of carbon atoms per molecule. These latter processes (isomerization processes) essentially change the shape of the molecule(s) and are used to improve the quality of the product (Chapter 18).

Nevertheless, the chemistry of conversion process may be quite complex (King et al., 1973), and an understanding of the chemistry involved in the conversion of a crude oil to a variety of products is essential to an understanding of refinery operations. It is therefore the purpose of this chapter to serve as an introduction to the chemistry involved in these conversion processes so that the subsequent chapters dealing with refining (Chapter 14 through Chapter 17) are easier to visualize and understand. However, understanding refining chemistry from the behavior of model compounds under refining conditions is not as straightforward as it may appear (Ebert et al., 1987).

The complexity of the individual reactions occurring in an extremely complex mixture and the interference of the products with those from other components of the mixture is unpredictable. Or the interference of secondary and tertiary products with the course of a reaction and, hence, with the formation of primary products may also be cause for concern. Hence, caution is advised when applying the data from model compound studies to the behavior of petroleum, especially the molecularly complex heavy oils. These have few, if any, parallels in organic chemistry.

15.2 CRACKING

15.2.1 THERMAL CRACKING

The term **cracking** applies to the decomposition of petroleum constituents that is induced by elevated temperatures ($>350°C$, $>660°F$), whereby the higher molecular weight constituents of petroleum are converted to lower molecular weight products. Cracking reactions involve carbon–carbon bond rupture and are thermodynamically favored at high temperature (Egloff, 1937).

Thus, cracking is a phenomenon by which higher boiling (higher molecular weight) constituents in petroleum are converted into lower boiling (lower molecular weight) products. However, certain products may interact with one another to yield products with higher molecular weights than the constituents of the original feedstock. Some of the products are expelled from the system as, say, gases, gasoline-range materials, kerosene-range materials, and the various intermediates that produce other products such as coke. Materials that have boiling ranges higher than gasoline and kerosene may (depending upon the refining options) be referred to as **recycle stock**, which is recycled in the cracking equipment until conversion is complete.

Two general types of reaction occur during cracking:

1. The decomposition of large molecules into small molecules (primary reactions):

$$CH_3 \, CH_2 \, CH_2 \, CH_3 \rightarrow CH_4 + CH_3 CH = CH_2$$
Butane   Methane   Propene
$$CH_3 \, CH_2 \, CH_2 \, CH_3 \rightarrow CH_3 CH_3 + CH_2 = CH_2$$
Butane   Ethane   Ethylene

2. Reactions by which some of the primary products interact to form higher molecular weight materials (secondary reactions):

$$CH_2 = CH_2 + CH_2 = CH_2 \rightarrow CH_3 \, CH_2 \, CH = CH_2$$

or

$$R \cdot CH = CH_2 + R^1 \cdot CH = CH_2 \rightarrow \text{Cracked residuum} + \text{Coke} + \text{Other products}$$

Thermal cracking is a free radical chain reaction; a free radical is an atom or group of atoms possessing an unpaired electron. Free radicals are very reactive, and it is their mode of reaction that actually determines the product distribution during thermal cracking. The free radical reacts with a hydrocarbon by abstracting a hydrogen atom to produce a stable end product and a new free radical. Free radical reactions are extremely complex, and it is hoped that these few reaction schemes illustrate potential reaction pathways. Any of the preceding reaction types are possible, but it is generally recognized that the prevailing conditions and those reaction sequences that are thermodynamically favored determine the product distribution.

One of the significant features of hydrocarbon free radicals is their resistance to **isomerization**, for example, migration of an alkyl group and, as a result, thermal cracking does not produce any degree of branching in the products other than that already present in the feedstock.

Data obtained from the thermal decomposition of pure compounds indicate certain decomposition characteristics that permit predictions to be made of the product types that arise from the thermal cracking of various feedstock. For example, normal paraffins are believed to form, initially, higher molecular weight material, which subsequently decomposes as the reaction progresses. Other paraffinic materials and α (terminal) olefins are produced. An increase in pressure inhibits the formation of low-molecular weight gaseous products and therefore promotes the formation of higher molecular weight materials.

Branched paraffins react somewhat differently to the normal paraffins during cracking processes and produce substantially higher yields of olefins with one less carbon atom than the parent hydrocarbon. Cycloparaffins (naphthenes) react differently to their noncyclic counterparts and are somewhat more stable. For example, cyclohexane produces hydrogen, ethylene, butadiene, and benzene: alkyl-substituted cycloparaffins decompose by means of scission of the alkyl chain to produce an olefin and a methyl or ethyl cyclohexane.

The aromatic ring is considered fairly stable at moderate cracking temperatures (350°C to 500°C, 660°F to 930°F). Alkylated aromatics, like the alkylated naphthenes, are more prone to dealkylation that to ring destruction. However, ring destruction of the benzene derivatives occurs above 500°C (930°F), but condensed aromatics may undergo ring destruction at somewhat lower temperatures (450°C, 840°F).

15.2.2 CATALYTIC CRACKING

Catalytic cracking is the thermal decomposition of petroleum constituents' hydrocarbons in the presence of a catalyst (Pines, 1981). Thermal cracking has been superseded by catalytic cracking as the process for gasoline manufacture. Indeed, gasoline produced by catalytic cracking is richer in branched paraffins, cycloparaffins, and aromatics, all of which serve to increase the quality of the gasoline. Catalytic cracking also results in production of the maximum amount of butenes and butanes (C_4H_8 and C_4H_{10}), rather than ethylene and ethane (C_2H_4 and C_2H_6).

Catalytic cracking processes evolved in the 1930s, from research on petroleum and coal liquids. The petroleum work came to fruition with the invention of acid cracking. The work to produce liquid fuels from coal, most notably in Germany, resulted in metal sulfide hydrogenation catalysts. In the 1930s, a catalytic cracking catalyst for petroleum that used solid acids as catalysts was developed using acid-treated clays.

Clays are a family of crystalline aluminosilicate solids, and the acid treatment develops acidic sites by removing aluminum from the structure. The acid sites also catalyze the formation of coke, and Houdry developed a moving-bed process that continuously removed the cooked beads from the reactor for regeneration by oxidation with air (McEvoy, 1996).

Although thermal cracking is a free radical (neutral) process, catalytic cracking is an ionic process involving carbonium ions, which are hydrocarbon ions having a positive charge on a carbon atom. The formation of carbonium ions during catalytic cracking can occur by:

1. Addition of a proton from an acid catalyst to an olefin
2. Abstraction of a hydride ion (H^-) from a hydrocarbon by the acid catalyst or by another carbonium ion

However, carbonium ions are not formed by cleavage of a carbon–carbon bond. In essence, the use of a catalyst permits alternate routes for cracking reactions, usually by lowering the free energy of activation for the reaction. The acid catalysts first used in catalytic cracking were amorphous solids composed of approximately 87% silica (SiO_2) and 13% alumina (Al_2O_3) and were designated low-alumina catalysts. However, this type of catalyst is now being replaced by crystalline aluminosilicates (zeolites) or molecular sieves.

The first catalysts used for catalytic cracking were acid-treated clays, formed into beads. In fact, clays are still employed as catalyst in some cracking processes (Chapter 18).

Clays are a family of crystalline aluminosilicate solids, and the acid treatment develops acidic sites by removing aluminum from the structure. The acid sites also catalyze the formation of coke, and the development of a moving-bed process that continuously removed the cooked beads from the reactor reduced the yield of coke; clay regeneration was achieved by oxidation with air (McEvoy, 1996).

Clays are natural compounds of silica and alumina, containing major amounts of the oxides of sodium, potassium, magnesium, calcium, and other alkali and alkaline earth metals. Iron and other transition metals are often found in natural clays, substituted for the aluminum cations. Oxides of virtually every metal are found as impurity deposits in clay minerals.

Clays are layered crystalline materials. They contain large amounts of water within and between the layers (Keller, 1985). Heating the clays above 100°C can drive out some or all of this water; at higher temperatures, the clay structures themselves can undergo complex solid-state reactions. Such behavior makes the chemistry of clays a fascinating field of study in its own right. Typical clays include kaolinite, montmorillonite, and illite (Keller, 1985). They are found in most natural soils and in large, relatively pure deposits, from which they are mined for applications ranging from adsorbents to paper making.

Once the carbonium ions are formed, the modes of interaction constitute an important means by which product formation occurs during catalytic cracking. For example, isomerization either by hydride ion shift or by methyl group shift, both of which occur readily. The trend is for stabilization of the carbonium ion by movement of the charged carbon atom toward the center of the molecule, which accounts for the isomerization of α-olefins to internal olefins when carbonium ions are produced. Cyclization can occur by internal addition of a carbonium ion to a double bond which, by continuation of the sequence, can result in aromatization of the cyclic carbonium ion.

Like the paraffins, naphthenes do not appear to isomerize before cracking. However, the naphthenic hydrocarbons (from C_9 upward) produce considerable amounts of aromatic hydrocarbons during catalytic cracking. Reaction schemes similar to that outlined here provide possible routes for the conversion of naphthenes to aromatics. Alkylated benzenes undergo nearly quantitative dealkylation to benzene without apparent ring degradation below 500°C (930°F). However, polymethlybenzenes undergo disproportionation and isomerization with very little benzene formation.

Catalytic cracking can be represented by simple reaction schemes. However, questions have arisen as to how the cracking of paraffins is initiated. Several hypotheses for the initiation step in catalytic cracking of paraffins have been proposed (Cumming and Wojciechowski, 1996).

The Lewis site mechanism is the most obvious, as it proposes that a carbenium ion is formed by the abstraction of a hydride ion from a saturated hydrocarbon by a strong Lewis acid site: a tricoordinated aluminum species. On Brønsted sites a carbenium ion may be readily formed from an olefin by the addition of a proton to the double bond or, more rarely, via the abstraction of a hydride ion from a paraffin by a strong Brønsted proton. This latter process requires the formation of hydrogen as an initial product. This concept was, for various reasons that are of uncertain foundation, often neglected.

It is therefore not surprising that the earliest cracking mechanisms postulated that the initial carbenium ions are formed only by the protonation of olefins generated either by thermal cracking or present in the feed as an impurity. For a number of reasons this proposal was not convincing, and in the continuing search for initiating reactions it was even proposed that electrical fields associated with the cations in the zeolite are responsible for the polarization of reactant paraffins, thereby activating them for cracking. More recently, however, it has been convincingly shown that a penta-coordinated carbonium ion can be formed on the alkane itself by protonation, if a sufficiently strong Brønsted proton is available (Cumming and Wojciechowski, 1996).

Coke formation is considered, with just cause as a malignant side reaction of normal carbenium ions. However, while chain reactions dominate events occurring on the surface, and produce the majority of products, certain less desirable bimolecular events have a finite chance of involving the same carbenium ions in a bimolecular interaction with one another. Of these reactions, most will produce a paraffin and leave carbine- or carboid-type species (Chapter 1) on the surface. This carbine- or carboid-type species can produce other products, but the most damaging product will be one which remains on the catalyst surface and cannot be desorbed and results in the formation of coke, or remains in a noncoke form but effectively blocks the active sites of the catalyst.

A general reaction sequence for coke formation from paraffins involves oligomerization, cyclization, and dehydrogenation of small molecules at active sites within zeolite pores:

$$\text{Alkanes} \rightarrow \text{Alkenes}$$
$$\text{Alkenes} \rightarrow \text{Oligomers}$$
$$\text{Oligomers} \rightarrow \text{Naphthenes}$$
$$\text{Naphthenes} \rightarrow \text{Aromatics}$$
$$\text{Aromatics} \rightarrow \text{Coke}$$

Whether or not these are the true steps to coke formation can only be surmised. The problem with this reaction sequence is that it ignores sequential reactions in favor of consecutive reactions. And it must be accepted that the chemistry leading up to coke formation is a complex process, consisting of many sequential and parallel reactions.

There is a complex and little understood relationship between coke content, catalyst activity, and the chemical nature of coke. For instance, the atomic hydrogen or carbon ratio of coke depends on how the coke was formed; its exact value will vary from system to system (Cumming and Wojciechowski, 1996). And it seems that catalyst decay is not related in any simple way to the hydrogen-to-carbon atomic ratio of coke, or to the total coke content of the catalyst, or any simple measure of coke properties. Moreover, despite many and varied attempts, there is currently no consensus as to the detailed chemistry of coke formation. There is, however, much evidence and good reason to believe that catalytic coke is formed from carbenium ions which undergo addition, dehydrogenation and cyclization, and elimination side reactions in addition to the main-line chain propagation processes (Cumming and Wojciechowski, 1996).

15.2.3 DEHYDROGENATION

The common primary reactions of pyrolysis are dehydrogenation and carbon bond scission. The extent of one or the other varies with the starting material and operating conditions, but because of its practical importance, methods have been found to increase the extent of dehydrogenation and, in some cases, to render it almost the only reaction.

Dehydrogenation is essentially the removal of hydrogen from the parent molecule. For example, at 550°C (1025°F) n-butane loses hydrogen to produce butene-1 and butene-2. The development of selective catalysts, such as chromic oxide (chromia, Cr_2O_3) on alumina (Al_2O_3) has rendered the dehydrogenation of paraffins to olefins particularly effective, and the formation of higher molecular weight material is minimized.

Naphthenes are somewhat more difficult to dehydrogenate, and cyclopentane derivatives form only aromatics if a preliminary step to form the cyclohexane structure can occur. Alkyl derivatives of cyclohexane usually dehydrogenate at 480°C to 500°C (895°F to 930°F), and polycyclic naphthenes are also quite easy to dehydrogenate thermally. In the presence of catalysts, cyclohexane and its derivatives are readily converted into aromatics; reactions of this type are prevalent in catalytic cracking and reforming. Benzene and toluene are prepared by the catalytic dehydrogenation of cyclohexane and methylcyclohexane respectively.

Polycyclic naphthenes can also be converted to the corresponding aromatics by heating at 450°C (840°F) in the presence of a chromia–alumina ($Cr_2O_3-Al_2O_3$) catalyst.

Alkylaromatics also dehydrogenate to various products. For example, styrene is prepared by the catalytic dehydrogenation of ethylbenzene.

Other alkylbenzenes can be dehydrogenated similarly; iso-propyl benzene yields α-methyl styrene.

15.2.4 DEHYDROCYCLIZATION

Catalytic aromatization involving the loss of 1 mol of hydrogen followed by ring formation and further loss of hydrogen has been demonstrated for a variety of paraffins (typically n-hexane and n-heptane). Thus, n-hexane can be converted to benzene, heptane is converted to toluene, and octane is converted to ethyl benzene and o-xylene. Conversion takes place at low pressures, even atmospheric, and at temperatures above 300°C (570°F), although 450°C to 550°C (840°F to 1020°F) is the preferred temperature range.

The catalysts are metals (or their oxides) of the titanium, vanadium, and tungsten groups and are generally supported on alumina; the mechanism is believed to be dehydrogenation of the paraffin to an olefin, which in turn is cyclized and dehydrogenated to the aromatic hydrocarbon. In support of this, olefins can be converted to aromatics much more easily that the corresponding paraffins.

15.3 HYDROGENATION

The purpose of hydrogenating petroleum constituents is (1) to improve existing petroleum products or develop new products or even new uses, (2) to convert inferior or low-grade materials into valuable products, and (3) to transform higher molecular weight constituents into liquid fuels.

The distinguishing feature of the hydrogenating processes is that, although the composition of the feedstock is relatively unknown and a variety of reactions may occur simultaneously, the final product may actually meet all the required specifications for its particular use (Furimsky, 1983; Speight, 2000).

Hydrogenation processes (Chapter 20 and 21) for the conversion of petroleum and petroleum products may be classified as destructive and nondestructive. The former (*hydrogenolysis* or **hydrocracking**) is characterized by the rupture of carbon–carbon bonds and is accompanied by hydrogen saturation of the fragments to produce lower boiling products. Such treatment requires rather high temperatures and high hydrogen pressures, the latter to minimize coke formation. Many other reactions, such as isomerization, dehydrogenation, and cyclization, can occur under these conditions (Dolbear et al., 1987).

On the other hand, nondestructive, or simple, hydrogenation is generally used for the purpose of improving product (or even feedstock) quality without appreciable alteration of the boiling range. Treatment under such mild conditions is often referred to as **hydrotreating** or **hydrofining** and is essentially a means of eliminating nitrogen, oxygen, and sulfur as ammonia, water, and hydrogen sulfide, respectively.

15.3.1 HYDROCRACKING

Hydrocracking (Chapter 21) is a thermal process (>350°C, >660°F) in which hydrogenation accompanies cracking. Relatively high pressure (100 to 2000 psi) is employed, and the overall result is usually a change in the character or quality of the products.

The wide range of products possible from hydrocracking is the result of combining catalytic cracking reactions with hydrogenation. The reactions are catalyzed by dual-function catalysts in which the cracking function is provided by silica–alumina (or zeolite) catalysts, and platinum, tungsten oxide, or nickel provides the hydrogenation function.

Essentially, all the initial reactions of catalytic cracking occur, but some of the secondary reactions are inhibited or stopped by the presence of hydrogen. For example, the yields of olefins and the secondary reactions that result from the presence of these materials are substantially diminished and branched-chain paraffins undergo demethanation. The methyl groups attached to secondary carbons are more easily removed than those attached to tertiary carbon atoms, whereas methyl groups attached to quaternary carbons are the most resistant to hydrocracking.

The effect of hydrogen on naphthenic hydrocarbons is mainly that of ring scission followed by immediate saturation of each end of the fragment produced. The ring is preferentially broken at favored positions, although generally all the carbon–carbon bond positions are attacked to some extent. For example, methyl-cyclopentane is converted (over a platinum-carbon catalyst) to 2-methylpentane, 3-methylpentane, and *n*-hexane.

Aromatic hydrocarbons are resistant to hydrogenation under mild conditions, but under more severe conditions, the main reactions are conversion of the aromatic to naphthenic rings and scissions within the alkyl side chains. The naphthenes may also be converted to paraffins. However, polynuclear aromatics are more readily attacked than the single-ring compounds, the reaction proceeding by a stepwise process in which one ring at a time is saturated and then opened. For example, naphthalene is hydrocracked over a molybdenum oxide–molecular catalyst to produce a variety of low weight paraffins ($\leq C_6$).

15.3.2 HYDROTREATING

It is generally recognized that the higher the hydrogen content of a petroleum product, especially the fuel products, the better is the quality of the product. This knowledge has stimulated the use of hydrogen-adding processes in the refinery.

Thus, hydrotreating (i.e., hydrogenation without simultaneous cracking) (Chapter 20) is used for saturating olefins or for converting aromatics to naphthenes as well as for heteroatom removal. Under atmospheric pressure, olefins can be hydrogenated up to about 500°C (930°F), but beyond this temperature dehydrogenation commences. Application of pressure and the presence of catalysts make it possible to effect complete hydrogenation at room or even cooler temperature; the same influences are helpful in minimizing dehydrogenation at higher temperatures.

A wide variety of metals are active hydrogenation catalysts; those of most interest are nickel, palladium, platinum, cobalt, iron, nickel-promoted copper, and copper chromite. Special preparations of the first three are active at room temperature and atmospheric pressure. The metallic catalysts are easily poisoned by sulfur-containing and arsenic-containing compounds, and even by other metals. To avoid such poisoning, less effective but more resistant metal oxides or sulfides are frequently employed, generally those of tungsten, cobalt, chromium, or molybdenum.

Alternatively, catalysts poisoning can be minimized by mild hydrogenation to remove nitrogen, oxygen, and sulfur from feedstock in the presence of more resistant catalysts, such as cobalt—molybdenum–alumina (Co–Mo–Al$_2$O$_3$). The reactions involved in nitrogen removal are somewhat analogous to those of the sulfur compounds and follow a stepwise mechanism to produce ammonia and the relevant substituted aromatic compound.

15.4 ISOMERIZATION

The importance of isomerization in petroleum-refining operations is twofold. First, the process is valuable in converting *n*-butane into *iso*-butane, which can be alkylated to liquid hydrocarbons in the gasoline boiling range. Second, the process can be used to increase the octane number of the paraffins, boiling in the gasoline boiling range, by converting some of the *n*-paraffins present into *iso*-paraffins.

The process involves contact of the hydrocarbon and a catalyst under conditions favorable to good product recovery (Chapter 25). The catalyst may be aluminum chloride promoted with hydrochloric acid or a platinum-containing catalyst. Both are very reactive and can lead to undesirable side reactions along with isomerization. These side reactions include disproportionation and cracking, which decrease the yield and produce olefinic fragments that may combine with the catalyst and shorten its life. These undesired reactions are controlled by such techniques as the addition of inhibitors to the hydrocarbon feed or by carrying out the reaction in the presence of hydrogen.

Paraffins are readily isomerized at room temperature, and the reaction is believed to occur by the formation and rearrangement of carbonium ions. The chain-initiating ion R$^+$ is formed by the addition of a proton from the acid catalyst to an olefin molecule, which may be added, present as an impurity, or formed by dehydrogenation of the paraffin.

Except for butane, the isomerization of paraffins is generally accompanied by side reactions involving carbon–carbon bond scissions when catalysts of the aluminum halide type are used. Products boiling both higher and lower than the starting material are formed, and the disproportionation reactions that occur with the pentanes and higher paraffins (>C$_5$) are caused by unpromoted aluminum halide. A substantial pressure of hydrogen tends to minimize these side reactions.

The ease of paraffin isomerization increases with molecular weight, but the extent of disproportionation reactions also increases. Conditions can be established under which isomerization takes place only with the butanes, but this is difficult for the pentanes and higher hydrocarbons. At 27°C (81°F) over aluminum bromide (AlBr$_3$), the equilibrium mixture of *n*-pentane and *iso*-pentane, contains over 70% of the branched isomer; at 0°C (32°F) approximately 90% of the branched isomer is present. Higher and lower boiling hydrocarbon products, hexanes, heptanes, and *iso*-butane are also formed in side reactions even at 0°C (32°F) and in increased amounts when the temperature is raised. Although the thermodynamic conditions are favorable, neo-pentane [C(CH$_3$)$_4$] does not appear to isomerize under these conditions.

Olefins are readily isomerized; the reaction involves either movement of the position of the double bond (hydrogen-atom shift) or skeletal alteration (methyl group shift). The double-bond shift may also include a reorientation of the groups around the double bond to bring about a cis–trans isomerization. Thus, 1-butene is isomerized to a mixture of *cis*- and *trans*-2-butene. *Cis* (same side) and *trans* (opposite side) refer to the spatial arrangement of the methyl groups with respect to the double bond.

Olefins having a terminal double bond are the least stable. They isomerize more rapidly than those in which the double bond carries the maximum number of alkyl groups.

Naphthenes can isomerize in various ways; for example, in the case of cyclopropane (C_3H_6) and cyclobutane (C_4H_8), ring scission can occur to produce an olefin. Carbon–carbon rupture may also occur in any side chains to produce polymethyl derivatives, whereas cyclopentane (C_5H_{10}) and cyclohexane (C_6H_{12}) rings may expand and contract, respectively.

The isomerization of alkylaromatics may involve changes in the side-chain configuration, disproportionation of the substituent groups, or their migration about the nucleus. The conditions needed for isomerization within attached long side chains of alkylbenzenes and alkyl-naphthalenes are also those for the scission of such groups from the ring. Such isomerization, therefore, does not take place unless the side chains are relatively short. The isomerization of ethylbenzene to xylenes, and the reverse reaction, occurs readily.

Disproportionation of attached side chains is also a common occurrence; higher and lower alkyl substitution products are formed. For example, xylenes disproportionate in the presence of hydrogen fluoride–boron trifluoride or aluminum chloride to form benzene, toluene, and higher alkylated products; ethylbenzene in the presence of boron trifluoride forms a mixture of benzene and 1,3-diethylbenzene.

15.5 ALKYLATION

Alkylation in the petroleum industry refers to a process for the production of high-octane motor fuel components by the combination of olefins and paraffins. The reaction of *iso*-butane with olefins, using an aluminum chloride catalyst, is a typical alkylation reaction.

In acid-catalyzed alkylation reactions, only paraffins with tertiary carbon atoms, such as *iso*-butane and *iso*-pentane react with the olefin. Ethylene is slower to react that the higher olefins. Olefins higher than propene may complicate the products by engaging in hydrogen exchange reactions.

Cycloparaffins, especially those containing tertiary carbon atoms, are alkylated with olefins in a manner similar to the *iso*-paraffins; the reaction is not as clean, and the yields are low because of the several side reactions that take place.

Aromatic hydrocarbons are more easily alkylated than the *iso*-paraffins by olefins. Cumene (*iso*-propylbenzene) is prepared by alkylating benzene with propene over an acid catalyst. The alkylating agent is usually an olefin, although cyclopropane, alkyl halides, aliphatic alcohols, ethers, and esters may also be used. The alkylation of aromatic hydrocarbons is presumed to occur through the agency of the carbonium ion.

Thermal alkylation is also used in some plants, but like thermal cracking, it is presumed to involve the transient formation of neutral free radicals and therefore tends to be less specific in production distribution.

15.6 POLYMERIZATION

Polymerization is a process in which a substance of low molecular weight is transformed into one of the same composition, but of higher molecular weight while maintaining the atomic arrangement present in the basic molecules. It has also been described as the successive addition of one molecule to another by means of a functional group, such as that present in an aliphatic olefin.

In the petroleum industry, polymerization is used to indicate the production of, say, gasoline components that fall into a specific (and controlled) molecular weight range, hence the term **polymer gasoline**. Further, it is not essential that only one type of monomer be involved:

$$CH_3 \cdot CH = CH_2 + CH_2 = CH_2 \rightarrow CH_3 \cdot CH_2 \cdot CH_2 \cdot CH = CH_2$$

This type of reaction is correctly called copolymerization, but polymerization in the true sense of the word is usually prevented, and all attempts are made to terminate the reaction at the dimer or trimer (three monomers joined together) stage. It is the four- to twelve-carbon compounds that are required as the constituents of liquid fuels. However, in the petrochemical section of the refinery, polymerization, which results in the production of, say, polyethylene, is allowed to proceed until materials of the required high molecular weight have been produced.

15.7 PROCESS CHEMISTRY

In a mixture as complex as petroleum, the reaction processes can only be generalized because of difficulties in analyzing not only the products but also the feedstock as well as the intricate and complex nature of the molecules that make up the feedstock. The formation of coke from the higher molecular weight and polar constituents of a given feedstock is detrimental to process efficiency and to catalyst performance (Speight, 1987; Dolbear, 1998).

Refining the constituents of heavy oil and bitumen has become a major issue in modern refinery practice. The limitations of processing heavy oils and residua depend to a large extent on the amount of higher molecular weight constituents (i.e., asphaltenes) present in the feedstock (Ternan, 1983; Speight, 1984; LePage and Davidson, 1986; Schabron and Speight, 1997; Speight, 2000) that are responsible for high yields of thermal and catalytic coke (Chapter 9).

15.7.1 THERMAL CHEMISTRY

When petroleum is heated to temperatures in excess of 350°C (660°F), the rate of thermal decomposition of the constituents increases significantly. The higher the temperature, the shorter the time to achieve a given conversion, and the severity of the process conditions is a combination of residence time of the crude oil constituents in the reactor and the temperature needed to achieve a given conversion.

Thermal conversion does not require the addition of a catalyst. This approach is the oldest technology available for residue conversion, and the severity of thermal processing determines the conversion and the product characteristics. As the temperature and residence time are increased, the primary products undergo further reaction to produce various secondary products, and so on, with the ultimate products (coke and methane) being formed at extreme temperatures of approximately 1000°C (1830°F).

The thermal decomposition of petroleum asphaltenes has received some attention (Magaril and Aksenova, 1968; Magaril and Ramazaeva, 1969; Magaril and Aksenova, 1970; Magaril et al., 1970; Magaril et al., 1971; Schucker and Keweshan, 1980; Shiroto et al., 1983). Special attention has been given to the nature of the volatile products of asphaltene decomposition, mainly because of the difficulty of characterizing the nonvolatile coke.

The organic nitrogen originally in the asphaltenes invariably undergoes thermal reaction to concentrate in the nonvolatile coke (Speight, 1970, 1989; Vercier, 1981) (Chapter 10). Thus, although asphaltenes produce high yields of thermal coke, little is known of the actual chemistry of coke formation. In a more general scheme, the chemistry of asphaltene **coking** has been suggested to involve the thermolysis of thermally labile bonds to form reactive species that then react with each other (condensation) to form coke. In addition, the highly aromatic and highly polar (refractory) products separate from the surrounding oil medium as an insoluble phase and proceed to form coke.

It is also interesting to note that although the aromaticity of the asphaltene constituents is approximately equivalent to the yield of thermal coke (Figure 15.1), not all the original aromatic carbon in the asphaltene constituents forms coke. Volatile aromatic species are eliminated during thermal decomposition, and it must be assumed that some of the original aliphatic carbon plays a role in coke formation.

Various patterns of thermal behavior have been observed for the constituents of petroleum feedstock (Table 15.1). Since the chemistry of thermal and catalytic cracking has been studied and well resolved, there has been a tendency to focus on the refractory (nonvolatile) constituents. These constituents of petroleum generally produce coke in yields varying from almost zero to more than 60% by weight (Figure 15.2). As an aside, it should also be noted that the differences in thermal behavior of the different subfractions of the asphaltene fraction detract from the concept of average structure. However, the focus of thermal studies has been, for obvious reasons, on the asphaltene constituents that produce thermal coke in amounts varying from approximately 35% by weight to approximately 65% by weight. Petroleum mapping techniques often show the nonvolatile constituents, specifically the asphaltene constituents and the resin constituents, producing coke while the volatile constituents produce distillates. It is often ignored that the asphaltene constituents also produce high yields (35% to 65% by weight) of volatile thermal products which vary from condensable liquids to gases.

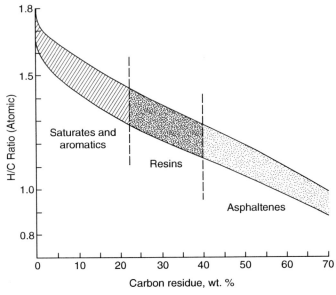

FIGURE 15.1 Yields of thermal coke (as determined by the Conradson carbon residue test method) for various petroleum fractions.

TABLE 15.1
General Indications of Feedstock Cracking

Feedstock Type	Characterization Factor[a], K	Naphtha Yield vol.%	Coke Yield wt.%	Relative Reactivity (Relative Crackability)
Aromatic	11.0 (1)	35.0	13.5	Refractory
Aromatic	11.2 (2)	49.6	12.5	Refractory
Aromatic	11.2 (1)	37.0	11.5	Refractory
Aromatic–naphthenic	11.4 (2)	47.0	9.1	Intermediate
Aromatic–naphthenic	11.4 (1)	39.0	9.0	Intermediate
Naphthenic	11.6 (2)	45.0	7.1	Intermediate
Naphthenic	11.6 (1)	40.0	7.2	Intermediate
Naphthenic–paraffinic	11.8 (2)	43.0	5.3	High
Naphthenic–paraffinic	11.8 (1)	41.0	6.0	High
Naphthenic–paraffinic	12.0 (2)	41.5	4.0	High
Naphthenic–paraffinic	12.0 (1)	41.5	5.3	High
Paraffinic	12.2 (2)	40.0	3.0	High

[a](1) Cycle oil or cracked feedstocks, 60% conversion; (2) Straight run or uncracked feedstocks, 60% conversion.

It has been generally thought that the chemistry of coke formation involves immediate condensation reactions to produce higher molecular weight, condensed aromatic species. And there is the claim that coking is a bimolecular process. However, more recent approaches to the chemistry of coking render the bimolecular process debatable. The rate of decomposition will vary with the nature of the individual constituents thereby giving rise to the perception of

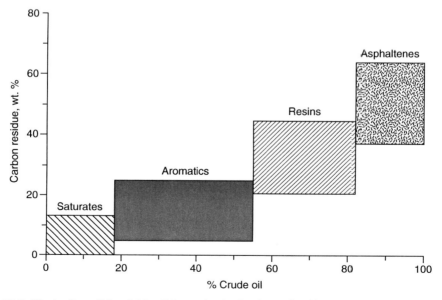

FIGURE 15.2 Illustration of the yields of thermal coke (as determined by the Conradson carbon residue test method) from fractions and subfractions of one petroleum.

second order or even multi-order kinetics. The initial reactions of asphaltene constituents involves thermolysis of pendant alkyl chains to form lower molecular weight higher polar species (**carbenes** and **carboids**) which then react to form coke. Indeed, as opposed to the bimolecular approach, the initial reactions in the coking of petroleum feedstocks that contain asphaltene constituents appear to involve unimolecular thermolysis of asphaltene aromatic–alkyl systems to produce volatile species (paraffins and olefins) and nonvolatile species (aromatics) (Figure 15.3) (Speight, 1987; Roberts, 1989; Schabron and Speight, 1997).

Thermal studies using model compounds confirm that volatility of the fragments is a major influence in carbon residue formation and a pendant-core model for the high molecular weight constituents of petroleum has been proposed (Wiehe, 1994). In such a model, the scission of alkyl side chains occurs, thereby leaving a polar core of reduced volatility that commences to produce a carbon residue (Speight, 1994; Wiehe, 1994). In addition, the pendant-core model also suggests that even one-ring aromatic cores can produce a carbon residue if multiple bonds need to be broken before a core can volatilize (Wiehe, 1994).

In support of the participation of asphaltenes in sediment or coke formation, it has been reported that the formation of a coke-like substance during heavy oil upgrading is dependant upon several factors (Storm et al., 1997):

1. Degree of polynuclear condensation in the feedstock
2. Average number of alkyl groups on the polynuclear aromatic systems
3. Ratio of heptane-insoluble material to the pentane-insoluble and heptane-soluble fraction
4. Hydrogen-to-carbon atomic ratio of the pentane-insoluble and heptane-soluble fraction

These findings correlate quite well with the proposed chemistry of coke or sediment formation during the processing of heavy feedstocks and even offer some predictability, since the characteristics of the whole feedstocks are evaluated.

Nitrogen species also appear to contribute to the pattern of the thermolysis. For example, the hydrogen or carbon–carbon bonds adjacent to a ring nitrogen undergo thermolysis quite readily, as if promoted by the presence of the nitrogen atom (Fitzer et al., 1971;). If it can be

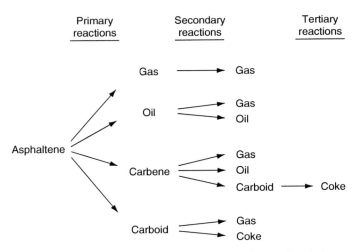

FIGURE 15.3 Multi-reaction sequence for the thermal decomposition of asphaltene constituents.

assumed that heterocyclic nitrogen plays a similar role in the thermolysis of asphaltenes, the initial reactions therefore will involve thermolysis of aromatic–alkyl bonds that are enhanced by the presence of heterocyclic nitrogen. An ensuing series of secondary reactions, such as aromatization of naphthenic species and condensation of the aromatic ring systems, then leads to the production of coke. Thus, the initial step in the formation of coke from asphaltenes is the formation of volatile hydrocarbon fragments and nonvolatile heteroatom-containing systems.

It has been reported that as the temperature of a 1-methylnaphthalene is raised from 100°C (212°F) to 400°C (750°F) there is a progressive decrease in the size of the asphaltenes particle (Thiyagarajan et al., 1995). Further, there is also the inference that the structural integrity of the asphaltene particle is compromised and that irreversible thermochemistry has occurred. Indeed, that is precisely what is predicted and expected from the thermal chemistry of asphaltenes and molecular weight studies of asphaltenes.

An additional corollary to this work is that conventional models of petroleum asphaltenes (which, despite evidence to the contrary, invoked the concept of a large polynuclear aromatic system) offer little, if any, explanation of the intimate events involved in the chemistry of coking. Models that invoke the concept of asphaltenes as a complex solubility class with molecular entities composed of smaller polynuclear aromatic systems (Chapter 11) are more in keeping with the present data.

Little has been acknowledged here of the role of low-molecular-weight polar species (resins) in coke formation. However, it is noteworthy that the resins are presumed to be lower molecular weight analogs of the asphaltenes. This being the case, similar reaction pathways may apply.

Thus, it is now considered more likely that molecular species, within the asphaltene fraction which contains nitrogen and other heteroatoms (and have lower volatility than the pure hydrocarbons), are the prime movers in the production of coke (Speight, 1987). Such species, containing various polynuclear aromatic systems, can be denuded of the attendant hydrocarbon moieties and are undoubtedly insoluble (Bjorseth, 1983; Dias, 1987, 1988) in the surrounding hydrocarbon medium. The next step is gradual carbonization of such entities to form coke (Cooper and Ballard, 1962; Magaril and Aksenova, 1968; Magaril and Ramzaeva, 1969; Magaril et al., 1970).

Thermal processes (such as **visbreaking** and coking) are the oldest methods for crude oil conversion and are still used in modern refineries. The thermal chemistry of petroleum constituents has been investigated for more than five decades, and the precise chemistry of the lower molecular weight constituents has been well defined because of the bountiful supply of pure compounds. The major issue in determining the thermal chemistry of the nonvolatile constituents is, of course, their largely unknown chemical nature and, therefore the inability to define their thermal chemistry with any degree of accuracy. Indeed, it is only recently that some light has been cast on the thermal chemistry of the nonvolatile constituents.

Thus, the challenges facing process chemistry and physics are determining (1) the means by which petroleum constituents thermally decompose, (2) the nature of the products of thermal decomposition, (3) the subsequent decomposition of the primary thermal products, (4) the interaction of the products with each other, (5) the interaction of the products with the original constituents, and (6) the influence of the products on the composition of the liquids.

When petroleum is heated to temperatures over approximately 410°C (770°F), the thermal or free radical reactions start to crack the mixture at significant rates. Thermal conversion does not require the addition of a catalyst; therefore, this approach is the oldest technology available for residue conversion. The severity of thermal processing determines the conversion and the product characteristics.

Asphaltene constituents are major components of residua and heavy oils and their thermal decomposition has been the focus of much attention (Wiehe, 1993a, b and references cited therein; Gray, 1994 and references cited therein; Speight, 1994 and references cited therein). The thermal decomposition not only produces high yields (40 wt.%) of coke but also, optimistically and realistically, produces equally high yields of volatile products (Speight, 1970). Thus, the challenge in studying the thermal decomposition of asphaltenes is to decrease the yields of coke and increase the yields of volatile products.

Several chemical models describe the thermal decomposition of asphaltenes (Wiehe, 1993a, b; Gray, 1994; Speight, 1994; and references cited therein). Using these available asphaltene models as a guide, the prevalent thinking is that the asphaltene nuclear fragments become progressively more polar as the paraffinic fragments are stripped from the ring systems by scission of the bonds (preferentially) between the carbon atoms alpha and beta to the aromatic rings.

The higher polarity polynuclear aromatic systems that have been denuded of the attendant hydrocarbon moieties are somewhat less soluble in the surrounding hydrocarbon medium than their parent systems (Bjorseth, 1983; Dias, 1987, 1988). Two factors are operative in determining the solubility of the polynuclear aromatic systems in the liquid product. The alkyl moieties that have a solubilizing effect have been removed and there is also enrichment of the liquid medium in paraffinic constituents. Again, there is an analogy with the deasphalting process (Chapter 7, Chapter 10, and Chapter 19), except that the paraffinic material is a product of the thermal decomposition of the asphaltene molecules and is formed in situ rather than being added separately.

The coke has a lower hydrogen-to-carbon atomic ratio than the hydrogen-to-carbon ratio of any of the constituents present in the original crude oil. The hydrocarbon products may have a higher hydrogen-to-carbon atomic ratio than the hydrogen-to-carbon ratio of any of the constituents present in the original crude oil or hydrogen-to-carbon atomic ratios at least equal to those of many of the original constituents. It must also be recognized that the production of coke and volatile hydrocarbon products is accompanied by a shift in the hydrogen distribution.

Mild-severity and high-severity processes are frequently used for the processing of residue fractions, whereas conditions similar to those of ultrapyrolysis (high temperature and very short residence time) are used commercially only for cracking ethane, propane, butane, and light distillate feeds to produce ethylene and higher olefins.

The formation of solid sediments, or coke, during thermal processes is a major limitation on processing. Further, the presence of different types of solids shows that solubility controls the formation of solids. And the tendency for solid formation changes in response to the relative amounts of the light ends, middle distillates, and residues and to their changing chemical composition during the process (Gray, 1994). In fact, the prime mover in the formation of incompatible products during the processing of feedstocks containing asphaltenes is the nature of the primary thermal decomposition products, particularly those designated as carbenes and carboids (Chapter 1) (Speight, 1987, 1992; Wiehe, 1992, 1993a, b;).

Coke formation during the thermal treatment of petroleum residua is postulated to occur by a mechanism that involves the liquid–liquid **phase separation** of reacted asphaltenes (which may be carbenes) to form a phase that is lean in abstractable hydrogen. The unreacted asphaltenes were found to be the fraction with the highest rate of thermal reaction but with the least extent of reaction. This not only described the appearance and disappearance of asphaltene constituents, but also quantitatively described the variation in molecular weight and hydrogen content of the asphaltenes with reaction time. Thus, the main features of coke formation are:

1. Induction period prior to coke formation
2. Maximum concentration of asphaltene constituents in the reacting liquid
3. Decrease in the asphaltene concentration that parallels the decrease in heptane-soluble material
4. High reactivity of the unconverted asphaltene constituents

The induction period has been observed experimentally by many previous investigators (Levinter et al., 1966, 1967; Magaril and Aksenova, 1968; Magaril and Aksenova, 1970; Valyavin et al., 1979; Takatsuka et al., 1989a) and makes visbreaking and the Eureka processes possible. The postulation that coke formation is triggered by the phase separation of asphaltenes (Magaril et al., 1971) led to the use of linear variations of the concentration of each fraction with reaction time, resulting in the assumption of zero-order kinetics rather than first-order kinetics. More recently (Yan, 1987), coke formation in visbreaking was described as resulting from a phase separation step, but the phase separation step was not included in the resulting kinetic model for coke formation.

This model represents the conversion of asphaltenes over the entire temperature range and of heptane-soluble materials in the coke induction period as first-order reactions. The data also show that the four reactions give simultaneously lower aromatic and higher aromatic products, on the basis of other evidence (Wiehe, 1992). Also, the previous work showed that residua fractions can be converted without completely changing solubility classes (Magaril et al., 1971) and that coke formation is triggered by the phase separation of converted asphaltenes.

The maximum solubility of these product asphaltenes is proportional to the total heptane-soluble materials, as suggested by the observation that the decrease in asphaltenes parallels the decrease of heptane-soluble materials. Finally, the conversion of the insoluble product asphaltenes into toluene-insoluble coke is pictured as producing a heptane-soluble by-product, which provides a mechanism for the heptane-soluble conversion to deviate from first-order behavior, once coke begins to form.

In support of this assumption, it is known (Langer et al., 1961) that partially hydrogenated refinery process streams provide abstractable hydrogen and as a result, inhibit coke formation during residuum thermal conversion. Thus, the heptane-soluble fraction of a residuum which contains naturally occurring, partially hydrogenated aromatics can provide abstractable hydrogen during thermal reactions.

As the conversion proceeds, the concentration of asphaltene cores continues to increase and the heptane-soluble fraction continues to decrease until the solubility limit, S_L is reached. Beyond the solubility limit, the excess asphaltene cores, $A_{ex}*$, phase separate to form a second liquid phase that is lean in abstractable hydrogen. In this new phase, asphaltene radical–asphaltene radical recombination is quite frequent, causing a rapid reaction to form solid coke and a byproduct of a heptane-soluble core.

The asphaltene concentration varies little in the coke induction period (Wiehe, 1993a, b), but then decreases once coke begins to form. Observing this, it might be concluded that asphaltenes are unreactive, but it is the high reactivity of the asphaltenes down to the asphaltene core that offsets the generation of asphaltene cores from the heptane-soluble materials to keep the overall asphaltene concentration nearly constant.

Previously, it was demonstrated (Schucker and Keweshan, 1980; Savage et al., 1988) that the hydrogen-to-carbon atomic ratio of the asphaltenes decreases rapidly with reaction time for asphaltene thermolysis and then approaches an asymptotic limit at long reaction times, which provides qualitative evidence for asphaltene cracking down to a core.

The measurement of the molecular weight of petroleum asphaltenes is known to give different values depending on the technique, the solvent and the temperature (Chapter 10)

(Dickie and Yen, 1967; Moschopedis et al., 1976; Speight et al., 1985). As shown by small-angle x-ray (Kim and Long, 1979) and neutron (Overfield et al., 1989) scattering, this is because asphaltenes tend to self-associate and form aggregates.

Thus, coke formation is a complex process involving both chemical reactions and thermodynamic behavior. Reactions that contribute to this process are cracking of side chains from aromatic groups, dehydrogenation of naphthenes to form aromatics, condensation of aliphatic structures to form aromatics, condensation of aromatics to form higher fused-ring aromatics, and dimerization or oligomerization reactions. Loss of side chains always accompanies thermal cracking, and dehydrogenation and condensation reactions are favored by hydrogen deficient conditions.

The importance of solvents in coking has been recognized for many years (e.g., Langer et al., 1961), but their effects have often been ascribed to hydrogen donor reactions rather than phase behavior. The separation of the phases depends on the solvent characteristics of the liquid. Addition of aromatic solvents suppresses phase separation, whereas paraffins enhance separation. Microscopic examination of coke particles often shows evidence for the presence of mesophase, spherical domains that exhibit the anisotropic optical characteristics of liquid crystals.

This phenomenon is consistent with the formation of a second liquid phase; the mesophase liquid is denser than the rest of the hydrocarbon, has a higher surface tension, and probably wets metal surfaces better than the rest of the liquid phase. The mesophase characteristic of coke diminishes as the liquid phase becomes more compatible with the aromatic material.

The phase separation phenomenon that is the prelude to coke formation can also be explained by use of the solubility parameter, δ, for petroleum fractions and for the solvents (Yen, 1984; Speight, 1994) (Chapter 11). As an extension of this concept, there is sufficient data to draw a correlation between the atomic hydrogen to carbon ratio and the solubility parameter for hydrocarbons and the constituents of the lower boiling fractions of petroleum (Speight, 1994). Recognition that hydrocarbon liquids can dissolve polynuclear hydrocarbons, a case in which there is usually less than a three-point difference between the lower solubility parameter of the solvent and the higher solubility parameter of the solute. Thus, a parallel, or a near-parallel line can be assumed that allows the solubility parameter of the asphaltenes and resins to be estimated.

By this means, the solubility parameter of asphaltenes can be estimated to fall in the range of 9 to 12, which is in keeping with the asphaltenes being composed of a mixture of different compound types with an accompanying variation in polarity. Removal of alkyl side chains from the asphaltenes decreases the hydrogen-to-carbon atomic ratio (Wiehe, 1993a, b; Gray, 1994) and increases the solubility parameter, thereby bringing about a concurrent decrease of the asphaltene product in the hydrocarbon solvent.

In fact, on the molecular weight polarity diagram for asphaltenes, carbenes and carboids can be shown as lower molecular weight, highly polar entities in keeping with molecular fragmentation models (Speight, 1994). If this increase in polarity and solubility parameters (Mitchell and Speight, 1973) is too drastic relative to the surrounding medium (Figure 15.4), phase separation will occur Further, the available evidence favors a multi-step mechanism rather than a stepwise mechanism (Figure 15.5) as the means by which the thermal decomposition of petroleum constituents occurs (Speight, 1997).

Any chemical or physical interactions (especially thermal effects) that cause a change in the solubility parameter of the solute relative to that of the solvent will also cause **incompatibility** be it called **instability**, phase separation, sediment formation, or sludge formation.

Instability or incompatibility (Chapter 13) resulting in the separation of solids during refining, can occur during a variety of processes, either by intent (such as in the deasphalting process) or inadvertently, when the separation is detrimental to the process. Thus, separation

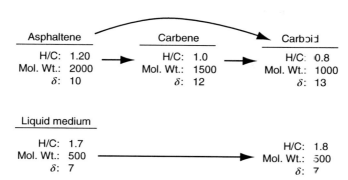

FIGURE 15.4 Illustration of the changes in the solubility parameter of the various fractions of petroleum during thermal treatment.

of solids occurs whenever the solvent characteristics of the liquid phase are no longer adequate to maintain polar or high molecular weight material in solution. Examples of such occurrences are:

1. Asphaltene separation, which occurs when the paraffin content or character of the liquid medium increases (Chapter 10)
2. Wax separation, which occurs when there is a drop in temperature or the aromatic content or character of the liquid medium increases
3. Sludge or sediment formation in a reactor, which occurs when the solvent characteristics of the liquid medium change so that asphalt or wax materials separate

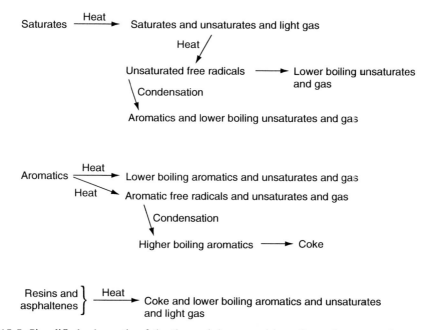

FIGURE 15.5 Simplified schematic of the thermal decomposition of petroleum constituents.

4. Coke formation, which occurs at high temperatures and commences when the solvent power of the liquid phase is not sufficient to maintain the coke precursors in solution (Chapter 10 and Chapter 11)
5. Sludge or sediment formation in fuel products which occurs because of the interplay of several chemical and physical factors

This mechanism also appears to be operable during residua hydroconversion, which has included a phase-separation step (the formation of dry sludge) in a kinetic model but this was not included as a preliminary step to coke formation in a thermal cracking model (Takatsuka et al., 1989a, b; Speight, 2004).

15.7.2 HYDROCONVERSION CHEMISTRY

There have also been many attempts to focus attention on the asphaltenes during hydro-cracking studies. The focus has been on the macromolecular changes that occur by investigation of the changes to the generic fractions, that is, the asphaltenes, the resins, and the other fractions that make up such a feedstock (Drushel, 1972). In terms of hydroprocessing, the means by which asphaltene constituents are desulfurized, as one step of a hydrocracking operation, is also suggested as part of the process. This concept can then be taken one step further to show the dealkylation of the aromatic systems as a definitive step in the hydro-cracking process (Speight, 1987).

When catalytic processes are employed, complex molecules (such as those that may be found in the original asphaltene fraction) or those formed during the process, are not sufficiently mobile (or are too strongly adsorbed by the catalyst) to be saturated by the hydrogenation components. Hence, these molecular species continue to condense and eventually degrade to coke. These deposits deactivate the catalyst sites and eventually interfere with the process.

Several noteworthy attempts have been made to focus attention on the asphaltene constituents during hydroprocessing studies. The focus has been on the macromolecular changes that occur by investigation of the changes in the generic fractions, i.e., the asphaltene constituents, the resin constituents, and the other fractions that make up such a feedstock. This option suggests that the overall pathway by which hydrotreating and hydrocracking of heavy oils and residua occur involves a stepwise mechanism:

Asphaltene constituents → Polar aromatics (resin-type components)
Polar aromatics   → Aromatics
Aromatics   → Saturates

A direct step from either the asphaltene constituents or the resin constituents to the saturates, is not considered a predominant pathway for hydroprocessing.

The means by which asphaltenes are desulfurized, as one step of a hydrocracking operation, is also suggested as part of this process. This concept can then be taken one step further to show the dealkylation of the aromatic systems as a definitive step in the hydrocracking process (Speight, 1987). It is also likely that molecular species within the asphaltene fractions that contains nitrogen and other heteroatoms, and have lower volatility than their hydrocarbon analogs, are the prime movers in the production of coke (Speight, 1987).

When catalytic processes are employed, complex molecules such as those that may be found in the original asphaltene fraction or those or formed during the process, are not sufficiently mobile (or are too strongly adsorbed by the catalyst) to be saturated by the hydrogenation components and, hence, continue to condense and eventually degrade to coke. These deposits deactivate the catalyst sites and eventually interfere with the hydropro-cess.

A convenient means of understanding the influence of feedstock on the hydrocracking process is through the study of the hydrogen content (hydrogen-to-carbon atomic ratio) and molecular weight (carbon number) of the feedstocks and products. Such data show the extent to which the carbon number must be reduced and the relative amount of hydrogen that must be added to generate the desired lower molecular weight, hydrogenated products. In addition, it is possible to use data for hydrogen usage in residuum processing, where the relative amount of hydrogen consumed in the process can be shown to be dependent upon the sulfur content of the feedstock.

15.7.3 Chemistry in the Refinery

Thermal cracking processes are commonly used to convert petroleum residua into distillable liquid products, although thermal cracking processes as used in the early refineries are no longer in use. Examples of modern thermal cracking processes are visbreaking and coking (**delayed coking**, **fluid coking**, and **flexicoking**) (Chapter 14). In all of these processes, the simultaneous formation of sediment or coke limits the conversion to usable liquid products. However, for the purposes of this section, the focus will be on the visbreaking and hydro-cracking processes. The coking processes, in which the reactions are taken to completion with the maximum yields of products, are not a part of this discussion.

15.7.3.1 Visbreaking

To study the thermal chemistry of petroleum constituents, it is appropriate to select the visbreaking process (a **carbon rejection** process) and the hydrocracking process (a **hydrogen addition** process) as used in a modern refinery (Chapter 14). The processes operate under different conditions (Figure 15.6) and have different levels of conversion (Figure 15.7) and, although they do offer different avenues for conversion, these processes are illustrative of the thermal chemistry that occurs in refineries.

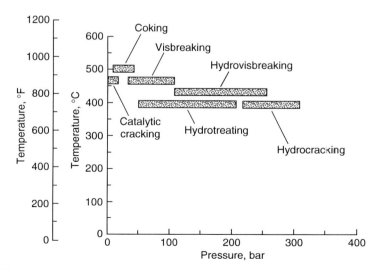

FIGURE 15.6 Temperature and pressure ranges for various processes.

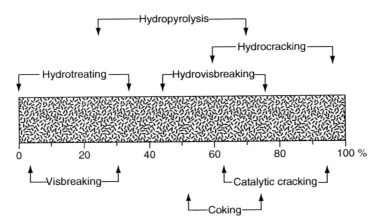

FIGURE 15.7 Feedstock conversion in various processes.

The visbreaking process (Chapter 17) is primarily a means of reducing the viscosity of heavy feedstocks by controlled thermal decomposition insofar as the hot products are quenched before complete conversion can occur (Speight and Ozum, 2002). However, the process is often plagued by sediment formation in the products. This sediment, or sludge, must be removed if the products are to meet fuel oil specifications.

The process (Figure 15.8) uses the mild thermal cracking (partial conversion) as a relatively low-cost and low-severity approach to improving the viscosity characteristics of the residue without attempting significant conversion to distillates. Low residence times are required to avoid coking reactions, although additives can help to suppress coke deposits on the tubes of the furnace (Allan et al., 1983).

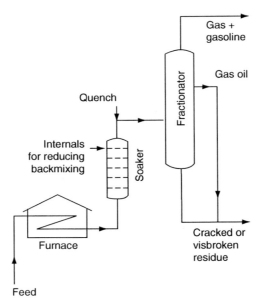

FIGURE 15.8 The visbreaking process using a soaker. (From OSHA Technical Manual, Petroleum Refining processes.)

A visbreaking unit consists of a reaction furnace, followed by quenching with a recycled oil, and fractionation of the product mixture. All of the reaction in this process occurs as the oil flows through the tubes of the reaction furnace. The severity is controlled by the flow rate through the furnace and the temperature; typical conditions are 475°C to 500°C (885°F to 930°F) at the furnace exit with a residence time of 1 to 3 min, with operation for 3 to 6 month on stream (continuous use) is possible before the furnace tubes must be cleaned and the coke removed (Gary and Handwerk, 1984). The operating pressure in the furnace tubes can range from 0.7 to 5 MPa depending on the degree of vaporization and the residence time desired. For a given furnace tube volume, a lower operating pressure will reduce the actual residence time of the liquid phase.

The reduction in viscosity of the unconverted residue tends to reach a limiting value with conversion, although the total product viscosity can continue to decrease (Figure 15.9). Conversion of residue in visbreaking follows first-order reaction kinetics (Henderson and Weber, 1965). The minimum viscosity of the unconverted residue can lie outside the range of allowable conversion if sediment begins to form (Rhoe and de Blignieres, 1979). When pipelining of the visbreaker product is a process objective, a diluent such as gas condensate can be added to achieve further reduction in viscosity.

The high viscosity of the heavier feedstocks and residua is thought to be due to the entanglement of the high molecular weight components of the oil and the formation of ordered structures in the liquid phase. Thermal cracking at low conversion can remove side chains from the asphaltenes and break bridging aliphatic linkages. A 5% to 10% conversion of atmospheric residue to naphtha is sufficient to reduce the entanglements and structures in the liquid phase and give at least a five-fold reduction in viscosity.

The stability of visbroken products is also an issue that might be addressed at this time. Using this simplified model, visbroken products might contain polar species that have been denuded of some of the alkyl chains and which, on the basis of solubility, might be more rightly called carbenes and carboids, but an induction period is required for phase separation or agglomeration to occur. Such products might initially be soluble in the liquid phase, but after the induction period, cooling, and diffusion of the products, incompatibility (phase separation, sludge formation, agglomeration) occurs.

On occasion, higher temperatures are employed in various reactors as it is often assumed that, if no side reactions occur, longer residence times at a lower temperature are equivalent to

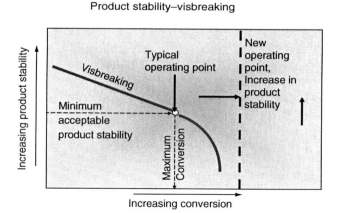

FIGURE 15.9 Representation of the break point above which maximum conversion is assured but product stability (i.e., inhibition of sediment formation) is less certain. (From Universal Oil Products [UOP].)

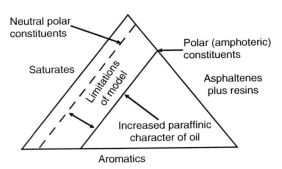

FIGURE 15.10 The limitations of the visbreaking process when predictions are based on average parameters for high-asphaltene feedstocks.

shorter residence times at a higher temperature. However, this assumption does not acknowledge the change in thermal chemistry that can occur at the higher temperatures, irrespective of the residence time. Thermal conditions can, indeed, induce a variety of different reactions in crude oil constituents, so that selectivity for a given product may change considerably with temperature. The onset of secondary, tertiary, and even quaternary reactions under the more extreme high-temperature conditions can convert higher molecular weight constituents of petroleum to low-boiling distillates, butane, propane, ethane, and (ultimately) methane. Caution is advised in the use of extreme temperatures.

Obviously, the temperature and residence time of the asphaltene constituents in the reactor are key to the successful operation of a visbreaker. Visbreakers must operate in temperature and residence time regimes that do not promote the formation of sediment (often referred to as coke). However, as already noted, there is a break point above which conversion might be increased, but the possibility of sediment deposition increases (Figure 15.9). At the temperatures and residence times outside of the most beneficial temperature and residence time regimes, thermal changes to the asphaltene constituents cause phase separation of a solid product that then progresses to coke. Further, it is in such operations that models derived from average parameters can be ineffective and misleading.

For example the amphoteric constituents of the asphaltene (Chapter 11) are more reactive than the less polar constituents. The thermal products from the amphoteric constituents form first and will separate out from the reaction matrix before other products (Figure 15.10). Under such conditions, models based on average structural parameters or on average properties will not predict early phase separation to the detriment of the product and the process as a whole.

Knowing the actual nature of the subtypes of the asphaltene constituents is obviously beneficial and will allow steps to be taken to correct any such unpredictable occurrence. Indeed, the concept of hydrovisbreaking (visbreaking in the presence of hydrogen) could be of valuable assistance when high asphaltene content feedstocks are used.

15.7.3.2 Hydroprocessing

Hydrotreating (Chapter 16) is the (relatively) low temperature removal of heteroatomic species by treatment of a feedstock or product in the presence of hydrogen (Chapter 20). Hydrocracking (Figure 15.11) is the thermal decomposition of a feedstock in which carbon–carbon bonds are cleaved in addition to the removal of heteroatomic species (Chapter 21).

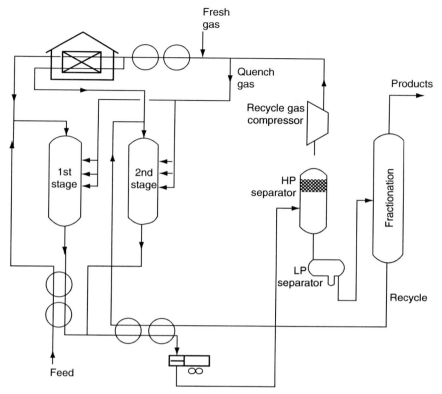

FIGURE 15.11 A two-stage hydrocracking unit. (From OSHA Technical Manual, Section IV, Chapter 2, Petroleum Refining Processes.)

The presence of hydrogen changes the nature of the products (especially the decreasing coke yield) by preventing the buildup of precursors that are incompatible in the liquid medium and form coke (Magaril and Aksenova, 1968, 1970; Magaril and Ramazaeva, 1969; Magaril et al., 1970; Speight and Moschopedis, 1979). In fact, the chemistry involved in the reduction of asphaltenes to liquids using models in which the polynuclear aromatic system borders on graphitic is difficult to visualize, let alone justify (Chapter 10). However, the paper chemistry derived from the use of a molecularly designed model composed of smaller polynuclear aromatic systems is much easier to visualize (Speight, 1994). But precisely how asphaltenes react with the catalysts is open to much more speculation.

In contrast to the visbreaking process, in which the general principle is the production of products for use as fuel oil, the hydroprocessing is employed to produce a slate of products for use as liquid fuels. Nevertheless, the decomposition of asphaltenes is, again, an issue, and just as models consisting of large polynuclear aromatic systems are inadequate to explain the chemistry of visbreaking, they are also of little value for explaining the chemistry of hydrocracking.

Deposition of solids or incompatibility is still possible when asphaltenes interact with catalysts, especially acidic support catalysts, through the functional groups, e.g., the basic nitrogen species just as they interact with adsorbents. And there is a possibility for interaction of the asphaltene with the catalyst through the agency of a single functional group in which the remainder of the asphaltene molecule remains in the liquid phase. There is also a less desirable option in which the asphaltene reacts with the catalyst at several points of contact, causing immediate incompatibility on the catalyst surface.

There is evidence to show that during the early stages of the hydrotreating process, the chemistry of the asphaltene constituents follows the same routes as thermal chemistry (Ancheyta et al., 2006). Thus, initially there is an increase in the amount of asphaltene constituents followed by a decrease, indicating that in the early stages of the process, resin constituents are being converted to asphaltene material by aromatization and by some dealkylation. In addition, aromatization and dealkylation of the original asphaltene constituents yields asphaltene products that are of higher polarity and lower molecular weight than the original asphaltene constituents. Analogous to the thermal processes, this produces an overall asphaltene fraction that is a more polar material and also of lower molecular weight. As the hydrotreating process proceeds, the amount of asphaltene constituents precipitated decreases due to conversion of the asphaltene constituents to products. At more prolonged on-stream times there is a steady increase in the yield of the asphaltene constituents. This is accompanied by a general increase in the molecular weight of the precipitated material.

These observations are in keeping with those for the thermal reactions of asphaltene constituents in the absence of hydrogen, where the initial events are a reduction in the molecular weight of the asphaltene constituents, leading to lower molecular weight by more polar products that are derived from the asphaltene constituents, but are often referred to as carbenes and carboids. As the reaction progresses, these derived products increase in molecular weight and eventually become insoluble in the reaction medium, deposit on the catalyst, and form coke.

As predicted from the chemistry of the thermal reactions of the asphaltene constituents, there is a steady increase in aromaticity (reflected as a decrease in the hydrogen to carbon atomic ratio) with on-stream time. This is due to (1) aromatization of the naphthene ring system that is present in asphaltene constituents, (2) cyclodehydrogenation of alkyl chains to form other naphthene ring systems (3) dehydrogenation of the new naphthene ring systems to form more aromatic rings, and (4) dealkylation of aromatic ring systems.

As the reaction progresses, the aromatic carbon atoms in the asphaltene constituents show a general increase and the degree of substitution of the aromatic rings decreases. Again, this is in keeping with the formation of products from the original asphaltene constituents (carbenes, carboids, and eventually coke) that have an increased aromaticity and decreased number of alkyl chains, as well as a decrease in the alkyl chain length. Thus, as the reaction progresses with increased on-stream time, new asphaltene constituents are formed that, relative to the original asphaltene constituents, the new species have increased aromaticity coupled with a lesser number of alkyl chains that are shorter than the original alky chains.

It may be that the chemistry of hydrocracking has to be given serious reconsideration insofar as the data show that the initial reactions of the asphaltene constituents appear to be the same as the reactions under thermal conditions where hydrogen is not present. Re-thinking of the process conditions and the potential destruction of the catalyst by the deposition of carbenes and carboids require further investigation of the chemistry of asphaltene hydrocracking.

If these effects are prevalent during the hydrocracking of high-asphaltene feedstocks, the option may be to hydrotreat the feedstock first and then to hydrocrack the hydrotreated feedstock. There are indications that such hydrotreatment can (at some obvious cost) act beneficially in the overall conversion of the feedstocks to liquid products.

REFERENCES

Allan, D.E., Martinez, C.H., Eng, C.C., and Barton, W.J. 1983. *Chem. Eng. Progr.* 79(1): 85.
Ancheyta, J., Centeno, G., Trejo, F., Betancourt, G., and Speight, J.G. 2006. *Catal. Today.* In press.
Bjorseth, A. 1983. *Handbook of Polycyclic Aromatic Hydrocarbons.* Marcel Dekker Inc., New York.

Boduszynski, M.M. 1987. *Energy Fuels* 1: 2.

Cooper, T.A., and Ballard, W.P. 1962. *Advances in Petroleum Chemistry and Refining*, Volume 6. K.A. Kobe and J.J. McKetta, eds. Interscience, New York. Chapter 4

Cumming, K.A. and Wojciechowski, B.W. 1996. *Catal. Rev. Sci. Eng.* 38: 101.

Dias, J.R. 1987. *Handbook of Polycyclic Hydrocarbons. Part A. Benzenoid Hydrocarbons*. Elsevier, New York.

Dias, J.R. 1988. *Handbook of Polycyclic Hydrocarbons, Part B. Polycyclic Isomers and Heteroatom Analogs of Benzenoid Hydrocarbons*. Elsevier, New York.

Dickie, J.P. and Yen, T.F. 1967. *Anal. Chem.* 39: 1847.

Dolbear, G.E. 1998. *Petroleum Chemistry and Refining*. J.G. Speight, ed. Taylor & Francis Publishers, Washington, DC. Chapter 7.

Dolbear, G.E., Tang, A., and Moorehead, E.L. 1987. *Fuel* 66: 267.

Drushel, H.V. 1972. Preprints. *Div. Petrol. Chem. Am. Chem. Soc.* 17(4): F92.

Ebert, L.B., Mills, D.R., and Scanlon, J.C. 1987. Preprints. *Div. Petrol. Chem. Am. Chem. Soc.* 32(2): 419.

Egloff, G. 1937. *The Reactions of Pure Hydrocarbons*. Reinhold, New York.

Eliel, E. and Wilen, S. 1994. *Stereochemistry of Organic Compounds*. John Wiley & Sons Inc., New York.

Fabuss, B.M., Smith, J.O., and Satterfield, C.N. 1964. *Advances in Petroleum Chemistry and Refining*, VanNostrand, New York. Volume IX.

Fitzer, E., Mueller, K., and Schaefer, W. 1971. *Chem. Phys. Carbon* 7: 237.

Furimsky, E. 1983. *Erdol. Kohle* 36: 518.

Gary, J.G. and Handwerk, G.E. 1984. *Petroleum Refining: Technology and Economics,* 2nd edn. Marcel Dekker Inc., New York.

Gray, M.R. 1994. *Upgrading Petroleum Residues and Heavy Oils*. Marcel Dekker Inc., New York.

Henderson, J.H. and Weber, L. 1965. *J. Can. Pet. Tech.* 4: 206.

Hurd, C.D. 1929. *The Pyrolysis of Carbon Compounds*. The Chemical Catalog Company Inc., New York.

Jones, D.S.J. 1995. *Elements of Petroleum Processing*. John Wiley & Sons Inc., New York.

Keller, W.D. 1985. Clays. *Kirk Othmer Concise Encyclopedia of Chemical Technology*. M. Grayson, ed. Wiley Interscience, New York. p. 283.

Kim, H. and Long, R.B. 1979. *Ind. Eng. Chem. Fundam.* 18: 60.

King, P.J., Morton, F., and Sagarra, A. 1973. *Modern Petroleum Technology*. G D. Hobson and W. Pohl eds. Applied Science Publishers, Barking, Essex, UK.

Langer, A.W., Stewart, J., Thompson, C.E., White, H.T., and Hill, R.M. 1961 *Ind. Eng. Chem.* 53: 27.

Laszlo, P. 1995. *Organic Reactions: Logic and Simplicity*. John Wiley & Sons Inc., New York.

LePage, J.F. and Davidson, M. 1986. *Rev. Inst. Franç. Petrol.* 41: 131.

Levinter, M.E., Medvedeva, M.I., Panchenkov, G.M., Aseev, Y.G., Nedoshivir, Y.N., Finkelshtein, G.B., and Galiakbarov, M.F. 1966. *Khim. Tekhnol. Topl. Masel.* 9: 31.

Levinter, M.E., Medvedeva, M.I., Panchenkov, G.M., Agapov, G.I., Galiakbarov, M.F., and Galikeev, R.K. 1967. *Khim. Tekhol. Topl. Masel.* 4: 20.

Magaril, R.A. and Aksenova, E.L. 1968. *Int. Chem. Eng.* 8: 727.

Magaril, R.Z. and Aksenova, E.I. 1970. *Khim. Tekhnol. Topl. Masel.* 7: 22.

Magaril, R.A. and Ramazaeva, L.F. 1969. *Izv. Vyssh. Ucheb. Zaved. Neft Gaz.* 12(1): 61.

Magaril, R.L., Ramazaeva, L.F., and Askenova, E.I. 1970. *Khim. Tekhnol. Topliv Masel.* 15(3): 15.

Magaril, R.Z., Ramazeava, L.F., and Aksenova, E.I. 1971. *Int. Chem. Eng.* 11: 250.

Masel, R.I. 1995. *Principles of Adsorption and Reaction on Solid Surfaces*. John Wiley & Sons Inc., New York.

McEvoy, J. 1996. *Chemtech.* 26(2): 6.

Mitchell, D.L. and Speight, J.G. 1973. *Fuel* 52: 149.

Moschopedis, S.E., Fryer, J.F., and Speight, J.G. 1976. *Fuel* 55: 227.

Overfield, R.E., Sheu, E.Y., Sinha, S.K., and Liang, K.S. 1989. *Fuel Sci. Technol. Int.* 7: 611.

Pines, H. 1981. *The Chemistry of Catalytic Hydrocarbon Conversions*. Academic Press, New York.

Rhoe, A. and de Blignieres, C. 1979. *Hydrocarbon Process.* 58(1): 131.

Roberts, I. 1989. Preprints. *Div. Petrol. Chem. Am. Chem. Soc.* 34(2): 251.

Samorjai, G.A. 1994. *Introduction to Surface Chemistry and Catalysis.* John Wiley & Sons Inc., New York.

Savage, P.E., Klein, M.T., and Kukes, S.G. 1988. *Energy Fuels* 2: 619.

Schabron, J.F. and Speight, J.G. 1997. *Revue Institut Français de Pétrole.* 52(1): 73–85.

Schucker, R.C. and Keweshan, C.F. 1980. Preprints. *Div. Fuel Chem. Am. Chem. Soc.* 25: 155.

Shiroto, Y., Nakata, S., Fukul, Y., and Takeuchi, C. 1983. *Ind. Eng. Chem. Process Design Dev.* 22: 248.

Smith, M.B. 1994. *Organic Synthesis.* McGraw-Hill Inc., New York.

Speight, J.G. 1970. *Fuel* 49: 134.

Speight, J.G. 1984. *Catalysis on the Energy Scene.* S. Kaliaguine and A. Mahay, eds. Elsevier, Amsterdam.

Speight, J.G. 1987 Preprints. *Div. Petrol. Chem. Am. Chem. Soc.* 32(2): 413.

Speight, J.G. 1989. *Neftekhimiya* 29: 723.

Speight, J.G. 1992. *Proceedings.* 4th International Conference on the Stability and Handling of Liquid Fuels. U.S. Department of Energy (DOE/CONF-911102). p. 169.

Speight, J.G. 1994. *Asphalts and Asphaltenes, 1.* T.F. Yen and G.V. Chilingarian, eds. Elsevier, Amsterdam. Chapter 2.

Speight, J.G. 1997. *Petroleum Chemistry and Refining.* J.G. Speight, ed. Taylor & Francis Publishers, Washington, DC. Chapter 5.

Speight, J.G. 2000. *The Desulfurization of Heavy Oils and Residua,* 2nd edn. Marcel Dekker Inc., New York.

Speight, J.G. 2004. *Catal. Today* 98(1–2): 55–60.

Speight, J.G. and Moschopedis, S.E. 1979. *Fuel Process. Technol.* 2: 295.

Speight, J.G. and Ozum, B. 2002. *Petroleum Refining Processes.* Marcel Dekker Inc., New York.

Speight, J.G., Wernick, D.L., Gould, K.A., Overfield, R.E., Rao, B.M.L., and Savage, D.W. 1985. *Rev. Inst. Franç. Petrol.* 40: 51.

Storm, D.A., Decanio, S.J., Edwards, J.C., and Sheu, E.Y. 1997. *Pet. Sci. Technol.* 15: 77.

Takatsuka, T., Kajiyama, R., Hashimoto, H., Matsuo, I., and Miwa, S.A. 1989a. *J. Chem. Eng. Japan* 22: 304.

Takatuska, T., Wada, Y., Hirohama, S., and Fukui, Y.A. 1989b. *J. Chem. Eng. Japan* 22: 298.

Ternan, M. 1983. *Can. J. Chem. Eng.* 61: 133, 689.

Thiyagarajan, P., Hunt, J.E., Winans, R.E., Anderson, K.B., and Miller, J.T. 1995. *Energy Fuels* 9: 829.

Valyavin, G.G., Fryazinov, V.V., Gimaev, R.H., Syunyaev, Z.I., Vyatkin, Y.L., and Mulyukov, S.F. 1979. *Khim. Tekhol. Topl. Masel.* 8: 8.

Vercier, P. 1981. *The Chemistry of Asphaltenes.* J.W. Bunger and N.C. Li, eds. Advances in Chemistry Series No. 195. American Chemical Society, Washington, DC.

Wiehe, I.A. 1992. *Ind. Eng. Chem. Res.* 31: 530.

Wiehe, I.A. 1993a. Preprints. *Div. Petrol. Chem. Am. Chem. Soc.* 38: 428.

Wiehe, I.A. 1993b. *Ind. Eng. Chem. Res.* 32: 2447.

Wiehe, I.A. 1994. *Energy Fuels* 8: 536.

Yan, T.Y. 1987. Preprints. *Div. Petrol. Chem. Am. Chem. Soc.* 32: 490.

Yen, T.F. 1984. *The Future of Heavy Crude Oil and Tar Sands.* R.F. Meyer, J.C. Wynn, and J.C. Olson, eds. McGraw-Hill, New York.

16 Distillation

16.1 INTRODUCTION

Petroleum in the unrefined state is of limited value and of limited use. Refining is required to produce the products that are attractive to the market place. Thus, petroleum refining is a series of steps by which the crude oil is converted into saleable products in the desired qualities and in the amounts dictated by the market (Priestley, 1973). In fact, a refinery is essentially a group of manufacturing plants that vary in number depending on the variety of products produced; processes are selected and products manufactured to give a balanced operation.

Most petroleum products, including kerosene, fuel oil, **lubricating oil**, and wax, are fractions of petroleum that have been treated to remove undesirable components. Other products, for example gasoline, aromatic solvents, and even asphalt, may be partly or totally synthetic in that they have compositions that are impossible to achieve by direct separation of these materials from crude oil. They result from chemical processes that change the molecular nature of selected portions of crude oil; in other words, they are the products of refining or they are refined products (Nelson, 1958; Bland and Davidson, 1967).

The petroleum refinery of the twenty-first century is a much more complex operation (Chapter 14) than those refineries of 100 to 120 years ago. Early refineries were predominantly distillation units, perhaps with ancillary units to remove objectionable odors from the various product streams. The refinery of the 1930s was somewhat more complex; it was essentially a distillation unit, but at this time **cracking** and coking units were starting to appear in the scheme of refinery operations. These units were not what we imagine today as a cracking and coking unit, but were the forerunners of today's units. Also at this time, asphalt was becoming a recognized petroleum product. Finally, current refineries are a result of major evolutionary trends and are highly complex operations. Most of the evolutionary adjustments to refineries have occurred during the decades since the commencement of World War II. In the petroleum industry, as in many other industries, supply and demand are key factors in efficient and economic operation. Innovation is also a key (Table 13.1).

A refinery is essentially a group of manufacturing plants (Chapter 14) that vary in number with the variety of products produced. Refinery processes must be selected and products manufactured to give a balanced operation: that is, crude oil must be converted into products according to the rate of sale of each. For example, the manufacture of products from the lower boiling portion of petroleum automatically produces a certain amount of higher boiling components. If the latter cannot be sold as, say, heavy fuel oil, they accumulate until refinery storage facilities are full. To prevent the occurrence of such a situation, the refinery must be flexible and able to change operations as needed. This usually means more processes to accommodate the ever-changing demands of the market (Hobson and Pohl, 1973). This could be reflected in the inclusion of a cracking process to change an excess of heavy fuel oil into more gasoline with coke as the residual product or inclusion of a **vacuum distillation** process to separate the heavy oil into lubricating oil stocks and asphalt.

In addition, a refinery must include the following (Kobe and McKetta, 1958):

1. All necessary nonprocessing facilities
2. Adequate tank capacity for storing crude oil, intermediate, and finished products
3. Dependable source of electrical power
4. Material-handling equipment
5. Workshops and supplies for maintaining a continuous 24 h/d, 7 d/week operation
6. Waste-disposal and water-treating equipment
7. Product-blending facilities

Distillation has remained a major refinery process and a process to which crude oil is subjected to. A multitude of separations are accomplished by distillation, but its most important and primary function in the refinery is its use for the separation of crude oil into component fractions (Gruse and Stevens, 1960). Thus, it is possible to obtain products ranging from gaseous materials taken off the top of the distillation column to a non-volatile atmospheric **residuum (bottoms, reduced crude)** with correspondingly lower-boiling materials (gas, gasoline, naphtha, kerosene, and gas oil) taken off at intermediate points (Gary and Handwerk, 1994).

The reduced crude may then be processed by vacuum or steam distillation to separate the high-boiling lubricating oil fractions without the danger of decomposition, which occurs at high (>350°C, 660°F) temperatures (Chapter 14). Indeed, atmospheric distillation may be terminated with a lower boiling fraction (boiling cut), if it is thought that vacuum or steam distillation will yield a better quality product or if the process appears to be economically more favorable.

It should be noted at this point that not all crude oils yield the same distillation products. In fact, the nature of the crude oil dictates the processes that may be required for refining. Petroleum can be classified according to the nature of the distillation residue, which in turn depends on the relative content of hydrocarbon types: paraffins, naphthenes, and aromatics. The majority of crude oils fall into one of the following classifications (Chapter 2):

1. *Asphalt-base crude oil* contains very little paraffin wax and a residue primarily asphaltic (predominantly condensed aromatics); sulfur, oxygen, and nitrogen contents are often relatively high. Light and intermediate fractions have high percentages of naphthenes. These crude oils are particularly suitable for making high-quality gasoline, machine lubricating oils, and asphalt.
2. *Paraffin-base crude oil* contains very little asphaltic materials and is a good source of paraffin wax, quality motor lubricating oils, and high-grade kerosene. Paraffin-base crude oil usually has a lower heteroatom content than asphalt-base crude oil.
3. *Mixed-base crude oil* contains considerable amounts of both wax and asphalt. Virtually all products can be obtained, although at lower yields than from the other two classes.

For example, a paraffin-base crude oil produces distillation cuts with higher proportions of paraffins than an asphalt-base crude. The converse is also true; that is, an asphalt-base crude oil produces materials with higher proportions of cyclic compounds. A paraffin-base crude oil yields wax distillates rather than the lubricating distillates produced by the naphthenic-base crude oils. The residuum from paraffin-base petroleum is referred to as cylinder stock rather than asphaltic bottoms, which is the name often given to the residuum from the distillation of naphthenic crude oil. It is emphasized that, in these cases, it is not a matter of the use of archaic terminology, but a reflection of the nature of the product and the petroleum from which it is derived.

Petroleum refining, as we know it, is a very recent science and for the purposes of this chapter will be acknowledged as such. Many innovations have evolved during the twentieth century and it is the purpose of the present chapter to illustrate the evolution of petroleum refining from the early processes to those in use at the present day.

16.2 PRETREATMENT

Even though distillation is, to all appearances, the first step in crude oil refining, it should be recognized that crude oil that is contaminated by salt water, either from the well or during transportation to the refinery, must be treated to remove the emulsion. If salt water is not removed, the materials of construction of the heater tubes and column intervals are exposed to chloride ion attack and the corrosive action of hydrogen chloride, which may be formed at the temperature of the column feed.

Various methods of pretreatment are open to the petroleum refiner, but three general approaches have been taken to the desalting of crude petroleum (Figure 16.1). Numerous variations of each type have been devised, but the selection of a particular process depends on the type of salt dispersion and the properties of the crude oil.

For example, desalting operations (Burris, 1992) are necessary to remove salt from the brines that are present with the crude oil after recovery. The salt or brine suspensions may be removed from crude oil by heating (90°C–150°C, 200°F–300°F) under pressure (50–250 psi) that is sufficient to prevent vapor loss and then allowing the material to settle in a large vessel. Alternatively, coalescence is aided by passage through a tower packed with sand, gravel, and the like.

Emulsions may also be broken by the addition of treating agents, such as soaps, fatty acids, sulfonates, and long-chain alcohols. When a chemical is used for emulsion breaking during desalting, it may be added at one or more of three points in the system. First, it may be added to the crude oil before it is mixed with fresh water. Second, it may be added to the fresh water before mixing with the crude oil. Third, it may be added to the mixture of crude oil and water. A high-potential field across the settling vessel also aids coalescence and breaks emulsions, in which case dissolved salts and impurities are removed with the water.

Finally, flashing the crude oil feed can reduce corrosion in the principal distillation column. The temperature of the feed is raised by heat exchange with the products from the distillation stages and fed to a flash tower at a pressure of around 30 to 45 psi. Dissolved hydrogen sulfide

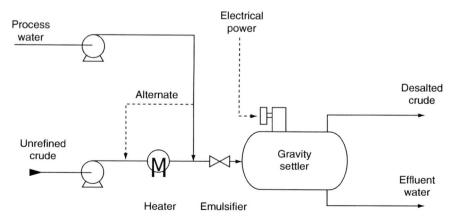

FIGURE 16.1 An electrostatic desalting unit. (From OSHA Technical Manual, SectionIV, Chapter 2, Petroleum Refining Processes.)

may thus be removed before reaching the atmospheric column, which would otherwise be exposed to the corrosive attack of hydrogen sulfide at an elevated temperature and in the presence of steam.

16.3 ATMOSPHERIC AND VACUUM DISTILLATION

Distillation columns are the most commonly used separation units in a refinery. Operation is based on the difference in boiling temperatures of the liquid mixture components, and on recycling counter-current gas–liquid flow. The properly organized temperature distribution up the column results in different mixture compositions at different heights. While multi-component inter-phase mass transfer is a common phenomenon for all column types, the flow regimes are very different, depending on the internal elements used. The two main types are a tray column and a packed column, the latter equipped with either random or structured packing. Different types of distillation columns are used for different processes, depending on the desired liquid holdup, capacity (flow rates), and pressure drop, but each column is a complex unit, combining many structural elements.

The tray column typically combines the open channel flow, with weirs, **downcomers** and heat exchangers. Free surface flow over the tray is disturbed by gas bubbles coming through the perforated tray, and possible leakage of liquid dropping through the upper tray.

A packed column is similar to a trickle-bed reactor, where liquid film flows down over the packing surface in contact with the upward gas flow. A small fragment of packing geometry can be accurately analyzed assuming the periodic boundary conditions, which allows calibration of the porous media model for a big packing segment.

In early refineries, distillation was the primary means by which products were separated from crude petroleum. As the technologies for refining evolved into the twentieth century, refineries became much more complex, but distillation remained the prime means by which petroleum is refined. Indeed, the distillation section of a modern refinery is the most flexible unit in the refinery since conditions can be adjusted to process a wide range of refinery feedstocks from the lighter crude oils to the heavier, more viscous crude oils. However, the maximum permissible temperature (in the vaporizing furnace or heater) to which the feedstock can be subjected is 350°C (660°F). The rate of thermal decomposition increases markedly above this temperature; if decomposition occurs within a distillation unit, it can lead to coke deposition in the heater pipes or in the tower itself with the resulting failure of the unit.

Of all the units in a refinery, the distillation section, comprising the atmospheric unit (Figure 16.2) and the vacuum unit (Figure 16.3), is required to have the greatest flexibility in terms of variable quality of feedstock and range of product yields (Figure 16.4). The maximum permissible temperature of the feedstock in the vaporizing furnace is the factor limiting the range of products in a single-stage (atmospheric) column. Thermal decomposition or cracking of the constituents begins as the temperature of the oil approaches 350°C (660°F) and the rate increases markedly above this temperature. This thermal decomposition is generally regarded as being undesirable because the coke-like material produced, tends to be deposited on the tubes with consequent formation of hot spots and eventual failure of the affected tubes. In the processing of lubricating oil stocks, an equally important consideration in the avoidance of these high temperatures is the deleterious effect on the lubricating properties. However, there are occasions when cracking distillation might be regarded as beneficial and the still temperature will be adjusted accordingly. In such a case, the products will be named accordingly using the prefix cracked, e.g., cracked residuum in which case the term **pitch** (Chapter 1) is applied.

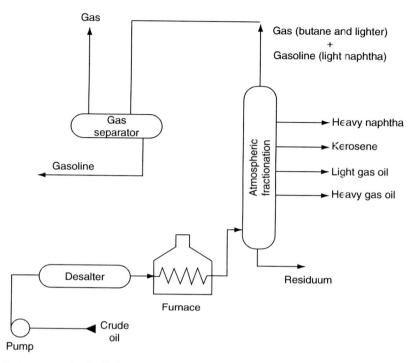

FIGURE 16.2 An atmospheric distillation unit. (From OSHA Technical Manual, Section IV, Chapter 2, Petroleum Refining Processes.)

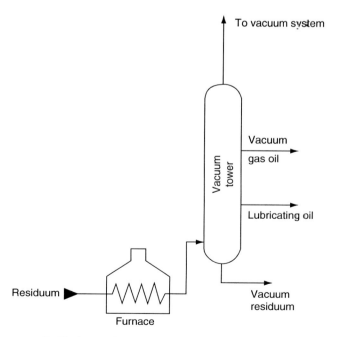

FIGURE 16.3 A vacuum distillation unit. (From OSHA Technical Manual, Section IV, Chapter 2, Petroleum Refining Processes.)

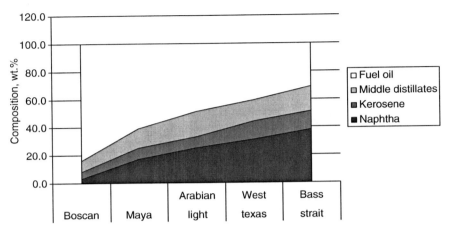

FIGURE 16.4 Variation of distillate yields and distillate composition for different feedstocks. (From Speight J.G. 2002. *Handbook of Petroleum Product Analysis*. John Wiley & Sons Inc., NJ. With permission.)

16.3.1 ATMOSPHERIC DISTILLATION

The present-day petroleum distillation unit is, in fact, a collection of distillation units that enable a fairly efficient degree of fractionation to be achieved. In contrast to the early units, which consisted of separate stills, a tower is used in the modern-day refinery.

It is common practice to use furnaces to heat the feedstock only when distillation temperatures above 205°C (400°F) are required. Lower temperatures (such as that used in the redistillation of naphtha and similar low-boiling products) are provided by heat exchangers and steam reboilers.

The feed to a fractional distillation tower is heated by flow-through pipe arranged within a large furnace. The heating unit is known as a pipestill heater or pipestill furnace, and the heating unit and the fractional distillation tower make up the essential parts of a distillation unit or **pipestill**. The pipestill furnace heats the feed to a predetermined temperature, usually a temperature at which a calculated portion of the feed changes into vapor. The vapor is held under pressure in the pipestill furnace, until it discharges as a foaming stream into the fractional distillation tower. Here, the vapors pass up the tower to be fractionated into gas oil, kerosene, and naphtha while the nonvolatile or liquid portion of the feed descends to the bottom of the tower to be pumped away as a bottom product.

Pipestill furnaces vary greatly in size, shape, and interior arrangement and can accommodate 25,000 bbl or more of crude petroleum per day. The walls and ceiling are insulated with firebrick, and gas or oil burners are inserted through one or more walls. The interior of the furnace is partially divided into two sections: a smaller convection section where the oil first enters the furnace and a larger section into which the burners discharge and where the oil reaches its highest temperature.

Heat exchangers are also used to preheat the feedstock before it enters the furnace. These exchangers are bundles of tubes arranged within a shell so that a stream passes through the tubes in the opposite direction of a stream passing through the shell. Thus cold crude oil, by passing through a series of heat exchangers where hot products from the distillation tower are cooled, before entering the furnace and saving of heat in this manner, may be a major factor in the economical operation of refineries.

Steam reboilers may take the form of a steam coil at the bottom of the fractional distillation tower or in a separate vessel. In the latter case, the bottom product from the tower enters the reboiler where part is vaporized by heat from the steam coil. The hot vapor is directed back to the bottom of the tower and provides part of the heat needed to operate the tower. The nonvolatile product leaves the reboiler and passes through a heat exchanger, where its heat is transferred to the feed to the tower. Steam may also be injected into a fractional distillation tower, not only to provide heat but also to induce boiling to take place at lower temperatures. Reboilers generally increase the efficiency of fractionation, but a satisfactory degree of separation can usually be achieved more conveniently by the use of a **stripping** section.

The stripping operation (please see Section 16.5.1, below) occurs in that part of the tower below the point at which the feed is introduced. The more volatile components are stripped from the descending liquid. Above the feed point (the rectifying section), the concentration of the less volatile component in the vapor is reduced.

The tower is divided into a number of horizontal sections by metal trays or plates, and each is the equivalent of a still. These force a rising vapor to pass though a pool of descending liquid. Therefore, the more trays, the more redistillation, and hence the better is the fractionation or separation of the mixture fed into the tower. A tower for fractionating crude petroleum may be 13 ft. in diameter and 85 ft. high, but a tower stripping unwanted volatile material from gas oil may be only 3 or 4 ft. in diameter and 10 ft. high. Towers concerned with the distillation of liquefied gases are only a few feet in diameter but may be up to 200 ft. in height. A tower used in the fractionation of crude petroleum may have from 16 to 28 trays, but one used in the fractionation of liquefied gases may have 30–100 trays. The feed to a typical tower enters the vaporizing or flash zone, an area without trays. The majority of the trays are usually located above this area. The feed to a bubble tower, however, may be at any point from top to bottom with trays above and below the entry point, depending on the kind of feedstock and the characteristics desired in the products.

Liquid collects on each tray to a depth of, say, several inches and the depth controlled by a dam or weir. As the liquid level rises, excess liquid spills over the weir into a channel (downspout), which carries the liquid to the tray below.

The temperature of the trays is progressively cooler from bottom to top (Figure 16.5). The bottom tray is heated by the incoming heated feedstock, although in some instances a steam coil (reboiler) is used to supply additional heat. As the hot vapors pass upward in the tower, condensation occurs onto the trays, until refluxing (simultaneous boiling of a liquid and

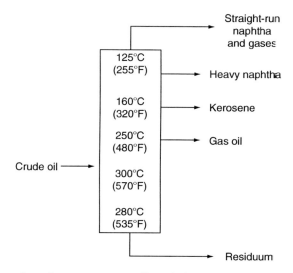

FIGURE 16.5 Representation of temperature profiles within an atmospheric distillation tower.

condensing of the vapor) occurs on the trays. Vapors continue to pass upward through the tower, whereas the liquid on any particular tray spills onto the tray below, and so on, until the heat at a particular point is too intense for the material to remain liquid. It then becomes vapor and joins the other vapors passing upward through the tower. The whole tower thus simulates a collection of several (or many) stills, with the composition of the liquid, at any one point or on any one tray, remaining fairly consistent. This allows part of the refluxing liquid to be tapped off at various points as side-stream products. Thus, in the distillation of crude petroleum, light naphtha and gases are removed as vapor from the top of the tower, heavy naphtha, kerosene, and gas oil are removed as side-stream products, and reduced crude is taken from the bottom of the tower.

The efficient operation of the distillation, or fractionating, tower requires the rising vapors to mix with the liquid on each tray. This is usually achieved by installing a short chimney on each hole in the plate and a cap with a serrated edge (**bubble cap**, hence bubble-cap tower) over each chimney (Figure 16.6). The cap forces the vapors to go below the surface of the liquid and to bubble up through it. Since the vapors may pass up the tower at substantial velocities, the caps are held in place by bolted steel bars.

Perforated trays are also used in fractionating towers. This tray is similar to the bubble-cap tray but has smaller holes (~1/4 in., 6 mm, versus 2 in., 50 mm). The liquid spills back to the tray below through weirs and is actually prevented from returning to the tray below through the holes by the velocity of the rising vapors. Needless to say, a minimum vapor velocity is required to prevent return of the liquid through the perforations.

In simple refineries, cut points can be changed slightly to vary yields and balance products, but the more common practice is to produce relatively narrow fractions and then process (or blend) to meet product demand. Since all these primary fractions are equilibrium mixtures, they all contain some proportion of the lighter constituents, characteristic of a lower boiling fraction and so are stripped of these constituents, or stabilized, before further processing or storage. Thus, gasoline is stabilized to controlled butanes–pentanes content, and the overhead may be passed to superfractionators, towers with a large number of plates that can produce nearly pure C_1–C_4 hydrocarbons (methane to butanes, CH_4 to C_4H_{10}), the successive columns termed *deethanizers*, *depropanizers*, *debutanizers*, and so on.

Kerosene and gas oil fractions are obtained as side-stream products from the atmospheric tower (primary tower), and these are treated in stripping columns (i.e., vessels of a few bubble trays) into which steam is injected, and the volatile overhead from the stripper is returned to

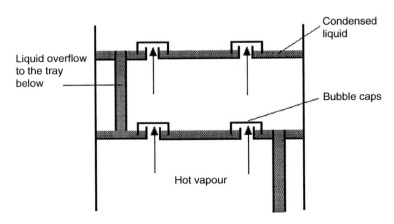

FIGURE 16.6 A bubble cap tray.

the primary tower. Steam is usually introduced by the stripping section of the primary column to lower the temperature at which fractionation of the heavier ends of the crude can occur.

The specifications for most petroleum products make it extremely difficult to obtain marketable material by distillation only. In fact, the purpose of atmospheric distillation is considered the provision of fractions that serve as feedstock for intermediate refining operations and for blending. Generally, this is carried out at atmospheric pressure, although light crude oils may be topped at an elevated pressure and the residue then distilled at atmospheric pressure.

The topping operation differs from normal distillation procedures, insofar as most of the heat is directed to the feed stream rather than by reboiling the material in the base of the tower. In addition, products of volatility intermediate between that of the overhead fractions and bottoms (residua) are withdrawn as side-stream products. Further, steam is injected into the base of the column and the side-stream strippers to adjust and control the initial boiling range (or point) of the fractions.

Topped crude oil must always be stripped with steam to elevate the flash point or to recover the final portions of gas oil. The composition of the topped crude oil is a function of the temperature of the vaporizer (or flasher). In addition, the properties of the residuum are highly dependent upon the extent of volatiles removal, either by atmospheric distillation or by vacuum distillation (Table 16.1).

16.3.2 VACUUM DISTILLATION

The boiling range of the highest boiling fraction that can be produced at atmospheric pressure is limited by the temperature at which the residue starts to decompose or crack. If the stock is

TABLE 16.1
Properties of Various Residua

Feedstock	Gravity API	Sulfur wt.%	Nitrogen wt.%	Nickel ppm	Vanadium ppm	Asphaltenes (Heptane) wt.%	Carbon Residue (Conradson) wt.%
Arabian light, >650°F	17.7	3.0	0.2	10.0	26.0	1.8	7.5
Arabian light, >1050°F	8.5	4.4	0.5	24.0	66.0	4.3	14.2
Arabian heavy, > 650°F	11.9	4.4	0.3	27.0	103.0	8.0	14.0
Arabian heavy, >1050°F	7.3	5.1	0.3	40.0	174.0	10.0	19.0
Alaska, North slope, >650°F	15.2	1.6	0.4	18.0	30.0	2.0	8.5
Alaska, North slope, >1050°F	8.2	2.2	0.6	47.0	82.0	4.0	18.0
Lloydminster (Canada), >650°F	10.3	4.1	0.3	65.0	141.0	14.0	12.1
Lloydminster (Canada), >1050°F	8.5	4.4	0.6	115.0	252.0	18.0	21.4
Kuwait, >650°F	13.9	4.4	0.3	14.0	50.0	2.4	12.2
Kuwait, >1050°F	5.5	5.5	0.4	32.0	102.0	7.1	23.1
Tia Juana, >650°F	17.3	1.8	0.3	25.0	185.0		9.3
Tia Juana, >1050°F	7.1	2.6	0.6	64.0	450.0		21.6
Taching, >650°F	27.3	0.2	0.2	5.0	1.0	4.4	3.8
Taching, >1050°F	21.5	0.3	0.4	9.0	2.0	7.6	7.9
Maya, >650°F	10.5	4.4	0.5	70.0	370.0	16.0	15.0

required for the manufacture of lubricating oils, further fractionation without cracking may be desirable, and this may be achieved by distillation under vacuum.

Vacuum distillation evolved because of the need to separate the less volatile products, such as lubricating oils, from the petroleum without subjecting these high-boiling products to cracking conditions. The boiling range of the highest boiling fraction obtainable at atmospheric pressure is limited by the temperature (ca. 350°C; ca. 660°F) at which the residue starts to decompose or crack, unless cracking distillation is preferred. When the feedstock is required for the manufacture of lubricating oils, further fractionation without cracking is desirable and this can be achieved by distillation under vacuum (reduced pressure) conditions.

The distillation of high-boiling lubricating oil stocks may require pressures as low as 15 to 30 mm Hg (0.29 to 0.58 psi), but operating conditions are more usually 50 to 100 mm Hg (0.97 to 1.93 psi). Volumes of vapor at these pressures are large and pressure drops must be small to maintain control, so vacuum columns are necessarily of large diameter. Differences in vapor pressure of different fractions are relatively larger than for lower boiling fractions, and relatively few plates are required. Under these conditions, heavy gas oil may be obtained as an overhead product at temperatures of about 150°C (300°F). Lubricating oil fractions may be obtained as side-stream products at temperatures of 250°C–350°C (480°F–660°F). The feedstock and residue temperatures are kept below the temperature of 350°C (660°F), above which the rate of thermal decomposition increases (Chapter 12) and cracking occurs. The partial pressure of the hydrocarbons is effectively reduced yet further by the injection of steam. The steam added to the column, principally for the stripping of asphaltic constituents at the base of the column, is superheated in the convection section of the heater.

At the point where the heated feedstock is introduced in the vacuum column (the flash zone) the temperature should be high and the pressure as low as possible to obtain maximum distillate yield. The flash temperature is restricted to about 420°C (790°F), however, in view of the cracking tendency of the feedstock constituents. Vacuum is maintained with vacuum ejectors and lately also with liquid ring pumps. In the older type high vacuum units, the required low hydrocarbon partial pressure in the flash zone could not be achieved without the use of lifting steam that acts in a similar manner as the stripping steam of atmospheric distillation units. Units of this type of units is called wet units. One of the latest developments in vacuum distillation has been the deep vacuum flashers, in which no steam is required. These dry units operate at very low flash zone pressures and low pressure drops over the column internals. For that reason, the conventional reflux sections with fractionation trays have been replaced by low pressure-drop spray sections. Cooled reflux is sprayed via a number of specially designed spray nozzles in the column countercurrent to the up-flowing vapor. This spray of small droplets comes into close contact with the hot vapor, resulting in good heat and mass transfer between the liquid and vapors phase.

When trays similar to those used in the atmospheric column are used in vacuum distillation, the column diameter may be extremely high, up to 45 ft. To maintain low pressure drops across the trays, the liquid seal must be minimal. The low holdup and the relatively high viscosity of the liquid limit the tray efficiency, which tends to be much lower than in the atmospheric column. The vacuum is maintained in the column by removing the noncondensable gas that enters the column by way of the feed to the column or by leakage of air.

The fractions obtained by vacuum distillation of reduced crude depend on whether the run is designed to produce lubricating or vacuum gas oils. In the former case, the fractions include:

1. *Heavy gas oil*, an overhead product and is used as catalytic cracking stock or, after suitable treatment, a light lubricating oil
2. *Lubricating oil* (usually three fractions: light, intermediate, and heavy), obtained as a side-stream product
3. *Residuum*, the nonvolatile product that may be used directly as asphalt or converted to asphalt

The residuum may also be used as a feedstock for a coking operation or blended with gas oils to produce a heavy fuel oil. However, if the reduced crude is not required as a source of lubricating oils, the lubricating and heavy gas oil fractions are combined or, more likely, removed from the residuum as one fraction and used as a catalytic cracking feedstock.

The continued use of atmospheric and vacuum distillation has been a major part of refinery operations during this century and no doubt will continue to be employed, at least into the beginning decades of the twenty-first century, as the primary refining operation.

Three types of high-vacuum units for long residue upgrading have been developed for commercial application: (1) feedstock preparation units, (2) lube oil high-vacuum units, and (3) high-vacuum units for asphalt production.

The feedstock preparation units make a major contribution to deep conversion upgrading and produce distillate feedstocks for further upgrading in catalytic crackers, hydrocracking units, and coking units. To obtain an optimum waxy distillate quality, a wash oil section is installed between the feed flash zone and waxy distillate draw-off. The wash oil produced is used as a fuel component or recycled to feed. The flashed residue (short residue) is cooled by heat exchange against long residue feed. A slipstream of this cooled short residue is returned to the bottom of the high-vacuum column as quench to minimize cracking (maintain low bottom temperature).

Lube oil high-vacuum units are specifically designed to produce high-quality distillate fractions for lube oil manufacturing. Special precautions are therefore taken to prevent thermal degradation of the distillates produced. The units are of the wet type. Normally, three sharply fractionated distillates are produced (spindle oil, light machine oil and medium machine oil). Cut points between those fractions are typically controlled on their viscosity quality. Spindle oil and light machine oil are subsequently steam-stripped in dedicated strippers. The distillates are further processed to produce lubricating base oil. The short residue is normally used as feedstock for the solvent deasphalting process to produce deasphalted oil, an intermediate for bright stock manufacturing. High-vacuum units for asphalt production are designed to produce straight-run asphalt and feedstocks for residuum blowing to produce **blown asphalt** that meets specifications. In principle, these units are designed on the same basis as feed preparation units, which may also be used to provide feedstocks for asphalt manufacturing.

Deep cut vacuum distillation, which involves a revamp of the vacuum distillation unit to cut deeper into the residue, is one of the first options available to the refiner. In addition to the limits of the major equipment, other constraints include: (1) the Vacuum gas oil (VGO) quality specification required by downstream conversion units, (2) the minimum flash zone pressure achievable, and (3) the maximum heater outlet temperature achievable without excessive cracking. These constraints typically limit the cut point (TBP) from 560°C to 590°C (1040°F to 1100°F), although units are designed for cut points (TBP) as high as 627°C (1160°F).

16.4 EQUIPMENT

16.4.1 COLUMNS

Distillation columns (distillation towers) are made up of several components, each of which is used either to transfer heat energy or enhance material transfer. A typical distillation column consists of several major parts:

1. Vertical shell where separation of the components is carried out
2. Column internals such as trays, or plates, or packings that are used to enhance component separation
3. Reboiler to provide the necessary vaporization for the distillation process

4. Condenser to cool and condense the vapor leaving the top of the column
5. Reflux drum to hold the condensed vapor from the top of the column, so that liquid (reflux) can be recycled back to the column

The vertical shell houses the column internals and together with the condenser and reboiler constitutes a distillation column (Figure 16.7).

In a petroleum distillation unit, the feedstock liquid mixture is introduced, usually somewhere near the middle of the column, to a tray known as the feed tray. The feed tray divides the column into a top (enriching, rectification) section and a bottom (stripping) section. The feed flows down the column where it is collected at the bottom in the reboiler. Heat is supplied to the reboiler to generate vapor. The source of heat input can be any suitable fluid, although in most chemical plants this is normally steam. In refineries, the heating source may be the output streams of other columns. The vapor raised in the reboiler is reintroduced into the unit at the bottom of the column. The liquid removed from the reboiler is known as the bottoms.

The vapor moves up the column, and as it exits the top of the unit, it is cooled by a condenser. The condensed liquid is stored in a holding vessel known as the reflux drum. Some of this liquid is recycled back to the top of the column and this is called the reflux. The condensed liquid that is removed from the system is known as the distillate or top product. Thus, there are internal flows of vapor and liquid within the column as well as external flows of feeds and product streams, into and out of the column.

The column is divided into a number of horizontal sections by metal trays or plates, and each is the equivalent of a still. The more trays, the more redistillation, and hence the better is the fractionation or separation of the mixture fed into the tower. A tower for fractionating crude petroleum may be 13 ft. in diameter and 85 ft. high, according to a general formula:

$$c = 220d^2r$$

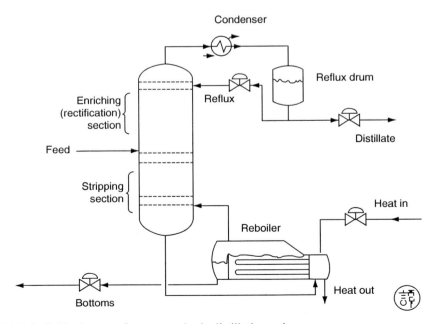

FIGURE 16.7 Individual parts of an atmospheric distillation column.

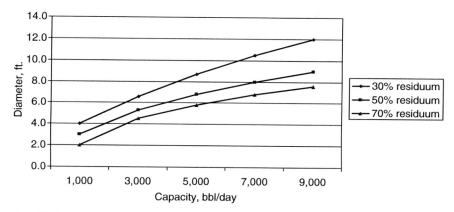

FIGURE 16.8 Variation of column (tower) capacity with diameter according to the amount of residuum in the feedstock.

where *c* is the capacity in bbl/day, *d* is the diameter in feet, and *r* is the amount of residuum expressed as a fraction of the feedstock (Figure 16.8) (Nelson, 1943).

A tower stripping unwanted volatile material from gas oil may be only 3 or 4 ft. in diameter and 10 ft. high with less than 20 trays. Towers concerned with the distillation of liquefied gases are only a few feet in diameter, but may be up to 200 ft. in height. A tower used in the fractionation of crude petroleum may have from 16 to 28 trays, but the ones used in the fractionation (superfractionation) of liquefied gases may have 30–100 trays. The feed to a typical tower enters the vaporizing or flash zone, an area without trays. The majority of the trays are usually located above this area. The feed to a bubble tower, however, may be at any point from top to bottom with trays above and below the entry point, depending on the kind of feedstock and the characteristics desired in the products.

16.4.2 PACKINGS

The packing in a distillation column creates a surface for the liquid to spread on, thereby providing a high surface area for mass transfer between the liquid and the vapor.

16.4.3 TRAYS

Usually, trays are horizontal, flat, specially prefabricated metal sheets, which are placed at a regular distance in a vertical cylindrical column. Trays have two main parts: (1) the part where vapor (gas) and liquid are being contacted—the contacting area and (2) the part where vapor and liquid are separated, after having been contacted—the downcomer area.

Classification of trays is based on: (1) the type of plate used in the contacting area, (2) the type and number of downcomers making up the downcomer area, (3) the direction and path of the liquid flowing across the contacting area of the tray, (4) the vapor (gas) flow direction through the (orifices in) the plate, and (5) the presence of baffles, packing or other additions to the contacting area to improve the separation performance of the tray.

Common plate types, for use in the contacting area are: (1) bubble cap tray in which caps are mounted over risers fixed on the plate (Figure 16.6). The caps come in a wide variety of sizes and shapes (round, square, and rectangular (tunnel)), (2) sieve trays come with different hole shapes (round, square, triangular, rectangular (slots), star), various hole sizes (from about 2 to 25 mm) and several punch patterns (triangular, square, rectangular), and (3) the valve tray which is also available in a variety of valve shapes (round, square, rectangular, triangular), valve sizes, valve weights (light and heavy), orifice sizes and either as fixed or floating valves.

Trays usually have one or more downcomers. The type and number of downcomers used mainly depends on the amount of downcomer area required to handle the liquid flow. Single pass trays are trays with one downcomer delivering the liquid from the next higher tray, a single bubbling area across which the liquid passes to contact the vapor and one downcomer for the liquid to the next lower tray.

Trays with multiple downcomers, and hence multiple liquid passes, can have a number of layout geometries. The downcomers may extend, in parallel, from wall to wall, as in. The downcomers may be rotated 90 (or 180) degrees on successive trays. The downcomer layout pattern determines the liquid flow path arrangement and liquid flow direction in the contacting area of the trays.

Giving a preferential direction to the vapor flowing through the orifices in the plate will induce the liquid to flow in the same direction. In this way, liquid flow rate and flow direction, as well as liquid height, can be manipulated. The presence of baffles, screen mesh or demister mats, loose or restrained dumped packing and the addition of other devices in the contacting area can be beneficial for improving the contacting performance of the tray; *viz.* its separation efficiency.

The most important parameter of a tray is its separation performance and four parameters are of importance in the design and operation of a tray-column: (1) the level of tray efficiency in the normal operating range, (2) the vapor rate at the upper limit, i.e., the maximum vapor load, (3) the vapor rate at the lower limit, i.e., the minimum vapor load and (4) the tray pressure drop.

The separation performance of a tray is the basis of the performance of the column as a whole. The primary function of, for instance, a distillation column is the separation of a feed stream into (at least) one top product stream and one bottom product stream. The quality of the separation performed by a column can be judged from the purity of the top and bottom product streams. The specification of the impurity levels in the top and bottom streams and the degree of recovery of pure products set the targets for a successful operation of a distillation column. It is evident that tray efficiency is influenced by (1) the specific component under consideration (this holds specially for multi-component. systems in which the efficiency can be different for each component, because of different diffusivities, diffusion interactions, and different stripping factors, and (2) the vapor flow rate; usually increasing the flow rate increases the effective mass transfer rate, while it decreases the contact time, at the same time. These counteracting effects lead to a roughly constant efficiency value, for a tray in its normal operating range. Upon approaching the lower operating limit, a tray starts weeping and loses efficiency.

16.5 OTHER PROCESSES

Atmospheric distillation and vacuum distillation provide the primary fractions from crude oil to use as feedstocks for other refinery processes, for conversion into products. Many of these subsequent processes involve fractional distillation and some of the procedures are so specialized and used with such frequency that they are identified by name.

16.5.1 STRIPPING

Stripping is a fractional distillation operation carried out on each side-stream product immediately after it leaves the main distillation tower. Since perfect separation is not accomplished in the main tower, unwanted components are mixed with those of the side-stream product. The purpose of stripping is to remove the more volatile components and thus reduce the flash point of the side-stream product. Thus, a side-stream product enters at the top tray of a stripper, and as it spills down four to six trays, steam injected into the bottom of the stripper removes the volatile components. The steam and volatile components leave the top of the stripper to return to the main tower. The stripped side-stream product leaves at the bottom and, after being cooled in a heat exchanger, goes to storage. Since strippers are short, they are arranged one above another in a single tower; each stripper, however, operates as a separate unit.

16.5.2 RERUNNING

Rerunning is a general term covering the redistillation of any material and indicating, usually, that a large part of the material is distilled overhead. Stripping, in contrast, removes only a relatively small amount of material as an overhead product. A rerun tower may be associated with a crude distillation unit that produces wide boiling range naphtha as an overhead product. By separating the wide-cut fraction into light and heavy naphtha, the rerun tower acts in effect as an extension of the crude distillation tower.

The product from chemical treating process of various fractions may be rerun to remove the treating chemical or its reaction products. If the volume of material being processed is small, a **shell still** may be used instead of a continuous fractional distillation unit. The same applies to gas oils and other fractions from which the front end or tail must be removed for special purposes.

16.5.3 STABILIZATION AND LIGHT END REMOVAL

The gaseous and more volatile liquid hydrocarbons produced in a refinery are collectively known as **light hydrocarbons** or **light ends**. Light ends are produced in relatively small quantities from crude petroleum and in large quantities when gasoline is manufactured by cracking and re-forming. When a naphtha or gasoline component at the time of its manufacture is passed through a condenser, most of the light ends do not condense and are withdrawn and handled as a gas. A considerable part of the light ends, however, can remain dissolved in the condensate, thus forming a liquid with a high vapor pressure.

Liquids with high vapor pressures may be stored in refrigerated tanks or in tanks capable of withstanding the pressures developed by the gases dissolved in the liquid. The more usual procedure, however, is to separate the light ends from the liquid by a distillation process, generally known as **stabilization**. Enough of the light ends are removed to make a stabilized liquid, that is, a liquid with a low enough vapor pressure to permit its storage in ordinary tanks without loss of vapor. The simplest stabilization process is a stripping process. Light naphtha from a crude tower, for example, may be pumped into the top of a tall, small-diameter fractional distillation tower operated under a pressure of 50–80 psi. Heat is introduced at the bottom of the tower by a steam reboiler. As the naphtha cascades down the tower, the light ends separate and pass up the tower to leave as an overhead product. Since reflux is not used, considerable amounts of liquid hydrocarbons pass overhead with the light ends.

Stabilization is usually a more precise operation than that just described. An example of more precise stabilization can be seen in the handling of the mixture of hydrocarbons produced by cracking. The overhead from the atmospheric distillation tower that fractionates the cracked mixture consists of light ends and cracked gasoline with light ends dissolved in it. If the latter is pumped to the usual type of tank storage, the dissolved gases cause the gasoline to boil, with consequent loss of the gases and some of the liquid components. To prevent this, the gasoline and the gases dissolved in it are pumped to a stabilizer, maintained under a pressure of approximately 100 psi and operated with reflux. This fractionating tower makes a cut between the highest boiling gaseous component (butane) and the lowest boiling liquid component (pentane). The bottom product is thus a liquid free of all gaseous components, including butane; hence the fractionating tower is known as a **debutanizer**. The debutanizer bottoms (gasoline constituents) can be safely stored, whereas the overhead from the debutanizer contains the butane, propane, ethane, and methane fractions. The butane fraction, which consists of all the hydrocarbons containing four carbon atoms, is particularly needed to give easy starting characteristics to motor gasoline. It must be separated from the other gases and blended with motor gasoline in amounts that vary with the season; more in the winter and less in the summer. Separation of the butane fraction is effected by another distillation in a fractional distillation tower called a **depropanizer**, since its purpose is to separate propane and the lighter gases from the butane fraction.

The depropanizer is very similar to the debutanizer, except that it is smaller in diameter because of the smaller volume being distilled and is taller because of the larger number of trays required to make a sharp cut between the butane and propane fractions. Since the normally gaseous propane must exist as a liquid in the tower, a pressure of 200 psi is maintained. The bottom product, known as the butane fraction (also known as stabilizer bottoms or refinery casinghead) is a high-vapor-pressure material that must be stored in refrigerated tanks or pressure tanks. The depropanizer overhead, consisting of propane and lighter gases, is used as a petrochemical feedstock or as a refinery fuel gas, depending on the composition.

A **depentanizer** is a fractional distillation tower that removes the pentane fraction from a debutanized (butane-free) fraction. Depentanizers are similar to debutanizers and have been introduced recently to segregate the pentane fractions from cracked gasoline and reformate. The pentane fraction, when added to a premium gasoline, makes this gasoline extraordinarily responsive to the demands of an engine accelerator.

The gases produced as overhead products from crude distillation, stabilization, and depropanizer units may be delivered to a gas absorption plant for the recovery of small amounts of butane and higher boiling hydrocarbons. The gas absorption plant consists essentially of two towers. One tower is the absorber where the butane and higher boiling hydrocarbons are removed from the lighter gases.

This is done by spilling a light oil (**lean oil**) down the absorber over trays similar to those in a fractional distillation tower. The gas mixture enters at the bottom of the tower and rises to the top. As it does this, it contacts the lean oil, which absorbs the butane and higher boiling hydrocarbons, but not the lower boiling hydrocarbons. The latter leaves the top of the absorber as dry gas. The lean oil that has become enriched with butane and higher boiling hydrocarbons is now termed **fat oil**. This is pumped from the bottom of the absorber into the second tower, where fractional distillation separates the butane and higher boiling hydrocarbons as an overhead fraction and the oil, once again lean oil, as the bottom product.

The condensed butane and higher boiling hydrocarbons are included with the refinery casinghead bottoms or stabilizer bottoms. The dry gas is frequently used as fuel gas for refinery furnaces. It contains propane and propylene, however, which may be required for liquefied petroleum gas for the manufacture of polymer gasoline or petrochemicals. Separation of the propane fraction (propane and propylene) from the lighter gases is accomplished by further distillation in a fractional distillation tower, similar to those previously described and particularly designed to handle liquefied gases. Further separation of hydrocarbon gases is required for petrochemical manufacture.

16.5.4 SUPERFRACTIONATION

The term *superfractionation* is sometimes applied to a highly efficient fractionating tower used to separate ordinary petroleum products. For example, to increase the yield of furnace fuel oil, heavy naphtha may be redistilled in a tower that is capable of making a better separation of the naphtha and the fuel oil components. The latter, obtained as a bottom product, is diverted to furnace fuel oil.

Fractional distillation as normally carried out in a refinery does not completely separate one petroleum fraction from another. One product overlaps another, depending on the efficiency of the fractionation, which in turn depends on: the number of trays in the tower, the amount of reflux used, and the rate of distillation. Kerosene, for example, normally contains a small percentage of hydrocarbons that (according to their boiling points) belong in the naphtha fraction and a small percentage that should be in the gas oil fraction. Complete separation is not required for the ordinary uses of these materials, but certain materials, such as solvents for particular purposes (hexane, heptane, and aromatics), are required as essentially pure compounds. Since they occur in mixtures of hydrocarbons they must be separated by distillation, with no overlap of one hydrocarbon with another. This requires highly efficient fractional distillation towers, specially designed for the purpose and referred to as superfractionators. Several towers with 50–100 trays operated with a high reflux ratio may be required to separate a single compound with the necessary purity.

16.5.5 Azeotropic Distillation

Azeotropic distillation is the use of a third component to separate two close-boiling components by means of the formation of an azeotropic mixture between one of the original components and the third component to increase the difference in the boiling points and facilitates separation by distillation.

All compounds have definite boiling temperatures, but a mixture of chemically dissimilar compounds sometimes causes one or both of the components to boil at a temperature other than that expected. For example, benzene boils at 80°C (176°F), but if it is mixed with hexane, it distills at 69°C (156°F). A mixture that boils at a temperature lower than the boiling point of either of the components is called an azeotropic mixture.

Two main types of azeotropes exist, i.e., the homogeneous azeotrope, where a single liquid phase is in the equilibrium with a vapor phase; and the heterogeneous azeotropes, where the overall liquid composition which form two liquid phases, is identical to the vapor composition. Most methods of distilling azeotropes and low relative volatility mixtures rely on the addition of specially chosen chemicals to facilitate the separation.

The five methods for separating azeotropic mixtures are:

1. *Extractive distillation* and homogeneous azeotropic distillation, where the liquid separating agent is completely miscible.
2. *Heterogeneous azeotropic distillation*, or more commonly, azeotropic distillation where the liquid separating agent (the entrainer) forms one or more azeotropes with the other components in the mixture and causes two liquid phases to exist over a wide range of compositions. This immiscibility is the key to making the distillation sequence work.
3. *Distillation using ionic salts*. The salts dissociate in the liquid mixture, and alter the relative volatilities sufficiently so that the separation becomes possible.
4. *Pressure-swing distillation*, where a series of columns operating at different pressures are used to separate binary azeotropes, which change appreciably in composition over a moderate pressure range or where a separating agent that forms a pressure-sensitive azeotrope is added to separate a pressure-insensitive azeotrope.
5. *Reactive distillation*, where the separating agent reacts preferentially and reversibly with one of the azeotropic constitutes. The reaction product is then distilled from the non-reacting components and the reaction is reversed to recover the initial component.

In simple distillation, a multi-component liquid mixture is slowly boiled in a heated zone and the vapors are continuously removed as they form and, at any instant in time, the vapor is in equilibrium with the liquid remaining on the still. Because the vapor is always richer in the more volatile components than the liquid, the liquid composition changes continuously with time, becoming more and more concentrated in the least volatile species. A simple distillation residue curve (Figure 16.9) is a means by which the changes in the composition of the liquid residue curve on the pot changes over time. A residue curve map is a collection of the liquid residue curves originating from different initial compositions. Residue curve maps contain the same information as phase diagrams, but represent this information in a way that is more useful for understanding how to synthesize a distillation sequence to separate a mixture.

All of the residue curves originate at the light (lowest boiling) pure component in a region, move towards the intermediate boiling component, and end at the heavy (highest boiling) pure component in the same region. The lowest temperature nodes are termed as *unstable nodes*, as all trajectories leave from them; while the highest temperature points in the region are termed *stable nodes*, as all trajectories ultimately reach them. The point that the trajectories approach from one direction and end in a different direction (as always is the point of intermediate boiling component) are termed *saddle point*. Residue curves that divide the composition space into different distillation regions are called distillation boundaries.

Many different residue curve maps are possible when azeotropes are present. Ternary mixtures containing only one azeotrope may exhibit six possible residue curve maps that differ by the binary pair forming the azeotrope and by whether the azeotrope is at minimum or maximum boiling. By identifying the limiting separation achievable by distillation, residue curve maps are also useful in synthesizing separation sequences combining distillation with other methods.

However, the separation of components of similar volatility may become economical if an entrainer can be found that effectively changes the relative volatility. It is also desirable that the entrainer be reasonably cheap, stable, nontoxic, and readily recoverable from the components. In practice, it is probably this last criterion that severely limits the application of extractive and azeotropic distillation. The majority of successful processes, in fact, are those in which the entrainer and one of the components separate into two liquid phases on cooling, if direct recovery by distillation is not feasible.

A further restriction in the selection of an azeotropic entrainer is that the boiling point of the entrainer be in the range 10°C to 40°C (18°F to 72°F) below that of the components. Thus, although the entrainer is more volatile than the components and distills off in the overhead product, it is present in a sufficiently high concentration in the rectification section of the column.

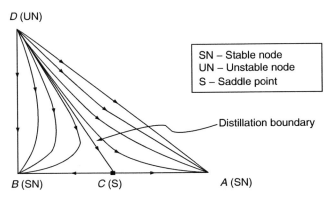

FIGURE 16.9 A residue curve map.

16.5.6 EXTRACTIVE DISTILLATION

Extractive distillation is the use of a third component to separate two close-boiling components in which one of the original components in the mixture is extracted by the third component and retained in the liquid phase to facilitate separation by distillation.

Using acetone-water as an extractive solvent for butanes and butenes, butane is removed as overhead from the extractive distillation column with acetone-water charged at a point close to the top of the column. The bottoms product of butenes and the extractive solvent are fed to a second column where the butenes are removed as overhead. The acetone-water solvent from the base of this column is recycled to the first column. Extractive distillation may also be used for the continuous recovery of individual aromatics, such as benzene, toluene, or xylenes, from the appropriate petroleum fractions. Prefractionation concentrates a single aromatic cut into a close-boiling cut, after which the aromatic concentrate is distilled with a solvent (usually phenol) for benzene or toluene recovery. Mixed cresylic acids (cresols and methylphenols) are used as the solvent for xylene recovery.

Extractive distillation is successful because the solvent is specially chosen to interact differently with the components of the original mixture, thereby altering their relative volatilities. Because these interactions occur predominantly in the liquid phase, the solvent is continuously added near the top of the extractive distillation column so that an appreciable amount is present in the liquid phase on all of the trays below. The mixture to be separated is added through the second feed point further down the column. In the extractive column, the component with the greater volatility, not necessarily the component with the lowest boiling point, is taken overhead as a relatively pure distillate. The other component leaves with the solvent via the column bottoms. The solvent is separated from the remaining components in a second distillation column and then recycled back to the first column.

One of the most important steps in developing a successful (economical) extractive distillation sequence is selecting a good solvent. In general, selection criteria for the solvent include the following:

1. Should enhance significantly the natural relative volatility of the key component
2. Should not require an excessive ratio of solvent to nonsolvent (because of handling cost in the column and auxiliary equipment)
3. Should remain soluble in the feed components and should not lead to the formation of two phases
4. Should be easily separable from the bottom product
5. Should be inexpensive and readily available
6. Should be stable at the temperature of the distillation and solvent separation
7. Should be nonreactive with the components in the feed mixture
8. Should have a low latent heat
9. Should be noncorrosive and nontoxic

No single solvent or solvent mixture satisfies all of the criteria for use in extractive distillation. However, the following solvent selection criteria assist in choosing the best possible solvent:

1. Screen by functional group or chemical family.
   a. Select candidate solvent from the high boiling homologous series of both light and heavy key components.
   b. Select candidate solvents from groups that tend to give positive or no deviations from Raoult's law for the key component desire in the distillate, and negative or no deviations for the other key.

    c. Select solvents that are likely to cause the formation of hydrogen bonds with the key component to be removed in the bottoms, or disruption of hydrogen bonds with the key to be removed in the distillate. Formation and disruption of hydrogen bonds are often associated with strong negative and positive deviations respectively, from Raoult's Law.

    d. Select candidate solvents from chemical groups that tend to show higher polarity than one key component or lower polarity than the other key.

2. Identify the individual candidate solvents.

    a. Select only candidate solvents that boil at least 30°C–40°C above the key components to ensure that the solvent is relatively nonvolatile and remains largely in the liquid phase. With this boiling point difference, the solvent should also not form azeotropes with the other components.

    b. Rank the candidate solvents according to their selectivity at infinite dilution.

    c. Rank the candidate solvents by the increase in relative volatility caused by the addition of the solvent.

Residue curve maps are of limited use at the preliminary screening stage because there is usually insufficient information available to sketch the them, but they are valuable and should be sketched or calculated as part of the second stage of the solvent selection.

In general, none of the fractions or combinations of fractions separated from crude petroleum is suitable for immediate use as petroleum products. Each fraction must be separately refined by processes that vary with the impurities in the fraction and the properties required in the finished product (Chapter 20 and Chapter 24). The simplest treatment is the washing of a fraction with a lye solution to remove sulfur compounds. The most complex is the series of treatments—solvent treating, dewaxing, clay treating or hydrorefining, and blending—required to produce lubricating oils. On rare occasions, no treatment of any kind is required. Some crude oils yield a light gas oil fraction that is suitable as furnace fuel oil or as a diesel fuel.

16.5.7 PROCESS OPTIONS FOR HEAVY FEEDSTOCKS

In order to further distill the residuum or topped crude from the atmospheric tower at higher temperatures, reduced pressure is required to prevent thermal cracking and the process takes place in one or more vacuum distillation towers. The principles of vacuum distillation resemble those of fractional distillation and, except that, larger-diameter columns are used to maintain comparable vapor velocities at the reduced pressures, the equipment is also similar. The internal designs of some vacuum towers are different from atmospheric towers, in that random packing and demister pads are used instead of trays. A typical first-phase vacuum tower may produce gas oil, lubricating-oil base stock, and a heavy residuum for propane **deasphalting**. A second-phase tower operating at lower vacuum may distill surplus residuum from the atmospheric tower (which is not used for lube-stock processing) and surplus residuum from the first vacuum tower not used for deasphalting.

Vacuum towers are typically used to separate catalytic cracking feedstock from surplus residuum and heavy oil and tar sand bitumen have fewer components distilling at atmospheric pressure and under vacuum than conventional petroleum. Nevertheless, some heavy oil still passes through the distillation stage of a refinery before further processing is undertaken. In addition, a vacuum tower has recently been installed at the Syncrude, Canada plant to offer an additional process option for upgrading tar sand bitumen (Speight, 2005 and references cited therein). The installation of such a tower as a means of refining heavy feedstocks (with the possible exception of the residua that are usually produced through a vacuum tower) is a question of economics and the ultimate goal of the refinery in terms of product slate. After distillation, the residuum from the heavy oil might pass to a cracking unit, such as visbreaking or coking, to produce saleable products. Catalytic cracking of the residuum or the whole heavy oil is also an option, but success of the process is highly dependent on the constituents of the feedstock and their interaction with the catalyst.

The development of the catalytic or reactive distillation which unites in the same equipment catalyst and distillation devices, finds its main applications for reversible reactions, such as methyl tetrabutyl ether (MTBE), ethyl tributyl ether (ETBE) synthesis, so as to shift an unfavorable equilibrium by continuous reaction product withdrawal (DeCroocq, 1997). But catalytic distillation can also provide several advantages in the selective hydrogenation of C_3, C_4, and C_5 cuts for petrochemistry. Inserting the catalyst in the fractionation column improves mercaptans removal, catalyst fouling resistance, and selective hydrogenation performances by modifying the reaction mixture composition along the column.

Thus, there is the potential for applying a related concept to the deep distillation of heavy oil.

REFERENCES

Bland, W.F. and Davidson, R.L. 1967. *Petroleum Processing Handbook*. McGraw-Hill, New York.

Burris, D.R. 1992. *Petroleum Processing Handbook*. J.J. McKetta, ed Marcel Dekker Inc., New York. p. 666.

DeCroocq, D. 1997. *Rev. Inst. Franç. Pétrol.* 52(5): 469–489.

Gary, J.H. and Handwerk, G.L. 1994. *Petroleum Refining: Technology and Economics*. 4th edn. Marcel Dekker Inc., New York.

Gruse, W.A. and Stevens, D.R. 1960. *Chemical Technology of Petroleum*. McGraw-Hill, New York.

Hobson, G.D. and Pohl, W. 1973. *Modern Petroleum Technology*. Applied Science Publishers, Barking, Essex, UK.

Kobe, K.A. and McKetta, J.J. 1958. *Advances in Petroleum Chemistry and Refining*. Interscience, New York.

Nelson, W.L. 1943. *Oil Gas J.* 41(16): 72.

Nelson, W.L. 1958. *Petroleum Refinery Engineering*. McGraw-Hill, New York. p. 226 et seq.

Priestley, R. 1973. *Modern Petroleum Technology*. G.D. Hobson and W. Pohl, eds. Applied Science Publishers, Barking, Essex, UK.

Speight, J.G. 2005. Natural Bitumen (Tar Sands) and Heavy Oil. *Coal, Oil Shale, Natural Bitumen, Heavy Oil and Peat*, from *Encyclopedia of Life Support Systems (EOLSS)*, Developed under the Auspices of the UNESCO, EOLSS Publishers, Oxford, UK, [http://www.eolss.net].

17 Thermal Cracking

17.1 INTRODUCTION

Distillation (Chapter 16) has remained a major refinery process and a process to which just about every crude oil that enters the refinery is subjected (Speight and Ozum, 2002, and references cited therein). However, not all crude oils yield the same distillation products. In fact, the nature of the crude oil dictates the processes that may be required for refining. And balancing product yield with demand is a necessary part of refinery operations.

After 1910, the demand for automotive fuel began to outstrip the market requirements for kerosene, and refiners were pressed to develop new technologies to increase gasoline yields. The earliest process, called **thermal cracking**, consisted of heating heavier oils (for which there was a low market requirement) in pressurized reactors and thereby cracking, or splitting, their large molecules into the smaller ones that form the lighter, more valuable fractions such as gasoline, kerosene, and light industrial fuels.

Gasoline manufactured by the cracking process performed better in automobile engines than gasoline derived from distillation of unrefined petroleum. The development of more powerful aircraft engines in the late 1930s gave rise to a need to increase the combustion characteristics of gasoline and spurred the development of lead-based fuel additives to improve engine performance.

During the 1930s and World War II, improved refining processes involving the use of catalysts led to further improvements in the quality of transportation fuels and further increased their supply. These improved processes, including catalytic cracking of heavy oils (Chapter 23), alkylation (Chapter 23), polymerization (Chapter 23), and isomerization (Chapter 23), enabled the petroleum industry to meet the demands of high-performance combat aircraft and, after the war, to supply increasing quantities of transportation fuels.

The 1950s and 1960s brought a large-scale demand for jet fuel and high-quality lubricating oils. The continuing increase in demand for petroleum products also heightened the need to process a wider variety of crude oils into high-quality products. Catalytic reforming of naphtha (Chapter 23) replaced the earlier thermal reforming process and became the leading process for upgrading fuel qualities to meet the needs of higher-compression engines. Hydrocracking, a catalytic cracking process conducted in the presence of hydrogen (Chapter 21), was developed as a versatile manufacturing process for increasing the yields of either gasoline or jet fuels.

Balancing product yield and market demand, without the manufacture of large quantities of fractions with low commercial value, has long required processes for the conversion of hydrocarbons of one molecular weight range or structure into some other molecular weight range or structure. Basic processes for this are still the so-called cracking processes, in which relatively high boiling constituents are cracked (thermally decomposed) into lower molecular weight, lower boiling molecules, although reforming, alkylation, polymerization, and hydrogen-refining processes have wide applications in producing premium-quality products (Corbett, 1990; Trash, 1990).

It is generally recognized that the most important part of any refinery is its gasoline (and liquid fuels) manufacturing facilities; other facilities are added to manufacture additional products, as indicated by technical feasibility and economic gain. More equipments are used in the manufacture of gasoline, the equipments are more elaborate, and the processes are more complex than those for any other product. Among the processes that have been used for liquid fuels production are thermal cracking, catalytic cracking, thermal reforming, catalytic reforming, polymerization, alkylation, coking, and distillation of fractions directly from petroleum. Each of these processes may be carried out in a number of ways, which differ in details of operation, or essential equipment, or both (Bland and Davidson, 1967).

When kerosene was the major product, gasoline was the portion of crude petroleum too volatile to be included in kerosene. The first refiners had no use for it and often dumped an accumulation of gasoline into a nearby stream or river. As the demand for gasoline increased, more and more of the lighter kerosene components were included in gasoline, but the maximum suitable portion depended on the kind of crude oil, and rarely exceeded 20% of the crude oil. To increase the supply of gasoline, more crude oil was run to the stills, resulting in overproduction of the heavier petroleum fractions, including kerosene. The problem of how to get more gasoline from less crude oil was solved in 1913, by the use of cracking in which fractions heavier than gasoline were converted into gasoline (Purdy, 1958).

Thermal processes are essentially processes that decompose, rearrange, or combine hydrocarbon molecules by the application of heat. The major variables involved are feedstock type, time, temperature, and pressure and, as such, are usually considered in promoting cracking (thermal decomposition) of the heavier molecules to lighter products and in minimizing coke formation.

The origins of cracking are unknown. There are records that illustrate the use of naphtha in Greek fire almost 2000 years ago (Chapter 1), but whether the naphtha was produced naturally by distillation or by cracking distillation is not clear. Cracking was used commercially in the production of oils from coal and shales, before the beginning of the modern petroleum industry. The ensuing discovery that the higher boiling materials could be decomposed to lower molecular weight products was used to increase the production of kerosene and was called cracking distillation (Kobe and McKetta, 1958). Thus, a batch of crude oil was heated until most of the kerosene was distilled from it and the overhead material became dark in color. At this point, the still fires were lowered, the rate of distillation decreased, and the heavy oils were held in the hot zone, during which time some of the large hydrocarbons were decomposed to yield lower molecular weight (lower boiling) products. After a suitable time, the still fires were increased and the distillation continued in the normal way. The overhead product, however, was light oil suitable for kerosene, instead of the heavy oil that would otherwise have been produced.

The precise origins of the modern version of cracking distillation, as applied in the modern petroleum industry, are unknown. It is rumored that, in 1861, a stillman had to leave his charge for a longer time than he intended (the reason is not known), during which time the still overheated. When he returned, he noticed that the distillate in the collector was much more volatile than anticipated at that particular stage of the distillation. Further investigation led to the development of cracking distillation (i.e., thermal degradation with the simultaneous production of distillate). However, before giving too much credit to the absence of a stillman, it is essential to recognize that the production of a volatile product by the destructive distillation of wood and coal was known for many years, if not decades or centuries, before the birth of the modern petroleum industry. Indeed, the production of spirits of fire (i.e., naphtha, the flammable constituent of Greek fire) was known from early times. The occurrence of bitumen at Hit (Mesopotamia) that was used as a mastic by the Assyrians was further developed for use in warfare, through the production of naphtha by destructive distillation.

When petroleum fractions are heated to temperatures over 350°C (660°F), the rates of thermal decomposition proceed at significant rates (Chapter 15). Thermal decomposition does not require the addition of a catalyst. Therefore, this approach is the oldest technology available for residue conversion. The severity of thermal processing determines the conversion and the product characteristics. Thermal treatment of residues ranges from mild treatment for reduction of viscosity to ultrapyrolysis (high-temperature cracking at very short residence time) for better conversion to overhead products (Hulet et al., 2005). A higher temperature requires a shorter time to achieve a given conversion but, in many cases, there is a change in the chemistry of the reaction. The severity of the process conditions is the combination of reaction time and temperature to achieve a given conversion.

Sufficiently high temperatures convert oils entirely to gases and coke; cracking conditions are controlled to produce as much as possible of the desired product, which is usually gasoline, but may be cracked gases for petrochemicals or a lower viscosity oil for use as a fuel oil. The feedstock, or cracking stock, may be almost any fraction obtained from crude petroleum, but the greatest amount of cracking is carried out on **gas oils**, a term that refers to the portion of crude petroleum boiling between the fuel oils (kerosene or stove oil) and the residuum. Residua are also cracked, but the processes are somewhat different from those used for gas oils.

Cracking, as carried out to produce gasoline, breaks up high molecular weight species into fragments of various sizes. The smallest fragments are usually the hydrocarbon gases; the larger fragments are hydrocarbons that boil in the gasoline range. Some of the intermediate fragments combine to form molecules larger than those in the feedstock, cracked residua, and coke. Consequently, a series of hydrocarbons with a boiling range similar to that of crude oil is created by cracking, but this material is quite different from crude oil. It contains much more hydrocarbon material boiling in the gasoline range, but usually no fraction suitable for asphalt. It does contain gas oils and **residual oils** suitable for light and **heavy fuel oils** and a much larger proportion of gases than is associated with crude petroleum, as delivered to a refinery. In addition, olefins will also be present that were not present in the original crude oil.

Thus, thermal conversion processes are designed to increase the yield of lower boiling products obtainable from petroleum, either directly (by means of the production of gasoline components from higher-boiling feedstocks) or indirectly (by the production of olefins and the like, which are precursors of the gasoline components). These processes may also be characterized by the physical state (liquid or vapor phase) in which the decomposition occurs. The state depends on the nature of the feedstock as well as conditions of pressure and temperature (Nelson, 1976; Vermillion and Gearhart, 1983; Trimm, 1984; Thomas et al., 1989; Speight and Ozum, 2002).

From the chemical viewpoint, the products of cracking are very different from those obtained directly from crude petroleum. When a twelve-carbon atom hydrocarbon, typical of a straight-run **gas oil** is cracked, there are several potential reactions that can occur that lead to a variety of products, for example:

$$CH_3(CH_2)_{10}CH_3 \rightarrow CH_3(CH_2)_8CH_3 + CH_2 = CH_2$$
$$CH_3(CH_2)_{10}CH_3 \rightarrow CH_3(CH_2)_7CH_3 + CH_2 = CHCH_3$$
$$CH_3(CH_2)_{10}CH_3 \rightarrow CH_3(CH_2)_6CH_3 + CH_2 = CHCH_2CH_3$$
$$CH_3(CH_2)_{10}CH_3 \rightarrow CH_3(CH_2)_5CH_3 + CH_2 = CH(CH_2)_2CH_3$$
$$CH_3(CH_2)_{10}CH_3 \rightarrow CH_3(CH_2)_4CH_3 + CH_2 = CH(CH_2)_3CH_3$$
$$CH_3(CH_2)_{10}CH_3 \rightarrow CH_3(CH_2)_3CH_3 + CH_2 = CH(CH_2)_4CH_3$$
$$CH_3(CH_2)_{10}CH_3 \rightarrow CH_3(CH_2)_2CH_3 + CH_2 = CH(CH_2)_5CH_3$$
$$CH_3(CH_2)_{10}CH_3 \rightarrow CH_3CH_2CH_3 + CH_2 = CH(CH_2)_6CH_3$$
$$CH_3(CH_2)_{10}CH_3 \rightarrow CH_3CH_3 + CH_2 = CH(CH_2)_7CH_3$$
$$CH_3(CH_2)_{10}CH_3 \rightarrow CH_4 + CH_2 = CH(CH_2)_8CH_3$$

The products are dependent on temperature and residence time, and these simple reactions shown, do not take into account the potential for isomerization of the products or secondary, and even tertiary, reactions that (and do) occur.

The hydrocarbons with the least thermal stability are the paraffins, and the olefins produced by the cracking of paraffins are also reactive. Cycloparaffins (naphthenes) are less easily cracked, their stability depending mainly on any side chains present, but ring splitting may occur, and dehydrogenation can lead to the formation of unsaturated naphthenes and aromatics. Aromatics are the most stable (refractory) hydrocarbons, the stability depending on the length and the stability of side chains. Very severe thermal cracking of high-boiling feedstocks can result in condensation reactions of ring compounds, yielding a high proportion of coke (Speight, 1986).

The higher-boiling oils produced by cracking are light and heavy gas oils as well as a residual oil, which in the case of thermal cracking is usually (erroneously) called **tar**, and in the case of catalytic cracking is called cracked fractionator bottoms. The residual oil may be used as heavy fuel oil, and gas oil from catalytic cracking are suitable as domestic fuel oil and industrial fuel oil or as **diesel fuel**, if blended with straight-run gas oils. Gas oils from thermal cracking must be mixed with straight-run (distilled) gas oils, before they become suitable for domestic fuel oils and diesel fuels.

The gas oil produced by cracking is, in fact, a further important source of gasoline. In a once-through cracking operation, all the cracked material is separated into products and may be used as such. However, cracked gas oil is more resistant to cracking (more refractory) than straight-run gas oil, but can still be cracked to produce gasoline. This is done in a recycling operation in which the cracked gas oil is combined with fresh feed for another trip through the cracking unit. The operation may be repeated until the cracked gas oil is almost completely decomposed (cracking to extinction) by recycling (recycling to extinction) the higher-boiling product, but it is more usual to withdraw part of the cracked gas oil from the system, according to the need for fuel oils. The extent to which recycling is carried out affects the amount or yield of cracked gasoline resulting from the process.

The gases formed by cracking are particularly important because of their chemical properties and quantity. Only relatively small amounts of paraffinic gases are obtained from crude oil, and these are chemically inactive. Cracking produces both paraffinic gases (e.g., propane, C_3H_8) and olefinic gases (e.g., propene, C_3H_6); the latter are used in the refinery as the feed for polymerization plants, where high-octane polymer gasoline is made. In some refineries, the gases are used to make **alkylate**, a high-octane component for aviation gasoline and motor gasoline. In particular, the cracked gases are the starting points for many petrochemicals (Chapter 29).

In summary, the cracking of petroleum constituents can be visualized as a series of thermal conversions (Chapter 15). The reactions involve the formation of transient free radical species that may react further in several ways to produce the observed product slate. Because of this, the slate of products from thermal cracking is considered difficult to predict (Germain, 1969).

The available data suggest that thermal conversion (leading to coke formation) is a complex process involving both chemical reactions and thermodynamic behavior (Chapter 15), and can be summarized as follows:

1. Thermal reactions of crude oil constituents result in the formation of volatile products.
2. Thermal reactions of crude oil constituents also result in the formation of high molecular weight and high-polarity aromatic components.
3. Once the concentration of the high molecular weight high-polarity material reaches a critical concentration, phase separation occurs, giving a denser, aromatic, liquid phase.

Reactions that contribute to this process are cracking of side chains from aromatic groups, dehydrogenation of naphthenes to form aromatics, condensation of aliphatic structures to form aromatics, condensation of aromatics to form higher fused-ring aromatics, and dimerization or oligomerization reactions. Loss of side chains always accompanies thermal cracking, while dehydrogenation and condensation reactions are favored by hydrogen-deficient conditions. Formation of oligomers is enhanced by the presence of olefins or diolefins, which are products of cracking. The condensation and oligomerization reactions are also enhanced by the presence of Lewis acids, for example, aluminum chloride ($AlCl_3$).

The importance of solvents to mitigate coke formation has been recognized for many years, but their effects have often been ascribed to hydrogen-donor reactions rather than phase behavior. The separation of the phases depends on the solvent characteristics of the liquid. Addition of aromatic solvents will suppress phase separation (Chapter 15), while paraffins will enhance separation. Microscopic examination of coke particles often shows evidence for the presence of a **mesophase**; spherical domains that exhibit the anisotropic optical characteristics of liquid crystal. This phenomenon is consistent with the formation of a second liquid phase; the mesophase liquid is denser than the rest of the hydrocarbon, has a higher surface tension, and likely wets metal surfaces better than the rest of the liquid phase. The mesophase characteristic of coke diminishes as the liquid phase becomes more compatible with the aromatic material.

From this mechanism, the following trends for coke yield production in thermal processes are anticipated:

1. Higher molecular weight fractions should give more coke (Chapter 15).
2. Coke formation depends on phase incompatibility (Chapter 13 and Chapter 15).
3. Acidic contaminants (such as clay) in a feedstock may promote coking.
4. Higher asphaltene content in a feed will, in general, correlate with higher coke yield (Chapter 15) (Schabron and Speight, 1997).
5. Coke may not form immediately if the point of incipient flocculation of the coke precursor is not exceeded, so that an induction time is observed (Magaril and Aksenova, 1968, 1970; Magaril and Ramazaeva, 1969; Magaril et al., 1970, 1971; Savage et al., 1988; Speight, 1992, 1994; Wiehe, 1993).
6. Phase separation may be very sensitive to surface chemistry, hydrodynamics, and surface-to-volume ratio, similar to other processes that require nucleation.

Putting this chemical information in perspective allows an understanding of the pathways by which the various thermal processes proceed and also the chemical pathways by which excessive yields of coke can be recorded (Chapter 15).

17.2 EARLY PROCESSES

As the demand for gasoline increased with the onset of World War I and the ensuing 1920s, more crude oil had to be distilled not only to meet the demand for gasoline but also to reduce the overproduction of the heavier petroleum fractions, including kerosene. The problem of how to produce more gasoline from less crude oil was solved in 1913 by the incorporation of cracking units into refinery operations, in which fractions heavier than gasoline were converted into gasoline by thermal decomposition. The early (pre-1940) processes employed for gasoline manufacture were processes in which the major variables involved were feedstock type, time, temperature, and pressure, which need to be considered to achieve the cracking of the feedstock to lighter products with minimal coke formation.

One of the earliest processes used in the petroleum industry, after distillation, is the noncatalytic conversion of higher-boiling petroleum stocks into lower-boiling products, known as thermal cracking.

The yields of kerosene products were usually markedly increased by means of cracking distillation, but the technique was not suitable for gasoline production. As the need for gasoline arose, the necessity of prolonging the cracking process became apparent, and a process known as pressure cracking evolved. Pressure cracking was a batch operation in which some 200 bbl gas oil was heated to about 425°C (800°F) in stills (shell stills), especially reinforced to operate at pressures as high as 95 psi. The gas oil was retained in the reactor under maximum pressure for 24 h. Distillation was then started and during the next 48 h, 70 to 100 bbl of a lighter distillate was obtained that contained the gasoline components. This distillate was treated with sulfuric acid to remove unstable gum-forming components and then redistilled to produce cracked gasoline (boiling range <205°C, <400°F) and a residual fuel oil (Stephens and Spencer, 1956).

The Burton cracking process for the large-scale production of cracked gasoline was first used in 1912. The process employed batch distillation in horizontal shell stills and operated at about 400°C (ca. 750°F) and 75 to 95 psi. It was, in fact, the first successful method of converting heavier oils into gasoline. Nevertheless, heating a bulk volume of oil was soon considered cumbersome, and during the years 1914 to 1922, a number of successful continuous cracking processes were developed. In these processes, gas oil was continuously pumped through a unit that heated the gas oil to the required temperature, held it for a time under pressure, and then discharged the cracked material into distillation equipment where it was separated into gases, gasoline, gas oil, and tar.

The tube and tank cracking process is typical of the early continuous cracking processes. Gas oil, preheated by exchange with the hot products of cracking, was pumped into the cracking coil, which consisted of several hundred feet of very strong pipe that lined the inner walls of a furnace, where oil or gas burners raised the temperature of the gas oil to 425°C (800°F). The hot gas oil passed from the cracking coil into a large reaction chamber (soaker), where the gas oil was held under these temperature and pressure conditions, long enough for the cracking reactions to be completed. The cracking reactions formed coke, which over the course of several days filled the soaker. The gas oil stream was then switched to a second soaker, and drilling operations similar to those used in drilling an oil well cleaned out the first soaker.

The cracked material (other than coke) left the on-stream soaker to enter an evaporator (tar separator) maintained under a much lower pressure than the soaker, where, because of the lower pressure, all the cracked material, except the tar, became vaporized. The vapor left the top of the separator, where it was distilled into separate fractions gases, gasoline, and gas oil. The tar that was deposited in the separator was pumped out for use as an asphalt or as a heavy fuel oil.

Early in the development of tube-and-tank thermal cracking, it was found that adequate yields of gasoline could not be obtained by one passage of the stock through the heating coil. Attempts to increase the conversion in one pass brought about undesirable high yields of gas and coke. It was better to crack to a limited extent, remove the products, and recycle the rest of the oil (or a distilled fraction free of tar) for repeated partial conversion. The high-boiling constituents, once exposed to cracking, were so changed in composition as to be more refractory than the original feedstock.

As refining technology evolved throughout the twentieth century, the feedstocks for cracking processes became the residuum or gas oil from a distillation unit. In addition, the residual oils produced as the end products of distillation processes, and even some of the higher boiling crude oil, often contain substantial amounts of asphaltic materials, which preclude use of the residuum as fuel oils or lubricating stocks (Speight, 2000). However, subjecting these residua directly to thermal processes has become economically advantageous, since, on the one hand, the end result is the production of lower-boiling products. On the other hand, the asphaltene and the resin constituents that are concentrated in residua are regarded as the precursors to coke.

As the thermal processes evolved and catalysts were employed with more frequency, poisoning of the catalyst with a concurrent reduction in the lifetime of the catalyst became a major issue for refiners. To avoid catalyst poisoning, it became essential that as much of the nitrogen and metals (such as vanadium and nickel) as possible should be removed from the feedstock. The majority of the heteroatoms (nitrogen, oxygen, and sulfur) and the metals are contained in, or associated with, the residuum. It became necessary that this fraction be removed from cracking feedstocks and a number of thermal processes such as flash distillation, vacuum flashing, visbreaking, and coking, also came into wider usage by refiners, and were directed at upgrading feedstocks by removal of the nonvolatile fraction (asphaltene plus resin fractions).

17.3   COMMERCIAL PROCESSES

As crude oil slates are expected to continue to grow heavier with higher sulfur contents, environmental restrictions are expected to significantly reduce the demand for high-sulfur residual fuel oil. Light sweet crude oils will continue to be available and in even greater demand than they are today. Refiners will be faced with the choice of purchasing light sweet crude oils at a premium price, or adding bottom–of–the–barrel upgrading capability, through additional new investment, to reduce the production of high-sulfur residual fuel oil and increase the production of low-sulfur distillate transportation fuels. There are a number of different processing technologies available to meet the refiners' differing needs. In this Chapter, we examine the various options available and the variables that influence the choice of one technology over another.

There are several pressures that are expected to push refiners toward increased bottom–of–the–barrel upgrading: (1) increasing low-sulfur distillate fuel demand which will require increased refining capacity, (2) heavier crude oil slates which will result in a greater high-sulfur residual fuel oil production, if conversion is not added, and (3) environmental restrictions which will result in reduced demand for high-sulfur residual fuel oil.

One answer to the need for conversion of the heavier feedstocks is the installation and use of one or more thermal cracking process.

Thermal cracking processes offer attractive methods of conversion of heavy feedstocks because they enable low operating pressure, while involving high operating temperature, without requiring expensive catalysts. Currently, the widest operated residuum conversion processes are **visbreaking** (*q.v.*) and **delayed coking** (*q.v.*). And, these are still attractive processes for refineries from an economic point of view (Bland and Davidson, 1967; Dickenson et al., 1997).

The majority of regular thermal cracking processes use temperatures of 455°C to 540°C (850°F to 1005°F) and pressures of 100 to 1000 psi. The Dubbs process, to which a reduced crude is charged, may be taken as a typical application of conventional thermal cracking. The feedstock (reduced crude) is preheated by direct exchange with the cracked products in the fractionating columns. Cracked gasoline and middle distillate fractions are removed from the upper section of the column. Light and heavy distillate fractions are removed from the lower section and are pumped to separate heaters. Higher temperatures are used to crack the more refractory fractions. The streams from the heaters are combined and sent to a soaking chamber, where additional time is provided to complete the cracking reactions. The cracked products are then separated in a low-pressure flash chamber, where a heavy fuel oil is removed as bottoms. The remaining cracked products are sent to the fractionating columns.

Low pressures (<100 psi) and temperatures in excess of 500°C (930°F) tend to produce lower molecular weight hydrocarbons than those produced at higher pressures (400 to 1000 psi) and at temperatures below 500°C (930°F). The reaction time is also important; light feeds (gas oils) and recycle oils require longer reaction times than the readily cracked heavy residues. Mild cracking conditions (defined here as a low conversion per cycle) favor a high yield of gasoline components with low gas production and decreased coke production, but the gasoline quality is not high, whereas more severe conditions give increased gas and coke production and reduced gasoline yield (but of higher quality). With limited conversion per cycle, the heavier residues must be recycled. However, the recycled oils become increasingly refractory upon repeated cracking, and if they are not required as a fuel oil stock, they may be subjected to a coking operation to increase gasoline yield or refined by means of a hydrogen process.

Although new thermal cracking units are now under development for heavy feedstocks (Chapter 23), processes that can be regarded as having evolved from the original concept of thermal cracking are visbreaking and the various coking processes (Table 17.1). It is the purpose this chapter to present these processes in the light of their use in modern refineries and the information that should be borne in mind when considering and deciding upon the potential utility of any process presented throughout this and subsequent chapters.

17.3.1 Visbreaking

Visbreaking (viscosity reduction, viscosity breaking) is a relatively mild thermal cracking operation used to reduce the viscosity of residua to produce fuel oil that meets specifications (Ballard et al., 1992; Dominici and Sieli, 1997). Residua are sometimes blended with lighter heating oils to produce fuel oils of acceptable viscosity. By reducing the viscosity of the nonvolatile fraction, visbreaking reduces the amount of the more valuable light heating oil that is required for blending to meet the fuel oil specifications. The process is also used to reduce the pour point of a waxy residue.

The visbreaking process uses the approach of mild thermal cracking as a relatively low-cost and low-severity approach to improving the viscosity characteristics of the residue, without attempting significant conversion to distillates. Low residence times are required to avoid coke formation, although additives can help to suppress coke deposits on the tubes of the furnace.

TABLE 17.1
Comparison of Visbreaking with Delayed Coking and Fluid Coking

Thermal Cracking

Purpose: to produce volatile products of low-volatile or nonvolatile feedstocks (conversion is the prime purpose)

Cracking with simultaneous removal of distillate (semi-continuous)

Batch cracking (noncontinuous)

High conversion

Process configuration: various

Visbreaking

Purpose: to reduce viscosity of fuel oil to acceptable levels (conversion is not a prime purpose)

Mild (470°C to 495°C; 880°F to 920°F) heating at pressures of 50 to 200 psi

    Reactions quenched before going to completion

    Low conversion (10%) to products boiling less than 220°C (430°F)

Heated coil or drum (soaker)

Delayed Coking

Purpose: to produce maximum yields of distillate products

Moderate (480°C to 515°C; 900°F to 960°F) heating at pressures of 90 psi

    Reactions allowed to proceed to completion

    Complete conversion of the feedstock

Soak drums (845°F to 900°F) used in pairs (one on stream and one off stream being de-coked)

Coked until drum solid

Coke removed hydraulically from off-stream drum

Coke yield: 20% to 40% by weight (dependent upon feedstock)

Yield of distillate boiling below 220°C (430°F): ca. 30% (but feedstock dependent)

Fluid Coking

Purpose: to produce maximum yields of distillate products

Severe (480°C to 565°C; 900°F to 1050°F) heating at pressures of 10 psi

    Reactions allowed to proceed to completion

    Complete conversion of the feedstock

Oil contacts refractory coke

Bed fluidized with steam; heat dissipated throughout the fluid bed

Higher yields of light ends ($<C_5$) than delayed coking

Less coke make than delayed coking (for one particular feedstock)

Two visbreaking processes are commercially available: the soaker visbreaker and the coil visbreaker.

The soaker visbreaking process (Figure 17.1) achieves some conversion within the heater, but the majority of the conversion occurs in a reaction vessel or soaker that holds the two-phase effluent at an elevated temperature for a predetermined length of time. Soaker visbreaking is described as a low-temperature, high-residence-time route. Product quality and yields from the coil and the soaker drum design are essentially the same at a specified severity, being independent of visbreaker configuration. By providing the residence time required to achieve the desired reaction, the soaker drum design allows the heater to operate at a lower outlet temperature (thereby saving fuel), but there are disadvantages. The main disadvantage is the decoking operation of the heater and the soaker drum and, although decoking requirements of the soaker drum design are not as frequent as those of the coil-type design, the soaker design requires more equipment for coke removal and handling. The customary practice of removing coke from a drum is to cut it out with high-pressure water, thereby producing a significant amount of coke-laden water that needs to be handled, filtered, and then recycled for use again.

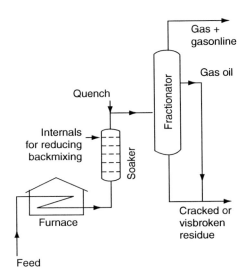

FIGURE 17.1 A soaker visbreaker. (From OSHA Technical Manual, Section IV, Chapter 2, Petroleum Refining Processes.)

The coil visbreaking process differs from soaker visbreaking, insofar as the coil process achieves conversion by high-temperature cracking within a dedicated soaking coil in the furnace. With conversion primarily achieved as a result of temperature and residence time, coil visbreaking is described as a high-temperature, short-residence-time route. The main advantage of the coil-type design is the two-zone fired heater that provides better control of the material being heated and, with the coil-type design, decoking of the heater tubes is accomplished more easily by the use of steam-air decoking.

In summary, the two different types of visbreaking (the soaker visbreaker and the coil visbreaker) are very similar. The only difference is that the soaker visbreaker works with a lower temperature with a higher residence time, and the coil visbreaker operates at a higher temperature with a lower residence time. The end result is that the soaker visbreaker has lower energy consumption for the same visbreaking severity.

Visbreaking conditions range from 455°C to 510°C (850°F to 950°F) at a short residence time and from 50 to 300 psi at the heating coil outlet. It is the short residence time that brings to visbreaking the concept of being a mild thermal reaction in contrast to, for example, the delayed coking process where residence times are much longer and the thermal reactions are allowed to proceed to completion. The visbreaking process uses a quench operation to terminate the thermal reactions. Liquid-phase cracking takes place under these low-severity conditions to produce some naphtha, as well as material in the kerosene and gas oil boiling range. The gas oil may be used as additional feed for catalytic cracking units, or as heating oil.

Thus, a crude oil residuum or tar sand bitumen is passed through a furnace where it is heated to a temperature of 480°C (895°F) under an outlet pressure of about 100 psi (Figure 17.1). The heating coils in the furnace are arranged to provide a soaking section of low heat density, where the charge remains until the visbreaking reactions are completed. The cracked products are then passed into a flash-distillation chamber. The overhead material from this chamber is then fractionated to produce naphtha as an overhead product and light gas oil. The liquid products from the flash chamber are cooled with a gas oil flux and then sent to a vacuum fractionator. This yields a heavy gas oil distillate and a residuum of reduced viscosity (Table 17.2).

A 5% to 10% conversion of atmospheric residua to naphtha is usually sufficient to afford at least an approximate fivefold reduction in viscosity. Reduction in viscosity is also accompanied by a reduction in the pour point. An alternative option is to use lower furnace temperatures and longer

TABLE 17.2
Examples of Product Yields and Properties from Visbreaking Athabasca Tar Sand Bitumen and Feedstocks Having a Similar API Gravity

	Arabian Light Vacuum Residuum	Arabian Light Vacuum Residuum	Iranian Light Vacuum Residue	Athabasca Bitumen
Feedstock				
API gravity	7.1	6.9	8.2	8.6
Carbon residue[a]	20.3		22.0	13.5
Sulfur, wt.%	4.0	4.0	3.5	4.8
Product yields,[b] *vol.%*				
Naphtha (<425°F, <220°C)	6.0	8.1	4.8	7.0
Light gas oil (425°F to 645°F, 220°C to 340°C)	16.0	10.5	13.1	21.0
Heavy gas oil (645°F to 1000°F, 340°C to 540°C)		20.8	—[b]	35.0
Residuum	76.0	60.5	79.9	34.0
API gravity	3.5	0.8	5.5	
Carbon residue[a]				
Sulfur, wt.%	4.7	4.6	3.8	

[a]Conradson.
[b]A blank product yield line indicates that the yield of the lower boiling product has been included in the yield of the higher boiling product.
Source: Speight, J.G. and Ozum, B. 2002. *Petroleum Refining Processes.* Marcel Dekker Inc., New York. With permission.

times, achieved by installing a soaking drum between the furnace and the fractionator. The disadvantage of this approach is the need to remove coke from the soaking drum.

The Shell soaker visbreaking process is claimed to be suitable for production of fuel oil by residuum (atmospheric residuum, vacuum residuum, or solvent deasphalter bottoms) viscosity reduction with maximum production of distillates. The basic configuration of the process includes the heater, soaker, and fractionator, and more recently, a vacuum flasher to recover more distillate products (Figure 17.2). The cut point of the heavy gas oil stream taken from the vacuum flasher is approximately 520°C (970°F).

In the process (Figure 17.2), the feedstock is pumped through preheat exchangers before entering the visbreaker heater, where the residue is heated to the required cracking temperature. In the convection section of the visbreaker heater, superheated steam is generated. Heater effluent is sent to the soaker drum, where most of the thermal cracking and viscosity reduction takes place under controlled conditions. The pressure in the soaker drum can be adjusted, which results in a change in residence time, and the amount of heavies that reside in the liquid phase, thereby providing the possibility to reach optimum selectivity. Soaker drum effluent is flashed and then quenched in the fractionator, and the flashed vapors are fractionated into gas, naphtha, gas oil, and visbreaker residue. The visbreaker residue is steam-stripped in the bottom of the fractionator and pumped through the cooling circuit to battery limits. Visbreaker gas oil, which is drawn off as a side stream, is steam-stripped, cooled, and sent to battery limits. Alternately, the gas oil fraction can be included with the visbreaker residue as cutter stock. Other cutter stocks, such as light cycle oil or heavy atmospheric gas oil, are typically added to the visbreaker residue and gas oil mixture to meet the desired fuel oil specification. Product yields (Table 17.3) are dependent on feed type and product specifications. The heavy gas oil stream for the visbreaker can be used as feedstock for a thermal distillate cracking unit or for a catalytic cracker for the production of lower-boiling distillate products.

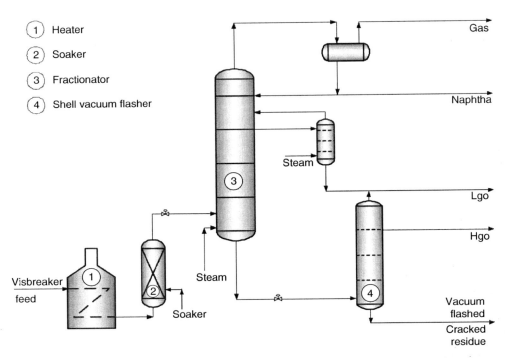

1. Heater
2. Soaker
3. Fractionator
4. Shell vacuum flasher

FIGURE 17.2 Shell soaker visbreaking technology.

Overall, the main limitation of the visbreaking process, and for that matter all thermal processes, is that the products can be unstable. Thermal cracking at low pressure gives olefins, particularly in the naphtha fraction. These olefins give a very unstable product, which tends to undergo secondary reactions to form gum and intractable residua.

Conversion of residua in visbreaking is complex, but generally follows first-order reaction kinetics. The reduction in viscosity of distillation residua tends to reach a limiting value with conversion, although the total product viscosity can continue to decrease. The minimum viscosity of the unconverted residue can lie outside the range of allowable conversion if sediment begins to form. When shipment of the visbreaker product by pipeline is the process objective, addition of a diluent such as gas condensate can be used to achieve a further reduction in viscosity.

Other variations of visbreaking technology include the Tervahl T and Tervahl H processes (*q.v.*). The Tervahl T alternative includes only the thermal section to produce a synthetic

TABLE 17.3

Typical Product Yields for Middle East Vacuum Residuum for the Shell Soaker Visbreaking Process

Product	Yield (% by Weight)	
Gas	2	
Naphtha	4	Endpoint 165°C
Light gas oil	12	Endpoint 350°C
Heavy gas oil	18	Endpoint 520°C
Vacuum flashed residuum	64	

crude oil with better transportability, by having reduced viscosity and greater stability. The Tervahl H alternative adds hydrogen that also increases the extent of the desulfurization and decreases the carbon residua.

17.3.2 Coking Processes

Coking is a thermal process for the continuous conversion of residua into lower-boiling products. The feedstock can be reduced crude, straight-run residua, or cracked residua, and the products are gases, naphtha, fuel oil, gas oil, and coke. The gas oil may be the major product of a coking operation, and serves primarily as a feedstock for catalytic cracking units. The coke obtained is usually used as fuel, but processing for specialty uses, such as electrode manufacture, production of chemicals, and metallurgical coke, is also possible and increases the value of the coke. For these uses, the coke may require treatment to remove sulfur and metal impurities. Further, the increasing attention paid to reducing atmospheric pollution has also served to direct some attention to coking, since the process not only concentrates such pollutants as feedstock sulfur in the coke, but usually yields products that can be conveniently subjected to desulfurization processes.

Coking processes generally utilize longer reaction times than thermal cracking processes. To accomplish this, drums or chambers (reaction vessels) are employed, but it is necessary to use two or more such vessels, so that coke removal can be accomplished in those vessels not on-stream, without interrupting the semi-continuous nature of the process.

Coking processes have the virtue of eliminating the residue fraction of the feed, at the cost of forming a solid carbonaceous product. The yield of coke in a given coking process tends to be proportional to the carbon residue of the feed (measured as the Conradson carbon residue; see Chapter 9).

The formation of large quantities of coke is a severe drawback, unless the coke can be put to use. Calcined petroleum coke can be used for making anodes for aluminum manufacture, and a variety of carbon or graphite products such as brushes for electrical equipment. These applications, however, require a coke that is low in mineral matter and sulfur.

If the feedstock produces a high sulfur-high ash-high vanadium coke, one option for use of the coke is combustion to produce process steam (and large quantities of sulfur dioxide, unless the coke is first gasified or the combustion gases are scrubbed). Another option is stockpiling.

For some feedstocks, particularly from heavy oil, the combination of poor coke properties for anode use, limits on sulfur dioxide emissions, and loss of liquid product volume has tended to relegate coking processes to a strictly secondary role in any new upgrading facility.

In spite of the various limitation outlined earlier, visbreaking has much potential and, in fact, remains an important, relatively inexpensive bottom-of-the-barrel upgrading process in many areas of the world. Most of the existing visbreakers are the soaker type, which utilize a soaker drum in conjunction with a fired heater to achieve conversion, and which reduce, the temperature required to achieve conversion while producing a stable residue product, thereby increasing the heater run length and reducing the frequency of unit shut down for heater decoking.

However, a recurring issue with the soaker visbreaker is the need to periodically decoke the soaker drum and the inability of the soaker process to easily adjust to changes in feedstock quality because of the need to fine-tune two process variables, temperature and residence time. Recent combination of the visbreaking technology and the addition of new coil visbreaker design features has provided the coil process with a competitive advantage over the traditional soaker visbreaker process. Limitations in heater run length are no longer a problem for the coil visbreaker. Advances in visbreaker coil heater design now allow for the isolation of one or more heater passes for decoking, eliminating the need to shut the entire visbreaker down for furnace decoking.

The higher heater outlet temperature specified for a coil visbreaker is now viewed as an important advantage of coil visbreaking. The higher heater outlet temperature is used to recover significantly higher quantities of heavy visbroken gas oil, through the use of the Wood's technology. This capability cannot be achieved with a soaker visbreaker, without the addition of a vacuum flasher.

17.3.2.1 Delayed Coking

Delayed coking has been selected by many refiners as their preferred choice for bottom–of–the–barrel upgrading, because of the process' inherent flexibility to handle even the heaviest of residues. The process provides, essentially, complete rejection of metals and precursors to term *coke* (precursors to Conradson carbon), while providing partial or complete conversion to naphtha and diesel. In the past, many cokers were designed to provide complete conversion of atmospheric residue to diesel and lighter, and today several cokers still operate in this mode. However, most recent cokers have been designed to minimize coke and produce a heavy coker gas oil (HCGO) that is catalytically upgraded. The economics of delayed coking is driven by the differential between transportation fuels and high-sulfur residual fuel oil. The yield slate for a delayed coker can be varied to meet a refiner's objectives through the selection of operating parameters. Coke yield and the conversion of heavy coker gas oil are reduced, as the operating pressure and recycle are reduced, and to a lesser extent, as temperature is increased.

Delayed coking is a semi-continuous (semi-batch) process in which the heated charge is transferred to large soaking (or coking) drums, which provide the long residence time needed to allow the cracking reactions to proceed to completion (McKinney, 1992; Feintuch and Negin, 1997; Hydrocarbon Processing, 2004).

The delayed coking process (Figure 17.3) is widely used for treating residua (Table 17.4). The process uses long reaction times in the liquid phase to convert the residue fraction of the

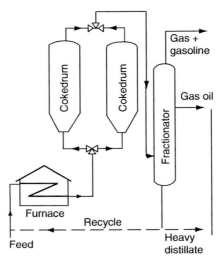

FIGURE 17.3 A delayed coker. (From OSHA Technical Manual. Petroleum Refining Processes.)

TABLE 17.4
Product Yields and Product Properties for Delayed Coking of Athabasca Tar Sand Bitumen and Similar Low API Gravity Feedstocks

	Kuwait Residuum	West Texas Residuum	Tia Juana Residuum	Alaska NS Residuum	Arabian Light Residuum	Athabasca Bitumen
Feedstock						
API gravity	6.7	8.9	8.5	7.4	6.9	7.3
Carbon residue[a]	19.8	17.8	22.0	18.1		17.9
Sulfur, wt.%	5.2	3.0	2.9	2.0	4.0	5.3
Product yields, vol.%						
Naphtha	26.7	28.9	25.6	12.5	19.1	20.3
(95°F to 425°F, 35°C to 220°C)						
Light gas oil	28.0	16.5	26.4	—[b]	—[b]	—[b]
(425°F to 645°F, 220°C to 340°C)						
Heavy gas oil	18.4	26.4	13.8	51.2	48.4	58.8
(645°F to 1000°F, 340°C to 540°C)						
Coke	30.2	28.4	33.0	27.2	32.8	21.0
Sulfur, wt.%	7.5	4.5		2.6	5.6	8.0

[a]Conradson.
[b]A blank product line indicates that the yield of the lower boiling product has been included in the yield of the higher boiling product.
Source: Speight, J.G. and Ozum, B. 2002. *Petroleum Refining Processes*. Marcel Dekker Inc., New York. With permission.

feed to gases, distillates, and coke. The condensation reactions that give rise to the highly aromatic coke product also tend to retain sulfur, nitrogen, and metals, so that the coke is enriched in these elements, relative to the feed.

In the process, the feedstock is introduced into the product fractionator, where it is heated, and the lighter fractions are removed as side-stream products. The fractionator bottoms, including a recycle stream of heavy products, are then heated in a furnace, whose outlet temperature varies from 480°C to 515°C (895°F to 960°F). The heated feedstock enters one of a pair of coking drums, where the cracking reactions continue. The cracked products leave as an overhead stream, and coke deposits form on the inner surface of the drum. To give continuous operation, two drums are used; while one is on steam, the other is cleaned. The temperature in the coke drum ranges from 415°C to 450°C (780°F to 840°F), at pressures from 15 to 90 psi.

Delayed coking units fractionate the coke drum overhead products into fuel gas (low molecular weight gases up to and including ethane), propane and propylene ($CH_3CH_2CH_3$ and $CH_3CH=CH_2$), butane and butene ($CH_3CH_2CH_2CH_3$ and $CH_3CH_2CH=CH_2$), naphtha, light gas oil, and heavy gas oil. Yields and product quality vary widely due to the broad range of feedstock types available for delayed coking units, and there is a decrease in overhead yield with the asphaltene content of the feedstock (Figure 17.4) (Schabron and Speight, 1997).

Overhead products go to the fractionator, where naphtha and heating oil fractions are recovered. The heavy recycle material is combined with preheated fresh feed and returned to the reactor.

A pair of coke drums is used so that while one drum is on stream, the other is cleaned, allowing continuous processing, and the drum operation cycle is typically 48 h. A coke drum

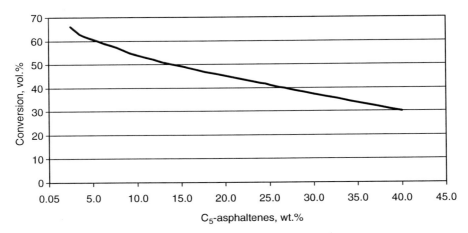

FIGURE 17.4 Variation of conversion (yield of overhead products) with asphaltene content. (From Speight, J.G. and Ozum, B. 2002. *Petroleum Refining Processes.* Marcel Dekker Inc., New York. With permission.)

is usually on-stream for about 24 h before becoming filled with porous coke, and the following procedure is used to remove the coke:

1. The coke deposit is cooled with water.
2. One of the heads of the coking drum is removed to permit the drilling of a hole through the center of the deposit.
3. A hydraulic cutting device, which uses multiple high-pressure water jets, is inserted into the hole and the wet coke is removed from the drum.

Normally, 24 h is required to complete the cleaning operation and to prepare the coke drum for subsequent use on-stream.

Delayed coking has an increasingly important role to play in the integration of modem petroleum refineries, because of its ability to convert heavy vacuum residues to lighter distillates and petroleum coke. The flexibility of operation inherent in delayed coking permits refiners to process a wide variety of crude oils, including those containing heavy, high-sulfur residua (Elliot, 1995).

The disadvantages of delayed coking are that it is a thermal cracking process and it is a more expensive process than solvent deasphalting (Chapter 19), although still less expensive than other conversion processes on heavier crude oil. One common misconception of delayed coking is that the product, coke, is a disadvantage. Although coke is a low-valued byproduct, compared to transportation fuels, there is a significant worldwide trade and demand even for high-sulfur petroleum coke from delayed cokers, as coke is a very economical fuel. However, for coke-burning power plants, this has required the installation of flue gas scrubbers and, in some of these plants, a circulating bed of limestone captures the sulfur.

Low-pressure coking is a process designed for a once-through, low-pressure operation. The process is similar to delayed coking, except that recycling is not usually practised and the coke chamber operating conditions are 435°C (815°F), 25 psi. Excessive coking is inhibited by the addition of water to the feedstock, in order to quench and restrict the reactions of the reactive intermediates.

High-temperature coking is a semi-continuous thermal conversion process designed for high-melting asphaltic residua that yields coke and gas oil as the primary products. The coke may be treated to remove sulfur to produce a low-sulfur coke ($\leq$5%), even though the feedstock contained as much as 5% w/w sulfur.

Thus, the feedstock is transported to the **pitch** accumulator, then to the heater (370°C, 700°F, 30 psi), and finally to the coke oven, where temperatures may be as high as 980°C to 1095°C (1800°F to 2000°F). Volatile materials are fractionated, and after the cycle is complete, coke is collected for sulfur removal and quenching before storage.

17.3.2.2 Fluid Coking

Fluid coking is a continuous process that uses the fluidized solids technique to convert residua, including vacuum residua and cracked residua, to more valuable products (Roundtree, 1997). The process is useful for processing of residua (Table 17.5). The yield of distillates from coking can be improved by reducing the residence time of the cracked vapors. In order to simplify handling of the coke product, and enhance product yields, fluidized-bed coking, or fluid coking, was developed in the mid-1950s.

The residuum feedstock is decomposed by spraying it into a fluidized bed of hot, fine coke particles, which permits the coking reactions to be conducted at higher temperatures and shorter contact times, than can be employed in delayed coking. Moreover, these conditions result in decreased yields of coke; greater quantities of more valuable liquid product are recovered in the fluid coking process.

Fluid coking uses two vessels, a reactor and a burner; coke particles are circulated between these to transfer heat (generated by burning a portion of the coke) to the reactor (Figure 17.5) (Blaser, 1992). The reactor holds a bed of fluidized coke particles, and steam is introduced at the bottom of the reactor to fluidize the bed. The feed coming from the bottom of a vacuum tower at, for example, 260°C to 370°C (500°F to 700°F), is injected directly into the reactor. The temperature in the coking vessel ranges from 480°C to 565°C (900°F to 1050°F), and the pressure is substantially atmospheric, so the incoming feed is partly vaporized and partly deposited on the fluidized coke particles. The material on the particle surface then cracks and vaporizes, leaving a residue that dries to form coke. The vapor products pass through cyclones that remove most of the entrained coke.

The vapor is discharged into the bottom of a scrubber, where the products are cooled to condense to a heavy tar containing the remaining coke dust, which is recycled to the coking reactor. The upper part of the scrubber tower is a fractionating zone, from which coker gas oil is withdrawn and then fed to a catalytic cracking unit; naphtha and gas are taken overhead to condensers.

In the reactor, the coke particles flow down through the vessel into a stripping zone at the bottom. Steam displaces the product vapors between the particles, and the coke then flows into a riser that leads to the burner. Steam is added to the riser to reduce the solids loading and to induce upward flow. The average bed temperature in the burner is 590°C to 650°C (1095°F to 1200°F), and air is added, as needed, to maintain the temperature by burning part of the product coke. The pressure in the burner may range from 5 to 25 psi. Flue gases from the burner bed pass through cyclones and discharge to the stack. Hot coke from the bed is returned to the reactor through a second riser assembly.

Coke is one of the products of the process, and it must be withdrawn from the system to keep the solids inventory from increasing. The net coke produced is removed from the burner bed through a quench elutriator drum, where water is added for cooling, and cooled coke is withdrawn and sent to storage. During the course of the coking reaction, the particles tend to grow in size. The size of the coke particles remaining in the system is controlled by a grinding system within the reactor.

The yields of products are determined by the feed properties, the temperature of the fluid bed, and the residence time in the bed. The use of a fluidized bed reduces the residence time of the vapor-phase products in comparison to delayed coking, which in turn reduces cracking

25

TABLE 17.5
Product Yields and Product Properties for Fluid Coking Low API Gravity Feedstocks

	LA Basin Vacuum Residuum	Kuwait Vacuum Residuum	Hawkins Vacuum Residuum	Tia Juana Vacuum Residuum	Arabian Heavy Vacuum Residuum	Iranian Heavy Vacuum Residuum	Bachaquero Vacuum Residuum	Zaca Vacuum Residuum
Feedstock								
API gravity	6.7	5.6	4.2	8.5	4.4	5.1	2.6	4.7
Carbon residue[a]	17.0	21.8	24.5	22.0	24.4	21.4	21.4	19.0
Sulfur, wt.%	2.1	5.5	4.3	2.9	5.3	3.4	3.7	7.8
Product yields, vol.%								
Naphtha (95°F to 425°F, 35°C to 220°C)	17.0	21.0	19.5	20.7	15.0	15.4	14.7	20.5
Gas oil (425°F to 1000°F, 220°C to 540°C)	62.0	48.0	53.0	48.3	47.7	55.1	48.3	61.0
Coke, wt.%	21.0	28.0	27.5	20.0	30.4	26.4	32.9	17.5

[a]Conradson.

Source: Speight, J.G. and Ozum, B. 2002. *Petroleum Refining Processes.* Marcel Dekker Inc., New York. With permission.

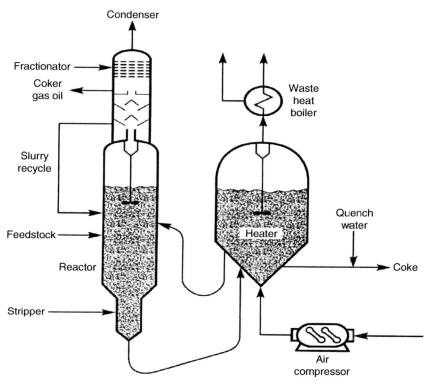

FIGURE 17.5 A fluid coker.

reactions. The yield of coke is thereby reduced, and the yield of gas oil and olefins increased. An increase of 5°C (9°F) in the operating temperature of the fluid-bed reactor typically increases gas yield by 1% by weight and naphtha by about 1% by weight.

The lower limit on operating temperature is set by the behavior of the fluidized coke particles. If the conversion to coke and light ends is too slow, the coke particles agglomerate in the reactor, a condition known as **bogging**.

The disadvantage of burning the coke to generate process heat is that sulfur from the coke is liberated as sulfur dioxide. The gas stream from the coke burner also contains carbon monoxide (CO), carbon dioxide (CO_2), and nitrogen (N_2). An alternate approach is to use a coke gasifier to convert the carbonaceous solids to a mixture of carbon monoxide (CO), carbon dioxide (CO_2), and hydrogen (H_2).

Currently, delayed coking and fluid coking are the processes of choice for conversion of Athabasca bitumen to liquid products. Both processes are termed the primary conversion processes for the tar sand plants in Ft. McMurray, Canada. The unstable liquid product streams are hydrotreated before recombining to the synthetic crude oil (Spragins, 1978; Speight, 1990).

17.3.2.3 Flexicoking

Flexicoking is a direct descendent of fluid coking and uses the same configuration as the fluid coker, but includes a gasification section in which excess coke can be gasified to produce refinery fuel gas (Figure 17.6) (Roundtree, 1997). The flexicoking process was designed during the late 1960s and the 1970s, as a means by which excess coke could be reduced in view of the gradual incursion of the heavier feedstocks in refinery operations. Such feedstocks are notorious for producing high yields of coke (>15% by weight) in thermal and catalytic operations.

In the process, excess coke is converted to a low-heating value gas in a fluid bed gasifier with steam and air. The air is supplied to the gasifier to maintain temperatures of 830°C to 1000°C (1525°F to 1830°F), but is insufficient to burn all of the coke. Under these reducing conditions, the sulfur in the coke is converted to hydrogen sulfide, which can be scrubbed from the gas prior to combustion. A typical gas product, after removal of hydrogen sulfide, contains carbon monoxide (CO, 18%), carbon dioxide (CO_2, 10%), hydrogen (H_2, 15%), nitrogen (N_2, 51%), water (H_2O, 5%), and methane (CH_4, 1%). The heater is located between the reactor and the gasifier, and it serves to transfer heat between the two vessels.

Yields of liquid products from flexicoking are the same as from fluid coking, because the coking reactor is unaltered. The main drawback of gasification is the requirement for a large additional reactor, especially if high conversion of the coke is required. Units are designed to gasify 60% to 97% of the coke from the reactor. Even with the gasifier, the product coke will contain more sulfur than the feed, which limits the attractiveness of even the most advanced of coking processes.

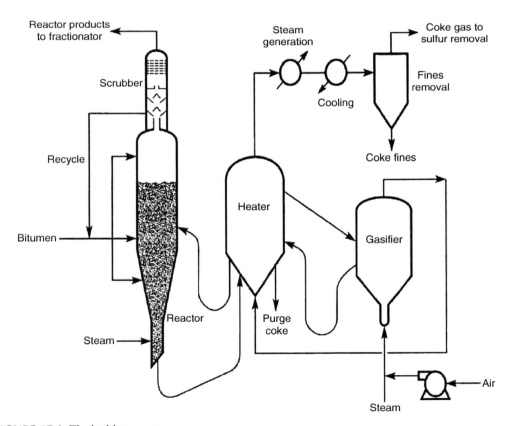

FIGURE 17.6 Flexicoking process.

17.3.3   PROCESS OPTIONS FOR HEAVY FEEDSTOCKS

17.3.3.1   Aquaconversion

The Aquaconversion process (Figure 17.7) is a hydrovisbreaking technology that uses cata-lyst-activated transfer of hydrogen from water added to the feedstock. Reactions that lead to coke formation are suppressed, and there is no separation of asphaltene-type material (Marzin et al., 1998). The important aspect of the Aquaconversion technology is that it does not produce any solid byproduct such as coke, nor does it require any hydrogen source or high-pressure equipment. In addition, the Aquaconversion process can be implanted in the production area, and thus, the need for external diluent and its transport over large distances is eliminated. Light distillates from the raw crude can be used as diluent for both the production and the desalting processes.

17.3.3.2   Asphalt Coking Technology (ASCOT) Process

The ASCOT process is a residual oil upgrading process that integrates the delayed coking process and the deep solvent deasphalting process (low energy deasphalting, LEDA) (Chapter 19) (Bonilla, 1985; Bonilla and Elliott, 1987;). Removing the deasphalted oil fraction prior to delayed coking has two benefits: (1) in the coking process, this fraction is thermally cracked to extinction, degrading this material as an fluid catalytic cracker (FCC) feedstock, and (2) in thermally cracking this material to extinction, a significant portion will convert to coke.

In the process, the vacuum residuum is brought to the desired extraction temperature and then sent to the extractor where solvent (straight-run naphtha, coker naphtha) flows upward, extracting soluble material from the down-flowing feedstock. The solvent-deasphalted phase leaves the top of the extractor and flows to the solvent recovery system, where the solvent is separated from the deasphalted oil and recycled to the extractor. The deasphalted oil is sent to the delayed coker, where it is combined with the heavy coker gas oil from the coker fractionator and sent to the heavy coker gas oil stripper, where low-boiling hydrocarbons are stripped off and returned to the fractionator. The stripped deasphalted oil and heavy coker gas oil mixture is removed from the bottom of the stripper and used to provide heat to the naphtha stabilizer–reboiler, before sending to battery limits as a cracking stock. The raffinate

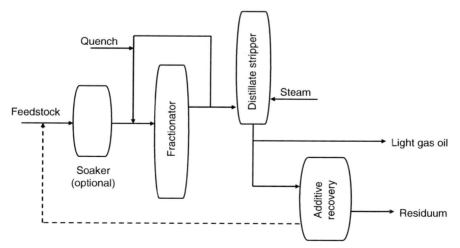

FIGURE 17.7 The Aquaconversion process. (From Speight, J.G. and Ozum, B. 2002. *Petroleum Refining Processes.* Marcel Dekker Inc., New York. With permission.)

phase containing the asphalt and some solvent, flows at a controlled rate from the bottom of the extractor, and is charged directly to the coking section.

The solvent contained in the asphalt and deasphalted oil is condensed in the fractionator overhead condensers, where it can be recovered and used as lean oil for a propane or butane recovery in the absorber, eliminating the need to recirculate lean oil from the naphtha stabilizer. The solvent introduced in the coker heater and coke drums results in a significant reduction in the partial pressure of asphalt feed, compared with a regular delayed coking unit. The low asphalt partial pressure results in low coke and high liquid yields in the coking reaction.

With the ASCOT process, there is a significant reduction in byproduct fuel, as compared to either solvent deasphalting or delayed coking (Table 17.6), and the process can be tailored to process a specific quantity, or process to a specific quality of cracking stock.

17.3.3.3 Comprehensive Heavy Ends Reforming Refinery (Cherry-P) Process

The Cherry-P process is a process for the conversion of heavy crude oil or residuum into distillate and a cracked residuum (Ueda, 1976, 1978). In this process, the principal aim is to upgrade heavy petroleum residues at conditions between those of conventional visbreaking and delayed coking. Although coal is added to the feedstock, it is mainly intended to act as a scavenger to prevent coke build up on the reactor wall.

TABLE 17.6
Comparison of Product Yields from the ASCOT Process with Product Yields from Delayed Coking and the LEDA Process

	ASCOT	Delayed Coking	LEDA
Feedstock			
API	2.8	2.8	2.8
Sulfur, wt.%	4.2	4.2	4.2
Nitrogen, wt.%	1.0	1.0	1.0
Carbon residue, wt.%	22.3	22.3	22.3
Products			
Naphtha, vol.%	7.7	19.4	
API	54.7		
Sulfur, wt.%	1.1		
Nitrogen, wt.%	0.1		
Gas oil, vol.%	69.9	51.8	
API	13.4		
Sulfur, wt.%	3.4		
Nitrogen, wt.%	0.5		
Coke, wt.%	25.0	32.5	
Sulfur, wt.%	5.8	5.7	
Nitrogen, wt.%	2.7	2.6	
Nickel, ppm	774.0	609.0	
Vanadium, ppm	2656.0	2083.0	
Deasphalted oil, vol.%			50.0
Asphalt, vol.%			50.0
Sulfur, wt.%			5.0
Nitrogen, wt.%			1.4
Nickel, ppm			365.0
Vanadium, ppm			1250.0

Source: Speight, J.G. and Ozum, B. 2002. *Petroleum Refining Processes*. Marcel Dekker Inc., New York. With permission.

In the process, the feedstock is mixed with coal powder in a slurry-mixing vessel, heated in the furnace, and fed to the reactor, where the feedstock undergoes thermal cracking reactions for several hours at a temperature higher than 400°C to 450°C (750°F to 840°F) and under pressure (70 to 290 psi). The residence time is in the range 1 to 5 h. No catalyst or hydrogen is added. Gas and distillate from the reactor are sent to a fractionator, and the cracked residuum residue is extracted out of the system after distilling low-boiling fractions by the flash drum and vacuum flasher to adjust its softening point. Distillable product yields of 44% by weight on total feed are reported (Table 17.7). Since this yield is obtained when using anthracite, the proportion that is derived from the coal is likely to be very low.

The distillates produced by this process are generally lower in the content of olefin hydrocarbons than the other thermal cracking processes, comparatively easy to desulfurize in hydrotreating units, and compatible with straight-run distillates.

17.3.3.4 Decarbonizing

The **decarbonizing** thermal process is designed to minimize coke and gasoline yields but, at the same time, to produce maximum yields of gas oil. Decarbonizing, in this sense of the term, should not be confused with **propane decarbonizing**, which is essentially a solvent deasphalting process (Chapter 19).

Thermal decarbonizing is essentially the same as the delayed coking process, but lower temperatures and pressures are employed. For example, pressures range from 10 to 25 psi, heater outlet temperatures may be 485°C (905°F), and coke drum temperatures may be of the order of 415°C (780°F).

TABLE 17.7
Feedstock and Product Data for the Cherry-P Process

Feedstock: Iranian Heavy Vacuum Residuum	
API	6.4
Sulfur, wt.%	3.1
Carbon residue, wt.%.	21.3
Products	
Naphtha, vol.%	9.3
API	58.9
Sulfur, wt.%	0.6
Kerosene, vol.%	9.2
API	40.4
Sulfur, wt.%	0.8
Gas oil, vol.%	17.9
API	32.5
Sulfur, wt.%	1.4
Vacuum gas oil, vol.%	7.7
API	20.5
Sulfur, wt.%	2.5
Residuum, wt.%	49.8
Sulfur, wt.%	3.9
Carbon residue, wt.%	42.2

Source: Speight, J.G. and Ozum, B. 2002. *Petroleum Refining Processes*. Marcel Dekker Inc., New York. With permission.

17.3.3.5 ET-II Process

The ET-II process is a thermal cracking process for the production of distillates and cracked residuum for use as metallurgical coke, and is designed to accommodate feedstocks such as heavy oils, atmospheric residua, and vacuum residua (Kuwahara, 1987). The distillate (referred to in the process as **cracked oil**) is suitable as a feedstock to hydrocracker and fluid catalytic cracking. The basic technology of the ET-II process is derived from that of the original Eureka process (*q.v.*).

In the process, the feedstock is heated up to 350°C (660°F) by passage through the preheater and fed into the bottom of the fractionator, where it is mixed with recycle oil and the high-boiling fraction of the cracked oil. The ratio of recycle oil to feedstock is within the range 0.1% to 0.3% by weight. The feedstock mixed with recycle oil is then pumped out and fed into the cracking heater, where the temperature is raised to approximately 490°C to 495°C (915°F to 925°F), and the outflow is fed to the stirred-tank reactor, where it is subjected to further thermal cracking. Both cracking and condensation reactions take place in the reactor.

The heat required for the cracking reaction is brought in by the effluent itself from the cracking heater, as well as by the superheated steam, which is heated in the convection section of the cracking heater and blown into the reactor bottom. The superheated steam reduces the partial pressure of the hydrocarbons in the reactor and accelerates the stripping of volatile components from the cracked residuum. This residual product is discharged through a transfer pump and transferred to a cooling drum, where the thermal cracking reaction is terminated by quenching with a water spray, after which it is sent to the pitch water slurry preparation unit.

The cracked oil and gas products, together with steam from the top of the reactor, are introduced into the fractionator, where the oil is separated into two fractions, cracked light oil and vacuum gas oil, and pitch (Table 17.8).

17.3.3.6 Eureka Process

The Eureka process is a thermal cracking process to produce a cracked oil and aromatic residuum from heavy residual materials (Table 17.9) (Aiba et al., 1981). In this process (Figure 17.8), the feedstock, usually a vacuum residuum, is fed to the preheater and then enters the bottom of the fractionator, where it is mixed with the recycle oil. The mixture is then fed to the reactor system that consists of a pair of reactors operating alternately. In the reactor, thermal cracking reaction occurs in the presence of superheated steam, which is injected to strip the cracked products out of the reactor and supply a part of the heat required for the cracking reaction. At the end of the reaction, the bottom product is quenched. The oil and gas products (and steam) pass from the top of the reactor to the lower section of the fractionator, where a small amount of entrained material is removed by a wash operation. The upper section is an ordinary fractionator, where the heavier fraction of cracked oil is drawn as a side stream. The process bottoms (pitch) can be used as boiler fuel, as partial oxidation feedstock for producing hydrogen and carbon monoxide, and as binder pitch for manufacturing metallurgical coke.

The process reactions proceed at lower cracked oil partial pressure by injecting steam into the reactor, keeping petroleum pitch in a homogeneous liquid state and, unlike a conventional delayed coker, a higher cracked oil yield can be obtained. A wide range of residua can be used as feedstock, such as atmospheric and vacuum residue of petroleum crude oils, various cracked residues, asphalt products from solvent deasphalting, and native asphalt. After hydrotreating, the cracked oil is used as feedstock for a fluid catalytic cracker or hydrocracker.

TABLE 17.8
Feedstock and Product Data for the ET-II Process

Feedstock: Mix of Bachaquero, Khafji, and Maya Vacuum Residua

API	6.7
Sulfur, wt.%	4.1
C$_7$-asphaltenes, wt.%	10.0
Carbon residue, wt.%.	22.4
Nickel, ppm	67.0
Vanadium, ppm	343.0
Products	
Light oil, wt.% (<315°C, <600°F)	28.6
API	41.3
Sulfur, wt.%	1.7
Nitrogen, wt.%	<0.1
Vacuum gas oil, vol.% (315°C to 540°C, 600°F to 1000°F)	32.3
API	18.3
Sulfur, wt.%	3.2
Nitrogen, wt.%	0.2
Carbon residue, wt.%	1.0
C$_7$-asphaltenes, wt.%	<0.1
Pitch, wt.%	32.8
Sulfur, wt.%	6.3
Nitrogen, wt.%	1.6
Carbon residue, wt.%	60 (estimated)
Nickel, ppm	204.0
Vanadium, ppm	1045.0

Source: Speight, J.G. and Ozum, B. 2002. *Petroleum Refining Processes*. Marcel Dekker Inc., New York.

The original Eureka process uses two batch reactors, while the newer ET II (*q.v.*) and the HSC process (*q.v.*), both employ continuous reactors.

17.3.3.7 Fluid Thermal Cracking (FTC) Process

The FTC process is a heavy oil and residuum upgrading process, in which the feedstock is thermally cracked to produce distillate and coke (Table 17.10). The coke is gasified to fuel gas (Miyauchi et al., 1981, 1987; Miyauchi and Ikeda, 1988).

The feedstock, mixed with recycle stock from the fractionator, is injected into the cracker and is immediately absorbed into the pores of the particles by capillary force and subjected to thermal cracking. In consequence, the surface of the noncatalytic particles is kept dry, and good fluidity is maintained allowing a good yield of, and selectivity for, middle distillate products. Hydrogen-containing gas from the fractionator is used for the fluidization in the cracker. Excessive coke, caused by the metals accumulated on the particle, is suppressed under the presence of hydrogen. The particles with deposited coke from the cracker are sent to the gasifier, where the coke is gasified and converted into carbon monoxide (CO), hydrogen (H$_2$), carbon dioxide (CO$_2$), and hydrogen sulfide (H$_2$S) with steam and air. Regenerated hot particles are returned to the cracker.

Excess coke formation due to the presence of nickel and vanadium in the feedstock is effectively avoided by the low pressure.

TABLE 17.9
Feedstock and Product Data for the Eureka Process

Feedstock: Middle East Vacuum Residuum Mix[a]	
API	7.6
Sulfur, wt.%	3.9
C_7-asphaltenes, wt.%	5.7
Carbon residue, wt.%	20.0
Nickel, ppm	136.0
Vanadium, ppm	202.0
Products	
Light oil, wt.% (C5–240°C, C5–465°F)	14.9
API	53.0
Sulfur, wt.%	1.1
Nitrogen, wt.%	<0.1
Gas oil, wt.% (240°C to 540°C, 465°F to 1000°F)	50.7
API	21.3
Sulfur, wt.%	2.7
Nitrogen, wt.%	0.3
Pitch, wt.% (>540°C, >1000°F)	29.6
Sulfur, wt.%	5.7
Nitrogen, wt.%	1.2
Nickel, ppm	487.0
Vanadium, ppm	688.0

[a] >500°C, >930°F, not defined by name.
Source: Speight, J.G. and Ozum, B. 2002. *Petroleum Refining Processes*. Marcel Dekker Inc., New York. With permission.

17.3.3.8 High Conversion Soaker Cracking (HSC) Process

The HSC process is a cracking process designed for moderate conversion, higher than visbreaking but lower than coking (Table 17.11) (Watari et al., 1987; Washimi, 1989). The process (Figure 17.9) features less gas-make and a higher yield of distillate, with compared other thermal cracking processes. The process can be used to convert a wide range of

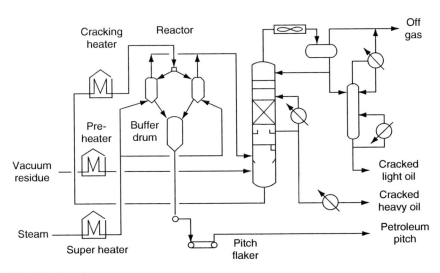

FIGURE 17.8 The Eureka process.

TABLE 17.10
Feedstock and Product Data for FTC Process

Feedstock: Bachaquero, Khafji, and Maya Vacuum Residua Mix

API	6.4
Sulfur, wt.%	4.5
Carbon residue, wt.%.	21.9
Nickel + vanadium, ppm	500.0
Products	
Naphtha, wt.% (C5–150°C, C5–300°F)	14.9
API	57.1
Sulfur, wt.%	0.4
Middle distillate, wt.% (150°C to 310°C, 300°F to 590°F)	34.4
API	39.5
Sulfur, wt.%	1.4
Heavy distillate, wt.% (310°C to 525°C, 590°F to 975°F)	21.8
API	18.7
Sulfur, wt.%	3.1
Nitrogen, wt.%	0.2
Coke	19.0 (estimate)

Source: Speight, J.G. and Ozum, B. 2002. *Petroleum Refining Processes*. Marcel Dekker Inc., New York. With permission.

TABLE 17.11
Feedstock and Product Data for the HSC Process

	Iranian Heavy Vacuum Residuum	Maya Vacuum Residuum
Feedstock		
API	5.7	2.2
Sulfur, wt.%	4.8	5.1
Nitrogen, wt.%	0.6	0.8
C_7-asphaltenes, wt.%	11.3	19.0
Carbon residue, wt.%.	22.6	28.7
Nickel, ppm	69.0	121.0
Vanadium, ppm	205.0	649.0
Products		
Naphtha, wt.% (C5–200°C, C5–390°F)	6.3	3.8
API	54.4	51.5
Sulfur, wt.%	1.1	1.1
Light gas oil, wt.% (200°C to 350°C, 390°F to 660°F)	15.0	12.3
API	30.2	29.5
Sulfur, wt.%	2.6	3.1
Nitrogen, wt.%	0.1	<0.1
Heavy gas oil, wt.% (350°C to 520°C, 660°F to 970°F)	32.2	18.6
API	16.4	16.5
Sulfur, wt.%	3.5	3.7
Nitrogen, wt.%	0.3	0.3
Vacuum residue, wt.% (>520°C, > 970°F)	43.2	62.0
Sulfur, wt.%	5.8	5.6
Nitrogen, wt.%	1.0	1.2
Carbon residue, wt.%	49.2	49.2
Nickel, ppm	148.0	148.0
Vanadium, ppm	453.0	453.0

Source: Speight, J.G. and Ozum, B. 2002. *Petroleum Refining Processes*. Marcel Dekker Inc., New York. With permission.

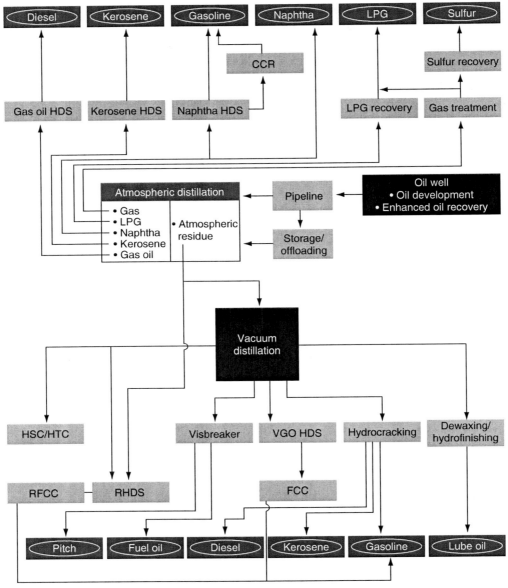

FIGURE 17.9 The relative placement of the HSC process and the related HTC process in a heavy ends section of the refinery.

feedstocks with high sulfur and metals content, including heavy oils, oil sand bitumen, residua, and visbroken residua. As a note of interest, the HSC process employs continuous reactors, whereas the original Eureka process (*q.v.*) uses two batch reactors.

The preheated feedstock enters the bottom of the fractionator, where it is mixed with the recycle oil. The mixture is pumped up to the charge heater and fed to the soaking drum (ca. atmospheric pressure, steam injection at the top and the bottom), where sufficient residence time is provided to complete the thermal cracking. In the soaking drum, the feedstock and some product flow downward, passing through a number of perforated plates, while steam with cracked gas and distillate vapors flow through the perforated plates counter-currently.

The volatile products from the soaking drum enter the fractionator, where the distillates are fractionated into desired product oil streams, including a heavy gas oil fraction. The cracked gas product is compressed and used as refinery fuel gas after sweetening. The cracked oil product after hydrotreating is used as fluid catalytic cracking or hydrocracker feedstock. The residuum is suitable for use as boiler fuel, road asphalt, binder for the coking industry, and as a feedstock for partial oxidation.

17.3.3.9 Mixed-Phase Cracking

Mixed-phase cracking (also called liquid-phase cracking) is a continuous thermal decomposition process for the conversion of heavy feedstocks to products boiling in the gasoline range. The process generally employs rapid heating of the feedstock (kerosene, gas oil, reduced crude, or even whole crude), after which it is passed to a reaction chamber and then to a separator where the vapors are cooled. Overhead products from the flash chamber are fractionated to gasoline components and recycle stock, and flash chamber bottoms are withdrawn as a heavy fuel oil. Coke formation, which may be considerable at the process temperatures (400°C to 480°C, 750°F to 900°F), is minimized by use of pressures in excess of 350 psi.

17.3.3.10 Naphtha Cracking

The thermal cracking of naphtha involves the upgrading of low-octane fractions of catalytic naphtha to higher-quality material. The process is designed, in fact, to upgrade the heavier portions of naphtha, which contain virgin feedstock, and to remove naphthenes, as well as paraffins. Some heavy aromatics are produced by condensation reactions, and substantial quantities of olefins occur in the product streams.

17.3.3.11 Selective Cracking

Selective cracking is a thermal conversion process that utilizes different conditions, depending on feedstock composition. For example, a heavy oil may be cracked at 494°C to 515°C (920°F to 960°F) and 300 to 500 psi a lighter gas oil may be cracked at 510°C to 530°C (950°F to 990°F) and 500 to 700 psi (Moschopedis et al., 1998; Speight and Ozum, 2002).

Each feedstock has its own particular characteristics that dictate the optimum conditions of temperature and pressure for maximum yields of the products. These factors are utilized in selective combination of cracking units, in which the more refractory feedstocks are cracked for longer periods of time or at higher temperatures than the less-stable feedstocks, which are cracked at lower temperatures.

The process eliminates the accumulation of stable low-boiling material in the recycle stock and also minimizes coke formation from high-temperature cracking of the higher-boiling material. The end result is the production of fairly high yields of gasoline, middle distillates, and olefin gases.

17.3.3.12 Shell Thermal Cracking

The Shell thermal distillate cracking unit is based on the principle of converting high-boiling feedstocks to lower-boiling products (Table 17.12) (Figure 17.10). Thermal cracking of heavy

TABLE 17.12
Typical Product Yields from the Shell Thermal Distillate Cracking Process

Product	Yield (% by Weight)	
Gas	9	
Naphtha	17	Endpoint 165°C
Gas oil	54	Endpoint 350°C
Residuum	20	

gas oil takes place in the liquid phase in a furnace, at elevated pressure and temperature, and the products are residuum and distillate products. The majority of the heavy gas oil is converted to light gas oil.

In the process (Figure 17.11), heavy gas oil from the atmospheric distillation unit, or vacuum gas oil from the vacuum distillation unit, is sent to a surge drum. Liquid from this drum is pumped to the distillate heater, which typically operates at a temperature of approximately 490°C (915°C) and at a pressure of 290 psi. Under these conditions, the cracking reactions take place in the liquid phase.

Fluid from the distillate heater is then routed to the combi tower, where separation is achieved between residue, gas oil, and lighter products. In addition, a heavy gas oil fraction is taken from the combi tower, returned to the surge drum, and then recycled through the distillate heater. The bottom product of the combi tower is routed to a vacuum flasher, where heavy gas oil is recovered from the residuum stream and routed back to the distillate heater. The vacuum-flashed residuum from the vacuum flasher can be routed to fuel oil blending, or can be used internally as refinery fuel. The recycling of heavy gas oil from both the vacuum flasher and the combi tower to the distillate heater means that all of the heavy gas oil is converted. Light gas oil from the combi tower is first stripped, and is then routed to a hydrotreater. Alternatively, the light gas oil can be used as cutter stock.

The feedstock and product requirements of the thermal distillate cracking process are flexible, and the process has the capability to optimize conversion through adjustment of the heavy gas oil recycle rate.

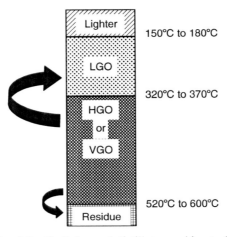

FIGURE 17.10 Basic principle of the Shell thermal distillate cracking technology.

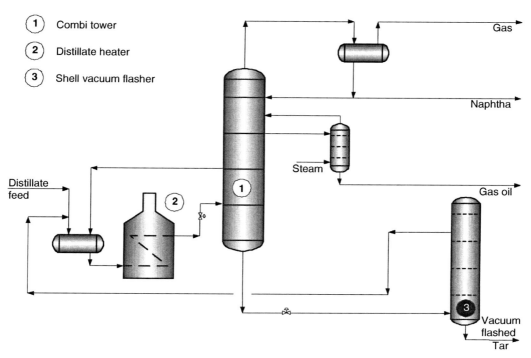

FIGURE 17.11 The Shell thermal distillate cracking technology.

17.3.3.13 Tervahl T Process

The Tervahl process (Figure 17.12) offers two options that cater to the feedstock and the desired products.

In the Tervahl T process (LePage et al., 1987; Peries et al., 1988), the feedstock is heated to the desired temperature using the coil heater, and heat recovered in the stabilization section

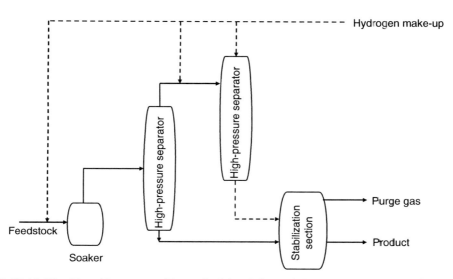

FIGURE 17.12 The Tervahl process. (From Speight, J.G. and Ozum, B. 2002. *Petroleum Refining Processes.* Marcel Dekker Inc., New York. With permission.)

TABLE 17.13
**Feedstock and Product Data for the Tervahl T
and Tervahl H Processes**

Feedstock: Boscan Heavy Crude Oil

		Process	
		Tervahl T	**Tervahl H**
API	10.5		
Distillate, wt.% (<500°C, <930°F)	35.5		
Product			
API		11.7	14.8
Distillate, wt.% (<500°C, <930°F)		52.5	55.3

Source: Speight, J.G. and Ozum, B. 2002. *Petroleum Refining Processes.* Marcel Dekker Inc., New York. With permission.

and held for a specified residence time in the soaking drum. The soaking drum effluent is quenched and sent to a conventional stabilizer or fractionator, where the products are separated into the desired streams (Table 17.13). The gas produced from the process is used for fuel.

In the related Tervahl H process (a hydrogenation process but covered here for convenient comparison with the Tervahl T process), the feedstock and hydrogen-rich stream are heated using heat recovery techniques and a fired heater, and held in the soak drum as in the Tervahl T process. The gas and the oil from the soaking drum effluent are mixed with recycle hydrogen and separated in the hot separator, where the gas is cooled, passed through a separator, and recycled to the heater and soaking drum effluent. The liquids from the hot and cold separator are sent to the stabilizer section, where purge gas and synthetic crude are separated. The gas is used as fuel, and the synthetic crude can now be transported or stored.

17.3.3.14 Vapor-Phase Cracking

Vapor-phase cracking is a high-temperature (545°C to 595°C, 1000°F to 1100°F), low-pressure (<50 psi) thermal conversion process that favors dehydrogenation of feedstock (gaseous hydrocarbons to gas oils) components to olefins and aromatics. Coke is often deposited in heater tubes, causing shutdowns. Relatively large reactors are required for these units.

REFERENCES

Aiba, T., Kaji, H., Suzuki, T., and Wakamatsu, T. 1981. *Chem. Eng. Progr.* February: 37.
Ballard, W.P., Cottington, G.I., and Cooper, T.A. 1992. *Petroleum Processing Handbook.* J.J. McKetta (ed.). Marcel Dekker Inc., New York. p. 309.
Bland, W.F. and Davidson, R.L. 1967. *Petroleum Processing Handbook.* McGraw-Hill, New York.
Blaser, D.E. 1992. *Petroleum Processing Handbook.* J.J. McKetta (ed.). Marcel Dekker Inc., New York. p. 255.
Bonilla, J. 1985. *Energy Prog* 5(4): 239–244.
Bonilla, J. and Elliott, J.D. 1987. United States Patent 4,686,027. August 11.
Corbett, R.A. 1990. *Oil Gas J.* 88(13): 49.
Dickenson, R.L., Biasca, F.E., Schulman, B.L., and Johnson, H.E. 1997. *Hydrocarbon Process.* 76(2): 57.
Dominici, V.E. and Sieli, G.M. 1997. *Handbook of Petroleum Refining Processes.* R.A. Meyers (ed.)., McGraw-Hill, New York. Chapter 12.3.

Elliot, J.D. 1995. *Khim. Tekhnol. Topl. Masel* 2: 9–17.

Feintuch, H.M. and Negin, K.M. 1997. *Handbook of Petroleum Refining Processes*. R.A. Meyers (ed.). McGraw-Hill, New York. Chapter 12.2.

Germain, J E. 1969. *Catalytic Conversion of Hydrocarbons*. Academic Press, New York.

Hulet, C., Briens, C., Berruti, F., and Chan, E.W. 2005. *Int. J. Reactor Eng.* 3: R1. *Hydrocarbon Process.* 75(11): 89.

Kobe, K.A. and McKetta, J.J. 1958. *Advances in Petroleum Chemistry and Refining*. Interscience, New York.

Kuwahara, I. 1987. *Kagaku Kogaku* 51: 1.

LePage, J. F., Morel, F., Trassard, A.M. and Bousquet, J. 1987. Preprints *Div. Fuel Chem.* 32: 470.

Magaril, R.A. and Aksenova, E.L. 1968. *Int. Chem. Eng.* 8: 727.

Magaril, R.Z. and Aksenova, E.I. 1970. *Khim. Tekhnol. Topl. Masel* 7: 22.

Magaril, R.A. and Ramazaeva, L.F. 1969. *Izv. Vyssh. Ucheb. Zaved. Neft Gaz.* 12(1): 61.

Magaril, R.L., Ramazaeva, L.F., and Askenova, E.I. 1970. *Khim. Tekhnol. Topliv Masel.* 15(3): 15.

Magaril, R.Z., Ramazeava, L.F., and Aksenora, E.I. 1971. *Int. Chem. Eng.* 11: 250.

Marzin, R., Pereira, P., McGrath, M.J., Feintuch, H.M., and Thompson, G. 1998. *Oil Gas J.* 97(44): 79.

McKinney, J.D. 1992. *Petroleum Processing Handbook*. McKetta, J.J. ed. Marcel Dekker Inc., New York. p. 245.

Miyauchi, T., Furusaki, S., and Morooka, Y. 1981. *Advances in Chemical Engineering*. Academic Press Inc., New York. Chapter 11.

Miyauchi, T., Tsutsui, T., and Nozaki, Y. 1987. A new fluid thermal cracking process for upgrading resid. Paper 65B. *Proceedings*. Spring National Meeting, American Institute of Chemical Engineers, Houston. March 29.

Miyauchi, T. and Ikeda, Y. 1988. United States Patent 4,722,378.

Moschopedis, S.E., Ozum, B., and Speight, J.G. 1998. *Rev. Process Chem. Eng.* 1(3): 201–259.

Nelson, W.L. 1976. *Oil Gas J.* 74(21): 60.

Peries, J.P., Quignard, A., Farjon C., and Laborde, M. 1988. *Rev. Inst. Franç. Pétrol.* 43(6): 847–853.

Purdy, G.A. 1958. *Petroleum: Prehistoric to Petrochemicals*. Copp Clark, Toronto, Canada.

Roundtree, E.M. 1997. *Handbook of Petroleum Refining Processes*. R.A. Meyers (ed.). McGraw-Hill, New York. Chapter 12.1.

Schabron, J.F. and Speight, J.G. 1997. *Rev. Inst. Franç. Pétrol.* 52(1): 73–85.

Speight, J.G. 1986. Upgrading Heavy Feedstocks. *Annu. Rev. Energ.* 11: 253.

Speight, J.G. 1990. *Fuel Science and Technology Handbook*. J.G. Speight (ed.). Marcel Dekker Inc., New York. Part II.

Speight, J.G. 1992. *Proceedings*. 4th International Conference on the Stability and Handling of Liquid Fuels. U.S. Department of Energy (DOE/CONF-911102). p. 169.

Speight, J.G. 1994. *Asphalts and Asphaltenes, 1*. T.F. Yen and G.V. Chilingarian (eds.). Elsevier, Amsterdam. Chapter 2.

Speight, J.G. 2000. *The Desulfurization of Heavy Oils and Residua*, 2nd Edition. Marcel Dekker Inc., New York.

Speight, J.G. and Ozum, B. 2002. *Petroleum Refining Processes*. Marcel Dekker Inc., New York.

Spragins, F.K. 1978. *Bitumens, Asphalts, and Tar Sands*. Chilingarian, G.V. and T.F. Yen (eds.). Elsevier, Amsterdam. p. 92.

Stephens, M.M. and Spencer, O.F. 1956. *Petroleum Refining Processes*. Penn State University Press, University Park, PA.

Thomas, M., Fixari, B., Le Perchec, P., Princic, Y., and Lena, L. 1989. *Fuel* 68: 318.

Trash, L.A. 1990. *Oil Gas J.* 88(13): 77.

Trimm, D.L. 1984. *Chem. Eng. Proc.* 18: 137.

Ueda, K. 1976. *J. Japan Petrol. Inst.* 19(5): 417.

Ueda, H. 1978. *J. Fuel Soc. Jpn.* 57: 963.

Vermillion, W.L. and Gearhart, W. 1983. *Hydrocarbon Process.* 62(9): 89.

Washimi, K. 1989. *Hydrocarbon Process.* 68(9): 69.

Watari, R., Shoji, Y., Ishikawa, T., Hirotani, H., and Takeuchi, T. 1987. Annual Meeting. National Petroleum Refiners Association, San Antonio, TX. Paper AM-87-43.

18 Catalytic Cracking

18.1 INTRODUCTION

Catalytic cracking is a conversion process that can be applied to a variety of feedstocks ranging from gas oil to heavy oil (Speight and Ozum, 2002). Fluid catalytic cracking units are currently in place in approximately 400 refineries around the world and the units are considered to be one of the most important achievements of the twentieth century.

Catalytic cracking is basically the same as thermal cracking, but it differs by the use of a catalyst that is not (in theory) consumed in the process and it is one of several practical applications used in a refinery that employs a catalyst to improve process efficiency. The original incentive to develop cracking processes arose from the need to increase gasoline supplies and to increase the octane number of gasoline, while maintaining yield from high-boiling stocks using catalysts (Chapter 1) (Germain, 1969). As cracking could virtually double the volume of gasoline from a barrel of crude oil, the purpose of cracking was wholly justified.

This history of catalytic cracking started in the early part of the twentieth century (Table 18.1). In the 1930s, thermal cracking units produced about half the total gasoline manufactured, the octane number of which was about 70 compared to 60 for straight-run gasoline. These were usually blended together with light ends and sometimes with **polymer gasoline** and **reformate** to produce gasoline base stock with an octane number of about 65. The addition of **tetraethyl lead** (ethyl fluid) increased the octane number to about 70 for the regular grade gasoline and 80 for premium grade gasoline. The thermal **reforming** and **polymerization** processes that were developed during the 1930s could be expected to further increase the octane number of gasoline to some extent, but something new was needed to break the octane barrier that threatened to stop the development of more powerful automobile engines. In 1936, a new cracking process opened the way to higher-octane gasoline; this process was catalytic cracking. Since that time, the use of catalytic materials in the petroleum industry has spread to other processes (Bradley et al., 1989).

The last 70 years have seen substantial advances in the development of catalytic processes (Luckenbach et al., 1992). This has involved not only rapid advances in the chemistry and physics of the catalysts themselves but also major engineering advances in reactor design, for example, the evolution of the design of the catalyst beds from **fixed-beds** to **moving-beds** to fluidized-beds, as well as the different designs of the fluidized-bed reactors currently in use. Catalyst chemistry and physics and bed design have allowed major improvements in process efficiency and product yields (Sadeghbeigi, 1995).

An important purpose evolved from the ability of cracked gasoline to resist detonation that is the cause of knocking in automobile engines. In the early days of automobile engine development, straight-run gasoline had a lower end point (165°C, 330°F) than that of cracked gasoline (205°C, 400°F) and, further, did not color when exposed to sunlight. Similar exposure caused the cracked gasoline to turn brown, and hence straight-run gasoline was regarded as the premium material.

521

TABLE 18.1
Brief History of Catalytic Cracking

1915: Batch reactor catalytic cracking to produce light distillates (1915)
  • Catalyst: aluminum chloride (AlCl$_3$)—a Lewis acid, electron acceptor
  • Alkane—electron (abstracted by AlCl$_3$) to produce a carbocation (C$^+$)
  • Ionic chain reactions to crack long chains

1936: Houdry process
  • Continuous feedstock flow with multiple fixed-bed reactors
  • Cracking/catalyst regeneration cycles
  • Catalyst: clays, natural alumina/silica particles

1942: Thermoform catalytic cracking
  • Continuous feedstock flow with moving-bed catalysts
  • Catalyst: synthetic alumina/silica particles
  • Higher thermal efficiency by process integration

1942: Fluid catalytic cracking (FCC) (1942)
  • Continuous feedstock flow with fluidized-bed catalysts

1965: Promising new catalysts
  • Synthetic alumina/silica and zeolite catalysts

Gasoline is now produced, for the most part, by catalytic cracking processes or by reforming processes. In catalytic cracking, the petroleum (or petroleum-derived feedstock) is fed into a reaction vessel containing a catalyst. In reforming processes (Chapter 23), naphtha (refined or unrefined) that may have been produced by catalytic cracking of higher molecular weight feedstocks is heated with hydrogen in the presence of a catalyst. Reforming causes a rearrangement of the structures of the molecular constituents and creates a gasoline product.

Catalytic cracking has a number of advantages over thermal cracking (Table 18.2) (see also Avidan et al., 1990). Gasoline produced by catalytic cracking has a higher octane number and consists largely of *iso*-paraffins and aromatics. The *iso*-paraffins and aromatic hydrocarbons have high octane numbers and greater chemical stability than mono-olefins (R$-$CH=CH$_2$ or R$-$CH=CH$-$R') and diolefins (R$-$CH=CH$-$CH=CH$_2$ or R$-$CH=CH(CH$_2$)$_n$ CH=CH$-$R'). The olefins and diolefins are present in much greater quantities in gasoline produced by thermal cracking processes. Further, substantial quantities of olefin gases and smaller quantities of methane (CH$_4$), ethane (CH$_3$CH$_3$), and ethylene (CH$_2$=CH$_2$) are produced by catalytic cracking. The olefin gases are suitable for polymer gasoline manufacture (Chapter 23).

Sulfur compounds are changed in such a way that the sulfur content of gasoline produced by catalytic cracking gasoline is lower than the sulfur content of gasoline produced by thermal cracking. Catalytic cracking produces less residuum and more of the useful gas oil constituents than thermal cracking. Finally, the process has considerable flexibility, permitting the manufacture of both motor gasoline and aviation gasoline and a variation in the gas oil production to meet changes in the fuel oil market (Speight, 1986).

The feedstocks for catalytic cracking can be any one (or blends) of the flowing: (1) straight-run gas oil, (2) vacuum gas oil, (3) atmospheric residuum, and (4) vacuum residuum. If blends of the above feedstocks are employed, compatibility of the constituents of the blends must be assured under reactor to conditions or excessive coke will be laid down on to the catalyst. In addition, there are several pretreatment options for the feedstocks for catalytic cracking units and these are: (1) deasphalting to prevent excessive coking on catalyst surfaces, (2) demetallation, i.e., removal of nickel, vanadium, and iron to prevent catalyst deactivation,

---

TABLE 18.2
Comparison of Thermal Cracking and Catalytic Cracking

Thermal cracking
- No catalyst
- Higher temperature
- Higher pressure
- Free radical reaction mechanisms
- Moderate thermal efficiency
- No regeneration of catalyst needed
- Moderate yields of gasoline and other distillates
- Gas yields feedstock dependent
- Low-to-moderate product selectivity
- Alkanes produced but feedstock dependent yields
- Low octane number gasoline
- Some chain-branching in alkanes
- Low to moderate yield of C_4 olefins
- Low to moderate yields of aromatics

Catalytic cracking
- Uses a catalyst
- Lower temperature
- Lower pressure
- More flexible in terms of product slate
- Ionic reaction mechanisms
- High thermal efficiency
- Good integration of cracking and regeneration
- High yields of gasoline and other distillates
- Low gas yields
- High product selectivity
- Low *n*-alkane yields
- High octane number
- Chain-branching and high yield of C_4 olefins
- High yields of aromatics

---

(3) use of a short residence time as a means of preparing the feedstock, and (4) hydrotreating or mild hydrocracking to prevent excessive coking in the fluid catalytic cracking unit (Bartholic, 1981a,b; Speight, 2000, 2004; Speight and Ozum, 2002; and references cited therein).

In addition, the goal is to reduce the sulfur content of the naphtha and there are a number of process options for reducing the sulfur level in fluid catalytic cracker naphtha, the two options that are generating the most interest are: (1) hydrotreat the feedstock to the catalytic cracking unit; and (2) hydrotreat the naphtha from the catalytic cracker.

Hydrotreating the fluid catalytic cracker feed improves naphtha yield (Table 18.3) and quality and reduces the sulfur oxide (SO_x) emissions from the catalytic cracker unit, but it is typically a high-pressure process and, further, manipulation of feedstock sulfur alone may not be sufficient to meet future gasoline performance standards. Refineries wishing to process heavier crude oil may only have the option to desulfurize the resulting higher-sulfur naphtha. Hydrodesulfurization of catalytic cracker naphtha is a low-pressure process. Obviously, the selection of an optimum hydrotreating process option for reducing sulfur in catalytic cracker naphtha is determined by economic factors specific to a refinery and to the feedstock. Hydrotreating catalytic cracker feedstock can be very profitable for a refiner, despite its large capital investment. By taking advantage of feedstock hydrotreating, margins can be

TABLE 18.3
Feedstock and Product Data for the Fluid Catalytic Process with and without Feedstock Hydrotreating

	No Pretreatment	With Hydrotreatment
Feedstock ($>370°C$, $>700°F$)		
API	15.1	20.1
Sulfur, wt.%	3.3	0.5
Nitrogen, wt.%	0.2	0.1
Carbon residue, wt.%.	8.9	4.9
Nickel + vanadium, ppm	51.0	7.0
Products		
Naphtha (C5–221°C, C5–430°F), vol.%	50.6	58.0
Light cycle oil (221–360°C, 430–680°F), vol.%	21.4	18.2
Residuum ($>360°C$, $>680°F$), wt.%	9.7	7.2
Coke, wt.%	10.3	7.0

Source: Speight, J.G., and Ozum, B. 2002. *Petroleum Refining Processes*. Marcel Dekker Inc., New York. With permission.

optimized by considering the feedstock hydrotreater unit, the fluid catalytic cracker, and any post-catalytic cracker hydrotreaters as one integrated upgrading step.

Catalytic cracking in the usual commercial process involves contacting a feedstock (usually a gas oil fraction) with a catalyst under suitable conditions of temperature, pressure, and residence time. By this means, a substantial part ($>50\%$) of the feedstock is converted into gasoline and lower-boiling products, usually in a single-pass operation (Bland and Davidson, 1967). However, during the cracking reaction, carbonaceous material is deposited on the catalyst, which markedly reduces its activity, and removal of the deposit is necessary. The carbonaceous deposit arises from the thermal decomposition of high molecular weight polar species (Chapter 10) in the feedstock. Removal of the deposit from the catalyst is usually accomplished by burning in the presence of air until catalyst activity is reestablished.

The reactions that occur during catalytic cracking are complex (Germain, 1969), but there is a measure of predictability now that catalyst activity is better understood. The major catalytic cracking reaction exhibited by paraffins is carbon–carbon bond scission into a lighter paraffin and olefin. Bond rupture occurs at certain definite locations on the paraffin molecule, rather than randomly as in thermal cracking. For example, paraffins tend to crack toward the center of the molecule, the long chains cracking in several places simultaneously (Chapter 15). Normal paraffins usually crack at γ carbon–carbon bonds or still nearer the center of the molecule. On the other hand, *iso*-paraffins tend to rupture between carbon atoms that are, respectively β and γ to a tertiary carbon. In either case, catalytic cracking tends to yield products containing three or four carbon atoms rather than the one- or two-carbon atom molecules produced in thermal cracking.

As in thermal cracking (Chapter 15 and Chapter 17), large molecules usually crack more readily than small molecules. However, paraffins with more than six carbon atoms may also undergo rearrangement of their carbon skeletons before cracking, and a minor amount of **dehydrocyclization** also occurs, yielding aromatics and hydrogen.

Olefins are the most reactive class of hydrocarbons in catalytic cracking and tend to crack from 1,000 to 10,000 times faster than in thermal processes. Severe cracking conditions destroy olefins almost completely, except for those in the low-boiling gasoline and gaseous hydrocarbon range and, as in the catalytic cracking of paraffins, *iso*-olefins crack more readily than *n*-olefins.

Olefins also tend to undergo rapid **isomerization** and yield mixtures with an equilibrium distribution of double-bond positions. In addition, the chain-branching isomerization of olefins is fairly rapid and often reaches equilibrium. These branched-chain olefins can then undergo **hydrogen transfer** reactions with naphthenes and other hydrocarbons. Other olefin reactions include polymerization as well as condensation to yield aromatic molecules, which in turn may be the precursors of coke formation.

Cycloparaffins (naphthenes) catalytically crack more readily than paraffins, but not as readily as olefins. Naphthene cracking occurs by both ring and chain rupture and yields olefins and paraffins, but formation of methane and the C_2 hydrocarbons [ethane, CH_3CH_3, ethylene, $CH_2{=}CH_2$, and acetylene, $CH{\equiv}(CH)$] is relatively minor if present at all.

Aromatic hydrocarbons exhibit wide variations in their susceptibility to catalytic cracking. The benzene ring is relatively inert, and condensed-ring compounds, such as naphthalene, anthracene, and phenanthrene, crack very slowly. When these aromatics crack, a substantial part of their **conversion** is reflected in the amount of coke deposited on the catalyst. Alkyl-benzenes with attached groups of C_2 or larger, primarily form benzene and the corresponding olefins, and heat sensitivity increases as the size of the alkyl group increases.

18.2 EARLY PROCESSES

In general, catalytic cracking may be regarded as the modern method for converting high-boiling petroleum fractions, such as gas oil, into high-quality gasoline and other added value products.

In the 1930s, thermal cracking units produced approximately 50% of the total gasoline. The octane number of this gasoline was about 70 compared to 60 for straight-run (distilled) gasoline. The thermal reforming and polymerization processes that were developed during the 1930s could be expected to further increase the octane number of gasoline to some extent, but an additional innovation was needed to increase the octane number of gasoline to enhance the development of more powerful automobile engines.

In 1936, a new cracking process opened the way to higher-octane gasoline; this process was catalytic cracking. The process is basically the same as thermal cracking but differs by the use of a catalyst, which is not (in theory) consumed in the process and directs the course of the cracking reactions to produce more of the desired higher-octane hydrocarbon products.

18.3 COMMERCIAL PROCESSES

Catalytic cracking is another innovation that truly belongs to the twentieth century. It is the modern method for converting high-boiling petroleum fractions, such as gas oil, into gasoline and other low-boiling fractions. The several processes currently employed in catalytic cracking differ mainly in the method of catalyst handling, although there is an overlap with regard to catalyst type and the nature of the products. The catalyst, which may be an activated natural or synthetic material, is employed in bead, pellet, or microspherical form and can be used as *fixed-bed*, *moving-bed*, or **fluid-bed** configurations.

The fixed-bed process was the first to be used commercially and uses a static bed of catalyst in several reactors, which allows a continuous flow of feedstock to be maintained. Thus, the cycle of operations consists of (1) flow of feedstock through the catalyst bed, (2) discontinuance of feedstock flow and removal of coke from the catalyst by burning, and (3) insertion of the reactor on-stream.

The moving-bed process uses a reaction vessel in which cracking takes place and a kiln in which the spent catalyst is regenerated, catalyst movement between the vessels is provided by various means.

The fluid-bed process differs from the fixed-bed and moving-bed processes insofar as the powdered catalyst is circulated essentially as a fluid with the feedstock (Sadeghbeigi, 1995). The several fluid catalytic cracking processes in use differ primarily in mechanical design (see, for example, Hemler, 1997; Hunt, 1997; Johnson and Niccum, 1997; Ladwig, 1997). Side-by-side, reactor–**regenerator** configuration of the reactor, either above or below the regenerator is the main mechanical variation. From a flow standpoint, all fluid catalytic cracking processes contact the feedstock (and any recycle streams) with the finely divided catalyst in the reactor.

Feedstocks may range from naphtha to atmospheric residuum (**reduced crude**). Feed preparation (to remove metallic constituents and high molecular weight nonvolatile materials) is usually carried out through any one of the following ways: coking, propane deasphalting, furfural extraction, vacuum distillation, viscosity breaking, thermal cracking, and hydrodesulfurization (see, for example, Speight, 2000).

The major process variables are temperature, pressure, catalyst-to-feedstock ratio (ratio of the weight of catalyst entering the reactor per hour to the weight of feedstock charged per hour), and space velocity (weight or volume of the feedstock charged per hour per weight or volume of catalyst in the reaction zone). Wide flexibility in product distribution and quality is possible through control of these variables, along with internal recycling. Increased conversion can be obtained by applying higher temperature or higher pressure. Alternatively, lower space velocity and higher catalyst to feedstock ratio will also contribute to an increased conversion.

When cracking is conducted in a single stage, the more reactive hydrocarbons may be cracked, with a high conversion to gas and coke, in the reaction time necessary for reasonable conversion of the more refractory hydrocarbons. However, in a two-stage process, gas and gasoline from a short-reaction-time, high-temperature cracking operation are separated before the main cracking reactions take place in a second-stage reactor.

18.3.1 FIXED-BED PROCESSES

Historically, the Houdry fixed-bed process, which went on-stream in June 1936, was the first of the modern catalytic cracking processes. It was preceded only by the McAfee batch process, which employed a metal halide (aluminum chloride) catalyst, but has long since lost any commercial significance.

In the fixed-bed process, the catalyst in the form of small lumps or pellets was made up in layers or beds in several (four or more) catalyst-containing drums called converters. Feedstock vaporized at about 450°C (840°F) and less than 7 to 15 psi pressure, and passed through one of the converters where the cracking reactions took place. After a short time, deposition of coke on the catalyst rendered it ineffective, and using a synchronized valve system, the feed stream was turned into a neighboring converter while the catalyst in the first converter was regenerated by carefully burning the coke deposits with air. After about 10 min, the catalyst was ready to go on-stream again.

Fixed-bed processes have now generally been replaced by moving-bed or fluid-bed processes.

18.3.2 FLUID-BED PROCESSES

18.3.2.1 Fluid-Bed Catalytic Cracking

The fluid catalytic cracking process (Figure 18.1) (Bartholic and Haseltine, 1981) is the most widely used process and is characterized by the use of a finely powdered catalyst that is moved through the reactor (Figure 18.2) and flow patterns may vary depending upon the precise configuration of the reactor. The catalyst particles are of such a size that when aerated with air or hydrocarbon vapor, the catalyst behaves like a liquid and can be moved through pipes.

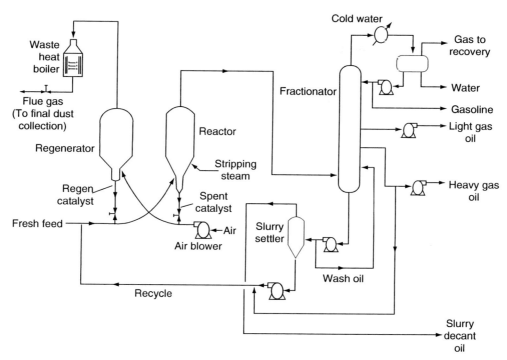

FIGURE 18.1 The fluid catalytic cracking process.

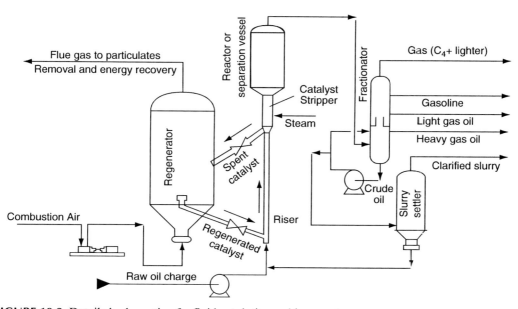

FIGURE 18.2 Detailed schematic of a fluid catalytic cracking reactor.

Thus, vaporized feedstock and fluidized catalyst flow together into a reaction chamber where the catalyst, still dispersed in the hydrocarbon vapors, forms beds in the reaction chamber and the cracking reactions take place. The cracked vapors pass through cyclones located in the top of the reaction chamber, and the catalyst powder is thrown out of the vapors by centrifugal force. The cracked vapors then enter the bubble towers where fractionation into cracked light and heavy gas oils, cracked gasoline, and cracked gases takes place.

As the catalyst in the reactor becomes contaminated with coke, the catalyst is continuously withdrawn from the bottom of the reactor and lifted by means of a stream of air into a regenerator where the coke is removed by controlled burning. The regenerated catalyst then flows to the fresh feed line, where the heat in the catalyst is sufficient to vaporize the fresh feed before it reaches the reactor, where the temperature is about 510°C (950°F).

18.3.2.2 Model IV Fluid-Bed Catalytic Cracking Unit

This unit involves a procedure in which the catalyst is transferred between the reactor and regenerator by means of U bends and the catalyst flow rate can be varied in relation to the amount of air injected into the spent-catalyst U bend. Regeneration air, other than that used to control circulation, enters the regenerator through a grid, and the reactor and regenerator are mounted side by side.

The model IV low-elevation design was preceded by the model III (1947) balanced-pressure design, the model II (1944) downflow design, and the original model I (1941) upflow design. The first commercial model IV installation in the United States was placed on-stream in 1952.

18.3.2.3 Orthoflow Fluid-Bed Catalytic Cracking

This process uses the unitary vessel design, which provides a straight-line flow of catalyst and thereby minimizes the erosion encountered in pipe bends. Commercial orthoflow designs are of three types: models A and C, with the regenerator beneath the reactor, and model B, with the regenerator above the reactor. In all cases, the catalyst-stripping section is located between the reactor and the regenerator. All designs employ the heat-balanced principle incorporating fresh feed—recycle feed cracking.

18.3.2.4 Shell Two-Stage Fluid-Bed Catalytic Cracking

Two-stage fluid catalytic cracking was devised to permit greater flexibility in shifting product distribution when dictated by demand. Thus, feedstock is first contacted with cracking catalyst in a **riser** reactor, that is, a pipe in which fluidized catalyst and vaporized feedstock flow concurrently upward, and the total contact time in this first stage is of the order of seconds. High temperatures, 470°C to 565°C (875°F to 1050°F), are employed to reduce undesirable coke deposits on catalyst without destruction of gasoline by secondary cracking. Other operating conditions in the first stage are a pressure of 16 psi and a catalyst-to-feedstock ratio of 3:1 to 50:1, and volume conversion ranges between 20% and 70% have been recorded.

All or part of the unconverted or partially converted gas oil product from the first stage is then cracked further in the second-stage fluid-bed reactor. Operating conditions are 480°C to 540°C (900°F to 1000°F) and 16 psi with a catalyst-to-oil ratio of 2:1 to 12:1. Conversion in the second stage varies between 15% and 70%, with an overall conversion range of 50% to 80%.

18.3.2.5 Universal Oil Products (UOP) Fluid-Bed Catalytic Cracking

This process is adaptable to the needs of both large and small refineries. The major distinguishing features of the process are: (1) elimination of the air riser with its attendant large expansion joints, (2) elimination of considerable structural steel supports, and (3) reduction in regenerator and in air-line size through use of 15 to 18 psi pressure operation.

18.3.3 MOVING-BED PROCESSES

18.3.3.1 Airlift Thermofor Catalytic Cracking (Socony Airlift TCC Process)

This process is a moving-bed, reactor-over-generator continuous process for conversion of heavy gas oils into lighter high-quality gasoline and middle distillate fuel oils. Feed preparation may consist of flashing in a separator to obtain vapor feed, and the separator bottoms may be sent to a vacuum tower from which the liquid feed is produced.

The gas oil vapor-liquid flows downward through the reactor concurrently with the regenerated synthetic bead catalyst. The catalyst is purged by steam at the base of the reactor and gravitates into the kiln, or regeneration is accomplished by the use of air injected into the kiln. Approximately 70% of the carbon on the catalyst is burned in the upper kiln burning zone and the remainder in the bottom-burning zone. Regenerated, cooled catalyst enters the lift pot, where low-pressure air transports it to the surge hopper above the reactor for reuse.

18.3.3.2 Houdresid Catalytic Cracking

Houdresid catalytic cracking is a process that uses a variation of the continuously moving catalyst bed designed to obtain high yields of high-octane gasoline and light distillate from reduced crude charge.

Residua, ranging from atmospheric residue to vacuum residua including residua high in sulfur or nitrogen, can be used as the feedstock and the catalyst is synthetic or natural. Although the equipment employed is similar in many respects to that used in Houdriflow units, novel process features modify or eliminate the adverse effects, and catalyst and product selectivity usually results when metals (such as nickel, vanadium copper, and iron) are present in the fuel. The Houdresid catalytic reactor and catalyst-regenerating kiln are contained in a single vessel. Fresh feed plus recycled gas oil are charged to the top of the unit in a partially vaporized state and mixed with steam.

18.3.3.3 Houdriflow Catalytic Cracking

This is a continuous, moving-bed process employing an integrated single vessel for the reactor and regenerator kiln. The sweet or sour feedstock can be any fraction of the crude boiling between naphtha and atmospheric residua. The catalyst is transported from the bottom of the unit to the top in a gas lift employing compressed flue gas and steam.

The reactor feed and catalyst pass concurrently through the reactor zone to a disengager section, in which vapors are separated and directed to a conventional fractionation system. The spent catalyst, which has been steam purged of residual oil, flows to the kiln for regeneration, after which steam and flue gas are used to transport the catalyst to the reactor.

18.3.3.4 Suspensoid Catalytic Cracking

Suspensoid cracking was developed from the thermal cracking process carried out in tube and tank units (Chapter 14). Small amounts of powdered catalyst or a mixture with the feedstock are pumped through a cracking coil furnace. Cracking temperatures are 550°C to 610°C (1025°F to 1130°F), with pressures of 200 to 500 psi. After leaving the furnace, the cracked material enters a separator where the catalyst and high boiling entrained material (**tar**) are left behind. The cracked vapors enter a bubble tower where they are separated into two parts, gas oil and pressure distillate. The latter is separated into gasoline and gases. The spent catalyst is filtered from the organic carry-over, which is used as a heavy industrial fuel oil.

The process is actually a compromise between catalytic and thermal cracking. The main effect of the catalyst is to allow a higher cracking temperature and to assist mechanically in keeping coke from accumulating on the walls of the tubes. The normal catalyst employed is spent clay (2 to 10 Lb per barrel of feed) that is obtained from the contact filtration of lubricating oils (Chapter 24).

18.3.4 PROCESS OPTIONS FOR HEAVY FEEDSTOCKS

The processes described later are the evolutionary offspring of the fluid catalytic cracking and the residuum catalytic cracking processes. Some of these newer processes use catalysts with different silica/alumina ratios as acid support of metals such as Mo, Co, Ni, and W. In general, the first catalyst used to remove metals from oils was the conventional hydrodesulfurization (HDS) catalyst. Diverse natural minerals are also used as raw material for elaborating catalysts addressed to the upgrading of heavy fractions. Among these minerals are: clays; manganese nodules; bauxite activated with vanadium (V), nickel (Ni), chromium (Cr), iron (Fe), and cobalt (Co), as well as iron laterites, sepiolites; and mineral nickel and transition metal sulfides supported on silica and alumina. Other kinds of catalysts, such as vanadium sulfide, are generated **in situ**, possibly in colloidal states.

In the past decades, in spite of the difficulty of handling heavy feedstocks, residuum fluidized catalytic cracking (RFCC), has evolved to become a well-established approach for converting a significant portion of the heavier fractions of the crude barrel into a high-octane gasoline blending component. Residuum fluidized catalytic cracking, which is an extension of conventional fluid catalytic cracking technology for applications involving the conversion of highly contaminated residua, has been commercially proven on feedstocks ranging from gas oil-residuum blends to atmospheric residua, as well as blends of atmospheric and vacuum residua blends. In addition to high gasoline yields, the residuum fluidized catalytic cracking unit also produces gaseous, distillate, and fuel oil-range products.

The product quality from the residuum fluidized catalytic cracker is directly affected by its feedstock quality. In particular, and unlike hydrotreating, the residuum fluidized catalytic cracking redistributes sulfur among the various products, but does not remove sulfur from the products unless, of course, one discount the sulfur that is retained by the coke formed on the catalyst. Consequently, tightening product specifications have forced refiners to hydrotreat some, or all, of the products from the RFCC unit. Similarly, in the future, the emissions of sulfur oxides (SO_x) from an RFCC may become more of an obstacle for residue conversion projects. For these reasons, a point can be reached where the economic operability of the unit can be sufficient to justify hydrotreating the feedstock to the cat cracker.

As an integrated conversion block, residue hydrotreating and residuum fluidized catalytic cracking complement each other and can offset many of the inherent deficiencies related to residue conversion.

18.3.4.1 Asphalt Residual Treating (ART) Process

The ART process is a process for increasing the production of transportation fuels and reduces heavy fuel oil production, without hydrocracking (Table 18.4).

In the process, the preheated feedstock (which may be whole crude, atmospheric residuum, vacuum residuum, or tar sand bitumen) is injected into a stream of fluidized, hot catalyst (trade name: ArtCat). Complete mixing of the feedstock with the catalyst is achieved in the contactor, which is operated within a pressure—temperature envelope to ensure selective

TABLE 18.4
Feedstock and Product Data for the ART Process

	Mixed Residua[a]
Feedstock	
API	14.9
Sulfur, wt.%	4.1
Nitrogen, wt.%	0.3
C_5-asphaltenes	12.4
Carbon residue, wt.%	15.8
Nickel, ppm	52.0
Vanadium, ppm	264.0
Products	
Naphtha, vol.%	62.1
No. 2 fuel oil, vol.%	35.5
No. 6 fuel oil, vol.%	2.3

[a]Blend of Arabian light vacuum residuum plus Arabian heavy vacuum residuum.
Source: Speight, J.G., and Ozum, B. 2002. *Petroleum Refining Processes*. Marcel Dekker Inc., New York. With permission.

vaporization. The vapor and the contactor effluent are quickly and efficiently separated from each other and entrained hydrocarbons are stripped from the contaminant (containing spent solid) in the stripping section. The contactor vapor effluent and vapor from the stripping section are combined and rapidly quenched in a quench drum to minimize product degradation. The cooled products are then transported to a conventional fractionator that is similar to that found in a fluid catalytic cracking unit. Spent solid from the stripping section is transported to the combustor bottom zone for carbon burn-off.

In the combustor, coke is burned from the spent solid that is then separated from combustion gas in the surge vessel. The surge vessel circulates regenerated catalyst streams to the contactor inlet for feed vaporization, and to the combustor bottom zone for premixing.

The components of the combustion gases include carbon dioxide (CO_2), nitrogen (N_2), oxygen (O_2), sulfur oxides (SO_x), and nitrogen oxides (NO_x) that are released from the catalyst with the combustion of the coke in the combustor. The concentration of sulfur oxides in the combustion gas requires treatment for their removal.

18.3.4.2 Residue Fluid Catalytic Cracking (HOC) Process

This process is a version of the fluid catalytic cracking process that has been adapted to conversion of residua that contain high amounts of metal and asphaltene constituents.

In the process, a residuum is desulfurized and the nonvolatile fraction from the hydrodesulfurization unit is charged to the RFCC unit. The reaction system is an external vertical riser terminating in a closed cyclone system. Dispersion steam in amounts higher than that used for gas oils is used to assist in the vaporization of any volatile constituents of heavy feedstocks.

A two-stage stripper is utilized to remove hydrocarbons from the catalyst. Hot catalyst flows at low velocity in dense phase through the catalyst cooler and returns to the regenerator. Regenerated catalyst flows to the bottom of the riser to meet the feed.

The coke deposited on the catalyst is burned off in the regenerator along with the coke formed during the cracking of the gas oil fraction. If the feedstock contains high proportions of metals, control of the metals on the catalyst requires excessive amounts of catalyst withdrawal and fresh catalyst addition. This problem can be addressed by feedstock pretreatment.

The feedstocks for the process are rated on the basis of carbon residue and content of metals. Thus, good quality feedstocks have less than 5% by weight carbon residue and less than 10 ppm metals. Medium quality feedstocks have greater than 5% but less than 10% by weight carbon residue and greater than 10 but less than 30 ppm metals. Poor quality feedstocks have greater than 10% but less than 20% by weight carbon residue and greater than 30 but less than less than 150 ppm metals. Finally, bad quality feedstocks have greater than 20% by weight carbon residue and greater than 150 ppm metals. One might question the value of this rating of the feedstocks for the HOC process as these feedstock ratings can apply to virtually any fluid catalytic cracking processes.

18.3.4.3 Heavy Oil Treating (HOT) Process

The HOT process is a catalytic cracking process for upgrading heavy feedstocks such as topped crude oils, vacuum residua, and solvent deasphalted bottoms (Table 18.5) using a fluidized bed of iron ore particles.

The main section of the process consists of three fluidized reactors and separate reactions take place in each reactor (cracker, regenerator, and desulfurizer):

Fe_3O_4 + asphaltene constituents → coke/Fe_3O_4 + Oil + Gas (in the *cracker*)

$$3FeO + H_2O \rightarrow Fe_3O_4 + H_2 \text{(in the cracker)}$$

$$Coke/Fe_3O_4 + O^2 \rightarrow 3FeO + CO + CO_2 \text{(in the regenerator)}$$

$$FeO + SO_2 + 3CO \rightarrow FeS + 3CO_2 \text{(in the regenerator)}$$

$$3FeS + 5O_2 + Fe_3O_4 + 3SO_2 \text{(in the desulfurizer)}$$

In the cracker, heavy oil cracking and the steam-iron reaction take place simultaneously, under conditions similar or usual to thermal cracking. Any unconverted feedstock is recycled to the cracker from the bottom of the scrubber. The scrubber effluent is separated into hydrogen gas, liquefied petroleum gas (LPG), and liquid products that can be upgraded by conventional technologies to priority products.

In the regenerator, coke deposited on the catalyst is partially burned to form carbon monoxide in order to reduce iron tetroxide and to act as a heat supply. In the desulfurizer, sulfur in the solid catalyst is removed and recovered as molten sulfur in the final recovery stage.

TABLE 18.5
Feedstock and Product Data for the HOT Process

	Arabian Light Vacuum Residuum
Feedstock	
API	7.1
Sulfur, wt.%	4.2
Carbon residue, wt.%	21.6
Products	
Light naphtha (C5–180°C, C5–355°F), vol.%	15.2
Heavy naphtha (180–230°C, 355–445°F), vol.%	8.2
Light gas oil (230–360°C, 445–680°F), vol.%	2.3
Heavy gas oil (360–510°C, 680–950°F), vol.%	28.2

Source: Speight, J.G. and Ozum, B. 2002. *Petroleum Refining Processes*. Marcel Dekker Inc., New York. With Permission.

18.3.4.4 R2R Process

The R2R process is a fluid catalytic cracking process for conversion of heavy feedstocks (Table 18.6) (Heinrich and Mauleon, 1994; Inai, 1994).

In the process, the feedstock is vaporized upon contacting hot regenerated catalyst at the base of the riser and lifts the catalyst into the reactor vessel separation chamber (Figure 18.3), where rapid disengagement of the hydrocarbon vapors from the catalyst is accomplished by both, a special solids separator and cyclones. The bulk of the cracking reactions takes place at the moment of contact and continues as the catalyst and hydrocarbons travel up the riser. The reaction products, along with a minute amount of entrained catalyst, then flow to the fractionation column. The stripped spent catalyst, deactivated with coke, flows into the Number 1 regenerator.

Partially regenerated catalyst is pneumatically transferred via an air riser to the Number 2 regenerator, where the remaining carbon is completely burned in a dryer atmosphere. This regenerator is designed to minimize catalyst inventory and residence time at high temperature, while optimizing the coke-burning rate. Flue gases pass through external cyclones to a waste heat recovery system. Regenerated catalyst flows into a withdrawal well and after stabilization is charged back to the oil riser.

18.3.4.5 Reduced Crude Oil Conversion (RCC) Process

In recent years, because of a trend for low-boiling products, most refineries perform the operation by partially blending residua into vacuum gas oil. However, conventional fluid catalytic cracking processes have limits in residue processing, so residue fluid catalytic cracking processes have lately been employed, one after another. Because the residue fluid catalytic cracking process enables efficient gasoline production directly from residues, it will play the most important role as a residue cracking process, along with the residue hydroconversion process. Another role of the residuum fluid catalytic cracking process is to generate high-quality gasoline blending stock and petrochemical feedstock. Olefins (propene, butenes, and pentenes) serve as feed for alkylating processes, for polymer gasoline, as well as for **additives** for reformulated gasoline.

TABLE 18.6
Feedstock and Product Data for the R2R Process

	Atmospheric Residuum + Vacuum Gas Oil[a]	Atmospheric Residuum[a]	Atmospheric Residuum[a]	Hydrotreated Atmospheric Residuum[a]	Hydrotreated Atmospheric Residuum
Feedstock					
API	28.4	26.4	22.6	20.7	20.7
Carbon residue, wt.%	0.2	6.4	5.5	4.8	4.8
Nickel + vanadium, ppm	1.0	22.0	34.0	20.0	20.0
Products					
Naphtha, vol.%	63.4	59.5	58.0	60.4	49.6
Distillate, vol.%	16.6	14.1	16.3	17.7	29.7
Heavy gas oil/residuum, vol.%	3.9	6.7	9.8	6.8	12.5
Coke, wt.%	4.6	7.4	6.9	7.2	6.7

[a]Processed for maximum gasoline production.
Source: Speight, J.G. and Ozum, B. 2002. *Petroleum Refining Processes*. Marcel Dekker Inc., New York. With permission.

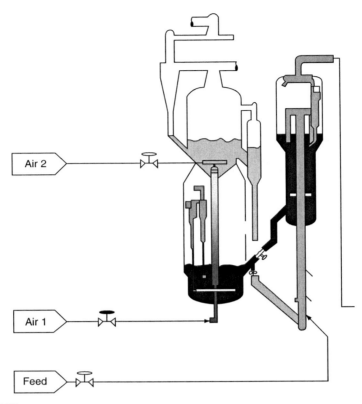

FIGURE 18.3 R2R reactor.

In the RCC process (Table 18.7), the clean regenerated catalyst enters the bottom of the reactor riser where it contacts low-boiling hydrocarbon lift gas that accelerates the catalyst up the riser prior to feed injection. At the top of the lift gas zone, the feed is injected through a series of nozzles located around the circumference of the reactor riser.

The catalyst and oil disengaging system is designed to separate the catalyst from the reaction products and then rapidly remove the reaction products from the reactor vessel. Spent catalyst from the reaction zone is first steam stripped to remove adsorbed hydrocarbon, and then routed to the regenerator. In the regenerator, all the carbonaceous deposits are removed from the catalyst by combustion, restoring the catalyst to an active state with a very low carbon content. The catalyst is then returned to the bottom of the reactor riser at a controlled rate to achieve the desired conversion and selectivity to the primary products.

18.3.4.6 Shell FCC Process

The Shell FCC process is designed to maximize the production of distillates from residua (Table 18.8).

In the process, the preheated feedstock (vacuum gas oil, atmospheric residuum) is mixed with the hot regenerated catalyst. After reaction in a riser, volatile materials and catalyst are separated, after which the spent catalyst is immediately stripped of entrained and adsorbed hydrocarbons in a very effective multistage stripper. The stripped catalyst gravitates through a short standpipe into a single-vessel, simple, reliable, and yet efficient catalyst regenerator.

TABLE 18.7
Feedstock and Product Data for the RCC Process

	Residuum[a]	Residuum[a]	Residuum[a]
Feedstock			
API	22.8	21.3	19.2
Sulfur, wt.%			
Nitrogen, wt.%			
Carbon residue, wt.%	0.2	6.4	5.5
Nickel, ppm	1.0	22.0	34.0
Vanadium, ppm			
Products			
Gasoline C5–221°C, C5–430°F), vol.%	59.1	56.6	55.6
Light cycle oil (221–322°C, 430–610°F), vol.%	16.3	15.4	16.3
Gas oil (>322°C, >610°F), vol.%	6.2	9.0	9.6
Coke, wt.%	8.4	9.1	10.8

[a]Unspecified.
Source: Speight, J.G. and Ozum, B. 2002. *Petroleum Refining Processes*. Marcel Dekker Inc., New York. With permission.

Regenerative flue gas passes via a cyclone and swirl tube combination to a power recovery turbine. From the expander turbine, the heat in the flue gas is further recovered in a waste heat boiler. Depending on the environmental conservation requirements, a deNO$_x$ing, deSO$_x$ing, and particulate emission control device can be included in the flue gas train.

There is a claim that feedstock pretreatment of bitumen (by hydrogenation) prior to fluid catalytic cracking (or for that matter any catalytic cracking process) can result in enhanced yield of naphtha. It is suggested that mild hydrotreating be carried out upstream of a fluid catalytic cracking unit to provide an increase in yield and quality of distillate products. This is in keeping with earlier work, where mild hydrotreating of bitumen was reported to produce low-sulfur liquids that would be amenable to further catalytic processing (Figure 18.4).

TABLE 18.8
Feedstock and Product Data for the Shell FCC Process

	Residuum[a]	Residuum[a]
Feedstock		
API	18.2	13.4
Sulfur, wt.%	1.1	1.3
Carbon residue, wt.%	1.2	4.7
Products		
Gasoline (C5–221°C, C5–430°F), wt.%	49.5	46.2
Light cycle oil (221–370°C, 430–700°F), wt.%	20.1	19.1
Heavy cycle oil (>370°C, >700°F), wt.%	5.9	10.8
Coke, wt.%	5.9	7.6

[a]Unspecified.
Source: Speight, J.G. and Ozum, B. 2002. *Petroleum Refining Processes*. Marcel Dekker Inc., New York. With permission.

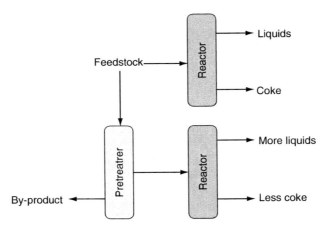

FIGURE 18.4 Potential pretreating scheme for catalytic cracker feedstocks.

18.3.4.7   S&W Fluid Catalytic Cracking Process

The S&W FCC process is also designed to maximize the production of distillates from residua (Table 18.9).

In the S&W FCC process, the heavy feedstock is injected into a stabilized, upward flowing catalyst stream, whereupon the feedstock–steam–catalyst mixture travels up the riser and is separated by a high-efficiency inertial separator. The product vapor goes overhead to the main fractionator.

The spent catalyst is immediately stripped in a staged, baffled stripper to minimize hydrocarbon carryover to the regenerator system. The first regenerator (650°C to 700°C, 1200°F to 1290°F) burns 50% to 70% of the coke in an incomplete carbon monoxide combustion mode running counter-currently. This relatively mild, partial regeneration step minimizes the significant contribution of hydrothermal catalyst deactivation. The remaining coke is burned in

TABLE 18.9
Feedstock and Product Data for the S&W FCC Process

	Residuum[a]	Residuum[a]
Feedstock		
API	24.1	22.3
Sulfur, wt.%	0.8	1.0
Carbon residue, wt.%	4.4	6.5
Products		
Naphtha, vol.%	61.5	60.2
Light cycle oil, vol.%	16.6	17.5
Heavy cycle oil, vol.%	5.6	6.6
Coke, wt.%	7.1	7.8
Conversion, vol.%	77.7	75.9

[a]Unspecified.
Source: Speight, J.G. and Ozum, B. 2002. *Petroleum Refining Processes*. Marcel Dekker Inc., New York. With permission.

the second regenerator (about 775°C, 1425°F) with an extremely low steam content. Hot clean catalyst enters a withdrawal well that stabilizes its fluid qualities prior to being returned to the reaction system.

18.4 CATALYSTS

Cracking crude oil fractions occur over many types of catalytic materials, but high yields of desirable products are obtained with hydrated aluminum silicates. These may be either activated (acid-treated natural clays of the bentonite type) or synthesized silica–alumina or silica–magnesia preparations. Their activity to yield essentially the same products may be enhanced to some extent by the incorporation of small amounts of other materials, such as the oxides of zirconium (zirconia, ZrO_2), boron (boria, B_2O_3, which has a tendency to volatilize away on use), and thorium (thoria, ThO_2). Both the natural and the synthetic catalysts can be used as pellets or beads, and also in the form of powder; in either case, replacements are necessary because of attrition and gradual loss of efficiency (DeCroocq, 1984; Thakur, 1985; Le Page et al., 1987).

The catalysts must be stable to physical impact loading and thermal shocks and must withstand the action of carbon dioxide, air, nitrogen compounds, and steam. They should also be resistant to sulfur compounds; the synthetic catalysts and certain selected clays appear to be better in this regard than average untreated natural catalysts. The silica–alumina catalysts are reported to give the highest-octane gasoline, and silica–magnesia the largest yields, with the natural clays falling between them.

Neither silica (SiO_2) nor alumina (Al_2O_3) alone is effective in promoting catalytic cracking reactions. In fact, they (and also activated carbon) promote decomposition of hydrocarbon constituents that match the thermal decomposition patterns. Mixtures of anhydrous silica and alumina ($SiO_2 \cdot Al_2O_3$) or anhydrous silica with hydrated alumina ($2SiO_2 \cdot 2Al_2O_3 \cdot 6H_2O$) are also essentially not effective. A catalyst with appreciable cracking activity is obtained only when prepared from hydrous oxides followed by partial dehydration (**calcining**). The small amount of water remaining is necessary for proper functioning.

The catalysts are porous and highly adsorptive, and their performance is affected markedly by the method of preparation. Two catalysts that are chemically identical but have pores of different size and distribution may have different activity, selectivity, temperature coefficient of reaction rate, and response to poisons. The intrinsic chemistry and catalytic action of a surface may be independent of pore size, but small pores appear to produce different effects because of the manner and time in which hydrocarbon vapors are transported into and out of the interstices.

Commercial synthetic catalysts are amorphous and contain more silica than is called for by the preceding formulas; they are generally composed of 10% to 15% alumina (Al_2O_3) and 85% to 90% silica (SiO_2). The natural materials, montmorillonite, a nonswelling bentonite, and halloysite, are hydrosilicates of aluminum, with a well-defined crystal structure and approximate composition of Al_2O_3 $4Si_2O \cdot xH_2O$. Some of the newer catalysts contain up to 25% alumina and are reputed to have a longer active life.

Commercially used cracking catalysts are insulator catalysts possessing strong acidic properties. They function as catalysts by altering the cracking process mechanisms through an alternative mechanism involving chemisorption by proton donation and **desorption**, resulting in cracked oil and theoretically restored catalyst. Thus, it is not surprising that all cracking catalysts are poisoned by proton-accepting vanadium.

The catalyst to oil volume ratios range from 5:1 to 30:1 for the different processes, although most processes are operated to 10:1. However, for moving-bed processes, the catalyst-to-oil volume ratios may be substantially lower than 10:1.

18.4.1 CATALYST TREATMENT

The latest technique developed by the refining industry to increase gasoline yield and quality is to treat the catalysts from the cracking units to remove metal poisons that accumulate on the catalyst (Gerber et al., 1999). Nickel, vanadium, iron, and copper compounds contained in catalytic cracking feedstocks are deposited on the catalyst during the cracking operation, thereby adversely affecting both catalyst activity and selectivity. Increased catalyst metal contents affect catalytic cracking yields by increasing coke formation, decreasing gasoline and butane and butylene production, and increasing hydrogen production.

The recent commercial development and adoption of cracking catalyst-treating processes definitely improve the overall catalytic cracking process economics.

18.4.1.1 Demet

A cracking catalyst is subjected to two pretreatment steps. The first step affects vanadium removal; the second, nickel removal, to prepare the metals on the catalyst for chemical conversion to compounds (chemical treatment step) that can be readily removed through water washing (catalyst wash step). The treatment steps include use of a sulfurous compound followed by chlorination with an anhydrous chlorinating agent (e.g., chlorine gas) and washing with an aqueous solution of a chelating agent [e.g., citric acid, $HO_2CCH_2C(OH)$ $(CO_2H)CH_2CO_2H$, 2-hydroxy-1,2,3-propanetricarboxylic acid]. The catalyst is then dried and further treated before returning to the cracking unit.

18.4.1.2 Met-X

This process consists of cooling, mixing, and ion-exchange separation, filtration, and resin regeneration. Moist catalyst from the filter is dispersed in oil and returned to the cracking reactor in a slurry. On a continuous basis, the catalyst from a cracking unit is cooled and then transported to a stirred reactor and mixed with an ion-exchange resin (introduced as slurry). The catalyst-resin slurry then flows to an elutriator for separation. The catalyst slurry is taken overhead to a filter, and the wet filter cake is slurried with oil and pumped into the catalytic cracked feed system. The resin leaves the bottom of the elutriator and is regenerated before returning to the reactor.

18.5 PROCESS PARAMETERS

Catalytic cracking is endothermic and, that being the case, heat is absorbed by the reactions and the temperature of reaction mixture declines as the reactions proceed and a source of heat for the process is required. This heat comes from combustion of coke formed in the process. Coke is one of the important, though undesirable, products of cracking as it forms on the surface and in the pores of the catalyst during the cracking process, covering active sites and deactivating the catalysts. During regeneration, this coke is burned off the catalyst to restore catalytic activity and, like all combustion processes, the process is exothermic, liberating heat.

Most fluid catalytic cracking units are operated to maximize conversion to gasoline and LPG. This is particularly true when building gasoline inventory for peak season demand or reducing clarified oil yield due to low market demand. Maximum conversion of a specific feedstock is usually limited by both FCCU design constraints (i.e., regenerator temperature, wet gas capacity, etc.) and the processing objectives. However, within these limitations, the FCCU operator has many operating and catalyst property variables to select from to achieve maximum conversion.

Conversion usually refers to the amount of fresh feedstock cracked to gasoline and lighter products and coke:

$$\text{Conversion} = 100 - (\text{LCO} + \text{HCO} + \text{CO})$$

LCO is light cycle oil, HCO is heavy cycle oil, and CO is clarified oil; all are expressed as a % of the fresh feedstock.

A low conversion operation for maximum production of light cycle oil is typically 40% to 60%, whereas a high conversion operation for maximum gasoline production is 70% to 85%. Again, the range is dependent on the character of the feedstock. Each fluid catalytic cracking unit that is operated for maximum conversion at constant fresh feed quality has an optimum conversion point, beyond which a further increase in conversion reduces gasoline yield and increases the yield of liquefied petroleum gas (Figure 18.5) and the optimum conversion point is referred to as the overcracking point.

18.5.1 REACTOR

The three main components of a fluid catalytic cracking unit are: (1) the reactor, (2) the stripper, and (3) the regenerator.

In the unit, the catalyst and the feed and product hydrocarbons are lifted up the riser pipe to the reactor, where the predominately endothermic cracking processes take place. As the reactions are endothermic, reaction temperature declines from bottom to top. At the top, the mixture enters a solid–gas separator, and the product vapors are led away. Cracked gases

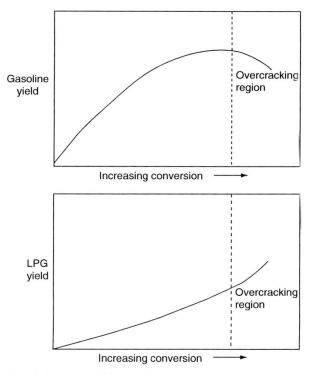

FIGURE 18.5 Illustration of the overcracking region.

are separated and fractionated; the catalyst and residue, together with recycle oil from a second-stage fractionator, pass to the main reactor for further cracking. The products of this second-stage reaction are gas, gasoline and gas oil streams, and recycle oil.

The coked catalyst enters the stripper where steam is added and unreacted-reacted hydrocarbons adsorbed on the catalyst are released. The stripped catalyst is then directed into the regenerator, where air is added and the combustion of coke on the catalyst (and any hydrocarbons still adsorbed, which were not stripped) occurs with the liberation of heat. Regenerator temperatures are typically 705°C to 760°C (1300°F to 1400°F). Heat exchangers and the circulating catalyst capture the heat evolved during regeneration to be used in preheating the reactor feed to appropriate cracking temperatures [usually in the range, 495°C to 550°C (925°F to 1020°F].

During operations, the entire catalyst inventory is continually circulated through the unit. Catalyst residence time in the riser reactor section is typically 1 to 3 sec (with current trends to even shorter residence times), and the entire reactor–stripper–regenerator cycle is less than 10 minutes. To achieve cycle times of this order, catalyst circulation rates as high as 1 ton per sec in large units are required. To withstand such movement, the catalyst must be sufficiently robust to withstand the operational stress.

Process temperatures are high, coke is repeatedly deposited and burned off, and the catalyst particles are kept moving at high speed through steel reactors and pipes. Contact between the catalyst particles, the metal walls, and inter-particle contact are unavoidable. Thus, catalyst loss from the unit caused by poor attrition resistance can be a serious problem, as the quantities lost must be replaced by fresh catalyst additions to maintain constant unit performance. Catalyst manufacturers work hard to prevent inordinate losses due to attrition, and refineries keep a close watch on catalyst quality to be sure the produce conforms to their specifications. Therefore, the robustness of the catalyst is carefully monitored and controlled to a high attrition resistance that is determined by rigorous test methods that place a semi-quantitative evaluation on attrition resistance, which is generally related to breakdown with time in commercial units.

As intimated earlier, in some units cracking does not always take place in the reactor and reaction often occurs in the vertical or upward sloped pipe called the riser (giving credence to the name riser reactor and riser pipe cracking) forming products, including coke (e.g., Bartholic, 1989). Preheated feedstock is sprayed into the base of the riser via feed nozzles, where it contacts extremely hot fluidized catalyst at 1230°F to 1400°F (665°C to 760°C). The hot catalyst vaporizes the feed and catalyzes the cracking reactions that break down the high molecular weight oil into lighter components including liquefied petroleum gas, constituents, gasoline, and diesel. The catalyst–hydrocarbon mixture flows upward through the riser for just a few seconds and then the mixture is separated via cyclones. The catalyst-free hydrocarbons are routed to a main fractionator for separation into fuel gas, propane and butanes, gasoline, light cycle oils used in diesel and jet fuel, and heavy fuel oil.

18.5.2 COKING

The formation of coke deposits has been observed in virtually every unit in operation and the deposits can be very thick, thicknesses up to 4 ft have been reported (McPherson, 1984). Coke has been observed to form where condensation of hydrocarbon vapors occurs. The reactor walls and plenum offer a colder surface where hydrocarbons can condense. Higher boiling constituents in the feedstock may be very close to their dew point, and they will readily condense and form coke nucleation sites on even slightly cooler surfaces.

Unvaporized feed droplets readily collect to form coke precursors on any available surface as the high boiling feedstock constituents do not vaporize at the mixing zone of the riser.

Thus, it is not surprising that residuum processing makes this problem even worse. Low residence time cracking also contributes to coke deposits, as there is less time for heat to transfer to feed droplets and vaporize them. This is an observation in line with the increase in coking when short contact time riser crackers (*q.v.*) were replacing the longer residence time fluid-bed reactors.

Higher boiling feedstocks that have high aromaticity result in higher yields of coke. Further, polynuclear aromatics and aromatics containing heteroatoms (i.e., nitrogen, oxygen, and sulfur) are more facile coke makers than simpler aromatics (Appleby et al., 1976; Speight, 1987). However, feed quality alone is not a foolproof method of predicting where coking will occur. However, it is known that feedstock hydrotreaters rarely have coking problems. The hydrotreating step mitigates the effect of the coke formers and coke formation is diminished.

The recognition that significant post riser cracking occurs in commercial catalytic cracking units and results in substantial production of dry gas and other low valued products (Avidan and Krambeck, 1990). There are two mechanisms by which this post-riser cracking occurs, thermal and dilute phase catalytic cracking.

Thermal cracking results from extended residence times of hydrocarbon vapors in the reactor disengaging area, and leads to high dry gas yields via nonselective free radical cracking mechanisms. On the other hand, dilute phase catalytic cracking results from extended contact between catalyst and hydrocarbon vapors downstream of the riser. Although much of this undesirable cracking was eliminated in the transition from bed to riser cracking, there is still a substantial amount of nonselective cracking occurring in the dilute phase due to the significant catalyst holdup.

Many catalytic cracking units are equipped with advanced riser termination systems to minimize post-riser cracking (Long et al., 1993). However, due to the complexity and diversity of catalytic cracking units, there are many variations of these systems and many such as closed cyclones and many designs are specific to the unit configuration, but all serve the same fundamental purpose of reducing the undesirable post-riser reactions. Further, there are many options for taking advantage of reduced post-riser cracking to improve yields. A combination of higher reactor temperature, higher catalyst-to-oil ratio, higher feed rate, and poorer quality feed is typically employed. Catalyst modification is also appropriate, and typical catalyst objectives such as low coke and dry gas selectivity are reduced in importance due to the process changes, whereas other features such as activity stability and bottoms cracking selectivity become more important for the new unit constraints.

Certain catalyst types seem to increase coke deposit formation. For example, those catalysts (some rare earth zeolites) that tend to form aromatics from naphthenes as a result of secondary hydrogen transfer reactions, and the catalysts that contribute to coke formation indirectly, because the products that they produce have a greater tendency to be coke precursors. In addition, high zeolite content, low surface areas cracking catalysts are less efficient at heavy oil cracking than many amorphous catalysts because the nonzeolite catalysts contain a matrix that is better able to crack heavy oils and convert the coke precursors. The active matrix of some modern catalysts serves the same function.

Once coke is formed, it is a matter of where it will appear. Coke deposits are most often found in the reactor (or disengager), transfer line, and slurry circuit and cause major problems in some units such as increased pressure drops, when a layer of coke reduces the flow through a pipe, or plugging, when chunks of coke spall off and block the flow completely. Deposited coke is commonly observed in the reactor as a black deposit on the surface of the cyclone barrels, reactor dome, and walls. Coke is also often deposited on the cyclone barrels, 180° away from the inlet. Coking within the cyclones can be potentially very troublesome, as any coke spalls going down into the dipleg could restrict catalyst flow or jam the flapper valve. Either situation reduces cyclone efficiency and can increase catalyst losses from the reactor.

Coke formation also occurs at nozzles, which can increase the nozzle pressure drop. It is possible for steam or instrument nozzles to be plugged completely, a serious problem in the case of unit instrumentation.

Coking in the transfer line between the reactor and main fractionator is also common, especially at the elbow where it enters the fractionator. Transfer line coking causes pressure drop and spalling and can lead to reduced throughput. Further, any coke in the transfer line that spalls off can pass through the fractionator into the circulating slurry system where it is likely to plug up exchangers, resulting in lower slurry circulation rates and reduced heat removal. Pressure balance is obviously affected if the reactor has to be run at higher pressures to compensate for transfer line coking. On units where circulation is limited by low slide valve differentials, coke laydown may then indirectly reduce catalyst circulation. The risk of a flow reversal is also increased. In units with reactor grids, coking increases grid pressure drop, which can directly affect the catalyst circulation rate.

Shutdowns and startups can aggravate problems due to coking. The thermal cycling leads to differential expansion and contraction between the coke and the metal wall that will often cause the coke to spall in large pieces. Another hazard during shutdowns is the possibility of an internal fire when the unit is opened up to the atmosphere. Proper shutdown procedures, which ensure that the internals have sufficiently cooled before air enters the reactor, will eliminate this problem. In fact, the only defense against having coke plugging problems during startup is to thoroughly clean the unit during the turnaround and remove all the coke. If there are strainers on the lines they will have to be cleaned frequently.

The two basic principles to minimize coking are to avoid dead spots and prevent heat losses. An example of minimizing dead spots is using purge steam to sweep out stagnant areas in the disengager system. The steam prevents collection of high-boiling condensable products in the cooler regions. Steam also provides a reduced partial pressure or steam distillation effect on the high boiling constituents and causes enhanced vaporization at lower temperatures. Steam for purging should preferably be superheated, as medium-pressure low-velocity steam in small pipes with high heat losses is likely to be very wet at the point of injection and will cause more problems. Cold spots are often caused by heat loss through the walls, in which case, increased thermal resistance might help reduce coking. The transfer line, being a common source of coke deposits, should be as heavily insulated as possible, provided stress-related problems have been taken into consideration.

In some cases, changing the catalyst type or the use of an additive (q.v.) can alleviate coking problems. The catalyst types that appear to result in the least coke formation (not delta coke or catalytic coke) contain low or zero earth zeolites with moderate matrix activities. Eliminating heavy recycle streams can lead to reduced coke formation. As clarified oil is a desirable feedstock to make needle coke in a coker, then it must also be a potential coke maker in the disengager.

One of the trends in recent years has been to improve product yields by means of better feed atomization. The ultimate objective is to produce an oil droplet, small enough so that a single particle of catalyst will have sufficient energy to vaporize it. This has the double benefit of improving cracking selectivity and reducing the number of liquid droplets that can collect to form coke nucleation sites.

18.5.3 CATALYST VARIABLES

The primary variables available to the operation of fluid catalytic cracking units for maximum unit conversion for a given feedstock quality include catalytic variables such as: (1) catalyst activity and (2) catalyst design, which includes availability of cracking sites and the presence of carbon on the regenerated catalyst.

The equilibrium catalyst activity, as measured by a microactivity test (MAT), is a measure of the availability of zeolite and active matrix cracking sites for conversion. Therefore, an increase in the unit activity can effect an increase in conversion and activity is increased by one, or a combination of: (1) increased fresh catalyst addition rate, (2) increased fresh catalyst zeolite activity, (3) increased fresh catalyst matrix activity, (4) addition of catalyst additives to trap or passivate the deleterious effects of feed nitrogen, alkalis (i.e., calcium and sodium), vanadium, and other feed metal contaminants, and (5) increased fresh catalyst matrix surface area to trap or remove feedstock contaminants.

In general, a two-digit increase in the activity as determined by the microactivity test activity appears to coincide with a 1% absolute increase in conversion. The increased matrix surface area improves conversion by providing more amorphous sites for cracking high boiling range compounds in the feedstock, which cannot be cracked by the zeolite. Increased zeolite, on the other hand, provides the necessary acid cracking sites for selectively cracking the amorphous cracked high boiling compounds and lighter boiling compounds.

In addition to zeolite and matrix activity, many of the catalyst's physical and chemical properties (catalyst design) contribute to increased conversion through selectivity differences. These include zeolite type, pore size distribution, relative matrix to total surface area, and chemical composition.

Increasing the concentration of catalyst in the reactor, often referred to as catalyst to oil ratio, will increase the availability of cracking for maximum conversion, assuming the unit is not already operating at a catalyst circulation limit. This can be achieved by increasing reactor heat load or switching to a lower coke selective (i.e., lower delta coke) catalyst. Reactor heat load can be raised by increased reactor temperature or lower feed preheat temperature. This, in turn, increases the catalyst-to-oil ratio, to maintain the unit in heat balance.

The lower the carbon on regenerated catalyst, the higher the availability of cracking sites, as less coke is blocking acid cracking sites. The carbon on the regenerated catalyst is reduced by increasing regeneration efficiency through the use of carbon monoxide oxidation promoters. Carbon on the regenerated catalyst can also be reduced by more efficient air and spent catalyst contact. Increased regenerator bed levels also reduce the amount of carbon on the regenerated catalyst through increased residence time, but this must be traded off with reduced dilute phase disengager residence time and the possibility of increased catalyst losses.

18.5.4 PROCESS VARIABLES

As already noted, there are primary variables available to the operation of FCCUs for maximum unit conversion for a given feedstock quality, which can be divided into two groups: catalytic variables and process variables. In addition to the catalyst variables (*q.v.*) there are also process variables that include (1) pressure, (2) reaction time, (5) reactor temperature. Higher conversion and coke yield are thermodynamically favored by higher pressure. However, pressure is usually varied over a very narrow range due to limited air blower horsepower. Conversion is not significantly affected by unit pressure as a substantial increase in pressure is required to significantly increase conversion.

An increase in reaction time available for cracking also increases conversion. Fresh feed rate, riser steam rate, recycle rate, and pressure are the primary operating variables that affect reaction time for a given unit configuration. Conversion varies inversely with these stream rates due to limited reactor size available for cracking. Conversion has been increased by a decrease in rate in injection of fresh feedstock. Under these circumstances, overcracking of gasoline to liquefied petroleum gas and to dry gas may occur due to the increase in reactor residence time. One approach to offset any potential gasoline overcracking is to add additional riser steam to lower hydrocarbon partial pressure for more selective cracking.

Alternatively, an operator may choose to lower reactor pressure or increase the recycle rate to decrease residence time. Gasoline overcracking may be controlled by reducing the availability of catalytic cracking sites by lowering the catalyst-to-oil ratio.

Increased reactor temperature increases feedstock conversion, primarily through a higher rate of reaction for the endothermic cracking reaction and also through increased catalyst-to-oil ratio. A 10°F increase in reactor temperature can increase conversion by 1% to 2% absolute but again, this is feedstock dependent. Higher reactor temperature also increases the amount of olefins in gasoline and in the gases. This is due to the higher rate of primary cracking reactions relative to secondary hydrogen transfer reactions, which saturate olefins in the gasoline boiling range and lower gasoline octane.

However, these variables are not always available for maximizing conversion, as most FCCUs operate at an optimum conversion level corresponding to a given feed rate, feed quality, set of processing objectives, and catalyst at one or more unit constraints (e.g., wet gas compressor capacity, fractionation capacity, air blower capacity, reactor temperature, regenerator temperature, catalyst circulation). Once the optimum conversion level is found, there are very few additional degrees of freedom for changing the operating variables.

18.5.5 Additives

In addition to the cracking catalyst described earlier, a series of additives has been developed, which catalyze or otherwise alter the primary catalyst's activity or selectivity or act as pollution control agents. Additives are most often prepared in microspherical form to be compatible with the primary catalysts and are available separately in compositions that (1) enhance gasoline octane and light olefin formation, (2) selectively crack heavy cycle oil, (3) passivate vanadium and nickel present in many heavy feedstocks, (4) oxidize coke to carbon dioxide, and (5) reduce sulfur dioxide emissions.

Both vanadium and nickel deposit on the cracking catalyst and are extremely deleterious when present in excess of 3000 ppm on the catalyst. Formulation changes to the catalyst can improve tolerance to vanadium and nickel, but the use of additives that specifically passivate either metal is often preferred.

REFERENCES

Appleby, W., Gibson, J., and Good, G. 1976. *Ind. Eng. Chem. Process Design Dev.* 1: 102.

Avidan, A.A. and Krambeck, F.J. 1990. FCC closed cyclone system eliminates post riser cracking. *Proceedings.* Annual Meeting. National Petrochemical and Refiners Association.

Bartholic, D.B. 1981a. United States Patent 4,243,514. Preparation of FCC Charge from Residual Fractions. January 6.

Bartholic, D.B. 1981b. United States Patent 4,263,128. Upgrading petroleum and residual fractions thereof. April 21.

Bartholic, D.B. 1989. United States Patent 4,804,459. Process for upgrading tar sand bitumen. February 14.

Bartholic, D.B. and Haseltine, R.P. 1981. *Oil Gas J.* 79(45): 242.

Bland, W.F. and Davidson, R.L. 1967. *Petroleum Processing Handbook.* McGraw-Hill, New York.

Bradley, S.A., Gattuso, M.J., and Bertolacini, R.J. 1989. *Characterization and Catalyst Development.* Symposium Series No. 411. American Chemical Society, Washington, DC.

DeCroocq, D. 1984. *Catalytic Cracking of Heavy Petroleum Hydrocarbons.* Editions Technip, Paris.

Gerber, M.A., Fulton, J.L., Frye, J.G., Silva, L.J., Bowman, L.E., and Wai, C.M. 1999. *Regeneration of Hydrotreating and FCC Catalysts.* Report No. PNNL-13025. U.S. Department of Energy Contract No. DE-AC06-76RLO 1830. Pacific Northwest National Laboratory, Richland, WA.

Germain, G.E. 1969. *Catalytic Conversion of Hydrocarbons.* Academic Press, New York.

Heinrich, G. and Mauleon, J-L. 1994. *Rev. Inst. Franç. Pétrol.* 49(5): 509–520.

Inai, K. 1994. *Rev. Inst. Franç. Pétrol.* 49(5): 521–527.

Hemler, C.L. 1997. *Handbook of Petroleum Refining Processes.* Meyers, R.A., ed. McGraw-Hill, New York. Chapter 3.3.

Hunt, D.A. 1997. *Handbook of Petroleum Refining Processes.* Meyers, R.A., ed. McGraw-Hill, New York. Chapter 3.5.

Johnson, T.E. and Niccum, P.K. 1997. *Handbook of Petroleum Refining Processes.* Meyers, R.A., ed. McGraw-Hill, New York. Chapter 3.2.

Ladwig, P.K. 1997. *Handbook of Petroleum Refining Processes.* Meyers, R.A., ed. McGraw-Hill, New York. Chapter 3.1.

Le Page, J.F., Cosyns, J., Courty, P., Freund, E., Franck, J.P., Jacquin, Y., Juguin, B., Marcilly, C., Martino, G., Miguel, J., Montarnal, R., Sugier, A., and von Landeghem, H. 1987. *Applied Heterogeneous Catalysis.* Editions Technip, Paris.

Long, S.L., Johnson, A.R., and Dharia, D. 1993. Advances in residual oil FCC. *Proceedings.* Annual Meeting. National Petrochemical and Refiners Association. Paper AM-93-50

Luckenbach, E.C., Worley, A.C., Reichle, A.D., and Gladrow, E.M. 1992. *Petroleum Processing Handbook.* McKetta, J.J., ed. Marcel Dekker Inc., New York. p. 349.

McPherson, L.J. 1984. *Oil Gas J.* (September 10): 139.

Sadeghbeigi, R. 1995. *Fluid Catalytic Cracking: Design, Operation, and Troubleshooting of FCC Facilities.* Gulf Publishing Company, Houston, TX.

Speight, J.G. 1986. *Ann. Rev. Energy* 11: 253.

Speight, J.G. 1987. Preprints, *Am. Chem. Soc. Div. Petrol. Chem.* 32(2): 413.

Speight J.G. 2000. *The Desulfurization of Heavy Oils and Residua.* 2nd Edition. Marcel Dekker Inc., New York.

Speight, J.G. 2004. *Catalysis Today* 98(1–2): 55–60.

Speight, J.G. and Ozum, B. 2002. *Petroleum Refining Processes.* Marcel Dekker Inc., New York.

Thakur, D.S. 1985. *Appl. Catalysis* 15: 197.

19 Deasphalting and Dewaxing Processes

19.1 INTRODUCTION

Solvent deasphalting processes are a major part of refinery operations (Bland and Davidson, 1967; Hobson and Pohl, 1973; Gary and Handwerk, 1994; Speight and Ozum, 2002), and are not often appreciated for the tasks for which they are used. In the solvent deasphalting processes, an alkane is injected into the feedstock to disrupt the dispersion of components and cause the polar constituents to precipitate. Propane (or sometimes propane and butane mixtures) is extensively used for deasphalting and produces a deasphalted oil (DAO) and propane deasphalter asphalt (PDA or PD tar) (Dunning and Moore, 1957). Propane has unique solvent properties; at lower temperatures (38°C to 60°C; 100°F to 140°F), paraffins are very soluble in propane and at higher temperatures (about 93°C; 200°F), all hydrocarbons are almost insoluble in propane.

A solvent deasphalting unit processes the residuum from the vacuum distillation unit and produces deasphalted oil (DAO), used as feedstock for a fluid catalytic cracking unit, and the asphaltic residue (deasphalter tar, deasphalter bottoms) which, as a residual fraction, can only be used to produce asphalt or as a blend stock or visbreaker feedstock for low-grade fuel oil. Solvent deasphalting processes have not realized their maximum potential. With on-going improvements in energy efficiency, such processes would display its effects in combination with other processes. **Solvent deasphalting** allows removal of sulfur and nitrogen compounds, as well as metallic constituents, by balancing yield with the desired feedstock properties (Ditman, 1973).

19.2 COMMERCIAL PROCESSES

Petroleum processing normally involves separation into various fractions that require further processing in order to produce marketable products. The initial separation process is distillation (Chapter 14), in which crude oil is separated into fractions of increasingly higher boiling range. Since petroleum fractions are subject to thermal degradation, there is a limit to the temperatures that can be used in simple separation processes. The crude cannot be subjected to temperatures much above 395°C (740°F), irrespective of the residence time, without encountering some thermal cracking. Therefore, to separate the higher molecular weight and higher boiling fractions from crude, special processing steps must be used.

Thus, although a crude oil might be subjected to atmospheric distillation and vacuum distillation, there may still be some valuable oils left in the vacuum-residuum. These valuable oils are recovered by solvent extraction, and the first application of solvent extraction in refining was the recovery of heavy lube oil base stocks by propane (C_3H_8) deasphalting. In order to recover more oil from vacuum-reduced crude, mainly for catalytic cracking feedstocks, higher molecular weight solvents such as butane (C_4H_{10}), and even pentane (C_5H_{12}), have been employed.

19.2.1 Deasphalting Process

The deasphalting process is a mature process but, as refinery operations evolve, it is necessary to include a description of the process here, so that the new processes might be compared with new options that also provide for deasphalting various feedstocks. Indeed, several of these options, such as the ROSE process, have been on-stream for several years and are included here for this reason. Thus, this section provides a one-stop discussion of solvent recovery processes and their integration into refinery operations.

The separation of residua into oil and asphalt fractions was first performed on a production scale by mixing the vacuum residuum with propane (or mixtures of normally gaseous hydrocarbons) and continuously decanting the resulting phases in a suitable vessel. Temperature was maintained within about 55°C (100°F) of the critical temperature of the solvent, at a level that would regulate the yield and properties of the deasphalted oil in solution and that would reject the heavier undesirable components as asphalt.

Currently, deasphalting and delayed coking are used frequently for residuum conversion. The high demand for petroleum coke, mainly for use in the aluminum industry, has made delayed coking a major residuum conversion process. However, many crude oils will not produce coke that meets the sulfur and metals specifications for aluminum electrodes, and coke gas oils are less desirable feedstocks for fluid catalytic cracking than virgin gas oils. In comparison, the solvent deasphalting process can apply to most vacuum residua. The deasphalted oil is an acceptable feedstock for both fluid catalytic cracking and, in some cases, hydrocracking. Since it is relatively less expensive to desulfurize the deasphalted oil than the heavy vacuum residuum, the solvent deasphalting process offers a more economical route for disposing of vacuum residuum from high sulfur crude. However, the question of disposal of the asphalt remains. Use as a road asphalt is common and as a refinery fuel is less common, since expensive stack gas clean-up facilities may be required when used as fuel.

In the process (Figure 19.1), the feedstock is mixed with dilution solvent from the solvent accumulator and then cooled to the desired temperature before entering the extraction **tower**. Because of its high viscosity, the charge oil can neither be cooled easily to the required temperature nor will it mix readily with solvent in the extraction tower. By adding a relatively small portion of solvent upstream of the charge cooler (insufficient to cause phase separation), the viscosity problem is avoided.

The feedstock, with a small amount of solvent, enters the extraction tower at a point about two-thirds up the column. The solvent is pumped from the accumulator, cooled, and enters near the bottom of the tower. The extraction tower is a multistage contactor, normally equipped with baffle trays and the heavy oil flows downward, while the light solvent flows upward. As the extraction progresses, the desired oil goes to the solvent and the asphalt separates and moves toward the bottom. As the extracted oil and solvent rise in the tower, the temperature is increased in order to control the quality of the product by providing adequate reflux for optimum separation. Separation of oil from asphalt is controlled by maintaining a temperature gradient across the extraction tower and by varying the solvent to oil ratio. The tower top temperature is regulated by adjusting the feed inlet temperature and the steam flow to the heating coils at the top of the tower. The temperature at the bottom of the tower is maintained at the desired level by the temperature of the entering solvent. The deasphalted oil–solvent mixture flows from the top of the tower under pressure control to a kettle-type evaporator heated by low-pressure steam. The vaporized solvent flows through the condenser into the solvent accumulator.

The liquid phase flows from the bottom of the evaporator, under level control, to the deasphalted oil flash tower where it is re-boiled by means of a fired heater. In the flash tower,

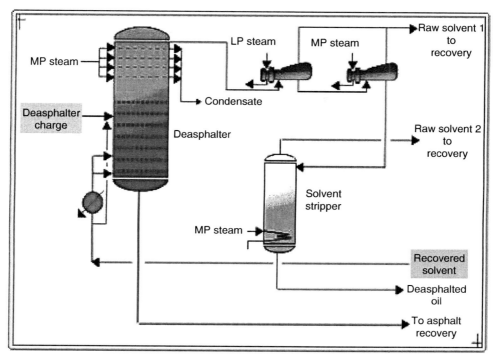

FIGURE 19.1 A deasphalting unit. (From OSHA Technical Manual, Section IV, Chapter 2, Petroleum Refining Processess.)

most of the remaining solvent is vaporized and flows overhead, joining the solvent from the low-pressure steam evaporator. The deasphalted oil, with relatively minor solvent, flows from the bottom of the flash tower under level control to a steam stripper operating at essentially atmospheric pressure. Superheated steam is introduced into the lower portion of the tower. The remaining solvent is stripped out and flows overhead with the steam through a condenser into the compressor suction drum, where the water drops out. The water flows from the bottom of the drum under level control to appropriate disposal.

The asphalt–solvent mixture is pressured from the extraction tower bottom on flow control to the asphalt heater and on to the asphalt flash drum, where the vaporized solvent is separated from the asphalt. The drum operates essentially at the solvent condensing pressure, so that the overhead vapors flow directly through the condenser into the solvent accumulator. Hot asphalt with a small quantity of solvent flows from the asphalt flash drum bottom, under level control, to the asphalt stripper that is operated at near atmospheric pressure. Superheated steam is introduced into the bottom of the stripper. The steam and solvent vapors pass overhead, join the deasphalted oil stripper overhead, and flow through the condenser into the compressor suction drum. The asphalt is pumped from the bottom of the stripper under level control to storage.

The propane deasphalting process is similar to solvent extraction, in that a packed or baffled extraction tower or rotating disc contactor is used to mix the oil feed stocks with the solvent. In the tower method, 4 to 8 volumes of propane are fed to the bottom of the tower for every volume of feed flowing down from the top of the tower. The oil, which is more soluble in the propane, dissolves and flows to the top. The asphaltene and resins flow to the bottom of the tower, where they are removed in a propane mix. Propane is recovered from the two streams through two-stage flash systems, followed by steam stripping in which propane is condensed and removed by cooling at high pressure in the first stage and at low

TABLE 19.1
Feedstock and Product Data for the Deasphalting Process

Crude Source	Arab	West Texas	California	Canadian	Kuwait	Kuwait
Feedstock						
Crude, vol%	23.0	29.2	20.0	16.0	22.2	32.3
Gravity, °API	6.8	12.0	6.3	9.6	5.6	8.1
Conradson carbon, wt%	15.0	12.1	22.2	18.9	24.0	19.7
SUS at 210°F	75,000	526	9600	1740	14,200	3270
Metals, wppm						
Ni	73.6	16.0	139	46.6	29.9	29.7
V	365.0	27.6	136	30.9	110.0	89
Cu + Fe	15.5	14.8	94	40.7	13.7	7.5
Deasphalted oil						
Vol% feed	49.8	66.0	52.8	67.8	45.6	54.8
Gravity, °API	18.1	19.6	18.3	17.8	16.2	17.1
Conradson carbon, wt%	5.9	2.2	5.3	5.4	4.5	5.4
SUS at 210°F	615	113	251	250	490	656
Metals, wppm						
Ni	3.5	1.0	8.1	3.9	0.9	0.6
V	12.4	1.3	2.3	1.4	0.7	4.0
Cu + Fe	0.2	0.8	3.5	0.2	0.8	0.8
Asphalt						
Vol% feed	50.2	34.0	47.2	32.2	54.4	45.2
Gravity, °API	−1.3	−0.9	−5.1	−5.1	−1.3	−2.0

Source: Speight, J.G. and Ozum, B. 2002. Petroleum Refining Processes. Marcel Dekker Inc., New York. With permission.

pressure in the second stage. The asphalt recovered can be blended with other asphalts or heavy fuels, or can be used as feed to the coker.

The yield of deasphalted oil varies with the feedstock (Table 19.1; Figure 19.2), but the deasphalted oil does make less coke and more distillate than the feedstock. Therefore,

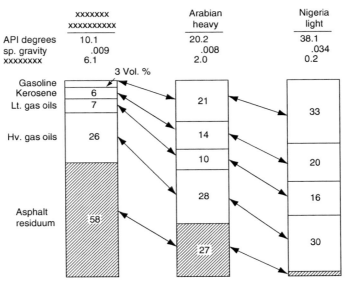

FIGURE 19.2 Variation of composition of selected crude oils.

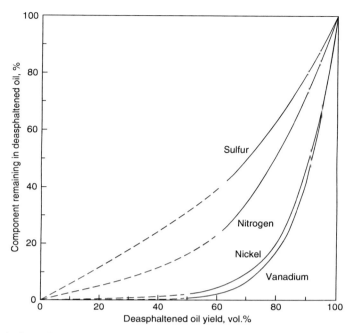

FIGURE 19.3 Variation of deasphalted oil properties with yield.

the process parameters for a deasphalting unit must be selected with care, according to the nature of the feedstock and the desired final products. The metals content of the deasphalted oil is relatively low and the nitrogen and sulfur contents in the deasphalted oil are also related to the yield of deasphalted oil yield (Figure 19.3). The character of the deasphalting process is a molecular weight separation in which the solvent takes a cross-cut across the feedstock, effecting separation by molecular weight and by polarity (Figure 19.4).

Further to the selection of the process parameters, the choice of solvent is vital to the flexibility and performance of the unit. The solvent must be suitable, not only for

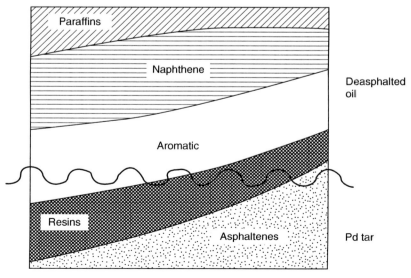

FIGURE 19.4 Illustration of the deasphalting process on the basis of molecular weight and polarity (PD, propane deasphalter tar or bottoms).

the extraction of the desired oil fraction, but also for control of the yield and quality of the deasphalted oil at temperatures that are within the operating limits. If the temperature is too high (i.e., close to the critical temperature of the solvent), the operation becomes unreliable in terms of product yields and character. If the temperature is too low, the feedstock may be too viscous and have an adverse effect on contact with the solvent in the tower.

Liquid propane is by far the most selective solvent among the light hydrocarbons used for deasphalting. At temperatures ranging from 38°C to 65°C (100°F to 150°F), most hydrocarbons are soluble in propane, while asphaltic and resinous compounds are not, thereby allowing rejection of these compounds, resulting in a drastic reduction (relative to the feedstock) of the nitrogen content and the metals in the deasphalted oil. Although the deasphalted oil from propane deasphalting has the best quality, the yield is usually less than the yield of deasphalted oil produced using a higher molecular weight (higher boiling) solvent.

The ratios of propane to oil required vary from 6 to 1, to 10 to 1 by volume, with the ratio occasionally being as high as 13 to 1. Since the critical temperature of propane is 97°C (206°F), this limits the extraction temperature to about 82°C (180°F). Therefore, propane alone may not be suitable for high viscosity feedstocks because of the relatively low operating temperature.

Iso-butane and *n*-butane are more suitable for deasphalting high viscosity feedstocks, since their critical temperatures are higher (134°C, 273°F, and 152°C, 306°F, respectively) than that of the critical temperature of propane. Higher extraction temperatures can be used to reduce the viscosity of the heavy feed and to increase the transfer rate of oil to solvent. Although *n*-pentane is less selective for metals and carbon residue removal, it can increase the yield of deasphalted oil from a heavy feed by a factor of 2 to 3 over propane (Speight, 2000, Speight and Ozum 2002). However, if the content of the metals and carbon residue of the pentane-deasphalted oil is too high (defined by the ensuing process), the deasphalted oil may be unsuitable as a cracking feedstock. In certain cases, the nature of the cracking catalyst may dictate that the pentane-deasphalted oil be blended with vacuum gas oil that, after further treatment such as hydrodesulfurization, produces a good cracking feedstock.

Solvent composition is an important process variable for deasphalting units. The use of a single solvent may (depending on the nature of the solvent) limit the range of feedstocks that can be processed in a deasphalting unit. When a deasphalting unit is required to handle a variety of feedstocks and produce various yields of deasphalted oil (as is the case in these days of variable feedstock quality), a dual solvent may be the only option to provide the desired flexibility. For example, a mixture of propane and *n*-butane might be suitable for feedstocks that vary from vacuum residua to both the heavy resid to heavy gas oils that contain asphaltic materials. Adjusting the solvent composition allows the most desirable product quantity and quality within the range of temperature control.

Besides the solvent composition, the solvent to oil ratio also plays an important role in a deasphalting operation. Solvent to oil ratios vary considerably and are governed by feedstock characteristics and desired product qualities and, for each individual feedstock, there is a minimum operable solvent to oil ratio. Generally, increasing the solvent to oil ratio almost invariably results in improving the deasphalted oil quality at a given yield, but other factors must also be taken into consideration and (generalities aside) each plant and feedstock will have an optimum ratio.

The main consideration in the selection of the operating temperature is its effect on the yield of deasphalted oil. For practical applications, the lower limits of operable temperature are set by the viscosity of the oil-rich phase. When the operating temperature is near the critical temperature of the solvent, control of the extraction tower becomes difficult, since the rate of change of solubility with temperature becomes very large at conditions close to the critical point of the solvent. Such changes in solubility cause large amounts of oil to transfer between the solvent-rich and the oil-rich phases that, in turn, cause **flooding** and uncontrollable changes in product quality. To mitigate such effects, the upper limits of operable temperatures must lie below the critical temperature of the solvent in order to ensure good control of the product quality and to maintain a stable condition in the extraction tower.

The temperature gradient across the extraction tower influences the sharpness of separation of the deasphalted oil and the asphalt, because of internal reflux that occurs when the cooler oil and solvent solution in the lower section of the tower attempt to carry a large portion of oil to the top of the tower. When the oil or solvent solution reaches the steam-heated, higher-temperature area near the top of the tower, some oil of higher molecular weight in the solvent solution is rejected, because the oil is less soluble in solvent at higher temperature. The heavier oil (rejected from the solution at the top of the tower) attempts to flow downward and causes the internal reflux. In fact, generally, the greater the temperature difference between the top and the bottom of the tower, the greater will be the internal reflux and the better will be the quality of the deasphalted oil. However, too much internal reflux can cause tower flooding and jeopardize the process.

The process pressure is usually not considered to be an operating variable, since it must be higher than the vapor pressure of the solvent mixture at the tower operating temperature to maintain the solvent in the liquid phase. The tower pressure is usually only subject to change when there is a need to change the solvent composition or the process temperature.

Proper contact and distribution of the oil and solvent in the tower are essential to the efficient operation of any deasphalting unit. In early units, days, mixer-settlers were used as contactors, but proved to be less efficient than the countercurrent contacting devices. Packed towers are difficult to operate in this process because of the large differences in viscosity and density between the asphalt phase and the solvent-rich phase.

The extraction tower for solvent deasphalting consists of two contacting zones: (1) a rectifying zone above the oil feed and (2) a stripping zone below the oil feed. The rectifying zone contains some elements designed to promote contacting and to avoid channeling. Steam-heated coils are provided to raise the temperature sufficiently to induce an oil-rich reflux in the top section of the tower. The stripping zone has disengaging spaces at the top and the bottom, and consists of contacting elements between the oil inlet and the solvent inlet.

A countercurrent tower with static baffles is widely used in solvent deasphalting service. The baffles consist of fixed elements formed of expanded metal gratings in groups of two or more to provide maximum change of direction without limiting capacity. The rotating disk contactor has also been employed and consists of disks connected to a rotating shaft that are used in place of the static baffles in the tower. The rotating element is driven by a variable speed drive at either the top or the bottom of the column, and operating flexibility is provided by controlling the speed of the rotating element and, thus, the amount of mixing in the contactor.

In the deasphalting process, the solvent is recovered for circulation, and the efficient operability of a deasphalting unit is dependent on the design of the solvent recovery system.

The solvent may be separated from the deasphalted oil in several ways, such as, conventional evaporation or the use of a flash tower. Irrespective of the method of solvent recovery from the deasphalted oil, it is usually the most efficient to recover the solvent at a temperature close to the extraction temperature. If a higher temperature for solvent recovery is used, heat is wasted in the form of high vapor temperature and, conversely, if a lower temperature is used, the solvent must be reheated, thereby requiring additional energy input. The solvent recovery pressure should be low enough to maintain a smooth flow under pressure from the extraction tower.

The asphalt solution from the bottom of the extraction tower usually contains less than an equal volume of solvent. A fired heater is used to maintain the temperature of the asphalt solution well above the foaming level and to keep the asphalt phase in a fluid state. A flash drum is used to separate the solvent vapor from asphalt, with the design being such as to prevent carryover of asphalt into the solvent outlet line and to avoid fouling the downstream solvent condenser. The solvent recovery system from asphalt is not usually subject to the same degree of variations as the solvent recovery system for the deasphalted oil, and operation at constant temperature and pressure with a separate solvent condenser and accumulator is possible.

Asphalt from different crude oils varies considerably, but the viscosity is often too high for fuel oil although, in some cases they can be blended with refinery cutter stocks to make No. 6 fuel oil. When the sulfur content of the original residuum is high, even the blend fuel oil will not be able to meet the sulfur specification of fuel oil, unless stack gas clean up is available.

The deasphalted oil and solvent asphalt are not finished products and require further processing or blending, depending on the final use. Manufacture of **lubricating oil** is one possibility, and the deasphalted oil may also be used as a catalytic cracking feedstock or it may be desulfurized. It is perhaps these last two options that are more pertinent to the present text and future refinery operations.

Briefly, catalytic cracking or hydrodesulfurization of atmospheric and vacuum residua from high-sulfur or high-metal crude oil is, theoretically, the best way to enhance their value. However, the concentrations of sulfur (in the asphaltene fraction) in the residua can severely limit the performance of cracking catalysts and hydrodesulfurization catalysts (Speight, 2000). Both processes generally require tolerant catalysts, as well as (in the case of hydrodesulfurization), high hydrogen pressure, low space velocity, and high hydrogen recycle ratio.

For both processes, the advantage of using the deasphalting process to remove the troublesome compounds becomes obvious. The deasphalted oil, with no asphaltene constituents and low metal content, is easier to process than the residua. Indeed, in the hydrodesulfurization process, the deasphalted oil may consume only 65% of the hydrogen required for direct hydrodesulfurization of topped crude oil.

As always, the use of the material rejected by the deasphalting unit remains an issue. It can be used (apart from its use for various types of asphalt) as feed to a partial oxidation unit to make a hydrogen-rich gas for use in hydrodesulfurization processes and hydrocracking processes. Alternatively, the asphalt may be treated in a visbreaker to reduce its viscosity, thereby minimizing the need for cutter stock to be blended with the solvent asphalt for making fuel oil. Or, hydrovisbreaking offers an option of converting the asphalt to feedstock for other conversion processes.

19.2.2 PROCESS OPTIONS FOR HEAVY FEEDSTOCKS

Solvent deasphalting is a separation process that represents a further step in the minimization of residual byproduct fuel. However, solvent deasphalting processes, far from realizing their maximum potential for heavy feedstocks, are now under further investigation and, with on-going improvements in energy efficiency, such processes are starting to display maximum benefits when used in combination with other processes. The process takes advantage of the fact that maltenes are more soluble in light paraffinic solvents than asphaltene constituents. This solubility increases with solvent molecular weight and decreases with temperature (Girdler, 1965; Mitchell and Speight, 1973). As with vacuum distillation, there are constraints with respect to how deep a solvent deasphalting unit can cut into the residue or how much DAO can be produced. In the case of solvent deasphalting, the constraint is usually related to deasphalted oil quality specifications required by downstream conversion units.

However, solvent deasphalting has the flexibility to produce a wide range of deasphalted oil that matches the desired properties. The process has very good selectivity for asphaltene constituents (and, to a lesser extent, resin constituents) as well as metals rejection. There is also some selectivity for rejection of carbon residue precursors, but there is less selectivity for sulfur-containing and nitrogen-containing constituents. The process is best suited for the more paraffinic vacuum residua with a somewhat lower efficiency, when applied to high-asphaltene residua that contain high proportions of metals and coke-forming constituents. The advantages and disadvantages of the process are that it performs no conversion and produces a very high-viscosity byproduct deasphalter bottoms and, where high-quality deasphalted oil is required, the process is limited in the quality of feedstock that can be economically processed. In those situations where there is a ready outlet for use for the bottoms, solvent deasphalting is an attractive option for treating heavy feedstocks. One such situation is the cogeneration of steam and power, both to supply the refiner's needs and for export to nearby users.

19.2.2.1 Deep Solvent Deasphalting Process

The deep solvent deasphalting process is an application of the low energy deasphalting (LEDA) process (Table 19.2; Figure 19.5) (RAROP, 1991, p. 91; *Hydrocarbon Processing*, 1998, p. 67) that is used to extract high-quality lubricating oil bright stock or prepare catalytic cracking feeds, hydrocracking feeds, hydrodesulfurization unit feeds, and asphalt from vacuum residue materials. The LEDA process uses a low-boiling hydrocarbon solvent, specifically formulated to ensure the most economical deasphalting design for each operation. For example, a propane solvent may be specified for a low deasphalted oil yield operation, while a higher-boiling solvent, such as pentane or hexane, may be used to obtain a high deasphalted oil yield from a vacuum residuum (Table 19.3). The deep deasphalting process can be integrated with a delayed coking operation (*q.v.*, ASCOT process). In this case, the solvent can be low-boiling naphtha (Table 19.4).

Low-energy deasphalting operations are usually carried out in a rotating disc contactor (RDC) that provides more extraction stages than a mixer-settler or baffle type column. Although not essential to the process, the rotating disc contactor provides higher-quality deasphalted oil at the same yield, or higher yields at the same quality. The low-energy solvent deasphalting process selectively extracts more paraffinic components from vacuumresidua

TABLE 19.2
Feedstock and Product Data for the LEDA Process

	Residuum[a]	Residuum[a]
Feedstock		
API	6.5	6.5
Sulfur, wt.%	3.0	3.0
Carbon residue, wt.%	21.8	21.8
Nickel, ppm	46.0	46.0
Vanadium, ppm	125.0	125.0
Products		
Deasphalted oil, vol.%	53.0	65.0
API	17.6	15.1
Sulfur, wt.%	1.9	2.2
Carbon residue, wt.%	3.5	6.2
Nickel, ppm	1.8	4.5
Vanadium, ppm	3.4	0.3

[a]Unspecified.
Source: Speight, J.G. and Ozum, B. 2002. *Petroleum Refining Processes*. Marcel Dekker Inc., New York. With permission.

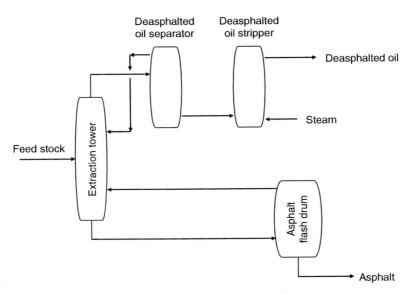

FIGURE 19.5 The LEDA process. (From Speight, J.G., and Ozum, B. 2002. *Petroleum Refining Processes.* Marcel Dekker Inc., New York. With permission.)

TABLE 19.3
Feedstock and Product Data for the ASCOT Process (Compared to Delayed Coking and the LEDA Process)

	ASCOT	Delayed Coking	LEDA
Feedstock			
API	2.8	2.8	2.8
Sulfur, wt.%	4.2	4.2	4.2
Nitrogen, wt.%	1.0	1.0	1.0
Carbon residue, wt.%	22.3	22.3	22.3
Products			
Naphtha, vol.%	7.7	19.4	
API	54.7		
Sulfur, wt.%	1.1		
Nitrogen, wt.%	0.1		
Gas oil, vol.%	69.9	51.8	
API	13.4		
Sulfur, wt.%	3.4		
Nitrogen, wt.%	0.5		
Coke, wt.%	25.0	32.5	
Sulfur, wt.%	5.8	5.7	
Nitrogen, wt.%	2.7	2.6	
Nickel, ppm	774.0	609.0	
Vanadium, ppm	2656.0	2083.0	
Deasphalted oil, vol.%			50.0
Asphalt, vol.%			50.0
Sulfur, wt.%			5.0
Nitrogen, wt.%			1.4
Nickel, ppm			365.0
Vanadium, ppm			1250.0

Source: Speight, J.G. and Ozum, B. 2002. *Petroleum Refining Processes.* Marcel Dekker Inc., New York. With permission.

TABLE 19.4
Feedstock and Product Data for the Demex Process

	Vacuum Residuum[a]	Vacuum Residuum[a]	Arabian Light Vacuum Residuum	Arabian Light Vacuum Residuum	Arabian Heavy Vacuum Residuum	Arabian Heavy Vacuum Residuum
Feedstock						
API	7.2	7.2	6.9	6.9	3.0	3.0
Sulfur, wt.%	4.0	4.0	4.0	4.0	6.0	6.0
Nitrogen, wt.%	0.3	0.3	0.3	0.3	0.5	0.5
Carbon residue, wt.%	20.8	20.8	20.8	20.8	27.7	27.7
Nickel, ppm			23.0	23.0	64.0	64.0
Vanadium, ppm			75.0	75.0	205.0	205.0
Nickel + vanadium, ppm	98.0	98.0				
C_6-asphaltenes, wt.%	10.0	10.0				
C_7-asphaltenes, wt.%			10.0	10.0	15.0	15.0
Products						
Demetallized oil, vol.%	56.0	78.0	40.0	60.0	30.0	55.0
API	16.0	12.0	18.9	15.3	16.3	12.0
Sulfur, wt.%	2.7	3.3	2.3	2.8	3.5	4.3
Nitrogen, wt.%	0.1	0.2	0.1	0.2	0.1	0.2
Carbon residue, wt.%	5.6	10.7	2.9	6.4	4.8	10.1
Nickel + vanadium, ppm	6.0	19.0	2.5	7.2	16.0	38.0
C_6-asphaltenes, wt.%	<0.1	<0.1				
Pitch, vol.%	44.0	22.0				
API	<0.0	<0.0	<0.0	<0.0	<0.0	<0.0
Sulfur, wt.%	5.4	6.3	5.0	5.5	6.9	7.8
Nickel + vanadium, ppm	201.0	341.0	154.0	216.0	364.0	515.0

[a]Unspecified.
Source: Speight, J.G. and Ozum, B. 2002. *Petroleum Refining Processes.* Marcel Dekker Inc., New York. With permission.

while rejecting the condensed ring aromatics. As expected, deasphalted oil yields vary as a function of solvent type and quantity, and feedstock properties (Figure 19.2).

In the process, vacuum residue feed is combined with a small quantity of solvent to reduce its viscosity and cooled to a specific extraction temperature before entering the rotating disc contactor. Recovered solvent from the high-pressure and the low-pressure solvent receivers are combined, adjusted to a specific temperature by the solvent heater-cooler, and injected into the bottom section of the rotating disc contactor. Solvent flows upward, extracting the paraffinic hydrocarbons from the vacuum residuum, which is flowing downward through the rotating disc contactor.

Steam coils at the top of the tower maintain the specified temperature gradient across the rotating disc contactor. The higher temperature in the top section of the rotating disc contactor results in separation of the less-soluble heavier material from the deasphalted oil mix and provides internal reflux, which improves the separation. The deasphalted oil mix leaves the top of the rotating disc contactor tower. It flows to an evaporator, where it is heated to vaporize a portion of the solvent. It then flows into the high-pressure flash tower, where high-pressure solvent vapors are taken overhead.

The deasphalted oil mix from the bottom of this tower flows to the pressure vapor heat exchanger, where additional solvent is vaporized from the deasphalted oil mix by condensing high-pressure flash. The high-pressure solvent, totally condensed, flows to the high-pressure solvent receiver. Partially vaporized, the deasphalted oil mix flows from the pressure vapor heat exchanger to the low-pressure flash tower, where low-pressure solvent vapor is taken overhead, condensed, and collected in the low-pressure solvent receiver. The deasphalted oil mix flows down the low-pressure flash tower to the reboiler, where it is heated, and then to the deasphalted oil stripper, where the remaining solvent is stripped overhead with superheated steam. The deasphalted oil product is pumped from the stripper bottom and is cooled, if required, before flowing to battery limits.

The raffinate phase containing asphalt and small amount of solvent, flows from the bottom of the rotating disc contactor to the asphalt mix heater. The hot, two-phase asphalt mix from the heater is flashed in the asphalt mix flash tower, where solvent vapor is taken overhead, condensed, and collected in the low-pressure solvent receiver. The remaining asphalt mix flows to the asphalt stripper, where the remaining solvent is stripped overhead with super-heated steam. The asphalt stripper overhead vapors are combined with the overhead from the deasphalted oil stripper, condensed, and collected in the stripper drum. The asphalt product is pumped from the stripper and is cooled by generating low-pressure steam.

19.2.2.2 Demex Process

The Demex process is a solvent extraction, demetallizing process that separates high metal vacuum residuum into demetallized oil of relatively low metal content and asphaltene of high metal content (Table 19.4) (Salazar, 1986; RAROP, 1991, p. 93; Houde, 1997) (Chapter 18). The asphaltene and condensed aromatic contents of the demetallized oil are very low. The demetallized oil is a desirable feedstock for fixed-bed hydrodesulfurization and, in cases where the metals content and carbon residue are sufficiently low, is a desirable feedstock for fluid catalytic cracking and hydrocracking units. Overall, the Demex process is an extension of the propane deasphalting process and employs a less selective solvent to recover not only the high-quality oils, but also higher molecular weight aromatics and other constituents present in the feedstock. Further, the Demex process requires a much less solvent circulation in achieving its objectives, thus, reducing the utility costs and unit size significantly. The process selectively rejects asphaltenes, metals, and high molecular weight aromatics from vacuum residua. The resulting demetallized oil can then be combined with vacuum gas oil to give a greater availability of acceptable feed to subsequent conversion units.

In the process, the vacuum residuum feedstock, mixed with Demex solvent recycling from the second stage, is fed to the first stage extractor. The pressure is kept high enough to maintain the solvent in the liquid phase. The temperature is controlled by the degree of cooling of the recycle solvent. The solvent rate is set near the minimum required to ensure the desired separation occurs. Asphaltene constituents are rejected in the first stage. Some resins are also rejected to maintain sufficient fluidity of the asphaltene for efficient solvent recovery. The asphaltene is heated and steam stripped to remove solvent. The first stage overhead is heated by an exchange with hot solvent. The increase in temperature decreases the solubility of resins and high molecular weight aromatics (Mitchell and Speight, 1973). These precipitate in the second stage extractor. The bottom stream of this second stage extractor is recycled to the first stage. A portion of this stream can also be drawn as a separate product.

The overhead from the second stage is heated by an exchange with hot solvent. The fired heater further raises the temperature of the solvent and demetallized oil mixture to a point above the critical temperature of the solvent. This causes the demetallized oil to separate. It is then flashed and steam-stripped to remove all traces of solvent. The vapor streams from the demetallized oil and asphalt strippers are condensed, dewatered, and pumped up to process pressure for recycle. The bulk of the solvent goes overhead in the supercritical separator. This hot solvent stream is then effectively used for process heat exchange. The subcritical solvent recovery techniques, including multiple effect systems, allow much less heat recovery. Most of the low-grade heat in the solvent vapors from the subcritical flash vaporization must be released to the atmosphere requiring additional heat input to the process.

19.2.2.3 MDS Process

The MDS process is a technical improvement of the solvent deasphalting process, particularly effective for upgrading heavy crude oils (Table 19.5) (Kashiwara, 1980; RAROP, 1991, p. 95). Combined with hydrodesulfurization, the process is fully applicable to the feed preparation for fluid catalytic cracking and hydrocracking. The process is capable of using a variety of feedstocks, including atmospheric and vacuum residua derived from various crude oils, tar sand bitumen, and nonvolatile products from a visbreaker.

In the process, the feed and the solvent are mixed and fed to the deasphalting tower. Deasphalting extraction proceeds in the upper half of the tower. After the removal of the asphalt, the mixture of deasphalted oil and solvent flows out of the tower through the tower top. Asphalt flows downward to come in contact with a countercurrent of rising solvent. The contact eliminates oil from the asphalt, and the asphalt then accumulates on the bottom. Deasphalted oil containing solvent is heated through a heating furnace, and fed to the deasphalted oil flash tower, where most of the solvent is separated under pressure. Deasphalted oil still containing a small amount of solvent is again heated and fed to the stripper, where the remaining solvent is completely removed.

Asphalt is withdrawn from the bottom of the extractor. Since this asphalt contains a small amount of solvent, it is heated through a furnace and fed to the flash tower to remove most of the solvent. Asphalt is then sent to the asphalt stripper, where the remaining portion of solvent is completely removed.

Solvent recovered from the deasphalted oil and asphalt flash towers is cooled and condensed into liquid, and sent to a solvent tank. The solvent vapor leaving both strippers is cooled to remove water and compressed for condensation. The condensed solvent is then sent to the solvent tank for further recycling.

TABLE 19.5
Feedstock and Product Data for the MDS Process

	Iranian Heavy Atmospheric Residuum	Kuwait Atmospheric Residuum	Khafji Vacuum Residuum
Feedstock			
API	17.0	16.4	5.2
Sulfur, wt.%	2.7	3.7	5.2
Carbon residue, wt.%	9.1	9.4	21.9
Nickel, ppm	40.0	14.0	49.0
Vanadium, ppm	130.0	48.0	140.0
Products			
Deasphalted oil, vol.%	93.4	93.8	72.4
API	19.0	16.4	11.3
Sulfur, wt.%	2.4	3.7	4.3
Carbon residue, wt.%	5.9	9.4	10.9
Nickel, ppm	18.0	14.0	6.0
Vanadium, ppm	53.0	48.0	28.0
Asphalt, vol.%	6.6	6.2	27.6
API	<0.0	<0.0	<0.0
Sulfur, wt.%	5.4	7.2	7.3
Carbon residue, wt.%			49.3
Nickel, ppm	320.0	113.0	150.0
Vanadium, ppm	1010.0	425.0	400.0

Source: Speight, J.G. and Ozum, B. 2002. *Petroleum Refining Processes*. Marcel Dekker Inc., New York. With permission.

19.2.2.4   Residuum Oil Supercritical Extraction (ROSE) Process

The ROSE process is a solvent deasphalting process with minimum energy consumption using a super-critical solvent recovery system, and the process is of value in obtaining oils for further processing (Gearhart and Gatwin, 1976; Gearhart, 1980; RAROP, 1991, p. 97; Low et al., 1995, *Hydrocarbon Processing*, 1996; Northrup and Sloan, 1996).

The process uses supercritical solvents and is a natural progression from propane deasphalting and allows the separation of residua into their base components (asphaltene constituents, resin constituents, and oil constituents) for recombination to optimum properties. Propane, butane, and pentane may be used as the solvent, depending on the feedstock and the desired compositions. A mixer is used to blend residue with liquefied solvent at elevated temperature and pressure. The blend is pumped into the first stage separator where, through counter current flow of solvent, the asphaltene constituents are precipitated, separated, and stripped of solvent by steam. The overhead solution from the first tower is taken to a second stage where it is heated to a higher temperature. This causes the resin constituents to separate. The final material is taken to a third stage and heated to a supercritical temperature. This makes the oils insoluble and separation occurs. This process is very flexible and allows precise blending to required compositions.

In the process (Figure 19.6), the residuum is mixed with several-fold volume of a low-boiling hydrocarbon solvent and passed into the asphaltene separator vessel. Asphaltenes rejected by the solvent are separated from the bottom of the vessel and are further processed by heating and steam stripping to remove a small quantity of dissolved solvent. The solvent-free asphaltenes are sent to a section of the refinery for further processing. The main flow,

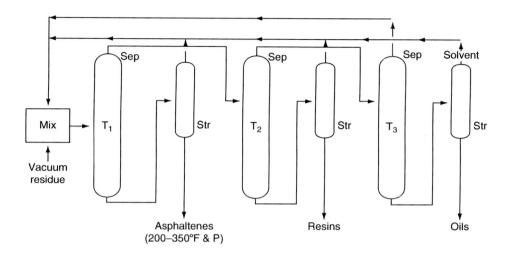

Sep = Separator
Str = Stripping
Sp = Softening point

FIGURE 19.6 The ROSE process.

solvent and extracted oil, passes overhead from the asphaltene separator through a heat exchanger and heater into the oil separator, where the extracted oil is separated without solvent vaporization. The solvent, after heat exchange, is recycled to the process. The small amount of solvent contained in the oil is removed by steam stripping, and the resulting vaporized solvent from the strippers is condensed and returned to the process. Product oil is cooled by heat exchange before being pumped to storage or further processing.

The deasphalting efficiency in processes using propane is of the order of 75% to 83%, with an overall deasphalted oil recovery yield of the order of 50%.

19.2.2.5 Solvahl Process

The Solvahl process is a solvent deasphalting process for application to vacuum residua (Table 19.6) (RAROP, 1991, p. 9; Billon et al., 1994).

The process was developed to give maximum yields of deasphalted oil, while eliminating asphaltenes and reducing metals content to a level compatible with the reliable operation of downstream units (Peries et al., 1995; *Hydrocarbon Processing*, 1996).

19.2.2.6 Lube Deasphalting

Other facilities incorporate lube deasphalting to process vacuum residuum into lube oil base stocks. Propane deasphalting is most commonly used to remove asphaltene constituents and resins, which contribute an undesirable dark color to the lube base stocks. This process typically uses baffle towers or rotating disk contactors to mix the propane with the feed. Solvent recovery is accomplished with evaporators, and supercritical solvent recovery processes are also used in some deasphalting units.

The Duo-Sol process is used to deasphalt and extract lubricating oil feedstocks. Propane is used as the deasphalting solvent, and a mixture of phenol and cresylic acids (cresols, hydroxytoluenes) are used as the extraction solvent. The extraction is conducted in a series of batch extractors followed by solvent recovery in multistage flash distillation and stripping towers.

TABLE 19.6
Feedstock and Product Data for the Solvahl Process

	Arabian Light Vacuum Residuum
Feedstock	
API	9.6
Sulfur, wt.%	4.1
Nitrogen, wt.%	0.3
C_7-asphaltenes, wt.%	4.2
Carbon residue, wt.%	16.4
Nickel, ppm	19.0
Vanadium, ppm	61.0
Products	
C_4-deasphalted oil, wt.%	70.1
API	16.0
Sulfur, wt.%	3.3
Nitrogen, wt.%	0.2
C_7-asphaltenes, wt.%	<0.1
Carbon residue, wt.%	5.3
Nickel, ppm	2.0
Vanadium, ppm	3.0
C_5-deasphalted oil, wt.%	85.5
API	13.8
Sulfur, wt.%	3.7
Nitrogen, wt.%	0.2
C_7-asphaltenes, wt.%	<0.1
Carbon residue, wt.%	7.9
Nickel, ppm	7.0
Vanadium, ppm	16.0

Source: Speight, J.G. and Ozum, B. 2002. *Petroleum Refining Processes*. Marcel Dekker Inc., New York. With permission.

19.3 DEWAXING PROCESSES

Paraffinic crude oils often contain microcrystalline or paraffin waxes. The crude oil may be treated with a solvent such as methyl-ethyl-ketone (MEK) to remove this wax before it is processed. This is not a common practice, however, and **solvent dewaxing** processes are designed to remove wax from lubricating oils to give the product good fluidity characteristics at low temperatures (e.g., low pour points) rather than from the whole crude oil. The mechanism of solvent dewaxing involves either the separation of wax as a solid that crystallizes from the oil solution at low temperature or the separation of wax as a liquid that is extracted at temperatures above the melting point of the wax through preferential selectivity of the solvent. However, the former mechanism is the usual basis for commercial dewaxing processes.

In the 1930s, two types of stocks, naphthenic and paraffinic, were used to make motor oils. Both types were solvent extracted to improve their quality, but in the high-temperature conditions encountered in service, the naphthenic type could not stand up as well as the paraffinic type. Nevertheless, the naphthenic type was the preferred oil, particularly in cold weather, because of its fluidity at low temperatures. Previous to 1938, the highest-quality lubricating oils were of the naphthenic type and were phenol treated to pour points of −40°C to −7°C (−40°F to 20°F), depending on the viscosity of the oil. Paraffinic oils were also available and could be phenol treated to higher-quality oil, but their wax content was so high that the oils were solid at room temperature.

The lowest viscosity paraffinic oils were dewaxed by the cold press method to produce oils with a pour point of 2°C (35°F). The light paraffin distillate oils contained a paraffin wax that crystallized into large crystals when chilled, and could thus readily be separated from the oil by the cold press filtration method. The more viscous paraffinic oils (intermediate and heavy paraffin distillates) contained amorphous or **microcrystalline waxes**, which formed small crystals that plugged the filter cloths in the cold press and prevented filtration. Because the wax could not be removed from intermediate and heavy paraffin distillates, the high-quality, high-viscosity lubricating oils in them could not be used except as cracking stock.

Methods were therefore developed to dewax these high-viscosity paraffinic oils. The methods were essentially alike in that the waxy oil was dissolved in a solvent that would keep the oil in solution; the wax separated as crystals when the temperature was lowered. The processes differed chiefly in the use of the solvent. Commercially used solvents were naphtha, propane, sulfur dioxide, acetone–benzene, trichloroethylene, ethylene dichloride–benzene (Barisol), methyl ethyl ketone–benzene (**benzol**), methyl-n-butyl ketone, and methyl-n-propyl ketone.

In the first solvent dewaxing process (developed in 1924), the waxy oil was mixed with naphtha and filter aid (Fuller's earth or diatomaceous earth). The mixture was chilled and filtered, and the filter aid assisted in building a wax cake on the filter cloth. This process is now obsolete, and most of the modern dewaxing processes use a mixture of methyl ethyl ketone and benzene. Other ketones may be substituted for dewaxing, but regardless of what ketone is used, the process is generally known as ketone dewaxing.

The process, as now practised, involves mixing the feedstock with 1 to 4 times its volume of the ketone (Figure 19.7) (Scholten, 1992). The mixture is then heated until the oil is in solution and the solution is chilled at a slow, controlled rate in double-pipe, scraped-surface exchangers. Cold solvent, such as filtrate from the filters, passes through the 2 inch annular space between the inner and the outer pipes and chills the waxy oil solution flowing through the inner 6 inch pipe.

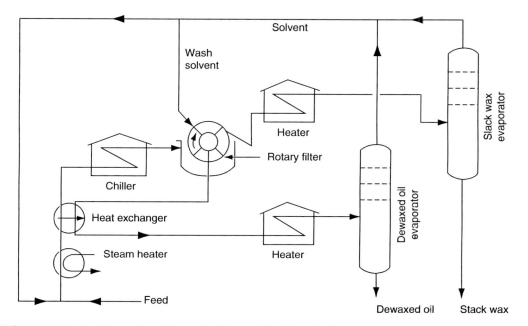

FIGURE 19.7 A solvent dewaxing unit. (From OSHA Technical Manual, Section IV, Chapter 2, Petroleum Refining Processes.)

To prevent wax from depositing on the walls of the inner pipe, blades or scrapers extending the length of the pipe are fastened to a central rotating shaft to scrape off the wax. Slow chilling reduces the temperature of the waxy oil solution to 2°C (35°F), and then faster chilling reduces the temperature to the approximate pour point required in the dewaxed oil. The waxy mixture is pumped to a filter case into which the bottom half of the drum of a rotary vacuum filter dips. The drum (8 feet in diameter, 14 feet long), covered with filter cloth, rotates continuously in the filter case. Vacuum within the drum sucks the solvent and the oil dissolved in the solvent through the filter cloth and into the drum. Wax crystals collect on the outside of the drum to form a wax cake, and as the drum rotates, the cake is brought above the surface of the liquid in the filter case and under sprays of ketone that wash oil out of the cake and into the drum. A knife-edge scrapes off the wax, and the cake falls into the conveyor and is moved from the filter by the rotating scroll.

The recovered wax is actually a mixture of wax crystals with a little ketone and oil, and the filtrate consists of the dewaxed oil dissolved in a large amount of ketone, which is removed from both by distillation. But before the wax is distilled, it is de-oiled, mixed with more cold ketone, and pumped to a pair of rotary filters in series, where further washing with cold ketone produces a wax cake that contains very little oil. The de-oiled wax is melted in heat exchangers and pumped to a distillation tower operated under vacuum, where a large part of the ketone is evaporated or flashed from the wax. The rest of the ketone is removed by heating the wax and passing it into a fractional distillation tower operated at atmospheric pressure and then into a stripper where steam removes the last traces of ketone.

An almost identical system of distillation is used to separate the filtrate into dewaxed oil and ketone. The ketone from both the filtrate and the wax slurry is reused. Clay treatment, or hydrotreating, finishes the dewaxed oil, as previously described. The wax (**slack wax**), even though it contains essentially no oil as compared to 50% in the slack wax obtained by cold pressing, is the raw material for either sweating or wax recrystallization, which subdivides the wax into a number of wax fractions with different melting points (Chapter 26).

Solvent dewaxing can be applied to light, intermediate, and heavy lubricating oil distillates, but each distillate produces a different kind of wax. Each of these waxes is actually a mixture of a number of waxes. For example, the wax obtained from light paraffin distillate consists of a series of paraffin waxes that have melting points in the range of 30°C to 70°C (90°F to 160°F), which are characterized by a tendency to harden into large crystals. However, heavy paraffin distillate yields a wax composed of a series of waxes with melting points in the range of 60°C to 90°C (140°F to 200°F), which harden into small crystals from which they derive the name of microcrystalline wax or microwax. On the other hand, intermediate paraffin distillates contain paraffin waxes and waxes intermediate in properties between paraffin and microwax.

Thus, the solvent dewaxing process produces three different slack waxes (also known as crude waxes or raw waxes), depending on whether light, intermediate, or heavy paraffin distillate is processed. The slack wax from heavy paraffin distillate may be sold as dark raw wax, and the wax from intermediate paraffin distillate as pale raw wax. The latter is treated with lye and clay to remove odor and improve color.

There are several processes in use for solvent dewaxing, but all have the same general steps, which are (1) contacting the feedstock with the solvent, (2) precipitating the wax from the mixture by chilling, and (3) recovering the solvent from the wax and dewaxed oil for recycling. The processes use benzene–acetone (solvent dewaxing), propane (**propane dewaxing**), trichloroethylene (**separator-Nobel dewaxing**), ethylene dichloride–benzene (Barisol dewaxing), and urea (**urea dewaxing**), as well as liquid sulfur dioxide–benzene mixtures.

Urea dewaxing (Chapter 21) (Scholten, 1992) is worthy of further mention insofar as the process is highly selective and, in contrast to the other dewaxing techniques, can be achieved without the use of refrigeration. However, the process cannot compete economically with the solvent dewaxing processes for treatment of the heavier lubricating oils. But when it is applied to the lighter materials that already may have been subjected to a solvent dewaxing operation, products are obtained that may be particularly useful as refrigerator oils, transformer oils, and the like.

The process description is essentially the same as that used for solvent dewaxing with the omission of the chilling stage and the insertion of a contactor, where the feedstock and the urea (with a solvent) are thoroughly mixed before filtration. The solvents are recovered from the dewaxed oil by evaporation, and the urea complex is decomposed in a urea recovery system.

Residual lubricating oils, such as cylinder oils and bright stocks, are made from paraffinic or mixed-base reduced crude oils and contain waxes of the microcrystalline type. Removal of these waxes from reduced crude produces **petrolatum**, a grease-like material that is known in a refined form as **petroleum jelly**. This material can be separated from reduced crude in several ways. The original method was cold settling, whereby reduced crude was dissolved in a suitable amount of naphtha and allowed to stand over winter until the microwax settled out. This method is still used, but the reduced crude naphtha solution is held in refrigerated tanks until the petrolatum settles out. The supernatant naphtha–oil layer is pumped to a still where the naphtha is removed, leaving cylinder stock that can be further treated to produce bright stock. The petrolatum layer is also distilled to remove naphtha and may be clay treated, or acid and clay treated to improve the color.

Another method of separating petrolatum from reduced crude is centrifuge dewaxing. In this process, the reduced crude is dissolved in naphtha and chilled to $-18°C$ ($0°F$) or lower, which causes the wax to separate. The mixture is then fed to a battery of centrifuges where the wax is separated from the liquid. However, the centrifuge method has now been largely displaced by solvent dewaxing methods and by more modern methods of wax removal.

There are also later-generation dewaxing processes that are being brought on-stream in various refineries (Hargrove, 1992; Genis, 1997). For example, British Petroleum has developed a hydrocatalytic dewaxing process that is reputed to overcome some of the disadvantages of the solvent dewaxing processes. In the process of dewaxing, waxy molecules are removed from heavy distillate fuel cuts or lube distillates.

Catalytic dewaxing is a hydrocracking process and is therefore operated at elevated temperatures ($280°C$ to $400°C$, $550°F$ to $750°F$) and pressures (300 to 1500 psi). However, the conditions for a particular dewaxing operation depend upon the nature of the feedstock and the product pour point required. The catalyst employed for the process is a mordenite-type catalyst that has the correct pore structure to be selective for n-paraffin cracking. Platinum on the catalyst serves to hydrogenate the reactive intermediates, so that further paraffin degradation is limited to the initial thermal reactions. The process has been employed to successfully dewax a wide range of naphthenic feedstocks (Hargrove et al., 1979), but it may not be suitable to replace solvent dewaxing in all cases. The process has the flexibility to fit into normal refinery operations and can be adapted for prolonged periods on-stream.

Other processes include the Exxon Mobil distillate dewaxing (MDDW) process (Smith et al., 1980; Safre, 2003), dewaxing is achieved by selective cracking, where the long paraffin chains are cracked to form shorter chains using a shape-selective zeolite that rejects ring compounds and *iso*-paraffins. In a related process (MIDW process), the paraffins are selectively isomerized using low-pressure conditions (Smith et al., 1980; Safre, 2003). This process also uses a zeolite catalyst to convert low-quality gas oil into diesel fuel.

In the process, the proprietary catalyst can be reactivated to fresh activity by relatively mild nonoxidative treatment. Of course, the time allowed between reactivation is a function of the feedstock, but after numerous reactivations it is possible that there will be coke buildup on the catalyst. The process can be used to dewax a full range of lubricating base stocks and, as such, has the potential to completely replace solvent dewaxing or can even be used in combination with solvent dewaxing. This latter option, of course, serves to de-bottleneck existing solvent dewaxing facilities.

Both the catalytic dewaxing processes have the potential to change conventional thoughts about dewaxing, insofar as they are not solvent processes and may be looked upon (more correctly) as thermal processes rather than treatment processes. However, both provide viable alternatives to the solvent processes and offer a further advance in the science and technology of refinery operations.

Catalytic dewaxing yields various grades of lube oils and fuel components suitable for extreme winter conditions. Paraffinic (waxy) components that precipitate out at low temperatures are removed. In the UOP catalytic dewaxing process, the first stage saturates olefins and desulfurizes and denitrifies the feed via hydrotreating (Genis, 1997). In the second stage, a dual-function, nonnoble metal zeolite catalyst selectively adsorbs and then selectively hydrocracks the normal and near-normal long-chain paraffins to form shorter-chain (nonwaxy) molecules.

Alternatively, in the Chevron isodewaxing process, the dewaxing results from isomerizing the linear paraffins to branched paraffins by using a molecular sieve catalyst containing platinum (Miller, 1994a, b). In the isodewaxing process, which is followed by a hydrofinishing step (Figure 19.8), waxy feedstock from a hydrocracker or hydrotreater, together with the hydrogen-containing gas, is heated and fed to the isodewaxing reactor. The conditions in the reactor cause isomerization of n-paraffins to iso-paraffins, and other paraffins are cracked to highly saturated low boiling products such as jet fuel and diesel fuel. The effluent from the isodewaxing reactor is then sent to the hydrofinisher where hydrofinishing, including aromatics saturation, provides the product. The catalysts used in the isodewaxing and hydrofinishing units are selective for dewaxing and hydrogenation. The catalysts are at their maximum efficiency with low sulfur and low nitrogen feedstocks. The process generally uses a high-pressure recycle loop. Because of the conversion of wax constituents to other usable products, the process has obvious benefits over solvent dewaxing, insofar as quality of the product is increased.

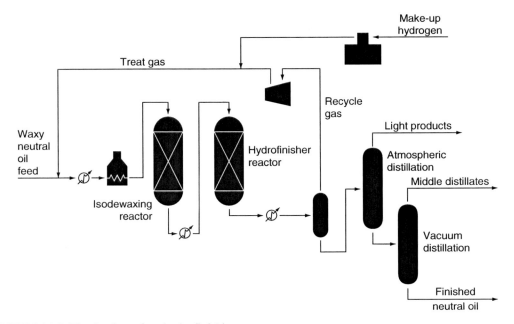

FIGURE 19.8 The Isodewaxing–hydrofinishing process.

REFERENCES

Billon, A., Morel, F., Morrisson, M.E., and Peries, J.P. 1994. *Rev. Inst. Franç. Pétrol.* 49(5): 495–507.

Bland, W.F. and Davidson, R.L. 1967. *Petroleum Processing Handbook*. McGraw-Hill, New York.

Ditman, J.G. 1973. *Hydrocarbon Process.* 52(5): 110.

Dunning, H.N. and Moore, J.W. 1957. *Petrol. Refin.* 36(5): 247–250.

Gary, J.H. and Handwerk, G.E. 1994. *Petroleum Refining: Technology and Economics,* 4th edn. Marcel Dekker Inc., New York.

Gearhart, J.A. 1980. *Hydrocarbon Process.* 59(5): 150.

Gearhart, J.A. and Gatwin, L. 1976. *Hydrocarbon Process.* (May): 125–128.

Genis, O. 1997. *Handbook of Petroleum Refining Processes*. R.A. Meyers, ed. McGraw-Hill, New York. Chapter 8.5.

Girdler, R.B. 1965. *Proc. Assoc. Asphalt Paving Technol.* 34: 45.

Hargrove, J.D. 1992. *Petroleum Processing Handbook*. J.J. McKetta, ed. Marcel Dekker Inc., New York. p. 558.

Hargrove, J.D., Elkes, G.J., and Richardson, A.H. 1979. *Oil Gas J.* 77(3): 103.

Hobson, G.D. and Pohl, W. 1973. *Modern Petroleum Technology*. Applied Science Publishers, Barking, Essex, UK.

Houde, E.J. 1997. *Handbook of Petroleum Refining Processes,* 2nd edn. R.A. Meyers, ed. McGraw-Hill, New York. Chapter 10.4.

Hydrocarbon Processing. 1996. 75(11): 89 *et seq.*

Hydrocarbon Processing. 1998. 77(11): 53 *et seq.*

Kashiwara, H. 1980. *Kagaku Kogaku*. (7): 44.

Low, J.Y., Hood, R.L. and Lynch, K.Z. 1995. Preprints. *Div. Petrol. Chem. Am. Chem. Soc.* 40: 780.

Miller, S.J. 1994a. *Studies in Surface Science and Catalysis*. J. Weitkamp, ed. Elsevier, Amsterdam. 84C: 2319–2326.

Miller, S.J. 1994b. *Micropor. Mater.* 2: 439–450.

Mitchell. D.L. and Speight, J.G. 1973. *Fuel* 52: 149.

Northrup, A.H., and Sloan, H.D. 1996. Annual Meeting. National Petroleum Refiners Association. Houston, TX. Paper AM-96–55.

Peries, J.P., Billon, A., Hennico, A., Morrison, E., and Morel, F., 1995. *Proceedings*. 6th UNITAR International Conference on Heavy Crude and Tar Sand. Volume 2. p. 229.

RAROP 1991. *RAROP Heavy Oil Processing Handbook*. Research Association for Residual Oil Processing. T. Noguchi (Chairman). Ministry of Trade and International Industry (MITI), Tokyo, Japan.

Safre, A. 2003. *Proceedings*. 4th International Conference on Oil Refining and Petrochemicals in the Middle East. Abu Dhabi. January 28.

Salazar, J.R. 1986. *Handbook of Petroleum Refining Processes*. R.A. Meyers, ed. McGraw-Hill, New York. Chapter 8.5.

Scholten G.G. 1992. *Petroleum Processing Handbook*. J.J. McKetta, ed. Marcel Dekker Inc., New York. p. 565.

Smith, K.W., Starr, W.C., and Chen, N.Y. 1980. *Oil Gas J.* 78(21): 75.

Speight, J.G. 2000. *The Desulfurization of Heavy Oils and Residua*. 2nd edn. Marcel Dekker Inc., New York.

Speight, J.G. and Ozum, B. 2002. *Petroleum Refining Processes*. Marcel Dekker Inc., New York.

20 Hydrotreating and Desulfurization

20.1 INTRODUCTION

Product improvement is the treatment of petroleum products to ensure that they meet utility and performance specifications. As already noted, product improvement usually involves changes in molecular shape (**reforming** and **isomerization**) or change in molecular size (**alkylation** and **polymerization**) (Chapter 23), and **hydrotreating** can play a major role in product improvement.

The use of hydrogen in thermal processes is perhaps the single most significant advance in refining technology during the twentieth century (Bridge et al., 1981; Dolbear, 1998). In fact, hydrogenation processes are the principal processes used in the manufacture of gasoline (Figure 20.1). Indeed, with the influx of heavier feedstocks into refineries, **hydroprocessing** will assume a greater role in the refinery of the future. The process uses the principle that the presence of hydrogen during a thermal reaction of a petroleum feedstock terminates many of the coke-forming reactions and enhances the yields of the lower boiling components, such as gasoline, kerosene, and jet fuel.

Hydrogenation processes for the conversion of petroleum fractions and petroleum products may be classified as destructive and nondestructive. Destructive hydrogenation (hydrogenolysis or **hydrocracking**) is characterized by the cleavage of carbon–carbon linkages accompanied by hydrogen saturation of the fragments to produce lower-boiling products. Such treatment requires severe processing conditions and the use of high hydrogen pressures to minimize polymerization and condensation that lead to coke formation. Many other reactions, such as isomerization, dehydrogenation, and cyclization, occur under the drastic conditions employed.

Furthermore, a growing dependence on high-heteroatom heavy oils and residua has emerged as a result of continuing decreasing availability of conventional crude oil, owing to the depletion of reserves in various parts of the world. Thus, the ever-growing tendency to convert as much as possible of lower-grade feedstocks to liquid products is causing an increase in the total sulfur content in refined products. Refiners must, therefore, continue to remove substantial portions of sulfur from the lighter products, but residua and heavy crude oil pose a particularly difficult problem. Indeed, it is now clear that there are other problems involved in the processing of the heavier feedstocks and that these heavier feedstocks, which are gradually emerging as the liquid fuel supply of the future, need special attention.

There are several valid reasons for removing heteroatoms from petroleum fractions. These include:

1. Reduction, or elimination, of corrosion during refining, handling, or use of the various products

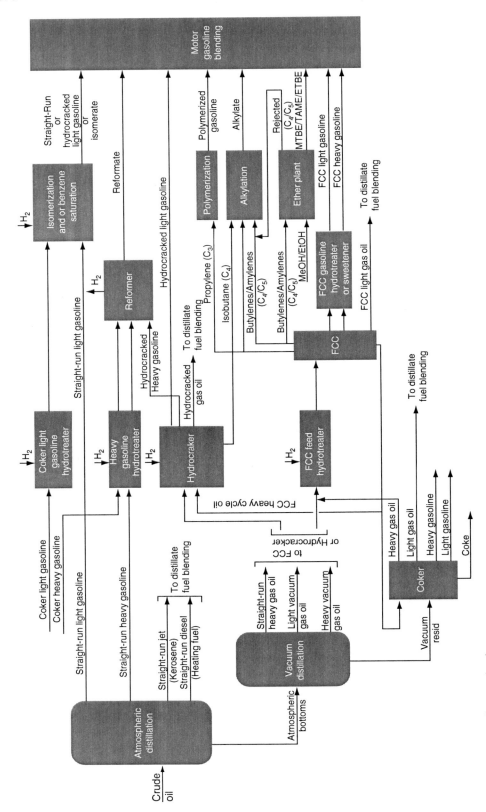

FIGURE 20.1 Gasoline production.

2. Production of products having an acceptable odor and specification
3. Increasing the performance (and stability) of gasoline
4. Decreasing smoke formation in kerosene
5. Reduction of heteroatom content in fuel oil to a level that improves burning characteristics and is environmentally acceptable

Heteroatom removal, as practiced in various refineries, can take several forms (Speight, 2000) such as concentration in residua during distillation, concentration in coke during coking (Chapter 16), hydroprocessing (Chapter 21), or chemical removal (acid treating, caustic treating, i.e., sweetening or finishing processes) (Chapter 23). Nevertheless, the removal of heteroatom from petroleum feedstocks is almost universally accomplished by the catalytic reaction of hydrogen with the feedstock constituents. However, there are certain other refinery processes that are adaptable to residua and heavy oils and that may be effective for reducing, but not necessarily effective for removing completely, the heteroatom content.

Thus, catalytic hydrotreating is a hydrogenation process used to remove about 90% of contaminants such as nitrogen, sulfur, oxygen, and metals from liquid petroleum fractions. These contaminants, if not removed from the petroleum fractions as they travel through the refinery processing units, can have detrimental effects on the equipment, the catalysts, and the quality of the finished product. Typically, hydrotreating is done prior to processes, such as catalytic reforming, so that the catalyst is not contaminated by untreated feedstock. Hydrotreating is also used prior to catalytic cracking to reduce sulfur and improve product yields, and to upgrade middle-distillate petroleum fractions into finished kerosene, diesel fuel, and heating fuel oils. In addition, hydrotreating converts olefins and aromatics to saturated compounds.

Hydrotreating (variously referred to as hydroprocessing to avoid any confusion with those processes that are referred to as hydrotreating processes) is a refining process in which the feedstock is treated with hydrogen at temperature and under pressure in which hydrocracking (thermal decomposition in the presence of hydrogen) is minimized (Figure 20.2 and Figure 20.3) (Bridge, 1997). The usual goal of hydrotreating is to hydrogenate olefins and to

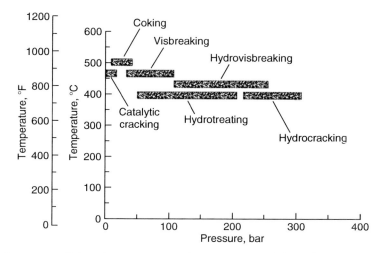

FIGURE 20.2 Process conditions (temperature and pressure ranges) for refinery processes.

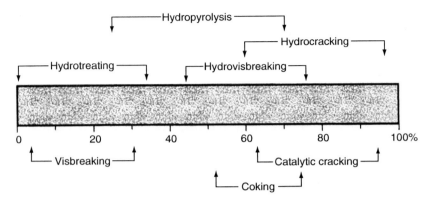

FIGURE 20.3 Conversion (based on feedstock) to liquids for refinery processes.

remove heteroatoms, such as sulfur, and to saturate aromatic compounds and olefins (Meyers, 1997; Speight, 2000).

On the other hand, hydrocracking is a process in which thermal decomposition is extensive, and the hydrogen assists in the removal of the heteroatoms, as well as mitigating the coke formation that usually accompanies thermal cracking of high molecular weight polar constituents.

The major differences between hydrotreating and hydrocracking are the time at which the feedstock remains at reaction temperature and the extent of the decomposition of the non-heteroatom constituents and products. The upper limits of hydrotreating conditions may overlap with the lower limits of hydrocracking conditions. Where the reaction conditions overlap, feedstocks to be hydrotreated will generally be exposed to the reactor temperature for shorter periods; hence the reason why hydrotreating conditions may be referred to as mild. All is relative.

Hydrotreating (nondestructive hydrogenation) is generally used for the purpose of improving product quality without appreciable alteration of the boiling range. Mild processing conditions are employed so that only the more unstable materials are attacked. Thus, nitrogen, sulfur, and oxygen compounds undergo hydrogenolysis to split ammonia, hydrogen sulfide, and water, respectively. Olefins are saturated, and unstable compounds, such as diolefins, which might lead to the formation of gums or insoluble materials, are converted to more stable compounds. Heavy metals present in the feedstock are also usually removed during hydrogen processing (Scott and Bridge, 1971; Aalund, 1975).

In addition to hydrodesulfurization (HDS) and hydrodenitrogenation (HDN), the removal of aromatic constituents from some product streams has also become essential. The high aromatic content in diesel fuel has been recognized both to lower the fuel quality and to contribute significantly to the formation of undesired emissions in exhaust gases (Stanislaus and Cooper, 1994). Indeed, as a result of the stringent environmental regulations, processes for aromatic reduction in middle distillates have received considerable attention in recent years. Studies have shown that existing middle distillate hydrotreaters designed to reduce sulfur and nitrogen levels would lower the diesel aromatics only marginally (Stanislaus and Cooper, 1994).

There is a rough correlation between the quality of petroleum products and their hydrogen content (Dolbear, 1998). It is a fact that desirable aviation gasoline, kerosene, diesel fuel, and lubricating oil are made up of hydrocarbons containing high proportions of hydrogen. In addition, it is usually possible to convert olefins and higher molecular weight constituents to paraffins and monocyclic hydrocarbons by **hydrogen-addition** processes. These facts have for many years encouraged attempts to employ hydrogenation for refining operations; despite considerable technical success, such processes were not economically possible until hydrogen became available as a result of the rise of reforming, which converts naphthenes to aromatics with the release of hydrogen.

Unsaturated compounds, such as olefins, are not indigenous to petroleum and are produced during cracking processes and need to be removed from product streams because of the tendency of unsaturated compounds and heteroatomic polar compounds to form gum and sediment (Chapter 22). On the other hand, aromatic compounds are indigenous to petroleum and some may be formed during cracking reactions. The most likely explanation is that the aromatic compounds present in product streams are related to the aromatic compounds originally present in petroleum but now having shorter alkyl side-chains. Thus, in addition to olefins, product streams will contain a range of aromatic compounds that have to be removed to enable many of the product streams to meet product specifications.

Of the aromatic constituents, the polycyclic aromatics are first partially hydrogenated before the cracking of the aromatic nucleus takes place. The sulfur and nitrogen atoms are converted to hydrogen sulfide and ammonia, but a more important role of the hydrogenation is probably to hydrogenate the coke precursors rapidly and prevent their conversion to coke.

Hydrotreating catalysts are usually cobalt plus molybdenum or nickel plus molybdenum (in the sulfide) forms, impregnated on an alumina base (Topsøe and Clausen, 1984). The hydrotreating operating conditions (1000 to 2000 psi hydrogen and about 370°C, 700°F) are such that appreciable hydrogenation of aromatics does not occur. The desulfurization reactions are invariably accompanied by small amounts of hydrogenation and hydrocracking, the extent of which depends on the nature of the feedstock and the severity of desulfurization.

One of the problems in the processing of high-sulfur and high-nitrogen feeds is the large quantities of hydrogen sulfide (H_2S) and ammonia (NH_3) that are produced. Substantial removal of both compounds from the recycle gas can be achieved by the injection of water in which, under the high-pressure conditions employed, both hydrogen sulfide and ammonia are very soluble compared with hydrogen and hydrocarbon gases. The solution is processed in a separate unit for the recovery of anhydrous ammonia and hydrogen sulfide.

Hydrotreating is carried out by charging the feed to the reactor, together with a portion of all the hydrogen produced in the catalytic reformer. Suitable catalysts are tungsten–nickel sulfide, cobalt–molybdenum–alumina, nickel oxide–silica–alumina, and platinum–alumina. Most processes employ cobalt–molybdenum catalysts, which generally contain about 10% by weight molybdenum oxide and less than 1% by weight cobalt oxide supported on alumina. The temperatures employed are in the range of 300°C to 345°C (570°F to 850°F), and the hydrogen pressures are about 500 to 1000 psi.

The reaction generally takes place in the vapor phase but, depending on the application, may be a mixed-phase reaction. The reaction products are cooled in a heat exchanger and conducted to a high-pressure separator, where hydrogen gas is separated for recycling. Liquid products from the high-pressure separator flow to a low-pressure separator (stabilizer), where dissolved light gases are removed. The product may then be fed to a reforming or cracking unit if desired.

Generally, it is more economical to hydrotreat high-sulfur feedstocks before catalytic cracking than to hydrotreat the products after catalytic cracking. The advantages are:

1. Products require less finishing.
2. Sulfur is removed from the catalytic cracking feedstock, and corrosion is reduced in the cracking unit.
3. Coke formation during cracking is reduced and higher conversions result.
4. Catalytic cracking quality of the gas oil fraction is improved.

Although hydrocracking will occur during hydrotreating, attempts are made to minimize such effects, but the degree of cracking is dependent on the nature of the feedstock. For example, decalin (decahydronaphthalene) cracks more readily than the corresponding paraffin analog, n-decane [CH$_3$ (CH$_2$)$_8$ CH$_3$], to give higher *iso*-paraffin to n-paraffin product ratios than those obtained from the paraffin. A large yield of single-ring naphthenes is also produced, and these are resistant to further hydrocracking and contain a higher than equilibrium ratio of methylcyclopentane to cyclohexane.

When applied to residua, the hydrotreating processes can be used for processes such as:

1. Fuel oil desulfurization
2. Residuum hydrogenation (with HDS, HDN, and partial conversion) to produce products suitable as feedstocks for other processes, such as catalytic cracking

One of the chief problems with hydroprocessing residua is the deposition of metals, in particular vanadium, on the catalyst. It is not possible to remove vanadium from the catalyst, which must therefore be replaced when deactivated, and the time taken for catalyst replacement can significantly reduce the unit time efficiency. Fixed-bed catalysts tend to plug owing to solids in the feed or carbon deposits when processing residual feeds. As mentioned previously, the highly exothermic reaction at high conversion presents complex reactor design problems in heat removal and temperature control.

The problems encountered in hydrotreating heavy feedstocks can be directly equated to the amount of complex, higher-boiling constituents that may require pretreatment (Speight and Moschopedis, 1979; Reynolds and Beret, 1989). Processing these feedstocks is not merely a matter of applying know-how derived from refining conventional crude oils but requires knowledge of the composition (Chapter 7 and Chapter 8). The materials are not only complex in terms of the carbon number and boiling point ranges but also because a large part of this envelope falls into the range of model compounds, and very little is known about their properties. It has also been established that the majority of the higher molecular weight materials produce coke (with some liquids), while the majority of the lower molecular weight constituents produce liquids (with some coke). It is to both of these trends that hydrocracking is aimed.

It is the physical and chemical composition of a feedstock that plays a large part not only in determining the nature of the products that arise from refining operations but also in determining the precise manner by which a particular feedstock should be processed (Speight, 1986). Furthermore, it is apparent that the conversion of heavy oils and residua requires new lines of thought to develop suitable processing scenarios (Celestinos et al., 1975). Indeed, the use of thermal (**carbon rejection**) processes and of hydrothermal (hydrogen addition) processes, which were inherent in the refineries designed to process lighter feedstocks, has been a particular cause for concern. This has brought about the evolution of processing schemes that accommodate the heavier feedstocks (Wilson, 1985; Boening et al., 1987; Corbett, 1989; Khan, 1997).

The choice of processing schemes for a given hydrotreating application depends upon the nature of the feedstock as well as the product requirements (Murphy and Treese, 1979; Aalund, 1981; Nasution, 1986; Suchanek and Moore, 1986). For higher-boiling feedstocks, the process is usually hydrocracking and can be simply illustrated as a single-stage or as a two-stage operation (Figure 20.4). Variations to the process are feedstocks dependent.

For example, the single-stage process can be used to produce gasoline but is more often used to produce middle distillate from heavy vacuum gas oils. The two-stage process was developed primarily to produce high yields of gasoline from straight-run gas oil, and the first stage may actually be a purification step to remove sulfur-containing (as well as

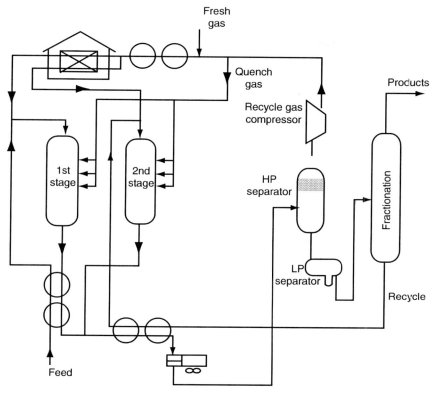

FIGURE 20.4 A single-stage or two-stage (optional) hydrocracking unit. (From OSHA Technical Manual, Section IV, Chapter 2, Petroleum Refining Processes.)

nitrogen-containing) organic materials. Both processes use an extinction-recycling technique to maximize the yields of the desired product. Significant conversion of heavy feedstocks can be accomplished by hydrocracking at high severity (Howell et al., 1985). For some applications, the products boiling up to 340°C (650°F) can be blended to give the desired final product.

For lower-boiling feedstocks, the commercial processes for treating or finishing petroleum fractions with hydrogen all operate in essentially the same manner. The feedstock is heated and passed with hydrogen gas through a tower or reactor filled with catalyst pellets (Figure 20.5). The reactor is maintained at a temperature of 260°C to 425°C (500°F to 800°F) at pressures from 100 to 1000 psi, depending on the particular process, the nature of the feedstock, and the degree of hydrogenation required. After leaving the reactor, the excess hydrogen is separated from the treated product and recycled through the reactor after removal of hydrogen sulfide. The liquid product is passed into a stripping tower where steam removes dissolved hydrogen and hydrogen sulfide, and after cooling the product is run to finished product storage or, in the case of feedstock preparation, pumped to the next processing unit.

Excessive contact time or temperature will create coking. Precautions need to be taken when unloading coked catalyst from the unit to prevent iron sulfide fires. The coked catalyst should be cooled to below 49°C (<120°F) before removal, or dumped into nitrogen-blanketed bins where it can be cooled before further handling. Antifoam additives may be used to prevent catalyst poisoning from silicone carryover in the coker feedstock. There is a potential for exposure to hydrogen sulfide or hydrogen gas in the event of a release, or to ammonia,

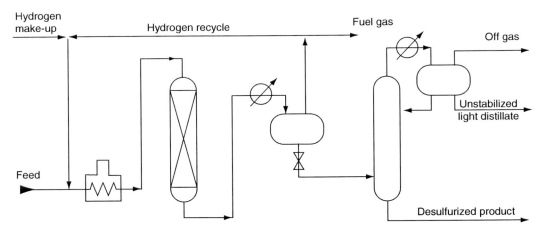

FIGURE 20.5 Distillate hydrotreating.

should a sour-water leak or spill occur. Phenol also may be present if high boiling-point feedstocks are processed.

Attention must also be given to the coke mitigation aspects of hydrotreating as a preliminary treatment option of feedstocks for other processes, especially heavier feedstocks. Although the visbreaking process (Chapter 17) reduces the viscosity of residua and partially converts the residue to lighter hydrocarbons and coke, the process can also be used to remove the undesirable higher molecular weight polar constituents before sending the visbroken feedstock to a catalytic cracking unit. The solvent deasphalting process (Chapter 19) separates the higher-value liquid product (DAO), using a light paraffinic solvent, from low-value asphaltene-rich pitch stream. Various residuum hydrotreating (in fact, hydrocracking) processes (Chapter 21), in which the feedstock is processed under high temperature and pressure, using a robust catalyst to remove sulfur, metals, condensed aromatic, or nitrogen, and increase the residue's hydrogen content to a desired degree are also available. However, an increased number of options are becoming available, in which the residuum is first hydrotreated (under milder conditions to remove heteroatoms and mitigate the effects of the asphaltene constituents and resin constituents) before sending the hydrotreated product to, for example, a fluid catalytic cracking unit.

In such cases, it is even more important that particular attention must be given to hydrogen management and promoting HDS and HDN (even fragmentation) of asphaltene and resin constituents, thereby producing a product that may be suitable as a feedstock for catalytic cracking with reduced catalyst destruction. The presence of a material with good solvating power to assist in the hydrotreating process is preferred. In this respect, it is worth noting the reappearance of donor solvent processing of heavy feedstocks (Vernon et al., 1984) that has its roots in the older hydrogen donor diluent visbreaking process (Carlson et al., 1958; Langer et al., 1962; Bland and Davidson, 1967).

However, it must be forgotten that product distribution and quality vary considerably, depending upon the nature of the feedstock constituents as well as on the process. In modern refineries, hydrocracking is one of several process options that can be applied to the production of liquid fuels from the heavier feedstocks (see also Chapter 23). A most important aspect of the modern refinery operation is the desired product slate, which dictates the matching of a process with any particular feedstock to overcome differences in feedstock composition.

Hydrogen consumption is also a parameter that varies with feedstock composition thereby indicating the need for a thorough understanding of the feedstock constituents if the process is to be employed to maximum efficiency. A convenient means of understanding the influence of feedstock on the hydrotreating process is through a study of the hydrogen content (H/C atomic ratio) and molecular weight (carbon number) of the various feedstocks or products. It is also possible to use data for hydrogen usage in residuum processing, where the relative amount of hydrogen consumed in the process can be shown to be dependent upon the sulfur content of the feedstock (Bridge et al., 1975, 1981).

Hydrotreating processes differ depending upon the feedstock available and the catalysts used. Hydrotreating can be used to improve the burning characteristics of distillates such as kerosene. Hydrotreatment of a kerosene fraction can convert aromatics into naphthenes, which are cleaner burning compounds.

Lube-oil hydrotreating uses catalytic treatment of the oil with hydrogen to improve product quality. The objectives in mild lube hydrotreating include saturation of olefins and improvements in the color, odor, and acid nature of the oil. Mild lube hydrotreating also may be used following solvent processing. Operating temperatures are usually below 315°C (600°F) and operating pressures below 800 psi. Severe lube hydrotreating, at temperatures in the 315°C to 400°C (600°F to 750°F) range and hydrogen pressures up to 3000 psi, is capable of saturating aromatic rings, along with sulfur and nitrogen removal, to impart specific properties not achieved at mild conditions.

Hydrotreating can also be employed to improve the quality of pyrolysis gasoline, a byproduct from the manufacture of ethylene. Traditionally, the outlet for pyrolysis-produced gasoline has been motor gasoline blending, a suitable route in view of its high octane number. However, only small portions can be blended untreated owing to the unacceptable odor, color, and gum-forming tendencies of this material. The quality of pyrolysis gasoline, which is high in olefin content, can be satisfactorily improved by hydrotreating, whereby conversion of olefins into mono-olefins provides an acceptable product for motor gas blending.

20.2 PROCESS PARAMETERS AND REACTORS

All HDS processes react hydrogen with a hydrocarbon feedstock to produce hydrogen sulfide and a desulfurized hydrocarbon product (Figure 20.5). The feedstock is preheated and mixed with hot recycle gas containing hydrogen, and the mixture is passed over the catalyst in the reactor section at temperatures between 290°C and 445°C (550°F and 850°F) and pressures between 150 and 3000 psig (1 to 20.7 MPa gauge) (Table 20.1). The reactor effluent is then cooked by heat exchange, and desulfurized liquid hydrocarbon product and recycle gas are

TABLE 20.1
Process Parameters for Hydrodesulfurization

Parameter	Naphtha	Residuum
Temperature (°C)	300 to 400	340 to 425
Pressure (atm.)	35 to 70	55 to 170
LHSV	4.0 to 10.0	0.2 to 1.0
H_2 recycle rate (scf/bbl)	400 to 1000	3000 to 5000
Catalyst life (years)	3.0 to 10.0	0.5 to 1.0
Sulfur removal (%)	99.9	85.0
Nitrogen removal (%)	99.5	40.0

Source: Speight, J.G. and Ozum, B. 2002. *Petroleum Refining Processes*. Marcel Dekker Inc., New York. With permission.

separated at essentially the same pressure as used in the reactor. The recycle gas is then scrubbed and purged of the hydrogen sulfide and light hydrocarbon gases, mixed with fresh hydrogen makeup, and preheated prior to mixing with hot hydrocarbon feedstock.

The recycle gas scheme is used in the HDS process to minimize the physical losses of expensive hydrogen. HDS reactions require a high hydrogen partial pressure in the gas phase to maintain high desulfurization reaction rates and to suppress carbon laydown (catalyst deactivation). The high hydrogen partial pressure is maintained by supplying hydrogen to the reactors at several times the chemical hydrogen consumption rate. The majority of the unreacted hydrogen is cooled to remove hydrocarbons, recovered in the separator, and recycled for further utilization. Hydrogen is physically lost in the process by solubility in the desulfurized liquid hydrocarbon product, and from losses during the scrubbing or purging of hydrogen sulfide and light hydrocarbon gases from the recycle gas.

The operating conditions in distillate HDS are dependent upon the feedstock as well as the desired degree of desulfurization or quality improvement. Kerosene and light gas oils are generally processed at mild severity and high throughput, whereas light catalytic cycle oils and thermal distillates require slightly more severe conditions. Higher-boiling distillates, such as vacuum gas oils and lube oil extracts, require the most severe conditions.

The principal variables affecting the required severity in distillate desulfurization are: (1) hydrogen partial pressure, (2) space velocity, (3) reaction temperature, and (4) feedstock properties.

20.2.1 HYDROGEN PARTIAL PRESSURE

The important effect of hydrogen partial pressure is the minimization of coking reactions. If the hydrogen pressure is too low for the required duty at any position within the reaction system, premature aging of the remaining portion of the catalyst will be encountered. In addition, the effect of hydrogen pressure on desulfurization varies with feed boiling range. For a given feed, there exists a threshold level above which hydrogen pressure is beneficial to the desired desulfurization reaction. Below this level, desulfurization drops off rapidly as hydrogen pressure is reduced.

20.2.2 SPACE VELOCITY

As the space velocity is increased, desulfurization is decreased, but increasing the hydrogen partial pressure and the reactor temperature can offset the detrimental effect of increasing space velocity.

20.2.3 REACTION TEMPERATURE

A higher reaction temperature increases the rate of desulfurization at constant feed rate, and the start-of-run temperature is set by the design desulfurization level, space velocity, and hydrogen partial pressure. The capability to increase temperature as the catalyst deactivates is built into most process or unit designs. Temperatures of 415°C (780°F) and above result in excessive coking reactions and higher-than-normal catalyst aging rates. Therefore, units are designed to avoid the use of such temperatures for any significant part of the cycle life.

20.2.4 CATALYST LIFE

Catalyst life depends on the charge stock properties and the degree of desulfurization desired. The only permanent poisons to the catalyst are metals in the feedstock that deposit on the catalyst, usually quantitatively, causing permanent deactivation as they accumulate.

However, this is usually of little concern, except when deasphalted oils are used as feedstocks, as most distillate feedstocks contain low amounts of metals. Nitrogen compounds are a temporary poison to the catalyst, but there is essentially no effect on catalyst aging, except that caused by a higher temperature requirement to achieve the desired desulfurization. Hydrogen sulfide can be a temporary poison in the reactor gas, and recycle gas scrubbing is employed to counteract this condition.

Providing that pressure drop buildup is avoided, cycles of 1 year or more and ultimate catalyst life of 3 years or more can be expected. The catalyst employed can be regenerated by normal steam–air or recycle combustion gas–air procedures. The catalyst is restored to near fresh activity by regeneration during the early part of its ultimate life. However, permanent deactivation of the catalyst occurs slowly during usage and repeated regenerations, so replacement becomes necessary.

20.2.5 FEEDSTOCK EFFECTS

The nature of the **feedstock properties**, especially the feed boiling range, has a definite effect on the ultimate design of the desulfurization unit and process flow. In agreement, there is a definite relationship between the percent by weight sulfur in the feedstock and the hydrogen requirements.

In addition, the reaction rate constant in the kinetic relationships decreases rapidly with increasing average boiling point in the kerosene and light gas oil range but much more slowly in the heavy gas oil range. This is attributed to the difficulty in removing sulfur from ring structures present in the entire heavy gas oil boding range.

The HDS of light (low-boiling) distillate (naphtha or kerosene) is one of the more common catalytic HDS processes, as it is usually used as a pretreatment of such feedstocks prior to deep HDS or prior to catalytic reforming (Datsevitch et al., 2003). This is similar to the concept of pretreating residua prior to hydrocracking to improve the quality of the products (Chapter 21). HDS of such feedstocks is required because sulfur compounds poison the precious metal catalysts used in reforming, and desulfurization can be achieved under relatively mild conditions and is near quantitative (Table 20.2). If the feedstock arises from a cracking operation, HDS will be accompanied with some degree of saturation, resulting in increased hydrogen consumption.

The HDS of low-boiling (naphtha) feedstocks is usually a gas-phase reaction and may employ the catalyst in fixed beds, and (with all of the reactants in the gaseous phase) only minimal diffusion problems are encountered within the catalyst pore system. It is,

TABLE 20.2
Hydrodesulfurization of Various Naphtha Fractions

	Boiling Range		Sulfur	Desulfurization
	°C	°F	wt.%	%
Feedstock				
Visbreaker naphtha	65 to 220	150 to 430	1.00	90
Visbreaker-coker naphtha	65 to 220	150 to 430	1.03	85
Straight-run naphtha	85 to 170	185 to 340	0.04	99
Catalytic naphtha (light)	95 to 175	200 to 350	0.18	89
Catalytic naphtha (heavy)	120 to 225	250 to 440	0.24	71
Thermal naphtha (heavy)	150 to 230	300 to 450	0.28	57

Source: Speight, J.G. and Ozum, B. 2002. *Petroleum Refining Processes*. Marcel Dekker Inc., New York. With permission.

however, important that the feedstock be completely volatile before entering the reactor, as there may be the possibility of pressure variations (leading to less satisfactory results) if some of the feedstock enters the reactor in the liquid phase and is vaporized within the reactor.

In applications of this type, the sulfur content of the feedstock may vary from 100 ppm to 1%, and the necessary degree of desulfurization to be effected by the treatment may vary from as little as 50% to more than 99%. If the sulfur content of the feedstock is particularly low, it will be necessary to presulfide the catalyst. For example, if the feedstock only has 100 to 200 ppm sulfur, several days may be required to sulfide the catalyst as an integral part of the desulfurization process even with complete reaction of all of the feedstock sulfur to, say, cobalt and molybdenum (catalyst) sulfides. In such a case, presulfiding can be conveniently achieved by addition of sulfur compounds to the feedstock or by addition of hydrogen sulfide to the hydrogen.

Generally, HDS of naphtha feedstocks to produce catalytic reforming feedstocks is carried to the point where the desulfurized feedstock contains less than 20 ppm sulfur. The net hydrogen produced by the reforming operation may actually be sufficient to provide the hydrogen consumed in the desulfurization process.

The HDS of middle distillates is also an efficient process, and applications include predominantly the desulfurization of kerosene, diesel fuel, jet fuel, and heating oils that boil over the general range of 250°C to 400°C (480°F to 750°F). However, with this type of feedstock, hydrogenation of the higher-boiling catalytic cracking feedstocks has become increasingly important, where HDS is accomplished alongside the saturation of condensed-ring aromatic compounds as an aid to subsequent processing.

Under the relatively mild processing conditions used for the HDS of these particular feedstocks, it is difficult to achieve complete vaporization of the feed. Process conditions may dictate that only part of the feedstock is actually in the vapor phase and that sufficient liquid phase is maintained in the catalyst bed to carry the larger molecular constituents of the feedstock through the bed. If the amount of liquid phase is insufficient for this purpose, molecular stagnation (leading to carbon deposition on the catalyst) will occur.

HDS of middle distillates causes a more marked change in the specific gravity of the feedstock, and the amount of low-boiling material is much more significant when compared with the naphtha-type feedstock. In addition, the somewhat more severe reaction conditions (leading to a designated degree of hydrocracking) also cause an overall increase in hydrogen consumption when middle distillates are employed as feedstocks in place of the naphtha.

High-boiling distillates, such as the atmospheric and vacuum gas oils, are not usually produced as a refinery product but merely serve as feedstocks to other processes for conversion to lower-boiling materials. For example, gas oils can be desulfurized to remove more than 80% of the sulfur originally in the gas oil with some conversion of the gas oil to lower-boiling materials. The treated gas oil (which has a reduced carbon residue as well as lower sulfur and nitrogen contents relative to the untreated material) can then be converted to lower-boiling products in, say, a catalytic cracker where an improved catalyst life and volumetric yield may be noted.

The conditions used for the HDS of gas oil may be somewhat more severe than the conditions employed for the HDS of middle distillates with, of course, the feedstock in the liquid phase.

In summary, the HDS of the low-, middle-, and high-boiling distillates can be achieved quite conveniently using a variety of processes. One major advantage of this type of feedstock is that the catalyst does not become poisoned by metal contaminants in the feedstock, as only negligible amounts of these contaminants will be present. Thus, the catalyst may be regenerated several times, and on-stream times between catalyst regeneration (while varying with the process conditions and application) may be of the order of 3 to 4 years.

20.2.6 REACTORS

20.2.6.1 Downflow Fixed-Bed Reactor

The reactor design commonly used in HDS of distillates is the fixed-bed reactor design in which the feedstock enters at the top of the reactor and the product leaves at the bottom of the reactor (Figure 20.5). The catalyst remains in a stationary position (fixed-bed) with hydrogen and petroleum feedstock passing in a downflow direction through the catalyst bed. The HDS reaction is exothermic and the temperature rises from the inlet to the outlet of each catalyst bed. With a high hydrogen consumption and subsequent large temperature rise, the reaction mixture can be quenched with cold recycled gas at intermediate points in the reactor system. This is achieved by dividing the catalyst charge into a series of catalyst beds, and the effluent from each catalyst bed is quenched to the inlet temperature of the next catalyst bed.

The extent of desulfurization is controlled by raising the inlet temperature in each catalyst bed to maintain constant catalyst activity over the course of the process. Fixed-bed reactors are mathematically modeled as plug-flow reactors with very little back mixing in the catalyst beds. The first catalyst bed is poisoned with vanadium and nickel at the inlet to the bed and may be a cheaper catalyst (**guard bed**). As the catalyst is poisoned in the front of the bed, the temperature exotherm moves down the bed, and the activity of the entire catalyst charge declines, thus requiring a raise in the reactor temperature over the course of the process sequence. After catalyst regeneration, the reactors are opened and inspected, and the high metal content catalyst layer at the inlet to the first bed may be discarded and replaced with fresh catalyst. The catalyst loses activity after a series of regenerations and, consequently, after a series of regenerations, it is necessary to replace the complete catalyst charge. In the case of very high metal content feedstocks (such as residua), it is often necessary to replace the entire catalyst charge rather than to regenerate it. This is due to the fact that the metal contaminants cannot be removed by economical means during rapid regeneration, and the metals have been reported to interfere with the combustion of carbon and sulfur, catalyzing the conversion of sulfur dioxide (SO_2) to sulfate (SO_4^{-2}) that has a permanent poisoning effect on the catalyst.

Fixed-bed HDS units are generally used for distillate HDS and may also be used for residuum HDS but require special precautions in processing. The residuum must undergo a two-stage electrostatic desalting so that salt deposits do not plug the inlet to the first catalyst bed and the residuum must be low in vanadium and nickel content to avoid plugging the beds with metal deposits. Hence, the need for a guard bed in residuum HDS reactors.

During the operation of a fixed-bed reactor, contaminants entering with fresh feed are filtered out and fill the voids between catalyst particles in the bed. The buildup of contaminants in the bed can result in the channeling of reactants through the bed and reducing the HDS efficiency. As the flow pattern becomes distorted or restricted, the pressure drop throughout the catalyst bed increases. If the pressure drop becomes high enough physical damage to the reactor internals can result. When high-pressure drops are observed throughout any portion of the reactor, the unit is shut down and the catalyst bed is skimmed and refilled.

With fixed-bed reactors, a balance must be reached between reaction rate and pressure drop across the catalyst bed. As catalyst particle size is decreased, the desulfurization reaction rate increases but so does the pressure drop across the catalyst bed. Expanded-bed reactors do not have this limitation, and small 1/32 inch (0.8 mm) extrudate catalysts or fine catalysts may be used without increasing the pressure drop.

20.2.6.2 Upflow Expanded-Bed Reactor

Expanded-bed reactors are applicable to distillates, but are commercially used for very heavy, high metals, and dirty feedstocks having extraneous fine solids material. They operate in such a way that the catalyst is in an expanded state so that the extraneous solids pass through the catalyst bed without plugging. They are isothermal, which conveniently handles the high-temperature exotherms associated with high hydrogen consumptions. As the catalyst is in an expanded state of motion, it is possible to treat the catalyst as a fluid and to withdraw and add catalyst during operation.

Expanded beds of catalyst are referred to as particulate fluidized, insofar as the feed-stock and hydrogen flow upward through an expanded bed of catalyst with each catalyst particle in independent motion. Thus, the catalyst migrates throughout the entire reactor bed. Expanded-bed reactors are mathematically modeled as back-mix reactors, with the entire catalyst bed at one uniform temperature. Spent catalyst may be withdrawn and replaced with fresh catalyst on a daily basis. Daily catalyst addition and withdrawal eliminate the need for costly shutdowns to change out catalyst and also result in a constant equilibrium catalyst activity and product quality. The catalyst is withdrawn daily and has a vanadium, nickel, and carbon content that is representative on a macro-scale of what is found throughout the entire reactor. On a micro-scale, individual catalyst particles have ages from that of fresh catalyst to as old as the initial catalyst charge to the unit, but the catalyst particles of each age group are so well dispersed in the reactor that the reactor contents appear uniform.

In the unit, the feedstock and hydrogen recycle gas enter the bottom of the reactor, pass up through the expanded catalyst bed, and leave from the top of the reactor. Commercial expanded-bed reactors normally operate with 1/32 inch (0.8 mm) extrudate catalysts that provide a higher rate of desulfurization than the larger catalyst particles used in fixed-bed reactors. With extrudate catalysts of this size, the upward liquid velocity based on fresh feedstock is not sufficient to keep the catalyst particles in an expanded state. Therefore, for each part of the fresh feed, several parts of product oil are taken from the top of the reactor, recycled internally through a large vertical pipe to the bottom of the reactor, and pumped back up through the expanded catalyst bed. The amount of catalyst bed expansion is controlled by the recycle of product oil back up through the catalyst bed.

The expansion and turbulence of gas and oil passing upward through the expanded catalyst bed are sufficient to cause almost complete random motion in the bed (particulate fluidized). This effect produces the isothermal operation. It also causes almost complete back-mixing. Consequently, to effect near-complete sulfur removal (over 75%), it is necessary to operate with two or more reactors in series. The ability to operate at a single temperature throughout the reactor or reactors, and to operate at a selected optimum temperature rather than an increasing temperature from the start to the end of the run, results in more effective use of the reactor and catalyst contents. When all these factors are put together, that is, use of a smaller catalyst particle size, isothermal, fixed temperature throughout run, back-mixing, daily catalyst addition, and constant product quality, the reactor size required for an expanded bed is often smaller than that required for a fixed bed to achieve the same product goals. This is generally true when the feeds have high initial boiling points or the hydrogen consumption is very high.

20.2.6.3 Demetallization Reactor (Guard Bed Reactor)

Feedstocks that have relatively high metals contents (>300 ppm) substantially increase catalyst consumption because the metals poison the catalyst, thereby requiring frequent catalyst replacement. The usual desulfurization catalysts are relatively expensive for these consumption rates, but there are catalysts that are relatively inexpensive and can be used in the first reactor to remove a large percentage of the metals. Subsequent reactors downstream of the first reactor would use normal HDS catalysts. As the catalyst materials are proprietary, it is not possible to identify them here. However, it is understood that such catalysts contain little or no metal promoters, that is, nickel, cobalt, molybdenum. Metals removal on the order of 90% has been observed with these materials.

Thus, one method of controlling demetallization is to employ separate smaller guard reactors just ahead of the fixed-bed HDS reactor section. The preheated feed and hydrogen pass through the guard reactors that are filled with an appropriate catalyst for demetallization that is often the same as the catalyst used in the HDS section. The advantage of this system is that it enables replacement of the most contaminated catalyst (guard bed), where pressure drop is highest, without having to replace the entire inventory or shut down the unit. The feedstock is alternated between guard reactors while the catalyst in the idle guard reactor is being replaced.

When the expanded-bed design is used, the first reactor could employ a low-cost catalyst (5% of the cost of Co/Mo catalyst) to remove the metals, and subsequent reactors can use the more selective HDS catalyst. The demetallization catalyst can be added continuously without taking the reactor out of service and the spent demetallization catalyst can be loaded to more than 30% vanadium, which makes it a valuable source of vanadium.

20.3 COMMERCIAL PROCESSES

Hydrotreating technology is one of the most commonly used refinery processes, designed to remove contaminants such as sulfur, nitrogen, condensed ring aromatics, or metals. The feedstocks used in the process range from naphtha to vacuum resid, and the products in most applications are used as environmentally acceptable clean fuels. Hydrotreating technology has a long history of commercial application since the 1950s, with more than 500 licensed units placed in operation worldwide. Furthermore, recent regulatory requirements to produce ultra-low-sulfur diesel (ULSD) and gasoline have created a very dynamic market, as refiners must build new units or revamp their existing assets to produce the low-sulfur, even no-sulfur fuels.

The petroleum industry often employs a two-stage process in which the feedstock undergoes both hydrotreating and hydrocracking. In the first, or pretreating, stage the main purpose is conversion of nitrogen compounds in the feed to hydrocarbons and to ammonia by hydrogenation and mild hydrocracking. Typical conditions are 340°C to 390°C (650°F to 740°F), 150 to 2500 psi (1 to 17 MPa), and a catalyst contact time of 0.5 to 1.5 h; up to 1.5% w/w hydrogen is absorbed, partly by conversion of the nitrogen compounds but chiefly by aromatic compounds that are hydrogenated.

It is most important to reduce the nitrogen content of the product oil to less than 0.001% w/w (10 ppm). This stage is usually carried out with a bifunctional catalyst containing hydrogenation promoters, for example, nickel and tungsten or molybdenum sulfides, on an acid support, such as silica–alumina. The metal sulfides hydrogenate aromatics and nitrogen compounds and prevent deposition of carbonaceous deposits; the acid support accelerates nitrogen removal as ammonia by breaking carbon–nitrogen bonds. The catalyst is generally used as $1/8 \times 1/8$ inch (0.32×0.32 cm) or $1/16 \times 1/8$ inch (0.16×0.32 cm) pellets, formed by extrusion.

Most of the hydrotreating cracking is accomplished in the first stage. Ammonia and some gasoline are usually removed from the first-stage product, and then the remaining first-stage product, which is low in nitrogen compounds, is passed over the second-stage catalyst. Again, typical conditions are 300°C to 370°C (600°F to 700°F), 1500 to 2500 psi (10 to 17 MPa) hydrogen pressure, and 0.5 to 1.5 h contact time; 1% to 1.5% w/w hydrogen may be absorbed. Conversion to gasoline or jet fuel is seldom complete in one contact with the catalyst, so the lighter oils are removed by distillation of the products, and the heavier, higher-boiling product combined with fresh feed and recycled over the catalyst until it is completely converted.

20.3.1 AUTOFINING

The autofining process differs from other hydrorefining processes, in that an external source of hydrogen is not required. Sufficient hydrogen to convert sulfur to hydrogen sulfide is obtained by dehydrogenation of naphthenes in the feedstock.

The processing equipment is similar to that used in **hydrofining** (see later). The catalyst is cobalt oxide and molybdenum oxide on alumina, and operating conditions are usually 340°C to 425°C (650°F to 800°F) at pressures of 100 to 200 psi. Hydrogen formed by dehydrogenation of naphthenes in the reactor is separated from the treated oil and is then recycled through the reactor. The catalyst is regenerated with steam and air at 200 to 1000 h intervals, depending on whether light or heavy feedstocks have been processed. The process is used for the same purpose as hydrofining but is limited to fractions with end points no higher than 370°C (700°F).

20.3.2 FERROFINING

This mild hydrogen-treating process was developed to treat distilled and solvent-refined lubricating oils. The process eliminates the need for acid and clay treatment. The catalyst is a three-component material on an alumina base with low hydrogen consumption and a life expectancy of 2 years or more. Process operations include heating the hydrogen–oil mixture and charging to a downflow catalyst-filled reactor. Separation of oil and gas is a two-stage operation, whereby gas is removed to the fuel system. The oil is then steam stripped to control the flash point and dried in vacuum, and a final filtering step removes catalyst fines.

20.3.3 GULF-HDS

This is a regenerative fixed-bed process to upgrade petroleum residues by catalytic hydrogenation to refined heavy fuel oils or to high-quality catalytic charge stocks. Desulfurization and quality improvement are the primary purposes of the process, but if the operating conditions and catalysts are varied, light distillates can be produced and the viscosity of heavy material can be lowered. Long on-stream cycles are maintained by reducing random hydrocracking reactions to a minimum, and whole crude oils, virgin, or cracked residua may serve as feedstock.

The catalyst is a metallic compound supported on pelletized alumina and may be regenerated in situ with air and steam or flue gas through a temperature cycle of 400°C to 650°C (750°F to 1200°F). On-stream cycles of 4 to 5 months can be obtained at desulfurization levels of 65% to 75%, and catalyst life may be as long as 2 years.

20.3.4 HYDROFINING

Hydrofining is a process used for reducing the sulfur content of feedstocks by treating the feedstock in the presence of a catalyst. Thus, hydrofining may be used to upgrade straight run, cracked, or coker-derived naphtha and distillate boiling-range streams, thus increasing the supply of catalytic reformer feedstock, solvent naphtha, and other naphtha-type materials. The sulfur content of kerosene can be reduced with improved color, odor, and wick-char characteristics. The tendency of kerosene to form smoke is not affected as aromatics, which cause smoke, are not affected by the mild hydrofining conditions. Cracked gas oil having a high sulfur content can be converted to excellent fuel oil and diesel fuel by reduction of sulfur content and by the elimination of components that form gum and carbon residues.

This process can be applied to lubricating oil, naphtha, and gas oil. The feedstock is heated in a furnace and passed with hydrogen through a reactor containing a suitable metal oxide catalyst, such as cobalt and molybdenum oxides or alumina. Hydrogen is obtained from catalytic reforming units. Reactor operating conditions range from 205°C to 425°C (400°F to 800°F) and from 50 to 800 psi, depending on the kind of feedstock and the degree of treatment required. Higher-boiling feedstocks, high sulfur content, and maximum sulfur removal require higher temperatures and pressures (*Hydrocarbon Processing, 1974*).

After passing through the reactor, the treated oil is cooled and separated from the excess hydrogen, which is recycled through the reactor. The treated oil is pumped to a stripper tower, where hydrogen sulfide formed by the hydrogenation reaction is removed by steam, vacuum, or flue gas and the finished product leaves the bottom of the stripper tower. The catalyst is not usually regenerated; it is replaced after about a year's use.

Lube oil hydrofining is a catalytic technology to prepare lube base stocks for further processing or it may be used as a base-stock finishing step. The process is usually integrated Exol N technology, into an Exolfining configuration, to treat the waxy raffinate from extraction upstream from the lube dewaxing unit. It saturates multi-ring aromatics, destroys acids, and removes sulfur, nitrogen, and color bodies to improve the color, and the color and oxidation stability of base stocks. As a finishing step, the lube hydrofining process may be used to treat a dewaxed oil product with similar color and oxidation stability benefits. Lube hydrofining, when used on lube distillate feedstocks, can enhance the ensuing solvent extraction performance.

Lube hydrofining technology generally produces a product with (1) improved color, (2) improved oxidation stability, (3) improved antioxidant response, (4) better odor, (5) better surface properties, (6) optimal sulfur reduction, (7) significant viscosity index (VI) increase at high desulfurization levels, and (8) reduction in toxicity.

Generally, the lube hydrofining process (Figure 20.6) is suitable for feedstocks from all crude sources, and the process conditions are moderate. In an Exolfining N configuration, the lube hydrofining step is fully heat integrated with the solvent extraction step. The feedstock to the lube hydrofining reactor is the hot bottom stream from the raffinate recovery tower. In a stand-alone configuration, the reactor feedstock is typically brought to reaction temperature by a combination of heat exchange with the hot hydrofinished product and furnace preheater. The fixed-bed reactor contains a hydrofinishing catalyst and once-through or recycle hydrogen treat gas may be used. The reactor effluent is flashed to recover the unreacted hydrogen treat gas at high pressure as well as to separate the hydrogen sulfide and ammonia resulting from the hydrofinishing reactions. The hydrofinished oil is then steam stripped and dried under vacuum. As an additional option, the steam stripper tower may be designed to correct the volatility of the hydrofinished oil product.

Wax hydrofining is used to wax from a lube oil plant to produce a food-grade wax that can be used for wax paper, fruit coatings, paper cups, as a seal for medicines, and for similar

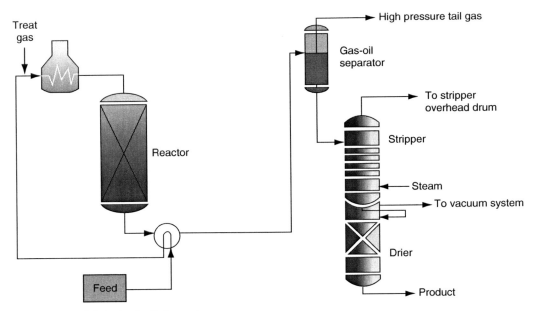

FIGURE 20.6 A lube hydrofining unit.

applications where high purity is required. Wax hydrofining is the final critical step for processing low oil content waxes from dewaxing–deoiling to produce wax with low odor, improved color, and stability. This technology is applicable to either crystalline wax from lube distillates or microcrystalline wax from deasphalted oils. The wax hydrofining option processes low oil content (<1% by weight) wax from conventional solvent lube plant dewaxing–deoiling units under moderate hydroprocessing conditions.

Thus, liquefied wax and hydrogen are preheated to reaction temperature in a conventional coil furnace and fed to a fixed-bed reactor containing a supported hydrotreating catalyst. Sulfur and nitrogen compounds are converted to hydrogen sulfide and ammonia, multi-ring aromatics are saturated, and trace acids and solvent residues are removed. The reactor effluent is flashed to separate hydrogen, hydrogen sulfide, and ammonia, and the wax is steam stripped and dried to improve volatility and color. Both crystalline and microcrystalline waxes can be processed in the same unit, although the types of oil molecules in microcrystalline waxes typically require more severe operating conditions. SCANfining (*q.v.*) is the next step up from hydrofining, and is used to meet environmental requirements for sulfur in gasoline using a proprietary RT-225 catalyst to treat gasoline from a fluid catalytic cracking unit. The process is specially designed to achieve high selectivity for sulfur removal without excessive olefin saturation and octane loss. Along similar paths, the diesel oil deep desulfurization (DODD) process for removing sulfur to very low levels and the GO-fining and residfining processes are used for upgrading gas oils and residua, respectively, for feedstocks to a catalytic cracking unit and also to use as low sulfur fuel oil.

Finally, as an additional option, most feedstocks to a fluid catalytic cracking unit are high sulfur with a high ratio of residue blended. This type of feedstocks results in high-sulfur high-olefin gasoline. This kind of gasoline can be treated in a hydrofiner, and the hydrofined naphtha is of good quality containing more saturated hydrocarbons, through which the high sulfur content problem of gasoline is resolved. Typical operating conditions are: 280°C to 320°C (535°F to 610°F), 290 to 1160 psi with a liquid hourly space velocity (LHSV) in the range 1.5 to −3.0 h^{-1} and a hydrogen–oil ratio (v/v) of 200 to 400.

The catalysts for hydrofining typically contain tungsten oxide or molybdenum oxide, nickel oxide and cobalt oxide supported on an alumina carrier. Catalysts with a lower metal content may exhibit higher activity at low temperature (Xia et al., 2000; Tanaka, 2004; Kumagai et al., 2005).

20.3.5 ISOMAX

The isomax process is a two-stage, fixed-bed catalyst system that operates under hydrogen pressures from 500 to 1500 psi in a temperature range of 205°C to 370°C (400°F to 700°F), for example, with middle distillate feedstocks. Exact conditions depend on the feedstock and product requirements, and hydrogen consumption is of the order of 1000 to 1600 ft^3 bbl^{-1} of feed processed. Each stage has a separate hydrogen recycling system. Conversion may be balanced to provide products for variable requirements, and recycling can be taken to extinction if necessary. Fractionation can also be handled in a number of different ways to yield desired products.

20.3.6 ULTRAFINING

Ultrafining is a regenerative, fixed-bed, catalytic process to desulfurize and hydrogenate refinery stocks from naphtha up to and including lubricating oil. The catalyst is cobalt–molybdenum on alumina and may be regenerated in situ using an air–stream mixture. Regeneration requires 10 to 20 h, and may be repeated 50 to 100 times for a given batch of catalyst; catalyst life is 2 to 5 years, depending on feedstock.

20.3.7 UNIFINING

This is a regenerative, fixed-bed, catalytic process to desulfurize and hydrogenate refinery distillates of any boiling range. Contaminating metals, nitrogen compounds, and oxygen compounds are eliminated, along with sulfur. The catalyst is a cobalt–molybdenum–alumina type that may be regenerated in situ with steam and air.

20.3.8 UNIONFINING

Unionfining process is a catalytic hydrotreating technology applied to produce low-sulfur, color-stable diesel fuel. The process is also used to reduce aromatics in diesel and to improve the diesel cetane number. In addition, the process can be used to prepare naphtha for catalytic reformers and to refine coker-naphtha.

In the process (Figure 20.7), the reactor feed is heat exchanged with the reactor effluent, and the reactor inlet temperature is controlled by charge heater firing. As hydrotreating is an exothermic reaction—especially when feedstocks are unsaturated—quench sections may be used to cool the reaction fluids and at the same time redistribute vapors and liquids between catalyst beds. Wash water is added to the cooled reactor effluent upstream of the final cooler to minimize corrosion and prevent deposits of ammonium salts. The cooled stream enters a cold high-pressure separator to separate out the hydrogen recycle that is recycled back to the reactor. Desulfurized product is recovered by stripping off the light ends with steam.

20.3.9 PROCESS OPTIONS FOR HEAVY FEEDSTOCKS

The major goal of residuum hydroconversion is cracking of residua with desulfurization, metal removal, denitrogenation, and asphaltene conversion. The residuum hydroconversion

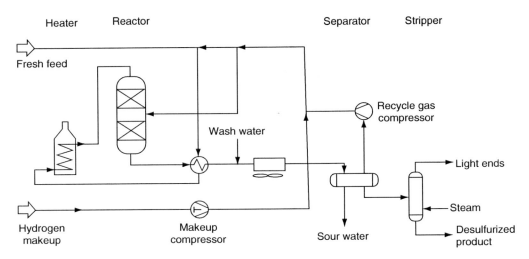

FIGURE 20.7 The unionfining process.

process offers production of kerosene and gas oil, and production of feedstocks for hydro-cracking, **fluid catalytic cracking**, and petrochemical applications. The processes that follow are listed in alphabetical order with no other preference in mind.

Residue hydrotreating is another method for reducing high-sulfur residual fuel oil yields. This technology was originally developed to reduce the sulfur content of atmospheric residues to produce specification low-sulfur residual fuel oil. Changes in crude oil quality and product demand, however, have shifted the commercial importance of this technology to include pretreating conversion unit feedstocks to minimize catalyst replacement costs, and coker feedstocks to reduce the yield and increase the quality of the byproduct coke fraction. Although residue hydrotreaters are capable of processing feedstocks having a wide range of contaminants, the feedstock's organometallic and asphaltene components typically determine its processability. Economics generally tend to limit residue hydrotreating applications to feedstocks with limitations (dictated by the process catalyst) on the content of nickel plus vanadium.

In many cases, application of hydrotreating technology to heavy feedstocks may also cause some cracking and, by inference, application of hydrocracking to heavy feedstocks will also cause desulfurization and denitrogenation. Rather than promote unnecessary duplication of the process description, two hydrotreating processes (the RDS and VRDS process options are described together as subcategories of one process) are described here with the reminder appearing elsewhere (Chapter 21) as the other processes are more amenable to hydrocracking than to hydrotreating.

20.3.9.1   Residuum Desulfurization and Vacuum Residuum Desulfurization Process

Residuum hydrotreating processes have two definite roles: (1) desulfurization to supply low-sulfur fuel oils and (2) pretreatment of feed residua for residuum fluid catalytic cracking processes. The main goal is to remove sulfur, metal, and asphaltene contents from residua and other heavy feedstocks to a desired level. On the other hand, the major goal of residuum hydroconversion is cracking of residua with desulfurization, metal removal, denitrogenation, and asphaltene conversion. The residuum hydroconversion process offers production of kerosene and gas oil, and production of feedstocks for hydrocracking, fluid catalytic cracking, and petrochemical applications.

The residuum desulfurizer/vacuum residuum desulfurizer (RDS/VRDS) process is (like the residfining process, *q.v.*) a hydrotreating process that is designed to hydrotreat vacuum gas oil, atmospheric residuum, or vacuum residuum to remove sulfur metallic constituents, while part of the feedstock is converted to lower-boiling products. In the case of residua, the asphaltene content is reduced (*Hydrocarbon Processing*, 1996).

The process consists of a once-through operation of hydrocarbon feed contacting graded catalyst systems designed to maintain activity and selectivity in the presence of deposited metals. Process conditions are designed for a 6 month to 1 year operating cycle between catalyst replacements. The process is ideally suited to produce feedstocks for residuum fluid catalytic crackers or delayed coking units to achieve minimal production of residual products in a refinery.

The major product of the processes is a low-sulfur fuel oil, and the amount of gasoline and middle distillates is maintained at a minimum to conserve hydrogen. The basic elements of each process are similar and consist of a once-through operation of the feedstock coming into contact with hydrogen and the catalyst in a downflow reactor that is designed to maintain activity and selectivity in the presence of deposited metals. Moderate temperatures and pressures are employed to reduce the incidence of hydrocracking and, hence, minimize production of low-boiling distillates (Table 16.11). The combination of a desulfurization step and a VRDS is often seen as an attractive alternate to the atmospheric RDS. In addition, either the RDS option or the VRDS option can be coupled with other processes (such as delayed coking, fluid catalytic cracking, and solvent deasphalting) to achieve the most optimum refining performance.

20.3.9.2 Residfining Process

The residfining process is a catalytic fixed-bed process for the desulfurization and demetallization of residua. The process can also be used to pretreat residua to suitably low contaminant levels prior to catalytic cracking. In the process, liquid feed to the unit is filtered, pumped to pressure, preheated, and combined with treat gas prior to entering the reactors. A small guard reactor would typically be employed to prevent plugging/fouling of the main reactors. Provisions are employed to periodically remove the guard while keeping the main reactors on-line. The temperature rise associated with the exothermic reactions is controlled utilizing either a gas or liquid quench. A train of separators is employed to separate the gas and liquid products. The recycle gas is scrubbed to remove ammonia and H_2S. It is then combined with fresh makeup hydrogen before being reheated and recombined with fresh feed. The liquid product is sent to a fractionator where the product is fractionated.

The different catalysts allow other minor differences in operating conditions and peripheral equipment. Primary differences include the use of higher-purity hydrogen makeup gas (usually 95% or greater), inclusion of filtration equipment in most cases, and facilities to upgrade the off-gases to maintain higher concentration of hydrogen in the recycle gas. Most of the processes utilize downflow operation over fixed-bed catalyst systems but exceptions to this are the H-Oil and LC-Fining processes (which are predominantly conversion processes) that employ upflow designs and ebullating catalyst systems with continuous catalyst removal capability, and the Shell Process (a conversion process) that may involve the use of a bunker flow reactor ahead of the main reactors to allow periodic changeover of catalyst.

The primary objective in most of the residue desulfurization processes is to remove sulfur with minimum consumption of hydrogen. Substantial percentages of nitrogen, oxygen, and metals are also removed from the feedstock. However, complete elimination of other reactions is not feasible and, in addition, hydrocracking, thermal cracking, and aromatic saturation reactions occur to some extent. Certain processes, that is, H-Oil (Chapter 21) using a single-stage or a two-stage reactor and LC-Fining (Chapter 21) using an expanded reactor, can be designed to accomplish greater amounts of hydrocracking to yield larger quantities of lighter distillates at the expense of desulfurization.

Removal of nitrogen is much more difficult than removal of sulfur. For example, nitrogen removal may be only about 25% to 30% when sulfur removal is at a 75% to 80% level. Metals are removed from the feedstock in substantial quantities and are mainly deposited on the catalyst surface and exist as metal sulfides at processing conditions. As these deposits accumulate, the catalyst pores eventually become blocked and inaccessible; thus catalyst activity is lost.

Desulfurization of residua is considerably more difficult than desulfurization of distillates (including vacuum gas oil), because many more contaminants are present and very large, complex molecules are involved. The most difficult portion of feed in residue desulfurization is the asphaltene fraction that forms coke readily and it is essential that these large molecules be prevented from condensing with each other to form coke, which deactivates the catalyst. This is accomplished by selection of proper catalysts, use of adequate hydrogen partial pressure, and assuring intimate contact of the hydrogen-rich gases and oil molecules in the process design.

20.4 CATALYSTS

HDS catalysts consist of metals impregnated on a porous alumina support. Almost all of the surface area is found in the pores of the alumina (200 to 300 m^2/g) and the metals are dispersed in a thin layer over the entire alumina surface within the pores (Table 20.3). This type of catalyst does display a huge catalytic surface for a small weight of catalyst.

TABLE 20.3
Composition and Properties of Hydrotreating Catalysts

	Range
Composition[a]	
Active phases (% by weight)	
MoO$_3$	13 to 20
CoO	2.5 to 3.5
NiO	2.5 to 3.5
Promoters (% by weight)	
SiO	1.0 to 10.0
Properties	
Surface area (m^2/gm)	150 to 500
Pore volume (cc/gm)	0.2 to 0.8
Pore diameter	
Mesopores (nm)	3.0 to 50.0
Macropores (nm)	100 to 5000
Extrudable diameter (mm)	0.8 to 4.0
Extrudable length/diameter (mm)	2.0 to 4.0
Bulk density (kg/m^3)	500 to 1000

[a]Catalyst is typically composed of active phases, promoters, and a gamma-alumina carrier.
Source: Speight, J.G. and Ozum, B. 2002. *Petroleum Refining Processes*. Marcel Dekker Inc., New York. With permission.

Cobalt (Co), molybdenum (Mo), and nickel (Ni) are the most commonly used metals for desulfurization catalysts. The catalysts are manufactured with the metals in an oxide state. In the active form, they are in the sulfide state, which is obtained by sulfiding the catalyst either prior to use or with the feed during actual use. Any catalyst that exhibits hydrogenation activity will catalyze HDS to some extent. However, the Group VIB metals (chromium, molybdenum, and tungsten) are particularly active for desulfurization, especially when promoted with metals from the iron group (iron, cobalt, nickel).

Furthermore, the increasing importance of HDS and HDN in petroleum processing to produce clean-burning fuels has led to a surge of research on the chemistry and engineering of heteroatom removal, with sulfur removal being the most prominent area of research. Most of the earlier works are focused on (1) catalyst characterization by physical methods, (2) low-pressure reaction studies of model compounds having relatively high reactivity, (3) process development, or (4) on cobalt–molybdenum (Co–Mo) catalysts, nickel–molybdenum catalysts (Ni–Mo), or nickel–tungsten (Ni–W) catalysts supported on alumina, often doped by fluorine or phosphorus.

HDS and demetallization occur simultaneously on the active sites within the catalyst pore structure. Sulfur and nitrogen occurring in residua are converted to hydrogen sulfide and ammonia in the catalytic reactor, and these gases are scrubbed out of the reactor effluent gas stream. The metals in the feedstock are deposited on the catalyst in the form of metal sulfides, and cracking of the feedstock to distillate produces a laydown of carbonaceous material on the catalyst; both events poison the catalyst and activity or selectivity suffers. The deposition of carbonaceous material is a fast reaction that soon equilibrates to a particular carbon level and is controlled by hydrogen partial pressure within the reactors. On the other hand, metal deposition is a slow reaction that is directly proportional to the amount of feedstock passed over the catalyst.

The need to develop catalysts that can carry out deep HDS and deep HDN has become even more pressing in view of recent environmental regulations limiting the amount of sulfur and nitrogen emissions. The development of a new generation of catalysts to achieve this objective of low nitrogen and sulfur levels in the processing of different feedstocks presents an interesting challenge for catalyst development.

Basic nitrogen-containing compounds in a feed diminish the cracking activity of hydro-cracking catalysts. However, zeolite catalysts can operate in the presence of substantial concentrations of ammonia, in marked contrast to silica–alumina catalysts, which are strongly poisoned by ammonia. Similarly, sulfur-containing compounds in a feedstock adversely affect the noble metal hydrogenation component of hydrocracking catalysts. These compounds are hydrocracked to hydrogen sulfide, which converts the noble metal to the sulfide form. The extent of this conversion is a function of the hydrogen and hydrogen sulfide partial pressures.

Removal of sulfur from the feedstock results in a gradual increase in catalyst activity, returning almost to the original activity level. As with ammonia, the concentration of the hydrogen sulfide can be used to control precisely the activity of the catalyst. Nonnoble metal-loaded zeolite catalysts have an inherently different response to sulfur impurities, as a minimum level of hydrogen sulfide is required to maintain the nickel–molybdenum and nickel–tungsten in the sulfide state.

HDN is more difficult to accomplish than HDS, but the relatively smaller amounts of nitrogen-containing compounds in conventional crude oil (Chapter 6 and Chapter 7) made this of little concern to refiners. However, the trend to heavier feedstocks in refinery operations, which are richer in nitrogen than the conventional feedstocks, has increased the awareness of refiners to the presence of nitrogen compounds in crude feedstocks. For the most part, however, HDS catalyst technology has been used to accomplish HDN

(Topsøe and Clausen, 1984), although such catalysts are not ideally suited to nitrogen removal (Katzer and Sivasubramanian, 1979). However, in recent years, the limitations of HDS catalysts when applied to HDN have been recognized, and there are reports of attempts to manufacture catalysts more specific to nitrogen removal (Ho, 1988).

The character of the hydrotreating processes is chemically very simple as they essentially involve removal of sulfur and nitrogen as hydrogen sulfide and ammonia, respectively:

$$R-S-R' + H_2 \rightarrow RH + R'H + H_2S$$
$$R-N(R')-R'' + 3H_2 \rightarrow RH + R'H + R'H + 2NH_3$$

However, nitrogen is the most difficult contaminant to remove from feedstocks, and processing conditions are usually dictated by the requirements for nitrogen removal.

In general, any catalyst capable of participating in hydrogenation reactions may be used for HDS. The sulfides of hydrogenating metals are particularly used for HDS, and catalysts containing cobalt, molybdenum, nickel, and tungsten are widely used on a commercial basis (Kellett et al., 1980).

Hydrotreating catalysts are usually cobalt–molybdenum catalysts and under the conditions whereby nitrogen removal is accomplished, desulfurization usually occurs as well as oxygen removal. Indeed, it is generally recognized that the fullest activity of the hydrotreating catalyst is not reached until some interaction with the sulfur (from the feedstock) has occurred, with part of the catalyst metals converted to the sulfides. Too much interaction may of course lead to catalyst deactivation.

The poisoning effect of nitrogen can be offset to a certain degree by operation at a higher temperature. However, the higher temperature tends to increase the production of material in the methane (CH_4) to butane (C_4H_{10}) range and decrease the operating stability of the catalyst so that it requires more frequent regeneration. Catalysts containing platinum or palladium (approximately 0.5% wet) on a zeolite base appear to be somewhat less sensitive to nitrogen than are nickel catalysts, and successful operation has been achieved with feedstocks containing 40 ppm nitrogen. This catalyst is also more tolerant of sulfur in the feed, which acts as a temporary poison, the catalyst recovering its activity when the sulfur content of the feed is reduced.

On such catalysts as nickel or tungsten sulfide on silica–alumina, isomerization does not appear to play any part in the reaction, as uncracked normal paraffins from the feedstock tend to retain their normal structure. Extensive splitting produces large amounts of low molecular weight ($C_3–C_6$) paraffins, and it appears that a primary reaction of paraffins is catalytic cracking followed by hydrogenation to form iso-paraffins. With catalysts of higher hydrogenation activity, such as platinum on silica–alumina, direct isomerization occurs. The product distribution is also different, and the ratio of low molecular weight to intermediate molecular weight paraffins in the breakdown product is reduced.

In addition to the chemical nature of the catalyst, the physical structure of the catalyst is also important in determining the hydrogenation and cracking capabilities, particularly for heavy feedstocks (van Zijll Langhout et al., 1980; Fischer and Angevine, 1986; Kobayashi et al., 1987; Kang et al., 1988). When gas oils and residua are used, the feedstock is present as liquids under the conditions of the reaction. Additional feedstock and the hydrogen must diffuse through this liquid before reaction can take place at the interior surfaces of the catalyst particle.

At high temperatures, reaction rates can be much higher than diffusion rates, and concentration gradients can develop within the catalyst particle. Therefore, the choice of catalyst porosity is an important parameter. When feedstocks are to be hydrocracked to liquefied petroleum gas and gasoline, pore diffusion effects are usually absent. High surface area (about 300 m^2/g) and low- to moderate-porosity (from 12 Å pore diameter with crystalline acidic components to 50 Å or more with amorphous materials) catalysts are used. With reactions involving high molecular weight feedstocks, pore diffusion can exert a large influence, and catalysts with pore diameters greater than 80 Å are necessary for more efficient conversion (Scott and Bridge, 1971).

Catalyst operating temperature can influence reaction selectivity as the activation energy for hydrotreating reactions is much lower than for hydrocracking reaction. Therefore, raising the temperature in a residuum hydrotreater increases the extent of hydrocracking relative to hydrotreating, which also increases the hydrogen consumption (Bridge et al., 1975, 1981).

Aromatic hydrogenation in petroleum refining may be carried out over supported metal or metal sulfide catalysts, depending on the sulfur and nitrogen levels in the feedstock. For hydrorefining of feedstocks that contain appreciable concentrations of sulfur and nitrogen, sulfided nickel–molybdenum (Ni–Mo), nickel–tungsten (Ni–W), or cobalt–molybdenum (Co–Mo) on alumina (γ–Al$_2$O$_3$) catalysts are generally used, whereas supported noble metal catalysts have been used for sulfur- and nitrogen-free feedstocks. Catalysts containing noble metals on Y-zeolites have been reported to be more sulfur tolerant than those on other supports (Jacobs, 1986). Within the series of cobalt-promoted or nickel-promoted group VI metal (Mo or W) sulfides supported on γ–Al$_2$O$_3$, the ranking for hydrogenation is:

$$Ni-W > Ni-Mo > Co-Mo > Co-W$$

Nickel–tungsten (Ni–W) and nickel–molybdenum (Ni–Mo) on Al$_2$O$_3$ catalysts are widely used to reduce sulfur, nitrogen, and aromatics levels in petroleum fractions by hydrotreating.

Molybdenum sulfide (MoS$_2$), usually supported on alumina, is widely used in petroleum processes for hydrogenation reactions. It is a layered structure that can be made much more active by addition of cobalt or nickel. When promoted with cobalt sulfide (CoS), making what is called cobalt–moly catalysts, it is widely used in HDS processes. The nickel sulfide (NiS)-promoted version is used for HDN as well as HDS. The closely related tungsten compound (WS$_2$) is used in commercial hydrocracking catalysts. Other sulfides (iron sulfide, FeS, chromium sulfide, Cr$_2$S$_3$, and vanadium sulfide, V$_2$S$_5$) are also effective and used in some catalysts. A valuable alternative to the base metal sulfides is palladium sulfide (PdS). Although it is expensive, PdS forms the basis for several very active catalysts.

The life of a catalyst used to hydrotreat petroleum residua is dependent on the rate of carbon deposition and the rate at which organometallic compounds decompose and form metal sulfides on the surface. Several different metal complexes exist in the asphaltene fraction of the residuum, and an explicit reaction mechanism of decomposition that would be a perfect fit for all of the compounds is not possible. However, in general terms, the reaction can be described as hydrogen (A) dissolved in the feedstock contacting an organometallic compound (B) at the surface of the hydrotreating catalyst and producing a metal sulfide (C) and a hydrocarbon (D):

$$A + B \rightarrow C + D$$

Different rates of reaction may occur with various types and concentrations of metallic compounds. For example, a medium metal-content feedstock will generally have a lower rate of demetallization compared with high-metal content feedstock. And, although individual organometallic compounds decompose according to both first- and second-order rate expressions, for reactor design, a second-order rate expression is applicable to the decomposition of a residuum as a whole.

Obviously, the choice of the hydrogenation catalyst depends on what the catalyst designer wishes to accomplish. In catalysts to make gasoline, for instance, vigorous cracking is needed to convert a large fraction of the feed to the kinds of molecules that will make a good gasoline blending stock. For this vigorous cracking, a vigorous hydrogenation component is needed. As palladium is the most active catalyst for this, the extra expense is warranted. On the other hand, many refiners wish only to make acceptable diesel, a less demanding application. For this, the less-expensive molybdenum sulfides are adequate.

20.5 BIODESULFURIZATION

Refiners are being continually challenged to produce products with ever-decreasing levels of sulfur. At the same time, the supplies of light, low-sulfur crude oil that favor distillate production are limited and even decreasing. Generally, the sulfur content of petroleum continues to rise (Swain, 1991, 1993, 1997, 2000) with the accompanying decrease in API gravity and an increase in the proportion of residua in the crude oil. These factors require the crude oil to be processed more severely to produce gasoline and other transportation fuels. Thus, many refineries are now configured for maximum gasoline production that also includes increasingly processing highly aromatic distillate byproducts, such as light cycle oil, for the additional feedstock to produce more distillate.

In microbial-enhanced oil recovery processes (Chapter 6), microbial technology is exploited in oil reservoirs to improve recovery. In the process, injected nutrients, together with indigenous or added microbes, promote in situ microbial growth and generation of products that mobilize additional oil and move it to producing wells through reservoir repressurization, interfacial tension/oil viscosity reduction, and selective plugging of the most permeable zones. Alternatively, the oil-mobilizing microbial products may be produced by fermentation and injected into the reservoir.

Biocatalyst desulfurization of petroleum distillates is one of the number of possible modes of applying biologically based processing to the needs of the petroleum industry (McFarland et al., 1998; Setti et al., 1999). In addition, *Mycobacterium goodie* has been found to desulfurize benzothiophene (Li et al., 2005). The desulfurization product was identified as α-hydroxystyrene. This strain appeared to have the ability to remove organic sulfur from a broad range of sulfur species in gasoline. When straight-run gasoline containing various organic sulfur compounds was treated with immobilized cells of *Mycobacterium goodie* for 24 h at 40°C (104°F), the total sulfur content significantly decreased, from 227 to 71 ppm at 40°C. Furthermore, when immobilized cells were incubated at 40°C (104°F) with *Mycobacterium goodie*, the sulfur content of the gasoline decreased from 275 to 54 ppm in two consecutive reactions.

A dibenzothiophene-degrading bacterial strain, *Nocardia* sp., was able to convert dibenzothiophene to 2-hydroxybiphenyl as the end metabolite through a sulfur-specific pathway (Chang et al., 1998). Other organic sulfur compounds, such as thiophene derivatives, thiazole derivatives, sulfides, and disulfides were also desulfurized by *Nocardia* sp. When a sample in which dibenzothiophene was dissolved in hexadecane and treated with growing cells was used, the dibenzothiophene was desulfurized in approximately 80 h.

The soil-isolated strain microcobe identified as *Rhodococcus erythropolis* can efficiently desulfurize benzonaphthothiophene (Yu et al., 2006). The desulfurization product was α-hydroxy-ß-phenyl-naphthalene. Resting cells were able to desulfurize diesel oil (total organic sulfur, 259 ppm) after HDS and the sulfur content of diesel oil was reduced by 94.5% after 24 h at 30°C (86°F). Biodesulfurization of crude oils was also investigated, and after 72 h at 30°C (86°F), 62.3% of the total sulfur content in Fushun crude oil (initial total sulfur content, 3210 ppm) and 47.2% of the sulfur in Sudanese crude oil (initial total sulfur, 1237 ppm) were removed (see also Abbad-Andaloussi et al., 2003, and references cited therein).

Heavy crude oil recovery, facilitated by microorganisms, was suggested in the 1920s and received growing interest in the 1980s as microbial-enhanced oil recovery. However, such projects have been slow to get underway, although in situ biosurfactant and bio-polymer applications continue to garner interest (Van Hamme et al., 2003, and references cited therein). In fact, studies have been carried out on biological methods of removing heavy metals such as nickel and vanadium from petroleum distillate fractions, coal-derived liquid shale, bitumen, and synthetic fuels . However, further characterization on the biochemical mechanisms and bioprocessing issues involved in petroleum upgrading are required to develop reliable biological processes.

For upgrading options, the use of microbes has to show a competitive advantage of enzyme over the tried-and-true chemical methods prevalent in the industry. Currently, the range of reactions using microbes is large but is usually related to production of bioactive compounds or precursors. But the door is not closed, and the issues of biodesulfurization and bioupgrad-ing remain open for the challenge of bulk petroleum processing. These drawbacks limit the applicability of this technology to speciality chemicals and steer it away from bulk petroleum processing.

Biodesulfurization is, therefore, another technology to remove sulfur from the feedstock. However, several factors may limit the application of this technology. Many ancillary processes novel to petroleum refining would be needed, including a biocatalyst fermentor to regenerate the bacteria. The process is also sensitive to environmental conditions such as sterilization, temperature, and residence time of the biocatalyst. Finally, the process requires the existing hydrotreater to continue in operation to provide a lower sulfur feedstock to the unit and is more costly than conventional hydrotreating. Nevertheless, the limiting factors should not stop the investigations of the concept, and work should be continued with success in mind.

Once the concept has been proven on the scale that a refiner would require, the successful microbial technology will most probably involve a genetically modified bacterial strain for (1) upgrading distillates and other petroleum fractions in refineries, (2) upgrading crude petrol-eum upstream, and (3) dealing with environmental problems that face industry, especially in areas related to spillage of crude oil and products. These developments are part of a wider trend to use bioprocessing to make products and do many of the tasks that are accomplished currently by conventional chemical processing. If commercialized for refineries, however, biologically based approaches will be at scales and with economic impacts beyond anything previously seen in industry.

In addition, the successful biodesulfurization process will, most likely, be based on naturally occurring aerobic bacteria that can remove organically bound sulfur in hetero-cyclic compounds without degrading the fuel value of the hydrocarbon matrix. Because of the susceptibility of bacteria to heat, the process will need to operate at temperatures and pressures close to ambient and also use air to promote sulfur removal from the feedstock.

20.6 GASOLINE AND DIESEL FUEL POLISHING

Briefly, the two desulfurization processes used for fuels purification (desulfurization) are (1) sweetening (Chapter 24) and (2) hydrotreating (Chapter 20). Sweetening is effective only against mercaptans, which are the predominant species in light gasoline. Hydrotreating is effective against all sulfur species and is more widely used.

In the sweetening process, a light naphtha stream is washed with amine to remove hydrogen sulfide and then reacted with caustic, which promotes the conversion of mercaptans to disulfides.

$$R–SH \rightarrow RSSR$$

The disulfides can subsequently be extracted and removed in what is referred to as extractive sweetening.

In the hydrotreating process, the feed is reacted with hydrogen, in the presence of a solid catalyst. The hydrogen removes sulfur by conversion to hydrogen sulfide, which is subsequently separated and removed from the reacted stream. As the reaction is favored by both temperature and pressure, hydrotreaters are typically designed and operated at approximately 370°C (700°F) and 1000 to 2000 psi hydrogen. The lower ends of the ranges typically apply to gasoline desulfurization, whereas gas oil desulfurization requires a more severe operation.

Hydrogen is provided in the form of treating gas at a purity that is typically around on the order of 90% by volume, although gas with as little as 60% by volume hydrogen is reputed to be used. Hydrogen is produced by catalytic reformers or hydrogen-generation units (Chapter 22) and distributed to the hydrotreaters through a refinery-wide network.

In a hydrotreating unit, feed and treating gas are combined and brought to the reaction temperature and pressure, prior to entering the reactor. The reactor is a vessel preloaded with solid catalyst, which promotes the reaction. The catalyst is slowly deactivated by the continuous exposure to high temperatures and by the formation of a coke layer on its surface. Refineries have to shut down the units periodically and regenerate or replace the catalyst.

The severity of operation of an existing unit can be increased by increasing the reaction temperature but there is a negative impact on the life of the catalyst. The severity of operation can also be increased by increasing the catalyst volume of the unit. In this case, the typical solution is to add a second reactor identical to the existing one, doubling the reactor volume. The pressure of an existing unit cannot be changed to increase its severity, because the pressure is related to the material of construction and the thickness of metal surfaces. If higher pressure is required, the typical solution is to install a new unit and use the existing one for a less severe service.

In contrast, there are also processes that do not require either of the above technologies.

Biodesulfurization (q.v.) is only one of several concepts by which gasoline and diesel fuel might be polished, that is, sulfur removed to an extremely low level, if not to a zero level. At this point, it is pertinent that a brief review of the potential methods for fuel polishing should be introduced.

One new technology is the use of adsorption by metal oxides, in which the oxides either react by physical adsorption or by chemical adsorption, insofar as adsorption followed by chemical reaction is promoted. The major distinction of this type of process from conventional hydrotreating is that the sulfur in the sulfur-containing compounds adsorbs to the catalyst after the feedstock–hydrogen mixture interacts with the catalyst. The catalyst does need to be regenerated constantly.

Another option involves sulfur oxidization in which a petroleum and water emulsion is reacted with hydrogen peroxide (or another oxidizer) to convert the sulfur in sulfur-containing compounds to sulfones. The sulfones are separated from the hydrocarbons for post-processing. The major advantages of this new technology include low reactor temperatures and pressures, short residence time, no emissions, and no hydrogen requirement. The technology preferentially treats dibenzothiophene derivatives, one of several streams that are most difficult to desulfurize.

One way to add to the supply of ultra-low sulfur fuels is to turn to a nonoil-based diesel. The **Fischer-Tropsch process**, for example, can be used to convert natural gas to a synthetic, sulfur-free diesel fuel. Commercial viability of gas-to-liquids projects depends (in addition to capital costs) on the market for petroleum products and possible price premiums for gas-to-liquids fuels as well as the value of any byproduct.

A second way to avoid desulfurization is with biodiesel made from vegetable oil or animal fats. Although other processes are available, most biodiesel is made with a base-catalyzed reaction. In the process, fat or oil is reacted with an alcohol, such as methanol, in the presence of a catalyst to produce glycerin and methyl esters or biodiesel. The methanol is charged in excess to assist in quick conversion and recovered for reuse. The catalyst, usually sodium or potassium hydroxide, is mixed with the methanol. Biodiesel is a strong solvent and can dissolve paint as well as deposits left in fuel lines by petroleum-based diesel, sometimes leading to engine problems. Biodiesel also freezes at a higher temperature than petroleum-based diesel.

REFERENCES

Aalund, L.R. 1975. *Oil Gas J*. 77(35): 339.

Aalund, L.R. 1981. *Oil Gas J*. 79(13): 70; and 79(37): 69.

Abbad-Andaloussi, S., Warzywoda, M., and Monot, F. 2003. *Revue Institut Français Du Pétrole*. 58(4): 505–513.

Bland, W.F. and Davidson, R.L. 1967. *Petroleum Processing Handbook*. McGraw-Hill, New York.

Boening, L.G., McDaniel, N.K., Petersen, R.D., and van Dreisen, R.P. 1987. *Hydrocarbon Process*. 66(9): 59.

Bridge, A.G. 1997. In *Handbook of Petroleum Refining Processes*. R.A. Meyers, ed. McGraw-Hill, New York. Chapter 14.1.

Bridge, A.G., Reed, E.M., and Scott, J.W. 1975. Paper presented at the API Midyear Meeting, May.

Bridge, A.G., Gould, G.D., and Berkman, J.F. 1981. *Oil Gas J*. 79(3): 85.

Carlson, C.S., Langer, A.W., Stewart, J., and Hill, R.M. 1958. *Ind. Eng. Chem*. 50: 1067.

Celestinos, J.A., Zermeno, R.G., Van Dreisen, R.P., and Wysocki, E.D. 1975. *Oil Gas J*. 73(48): 127.

Chang, J.H., Rhee, S.K., Chang, Y.K., and Chang H.N. 1998. *Biotechnol. Prog*. 14(6): 851–855.

Corbett, R.A. 1989. *Oil Gas J*. 87(26): 33.

Datsevitch, L., Gudde, N.J., Jess, A., and Schmitz, C. 2003. Proceedings. DGMK Conference on Innovation in the Manufacture and Use of Hydrogen. Dresden, Germany. October 15–17. p. 321.

Dolbear, G.E. 1998. In *Petroleum Chemistry and Refining*. J.G. Speight, ed. Taylor & Francis, Washington, DC. Chapter 7.

Fischer, R.H. and Angevine, P.V. 1986. *Appl. Catal*. 27: 275.

Ho, T.C. 1988. *Catal. Rev. Sci. Eng*. 30: 117.

Howell, R.L., Hung, C., Gibson, K.R., and Chen, H.C. 1985. *Oil Gas J*. 83(30): 121.

Hydrocarbon Processing. 1974. 53(9): 138 and 149.

Hydrocarbon Processing. 1978. 57(9): 133.

Hydrocarbon Processing. 1996.

Jacobs, P.A. 1986. In *Metal Clusters in Catalysis, Studies in Surface Science and Catalysis*. B.C. Gates, ed. Elsevier, Amsterdam. 29: 357.

Kang, B.C., Wu, S.T., Tsai, H.H., and Wu, J.C. 1988. *Appl. Catal*. 45: 221.

Katzer, J.R. and Sivasubramanian, R. 1979. *Catal. Rev. Sci. Eng*. 20: 155.

Kellett, T.F., Trevino, C.A., and Sartor, A.F. 1980. *Oil Gas J*. 78(18): 244.

Khan, M.R. 1997. In *Petroleum Chemistry and Refining*. J.G. Speight, ed. Taylor & Francis, Washington, DC. Chapter 6.

Kobayashi, S., Kushiyama S., Aizawa, R., Koinuma, Y., Inoue, K., Shmizu, Y., and Egi, K. 1987. *Ind. Eng. Chem. Res*. 26: 2241 and 2245.

Kumagai, H., Koyama, H., Nakamura, K., Igarashi, N., Mori, M., and Tsukada, T. 2005. *Hydrofining catalyst and hydrofining process*. United States Patent 6,858,132. February 22.

Langer, A.W., Stewart, J., Thompson, C.E., White, H.Y., and Hill, R.M. 1962. *Ind. Eng. Chem. Proc. Design Dev.* 1: 309.

Li, F., Xu, P., Feng, J., Meng, L., Zheng, Y., Luo, L., and Ma, C. 2005. *Appl. Environ Microbiol.* 71(1): 276–281.

McFarland, B.L., Boron, D.J., Deever, W., Meyer, J.A., Johnson, A.R., and Atlas, R.M. 1998. *Crit. Rev. Microbiol.* 24: 99–147.

Meyers, R.A. (ed.). 1997. *Handbook of Petroleum Refining Processes*. 2nd edn. New York.

Murphy, J.R. and Treese, S.A. 1979. *Oil Gas J.* 77(26): 135.

Nasution, A.S. 1986. Preprints. *Div. Petrol. Chem. Am. Chem. Soc.* 31(3): 722.

Reynolds, J.G. and Beret, S. 1989. *Fuel Sci. Technol. Int.* 7: 165.

Scott, J.W. and Bridge, A.G. 1971. In *Origin and Refining of Petroleum*. H.G. McGrath and M.E. Charles, eds. Advances in Chemistry Series 103. American Chemical Society, Washington, DC, p. 113.

Setti, L., Farinelli, P., Di Martino, S., Frassinetti, S., Lanzarini, G., and Pifferia, P.G. 1999. *Appl. Microbiol. Biotechnol.* 52: 111–117.

Speight, J.G. 1986. *Ann. Rev. Energy.* 11: 253.

Speight, J.G. 2000. *The Desulfurization of Heavy Oils and Residua*. 2nd edn. Marcel Dekker Inc., New York.

Speight, J.G. and Moschopedis, S.E. 1979. *Fuel Process. Technol.* 2: 295.

Stanislaus, A. and Cooper, B.H. 1994. *Catal. Rev.—Sci. Eng.* 36(1): 75.

Suchanek, A.J. and Moore, A.S. 1986. *Oil Gas J.* 84(31): 36.

Swain, E.J. 1991. *Oil Gas J.* 89(36): 59.

Swain, E.J. 1993. *Oil Gas J.* 91(9): 62.

Swain, E.J. 1997. *Oil Gas J.* 95(45): 79.

Swain, E.J. 2000. *Oil Gas J.* March 13.

Tanaka, H. 2004. *Process for producing hydrofining catalyst*. United States Patent 6,689,712. February 10.

Topsøe, H. and Clausen, B.S. 1984. *Catal. Rev. Sci. Eng.* 26: 395.

Van Hamme, J.D., Singh, A., and Ward, O.P. 2003. *Microbiol. Mol. Biol. Rev.* 67(4): 503–549.

Van Zijll Langhout, W.C. Ouwerkerk, C., and Pronk, K.M.A. 1980. *Oil Gas J.* 78(48): 120.

Vernon, L.W., Jacobs, F.E., and Bauman, R.F. 1984. United States Patent 4,425,224.

Wilson, J. 1985. *Energy Proc.* 5: 61.

Yu, B., Xu, P., Shi, Q., and Ma, C. 2006. *Appl. Environ. Microbiol.* 72: 54–58.

Xia, G., Zhu, M., Min, E., Shi, Y., L., M., Nie, H., Tao, Z., Huang, H., Zhang, R., Li, J., Wang, Z., and Ran, G. 2000. United States Patent 6,037,306. March 14.

21 Hydrocracking

21.1 INTRODUCTION

Hydrocracking is a refining technology that, like hydrotreating (Chapter 20), also falls under the general umbrella of **hydroprocessing**. The outcome is the conversion of a variety of feedstocks to a range of products; units to accomplish this goal can be found at various points in a refinery (Chapter 14).

The history of the process goes back to the late 1920s when it was realized that there was a need for gasoline of a higher quality than that obtained by catalytic cracking, which led to the development of the hydrocracking process. One of the first plants to use hydrocracking and commissioned for the commercial hydrogenation of brown coal was at Leuna in Germany. Tungsten sulfide was used as a catalyst in this one-stage unit, in which high reaction pressures, 2900 to 4350 psi, were applied. The catalyst displayed a very high hydrogenation activity: the aromatic feedstock, coal, and heavy fractions of oil, containing sulfur, nitrogen, and oxygen, were virtually completely converted into paraffins and iso-paraffins. In 1939, the Imperial Chemical Industries in Britain developed the second-stage catalyst for a plant that contributed largely to Britain's supply of aviation gasoline in the subsequent years.

During World War II, two-stage processes were applied on a limited scale in Germany, Britain, and the United States. In Britain, feedstocks were creosote from coal tar and gas oil from petroleum. In the United States, Standard Oil of New Jersey operated a plant at Baton Rouge, producing gasoline from a Venezuelan kerosene/light gas oil fraction. Operating conditions in those units were comparable: approximate reaction temperature was 400°C (750°F) and reaction pressures of 2900 to 4350 psi. After the war, commercial hydrocracking was very expensive but by the end of the 1950s, the process had become economical. The development of improved catalyst made it possible to operate the process at considerably lower pressure, namely 1000 to 2200 psi. This, in turn, resulted in a reduction in equipment wall thickness, whereas simultaneously, advances were made in mechanical engineering, especially in the field of reactor design and heat transfer. These factors, together with the availability of relatively low-cost hydrogen from the steam reforming process, brought hydrocracking back on the refinery scene. The first units of the second generation were built in the United States to meet the demand for conversion of surplus fuel oil in the gasoline-oriented refineries.

The older hydrogenolysis type of hydrocracking practiced in Europe during and after World War II used tungsten sulfide (WS_2) or molybdenum sulfide (MoS) as catalysts. These processes required high reaction temperatures and operating pressures, sometimes in excess of about 3000 psi (20,684 kPa) for continuous operation. The modern hydrocracking processes were initially developed for converting refractory feedstocks to gasoline and jet fuel, process and catalyst improvements, and modifications have made it possible to yield products from gases and naphtha to furnace oils and catalytic cracking feedstocks. The **zeolites** most frequently used in commercial **hydrocracking catalysts** are partially dealuminated and low-sodium, or high-silica, type Y zeolites in hydrogen or rare-earth forms. Other zeolites and mixtures of zeolites are also used. The zeolites are often imbedded in a high-surface area amorphous matrix, which serves as a binder. The metals can reside inside the zeolite and on the amorphous matrix.

Thus, hydrocracking is a more recent process development compared with the older thermal cracking, visbreaking, and coking. In fact, the use of hydrogen in thermal processes is perhaps the single most significant advance in refining technology during the twentieth century and the ability of refiners to cope with the renewed trend toward distillate production from heavier feedstocks with low atomic hydrogen/carbon ratios has created a renewed interest in hydrocracking (Bridge et al., 1981; Scherzer and Gruia, 1996; Dolbear, 1998). Without the required conversion units, heavier crude oils produce in lower yields of naphtha and middle distillate. To maintain the current gasoline and middle distillate production levels, additional conversion capacity is required because of the differential in the amount of distillates produced from light crude oil and the distillate products produced from heavier crude oil.

The concept of hydrocracking allows the refiner to produce products having a lower molecular weight with higher hydrogen content and a lower yield of **coke**. In summary, hydrocracking facilities add flexibility to refinery processing and to the product slate. Hydrocracking is more severe than hydrotreating (Chapter 20), there being the intent, in hydrocracking processes, to convert the feedstock to lower-boiling products rather than to treat the feedstock for heteroatom and metals removal only. Process parameters (Figure 21.1 and Figure 21.2) emphasize the relatively severe nature of the hydrocracking process.

Hydrocracking is an extremely versatile process that can be used in many different ways, and one of the advantages of hydrocracking is its ability to break down high-boiling aromatic stocks produced by catalytic cracking or coking. To take full advantage of hydrocracking, the process must be integrated in the refinery with other process units (Chapter 14). In gasoline production, for example, the hydrocracker product must be further processed in a catalytic reformer as it has a high naphthene content and relatively low octane number. The high naphthene content makes the hydrocracker gasoline an excellent feed for catalytic reforming, and good yields of high octane number gasoline can be obtained.

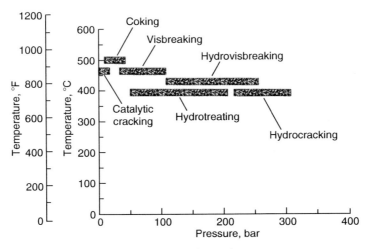

FIGURE 21.1 Temperature and pressure parameters for various processes.

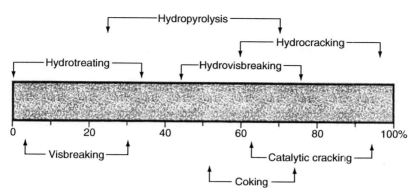

FIGURE 21.2 Feedstock conversion in various processes.

If high molecular weight petroleum fractions are pyrolyzed, that is, if no hydrogenation occurs, progressive cracking and condensation reactions generally lead to the final products. These products are usually:

1. Gaseous and low-boiling liquid compounds of high hydrogen content
2. Liquid material of intermediate molecular weight with a hydrogen–carbon atomic ratio differing more or less from that of the original feedstock, depending on the method of operation
3. Material of high molecular weight, such as coke, possessing a lower hydrogen–carbon atomic ratio than the starting material

Highly aromatic or refractory recycle stocks or gas oils that contain varying proportions of highly condensed aromatic structures (e.g., naphthalene and phenanthrene) usually crack, in the absence of hydrogen, to yield intractable residues and coke.

An essential difference between **pyrolysis** (thermal decomposition, usually in the absence of any added agent) and hydrogenolysis (thermal decomposition in the presence of hydrogen and a catalyst or a hydrogen-donating solvent) of petroleum is that in pyrolysis a certain amount of polymerized heavier products, like cracked residuum and coke, is always formed along with the light products, such as gas and gasoline. During hydrogenolysis (destructive hydrogenation) polymerization may be partly or even entirely prevented so that only light products are formed. The prevention of coke formation usually results in an increased distillate (e.g., gasoline) yield. The condensed type of molecule, such as naphthalene or phenanthrene, is one that is closely associated with the formation of coke, but in an atmosphere of hydrogen and in contact with catalysts, these condensed molecules are converted into lower molecular weight saturated compounds that boil within the gasoline range.

The mechanism of hydrocracking is basically similar to that of catalytic cracking, but with concurrent hydrogenation. The catalyst assists in the production of carbonium ions via olefin intermediates and these intermediates are quickly hydrogenated under the high hydrogen partial pressures employed in hydrocracking. The rapid hydrogenation prevents adsorption of olefins on the catalyst and, hence, prevents their subsequent dehydrogenation, which ultimately leads to coke formation so that long on-stream times can be obtained without the necessity of catalyst regeneration.

One of the most important reactions in hydrocracking is the partial hydrogenation of polycyclic aromatics followed by rupture of the saturated rings to form substituted monocyclic aromatics. The side chains may then be split off to give *iso*-paraffins. It is desirable to avoid excessive hydrogenation activity of the catalyst so that the monocyclic aromatics become hydrogenated to naphthenes; furthermore, repeated hydrogenation leads to loss in octane number, which increases the catalytic reforming required to process the hydrocracked naphtha.

Side chains of three or four carbon atoms are easily removed from an aromatic ring during catalytic cracking, but the reaction of aromatic rings with shorter side chains appears to be quite different. For example, hydrocracking single-ring aromatics containing four or more methyl groups produces largely *iso*-butane and benzene. It may be that successive isomerization of the feed molecule adsorbed on the catalyst occurs until a four-carbon side chain is formed, which then breaks off to yield *iso*-butane and benzene. Overall, coke formation is very low in hydrocracking as the secondary reactions and the formation of the precursors to coke are suppressed as the hydrogen pressure is increased.

When applied to residua, the hydrocracking process can be used for processes such as

1. Fuel oil desulfurization
2. Residuum conversion to lower boiling distillates

The products from hydrocracking are composed of either saturated or aromatic compounds; no olefins are found. In making gasoline, the lower paraffins formed have high octane numbers; for example, the five- and six-carbon number fractions have leaded research octane numbers of 99 to 100. The remaining gasoline has excellent properties as a feed to catalytic reforming, producing a highly aromatic gasoline that is capable of a high octane number. Both types of gasoline are suitable for premium-grade motor gasoline. Another attractive feature of hydrocracking is the low yield of gaseous components, such as methane, ethane, and propane, which are less desirable than gasoline. When making jet fuel, more hydrogenation activity of the catalysts is used, as jet fuel contains more saturates than gasoline.

Like many refinery processes, the problems encountered in hydrocracking heavy feedstocks can be directly equated to the amount of complex, higher boiling constituents that may require pretreatment (Speight and Moschopedis, 1979; Reynolds and Beret, 1989; Gray, 1994; Speight, 2000; Speight and Ozum, 2002). Processing these feedstocks is not merely a matter of applying know-how derived from refining conventional crude oils but requires knowledge of composition and properties (Chapter 7 through Chapter 9). The materials are not only complex in terms of the carbon number and boiling point ranges but also because a large part of this envelope falls into a range of model compounds where very little is known about the properties. It is also established that the majority of the higher molecular weight materials produce coke (with some liquids) but the majority of the lower molecular weight constituents produce liquids (with some coke). It is to both of these trends that hydrocracking is aimed.

It is the physical and chemical composition of a feedstock that plays a large part not only in determining the nature of the products that arise from refining operations but also in determining the precise manner by which a particular feedstock should be processed (Speight, 1986). Furthermore, it is apparent that the conversion of heavy oils and residua requires new lines of thought to develop suitable processing scenarios (Celestinos et al., 1975). Indeed, the use of thermal (**carbon rejection**) processes and of hydrothermal (**hydrogen addition**) processes, which were inherent in the refineries designed to process lighter feedstocks, has been a particular cause for concern. This has brought about the evolution of processing schemes that accommodate the heavier feedstocks (Chapter 23) (Wilson, 1985; Boening et al., 1987; Corbett, 1989; Khan and Patmore, 1998). As a point of reference, an example of the former option is the delayed coking process in which the feedstock is converted to overhead with the concurrent deposition of coke, for example, that used by Suncor, Inc., at their oil sands plant (Speight, 1990).

The hydrogen addition concept is illustrated by the hydrocracking process in which hydrogen is used in an attempt to stabilize the reactive fragments produced during the cracking, thereby decreasing their potential for recombination to heavier products and ultimately to coke. The choice of processing schemes for a given hydrocracking application depends on the nature of the feedstock as well as the product requirements (Murphy and Treese, 1979; Aalund, 1981; Nasution, 1986; Suchanek and Moore, 1986). The process can be simply illustrated as a single-stage or as a two-stage operation (Figure 21.3).

The single-stage process can be used to produce gasoline but is more often used to produce middle distillate from heavy vacuum gas oils (VGO). The two-stage process was developed primarily to produce high yields of gasoline from straight-run gas oil, and the first stage may actually be a purification step to remove sulfur-containing (as well as nitrogen-containing) organic materials. In terms of sulfur removal, it appears that nonasphaltene sulfur in nonasphaltene constituents may be removed before the more refractory sulfur in asphaltene constituents (Figure 21.4), thereby requiring thorough desulfurization. This is a good reason for processes to use an extinction–recycling technique to maximize desulfurization and the yields of the desired product. Significant conversion of heavy feedstocks can be accomplished by hydrocracking at high severity (Howell et al., 1985). For some applications, the products boiling up to 340°C (650°F) can be blended to give the desired final product.

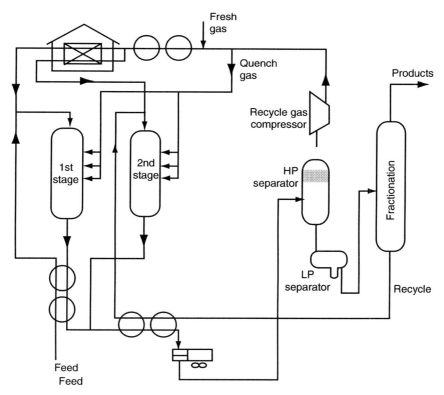

FIGURE 21.3 A single-stage or two-stage (optional) hydrocracking unit. (From OSHA Technical Manual, Section IV, Chapter 2, Petroleum Refining Processes.)

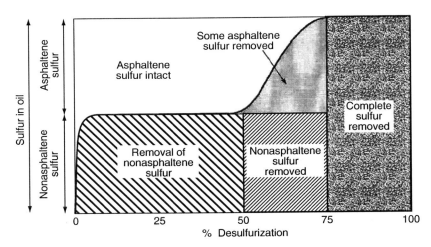

FIGURE 21.4 Trends for sulfur removal from crude oil.

Hydrocracking is similar to catalytic cracking, with hydrogenation superimposed and the reactions taking place either simultaneously or sequentially. Hydrocracking was initially used to upgrade low-value distillate feedstocks, such as cycle oils (highly aromatic products from a catalytic cracker that usually are not recycled to extinction for economic reasons), thermal and coker gas oils, and heavy-cracked and straight-run naphtha. These feedstocks are difficult to process by either catalytic cracking or reforming, as they are usually characterized by a high polycyclic aromatic content or by high concentrations of the two principal catalyst poisons, sulfur and nitrogen compounds (Bland and Davidson, 1967).

While whole families of catalysts are required depending on the feed available and the desired product slate or product character, the number of process stages is also important for the choice of catalysts. Generally, the refinery uses one of three options. Thus, depending on the feedstock being processed and the type of plant design employed (single-stage or two-stage), flexibility can be provided to vary product distribution among the end products.

Fundamentally, the trend toward lower API gravity feedstocks is related to an increase in the hydrogen/carbon atomic ratio of crude oils because of higher content of residuum. This can be overcome by upgrading methods that lower this ratio by adding hydrogen, rejecting carbon, or using a combination of both methods.

Though several technologies exist to upgrade heavy feedstocks (Chapter 23), selection of the optimum process units is very much dependent on each refiner's needs and goals, with the market pull being the prime motivator. Furthermore, processing options systems to dig deeper into the barrel by converting more of the higher boiling materials to distillable products should not only be cost effective and reliable, but also be flexible.

Hydrocracking adds that flexibility and offers the refiner a process that can handle varying feeds and operate under diverse process conditions. Utilizing different types of catalysts can modify the product slate produced. Reactor design and the number of processing stages play an important role in this flexibility.

Finally, a word about conversion measures for upgrading processes. Such measures are necessary for any conversion process but more particularly for hydrocracking processes where hydrogen management is an integral, and essential, part of process design.

The objective of any upgrading process is to convert heavy feedstock into marketable products by reducing their heteroatom (nitrogen, oxygen, and sulfur) and metal contents, modifying the structures of the asphaltene constituents (reducing coke precursors), and converting the high molecular weight polar species large molecules into lower molecular weight and lower boiling hydrocarbon products. Upgrading processes are evaluated on the basis of liquid yield (i.e., naphtha, distillate, and gas oil), heteroatom removal efficiency, feedstock conversion (FC), carbon mobilization (CM), and hydrogen utilization (HU), along with other process characteristics. Definitions of FC, CM, and HU are

$$FC = (Feedstock_{IN} - Feedstock_{OUT})/Feedstock_{IN} \times 100$$
$$CM = Carbon_{LIQUIDS}/Carbon_{FEEDSTOCK} \times 100$$
$$HU = Hydrogen_{LIQUIDS}/Hydrogen_{FEEDSTOCK} \times 100$$

High carbon mobilization (CM <100%) and high hydrogen utilization correspond to high feedstock conversion processes involving hydrogen addition such as hydrocracking. As hydrogen is added, hydrogen utilization can be greater than 100%. These tasks can be achieved by using thermal or catalytic processes. Low carbon mobilization and low hydrogen utilization correspond to low feedstock conversion such as coking (carbon rejection) processes (Chapter 17). Maximum efficiency from an upgrading process can be obtained by maximizing the liquid yield and its quality by minimizing the gas (C_1–C_4) yield, simultaneously. Under these operating conditions the hydrogen consumption would be the most efficient, that is, hydrogen is consumed to increase the liquid yield and its quality (Towler et al., 1996; Speight, 2000). Several process optimization models can be formulated if the reaction kinetics is known.

21.2 COMMERCIAL PROCESSES

Refiners are continuously faced with trends toward increased conversion, better product qualities, and more rapidly changing product patterns. Various processes are available that can meet the requirements to a greater or lesser degree: coking, visbreaking/thermal cracking, catalytic cracking, and hydrocracking.

The type of process applied and the complexity of refineries in various parts of the world are determined to a greater extent by the product distribution required. As a consequence, the relative importance of the above process in traditionally fuel-oil dominated refineries such as those in Western Europe will be quite different from those of gasoline-oriented refineries in, for instance, the United States.

A particular feature of the hydrocracking process, as compared with its alternatives, is its flexibility with respect to product outturn and the high quality of its products. In the areas where quantitative imbalance exists of lighter products, middle distillates and fuel, hydrocracking is a most suitable process for correction. Moreover, the hydrocracker does not yield any coke or pitch byproduct: the entire feedstock is converted into the required product range, an important consideration in a situation of limited crude oil availability. The development of the low-pressure catalytic reforming process, which produces relatively cheap, high-quality hydrogen, has contributed substantially to the economic viability of hydrocracking. On the whole, hydrocracking can handle a wider range of feedstock than catalytic cracking, although the latter process has seen some recent catalyst developments that have narrowed the gap. There are also examples where hydrocracking is complementary rather than alternative to the other conversion process; an example, cycle oils, which cannot be recycled to extinction in the catalytic cracker, can be processed in the hydrocracker.

21.2.1 Process Design

All hydrocracking processes are characterized by the fact that in a catalytic operation under relatively high hydrogen pressure a heavy oil fraction is treated to give products of lower molecular weight.

Most hydrocracking units use fixed beds of catalyst with downflow of reactants. The H-Oil process developed by Hydrocarbon Research Corp and Cities Service R&D employs an ebullient bed reactor in which the beds of particulate catalyst are maintained in an ebullient or fluidized condition with up-flowing reactants.

When the processing severity in a hydrocracker is increased, the first reaction that occurs leads to saturation of olefins present in feedstock. Next, desulfurization, denitrogenation, and deoxygenation reactions occur. These reactions constitute treating steps during which, in most cases, only limited cracking takes place. When the severity is increased further, hydro-cracking reaction is initiated. This proceeds at various rates with the formation of intermediate products (e.g., saturation of aromatics), which are subsequently cracked into lighter products.

21.2.1.1 Single-Stage and Two-Stage Options

The most common form of hydrocracking process is a two-stage operation (Figure 21.3). This flow scheme has been very popular as it can be used to maximize the yield of transportation fuels and has the flexibility to produce gasoline, naphtha, jet fuel, or diesel fuel to meet seasonal swings in product demand.

When the treating step is combined with the cracking reaction to occur in one reactor, the process is called a single-stage process. In this simplest of the hydrocracker configuration, the layout of the reactor section generally resembles that of the hydrotreating unit. This configuration will find application in cases where only moderate degree of conversion (say 60% or less) is required. It may also be considered if full conversion, but with a limited reduction in molecular weight, is aimed at. An example is the production of middle distillates from heavy distillate oils. The catalyst used in a single-stage process comprises a hydrogenation function in combination with a strong cracking function. Sulfided metals such as cobalt, molybdenum, and nickel provide the hydrogenation function. An acidic support, usually alumina, attends to the cracking function. Nitrogen compounds and ammonia produced by hydrogenation interfere with the acidic activity of the catalyst. In cases where high or full conversion is required, the reaction temperatures and run lengths of interest in commercial operation can no longer be adhered to. And moreover, conversion asymptotes out with increasing hydrogen pressure (Figure 21.5), so more hydrogen in the reactor is not the answer. In fact, it becomes necessary to switch to a different reactor bed system or to a multistage process, in which the cracking reaction mainly takes place in an added reactor.

With regard to the adverse effect of ammonia and nitrogen compounds on catalyst activity, two versions of the multistage hydrocracker have been developed: the two-stage hydrocracker and the series flow hydrocracker.

In the first type, the undesirable compounds are removed from the unconverted hydrocarbons before the latter are charged to the cracking reactor. This type is called the two-stage process. The other variety is often referred to as the series-flow hydrocracker. This type uses a catalyst with an increased tolerance toward nitrogen, both as ammonia and in organic form.

In the two-stage configuration, fresh feed is preheated by heat exchange with effluent from the first reactor. It is combined with part of a not fresh gas–recycle gas mixture and passes through a first reactor for desulfurnation/denitrogenation step. These reactions, as well as those of hydrocracking, which occur to a limited extent in the first reactor, are exothermic. The catalyst inventory is therefore divided among a number of fixed beds.

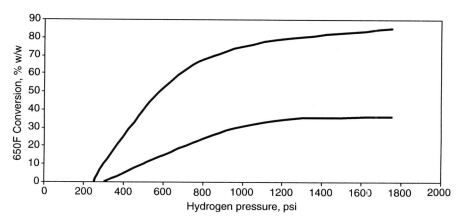

FIGURE 21.5 Relationship of conversion to hydrogen pressure.

Reaction temperatures are controlled by introducing part of the recycle gas as a quench medium between beds. The ensuing liquid is fractionated to remove the product made in the first reactor. Unconverted material, with low nitrogen content and free of ammonia, is taken as a bottom stream from the fractionation section. After heat exchange with reactor effluent and mixing with heated recycle gas, it is sent to the second reactor. Here, most of the hydrocracking reactions occur. Strongly acidic catalyst with a relatively low hydrogenation activity (metal sulfides on, e.g., amorphous silica–alumina) is usually applied. As in the first reactor, the exothermic nature of the process is controlled by recycle gas as the quench medium in the catalyst beds. Effluent from the second reactor is cooled and joins first stage effluent for separation from recycle gas and fractionation. The part of the second reactor feed that has remained unconverted is recycled to the reactor. Feedstock is thereby totally converted to the product boiling range.

In the series-flow configuration, the principal difference is the elimination of first stage cooling and gas or liquid separation and the ammonia removal step. The effluent from the first stage is mixed with more recycle gas and routed direct to the inlet of the second reactor. In contrast with the amorphous catalyst of the two-stage process, the second reactor in the series flow generally has a zeolite catalyst, based on crystalline silica–alumina. As in the two-stage process, material not converted to the product boiling range is recycled from the fractionation section.

Both two-stage and series flow hydrocracking are flexible processes: they may yield, in one mode of operation, only naphtha and lighter products and, in a different mode, only gas oil and lighter products. In the naphtha mode, both configurations have comparable yield patterns. In modes for heavier products, kerosene and gas oil, the two-stage process is more selective because product made in the first reactor is removed from the second reactor feed. In the series flow operation this product is partly cracked into lighter products in the second reactor.

Hydrocracking reactor stages usually have similar process flow schemes. The oil feed is combined with a preheated mixture of makeup hydrogen and hydrogen-rich recycle gas, and heated to reactor inlet temperature via a feed-effluent exchanger and a reactor charge heater. The reactor charge heater design philosophy is based on many years of safe operation with such two-phase furnaces. The feed-effluent exchangers take advantage of special high-pressure exchanger design features developed by Chevron engineers to give leak-free end closures. From the charge heater, the partially vaporized feed enters the top of the reactor.

The catalyst is loaded in separate beds in the reactor with facilities between the beds for quenching the reaction mix and ensuring good flow distribution through the catalyst.

The reactor effluent is cooled through a variety of heat exchangers including the feed-effluent exchanger and one or more air coolers. Deaerated condensate is injected into the first-stage reactor effluent before the final air cooler to remove ammonia and some of the hydrogen sulfide. This prevents solid ammonium bisulfide from depositing in the system. Expertise in the field of materials selection for hydrocracker cooling trains is quite important for proper design.

The reactor effluent leaving the air cooler is separated into hydrogen-rich recycle gas, a sour water stream, and a hydrocarbon liquid stream in the high-pressure separator. The sour water effluent stream is then sent to a plant for ammonia recovery and for purification so water can be recycled back to the hydrocracker. The hydrocarbon-rich stream is pressure reduced and fed to the distillation section after light products are flashed off in a low-pressure separator. The hydrogen-rich gas stream from the high-pressure separator is recycled back to the reactor feed by using a recycle compressor. Sometimes with sour feeds, the first-stage recycle gas is scrubbed with an amine system to remove hydrogen. If the feed sulfur level is high, this option can improve the performance of the catalyst and result in less costly materials of construction.

The distillation section consists of a hydrogen sulfide (H_2S) stripper and a recycle splitter. This latter column separates the product into the desired cuts. The column bottoms stream is recycled back to the second-stage feed. The recycle cut point is changed depending on the light products needed. It can be as low as 160°C (320°F) if naphtha production is maximized (for aromatics) or as high as 380°C (720°F) if a low pour point diesel is needed. Between these two extremes a recycle cut point of 260°C to 285°C (500°F to 550°F) results in high yields of high smoke point and low freezing point jet fuel.

Overall, a single-stage once-through (SSOT) unit resembles the first stage of the two-stage plant. This type of hydrocracker usually requires the least capital investment. The feedstock is not completely converted to lighter products. For this application the refiner must have a demand for highly refined heavy oil. In many refining situations, such an oil product can be used as lube oil plant feed or as FCC plant feed or in low sulfur oil blends or as ethylene plant feed. It also lends itself to stepwise construction of a future two-stage hydrocracker for full feed conversion.

A single-stage recycle (SSREC) unit converts heavy oil completely into light products with a flow scheme resembling the second stage of the two-stage plant. Such a unit maximizes the yield of naphtha, jet fuel, or diesel depending on the recycle cut point used in the distillation section. This type of unit is more economical than the more complex two-stage unit when plant design capacity is less than about 10,000 to 15,000 bbl/day. Commercial SSREC plants have operated to produce low pour point diesel fuel from waxy Middle East VGOs. Recent emphasis has been placed on the upgrading of lighter gas oils into jet fuels.

Building on the theme of one- or two-stage hydrocracking, the once-through partial conversion (OTPC) concept evolved. This concept offers the means to convert heavy VGO feed into high-quality gasoline, jet fuel, and diesel products by a partial conversion operation. The advantage is lower initial capital investment and also lower utilities consumption than a plant designed for total conversion. Because total conversion of the higher molecular weight compounds in the feedstock is not required, once-through hydrocracking can be carried out at lower temperatures and in most cases at lower hydrogen partial pressures than in recycle hydrocracking, where total conversion of the feedstock is normally an objective.

Proper selection of the types of catalysts employed can even permit partial conversion of heavy gas oil feeds to diesel and lighter products at the low hydrogen partial pressures for which gas oil hydrotreaters are normally designed. This so-called mild hydrocracking has been attracting a great deal of interest from refiners who have existing hydrotreaters and wish to increase their refinery's conversion of fuel oil into lower boiling higher value products.

Recycle hydrocracking plants are designed to operate at hydrogen partial pressures from about 1200 to 2300 psi (8274 to 15,858 kPa) depending on the type of feed being processed. Hydrogen partial pressure is set in the design in part depending on required catalyst cycle length, but also to enable the catalyst to convert high molecular weight polynuclear aromatic and naphthene compounds that must be hydrogenated before they can be cracked. Hydrogen partial pressure also affects the properties of the hydrocracked products that depend on hydrogen uptake, such as jet fuel aromatics content and smoke point and diesel cetane number. In general, the higher the feed endpoint, the higher the required hydrogen partial pressure necessary to achieve satisfactory performance of the plant.

OTPC hydrocracking of a given feedstock may be carried out at hydrogen partial pressures significantly lower than required for recycle total conversion hydrocracking. The potential higher catalyst deactivation rates experienced at lower hydrogen partial pressures can be offset by using higher activity catalysts and designing the plant for lower catalyst space velocities. Catalyst deactivation is also reduced by the elimination of the recycle stream. The lower capital cost resulting from the reduction in plant operating pressure is much more significant than the increase resulting from the possible additional catalyst requirement and larger volume reactors.

Additional capital cost savings from once-through hydrocracking result from the reduced overall required hydraulic capacity of the plant for a given fresh feed rate as a result of the elimination of a recycle oil stream. Hydraulic capacity at the same fresh feed rate is 30% to 40% lower for a once-through plant compared to one designed for recycle. Utilities savings for a once-through vs. recycle operation arise from lower pumping and compression costs as a result of the lower design pressure possible and also lower hydrogen consumption. Additional savings are realized as a result of the lower oil and gas circulation rates required, as recycle of oil from the fractionator bottoms is not necessary. Lower capital investment and operating costs are obvious advantages of once-through hydrocracking compared with a recycle design. This type of operation may be adaptable for use in an existing gas oil hydrotreater or atmospheric resid desulfurization plant. The change from hydrotreating to hydrocracking service will require some modifications and capital expenditure, but in most cases these changes will be minimal.

The fact that unconverted oil is produced by the plant is not necessarily a disadvantage. The unconverted oil produced by once-through hydrocracking is a high quality, low sulfur and nitrogen material that is an excellent feedstock for a **fluid catalytic cracking** or ethylene pyrolysis furnace or as a source of high viscosity index lube oil base stock. The properties of the oil are a function of the degree of conversion and other plant operating conditions.

One disadvantage of once-through hydrocracking compared with a recycle operation is a somewhat reduced flexibility for varying the ratio of gasoline to middle distillate that is produced. A greater quantity of naphtha can be produced by increasing conversion and production of jet fuel, moreover diesel can also be increased. But selectivity for higher boiling products is also a function of conversion. Selectivity decreases as once-through conversion increases. If conversion is increased too much, the yield of desired product will decrease, accompanied by an increase in light ends and gas production. Higher yields of gasoline or jet fuel and diesel are possible from a recycle than from a once-through operation.

Middle distillate products made by once-through hydrocracking are generally higher in aromatics content of poorer burning quality than those produced by recycle hydrocracking. However, the quality is generally better than produced by catalytic cracking or from straight run. Middle distillate product quality improves as the degree of conversion increases and as hydrogen partial pressure is increased.

The hydrocracking process employs high-activity catalysts that produce a significant yield of light products. Catalyst selectivity to middle distillate is a function of both the conversion level and operating temperature, with values in excess of 90% reported in commercial operation. In addition to the increased hydrocracking activity of the catalyst, percentage desulfurization and denitrogenation at start-of-run conditions are also substantially increased. End-of-cycle is reached when product sulfur has risen to the level achieved in conventional VGO hydrodesulfurization (HDS) process.

An important consideration, however, is that commercial hydrocracking units are often limited by design constraints of existing VGO hydrotreating units. Thus, the proper choice of catalyst(s) is critical when searching for optimum performance. Typical commercial distillate hydrocracking (DHC) catalysts contain both the hydrogenation (metal) and cracking (acid sites) functions required for service in existing desulfurization units.

21.2.2 PROCESS OPTIONS FOR HEAVY FEEDSTOCKS

The major goal of residuum hydroconversion is cracking of residua with desulfurization, metal removal, denitrogenation, and asphaltene conversion. The residuum hydroconversion process offers production of kerosene and gas oil, and production of feedstocks for hydrocracking, fluid catalytic cracking, and petrochemical applications. The processes that follow are listed in alphabetical order with no other preference in mind.

21.2.2.1 Asphaltenic Bottom Cracking (ABC) Process

The ABC process can be used for distillate production (Table 21.1), hydrodemetallization (HDM), asphaltene cracking, and moderate HDS as well as sufficient resistance to coke fouling and metal deposition using such feedstocks as vacuum residua, thermally cracked residua, solvent deasphalted bottoms, and bitumen with fixed catalyst beds.

The process can be combined with (1) solvent deasphalting for complete or partial conversion of the residuum or (2) HDS to promote the conversion of residue, to treat feedstock with high metals and to increase catalyst life, or (3) **hydrovisbreaking** to attain high conversion of residue with favorable product stability.

In the process, the feedstock is pumped up to the reaction pressure and mixed with hydrogen. The mixture is heated to the reaction temperature in the charge heater after a heat exchange and fed to the reactor.

In the reactor, HDM and subsequent asphaltene cracking with moderate HDS take place simultaneously under conditions similar to residuum HDS. The reactor effluent gas is cooled, cleaned up, and recycled to the reactor section, whereas the separated liquid is distilled into distillate fractions and vacuum residue that is further separated by deasphalting (Chapter 19) into deasphalted oil and asphalt using butane or pentane.

In the case of ABC-hydrodesulfurization catalyst combination, the ABC catalyst is placed upstream of the HDS catalyst and can be operated at a higher temperature than the HDS catalyst under conventional residuum HDS conditions. In the VisABC process, a soaking drum is provided after the heater, when necessary. Hydrovisbroken oil is first stabilized by the ABC catalyst through hydrogenation of coke precursors and then desulfurized by the HDS catalyst.

21.2.2.2 CANMET Hydrocracking Process

The CANMET hydrocracking process for heavy oils, atmospheric residua, and vacuum residua (Pruden, 1978; Waugh, 1983; Pruden et al., 1993). Energizing hydrocracking technology, such as that developed by the CANMET process shows promise for some applications. The scheme is a high conversion, high demetallization, residuum hydrocracking process which,

TABLE 21.1
Feedstock and Product Adapt for the ABC Process

	California Atmospheric Residuum	Vacuum Residuum			
		Arabian Light	Arabian Light	Arabian Heavy	Cerro Negro
Feedstock					
API	6.1	5.5	7.0	5.1	1.7
Sulfur, wt.%	6.5	4.4	4.0	5.3	4.3
Carbon residue, wt.%.	17.5	24.6	20.8	23.3	23.6
C7-asphaltenes	16.2	8.9	7.0	13.1	19,819.8
Nickel, ppm	150.0	30.0	223.0	52.0	150.0
Vanadium, ppm	380.0	90.0	76.0	150.0	640.0
Products					
Naphtha, vol.%	12.6	9.3	6.5	7.7	15.1
API	56.7	58.7	57.2	57.2	54.7
Distillate, vol.%	23.0	20.1	16.0	19.8	21.3
API	34.6[a]	33.2	34.2[a]	34.2[a]	32.5[a]
VGO, vol.%	38.4	33.1	34.3	38.1	32.8
API	23.1	20.3	24.7	21.6	15.4
Sulfur, wt.%	0.4	0.4	0.2	1.7	0.5
Vacuum residuum, vol.%	29.2	40.8	46.2	37.9	34.7
API	5.4	7.5	10.6	7.8	<0.0
Sulfur, wt.%	2.4	1.2	0.6	1.7	2.2
Carbon residue, wt.%	26.0	28.6	13.6	26.5	13.6
C7-asphaltenes, wt.%		12.0			
Nickel, ppm	100.0		9.0	45.0	117.0
Vanadium, ppm	180.0		11.0	75.0	371.0
Conversion	60.0		55.0	60.0	60.0

[a]Estimated.

Source: Speight, J.G. and Ozum, B. 2002. *Petroleum Refining Processes.* Marcel Dekker Inc., New York. With permission.

using an additive to inhibit coke formation, achieves conversion of high boiling point hydrocarbons into lighter products. It was initially developed to upgrade heavy oil and tar sand bitumen as well as residua. The process does not use a catalyst but employs a low-cost additive to inhibit coke formation and allow high conversion of heavy feedstocks (such as heavy oil and bitumen) into lower boiling products using a single reactor (Table 21.2). The process is unaffected by high levels of feed contaminants such as sulfur, nitrogen, and metals. Conversion of over 90% of the $525°C^+$ ($975°F^+$) fraction into distillates has been attained.

In the process, the feedstock and recycle hydrogen gas are heated to reactor temperature in separate heaters. A small portion of the recycle gas stream and the required amount of additive are routed through the oil heater to prevent coking in the heater tubes. The outlet streams from both heaters are fed to the bottom of the reactor.

The vertical reactor vessel is free of internal equipment and operates in a three-phase mode. The solid additive particles are suspended in the primary liquid hydrocarbon phase through which the hydrogen and product gases flow rapidly in bubble form. The reactor exit stream is quenched with cold recycle hydrogen prior to the high-pressure separator. The heavy liquids are further reduced in pressure to a hot medium pressure separator and from there to fractionation. The spent additive leaves with the heavy fraction and remains in the unconverted vacuum residue.

TABLE 21.2
Feedstock and Product Data for the CANMET Process

Feedstock[a]	
API gravity	4.4
Sulfur, wt.%	5.1
Nitrogen, wt %	0.6
Asphaltenes, wt.%	15.5
Carbon residue, wt.%	20.6
Metals, ppm	
Ni	80
V	170
Residuum (>525°C, >975°F) wt.%	
Products[b], *wt.%*	
Naphtha (C5–204°C, 400°F)	19.8
Nitrogen, wt.%	0.1
Sulfur, wt.%	0.6
Distillate (204–343°C, 400–650°F)	33.5
Nitrogen, wt.%	0.4
Sulfur, wt.%	1.8
VGO (343–534°C, 650–975°F)	28.5
Nitrogen, wt.%	0.6
Sulfur, wt.%	2.3
Residuum (>534°C, >975°F)	4.5
Nitrogen, wt.%	1.6
Sulfur, wt.%	3.1

[a]Cold Lake (Canada) heavy oil vacuum residuum.
[b]Residuum, 93.5% by weight.
Source: Speight, J.G. and Ozum, B. 2002. *Petroleum Refining Processes*. Marcel
Dekker Inc., New York. With permission.

The vapor stream from the hot high-pressure separator is cooled stepwise to produce middle distillate and naphtha that are sent to fractionation. High-pressure purge of low-boiling hydrocarbon gases is minimized by a sponge oil circulation system. Product naphtha will be hydrotreated and reformed, light gas oil will be hydrotreated and sent to the distillate pool, the heavy gas oil will be processed in the FCC, and the pitch will be sold.

The additive, prepared from iron sulfate [$Fe_2(SO_4)_3$], is used to promote hydrogenation and effectively eliminate coke formation. The effectiveness of the dual-role additive permits the use of operating temperatures that give high conversion in a single-stage reactor. The process maximizes the use of reactor volume and provides a thermally stable operation with no possibility of temperature runaway.

The process also offers the attractive option of reducing the coke yield by slurrying the feedstock with less than 10 ppm of catalyst (molybdenum naphthenate) and sending the slurry to a hydroconversion zone to produce low-boiling products (Kriz and Ternan, 1994).

21.2.2.3 H-Oil Process

The H-Oil process (Hydrocarbon Processing, 1998, p. 86) is a catalytic process that uses a single-stage, two-stage, or three-stage ebullated-bed reactor in which, during the reaction, considerable hydrocracking takes place (Figure 21.6). The process is designed for hydrogenation of residua

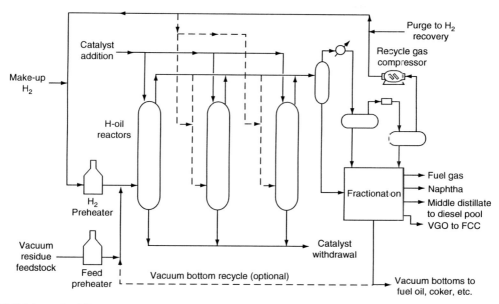

FIGURE 21.6 H-Oil process.

and other high feedstocks in an ebullated bed reactor to produce upgraded petroleum products (Hydrocarbon Processing, 1998). The process is able to convert all types of feedstocks to either distillate products as well as to desulfurize and demetallize residues for feed to coking units or residue fluid catalytic cracking units, for production of low sulfur fuel oil, or for production to asphalt blending. A modification of the H-Oil process (Hy-C Cracking process) converts high-boiling distillates to middle distillates and kerosene (Table 21.3).

A wide variety of process options can be used with the H-Oil process depending on the specific operation. In all cases, a catalytic ebullated-bed reactor system is used to provide an efficient hydroconversion. The system insures uniform distribution of liquid, hydrogen-rich gas and catalyst across the reactor. The ebullated-bed system operates under essentially isothermal conditions, exhibiting little temperature gradient across the bed (Kressmann et al., 2000). The heat of reaction is used to bring the feed oil and hydrogen up to reactor temperature.

In the process, the feedstock (a vacuum residuum) is mixed with recycle vacuum residue from downstream fractionation, hydrogen-rich recycle gas, and fresh hydrogen. This combined stream is fed into the bottom of the reactor whereby the upward flow expands the catalyst bed. The mixed vapor liquid effluent from the reactor goes either to the flash drum for phase separation or to the next reactor. A portion of the hydrogen rich gas is recycled to the reactor. The product oil is cooled and stabilized and the vacuum residue portion is recycled to increase conversion.

A catalyst of small particle size can be used, providing efficient contact among gas, liquid, and solid with good mass and heat transfer. Part of the reactor effluent is recycled back through the reactors for temperature control and to maintain the requisite liquid velocity. The entire bed is held within a narrow temperature range, which provides essentially an isothermal operation with an exothermic process. Owing to the movement of catalyst particles in the liquid–gas medium, deposition of tar and coke is minimized and fine solids entrained in the feed do not lead to reactor plugging. The catalyst can also be added and withdrawn from the reactor without destroying the continuity of the process. The reactor effluent is cooled by exchange and separates into vapor and liquid. After scrubbing in a lean oil absorber,

TABLE 21.3
Feedstock and Product Data for the H-Oil Process

| | Arabian Medium Vacuum Residuum | | Athabasca Bitumen |
	65%[a]	90%[a]	
Feedstock			
API gravity	4.9	4.9	8.3
Sulfur, wt.%	5.4	5.4	4.9
Nitrogen, wt %			0.5
Carbon residue, wt.%			
Metals, ppm (Ni and V)	128.0	128.0	
Residuum (>525°C, >975°F) wt.%			50.3
Products[b], wt.%			
Naphtha (C5–204°C, 400°F)	17.6	23.8	16.0
Sulfur, wt.%			1.0
Distillate (204–343°C, 400–650°F)	22.1	36.5	43.0
Sulfur, wt.%			2.0
VGO (343–534°C, 650–975°F)	34.0	37.1	26.4
Sulfur, wt.%			3.5
Residuum (>534°C, >975°F)	33.2	9.5	16.0
Sulfur, wt.%			5.7

[a] % conversion.
[b] % desulfurization.
Source: Speight, J.G. and Ozum, B. 2002. *Petroleum Refining Processes*. Marcel Dekker Inc., New York. With permission.

hydrogen is recycled and the liquid product is either stored directly or fractionated before storage and blending.

H-Oil and LC-fining technologies are often practiced commercially at about the 85% conversion level.

21.2.2.4   Hydrovisbreaking (HYCAR) Process

Briefly, hydrovisbreaking, a noncatalytic process, is conducted under similar conditions to visbreaking (Chapter 17) and involves treatment with hydrogen under mild conditions. The presence of hydrogen leads to more stable products (lower **flocculation threshold**) than can be obtained with straight visbreaking, which means that higher conversions can be achieved, producing a lower viscosity product.

The HYCAR process is composed fundamentally of three divisions (1) visbreaking, (2) HDM, and (3) hydrocracking. In the visbreaking section, the heavy feedstock (e.g., vacuum residuum or bitumen) is subjected to moderate thermal cracking whereas no coke formation is induced. The visbroken oil is fed to the demetallization reactor in the presence of catalysts, which provide sufficient pore for diffusion and adsorption of high molecular weight constituents. The product from this second stage proceeds to the hydrocracking reactor, where desulfurization and denitrogenation take place along with hydrocracking.

21.2.2.5   Hyvahl F Process

The process is used to hydrotreat atmospheric and vacuum residua to convert the feedstock to naphtha and middle distillates (Table 21.4) (Peries et al., 1988; Billon et al., 1994; Hydrocarbon Processing, 1998).

TABLE 21.4
Feedstock and Product Data for the Hyvahl Process

	Iranian Crude Oil (Topped) >360°C; >680°F	Hyvahl F (Once-Through)	Hyvahl F plus R2R
Feedstock			
API	15.2		
Sulfur, wt.%	2.6		
Nitrogen, wt.%	0.4		
Carbon residue, wt %.	9.4		
C7-asphaltenes	2.9		
Nickel + vanadium, ppm	191.0		
Products			
Naphtha, wt.%		4.0	48.0
Distillate/gas oil/VGO, wt.%		24.5	17.5
Vacuum residuum, wt.%		67.5	8.4
Coke			6.4

Source: Speight, J.G. and Ozum, B. 2002. *Petroleum Refining Processes*. Marcel Dekker Inc., New York. With permission.

The main features of this process are its dual catalyst system and its fixed bed swing-reactor concept. The first catalyst has a high capacity for metals (up to 100% by weight of new catalyst) and is used for both HDM and most of the conversion. This catalyst is resistant to fouling, coking, and plugging by asphaltene constituents (as well as by reacted asphaltene constituents) and shields the second catalyst from the same. Protected from metal poisons and deposition of coke-like products, the highly active second catalyst can carry out its deep HDS and refining functions. Both catalyst systems use fixed beds that are more efficient than moving beds and are not subject to attrition problems.

The swing-reactor design reserves two of the HDM reactors for use as guard reactors: one of them can be removed from service for catalyst reconditioning and put on standby whereas the rest of the unit continues to operate. More that 50% of the metals are removed from the feed in the guard reactors.

In the process, the preheated feedstock enters one of the two guard reactors where a large proportion of the nickel and vanadium is adsorbed and hydroconversion of the high molecular weight constituents commences. Meanwhile, the second guard reactor catalyst undergoes a reconditioning process and then is put on standby. From the guard reactors, the feedstock flows through a series of HDM reactors that continue the metals removal and the conversion of heavy ends.

The next processing stage, HDS, is where most of the sulfur, some of the nitrogen, and the residual metals are removed. A limited amount of conversion also takes place. From the final reactor, the gas phase is separated, hydrogen is recirculated to the reaction section, and the liquid products are sent to a conventional fractionation section for separation into naphtha, middle distillates, and heavier streams.

21.2.2.6 IFP Hydrocracking Process

The process features a dual catalyst system: the first catalyst is a promoted nickel–molybdenum amorphous catalyst. It acts to remove sulfur and nitrogen and hydrogenate aromatic rings. The second catalyst is a zeolite that finishes the hydrogenation and promotes the hydrocracking reaction.

TABLE 21.5
Feedstock and Product Data for the IFP Process

	Vacuum Gas Oil (350–570°C; 660–1060°F)
Feedstock	
API gravity	20.6
Sulfur, wt.%	2.3
Nitrogen, wt %	1.0
Carbon residue, wt.%	0.5
Products, wt.%	
Light naphtha	6.0
Sulfur, ppm	<1
Distillate (heavy naphtha, jet fuel, diesel fuel)	91.0
Sulfur, ppm	<50

Source: Speight, J.G. and Ozum, B. 2002. *Petroleum Refining Processes*. Marcel Dekker Inc., New York. With permission.

In the two-stage process, feedstock and hydrogen are heated and sent to the first reaction stage where conversion to products occurs (RAROP, 1991, p. 85). The reactor effluent phases are cooled and separated and the hydrogen-rich gas is compressed and recycled. The liquid leaving the separator is fractionated, the middle distillates and lower-boiling streams (Table 21.5) are sent to storage, and the high-boiling stream is transferred to the second reactor section and then recycled back to the separator section.

In the single-stage process, the first reactor effluent is sent directly to the second reactor, followed by the separation and fractionation steps. The fractionator bottoms are recycled to the second reactor or sold.

21.2.2.7 Isocracking Process

The process has been applied commercially in the full range of process flow schemes: single-stage, once-through liquid; single-stage, partial recycle of heavy oil; single-stage extinction recycle of oil (100% conversion); and two-stage extinction recycle of oil (Bridge, 1997; Hydrocarbon Processing, 1998, p. 84). The preferred flow scheme will depend on the feed properties, the processing objectives, and, to some extent, the specified feed rate.

The process uses multibed reactors and, in most applications, a number of catalysts are used in a reactor. The catalysts are dual functional being a mixture of hydrous oxides (for cracking) and heavy metal sulfides (for hydrogenation) (Bridge, 1997). The catalysts are used in a layered system to optimize the processing of the feedstock that undergoes changes in its properties along the reaction pathway (Table 21.6). In most commercial isocracking units, the entire fractionator bottom fractions are recycled or it is drawn as heavy product, depending on whether the low-boiling or high-boiling products are of greater value. If the low-boiling distillate products (naphtha or naphtha/kerosene) are the most valuable products, the higher boiling point distillates (like diesel) can be recycled to the reactor for conversion rather than drawn as a product (RAROP, 1991, p. 83). Product distribution depends on the mode of operation.

Heavy feedstocks have been used in the process and the product yield is very much dependent on the catalyst and the process parameters (Bridge, 1997).

TABLE 21.6
Feedstock and Product Data for the Isocracking Process

	Vacuum Gas Oil (360–530°C; 630–985°F)
Feedstock	
API gravity	22.6
Sulfur, wt.%	2.2
Nitrogen, wt.%	0.6
Carbon residue, wt.%	0.3
Metals	
Ni	0.1
V	0.3
Products, wt.%	
Naphtha (C5–124°C, C5–255°F)	15.9
Sulfur, ppm	<2
Distillate (124–295°C, 255–565°F)	51.6
Sulfur, ppm	<5
Heavy distillate (295–375°C, 565–705°F)	42.3
Sulfur, ppm	<5

Source: Speight, J.G. and Ozum, B. 2002. *Petroleum Refining Processes.* Marcel Dekker Inc., New York. With permission.

21.2.2.8 LC-Fining Process

The LC-Fining process is a hydrocracking process capable of desulfurizing, demetallizing, and upgrading a wide spectrum of heavy feedstocks by means of an expanded bed reactor (Van Driesen, et al., 1979; Fornoff, 1982; Bishop, 1990; RAROP, 1991, p. 61; Reich et al., 1993; Hydrocarbon Processing, 1998, p. 82). Operating with the expanded bed allows the processing of heavy feedstocks, such as atmospheric residua, vacuum residua, and oil sand bitumen. The catalyst in the reactor behaves like fluid that enables the catalyst to be added to and withdrawn from the reactor during operation. The reactor conditions are near isothermal because the heat of reaction is absorbed by the cold fresh feed immediately owing to thorough mixing.

In the process (Figure 21.7), the feedstock and hydrogen are heated separately and then passed upward in the hydrocracking reactor through an expanded bed of catalyst (Figure 21.8). Reactor products flow to the high pressure–high temperature separator. Vapor effluent from the separator is let down in pressure, and then goes to the heat exchange and thence to a section for the removal of condensable products, and purification (Table 21.7).

Liquid is let down in pressure and passes to the recycle stripper. This is a most important part of the high conversion process. The liquid recycle is prepared to the proper boiling range for return to the reactor. In this way the concentration of bottoms in the reactor, and therefore the distribution of products, can be controlled. After the stripping, the recycle liquid is then pumped through the coke precursor removal step where high molecular weight constituents are removed. The clean liquid recycle then passes to the suction drum of the feed pump. The product from the top of the recycle stripper goes to fractionation and any heavy oil product is directed from the stripper bottoms pump discharge.

The residence time in the reactor is adjusted to provide the desired conversion levels. Catalyst particles are continuously withdrawn from the reactor, regenerated, and recycled

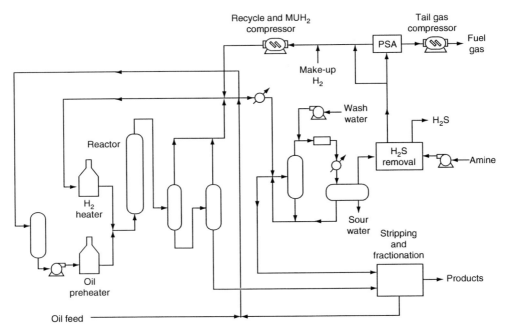

FIGURE 21.7 LC-Fining process.

back into the reactor, which provides the flexibility to process a wide range of heavy feedstock such as atmospheric and vacuum tower bottoms, coal-derived liquids and bitumen. An internal liquid recycle is provided with a pump to expand the catalyst bed, continuously. As

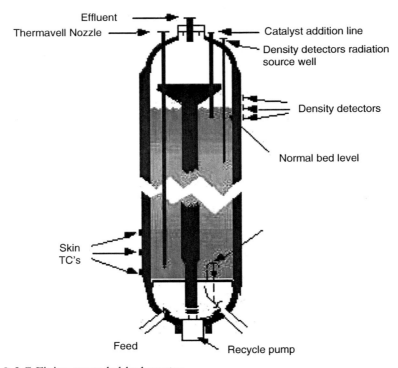

FIGURE 21.8 LC-Fining expanded-bed reactor.

TABLE 21.7
Feedstock and Product Data for the LC-Fining Process

	Kuwait Atmospheric Residuum	Gach Saran Vacuum Residuum	Arabian Heavy Vacuum Residuum	Athabasca Bitumen
Feedstock				
API gravity	15.0	6.1	7.5	9.1
Sulfur, wt.%	4.1	3.5	4.9	5.5
Products, wt.%				
Naphtha (C5–205°C, C5–400°F)	2.5	9.7	14.3	11.9
Sulfur, wt.%				1.1
Distillate (205–345°C, 400–650°F)	22.7	14.1	26.5	37.7
Sulfur, wt.%				0.7
Heavy distillate (345–525°C, 650–975°F)	34.7	24.1	31.1	30
Sulfur, wt.%				1.1
Residuum (>525°C, >975°F)	35.5	47.5	21.3	12.9
Sulfur, wt.%				3.4

Source: Speight, J.G. and Ozum, B. 2002. *Petroleum Refining Processes*. Marcel Dekker Inc., New York. With permission.

a result of expanded bed operating mode, small pressure drops and isothermal operating conditions are accomplished. Small diameter extruded catalyst particles as small as 0.8 mm (1/32 in.) can be used in this reactor.

Although the process may not be the means by which direct conversion of the bitumen to a synthetic crude oil would be achieved it does nevertheless offer an attractive means of bitumen conversion. Indeed, the process would play the part of the primary conversion process from which liquid products would accrue—these products would then pass to a secondary upgrading (hydrotreating) process to yield a synthetic crude oil.

21.2.2.9 MAKfining Process

The process uses a multiple catalyst system in multibed reactors that include quench and redistribution system internals (Hunter et al., 1997; Hydrocarbon Processing, 1998, p. 86).

In the process (Figure 21.9), the feedstock and recycle gas are preheated and brought into contact with the catalyst in a downflow fixed-bed reactor. The reactor effluent is sent to high- and low-temperature separators. Product recovery is a stripper/fractionator arrangement. Typical operating conditions in the reactors are 370°C to 425°C (700°F to 800°F) (single-pass) and 370°C to 425°C (700°F to 800°F) (recycle) with pressures of 1000 to 2000 psi (6895 to 13,790 kPa) (single-pass) and 1500 to 3000 psi (10,342 to 20,684 kPa) (recycle). Product (Table 21.8) yields depend on the extent of the conversion.

21.2.2.10 Microcat-RC Process

The Microcat-RC process (also referred to as the M-Coke process) is a catalytic ebullated-bed hydroconversion process that is similar to Residfining (*q.v.*) and which operates at relatively moderate pressures and temperatures (Bearden and Aldridge, 1981; Bauman et al., 1993). The catalyst particles, containing a metal sulfide in a carbonaceous matrix formed within the process, are uniformly dispersed throughout the feed. Because of their ultra small size (10^{-4} in. diameter)

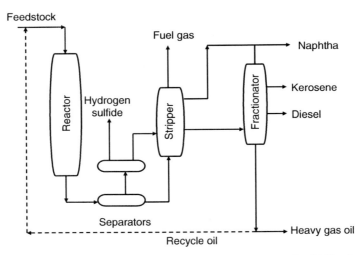

FIGURE 21.9 MAKfining process. (From Speight, J.G. and Ozum, B. 2002. *Petroleum Refining Processes*. Marcel Dekker Inc., New York. With permission.)

there are typically several orders of magnitude more of these microcatalyst particles per cubic centimeter of oil than is possible in other types of hydroconversion reactors using conventional catalyst particles. This results in smaller distances between particles and less time for a reactant molecule or intermediate to find an active catalyst site. Because of their physical structure, micro-catalysts suffer none of the pore-plugging problems that plague conventional catalysts suffer.

In the process, fresh vacuum residuum, microcatalyst, and hydrogen are fed to the hydro-conversion reactor. Effluent is sent to a flash separation zone to recover hydrogen, gases, and liquid products, including naphtha, distillate, and gas oil (Table 21.9). The residuum from the flash step is then fed to a vacuum distillation tower to obtain a $565°C^-$ ($1050°F^-$) product oil and a $565°C^+$ ($1050°F^+$) bottoms fraction that contains unconverted feed, microcatalyst, and essentially all the feed metals.

Hydrotreating facilities may be integrated with the hydroconversion section or built on a stand-alone basis, depending on product quality objectives and owner preference.

TABLE 21.8
Feedstock and Product Data for the MAKfining Process

	AL/AH[a] Vacuum	Gas Oil	AL/AH[a] Light Cycle Oil
Feedstock			
API gravity	20.2	20.2	19.0
Sulfur, wt.%	2.9	2.9	1.0
Nitrogen, wt.%	0.9	0.9	0.6
Products, vol.%			
Naphtha	12.9	22.6	54.0
Kerosene	14.1	24.5	
Diesel	31.8	32.5	54.3
Light gas oil	50.0	30.0	
Conversion, %	50.0	70.0	50.0

[a]AL/AH: Arabian light crude oil blended with Arabian heavy crude oil.
Source: Speight, J.G. and Ozum, B. 2002. *Petroleum Refining Processes*. Marcel Dekker Inc., New York. With permission.

TABLE 21.9
Feedstock and Product Data for the Microcat Process

	Cold Lake Heavy Oil Vacuum Residuum
Feedstock	
API gravity	4.4
Sulfur, wt.%	
Nitrogen, wt.%	
Metals (Ni + V), ppm	480.0
Carbon residue, wt.%	24.4
Products, vol.%	
Naphtha (C5–177°C, C5–350°F)	17.2
Distillate (177–343°C, 350–650°F)	63.6
Gas oil (343–566°C, 650–1050°F)	21.9
Residuum (>566°C, >1050°F)	2.1

Source: Speight, J.G. and Ozum, B. 2002. *Petroleum Refining Processes*. Marcel Dekker Inc., New York. With permission.

21.2.2.11 Mild Hydrocracking Process

The mild hydrocracking process uses operating conditions similar to those of a VGO desulfurizer to convert a VGO to significant yields of lighter products. Consequently, the flow scheme for a mild hydrocracking unit is virtually identical to that of a VGO desulfurizer.

For example, in a simplified process for VGO desulfurization, the VGO feedstock is mixed with hydrogen make-up gas and preheated against reactor effluent. Further preheating to reaction temperature is accomplished in a fired heater. The hot feed is mixed with recycle gas before entering the reactor. The temperature rises across the reactor due to the exothermic heat of reaction. Catalyst bed temperatures are usually controlled by using multiple catalyst beds and by introducing recycle gas as an interbed quench medium. Reactor effluent is cooled against incoming feed and air or water before entering the high-pressure separator. Vapors from this separator are scrubbed to remove hydrogen sulfide (H_2S) before compression back to the reactor as recycle and quench. A small portion of these gases is purged to fuel gas to prevent buildup of light ends. Liquid from the high-pressure separator is flashed into the low-pressure separator. Sour flash vapors are purged from the unit. Liquid is preheated against stripper bottoms and in a feed heater before steam stripping in a stabilizer tower. Water wash facilities are provided upstream of the last reactor effluent cooler to remove ammonium salts produced by denitrogenation of the VGO feedstock.

Variation of this process leads to the hot separator design. The process flow scheme is identical to that described earlier up to the reactor outlet. After initial reactor effluent cooling against incoming VGO feed and make-up hydrogen, a hot separator is installed. Hot liquid is routed directly to the product stabilizer. Hot vapors are further cooled by air or water before entering the cold separator. This arrangement reduces the stabilizer feed preheat duty and the effluent cooling duty by routing hot liquid direct to the stripper tower.

The conditions for mild hydrocracking are typical of many low-pressure desulfurization units that for hydrocracking units, in general, are marginal in pressure and hydrogen oil ratio capabilities. For hydrocracking, to obtain satisfactory run lengths (approximately 11 months), reduction in feed rate or addition of an extra reactor may be necessary. In most cases, as the product slate will be lighter than for normal desulfurization service only, changes in the fractionation system may be necessary. When these limitations can be tolerated, the product value from mild hydrocracking vs. desulfurization can be greatly enhanced.

In summary, the so-called mild hydrocracking process is a simple form of hydrocracking. The hydrotreaters designed for VGO desulfurization and catalytic cracker feed pretreatment are converted to once-through hydrocracking units and, because existing units are being used, the hydrocracking is often carried out under nonideal hydrocracking conditions.

21.2.2.12 MRH Process

The MRH process is a hydrocracking process designed to upgrade heavy feedstocks containing large amount of metals and asphaltene, such as vacuum residua and bitumen, and to produce mainly middle distillates (Table 21.10) (Sue, 1989; RAROP, 1991, p. 65). The reactor is designed to maintain a mixed three-phase slurry of feedstock, fine powder catalyst and hydrogen, and to promote effective contact.

In the process, a slurry consisting of heavy oil feedstock and fine powder catalyst is preheated in a furnace and fed into the reactor vessel. Hydrogen is introduced from the

TABLE 21.10
Feedstock and Product Data for the MRH Process

	Arabian Heavy Vacuum Residuum	Athabasca (Canada) Bitumen
Feedstock		
API gravity	5.9	10.2
Sulfur, wt.%	5.1	4.3
Nitrogen, wt.%	0.3	0.4
C7-asphaltenes, wt.%	11.4	8.1
Metals, ppm		
Nickel	41.0	85.0
Vanadium	127.0	182.0
Carbon residue, wt.%	21.7	13.3
Distillation profile, vol.%		
Naphtha	0.0	2.2
Kerosene	0.0	5.3
Light gas oil	0.0	12.1
VGO	4.0	31.8
Vacuum residuum	96.0	48.6
Atmospheric residuum	100.0	80.4
Products, wt.%		
Naphtha	13.0	12.0
Sulfur	0.2	1.1
Nitrogen	0.03	0.05
Kerosene	6.0	11.0
Sulfur	1.0	1.2
Nitrogen	0.06	0.08
Light gas oil	17.0	29.0
Sulfur	2.5	2.2
Nitrogen	0.06	0.11
Atmospheric residuum	55.0	41.0
Sulfur	3.8	3.8

Source: Speight, J.G. and Ozum, B. 2002. *Petroleum Refining Processes*. Marcel Dekker Inc., New York. With permission.

bottom of the reactor and flows upward through the reaction mixture, maintaining the catalyst suspension in the reaction mixture. Cracking, desulfurization, and demetallization reactions take place via thermal and catalytic reactions. In the upper section of the reactor, vapor is disengaged from the slurry, and hydrogen and other gases are removed in a high-pressure separator. The liquid condensed from the overhead vapor is distilled and then flows out to the secondary treatment facilities.

From the lower section of the reactor, bottom slurry oil (SLO) that contains catalyst, uncracked residuum, and a small amount of VGO fraction are withdrawn. VGO is recovered in the slurry separation section, and the remaining catalyst and coke are fed to the regenerator.

Product distribution focuses on middle distillates with the process focused as a residuum processing unit and inserted into a refinery just downstream from the vacuum distillation unit.

21.2.2.13 RCD Unibon (BOC) Process

The RCD Unibon (BOC) process is a process to upgrade vacuum residua (RAROP, 1991, p. 67; Thompson, 1997; Hydrocarbon Processing, 1998). There are several possible flow scheme variations involved for the process. It can operate as an independent unit or be used in conjunction with a thermal conversion unit. In this configuration, hydrogen and a vacuum residuum are introduced separately to the heater, and mixed at the entrance to the reactor. To avoid thermal reactions and premature coking of the catalyst, temperatures are carefully controlled and conversion is limited to approximately 70% of the total projected conversion. The removal of sulfur, heptane-insoluble materials, and metals is accomplished in the reactor. The effluent from the reactor is directed to the hot separator. The overhead vapor phase is cooled, condensed, and the separated hydrogen is recycled to the reactor.

Liquid product goes to the thermal conversion heater where the remaining conversion of nonvolatile materials occurs. The heater effluent is flashed and the overhead vapors are cooled, condensed, and routed to the cold flash drum. The bottoms liquid stream then goes to the vacuum column where the gas oils are recovered for further processing, and the residuals are blended into the heavy fuel oil pool.

21.2.2.14 Residfining Process

Residfining is a catalytic fixed-bed process for the desulfurization and demetallization of atmospheric and vacuum residua (RAROP, 1991, p. 69; Hydrocarbon Processing, 1998). The process can also be used to pretreat residua to suitably low contaminant levels prior to catalytic cracking.

In the process, liquid feedstock to the unit is filtered, pumped to pressure, preheated, and combined with treat gas prior to entering the reactors. A small guard reactor would typically be employed to prevent plugging/fouling of the main reactors. Provisions are employed to periodically remove the guard whereas keeping the main reactors online. The temperature rise associated with the exothermic reactions is controlled using either a gas- or liquid-quench. A train of separators is employed to separate the gas and liquid products. The recycle gas is scrubbed to remove ammonia and H_2S. It is then combined with fresh makeup hydrogen before it is reheated and recombined with fresh feed. The liquid product is sent to a fractionator where the product is fractionated.

Residfining is an option that can be used to reduce the sulfur, to reduce metals and coke-forming precursors, and to accomplish some conversion to lower-boiling products as a feed pretreat step ahead of a fluid catalytic cracking unit. There is also a hydrocracking option where substantial conversion of the resid occurs.

21.2.2.15 Residue Hydroconversion (RHC) Process

The residue hydroconversion process is a high-pressure fixed-bed trickle-flow hydrocatalytic process (RAROP, 1991, p. 71). The feedstock can be desalted atmospheric or vacuum residue (Table 21.11).

The reactors are of multibed design with interbed cooling and the multicatalyst system can be tailored according to the nature of the feedstock and the target conversion. For residua with high metal content, a HDM catalyst is used in the front-end reactor(s), which excels in its high metal uptake capacity and good activities for metal removal, asphaltene conversion, and residue cracking. Downstream of the demetallization stage, one or more hydroconversion stages, with optimized combination of catalysts' hydrogenation function and texture, is used to achieve desired catalyst stability and activities for denitrogenation, desulfurization, and heavy hydrocarbon cracking. A guard reactor may be employed to remove contaminants that promote plugging or fouling of the main reactors with periodic removal of the guard reactor whereas keeping the main reactors online.

21.2.2.16 Tervahl-H Process

In the Tervahl-H process, the feedstock and hydrogen-rich stream are heated using heat recovery techniques and fired heater and held in the soak drum as in the Tervahl-T process. The gas and oil from the soaking drum effluent are mixed with recycle hydrogen and separated in the hot separator where the gas is cooled, passed through a separator and recycled to the heater and soaking drum effluent. The liquids from the hot and cold separator are sent to the stabilizer section where purge gas and synthetic crude are separated. The gas is used as fuel and the synthetic crude can now be transported or stored.

In the related Tervahl-T process (a thermal process but covered here for convenient comparison with the Tervahl-T process), the feedstock is heated to the desired temperature using the coil heater and heat recovered in the stabilization section and held for a specified residence time in the soaking drum. The soaking drum effluent is quenched and sent to a conventional stabilizer or fractionator where the products are separated into the desired streams. The gas produced from the process is used for fuel.

TABLE 21.11
Feedstock and Product Data for the RHC Process

	Unspecified Residuum
Feedstock	
API gravity	24.4
Sulfur, wt.%	0.1
Nitrogen, wt.%	0.1
Carbon residue, wt.%	0.3
Products, vol.%	
Naphtha	4.5
Light gas oil	19.4
VGO	77.10
Sulfur, wt.%	0.01
Carbon residue, wt.%	0.20

Source: Speight, J.G. and Ozum, B. 2002. *Petroleum Refining Processes*. Marcel Dekker Inc., New York. With permission.

21.2.2.17 Unicracking Process

Unicracking is a fixed-bed catalytic process that employs a high-activity catalyst with a high tolerance for sulfur and nitrogen compounds and can be regenerated (Reno, 1997). The design is based on a single-stage or a two-stage system with provisions to recycle to extinction (RAROP, 1991, p. 79).

In the process, a two-stage reactor system receives untreated feed, makeup hydrogen, and a recycle gas at the first stage, in which gasoline conversion may be as high as 60% by volume. The reactor effluent is separated to recycle gas, liquid product, and unconverted oil (Table 21.12). The second-stage oil may be either once-through or recycle cracking; feed to the second stage is a mixture of unconverted first-stage oil and second-stage recycle. The process operates satisfactorily for a variety of feedstocks that vary in sulfur content from about 1.0% by weight to about 5% by weight. The rate of desulfurization is dependent on the sulfur content of the feedstock as is catalyst life, product sulfur, and hydrogen consumption (Speight, 2000 and references cited therein).

In the process, the feedstock and hydrogen-rich recycle gas are preheated, mixed, and introduced into a guard reactor that contains a relatively small quantity of the catalyst. The guard chamber removes particulate matter and residual salt from the feed. The effluent from the guard chamber flows down through the main reactor, where it contacts one or more catalysts designed for removal of metals and sulfur. The catalysts, which induce desulfurization, denitrogenation, and hydrocracking, are based on both amorphous and molecular-sieve containing supports. The product from the reactor is cooled, separated

TABLE 21.12
Feedstock and Product Data for the Unicracking Process

	Atmospheric Residuum				
	Alaska North Slope	Gach Saran	Kuwait	Kuwait	California
Feedstock					
API gravity	15.2	16.3	16.7	14.4	9.9
Sulfur, wt.%	1.7	2.4	3.8	4.2	4.5
Nitrogen, wt.%	0.4	0.4	0.2	0.2	0.4
Metals (Ni + V), ppm	44.0	220.0	46.0	66.0	213.0
Carbon residue, wt.%	8.4	8.5	8.5	10.0	13.6
Products, vol.%					
Naphtha (<185°C, <365°F)	0.8	1.8	1.1		
Naphtha (C5–205°C, C5–400°F)				2.1	4.2
Light gas oil				11.1	18.0
Residuum (>345°C, >650°F)				89.5	81.6
API				22.7	21.6
Sulfur, wt.%				<0.3	<0.3
Nitrogen, wt.%				<0.2	<0.2
Metals (Ni + V), ppm				<25	<5
Residuum (>185°C, >365°F)	100.5	100.4	100.4		
API	19.8	22.8	24.4		
Sulfur, wt.%	0.3	0.3	0.3		
Nitrogen, wt.%	0.3	0.3	0.1		
Metals (Ni + V), ppm	14.0	55.0	15.0		
Carbon residue, wt.%	4.0	4.0	3.0		

Source: Speight, J.G. and Ozum, B. 2002. *Petroleum Refining Processes*. Marcel Dekker Inc., New York. With permission.

from hydrogen-rich recycle gas, and either stripped to meet fuel oil flash point specifications, or fractionated to produce distillate fuels, upgraded VGO, and upgraded vacuum residuum. Recycle gas, after hydrogen sulfide removal, is combined with makeup gas and returned to the guard chamber and main reactors.

The most commonly implemented configuration is a single-stage Unicracking design, where the fresh feed and recycle oil are converted in the same reaction stage. This configuration simplifies the overall unit design by reducing the quantity of equipment in high-pressure service and keeping high-pressure equipment in a single train. The two-stage design has a separation system in each reaction stage. However, the optimum flow scheme depends on feedstock capacity and product slate objectives.

The high efficiency of the process is due to the excellent distribution of the feedstock and hydrogen that occurs in the reactor where a proprietary liquid distribution system is employed. In addition, the process catalyst (also proprietary) was designed for the desulfurization of residua and is not merely an upgraded gas oil hydrotreating catalyst as often occurs in various processes. It is a cobalt–molybdena–alumina catalyst with a controlled pore structure that permits a high degree of desulfurization and, at the same time, minimizes any coking tendencies.

The process uses base-metal or noble-metal hydrogenation-activity promoters impregnated on combinations of zeolites and amorphous-aluminosilicates for cracking activity (Reno, 1997). The specific metals chosen and the proportions of the metals, zeolite, and nonzeolite aluminosilicates are optimized for the feedstock and desired product balance. This is effective in the production of clean fuels, especially for cases where a partial conversion Unicracking unit and a fluid catalytic cracking unit are integrated.

The Unicracking process converts feedstocks into lower molecular weight products that are more saturated than the feed. Feedstocks include atmospheric gas oil, VGO, fluid catalytic cracking/resid catalytic cracking cycle oil, coker gas oil, deasphalted oil, and naphtha. Hydrocracking catalysts promote sulfur and nitrogen removal, aromatic saturation, and molecular weight reduction. All these reactions consume hydrogen and as a result, the volume of recovered liquid product normally exceeds the feedstock. Many units are operated to make naphtha (for petrochemical or motor-fuel use) as a primary product.

Unicracking catalysts are designed to function in the presence of hydrogen sulfide (H_2S) and ammonia (NH_3). This gives rise to an important difference between Unicracking and other hydrocracking processes: the availability of a single-stage design. In a single-stage unit, the absence of a stripper between treating and cracking reactors reduces investment costs by making use of a common recycle gas system. Process objectives determine catalyst selection for a specific unit. Product from the reactor section is condensed, separated from hydrogen-rich gas, and fractionated into desired products. Unconverted oil is recycled or used as lube stock, fluid catalytic cracking feedstock, or ethylene plant feedstock.

The Advanced Partial Conversion Unicracking (APCU) process is a recent advancement in the area of ultra low-sulfur diesel (ULSD) production and feedstock pretreatment for catalytic cracking units. At low conversions (20% to 50%) and moderate pressure, the APCU technology provides an improvement in product quality compared with traditional mild hydrocracking. In the process, high sulfur feeds such as VGO and heavy cycle gas oil are mixed with a heated hydrogen-rich recycle gas stream and passed over consecutive beds of high activity pretreat catalyst and distillate selective unicracking catalyst. This combination of catalysts removes refractory sulfur and nitrogen contaminants, saturates polynuclear aromatic compounds and converts a portion of the feed to ultra low sulfur diesel fuel. The hydrocracked products and desulfurized feedstock for a fluid catalytic cracking unit are separated at reactor pressure in an enhanced hot separator. The overhead products for the separator are immediately hydrogenated in the integrated finishing reactor.

As pretreatment severity is increased, conversion is increased in the fluid catalytic cracker, and both gasoline and alkylate octane-barrel output per barrel of cat cracker feedstock increases. APCU units can be customized to achieve maximum octane-barrel production in the catalytic cracker.

Another development in the unicracking family is the HyCycle Unicracking technology that is designed to maximize diesel production for full conversion applications.

21.2.2.18 Veba Combi Cracking Process

The Veba Combi Cracking (VCC) process is a thermal hydrocracking/hydrogenation process for converting residua and other heavy feedstocks (Table 21.13) (Niemann et al., 1988; RAROP, 1991, p.81; Wenzel and Kretsmar, 1993; Hydrocarbon Processing, 1998, p. 88). The process is based on the Bergius–Pier technology that was used for coal hydrogenation in Germany up to 1945. The heavy feedstock is hydrogenated (hydrocracked) using a commercial catalyst and liquid-phase hydrogenation reactor operating at 440°C to 485°C (825°F to 905°F) and 2175 to 4350 psi (14,996 to 29,993 kPa) pressure. The product obtained from the reactor is fed into the hot separator operating at temperatures slightly below the reactor temperature. The liquid and solid materials are fed into a vacuum distillation column, the gaseous products are fed into a gas-phase hydrogenation reactor operating at an identical pressure (Graeser and Niemann, 1982, 1983). This high-temperature, high-pressure coupling of the reactor products with further hydrogenation provides a specific process economics.

In the process, the residue feed is slurried with a small amount of finely powdered additive and mixed with hydrogen and recycle gas prior to preheating. The feed mixture is routed to the liquid phase reactors. The reactors are operated in an upflow mode and arranged in series. In a once-through operation conversion rates of >95% are achieved. Substantial conversion of asphaltene constituents, desulfurization, and denitrogenation take place at high levels of residue conversion. Temperature is controlled by a recycle gas quench system.

The flow from the liquid phase hydrogenation reactors is routed to a hot separator, where gases and vaporized products are separated from unconverted material. A vacuum flash recovers distillates in the hot separator bottom product.

TABLE 21.13
Feedstock and Product Data for the Veba Combi Cracking Process

	Arabian Heavy Vacuum Residuum[a]
Feedstock	
API gravity	3.4
Sulfur, wt.%	5.5
C7-asphaltenes, wt.%	13.5
Metals (Ni + V), ppm	230.0
Carbon residue, wt.%	8.4
Products, vol.%	
Naphtha (<170°C, <340°F)	26.9
Middle distillate (170–370°C, 340–700°F)	36.5
Gas oil (>370°C, 700°F)	19.9

[a] >550°C, >1025°F; conversion, 95%.
Source: Speight, J.G. and Ozum, B. 2002. *Petroleum Refining Processes*. Marcel Dekker Inc., New York. With permission.

The hot separator top product, together with recovered distillates and straight-run distillates enters the gas phase hydrogenation reactor. The gas phase hydrogenation reactor operates at the same pressure as the liquid phase hydrogenation reactor and contains a fixed bed of commercial hydrotreating catalyst. The operation temperature (340°C to 420°C) is controlled by a hydrogen quench. The system operates in a trickle flow mode, which may not be efficient for some heavy feedstocks. The separation of the synthetic crude from associated gases is performed in a cold separator system. The synthetic crude may be sent to stabilization and fractionation units as required. The gases are sent to a lean oil scrubbing system for contaminant removal and are recycled.

21.3 CATALYSTS

The early 1960s saw an increasing demand for high-octane gasoline for the high-compression-ratio engines in new high-performance cars. Demand also grew for diesel fuel for diesel-electric locomotives, and low-freeze-point jet fuel. These needs were met by rapid growth in hydrocracking of the more-refractory crude fractions that were not converted to gasoline and lighter products in the catalytic cracking units. This growth in demand was accompanied by the pioneering development of new zeolite-based hydrocracking catalysts that had improved activity and selectivity. The need to develop catalysts that can carry out cracking and hydrogenation has become even more pressing in view of recent environmental regulations limiting the amount of sulfur and nitrogen emissions. The development of a new generation of catalysts to achieve this objective of low nitrogen and sulfur levels in the processing of different feedstocks presents an interesting challenge for catalyst development.

The deposition of coke and metals on to the catalyst diminishes the cracking activity of hydrocracking catalysts. Basic nitrogen plays a major role because of the susceptibility of such compounds for the catalyst and their predisposition to form coke (Speight, 2000; Speight and Ozum, 2002). The reactions of hydrocracking require a dual-function catalyst with high cracking and hydrogenation activities (Katzer and Sivasubramanian, 1979; Ho, 1988). The catalyst base, such as acid-treated clay, usually supplies the cracking function or alumina or silica–alumina that is used to support the hydrogenation function supplied by metals, such as nickel, tungsten, platinum, and palladium. These highly acid catalysts are very sensitive to nitrogen compounds in the feed, which break down the conditions of reaction to give ammonia and neutralize the acid sites. As many heavy gas oils contain substantial amounts of nitrogen (up to approximately 2500 ppm) a purification stage is frequently required. Denitrogenation and desulfurization can be carried out using cobalt–molybdenum or nickel–cobalt–molybdenum on alumina or silica–alumina.

Hydrocracking catalysts typically contain separate hydrogenation and cracking functions. Palladium sulfide and promoted group VI sulfides (nickel molybdenum or nickel tungsten) provide the hydrogenation function. These active compositions saturate aromatics in the feed, saturate olefins formed in the cracking, and protect the catalysts from poisoning by coke. Zeolites or amorphous silica–alumina provide the cracking functions. The zeolites are usually type Y (faujasite), ion exchanged to replace sodium with hydrogen and make up 25% to 50% of the catalysts. Pentasils (silicalite or ZSM-5) may be included in dewaxing catalysts.

Hydrocracking catalysts, such as nickel (5% by weight) on silica–alumina, work best on feedstocks that have been hydrofined to low nitrogen and sulfur levels. The nickel catalyst then operates well at 350°C to 370°C (660°F to 700°F) and a pressure of about 1500 psi to give good conversion of feed to lower boiling liquid fractions with minimum saturation of single-ring aromatics and a high *iso*-paraffin to *n*-paraffin ratio in the lower molecular weight paraffins.

The poisoning effect of nitrogen can be offset to a certain degree by operation at a higher temperature. However, the higher temperature tends to increase the production of material in the methane (CH_4) to butane (C_4H_{10}) range and decrease the operating stability of the catalyst so that it requires more frequent regeneration. Catalysts containing platinum or palladium (approximately 0.5% wet) on a zeolite base appear to be somewhat less sensitive to nitrogen than are nickel catalysts, and successful operation has been achieved with feedstocks containing 40 ppm nitrogen. This catalyst is also more tolerant of sulfur in the feed, which acts as a temporary poison, the catalyst recovering its activity when the sulfur content of the feed is reduced.

On such catalysts as nickel or tungsten sulfide on silica–alumina, isomerization does not appear to play any part in the reaction, as uncracked normal paraffins from the feedstock tend to retain their normal structure. Extensive splitting produces large amounts of low-molecular-weight (C_3–C_6) paraffins, and it appears that a primary reaction of paraffins is catalytic cracking followed by hydrogenation to form *iso*-paraffins. With catalysts of higher hydrogenation activity, such as platinum on silica–alumina, direct isomerization occurs. The product distribution is also different, and the ratio of low- to intermediate-molecular-weight paraffins in the breakdown product is reduced.

In addition to the chemical nature of the catalyst, the physical structure of the catalyst is also important in determining the hydrogenation and cracking capabilities, particularly for heavy feedstocks (van Zijll Langhout et al., 1980; Fischer and Angevine, 1986; Kobayashi et al., 1987; Kang et al., 1988). When gas oils and residua are used, the feedstock is present as liquids under the conditions of the reaction. Additional feedstock and the hydrogen must diffuse through this liquid before reaction can take place at the interior surfaces of the catalyst particle.

At high temperatures reaction rates can be much higher than diffusion rates and concentration gradients can develop within the catalyst particle. Therefore, the choice of catalyst porosity is an important parameter. When feedstocks are to be hydrocracked to liquefied petroleum gas and gasoline, pore diffusion effects are usually absent. High surface area (about 300 m²/g) and low to moderate porosity (from 12 Å pore diameter with crystalline acidic components to 50 Å or more with amorphous materials) catalysts are used. With reactions involving high-molecular-weight feedstocks, pore diffusion can exert a large influence, and catalysts with pore diameters greater than 80 Å are necessary for more efficient conversion (Scott and Bridge, 1971).

Catalyst operating temperature can influence reaction selectivity as the activation energy for hydrotreating reactions is much lower than for hydrocracking reaction. Therefore, raising the temperature in a residuum hydrotreater increases the extent of hydrocracking relative to hydrotreating, which also increases the hydrogen consumption (Bridge et al., 1975, 1981).

Molybdenum sulfide (MoS_2), usually supported on alumina, is widely used in petroleum processes for hydrogenation reactions. It is a layered structure that can be made much more active by the addition of cobalt or nickel. When promoted with cobalt sulfide (CoS), making what is called cobalt-moly catalysts, it is widely used in HDS processes. The nickel sulfide (NiS)-promoted version is used for HDN as well as HDS. The closely related tungsten compound (WS_2) is used in commercial hydrocracking catalysts. Other sulfides (iron sulfide, FeS, chromium sulfide, Cr_2S_3, and vanadium sulfide, V_2S_5) are also effective and used in some catalysts. A valuable alternative to the base metal sulfides is palladium sulfide (PdS). Although it is expensive, PdS forms the basis for several very active catalysts.

Clays have been used as cracking catalysts particularly for heavy feedstocks (Chapter 23), and have also been explored in the demetallization and upgrading of heavy crude oil (Rosa-Brussin, 1995). The results indicated that the catalyst prepared was mainly active toward demetallization and conversion of the heaviest fractions of crude oils.

The cracking reaction results from attack of a strong acid on a paraffinic chain to form a carbonium ion (a carbon cation, e.g., R^+) (Dolbear, 1998). Strong acids come in two fundamental types, **Brønsted and Lewis acids. Brønsted acids** are the familiar proton-containing acids; Lewis acids are a broader class including inorganic and organic species formed by positively charged centers. Both kinds have been identified on the surfaces of catalysts; sometimes both kinds of sites occur on the same catalyst. The mixture of Brønsted and Lewis acids sometimes depends on the level of water in the system.

Examples of Brønsted acids are the familiar proton-containing species such as sulfuric acid (H_2SO_4). Acidity is provided by the very active hydrogen ion (H^+), which has a very high positive charge density. It seeks out centers of negative charge such as the pi electrons in aromatic centers. Such reactions are familiar to organic chemistry students, who are taught that bromination of aromatics takes place by attack of the bromonium ion (Br^+) on such a ring system. The proton in strong acid systems behaves in much the same way, adding to the pi electrons and then migrating to a site of high electron density on one of the carbon atoms.

In reactions with hydrocarbons, both Lewis and Brønsted acids can catalyze cracking reactions. For example, the proton in Brønsted acids can add to an olefinic double bond to form a carbon cation. Similarly, a Lewis acid can abstract a hydride from the corresponding paraffin to generate the same intermediate (Dolbear, 1998). Although these reactions are written to show identical intermediates in the two reactions, in real catalytic systems the intermediates would be different. This is because the carbon cations would probably be adsorbed on surface sites that would be different in the two kinds of catalysts.

Zeolites and amorphous silica–alumina provide the cracking function in hydrocracking catalysts (Sherman, 1998). Both have similar chemistry at the molecular level, but the crystalline structure of the zeolites provides higher activities and controlled selectivity not found in the amorphous materials. Zeolites (Greek: *zeo*, to boil, and *lithos*, stone) consist primarily of silicon, aluminum, and oxygen and host an assortment of other elements. In addition, zeolites are highly porous crystals veined with submicroscopic channels. The channels contain water (hence the bubbling at high temperatures), which can be eliminated by heating (combined with other treatments) without altering the crystal structure (Occelli and Robson, 1989).

Typical naturally occurring zeolites include analcite (also called analcime) Na ($AlSi_2O_6$), and **faujasite** $Na_2Ca(AlO_2)_2$ ($SiO_2)_4 \cdot H_2O$ that is the structural analog of the synthetic zeolite X and zeolite Y. Sodalite ($Na_8[(Al_2O_2)_6(SiO_2)_6]Cl_2$) contains the truncated octahedral structural unit known as the sodalite cage that is found in several zeolites. The corners of the faces of the cage are defined by either four or six Al/Si atoms, which are joined together through oxygen atoms. The zeolite-structure is generated by joining sodalite cages through the four-Si/Al rings, so enclosing a cavity or super cage bounded by a cube of eight sodalite cages and readily accessible through the faces of that cube (channels or pores). Joining sodalite cages together through the six-Si/Al faces generates the structural frameworks of faujasite, zeolite X, and zeolite Y. In zeolites, the effective width of the pores is usually controlled by the nature of the cation (M^+ or M^{2+}).

Natural zeolites form hydrothermally (e.g., by the action of hot water on volcanic ash or lava), and synthetic zeolites can be made by mixing solutions of aluminates and silicates and maintaining the resulting gel at temperatures of 100°C (212°F) or higher for appropriate periods (Swaddle, 1997). Zeolite-A can form at temperatures below 100°C (212°F), but most zeolite syntheses require hydrothermal conditions (typically 150°C/300°F at the appropriate pressure). The reaction mechanism appears to involve dissolution of the gel and precipitation as the crystalline zeolite and the identity of the zeolite produced depends on the composition of the solution. Aqueous alkali metal hydroxide solutions favor zeolites with relatively high aluminum contents, whereas the presence of organic molecules such as amines or alcohols favors highly siliceous zeolites such as silicalite or ZSM-5. Various tetra-alkyl ammonium cations favor the formation of certain specific zeolite structures and are known as template ions, although it should not be supposed that the channels and cages form simply by the wrapping of aluminosilicate fragments around suitably shaped cations.

Zeolite catalysts have also found use in the refining industry during the last two decades. Like the silica–alumina catalysts, zeolites also consist of a framework of tetrahedral usually with a silicon atom or an aluminum atom at the center. The geometric characteristics of the zeolites are responsible for their special properties, which are particularly attractive to the refining industry (DeCroocq, 1984). Specific zeolite catalysts have shown up to 10,000 times more activity than the so-called conventional catalysts in specific cracking tests. The mordenite-type catalysts are particularly worthy of mention as they have shown up to 200 times greater activity for hexane cracking in the temperature range 360°C to 400°C (680°F to 750°F).

Other zeolite catalysts have also shown remarkable adaptability to the refining industry. For example, the resistance to deactivation of the type Y zeolite catalysts containing either noble or nonnoble metals is remarkable, and catalyst life of up to 7 years has been obtained commercially in processing heavy gas oils in the Unicracking-JHC processes. Operating life depends on the nature of the feedstock, the severity of the operation, and the nature and extent of operational upsets. Gradual catalyst deactivation in commercial use is counteracted by incrementally raising the operating temperature to maintain the required conversion per pass. The more active a catalyst, the lower is the temperature required. When processing for gasoline, lower operating temperatures have the additional advantage that less of the feedstock is converted to *iso*-butane.

Any given zeolite is distinguished from other zeolites by structural differences in its unit cell, which is a tetrahedral structure arranged in various combinations. Oxygen atoms establish the four vertices of each tetrahedron, which are bound to, and enclose, either a silicon (Si) or an aluminum (Al) atom. The vertex oxygen atoms are each shared by two tetrahedrons, so that every silicon atom or aluminum atom within the tetrahedral cage is bound to four neighboring caged atoms through an intervening oxygen. The number of aluminum atoms in a unit cell is always smaller than, or at most equal to, the number of silicon atoms because two aluminum atoms never share the same oxygen.

The aluminum is actually in the ionic form and can readily accommodate electrons donated from three of the bound oxygen atoms. The electron donated by the fourth oxygen imparts a negative, or anionic, charge to the aluminum atom. This negative charge is balanced by a cation from the alkali metal or the alkaline earth groups of the periodic table. Such cations are commonly sodium, potassium, calcium, or magnesium. These cations play a major role in many zeolite functions and help to attract polar molecules, such as water. However, the cations are not part of the zeolite framework and can be exchanged for other cations without any effect on crystal structure.

Zeolites provide the cracking function in many hydrocracking catalysts, as they do in fluid catalytic cracking catalysts. The zeolites are crystalline aluminosilicates, and in almost all commercial catalysts today, the zeolite used is faujasite. Pentasil zeolites, including silicalite and ZSM-5, are also used in some catalysts for their ability to crack long chain paraffins selectively. Typical levels are 25% to 50% by wt. zeolite in the catalysts, with the remainder being the hydrogenation component and a silica (SiO_2) or alumina (Al_2O_3) binder. Exact recipes are guarded as trade secrets.

Crystalline zeolite compounds provide a broad family of solid acid catalysts. The chemistry and structures of these solids are beyond the scope of this book. What is important here is that the zeolites are not acidic as crystallized. They must be converted to acidic forms by ion exchange processes. In the process of doing this conversion, the chemistry of the crystalline structure is often changed. This complication provides tools for controlling the catalytic properties, and much work has been done on understanding and applying these reactions as a way to make catalysts with higher activities and more desirable selectivity.

As an example, the zeolite faujasite crystallizes with the composition SiO_2 $(NaAlO_2)_x$ $(H_2O)_y$. The ratio of silicon to aluminum, expressed here by the subscript x, can be varied in the crystallization from 1 to greater than 10. What does not vary is the total number of silicon and aluminum atoms per unit cell, 192. For legal purposes to define certain composition of matter patents, zeolites with a ratio of 1 to 1.5 are called type X; those with ratio greater than 1.5 are type Y.

Both silicon and aluminum in zeolites are found in tetrahedral oxide sites. The four oxides are shared with another silicon or aluminum (except that two aluminum ions are never found in adjacent, linked tetrahedral). Silicon with a plus four charge balances exactly half of the charge of the oxide ions it is linked to; as all of the oxygen atoms are shared, silicon balances all of the charge around it and is electrically neutral. Aluminum, with three positive charges, leaves one charge unsatisfied. Sodium neutralizes this charge.

The sodium, as expected from its chemistry, is not linked to the oxides by covalent bonds as the silicon and aluminum are. The attraction is simply ionic, and sodium can be replaced by other cations by ion exchange processes. In extensive but rarely published experiments, virtually every metallic and organic cation has been exchanged into zeolites in studies by catalyst designers.

The most important ion exchanged for sodium is the proton. In the hydrogen ion form, faujasite zeolites are very strong acids, with strengths approaching that of oleum. Unfortunately, direct exchange using mineral acids such as hydrochloric acid is not practical. The acid tends to attack the silica alumina network, in the same way that strong acids attack clays in the activation processes developed by Houdry. The technique adopted to avoid this problem is indirect exchange, beginning with exchange of ammonium ion for the sodium. When heated to a few hundred degrees, the ammonium decomposes, forming gaseous ammonia and leaving behind a proton:

$$R^-NH_4^+ \rightarrow R^-H^+ + NH_3 \uparrow$$

The step is accompanied by a variety of solid-state reactions that can change the zeolite structure in subtle but important ways. This chemistry and the related structural alterations have been described in many articles.

Although zeolites provided a breakthrough that allowed catalytic hydrocracking to become commercially important, continued advances in the manufacture of amorphous silica alumina made these materials competitive in certain kinds of applications. This was important, because patents controlled by Unocal and Exxon dominated the application of zeolites in this area.

Typical catalysts of this type contain 60 to 80 wt.% of the silica alumina, with the remainder as the hydrogenation component. The compositions of these catalysts are closely held secrets. Over the years, broad ranges of silica/alumina molar ratios have been used in various cracking applications, but silica is almost always in excess for high acidity and stability. A typical level might be 25 wt.% alumina (Al_2O_3).

Amorphous silica–alumina is made by a variety of precipitation techniques. The whole class of materials traces its beginnings to silica gel technology, in which sodium silicate is acidified to precipitate the hydrous silica–alumina sulfate; sulfuric acid is used as some or all of the acid for this precipitation, and a mixed gel is formed. The properties of this gel, including acidity and porosity, can be varied by changing the recipe—concentrations, order of addition, pH, temperature, aging time, and the like. The gels are isolated by filtration and washed to remove sodium and other ions.

Careful control of the precipitation allows the pore size distributions of amorphous materials to be controlled but the distributions are still much broader than those in the zeolites. This limits the activity and selectivity. One effect of the reduced activity has been that these materials have been applied only in making middle distillates: diesel and turbine fuels. At higher process severities, the poor selectivity results in production of unacceptable amounts of methane (CH_4) to butane (C_4H_{10}) hydrocarbons.

Hydrocarbons, especially aromatic hydrocarbons, can react in the presence of strong acids to form coke. This coke is a complex polynuclear aromatic material that is low in hydrogen. Coke can deposit on the surface of a catalyst, blocking access to the active sites and reducing the activity of the catalyst. Coke poisoning is a major problem in fluid catalytic cracking catalysts, where coked catalysts are circulated to a fluidized bed combustor to be regenerated. In hydrocracking, coke deposition is virtually eliminated by the catalyst's hydrogenation function.

However, the product referred to as coke is not a single material. The first products deposited are tarry deposits that can, with time and temperature, continue to polymerize. Acid catalyzes these polymerizations. The stable product would be graphite, with very large aromatic sheets and no hydrogen. This product forms only with very high temperature aging, far more severe than that found in a hydrocracker. The graphitic material is both more thermodynamically stable and less kinetically reactive. This kinetic stability results from the lack of easily hydrogenated functional groups.

In a well-designed hydrocracking system, the hydrogenation function adds hydrogen to the tarry deposits. This reduces the concentration of coke precursors on the surface. There is, however, a slow accumulation of coke that reduces activity over a 1 to 2-year period. Refiners respond to this slow reduction in activity by raising the average temperature of the catalyst bed to maintain conversions. Eventually, however, an upper limit to the allowable temperature is reached and the catalyst must be removed and regenerated.

Catalysts carrying coke deposits can be regenerated by burning off the accumulated coke. This is done by service in rotary or similar kilns rather than leaving catalysts in the hydrocracking reactor, where the reactions could damage the metals in the walls. Removing the catalysts also allows inspection and repair of the complex and expensive reactor internals, discussed later. Regeneration of a large catalyst charge can take weeks or months, so refiners may own two catalyst loads, one in the reactor, one regenerated and ready for reload.

The thermal reactions also convert the metal sulfide hydrogenation functions to oxides and may result in agglomeration. Excellent progress has been made since the 1970s in regenerating hydrocracking catalysts; similar regeneration of hydrotreating catalysts is widely practiced.

After combustion to remove the carbonaceous deposits, the catalysts are treated to disperse active metals. Vendor documents claim more than 95% recovery of activity and selectivity in these regenerations. Catalysts can undergo successive cycles of use and regeneration, providing long functional life with these expensive materials.

Hydrocracking allows refiners the potential to balance fuel oil supply and demand by adding VGO cracking capacity. Situations where this is the case include (1) refineries with no existing VGO cracking capacity, (2) refineries with more VGO available than VGO conversion capacity, (3) refineries where addition of vacuum residuum conversion capacity has resulted in production of additional feedstocks boiling in the VGO range (e.g., coker gas oil), and (4) refineries that have one of the two types of VGO conversion units but could benefit from adding the second type. In some cases, a refiner might add both gas oil cracking and residuum conversion capacity simultaneously.

Those refiners who choose gas oil cracking as part of their strategy for balancing residual fuel oil supply and demand must decide whether to select a hydrocracker or a fluid catalytic cracking unit. Although the two processes have been compared vigorously over the years, neither process has evolved to be the universal choice for gas oil cracking. Both processes have their advantages and disadvantages, and process selection can be properly made only after careful consideration of many case-specific factors. Among the most important factors are (1) product slate required, (2) amount of flexibility required to vary the product slate, (3) product quality (specifications) required, and (4) the need to integrate the new facilities in a logical and cost-effective way with any existing facility.

As illustrated earlier for various forms of more conventional hydrocracking, the type of catalyst used can influence the product slate obtained. For example, for a mild hydro-cracking operation at constant temperature, the selectivity of the catalyst varies from about 65% to about 90% by volume. Indeed, several catalytic systems have now been developed with a group of catalysts specifically for mild hydrocracking operations. Depending on the type of catalyst, they may be run as a single catalyst or in conjunction with a hydrotreating catalyst.

Precious metal catalysts, particularly catalysts incorporating platinum or platinum and palladium, are used in the latter stages of deep-desulfurization process. They have excellent performance in hydrogenation of monocyclic aromatic hydrocarbons and are likely to become more and more important in the years ahead and refiners seek to hydrocrack higher amounts of the heavy feedstocks. Current efforts are seeking to produce such catalysts with increased hydrogenation activity as well as resistance to sulfur and nitrogen poisoning while balancing these characteristics. In addition, work is done to assign appropriate cracking activity to match specific applications using inorganic (composite) oxides, such as amorphous alumina, silica alumina, or crystalline silica alumina, as carriers and optimization of the amount of the precious metals and highly dispersed metal impregnation methods.

The cobalt–molybdenum, nickel–molybdenum, and precious metal catalysts are also available for use in deep desulfurization and aromatics hydrogenation. These catalysts should be helpful in producing diesel fuel with a sulfur content of 50 ppm or less only by substituting catalysts.

Catalysts used in residuum upgrading processes typically use an association of several kinds of catalysts, each of them playing a specific and complementary role (Kressmann et al., 1998). The first major function to be performed is HDM. Therefore, the HDM catalyst must desegregate asphaltene constituents and remove as much metals (nickel and vanadium) as possible. One catalyst in particular has been developed by optimizing the support pore structure and acidity (Toulhoat et al., 1990). This catalyst allows a uniform distribution of metals deposited and therefore a high metal retention capacity is reached. A specific HDS catalyst can be placed downstream the HDM catalyst and the main function of such positioning is to desulfurize the already deeply demetallized feedstock as well as to reduce coke precursors (Radler, 1999). Thus, the main function the HDS catalyst is not the same function as that of the HDM catalyst. In addition, for fixed-bed processes, swing guard reactors may be used to improve the protection of downstream catalysts and increase the unit cycle length. For example, the Hyvahl Process (q.v.) includes two swing guard reactors followed by conventional HDM and HDS reactors (DeCroocq, 1997). The HDM catalyst in the guard reactors may be replaced during unit operation and the total catalyst amount is replaced at the end of a cycle.

REFERENCES

Aalund, L.R. 1981. *Oil Gas J.* 79(13): 70 and 79(37): 69.

Bauman, R.F., Aldridge, C.L., Bearden, R., Jr., Mayer, F.X., Stuntz, G.F., Dowdle, L.D., and Fiffron, E. 1993. Preprints. *Oil Sands—Our Petroleum Future.* Alberta Research Council, Edmonton, Alberta, Canada. p. 269.

Bearden, R. and Aldridge, C.L. 1981. *Energy Progr.* 1: 44.

Billon, A., Morel, F., Morrison, M.E., and Peries, J.P. 1994. *Revue Institut Français du Pétrole.* 49(5): 495–507.

Bishop, W. 1990. *Proceedings.* Symposium on Heavy Oil: Upgrading to Refining. Canadian Society for Chemical Engineers. p. 14.

Bland, W.F. and Davidson, R.L. 1967. *Petroleum Processing Handbook.* McGraw-Hill, New York.

Boening, L.G., McDaniel, N.K., Petersen, R.D., and van Driesen, R.P. 1987. *Hydrocarb. Process.* 66(9): 59.

Bridge, A.G. 1997. *Handbook of Petroleum Refining Processes,* 2nd Edn. R.A. Meyers, ed. McGraw-Hill, New York. Chapter 7.2.

Bridge, A.G., Reed, E.M., and Scott, J.W. 1975. Paper presented at the API Midyear Meeting, May.

Bridge, A.G., Gould, G.D., and Berkman, J.F. 1981. *Oil Gas J.* 79(3): 85.

Celestinos, J.A., Zermeno, R.G., Van Dreisen, R.P., and Wysocki, E.D. 1975. *Oil Gas J.* 73(48): 127.

Corbett, R.A. 1989. *Oil Gas J.* 87(26): 33.

DeCroocq, D. 1984. *Catalytic Cracking of Heavy Petroleum Hydrocarbons.* Editions Technip, Paris.

DeCroocq, D. 1997. *Rev. Inst. Franç. Pétrol.* 52(5): 469–489.

Dolbear, G.E. 1998. *Petroleum Chemistry and Refining.* J.G. Speight, ed. Taylor & Francis. Washington, DC. Chapter 7.

Fischer, R.H. and Angevine, P.V. 1986. *Appl. Catal.* 27: 275.

Fornoff, L.L. 1982. *Proceedings.* Second International Conference on the Future of Heavy Crude and Tar Sands, Caracas, Venezuela.

Graeser, U. and Niemann, K. 1982. *Oil Gas J.* 80(12): 121.

Graeser, U. and Niemann, K. 1983, Preprints. *Am. Chem. Soc. Div. Petrol. Chem.* 28(3): 675.

Gray, M.R. 1994. *Upgrading Petroleum Residues and Heavy Oils.* Marcel Dekker Inc., New York.

Ho, T.C. 1988. *Catal. Rev. Sci. Eng.* 30: 117.

Howell, R.L., Hung, C., Gibson, K.R., and Chen, H.C. 1985. *Oil Gas J.* 83(30): 121.

Hunter, M.G., Pasppal, D.A., and Pesek, C.L. 1997. *Handbook of Petroleum Refining Processes,* 2nd Edn. R.A. Meyers, ed. McGraw-Hill, New York. Chapter 7.1.

Hydrocarbon Processing. 1998. Refining Processes 77(11): 53.

Kang, B.C., Wu, S.T., Tsai, H.H., and Wu, J.C. 1988. *Appl. Catal.* 45: 221.

Katzer, J.R. and Sivasubramanian, R. 1979. *Catal. Rev. Sci. Eng.* 20: 155.

Khan, M.R. and Patmore, D.J. 1998. *Petroleum Chemistry and Refining.* J.G. Speight, ed. Taylor & Francis, Washington, DC. Chapter 6.

Kobayashi, S., Kushiyama S., Aizawa, R., Koinuma, Y., Inoue, K., Shmizu, Y., and Egi, K. 1987. *Ind. Eng. Chem. Res.* 26: 2241 and 2245.

Kressmann, S., Morel, F., Harlé, V., and Kasztelan, S. 1998. *Catal. Today* 43: 203–215.

Kressmann, S., Harlé, V., Kasztelan, S., Guibard, I., Tromeur, P., and Morel, F. 1999. *Abstracts.* National Meeting, American Chemical Society, New Orleans., 22–26 August.

Kressmann, S., Boyer, C., Colyar, J.J., Schweitzer, J.M., and Viguié, J.C. 2000. *Rev. Inst. Franç. Pétrol.* 55: 397–406.

Kriz, J.F. and Ternan, M. 1994. United States Patent 5,296,130. March 22.

Murphy, J.R. and Treese, S.A. 1979. *Oil Gas J.* 77(26): 135.

Nasution, A.S. 1986. Preprints. *Div. Petrol. Chem. Am. Chem. Soc.* 31(3): 722.

Niemann, K., Kretschmar, K., Rupp, M., and Merz, L. 1988. *Proceedings.* 4th UNITAR/UNDP International Conference on Heavy Crude and Tar Sand. Edmonton, Alberta, Canada. Volume 5, p. 225.

Occelli, M.L. and Robson, H.E. 1989. *Zeolite Synthesis.* Symposium Series No. 398. American Chemical Society, Washington, DC.

Peries, J.P., Quignard, A., Farjon, C., and Laborde, M. 1988. *Rev. Inst. Franç. Pétrol.* 43(6): 847–853.

Pruden, B.B. 1978. *Can. J. Chem. Eng.* 56: 277.

Pruden, B.B., Muir, G., and Skripek, M. 1993. Preprints. *Oil Sands—Our Petroleum Future.* Alberta Research Council, Edmonton, Alberta, Canada. p. 277.

Radler, M. 1999. *Oil Gas J.* 97(51): 445.

RAROP. 1991. *Heavy Oil Processing Handbook.* Y. Kamiya, ed. Research Association for Residual Oil Processing, Agency of Natural Resources and Energy, Ministry of International Trade and Industry, Tokyo, Japan.

Reich, A., Bishop, W., and Veljkovic, M. 1993. Preprints. *Oil Sands—Our Petroleum Future.* Alberta Research Council, Edmonton, Alberta, Canada. p. 216.

Reno, M. 1997. *Handbook of Petroleum Refining Processes,* 2nd Edn. R.A. Meyers, ed. McGraw-Hill, New York. Chapter 7.3.

Reynolds, J.G. and Beret, S. 1989. *Fuel Sci. Technol. Int.* 7: 165.

Rosa-Brussin, M.F. 1995. *Catal. Rev. Sci. Eng.* 37(1): 1.

Scherzer, J. and Gruia, A.J. 1996. *Hydrocracking Science and Technology.* Marcel Dekker Inc., New York.

Scott, J.W. and Bridge, A.G. 1971. *Origin and Refining of Petroleum.* H.G. McGrath and M.E. Charles, ed. Advances in Chemistry Series 103. American Chemical Society, Washington, DC. p. 113.

Sherman, J.D. 1998. Synthetic zeolites and other microporous oxide molecular sieves. *Proceedings.* Colloquium on Geology, Mineralogy, and Human Welfare. National Academy of Sciences, Irvine, CA.

Speight, J.G. 1986. *Ann. Rev. Energy* 11: 253.

Speight, J.G. 1990. *Fuel Science and Technology Handbook.* J.G. Speight, ed. Marcel Dekker Inc., New York. Chapters 12–16.

Speight, J.G. 2000. *The Desulfurization of Heavy Oils and Residua,* 2nd Edn. Marcel Dekker Inc., New York.

Speight, J.G. and Moschopedis, S.E. 1979. *Fuel Process. Technol.* 2: 295.

Speight, J.G. and Ozum, B. 2002. *Petroleum Refining Processes.* Marcel Dekker Inc., New York.

Suchanek, A.J. and Moore, A.S. 1986. *Oil Gas J.* 84(31): 36.

Sue, H. 1989. *Proceedings.* 4th UNITAR/UNDP Conference on Heavy Oil and Tar Sands. 5: 117.

Swaddle, T.W. 1997. *Inorganic Chemistry.* Academic Press Inc., New York.

Thompson, G.J. 1997. *Handbook of Petroleum Refining Processes.* R.A. Meyers, ed. McGraw-Hill, New York. Chapter 8.4.

Toulhoat, H., Szymanski, R., and Plumail, J.C. 1990. *Catal. Today* 7: 531.

Towler, G.P., Mann, R., Serriere, A.J.L., and Gabaude, C.M.D. 1996. *Ind. Eng. Chem. Res.* 35: 278.

Van Driesen, R.P., Caspers, J., Campbell, A.R., and Lunin, G. 1979. *Hydrocarbon Process.* 58(5): 107.

Van Zijll Langhout, W.C., Ouwerkerk, C., and Pronk, K.M.A. 1980. *Oil Gas J.* 78(48): 120.

Waugh, R.J. 1983. Annual Meeting. National Petroleum Refiners Association. San Francisco, CA.

Wenzel, F. and Kretsmar, K. 1993. Preprints. *Oil Sands—Our Petroleum Future.* Alberta Research Council, Edmonton, Alberta, Canada. p. 248.

Wilson, J. 1985. *Energy Proc.* 5: 61.

22 Hydrogen Production

22.1 INTRODUCTION

Throughout the previous chapters there have been several references or acknowledgments to a very important property of petroleum and petroleum products. And that is the hydrogen content or the use of hydrogen during refining in **hydrotreating** processes, such as desulfurization (Chapter 20) and in hydroconversion processes, such as **hydrocracking** (Chapter 21). Although the hydrogen recycle gas may contain up to 40% by volume of other gases (usually hydrocarbons), hydrotreater catalyst life is a strong function of hydrogen partial pressure. Optimum hydrogen purity at the reactor inlet extends catalyst life by maintaining desulphurization kinetics at lower operating temperatures and reducing carbon laydown. Typical purity increases resulting from hydrogen purification equipment and increased hydrogen sulfide removal as well as tuning hydrogen circulation and purge rates, extending catalyst life up to about 25%.

As hydrogen use has become more widespread in refineries, hydrogen production has moved from the status of a high-tech specialty operation to an integral feature of most refineries (Raissi, 2001). This has been made necessary by the increase in hydrotreating and hydrocracking, including the treatment of progressively heavier feedstocks (Chapter 21). In fact, the use of hydrogen in thermal processes is perhaps the single most significant advance in refining technology during the twentieth century (Bridge et al., 1981; Scherzer and Gruia, 1996; Bridge, 1997; Dolbear, 1998). The continued increase in hydrogen demand over the last several decades is a result of the conversion of petroleum to match changes in product slate and the supply of heavy, high-sulfur oil, and to make lower-boiling, cleaner, and more saleable products. There are also many reasons other than product quality for using hydrogen in processes along with the need to add hydrogen at relevant stages of the refining process and, most important according to the availability of hydrogen (Bezler, 2003; Miller and Penner, 2003; Ranke and Schödel, 2003)).

As an example of the popularity or necessity of using hydrotreating (hydrogen addition), data (Radler, 1999) show that of a total worldwide daily refinery capacity of approximately 81,500,000 bbl of oil, approximately 4,000,000 bbl/d is dedicated to catalytic hydrocracking, 8,500,000 bbl/d is dedicated to catalytic hydrorefining and approximately 28,100,000 bbl/d is dedicated to catalytic hydrotreating. On a national (United States) basis, the corresponding data is approximately 1,400,000 bbl/d for hydrocracking, approximately 1,780,000 bbl/d for hydrorefining, and approximately 1,900,000 bbl/d for catalytic hydrotreating out of a daily total refining capacity of approximately 16,500,000 bbl/d.

Thus, the production of hydrogen for refining purposes requires a major effort by refiners. In fact, the trend to increase the number of hydrogenation (hydrocracking and hydrotreating) processes in refineries coupled with the need to process the heavier oils, which require substantial quantities of hydrogen for upgrading because of the increased use of hydrogen in hydrocracking processes, has resulted in vastly increased demands for this gas. The hydrogen demands can be estimated to a very rough approximation using API gravity and the extent of the reaction, particularly the hydrodesulfurization reaction (Speight, 2000; Speight and Ozum, 2002). But accurate estimation requires equivalent process parameters and a thorough understanding of the nature of each process. Thus, as hydrogen production grows, a better understanding of the capabilities and requirements of a hydrogen plant becomes even more important to overall refinery operations as a means of making the best use of hydrogen supplies in the refinery.

The chemical nature of the crude oil used as the refinery feedstock has always played a major role in determining the hydrogen requirements of that refinery. For example, the lighter, more paraffinic crude oils will require somewhat less hydrogen for upgrading to, say, a gasoline product than a heavier more asphaltic crude oil (Speight, 2000). It follows that the hydrodesulfurization of heavy oils and residua (which, by definition, is a hydrogen-dependent process) needs substantial amounts of hydrogen as part of the processing requirements.

In general, considerable variation exists from one refinery to another in the balance between hydrogen produced and hydrogen consumed in the refining operations. However, what is more pertinent to the present text is the excessive amounts of hydrogen that are required for hydroprocessing operations, whether these be hydrocracking or somewhat milder hydrotreating processes. For effective hydroprocessing, a substantial hydrogen partial pressure must be maintained in the reactor and, to meet this requirement, an excess of hydrogen above that actually consumed by the process must be fed to the reactor. Part of the hydrogen requirement is met by recycling a stream of hydrogen-rich gas. However, the need still remains to generate hydrogen as a make up material to accommodate the process consumption of 500 scf/bbl to 3000 scf/bbl depending upon whether the heavy feedstock is being subjected to a predominantly hydrotreating (hydrodesulfurization) or to a predominantly hydrocracking process.

In some refineries, the hydrogen needs can be satisfied by hydrogen recovery from catalytic reformer product gases, but other external sources are also required. However, for the most part, many refineries now require on-site hydrogen production facilities to supply the gas for their own processes. Most of this nonreformer hydrogen is manufactured either by steam–methane reforming or by oxidation processes. However, other processes, such as steam–methanol interaction or ammonia dissociation, may also be used as sources of hydrogen. Electrolysis of water produces high-purity hydrogen, but the power costs may be prohibitive.

An early use of hydrogen in refineries was in naphtha hydrotreating, as a feed pretreatment for catalytic reforming (which in turn was producing hydrogen as a by-product). As environmental regulations tightened, the technology matured and heavier streams were hydrotreated. Thus in the early refineries, the hydrogen for hydroprocesses was provided as a result of catalytic reforming processes in which dehydrogenation (Chapter 23) is a major chemical reaction and, as a consequence, hydrogen gas is produced. The light ends from the catalytic reformer contain a high ratio of hydrogen to methane so the stream is freed from ethane or propane to get a high concentration of hydrogen in the stream.

The hydrogen is recycled though the reactors where the reforming takes place to provide the atmosphere necessary for the chemical reactions and also prevents the carbon from being deposited on the catalyst, thus extending its operating life. An excess of hydrogen above whatever is consumed in the process is produced, and, as a result, catalytic reforming processes are unique in that they are the only petroleum refinery processes to produce hydrogen as a byproduct. However, as refineries and refinery feedstocks evolved during the last four decades, the demand for hydrogen has increased and reforming processes are no longer capable of providing the quantities of hydrogen necessary for feedstock hydrogenation. Within the refinery, other processes are used as sources of hydrogen. Thus, the recovery of hydrogen from the byproducts of the coking units, visbreaker units, and catalytic cracking units is also practiced in some refineries.

In coking units and visbreaker units, heavy feedstocks are converted to petroleum coke, oil, light hydrocarbons (benzene, naphtha, liquefied petroleum gas) and gas (Chapter 17). Depending on the process, hydrogen is present in a wide range of concentrations. As coking processes need gas for heating purposes, adsorption processes are best suited to recover the hydrogen because they feature a very clean hydrogen product and an off-gas suitable as fuel.

Catalytic cracking is the most important process step for the production of light products from gas oil and increasingly from vacuum gas oil and heavy feedstocks (Chapter 18). In catalytic cracking the molecular mass of the main fraction of the feed is lowered, while another part is converted to coke which is deposited on the hot catalyst. The catalyst is regenerated in one or two stages by burning the coke off with air that also provides the energy for the endothermic cracking process. In the process, paraffins and naphthenes are cracked to olefins and to alkanes with shorter chain length, mono-aromatic compounds are dealkylated without ring cleavage, di-aromatics and poly-aromatics are dealkylated and converted to coke. Hydrogen is formed in the last type of reaction, whereas the first two reactions produce light hydrocarbons and therefore require hydrogen. Thus, a catalytic cracker can be operated in such a manner that enough hydrogen for subsequent processes is formed.

In reforming processes, naphtha fractions are reformed to improve the quality of gasoline (Speight, 2000; Speight and Ozum, 2002). The most important reactions occurring during this process are the dehydrogenation of naphthenes to aromatics. This reaction is endothermic and is favored by low pressures and the reaction temperature lies in the range of 300°C to 450°C (570°F to 840°F). The reaction is performed on platinum catalysts, with other metals, for example, rhenium, as promoters.

Hydrogen is generated in a refinery by the catalytic reforming process, but there may not always be the need to have a catalytic reformer as part of the refinery sequence. Nevertheless, assuming that a catalytic reformer is part of the refinery sequence, the hydrogen production from the reformer usually falls well below the amount required for hydroprocessing purposes. For example, in a 100,000 bbl/d hydrocracking refinery, assuming intensive reforming of hydrocracked gasoline, the hydrogen requirements of the refinery may still fall some 500 to 900 scf/bbl of crude charge below that necessary for the hydrocracking sequences. Consequently, an external source of hydrogen is necessary to meet the daily hydrogen requirements of any process where the heavier feedstocks are involved.

The trend to increase the number of hydrogenation (hydrocracking or hydrotreating) processes in refineries (Dolbear, 1998) coupled with the need to process the heavier oils, which require substantial quantities of hydrogen for upgrading, has resulted in vastly increased demands for this gas.

Hydrogen has historically been produced during catalytic reforming processes as a byproduct of the production of the aromatic compounds used in gasoline and in solvents. As reforming processes changed from fixed-bed to cyclic to continuous regeneration, process pressures have dropped and hydrogen production per barrel of **reformate** has tended to increase. However, hydrogen production as a byproduct is not always adequate to the needs of the refinery and other processes are necessary. Thus, hydrogen production by steam reforming or by partial oxidation of residua has also been used, particularly where heavy oil is available. Steam reforming is the dominant method for hydrogen production and is usually combined with pressure-swing adsorption (PSA) to purify hydrogen to greater than 99% by volume (Bandermann and Harder, 1982).

The gasification of residua and coke to produce hydrogen and power may become an attractive option for refiners (Campbell, 1997; Dickenson et al., 1997; Fleshman, 1997; Gross and Wolff, 2000). The premise that the gasification section of a refinery will be the garbage can for deasphalter residues, high-sulfur coke, as well as other refinery wastes is worthy of consideration.

Several other processes are available for the production of the additional hydrogen that is necessary for the various heavy feedstock hydroprocessing sequences (Chapter 23), and it is the purpose of the present chapter to present a general description of these processes. In general, most of the external hydrogen is manufactured by steam–methane reforming or by oxidation processes. Other processes such as ammonia dissociation, steam–methanol interaction, or electrolysis are also available for hydrogen production, but economic factors and feedstock availability assist in the choice between processing alternatives.

The processes described in this chapter are those gasification processes that are often referred as the garbage disposal units of the refinery. Hydrogen is produced for use in other parts of the refinery as well as for energy and it is often produced from process byproducts that may not be of any use elsewhere. Such byproducts might be the highly aromatic, heteroatom, and metal-containing reject from a deasphalting unit or from a mild hydrocracking process. However attractive this may seem, there will be the need to incorporate a gas cleaning operation to remove any environmentally objectionable components from the hydrogen gas.

22.2 PROCESSES REQUIRING HYDROGEN

22.2.1 HYDROTREATING

Catalytic hydrotreating is a hydrogenation process used to remove about 90% of contaminants such as nitrogen, sulfur, oxygen, and metals from liquid petroleum fractions (Chapter 20). These contaminants, if not removed from the petroleum fractions as they travel through the refinery processing units, can have detrimental effects on the equipment, the catalysts, and the quality of the finished product. Typically, hydrotreating is done prior to processes such as catalytic reforming so that the catalyst is not contaminated by untreated feedstock. Hydrotreating is also used prior to catalytic cracking to reduce sulfur and improve product yields, and to upgrade middle-distillate petroleum fractions into finished kerosene, diesel fuel, and heating fuel oils. In addition, hydrotreating converts olefins and aromatics to saturated compounds.

In a typical catalytic hydrodesulfurization unit, the feedstock is deaerated and mixed with hydrogen, preheated in a fired heater (315°F to 425°F; 600°F to 800°F) and then charged under pressure (upto 1000 psi) through a fixed-bed catalytic reactor (Figure 22.1). In the reactor, the sulfur and nitrogen compounds in the feedstock are converted into hydrogen sulfide and ammonia. The reaction products leave the reactor and after cooling to a low temperature enter a liquid or gas separator. The hydrogen-rich gas from the high-pressure separation is recycled to combine with the feedstock, and the low-pressure gas stream rich in hydrogen sulfide is sent to a gas treating unit where the hydrogen sulfide is removed. The clean gas is then suitable as fuel for the refinery furnaces. The liquid stream is the product from hydrotreating and is normally sent to a stripping column for removal of hydrogen sulfide and other undesirable components. In cases where steam is used for stripping, the product is sent to a vacuum drier for removal of water. Hydrodesulfurized products are blended or used as catalytic reforming feedstock.

Hydrotreating processes differ depending upon the feedstock available and catalysts used. Hydrotreating can be used to improve the burning characteristics of distillates such as kerosene. Hydrotreatment of a kerosene fraction can convert aromatics into naphthenes, which are cleaner-burning compounds. Lube-oil hydrotreating uses catalytic treatment of the oil with hydrogen to improve product quality. The objectives in mild lube hydrotreating include saturation of olefins and improvements in color, odor, and acid nature of the oil. Mild lube hydrotreating may also be used following solvent processing. Operating temperatures are usually below 315°C (600°F) and operating pressures below 800 psi. Severe lube hydrotreating,

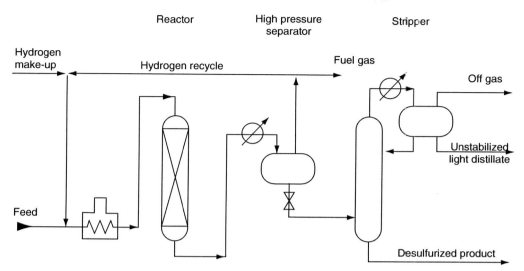

FIGURE 22.1 Distillate hydrodesulfurization. (From OSHA Technical Manual, Section IV, Chapter 2, Petroleum Refining Processes)

at temperatures in the 315°C to 400°C (600°F to 750°F) range and hydrogen pressures up to 3000 psi, is capable of saturating aromatic rings, along with sulfur and nitrogen removal, to impart specific properties not achieved at mild conditions.

Hydrotreating can also be employed to improve the quality of pyrolysis gasoline (pygas), a byproduct from the manufacture of ethylene. Traditionally, the outlet for pygas has been motor gasoline blending, a suitable route in view of its high octane number. However, only small portions can be blended untreated owing to the unacceptable odor, color, and gum-forming tendencies of this material. The quality of pygas, which is high in di-olefin content, can be satisfactorily improved by hydrotreating, whereby conversion of di-olefins into mono-olefins provides an acceptable product for motor gas blending.

22.2.2 HYDROCRACKING

Hydrocracking is a two-stage process combining catalytic cracking and hydrogenation, wherein heavier feedstocks are cracked in the presence of hydrogen to produce more desirable products. Hydrocracking also produces relatively large amounts of *iso*-butane for alkylation feedstock and the process also performs isomerization for pour-point control and smoke-point control, both of which are important in high-quality jet fuel.

Hydrocracking employs high pressure, high temperature, and a catalyst. Hydrocracking is used for feedstocks that are difficult to process by either catalytic cracking or reforming, as these feedstocks are characterized usually by high polycyclic aromatic content and high concentrations of the two principal catalyst poisons, sulfur and nitrogen compounds. The hydrocracking process depends largely on the nature of the feedstock and the relative rates of the two competing reactions, hydrogenation and cracking. Heavy aromatic feedstock is converted into lighter products under a wide range of very high pressures (1000 to 2000 psi) and fairly high temperatures (400°C to 815°C; 750°F to 1500°F), in the presence of hydrogen and special catalysts. When the feedstock has a high paraffinic content, the primary function of hydrogen is to prevent the formation of polycyclic aromatic compounds. Another important role of hydrogen in the hydrocracking process is to reduce tar formation and

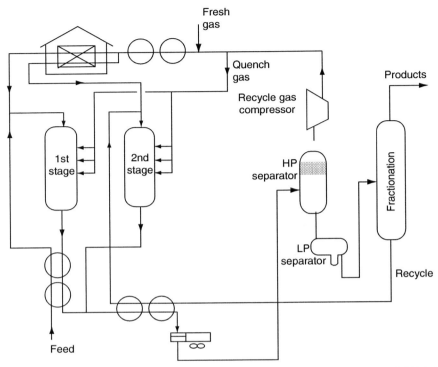

FIGURE 22.2 A single-stage or two-stage (optional) hydrocracking unit. (From OSHA Technical Manual, Section IV, Chapter 2, Petroleum Refining Processes.)

prevent buildup of coke on the catalyst. Hydrogenation also serves to convert sulfur and nitrogen compounds present in the feedstock to hydrogen sulfide and ammonia.

In the first stage of the process (Figure 22.2), preheated feedstock is mixed with recycled hydrogen and sent to the first-stage reactor, where catalysts convert sulfur and nitrogen compounds to hydrogen sulfide and ammonia. Limited hydrocracking also occurs. After the hydrocarbon leaves the first stage, it is cooled and liquefied and run through a hydrocarbon separator. The hydrogen is recycled to the feedstock. The liquid is charged to a fractionator. Depending on the products desired (gasoline components, jet fuel, and gas oil), the fractionator is run to cut out some portion of the first stage reactor outturn. Kerosene-range material can be taken as a separate side-draw product or included in the fractionator bottoms with the gas oil. The fractionator bottoms are again mixed with a hydrogen stream and charged to the second stage. As this material has already been subjected to some hydrogenation, cracking, and reforming in the first stage, the operations of the second stage are more severe (higher temperatures and pressures). Like the outturn of the first stage, the second stage product is separated from the hydrogen and charged to the fractionator.

22.3 FEEDSTOCKS

The most common, and perhaps the best, feedstocks for steam reforming are low boiling saturated hydrocarbons that have a low sulfur content; including **natural gas**, **refinery gas**, liquefied petroleum gas (LPG), and low-boiling naphtha.

Natural gas is the most common feedstock for hydrogen production as it meets all the requirements for reformer feedstock. Natural gas typically contains more than 90% methane and ethane with only a few percent of propane and higher boiling hydrocarbons (Speight, 1993). Natural gas may (or most likely will) contain traces of carbon dioxide with some nitrogen and other impurities. Purification of natural gas, before reforming, is usually relatively straightforward (Speight, 1993). Traces of sulfur must be removed to avoid poisoning the reformer catalyst; zinc oxide treatment in combination with hydrogenation is usually adequate.

Light refinery gas, containing a substantial amount of hydrogen, can be an attractive steam reformer feedstock as it is produced as a byproduct. Processing of refinery gas will depend on its composition, particularly the levels of olefins and of propane and heavier hydrocarbons. Olefins, that can cause problems by forming coke in the reformer, are converted to saturated compounds in the hydrogenation unit. Higher boiling hydrocarbons in refinery gas can also form coke, either on the primary reformer catalyst or in the preheater. If there is more than a few percent of C_3 and higher compounds, a promoted reformer catalyst should be considered, to avoid carbon deposits.

Refinery gas from different sources varies in suitability as hydrogen plant feed. Catalytic reformer off-gas (Chapter 24), for example, is saturated, very low in sulfur, and often has high hydrogen content. The process gases from a coking unit (Chapter 17) or from a fluid catalytic cracking unit (Chapter 18) are much less desirable because of the content of unsaturated constituents. In addition to olefins, these gases contain substantial amounts of sulfur that must be removed before the gas is used as feedstock. These gases are also generally unsuitable for direct hydrogen recovery, as the hydrogen content is usually too low. Hydrotreater off-gas lies in the middle of the range. It is saturated, so it is readily used as hydrogen plant feed. The content of hydrogen and heavier hydrocarbons depends to a large extent on the upstream pressure. Sulfur removal will generally be required.

22.4 PROCESS CHEMISTRY

Before the feedstock is introduced to a process, there is the need for application of a strict feedstock purification protocol. Prolonging catalyst life in hydrogen production processes is attributable to effective feedstock purification, particularly sulfur removal. A typical natural gas or other light hydrocarbon feedstock contains traces of hydrogen sulfide and organic sulfur.

To remove sulfur compounds, it is necessary to hydrogenate the feedstock to convert the organic sulfur to hydrogen that is then reacted with zinc oxide (ZnO) at approximately 370°C (700°F) that results in the optimal use of the zinc oxide as well as ensuring complete hydrogenation. Thus, assuming assiduous feedstock purification and removal of all of the objectionable contaminants, the chemistry of hydrogen production can be defined.

In steam reforming, low-boiling hydrocarbons such as methane are reacted with steam to form hydrogen:

$$CH_4 + H_2O \rightarrow 3H_2 + CO, \quad \Delta H_{298K} = +97,400 \text{ Btu/lb}$$

H is the heat of reaction. A more general form of the equation that shows the chemical balance for higher-boiling hydrocarbons is:

$$C_nH_m + nH_2O \rightarrow (n + m/2)H_2 + nCO$$

The reaction is typically carried out at approximately 815°C (1500°F) over a nickel catalyst packed into the tubes of a reforming furnace. The high temperature also causes the hydrocarbon feedstock to undergo a series of cracking reactions, plus the reaction of carbon with steam:

$$CH_4 \rightarrow 2H_2 + C$$
$$C + H_2O \rightarrow CO + H_2$$

Carbon is produced on the catalyst at the same time that hydrocarbon is reformed to hydrogen and carbon monoxide. With natural gas or similar feedstock, reforming predominates and the carbon can be removed by reaction with steam as fast as it is formed. When higher boiling feedstocks are used, the carbon is not removed fast enough and builds up thereby requiring catalyst regeneration or replacement. Carbon buildup on the catalyst (when high-boiling feedstocks are employed) can be avoided by addition of alkali compounds, such as potash, to the catalyst thereby encouraging or promoting the carbon–steam reaction.

However, even with an alkali-promoted catalyst, feedstock cracking limits the process to hydrocarbons with a boiling point less than 180°C (350°F). Natural gas, propane, butane, and light naphtha are most suitable. Prereforming, a process that uses an adiabatic catalyst bed operating at a lower temperature, can be used as a pretreatment to allow heavier feedstocks to be used with lower potential for carbon deposition (coke formation) on the catalyst.

After reforming, the carbon monoxide in the gas is reacted with steam to form additional hydrogen (the water–gas shift reaction):

$$CO + H_2O \rightarrow CO_2 + H_2$$
$$\Delta H_{298K} = -16,500 \text{ Btu/lb}$$

This leaves a mixture consisting primarily of hydrogen and carbon monoxide that is removed by conversion to methane:

$$CO + 3H_2O \rightarrow CH_4 + H_2O$$
$$CO_2 + 4H_2 \rightarrow CH_4 + 2H_2O$$

The critical variables for steam reforming processes are (1) temperature, (2) pressure, and (3) the steam/hydrocarbon ratio. Steam reforming is an equilibrium reaction, and conversion of the hydrocarbon feedstock is favored by high temperature, which in turn requires higher fuel use. Because of the volume increase in the reaction, conversion is also favored by low pressure, which conflicts with the need to supply the hydrogen at high pressure. In practice, materials of construction limit temperature and pressure.

On the other hand, and in contrast to reforming, shift conversion is favored by low temperature. The gas from the reformer is reacted over iron oxide catalyst at 315°C to 370°C (600°F to 700°F) with the lower limit being dictated by activity of the catalyst at low temperature.

Hydrogen can also be produced by partial oxidation (POX) of hydrocarbons in which the hydrocarbon is oxidized in a limited or controlled supply of oxygen:

$$2CH_4 + O_2 \rightarrow CO + 4H_2$$
$$\Delta H_{298K} = -10,195 \text{ Btu/lb}$$

The shift reaction also occurs and a mixture of carbon monoxide and carbon dioxide is produced in addition to hydrogen. The catalyst tube materials do not limit the reaction temperatures in partial oxidation processes and higher temperatures may be used that enhance the conversion of methane to hydrogen. Indeed, much of the design and operation of hydrogen plants involves protecting the reforming catalyst and the catalyst tubes because of the extreme temperatures and the sensitivity of the catalyst. In fact, minor variations in feedstock composition or operating conditions can have significant effects on the life of the catalyst or the reformer itself. This is particularly true of changes in molecular weight of the feed gas, or poor distribution of heat to the catalyst tubes.

As the high temperature takes the place of a catalyst, partial oxidation is not limited to the lower-boiling feedstocks that are required for steam reforming. Partial oxidation processes were first considered for hydrogen production because of expected shortages of lower boiling feedstocks and the need to have available a disposal method for higher-boiling, high-sulfur streams such as asphalt or petroleum coke.

Catalytic partial oxidation, also known as auto-thermal reforming, reacts oxygen with a light feedstock and by passing the resulting hot mixture over a reforming catalyst. The use of a catalyst allows the use of lower temperatures than in noncatalytic partial oxidation and which causes a reduction in oxygen demand.

The feedstock requirements for catalytic partial oxidation processes are similar to the feedstock requirements for steam reforming, and light hydrocarbons from refinery gas to naphtha are preferred. Oxygen substitutes for much of the steam in preventing coking and a lower steam/carbon ratio is required. In addition, because a large excess of steam is not required, catalytic partial oxidation produces more carbon monoxide and less hydrogen than steam reforming. Thus, the process is more suited to situations where carbon monoxide is the more desirable product such as, for example, as synthesis gas for chemical feedstocks.

22.5 COMMERCIAL PROCESSES

In spite of the use of low-quality hydrogen (that contains up to 40% by volume hydrocarbon gases), a high-purity hydrogen stream (95%–99% by volume hydrogen) is required for hydrodesulfurization, hydrogenation, hydrocracking, and petrochemical processes. Hydrogen, produced as a byproduct of refinery processes (principally hydrogen recovery from catalytic reformer product gases), often is not enough to meet the total refinery requirements, necessitating the manufacturing of additional hydrogen or obtaining supply from external sources.

Catalytic reforming remains an important process used to convert low-octane naphtha into high-octane gasoline blending components called reformate. Reforming represents the total effect of numerous reactions such as cracking, polymerization, dehydrogenation, and isomerization taking place simultaneously. Depending on the properties of the naphtha feedstock (as measured by the paraffin, olefin, naphthene, and aromatic content) and catalysts used, reformate can be produced with very high concentrations of toluene, benzene, xylene, and other aromatics useful in gasoline blending and petrochemical processing. Hydrogen, a significant byproduct, is separated from reformate for recycling and use in other processes.

A catalytic reformer comprises a reactor section and a product-recovery section. More or less standard is a feed preparation section in which, by combination of hydrotreatment and distillation, the feedstock is prepared to specification. Most processes use platinum as the active catalyst. Sometimes platinum is combined with a second catalyst (bimetallic catalyst) such as rhenium or another noble metal. There are many different commercial catalytic reforming processes (Chapter 24) including platforming (Figure 22.3), powerforming, ultraforming, and Thermofor catalytic reforming. In the platforming process, the first step is preparation of the naphtha feed to remove impurities from the naphtha and reduce catalyst degradation. The naphtha feedstock is then mixed with hydrogen, vaporized, and passed through a series of alternating furnace and fixed-bed reactors containing a platinum catalyst. The effluent from the last reactor is cooled and sent to a separator to permit removal of the hydrogen-rich gas stream from the top of the separator for recycling. The liquid product from the bottom of the separator

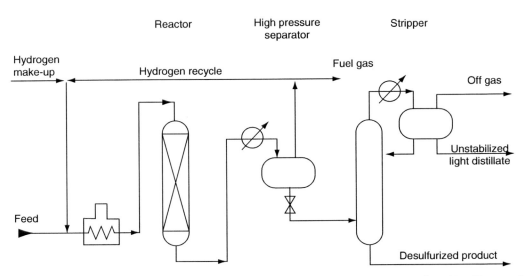

FIGURE 22.3 The platforming process. (From OSHA Technical Manual, Section IV, Chapter 2, Petroleum Refining Processes.)

is sent to a fractionator called a stabilizer (butanizer) and the bottom product (reformate) is sent to storage and butanes and lighter gases pass overhead and are sent to the saturated gas plant.

Some catalytic reformers operate at low pressure (50 to 200 psi), and others operate at high pressures (up to 1000 psi). Some catalytic reforming systems continuously regenerate the catalyst in other systems. One reactor at a time is taken off-stream for catalyst regeneration, and some facilities regenerate all of the reactors during turnarounds. Operating procedures should be developed to ensure control of hot spots during start-up. Safe catalyst handling is very important and care must be taken not to break or crush the catalyst when loading the beds, as the small fines will plug up the reformer screens. Precautions against dust when regenerating or replacing catalyst should also be considered and a water wash should be considered where stabilizer fouling has occurred due to the formation of ammonium chloride and iron salts. Ammonium chloride may form in pretreater exchangers and cause corrosion and fouling. Hydrogen chloride from the hydrogenation of chlorine compounds may form acid or ammonium chloride salt.

22.5.1 HEAVY RESIDUE GASIFICATION AND COMBINED CYCLE POWER GENERATION

Heavy residua are gasified and the produced gas is purified to clean fuel gas (Gross and Wolff, 2000). As an example, solvent deasphalter residuum is gasified by partial oxidation method under pressure of about 570 psi (3923 kPa) and at a temperature between 1300°C and 1500°C (2370°F and 2730°F). The high temperature generated gas flows into the specially designed waste heat boiler, in which the hot gas is cooled and high pressure saturated steam is generated. The gas from the waste heat boiler is then heat exchanged with the fuel gas and flows to the carbon scrubber, where unreacted carbon particles are removed from the generated gas by water scrubbing.

The gas from the carbon scrubber is further cooled by the fuel gas and boiler feed water and led into the sulfur compound removal section, where hydrogen sulfide (H_2S) and carbonyl sulfide (COS) are removed from the gas to obtain clean fuel gas. This clean fuel gas is heated with the hot gas generated in the gasifier and finally supplied to the gas turbine at a temperature of 250°C to 300°C (480°F to 570°F).

The exhaust gas from the gas turbine having a temperature of about 550°C to 600°C (1020°F to 1110°F) flows into the heat recovery steam generator consisting of five heat exchange elements. The first element is a superheater in which the combined stream of the high pressure saturated steam generated in the waste heat boiler and in the second element (high pressure steam evaporator) is super heated. The third element is an economizer, the fourth element is a low pressure steam evaporator, and the final or the fifth element is a de-aerator heater. The off-gas from heat recovery steam generator having a temperature of about 130°C is emitted into the air via stack.

To decrease the nitrogen oxide (NO_x) content in the flue gas, two methods can be applied. The first method is the injection of water into the gas turbine combustor. The second method is to selectively reduce the nitrogen oxide content by injecting ammonia gas in the presence of de-NO_x catalyst that is packed in a proper position of the heat recovery steam generator. The latter is more effective than the former to lower the nitrogen oxide emissions to the air.

22.5.2 HYBRID GASIFICATION PROCESS

In the hybrid gasification process, a slurry of coal and residual oil is injected into the gasifier where it is pyrolyzed in the upper part of the reactor to produce gas and chars. The chars produced are then partially oxidized to ash. The ash is removed continuously from the bottom of the reactor.

In this process, coal and vacuum residue are mixed together into slurry to produce clean fuel gas. The slurry fed into the pressurized gasifier is thermally cracked at a temperature of 850°C to 950°C (1560°F to 1740°F) and is converted into gas, tar, and char. The mixture of oxygen and steam in the lower zone of the gasifier gasify the char. The gas leaving the gasifier is quenched to a temperature of 450°C (840°F)in the fluidized bed heat exchanger, and is then scrubbed to remove tar, dust, and steam at around 200°C (390°F).

The coal and residual oil slurry is gasified in the fluidized-bed gasifier. The charged slurry is converted to gas and char by thermal cracking reactions in the upper zone of the fluidized bed. The produced char is further gasified with steam and oxygen that enter the gasifier just below the fluidizing gas distributor. Ash is discharged from the gasifier and indirectly cooled with steam and then discharged into the ash hopper. It is burned with an incinerator to produce process steam. Coke deposited on the silica sand is removed in the incinerator.

22.5.3 HYDROCARBON GASIFICATION

The gasification of hydrocarbons to produce hydrogen is a continuous, noncatalytic process that involves partial oxidation of the hydrocarbon. Air or oxygen (with steam or carbon dioxide) is used as the oxidant at 1095°C to 1480°C (2000°F to 2700°F). Any carbon produced (2% to 3% by weight of the feedstock) during the process is removed as a slurry in a carbon separator and pelletized for use either as a fuel or as raw material for carbon-based products.

22.5.4 HYPRO PROCESS

The hypro process is a continuous catalytic method for hydrogen manufacture from natural gas or from refinery effluent gases. The process is designed to convert natural gas:

$$CH_4 \rightarrow C + 2H_2$$

Hydrogen is recovered by phase separation to yield hydrogen of about 93% purity; the principal contaminant is methane.

22.5.5   PYROLYSIS PROCESSES

There has been a recent interest in the use of pyrolysis processes to produce hydrogen. Specifically, the interest has focused on the pyrolysis of methane (natural gas) and hydrogen sulfide.

Natural gas is readily available and offers a relatively rich stream of methane with lower amounts of ethane, propane, and butane also present. The thermocatalytic decomposition of natural gas hydrocarbons offers an alternate method for the production of hydrogen (Uemura et al., 1999; Weimer et al., 2000; Dahl and Weimer, 2001):

$$C_nH_m \rightarrow nC + (m/2)H_2$$

If a hydrocarbon fuel such as natural gas (methane) is to be used for hydrogen production by direct decomposition, then the process that is optimized to yield hydrogen production may not be suitable for production of high quality carbon black byproduct intended for the industrial rubber market. Moreover, it appears that the carbon produced from high-temperature (850°C to 950°C; 1560°F to 1740°F) direct thermal decomposition of methane is soot-like material with high tendency for catalyst deactivation (Murata, 1997). Thus, if the object of methane decomposition is hydrogen production, the carbon by-product may not be marketable as high-quality carbon black for rubber and tire applications.

The production of hydrogen by direct decomposition of hydrogen sulfide has been studied extensively and a process proposed (Figure 22.4) (Clark and Wassink, 1990; Zaman and Chakma, 1995; Donini, 1996; Luinstra, 1996).

Hydrogen sulfide decomposition is a highly endothermic process and equilibrium yields are poor (Clark et al., 1995). At temperatures less than 1500°C (2730°F), the thermodynamic equilibrium is unfavorable toward hydrogen formation. However, in the presence of catalysts

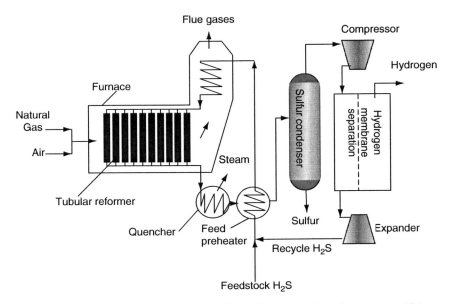

FIGURE 22.4 Simplified schematic for the production of hydrogen from hydrogen sulfide.

such as platinum–cobalt (at 1000°C; 1830°F), disulfides of molybdenum or tungsten Mo or W at 800°C (1470°F) (Kotera et al., 1976), or other transition metal sulfides supported by alumina (at 500°C to 800°C; 930°F to 1470°F), decomposition of hydrogen sulfide proceeds rapidly (Kiuchi, 1982; Bishara et al., 1987; Al-Shamma and Naman, 1989; Clark and Wassink, 1990; Megalofonos and Papayannakos, 1997; Arild, 2000; Raissi, 2001). In the temperature range of about 800°C to 1500°C (1470°F to 2730°F), thermolysis of hydrogen sulfide can be treated simply:

$$H_2S \rightarrow H_2 + 1/xS_x, \qquad \Delta H_{298K} = +34,300 \text{ Btu/lb}$$

where $x = 2$. Outside this temperature range, multiple equilibria may be present depending on temperature, pressure, and relative abundance of hydrogen and sulfur (Clark, 1990).

Above approximately 1000°C (1830°F), there is a limited advantage in using catalysts as the thermal reaction proceeds to equilibrium very rapidly (Clark and Wassink, 1990). The hydrogen yield can be doubled by preferential removal of either H_2 or sulfur from the reaction environment, thereby shifting the equilibrium. The reaction products must be quenched quickly after leaving the reactor to prevent reversible reactions.

22.5.6 SHELL GASIFICATION (PARTIAL OXIDATION) PROCESS

The shell gasification process is a flexible process for generating synthesis gas, principally hydrogen and carbon monoxide, for the ultimate production of high-purity high pressure hydrogen, ammonia, methanol, fuel gas, town gas or reducing gas by reaction of gaseous or liquid hydrocarbons with oxygen, air, or oxygen-enriched air.

The most important step in converting heavy residue to industrial gas is the partial oxidation of the oil using oxygen with the addition of steam. The gasification process takes place in an empty, refractory-lined reactor at temperatures of about 1400°C (2550°F) and pressures between 29 to 1140 psi (196 to 7845 kPa). The chemical reactions in the gasification reactor proceed without catalyst to produce gas containing carbon amounting to some 0.5% to 2% by weight, based on the feedstock. The carbon is removed from the gas with water, extracted in most cases with feed oil from the water and returned to the feed oil. The high reformed gas temperature is utilized in a waste heat boiler for generating steam. The steam is generated at 850 to 1565 psi (5884 to 10787 kPa). Some of this steam is used as process steam and for oxygen and oil preheating. The surplus steam is used for energy production and heating purposes.

22.5.7 STEAM–METHANE REFORMING

Steam–methane reforming is the benchmark process that has been employed over a period of several decades for hydrogen production. The process involves reforming natural gas in a continuous catalytic process in which the major reaction is the formation of carbon monoxide and hydrogen from methane and steam:

$$CH_4 + H_2O = CO + 3H_2, \qquad \Delta H_{298K} = +97,400 \text{ Btu/lb}$$

Higher molecular weight feedstocks can also be reformed to hydrogen:

$$C_3H_8 + 3H_2O \rightarrow 3CO + 7H_2$$

That is,

$$C_nH_m + nH_2O \rightarrow nCO + (0.5m + n)H_2$$

In the actual process, the feedstock is first desulfurized by passage through activated carbon, which may be preceded by caustic and water washes. The desulfurized material is then mixed with steam and passed over a nickel-based catalyst (730°C to 845°C, 1350°F to 1550°F and 400 psi, 2758 kPa). Effluent gases are cooled by the addition of steam or condensate to about 370°C (700°F), at which point carbon monoxide reacts with steam in the presence of iron oxide in a shift converter to produce carbon dioxide and hydrogen:

$$CO + H_2O = CO_2 + H_2, \qquad \Delta H_{298K} = -41.16 \text{ kJ/mol}$$

The carbon dioxide is removed by amine washing; the hydrogen is usually a high-purity (>99%) material.

As the presence of any carbon monoxide or carbon dioxide in the hydrogen stream can interfere with the chemistry of the catalytic application, a third stage is used to convert these gases to methane:

$$CO + 3H_2 \rightarrow CH_4 + H_2O$$
$$CO_2 + 4H_2 \rightarrow CH_4 + 2H_2O$$

For many refiners, sulfur-free natural gas (CH_4) is not always available to produce hydrogen by this process. In that case, higher-boiling hydrocarbons (such as propane, butane, or naphtha) may be used as the feedstock to generate hydrogen (*q.v.*).

The net chemical process for steam methane reforming is then given by:

$$CH_4 + 2H_2O \rightarrow CO_2 + 4H_2, \qquad \Delta H_{298K} = +165.2 \text{ kJ/mol}$$

Indirect heating provides the required overall endothermic heat of reaction for the steam-methane reforming.

One way of overcoming the thermodynamic limitation of steam reforming is to remove either hydrogen or carbon dioxide as it is produced, hence shifting the thermodynamic equilibrium toward the product side. The concept for sorption-enhanced methane steam reforming is based on in-situ removal of carbon dioxide by a sorbent such as calcium oxide (CaO).

$$CaO + CO_2 \rightarrow CaCO_3$$

Sorption enhancement enables lower reaction temperatures, which may reduce catalyst coking and sintering, while enabling use of less expensive reactor wall materials. In addition, heat release by the exothermic carbonation reaction supplies most of the heat required by the endothermic reforming reactions. However, energy is required to regenerate the sorbent to its oxide form by the energy-intensive calcination reaction, that is,

$$CaCO_3 \rightarrow CaO + CO_2$$

Use of a sorbent requires either that there be parallel reactors operated alternatively and out of phase in reforming and sorbent regeneration modes, or that sorbent be continuously transferred between the reformer/carbonator and regenerator/calciner (Balasubramanian et al., 1999; Hufton et al., 1999).

In autothermal (or secondary) reformers, the oxidation of methane supplies the necessary energy and is carried out either simultaneously or in advance of the reforming reaction (Brandmair et al., 2003; Ehwald et al., 2003; Nagaoka et al., 2003). The equilibrium of the methane steam reaction and the water–gas shift reaction determines the conditions for optimum hydrogen yields. The optimum conditions for hydrogen production require: high temperature at the exit of the reforming reactor (800°C to 900°C; 1470°F to 1650°F), high excess of steam (molar steam-to-carbon ratio of 2.5 to 3) and relatively low pressures (below 450 psi). Most commercial plants employ supported nickel catalysts for the process.

The steam–methane reforming process described briefly above would be an ideal hydrogen production process if it was not for the fact that large quantities of natural gas, a valuable resource, are required as both feed gas and combustion fuel. For each mole of methane reformed, more than one mole of carbon dioxide is co-produced and must be disposed. This can be a major issue as it results in the same amount of greenhouse gas emission as would be expected from direct combustion of natural gas or methane. In fact, the production of hydrogen as a clean burning fuel by way of steam reforming of methane and other fossil-based hydrocarbon fuels is not in environmental balance if in the process, carbon dioxide and carbon monoxide are generated and released into the atmosphere, although alternate scenarios are available (Gaudernack, 1996). Moreover, as the reforming process is not totally efficient, some of the energy value of the hydrocarbon fuel is lost by conversion to hydrogen but with no tangible environmental benefit, such as a reduction in emission of greenhouse gases. Despite these apparent shortcomings, the process has the following advantages: (1) produces 4 moles of hydrogen for each mole of methane consumed, (2) feedstocks for the process (methane and water are readily available, (3) the process is adaptable to a wide range of hydrocarbon feedstocks, (4) operates at low pressures, less than 450 psi, (5) requires a low steam/carbon ratio (2.5 to 3), (6) good utilization of input energy (reaching 93%), (7) can use catalysts that are stable and resist poisoning, and (8) good process kinetics.

Liquid feedstocks, either liquefied petroleum gas or naphtha (*q.v.*), can also provide backup feed, if there is a risk of natural gas curtailments. The feed handling system needs to include a surge drum, feed pump, vaporizer (usually steam-heated) followed by further heating before desulfurization. The sulfur in liquid feedstocks occurs as mercaptans, thiophene derivatives, or higher boiling compounds. These compounds are stable and will not be removed by zinc oxide, therefore a hydrogenation unit will be required. In addition, as with refinery gas, olefins must also be hydrogenated if they are present.

The reformer will generally use a potash-promoted catalyst to avoid coke buildup from cracking of the heavier feedstock. If liquefied petroleum gas is to be used only occasionally, it is often possible to use a methane-type catalyst at a higher steam/carbon ratio to avoid coking. Naphtha will require a promoted catalyst unless a preformer is used.

22.5.8 STEAM–NAPHTHA REFORMING

Steam–naphtha reforming is a continuous process for the production of hydrogen from liquid hydrocarbons and is, in fact, similar to steam–methane reforming that is one of several possible processes for the production of hydrogen from low boiling hydrocarbons other than ethane (Muradov, 1998, 2000; Murata, et al; 1997; Brandmair et al., 2003; Find et al., 2003). A variety of naphtha-types in the gasoline boiling range may be employed, including feeds containing up to 35% aromatics. Thus, following pretreatment to remove sulfur compounds, the feedstock is mixed with steam and taken to the reforming furnace (675°C to 815°C, 1250°F to 1500°F, 300 psi, 2068 kPa), where hydrogen is produced.

22.5.9 SYNTHESIS GAS GENERATION

The synthesis gas generation process is a noncatalytic process for producing synthesis gas (principally hydrogen and carbon monoxide) for the ultimate production of high-purity hydrogen from gaseous or liquid hydrocarbons.

In this process, a controlled mixture of preheated feedstock and oxygen is fed to the top of the generator where carbon monoxide and hydrogen emerge as the products. Soot, produced in this part of the operation, is removed in a water scrubber from the product gas stream and is then extracted from the resulting carbon–water slurry with naphtha and transferred to a fuel oil fraction. The oil–soot mixture is burned in a boiler or recycled to the generator to extinction to eliminate carbon production as part of the process.

The soot-free synthesis gas is then charged to a shift converter where the carbon monoxide reacts with steam to form additional hydrogen and carbon dioxide at the stoichiometric rate of 1 mole of hydrogen for every mole of carbon monoxide charged to the converter.

The reactor temperatures vary from 1095°C to 1490°C (2000°F to 2700°F), while pressures can vary from approximately atmospheric pressure to approximately 2000 psi (13,790 kPa). The process has the capability of producing high-purity hydrogen although the extent of the purification procedure depends upon the use to which the hydrogen is to be put. For example, carbon dioxide can be removed by scrubbing with various alkaline reagents, while carbon monoxide can be removed by washing with liquid nitrogen or, if nitrogen is undesirable in the product, the carbon monoxide should be removed by washing with copper-amine solutions.

This particular partial oxidation technique has also been applied to a whole range of liquid feedstocks for hydrogen production. There is now serious consideration being given to hydrogen production by the partial oxidation of solid feedstocks such as petroleum coke (from both delayed and fluid-bed reactors), lignite, and coal, as well as petroleum residua.

The chemistry of the process, using naphthalene as an example, may be simply represented as the selective removal of carbon from the hydrocarbon feedstock and further conversion of a portion of this carbon to hydrogen:

$$C_{10}H_8 + 5O_2 \rightarrow 10CO + 4H_2$$
$$10CO + 10H_2O \rightarrow 10CO_2 + 10H_2$$

Although these reactions may be represented very simply using equations of this type, the reactions can be complex and result in carbon deposition on parts of the equipment thereby requiring careful inspection of the reactor.

22.5.10 TEXACO GASIFICATION (PARTIAL OXIDATION) PROCESS

The Texaco gasification process is a partial oxidation gasification process for generating synthetic gas, principally hydrogen and carbon monoxide. The characteristic of Texaco gasification process is to inject feedstock together with carbon dioxide, steam, or water into the gasifier. Therefore, solvent deasphalted residua, or petroleum coke rejected from any coking method can be used as feedstock for this gasification process. The gas produced from this gasification process can be used for the production of high-purity high-pressurized hydrogen, ammonia, and methanol. The heat recovered from the high temperature gas is used for the generation of steam in the waste heat boiler. Alternatively the less expensive quench type configuration is preferred when high pressure steam is not needed or when a high degree of shift is needed in the downstream CO converter.

In the process, the feedstock, together with the feedstock carbon slurry recovered in the carbon recovery section, is pressurized to a given pressure, mixed with high-pressure steam and then blown into the gas generator through the burner together with oxygen.

The gasification reaction is a partial oxidation of hydrocarbons to carbon monoxide and hydrogen:

$$C_xH_{2y} + x/2O_2 \rightarrow xCO + yH_2$$
$$C_xH_{2y} + xH_2O \rightarrow xCO + (x+y)H_2$$

The gasification reaction is instantly completed, thus producing gas mainly consisting of H_2 and CO ($H_2 + CO = > 90\%$). The high-temperature gas leaving the reaction chamber of the gas generator enters the quenching chamber linked to the bottom of the gas generator and is quenched from 200°C to 260°C (390°F to 500°F) with water.

22.6 CATALYSTS

Hydrogen plants are one of the most extensive users of catalysts in the refinery. Catalytic operations include hydrogenation, steam reforming, shift conversion, and methanation.

22.6.1 REFORMING CATALYSTS

The reforming catalyst is usually supplied as nickel oxide that, during startup, is heated in a stream of inert gas, then steam. When the catalyst is near the normal operating temperature, hydrogen or a light hydrocarbon is added to reduce the nickel oxide to metallic nickel.

The high temperatures (up to 870°C, 1600°F), and the nature of the reforming reaction require that the reforming catalyst be used inside the radiant tubes of a reforming furnace. The active agent in reforming catalyst is nickel, and normally the reaction is controlled both by diffusion and by heat transfer. Catalyst life is limited as much by physical breakdown as by deactivation.

Sulfur is the main catalyst poison and the catalyst poisoning is theoretically reversible with the catalyst being restored to near full activity by steaming. However, in practice, the deactivation may cause the catalyst to overheat and coke, to the point that it must be replaced. Reforming catalysts are also sensitive to poisoning by heavy metals, although these are rarely present in low-boiling hydrocarbon feedstocks and in naphtha feedstocks.

Coking deposition on the reforming catalyst and ensuing gloss of catalyst activity is the most characteristic issue that must be assessed and mitigated.

While methane-rich streams such as natural gas or light refinery gas are the most common feeds to hydrogen plants, there is often a requirement for a variety of reasons to process a variety of higher boiling feedstocks, such as liquefied petroleum gas and naphtha. Feedstock variations may also be inadvertent due, for example, to changes in refinery off-gas composition from another unit or because of variations in naphtha composition because of feedstock variance to the naphtha unit.

Thus, when using higher boiling feedstocks in a hydrogen plant, coke deposition on the reformer catalyst becomes a major issue. Coking is most likely in the reformer unit at the point where both temperature and hydrocarbon content are high enough. In this region, hydrocarbons crack and form coke faster; then the coke is removed by reaction with steam or hydrogen and when catalyst deactivation occurs, there is a simultaneous temperature increase with a concomitant increase in coke formation and deposition. In other zones, where the hydrocarbon-to-hydrogen ratio is lower, there is less risk of coking.

Coking depends to a large extent on the balance between catalyst activity and heat input with the more active catalysts producing higher yields of hydrogen at lower temperature thereby reducing the risk of coking. A uniform input of heat is important in this region of the reformer as any catalyst voids or variations in catalyst activity can produce localized hot spots leading to coke formation and reformer failure.

Coke formation results in hotspots in the reformer that increases pressure drop, reduces feedstock (methane) conversion, leading eventually to reformer failure. Coking may be partially mitigated by increasing the steam/feedstock ratio to change the reaction conditions but the most effective solution may be to replace the reformer catalyst with one designed for higher boiling feedstocks.

A standard steam–methane reforming catalyst uses nickel on an α-alumina ceramic carrier that is acidic in nature. Promotion of hydrocarbon cracking with such a catalyst leads to coke formation from higher boiling feedstocks. Some catalyst formulations use a magnesia/alumina (MgO/Al_2O_3) support that is less acidic than α-alumina that reduces cracking on the support and allows higher boiling feedstocks (such as liquefied petroleum gas) to be used.

Further resistance to coking can be achieved by adding an alkali promoter, typically some form of potash (KOH) to the catalyst. Besides reducing the acidity of the carrier, the promoter catalyzes the reaction of steam and carbon. While carbon continues to be formed, it is removed faster than it can build up. This approach can be used with naphtha feedstocks boiling point up to approximately 180°C (350°F). Under the conditions in a reformer, potash is volatile and it is incorporated into the catalyst as a more complex compound that slowly hydrolyzes to release potassium hydroxide (KOH). Alkali-promoted catalyst allows the use of a wide range of feedstocks but, in addition to possible potash migration, which can be minimized by proper design and operation, the catalyst is also somewhat less active than a conventional catalyst.

Another option to reduce coking in steam reformers is to use a prereformer in which a fixed-bed of catalyst, operating at a lower temperature, upstream of the fired reformer is used. In a prereformer, adiabatic steam-hydrocarbon reforming is performed outside the fired reformer in a vessel containing high nickel catalyst. The heat required for the endothermic reaction is provided by hot flue gas from the reformer convection section. As the feed to the fired reformer is now partially reformed, the steam methane reformer can operate at an increased feed rate and produce 8% to 10% additional hydrogen at the same reformer load. An additional advantage of the prereformer is that it facilitates higher mixed feed preheat temperatures and maintains relatively constant operating conditions within the fired reformer regardless of variable refinery off-gas feed conditions. Inlet temperatures are selected so that there is minimal risk of coking and the gas leaving the prereformer contains only steam, hydrogen, carbon monoxide, carbon dioxide, and methane. This allows a standard methane catalyst to be used in the fired reformer and this approach has been used with feedstocks up to light kerosene. As the gas leaving the prereformer poses reduced risk of coking, it can compensate to some extent for variations in catalyst activity and heat flux in the primary reformer.

22.6.2 Shift Conversion Catalysts

The second important reaction in a steam reforming plant is the shift conversion reaction:

$$CO + H_2O \rightarrow CO_2 + H_2$$

Two basic types of shift catalyst are used in steam reforming plants: iron/chrome high temperature shift catalysts, and copper/zinc low temperature shift catalysts.

High-temperature shift catalysts operate in the range of 315°C to 430°C (600°F to 800°F) and consist primarily of magnetite (Fe_3O_4) with three-valent chromium oxide (Cr_2O_3) added as a stabilizer. The catalyst is usually supplied in the form of ferric oxide (Fe_2O_3) and six-valent chromium oxide (CrO_3) and is reduced by the hydrogen and carbon monoxide in the shift feed gas as part of the start-op procedure to produce the catalyst in the desired form. However, caution is necessary since, if the steam/carbon ratio of the feedstock is too low and the reducing environment too strong, the catalyst can be reduced further, to metallic iron. Metallic iron is a catalyst for Fischer–Tropsch reactions and hydrocarbons will be produced.

Low-temperature shift catalysts operate at temperatures on the order of 205°C to 230°C (400°F to 450°F). Because of the lower temperature, the reaction equilibrium is more controllable and lower amounts of carbon monoxide are produced. The low-temperature shift catalyst is primarily used in wet scrubbing plants that use a methanation for final purification. PSA plants do not generally use a low-temperature because any unconverted carbon monoxide is recovered as reformer fuel. Low-temperature shift catalysts are sensitive to poisoning by sulfur and are sensitive to water (liquid) that can cause softening of the catalyst followed by crusting or plugging.

The catalyst is supplied as copper oxide (CuO) on a zinc oxide (ZnO) carrier and the copper must be reduced by heating it in a stream of inert gas with measured quantities of hydrogen. The reduction of the copper oxide is strongly exothermic and must be closely monitored.

22.6.3 Methanation Catalysts

In wet scrubbing plants, the final hydrogen purification procedure involved is by methanation in which the carbon monoxide and carbon dioxide are converted to methane:

$$CO + 3H_2O \rightarrow CH_4 + H_2O$$
$$CO_2 + 4H_2 \rightarrow CH_4 + 2H_2O$$

The active agent is nickel, on an alumina carrier.

The catalyst has a long life, as it operates under ideal conditions and is not exposed to poisons. The main source of deactivation is plugging from carryover of carbon dioxide from removal solutions.

The most severe hazard arises from high levels of carbon monoxide or carbon dioxide that can result from breakdown of the carbon dioxide removal equipment or from exchanger tube leaks that quench the shift reaction. The results of breakthrough can be severe, as the methanation reaction produces a temperature rise of 70°C (125°F) per one percent of carbon monoxide or a temperature rise of 33°C (60°F) per one percent of carbon dioxide. While the normal operating temperature during methanation is approximately 315°C (600°F), it is possible to reach 700°C (1300°F) in cases of major breakthrough.

22.7 HYDROGEN PURIFICATION

When the hydrogen content of the refinery gas is greater than 50% by volume, the gas should first be considered for hydrogen recovery, using a membrane (Brüschke, 2003) or PSA unit (Figure 22.5). The tail gas or reject gas that will still contain a substantial amount of hydrogen can then be used as steam reformer feedstock. Generally, the feedstock purification process uses three different refinery gas streams to produce hydrogen. First, high-pressure hydrocracker purge gas is purified in a membrane unit that produces hydrogen at medium pressure and is combined with medium pressure off-gas that is first purified in a PSA unit. Finally, low-pressure off-gas is compressed, mixed with reject gases from the membrane and PSA units, and used as steam reformer feed.

Various processes are available to purify the hydrogen stream but since the product streams are available as a wide variety of composition, flows, and pressures, the best method of purification will vary. And there are several factors that must also be taken into consideration in the selection of a purification method. These are:

1. Hydrogen recovery
2. Product purity

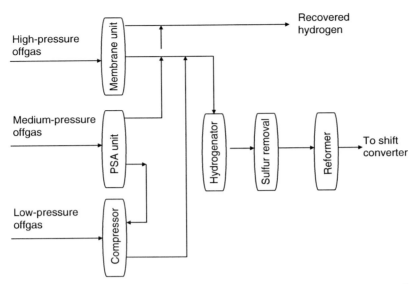

FIGURE 22.5 Hydrogen purification. (From Speight, J.G., and Ozum, B. 2002. *Petroleum Refining Processes*. Marcel Dekker Inc., New York. With permission.)

3. Pressure profile
4. Reliability

Cost, an equally important parameter is not considered here as the emphasis is on the technical aspects of the purification process.

22.7.1 WET SCRUBBING

Wet scrubbing systems, particularly amine or potassium carbonate systems, are used for removal of acid gases such as hydrogen sulfide or carbon dioxide (Speight, 1993; Dalrymple et al., 1994). Most systems depend on chemical reaction and can be designed for a wide range of pressures and capacities. They were once widely used to remove carbon dioxide in steam reforming plants, but have generally been replaced by PSA units except where carbon monoxide is to be recovered. Wet scrubbing is still used to remove hydrogen sulfide and carbon dioxide in partial oxidation plants.

Wet scrubbing systems remove only acid gases or heavy hydrocarbons but they do not remove methane or other hydrocarbon gases, hence have little influence on product purity. Therefore, wet scrubbing systems are most often used as a pretreatment step, or where a hydrogen-rich stream is to be desulfurized for use as fuel gas.

22.7.2 PRESSURE-SWING ADSORPTION UNITS

PSA units use beds of solid adsorbent to separate impurities from hydrogen streams leading to high-purity high-pressure hydrogen and a low-pressure tail gas stream containing the impurities and some of the hydrogen. The beds are then regenerated by depressuring and purging. Part of the hydrogen (up to 20%) may be lost in the tail gas.

PSA is generally the purification method of choice for steam reforming units because of its production of high purity hydrogen and is also used for purification of refinery off-gases, where it competes with membrane systems.

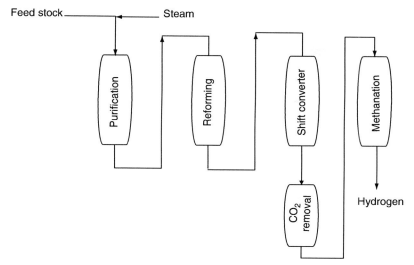

FIGURE 22.6 Hydrogen purification by wet scrubbing. (From Speight, J.G., and Ozum, B. 2002. *Petroleum Refining Processes*. Marcel Dekker Inc., New York. With permission.)

Many hydrogen plants that formerly used a wet scrubbing process (Figure 22.6) for hydrogen purification are now using *PSA* (Figure 22.7) for purification. The PSA process is a cyclic process that uses beds of solid adsorbent to remove impurities from the gas and generally produces a higher purity hydrogen (99.9 vol.% purity compared to less than 97 vol.% purity). The purified hydrogen passes through the adsorbent beds with only a tiny fraction absorbed and the beds are regenerated by depressurization followed by purging at low pressure.

When the beds are depressurized, a waste gas (or tail gas) stream is produced and consists of the impurities from the feed (carbon monoxide, carbon dioxide, methane, and nitrogen) plus some hydrogen. This stream is burned in the reformer as fuel and reformer operating conditions in a PSA plant are set so that the tail gas provides no more than about 85% of the reformer fuel. This gives good burner control because the tail gas is more difficult to burn than regular fuel gas and the high content of carbon monoxide can interfere with the stability of the flame. As the reformer operating temperature is increased, the reforming equilibrium shifts, resulting in more hydrogen and less methane in the reformer outlet and hence less methane in the tail gas.

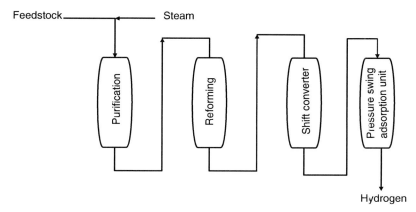

FIGURE 22.7 Hydrogen purification by pressure-swing adsorption. (From Speight, J.G., and Ozum, B. 2002. *Petroleum Refining Processes*. Marcel Dekker Inc., New York. With permission.)

22.7.3 MEMBRANE SYSTEMS

Membrane systems separate gases by taking advantage of the difference in rates of diffusion through membranes (Brüschke, 2003). Gases that diffuse faster (including hydrogen) become the permeate stream and are available at low pressure whereas the slower-diffusing gases become the nonpermeate and leave the unit at a pressure close to the pressure of the feedstock at entry point. Membrane systems contain no moving parts or switch valves and have potentially very high reliability. The major threat is from components in the gas (such as aromatics) that attack the membranes, or from liquids, which plug them.

Membranes arc fabricated in relatively small modules; for larger capacity more modules are added. Cost is therefore virtually linear with capacity, making them more competitive at lower capacities. The design of membrane systems involves a tradeoff between pressure drop (or diffusion rate) and surface area, as well as between product purity and recovery. As the surface area is increased, the recovery of fast components increases; however more of the slow components are recovered, which lowers the purity.

22.7.4 CRYOGENIC SEPARATION

Cryogenic separation units operate by cooling the gas and condensing some, or all, of the constituents for the gas stream. Depending on the product purity required, separation may involve flashing or distillation. Cryogenic units offer the advantage of being able to separate a variety of products from a single feed stream. One specific example is the separation of light olefins from a hydrogen stream.

Hydrogen recovery is in the range of 95%, with purity above 98% obtainable.

22.8 HYDROGEN MANAGEMENT

Many existing refinery hydrogen plants use a conventional process, which produces a medium-purity (94% to 97%) hydrogen product by removing the carbon dioxide in an absorption system and methanation of any remaining carbon oxides. Since the 1980s, most hydrogen plants are built with PSA technology to recover and purify the hydrogen to purities above 99.9%. As many refinery hydrogen plants utilize refinery off-gas feeds containing hydrogen, the actual maximum hydrogen capacity that can be synthesized via steam reforming is not certain as the hydrogen content of the off-gas can change due to operational changes in the hydrotreaters.

Hydrogen management has become a priority in current refinery operations and when planning to produce lower sulfur gasoline and diesel fuels (Zagoria et al., 2003; Davis and Patel, 2004). Along with increased hydrogen consumption for deeper hydrotreating, additional hydrogen is needed for processing heavier and higher sulfur crude slates. In many refineries, hydroprocessing capacity and the associated hydrogen network is limiting refinery throughput and operating margins. Furthermore, higher hydrogen purities within the refinery network are becoming more important to boost hydrotreater capacity, achieve product value improvements, and lengthen catalyst life cycles.

Improved hydrogen utilization and expanded or new sources for refinery hydrogen and hydrogen purity optimization are now required to meet the needs of the future transportation fuel market and the drive toward higher refinery profitability. Many refineries developing hydrogen management programs fit into the two general categories of either a catalytic reformer supplied network or an on-purpose hydrogen supply.

Some refineries depend solely on catalytic reformer(s) as their source of hydrogen for hydrotreating. Often, they are semi-regenerative reformers where off-gas hydrogen quantity, purity, and availability change with feed naphtha quality, as octane requirements change seasonally, and when the reformer catalyst progresses from start-of-run to end-of-run conditions and then goes offline for regeneration. Typically, during some portions of the year, refinery margins are reduced as a result of hydrogen shortages.

Multiple hydrotreating units compete for hydrogen—either by selectively reducing throughput, managing intermediate tankage logistics, or running the catalytic reformer suboptimally just to satisfy downstream hydrogen requirements. Part of the operating year still runs in hydrogen surplus, and the network may be operated with relatively low hydrogen utilization (consumption/production) at 70% to 80%. Catalytic reformer off-gas hydrogen supply may swing from 75% to 85% hydrogen purity. An hydrogen purity upgrade can be achieved through some hydrotreaters by absorbing heavy hydrocarbons. But without supplemental hydrogen purification, critical control of hydrogen partial pressure in hydroprocessing reactors is difficult, which can affect catalyst life, charge rates, and gasoline yields.

More complex refineries, especially those with hydrocracking units, also have on-purpose hydrogen production, typically with a steam methane reformer that utilizes refinery off-gas and supplemental natural gas as feedstock. The steam methane reformer plant provides the swing hydrogen requirements at higher purities (92% to more than 99% hydrogen) and serves an hydrogen network configured with several purity and pressure levels. Multiple purities and existing purification units allow for more optimized hydroprocessing operation by controlling hydrogen partial pressure for maximum benefit. Typical hydrogen utilization is 85% to 95%.

REFERENCES

Al-Shamma, L.M. and Naman, S.A. 1989. *Int. J. Hydrogen Energy* 14(3): 173–179.

Arild, V. 2000. Production of hydrogen and carbon with a carbon black catalyst. PCT Int. Appl. No. 0021878.

Balasubramanian, B., Ortiz, A.L., Kaytakoglu, S., and Harrison, D.P. 1999. *Chem. Eng. Sci.* 54: 3543–3552.

Bandermann, F. and Harder, K.B. 1982. *Int. J. Hydrogen Energy* 7(6): 471–475.

Bezler, J. 2003. *Proceedings.* DGMK Conference on Innovation in the Manufacture and Use of Hydrogen. Dresden, Germany. October 15–17. p. 65.

Bishara, A., Salman, O.S., Khraishi, N., and Marafi, A. 1987. *Int. J. Hydrogen Energy* 12(10): 679–685.

Brandmair, M., Find, J., and Lercher, J.A. 2003. *Proceedings.* DGMK Conference on Innovation in the Manufacture and Use of Hydrogen. Dresden, Germany. October 15–17. p. 273.

Bridge, A.G. 1997. In *Handbook of Petroleum Refining Processes*, 2nd edn. R.A. Meyers, ed. McGraw-Hill, New York. Chapter 14.1.

Bridge, A.G., Gould, G.D., and Berkman, J.F. 1981. *Oil Gas J.* 79(3): 85.

Brüschke, H. 2003. *Proceedings.* DGMK Conference on Innovation in the Manufacture and Use of Hydrogen. Dresden, Germany. October 15–17. p. 19.

Campbell, W.M. 1997. *Handbook of Petroleum Refining Processes*, 2nd edn. R.A. Meyers, ed. McGraw-Hill, New York. Chapter 6.1.

Clark, P.D. and Wassink B. 1990. *Alberta Sulfur Res. Quart. Bull.* 26(2/3/4): 1.

Clark, P.D., N.I. Dowling, J.B. Hyne, and D.L. Moon. 1995. *Alberta Sulfur Res. Quart. Bull.* 32(1): 11–28.

Dahl, J. and Weimer, A.W.. 2001. Preprints. *Div. Fuel Chem. Am. Chem.* Soc. 221.

Dalrymple, D.A., Trofe, T.W., and Leppin, D. 1994. *Oil Gas J.* May 23.

Davis, R.A. and Patel, N.M. 2004. *Petrol. Technol. Quart.* Spring: 28–35.

Dickenson, R.L., Biasca, F.E., Schulman, B.L., and Johnson, H.E. 1997. *Hydrocarbon Process.* 76(2): 57.

Dolbear, G.E. 1998. *Petroleum Chemistry and Refining.* J.G. Speight, ed. Taylor & Francis, Washington, DC. Chapter 7.

Donini, J.C. 1996. Separation and processing of hydrogen sulfide in the fossil fuel industry. Minimum Effluent Mills Symposium. pp. 357–363.

Ehwald, H., Kürschner, U., Smejkal, Q., and Lieske, H. 2003. *Proceedings*. DGMK Conference on Innovation in the Manufacture and Use of Hydrogen. Dresden, Germany. October 15–17. p. 345.

Find, J., Nagaoka, K., and Lercher, J.A.. 2003. *Proceedings*. DGMK Conference on Innovation in the Manufacture and Use of Hydrogen. Dresden, Germany. October 15–17. p. 257.

Fleshman, J.D. 1997. In *Handbook of Petroleum Refining Processes*, 2nd edn. R.A. Meyers, ed. McGraw-Hill, New York. Chapter 6.2.

Gaudernack, B. 1996. Hydrogen from natural gas without release of carbon dioxide into the atmosphere. Hydrogen Energy Prog. *Proceedings*. 11th World Hydrogen Energy Conference. Volume 1, pp. 511–523.

Gross, M. and Wolff, J. 2000. Gasification of residue as a source of hydrogen for the refining industry in India. *Proceedings*. Gasification Technologies Conference. San Francisco, CA. October 8–11.

Hufton, J.R., Mayorga, S., and Sircar, S. 1999. *AIChE J.* 45: 248–256.

Kiuchi, H. 1982. *Int. J. Hydrogen Energy* 7(6): 148–156.

Kotera, Y., Todo, N., and Fukuda, K. 1976. Process for production of hydrogen and sulfur from hydrogen sulfide as raw material. U.S. Patent No. 3,962,409.

Luinstra, E. 1996. Hydrogen from H2S-A review of the leading processes. *Proceedings*. 7th Sulfur Recovery Conference. Gas Research Institute, Chicago. pp. 149–165.

Megalofonos, S.K. and Papayannakos, N.G. 1997. *J. Appl. Catalysis A* 65(1–2): 249–258.

Miller G.Q. and Penner, D.W. 2003. *Proceedings*. DGMK Conference on innovation in the manufacture and use of hydrogen. Dresden, Germany. October 15–17. p. 7.

Muradov, N.Z. 1998. *Energy Fuels* 12(1): 41–48.

Muradov, N.Z. 2000. Thermocatalytic carbon dioxide-free production of hydrogen from hydrocarbon fuels. *Proceedings*. Hydrogen Program Review, NREL/CP-570-28890.

Murata, K., Ushijima H., and Fujita. K.1997. Process for producing hydrogen from hydrocarbon. United States Patent 5,650,132.

Nagaoka, K., Jentys, A., and Lecher, J.A. 2003. *Proceedings*. DGMK Conference on innovation in the manufacture and use of hydrogen. Dresden, Germany. October 15–17. p. 171.

Radler, M. 1999. *Oil Gas J.* 97(51): 445 *et seq*.

Raissi, A.T. 2001. Technoeconomic Analysis of Area II Hydrogen Production. Part 1. Proceedings. U.S. Doe Hydrogen Program Review Meeting, Baltimore, MD.Ranke, H. and Schödel, N. 2003. *Proceedings*. DGMK Conference on innovation in the manufacture and use of hydrogen. Dresden, Germany. October 15–17. p. 19.

Scherzer, J. and Gruia, A.J. 1996. *Hydrocracking Science and Technology*. Marcel Dekker Inc., New York.

Speight, J.G. 1993. *Gas Processing: Environmental Aspects and Methods*. Butterworth Heinemann, Oxford, UK.

Speight, J.G. 2000. *The Desulfurization of Heavy Oils and Residua*, 2nd edn. Marcel Dekker Inc., New York.

Speight, J.G. and Ozum, B. 2002. *Petroleum Refining Processes*. Marcel Dekker Inc., New York.

Uemura, Y., Ohe, H., Ohzuno, Y., and Hatate, Y. 1999. *Proc. Int. Conf. Solid Waste Technol. Mgmt.* 15: 5E/25–5E/30.

Weimer, A.W., Dahl, J., Tamburini, J., Lewandowski, A., Pitts, R., Bingham, C., and Glatzmaier, G.C. 2000. Thermal dissociation of methane using a solar coupled aerosol flow reactor. Proceedings. Hydrogen Program Review, NREL/CP-570–28890.

Zagoria, A., Huychke, R., and Boulter, P.H. 2003. *Proceedings*. DGMK Conference on innovation in the manufacture and use of hydrogen. Dresden, Germany. October 15–17. p. 95.

Zaman, J. and Chakma, A. 1995. *Fuel Process. Technol.* 41: 159–198.

23 Product Improvement

23.1 INTRODUCTION

As already noted (Chapter 1), petroleum and its derivatives have been used for millennia (Abraham, 1945; Forbes, 1958a,b, 1959; James and Thorpe, 1994) and petroleum is perhaps the most important raw material consumed in modern society. It provides not only raw materials for the ubiquitous plastics and other products, but also fuel for energy, industry, heating, and transportation. Thus, the use of petroleum and the development of related technology is not such a modern subject as we are inclined to believe. However, the petroleum industry is essentially a twentieth-century industry but to understand the evolution of the industry, it is essential to have a brief understanding of the first uses of petroleum.

From a chemical standpoint petroleum is an extremely complex mixture of hydrocarbon compounds, usually with minor amounts of nitrogen-, oxygen-, and sulfur-containing compounds as well as trace amounts of metal-containing compounds (Chapter 7) (Speight, 2001 and references cited therein; Speight, 2002 and references cited therein). In addition, the properties of petroleum vary widely (Chapter 1 and Chapter 9) and are not conducive to modern-day use. Thus, petroleum is not used in its raw state. A variety of processing steps are required to convert petroleum from its raw state to products that are usable in modern society.

The fuel products that are derived from petroleum supply more than half of the world's total supply of energy. Gasoline, kerosene, and diesel oil provide fuel for automobiles, tractors, trucks, aircraft, and ships. Fuel oil and natural gas are used to heat homes and commercial buildings, as well as to generate electricity. Petroleum products are the basic materials used for the manufacture of synthetic fibers for clothing and in plastics, paints, fertilizers, insecticides, soaps, and synthetic rubber. The uses of petroleum as a source of raw material in manufacturing are central to the functioning of modern industry.

For the purposes of terminology, it is preferable to subdivide petroleum and related materials into three major classes (Chapter 1):

1. Materials that are of natural origin
2. Materials that are manufactured
3. Materials that are integral fractions derived from the natural or manufactured products

The materials included in categories 1 and 2 are relevant here because of their participation in product streams. Straight-run constituents of petroleum (i.e., constituents distilled without change from petroleum) (Chapter 13) are used in products. Manufactured materials are produced by a variety of processes (Chapter 14 through Chapter 16) and are also used in product streams. Category 3 materials are usually those materials that are isolated from petroleum or from a product by use of a variety of techniques (Chapter 7) and are not included here.

The production of liquid product streams by distillation (Chapter 16) or by cracking processes (Chapter 17 through Chapter 18) is only the first of a series of steps that lead to the production of marketable liquid products. Several other unit processes are involved in the production of a final product. Such processes may be generally termed *secondary processes* or *product improvement processes* as they are not used directly on the crude petroleum but are used on primary product streams that have been produced from the crude petroleum (Bland and Davidson, 1967; Hobson and Pohl, 1973).

In addition, the term *product improvement* as used in this chapter includes processes such as **reforming** processes in which the molecular structure of the feedstock is reorganized. An example is the conversion (reforming, molecular rearrangement) of *n*-hexane to cyclohexane or cyclohexane to benzene. These processes reform or rearrange one particular molecular type to another thereby changing the properties of the product relative to the feedstock. Such processes are conducive to expansion of the utility of petroleum products and to sales.

It is, therefore, the purpose of this chapter to present the concepts behind these secondary processes with specific examples of those processes that have reached commercialization. It must be understood that the process examples presented here are only a selection of the total number available. The choice of a process for inclusion here was made to illustrate the different process types that are available.

23.2   REFORMING

When the demand for higher-octane gasoline developed during the 1930s, attention was directed to ways and means of improving the **octane number** of fractions within the boiling range of gasoline. Straight-run gasoline, for example, frequently had a low octane number and any process that would improve the octane number would aid in meeting the demand for higher quality (higher octane number) gasoline. Such a process, called **thermal reforming**, was developed and used widely but to a much lesser extent than thermal cracking.

Upgrading by reforming is essentially a treatment to improve a gasoline octane number and may be accomplished in part by an increase in the volatility (reduction in molecular size) or chiefly by the conversion of *n*-paraffins to *iso*-paraffins, olefins, and aromatics and the conversion of naphthenes to aromatics (Table 23.1). The nature of the final product is of course influenced by the source (and composition) of the feedstock. In thermal reforming, the reactions resemble the reactions that occur during gas oil cracking, that is, molecular size is reduced, and olefins and some aromatics are produced.

Gasoline has many specifications that must be satisfied before it can be sold on the market. The most widely recognized gasoline specification is the octane number, which refers to the percentage by volume of *iso*-octane in a mixture of *iso*-octane and heptane in a reference fuel that when tested in a laboratory engine, matches the antiknock quality, as measured for the fuel being tested under the same conditions. The octane number posted at the gasoline pump is actually the average of the research octane number (RON) and motor octane number (MON), commonly referred to as (R+M)/2. RON and MON are two different test methods that quantify the antiknock qualities of a fuel. As the MON is a test under more severe conditions than the RON test, for any given fuel, the RON is always higher than the MON.

Unfortunately, the desulfurized light and heavy naphtha fractions of crude oils have very low octane numbers. The heavy naphtha fraction is approximately 50 (R+M)/2. Reforming is the refinery process that reforms the molecular structure of the heavy naphtha to increase the percentage of high-octane components while reducing the percentage of low-octane components.

TABLE 23.1
Structure and Octane Numbers of Selected Hydrocarbons

Compound	n-Hexane	1-Hexene	Cyclohexane	Benzene
Formula	C_6H_{11}	C_6H_{12}	C_6H_{12}	C_6H_6
Structure	$CH_3(CH_2)_4CH_3$	$CH_2–CH_4(CH_2)_3CH_3$		
RON	25	76	83	123 (est.)
Compound	2,2,4-Trimethylpentane (Iso-octane)	2,4,4-Trimethyl-1-pentene (Iso-octane)	Cis 1,3-dimethyl-cyclohexane	1,3-Dimethylbenzene
Formula	C_8H_{18}	C_8H_{16}	C_8H_{16}	C_8H_{10}
Structure				
RON	100	106	72	118

When lead is phased out of gasoline (Chapter 26), the only way to produce high-octane-number gasolines is to use inherently high-octane hydrocarbons or to use alcohols (often referred to as **oxygenates**) (Table 23.2). The ether derivatives are also high-octane oxygenates (Table 23.3) and have been used widely as additives. The ethers may be produced at the refinery by reacting suitable alcohols such as methanol and ethanol with branched olefins from the fluid catalytic cracker, such as *iso*-butene and *iso*-pentene, under the influence of

TABLE 23.2
Oxygenates Allowed in Gasoline

Oxygenate	Maximum, Vol.%
Methanol	3
Ethanol	5
Iso-propyl alcohol	10
Iso-butyl alcohol	10
Tert-butyl alcohol	7
Ether (5 or more C atoms)	15
Other oxygenates	10

TABLE 23.3
Various Oxygenates Used in Gasoline

Name	Formula	Structure	Oxygen Content Mass %	Blending Research Octane Number (BRON)
Ethanol (EtOH)	C_2H_6O	CH_3CH_2OH	34.73	129
Methyl tert-butyl ether (MTBE)	$C_5H_{12}O$	$\begin{array}{c} CH_3 \\ | \\ CH_3O\;C\;CH_3 \\ | \\ CH_3 \end{array}$	18.15	118
Ethyl tert-butyl ether (ETBE)	$C_6H_{14}O$	$\begin{array}{c} CH_3 \\ | \\ CH_3CH_2O\;C\;CH_3 \\ | \\ CH_3 \end{array}$	15.66	119
Tert-amyl methyl ether (TAME)	$C_6H_{14}O$	$\begin{array}{c} CH_3 \\ | \\ CH_3O\;C\;CH_2CH_3 \\ | \\ CH_3 \end{array}$	15.66	112

acid catalysts. In the mid-1990s methyl-*t*-butyl ether (MTBE) (Table 23.3), made by etherification of *iso*-butene with methanol, became the predominant oxygenate used to meet reformulation requirements for adding oxygen to mitigate emissions from gasoline-powered vehicles.

Environmental issues involving MTBE have made it more desirable to dimerize *iso*-butene from the catalytic cracking unit rather than etherify it. Fortunately, *iso*-butene dimerization may be achieved with minimal modifications to existing MTBE plants and process conditions, using the same acidic catalysts. Where olefin levels are not restricted, the extra blending octane boost of the di-*iso*-butylene can be retained. Where olefin levels are restricted, the di-*iso*-butylene can be hydrotreated to produce a relatively pure *iso*-octane stream that can supplement **alkylate** for reducing olefins and aromatics in reformulated gasoline.

23.2.1 THERMAL REFORMING

Thermal reforming was a natural development from thermal cracking, as reforming is also thermal decomposition reaction. Cracking converts heavier oils into gasoline; reforming converts (reforms) gasoline into higher-octane gasoline. The equipment for thermal reforming is essentially the same as for thermal cracking, but higher temperatures are used (Nelson, 1958).

In carrying out thermal reforming, a feedstock, such as a 205°C (400°F) end-point naphtha or a straight-run gasoline, is heated to 510°C to 595°C (950°F to 1100°F) in a furnace much the same as a cracking furnace, with pressures from 400 to 1000 psi. As the heated naphtha leaves the furnace, it is cooled or quenched by the addition of cold naphtha. The quenched, reformed material then enters a fractional distillation tower where any heavy product is separated. The remainder of the reformed material leaves the top of the tower to be separated into gases and **reformate**. The higher octane number of the product (reformate) is due primarily to the cracking of longer chain paraffins into higher-octane olefins.

Thermal reforming is in general less effective than catalytic processes and has been largely supplanted. As it was practiced, a single-pass operation was employed at temperatures in the range of 540°C to 760°C (1000°F to 1140°F) and pressures in the range 500 to 1000 psi. Octane number improvement depended on the extent of conversion but was not directly proportional to the extent of cracking-per-pass.

The amount and quality of reformate is dependent on the temperature. A general rule is the higher the reforming temperature, the higher the octane number of the product but the yield of reformate is relatively low. For example, a gasoline with an octane number of 35 when reformed at 515°C (960°F) yields 92.4% of 56 octane reformate; when reformed at 555°C (1030°F) the yield is 68.7% of 83 octane reformate. However, high conversion is not always effective as coke production and gas production usually increase.

Modifications of the thermal reforming process caused by the inclusion of hydrocarbon gases with the feedstock are known as **gas reversion** and **polyforming**. Thus, olefinic gases produced by cracking and reforming can be converted into liquids boiling in the gasoline range by heating them under high pressure. As the resulting liquids (polymers) have high octane numbers, they increase the overall quantity and quality of gasoline produced in a refinery.

The gases most susceptible to conversion to liquid products are olefins with three and four carbon atoms. These are propylene ($CH_3 \cdot CH = CH_2$), which is associated with propane in the C_3 fraction, butylene ($CH_3 \cdot CH_2 \cdot CH = CH_2$ or $CH_3 \cdot CH = CH \cdot CH_3$) and *iso*-butylene [$(CH_3)_2C = CH_2$], which are associated with butane ($CH_3 \cdot CH_2 \cdot CH_2 \cdot CH_3$), and *iso*-butane [$(CH_3)_2CH \cdot CH_3$] in the C_4 fraction. When the C_3 and C_4 fractions are subjected to the temperature and pressure conditions used in thermal reforming, they undergo chemical reactions that result in a small yield of gasoline. When the C_3 and C_4 fractions are passed through a thermal reformer in admixture with naphtha, the process is called naphtha-gas reversion or naphtha polyforming.

These processes are essentially the same but differ in the manner in which the gases and naphtha are passed through the heating furnace. In gas reversion, the naphtha and gases flow through separate lines in the furnace and are heated independently of one another. Before leaving the furnace, both lines join to form a common soaking section where the reforming, **polymerization**, and other reactions take place. In naphtha reforming, the C_3 and C_4 gases are premixed with the naphtha and passed together through the furnace. Except for the gaseous components in the feedstock, both processes operate in much the same manner as thermal reforming and produce similar products.

23.2.2 CATALYTIC REFORMING

Like thermal reforming, **catalytic reforming** converts low-octane gasoline into high-octane gasoline (reformate) (Kelly et al., 1997). Although thermal reforming can produce reformate with a research octane number in the range of 65 to 80 depending on the yield, catalytic reforming produces reformate with octane numbers of the order of 90 to 95. Catalytic reforming is conducted in the presence of hydrogen over hydrogenation–dehydrogenation catalysts, which may be supported on alumina or silica–alumina. Depending on the catalyst, a definite sequence of reactions takes place, involving structural changes in the charge stock. The catalytic reforming process was commercially nonexistent in the United States before 1940. The process is really a process of the 1950s and showed phenomenal growth in the 1953–1959 period (Bland and Davidson, 1967; Riediger, 1971). As a result, thermal reforming is now somewhat obsolete (Schwarzenbek, 1971).

Catalytic reformer feeds are saturated (i.e., not olefinic) materials; in the majority of cases the feed may be a straight-run naphtha, but other by-product low-octane naphtha (e.g., coker naphtha) can be processed after treatment to remove olefins and other contaminants. Hydrocarbon naphtha that contains substantial quantities of naphthenes is also a suitable feed. The process uses a precious metal catalyst (platinum supported by an alumina base) in conjunction with very high temperatures to reform the paraffin and naphthene constituents into high-octane components. Sulfur is a poison to the reforming catalyst, which requires that virtually all the sulfur must be removed from the heavy naphtha through hydrotreating before reforming. Several different types of chemical reactions occur in the reforming reactors: paraffins are isomerized to branched chains and to a lesser extent to naphthenes, and naphthenes are converted to aromatics. Overall, the reforming reactions are endothermic. The resulting product stream (reformate) from catalytic reforming has a RON from 96 to 102 depending on the reactor severity and feedstock quality. The dehydrogenation reactions that convert the saturated naphthenes into unsaturated aromatics produce hydrogen, which is available for distribution to other refinery hydroprocesses.

The catalytic reforming process consists of a series of several reactors (Figure 23.1), which operate at temperatures of approximately 480°C (900°F). The hydrocarbons are reheated by direct-fired furnaces between the subsequent reforming reactors. As a result of the very high temperatures, the catalyst becomes deactivated by the formation of coke (i.e., essentially pure carbon) on the catalyst, which reduces the surface area available to contact with the hydrocarbons.

Catalytic reforming is usually carried out by feeding a naphtha (after pretreating with hydrogen if necessary) and hydrogen mixture to a furnace where the mixture is heated to the desired temperatures 450°C to 520°C (840°F to 965°F), and then passed through fixed-bed catalytic reactors at hydrogen pressures of 100 to 1000 psi. Normally, two (or more than one) reactors are used in series, and reheaters are located between adjoining reactors to compensate for the endothermic reactions taking place. Sometimes as many as four or five are kept on-stream in series while one or more is being regenerated. The on-stream cycle of any one reactor may vary from several hours to many days, depending on the feedstock and reaction conditions.

The product issuing from the last catalytic reactor is cooled and sent to a high-pressure separator where the hydrogen-rich gas (Table 23.4) is split into two streams: one stream goes to recycle, and the remaining portion represents excess hydrogen available for other uses. The excess hydrogen is vented from the unit and used in hydrotreating, as a fuel, or for manufacture of chemicals (e.g., ammonia). The liquid product (reformate) is stabilized (by

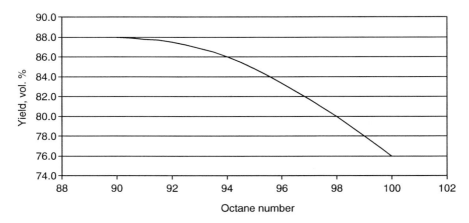

FIGURE 23.1 Octane number and reformate yield.

TABLE 23.4
Composition of Catalytic Reformer Product Gas

Constituent	% by Volume
Hydrogen	75 to 85
Methane	5 to 10
Ethane	5 to 10
Ethylene	0
Propane	5 to 10
Propylene	0
Butane	<5
Butylenes	0
Pentane plus	<2

removal of light ends) and used directly in gasoline or extracted for aromatic blending stocks for aviation gasoline.

The commercial processes available for use can be broadly classified as the moving-bed, **fluid-bed**, and **fixed-bed** types. The fluid-bed and moving-bed processes use mixed nonprecious metal oxide catalysts in units equipped with separate regeneration facilities. Fixed-bed processes use predominantly platinum-containing catalysts in units equipped for cycle, occasional, or no regeneration.

There are several types of catalytic reforming process configurations that differ in the manner that they accommodate the regeneration of the reforming catalyst. Catalyst regeneration involves burning off the coke with oxygen. The semi-regenerative process is the simplest configuration but does require that the unit be shut down for catalyst regeneration in which all reactors (typically four) are regenerated. The cyclic configuration uses an additional swing reactor that enables one reactor at a time to be taken off-line for regeneration while the other four remain in service. The continuous catalyst regeneration (CCR) configuration is the most complex configuration and enables the catalyst to be continuously removed for regeneration and replaced after regeneration. The benefits of more complex configurations are that operating severity may be increased as a result of higher catalyst activity but this does come at an increased capital cost for the process.

Although subsequent olefin reactions occur in thermal reforming, the product contains appreciable amounts of unstable unsaturated compounds. In the presence of catalysts and hydrogen (available from dehydrogenation reactions), hydrocracking of paraffins to yield two lower paraffins occurs. Olefins that do not undergo dehydrocyclization are also produced. The olefins are hydrogenated with or without isomerization, so that the end product contains only traces of olefins.

The addition of a hydrogenation–dehydrogenation catalyst to the system yields a dual-function catalyst complex. Hydrogen reactions—hydrogenation, dehydrogenation, dehydrocyclization, and hydrocracking take place on the one catalyst, and cracking, isomerization, and olefin polymerization take place on the acid catalyst sites.

Under the high-hydrogen partial pressure conditions used in catalytic reforming, sulfur compounds are readily converted into hydrogen sulfide, which, unless removed, builds up to a high concentration in the recycle gas. Hydrogen sulfide is a reversible poison for platinum and causes a decrease in the catalyst dehydrogenation and dehydrocyclization activities. In the first catalytic reformers the hydrogen sulfide was removed from the gas cycle stream by absorption in, for example, diethanolamine. Sulfur is generally removed from the feedstock by use of a conventional desulfurization over a cobalt–molybdenum catalyst. An additional benefit of desulfurization of the feed to a level of <5 ppm sulfur is the elimination of hydrogen sulfide (H_2S) corrosion problems in the heaters and reactors.

Organic nitrogen compounds are converted into ammonia under reforming conditions, and this neutralizes acid sites on the catalyst and thus represses the activity for isomerization, hydrocracking, and dehydrocyclization reactions. Straight-run materials do not usually present serious problems with regard to nitrogen, but feeds such as coker naphtha may contain around 50 ppm nitrogen and removal of this quantity may require high-pressure hydrogenation (800 to 1000 psi) over nickel–cobalt–molybdenum on an alumina catalyst.

The yield of gasoline of a given octane number and at given operating conditions depends on the hydrocarbon types in the feed. For example, high-naphthene stocks, which readily give aromatic gasoline, are the easiest to reform and give the highest gasoline yields. Paraffinic stocks, however, which depend on the more difficult isomerization, dehydrocyclization, and hydrocracking reactions, require more severe conditions and give lower gasoline yields than the naphthenic stocks. The end point of the feed is usually limited to about 190°C (375°F), partially because of increased coke deposition on the catalyst as the end point during processing at about 15°C (27°F). Limiting the feed end point avoids redistillation of the product to meet the gasoline end-point specification of 205°C (400°F), maximum.

Dehydrogenation is a main chemical reaction in catalytic reforming, and hydrogen gas is consequently produced in large quantities. The hydrogen is recycled through the reactors where the reforming takes place to provide the atmosphere necessary for the chemical reactions and also prevents the carbon from being deposited on the catalyst, thus extending its operating life. An excess of hydrogen above whatever is consumed in the process is produced, and as a result, catalytic reforming processes are unique in that they are the only petroleum refinery processes to produce hydrogen as a by-product.

23.2.2.1 Fixed-Bed Processes

Fixed-bed, continuous catalytic reforming can be classified by catalyst type: (1) cyclical regenerative with nonprecious metal oxide catalysts and (2) cyclic regenerative with platinum–alumina catalysts. Both types use swing reactors to regenerate a portion of the catalyst while the remainder stays on-stream.

The cyclic regenerative fixed-bed operation using a platinum catalyst is basically a low-pressure process (250 to 350 psi), which gives higher gasoline yields because of fewer hydrocracking reactions, as well as higher octane products from a given naphtha charge and better hydrogen yields because of more dehydrogenation and fewer hydrocracking reactions. The coke yield, with attendant catalyst deactivation, increases rapidly at low pressures.

23.2.2.1.1 Hydroforming

The hydroforming process made use of molybdena–alumina (MoO_2–Al_2O_3) catalyst pellets arranged in fixed beds; hence the process is known as fixed-bed hydroforming. The hydroformer had four reaction vessels or catalyst cases, two of which were regenerated; the other two were on the process cycle. Naphtha feed was preheated to 400°C to 540°C (900°F to 1000°F) and passed in series through the two catalyst cases under a pressure of 150 to 300 psi. Gas containing 70% hydrogen produced by the process was passed through the catalyst cases with the naphtha. The material leaving the final catalyst case entered a four-tower system where fractional distillation separated hydrogen-rich gas, a product (reformate) suitable for motor gasoline and an aromatic polymer boiling above 205°C (400°F).

After 4 to 16 h on process cycle, the catalyst was regenerated. This was done by burning carbon deposits from the catalyst at a temperature of 565°C (1050°F) by blowing air diluted with flue gas through the catalyst. The air also reoxidized the reduced catalyst (9% molybdenum oxide on activated alumina pellets) and removed sulfur from the catalyst.

TABLE 23.5
Feedstock and Product Data for the Houdriforming Process

Feedstock	
API	52.6
Boiling range	
°C	92–192
°F	197–377
Composition, vol.%	
Paraffins	53
Naphthenes	38
Aromatics	9
Product	
Research octane number	100
Composition, vol.%	
Paraffins	21
Naphthenes	2
Aromatics	77

Source: Speight, J.G., and Ozum, B. 2002. *Petroleum Refining Processes*. Marcel Dekker Inc., New York. With permission.

23.2.2.1.2   Iso-Plus Houdriforming

This is a combination process using a conventional Houdriformer operated at moderate severity for product production (Table 23.5), in conjunction with one of three possible alternatives:

1. Conventional catalytic reforming plus aromatic extraction and separate catalytic reforming of the aromatic raffinate
2. Conventional catalytic reforming plus aromatic extraction and recycling of the aromatic raffinate aligned to the reforming state
3. Conventional catalytic reforming followed by thermal reforming of the Houdriformer product and catalytic polymerization of the C_3 and C_4 olefins from thermal reforming

A typical feedstock for this type of unit is naphtha, and the use of a Houdry **guard bed** permits charging stocks of relatively high sulfur content.

23.2.2.1.3   Platforming

The first step in the **platforming** process (Figure 23.2) (Dachos et al., 1997) is preparation of the naphtha feed. For motor gasoline manufacture, the naphtha feed is distilled to separate a fraction boiling in the 120°C to 205°C (250°F to 400°F) range. As sulfur adversely affects the platinum catalyst, the naphtha fraction may be treated to remove sulfur compounds. Otherwise, the hydrogen-rich gas produced by the process, which is cycled through the catalyst cases, must be scrubbed free of its hydrogen sulfide content.

The prepared naphtha feed is heated to 455°C to 540°C (850°F to 1000°F) and passed into a series of three catalyst cases under a pressure of 200 to 1000 psi. Further heat is added to the naphtha between each of the catalyst cases in the series. The material from the final catalyst case is fractionated into a hydrogen-rich gas stream and a reformate stream (Table 23.6). The catalyst is composed of 1/8 in. pellets of alumina containing chlorine and about 0.5% platinum.

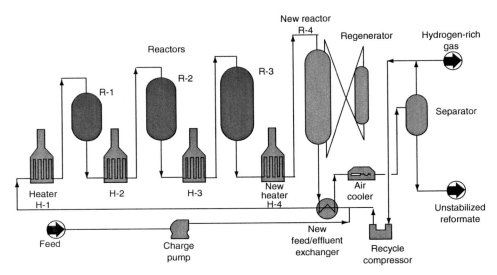

FIGURE 23.2 Schematic of a catalytic reforming process.

Each pound of catalyst reforms up to 100 bbl of naphtha before losing its activity. It is possible to regenerate the catalyst, but it is usual to replace the spent catalyst with a new catalyst.

Other fixed-bed processes include **catforming**, in which the catalyst is platinum (Pt), alumina (Al_2O_3), and silica–alumina (SiO_2–Al_2O_3) composition, which permits relatively

TABLE 23.6
Feedstock and Product Data for the Platforming Process

Feedstock			
API	59.0		
Boiling range			
°C	92–112		
°F	197–233		
Composition, vol.%			
Paraffins	69		
Naphthenes	21		
Aromatics	10		

Process Operation	**Semi-Regenerative**	**Continuous**	
Pressure			
psi	295	123	49
kPa	2040	850	340
Product			
Research octane number	100	100	100
Composition, vol.%			
C5+	70	78	82
Aromatics	46	55	58
Hydrogen, wt.%	2	3	4

Source: Speight, J.G., and Ozum, B. 2002. *Petroleum Refining Processes*. Marcel Dekker Inc., New York. With permission.

high space velocities and results in very high hydrogen purity. Regeneration to prolong catalyst life is practiced on a block-out basis with a dilute air in-stream mixture. In addition, Houdriforming is a process in which the catalyst may be regenerated, if necessary, on a block-out basis. A guard bed catalytic hydrogenation pretreating stage using the same Houdry catalyst as the Houdriformer reactors is available for high-sulfur feedstocks. Lead and copper salts are also removed under the mild conditions of the guard bed operation.

23.2.2.1.4 Powerforming

The cyclic **powerforming** process is based on frequent regeneration (carbon burn-off) and permits continuous operation. Reforming takes place in several (usually four or five) reactors and regeneration is carried out in the last (or swing) reactor. Thus, the plant need not be shut down to regenerate a catalyst reactor. The cyclic process assures a continuous supply of hydrogen gas for hydrorefining operations and tends to produce a greater yield of higher octane reformate (Table 23.7). Choice between the semi-regenerative process and the cyclic process depends on the size of plant required, type of feedstocks available, and the octane number needed in the product.

23.2.2.1.5 Rexforming

Rexforming is a combination process using platforming and aromatic extraction processes in which low-octane raffinate is recycled to the platformer. Operating temperatures may be as much as 27°C (50°F) lower than conventional platforming, and higher space velocities are used. A balance is struck between hydrocyclization and hydrocracking, excessive coke and gas formation thus being avoided. The glycol solvent in the aromatic extraction section is designed to extract low-boiling high-octane *iso*-paraffins as well as aromatics.

23.2.2.1.6 Selectoforming

The selectoforming process uses a fixed-bed reactor operating under a hydrogen partial pressure. Typical operating conditions depend on the process configuration but are in the ranges of 200 to 600 psi and 315°C to 450°C (600°F to 900°F). The catalyst used in the selectoforming process is nonnoble metal with low potassium content. As with the large-pore hydrocracking catalysts, the cracking activity increases with decreasing alkali metal content.

TABLE 23.7
Feedstock and Product Data for the Powerforming Process

Feedstock		
API	57.2	
Composition, vol.%		
Paraffins	57	
Naphthenes	30	
Aromatics	13	
Process Operation	**Semi-Regenerative**	**Cyclic**
Product		
Research octane number	99	101
Composition, vol.%		
C1–C4	13	11
C5+	79	79
Hydrogen, wt.%	2	3

Source: Speight, J.G., and Ozum, B. 2002. *Petroleum Refining Processes*. Marcel Dekker Inc., New York. With permission.

There are two configurations of the selectoforming process that are being used commercially. The first selectoformer was designed as a separate system and integrated with the reformer only to the extent of having a common hydrogen system. The reformer naphtha is mixed with hydrogen and passed into the reactor containing the shape-selective catalyst. The reactor effluent is cooled and separated into hydrogen, liquid petroleum, gas, and high-octane gasoline. The removal of *n*-paraffins reduces the vapor pressure of the reformate as these paraffins are in higher concentration in the front end of the feed. The separate selectoforming system has the additional flexibility of being able to process other refinery streams.

The second process modification is the terminal reactor system. In this system, the shape-selective catalysts replace all or part of the reforming catalyst in the last reforming reactor. Although this configuration is more flexible, the high reforming operating temperature causes butane and propane cracking and consequently decreases the liquid petroleum gas yield and generates higher ethane and methane production. The life of a selectoforming catalyst used in a terminal system is between 2 and 3 years, and regeneration only partially restores fresh catalytic activity.

23.2.2.2 Moving-Bed Processes

23.2.2.2.1 Hyperforming
Hyperforming is a moving-bed reforming process that uses catalyst pellets of cobalt molybdate with a silica-stabilized alumina base. In operation, the catalyst moves downward through the reactor by gravity flow and is returned to the top by means of a solids-conveying technique (hyperflow), which moves the catalyst at low velocities and with minimum attrition loss. Feedstock (naphtha vapor) and recycle gas flow upward, countercurrent to the catalyst, and regeneration of catalyst is accomplished in either an external vertical lift line or a separate vessel.

Hyperforming naphtha (65°C to 230°C, 150°F to 450°F) can result in improvement of the motor fuel component; in addition, sulfur and nitrogen removal is accomplished. Light gas oil stocks can also be charged to remove sulfur and nitrogen under mild hydrogenation conditions for the production of premium diesel fuels and middle distillates. Operating conditions in the reactor are 400 psi and 425°C to 480°C (800°F to 900°F), the higher temperature being employed for a straight-run naphtha feedstock; catalyst regeneration takes place at 510°C (950°F) and 415 psi.

23.2.2.2.2 Thermofor Catalytic Reforming (TCR)
This is also a moving-bed process that uses a synthetic bead coprecipitated chromia (CrO_2) and alumina (Al_2O_3) catalyst. Catalyst–naphtha ratios have little effect on product yield or quality when varied over a wide range. The catalyst flow downward through the reactor and the naphtha–recycle gas feed enters the center of the reactor. The catalyst is transported from the base of the reactor to the top of the regenerator by bucket-type elevators.

23.2.2.3 Fluid-Bed Processes

In catalytic reforming processes using a fluidized solids catalyst bed, continuous regeneration with a separate or integrated reactor is practiced to maintain catalyst activity by coke and sulfur removal. Cracked or virgin naphtha is charged with hydrogen-rich recycle gas to the reactor. A molybdena (Mo_2O_3, 10.0%) on alumina catalyst, not materially affected by normal amounts of arsenic, iron, nitrogen, or sulfur, is used. Operating conditions in the reactor are about 200 to 300 psi and 480°C to 950°C (900°F to 950°F).

Fluidized-bed operation with its attendant excellent temperature control prevents over- and under-reforming operations, resulting in more selectivity in the conditions needed for optimum yield of the desired product.

23.3   ISOMERIZATION

Catalytic reforming processes provide high-octane constituents in the heavier gasoline fraction, but the n-paraffin components of the lighter gasoline fraction, especially butane (C_4) to hexane (C_6), have poor octane ratings. The conversion of these n-paraffins to their isomers (isomerization) yields gasoline components of high octane rating in this lower boiling range. Conversion is obtained in the presence of a catalyst (aluminum chloride activated with hydrochloric acid), and it is essential to inhibit side reactions, such as cracking and olefin formation.

Various companies have developed and operated isomerization processes that increase the octane numbers of light naphtha from say, 70 or less to more than 80. In thermal catalytic alkylation ethylene or propylene is combined with iso-butane at 50°C to 280°C (125°F to 450°F) and 300 to 1000 psi in the presence of metal halide catalysts, such as aluminum chloride. Conditions are less stringent in catalytic alkylation; olefins (C_3, C_4, and C_5) are combined with iso-butane in the presence of an acid catalyst (sulfuric or hydrofluoric) at low temperatures (1°F to 40°F, 30°F to 105°F) and pressures from atmospheric to 150 psi.

In a typical process, naphtha is passed over an aluminum chloride catalyst at 120°C (250°F) and at a pressure of about 800 psi to produce the isomerate (Figure 23.3).

Isomerization, another innovation specific to recent times, found initial commercial applications during World War II for making high-octane aviation gasoline components and additional feed for alkylation units. The lowered alkylate demands in the post–World War II period caused a shutdown of the majority of the butane isomerization units. In recent years, the greater demand for high-octane motor fuel has resulted in the installation of new butane isomerization units.

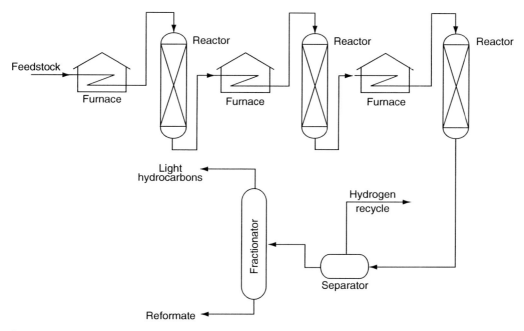

FIGURE 23.3 The platforming process.

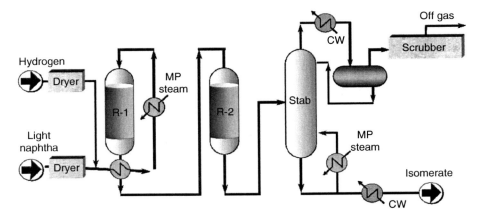

FIGURE 23.4 Isomerization using an aluminum chloride catalyst.

The earliest important process was the formation of *iso*-butane, which is required as an alkylation feed; the isomerization may take place in the vapor phase, with the activated catalyst supported on a solid phase, or in the liquid phase with a dissolved catalyst. Thus, a pure butane feed is mixed with hydrogen (to inhibit olefin formation) and passed to the reactor at 110°C to 170°C (230°F to 340°F) and 200 to 300 psi. The product is cooled and the hydrogen separated; the cracked gases are then removed in a stabilizer column. The stabilizer bottom product is passed to a superfractionator, and the *n*- and *iso*-butane are separated. With pentanes, the equilibrium is favorable at higher temperatures, and operating conditions of 300 to 1000 psi and 240°C to 500°C (465°F to 930°F) may be used.

Present isomerization applications are to provide additional feedstock for alkylation units or high-octane fractions for gasoline blending. Straight-chain paraffins (*n*-butane) (Figure 23.4) and mixtures of *n*-pentane and *n*-hexane (Figure 23.5) are converted to the respective *iso*-compounds by continuous catalytic (aluminum chloride and noble metals) processes. Natural gasoline or light straight-run gasoline can provide feed by first fractionating as a preparatory step. High volumetric yields (>95%) and 40% to 60% conversion per pass are characteristic of the isomerization reaction.

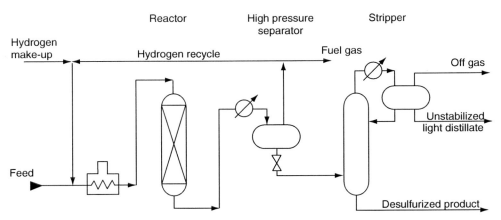

FIGURE 23.5 Butane isomerization. (From OSHA Technical Manual, Section IV, Chapter 2, Petroleum Refining Processes.)

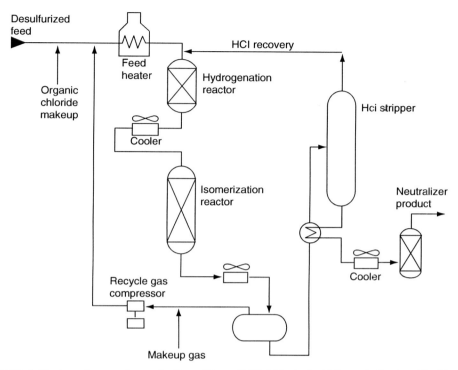

FIGURE 23.6 Pentane–hexane isomerization. (From OSHA Technical Manual, Section IV, Chapter 2, Petroleum Refining Processes.)

Nonregenerable aluminum chloride catalyst is employed with various carriers in a fixed-bed or liquid contactor. Platinum or other metal catalyst processes use a fixed-bed operation and can be regenerable or nonregenerable. The reaction conditions vary widely depending on the particular process and feedstock, 40°C to 480°C (100°F to 900°F) and 150 to 1000 psi; residence time in the reactor is 10 to 40 min.

23.3.1 BUTAMER PROCESS

The butamer process is designed to convert *n*-butane to *iso*-butane under mild operating conditions (Figure 23.6) (Cusher, 1997a). A platinum catalyst on a support is used in a fixed-bed reactor system. Using reformer off-gas can readily satisfy the low hydrogen requirement. The operation can be designed for once-through or recycle operation and is normally tied in with alkylation unit de-*iso*-butanizer operations to provide additional feed.

Butane feed is mixed with hydrogen, heated, and charged to the reactor at moderate pressure. The effluent is cooled before light gas separation and stabilization. The resultant butane mixture is then charged to a de-*iso*-butanizer to separate a recycle stream from the *iso*-butane product.

23.3.2 BUTOMERATE PROCESS

The butomerate process is specially designed to isomerize *n*-butane to produce additional alkylation feedstock. The catalyst contains a small amount of nonnoble hydrogenation metal on a high-surface-area support. The process operates with hydrogen recycle to eliminate coke deposition on the catalyst, but the isomerization reaction can continue for extended periods in the absence of hydrogen.

The feedstock should be dry and comparatively free of sulfur and water; the feed is heated, mixed with hydrogen, and conveyed to the reactor. Operating conditions range from 150°C to 260°C (300°F to 500°F) and 150 to 450 psi. The effluent is cooled and flashed, and the liquid product is stripped of light material.

23.3.3 HYSOMER PROCESS

The hysomer process uses hydrotreated feedstocks containing pentane(s) and hexane(s) without further pretreatment. Operating conditions are 400 to 450 psi hydrogen and about 290°C (550°F); it appears that the catalyst (zeolite) life is about 2 years.

The influence of sulfur on catalyst activity is minimal, and the catalyst can tolerate a permanent sulfur level of 10 ppm in the feedstock, but concentrations up to 35 ppm are not harmful. The process can operate at a water level of 50 ppm, and feedstocks having saturated water contents can be processed without a deleterious effect on either catalyst stability or conversion. A minimum quantity of water is essential for the activity of zeolite catalysts in this application.

23.3.4 ISO-KEL PROCESS

The iso-kel process is a fixed-bed, vapor-phase isomerization process that employs a precious metal catalyst and hydrogen. A wide variety of feedstocks, including natural gasoline, pentane, and hexane cuts, can be processed. Operating conditions include reactor temperatures and pressures from 345°C to 455°C (650°F to 550°F) and 350 to 600 psi.

23.3.5 ISOMATE PROCESS

The isomate process is a nonregenerative pentane and hexane or naphtha (C_6) isomerization process using an aluminum chloride–hydrocarbon complex catalyst with anhydrous hydrochloric acid as a promoter. Hydrogen partial pressure is maintained to suppress undesirable reactions (cracking and disproportionation) and retain catalyst activity. The feed is saturated with anhydrous hydrogen chloride in an absorber, then heated and combined with hydrogen and charged to the reactor (115°C, 240°F, and 700 psi). Catalyst is added to the reactor separately, and the reaction takes place in the liquid phase. The product is washed (caustic and water), acid stripped, and stabilized before going to storage.

23.3.6 ISOMERATE PROCESS

The isomerate process is a continuous isomerization process designed to convert pentanes and hexanes into highly branched isomers; a dual-function catalyst is used in a fixed-bed reactor system.

Operating conditions are mild, less than 750 psi and 400°C (750°F). Hydrogen is added to the feed along with recycle gas, and the usual operation includes fractionation facilities to allow the recycling of n-paraffins almost to extinction.

23.3.7 PENEX PROCESS

The penex process is a nonregenerative pentane(s) and hexane(s) isomerization process (Cusher, 1997b). The reaction takes place in the presence of hydrogen with a platinum catalyst, and the reactor conditions are selected so that catalyst life is long and regeneration is not required. The reactor temperatures range from 260°C to 480°C (500°F to 900°F) and pressures from 300 to 1000 psi.

The Penex process may be applied to many feedstocks by varying the fractionating system. Mixed feeds may be split into pentane and hexane fractions and respective iso-fractions separated from each. The system can also be operated in conjunction with reforming of the naphtha ($>C_7$) fraction.

23.3.8 PENTAFINING PROCESS

The **pentafining** process is a regenerable pentane isomerization process that uses platinum catalyst on a silica–alumina support. A number of process combinations are possible. For example, with natural gasoline and hydrogen as starting materials, pentanes are removed from the feedstock and the pentane fraction is passed to a low-pressure reformer. The pentane stream is split, and the *n*-pentane fraction is combined with a recycle stream and makeup hydrogen, and charged to the reactor (300 to 400 psi and 425°C to 480°C, 800°F to 900°F) where isomerization occurs.

$$CH_3CH_2CH_2CH_2CH_3 \rightarrow (CH_3)_2CHCH_2CH_3$$

n-pentane *iso*-pentane

Hydrogen is removed from the effluent, which is degassed and fractionated to separate *n*-pentane and *iso*-pentane (95% purity) fractions. The catalyst is regenerated at 260°C to 540°C (500°F to 1000°F) using a steam–air mixture.

23.4 ALKYLATION

Alkylation is the refinery process that provides an economically feasible outlet for several of the very light olefins produced from the catalytic cracking unit. Propylene, butylene, and pentylene (also known as amylene) products are available for alkylation. Propylene alkylation is the normal disposition for cat cracker product although some butylenes from a fluid catalytic cracking unit could be blended into gasoline. The high vapor pressure of butylenes prevents low cost butane blending stock from being blended into gasoline, thereby carrying a very high opportunity cost for this option.

In the alkylation process, the propylene, butylene, and pentylene are combined with *iso*-butane in the catalyzed alkylation reaction to produce branched, and saturated, seven, eight, or nine carbon molecules, respectively. The product (alkylate) consists of *iso*-heptane, *iso*-octane, and *iso*-nonane and is a low vapor pressure (relative to the feedstocks), very high-octane gasoline blending stock. The high-octane value makes alkylate an excellent blending stock for premium grades of gasolines. Furthermore, as alkylate contains no olefins, aromatics, or sulfur, it is also an excellent blending stock for use in reformulated gasoline.

The alkylation reaction is catalyzed by the presence of very strong acid, either sulfuric acid or hydrofluoric acid. Hydrofluoric acid exists in a vapor state at ambient conditions and this dictates that extreme precaution is necessary to ensure that this toxic substance is contained inside the process equipment. The alkylation reactors typically operate at temperatures of 2°C to 21°C (35°F to 70°F, maximum) to minimize polymerization of the olefins to form undesirable hydrocarbons for the sulfuric acid process. The hydrofluoric acid process, which is less sensitive to polymerization at warmer temperatures, typically operates at reactor temperatures of 21°C to 38°C (70°F–100°F). *Iso*-butane concentrations are maintained very high (i.e., at ratios of 4:1 or more above the reaction requirements) in the reactor vessels to ensure that all of the olefins are reacted. The reactor effluent is distilled to separate the propane, *iso*-butane, and alkylate boiling fractions. The propane is routed to propane product treating, the *iso*-butane is recycled back to the alkylation reactors, and the alkylate is routed to gasoline blending, or in some cases to additional solvents refinery processing.

The chemistry of the combination of olefins with paraffins to form higher *iso*-paraffins is simple:

$$(CH_3)_3CH + CH_2{=}CH_2 \rightarrow (CH_3)_3CHCH_2CH_3$$

As olefins are reactive (hence unstable) and are responsible for exhaust pollutants, their conversion to high-octane *iso*-paraffins is desirable when possible. In refinery practice, only *iso*-butane is alkylated by reaction with *iso*- or *n*-butene and *iso*-octane is the product. Although alkylation is possible without catalysts, commercial processes use sulfuric acid (Lerner, 1997) or hydrogen fluoride (Scheckler and Shah, 1997) as catalysts when the reactions can take place at low temperatures, minimizing undesirable side reactions, such as polymerization of olefins.

Alkylate is composed of a mixture of *iso*-paraffins that have octane numbers that vary with the olefins from which they were made. Butylenes produce the highest octane numbers, propylene the lowest, and pentylenes the intermediate. All alkylates, however, have high octane numbers (>87) and are particularly valuable because of these high octane numbers.

Alkylation development in petroleum processing in the late 1930s and during World War II were directed toward production of high-octane blending stock for aviation gasoline. The sulfuric acid process was introduced in 1938, and hydrogen fluoride alkylation was introduced in 1942. Rapid commercialization took place during the war to supply military needs, but many of these plants were shut down at the end of the war.

In the mid-1950s the demand for aviation gasoline started to decline, and motor gasoline quality requirements rose sharply. Whenever practical, refiners shifted the use of alkylate to premium motor fuel. Alkylate end point was increased for this service, and total alkylate was often used without rerunning. To help improve the economics of the alkylation process and also the sensitivity of the premium gasoline pool, additional olefins were gradually added to alkylation feed. New plants were built to alkylate propylene and the butylenes (butanes) produced in the refinery rather than the butane–butylene stream formerly used. More recently *n*-butane isomerization has been used to produce additional *iso*-butane for alkylation feed.

The alkylation reaction as practiced in petroleum refining is the union, through the agency of a catalyst, of an olefin (ethylene, $CH_2=CH_2$, propylene, $CH_3 \cdot CH=CH_2$, butene also called butylene, $CH_3 \cdot CH_2 \cdot CH=CH_2$, and amylene, $CH_3 \cdot CH_2 \cdot CH_2 \cdot CH=CH_2$) with *iso*-butane [$(CH_3)_3CH$] to yield high-octane branched-chain hydrocarbons in the gasoline boiling range. Olefin feedstock is derived from the gas make of a catalytic cracker; *iso*-butane is recovered from refinery gases or produced by catalytic butane isomerization.

Zeolite catalysts are also used for alkylation processes. For example, cumene (*iso*-propylbenzene) is produced by the alkylation of benzene by propylene (Wallace and Gimpel, 1997). The cumene can then be used for the production of phenol and acetone by means of oxidation processes.

23.4.1 CASCADE SULFURIC ACID ALKYLATION

This is a low-temperature process (Figure 23.7) employing concentrated sulfuric acid catalyst to react olefins with *iso*-butane to produce high-octane aviation or motor fuel blending stock. The olefin feed is split into equal streams and charged to the individual reaction zones of the cascade reactor. *Iso*-butane-rich recycle and refrigerant streams are introduced in the front of the reactor and passed through the reaction zones. The olefin is contacted with the *iso*-butane and acid in the reaction zones, which operate at 2°C to 7°C (35°F to 45°F) and 5 to 15 psi, after which vapors are withdrawn from the top of the reactor, compressed, and condensed. Part of this stream is sent to a de-propanizer to control propane concentration in the unit.

De-propanizer bottoms and the remainder of the stream are combined and returned to the reactor. Spent acid is withdrawn from the bottom of the settling zone; hydrocarbons spill over a baffle into a special withdrawal section and are hot-water washed with caustic addition for pH control before being successively de-propanized, de-*iso*-butanized, and de-butanized. Alkylate can then be taken directly to motor fuel blending or be rerun to produce aviation-grade blending stock.

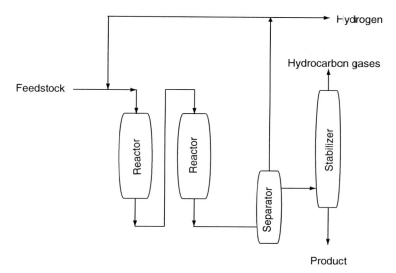

FIGURE 23.7 Butamer process. (From Speight, J.G., and Ozum, B. 2002. *Petroleum Refining Processes.* Marcel Dekker Inc., New York. With permission.)

23.4.2 HYDROGEN FLUORIDE ALKYLATION

This process (Figure 23.8) uses regenerable hydrofluoric acid as a catalyst to unite olefins with *iso*-butane to produce high-octane blending stock. The dried charge is intimately contacted in the reactor with acid at 20°C to 140°C (70°F to 100°F) and a high (15:1) *iso*-butane-olefin ratio. The mixture is separated in a settler and acid is returned to the reactor, but an acid sidestream must be continuously regenerated to 88% purity by fractionation to remove acid-soluble oils. The hydrocarbon fraction from the settler is de-*iso*-butanized, and alkylate is run to storage.

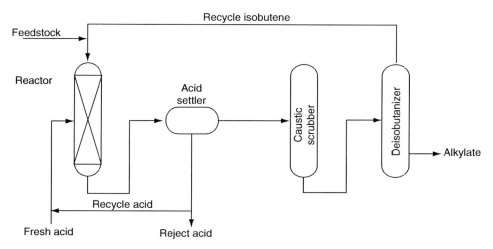

FIGURE 23.8 The sulfuric acid alkylation process. (From OSHA Technical Manual, Section IV, Chapter 2, Petroleum Refining Processes.)

23.5 POLYMERIZATION

Polymerization, as practiced in the petroleum industry, is a process that can claim to be the earliest to employ catalysts on a commercial scale. Catalytic polymerization came into use in the 1930s and was one of the first catalytic processes to be used in the petroleum industry.

In the usual industrial sense, polymerization is a process in which a substance of low molecular weight is transformed into one of the same composition but of higher molecular weight, maintaining the atomic arrangement present in the basic molecule. It has also been described as the successive addition of one molecule to another by means of a functional group, such as that present in an aliphatic olefin. In the petroleum industry, polymerization is the controlled process by which olefin gases are converted to liquid condensation products that may be suitable for gasoline (hence **polymer gasoline**, polymerate) or other liquid fuels.

The feedstock usually consists of propylene, propene ($CH_3 \cdot CH = CH_2$), and butylenes (butenes, various isomers of C_4H_8) from cracking processes or might even be selective olefins for dimer, trimer, or tetramer production:

$$nCH_2 = CH_2 \rightarrow H - (CH_2CH_2)_n - H$$

In this process, n is usually 2 (dimer), 3 (trimer), or 4 (tetramer); the molecular size of the product is limited to give products boiling in the gasoline range constituents. This is in contrast to polymerization that is carried out in the polymer industry where n may be on the order of several hundred. Thus, polymerization in the true sense of the word is usually prevented, and all attempts are made to terminate the reaction at the dimer or trimer (three monomers joined together) stage. The four-carbon to twelve-carbon compounds that are required as the constituents of liquid fuels are the prime products. However, in the petrochemical section of a refinery, polymerization, which results in the production of, for example, polyethylene, is allowed to proceed until the products having the required high molecular weight have been produced.

Polymerization may be accomplished thermally or in the presence of a catalyst at lower temperatures (York et al., 1997). Thermal polymerization is regarded not as effective as catalytic polymerization but has the advantage that it can be used to polymerize saturated materials that cannot be induced to react by catalysts. The process consists essentially of vapor-phase cracking of, say, propane and butane followed by prolonged periods at high temperature (510°C to 590°C, 950°F to 1100°F) for the reactions to proceed to near completion.

On the other hand, olefins can be conveniently polymerized by means of an acid catalyst. Thus, the treated olefin-rich feed stream is contacted with a catalyst (sulfuric acid, copper pyrophosphate, or phosphoric acid) at 150°C to 220°C (300°F to 425°F) and 150 to 1200 psi, depending on feedstock and product requirement. The reaction is exothermic, and the temperature is usually controlled by heat exchange. Stabilization and fractionation systems separate saturated and unreacted gases from the product. In both thermal and catalytic polymerization processes, the feedstock is usually pretreated to remove sulfur and nitrogen compounds.

23.5.1 THERMAL POLYMERIZATION

Thermal polymerization converts butanes and lighter gases into liquid condensation products. Olefins are produced by thermal decomposition and polymerized by heat and pressure. Thus, liquid feed under a pressure of 1200 to 2000 psi is pumped to a furnace heated to 510°C to 595°C (950°F to 1100°F), from which the various streams are separated by fractionation.

Thermal polymerization is regarded not as effective as catalytic polymerization but has the advantage that it can be used to polymerize saturated materials that cannot be induced to react by catalysts. The process consists essentially of vapor-phase cracking of, say, propane and butane followed by prolonged periods at high temperature (510°C to 595°C; 950°F to 1100°F) for the reactions to proceed to near completion.

23.5.2 SOLID PHOSPHORIC ACID CONDENSATION

Olefins can be conveniently polymerized by means of an acid catalyst (Figure 23.9). Thus, the treated olefin-rich feed stream is contacted with a catalyst (sulfuric acid, copper pyrophosphate, or phosphoric acid) at 150°C to 220°C (300°F to 425°F) and 150 to 1200 psi, depending on the feedstock and the desired product(s). The reaction is exothermic, and temperature is usually controlled by heat exchange. Stabilization and fractionation systems separate saturated and unreacted gases from the product. In both thermal and catalytic polymerization processes, the feedstock is usually pretreated to remove sulfur and nitrogen compounds.

This process converts propylene and butylene to high-octane gasoline or petrochemical polymers. The catalyst, pelleted kieselguhr (diatomaceous earth) impregnated with phosphoric acid, is used in either a chamber or tubular reactor. The exothermic reaction temperature is controlled by using saturates (separated from the effluent as recycle to the feed) as a quench liquid between the catalyst chamber beds. Tubular reactors are temperature-controlled by water or oil circulation around the catalyst tubes.

Reaction temperatures and pressures are 175°C to 225°C (350°F to 435°F) and 400 to 1200 psi. Olefins and aromatics may be united by alkylation for special applications at 205°C to 315°C (400°F to 600°F) and 400 to 900 psi, and a rerun column is required in addition to the usual fractionating.

23.5.3 BULK ACID POLYMERIZATION

This is a process to produce high-octane polymer gasoline from all types of light olefin feed, and the olefin concentration can be as high as 95%. Liquid phosphoric acid is used as the catalyst.

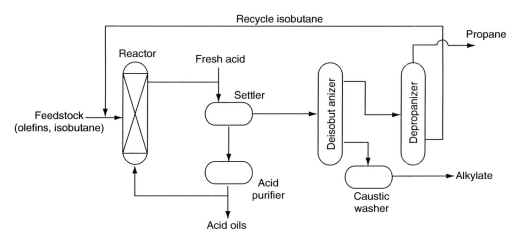

FIGURE 23.9 The hydrofluoric acid alkylation process. (From OSHA Technical Manual, Section IV, Chapter 2, Petroleum Refining Process.)

The olefin feed is washed (caustic and water) and then contacted thoroughly by liquid phosphoric acid in a small reactor. The effluent stream and the acid are separated in a settler, and acid is returned to the reactor through a cooler. Gasoline is first stabilized and washed with caustic before storage. The heat of reaction is removed by circulation through an exchanger before contact with the olefin feed, and catalyst activity is maintained by continuous addition of fresh acid and withdrawal of spent acid.

23.6 CATALYSTS

The various catalysts used in product improvement processes have already been mentioned in the context of a particular reaction or reactor. However, for convenience it is worth mentioning the salient facts about these catalysts under one particular heading.

23.6.1 REFORMING PROCESSES

The composition of a reforming catalyst is dictated by the composition of the reformer feedstock and the desired reformate. Reforming consists of two types of chemical reactions that are catalyzed by two different types of catalysts: (1) isomerization of straight-chain paraffins and isomerization (simultaneously with hydrogenation) of olefins to produce branched-chain paraffins and (2) dehydrogenation–hydrogenation of paraffins to produce aromatics and olefins to produce paraffins.

The composition of a reforming catalyst is dictated by the composition of the feedstock and the desired reformate. The catalysts used are principally molybdena–alumina (MoO_2–Al_2O_3), chromia–alumina (Cr_2O_3–Al_2O_3), or platinum (Pt) on a silica–alumina (SiO_2–Al_2O_3) or alumina (Al_2O_3) base. The nonplatinum catalysts are widely used in regenerative process for feeds containing, for example, sulfur, which poisons platinum catalysts, although pretreatment processes (e.g., hydrodesulfurization) may permit platinum catalysts to be employed.

The purpose of platinum on the catalyst is to promote dehydrogenation and hydrogenation reactions, that is, the production of aromatics, participation in hydrocracking, and rapid hydrogenation of carbon-forming precursors. For the catalyst to have an activity for isomerization of both paraffins and naphthenes—the initial cracking step of hydrocracking—and to participate in paraffin dehydrocyclization, it must have an acid activity. The balance between these two activities is most important in a reforming catalyst.

In the production of aromatics from cyclic saturated materials (naphthenes), it is important that hydrocracking be minimized to avoid loss of the desired product. Thus, the catalytic activity must be moderated relative to the case of gasoline production from a paraffinic feed, where dehydrocyclization and hydrocracking play an important part.

The acid activity can be obtained by means of halogens (usually fluorine or chlorine up to about 1% by weight in catalyst) or silica incorporated in the alumina base. The platinum content of the catalyst is normally in the range 0.3 to 0.8 wt.%. At higher levels there is some tendency to effect de-methylation and naphthene ring opening, which is undesirable; at lower levels the catalysts tend to be less resistant to poisons.

Most processes have a means of regenerating the catalyst as needed. The time between regeneration, which varies with the process, the severity of the reforming reactions, and the impurities of the feedstock, ranges from every few hours to several months. Several processes use a nonregenerative catalyst that can be used for a year or more, after which it is returned to the catalyst manufacturer for reprocessing. The processes that have moving beds of catalysts use continuous regeneration of the catalyst in separate regenerators.

The processes using bauxite (cycloversion) and clay (**isoforming**) differ from other catalytic reforming processes in that hydrogen is not formed and hence none is recycled through the reactors. As hydrogen is not involved in the reforming reactions, there is no limit to the amount of olefin that may be present in the feedstock. The Cycloversion process is also used as a catalytic cracking process and as a desulfurization process. The Isoforming process causes only a moderate increase in octane number.

23.6.2   ISOMERIZATION PROCESSES

During World War II, aluminum chloride was the catalyst used to isomerize butane, pentane, and hexane. Since then, supported metal catalysts have been developed for use in high-temperature processes that operate in the range of 370°C to 480°C (700°F to 900°F) and 300 to 750 psi; aluminum chloride plus hydrogen chloride is universally used for the low-temperature processes. However, aluminum chloride is volatile at commercial reaction temperatures and is somewhat soluble in hydrocarbons, and techniques must be employed to prevent its migration from the reactor. This catalyst is nonregenerable and is used in either a fixed-bed or liquid contactor.

23.6.3   ALKYLATION PROCESSES

Sulfuric acid, hydrogen fluoride, and aluminum chloride are the only catalysts used commercially. Sulfuric acid is used with propylene and higher-boiling feeds, but not with ethylene because it reacts to form ethyl hydrogen sulfate and a suitable catalyst contains a minimum of 85% titratable acidity. The acid is pumped through the reactor and forms an air emulsion with reactants; the emulsion is maintained at 50% acid. The rate of deactivation varies with the feed and *iso*-butane change rate. Butene feedstocks cause less acid consumption than propylene feeds.

Aluminum chloride is not widely used as an alkylation catalyst, but when employed hydrogen chloride is used as a promoter and water is injected to activate the catalyst. The form of catalyst is an aluminum chloride–hydrocarbon complex, and the aluminum chloride concentration is 63% to 84%.

Hydrogen fluoride is used for alkylation of higher boiling olefins. The advantage of hydrogen fluoride is that it is more readily separated and recovered from the resulting product. The usual concentration is 85% to 92% titratable acid, with about 1.5% water.

23.6.4   POLYMERIZATION PROCESSES

Phosphates are the principal catalysts for polymerization; the commercially used catalysts are liquid phosphoric acid, phosphoric acid on diatomaceous earth, copper pyrophosphate pellets, and phosphoric acid film on quartz (Figure 23.10). The latter is the least active but most used and easiest to regenerate simply by washing and recoating; the serious disadvantage is that residue must occasionally be burned off the support. The process using liquid phosphoric acid catalyst is far more responsive to attempts to raise production by increasing the temperature than the other processes.

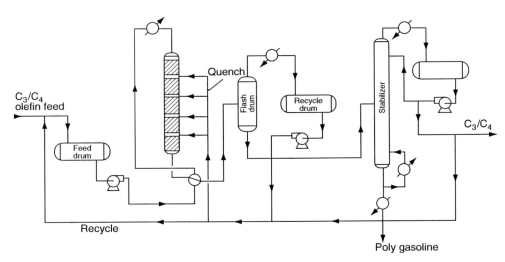

FIGURE 23.10 Polymerization process.

REFERENCES

Abraham, H. 1945. *Asphalts and Allied Substances*. Van Nostrand, New York.

Bland, W.F. and Davidson, R.L. 1967. *Petroleum Processing Handbook*. McGraw-Hill, New York.

Cusher, N.A. 1997a. *Handbook of Petroleum Refining Processes*. Meyers, R.A., ed. McGraw-Hill, New York. Chapter 9.2.

Cusher, N.A. 1997b. *Handbook of Petroleum Refining Processes*. Meyers, R.A., ed. McGraw-Hill, New York. Chapter 9.3.

Dachos, N., Kelly, A., Felch, D., and Reis, E. 1997. In *Handbook of Petroleum Refining Processes*. Meyers, R.A., ed. McGraw-Hill, New York. Chapter 4.1.

Forbes, R.J. 1958a. *A History of Technology*. Oxford University Press, Oxford, UK.

Forbes, R.J. 1958b. *Studies in Early Petroleum Chemistry*. E.J. Brill, Leiden, The Netherlands.

Forbes, R.J. 1959. *More Studies in Early Petroleum Chemistry*. E.J. Brill, Leiden, The Netherlands.

Hobson, G.D. and Pohl, W. 1973. *Modern Petroleum Technology*. Applied Science Publishers, Barking, Essex, England.

James, P. and Thorpe, N. 1994. *Ancient Inventions*. Ballantine Books, New York.

Kelly, A., Felch, D., and Reis, E. 1997. *Handbook of Petroleum Refining Processes*, 2nd Edition. Meyers, R.A., ed. McGraw-Hill, New York. Chapter 4.1.

Lerner, H. 1997. *Handbook of Petroleum Refining Processes*, 2nd Edition. Meyers, R.A., ed. McGraw-Hill, New York. Chapter 1.1.

Nelson, W.L. 1958. *Petroleum Refinery Engineering*. McGraw-Hill, New York.

Riediger, B. 1971. *The Refining of Petroleum*. Springer-Verlag, Heidelberg.

Scheckler, J.C. and Shah, B.R. 1997. *Handbook of Petroleum Refining Processes*, 2nd Edition. Meyers, R.A., ed. McGraw-Hill, New York. Chapter 1.4.

Schwarzenbek, E.F. 1971. *Origin and Refining of Petroleum*. Advances in Chemistry Series No. 103. McGrath, H.G., and Charles, M.E., eds. American Chemical Society, Washington, DC.

Speight, J.G. 2001. *Handbook of Petroleum Analysis*. John Wiley & Sons Inc., Hoboken, NJ.

Speight, J.G. 2002. *Handbook of Petroleum Product Analysis*. John Wiley & Sons Inc., Hoboken, NJ.

Wallace, J.W. and Gimpel, H.E. 1997. *Handbook of Petroleum Refining Processes*, 2nd Edition. Meyers, R.A., ed. McGraw-Hill, New York. Chapter 1.2.

York, D., Scheckler, J.C., and Tajbl, D.G. 1997. *Handbook of Petroleum Refining Processes*, 2nd Edition. Meyers, R.A., ed. McGraw-Hill, New York. Chapter 1.2.

24 Product Treating

24.1 INTRODUCTION

Fractions or streams produced by crude distillation (Chapter 16), cracking (Chapter 17 and Chapter 18), and other refinery processes (Chapter 14) (although usable in the refinery as process feedstocks) often contain small amounts of impurities that must be removed. Processes that remove these undesirable components are known as treating processes, and these processes are used not only to finish products for the market but also to prepare feedstocks for other processes (catalytic polymerization and **reforming**) (Chapter 23) in which catalysts would be harmed by impurities (Bland and Davidson, 1967).

The undesirable components in a petroleum fraction are referred to as impurities, but they are almost invariably either the normal constituents of crude oil or they are derivatives of the constituents of crude oil. The most common impurities are sulfur compounds, which are derived from the sulfur compounds that occur in crude oil (Table 24.1) (Chapter 7), such as sulfides (R–S–R′) and the foul-smelling **mercaptans** (R–SH), also called thiols. Oxygen compounds in the form of carboxylic acids (R–CO$_2$H) and phenols (Ar–OH, where Ar is an aromatic group) may also be present. Nitrogen-containing compounds derived from those that occur in crude oil (Table 24.2) (Chapter 7) are also present.

Sometimes olefins (R–CH=CH–R′) must be eliminated from a feedstock or aromatics removed from a solvent, and these olefins and aromatics are considered impurities. Similarly, polymerized material, asphaltic material, or resins may be impurities, depending on whether their presence in a finished product is harmful (Hobson and Pohl, 1973).

During processing (Chapter 14), much of the sulfur originally in the crude oil is converted to hydrogen sulfide (H$_2$S) and mercaptans (R–SH). Some sulfur may be retained in the coke that is formed. Processes that are primarily concerned with the removal of hydrogen sulfide and mercaptans are known as **sweetening** processes. Petroleum fractions containing mercaptans are readily recognized by their odor and are called sour. Fractions that are free of obnoxious sulfur compounds, either naturally or because of treatment, are called sweet, but a sweet fraction may contain sulfur compounds (e.g., sulfide R–S–R′ or disulfide R–S–S–R′) that have no odor. The more severe desulfurization of petroleum fractions (Speight, 2000) is not usually classed as a treating process but recognition has been given to processes such as hydrotreating (Chapter 20) and hydrocracking (Chapter 21) that contribute to product improvement.

The need to remove sulfur (desulfurization) from various feedstocks is of importance. Thus, desulfurization is a major refinery issue and is worthy of mention here. Indeed, considerable importance has been attached to the lowering of the sulfur level in distillates and residual stocks. For example, the stability of sulfur compounds is greatly reduced when they are heated in the presence of adsorptive catalysts, and this is employed in a number of desulfurization processes. The noncyclic sulfur compounds (mercaptans, R–SH, sulfides, R–S–R^1, and disulfides, R–S–S–R′) in straight-run distillates (e.g., naphtha) are readily

TABLE 24.1
Nomenclature and Types of Organic Sulfur Compounds

RSH	Thiols (Mercaptans)
RSR′	Sulfides
	Cyclic Sulfides
RSSR′	Disulfides
	Thiophene
	Benzothiophene
	Dibenzothiophene
	Naphtobenzothiophene

converted into hydrogen sulfide and olefins by contacting the vapors with clays, aluminum oxide, or alumina–silica (Al_2O_3–SiO_2) cracking catalysts (Chapter 18). These processes generally operate at about 345°C to 425°C (650°F to 800°F) and at a pressure of about 50 psi.

When hydrogen is added and a dehydrogenation catalyst, such as cobalt sulfide and molybdenum sulfide on alumina, is employed, extensive desulfurization or denitrogenation of a wider spectrum of heteroatomic compounds is brought about. Even the refractory cyclic nitrogen and sulfur compounds are decomposed and the heteroatom is released as its hydrogen analog (e.g., H_2S or NH_3).

Treatment processes for the removal of sulfur-containing and nitrogen-containing compounds are much less severe than the desulfurization and denitrogenation processes. In fact, it is generally recognized that the removal or conversion of sulfur and nitrogen compounds in distillates by treatment processes is usually limited to mercaptans and the lower molecular weight sulfur compounds. When there is more than a trace amount (>0.1%) of heteroatoms present, it is often more convenient and economical to resort to such methods as those thermal processes (e.g., hydroprocesses) that bring about a decrease in all types of heteroatomic compounds.

However, it must be added that the appearance of nitrogen compounds in distillates is theoretically possible but often does not occur. The nitrogen species in petroleum tend to concentrate in the nonvolatile species. Furthermore, thermal processes also tend to concentrate the nitrogen species in the coke. This is also true, but to a lesser extent, for the sulfur compounds. Sulfur tends to be ubiquitous throughout the boiling range of crude oil.

Some of the sulfur compounds in petroleum break down readily in the mild thermal treatments to which they are exposed in distillation. For others, extensive decomposition does not take place until the conditions necessary for cracking are reached. Some are refractory and survive even the more severe conditions employed in processes leading to tar and coke formation. The sulfur in the lower-boiling straight-run distillates is mainly in the form of mercaptans, sulfides, and disulfides, whereas thermally cracked distillates contain the more refractory thiophene-type sulfur, in addition to thiophenols (Ar–SH, where Ar is an aromatic group) that occur in catalytically cracked distillates.

The sulfur compounds that occur in light petroleum distillates are usually degradation products formed from higher molecular weight materials during distillation or cracking. In addition to traces of elementary sulfur, hydrogen sulfide, mercaptans, sulfides, disulfides, and thiophenes have also been identified in petroleum products. The removal of these various classes of sulfur compounds presents a series of individual problems.

The elemental sulfur contained in light distillates may be a product of oxidation (air) or conversion of hydrogen sulfide to sulfur and water. Sulfur is not generally affected by refining agents, except in doctor sweetening, and for this reason it is best to prevent its formation by separating the hydrogen sulfide from the distillate before refining treatments are made. If the distillate is heated, most of the sulfur appears as hydrogen sulfide and is removed in the stabilizing tower of the fractionating system; any remaining traces are easily removed by washing with aqueous alkali. For obvious reasons, hydrogen sulfide must be removed from refinery gases before the latter are cracked or polymerized to gasoline; hydrocarbon gases must also be essentially sulfur-free to make high-quality carbon blacks and certain chemical products. The sulfur compounds and elemental sulfur in crude oil, as well as their thermal decomposition products, are probably responsible for heavy corrosion losses in refinery equipment.

Choices of a treatment method depend on the amount and type of impurities in the fractions to be treated and the extent to which the process removes the impurities. Naturally occurring sweet kerosene, for example, may require only a simple treatment with alkali (lye) to remove hydrogen sulfide. If mercaptans are also present in the raw kerosene, a doctor treatment in addition to lye treatment is required, but poor-quality raw kerosene may require, in addition to these treatments, treatment with sulfuric acid and Fuller's earth. The lowest-quality raw kerosene requires treatment with strong sulfuric acid, neutralization with lye, and redistillation. As different fractions have the same impurities, the same treatment process may be used for several different products (Bland and Davidson, 1967).

There is a choice of several different treatment processes, but the primary purpose of the majority is the elimination of unwanted sulfur compounds. Some processes are limited to the conversion of certain sulfur compounds, as well as olefins, asphaltic materials, oxygen compounds, and nitrogen compounds. The various processes eliminate impurities by chemical reagents, by catalysts, and by adsorption on clays or similar materials (Chapter 8).

Some chemical treatment processes directly remove sour sulfur compounds and hence tend to supplant the older processes that sweeten by conversion of sulfur compounds. Caustic soda (lye) solutions have long been used to remove hydrogen sulfide and some of the lighter mercaptans. Caustic soda, however, does not remove the higher-boiling mercaptans. By combining caustic soda solution with various solubility promoters (methyl alcohol, cresols, naphthenic acids, and alkyl phenols), up to 99% of all mercaptans as well as oxygen and nitrogen compounds can be dissolved from petroleum fractions (Kalichevsky and Kobe, 1956).

Sulfuric acid (H_2SO_4) treatment has the longest history of any petroleum-treating method. Sulfuric acid was the preferred treating material for all petroleum fractions from naphtha to lubricating oil for a century and is still widely used. Modern use of sulfuric acid is only a fraction of its former use. Nevertheless, the removal or conversion of mercaptans can be effected by sulfuric acid or by anhydrous aluminum chloride (AlCl₃), but the chemistry of the sulfuric acid desulfurization is not fully understood. Sulfides and disulfides are removed by addition reactions with aluminum chloride or by solution with sulfuric acid. Cyclic sulfides, such as tetra- and penta-methylene sulfide as well as sulfones [R–S(=O)–R′] and sulfoxides [R–S(O)₂–R′] are also removed by solution in the acid.

A number of commonly used treatment processes (doctor treatment, copper chloride treatment, and sodium hypochlorite treatment) convert mercaptans into odorless disulfides, which remain dissolved in the petroleum fraction. If no harm will result from a small amount of disulfides, they may be left in finished products, such as kerosene or stove oil. When sulfur compounds cannot be tolerated, as in solvent naphtha, disulfides are removed in a bottom fraction obtained by redistilling the sweetened naphtha.

Hypochlorites, for example sodium hypochlorite (NaOCl, bleaching powder), oxidize mercaptans (R–SH) to disulfides (R–S–S–R′) and alkyl sulfides (thioethers) (R–S–R′) are oxidized to sulfoxides [R–S(O)₂–R′]. The lower molecular weight products may be water-soluble.

Mercaptans (e.g., R–CH₂·CH₂–SH) may also be removed from refinery products by a process in which they are converted to hydrogen sulfide (H₂S) and the corresponding olefin (R·CH=CH₂), either by heat alone or, more easily, in the presence of catalysts such as bauxite (crude alumina, Al₂O₃). However, thioethers may be produced if the catalyst contains zinc sulfide (ZnS) or cadmium sulfide (CdS).

A number of reagents have been found suitable for the efficient and inexpensive removal of hydrogen sulfide from gases. All use the principle of absorbing the hydrogen sulfide at a low temperature, after which any one of several treatment steps is used:

1. Air blowing (sodium carbonate process)
2. Heating to a higher temperature (phenolates, ethanolamines)
3. Treatment with solutions of olamines
4. Treatment with alkali salts of amino acids and potassium phosphate
5. Treatment by oxidizing the adsorbent solution to release elemental sulfur and to regenerate the absorbing agent (thioarsenates)

It is the purpose of this chapter to present an outline of the processes that are available for treatment of the various product streams to remove contaminants from streams that will be eventually used as stock for products. The processes outlined here are not usually shown on a refinery schematic but would be placed in the general area entitled finishing after the processes shown on a schematic and before the designation of the product.

24.2 COMMERCIAL PROCESSES

24.2.1 CAUSTIC PROCESSES

Treating of petroleum products by washing with solutions of alkali (caustic or lye) is almost as old as the petroleum industry itself. Early discoveries that product odor and color could be improved by removing organic acids (naphthenic acids and phenols) and sulfur compounds (mercaptans and hydrogen sulfide) led to the development of caustic washing.

Thus, it is not surprising that caustic soda washing (lye treatment) has been used widely on many petroleum fractions. In fact, it is sometimes used as a pretreatment for sweetening and other processes. The process consists of mixing a water solution of lye (sodium hydroxide or caustic soda) with a petroleum fraction. The treatment is carried out as soon as possible after the petroleum fraction is distilled, as contact with air forms free sulfur, which is very corrosive and difficult to remove. The lye reacts with any hydrogen sulfide present to form sodium sulfide, which is soluble in water.

24.2.1.1 Dualayer Distillate Process

The **Dualayer distillate process** is similar in character to the Duosol process in that it uses caustic solution and cresylic acid (cresol, methylphenol, $CH_3 \cdot C_6H_4 \cdot OH$) The process extracts organic acid substances (including mercaptans, R–SH) from cracked, or virgin, distillate fuels. In a typical operation, the Dualayer reagent is mixed with the distillate at about 55°C (130°F) and passed to the settler, where three layers separate with the aid of electrical coagulation. The product is withdrawn from the top layer; the Dualayer reagent is withdrawn from the bottom layer, relieved of excess water, fortified with additional caustic, and recycled.

24.2.1.2 Dualayer Gasoline Process

The **Dualayer gasoline process** is a modification of the **Dualayer distillate process** in that it is used to extract mercaptans from liquid petroleum gas, gasoline, and naphtha using the Dualayer reagents. Thus gasoline, free of hydrogen sulfide, is contacted with the Dualayer solution at 50°C (120°F) in at least two stages, after which the treated gasoline is washed and stored. The treating solution is diluted with water (60% to 70% of the solution volume) and stripped of mercaptans, gasoline, and excess water, and the correct amount of fresh caustic is added to obtain the regenerated reagent.

24.2.1.3 Electrolytic Mercaptan Process

The **electrolytic mercaptan process** employs aqueous solutions to extract mercaptans from refinery streams, and the electrolytic process is used to regenerate the solution. The charge stock is pre-washed to remove hydrogen sulfide and contacted countercurrently with the treating solution in a mercaptan extraction tower. The treated gasoline is stored; the spent solution is mixed with regenerated solution and oxygen. The mixture is pumped to the cell, where mercaptans are converted to disulfides that are separated from the regenerated solution.

24.2.1.4 Ferrocyanide Process

The **ferrocyanide process** is a regenerative chemical treatment for removing mercaptans from straight-run naphtha, as well as natural and recycle gasoline, using caustic-sodium ferrocyanide reagent.

For example, gasoline is washed with caustic to remove hydrogen sulfide and then washed countercurrently in a tower with the treating agent. The spent solution is mixed with fresh solution containing ferricyanide; the mercaptans are converted to insoluble disulfides and are removed by a countercurrent hydrocarbon wash. The solution is then recycled, and part of the ferrocyanide is converted to ferricyanide by an electrolyzer.

24.2.1.5 Lye Treatment

Lye treatment is carried out in continuous treaters, which essentially consist of a pipe containing baffles or other mixing devices into which the oil and lye solution are both pumped. The pipe discharges into a horizontal tank where the lye solution and oil separate. Treated oil is withdrawn from near the top of the tank; lye solution is withdrawn from the bottom and recirculated to mix with incoming untreated oil. A lye-treating unit may be incorporated as part of a processing unit, for example, the overhead from a bubble tower may be condensed, cooled, and passed immediately through a lye-treating unit. Such a unit is often referred to as a worm-end treater, as the unit is attached to the particular unit as a point beyond the cooling coil or cooling worm.

Caustic solutions ranging from 5% to 20% w/w are used at 20°C to 45°C (70°F to 110°F) and 5 to 40 psi. High temperatures and strong caustic are usually avoided because of the risk of color body formation and stability loss. Caustic-product treatment ratios vary from 1:1 to 1:10.

Spent lye is the term given to a lye solution in which about 65% of the sodium hydroxide content has been used by reaction with hydrogen sulfide, light mercaptans, organic acids, or mineral acids. A lye solution that is spent, as far as hydrogen sulfide is concerned, may still be used to remove mineral or organic acids from petroleum fractions. Lye solution spent by hydrogen sulfide is not regenerated, whereas blowing with steam can regenerate lye solution spent by mercaptans. This technique reforms sodium hydroxide and mercaptans from the spent lye. The mercaptans separate as a vapor and are normally destroyed by burning in a furnace. Spent lye can also be regenerated in a stripper tower with steam, and the overhead consists of steam and mercaptans, as well as the small amount of oil picked up by the lye solution during treatment. Condensing the overhead allows the mercaptans to separate from the water.

Nonregenerative caustic treatment is generally economically applied when the contaminating materials are low in concentration and waste disposal is not a problem. However, the use of nonregenerative systems is on the decline because of the frequently occurring waste disposal problems that arise from environmental considerations and because of the availability of numerous other processes that can effect more complete removal of contaminating materials.

24.2.1.6 Mercapsol Process

The **Mercapsol process** is another regenerative process for extracting mercaptans by means of sodium (or potassium) hydroxide, together with cresols, naphthenic acids, and phenol. Gasoline is contacted countercurrently with the mercapsol solution, and the treated product is removed from the top of the tower. Spent solution is stripped to remove gasoline, and the mercaptans are then removed by steam stripping.

24.2.1.7 Polysulfide Treatment

Polysulfide treatment is a nonregenerative chemical treatment process used to remove elemental sulfur from refinery liquids. Dissolving 1 Lb of sodium sulfide (Na_2S) and 0.1 Lb of elemental sulfur in a gallon of caustic solution prepare the polysulfide solution. The sodium sulfide can actually be prepared in the refinery by passing hydrogen sulfide, an obnoxious refinery by-product gas, through a caustic solution.

The solution is most active when the composition approximates Na_2S to Na_2S_3 but activity decreases rapidly when the composition approaches Na_2S_4. When the solution is discarded, a portion (about 20%) is retained and mixed with fresh caustic-sulfide solution, which eliminates the need to add free sulfur. Indeed, if the material to be treated contains hydrogen sulfide in addition to free sulfur, it is often necessary to simply add fresh caustic.

24.2.1.8 Sodasol Process

A lye solution removes only the lighter or lower-boiling mercaptans, but various chemicals can be added to the lye solution to increase its ability to dissolve the heavier mercaptans. The added chemicals are generally known as solubility promoters or solutizers. Several different solutizers have been patented and are used in processes that differ chiefly in the composition of the solutizers.

In the Sodasol process, the treating solution is composed of lye solution and alkyl phenols (acid oils), which occur in cracked naphtha and cracked gas oil and are obtained by washing cracked naphtha or cracked gas oil with the lye solution. The lye solution, with solutizers incorporated, is then ready to treat product streams, such as straight-run naphtha and gasoline. The process is carried out by pumping a sour stream up a treating tower counter-current to a stream of Sodasol solution that flows down the tower. As the two streams mix and pass, the solution removes mercaptans and other impurities, such as oxygen compounds (phenols and acids), as well as some nitrogen compounds.

The treated stream leaves the top of the tower; the spent Sodasol solution leaves the bottom of the tower to be pumped to the top of a regeneration tower, where mercaptans are removed from the solution by steam. The regenerated Sodasol solution is then pumped to the top of the treatment tower to treat more material. A variation of the Sodasol process is the Potasol process, which uses potassium hydroxide instead of lye (sodium hydroxide).

24.2.1.9 Solutizer Process

The **Solutizer process** is a regenerative process that uses such materials as potassium *iso*-butyrate and potassium alkylphenolate in strong aqueous potassium hydroxide to remove mercaptans. After removal of the mercaptans and recovery of the hydrocarbon stream, regeneration of the spent solution may be achieved by heating and steam blowing at 130°C (270°F) in a stripping column in which steam and mercaptans are condensed and separated.

On the other hand, the spent solution may be contacted with carbon dioxide air, after which the disulfides formed by oxidation of the mercaptans are extracted by a naphtha wash.

Air blowing in the presence of tannin (tannin Solutizer process) catalytically oxidizes mercaptans to the corresponding disulfides, but there may be reactions that can lead to reagent contamination.

24.2.1.10 Steam-Regenerative Caustic Treatment

Steam-regenerative caustic treatment is essentially directed toward removal of mercaptans from such products as light, straight-run gasoline. The caustic is regenerated by steam blowing in a stripping tower. The nature and concentration of the mercaptans to be removed dictate the quantity and temperature of the process. However, the caustic solution gradually deteriorates because of the accumulation of material that cannot be removed by stripping; the caustic quality must be maintained by either continuous or intermittent discard, and replacement, of a minimum amount of the operating solution.

24.2.1.11 Unisol Process

The **Unisol process** is a regenerative method for extracting not only mercaptans but also certain nitrogen compounds from sour gasoline or distillates. The gasoline, free of hydrogen sulfide, is washed countercurrently with aqueous caustic-methanol solution at about 40°C (100°F). The spent caustic is regenerated in a stripping tower (145°C to 150°C, 290°F to 300°F), where methanol, water, and mercaptans are removed.

24.2.2 Acid Processes

Treating petroleum products with acids is, like caustic treatment, a procedure that has been in use for a considerable time in the petroleum industry. Various acids, such as hydrofluoric acid, hydrochloric acid, nitric acid, and phosphoric acid, have been used in addition to the more commonly used sulfuric acid, but in most instances there is little advantage in using any acid other than sulfuric.

Until about 1930, acid treatment was almost universal for all types of refined petroleum products, especially for cracked gasoline, kerosene, and lubricating stocks. Cracked products were acid treated to stabilize against gum formation and color darkening (oxidation) and to reduce sulfur content if necessary. However, there were appreciable losses due to polymer formation (from olefins in cracked products) initiated by the sulfuric acid.

Other processes have now superseded the majority of the acid treatment processes. However, acid treatment has, to some extent, been continued for desulfurizing high-boiling fractions of cracked gasoline, for refining kerosene, for manufacture of low-cost lubricating oil, and for making such specialties as insecticide naphtha, pharmaceutical white oil, and insulating oil.

The reactions of sulfuric acid with petroleum fractions are complex. The undesirable components to be removed are generally present in small amounts; large excesses of acid are required for efficient removal, which may cause marked changes in the remainder of the hydrocarbon mixture.

Paraffin and naphthene hydrocarbons in their pure forms are not attacked by concentrated sulfuric acid at low temperatures and during the short time of conventional refining treatment, but solution of light paraffins and naphthenes in the acid sludge can occur. Fuming sulfuric acid (oleum) absorbs small amounts of paraffins when contact is induced by long agitation; the amount of absorption increases with time, temperature, concentration of the acid, and complexity of structure of the hydrocarbons. With naphthenes fuming sulfuric acid causes sulfonation as well as rupture of the ring.

The action of sulfuric acid on olefin hydrocarbons is very complex. The main reactions involve ester formation and polymerization. The esters formed by a reaction of sulfuric acid with olefins in cracked distillates are soluble in the acid phase but are also to some extent soluble in hydrocarbons, especially as the molecular weight of the olefin increases. The esters are usually difficult to hydrolyze with a view to removal by alkali washing. They are, however, unstable on standing for a long time, and products containing them (acid-treated cracked gasoline) may evolve sulfur dioxide and deposit intractable materials. The esters are quite unstable on heating, so that a redistilled, acid-treated cracked distillate usually requires alkali washing after the customary distillation.

Aromatics are not attacked by sulfuric acid to any great extent under ordinary refining conditions, unless they are present in high concentrations. However, if fuming acid is used or if the temperature is allowed to rise above normal, sulfonation may occur. When both aromatics and olefins are present, as in distillates from cracking units, alkylation can occur.

Thus, as indicated, acid treatment of cracked gasoline distillate brings about losses due to chemical reaction and polymerization of some of the olefins to constituents boiling above the gasoline range. This makes redistillation necessary, and such losses may total several percent, even when refrigeration is employed to maintain a low temperature.

Acid treatment of high-boiling distillates and residua presents different problems. Most of these contain at least a small proportion of dissolved or suspended asphaltic substances, and almost all the acid comes out as sludge (acid tar); its separation is aided by the addition of a little water or alkali solution. However, there may be obvious chemical changes, such as sulfur dioxide evolution, and washed (acid-free) sludge from the treatment of practically sulfur-free oils may contain up to 10% combined sulfur derived from the treating acid.

Although largely displaced for bulk production of both gasoline and lubricating oils, acid treatment still serves many special purposes. Paraffin distillates intended for **dewaxing** might receive light treatment to facilitate wax crystallization and refining, whereas insulating oils, refrigeration compressor oils, and white oils may be treated more severely.

The sludge produced on acid treatment of petroleum distillates, even gasoline and kerosene, is complex in nature. Esters and alcohols are present from reactions with olefins; sulfonation products from reactions with aromatics, naphthenes, and phenols; and salts from reactions with nitrogen bases. In addition, such materials as naphthenic acids, sulfur compounds, and asphaltic material are all retained by direct solution. To these constituents must be added the various products of oxidation–reduction reactions: coagulated resins, soluble hydrocarbons, water, and free acid.

Disposal of the sludge is difficult, as it contains unused free acid that must be removed by dilution and settling. The disposal is a comparatively simple process for the sludge resulting from treating gasoline and kerosene, the so-called light oils. The insoluble oil phase separates out as a mobile tar-like material, which can be mixed and burned without too much difficulty. Sludge from heavy oil and bitumen, however, separates out as granular semisolids, which presents considerable difficulty in handling.

In all cases, careful separation of reaction products is important for the recovery of well-refined materials. This may not be easy if the temperature has risen as a consequence of chemical reaction. This will result in a persistent dark color traceable to colloidally distributed reaction products. Separation may also be difficult at low temperature because of high viscosity of the stock, but this problem can be overcome by diluting with light naphtha or with propane.

When acid treatment cannot be applied continuously by mechanical agitation followed by effective separation, the older batch agitators are employed. These devices are vertical reactors holding up to several thousand barrels, provided with conical bottoms for sludge drainage. The contact time is difficult to control and may amount to several hours, but the separation of acid tar is desirable to avoid discoloration by resolution and to permit handling of the sludge before it becomes undesirably viscous. Breaking out of the suspended acid tar, often referred to as pepper sludge, is helped by adding a little water and agitating, and the subsequent separation of tar closely resembles the precipitation of a colloidal suspension. The sludge is allowed to settle, and the sour oil is washed with water, usually after transfer to another container, to avoid retention of acid tar in the system during the alkali washing that follows.

Sodium hydroxide solution (10% to 25% concentration) may be used for nonviscous products, but for viscous oils more dilute solutions are employed and only a very slight excess of alkali is used, but no attempt is made at its recovery. Emulsion breaking chemicals are sometimes required in alkali washing; the use of aqueous alcohol is customary when fuming acid has been employed, as for sulfonates and white oils. Final water washing followed by air blowing to dry the oils is the customary procedure.

24.2.2.1 Nalfining Process

The Nalfining process is a continuous process that employs acetic anhydride and a caustic rinse to convert contaminants into less objectionable, but oil-soluble, compounds. The anhydride is injected into the product stream, where it reacts with oxygen to form the ester, with sulfur to form the thiaester, with nitrogen to form substituted amides, and with complex organic impurities to form environmentally benign products. The caustic rinse neutralizes the potentially corrosive acetic acid.

24.2.2.2 Sulfuric Acid Treatment

Sulfuric acid treatment is a continuous or batch method that is used to remove sulfur compounds. The treatment will also remove asphaltic materials from various refinery stocks. The acid strength varies from fuming (>100%) to 80%; approximately 93% acid finds the most common use. The weakest suitable acid is used for each particular situation to reduce sludge formation from the aromatic and olefin hydrocarbons.

The use of strong acid dictates the use of a fairly low temperature (−4°C to 10°C, 25°F to 50°F), but higher temperatures (20°C to 55°C, 70°F to 130°F) are possible if the product is to be redistilled.

24.2.3 CLAY PROCESSES

Treating petroleum distillates and residua by passing them through materials possessing decolorizing power has been in operation for many years (Chapter 7). For example, various clays and similar materials are used to treat petroleum fractions to remove diolefins, asphaltic materials, resins, acids, and colored bodies. Cracked naphtha were frequently clay treated to remove diolefins that formed gums in gasoline. Other processes have now largely superseded this use of clay treatment, in particular by the use of inhibitors, which, added in small amounts to gasoline, prevent gums from forming. Nevertheless, clay treatment is still used as a finishing step in the manufacture of lubricating oils and waxes. The clay removes traces of asphaltic materials and other compounds that give oils and waxes unwanted odors and colors.

The original method of clay treatment was to percolate a petroleum fraction through a tower containing coarse clay pellets. As the clay absorbed impurities from the petroleum fraction, the clay became less effective. Removing it from the tower periodically restored the activity of the clay and the absorbed material was burnt under carefully controlled conditions so as not to sinter the clay. The percolation method of clay treatment was widely used for lubricating oils but has been largely replaced by clay contacting.

24.2.3.1 Alkylation Effluent Treatment

This is a continuous liquid percolation process in which reactor effluent is coalesced in a vessel containing glass wool and steel mesh and then charged, alternately, to two bauxite (medium mesh) towers. Regeneration of the bauxite is effected with a mixture of steam and gas.

24.2.3.2 Arosorb Process

The Arosorb process separates aromatics from various refinery streams by the use of fixed silica gel beds. The feedstock is preheated over activated alumina to remove water, as well as traces of nitrogen, oxygen, and sulfur compounds. The material is then passed into one of several gel cases and, after a suitable residence time (about 30 min), conveyed to a second gel case. The first container (case) is then fed with the desorbent (e.g., crude xylene stream) that removes the saturate compounds from the bed and then displaces the adsorbed aromatics (e.g., benzene and toluene).

24.2.3.3 Bauxite Treatment

This process is essentially the same as the previous process, except in this case a vaporized petroleum fraction is passed through beds of a porous mineral known as bauxite. The bauxite acts as a catalyst to convert many different sulfur compounds, in particular mercaptans, into hydrogen sulfide, which is subsequently removed by a lye treatment. Bauxite is used to treat gasoline, naphtha, and kerosene products that have unusually high mercaptan contents.

A typical bauxite treatment unit consists of a fire-heated coil, two bauxite treatment towers, a bubble tower, a superheater for steam and air, and the usual exchangers, coolers, and pumps. Naphtha, raw kerosene, or other stock to be treated is preheated in heat exchangers and passed through the heating coil where it is heated to 415°C (780°F). At this temperature the stock is superheated. The vaporized feed is then passed downward through one of two bauxite towers at a pressure of about 40 psi. Three beds of catalyst in the tower convert mercaptans to hydrogen sulfide to enter a continuous lye treatment unit where hydrogen sulfide is removed.

After a time the bauxite towers are switched, as the bauxite progressively loses its catalytic activity. The spent catalyst is restored to its original activity by regeneration. Regeneration involves bypassing superheated steam and air, carefully scheduled in proportions and rate, downward through the catalyst beds. The carbonaceous material that has accumulated on the catalyst is burned off. Combustion progresses downward through the beds; care is taken to prevent temperatures exceeding 595°C (1100°F), which would harm the bauxite. Air alone is finally used to burn away the last traces of carbonaceous material, after which the bauxite is ready for use again.

24.2.3.4 Continuous Contact Filtration Process

This is a continuous clay treatment process in which finely divided adsorbent is mixed with the charge stock and heated to 95°C to 175°C (200°F to 350°F). The slurry is then conveyed to a steam-stripping tower, after which it is cooled, vacuum filtered, and then vacuum stripped for further product specification control.

24.2.3.5 Cyclic Adsorption Process

The cyclic adsorption process is used for the separation of aromatics from petroleum product streams. Like the Arosorb process, the cyclic adsorption process employs fixed beds of silica gel. The various stages of the process are (1) extraction of the adsorbable material from the feedstock (refining), (2) concentration of the adsorbed phase (enriching), and (3) stripping for recovery of the extract and regeneration of the gel using a light gasoline or pentane fraction.

24.2.3.6 Gray Clay Treatment

This is a continuous vapor-phase process for selectively polymerizing the diolefin constituents of, and removing other gum-forming agents from, thermal gasoline. Two or more towers (10 ft in diameter and approximately 25 ft high) are used in parallel, and the hydrocarbon vapors are passed through the bed at temperatures (120°C to 245°C, 250°F to 475°F) just above the condensation point. The diolefins polymerize and drain from the base of the tower or they are separated from the gasoline by fractionation. Spent clay is either discarded or regenerated in kilns.

24.2.3.7 Percolation Filtration Process

This is a continuous-flow cyclic-regenerative liquid phase process in which oil is filtered through a bed (containing 10 to 50 tons of Fuller's earth) before storage. Two or more beds are used alternately on an operating and regenerating cycle. The spent clay is regenerated by washing it with naphtha, steaming, and burning.

24.2.3.8 Thermofor Continuous Percolation Process

The Thermofor continuous percolation process is a continuous regenerative process for stabilizing and decolorizing lubricants or waxes that have been distilled, solvent refined, or acid treated. The charge stock is heated to 50°C to 175°C (125°F to 350°F), injected into the base of a clay-filled tower, and allowed to percolate in countercurrent flow through the bed. Spent clay is continuously withdrawn from the base of the tower; regenerated clay is added to the top of the bed to maintain a constant level.

24.2.4 OXIDATIVE PROCESSES

Oxidative treatment processes are, in fact, processes that have been developed to convert the objectionable-smelling mercaptans to the less-objectionable disulfides by oxidation.

However, disulfides tend to reduce the tetraethyl lead susceptibility of gasoline, and recent trends are toward processes that are capable of completely removing the mercaptans.

24.2.4.1 Bender Process

The **Bender process** is a fixed-bed catalytic treatment method that employs a lead sulfide catalyst. Controlled amounts of sulfur, alkali, and air are added to the product stream, which is passed through lead sulfide catalyst beds.

In this method, the sulfur required to oxidize the mercaptides is also furnished by the air oxidation of lead sulfide:

$$PbS + 1/2O_2 \rightarrow PbO + S$$

$$PbO + 2NaOH \rightarrow Na_2PbO_2 + H_2O$$

$$Na2PbO_2 + 2RSH \rightarrow Pb(RS)_2 + 2NaOH$$

$$Pb(RS)_2 + S \rightarrow PbS + R_2S_2$$

When larger quantities of air must be supplied for treating gasoline of high mercaptan content, there is a tendency toward excessive plumbite formation and also excessive sulfur formation. In such cases, a controlled quantity of aqueous sodium sulfide is simultaneously added to reconvert the extra plumbite back to lead sulfide. The presence of the remaining extra sulfur is not desirable, and therefore it is advantageous to control the air oxidation carefully. The lead sulfide is essentially a catalyst, as only oxygen is consumed in the process; there is, however, a certain loss of alkali to sodium sulfate and thiosulfate.

24.2.4.2 Copper Sweetening Process

The oxidizing power of cupric (Cu^{2+}) salts is also used to convert mercaptans directly into disulfides; free sulfur is not employed, and polysulfides are not obtained. The process employs cupric chloride in the presence of strong salt solutions, which are generally made up by dissolving copper sulfate in an aqueous solution of sodium chloride.

$$4RSH + 2CuCl_2 \rightarrow R_2S_2 + 2CuSR + 4HCl$$

$$2CuSR + 2CuCl_2 \rightarrow R2S_2 + 4CuCl$$

$$4CuCl + 4HCl + O_2 \rightarrow 4CuCl_2 + H_2O$$

The cuprous chloride (CuCl) is soluble in the salt solution, and there is no precipitation. Under operating conditions, a certain amount of copper is retained by the sweetened petroleum fraction, probably as cuprous mercaptides or cuprous chloride-olefin addition products, but these can be removed by washing the material with aqueous sodium sulfide. Air blowing the cuprous chloride solution, after or during the sweetening operation, regenerates the cupric chloride. The copper chloride solution may be employed as such, or the sour fraction may be percolated through a porous mass saturated with the treating agent. Alternatively, the gasoline may be mixed with a solid carrier for the reagent, dispersed as a slurry.

Three methods of mechanical application of the copper chloride are used. If air will not cause the petroleum fraction to change color or form gum, a fixed-bed process may be used in which the sour material is passed through beds of an adsorbent that have been impregnated with cupric chloride. Air continuously regenerates the cupric chloride almost simultaneously with the sweetening reaction.

In the solution process, a solution of cupric chloride is continuously mixed in a centrifugal pump with the sour fraction. The mixture then enters a settling tank where the spent treatment solution separates from the petroleum liquid, and the treatment solution is withdrawn to a tank where blowing with air regenerates cupric chloride.

The slurry process makes use of clay or a similar material impregnated with cupric chloride. The clay is mixed with a small amount of, say, naphtha to form a slurry, which is pumped into the sour naphtha stream. Air or oxygen gas is added with the sour stream and continuously regenerates the cupric chloride. The treated material and clay slurry flow into a settling tank and separate, and then the clay slurry is recycled.

24.2.4.3 Doctor Process

The doctor process is a method of treating sour distillates (Brown, 1992) and consists of agitating the distillate with alkaline sodium plumbite (doctor solution) in the presence of a small amount of free sulfur. A black precipitate of lead sulfide is formed, and the material, which has improved odor, has been rendered sweet. The essential reactions of the doctor processes are as follows.

In practice, sour distillates are usually given an alkali wash before the doctor treatment to remove traces of hydrogen sulfide and some of the lower molecular weight mercaptans; this process has a marked effect in reducing the plumbite requirement. Slightly more sulfur than the theoretical amount is required as a result of the formation of complex lead intermediates. In the presence of lead mercaptides the extra sulfur acts to form alkyl polysulfides, which are chemically analogous to peroxide.

The precipitating effect is evidently a result of the presence of these polysulfides or possibly of sodium sulfide formed between mercaptans, sulfur, and the alkaline solution. The doctor solution leaving the reactor consists essentially of a mixture of lead sulfide in free alkali, containing emulsified hydrocarbons, and this spent solution is pumped to steam-heated vessels, where it is air blown for regeneration. Considerable amounts of sodium thiosulfate are also formed and the thiosulfate, in turn, may react with the alkali present to form sodium sulfite (Na_2SO_3) and sodium sulfide (Na_2S). The loss of lead is very low, and the main items of consumption are alkali and sulfur.

24.2.4.4 Hypochlorite Sweetening Process

This process employs sodium or calcium hypochlorite [$NaOCl$ or $Ca(OCl)_2$]. The principal reaction produces disulfides (R–S S–R') with some formation of sulfoxides ($R_2S=O$) and sulfonic acids (R–SO_2H). When hydrogen sulfide is present, an alkaline wash prevents the formation of elemental sulfur. An alkaline after-wash is frequently necessary to remove undesirable chlorinated products.

24.2.4.5 Inhibitor Sweetening Process

This is a continuous process that uses a phenylenediamine-type inhibitor, air, and caustic to sweeten low-mercaptan-content gasoline. The inhibitor and air are injected between the caustic washing stages, and the mercaptan disappearance may be attributed to reaction with the caustic and then to oxidation during both washing and storage. In the absence of caustic, excessive peroxide formation occurs, which leads to gasoline deterioration.

24.2.4.6 Merox Process

The Merox process is a combination process for mercaptan extraction and sweetening of gasoline or lower-boiling materials (Holbrook, 1997). The catalyst is a cobalt salt, which is insoluble in the oil and may be used in caustic solution or on a suitable solid support. Thus, gasoline is washed with alkali and contact by the catalyst and caustic in the extractor, air is then injected, and the treated product is stored in the regeneration step. Caustic is taken from the extractor and mixed with air in the oxidizer, after which disulfides and excess air are separated from the reagent in the disulfide separator. The regenerated caustic solution is recirculated to the top of the extractor.

24.2.5 SOLVENT PROCESSES

Solvent refining processes are of a physical nature rather than a chemical nature. The desirable constituents, as well as the undesirable constituents, of the mixture can be recovered unchanged and in the original state. In addition, the processes that use solvent as a means of refining are extremely versatile, insofar as both low-boiling and high-boiling fractions can be used as feedstocks. In general, the solvent processes can be classified as **deasphalting**, *solvent refining*, and *dewaxing*.

Deasphalting is usually applied to such materials as gas oils, lubricating stocks, residua, and even heavy crude oil before further treatment. The solvent separates the feedstock into constituents that are usually based on molecular size and polarity (Chapter 8) (Ditman, 1973; Mitchell and Speight, 1973; Gary and Handwerk, 2001; Long and Speight, 1998).

Solvent refining is usually applied to materials boiling in the lubricating oil range, and it is a means of selectively removing (by extraction) aromatics, naphthenes, or other constituents that adversely affect physical parameters, such as the viscosity index (Chapter 9). The feedstock to these solvent-refining processes has usually been through a **solvent deasphalting** treatment, and if desired, the effluent may be subjected to a dewaxing operation.

Dewaxing processes are applied to refined feedstocks before final treatment with clay. Wax in a finished lubricating oil adversely affects the properties of the oil, for example, by increasing the pour point and reducing the fluidity at low temperatures because of crystallization of the wax from the lubricating oil. Dewaxing operations are often combined with wax production when low-pour-point lubricating oil and high-melting-point wax can be produced simultaneously.

Feedstocks for cracking purposes are now usually the residuum or heavy distillate from a distillation sequence. Nevertheless, as the occasion demands, whole crude oil may also serve as the cracking feedstock, and when this is the case it is assumed that the crude has been subjected to one of the pretreatment processes (desalting) (Chapter 14 and Chapter 16) (Burris, 1992).

In addition, the residua produced as the end products of distillation processes (Chapter 16), and even some of the high-boiling gas oils, often contain substantial amounts of asphaltic materials, which preclude use of the residuum as fuel oils or lubricating stocks. However, subjecting residua directly to thermal processes is economically advantageous. On the one hand, the end result is the production of lower-boiling salable materials; on the other hand, the asphaltic materials in the residua are regarded as the coke-forming constituents and may even promote coke formation from other components of the residua.

It is therefore advantageous that these coke-forming constituents be kept at a minimum. This is especially true when the residuum is to be used as a feedstock for catalytic cracking, as the capacity of the unit is often limited by the rate at which carbon is burned from the catalyst in the regenerating chamber.

Furthermore, to avoid catalyst poisoning or reduction in catalyst activity, it is essential that as much of the nitrogen and metals (such as vanadium and nickel) as possible should be removed from the feedstock. It was noted earlier that the majority of the heteroatoms (nitrogen, oxygen, and sulfur) and the metals are contained in, or associated with, the asphaltic fraction. It is therefore often deemed necessary that this fraction be removed from cracking feedstocks (Hydrocarbon Processing, 1996).

The coke-forming tendencies of middle distillates, gas oils, lubricating oils, and residua can be reduced significantly through the precipitating, or extracting, action of solvents on the asphaltic and resinous materials that are present either in solution or in colloidal form. The solvents capable of being used in deasphalting operations may be divided into two major groups. The first group contains the low-molecular-weight liquid or liquefied hydrocarbons, of which propane is the most widely used. The second group contains such liquids as the alcohols and ethers, but in general these materials are not used to nearly the same extent as the hydrocarbons.

24.2.5.1 Deasphalting

Solvent deasphalting (Chapter 19) (Van Tine and Feintuch, 1997) provides an extension to vacuum distillation and is a later addition to the petroleum refinery. Before its use, many processes capable of removing asphaltic materials from feedstocks were employed in the form of distillation (atmospheric and vacuum), as well as clay and sulfuric acid treatment. In its present form, solvent deasphalting (when employed for feedstock preparation to catalytic cracking units) may be considered a competitor to vacuum distillation, visbreaking, and coking.

The highest-boiling lubricating oil fractions are valuable for the lubricating of engine parts, such as steam cylinders, which are subjected to high temperatures. These fractions and asphalt make up crude residua, and the most widely used method of separating high-boiling lubricating oil fractions from asphalt is vacuum distillation (Chapter 16). The highest-boiling lubricating oil fractions, however, cannot be separated from asphalt by distillation, because cracking temperatures must be used even with the highest possible vacuum.

One common method of obtaining the highest-boiling lubricating oil fractions is to treat asphaltic crude residues with sulfuric acid, which removes the asphaltic materials; another method of separating asphalt from residual lubricating oil fractions is deasphalting, also often referred to as **decarbonizing**.

Propane deasphalting focuses on the ability of liquefied petroleum gases to precipitate asphaltic materials from heavy feedstocks and residua whereas the lubricating oil components remain in solution. All liquefied hydrocarbon gases have this property to a marked extent, but propane is used to deasphalt residua and lubricating oils because of its relatively low cost and its ease of separation from lubricating oils. Propane can also be used to dewax lubricating oils, in which case the propane is the solvent for the oil and, by its evaporation, is the source of the needed refrigeration.

In propane deasphalting, the heavy feedstock and three to ten times its volume of liquefied propane are pumped together through a mixing device and then into a settling tank. The temperature is maintained between 27°C (80°F) and 71°C (160°F): the higher the temperature, the greater the tendency of asphaltic materials to separate. The propane is maintained in the liquid state by a pressure of about 200 psi. The asphalt settles in the settling tank and is pumped to an asphalt recovery unit, where propane is separated from the asphalt. The upper layer in the settling tank consists of lubricating oil dissolved in a large amount of propane, and while it is still dissolved in the propane, more impurities may be removed by sulfuric acid, or it may even be dewaxed. In the latter case, evaporation of some of the propane cools the mixture to a sufficiently low temperature ($-40°C$, $-40°F$) that wax crystals form. Filtration at low temperature separates wax from the liquid, and propane is separated from the oil–propane mixture in evaporators heated by steam. The last trace of propane is removed from the oil by steam in a stripper tower, and the propane is condensed and reused.

In place of a mixer and settling tank, most modern deasphalting plants use a countercurrent tower. Liquefied propane is pumped into the bottom of the tower to form a continuous phase, and lubricating stock or reduced crude (crude residuum) is pumped into the tower near the top. As the reduced crude descends, the oil components are dissolved and carried with the propane out the top of the tower and the asphaltic components are pumped from the bottom of the tower.

Propane fractionation is a continuous process for the segregation of vacuum distillation residua into two or more grades of lubricating oils. It is in fact an extension of the propane deasphalting process. The residuum is charged to the primary tower where it is contacted countercurrently with propane. The bottoms from the primary tower are then countercurrently extracted with propane in a secondary tower for production of a second propane soluble stream (bright stock) but at lower temperatures and pressures.

Propane decarbonizing is a solvent process usually used to recover catalytic cracking feed from residua. As butane, alone or with propane, can also be used as the solvent, the process is often called solvent decarbonizing. The process and equipment are essentially the same as those employed for the deasphalting of lubricating oils.

Apart from the additional use of butane, there is so little difference between the processes that the terms *deasphalting* and *decarbonizing* are often used interchangeably (Hydrocarbon Processing, 1978).

The Duosol process uses two solvents: one is propane and the other is a mixture of cresylic acid and phenol called selecto. A series of treatment compartments is used; propane is pumped into one end of the series and selecto into the other and the solvents pass each other countercurrently. Raw lubricating oil is pumped into the third compartment from the propane end. The propane dissolves the paraffinic oil and wax components, and the selecto dissolves naphthenic, unsaturated, asphaltic, and resinous components. The raw lubricating oil is thus divided into two parts that move in opposite directions through the treatment system, each carried by its particular solvent.

After leaving the treatment system each solvent stream passes through separate solvent recovery systems. Propane is separated from the treated oil in evaporators, and *selecto* is separated from the naphthenic and asphaltic extract by fractional distillation. The liquid products may be dewaxed, or if this is not required, the finishing step may be clay treatment.

24.2.5.2 Solvent Refining

Solvent refining (solvent treatment) is a widely used method of refining lubricating oils, as well as a host of other refinery stocks. The solvent processes yield products that meet the desired specifications by removing undesirable constituents (such as aromatics, naphthenes, and unsaturated compounds) from the charge material. There are, however, solvent refining processes in which the desirable constituents are aromatics and are extracted by the solvent from the petroleum fraction.

Nevertheless, the original object of solvent extraction, or solvent treatment, was to remove aromatic compounds from feedstocks, such as lubricating oils. Thus, a suitable solvent can convert inferior raw lubricating oil stocks into oils with as high a quality as desired and is higher than could ever be obtained with, say, sulfuric acid. In contrast to sulfuric acid treatment (which depends on chemical reaction for a large part of its effect), solvent treatment is a physical process in which undesirable olefin compounds, asphaltic compounds, aromatic compounds, and sulfur compounds are selectively dissolved in the solvent and removed with the solvent. After separation of the solvent by distillation from both the unwanted materials and the treated oil, the solvent can be reused.

The most widely used extraction solvents are phenol, furfural, and cresylic acid. The last is used with propane in the Duosol process. Other solvents less frequently used are liquid sulfur dioxide, nitrobenzene, and **chlorex** (2,2-dichloroethyl ether). All lubricating oil solvent extraction processes operate on the principle of mixing the oil with the solvent and then allowing the solvent to settle from the treated oil. The solvent carries the unwanted materials with it. The nature of the raw lubricating oil stock is important in determining the type of process to use and the extent to which treatment should be carried out. The greater is the extent of treatment, the smaller the yield of treated oil but the higher the quality.

It is possible to make oil of any desired quality from any raw lubricating oil stock. In fact, it is possible to over-treat lubricating oils, causing a loss in lubricating properties. White oils, for example, are lubricating oils that have been deliberately over-treated with very strong sulfuric acid to remove all traces of color. Such oils are so over-refined that they are not used as lubricating oils; their lubricating properties are inferior to oils refined specifically for use as lubricants. Thus, solvent extraction is a complex operation requiring the selective removal of only those components that reduce the lubricating qualities of the oil being treated.

The phenol treatment process is the most widely used. Phenol, also known as carbolic acid, is a poisonous solid that can cause serious flesh burns. It melts at 41°C (106°F) and boils at 183°C (361°F). In the phenol treatment process, phenol is used in the liquid state by maintaining the temperature at over 35°C (100°F).

In the process, raw lubricating oil is pumped into the bottom of the tower and phenol into the top; the phenol descends through the oil and is mixed with the oil by means of baffles or other mechanical mixing devices. The treated oil or raffinate leaves the top of the tower. The phenol that collects at the bottom of the tower contains the extract (aromatic, unsaturated, asphaltic, and sulfur compounds extracted from the raw oil) and is known as spent phenol.

The raffinate is heated to 260°C (500°F) and pumped to a pair of fractional vacuum distillation towers operated in series. Here the phenol in the raffinate is removed overhead. The bottom product from the second tower is the finished product.

The phenol is recovered from the spent phenol in a phenol recovery unit, which consists of another pair of fractional distillation towers and a stripper tower. The spent phenol is heated and passed into the primary tower, where process water is removed as an overhead product and some of the phenol recovered as a side-stream product. The bottom product from the primary tower is heated to about 240°C (460°F) and pumped into the secondary tower, where the majority of the remaining phenol is recovered as an overhead product. The extract (the bottom product) is heated to 345°C (650°F) and pumped into a vacuum stripper tower, where steam removes the phenol from the extract. The phenol obtained from the various distillation units is pumped to a heated storage tank for reuse. The extract may be used as a special-purpose product when high-molecular-weight aromatic hydrocarbons are needed; it may be added to heavy fuel oils, or it may be used as a cracking stock.

Liquid sulfur dioxide was first used commercially as an extraction solvent in 1909 to remove aromatic hydrocarbons from kerosene (**Edeleanu process**), when it was noted that aromatic and unsaturated hydrocarbons would dissolve in liquid sulfur dioxide but paraffinic and naphthenic hydrocarbons would not. The process was widely used for light distillates, such as kerosene, but was not used for heavier oils, such as lubricating oils, until about 1930.

The process is carried out in much the same way as phenol treatment, except that treatment takes place at about 0°C (32°F) and under sufficient pressure to maintain the sulfur dioxide as a liquid. The feedstock is contacted with liquid sulfur dioxide and the aromatic portion is concentrated in the extract phase, which is in turn contacted with a wash oil (such as kerosene) to remove the lower-boiling nonaromatic compounds. The streams are then stripped and fractionated to recover sulfur dioxide and the wash oil for recycling to the extraction tower.

Sulfur dioxide alone does not effectively dissolve the high-boiling aromatic hydrocarbons in lubricating oils and therefore is used in conjunction with benzene for treating lubricating oils. Mixing benzene with sulfur dioxide increases the solvent capacity but at the same time retains the selectivity for aromatic and nonparaffinic hydrocarbons. The percentage of benzene in the solvent mixture makes it possible to select the most advantageous treatment conditions for any particular feedstock to produce a product with the desired specifications.

Furfural is a heavy, straw-colored liquid that boils at 162°C (323°F). The process is carried out in the same manner as phenol treatment, but the raffinate or treated oil has so little furfural dissolved in it that distillation is not used to remove it; steam stripping is all that is required. The treatment (Figure 24.1) is conducted in counterflow towers, or multistage units, which normally operate at a temperature of 40°C to 120°C (100°F to 250°F). A high-temperature gradient in the treatment section permits a high yield of the refined oil.

Hydrogen fluoride treatment is a liquid–liquid extraction process removing sulfur and potential coke-forming materials from naphtha, middle distillates, and gas oil. The feedstock is contacted countercurrently with liquid hydrofluoric acid in an extraction tower, after which the overhead product is sent to a tower for removal of hydrogen fluoride. The solvent is recovered from the extract by use of evaporators and a stripper. The process is relatively insensitive to variations in temperature and pressure, but normally temperatures of 50°C (120°F) and pressures below 100 psi are employed.

The Udex extraction process is a liquid–liquid recovery method for the selective extraction of aromatic compounds from hydrocarbon mixtures by the use of a diethylene glycol (90% to 92%), water (8% to 10%), and solvent system. The feedstock, which may be pretreated to eliminate high conjugated dienes or alkenyl aromatics, rises countercurrent to the descending

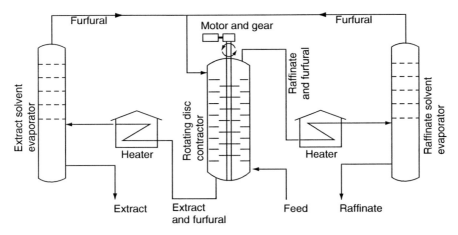

FIGURE 24.1 Aromatics extraction using furfural.

solvent. The stripper recovers the solvent for recycling; the overhead from the extractor is treated with clay and fractionated to separate the aromatics. The temperatures employed in the process are usually fairly low (120°C, 250°F); the pressures are just high enough to maintain liquid-phase conditions.

Sulfolane has been used primarily for the extraction of aromatics, such as benzene, toluene, and xylene(s). The chemicals usually arise from catalytic reforming (Chapter 23) of various feedstocks. Thus the feedstock is contacted with the solvent in an extractor, and the mixture is conveyed to an extractive distillation column. The bottom product from this column is a mixture of solvent and aromatic compounds. Vacuum and steam stripping remove the final traces of hydrocarbons in the bottom part of the recovery column.

24.2.5.3 Dewaxing

Dewaxing processes are designed to remove wax from lubricating oils to give the product good fluidity characteristics at low temperatures (e.g., low pour points). The mechanism of **solvent dewaxing** involves either the separation of wax as a solid that crystallizes from the oil solution at low temperature or the separation of wax as a liquid that is extracted at temperatures above the melting point of the wax through preferential selectivity of the solvent. However, the former mechanism is the usual basis for commercial dewaxing processes.

In the 1930s two types of stocks, naphthenic and paraffinic, were used to make motor oils. Both types were solvent extracted to improve their quality, but in the high-temperature conditions encountered in service, the naphthenic type could not stand up as well as the paraffinic type. Nevertheless, the naphthenic type was the preferred oil, particularly in cold weather, because of its fluidity at low temperatures. Before 1938 the highest quality lubricating oils were of the naphthenic type and were phenol treated to pour points of −40°C to −7°C (−40°F to 20°F), depending on the viscosity of the oil. Paraffinic oils were also available and could be phenol treated to a higher quality oil, but their wax content was so high that the oils were solid at room temperature.

The lowest viscosity paraffinic oils were dewaxed by the cold press method to produce oils with a pour point of 2°C (35°F). The light paraffin distillate oils contained a paraffin wax that crystallized into large crystals when chilled and could thus readily be separated from the oil by the cold press filtration method. The more viscous paraffinic oils (intermediate and heavy paraffin distillates) contained amorphous or microcrystalline waxes, which formed small crystals that plugged the filter cloths in the cold press and prevented filtration. As the wax could not be removed from intermediate and heavy paraffin distillates, the high-quality, high-viscosity lubricating oils in them could not be used except as cracking stock.

Methods were therefore developed to dewax these high-viscosity paraffinic oils. The methods were essentially alike in that the waxy oil was dissolved in a solvent that would keep the oil in solution; the wax separated as crystals when the temperature was lowered. The processes differed chiefly in the use of the solvent. Commercially used solvents were naphtha, propane, sulfur dioxide, acetone–benzene, trichloroethylene, ethylene dichloride–benzene (Barisol), methyl ethyl ketone–benzene (**benzol**), methyl-*n*-butyl ketone, and methyl-*n*-propyl ketone.

In the first solvent dewaxing process (developed in 1924), the waxy oil was mixed with naphtha and filter aid (Fuller's earth or diatomaceous earth). The mixture was chilled and filtered, and the filter aid assisted in building a wax cake on the filter cloth. This process is now obsolete, and most of the modern dewaxing processes use a mixture of methyl ethyl ketone and benzene. Other ketones may be substituted for dewaxing, but regardless of what ketone is used, the process is generally known as ketone dewaxing.

The process involves mixing the feedstock with one to four times its volume of the ketone (Scholten, 1992). The mixture is then heated until the oil is in solution. The solution is then chilled at a slow, controlled rate in double-pipe, scraped-surface exchangers. Cold solvent, such as filtrate from the filters, passes through the 2 in. annular space between the inner and outer pipes and chills the waxy oil solution flowing through the inner 6 in. pipe.

To prevent wax from depositing on the walls of the inner pipe, blades or scrapers extending the length of the pipe and fastened to a central rotating shaft scrape off the wax. Slow chilling reduces the temperature of the waxy oil solution to 2°C (35°F), and then faster chilling reduces the temperature to the approximate pour point required in the dewaxed oil. The waxy mixture is pumped to a filter case into which the bottom half of the drum of a rotary vacuum filter dips. The drum (8 ft in diameter, 14 ft long), covered with filter cloth, rotates continuously in the filter case. Vacuum within the drum sucks the solvent and the oil dissolved in the solvent through the filter cloth and into the drum. Wax crystals collect on the outside of the drum to form a wax cake, and as the drum rotates, the cake is brought above the surface of the liquid in the filter case and under sprays of ketone that wash oil out of the cake and into the drum. A knife edge scrapes off the wax, and the cake falls into the conveyor and is moved from the filter by the rotating scroll.

The recovered wax is actually a mixture of wax crystals with a little ketone and oil, and the filtrate consists of the dewaxed oil dissolved in a large amount of ketone. Ketone is removed from both by distillation, but before the wax is distilled, it is deoiled, mixed with more cold ketone, and pumped to a pair of rotary filters in series, where further washing with cold ketone produces a wax cake that contains very little oil. The deoiled wax is melted in heat exchangers and pumped to a distillation tower operated under vacuum, where a large part of the ketone is evaporated or flashed from the wax. The rest of the ketone is removed by heating the wax and passing it into a fractional distillation tower operated at atmospheric pressure and then into a stripper where steam removes the last traces of ketone.

An almost identical system of distillation is used to separate the filtrate into dewaxed oil and ketone. The ketone from both the filtrate and wax slurry is reused. Clay treatment or hydrotreating finishes the dewaxed oil as previously described. The wax (**slack wax**), even though it contains essentially no oil as compared to 50% in the slack wax obtained by cold pressing, is the raw material for either sweating or wax recrystallization, which subdivides the wax into a number of wax fractions with different melting points (Chapter 19).

Solvent dewaxing can be applied to light, intermediate, and heavy lubricating oil distillates, but each distillate produces a different kind of wax. Each of these waxes is actually a mixture of a number of waxes. For example, the wax obtained from light paraffin distillate consists of a series of paraffin waxes that have melting points in the range of 30°C to 70°C (90°F to 160°F), which are characterized by a tendency to harden into large crystals. However, heavy paraffin distillate yields a wax composed of a series of waxes with melting points in the range of 60°C to 90°C (140°F to 200°F), which harden into small crystals from which they derive the name of microcrystalline wax or microwax. On the other hand, intermediate paraffin distillates contain paraffin waxes and waxes intermediate in properties between paraffin and microwax.

Thus, the solvent dewaxing process produces three different slack waxes (also known as crude waxes or raw waxes) depending on whether light, intermediate, or heavy paraffin distillate is processed. The slack wax from heavy paraffin distillate may be sold as dark raw wax, the wax from intermediate paraffin distillate as pale raw wax. The latter is treated with lye and clay to remove odor and improve color.

There are several processes in use for solvent dewaxing, but all have the same general steps, which are (1) contacting the feedstock with the solvent, (2) precipitating the wax from the mixture by chilling, and (3) recovering the solvent from the wax and dewaxed oil for recycling. The processes use benzene–acetone (solvent dewaxing), propane (**propane dewaxing**), trichloroethylene (**separator–Nobel dewaxing**), ethylene dichloride–benzene (Barisol dewaxing), and urea (**urea dewaxing**), as well as liquid sulfur dioxide–benzene mixtures.

Urea dewaxing (Chapter 19) (Scholten, 1992) is worthy of further mention insofar as the process is highly selective and, in contrast to the other dewaxing techniques, can be achieved without the use of refrigeration. However, the process cannot compete economically with the solvent dewaxing processes for treatment of the heavier lubricating oils. But when it is applied to the lighter materials that already may have been subjected to a solvent dewaxing operation, products are obtained that may be particularly useful as refrigerator oils, transformer oils, and the like.

The process description is essentially the same as that used for solvent dewaxing, with the omission of the chilling stage and the insertion of a contactor where the feedstock and the urea (with a solvent) are thoroughly mixed before filtration. The solvents are recovered from the dewaxed oil by evaporation, and the urea complex is decomposed in a urea recovery system.

Residual lubricating oils, such as cylinder oils and bright stocks, are made from paraffinic or mixed-base reduced crude oils and contain waxes of the microcrystalline type. Removal of these waxes from reduced crude produces **petrolatum**, a grease-like material that is known in a refined form as **petroleum jelly**. This material can be separated from reduced crude in several ways. The original method was cold settling, whereby reduced crude was dissolved in a suitable amount of naphtha and allowed to stand over winter until the microwax settled out. This method is still used, but the reduced crude naphtha solution is held in refrigerated tanks until the petrolatum settles out. The supernatant naphtha-oil layer is pumped to a still where the naphtha is removed, leaving cylinder stock that can be further treated to produce bright stock. The petrolatum layer is also distilled to remove naphtha and may be clay treated or acid-and-clay treated to improve the color.

Another method of separating petrolatum from reduced crude is centrifuge dewaxing. In this process the reduced crude is dissolved in naphtha and chilled to $-18°C$ ($0°F$) or lower, which causes the wax to separate. The mixture is then fed to a battery of centrifuges where the wax is separated from the liquid. However, the centrifuge method has now been largely replaced by solvent dewaxing methods and by more modern methods of wax removal.

There is also later generation dewaxing processes that are being brought on-stream in various refineries (Hargrove, 1992; Genis, 1997). For example, British Petroleum has developed a hydrocatalytic dewaxing process that is reputed to overcome some of the disadvantages of the solvent dewaxing processes. The operating costs for the solvent dewaxing processes are reputed to be high because of the solvent cooling that is necessary and because the pour point that can be achieved is limited by the high cost of refrigerating to the very low temperatures.

Catalytic dewaxing is a hydrocracking process and is therefore operated at elevated temperatures ($280°C$ to $400°C$, $550°F$ to $750°F$) and pressures (300 to 1500 psi). However, the conditions for a particular dewaxing operation depend upon the nature of the feedstock and the product pour point required. The catalyst employed for the process is a mordenite-type catalyst that has the correct pore structure to be selective for n-paraffin cracking. Platinum on the catalyst serves to hydrogenate the reactive intermediates so that further paraffin degradation is limited to the initial thermal reactions. The process has been employed to successfully dewax a wide range of naphthenic feedstocks (Hargrove et al., 1979), but it may not be suitable to replace solvent dewaxing in all cases. The process has the flexibility to fit into normal refinery operations and can be adapted for prolonged periods on-stream.

Another catalytic dewaxing process has been developed by Mobil and also involves selective cracking of n-paraffins and those paraffins that may have minor branching in the chain (Smith et al., 1980). In the process, the proprietary catalyst can be reactivated to fresh activity by relatively mild nonoxidative treatment. Of course, the time allowed between reactivation is a function of the feedstock but after numerous reactivations it is possible that there will be coke buildup on the catalyst.

The process can be used to dewax a full range of lubricating base stocks and, as such, has the potential to completely replace solvent dewaxing or can even be used in combination with solvent dewaxing. This latter option, of course, serves to debottleneck existing solvent dewaxing facilities.

Both the catalytic dewaxing processes have the potential to change the conventional thoughts about dewaxing insofar as they are not solvent processes and may be looked upon (more correctly) as thermal processes rather than treatment processes. However, both provide viable alternatives to the solvent processes and offer a further advance in the science and technology of refinery operations.

REFERENCES

Bland, W.F. and Davidson, R.L. 1967. *Petroleum Processing Handbook*. McGraw-Hill, New York.

Brown, K.M. 1992. *Petroleum Processing Handbook*. J.J. McKetta, ed. Marcel Dekker Inc., New York. p. 736.

Burris, D.R. 1992. *Petroleum Processing Handbook*. J.J. McKetta, ed. Marcel Dekker Inc., New York. p. 666.

Ditman, J.G. 1973. *Hydrocarbon Process*. 52(5): 110.

Gary, J.H. and Handwerk, G.E. 2001. *Petroleum Refining: Technology and Economics*, 4th edn. Marcel Dekker Inc., New York.

Genis, O. 1997. *Handbook of Petroleum Refining Processes*. R.A. Meyers, ed. McGraw-Hill, New York. Chapter 8.5.

Hargrove, J.D. 1992. *Petroleum Processing Handbook*. J.J. McKetta, ed. Marcel Dekker Inc., New York. p. 558.

Hargrove, J.D., Elkes, G.J., and Richardson, A.H. 1979. *Oil Gas J.* 77(3): 103.

Hobson, G.D. and Pohl, W. 1973. *Modern Petroleum Technology*. Applied Science Publishers, Barking, Essex, England.

Holbrook, D.L. 1997. *Handbook of Petroleum Refining Processes*, 2nd Edition. R.A. Meyers, ed. McGraw-Hill, New York. Chapter 11.3.

Hydrocarbon Processing. 1996. November.

Kalichevsky, V.A. and Kobe, K.A. 1956. *Refining Petroleum with Chemicals*. Elsevier, New York.

Long, R.B. and Speight, J.G. 1998. *Petroleum Chemistry and Refining*. J.G. Speight, ed. Taylor & Francis Publishers, Washington, DC. Chapter 1.

Mitchell, D.L. and Speight, J.G. 1973. *Fuel* 52: 149.

Scholten, G.G. 1992. *Petroleum Processing Handbook*. J.J. McKetta, ed. Marcel Dekker Inc., New York. p. 565.

Smith, K.W., Starr, W.C., and Chen, N.Y. 1980. *Oil Gas J.* 78(21): 75.

Speight, J.G. 2000. *The Desulfurization of Heavy Oils and Residua*, 2nd Edition. Marcel Dekker Inc., New York.

Van Tine, F.M. and Feintuch, H.M. 1997. *Handbook of Petroleum Refining Processes*, 2nd edn. R.A. Meyers, ed. McGraw-Hill, New York. Chapter 10.2.

25 Gas Processing

25.1 INTRODUCTION

Gas processing (Mokhatab et al., 2006) consists of separating all of the various hydrocarbons and fluids from pure natural gas (Figure 25.1). Major transportation pipelines usually impose restrictions on the makeup of the natural gas that is allowed into the pipeline. That means that before the natural gas can be transported it must be purified. While the ethane, propane, butane, and pentanes must be removed from natural gas, this does not mean that they are all waste products. Gas processing is necessary to ensure that the natural gas intended for use is as clean and pure as possible, making it the clean burning and environmentally sound energy choice. Thus, natural gas, as it is used by consumers, is much different from the natural gas that is brought from underground up to the wellhead. Although the processing of natural gas is in many respects less complicated than the processing and refining of crude oil, it is equally as necessary before its use by end users. The natural gas used by consumers is composed almost entirely of methane. However, natural gas found at the wellhead, although still composed primarily of methane, is by no means as pure.

Raw natural gas comes from three types of wells: oil wells, gas wells, and condensate wells. Natural gas that comes from oil wells is typically termed *associated gas*. This gas can exist separate from oil in the formation (free gas), or dissolved in the crude oil (dissolved gas). Natural gas from gas and condensate wells, in which there is little or no crude oil, is termed nonassociated gas. Gas wells typically produce raw natural gas by itself, while condensate wells produce free natural gas along with a semi-liquid hydrocarbon condensate. Whatever the source of natural gas, once separated from crude oil (if present) it commonly exists in mixtures with other hydrocarbons; principally ethane, propane, butane, and pentanes. In addition, raw natural gas contains water vapor, hydrogen sulfide (H_2S), carbon dioxide, helium, nitrogen, and other compounds. In fact, associated hydrocarbons, known as **natural gas liquids** (NGLs) can be very valuable by-products of natural gas processing. NGLs include ethane, propane, butane, iso-butane, and natural gasoline that are sold separately and have a variety of different uses, including enhancing oil recovery in oil wells, providing raw materials for oil refineries or petrochemical plants, and as sources of energy.

Petroleum refining as it is currently known will continue at least for the next three decades. Various political differences have caused fluctuations in petroleum imports, and imports have continued to increase over the past several years (Chapter 3). It is also predictable that use of petroleum for the transportation sector will increase as increases in travel offset increased efficiency. As a consequence of this increase in use, petroleum will be the largest single source of carbon emissions from fuel. Acid gases corrode refining equipment, harm catalysts, pollute the atmosphere, and prevent the use of hydrocarbon components in petrochemical manufacture. When the amount of hydrogen sulfide is high, it may be removed from a gas stream and converted to sulfur or sulfuric acid. Some natural gases contain sufficient carbon dioxide to warrant recovery as dry ice (Bartoo, 1985).

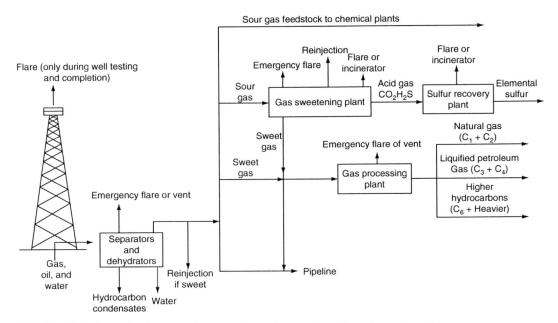

FIGURE 25.1 General scheme to show the flow of natural gas from the well to the consumer.

25.1.1 GAS STREAMS FROM CRUDE OIL

To process and transport associated dissolved natural gas, it must be separated from the oil in which it is dissolved. This separation of natural gas from oil is most often done using equipment installed at or near the wellhead.

The actual process used to separate oil from natural gas, as well as the equipment that is used, can vary widely. Although dry pipeline quality natural gas is virtually identical across different geographic areas, raw natural gas from different regions will vary in composition (Table 25.1) (Chapter 1) and therefore separation requirements may emphasize or de-emphasize the optional separation processes. In many instances, natural gas is dissolved

TABLE 25.1
Constituents of Natural Gas

Name	Formula	Vol.%
Methane	CH_4	>85
Ethane	C_2H_6	3–8
Propane	C_3H_8	1–5
Butane	C_4H_{10}	1–2
Pentane[a]	$C_5H_{12}{}^+$	1–5
Carbon dioxide	CO_2	1–2
Hydrogen sulfide	H_2S	1–2
Nitrogen	N_2	1–5
Helium	He	<0.5

[a] Pentane and higher molecular weight hydrocarbons, including benzene and toluene.

in oil underground primarily due to the formation pressure. When this natural gas and oil is produced, it is possible that it will separate on its own, simply due to decreased pressure; much like opening a can of soda pop allows the release of dissolved carbon dioxide. In these cases, separation of oil and gas is relatively easy, and the two hydrocarbons are sent in separate ways for further processing. The most basic type of separator is known as a conventional separator. It consists of a simple closed tank, where the force of gravity serves to separate the heavier liquids like oil, and the lighter gases, like natural gas.

In certain instances, however, specialized equipment is necessary to separate oil and natural gas. An example of this type of equipment is the low-temperature separator. This is most often used for wells producing high pressure gas along with light crude oil or condensate. These separators use pressure differentials to cool the wet natural gas and separate the oil and condensate. Wet gas enters the separator, being cooled slightly by a heat exchanger. The gas then travels through a high pressure liquid "knockout," which serves to remove any liquid into a low-temperature separator. The gas then flows into this low-temperature separator through a choke mechanism, which expands the gas as it enters the separator. This rapid expansion of the gas allows for the lowering of temperature in the separator. After liquid removal, the dry gas then travels back through the heat exchanger and is warmed by the incoming wet gas. By varying the pressure of the gas in various sections of the separator, it is possible to vary the temperature, which causes oil and some water to be condensed out of the wet gas stream. This basic pressure–temperature relationship can work in reverse as well, to extract gas from a liquid oil stream.

On the other hand, petroleum refining produces gas streams that contain substantial amounts of acid gases such as hydrogen sulfide and carbon dioxide. These gas streams are produced during initial distillation of the crude oil and during various conversion processes. Of particular interest is the hydrogen sulfide (H_2S) that arises from the hydrodesulfurization of feedstocks that contain organic sulfur:

$$[S]_{feedstock} + H_2 = H_2S + hydrocarbons$$

The terms **refinery gas** and *process gas* are also often used to include all of the gaseous products and by-products that emanate from a variety of refinery processes. There are also components of the gaseous products that must be removed prior to release of the gases to the atmosphere or prior to the use of gas in another part of the refinery, that is, as a fuel gas or as a process feedstock.

Petroleum refining involves, with the exception of heavy crude oil, primary distillation (Chapter 13) that results in separation into fractions differing in carbon number, volatility, specific gravity, and other characteristics. The most volatile fraction, that contains most of the gases that are generally dissolved in the crude, is referred to as **pipestill gas** or **pipestill light ends** and consists essentially of hydrocarbon gases ranging from methane to butane(s), or sometimes pentane(s).

The gas varies in composition and volume, depending on crude origin and on any additions to the crude made at the loading point. It is not uncommon to re-inject light hydrocarbons such as propane and butane into the crude oil before dispatch by tanker or pipeline. This results in a higher vapor pressure of the crude, but it allows one to increase the quantity of light products obtained at the refinery. As light ends in most petroleum markets command a premium, while in the oil field itself propane and butane may have to be re-injected or flared, the practice of spiking crude oil with liquefied petroleum gas is becoming fairly common.

In addition to the gases obtained by distillation of petroleum, more highly volatile products result from the subsequent processing of naphtha and middle distillate to produce gasoline. Hydrogen sulfide is produced in the desulfurization processes involving hydrogen treatment of naphtha, distillate, and residual fuel; and from the coking or similar thermal treatments of vacuum gas oils and residua (Chapter 16). The most common processing step in the production of gasoline is the catalytic reforming of hydrocarbon fractions in the heptane (C_7) to decane (C_{10}) range.

Additional gases are produced in **thermal cracking processes**, such as the coking or visbreaking processes (Chapter 14) for the processing of heavy feedstocks. In the visbreaking process, fuel oil is passed through externally fired tubes and undergoes liquid phase cracking reactions, which result in the formation of lighter fuel oil components. Oil viscosity is thereby reduced, and some gases, mainly hydrogen, methane, and ethane, are formed. Substantial quantities of both gas and carbon are also formed in coking (both fluid coking and delayed coking) in addition to the middle distillate and naphtha. When coking a residual fuel oil or heavy gas oil, the feedstock is preheated and contacted with hot carbon (coke), which causes extensive cracking of the feedstock constituents of higher molecular weight to produce lower molecular weight products ranging from methane, via liquefied petroleum gas(es) and naphtha, to gas oil and heating oil. Products from coking processes tend to be unsaturated and olefin components predominate in the tail gases from coking processes.

Another group of refining operations that contributes to gas production is that of the **catalytic cracking processes** (Chapter 15). These consist of fluid-bed catalytic cracking and there are many process variants (see also Chapter 18) in which heavy feedstocks are converted into cracked gas, liquefied petroleum gas, catalytic naphtha, fuel oil, and coke by contacting the heavy hydrocarbon with the hot catalyst. Both catalytic and thermal cracking processes, the latter being now largely used for the production of chemical raw materials, result in the formation of unsaturated hydrocarbons, particularly ethylene ($CH_2{=}CH_2$), but also propylene (propene, $CH_3CH{=}CH_2$), iso-butylene [iso-butene, $(CH_3)_2C{=}CH_2$] and the n-butenes ($CH_3CH_2CH{=}CH_2$, and $CH_3CH{=}CHCH_3$) in addition to hydrogen (H_2), methane (CH_4) and smaller quantities of ethane (CH_3CH_3), propane ($CH_3CH_2CH_3$), and butanes [$CH_3CH_2CH_2CH_3$, $(CH_3)_3CH$]. Diolefins such as butadiene ($CH_2{=}CHCH{=}CH_2$) are also present.

A further source of refinery gas is **hydrocracking**, a catalytic high-pressure pyrolysis process in the presence of fresh and recycled hydrogen (Chapter 16). The feedstock is again heavy gas oil or residual fuel oil, and the process is mainly directed at the production of additional middle distillates and gasoline. As hydrogen is to be recycled, the gases produced in this process again have to be separated into lighter and heavier streams; any surplus recycle gas and the liquefied petroleum gas from the hydrocracking process are both saturated.

In a series of **reforming processes** (Chapter 23), commercialized under names such as **Platforming**, (Chapter 23), paraffin and naphthene (cyclic nonaromatic) hydrocarbons are converted in the presence of hydrogen and a catalyst and are converted into aromatics, or isomerized to more highly branched hydrocarbons. Catalytic reforming processes thus not only result in the formation of a liquid product of higher octane number, but also produce substantial quantities of gases. The latter are rich in hydrogen, but also contain hydrocarbons from methane to butanes, with a preponderance of propane ($CH_3CH_2CH_3$), n-butane ($CH_3CH_2CH_2CH_3$), and iso-butane [$(CH_3)_3CH$].

The composition of the process gas varies in accordance with reforming severity and reformer feedstock. All catalytic reforming processes require substantial recycling of a hydrogen stream. Therefore, it is normal to separate reformer gas into a propane ($CH_3CH_2CH_3$) or a butane stream [$CH_3CH_2CH_2CH_3$ plus $(CH_3)_3CH$], which becomes a part of the refinery liquefied petroleum gas production, and a lighter gas fraction, part of which is recycled. In view of the excess of hydrogen in the gas, all products of catalytic reforming are saturated, and there are usually no olefin gases present in either gas stream.

Both hydrocracker gases and catalytic reformer gases are commonly used in catalytic desulfurization processes. In the latter, feedstocks ranging from light to vacuum gas oils are passed at pressures of 500–1000 psi (3.5–7.0×10^3 kPa) with hydrogen over a hydrofining catalyst. This results mainly in the conversion of organic sulfur compounds to hydrogen sulfide,

$$[S]_{\text{feedstock}} + H_2 = H_2S + \text{hydrocarbons}$$

This process also produces some light hydrocarbons by hydrocracking.

Thus, refinery gas streams, while ostensibly being hydrocarbon in nature, may contain large amounts of acid gases such as hydrogen sulfide and carbon dioxide. Most commercial plants employ hydrogenation to convert organic sulfur compounds into hydrogen sulfide. Hydrogenation is effected by means of recycled hydrogen-containing gases or external hydrogen over a nickel molybdate or cobalt molybdate catalyst.

The presence of impurities in gas streams may eliminate some of the sweetening processes, as some processes remove large amounts of acid gas but not to a sufficiently low concentration. On the other hand, there are those processes not designed to remove (or incapable of removing) large amounts of acid gases whereas they are capable of removing the acid gas impurities to very low levels when the acid gases are present only in low-to-medium concentration in the gas.

The processes that have been developed to accomplish gas purification vary from a simple once-through wash operation to complex multi-step recycling systems. In many cases, the process complexities arise because of the need for recovery of the materials used to remove the contaminants or even recovery of the contaminants in the original, or altered, form (Katz, 1959; Kohl and Riesenfeld, 1985; Newman, 1985).

From an environmental viewpoint, it is not the means by which these gases can be utilized that is of concern, but it is the effects of these gases on the environment when they are introduced into the atmosphere.

In addition to the corrosion of equipment of acid gases, the escape into the atmosphere of sulfur-containing gases can eventually lead to the formation of the constituents of acid rain, that is, the oxides of sulfur (SO_2 and SO_3). Similarly, the nitrogen-containing gases can also lead to nitrous and nitric acids (through the formation of the oxides NO_x, where $x = 1$ or 2) which are the other major contributors to acid rain. The release of carbon dioxide and hydrocarbons as constituents of refinery effluents can also influence the behavior and integrity of the ozone layer.

Finally, another acid gas, hydrogen chloride (HCl), although not usually considered to be a major emission, is produced from mineral matter and the brine that often accompany petroleum during production and is gaining increasing recognition as a contributor to acid rain. However, hydrogen chloride may exert severe local effects because it does not need to participate in any further chemical reaction to become an acid. Under atmospheric conditions that favor a buildup of stack emissions in the areas where hydrogen chloride is produced, the amount of hydrochloric acid in rain water could be quite high.

In summary, refinery processed gas, in addition to hydrocarbons, may contain other contaminants, such as carbon oxides (CO_x, where $x = 1$ or 2), sulfur oxides (SO_x, where $x = 2$ or 3), as well as ammonia (NH_3), mercaptans (R–SH), and carbonyl sulfide (COS). From an environmental viewpoint, petroleum processing can result in a variety of gaseous emissions. It is a question of degree insofar as the composition of the gaseous emissions may vary from process to process but the constituents are, in the majority of cases, the same.

25.1.2 Gas Streams from Natural Gas

Natural gas is also capable of producing emissions that are detrimental to the environment. While the major constituent of natural gas is methane, there are components such as carbon dioxide (CO_2), hydrogen sulfide (H_2S), and mercaptans (thiols; R–SH), as well as trace amounts of sundry other emissions such as carbonyl sulfide (COS). The fact that methane has a foreseen and valuable end-use makes it a desirable product, but in several other situations it is considered a pollutant, having been identified as a greenhouse gas.

A sulfur removal process must be very precise, as natural gas contains only a small quantity of sulfur-containing compounds that must be reduced several orders of magnitude. Most consumers of natural gas require less than 4 ppm in the gas.

A characteristic feature of natural gas that contains hydrogen sulfide is the presence of carbon dioxide (generally in the range of 1% to 4% by volume). In cases where the natural gas does not contain hydrogen sulfide, there may also be a relative lack of carbon dioxide.

In practice, heaters and scrubbers are installed, usually at or near the wellhead. The scrubbers serve primarily to remove sand and other large-particle impurities and the heaters ensure that the temperature of the gas does not drop too low. With natural gas that contains even low quantities of water, natural gas hydrates have a tendency to form when temperatures drop. These hydrates are solid or semi-solid compounds, resembling ice-like crystals. If the hydrates accumulate, they can impede the passage of natural gas through valves and gathering systems. To reduce the occurrence of hydrates, small natural gas-fired heating units are typically installed along the gathering pipe wherever it is likely that hydrates may form.

25.2 GAS CLEANING

The actual practice of processing natural gas to pipeline dry gas quality levels can be quite complex, but usually involves four main processes to remove the various impurities. Gas streams produced during petroleum and natural gas refining, while ostensibly being hydrocarbon in nature, may contain large amounts of acid gases such as hydrogen sulfide and carbon dioxide. Most commercial plants employ hydrogenation to convert organic sulfur compounds into hydrogen sulfide. Hydrogenation is effected by means of recycled hydrogen-containing gases or external hydrogen over a nickel molybdate or cobalt molybdate catalyst.

In summary, refinery process gas, in addition to hydrocarbons, may contain other contaminants, such as carbon oxides (CO_x, where $x = 1$ or 2), sulfur oxides (SO_x, where $x = 2$ or 3), as well as ammonia (NH_3), mercaptans (R–SH), and carbonyl sulfide (COS).

The presence of these impurities may eliminate some of the sweetening processes, as some processes remove large amounts of acid gas but not to a sufficiently low concentration. On the other hand, there are those processes not designed to remove (or incapable of removing) large amounts of acid gases whereas they are capable of removing the acid gas impurities to very low levels when the acid gases are present only in low-to-medium concentration in the gas (Katz, 1959).

The processes that have been developed to accomplish gas purification vary from a simple once-through wash operation to complex multi-step recycling systems (Speight, 1993). In many cases, the process complexities arise because of the need for recovery of the materials used to remove the contaminants or even recovery of the contaminants in the original, or altered, form (Katz, 1959; Kohl and Riesenfeld, 1985; Newman, 1985).

There are many variables in treating refinery gas or natural gas. The precise area of application, of a given process is difficult to define. Several factors must be considered:

1. Types of contaminants in the gas
2. Concentrations of contaminants in the gas
3. Degree of contaminant removal desired
4. Selectivity of acid gas removal required
5. Temperature of the gas to be processed

6. Pressure of the gas to be processed
7. Volume of the gas to be processed
8. Composition of the gas to be processed
9. Carbon dioxide–hydrogen sulfide ratio in the gas
10. Desirability of sulfur recovery due to process economics or environmental issues

In addition to hydrogen sulfide (H_2S) and carbon dioxide (CO_2), gas may contain other contaminants, such as mercaptans (RSH) and carbonyl sulfide (COS). The presence of these impurities may eliminate some of the sweetening processes as some processes remove large amounts of acid gas but not to a sufficiently low concentration. On the other hand, there are those processes that are not designed to remove (or are incapable of removing) large amounts of acid gases. However, these processes are also capable of removing the acid gas impurities to very low levels when the acid gases are there in low to medium concentrations in the gas.

Process selectivity indicates the preference with which the process removes one acid gas component relative to (or in preference to) another. For example, some processes remove both hydrogen sulfide and carbon dioxide; other processes are designed to remove hydrogen sulfide only. It is important to consider the process selectivity for, say, hydrogen sulfide removal compared to carbon dioxide removal that ensures minimal concentrations of these components in the product, thus the need for consideration of the carbon dioxide to hydrogen sulfide in the gas stream.

Gas processing involves the use of several different types of processes but there is always overlap between the various processing concepts. In addition, the terminology used for gas processing can often be confusing or misleading because of the overlap (Nonhebel, 1964; Curry, 1981; Maddox, 1982).

There are four general processes used for emission control (often referred to in another, more specific context as flue gas desulfurization): (1) **adsorption**; (2) absorption; (3) catalytic oxidation; and (4) thermal oxidation (Soud and Takeshita, 1994).

Adsorption is a physical–chemical phenomenon in which the gas is concentrated on the surface of a solid or liquid to remove impurities (Mantell, 1951). Usually, carbon is the adsorbing medium (Rook, 1994), which can be regenerated upon **desorption** (Fulker, 1972; Speight, 1993). The quantity of material adsorbed is proportional to the surface area of the solid and, consequently, adsorbents are usually granular solids with a large surface area per unit mass. Subsequently, the captured gas can be desorbed with hot air or steam either for recovery or for thermal destruction.

Adsorbers are widely used to increase a low gas concentration prior to incineration unless the gas concentration is very high in the inlet air stream. Adsorption also is employed to reduce problem odors from gases. There are several limitations to the use of adsorption systems, but it is generally felt that the major one is the requirement for minimization of particulate matter and condensation of liquids (e.g., water vapor) that could mask the adsorption surface and drastically reduce its efficiency (Table 25.2).

Absorption differs from adsorption, in that it is not a physical–chemical surface phenomenon, but an approach in which the absorbed gas is ultimately distributed throughout the absorbent (liquid). The process depends only on physical solubility and may include chemical reactions in the liquid phase (**chemisorption**). Common absorbing media used are water, aqueous amine solutions, caustic, sodium carbonate, and nonvolatile hydrocarbon oils, depending on the type of gas to be absorbed. Usually, the gas–liquid contactor designs that are employed are plate columns or packed beds.

Absorption is achieved by dissolution (a physical phenomenon) or by reaction (a chemical phenomenon) (Barbouteau and Dalaud, 1972; Ward, 1972). Chemical adsorption processes

TABLE 25.2
Comments on the Use of Adsorption Systems

1. Possibility of product recovery
2. Excellent control and response to process changes
3. No chemical-disposal problem when pollutant (product) is recovered and returned to process
4. Capability to remove gaseous or vapor contaminants from process streams to extremely low levels
5. Product recovery may require distillation or extraction
6. Adsorbent may deteriorate as the number of cycles increase
7. Adsorbent regeneration
8. Prefiltering of gas stream possibly required to remove particulate materials capable of plugging the adsorbent bed

adsorb sulfur dioxide onto a carbon surface where it is oxidized (by oxygen in the flue gas) and absorbs moisture to give sulfuric acid impregnated into and on the adsorbent.

As currently practiced, acid gas removal processes involve the selective absorption of the contaminants into a liquid, which is passed countercurrent to the gas. Then the absorbent is stripped of the gas components (regeneration) and recycled to the absorber. The process design will vary and, in practice, may employ multiple absorption columns and multiple regeneration columns (Table 25.3).

Liquid absorption processes (which usually employ temperatures below 50°C (120°F) are classified either as physical solvent processes or chemical solvent processes. The former processes employ an organic solvent, and absorption is enhanced by low temperatures, or high pressure, or both. Regeneration of the solvent is often accomplished readily (Staton et al., 1985). In chemical solvent processes, absorption of the acid gases is achieved mainly by use of alkaline solutions such as amines or carbonates (Kohl and Riesenfeld, 1985). Regeneration (desorption) can be brought about by the use of reduced pressures or high temperatures, whereby the acid gases are stripped from the solvent.

Solvents used for emission control processes should have (Speight, 1994):

1. High capacity for acid gas
2. Low tendency to dissolve hydrogen
3. Low tendency to dissolve low-molecular weight hydrocarbons
4. Low vapor pressure at operating temperatures to minimize solvent losses
5. Low viscosity
6. Low thermal stability
7. Absence of reactivity toward gas components

TABLE 25.3
Comments on the Use of Packed Column and Plate Column Absorption Systems

1. Relatively low pressure drop
2. Capable of achieving relatively high mass-transfer efficiencies
3. Increasing the height and type of packing or number of plates capable of improving mass transfer without purchasing a new piece of equipment
4. Ability to collect particulate materials as well as gases
5. May require water (or liquid) disposal

8. Low tendency for fouling
9. Low tendency for corrosion
10. Economically acceptable

Amine washing of gas emissions involves chemical reaction of the amine with any acid gases with the liberation of an appreciable amount of heat and it is necessary to compensate for the absorption of heat. Amine derivatives such as ethanolamine (monoethanolamine, MEA), diethanolamine (DEA), triethanolamine (TEA), methyldiethanolamine (MDEA), diisopropanolamine (DIPA), and diglycolamine (DGA) have been used in commercial applications (Table 25.4) (Katz, 1959; Jou et al., 1985; Kohl and Riesenfeld, 1985; Maddox et al., 1985; Polasek and Bullin, 1985; Pitsinigos and Lygeros, 1989; Speight, 1993).

The chemistry can be represented by simple equations for low partial pressures of the acid gases:

$$2RNH_2 + H_2S \rightarrow (RNH_3)_2S$$

$$2RNH_2 + CO_2 + H_2O \rightarrow (RNH_3)_2CO_3$$

At high acid gas partial pressure, the reactions will lead to the formation of other products:

$$(RNH_3)_2S + H_2S \rightarrow 2RNH_3HS$$

$$(RNH_3)_2CO_3 + H_2O \rightarrow 2RNH_3HCO_3$$

The reaction is extremely fast, the absorption of hydrogen sulfide being limited only by mass transfer; this is not so for carbon dioxide.

Regeneration of the solution leads to near complete desorption of carbon dioxide and hydrogen sulfide. A comparison between MEA, DEA, and DIPA shows that MEA is the cheapest of the three but shows the highest heat of reaction and corrosion; the reverse is true for DIPA.

Carbonate washing is a mild alkali process for emission control by the removal of acid gases (such as carbon dioxide and hydrogen sulfide) from gas streams (Speight, 1993) and uses the principle that the rate of absorption of carbon dioxide by potassium carbonate increases with temperature. It has been demonstrated that the process works best near the temperature of reversibility of the reactions:

$$K_2CO_3 + CO_2 + H_2O \rightarrow 2KHCO_3$$

$$K_2CO_3 + H_2S \rightarrow KHS + KHCO_3$$

Water washing, in terms of the outcome, is analogous to washing with potassium carbonate (Kohl and Riesenfeld, 1985), and it is also possible to carry out the desorption step by pressure reduction. The absorption is purely physical and there is also a relatively high absorption of hydrocarbons, which are liberated at the same time as the acid gases.

In chemical conversion processes, contaminants in gas emissions are converted to compounds that are not objectionable or that can be removed from the stream with greater ease than the original constituents. For example, a number of processes have been developed that remove hydrogen sulfide and sulfur dioxide from gas streams by absorption in an alkaline solution.

Catalytic oxidation is a chemical conversion process that is used predominantly for destruction of volatile organic compounds and carbon monoxide. These systems operate in

TABLE 25.4
Olamines Used for Gas Processing

Olamine	Formula	Derived Name	Molecular Weight	Specific Gravity	Melting Point, °C	Boiling Point, °C	Flash Point, °C	Relative Capacity %
Ethanolamine (monoethanolamine)	$HOC_2H_4NH_2$	MEA	61.08	1.01	10	170	85	100
Diethanolamine	$(HOC_2H_4)_2NH$	DEA	105.14	1.097	27	217	169	58
Triethanolamine	$(HOC_2H_4)_3NH$	TEA	148.19	1.124	18	335[a]	185	41
Diglycolamine (hydroxyethanolamine)	$H(OC_2H_4)_2NH_2$	DGA	105.14	1.057	-11	223	127	58
Diisopropanolamine	$(HOC_3H_6)_2NH$	DIPA	133.19	0.99	42	248	127	46
Methyldiethanolamine	$(HOC_2H_4)_2NCH_3$	MDEA	119.17	1.03	-21	247	127	51

[a]With decomposition

a temperature regime of 205°C to 595°C (400°F to 1100°F) in the presence of a catalyst. Without the catalyst, the system would require higher temperatures. Typically, the catalysts used are a combination of noble metals deposited on a ceramic base in a variety of configurations (e.g., honeycomb-shaped) to enhance good surface contact.

Catalytic systems are usually classified on the basis of bed types such as **fixed bed** (or packed bed) and **fluid bed** (fluidized bed). These systems generally have very high destruction efficiencies for most volatile organic compounds, resulting in the formation of carbon dioxide, water, and varying amounts of hydrogen chloride (from halogenated hydrocarbons). The presence in emissions of chemicals such as heavy metals, phosphorus, sulfur, chlorine, and most halogens in the incoming air stream act as poison to the system and can foul up the catalyst.

Thermal oxidation systems, without the use of catalysts, also involve chemical conversion (more correctly, chemical destruction) and operate at temperatures in excess of 815°C (1500°F), or 220°C to 610°C (395°F to 1100°F) higher than catalytic systems.

Historically, particulate matter control (dust control) (Mody and Jakhete, 1988) has been one of the primary concerns of industries, as the emission of particulate matter is readily observed through the deposition of fly ash and soot as well as in an impairment of visibility. Differing ranges of control can be achieved by use of various types of equipments. Upon proper characterization of the particulate matter emitted by a specific process, the appropriate piece of equipment can be selected, sized, installed, and performance tested. The general classes of control devices for particulate matter are as given in the following paragraphs.

Cyclone collectors are the most common of the inertial collector class. Cyclones are effective in removing coarser fractions of particulate matter. The particle-laden gas stream enters an upper cylindrical section tangentially and proceeds downward through a conical section. Particles migrate by centrifugal force generated by providing a path for the carrier gas to be subjected to a vortex-like spin. The particles are forced to the wall and are removed through a seal at the apex of the inverted cone. A reverse-direction vortex moves upward through the cyclone and discharges through a top center opening. Cyclones are often used as primary collectors because of their relatively low efficiency (50%–90% is usual). Some small-diameter high-efficiency cyclones are utilized. The equipment can be arranged either in parallel or in series to increase efficiency and decrease pressure drop. However, there are disadvantages that must be recognized (Table 25.5). These units for particulate matter operate by contacting the particles in the gas stream with a liquid. In principle the particles are incorporated in a liquid bath or in liquid particles that are much larger and therefore more easily collected.

Fabric filters are typically designed with nondisposable filter bags. As the dusty emissions flow through the filter media (typically cotton, polypropylene, Teflon, or fiberglass), particulate matter is collected on the bag surface as a dust cake. Fabric filters are generally classified

---

TABLE 25.5
Comments on the Use of Cyclone Collectors

1. Relatively low operating pressure drops (for degree of particulate removal obtained) in the range of approximately 2 to 6 inch water column
2. Dry collection and disposal
3. Relatively low overall particulate collection efficiencies, especially on particulates below 10 mm
4. Usually unable to process semi-solid (tacky) materials

TABLE 25.6
Comments on the Use of Fabric Filter Systems

1. High collection efficiency on both coarse and fine (submicrometer) particulates
2. Collected material recovered dry for subsequent processing of disposal
3. Corrosion and rusting of components usually not major issues
4. Relatively simple operation
5. Temperatures much in excess of 288°C (550°F) require special refractory materials
6. Potential for dust explosion hazard
7. Fabric life possible shortened at elevated temperatures and in the presence of acid or alkaline particulate or gas constituents
8. Hygroscopic materials, condensation of moisture, or tarry adhesive components possible causing crusty caking or plugging of the fabric or requiring special additives

on the basis of the filter bag cleaning mechanism employed. Fabric filters operate with collection efficiencies up to 99.9% although other advantages are evident. There are several issues that arise during use of such equipment (Table 25.6).

Wet scrubbers are devices in which a countercurrent spray liquid is used to remove particles from an air stream. Device configurations include plate scrubbers, packed beds, orifice scrubbers, venturi scrubbers, and spray towers, individually or in various combinations. Wet scrubbers can achieve high collection efficiencies at the expense of prohibitive pressure drops (Table 25.7).

Other methods include use of high-energy input venturi scrubbers or electrostatic scrubbers where particles or water droplets are charged, and flux force/condensation scrubbers where a hot humid gas is contacted with cooled liquid or where steam is injected into saturated gas. In the latter scrubber the movement of water vapor toward the cold water surface carries the particles with it (diffusiophoresis), while the condensation of water vapor on the particles causes the particle size to increase, thus facilitating collection of fine particles.

The foam scrubber is a modification of the wet scrubber in which the particle-laden gas is passed through a foam generator, where the gas and particles are enclosed by small bubbles of foam.

Electrostatic precipitators (Table 25.8) operate on the principle of imparting an electric charge to particles in the incoming air stream, which are then collected on an oppositely charged plate across a high-voltage field. Particles of high resistivity create the most difficulty in collection. Conditioning agents such as sulfur trioxide (SO_3) have been used to lower resistivity.

TABLE 25.7
Comments on the Use of Wet Scrubbers

1. No secondary dust sources
2. Ability to collect gases as well as particulates (especially sticky ones)
3. Ability to handle high-temperature, high-humidity gas streams
4. Ability to achieve high collection efficiencies on fine particulates (however, at the expense of pressure drop)
5. May be necessary for water disposal
6. Corrosion problems more severe than with dry systems
7. Potential for solids buildup at the wet–dry interface

TABLE 25.8
Comments on the Use of Electrostatic Precipitators

1. High particulate (coarse and fine) collection efficiencies
2. Dry collection and disposal
3. Operation under high pressure (to 150 lb/in.^{2}) or vacuum conditions
4. Operation at high temperatures (up to 704°C, 1300°F) when necessary
5. Relatively large gas flow rates capable of effective handling
6. Can be sensitive to fluctuations in gas-stream conditions
7. Explosion hazard when treating combustible gases/particulates
8. Ozone produced during gas ionization

Important parameters include design of electrodes, spacing of collection plates, minimization of air channeling, and collection-electrode rapping techniques (used to dislodge particles). Techniques under study include the use of high-voltage pulse energy to enhance particle charging, electron-beam ionization, and wide plate spacing. Electrical precipitators are capable of efficiencies >99% under optimum conditions, but performance is still difficult to predict in new situations.

25.3 WATER REMOVAL

Water is a common impurity in gas streams, and removal of water is necessary to prevent the condensation of water and the formation of ice or gas hydrates ($C_nH_{2n+2}\cdot xH_2O$). Water in the liquid phase causes corrosion or erosion problems in pipelines and equipment, particularly when carbon dioxide and hydrogen sulfide are present in the gas. The simplest method of water removal (refrigeration or cryogenic separation) is to cool the gas to a temperature at least equal to or (preferentially) below the dew point (Figure 25.2).

In addition to separating petroleum and some condensate from the wet gas stream, it is necessary to remove most of the associated water. Most of the liquid, free water associated with extracted natural gas is removed by simple separation methods at or near the wellhead.

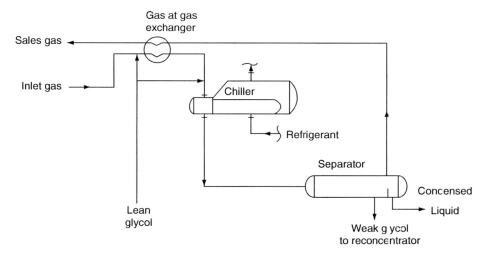

FIGURE 25.2 The refrigeration process using glycol.

However, the removal of the water vapor that exists in solution in natural gas requires a more complex treatment. This treatment consists of dehydrating the natural gas, which usually involves one of two processes: either absorption or adsorption.

Absorption occurs when the water vapor is taken out by a dehydrating agent. Adsorption occurs when the water vapor is condensed and collected on the surface.

In a majority of cases, cooling alone is insufficient and, for the most part, impractical for use in field operations. Other, more convenient, water removal options use (1) hygroscopic liquids (e.g., diethylene glycol (DEG) or triethylene glycol (TEG)) and (2) solid adsorbents or desiccants (e.g., alumina, silica gel, and molecular sieves). Ethylene glycol can be directly injected into the gas stream in refrigeration plants.

25.3.1 ABSORPTION

An example of absorption dehydration is known as glycol dehydration and DEG, the principal agent in this process, has a chemical affinity for water and removes water from the gas stream. In this process, a liquid desiccant dehydrator serves to absorb water vapor from the gas stream. Essentially, glycol dehydration involves using a glycol solution, usually either DEG or TEG, which is brought into contact with the wet gas stream in a contactor. The glycol solution will absorb water from the wet gas and, once absorbed, the glycol particles become heavier and sink to the bottom of the contactor where they are removed. The natural gas, having been stripped of most of its water content, is then transported out of the dehydrator. The glycol solution, bearing all of the water stripped from the natural gas, is put through a specialized boiler designed to vaporize only the water out of the solution. The boiling point differential between water (100°C, 212°F) and glycol (204°C, 400°F) makes it relatively easy to remove water from the glycol solution, allowing it be reused in the dehydration process.

As well as absorbing water from the wet gas stream, the glycol solution occasionally carries with it small amounts of methane and other compounds found in the wet gas. In the past, this methane was simply vented out of the boiler. In addition to losing a portion of the natural gas that was extracted, this venting contributes to air pollution and the greenhouse effect. In order to decrease the amount of methane and other compounds that are lost, flash tank separator-condensers work is to remove these compounds before the glycol solution reaches the boiler. Essentially, a flash tank separator consists of a device that reduces the pressure of the glycol solution stream, allowing the methane and other hydrocarbons to vaporize (flash). The glycol solution then travels to the boiler, which may also be fitted with air or water cooled condensers, which serve to capture any remaining organic compounds that may remain in the glycol solution. The regeneration (stripping) of the glycol is limited by temperature: DEG and TEG decompose at or before their respective boiling points. Such techniques as stripping of hot TEG with dry gas (e.g., heavy hydrocarbon vapors, the Drizo process) or vacuum distillation are recommended.

In practice, absorption systems recover 90% to 99% by volume of methane that would otherwise be flared into the atmosphere.

25.3.2 SOLID ADSORBENTS

Solid-adsorbent or solid-desiccant dehydration is the primary form of dehydrating natural gas using adsorption, and usually consists of two or more adsorption towers, which are filled with a solid desiccant. Typical desiccants include activated alumina or granular silica gel material. Wet natural gas is passed through these towers, from top to bottom. As the wet gas passes around the particles of desiccant material, water is retained on the surface of these desiccant particles. Passing through the entire desiccant bed, almost all of the water is adsorbed onto the desiccant material, leaving the dry gas to exit via the bottom of the tower.

Solid-adsorbent dehydrators are typically more effective than glycol dehydrators, and are usually installed as a type of straddle system along natural gas pipelines. These types of dehydration systems are best suited for large volumes of gas under very high pressure, and are thus usually located on a pipeline downstream of a compressor station. Two or more towers are required due to the fact that after a certain period of use, the desiccant in a particular tower becomes saturated with water. To regenerate the desiccant, a high-temperature heater is used to heat gas to a very high temperature. Passing this heated gas through a saturated desiccant bed vaporizes the water in the desiccant tower, leaving it dry and allowing for further natural gas dehydration.

Although two-bed adsorbent treaters have become more common (while one bed is removing water from the gas, the other undergoes alternate heating and cooling), on occasion, a three-bed system is used: one bed adsorbs, one is being heated, and one is being cooled. An additional advantage of the three-bed system is the facile conversion of a two-bed system so that the third bed can be maintained or replaced, thereby ensuring continuity of the operations, and reducing the risk of a costly plant shutdown.

Silica gel (SiO_2) and alumina (Al_2O_3) have good capacities for water adsorption (up to 8% by weight). Bauxite (crude alumina, Al_2O_3) adsorbs up to 6% by weight water, and molecular sieves adsorb up to 15% by weight water. Silica is usually selected for dehydration of sour gas because of its high tolerance to hydrogen sulfide and to protect molecular sieve beds from plugging by sulfur. Alumina **guard beds** (which serve as protectors by the act of attrition and may be referred to as an attrition catalyst) (Speight, 2000) may be placed ahead of the molecular sieves to remove the sulfur compounds. Downflow reactors are commonly used for adsorption processes, with an upward flow regeneration of the adsorbent and cooling in the same direction as adsorption.

25.3.3 USE OF MEMBRANES

Membrane separation processes are very versatile and are designed to process a wide range of feedstocks and offer a simple solution for removal and recovery of higher boiling hydrocarbons (NGLs) from natural gas (Figure 25.3) (Foglietta, 2004). The separation process is based

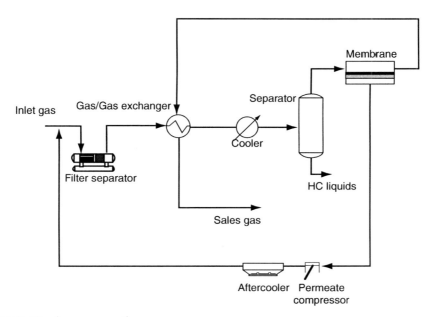

FIGURE 25.3 Membrane separation process.

on high-flux membranes that selectively permeate higher boiling hydrocarbons (compared to methane) and are recovered as a liquid after recompression and condensation. The residue stream from the membrane is partially depleted of higher boiling hydrocarbons, and is then sent to sales gas stream. Gas permeation membranes are usually made with vitreous polymers that exhibit good selectivity but, to be effective, the membrane must be very permeable with respect to the separation process.

25.4 LIQUIDS REMOVAL

Natural gas coming directly from a well contains many NGLs that are commonly removed. In most instances, NGLs have a higher value as separate products, and it is thus economical to remove them from the gas stream. The removal of NGLs usually takes place in a relatively centralized processing plant, and uses techniques similar to those used to dehydrate natural gas. Recovery of the liquid hydrocarbons can be justified either because it is necessary to make the gas saleable or because economics dictate this course of action. The justification for building a liquid recovery (or a liquid removal) plant depends on the price differential between the enriched gas (containing the higher molecular weight hydrocarbons) and lean gas with the added value of the extracted liquid.

There are two basic steps to the treatment of NGLs in the natural gas stream. First, the liquids must be extracted from the natural gas. Second, these NGLs must themselves be separated, down to their base components. These two processes account for approximately 90% of the total production of NGLs.

25.4.1 EXTRACTION

There are two principle techniques for removing NGLs from the natural gas stream: the absorption method and the cryogenic expander process.

In the process, a turboexpander is used to produce the necessary refrigeration and very low temperatures and high recovery of light components, such as ethane and propane, can be attained. The natural gas is first dehydrated using a molecular sieve followed by cooling (Figure 25.4). The separated liquid containing most of the heavy fractions is then demethanized,

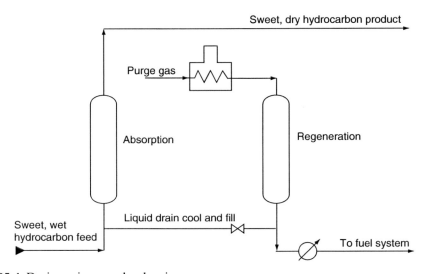

FIGURE 25.4 Drying using a molecular sieve.

and the cold gases are expanded through a turbine that produces the desired cooling for the process. The expander outlet is a two-phase stream that is fed to the top of the demethanizer column. This serves as a separator in which: (1) the liquid is used as the column reflux and the separator vapors combined with vapors stripped in the demethanizer are exchanged with the feed gas, and (2) the heated gas, which is partially recompressed by the expander compressor, is further recompressed to the desired distribution pressure in a separate compressor. This process allows for the recovery of about 90% to 95% by volume of the ethane originally in the gas stream. In addition, the expansion turbine is able to convert some of the energy released when the natural gas stream is expanded into recompressing the gaseous methane effluent, thus saving energy costs associated with extracting ethane.

The extraction of NGLs from the natural gas stream produces both cleaner, purer natural gas, as well as the valuable hydrocarbons that are the NGLs themselves.

25.4.2 ABSORPTION

The absorption method of extraction is very similar to using absorption for dehydration. The main difference is that, in the absorption of NGLs, absorbing oil is used as opposed to glycol. This absorbing oil has an affinity for NGLs in much the same manner as glycol has an affinity for water. Before the oil has picked up any NGLs, it is termed *lean* absorption oil.

The oil absorption process involves the countercurrent contact of the **lean** (or stripped) **oil** with the incoming wet gas with the temperature and pressure conditions programmed to maximize the dissolution of the liquefiable components in the oil. The rich absorption oil (sometimes referred to as **fat oil**), containing NGLs, exits the absorption tower through the bottom. It is now a mixture of absorption oil, propane, butanes, pentanes, and other higher boiling hydrocarbons. The **rich oil** is fed into lean oil stills, where the mixture is heated to a temperature above the boiling point of the NGLs but below that of the oil. This process allows for the recovery of around 75% by volume of the butanes, and 85% to 90% by volume of the pentanes and higher boiling constituents from the natural gas stream.

The basic absorption process above can be modified to improve its effectiveness, or to target the extraction of specific NGLs. In the refrigerated oil absorption method, where the lean oil is cooled through refrigeration, propane recovery can be upwards of 90% by volume and approximately 40% by volume of ethane can be extracted from the natural gas stream. Extraction of the other higher boiling NGLs can be close to 100% by volume using this process.

The advanced extraction technology (AET) process (Figure 25.5) for recovery of liquefied petroleum gas utilizes noncryogenic absorption to recover ethane, propane, and higher boiling constituents from natural gas streams. The absorbed gases in the rich solvent from the bottom of the absorber column are fractionated in the solvent regenerator column that separates gases (as an overhead fraction) and lean solvent (as a bottoms fraction). After heat recuperation, the lean solvent is presaturated with absorber overhead gases. The chilled solvent flows in the top of the absorber column. The separated gases are sent to storage. Depending upon the economics of ethane recovery, the operation of the plant can be switched on-line from ethane plus recovery to propane plus recovery without affecting the propane recovery levels. The AET liquefied petroleum gas plant uses lower boiling lean oils. For most applications there are no solvent make-up requirements.

25.4.3 FRACTIONATION OF NATURAL GAS LIQUIDS

After the NGLs have been removed from the natural gas stream, they must be broken down into their base components to be useful. That is, the mixed stream of different NGLs must be

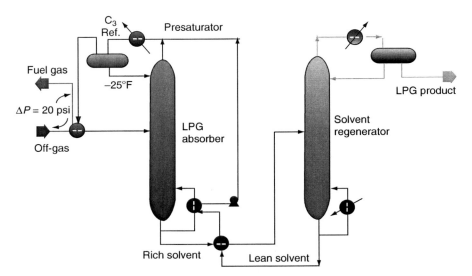

FIGURE 25.5 Schematic of the AET recovery plant.

separated. The process used to accomplish this task is called **fractionation**. Fractionation works based on the different boiling points of the different hydrocarbons in the NGLs stream. Essentially, fractionation occurs in stages consisting of the boiling off of hydrocarbons one by one. The name of a particular fractionator gives an idea as to its purpose, as it is conventionally named for the hydrocarbon that is boiled off. The entire fractionation process is broken down into steps, starting with the removal of the lighter NGLs from the stream. The particular fractionators are used in the following order: (1) de-ethanizer that separates the ethane from the stream of NGLs, (2) depropanizer that separates propane from the de-ethanized stream, (3) debutanizer that separates the butanes, leaving the pentanes and higher boiling hydrocarbons in the stream, (4) the butane splitter or de-isobutanizer that separates *iso*-butane and *n*-butane.

25.5 NITROGEN REMOVAL

Nitrogen may often occur in sufficient quantities in natural gas and, consequently, lower the heating value of the gas. Thus several plants for nitrogen removal from natural gas have been built, but it must be recognized that nitrogen removal requires liquefaction and fractionation of the entire gas stream, which may affect process economics. In many cases the nitrogen-containing natural gas is blended with a gas having a higher heating value and sold at a reduced price depending upon the thermal value (Btu/ft.^{3}).

25.6 ACID GAS REMOVAL

In addition to water and NGLs removal, one of the most important parts of gas processing involves the removal of hydrogen sulfide and carbon dioxide. Natural gas from some wells contains significant amounts of hydrogen sulfide and carbon dioxide and is usually referred to as sour gas. Sour gas is undesirable because the sulfur compounds it contains can be extremely harmful, even lethal, to breathe and the gas can also be extremely corrosive. The process for removing hydrogen sulfide from sour gas is commonly referred to as **sweetening** the gas.

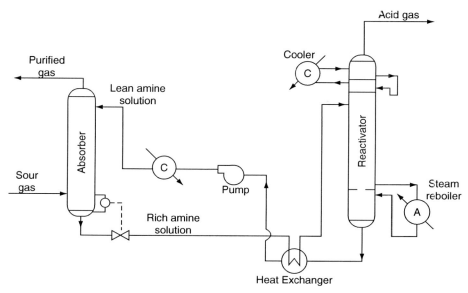

FIGURE 25.6 The amine or olamine process.

The primary process (Figure 25.6) for sweetening sour natural gas is quite similar to the processes of glycol dehydration and removal of NGLs by absorption. In this case, however, amine (olamine) solutions are used to remove the hydrogen sulfide (the amine process). The sour gas is run through a tower, which contains the olamine solution. There are two principle amine solutions used, MEA and DEA. Either of these compounds, in liquid form, will absorb sulfur compounds from natural gas as it passes through. The effluent gas is virtually free of sulfur compounds, and thus loses its sour gas status. Like the process for the extraction of NGLs and glycol dehydration, the amine solution used can be regenerated for reuse.

Although most sour gas sweetening involves the amine absorption process, it is also possible to use solid desiccants like iron sponge (*q.v.*) to remove hydrogen sulfide and carbon dioxide.

Treatment of gas to remove the acid gas constituents (hydrogen sulfide and carbon dioxide) is most often accomplished by contact of the natural gas with an alkaline solution. The most commonly used treating solutions are aqueous solutions of ethanolamine (Table 20.8) or alkali carbonates, although a considerable number of other treating agents have been developed in recent years. Most of these newer treating agents rely upon physical absorption and chemical reaction. When only carbon dioxide is to be removed in large quantities or when only partial removal is necessary, a hot carbonate solution or one of the physical solvents is the most economical selection.

The most well-known hydrogen sulfide removal process is based on the reaction of hydrogen sulfide with iron oxide (often also called the iron sponge process or the dry box method) in which the gas is passed through a bed of wood chips impregnated with iron oxide.

The iron oxide process is the oldest and still the most widely used batch process for sweetening natural gas and NGLs (Zapffe, 1963; Duckworth and Geddes, 1965; Anerousis and Whitman, 1984). The process was implemented during the nineteenth century. In the process (Figure 25.7), the sour gas is passed down through the bed. In the case where continuous regeneration is to be utilized a small concentration of air is added to the sour gas before it is processed. This air serves to continuously regenerate the iron oxide, which has

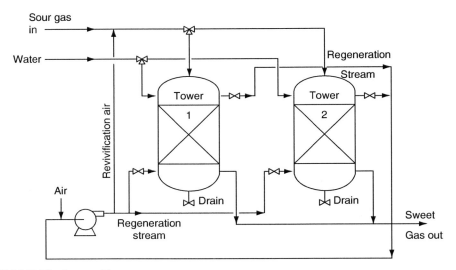

FIGURE 25.7 The iron oxide process.

reacted with hydrogen sulfide, which serves to extend the on-stream life of a given tower but probably serves to decrease the total amount of sulfur that a given weight of bed will remove.

The process is usually best applied to gases containing low to medium concentrations (300 ppm) of hydrogen sulfide or mercaptans. This process tends to be highly selective and does not normally remove significant quantities of carbon dioxide. As a result, the hydrogen sulfide stream from the process is usually of high purity. The use of iron sponge process for sweetening sour gas is based on adsorption of the acid gases on the surface of the solid sweetening agent followed by chemical reaction of ferric oxide (Fe_2O_3) with hydrogen sulfide:

$$2Fe_2O_3 + 6H_2S \rightarrow 2Fe_2S_3 + 6H_2O$$

The reaction requires the presence of slightly alkaline water and a temperature below 43°C (110°F) and bed alkalinity (pH +8 to 10) should be checked regularly, usually on a daily basis. The pH level is be maintained through the injection of caustic soda with water. If the gas does not contain sufficient water vapor, water may need to be injected into the inlet gas stream.

The ferric sulfide produced by the reaction of hydrogen sulfide with ferric oxide can be oxidized with air to produce sulfur and regenerate the ferric oxide:

$$2Fe_2S_3 + 3O_2 \rightarrow 2Fe_2O_3 + 6S$$
$$S_2 + 2O_2 \rightarrow 2SO_2$$

The regeneration step is exothermic and air must be introduced slowly, so that the heat of reaction can be dissipated. If air is introduced quickly the heat of reaction may ignite the bed. Some of the elemental sulfur produced in the regeneration step remains in the bed. After several cycles this sulfur will cake over the ferric oxide, decreasing the reactivity of the bed. Typically, after ten cycles the bed must be removed and a new bed introduced into the vessel.

The iron oxide process is one of several metal oxide-based processes that scavenge hydrogen sulfide and organic sulfur compounds (mercaptans) from gas streams through reactions with the solid-based chemical adsorbent (Kohl and Riesenfeld, 1985). They are typically nonregenerable, although some are partially regenerable, losing activity upon each regeneration cycle. Most of the processes are governed by the reaction of a metal oxide with hydrogen sulfide to form the metal sulfide. For regeneration, the metal oxide is reacted with oxygen to produce elemental sulfur and the regenerated metal oxide. In addition to iron oxide, the primary metal oxide used for dry sorption processes is zinc oxide.

In the zinc oxide process, the zinc oxide media particles are extruded cylinders 3 to 4 mm in diameter and 4 to 8 mm in length (Kohl and Nielsen, 1997) and react readily with the hydrogen sulfide:

$$ZnO + H_2S \rightarrow ZnS + H_2O$$

At increased temperatures (205°C to 370°C, 400°F to 700°F), zinc oxide has a rapid reaction rate, therefore providing a short mass transfer zone, resulting in a short length of unused bed and improved efficiency.

Removal of larger amounts of hydrogen sulfide from gas streams requires a continuous process, such as the Ferrox process or the Stretford process. The Ferrox process is based on the same chemistry as the iron oxide process except that it is fluid and continuous. The Stretford process employs a solution containing vanadium salts and anthraquinone disulfonic acid (Maddox, 1974). Most hydrogen sulfide removal processes return the hydrogen sulfide unchanged, but if the quantity involved does not justify installation of a sulfur recovery plant (usually a Claus plant) it is necessary to select a process that directly produces elemental sulfur.

The processes using ethanolamine and potassium phosphate are now widely used. The ethanolamine process, known as the **Girbotol process**, removes acid gases (hydrogen sulfide and carbon dioxide) from liquid hydrocarbons as well as from natural and from refinery gases. The Girbotol process uses an aqueous solution of ethanolamine ($H_2NCH_2CH_2OH$) that reacts with hydrogen sulfide at low temperatures and releases hydrogen sulfide at high temperatures. The ethanolamine solution fills a tower called an absorber through which the sour gas is bubbled. Purified gas leaves the top of the tower, and the ethanolamine solution leaves the bottom of the tower with the absorbed acid gases. The ethanolamine solution enters a reactivator tower where heat drives the acid gases from the solution. Ethanolamine solution, restored to its original condition, leaves the bottom of the reactivator tower to go to the top of the absorber tower, and acid gases are released from the top of the reactivator.

The process using potassium phosphate is known as phosphate desulfurization, and it is used in the same way as the Girbotol process to remove acid gases from liquid hydrocarbons as well as from gas streams. The treatment solution is a water solution of tripotassium phosphate (K_3PO_4), which is circulated through an absorber tower and a reactivator tower in much the same way as the ethanolamine is circulated in the Girbotol process; the solution is regenerated thermally.

Moisture may be removed from hydrocarbon gases at the same time as hydrogen sulfide is removed. Moisture removal is necessary to prevent harm to anhydrous catalysts and to prevent the formation of hydrocarbon hydrates (e.g., $C_3H_8 \cdot 18H_2O$) at low temperatures. A widely used dehydration and desulfurization process is the glycolamine process, in which the treatment solution is a mixture of ethanolamine and a large amount of glycol. The mixture is circulated through an absorber and a reactivator in the same way as ethanolamine is circulated in the Girbotol process. The glycol absorbs moisture from the hydrocarbon gas passing up the absorber; the ethanolamine absorbs hydrogen sulfide and carbon dioxide. The treated gas leaves the top of the absorber; the spent ethanolamine–glycol mixture enters the reactivator tower, where heat drives off the absorbed acid gases and water.

Other processes include the Alkazid *process* for removal of hydrogen sulfide and carbon dioxide using concentrated aqueous solutions of amino acids. The hot potassium carbonate process decreases the acid content of natural and refinery gas from as much as 50% to as low as 0.5% and operates in an unit similar to that used for amine treating. The Giammarco-Vetrocoke process is used for hydrogen sulfide and carbon dioxide removal. In the hydrogen sulfide removal section, the reagent consists of sodium or potassium carbonates containing a mixture of arsenites and arsenates; the carbon dioxide removal section utilizes hot aqueous alkali carbonate solution activated by arsenic trioxide or selenous acid or tellurous acid.

Molecular sieves are highly selective for the removal of hydrogen sulfide (as well as other sulfur compounds) from gas streams and over continuously high absorption efficiency. They are also an effective means of water removal and thus offer a process for the simultaneous dehydration and desulfurization of gas. Gas that has excessively high water content may require upstream dehydration, however (Rushton and Hayes, 1961).

The molecular sieve process is similar to the iron oxide process. Regeneration of the bed is achieved by passing heated clean gas over the bed. As the temperature of the bed increases, it releases the adsorbed hydrogen sulfide into the regeneration gas stream. The sour effluent regeneration gas is sent to a flare stack, and up to 2% of the gas seated can be lost in the regeneration process (Rushton and Hayes, 1961). A portion of the natural gas may also be lost by the adsorption of hydrocarbon components by the sieve.

In this process, unsaturated hydrocarbon components, such as olefins and aromatics, tend to be strongly adsorbed by the molecular sieve (Conviser, 1965). Molecular sieves are susceptible to poisoning by such chemicals as glycols and require thorough gas cleaning before the adsorption step. Alternatively, the sieve can be offered some degree of protection by the use of guard beds in which a less expensive catalyst is placed in the gas stream before contact of the gas with the sieve, thereby protecting the catalyst from poisoning. This concept is analogous to the use of guard beds or attrition catalysts in the petroleum industry (Speight, 2000).

25.7   ENRICHMENT

The purpose of enrichment is to produce natural gas for sale and enriched tank oil. The tank oil contains more light hydrocarbon liquids than natural petroleum, and the residue gas is drier (leaner, i.e., has lesser amounts of higher molecular weight hydrocarbons). Therefore, the process concept is essentially the separation of hydrocarbon liquids from the methane to produce a lean, dry gas.

Crude oil enrichment is used when there is no separate market for light hydrocarbon liquids or when the increase in API gravity of the crude provides a substantial increase in the price per unit volume as well as volume of the stock tank oil. A very convenient method of enrichment involves manipulation of the number and operating pressures of the gas–oil separators (traps). However, it must be recognized that alteration or manipulation of the separator pressure affects the gas compression operation as well as influences other processing steps.

One method of removing light ends involves the use of a pressure reduction (vacuum) system. Generally, stripping of light ends is achieved at low pressure, after which the pressure of the stripped crude oil is elevated so that the oil acts as an absorbent. The crude oil, which becomes enriched by this procedure, is then reduced to atmospheric pressure in stages or using fractionation (rectification).

25.8   FRACTIONATION

Fractionation processes are very similar to those processes classed as liquids removal processes but often appear to be more specific in terms of the objectives: hence the need to place the fractionation processes into a separate category. The fractionation processes are those processes that are used (1) to remove the more significant product stream first, or (2) to remove any unwanted light ends from the heavier liquid products.

In the general practice of natural gas processing, the first unit is a de-ethanizer followed by a depropanizer then by a debutanizer and, finally, a butane fractionator. Thus, each column can operate at a successively lower pressure, thereby allowing the different gas streams to flow from column to column by virtue of the pressure gradient, without the necessity of the use of pumps.

The purification of hydrocarbon gases by any of these processes is an important part of refinery operations, especially in regard to the production of liquefied petroleum gas (LPG). This is actually a mixture of propane and butane, which is an important domestic fuel, as well as an intermediate material in the manufacture of petrochemicals (Chapter 21). The presence of ethane in liquefied petroleum gas must be avoided because of the inability of this lighter hydrocarbon to liquefy under pressure at ambient temperatures and its tendency to register abnormally high pressures in the liquefied petroleum gas containers. On the other hand, the presence of pentane in liquefied petroleum gas must also be avoided, as this particular hydrocarbon (a liquid at ambient temperatures and pressures) may separate into a liquid state in the gas lines.

25.9 CLAUS PROCESS

The disposition of hydrogen sulfide, a toxic gas that originates in crude oils and is also produced in the coking, catalytic cracking, hydrotreating, and hydrocracking processes, is an issue with many refiners. Burning hydrogen sulfide as a fuel gas component or as a flare gas component is precluded by safety and environmental considerations, as one of the combustion products is the highly toxic sulfur dioxide (SO_2). As described above, hydrogen sulfide is typically removed from the refinery light ends gas streams through an olamine process after which application of heat regenerates the olamine and forms an acid gas stream. Following from this, the acid gas stream is treated to convert the hydrogen sulfide elemental sulfur and water. The conversion process utilized in most modern refineries is the Claus process, or a variant thereof.

The Claus process (Figure 25.8) involves combustion of approximately one-third of hydrogen sulfide to sulfur dioxide and then reaction of sulfur dioxide with the remaining hydrogen sulfide in the presence of a fixed bed of activated alumina, cobalt molybdenum catalyst resulting in the formation of elemental sulfur:

$$2H_2S + 3O_2 \rightarrow 2SO_2 + 2H_2O$$
$$2H_2S + SO_2 \rightarrow 3S + 2H_2O$$

Different process flow configurations are in use to achieve the correct hydrogen sulfide/sulfur dioxide ratio in the conversion reactors.

In a split-flow configuration, one-third split of the acid gas stream is completely combusted and the combustion products are then combined with the noncombusted acid gas upstream of the conversion reactors. In a once-through configuration, the acid gas stream is partially combusted by only providing sufficient oxygen in the combustion chamber to combust one-third of the acid gas. Two or three conversion reactors may be required depending on the level of hydrogen sulfide conversion required. Each additional stage provides incrementally less conversion than the previous stage.

Overall, conversion of 96% to 97% of hydrogen sulfide to elemental sulfur is achievable in the Claus process. If this is insufficient to meet air quality regulations, Claus process tail gas

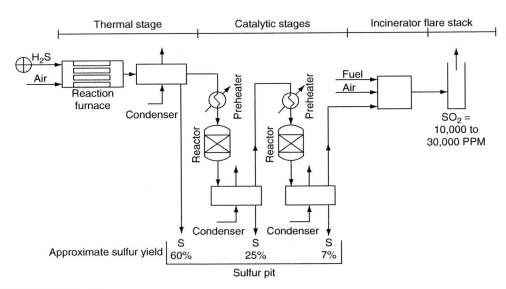

FIGURE 25.8 The Claus process.

treater is utilized to remove essentially the entire remaining hydrogen sulfide in the tail gas from the Claus unit. The tail gas treater may employ a proprietary solution to absorb the hydrogen sulfide followed by conversion to elemental sulfur.

The shell Claus off-gas treating (SCOT) unit is a most common type of tail gas unit and uses a hydrotreating reactor followed by amine scrubbing to recover and recycle sulfur, in the form of hydrogen, to the Claus unit (Nederland, 2004).

In the process (Figure 25.9), tail gas (containing hydrogen sulfide and sulfur dioxide) is contacted with hydrogen and reduced in a hydrotreating reactor to form hydrogen sulfide and water. The catalyst is typically cobalt or molybdenum on alumina. The gas is then cooled in a water contractor. The hydrogen sulfide-containing gas enters into an amine absorber that is

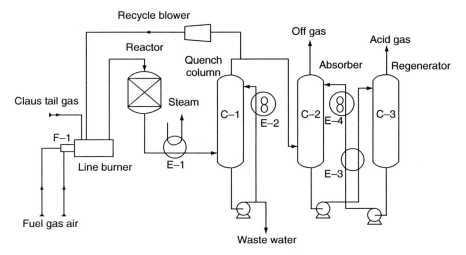

FIGURE 25.9 The SCOT process.

typically in a system segregated from the other refinery amine systems. The purpose of segregation is two-fold: (1) the tail gas treater frequently uses a different amine than the rest of the plant, and (2) the tail gas is frequently cleaner than the refinery fuel gas (in regard to contaminants) and segregation of the systems reduces maintenance requirements for the SCOT unit. Amines chosen for use in the tail gas system tend to be more selective for hydrogen sulfide and are not affected by the high levels of carbon dioxide in the off-gas.

The hydrotreating reactor converts sulfur dioxide in the off-gas to hydrogen sulfide that is then contacted with a Stretford solution (a mixture of vanadium salt, anthraquinone disulfonic acid, sodium carbonate, and sodium hydroxide) in a liquid-gas absorber. The hydrogen sulfide reacts stepwise with sodium carbonate and anthraquinone sulfonic acid to produce elemental sulfur, with vanadium serving as a catalyst. The solution proceeds to a tank where oxygen is added to regenerate the reactants. One or more froth or slurry tanks are used to skim the product sulfur from the solution, which is recirculated to the absorber.

Other tail gas treating processes include: (1) caustic scrubbing, (2) polyethylene glycol treatment, (3) Selectox process, and (4) sulfite/bisulfite tail gas treating.

REFERENCES

Anerousis, J.P. and Whitman, S.K. 1984. An Updated Examination of Gas Sweetening by the Iron Sponge Process. *Paper No. SPE 13280*. SPE Annual Technical Conference and Exhibition, Houston, TX. September.

Barbouteau, L. and Dalaud, R. 1972. *Gas Purification Processes for Air Pollution Control*. G. Nonhebel (ed.). Butterworth and Co., London, UK. Chapter 7.

Bartoo, R.K. 1985. *Acid and Sour Gas Treating Processes*. S.A. Newman (ed.). Gulf Publishing, Houston, TX.

Conviser, S.A. 1965. *Oil Gas J*. 63(49): 130.

Curry, R.N. 1981. *Fundamentals of Natural Gas Conditioning*. PennWell Publishing Co., Tulsa, OK.

Duckworth, G.L. and Geddes, J.H. 1965. *Oil Gas J*. 63(37): 94–96.

Foglietta, J.H. 2004. Dew Point Turboexpander Process: A Solution for High Pressure Fields. *Proceedings*. IAPG Gas Conditioning Conference, Neuquen, Argentina. October 18.

Fulker, R.D. 1972. *Gas Purification Processes for Air Pollution Control*. G. Nonhebel (ed.). Butterworth and Co., London, UK. Chapter 9.

Geist, J.M. 1985. *Oil Gas J*. 83(5): 56–60.

Jou, F.Y., Otto, F.D., and Mather, A.E. 1985. *Acid and Sour Gas Treating Processes*. S.A. Newman (ed.). Gulf Publishing Company, Houston, TX. Chapter 10.

Katz, D.K. 1959. *Handbook Of Natural Gas Engineering*. McGraw-Hill, New York.

Kohl, A.L. and Nielsen, R.B., 1997. *Gas Purification*. Gulf Publishing Company, Houston, TX.

Kohl, A.L. and Riesenfeld, F.C. 1985. *Gas Purification*, 4th Edition, Gulf Publishing Company, Houston, TX.

Maddox, R.N. 1974. *Gas and Liquid Sweetening*, 2nd Edition. Campbell Publishing Co., Norman, OK.

Maddox, R.N. 1982. *Gas Conditioning and Processing. Volume 4: Gas and Liquid Sweetening*. Campbell Publishing Co., Norman, OK.

Maddox, R.N., Bhairi, A., Mains, G.J., and Shariat, A. 1985. *Acid and Sour Gas Treating Processes*. S.A. Newman (ed.). Gulf Publishing Company, Houston, TX. Chapter 8.

Mantell, C.L. 1951. *Adsorption*. McGraw-Hill, New York.

Mody, V. and Jakhete, R. 1988. *Dust Control Handbook*. Noyes Data Corp., Park Ridge, NJ.

Mokhatab, S., Poe, W.A., and Speight, J.G. 2006. *Handbook of Natural Gas Transmission and Processing*. Elsevier, Amsterdam..

Nederland, J. 2004. *Sulphur*. University of Calgary, Calgary, Alberta, Canada. November.

Newman, S.A. 1985. *Acid and Sour Gas Treating Processes*. Gulf Publishing, Houston, TX.

Nonhebel, G. 1964. *Gas Purification Processes*. George Newnes Ltd., London, UK.

Pitsinigos, V.D. and Lygeros, A.I. 1989. *Hydrocarbon Process*. 58(4): 43.

Polasek, J. and Bullin, J. 1985. *Acid and Sour Gas Treating Processes*. S.A. Newman (ed.). Gulf Publishing Company, Houston, TX. Chapter 7.

Rook, R. 1994. *Chem. Process.* 57(11): 53.

Rushton, D.W. and Hayes, W. 1961. *Oil Gas J.* 59(38): 102.

Soud, H. and Takeshita, M. 1994. *FGD Handbook*. No. IEACR/65. International Energy Agency Coal Research, London, UK.

Speight, J.G. 1993. *Gas Processing: Environmental Aspects and Methods*. Butterworth Heinemann, Oxford, UK.

Speight, J.G. 1994. *The Chemistry and Technology of Coal*, 2nd Edition. Marcel Dekker Inc., New York.

Speight, J.G. 2000. *The Desulfurization of Heavy Oils and Residua*, 2nd Edition. Marcel Dekker Inc., New York.

Staton, J.S., Rousseau, R.W., and Ferrell, J.K. 1985. *Acid and Sour Gas Treating Processes*. S.A. Newman (ed.). Gulf Publishing Company, Houston, TX. Chapter 5.

Ward, E.R. 1972. *Gas Purification Processes for Air Pollution Control*. G. Nonhebel (ed.). Butterworth and Co., London, UK. Chapter 8.

Zapffe, F. 1963. *Oil Gas J.* 61(33): 103–104.

26 Products

26.1 INTRODUCTION

The constant demand for products such as liquid fuels is a major driving force behind the petroleum industry. Other products such as **lubricating oils**, waxes, and **asphalt** have also added to the popularity of petroleum as a major resource. There have, however, been many changes in emphasis on product demand since petroleum first came into use some five to six millennia ago (Table 26.1) (Chapter 1). It is these changes in product demand that have been largely responsible for the evolution of the industry from the demand for asphalt mastic used in ancient times to the current demand for **gasoline** and other liquid fuels.

The demand for petroleum products (Table 26.2)—particularly transportation fuels (gasoline and diesel) and petrochemical feedstocks (such as aromatics and olefins)—is increasing throughout the world. Traditional markets such as North America and Europe are experiencing moderate increase in demand, whereas emerging Asian markets such as India and China are witnessing a rapid surge. This has resulted in a squeeze on existing refineries, prompting a fresh technological approach to optimize efficiency and throughput. Major oil companies and technology suppliers and licensors are investing heavily to revamp their refining technologies in an effort to cater to the growing needs of customers.

A steady evolution in product specifications caused by an endless wave of fresh environmental regulations plays a major role in the development of petroleum refining technologies. In the United States and Europe, gasoline and diesel specifications have changed radically in the past decades and will continue to do so in the future. Currently, reducing the sulfur levels of finished products is the dominant objective. Sulfur is ubiquitous in petroleum and refiners are seeking technologies on how to achieve the mandated levels of sulfur in petroleum products.

As petroleum products are shipped worldwide, they need to comply with stringent environment-related regulations prevalent in specific countries. Japan and Singapore have already implemented strict legislation and many countries are likely to follow suit as they confront environmental issues such as smog. These changing rules also cause a negative impact on the market for heavy products such as fuel oil.

Refineries are eager to adapt to changing circumstances and are amenable to trying new technologies that are radically different in character. This is evident from the increasing use of ultrasonic technology and novel separation methods. Currently, they are also looking to exploit heavy (more viscous) crude oils, provided they have the refinery technology capable of handling them. Heavy crude oil, relative to light or conventional crude oil, is rich in higher boiling constituents (Figure 26.1) and transforming the heavier components into light fractions is thus a necessity. However, this requires technological changes—including more effective use of hydrogen within the refinery—and that increases operating costs.

Heavier crude oil could also be contaminated with sulfur and metal particles that must be removed to meet quality standards. "A deeper understanding of how catalysts work—both

TABLE 26.1
Petroleum and Derivatives Such as Asphalt Have Been Known and Used for Almost Six Thousand Years

3800 BC	First documented use of asphalt for caulking reed boats.
3500 BC	Asphalt used as cement for jewelry and for ornamental applications.
3000 BC	Use of asphalt as a construction cement by Sumerians; also believed to be used as a road material; asphalt used to seal bathing pool or water tank at Mohenjo Daro.
2500 BC	Use of asphalt and other petroleum liquids (oils) in the embalming process; asphalt believed to be widely used for caulking boats.
1500 BC	Use of asphalt for medicinal purposes and (when mixed with beer) as a sedative for the stomach; continued reference to use of asphalt liquids (oil) as illuminant in lamps.
1000 BC	Use of asphalt as a waterproofing agent by lake dwellers in Switzerland.
500 BC	Use of asphalt mixed with sulfur as an incendiary device in Greek wars; also use of asphalt liquid (oil) in warfare.
350 BC	Occurrence of flammable oils in wells in Persia.
300 BC	Use of asphalt and liquid asphalt as incendiary device (Greek fire) in warfare.
250 BC	Occurrences of asphalt and oil seepage in several areas of the Fertile Crescent (Mesopotamia); repeated use of liquid asphalt (oil) as an illuminant in lamps.
750 AD	Use in Italy of asphalt as a color in paintings.
950 AD	Report of destructive distillation of asphalt to produce distillate; reference to distillate as nafta (naft, naphtha).
1500 AD	Discovery of asphalt deposits in the Americas; first attempted documentation of the relationship of asphalt and naphtha (petroleum).
1600 AD	Asphalt used for a variety of tasks; relationship of asphalt to coal and wood tar studied; asphalt studied; used for paving; continued documentation of the use of naphtha as an illuminant and the production of naphtha from asphalt; importance of naphtha as fuel realized.
1859 AD	Discovery of petroleum in North America; birth of modern-day petroleum science and refining.

TABLE 26.2
Petroleum Products

Product	Lower Carbon Limit	Upper Carbon Limit	Lower Boiling Point (°C)	Upper Boiling Point (°C)	Lower Boiling Point (°F)	Upper Boiling Point (°F)
Refinery gas	C_1	C_4	−161	−1	−259	31
Liquefied petroleum gas	C_3	C_4	−42	−1	−44	31
Naphtha	C_5	C_{17}	36	302	97	575
Gasoline	C_4	C_{12}	−1	216	31	421
Kerosene/diesel fuel	C_8	C_{18}	126	258	302	575
Aviation turbine fuel	C_8	C_{16}	126	287	302	548
Fuel oil	C_{12}	$>C_{20}$	216	421	>343	>649
Lubricating oil	$>C_{20}$		>343		>649	
Wax	C_{17}	$>C_{20}$	302	>343	575	>649
Asphalt	$>C_{20}$		>343		>649	
Coke	C_{50}[a]		>1000[a]		>1832[a]	

[a] Carbon number and boiling point difficult to assess; inserted for illustrative purposes only.

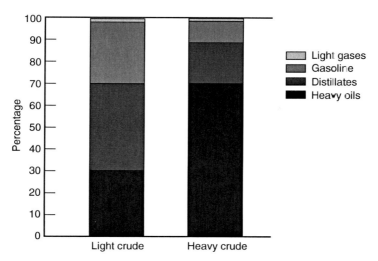

FIGURE 26.1 Variation in the distribution of products from light crude oil and heavy oil.

chemically and physically—is providing greater scope for technological improvements," says the analyst. "Nanotechnology and combinatorial chemistry are among the techniques that are likely to help push forward the frontiers of efficiency and selectivity," he says. Even though the nature of crude oil is changing, refineries are here to stay in the foreseeable future, as petroleum products satisfy wide-ranging energy requirements that are not fully catered to by **natural gas**, **liquefied petroleum gas** (LPG), or coal. At present, alternative energy schemes lack technological sophistication or economic sufficiency to be considered a substitute for petroleum products.

A major group of products from petroleum (petrochemicals) are the basis of a major industry. They are, in the strictest sense, different from petroleum products insofar as the petrochemicals are the basic building blocks of the chemical industry. They will not be considered here and will be dealt with separately in this work (Chapter 27).

Unlike processes, products are more difficult to place on an individual evolutionary scale. Processes changed and evolved to accommodate the demand for, say, higher octane fuel or longer-lasting asphalts or lower sulfur **coke**. In this section, a general overview of some petroleum products is presented to show the raison d'etre of the industry.

Petroleum products (in contrast to petrochemicals) are those bulk fractions that are derived from petroleum and have commercial value as a bulk product. In the strictest sense, petro-chemicals (Chapter 21) are also petroleum products but they are individual chemicals that are used as the basic building blocks of the chemical industry.

The use of petroleum and its products was established in pre-Christian times and is known largely through documentation by many of the older civilizations (Chapter 1) (Abraham, 1945; Forbes, 1958a,b, 1959; James and Thorpe, 1994). Thus, the use of petroleum and the development of related technology is not such a modern subject as we are inclined to believe. However, the petroleum industry is essentially a twentieth-century industry but to understand the evolution of the industry, it is essential to have a brief understanding of the first uses of petroleum (Chapter 1).

There have been many changes in emphasis on product demand since petroleum first came into use some five to six millennia ago (Chapter 1). It is these changes in product demand that have been largely responsible for the evolution of the industry, from the asphalt used in ancient times to the gasoline and other liquid fuels of today. The modern petroleum industry began in 1859 with the discovery and subsequent commercialization of petroleum in Pennsylvania. After completion of the first well (by Edwin Drake), the surrounding areas were immediately leased and extensive drilling took place. It is from this time onward that modern petroleum products began to evolve.

Petroleum is an extremely complex mixture of hydrocarbon compounds, usually with minor amounts of nitrogen-containing, oxygen-containing, and sulfur-containing compounds as well as trace amounts of metal-containing compounds (Chapter 6). In addition, the properties of petroleum vary widely (Chapter 1 and Chapter 9). Thus, petroleum is not used in its raw state. A variety of processing steps is required to convert petroleum from its raw state to products that have well-defined properties (Table 26.3).

The nomenclature of petroleum products is as diverse as the nomenclature applied to petroleum itself (Chapter 1). For example, several names can be applied to one product (Chapter 1) and caution is advised when attempting to define a petroleum product.

TABLE 26.3
Properties of Selected Petroleum Products

	Molecular Weight	Specific Gravity	Boiling Point (°F)	Ignition Temperature (°F)	Flash Point (°F)	Flammability Limits in Air (% v/v)
Benzene	78.1	0.879	176.2	1040	12	1.35–6.65
n-Butane	58.1	0.601	31.1	761	−76	1.86–8.41
iso-Butane	58.1		10.9	864	−117	1.80–8.44
n-Butene	56.1	0.595	21.2	829	Gas	1.98–9.65
iso-Butene	56.1		19.6	869	Gas	1.8–9.0
Diesel fuel	170–198	0.875			100–130	
Ethane	30.1	0.572	−127.5	959	Gas	3.0–12.5
Ethylene	28.0		−154.7	914	Gas	2.8–28.6
Fuel oil No. 1		0.875	304–574	410	100–162	0.7–5.0
Fuel oil No. 2		0.920		494	126–204	
Fuel oil No. 4	198.0	0.959		505	142–240	
Fuel oil No. 5		0.960			156–336	
Fuel oil No. 6		0.960			150	
Gasoline	113.0	0.720	100–400	536	−45	1.4–7.6
n-Hexane	86.2	0.659	155.7	437	−7	1.25–7.0
n-Heptane	100.2	0.668	419.0	419	25	1.00–6.00
Kerosene	154.0	0.800	304–574	410	100–162	0.7–5.0
Methane	16.0	0.553	−258.7	900–1170	Gas	5.0–15.0
Naphthalene	128.2		424.4	959	174	0.90–5.90
Neohexane	86.2	0.649	121.5	797	−54	1.19–7.58
Neopentane	72.1		49.1	841	Gas	1.38–7.11
n-Octane	114.2	0.707	258.3	428	56	0.95–3.2
iso-Octane	114.2	0.702	243.9	837	10	0.79–5.94
n-Pentane	72.1	0.626	97.0	500	−40	1.40–7.80
iso-Pentane	72.1	0.621	82.2	788	−60	1.31–9.16
n-Pentene	70.1	0.641	86.0	569	–	1.65–7.70
Propane	44.1		−43.8	842	Gas	2.1–10.1
Propylene	42.1		−53.9	856	Gas	2.00–11.1
Toluene	92.1	0.867	321.1	992	40	1.27–6.75
Xylene	106.2	0.861	281.1	867	63	1.00–6.00

The constant demand for products, such as liquid fuels, is the main driving force behind the petroleum industry (Guthrie, 1960; Hobson and Pohl, 1973). Other products, such as lubricating oils, waxes, and asphalt, have also added to the popularity of petroleum as a national resource. Indeed, fuel products that are derived from petroleum supply more than half of the world's total supply of energy. Gasoline, kerosene, and diesel oil provide fuel for automobiles, tractors, trucks, aircraft, and ships. Fuel oil and natural gas are used to heat homes and commercial buildings, as well as to generate electricity. Petroleum products are the basic materials used for the manufacture of synthetic fibers for clothing and in plastics, paints, fertilizers, insecticides, soaps, and synthetic rubber. The uses of petroleum as a source of raw material in manufacturing are central to the functioning of modern industry.

There is a myriad of other products that have evolved during the short span since the initiation of the petroleum industry. And the complexities of product composition have matched the evolution of the products (Hoffman, 1992). In fact, it is the complexity of product composition that had an adverse effect on product use. Product complexity has made the industry unique. Indeed, current analytical techniques that are accepted as standard methods, for example, the aromatics content of fuels (ASTM D-1319, ASTM D-2425, ASTM D-2549, ASTM D-2786, ASTM D-2789), as well as proton and carbon nuclear magnetic resonance methods, yield different information. Each method will yield the "% aromatics" in the sample but the data must be evaluated within the context of the method.

Product complexity, and the means by which the product is evaluated, has made the industry unique among industries. But product complexity has also brought to the fore issues such as instability and incompatibility. To understand the evolution of the products it is essential to have an understanding of the composition of the various products.

Product complexity becomes even more meaningful when various fractions from different types of crude oil as well as fractions from synthetic crude oil are blended with the corresponding petroleum stock. The implications for refining the fractions to salable products increase (Dooley et al., 1979). However, for the main part, the petroleum industry was inspired by the development of the automobile and the continued demand for gasoline and other fuels (Hoffman, 1992). Such a demand has been accompanied by the demand for other products: diesel fuel for engines, lubricants for engine and machinery parts, fuel oil to provide power for the industrial complex, and asphalt for roadways.

Unlike processes, products are more difficult to place on an individual evolutionary scale. Processes changed and evolved to accommodate the demand for, say, higher-octane fuels, longer-lasting asphalt, or lower sulfur coke. In this section, a general overview of some petroleum products is presented to show the raison d'être of the industry. Another consideration that must be acknowledged is the change in character and composition of the original petroleum feedstock (Chapter 7 and Chapter 8) (Long and Speight, 1998). In the early days of the petroleum industry several products were obtained by distillation and could be used without any further treatment. Nowadays the different character and composition of the petroleum dictates that any liquids obtained by distillation must go through one or more of the several available product improvement processes (Chapter 18). Such changes in feedstock character and composition have caused the refining industry to evolve in a direction such that changes in the petroleum can be accommodated.

It must also be recognized that adequate storage facilities for the gases, liquids, and solids that are produced during the refining operations are also an essential part of a refinery. Without such facilities, refineries would be incapable of operating efficiently.

The customary processing of petroleum does not usually involve the separation and handling of pure hydrocarbons. Indeed, petroleum-derived products are always mixtures: occasionally simple but more often very complex. Thus, for the purposes of this chapter, such materials as the gross fractions of petroleum (e.g., gasoline, **naphtha**, kerosene, and the like) that are usually obtained by distillation or refining are classed as petroleum products; asphalt and other solid products (e.g., wax) are also included in this division.

This type of classification separates this group of products from those obtained as petroleum chemicals (petrochemicals), for which the emphasis is on separation and purification of single chemical compounds, which are in fact starting materials for a host of other chemical products.

26.2   GASEOUS FUELS

Natural gas, which is predominantly methane, occurs in underground reservoirs separately or in association with crude oil (Chapter 3). The principal types of gaseous fuels are oil (distillation) gas, reformed natural gas, and reformed propane or LPG.

LPG is the term applied to certain specific hydrocarbons and their mixtures, which exist in the gaseous state under atmospheric ambient conditions but can be converted to the liquid state under conditions of moderate pressure at ambient temperature. These are the light hydrocarbon fractions of the paraffin series, derived from refinery processes, crude oil stabilization plants, and natural gas processing plants comprising propane ($CH_3CH_2CH_3$), butane ($CH_3CH_2CH_2CH_3$), *iso*-butane [$CH_3CH(CH_3)CH_3$] and to a lesser extent propylene ($CH_3CH=CH_2$), or butylene ($CH_3CH_2CH=CH_2$). The most common commercial products are propane, butane, or a mixture of the two (Table 26.4), and are generally extracted from natural gas or crude petroleum. Propylene and butylenes result from **cracking** other hydrocarbons in a petroleum refinery and are two important chemical feedstocks.

Mixed gas is a gas prepared by adding natural gas or LPG to a manufactured gas, giving a product of better utility and higher heat content or Btu value.

26.2.1   COMPOSITION

The principal constituent of natural gas is methane (CH_4). Other constituents are paraffinic hydrocarbons such as ethane (CH_3CH_3), propane ($CH_3CH_2CH_3$), and the butanes [$CH_3CH_2CH_2CH_3$ and $(CH_3)_3CH$]. Many natural gases contain nitrogen (N_2) as well as

TABLE 26.4
Properties of Propane and Butane

	Propane	Butane
Formula	C_3H_8	C_4H_{10}
Boiling point, °F	−44°	32°
Specific gravity—gas (air = 1.00)	1.53	2.00
Specific gravity—liquid (water = 1.00)	0.51	0.58
lb/gallon—liquid at 60°F	4.24	4.81
BTU/gallon—gas at 60°F	91,690	102,032
BTU/Lb—gas	21,591	21,221
BTU/ft³—gas at 60°F	2516	3280
Flash point, °F	−156	−96
Ignition temperature in air, °F	920–1020	900–1000
Maximum flame temperature in air, °F	3595	3615
Octane number (*iso*-octane = 100)	100+	92

carbon dioxide (CO_2) and hydrogen sulfide (H_2S). Trace quantities of argon, hydrogen, and helium may also be present. Generally, the hydrocarbons having a higher molecular weight than methane, carbon dioxide, and hydrogen sulfide are removed from natural gas before to its use as a fuel. Gases produced in a refinery contain methane, ethane, ethylene, propylene, hydrogen, carbon monoxide, carbon dioxide, and nitrogen, with low concentrations of water vapor, oxygen, and other gases.

26.2.2 Manufacture

Unless produced specifically as a product (e.g., LPG), the gaseous products of refinery operations are mixtures of various gases. Each gas is a by-product of a refining process. Thus, the compositions of natural, manufactured, and mixed gases can vary so widely, no single set of specifications could cover all situations.

26.2.3 Properties and Uses

As already noted, the compositions of natural, manufactured, and mixed gases can vary so widely, no single set of specifications could cover all situations. The requirements are usually based on performances in burners and equipment, on minimum heat content, and on maximum sulfur content. Gas utilities in most states come under the supervision of state commissions or regulatory bodies and the utilities must provide a gas that is acceptable to all types of consumers and that will give satisfactory performance in all kinds of consuming equipment. However, there are specifications for LPG (ASTM D-1835) that depend upon the required volatility.

As natural gas as delivered to pipelines has practically no odor, the addition of an odorant is required by most regulations so that the presence of the gas can be detected readily in case of accidents and leaks. This odorization is provided by the addition of trace amounts of some organic sulfur compounds to the gas before it reaches the consumer. The standard requirement is that a user will be able to detect the presence of the gas by odor when the concentration reaches 1% of gas in air. As the lower limit of flammability of natural gas is approximately 5%, this 1% requirement is essentially equivalent to one-fifth the lower limit of flammability. The combustion of these trace amounts of odorant does not create any serious problems of sulfur content or toxicity.

The different methods for gas analysis include absorption, distillation, combustion, mass spectroscopy, infrared spectroscopy, and gas chromatography (ASTM D-2163, ASTM D-2650, and ASTM D-4424). Absorption methods involve absorbing individual constituents one at a time in suitable solvents and recording of contraction in volume measured. Distillation methods depend on the separation of constituents by fractional distillation and measurement of the volumes distilled. In combustion methods, certain combustible elements are caused to burn to carbon dioxide and water, and the volume changes are used to calculate composition. Infrared spectroscopy is useful in particular applications. For the most accurate analyses, mass spectroscopy and gas chromatography are the preferred methods.

The specific gravity of product gases, including LPG, may be determined conveniently by a number of methods and a variety of instruments (ASTM D-1070, ASTM D-4891).

The heat value of gases is generally determined at constant pressure in a flow calorimeter in which the heat released by the combustion of a definite quantity of gas is absorbed by a measured quantity of water or air. A continuous recording calorimeter is available for measuring heat values of natural gases (ASTM D-1826).

The lower and upper limits of flammability of organic compounds (Table 26.5) indicate the percentage of combustible gas in air below which and above which flame will not propagate. When flame is initiated in mixtures having compositions within these limits, it will propagate

TABLE 26.5
Flammability Limits of Selected Organic Compounds

Compound	Limits of Flammability	
	Lower Vol.%	Upper Vol.%
Acetaldehyde	3.97	57.00
Acetic acid	5.40	20.00
Acetone	2.55	12.80
Acetylene	2.50	80.00
Allyl alcohol	2.50	18.00
Allyl bromide	4.36	7.25
Allyl chloride	3.28	11.15
n-Amyl acetate	1.10	7.50
n-Amyl alcohol	1.19	10.00
iso-Amyl alcohol	1.20	9.00
n-Amyl chloride	1.60	8.63
n-Amylene	1.42	8.70
Benzene	1.40	7.10
n-Butane	1.86	8.41
iso-Butane	1.80	8.44
Butene-1	1.65	9.95
Butene-2	1.75	9.70
n-Butyl acetate	1.39	7.55
n-Butyl alcohol	1.45	11.25
iso-Butyl alcohol	1.68	9.80
n-Butyl chloride	1.85	10.10
iso-Butyl chloride	2.05	8.75
Carbon disulfide	1.25	50.00
Crotonic aldehyde	2.12	15.50
Cyclohexane	1.26	7.75
Cyclopropane	2.40	10.40
n-Decane	0.77	5.35
Diethylamine	1.77	10.10
Diethyl ether	1.85	36.50
Diethyl peroxide	2.34	
Dimethylamine	2.80	14.40
2,3-Dimethylpentane	1.12	6.75
2,2-Dimethylpropane	1.38	7.50
1,4-Dioxane	1.97	22.25
Divinyl ether	1.70	27.00
Ethane	3.00	12.50
Ethyl acetate	2.18	11.40
Ethyl alcohol	3.28	18.95
Ethylamine	3.55	13.95
Ethyl bromide	6.75	11.25
Ethyl chloride	4.00	14.80
Ethylene	2.75	28.60
Ethylene dichloride	6.20	15.90
Ethylene oxide	3.00	80.00
Ethyl formate	2.75	16.40
Ethyl nitrate	3.80	
Ethyl nitrite	3.01	50.00

TABLE 26.5 (continued)

Compound	Limits of Flammability	
	Lower Vol.%	Upper Vol.%
Furfural	2.10	19.30
n-Heptane	1.10	6.70
n-Hexane	1.18	7.40
Methane	5.00	15.00
Methyl acetate	3.15	15.60
Methyl alcohol	6.72	36.50
Methylamine	4.95	20.75
Methyl bromide	13.50	14.50
Methyl iso-butyl ketone	1.35	7.60
Methyl chloride	8.25	18.70
Methylcyclohexane	1.15	6.70
Methyl ethyl ether	2.00	10.00
Methyl ethyl ketone	1.81	9.50
Methyl formate	5.05	22.70
Methyl iso-propyl ketone	1.55	8.15
n-Nonane	0.83	2.90
n-Octane	0.95	6.50
Paraldehyde	1.30	
n-Pentane	1.40	7.80
iso-Pentane (2-methylbutane)	1.32	7.60
Propane	2.12	9.35
n-Propyl acetate	1.77	8.00
Iso-Propyl acetate	1.78	7.80
n-Propyl alcohol	2.15	13.50
iso-Propyl alcohol	2.02	11.80
Propylamine	2.01	10.35
n-Propyl chloride	2.60	11.10
Propylene	2.00	11.20
Propylene dichloride	3.40	14.50
Propylene oxide	2.00	22.00
Pyridine	1.81	12.40
Toluene	1.27	6.75
Triethylamine	1.25	7.90
Trimethylamine	2.00	11.60
Turpentine	0.80	
Vinyl chloride	4.00	21.70
o-Xylene	1.00	6.00
m-Xylene	1.10	7.00
p-Xylene	1.10	7.00

and therefore the mixtures are flammable. Knowledge of flammable limits and their use in establishing safe practices in handling gaseous fuels is important, for example, when purging equipment used in gas service, in controlling factory or mine atmospheres, or in handling liquefied gases.

Many factors enter into the experimental determination of flammable limits of gas mixtures, including the diameter and length of the tube or vessel used for the test, the temperature and pressure of the gases, and the direction of flame propagation—upward or downward. For these and other reasons, great care must be used in the application of the data. In monitoring closed spaces where small amounts of gases enter the atmosphere, often the maximum concentration of the combustible gas is limited to one-fifth of the concentration of the gas at the lower limit of flammability of the gas–air mixture.

26.3 GASOLINE

Gasoline, also called gas (United States and Canada), or **petrol** (Great Britain) or **benzine** (Europe) is a mixture of volatile, flammable liquid hydrocarbons derived from petroleum and used as fuel for internal-combustion engines. It is also used as a solvent for oils and fats. Originally a by-product of the petroleum industry (kerosene being the principal product), gasoline became the preferred automobile fuel because of its high energy of combustion and capacity to mix readily with air in a carburetor.

Gasoline is a mixture of hydrocarbons that usually boil below 180°C (355°F) or, at most, below 200°C (390°F). The hydrocarbon constituents in this boiling range are those that have 4 to 12 carbon atoms in their molecular structure and fall into three general types: paraffins (including the cycloparaffins and branched materials), olefins, and aromatics.

Gasoline is still in great demand as a major product from petroleum. The network of interstate highways that links towns and cities in the United States are dotted with frequent service centers where motorists can obtain refreshment not only for themselves but also for their vehicles.

26.3.1 COMPOSITION

Gasoline is manufactured to meet specifications and regulations and not to achieve a specific distribution of hydrocarbons by class and size. However, chemical composition often defines properties. For example, volatility is defined by the individual hydrocarbon constituents and the lowest-boiling constituent(s) defines the volatile as determined by certain test methods.

Up to, and during, the first decade of the twentieth century, the gasoline produced was that which was originally present in crude oil or which could be condensed from natural gas. However, it was soon discovered that if the heavier portions of petroleum (such as the fraction that boiled higher than kerosene, e.g., gas oil) were heated to more severe temperatures, thermal degradation (or cracking) occurred to produce smaller molecules that were within the range suitable for gasoline. Therefore, gasoline that was not originally in the crude petroleum could be manufactured.

At first, cracked gasoline was regarded as an inferior product because of its comparative instability on storage but as more gasoline was required, the petroleum industry revolved around processes by which this material could be produced (e.g., catalytic cracking, thermal and catalytic reforming, **hydrocracking**, **alkylation**, and **polymerization**) and the problem of storage instability was addressed and resolved.

Automotive gasoline typically contains about almost 200 (if not several hundred) hydrocarbon compounds. The relative concentrations of the compounds vary considerably depending on the source of crude oil, refinery process, and product specifications. Typical hydrocarbon chain lengths range from C_4 through C_{12} with a general hydrocarbon distribution consisting of alkanes (4–8%), alkenes (2–5%), *iso*-alkanes 25–40%, cycloalkanes (3–7%), cycloalkenes (1–4%), and aromatics (20–50%). However, these proportions vary greatly.

The majority of the members of the paraffin, olefin, and aromatic series (of which there are about 500) boiling below 200°C (390°F) have been found in the gasoline fraction of petroleum. However, it appears that the distribution of the individual members of straight-run gasoline (i.e., distilled from petroleum without thermal alteration) is not even.

Highly branched paraffins, which are particularly valuable constituents of gasoline(s), are not usually the principal paraffinic constituents of straight-run gasoline. The more predominant paraffinic constituents are usually the normal (straight-chain) isomers, which may dominate the branched isomer(s) by a factor of 2 or more. This is presumed to indicate the tendency to produce long uninterrupted carbon chains during petroleum maturation rather than those in which branching occurs. However, this trend is somewhat different for the cyclic constituents of gasoline, that is, cycloparaffins (naphthenes) and aromatics. In these cases, the preference appears to be for several short side chains rather than one long substituent.

Gasoline can vary widely in composition: even those with the same **octane number** may be quite different, not only in the physical makeup but also in the molecular structure of the constituents. For example, the Pennsylvania petroleum is high in paraffins (normal and branched), but the California and Gulf Coast crude oils are high in cycloparaffins. Low-boiling distillates with high content of aromatic constituents (above 20%) can be obtained from some Gulf Coast and West Texas crude oils, as well as from crude oils from the Far East. The variation in aromatics content as well as the variation in the content of normal paraffins, branched paraffins, cyclopentanes, and cyclohexanes involve characteristics of any one individual crude oil and may in some instances be used for crude oil identification. Furthermore, straight-run gasoline generally shows a decrease in paraffin content with an increase in molecular weight, but the cycloparaffins (naphthenes) and aromatics increase with increasing molecular weight. Indeed, the hydrocarbon type variation may also vary markedly from process to process.

The reduction of the lead content of gasoline and the introduction of reformulated gasoline has been very successful in reducing automobile emissions (Wittcoff, 1987; Absi-Halabi et al., 1997). Further improvements in fuel quality have been proposed for the years 2000 and beyond. These projections are accompanied by a noticeable and measurable decrease in crude oil quality and the reformulated gasoline will help meet environmental regulations for emissions for liquid fuels (Chapter 28).

26.3.2 MANUFACTURE

Gasoline was at first produced by distillation, simply separating the volatile, more valuable fractions of crude petroleum. Later processes, designed to raise the yield of gasoline from crude oil, decomposed higher molecular weight constituents into lower molecular weight products by processes known as cracking (Chapter 14 and Chapter 15). And like typical gasoline, several processes produce the blending stocks for reformulated gasoline (Figure 26.2).

Up to and during the first decade of the present century, the gasoline produced was that originally present in crude oil or that which could be condensed from natural gas. However, it was soon discovered that if the heavier portions of petroleum (such as the fraction that boiled higher than kerosene, e.g., gas oil) were heated to more severe temperatures, thermal degradation (or cracking) occurred to produce smaller molecules within the range suitable for gasoline. Therefore, gasoline that was not originally in the crude petroleum could be manufactured.

Thermal cracking, employing heat and high pressures, was introduced in 1913 but was replaced after 1937 by catalytic cracking, the application of catalysts that facilitate chemical reactions producing more gasoline. Other methods used to improve the quality of gasoline and increase its supply include polymerization, alkylation, **isomerization**, and **reforming** (Chapter 18).

Polymerization is the conversion of gaseous olefins, such as propylene and butylene, into larger molecules in the gasoline range. Alkylation is a process combining an olefin and

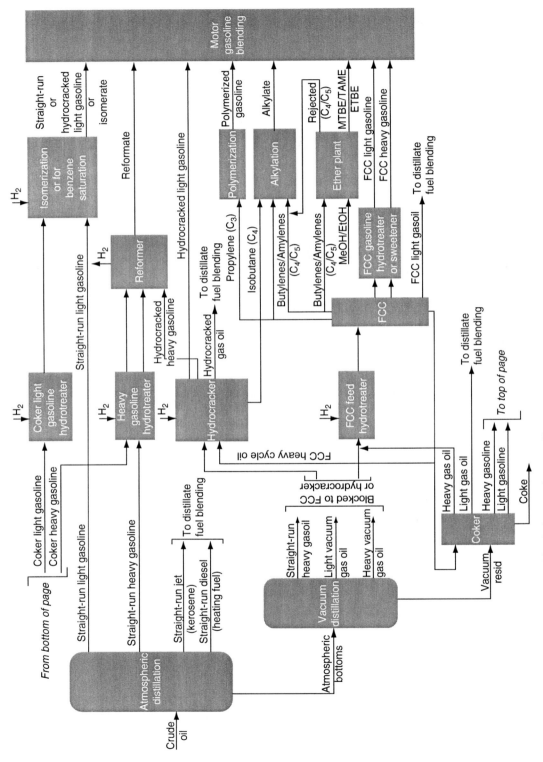

FIGURE 26.2 Refinery streams that are blended to produce gasoline.

paraffin such as *iso*-butane. Isomerization is the conversion of straight-chain hydrocarbons to branched-chain hydrocarbons. Reforming is the use of either heat or a catalyst to rearrange the molecular structure.

At first cracked gasoline was regarded as an inferior product because of its comparative instability on storage, and special techniques had to be developed to combat this marked instability. As more and better-quality gasoline was required, the petroleum industry revolved around processes by which this material could be produced. Such processes were catalytic cracking, thermal and catalytic reforming, hydrocracking, alkylation, and polymerization. The end product was the production of gasoline constituents with enhanced stability and performance.

However, despite the variations in the composition of the gasoline produced by the various available processes, this material is rarely if ever suitable for use as such. It is at this stage of a refinery operation that blending becomes important.

Aviation gasoline is a form of motor gasoline that has been especially prepared for use for aviation piston engines. It has an octane number suited to the engine, a freezing point of −60°C (−76°F), and a distillation range usually within the limits of 30°C to 180°C (86°F to 356°F) compared to −1°C to 200°C (30°F to 390°F) for automobile gasoline. The narrower boiling range ensures better distribution of the vaporized fuel through the more complicated induction systems of aircraft engines. Aircraft operate at altitudes at which the prevailing pressure is less than the pressure at the surface of the earth (pressure at 17,500 ft is 7.5 psi compared to 14.7 psi at the surface of the earth). Thus, the vapor pressure of aviation gasoline must be limited to reduce boiling in the tanks, fuel lines, and carburetors. Thus, aviation gasoline does not usually contain the gaseous hydrocarbons (butanes) that give automobile gasoline the higher vapor pressures.

Aviation gasoline is strictly limited regarding hydrocarbon composition. The important properties of the hydrocarbons are the highest octane numbers economically possible, boiling points in the limited temperature range of aviation gasoline, maximum heat contents per pound (high proportion of combined hydrogen), and high chemical stability to withstand storage. Aviation gasoline is composed of paraffins and *iso*-paraffins (50% to 60%), moderate amounts of naphthenes (20% to 30%), small amounts of aromatics (10%), and usually no olefins, whereas motor gasoline may contain up to 30% olefins and up to 40% aromatics.

Under conditions of use in aircraft, olefins have a tendency to form gum, cause preignition, and have relatively poor **antiknock** characteristics under lean mixture (cruising) conditions; for these reasons olefins are detrimental to aviation gasoline. Aromatics have excellent antiknock characteristics under rich mixture (takeoff) conditions, but are much like the olefins under lean mixture conditions; hence the proportion of aromatics in aviation gasoline is limited. Some naphthenes with suitable boiling temperatures are excellent aviation gasoline components but are not segregated as such in refinery operations. They are usually natural components of the straight-run naphtha (aviation base stocks) used in blending aviation gasoline. The lower-boiling paraffins (pentane and hexane), and both the high-boiling and low-boiling *iso*-paraffins (*iso*-pentane to *iso*-octane) are excellent aviation gasoline components. These hydrocarbons have high heat contents per pound and are chemically stable, and the *iso*-paraffins have high octane numbers under both lean and rich mixture conditions.

The manufacture of aviation gasoline is thus dependent on the availability and selection of fractions containing suitable hydrocarbons. The lower-boiling hydrocarbons are usually found in straight-run naphtha from certain crude petroleum. These fractions have high contents of *iso*-pentanes and *iso*-hexane and provide the needed volatility, as well as high octane number components. Higher-boiling *iso*-paraffins are provided by aviation alkylate, which consists mostly of branched octanes. Aromatics, such as benzene, toluene, and xylene, are obtained from catalytic reforming or a similar source.

To increase the proportion of higher-boiling octane components, such as aviation alkylate and xylenes, the proportion of lower-boiling components must also be increased to maintain the proper volatility. *Iso*-pentane and, to some extent, *iso*-hexane are the lower-boiling components used. *Iso*-pentane and *iso*-hexane may be separated from selected naphtha by superfractionators or synthesized from the normal hydrocarbons by isomerization. In general, most aviation gasoline are made by blending a selected straight-run naphtha fraction (aviation base stock) with *iso*-pentane and aviation alkylate.

26.3.3 PROPERTIES AND USES

Despite the diversity of the processes within a modern petroleum refinery, no single stream meets all the requirements of gasoline. Thus, the final step in gasoline manufacture is blending the various streams into a finished product (Figure 26.3). It is not uncommon for the finished gasoline to be made up of six or more streams (Figure 3.4) and several factors make this flexibility critical: (1) the requirements of the gasoline specification (ASTM D-4814) and the regulatory requirements and (2) performance specifications that are subject to local climatic conditions and regulations.

The early criterion for gasoline quality was Baumé (or API) gravity (Chapter 8). For example, a 70° API gravity gasoline contained fewer, if any, of the heavier gasoline constituents than a 60° API gasoline. Therefore, the 70° API gasoline was of a higher quality and, hence, economically more valuable gasoline. However, apart from being used as a rough estimation of quality (not only for petroleum products but also for crude petroleum), specific gravity is no longer of any significance as a true indicator of gasoline quality.

26.3.4 OCTANE NUMBERS

Gasoline performance and hence quality of an automobile gasoline is determined by its resistance to knock, for example, detonation or ping during service. The antiknock quality of the fuel limits the power and economy that an engine using that fuel can produce: the higher the antiknock quality of the fuel, the more the power and efficiency of the engine.

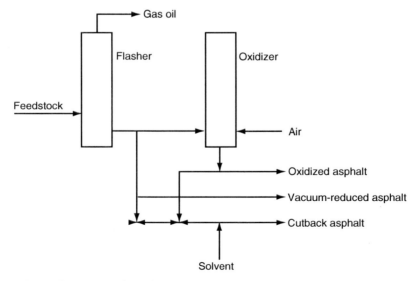

FIGURE 26.3 General representation of asphalt manufacture.

In the early days of automobile engine development, there was a demand for more powerful engines and knocking was not a problem. This demand for more powerful engines was met at first by adding more and larger pistons to the engines; some were even built with 16 cylinders. However, the practical limit to the size to which an engine could be built meant that other avenues to increased power had to be explored. The obvious method of getting more power from an engine of a given size was to increase its compression ratio (a measure of the extent to which the gasoline–air mixture is compressed in the cylinder of an engine). The more the mixture is compressed before ignition, the more power the engine can deliver, but the increased performance was accompanied by an increased tendency for detonation or knocking. Eventually, the cause of engine knock was traced to the gasoline. As some gasoline seemed to cause more knock than another, and as there was no suitable testing procedure, it was difficult to determine the relative antiknock characteristics of gasoline. It appeared, however, that cracked gasoline caused less knock than straight-run gasoline and, hence, the use of cracked gasoline as a fuel increased.

In 1922, tetraethyl lead was discovered to be an excellent antiknock material when added in small quantities to gasoline, and gasoline containing tetraethyl lead became widely available. However, the problem of how to increase the antiknock characteristics of cracked gasoline became acute in the 1930s. One feature of the problem concerned the need to measure the antiknock characteristics of gasoline accurately. This was solved in 1933 by the general use of a single-cylinder test engine, which allowed comparisons of the antiknock characteristics of gasoline to be made in terms of octane numbers. The octane numbers formed a scale ranging from 0 to 100: the higher the number, the greater the antiknock characteristics. In 1939, a second and less severe test procedure using the same test engine was developed, and results obtained by this test were also expressed in octane numbers.

Octane numbers are obtained by two test procedures; those obtained by the first method are called motor octane numbers (indicative of high-speed performance) (ASTM D-2700 and ASTM D-2723); and those obtained by the second method are called research octane numbers (indicative of normal road performance) (ASTM D-2699 and ASTM D-2722). Octane numbers quoted are usually, unless stated otherwise, research octane numbers.

In the test methods used to determine the antiknock properties of gasoline, comparisons are made with blends of two pure hydrocarbons, n-heptane and iso-octane (2,2,4-trimethyl-pentane). Iso-octane has an octane number of 100 and is high in its resistance to knocking; n-heptane is quite low (with an octane number of 0) in its resistance to knocking.

Extensive studies of the octane numbers of individual hydrocarbons (Table 26.5) have brought to light some general rules. For example, normal paraffins have the least desirable knocking characteristics, and these become progressively worse as the molecular weight increases. Iso-paraffins have higher octane numbers than the corresponding normal isomers, and the octane number increases as the degree of branching of the chain is increased. Olefins have markedly higher octane numbers than the related paraffins; naphthenes are usually better than the corresponding normal paraffins but rarely have very high octane numbers; aromatics usually have quite high octane numbers.

Blends of n-heptane and iso-octane thus serve as a reference system for gasoline and provide a wide range of quality used as an antiknock scale. The exact blend, which matches the antiknock resistance of the fuel under test, is found, and the percentage of iso-octane in that blend is termed the *octane number* of the gasoline. For example, gasoline with a knocking ability that matches that of a blend of 90% iso-octane and 10% n-heptane has an octane number of 90. However, many pure hydrocarbons and even commercial gasoline have antiknock quality above an octane number of 100. In this range it is common practice to extend the reference values by the use of varying amounts of tetraethyl lead in pure iso-octane.

With an accurate and reliable means of measuring octane numbers, it was possible to determine the cracking conditions—temperature, cracking time, and pressure—that caused increases in the antiknock characteristics of cracked gasoline. In general it was found that higher cracking temperatures and lower pressures produced higher octane gasoline, but unfortunately more gas, cracked residua, and coke were formed at the expense of the volume of cracked gasoline.

To produce higher-octane gasoline, cracking coil temperatures were pushed up to 510°C (950°F), and pressures dropped from 1000 to 350 psi. This was the limit of thermal cracking units, for at temperatures over 510°C (950°F) coke formed so rapidly in the cracking coil that the unit became inoperative after only a short time on-stream. Hence, it was at this stage that the nature of the gasoline-producing process was reexamined, leading to the development of other processes, such as reforming, polymerization, and alkylation (Chapter 18) for the production of gasoline components having suitably high octane numbers.

It is worthy of note here that the continued decline in petroleum reserves and the issue of environmental protection has emerged as of extreme importance in the search for alternatives to petroleum. In this light, **oxygenates**, either neat or as **additives** to fuels, appear to be the principal alternative fuel candidates beyond the petroleum refinery.

26.3.5 ADDITIVES

In the late twentieth century, the rising price of petroleum (and hence of gasoline) led to the increasing use of **gasohol**, which is a mixture of 90% unleaded gasoline and 10% ethanol (ethyl alcohol). Gasohol burns well in gasoline engines and is a desirable alternative fuel for certain applications because of the availability of ethanol, which can be produced from grains, potatoes, and certain other plant matter. Methanol and a number of other alcohols and ethers are considered high-octane enhancers of gasoline (Mills and Ecklund, 1987; Gray and Alson, 1989). They can be produced from various hydrocarbon sources other than petroleum and may also offer environmental advantages insofar as the use of oxygenates would presumably suppress the release of vehicle pollutants into the air.

During the manufacture and distribution of gasoline, it comes into contact with water and particulate matter and can become contaminated with such materials. Water is allowed to settle from the fuel in storage tanks and the water is regularly withdrawn and disposed of properly. Particulate matter is removed by filters installed in the distribution system. (ASTM D-4814, Appendix X6).

Adulteration differs from contamination insofar as unacceptable materials are deliberately added to gasoline for a variety of reasons not to be discussed here. Such activities may not only lower the octane number but will also adversely affect volatility, which in turn also affects performance. In some countries, dyes and markers are used to detect adulteration (e.g., ASTM D-86 distillation testing and ASTM D-2699/ASTM D-2700 octane number testing may be required to detect adulteration).

Additives are gasoline-soluble chemicals that are mixed with gasoline to enhance certain performance characteristics or to provide characteristics not inherent in the gasoline. Additives are generally derived from petroleum-based materials and their function and chemistry are highly specialized. They produce the desired effect at the parts-per-million (ppm) concentration range.

Oxidation inhibitors (antioxidants) are aromatic amines and hindered phenols that prevent gasoline components (particularly olefins) from reacting with oxygen in the air to form peroxides or **gums**. Peroxides can degrade antiknock quality, cause fuel pump wear, and attack plastic or elastomeric fuel system parts; soluble gums can lead to engine deposits and insoluble gums can plug fuel filters. Inhibiting oxidation is particularly important for fuels used in modern fuel-injected vehicles, as their fuel recirculation design may subject the fuel to more temperature and oxygen-exposure stress.

Corrosion inhibitors are carboxylic acids and carboxylates that prevent free water in the gasoline from rusting or corroding pipelines and storage tanks. Corrosion inhibitors are less important when the gasoline is in the vehicle. The metal parts in the fuel systems of today's vehicles are made of corrosion-resistant alloys or of steel coated with corrosion-resistant coatings. More plastic parts are replacing metals in the fuel systems and, in addition, service station systems and operations are designed to prevent free water from being delivered to a vehicle's fuel tank.

Demulsifiers are polyglycol derivatives that improve the water-separating characteristics of gasoline by preventing the formation of stable emulsions.

Antiknock compounds are compounds (such as tetraethyl lead) that increase the antiknock quality of gasoline. Gasoline containing tetraethyl lead was first marketed in the 1920s and the average concentration of lead in gasoline was gradually increased until it reached a maximum of about 2.5 g/gal (grams per gallon) in the late 1960s. After that, a series of events resulted in the use of less lead and EPA regulations required the phased reduction of the lead content of gasoline beginning in 1979. The EPA completely banned the addition of lead additives to on-road gasoline in 1996 and the amount of incidental lead may not exceed 0.05 g/gal.

Anti-icing additives are surfactants, alcohols, and glycols that prevent ice formation in the carburetor and fuel system. The need for this additive is being reduced as older-model vehicles with carburetors are replaced by vehicles with fuel injection systems.

Dyes are oil-soluble solids and liquids used to visually distinguish batches, grades, or applications of gasoline products. For example, gasoline for general aviation, which is manufactured to different and more exacting requirements, is dyed blue to distinguish it from motor gasoline for safety reasons.

Markers are a means of distinguishing specific batches of gasoline without providing an obvious visual clue. A refiner may add a marker to its gasoline so it can be identified as it moves through the distribution system.

Drag reducers are high-molecular-weight polymers that improve the fluid flow characteristics of low-viscosity petroleum products. Drag reducers lower pumping costs by reducing friction between the flowing gasoline and the walls of the pipe.

Oxygenates are carbon-, hydrogen-, and oxygen-containing combustible liquids that are added to gasoline to improve performance. The addition of oxygenates gasoline is not new as ethanol (ethyl alcohol or grain alcohol) has been added to gasoline for decades. Thus, **oxygenated gasoline** is a mixture of conventional hydrocarbon-based gasoline and one or more oxygenates. The current oxygenates belong to one of two classes of organic molecules: alcohols and ethers. The most widely used oxygenates in the United States are ethanol, methyl tertiary-butyl ether (MTBE), and tertiary-amyl methyl ether (TAME). Ethyl tertiary-butyl ether (ETBE) is another ether that could be used. Oxygenates may be used in areas of the United States where they are not required as long as concentration limits (as refined by environmental regulations) are observed.

Of all the oxygenates, MTBE is attractive for a variety of technical reasons. It has a low vapor pressure, can be blended with other fuels without phase separation, and has the desirable octane characteristics. If oxygenates achieve recognition as vehicle fuels, the biggest contributor will probably be methanol, the production of which is mostly from synthesis gas derived from methane (Chapter 23).

The higher alcohols also offer some potential as motor fuels. These alcohols can be produced at temperatures below 300°C (570°F) using copper oxide–zinc oxide–alumina catalysts promoted with potassium. *Iso*-butyl alcohol is of particular interest because of its high octane rating, which makes it desirable as a gasoline-blending agent. This alcohol can be reacted with methanol in the presence of a catalyst to produce MTBE. Although it is currently cheaper to make *iso*-butyl alcohol from *iso*-butylene, it can be synthesized from syngas with alkali-promoted zinc oxide catalysts at temperatures above 400°C (750°F).

26.4 SOLVENTS (NAPHTHA)

The term *petroleum solvent* describes the liquid hydrocarbon fractions obtained from petroleum and used in industrial processes and formulations. These fractions are also referred to as naphtha or as industrial naphtha. By definition the solvents obtained from the petrochemical industry such as alcohols, ethers, and the like are not included in this chapter. A refinery is capable of producing hydrocarbons of a high degree of purity and at the present time petroleum solvents covering a wide range of solvent properties including both volatile and high boiling qualities are aviailable.

Naphtha (often referred to as **naft** in the older literature) is actually a generic term applied to refined, partly refined, or unrefined petroleum products. In the strictest sense of the term, not less than 10% of the material should distill below 175°C (345°F); not less than 95% of the material should distill below 240°C (465°F) under standardized distillation conditions (ASTM D-86).

Naphtha has been available since the early days of the petroleum industry. Indeed, the infamous Greek fire documented as being used in warfare during the last three millennia is a petroleum derivative (Chapter 1). It was produced either by distillation of crude oil isolated from a surface seepage or (more likely) by destructive distillation of the bituminous material obtained from **bitumen** seepages (of which there were many known during the heyday of the civilizations of the Fertile Crescent) (Chapter 1). The bitumen obtained from the area of Hit (Tuttul) in Iraq (Mesopotamia) is an example of such an occurrence (Abraham, 1945; Forbes, 1958a).

Other petroleum products boiling within the naphtha boiling range include industrial spirit and white spirit.

Industrial spirit comprises liquids distilling between 30°C and 200°C (−1°F to 390°F), with a temperature difference between 5% volume and 90% volume distillation points, including losses, of not more than 60°C (140°F). There are several (up to eight) grades of industrial spirit, depending on the position of the cut in the distillation range defined earlier. On the other hand, white spirit is an industrial spirit with a flash point above 30°C (99°F) and has a distillation range from 135°C to 200°C (275°F to 390°F).

26.4.1 COMPOSITION

Naphtha is divided into two main types, aliphatic and aromatic. The two types differ in two ways: first, in the kind of hydrocarbons making up the solvent, and second, in the methods used for their manufacture. Aliphatic solvents are composed of paraffinic hydrocarbons and cycloparaffins (naphthenes), and may be obtained directly from crude petroleum by distillation. The second type of naphtha contains aromatics, usually alkyl-substituted benzene, and is very rarely, if at all, obtained from petroleum as straight-run materials.

Stoddard solvent is a petroleum distillate widely used as a dry cleaning solvent and as a general cleaner and degreaser. It may also be used as a paint thinner, as a solvent in some types of photocopier toners, in some types of printing inks, and in some adhesives. Stoddard solvent is considered to be a form of mineral spirits, white spirits, and naphtha but not all forms of mineral spirits, white spirits, and naphtha are considered to be Stoddard solvent. Stoddard solvent consists of linear alkanes (30–50%), branched alkanes (20–40%), cycloalkanes (30–40%), and aromatic hydrocarbons (10–20%). The typical hydrocarbon chain ranges from C_7 through C_{12} in length.

26.4.2 Manufacture

In general, naphtha may be prepared by any one of several methods, which include (1) fractionation of straight-run, cracked, and reforming distillates, or even fractionation of crude petroleum; (2) **solvent extraction**; (3) hydrogenation of cracked distillates; (4) polymerization of unsaturated compounds (olefins); and (5) alkylation processes. In fact, the naphtha may be a combination of product streams from more than one of these processes.

The more common method of naphtha preparation is distillation. Depending on the design of the distillation unit, either one or two naphtha steams may be produced: (1) a single naphtha with an end point of about 205°C (400°F) and similar to straight-run gasoline or (2) this same fraction divided into a light naphtha and a heavy naphtha. The end point of the light naphtha is varied to suit the subsequent subdivision of the naphtha into narrower boiling fractions and may be of the order of 120°C (250°F).

Before the naphtha is redistilled into a number of fractions with boiling ranges suitable for aliphatic solvents, it is usually treated to remove sulfur compounds, as well as aromatic hydrocarbons, which are present in sufficient quantity to cause an odor. Aliphatic solvents that are specially treated to remove aromatic hydrocarbons are known as deodorized solvents. Odorless solvent is the name given to heavy alkylate used as an aliphatic solvent, which is a by-product in the manufacture of aviation alkylate.

Sulfur compounds are most commonly removed or converted to a harmless form by chemical treatment with lye, doctor solution, copper chloride, or similar treating agents (Chapter 24). Hydrorefining processes (Chapter 20) are also often used in place of chemical treatment. Solvent naphtha is a solvent selected for low sulfur content, and the usual treatment processes, if required, remove only sulfur compounds. Naphtha with a small aromatic content has a slight odor, but the aromatic constituents increase the solvent power of the naphtha and there is no need to remove aromatics unless an odor-free solvent is specified.

Naphtha that is either naturally sweet (no odor), or has been treated until sweet, is subdivided into several fractions in efficient fractional distillation towers frequently called pipe stills, columns, and column steam stills (Chapter 14 and Chapter 16). A typical arrangement consists of primary and secondary fractional distillation towers and a stripper. Heavy naphtha, for example, is heated by a steam heater and passed into the primary tower, which is usually operated under vacuum. The vacuum permits vaporization of the naphtha at the temperatures obtainable from the steam heater.

The primary tower separates the naphtha into three parts:

1. High boiling material that is removed as a bottom product and sent to a cracking unit.
2. Side stream product of narrow boiling range that, after passing through the stripper, may be suitable for the aliphatic solvent Varsol.
3. Overhead product that is sent to the secondary (vacuum) tower where the overhead product from the primary tower is divided into an overhead and a bottom product in the secondary tower, which operates under a partial vacuum with steam injected into the bottom of the tower to assist in the fractionation. The overhead and bottom products are finished aliphatic solvents, or if the feed to the primary tower is light naphtha instead of heavy naphtha, other aliphatic solvents of different boiling ranges are produced.

Several methods, involving solvent extraction (Chapter 19) or destructive hydrogenation (hydrocracking) (Chapter 16) can accomplish the removal of aromatic hydrocarbons from naphtha. By this latter method, aromatic hydrocarbon constituents are converted into odorless, straight-chain paraffin hydrocarbons that are required in aliphatic solvents.

The **Edeleanu process** (Chapter 24) was originally developed to improve the burning characteristics of kerosene by extraction of the smoke-forming aromatic compounds. Thus it is not surprising that its use has been extended to the improvement of other products as well as to the segregation of aromatic hydrocarbons for use as solvents. Naphtha fractions rich in aromatics may be treated by the Edeleanu process for the purpose of recovering the aromatics, or the product stream from a catalytic reformer unit—particularly when the unit is operated to product maximum aromatics—may be Edeleanu treated to recover the aromatics. The other most widely used processes for this purpose are the **extractive distillation** process and the Udex processes. Processes such as the **Arosorb process** and cyclic adsorption processes are used to a lesser extent.

Extractive distillation (Chapter 16), is used to recover aromatic hydrocarbons from, say, reformate fractions in the following manner. By means of preliminary distillation in a 65-tray prefractionator, a fraction containing a single aromatic can be separated from reformate, and this aromatic concentrate is then pumped to an extraction distillation tower near the top and aromatic concentrate enters near the bottom. A reboiler in the extractive distillation tower induces the aromatic concentrate to ascend the tower, where it contacts the descending solvent.

The solvent removes the aromatic constituents and accumulates at the bottom of the tower; the nonaromatic portion of the concentrate leaves the top of the tower and may contain about 1% of the aromatics. The solvent and dissolved aromatics are conveyed from the bottom of the extractive distillation tower to a solvent stripper, where fractional distillation separates the aromatics from the solvent as an overhead product. The solvent is recirculated to the extractive distillation tower, whereas the aromatic stream is treated with sulfuric acid and clay to yield a finished product of high purity.

The Udex process (Chapter 24) is also employed to recover aromatic streams from reformate fractions. This process uses a mixture of water and diethylene glycol to extract aromatics. Unlike extractive distillation, an aromatic concentrate is not required and the solvent removes all the aromatics, which are separated from one another by subsequent fractional distillation.

The reformate is pumped into the base of an extractor tower. The feed rises in the tower countercurrent to the descending diethylene glycol–water solution, which extracts the aromatics from the feed. The nonaromatic portion of the feed leaves the top of the tower, and the aromatic-rich solvent leaves the bottom of the tower. Distillation in a solvent stripper separates the solvent from the aromatics, which are sulfuric acid and clay treated and then separated into individual aromatics by fractional distillation.

Silica gel (SiO_2) is an adsorbent for aromatics and has found use in extracting aromatics from refinery streams (Arosorb and cyclic adsorption processes) (Chapter 24). Silica gel is manufactured amorphous silica that is extremely porous and has the property of selectively removing and holding certain chemical compounds from mixtures. For example, silica gel selectively removes aromatics from a petroleum fraction, and after the nonaromatic portion of the fraction is drained from the silica gel, the adsorbed aromatics are washed from the silica gel by a stripper (or desorbent). Depending on the kind of feedstock, xylene, kerosene, or pentane may be used as the desorbent.

However, silica gel can be poisoned by contaminants, and the feedstock must be treated to remove water as well as nitrogen, oxygen, and sulfur-containing compounds by passing the feedstock through beds of alumina and other materials that remove impurities. The treated feedstock then enters one of several silica gel cases (columns) where the aromatics are adsorbed. The time required for adsorption depends on the nature of the feedstock; for example, reformate product streams have been known to require substantially less treatment time than kerosene fractions.

26.4.3 PROPERTIES AND USES

Generally, naphtha is valuable as a solvent because of its good dissolving power. The wide range of naphtha available, from the ordinary paraffin straight-run to the highly aromatic types, and the varying degree of volatility possible offer products suitable for many uses (Boenheim and Pearson, 1973; Hadley and Turner, 1973).

The main uses of naphtha fall into the general areas of (1) solvents (diluents) for paints, for example; (2) dry-cleaning solvents; (3) solvents for **cutback asphalt**; (4) solvents in the rubber industry; and (5) solvents for industrial extraction processes.

Turpentine, the older more conventional solvent for paints, has now been almost completely replaced with the discovery that the cheaper and more abundant petroleum naphtha is equally satisfactory. The differences in application are slight: naphtha causes a slightly greater decrease in viscosity when added to some paints than does turpentine, and depending on the boiling range, may also show difference in evaporation rate.

The boiling ranges of fractions that evaporate at rates permitting the deposition of good films have been fairly well established. Depending on conditions, products are employed as light as those boiling from 38°C to 150°C (100°F to 300°F) and as heavy as those boiling between 150°C and 230°C (300°F and 450°F). The latter are used mainly in the manufacture of backed and forced-drying products.

The solvent power required for conventional paint diluents is low and can be reached by distillates from paraffinic crude oils, which are usually recognized as the poorest solvents in the petroleum naphtha group. In addition to solvent power and correct evaporation rate, a paint thinner should also be resistant to oxidation, that is, the thinner should not develop bad color and odor during use. The thinner should be free of corrosive impurities and reactive materials, such as certain types of sulfur compounds, when employed with paints containing lead and similar metals. The requirements are best met by straight-run distillates from paraffinic crude oils that boil from 120°C to 205°C (250°F to 400°F). The components of enamels, varnishes, nitrocellulose lacquers, and synthetic resin finishes are not as soluble in paraffinic naphtha as the materials in conventional paints, and hence naphthenic and aromatic naphtha are favored for such uses.

Dry cleaning is a well-established industry, and the standardized requirements for the solvent are usually met by straight-run naphtha from a low-sulfur, suitably refined paraffinic crude oil. An aromatic hydrocarbon content is not desirable, as it may cause removal of dyes from fabrics or too efficient removal of natural oils from wool, for example. Such a product is usually high boiling and, hence, safe from fire risks, as well as stable enough for extensive reuse and reclaiming. It is especially important that dry-cleaning solvents leave no odor on the cloth, and for this reason (coupled with reuse and reclaiming) the solvents cannot be treated with sulfuric acid. The acid treatment leaves the oil with very small quantities of sulfonated hydrocarbons, which leave a residual odor on the cloth and render the solvent unstable when exposed to distillation temperatures.

Cutback asphalt is asphalt cement diluted with a petroleum distillate to make it suitable for direct application to road surfaces with little or no heating. Asphalt cement, in turn, is a combination of hard asphalt with a heavy distillate or with a viscous residuum of an asphaltic crude oil. The products are classified as rapid, medium, and slow curing, depending on the rate of evaporation of the solvent. A rapid-curing product may contain 40% to 50% of material distilling up to 360°C (680°F); a slow-curing mixture may have only 25% of such material. Gasoline naphtha, kerosene, and light fuel oils boiling from 38°C to 330°C (100°F to 30°F) are used in different products and for different purposes; the use may also dictate the nature of the asphaltic residuum that can be used for the asphalt.

Naphtha is used in the rubber industry for dampening the tread stocks of automobile tires during manufacture to obtain better adhesion between the units of the tire. They are also consumed extensively in making rubber cements (adhesives) or are employed in the fabrication of rubberized cloth, hot-water bottles, bathing caps, gloves, overshoes, and toys. These cements are solutions of rubber and were formerly made with benzene, but petroleum naphtha is now preferred because of its less toxic character.

Petroleum distillates are also added in amounts up to 25% and higher at various stages in the polymerization of butadiene-styrene to synthetic rubber. Those employed in oil extended rubber are of the aromatic type. These distillates are generally high-boiling fractions and preferably contain no wax, boil from 425°C to 510°C (800°F to 950°F), have characterization factors of 10.5 to 11.6 (Chapter 8), a viscosity index lower than 0, bromine numbers of 6 to 30, and API gravity of 3 to 24.

Naphtha is used for extraction on a fairly wide scale. It is supplied in extracting residual oil from castor beans, soybeans, cottonseed, and wheat germ and in the recovery of grease from mixed garbage and refuse. The solvent employed in these cases is a hexane cut, boiling from about 65°C to 120°C (150°F to 250°F). When the oils recovered are of edible grade or intended for refined purposes, stable solvents completely free of residual odor and taste are necessary, and straight-run streams from low-sulfur, paraffinic crude oils are generally satisfactory.

The recovery of wood resin by naphtha extraction of the resinous portions of dead trees of the resin-bearing varieties or stumps, for example, is also resorted to in the wood industry. The chipped wood is steamed to distill out the resinous products recoverable in this way and then extracted with a naphtha solvent, usually a well-refined, low-sulfur, paraffinic product boiling from, say, 95°C to 150°C (200°F to 300°F).

Petroleum distillates of various compositions and volatility are also employed as solvents in the manufacture of printing inks, leather coatings, diluents for dyes, and degreasing of wool fibers, polishes, and waxes, as well as rust-proofing and water-proofing compositions, mildew-proofing compositions, insecticides, and wood preservatives.

26.5 KEROSENE

Kerosene (kerosine), also called paraffin or paraffin oil, is a flammable pale-yellow or colorless oily liquid with a characteristic odor. It is obtained from petroleum and used for burning in lamps and domestic heaters or furnaces, as a fuel or fuel component for jet engines, and as a solvent for greases and insecticides.

Kerosene is intermediate in volatility between gasoline and gas/diesel oil. It is a medium oil distilling between 150°C and 300°C (300°F to 570°F). Kerosene has a flash point of about 25°C (77°F) and is suitable for use as an illuminant when burned in a wide lamp. The term *kerosene* is also often incorrectly applied to various fuel oils, but a fuel oil is actually any liquid or liquid petroleum product that produces heat when burned in a suitable container or that produces power when burned in an engine.

Kerosene was the major refinery product before the onset of the automobile age, but now *kerosene* can be termed one of several secondary petroleum products after the primary refinery product—gasoline. Kerosene originated as a straight-run petroleum fraction that boiled between approximately 205°C and 260°C (400°F to 500°F) (Walmsley, 1973). Some crude oils, for example, those from the Pennsylvania oil fields, contain kerosene fractions of very high quality, but other crude oils, such as those having an asphalt base, must be thoroughly refined to remove aromatics and sulfur compounds before a satisfactory kerosene fraction can be obtained.

Jet fuel comprises both gasoline and kerosene type jet fuels meeting specifications for use in aviation turbine power units and is often referred to as gasoline-type jet fuel and kerosene-type jet fuel.

Jet fuel is a light petroleum distillate that is available in several forms suitable for use in various types of jet engines. The major jet fuels used by the military are JP-4, JP-5, JP-6, JP-7, and JP-8. Briefly, JP-4 is a wide-cut fuel developed for broad availability. JP-6 has a higher cut than JP-4 and is characterized by fewer impurities. JP-5 is specially blended kerosene, and JP-7 is high flash point special kerosene used in advanced supersonic aircraft. JP-8 is kerosene modeled on Jet A-l fuel (used in civilian aircraft). From whatever data that are available, typical hydrocarbon chain lengths characterizing JP-4 range from C_4 to C_{16}. Aviation fuels consist primarily of straight and branched alkanes and cycloalkanes. Aromatic hydrocarbons are limited to 20% to 25% of the total mixture because they produce smoke when burned. A maximum of 5% alkenes is specified for JP-4. The approximate distribution by chemical class is: straight chain alkanes (32%), branched alkanes (31%), cycloalkanes (16%), and aromatic hydrocarbons (21%).

Gasoline-type jet fuel includes all light hydrocarbon oils for use in aviation turbine power units that distill between 100°C and 250°C (212°F to 480°F). It is obtained by blending kerosene and gasoline or naphtha in such a way that the aromatic content does not exceed 25% in volume. Additives can be included to improve fuel stability and combustibility.

Kerosene-type jet fuel is a medium distillate product that is used for aviation turbine power units. It has the same distillation characteristics and flash point as kerosene (between 150°C and 300°C, 300°F and 570°F, but not generally above 250°C, 480°F). In addition, it has particular specifications (such as freezing point) that are established by the International Air Transport Association (IATA).

26.5.1 COMPOSITION

Chemically, kerosene is a mixture of hydrocarbons; the chemical composition depends on its source, but it usually consists of about 10 different hydrocarbons, each containing from 10 to 16 carbon atoms per molecule; the constituents include *n*-dodecane (*n*-$C_{12}H_{26}$), alkyl benzenes, and naphthalene and its derivatives. Kerosene is less volatile than gasoline; it boils between about 140°C (285°F) and 320°C (610°F).

Kerosene, because of its use as a burning oil, must be free of aromatic and unsaturated hydrocarbons, as well as free of the more obnoxious sulfur compounds. The desirable constituents of kerosene are saturated hydrocarbons, and it is for this reason that kerosene is manufactured as a straight-run fraction, not by a cracking process.

Although the kerosene constituents are predominantly saturated materials, there is evidence for the presence of substituted tetrahydronaphthalene. Dicycloparaffins also occur in substantial amounts in kerosene. Other hydrocarbons with both aromatic and cycloparaffin rings in the same molecule, such as substituted indan, also occur in kerosene. The predominant structure of the dinuclear aromatics appears to be that in which the aromatic rings are condensed, such as naphthalene whereas the isolated two-ring compounds, such as biphenyl, are only present in traces, if at all.

26.5.2 MANUFACTURE

Kerosene was first manufactured in the 1850s from coal tar, hence the name coal oil is often applied to kerosene, but petroleum became the major source after 1859. From that time, the kerosene fraction is and has remained a distillation fraction of petroleum. However, the quantity and quality vary with the type of crude oil, and although some crude oils yield excellent kerosene quite simply, others produce kerosene that requires substantial refining.

Kerosene is now largely produced by cracking the less volatile portion of crude oil (Chapter 13) at atmospheric pressure and elevated temperatures (Chapter 14 and Chapter 15).

In the early days, the poorer-quality kerosene was treated with large quantities of sulfuric acid to convert it into a marketable product. However, this treatment resulted in high acid and kerosene losses, but the later development of the Edeleanu process (Chapter 19) overcame these problems.

Kerosene is a very stable product, and additives are not required to improve the quality. Apart from the removal of excessive quantities of aromatics by the Edeleanu process, kerosene fractions may need only a lye wash or a doctor treatment if hydrogen sulfide is present to remove mercaptans (Chapter 19).

26.5.3 Properties and Uses

Kerosene is by nature a fraction distilled from petroleum that has been used as a fuel oil from the beginning of the petroleum-refining industry. As such, low proportions of aromatic and unsaturated hydrocarbons are desirable to maintain the lowest possible level of smoke during burning. Although some aromatics may occur within the boiling range assigned to kerosene, excessive amounts can be removed by extraction; that kerosene is not usually prepared from cracked products almost certainly excludes the presence of unsaturated hydrocarbons.

The essential properties of kerosene are flash point, fire point, distillation range, burning, sulfur content, color, and cloud point. In the case of the flash point (ASTM D-56), the minimum flash temperature is generally placed above the prevailing ambient temperature; the fire point (ASTM D-92) determines the fire hazard associated with its handling and use.

The boiling range (ASTM D-86) is of lesser importance for kerosene than for gasoline, but it can be taken as an indication of the viscosity of the product, for which there is no requirement for kerosene. The ability of kerosene to burn steadily and cleanly over an extended period (ASTM D-187) is an important property and gives some indication of the purity or composition of the product.

The significance of the total sulfur content of a fuel oil varies greatly with the type of oil and the use to which it is put. Sulfur content is of great importance when the oil to be burned produces sulfur oxides that contaminate the surroundings. The color of kerosene is of little significance, but a product darker than usual may have resulted from contamination or aging, and in fact a color darker than specified (ASTM D-156) may be considered by some users as unsatisfactory. Finally, the cloud point of kerosene (ASTM D-2500) gives an indication of the temperature at which the wick may become coated with wax particles, thus lowering the burning qualities of the oil.

26.6 FUEL OIL

Fuel oil is classified in several ways but generally may be divided into two main types: distillate fuel oil and **residual fuel oil**. Distillate fuel oil is vaporized and condensed during a distillation process and thus has a definite boiling range and does not contain high-boiling constituents. A fuel oil that contains any amount of the residue from crude distillation of thermal cracking is a residual fuel oil. The terms *distillate fuel oil* and *residual fuel oil* are losing their significance, as fuel oil is now made for specific uses and may be either distillates or residuals or mixtures of the two. The terms *domestic fuel oil*, *diesel fuel oil*, and **heavy fuel oil** are more indicative of the uses of fuel oils.

Domestic fuel oil is fuel oil that is used primarily in the home. This category of fuel oil includes kerosene, stove oil, and furnace fuel oil; they are distillate fuel oils.

Diesel fuel oil is also a distillate fuel oil that distills between 180°C and 380°C (356°F to 716°F). Several grades are available depending on their uses: diesel oil for diesel compression ignition (cars, trucks, and marine engines) and light heating oil for industrial and commercial uses.

Heavy fuel oil comprises all residual fuel oils (including those obtained by blending). Heavy fuel oil constituents range from distillable constituents to residual (nondistillable) constituents that must be heated to 260°C (500°F) or more before they can be used. The kinematic viscosity is above 10 cst at 80°C (176°F). The flash point is always above 50°C (122°F) and the density is always higher than 0.900. In general, heavy fuel oil usually contains cracked residua, reduced crude, or cracking coil heavy product, which is mixed (cut back) to a specified viscosity with cracked gas oils and fractionator bottoms. For some industrial purposes in which flames or flue gases contact the product (ceramics, glass, heat treating, and open hearth furnaces) fuel oils must be blended to contain minimum sulfur contents, and hence low-sulfur residues are preferable for these fuels.

No. 1 fuel oil is a petroleum distillate that is one of the most widely used of the fuel oil types. It is used in atomizing burners that spray fuel into a combustion chamber where the tiny droplets burn while in suspension. It is also used as a carrier for pesticides, as a weed killer, as a mold release agent in the ceramic and pottery industry, and in the cleaning industry. It is found in asphalt coatings, enamels, paints, thinners, and varnishes. No. 1 fuel oil is a light petroleum distillate (straight-run kerosene) consisting primarily of hydrocarbons in the range $C_9–C_{16}$. Fuel oil No. 1 is very similar in composition to diesel fuel; the primary difference is in the additives.

No. 2 fuel oil is a petroleum distillate that may be referred to as domestic or industrial. The domestic fuel oil is usually lower boiling and a straight-run product. It is used primarily for home heating. Industrial distillate is a cracked product or a blend of both. It is used in smelting furnaces, ceramic kilns, and packaged boilers. No. 2 fuel oil is characterized by hydrocarbon chain lengths in the $C_{11}–C_{20}$ range. The composition consists of aliphatic hydrocarbons (straight chain alkanes and cycloalkanes) (64%), unsaturated hydrocarbons (alkenes) (1% to 2%), and aromatic hydrocarbons (including alkyl benzenes and 2-ring, 3-ring aromatics) (35%) but contains only low amounts of the polycyclic aromatic hydrocarbons (<5%;).

No. 6 fuel oil (also called **Bunker C oil** or residual fuel oil) is the residuum from crude oil after naphtha-gasoline, No. 1 fuel oil, and No. 2 fuel oil have been removed. No. 6 fuel oil can be blended directly to heavy fuel oil or made into asphalt. Residual fuel oil is more complex in composition and impurities than distillate fuels. Limited data are available on the composition of No. 6 fuel oil. Polycyclic aromatic hydrocarbons (including the alkylated derivatives) and metal-containing constituents are components of No. 6 fuel oil.

Stove oil, like kerosene, is always a straight-run fraction from suitable crude oils, whereas other fuel oils are usually blends of two or more fractions, one of which is usually cracked gas oil. The straight-run fractions available for blending into fuel oils are heavy naphtha, light and heavy gas oils, reduced crude, and pitch. Cracked fractions such as light and heavy gas oils from catalytic cracking, cracking coil tar, and fractionator bottoms from catalytic cracking, may also be used as blends to meet the specifications of the different fuel oils.

As the boiling ranges, sulfur contents, and other properties of even the same fraction vary from crude oil to crude oil and with the way the crude oil is processed, it is difficult to specify which fractions are blended to produce specific fuel oils. In general, however, furnace fuel oil is a blend of straight-run gas oil and cracked gas oil to produce a product boiling in the 175°C to 345°C (350°F to 50°F) range.

Diesel fuel oil is essentially the same as furnace fuel oil, but the proportion of cracked gas oil is usually less as the high aromatic content of the cracked gas oil reduces the cetane value of the diesel fuel.

Diesel fuels originally were straight-run products obtained from the distillation of crude oil. However, with the use of various cracking processes to produce diesel constituents, diesel fuels may also contain varying amounts of selected cracked distillates to increase the volume available for meeting the growing demand. Care is taken to select the cracked stocks in a manner that specifications are met as simply as possible.

Under the broad definition of diesel fuel, many possible combinations of characteristics (such as volatility, ignition quality, viscosity, gravity, stability, and other properties) exist. To characterize diesel fuels and thereby establish a framework of definition and reference, various classifications are used in different countries. An example is ASTM D-975 in the United States in which grades No. 1-D and 2-D are distillate fuels, the types most commonly used in high-speed engines of the mobile type, in medium-speed stationary engines, and in railroad engines. Grade 4-D covers the class of more viscous distillates and, at times, blends of these distillates with residual fuel oils. No. 4-D fuels are applicable for use in low and medium speed engines employed in services involving sustained load and predominantly constant speed.

Cetane number is a measure of the tendency of a diesel fuel to knock in a diesel engine. The scale is based upon the ignition characteristics of two hydrocarbons *n*-hexadecane (cetane) and 2,3,4,5,6,7,8-heptamethylnonane. Cetane has a short delay period during ignition and is assigned a cetane number of 100; heptamethylnonane has a long delay period and has been assigned a cetane number of 15. Just as the octane number is meaningful for automobile fuels, the cetane number is a means of determining the ignition quality of diesel fuels and is equivalent to the percentage by volume of cetane in the blend with heptamethylnonane, which matches the ignition quality of the test fuel (ASTM D-613).

The manufacture of fuel oils at one time largely involved using what was left after removing the desired products from crude petroleum. Now fuel oil manufacture is a complex matter of selecting and blending various petroleum fractions to meet definite specifications, and the production of a homogeneous, stable fuel oil requires experience backed by laboratory control.

26.7 LUBRICATING OIL

After kerosene, the early petroleum refiners wanted **paraffin wax** for the manufacture of candles, and lubricating oil was, at first, a by-product of wax manufacture. The preferred lubricants in the 1860s were lard oil, sperm oil, and tallow. The demand that existed for kerosene did not develop for petroleum-derived lubricating oils. In fact, oils were used to supplement the animal and vegetable oils used as lubricants. However, as the trend to heavier industry increased, the demand for mineral lubricating oils increased, and after the 1890s petroleum displaced animal and vegetable oils as the source of lubricants for most purposes.

Mineral oils are often used as lubricating oils but also have medicinal and food uses. A major type of hydraulic fluid is the mineral oil class of hydraulic fluids. The mineral-based oils are produced from heavy-end crude oil distillates. Hydrocarbon numbers ranging from C_{15} to C_{50} occur in the various types of mineral oils, with the heavier distillates having higher percentages of the higher carbon number compounds.

Crankcase oil (*motor oil*) may be either mineral-based or synthetic. The mineral-based oils are more widely used than the synthetic oils and may be used in automotive engines, railroad and truck diesel engines, marine equipment, jet and other aircraft engines, and most small 2- and 4-stroke engines. The mineral-based oils contain hundreds to thousands of hydrocarbon compounds, including a substantial fraction of nitrogen- and sulfur-containing compounds. The hydrocarbons are mainly mixtures of straight and branched chain hydrocarbons (alkanes), cycloalkanes, and aromatic hydrocarbons. Polynuclear aromatic hydrocarbons (and the alkyl derivatives) and metal-containing constituents are components of motor oils and crankcase oils, with the used oils typically having higher concentrations than the new unused oils. Typical carbon number chain lengths range from C_{15} to C_{50}.

26.7.1 COMPOSITION

Lubricating oil is distinguished from other fractions of crude oil by their usually high (>400°C, >750°F) boiling point, as well as their high viscosity. Materials suitable for the production of lubricating oils are comprised principally of hydrocarbons containing from 25 to 35 or even 40 carbon atoms per molecule, whereas residual stocks may contain hydrocarbons with 50 or more (up to 80 or so) carbon atoms per molecule. The composition of lubricating oil may be substantially different from the lubricant fraction from which it was derived, as wax (normal paraffins) is removed by distillation or refining by solvent extraction (Chapter 18) and adsorption preferentially removes nonhydrocarbon constituents as well as polynuclear aromatic compounds and the multiring cycloparaffins.

Normal paraffins up to C_{36} have been isolated from petroleum, but it is difficult to isolate any hydrocarbon from the lubricant fraction of petroleum. Various methods have been used in the analysis of products in the lubricating oil range, but the most successful procedure involves a technique based on the correlation of simple physical properties, such as refractive index, density, and molecular weight or viscosity (Chapter 9). Results are obtained in the form of carbon distribution and the methods may also be applied to oils that have not been subjected to extensive fractionation. Although there are relatively rapid methods of analysis, the lack of information concerning the arrangement of the structural groups within the component molecules is a major disadvantage.

Nevertheless, there are general indications that the lubricant fraction contains a greater proportion of normal and branched paraffins than the lower-boiling portions of petroleum. For the polycycloparaffin derivatives, a good proportion of the rings appear to be in condensed structures, and both cyclopentyl and cyclohexyl nuclei are present. The methylene groups appear principally in unsubstituted chains at least four carbon atoms in length, but the cycloparaffin rings are highly substituted with relatively short side chains.

Mono-, di-, and trinuclear aromatic compounds appear to be the main constituents of the aromatic portion, but material with more aromatic nuclei per molecule may also be present. For the dinuclear aromatics, most of the material consists of naphthalene types. For the trinuclear aromatics, the phenanthrene type of structure predominates over the anthracene type. There are also indications that the greater part of the aromatic compounds occurs as mixed aromatic-cycloparaffin compounds.

26.7.2 MANUFACTURE

Lubricating oil manufacture was well established by 1880, and the method depended on whether the crude petroleum was processed primarily for kerosene or for lubricating oils. Usually the crude oil was processed for kerosene, and primary distillation separated the crude into three fractions, naphtha, kerosene, and a residuum. To increase the production of kerosene the cracking distillation technique was used, and this converted a large part of the gas oils and lubricating oils into kerosene. The cracking reactions also produced coke products and asphalt-like materials, which gave the residuum a black color, and hence it was often referred to as **tar** (Chapter 1).

The production of lubricating oils is well established (Sequeira, 1992) and consists of four basic processes: (1) distillation (Chapter 13) to remove the lower-boiling and lower-molecular-weight constituents of the feedstock, (2) solvent refining, such as deasphalting (Chapter 19), and hydrogen treatment (Chapter 16) to remove the nonhydrocarbon constituents and to improve the feedstock quality, (3) dewaxing (Chapter 19) to remove the wax constituents and improve the low-temperature properties, (4) and clay treatment (Chapter 19) or hydrogen treatment (Chapter 16) to prevent instability (Chapter 22) of the product.

Chemical, solvent, and hydrogen refining processes have been developed and are used to remove aromatics and other undesirable constituents, and to improve the viscosity index and quality of lube base stocks. Traditional chemical processes that use sulfuric acid and clay refining have been replaced by solvent extraction/refining and hydrotreating, which are more effective, cost efficient, and generally more environmentally acceptable. Chemical refining is used most often for the reclamation of used lubricating oils or in combination with solvent or hydrogen refining processes for the manufacture of specialty lubricating oils and by-products.

26.7.2.1 Chemical Refining Processes

Acid-alkali refining, also called wet refining, is a process where lubricating oils are contacted with sulfuric acid followed by neutralization with alkali. Oil and acid are mixed and an **acid sludge** is allowed to coagulate. The **sludge** is removed or the oil is decanted after settling, and more acid is added and the process repeated.

Acid-clay refining, also called dry refining is similar to acid-alkali refining with the exception that clay and a neutralizing agent are used for neutralization. This process is used for oils that form emulsions during neutralization. Neutralization with aqueous and alcoholic caustic, soda ash lime, and other neutralizing agents is used to remove organic acids from some feedstocks. This process is conducted to reduce organic acid corrosion in downstream units or to improve the refining response and color stability of lube feedstocks.

26.7.2.2 Hydroprocessing

Hydroprocessing, which has been generally replaced with solvent refining, consists of lube hydrocracking as an alternative to solvent extraction, and hydrorefining to prepare specialty products or to stabilize hydrocracked base stocks. Hydrocracking catalysts consist of mixtures of cobalt, nickel, molybdenum, and tungsten on an alumina or silica–alumina-based carrier. Hydrotreating catalysts are proprietary but usually consist of nickel–molybdenum on alumina. The hydrocracking catalysts are used to remove nitrogen, oxygen, and sulfur, and convert polynuclear aromatics and polynuclear naphthenes to mononuclear naphthenes, aromatics, and iso-paraffins that are typically desired in lube base stocks. Feedstocks consist of unrefined distillates and deasphalted oils, solvent extracted distillates and deasphalted oils, cycle oils, hydrogen refined oils, and mixtures of these hydrocarbon fractions.

Lube hydrorefining processes are used to stabilize or improve the quality of lube base stocks from lube hydrocracking processes and for manufacture of specialty oils. Feedstocks are dependent on the nature of the crude source but generally consist of waxy or dewaxed-solvent extracted or hydrogen-refined paraffinic oils and refined or unrefined naphthenic and paraffinic oils from some selected crude oils.

26.7.2.3 Solvent Refining Processes

Feedstocks from solvent refining processes consist of paraffinic and naphthenic distillates, deasphalted oils, hydrogen refined distillates and deasphalted oils, cycle oils, and dewaxed oils. The products are refined oils destined for further processing or finished lube base stocks. The by-products are aromatic extracts that are used in the manufacture of rubber, carbon black, petrochemicals, catalytic cracking feedstock, fuel oil, or asphalt. The major solvents in use are N-methyl-2-pyrrolidone (NMP) and furfural, with phenol and liquid sulfur dioxide used to a lesser extent.

The solvents are typically recovered in a series of flash towers. Steam or inert gas strippers are used to remove traces of solvent, and a solvent purification system is used to remove water and other impurities from the recovered solvent.

Lube feedstocks typically contain increased wax content resulting from deasphalting and refining processes. These waxes are normally solid at ambient temperatures and must be removed to manufacture lube oil products with the necessary low-temperature properties.

Catalytic dewaxing and solvent dewaxing (the most prevalent) are processes currently in use. Older technologies include **cold settling**, pressure filtration, and centrifuge dewaxing.

26.7.2.4 Catalytic Dewaxing

As solvent dewaxing is relatively expensive for the production of low pour point oils, various catalytic dewaxing (selective hydrocracking) processes have been developed for the manufacture of lube oil base stocks. The basic process consists of a reactor containing a proprietary dewaxing catalyst followed by a second reactor containing a hydrogen finishing catalyst to saturate olefins created by the dewaxing reaction and to improve stability, color, and demulsibility of the finished lube oil.

26.7.2.5 Solvent Dewaxing

Solvent dewaxing (Chapter 19) consists of the following steps: crystallization, filtration, and solvent recovery. In the crystallization step, the feedstock is diluted with the solvent and chilled, solidifying the wax components. The filtration step removes the wax from the solution of dewaxed oil and solvent. Solvent recovery removes the solvent from the wax cake and filtrate for recycle by flash distillation and stripping. The major processes in use today are the ketone dewaxing processes. Other processes that are used to a lesser degree include the Di/Me Process and the propane dewaxing process. The most widely used ketone processes are the Texaco solvent dewaxing process and the Exxon Dilchill Process. Both processes consist of diluting the waxy feedstock with solvent while chilling at a controlled rate to produce a slurry. The slurry is filtered using rotary vacuum filters and the wax cake is washed with cold solvent. The filtrate is used to chill the feedstock and solvent mixture. The primary wax cake is diluted with additional solvent and filtered again to reduce the oil content in the wax. The solvent is recovered from the dewaxed oil and wax cake by flash vaporization and recycled back into the process.

The Texaco solvent dewaxing process (also called the MEK process) uses a mixture of MEK and toluene as the dewaxing solvent, and sometimes uses mixtures of other ketones and aromatic solvents. The Exxon Dilchill dewaxing process uses a direct cold solvent dilution-chilling process in a special crystallizer in place of the scraped surface exchangers used in the Texaco process. The Di/Me dewaxing process uses a mixture of dichloroethane and methylene dichloride as the dewaxing solvent. The propane dewaxing process is essentially the same as the ketone process except for the following: propane is used as the dewaxing solvent and higher pressure equipment is required, and chilling is done in evaporative chillers by vaporizing a portion of the dewaxing solvent. Although this process generates a better product and does not require crystallizers, the temperature differentia between the dewaxed oil and the filtration temperature is higher than for the ketone processes (higher energy costs), and dewaxing aids are required to get good filtration rates.

26.7.2.6 Finishing Processes

Hydrogen finishing processes have largely replaced acid and clay finishing processes. The hydrogen finishing processes are mild hydrogenation processes used to improve the color, odor, thermal, and oxidative stability, and demulsibility of lube base stocks.

The process consists of fixed-bed catalytic reactors that typically use a nickel–molybdenum catalyst to neutralize, desulfurize, and denitrify lube base stocks. These processes do not saturate aromatics or break carbon–carbon bonds as in other hydrogen finishing processes. Sulfuric acid treating is still used by some refiners for the manufacture of specialty oils and the reclamation of used oils. This process is typically conducted in batch or continuous processes similar to the chemical refining processes with the exception that the amount of acid used is much lower than that used in acid refining.

Clay contacting involves mixing the oil with fine bleaching clay at elevated temperature followed by separation of the oil and clay. This process improves the chemical, thermal, and color stability of the lube base stock, and is often combined with acid finishing. Clay percolation is a static bed absorption process used to purify, decolorize, and finish lube stocks and waxes. It is still used in the manufacture of refrigeration oils, transformer oils, turbine oils, **white oils**, and waxes.

26.7.2.7 Older Processes

Owing to cracking distillation in the primary distillation and the high temperatures used in the still, the paraffin distillate contained dark-colored, sludge-forming asphaltic materials. These undesirable materials were removed by treatment with sulfuric acid followed by lye washing. Then, to separate the wax from the acid-treated paraffin distillate, the latter was chilled and filtered. The chilled, semisolid paraffin distillate was then squeezed in canvas bags in a knuckle or rack press (similar to a cider press) so that the oil would filter through the canvas, leaving the wax crystals in the bag. Later developments saw chilled paraffin distillate filtered in hydraulically operated plate and frame presses, and the use of these continued almost to the present time.

The oil from the press was known as pressed distillate, which was subdivided into three fractions by redistillation. Two overhead fractions of increasing viscosity, the heavier with a Society of Automotive Engineers (SAE) (Chapter 8) viscosity of about 10, were called paraffin oils. The residue in the still (viscosity equivalent to a light SAE 30) was known as red oil. All three fractions were again acid and lye treated and then washed with water. The treated oils were pumped into shallow pans in the bleacher house, where air blowing through the oil and exposure to the sun through the glass roof of the bleacher house or pan removed cloudiness or made the oils bright.

Further treatment of the paraffin oil produced pale oil; thus if the paraffin oil was filtered through bone charcoal, Fuller's earth, clay, or similar absorptive material, the color was changed from a deep yellow to a pale yellow. The filtered paraffin oil was called pale oil to differentiate it from the nonfiltered paraffin oil, which was considered of lower quality.

The wax separated from paraffin distillate by **cold pressing** contained about 50% oil and was known as **slack wax**. The slack wax was melted and cast into cakes, which were again pressed in a hot or hard press. This squeezed more oil from the wax, which was known as scale wax. By a process known as **sweating**, the scale wax was subdivided into several paraffin waxes with different melting points.

In contrast, crude petroleum processed primarily as a source of lubricating oil was handled differently from crude oils processed primarily for kerosene. The primary distillation removed naphtha and kerosene fractions, but without using temperatures high enough to cause cracking. The yield of kerosene was thus much lower, but the absence of cracking reactions increased the yield of lubricating oil fractions. Furthermore, the residuum was distilled using steam, which eliminated the need for high distillation temperatures, and cracking reactions were thus prevented. Thus, various overhead fractions suitable for lubricating oils and known as neutral oils were obtained; many of these were so light that they did not contain wax and did not need dewaxing; the more viscous oils could be dewaxed by cold pressing.

If the wax in the residual oil could not be removed by cold pressing it was removed by cold settling. This involved admixture of the residual oil with a large volume of naphtha, which was then allowed to stand for as long as necessary in a tank exposed to low temperature, usually climatic cold (winter). This caused the waxy components to congeal and settle at the bottom of the tank. In the spring, the supernatant naphtha–oil mixture was pumped to a steam still, where the naphtha was removed as an overhead stream; the bottom product was known as steam-refined stock. If the steam-refined stock (**bright stock**) was filtered through charcoal or a similar filter material the improvement in color caused the oil to be known as bright stock. Mixtures of steam-refined stock with the much lighter paraffin, pale, red, and neutral oils produced oils of any desired viscosity.

The wax material that settled at the bottom of the cold settling tank was crude petrolatum. This was removed from the tank, heated, and filtered through a vessel containing clay, which changed its red color to brown or yellow. Further treatment with sulfuric acid produced white grades of **petrolatum**.

If the crude oil used for the manufacture of lubricating oils contained asphalt, it was necessary to acid treat the steam-refined oil before cold settling. Acid-treated, settled steam-refined stock was widely used as steam cylinder oils.

The crude oils available in North America until about 1900 were either paraffin base or mixed base; hence paraffin wax was always a component of the raw lubricating oil fraction. The mixed-base crude oils also contained asphalt, and this made acid treatment necessary in the manufacture of lubricating oils. However, the asphalt-base crude oils (also referred to as naphthene-base crude oils) that contained little or no wax yielded a different kind of lubricating oil. As wax was not present, the oils would flow at much lower temperatures than the oils from paraffin- and mixed-base crude oils even when the latter had been dewaxed. Hence, lubricating oils from asphalt-base crude oils became known as low cold-test oils; furthermore, these lubricating oils boiled at a lower temperature than oils of similar viscosity from paraffin-base crude oils. Thus higher–viscosity oils could be distilled from asphalt-base crude oils at relatively low temperatures, and the low cold-test oils were preferred because they left less carbon residue in gasoline engines.

The development of vacuum distillation led to a major improvement in both paraffinic and naphthenic (low cold-test) oils. By vacuum distillation the more viscous paraffinic oils (even oils suitable for bright stocks) could be distilled overhead and could be separated completely from residual asphaltic components. Vacuum distillation provided the means of separating more suitable lubricating oil fractions with predetermined viscosity ranges and removed the limit on the maximum viscosity that might be obtained in a distillate oil.

However, although vacuum distillation effectively prevented residual asphaltic material from contaminating lubricating oils, it did not remove other undesirable components. The naphthenic oils, for example, contained components (naphthenic acids) that caused the oil to form emulsions with water. In particular, naphthenic oils contained components that caused oil to thicken excessively when cold and become very thin when hot. The degree to which the viscosity of an oil is affected by temperature is measured on a scale that originally ranged from 0 to 100 and is called the viscosity index (Chapter 8). An oil that changes the least in viscosity when the temperature is changed has a high viscosity index. Naphthenic oils have viscosity indices of 35 or less, compared to 70 or more for paraffinic oils.

26.7.3 PROPERTIES AND USES

Lubricating oil may be divided into many categories according to the types of service they are intended to perform. However, there are two main groups: (1) oils used in intermittent service, such as motor and aviation oils, and (2) oils designed for continuous service, such as turbine oils. Lubricating oil is distinguished from other fractions of crude oil by a high (>400°C, >750°F) boiling point, as well as a high viscosity and, in fact, lubricating oil is identified by viscosity (Table 26.6).

This classification is based on the SAE (Society of Automotive Engineers) J 300 specification. The single grade oils (e.g., SAE 20, etc.) correspond to a single class and have to be selected according to engine manufacturer specifications, operating conditions, and climatic conditions. At −20°C (68°F), a multigrade lubricating oil such as SAE 10W-30 possesses the viscosity of a 10W oil and at 100°C (212°F) the multigrade oil possesses the viscosity of a SAE 30 oil.

Oils used in intermittent service must show the least possible change in viscosity with temperature; that is, their viscosity indices (Chapter 8) must be high. These oils must be changed at frequent intervals to remove the foreign matter collected during service. The stability of such oils is therefore of less importance than the stability of oils used in continuous service for prolonged periods without renewal.

Oils used in continuous service must be extremely stable, but their viscosity indices may be low because the engines operate at fairly constant temperature without frequent shutdown.

26.8 OTHER OIL PRODUCTS

Lubricating oil is not the only material manufactured from the high-boiling fraction of petroleum, there are several important types of oil that may be contained in the so-called lubricant fraction.

26.8.1 WHITE OIL

For many years much of the production of white oil originated from naphthenic stocks, but now white oils are prepared from paraffinic, mixed-base, or naphthenic fractions, depending on the final use of the oil. Naphthenic crude oils give products of high specific gravity and viscosity, desirable in pharmaceutical use, whereas paraffinic stocks produce oils of lighter gravity and lower viscosity suitable for lubrication purposes.

White oils generally fall into two classes: (1) those often referred to as technical white oils, which are employed for cosmetics, textile lubrication, insecticides, vehicles, paper impregnation, and so on; and (2) pharmaceutical white oils, which may be employed as laxatives or for the lubrication of food-handling machinery. The colorless character of these oils is important

TABLE 26.6
Viscosity of Various Lubricating Oil Grades

	Kinematic Viscosity, cs at 100°C (212°F)		Dynamic Viscosity, cp Maximum
	Minimum	Maximum	
SAE 10W	4.1		3500 at −20°C (−4°F)
SAE 15W	5.6		3500 at −15°C (5°F)
SAE 20W	5.6		4500 at −10°C (14°F)
SAE 25W	9.3		6000 at −5°C
SAE 20	5.6	9.3	
SAE 30	9.3	12.5	
SAE 40	12.5	16.3	
SAE 50	16.3	21.9	

in some cases, as it may indicate the chemically inert nature of the hydrocarbon constituents. Textile lubricants should be colorless to prevent the staining of light-colored threads and fabrics. Insecticide oils should be free of reactive (easily oxidized) constituents so as not to injure plant tissues when applied as sprays. Laxative oils should be free of odor, taste, and also hydrocarbons, which may react during storage and produce unwanted by-products. These properties are attained by the removal of nitrogen-containing, oxygen-containing, and sulfur-containing compounds, as well as reactive hydrocarbons by, say, sulfuric acid.

The crude or fraction chosen for refining may have been subjected to a preliminary refining with a differential solvent. The exact procedure for the acid treatment varies, but a preliminary acid treatment (chiefly for drying) may be followed by incremental addition of as much as 50% by volume of acid as strong as 20% fuming sulfuric acid. The sludge is promptly removed to limit oxidation-reduction reactions; the time, temperature, and method of application depend on the type of charge stock and the degree of refining desired. The product is neutralized with alkali and washed with ethyl or *iso*-propyl alcohol or acetone to remove the oil-soluble sulfonic **mahogany acids**; water-soluble green acids are recovered from the alkali washings. The treated oil is further refined and decolorized by adsorption, either by percolation or by contacting with clay.

It is evident that this sequence leaves only the most acid-resistant hydrocarbons behind, and as these are roughly the more generally stable compounds, the process is effective. The medicinal oils require a test showing minimal color change, but depending on its intended use, a technical oil showing rather marked color change may be satisfactory. The only further distinction between pharmaceutical and technical oils is that the high-quality medicinal oils are made as viscous as possible (250 to 350 SUS, Saybolt Universal Seconds, a measure of viscosity, see Chapter 8); the technical oils are likely to be made of the less viscous fractions.

26.8.2 INSULATING OIL

Petroleum oils for electrical insulation fall into two general classes: (1) those used in transformers, circuit breakers, and oil-filled cables; and (2) those employed for impregnating the paper covering of wrapped cables. The first are highly refined fractions of low viscosity and comparatively high boiling range and resemble heavy burning oils, such as mineral seal oil, or the very light lubricating fractions known as nonviscous neutral oils. The second are usually highly viscous products, often naphthenic distillates, and are not usually highly refined.

The insulating value of fresh transformer oils seems to vary little with chemical constitution, but physical purity, including freedom from water, is highly significant. A water content of 0.1% lowers an original dry insulating value by a factor of about 10; higher water content causes little additional change.

The deterioration of transformer oils in service is closely connected with oxidation by air, which brings on deposition of sludge and the development of acids, resulting in overheating and corrosion, respectively. The sludge formed is one of three types:

1. Sludge attributed to the direct oxidation of the hydrocarbon constituents to oil-insoluble products
2. Soap resulting from the reaction of acid products of oxidation with metals in the transformer
3. Carbon formed by any arc or corona discharge occurring in service

Testing transformer oils for suitability, in addition to the conventional inspection data (flash and pour points and viscosity, e.g.) and determination of insulating value, is concerned mostly with accelerated oxidation tests aimed at estimating probable life in service. Various procedures have been suggested. Nearly all involve heating the oil at a temperature near 120°C (250°F) in air or oxygen in the presence of copper as an oxidation catalyst. The changes then watched are color, interfacial tension, acidity development, sludge and water formation, steam emulsion number, and power factor.

26.8.3 INSECTICIDES

Petroleum oils, as such, usually applied in water-emulsion form, have marked killing power for certain species of insects. For many applications for which their own effectiveness is too slight, the oils serve as carriers for active poisons, as in the household and livestock sprays.

The most extensive use of petroleum itself as a killing agent is in fruit tree sprays. The spraying of swamp waters with an oil film as a method of mosquito control has also been practiced. The fruit tree spray oils are known to be elective in the control of scale insects, leaf rollers, red spiders, tree hoppers, mites, moth eggs, and aphids. Molecular weight and structure appear to be the factors determining the insecticidal power of these oils. Olefins and aromatics are both highly toxic to insects, but they also have a detrimental effect on the plant; thus spray oils generally receive some degree of refining, especially those of the summer oil type that come into contact with foliage.

Paraffins and naphthenes are the major components of the refined spray oils, and the former appear to be the more toxic. With both naphthenic and paraffinic hydrocarbons the insecticidal effect increases with molecular weight but becomes constant at about 350 for each; the maximum toxicity has also been attributed to that fraction boiling between 240°C and 300°C (465°F to 570°F) at 40 mm Hg pressure.

The physical properties of petroleum oils, such as their solvent power for waxy coatings on leaf surfaces and insect bodies, make them suitable as carriers for more active fungicides and insecticides. The additive substance may vary from fatty acids and soaps, the latter intended chiefly to affect favorably the spreading properties of the oil, to physiologically active compounds, such as pyrethrum, nicotine, rotenone, DDT, thiocyanates, methoxychlor, chlordane, lindane, and others. Solubility of the chlorine-containing insecticides is often aided by an accessory solvent rich in methylnaphthalene. The hydrocarbon-base solvent used in household insecticides is generally a high-flash (66°C, 150°F) 195°C to 250°C (380°F to 480°F) boiling naphtha that has been heavily treated with concentrated sulfur acid. Household and livestock sprays are also made up for application from aerosol containers, in which liquefied gases (generally dichlorodifluoromethane and trichloromonofluoromethane) are used as the propelling agents.

26.9 GREASE

Grease is lubricating oil to which a thickening agent has been added for the purpose of holding the oil to surfaces that must be lubricated. The most widely used thickening agents are soaps of various kinds, and grease manufacture is essentially the mixing of soaps with lubricating oils. Until a relatively short time ago, grease making was considered an art. To stir hot soap into hot oil is a simple business, but to do so in such a manner as to form a grease is much more difficult, and the early grease maker needed much experience to learn the essentials of the trade. Therefore, it is not surprising that grease making is still a complex operation. The signs that told the grease maker that the soap was cooked and that the batch of grease was ready to run have been replaced by scientific tests that follow the process of manufacture precisely.

The early grease makers made grease in batches in barrels or pans, and the batch method is still the chief method of making grease. Oil and soap are mixed in kettles that have double walls between which steam and water may be circulated to maintain the desired temperature. When temperatures higher than 150°C (300°F) are required, a kettle heated by a ring of gas burners is used. Mixing is usually accomplished in each kettle by horizontal paddles radiating from a central shaft.

The soaps used in grease making are usually made in the grease plant, usually in a grease making kettle. Soap is made by chemically combining a metal hydroxide with a fat or fatty acid:

$$R-CO_2H + NaOH \rightarrow R-CO_2^- Na^+ + H_2O$$
$$\text{Fatty acid} \qquad\qquad \text{Soap}$$

The most common metal hydroxides used for this purpose are calcium hydroxide, lye, lithium hydroxide, and barium hydroxide. Fats are chemical combinations of fatty acids and glycerin. If a metal hydroxide is reacted with a fat, a soap containing glycerin is formed. Frequently, a fat is separated into its fatty acid and glycerin components, and only the fatty acid portion is used to make soap. Commonly used fats for grease-making soaps are cottonseed oil, tallow, and lard. Among the fatty acids used are stearic acid (from tallow), oleic acid (from cottonseed oil), and animal fatty acids (from lard).

To make grease the soap is dispersed in the oil as fibers of such a size that it may be possible to detect them only by microscopy. The fibers form a matrix for the oil, and the type, amount, size, shape, and distribution of the soap fibers dictate the consistency, texture, and bleeding characteristics, as well as the other properties of grease. Greases may contain from 50% to 30% soap, and although the fatty acid influences the properties of a grease, the metal in the soap has the most important effect. For example, calcium soaps form smooth buttery greases that are resistant to water but are limited in use to temperatures under about 95°C (200°F).

Soda (sodium) salts form fibrous greases that disperse in water but can be used at temperatures well over 95°C (200°F). Barium and lithium soaps form greases similar to those from calcium soaps, but they can be used at both high temperatures and very low temperatures; hence barium and lithium soap greases are known as multipurpose greases.

The soaps may be combined with any lubricating oil from a light distillate to a heavy residual oil. The lubricating value of the grease is chiefly dependent on the quality and viscosity of the oil. In addition to soap and oil, greases may also contain various additives that are used to improve the ability of the grease to stand up under extreme bearing pressures, to act as a rust preventive, and to reduce the tendency of oil to seep or bleed from a grease. Graphite, mica, talc, or fibrous material may be added to grease that are used to lubricate rough machinery to absorb the shock of impact. Other chemicals can make grease more resistant to oxidation or modify the structure of the grease.

The older, more common method of grease making is the batch method, but grease is also made by a continuous method. The process involves soap manufacture in a series (usually three) of retorts. Soap-making ingredients are charged into one retort whereas soap is made in the second retort. The third retort contains finished soap, which is pumped through a mixing device where the soap and the oil are brought together and blended. The mixer continuously discharges finished grease into suitable containers.

26.9.1 Lime Soap

The soap for a lime (calcium) soap grease is made by mixing fats and an equal amount of lubricating oil with the proper amount of lime (calcium hydroxide) at a temperature of 175°C (350°F) in an enclosed, steam-heated, pressure vessel located above a grease kettle. Heating for 30 min at 35 psi forms the lime soap that is dropped into the grease kettle. Oil is run into the grease kettle while the rotating paddles stir the oil into the soap. If the soap–oil mixture cools to a temperature below 95°C (200°F), the soap separates into large masses. To prevent this a small quantity of water is added to the mixture when it is at 105°C (220°F), and the water combines with the soap to form a hydrate that disperses through the oil and produces a smooth grease. At about 80°C (180°F) the grease is run from the bottom of the kettle through screens into packages. If the water content of the grease is diminished by the use of the grease at too high a temperature, the soap separates from the oil and the grease is destroyed. To overcome this and to permit lime soap greases to be used at temperatures over 70°C (160°F), chemicals less affected by heat are now used in place of water.

26.9.2   Soda Soap

The soap for soda greases is made directly in a grease kettle. Thus fat and lye solution (sodium hydroxide) are mixed in the container for 2 h at about 150°C (300°F), and oil is then stirred into the soap. Water is not required to form a soda soap grease. Brick greases are soda soap greases with high soap contents that are cast in the form of bricks.

26.9.3   Lithium and Barium Soap

Lithium and barium soap greases combine the best properties of both lime and soda greases. They are water resistant, adhere to metal surfaces, and can be used at both high and low temperatures. They were first developed for aircraft use but are now widely used in automobile lubrication and in industry under the name of multipurpose greases, which have replaced a large part of the lime soap greases formerly made and have almost eliminated aluminum soap greases. Most lithium and barium soap greases are made in the same manner as soda soap greases but require mixing temperatures of about 205°C (400°F). The lubricating oil and fatty acid are run into the grease kettle and heated to 80°C (180°F); lithium hydroxide is then added, and the mixture is stirred and heated to 150°C (300°F), which forms the soap and evaporates the water. Alternatively, the grease may be prepared from the lubricating oil and lithium stearate.

26.9.4   Aluminum Soap

Aluminum soap is made from soda soap by addition to the soda soap of a water solution of aluminum sulfate. The aluminum replaces the sodium in the soap, and the sodium is removed with the water as sodium sulfate. The aluminum soap is then mixed with oil in the usual manner. Aluminum soap grease is a relatively fluid material and is generally used for slow-moving gears operated at low temperatures.

26.9.5   Cold Sett Grease

Cold sett grease is lime soap grease in which rosin oil, rather than a fat or fatty acid, is used. Rosin oil is obtained by distilling the rosin obtained from pine trees. The method of making sett greases is quite different from the methods used for other greases. Rosin oil is dissolved in the oil in a vessel, and a mixture of lime and water to which a small amount of rosin oil and mineral oil has been added is stirred into an emulsion in another vessel. Both vessels are at temperatures below 65°C (150°F), and running the materials continuously from both vessels into a small mixer forms the grease. The mixed materials flow directly into packages where the grease sets to a solid in a few minutes. Cold sett greases are water-resistant materials and are usually used to lubricate rough gears, wagon axles, and the like.

26.10 WAX

The use of paraffin wax in a historical sense is varied but for the purpose of this chapter can be taken to the eighteenth century (Burke, 1996).

At that time, documents were written or drawn on damp paper with special ink that included gum Arabic, which stayed moist for 24 h, during which copies could be made by pressing another smooth white sheet against the original and transferring the ink marks to the new sheet. Initially, the copier was not a success. Banks were opposed because they thought it would encourage forgery. Counting houses argued that it would be inconvenient when they were rushed, or working by candlelight. But by the end of the first year, Watt had sold 200 samples and had made a great impression with a demonstration at the houses of Parliament, causing such a stir that members had to be reminded they were in session. By 1785 the copier was in common use.

Then in 1823 Cyrus P. Dalkin of Concord, Massachusetts, improved on the technique by using two different materials whose effect on history was to be startling. By rolling a mixture of carbon black and hot paraffin wax onto the back of a sheet of paper, Dalkin invented carbon copies. The development lay relatively unnoticed until the 1868 balloon ascent by Lebbeus H. Rogers, the 21-year-old partner in a biscuit-and-greengrocery firm. His aerial event was being covered by the Associated Press, and in the local newspaper office after the flight. Rogers was interviewed by a reporter who was using the carbon paper developed by Dalkin. Impressed by what he saw, Rogers terminated his ballooning and biscuits efforts and started a business producing carbon paper for use in order books, receipt books, invoices, and the like. In 1873, he conducted a demonstration for the Remington typewriter company, and the new carbon paper became an instant success.

The paraffin wax Dalkin used, and which was therefore half responsible (together with carbon black) for changing the world of business, had originally been produced from oil shale rocks. After the discovery of petroleum in Pennsylvania, in 1857 (Chapter 1), paraffin oil was produced by distillation and was used primarily as an illuminant to make up for the dwindling supply of sperm-whale oil in a rapidly growing lamp market. Chilled down paraffin solidified into paraffin wax. Apart from its use in lighting, the wax was also used to preserve the crumbling Cleopatra's Needle obelisk in New York's Central Park

Petroleum wax is of two general types: (1) paraffin wax in petroleum distillates and (2) **microcrystalline wax** in petroleum residua. The melting point of wax is not directly related to its boiling point, because waxes contain hydrocarbons of a different chemical nature. Nevertheless, waxes are graded according to their melting point and oil content.

26.10.1 COMPOSITION

Paraffin wax is a solid crystalline mixture of straight-chain (normal) hydrocarbons ranging from C_{20} to C_{30} and possibly higher, that is, $CH_3(CH_2)_nCH_3$ where $n \geq 18$.

It is distinguished by its solid state at ordinary temperatures (25°C, 77°F) and low viscosity (35 to 45 SUS at 99°C, 210°F) when melted. However, in contrast to petroleum wax, petrolatum (**petroleum jelly**), although solid at ordinary temperatures, does in fact contain both solid and liquid hydrocarbons. It is essentially a low-melting, ductile, microcrystalline wax.

26.10.2 MANUFACTURE

Paraffin wax from a solvent dewaxing operation (Chapter 19) is commonly known as slack wax, and the processes employed for the production of waxes are aimed at deoiling the slack wax (petroleum wax concentrate).

Wax sweating was originally used in Scotland to separate wax fractions with various melting points from the wax obtained from shale oils. Wax sweating is still used to some extent but is being replaced by the more convenient wax recrystallization process. In wax sweating, a cake of slack wax is slowly warmed to a temperature at which the oil in the wax and the lower melting waxes become fluid and drip (or sweat) from the bottom of the cake, leaving a residue of higher melting wax. However, wax sweating can be carried out only when the residual wax consists of large crystals that have spaces between them, through which the oil and lower melting waxes can percolate; it is therefore limited to wax obtained from light paraffin distillate.

The amount of oil separated by sweating is now much smaller than it used to be owing to the development of highly efficient solvent dewaxing techniques (Chapter 19). In fact, wax sweating is now more concerned with the separation of slack wax into fractions with different melting points. A wax sweater consists of a series of about nine shallow pans arranged one above the other in a sweater house or oven, and each pan is divided horizontally by a wire screen. The pan is filled to the level of the screen with cold water. Molten wax is then introduced and allowed to solidify, and the water is then drained from the pan leaving the wax cake supported on the screen.

A single sweater oven may contain more than 600 barrels of wax, and steam coils arranged on the walls of the oven slowly heat the wax cakes, allowing oil and the lower melting waxes to sweat from the cakes and drip into the pans. The first liquid removed from the pans is called **foots oil**, which melts at 38°C (100°F) or lower, followed by interfoots oil, which melts in the range 38°C to 44°C (100°F to 112°F). Crude scale wax next drips from the wax cake and consists of wax fractions with melting points over 44°C (112°F).

When oil removal was an important function of sweating, the sweating operation was continued until the residual wax cake on the screen was free of oil. When the melting point of the wax on the screen has increased to the required level, allowing the oven to cool terminates sweating. The wax on the screen is a sweated wax with the melting point of a commercial grade of paraffin wax, which after a finished treatment becomes refined paraffinic wax. The crude scale wax obtained in the sweating operation may be recovered as such or treated to improve the color, in which case it is white crude scale wax. The crude scale wax and interfoots, however, are the sources of more waxes with lower melting points. The crude scale wax and interfoots are re-sweated several times to yield sweated waxes, which are treated to produce a series of refined paraffin waxes with melting points ranging from about 50°C to 65°C (125°F to 150°F).

Sweated waxes generally contain small amounts of unsaturated aromatic and sulfur compounds, which are the source of unwanted color, odor, and taste that reduce the ability of the wax to resist oxidation; the commonly used method of removing these impurities is clay treatment of the molten wax.

Wax recrystallization, like wax sweating, separates slack wax into fractions, but instead of using the differences in melting points, it makes use of the different solubility of the wax fractions in a solvent, such as the ketone used in the dewaxing process (Chapter 19). When a mixture of ketone and slack wax is heated, the slack wax usually dissolves completely, and if the solution is cooled slowly, a temperature is reached at which a crop of wax crystals is formed. These crystals will all be of the same melting point, and if they are removed by filtration, a wax fraction with a specific melting point is obtained. If the clear filtrate is further cooled, a second crop of wax crystals with a lower melting point is obtained. Thus by alternate cooling and filtration the slack wax can be subdivided into a large number of wax fractions, each with different melting points.

This method of producing wax fractions is much faster and more convenient than sweating and results in a much more complete separation of the various fractions. Furthermore, recrystallization can also be applied to the microcrystalline waxes obtained from intermediate and heavy paraffin distillates, which cannot be sweated. Indeed, the microcrystalline waxes have higher melting points and differ in their properties from the paraffin waxes obtained from light paraffin distillates; thus wax recrystallization makes new kinds of waxes available.

26.10.3 PROPERTIES AND USES

The melting point of paraffin wax (ASTM D-87) has both direct and indirect significance in most wax utilization. All wax grades are commercially indicated in a range of melting temperatures rather than at a single value, and a range of 1°C (2°F) usually indicates a good degree of refinement. Other common physical properties that help to illustrate the degree of refinement of the wax are color (ASTM D-156), oil content (ASTM D-721), API gravity (ASTM D-287), flash point (ASTM D-92), and viscosity (ASTM D-88 and ASTM D-445), although the last three properties are not usually given by the producer unless specifically requested.

Petroleum waxes (and petrolatum) find many uses in pharmaceuticals, cosmetics, paper manufacturing, candle making, electrical goods, rubber compounding, textiles, and many more too numerous to mention here. For additional information, more specific texts on petroleum waxes should be consulted.

26.11 ASPHALT

Asphalt is a major product of many petroleum refineries. Asphalt may be residual (straight-run) asphalt, which is made up of the nonvolatile hydrocarbons in the feedstock, along with similar materials produced by thermal alteration during the distillation sequences, or they may be produced by air blowing residua. Alternatively asphalt may be the residuum from a vacuum distillation unit. In either case, the properties of the asphalt are, essentially, the properties of the residuum (Table 26.7). If the properties are not suitable for the asphalt product to meet specifications, changing the properties by, for example, blowing is necessary.

26.11.1 COMPOSITION

Asphalt is the residue of mixed-base and asphalt-base crude oils. It cannot be distilled even under the highest vacuum, because the temperatures required to do this promote formation of coke. Asphalt have complex chemical and physical compositions that usually vary with the source of the crude oil and are considered dispersions of particles, called asphaltenes, in a high-boiling fluid composed of oil and resins (Traxler, 1961; Barth, 1962; Hoiberg, 1964; Broome, 1973; Broome and Wadelin, 1973) (Chapter 11).

The nature of the asphalt is determined by such factors as the nature of the medium (paraffinic or aromatic), as well as the nature and proportion of the asphaltenes and of the resins. The asphaltenes have been suggested to be lyophobic; the resins are lyophilic, and the interaction of the resins with the asphaltenes is responsible for asphaltene dispersion, which seems to exercise marked control on the nature of the asphalt. The asphaltenes vary in character but are of sufficiently high molecular weight to require dispersion as micelles, which are peptized by the resins. If the asphaltenes are relatively low in molecular weight, the resins plentiful, and the medium aromatic in nature, the result may be viscous asphalt without anomalous properties. If, however, the medium is paraffinic and the resins are scarce, and the asphaltenes are high in molecular (or micellar) weight (these conditions are encouraged by vacuum, steam reduction, or air blowing), the asphalt is of the gel type and exhibits the properties that accompany such structure. A high content of resins imparts to a product

TABLE 26.7
Properties of Atmospheric and Vacuum Residua

Feedstock	Gravity (API)	Sulfur (wt.%)	Nitrogen (wt.%)	Nickel (ppm)	Vanadium (ppm)	Asphaltenes (Heptane) (wt.%)	Carbon Residue (Conradson) (wt.%)
Arabian light, >650°F	17.7	3.0	0.2	10.0	26.0	1.8	7.5
Arabian light, >1050°F	8.5	4.4	0.5	24.0	66.0	4.3	14.2
Arabian heavy, >650°F	11.9	4.4	0.3	27.0	103.0	8.0	14.0
Arabian heavy, >1050°F	7.3	5.1	0.3	40.0	174.0	10.0	19.0
Alaska, north slope, >650°F	15.2	1.6	0.4	18.0	30.0	2.0	8.5
Alaska, north slope, >1050°F	8.2	2.2	0.6	47.0	82.0	4.0	18.0
Lloydminster (Canada), >650°F	10.3	4.1	0.3	65.0	141.0	14.0	12.1
Lloydminster (Canada), >1050°F	8.5	4.4	0.6	115.0	252.0	18.0	21.4
Kuwait, >650°F	13.9	4.4	0.3	14.0	50.0	2.4	12.2
Kuwait, >1050°F	5.5	5.5	0.4	32.0	102.0	7.1	23.1
Tia Juana, >650°F	17.3	1.8	0.3	25.0	185.0		9.3
Tia Juana, >1050°F	7.1	2.6	0.6	64.0	450.0		21.6
Taching, >650°F	27.3	0.2	0.2	5.0	1.0	4.4	3.8
Taching, >1050°F	21.5	0.3	0.4	9.0	2.0	7.6	7.9
Maya, >650°F	10.5	4.4	0.5	70.0	370.0	16.0	15.0

desirable adhesive character and plasticity; a high asphaltene content is usually responsible for the harder, more brittle, asphalt.

26.11.2 MANUFACTURE

Asphalt manufacture is, in essence, a matter of distilling everything possible from crude petroleum until a residue with the desired properties is obtained. This is usually done by stages (Figure 26.3); crude distillation at atmospheric pressure removes the lower-boiling fractions and yields reduced crude that may contain higher-boiling (lubricating) oils, asphalt, and even wax. Distillation of the reduced crude under vacuum removes the oils (and wax) as overhead products and the asphalt remains as a bottom (or residual) product. The majority of the polar functionality in the original crude oil tend to be nonvolatile and concentrate in the vacuum residuum (Figure 26.4). It is this concentration effect that confers upon asphalt some of its unique properties. At this stage the asphalt is frequently and incorrectly referred to as pitch and has a softening point related to the amount of oil removed: the more oil distilled from the residue, the higher the softening point (Corbett and Petrossi, 1978).

However, as there are wide variations in refinery operations and crude petroleum, asphalt with softening points ranging from 25°C to 55°C (80°F to 130°F) may be produced. Blending with higher and lower softening point asphalt may make asphalt of intermediate softening points. If lubricating oils are not required, the reduced crude may be distilled in a flash drum, which is similar to a bubble tower but has few, if any, trays. Asphalt descends to the base of the drum as the oil components pass out of the top of the drum. If the asphalt has a relatively low softening point, it can be hardened by further distillation with steam or by oxidation.

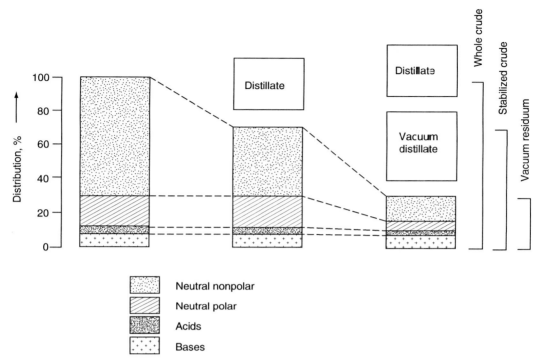

FIGURE 26.4 Accumulation of polar functions in residua during distillation.

Asphalt is also produced by propane deasphalting (Chapter 19) and the asphalt so produced may have a softening point of about 95°C (200°F). Softer grades are made by blending hard asphalt with the extract obtained in the solvent treatment of lubricating oils.

Soft asphalt can be converted into harder asphalt by oxidation, which promotes the formation of resins and asphaltenes from the lower-molecular-weight oil constituents:

$$\text{Oil constituents} \rightarrow \text{Resin constituents} \rightarrow \text{Asphaltene constituents}$$

The increase in the proportion of semisolid (resins) and solid (asphaltenes) constituents as a result of air blowing is accompanied by an increase in softening point with only a small loss in volume. A similar increase in softening point by removing oily constituents would cause a considerable decrease in volume. Oxidation is carried out by blowing air through asphalt heated to about 260°C (500°F) and is usually done in a tower (an oxidizer) equipped with a perforated pipe at the bottom through which the air is blown. The asphalt, in the batch mode or continuous mode, is heated until the oxidation reaction starts, but the reaction is exothermic, and the temperature is controlled by regulating the amount of air and by circulating oil or water through cooling coils within the oxidizer. Asphalt with softening points as high as 180°C (350°F) may be produced.

Asphalt, normally a liquid when applied at higher temperature, may be referred to as nonvolatile liquid asphalt, but semisolid or solid asphalt may be made liquid for easier handling by dissolving them in a solvent and are referred to as cutback asphalt. The asphalt and solvent (naphtha, kerosene, or gas oil) are heated to about 105°C (225°F) and passed together through a mechanical mixer, and the effluent then enters a horizontal tank, which is used for further mixing. Liquid asphalt is pumped from the top of the tank into a perforated pipe lying at the bottom of the tank, and circulation is continued until mixing is complete.

26.11.3 PROPERTIES AND USES

The use of asphalt (in many cases this was bitumen rather than a processed material; Chapter 1) goes back into antiquity. It was in fact the first petroleum derivative that was used extensively (Chapter 1). Nowadays, a good portion of the asphalt produced from petroleum is consumed in paving roads; the remainder is employed for roofing, paints, varnishes, insulating, rust-protective compositions, battery boxes, and compounding materials that go into rubber products, brake linings, and fuel briquettes. However, asphalt uses can be more popularly divided into use as road oils, cutback asphalt, asphalt emulsions, and solid asphalt. The properties of asphalt are defined by a variety of standard tests (Table 26.8) that can be used to define quality and viscosity specifications.

Road oils are, as the name implies, liquid asphaltic materials intended for easy application to earth roads; they do not provide a strong base or a hard surface but maintain a satisfactory passage for light traffic. Both straight-run and cracked residua have been employed successfully. Binding quality and adhesive character are important in governing the quality of the road produced; resistance to removal by emulsification has some influence on its permanence. Liquid road oils, cutbacks, and emulsions are of recent date, but the use of asphaltic solids for paving goes back to a European practice of about 1835. The asphaltic constituents employed may have softening points up to, say, 110°C (230°F).

Cutback asphalt are mixtures in which hard asphalt has been diluted with a lighter oil to permit application as a liquid without drastic heating. They are classified as rapid, medium, and slow curing, depending on the volatility of the diluent, which governs the rate of evaporation and consequent hardening.

An asphaltic material may be emulsified with water to permit application without heating. Such emulsions are normally of the oil-in-water type. They reverse or break on application to a stone or earth surface, so that the oil clings to the stone and the water disappears. In addition to their usefulness in road and soil stabilization, they are useful for paper impregnation and waterproofing. The emulsions are chiefly (1) the soap or alkaline type and (2) the neutral or clay type. The former break readily on contact, but the latter are more stable and probably lose water mainly by evaporation. Good emulsions must be stable during storage or freezing, suitably fluid, and amenable to control for speed of breaking.

26.12 COKE

Coke is the residue left by the destructive distillation of petroleum residua. That formed in catalytic cracking operations is usually nonrecoverable, as it is often employed as fuel for the process.

The composition of petroleum coke varies with the source of the crude oil, but in general, large amounts of high-molecular-weight complex hydrocarbons (rich in carbon but correspondingly poor in hydrogen) make up a high proportion. The solubility of petroleum coke in carbon disulfide has been reported to be as high as 50% to 80%, but this is in fact a misnomer, as the coke is the insoluble, honeycomb material that is the end product of thermal processes.

Petroleum coke is employed for a number of purposes, but its chief use is in the manufacture of carbon electrodes for aluminum refining, which requires a high-purity carbon low in ash and sulfur free; the volatile matter must be removed by calcining. In addition to its use as a metallurgical reducing agent, petroleum coke is employed in the manufacture of carbon

TABLE 26.8
Test Methods Used to Determine Asphalt Properties

Test	Organization/Number	Description
Adsorption	ASTM D-4469	Calculation of degree of adsorption of an asphalt by an aggregate.
Bond and adhesion	ASTM D-1191	Used primarily to determine whether an asphalt has bonding strength at low temperatures. See also ASTM D-3141 and ASTM D-5078.
Breaking point	IP 80	Indication of the temperature at which an asphalt possesses little or no ductility and would show brittle fracture conditions.
Compatibility	ASTM D-1370	Indicates whether asphalts are likely to be incompatible and disbond under stress. See also ASTM D-3407.
Distillation	ASTM D-402	Determination of volatiles content; applicable to road oils and cutback asphalts.
Ductility	ASTM D-113	Expressed as the distance in cm to which a standard briquet can be elongated before breaking; reflects cohesion and shear susceptibility.
Emulsified asphalts	ASTM D-244	Covers various tests for the composition, handling, nature and classification, storage, use, and specifications. See also ASTM D-977 and ASTM D-1187.
Flash point	ASTM D-92	Cleveland open cup method is commonly used; Tag open cup (ASTM D-3143) applicable to cutback asphalts.
Float test	ASTM D-139	Normally used for asphalts that are too soft for the penetration test.
Penetration	ASTM D-5	The extent to which a needle penetrates asphalt under specified conditions of load, time, and temperature; units are mm/10 measured from 0 to 300. See also ASTM D-243.
Sampling	ASTM D-140	Provides guidance for the sampling of asphalts.
Softening point	ASTM D-36	Ring and ball method; the temperature at which an asphalt attains a particular degree of softness under specified conditions; used to classify asphalt grades. See also ASTM D-2389.
Solubility in carbon disulfide	ASTM D-4	Determination of the carbon amount of carboids and carbenes and mineral matter; trichloroethylene and 1,1,1-trichloroethane have been used for this purpose. See also ASTM D-2042.
Specific gravity	ASTM D-70	See also ASTM D-3142.
Stain	ASTM D-1328	Measures the amount of stain on paper or other cellulosic materials.
Temperature–volume correction	ASTM D-1250	Allows the conversion of volumes of asphalts from one temperature to another. See also ASTM D-4311.
Thin film oven test	ASTM D-1754	Determines the hardening effect of heat and air on a film of asphalt. See also ASTM D-2872.
Viscosity	ASTM D-2170	A measure of resistance to flow. See also ASTM D-88 (now discontinued but a useful reference), ASTM D-1599, ASTM D-2171, ASTM D-2493, ASTM D-3205, ASTM D-3381, ASTM D-4402, and ASTM D-4957.
Water content	ASTM D-95	Determines the water content by distillation with a Dean and Stark receiver.
Weathering	ASTM D-529	Used for determining the relative weather resistance of asphalt. See also ASTM D-1669 and ASTM D-1670.

brushes, silicon carbide abrasives, and structural carbon (e.g., pipes and Rashig rings), as well as calcium carbide manufacture from which acetylene is produced:

$$\text{Coke} \rightarrow CaC_2$$
$$CaC_2 + H_2O \rightarrow HC \equiv CH$$

26.13 SULFONIC ACIDS

Sulfonic acids are produced when petroleum is treated with sulfuric acid. Sulfuric acid treating of petroleum distillates is generally applied to dissolve unstable or colored substances and sulfur compounds, as well as to precipitate asphaltic materials. When drastic conditions are employed, as in the treatment of lubricating fractions with large amounts of concentrated acid or when fuming acid is used in the manufacture of white oils, considerable quantities of petroleum sulfonic acids are formed. Extensive side reactions, mainly oxidation, also occur and increase with the proportion of sulfur trioxide in the acid.

Many of the lower paraffins are physically absorbed by concentrated and fuming sulfuric acids; chemical activity increases with rise in molecular weight, and compounds containing tertiary carbons are especially responsive. n-Hexane, n-heptane, and n-octane are essentially inactive in cold fuming acid; but at the boiling point of the hydrocarbons rapid sulfonation takes place to give mono- and disulfonic acids:

$$R{-}H + H_2SO_4 \rightarrow R{-}SO_3{-}H + H_2O$$
$$\qquad\text{Paraffin} \qquad\qquad \text{Sulfonic acid}$$

The five- and six-membered ring lower naphthene derivatives are stable to cold concentrated sulfuric acid, but fuming sulfuric acid reacts with cyclohexane to give mono- and dinaphthene and mono-aromatic sulfonic acids, along with products based on cyclic olefins formed through hydrogen-transfer reactions.

The action of sulfuric acid on hydrocarbons is indeed quite complex, but it is obvious that a reaction occurs readily with such compound types as aromatics and those tertiary carbon atoms in naphthenic rings that are both present in the lubricating fractions of petroleum. Ordinarily, a charge stock for sulfuric acid treatment will already have been refined by solvent extraction with, say, furfural (Chapter 19) to remove those more highly aromatic constituents. Thus the remaining hydrocarbons, which give higher yields of better sulfonated products, are those in which aromatic rings are entirely absent or are low in proportion relative to the naphthene rings and paraffinic chains, and hence the preferred sulfonic acids of commerce are probably naphthene sulfonic acids.

Sulfonic acids are also used as detergents made by the sulfonation of alkylated benzene. The number, size, and structure of the alkyl side chains are important in determining the performance of the finished detergent.

Two general methods are applied for the recovery of sulfonic acids from sulfonated oils and their sludge: (1) In one case the acids are selectively removed by adsorbents or by solvents (generally low-molecular-weight alcohols), and (2) in the other case the acids are obtained by salting out with organic salts or bases.

Petroleum sulfonic acids may be roughly divided into those soluble in hydrocarbons and those soluble in water. Because of their color, hydrocarbon-soluble acids are referred to as mahogany acids, and the water-soluble acids are referred to as green acids. The composition of each type varies with the nature of the oil sulfonated and the concentration of the acids produced. In general, those formed during light acid treatment are water soluble; oil-soluble acids result from more drastic sulfonation.

The salts of mixed petroleum sulfonic acids have many commercial applications. They find use as anticorrosion agents, leather softeners, and flotation agents and have been used in place of red oil (sulfonated castor oil) in the textile industry. Lead salts of the acids have been employed in greases as extreme pressure agents, and alkyl esters have been used as alkylating agents. The alkaline earth metal (Mg, Ca, and Ba) salts are used in detergent compositions for motor oils, and the alkali metal (K and Na) salts are used as detergents in aqueous systems.

26.14   ACID SLUDGE

Sludge produced during the use of sulfuric acid as a treating agent is mainly of two types: (1) sludge from light oils (gasoline and kerosene) and (2) sludge from lubricating stocks, medicinal oils, and the like. In the treatment of the latter oils it appears that the action of the acid causes precipitation of asphaltene constituents and resin constituents, as well as the solution of color-bearing and sulfur compounds. Sulfonation and oxidation-reduction reactions also occur but to a lesser extent as much of the acid can be recovered. In the desulfurization of cracked distillates, however, chemical interaction is more important, and polymerization, ester formation, aromatic-olefin condensation, and sulfonation also occur. Nitrogen bases are neutralized, and the acid dissolves naphthenic acids; thus the composition of the sludge is complex and depends largely on the oil treated acid strength and the temperature.

Sulfuric acid sludge from *iso*-paraffin alkylation and lubricating oil treatment are frequently decomposed thermally to produce sulfur dioxide (which is returned to the sulfuric acid plant) and sludge acid coke. The coke, in the form of small pellets, is used as a substitute for charcoal in the manufacture of carbon disulfide. Sulfuric acid coke is different from other petroleum coke in that it is pyrophoric in air and also reacts directly with sulfur vapors to form carbon disulfide.

26.15   PRODUCT BLENDING

The modern petroleum refinery consists of a very complex mix of high-technology processes that efficiently convert the wide array of crude oils into the hundreds of specification products we use daily. Each refinery has its own unique processing configuration as a result of the logistics and associated economics related to its specific crude oils and products markets. The refiner must continuously optimize the mix of product volumes and this is accomplished through executing decisions regarding parameters as varied as crude oil feedstock selection, adjustments in product cut-points, and reactor severities in individual processes. Additional options include changing the dispositions of intermediate product streams to alternative processing units, or alternative finished product blends.

In fact, many refinery products are typically the result of blending several component streams or blending stocks. In most cases, product blending is accomplished by controlling the volumes of blend stocks from individual component storage tanks that are mixed in the finished product storage tank. Samples of the finished blend are then analyzed by laboratory testing for all product specifications prior to shipping. Alternatively, in-line blending refers to pipeline shipments in which the finished product is actually blended directly into the product pipeline (as opposed to a standing product storage tank).

The most commonly recognized blending operations occur in the gasoline production section of the refinery. The various gasoline streams are such that specifications (dependent upon geographic location, environmental regulations, and weather patterns) can be met.

Gasoline blending involves combining of the components that make up motor gasoline. The components include the various hydrocarbon streams produced by distillation, cracking, reforming, and polymerization, tetraethyl lead, and identifying color dye, as well as other special-purpose components, such as solvent oil and anti-icing compounds. The physical process of blending the components is simple, but determination of how much of each component to include in a blend is much more difficult. The physical operation is carried out by simultaneously pumping all the components of a gasoline blend into a pipeline that leads to the gasoline storage, but the pumps must be set to deliver automatically the proper proportion of each component. Baffles in the pipeline are often used to mix the components as they travel to the storage tank.

Selection of the components and their proportions in a blend is the most complex problem in a refinery. Many different hydrocarbon streams may need to be blended to produce quality gasoline. Each property of each stream is a variable, and the effect on the product gasoline is considerable. For example, the low octane number of straight-run naphtha limits its use as a gasoline component, although its other properties may make it desirable. The problem is further complicated by changes in the properties of the component streams due to processing changes. For example, an increase in cracking temperature produces a smaller volume of higher octane cracked naphtha, but before this cracked naphtha can be included in a blend, adjustments must be made in the proportions of the other hydrocarbon components. Similarly, the introduction of new processes and changes in the specifications of the finished gasoline dictate reevaluation of the components that make up the gasoline (Gibbs, 1989).

Gasoline blending is not the only blending operation, other product blending operations are also in operation in a refinery. The applicable specifications vary by product but typically include properties pertinent to the behavior of the product in use. Many product specifications do not blend linearly by component volumes. In these circumstances, the finished blend properties are predicted using experience-based algorithms for the applicable blending components.

REFERENCES

Abraham, H. 1945. *Asphalt and Allied Substances*, 5th edn, Volume I. Van Nostrand Inc., New York. p. 1.

Absi-Halabi, M., Stanislaus, A., and Qabazard, H. 1997. *Hydrocarbon Process.* 76(2): 45.

Barth, E.J. 1962. *Asphalt: Science and Technology*. Gordon and Breach, New York.

Boenheim, A.F. and Pearson, A.J. 1973. *Modern Petroleum Technology*. G.D. Hobson and W. Pohl, eds. Applied Science Publishers Inc., Barking, Essex, UK. Chapter 19.

Broome, D.C. 1973. *Modern Petroleum Technology*. G.D. Hobson and W. Pohl, eds. Applied Science Publishers Inc., Barking, Essex, UK. Chapter 23.

Broome, D.C. and Wadelin, F.A. 1973. *Criteria for Quality of Petroleum Products*. J.P. Allinson, ed. Halsted Press, Toronto. Chapter 13.

Burke, J. 1996. *The Pinball Effect*. Little, Brown and Company, New York. p. 25 and 26.

Corbett, L.W. and Petrossi, V. 1978. *Ind. Eng. Chem. Prod. Res. Dev.* 17: 342.

Dooley, J.E., Lanning, W.C., and Thompson, C.J. 1979. *Refining of Synthetic Crudes*. M.L. Gorbaty and B.M. Harney, eds. Advances in Chemistry Series No. 179. American Chemical Society, Washington, DC. Chapter 1.

Forbes, R.J. 1958a. *A History of Technology*, Volume V. Oxford University Press, Oxford, UK. p. 102.

Forbes, R.J. 1958b. *Studies in Early Petroleum Chemistry*. E.J. Brill, Leiden, The Netherlands.

Forbes, R.J. 1959. *More Studies in Early Petroleum Chemistry*. E.J. Brill, Leiden, The Netherlands.

Gibbs, L.M. 1989. *Oil Gas J.* 87(17): 60.

Gray, C.L. and Alson, J.A. 1989. *Sci. Am.* 145(11): 108.

Guthrie, V. 1960. *Petrochemical Products Handbook*. McGraw-Hill. New York.

Hadley, D.J. and Turner, L. 1973. *Modern Petroleum Technology*. G.D. Hobson and W. Pohl, eds. Applied Science Publishers Inc., Barking, Essex, UK. Chapter 12.

Hobson, G.D. and Pohl, W. 1973. *Modern Petroleum Technology*. Applied Science Publishers, Barking, Essex, UK.

Hoffman, H.L. 1992. *Petroleum Processing Handbook*. J.J. McKetta, ed. Marcel Dekker Inc., New York. p. 2.

Hoiberg, A.J. 1964. *Bituminous Materials: Asphalts, Tar, and Pitches*. Interscience Publishers, New York.

James, P. and Thorpe, N. 1994. *Ancient Inventions*. New York. *Ballantine Books*.

Long, R.B. and Speight, J.G. 1998. *Petroleum Chemistry and Refining*. J.G. Speight, ed. Taylor & Francis Publishers, Washington, DC. Chapter 1.

Mills, G.A. and Ecklund, E.E. 1987. *Annu. Rev. Energy* 12: 47.

Sequeira, A. Jr. 1992. *Petroleum Processing Handbook*. J.J. McKetta, ed. Marcel Dekker Inc., New York. p. 634.

Traxler, R.N. 1961. *Asphalt: Its Composition, Properties, and Uses*. Reinhold Publishing Corp., New York.

Walmsley, A.G. 1973. *Modern Petroleum Technology*. G.D. Hobson and W. Pohl, eds. Applied Science Publishers Inc., Barking, Essex, UK. Chapter 17.

Wittcoff, H. 1987. *J. Chem. Educ.* 64: 773.

27 Petrochemicals

27.1 INTRODUCTION

Petroleum refining (Chapter 14) begins with the distillation or fractionation of crude oils into separate fractions of hydrocarbon groups. The resultant products (Chapter 26) are directly related to the characteristics of the crude oil being processed. Most of these products of distillation are further converted into more useable products by changing their physical and molecular structures through cracking, **reforming**, and other conversion processes. These products are subsequently subjected to various treatment and separation processes, such as extraction, hydrotreating, and sweetening, to produce finished products. Whereas the simplest refineries are usually limited to atmospheric and vacuum distillation, integrated refineries incorporate fractionation, conversion, treatment, and blending with lubricant, heavy fuels, and asphalt manufacturing; they may also include petrochemical processing.

It is during the refining process that other products are also produced. These products include the gases dissolved in the crude oil that are released during distillation as well as the gases produced during the various refining processes that provide fodder for the petrochemical industry.

The gas (often referred to as **refinery gas** or process gas) varies in composition and volume, depending on the origin of the crude oil and on any additions (i.e., other crude oils blended into the refinery feedstock) to the crude oil made at the loading point. It is not uncommon to re-inject light hydrocarbons such as propane and butane into the crude oil before dispatch by tanker or pipeline. This results in a higher vapor pressure of the crude, but it allows one to increase the quantity of light products obtained at the refinery. As light ends in most petroleum markets command a premium, while in the oil field itself propane and butane may have to be re-injected or flared, the practice of spiking crude oil with liquefied petroleum gas is becoming fairly common. These gases are recovered by distillation (Figure 27.1).

In addition to distillation (Chapter 16), gases that are produced in the various **thermal cracking processes** (Figure 27.2) are also available.

Thus, in processes such as coking or visbreaking (Chapter 17) a variety of gases are produced. Another group of refining operations that contributes to gas production is that of the **catalytic cracking** processes (Chapter 18). Both catalytic and thermal cracking processes result in the formation of unsaturated hydrocarbons, particularly ethylene ($CH_2=CH_2$), but also propylene (propene, $CH_3CH=CH_2$), *iso*-butylene [*iso*-butene, $(CH_3)_2C=CH_2$] and the *n*-butenes ($CH_3CH_2CH=CH_2$, and $CH_3CH=CHCH_3$) in addition to hydrogen (H_2), methane (CH_4) and smaller quantities of ethane (CH_3CH_3), propane ($CH_3CH_2CH_3$), and butanes [$CH_3CH_2CH_2CH_3$, $(CH_3)_3CH$]. Diolefins such as butadiene ($CH_2=CHCH=CH_2$) are also present. A further source of refinery gas is **hydrocracking**, a catalytic high-pressure pyrolysis process in the presence of fresh and recycled hydrogen (Chapter 16). The feedstock is again heavy gas oil or residual fuel oil, and the process is mainly directed at the production of additional middle distillates and gasoline. As hydrogen is

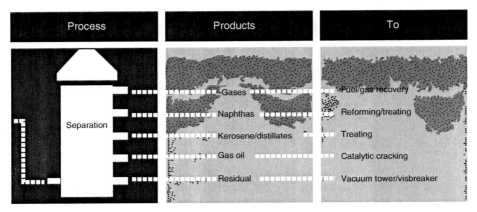

FIGURE 27.1 Gas recovery by distillation.

to be recycled, the gases produced in this process again have to be separated into lighter and heavier streams; any surplus recycle gas and the liquefied petroleum gas from the hydrocracking process are both saturated.

In a series of reforming processes (Chapter 23), commercialized under names such as **platforming**, paraffin, and naphthene, (cyclic nonaromatic) hydrocarbons are converted in the presence of hydrogen and a catalyst into aromatics, or isomerized to more highly branched hydrocarbons. Catalytic reforming processes thus not only result in the formation of a liquid product of higher octane number, but also produce substantial quantity of gases. The latter are rich in hydrogen, but also contain hydrocarbons from methane to butanes, with a preponderance of propane ($CH_3CH_2CH_3$), n-butane ($CH_3CH_2CH_2CH_3$), and *iso*-butane $[(CH_3)_3CH]$.

The composition of the process gas varies in accordance with the reforming severity and reformer feedstock. All catalytic reforming processes require substantial recycling of a hydrogen stream. Therefore, it is normal to separate reformer gas into a propane ($CH_3CH_2CH_3$) or a butane stream $[CH_3CH_2CH_2CH_3$ plus $(CH_3)_3CH]$, which becomes part of the refinery liquefied petroleum gas production, and a lighter gas fraction, part of which is recycled. In view of the excess of hydrogen in the gas, all products of catalytic reforming are saturated, and there are usually no olefin gases present in either gas stream.

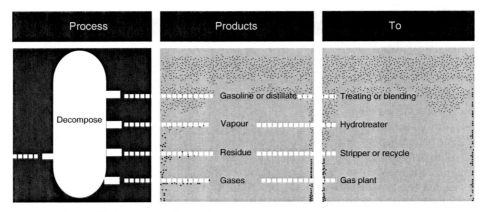

FIGURE 27.2 Gas prediction during thermal processes.

TABLE 27.1
Naphtha Production

Process	Primary Product	Secondary Process	Secondary Product
Atmospheric distillation	Naphtha		Light naphtha
			Heavy naphtha
	Gas oil	Catalytic cracking	Naphtha
	Gas oil	Hydrocracking	Naphtha
Vacuum distillation	Gas oil	Catalytic cracking	Naphtha
		Hydrocracking	Naphtha
	Residuum	Coking	Naphtha
		Hydrocracking	Naphtha

In many refineries, naphtha, in addition to other refinery gases, is also used as the source of petrochemical feedstocks. In the process, naphtha crackers convert naphtha feedstock (produced by various process) (Table 27.1) into ethylene, propylene, benzene, toluene, and xylene as well as other by-products in a two-step process of cracking and separating. In some cases, a combination of naphtha, gas oil, and liquefied petroleum gas may be used. The feedstock, typically naphtha, is introduced into the pyrolysis section where it is cracked in the presence of steam. The naphtha is converted into lower boiling fractions, primarily ethylene and propylene. The hot gas effluent from the furnace is then quenched to inhibit further cracking and to condense higher molecular weight products. The higher molecular weight products are subsequently processed into fuel oil, light cycle oil and pygas by-products. The pygas stream can then be fed to the aromatics plants for benzene and toluene production.

The cooled gases are then compressed, treated to remove acid gases, dried over a desiccant and fractionated into separate components at low temperature through a series of refrigeration processes (Chapter 25). Hydrogen and methane are removed by way of a compression/expansion process after which the methane is distributed to other processes as deemed appropriate or used as fuel gas. Hydrogen is collected and further purified in a pressure swing unit for use in the hydrogenation process (Chapter 20 and Chapter 21). Polymer grade ethylene and propylene are separated in the cold section after which the ethane and propane streams are recycled back to the furnace for further cracking while the mixed butane (C_4) stream is hydrogenated prior to recycling back to the furnace for further cracking.

The refinery gas (or the process gas) stream and the products of naphtha cracking are the source of a variety of petrochemicals. Thus, petrochemicals are chemicals derived from petroleum and natural gas and, for convenience of identification, petrochemicals can be divided into two groups: (1) primary petrochemicals (Figure 27.3) and (2) intermediates and derivatives.

Primary petrochemicals include: olefins (ethylene, propylene, and butadiene), aromatics (benzene, toluene, and xylenes), and methanol. Petrochemical intermediates are generally produced by chemical conversion of primary petrochemicals to form more complicated derivative products. Petrochemical derivative products can be made in a variety of ways: directly from primary petrochemicals; through intermediate products that still contain only carbon and hydrogen; and, through intermediates that incorporate chlorine, nitrogen or oxygen in the finished derivative. In some cases, they are finished products; in others, more steps are needed to arrive at the desired composition.

In the strictest sense, petrochemicals is any of a large group of chemicals manufactured from petroleum and natural gas as distinct from fuels and other products (Chapter 26); derived from

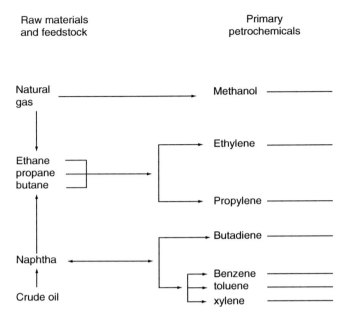

FIGURE 27.3 Raw materials and primary petrochemicals.

petroleum and natural gas and used for a variety of commercial purposes. The definition has been broadened to include the whole range of organic chemicals (Figure 27.4). In many instances, a specific chemical included among the petrochemicals may also be obtained from other sources, such as coal, coke, or vegetable products. For example, materials such as benzene and naphthalene can be made from either petroleum or coal, while ethyl alcohol may be of petrochemical or vegetable origin. This makes it difficult to categorize a specific substance as, strictly speaking, petrochemical or nonpetrochemical.

The chemical industry is in fact the chemical process industry by which a variety of chemicals are manufactured. The chemical process industry is, in fact, subdivided into other categories as follows:

1. Chemicals and allied products in which chemicals are manufactured from a variety of feedstocks and may then be put to further use
2. Rubber and miscellaneous products that focus on the manufacture of rubber and plastic materials
3. Petroleum refining and related industries which, on the basis of prior chapters in this text, are now self-explanatory

Thus, the petrochemical industry falls under the subcategory of petroleum and related industries.

The petroleum era was ushered in by the discovery of petroleum at Titusville, Pennsylvania in 1859. But the production of chemicals from natural gas and petroleum has been a recognized industry only since the early twentieth century. Nevertheless, the petrochemical industry has made quantum leaps in the production of a wide variety of chemicals (Kolb and Kolb, 1979; Chenier, 1992), which being based on starting feedstocks from petroleum are termed *petrochemicals*.

Thermal cracking processes (Chapter 17) developed for crude oil refining, starting in 1913 and continuing for the next two decades, were focused primarily on increasing the quantity

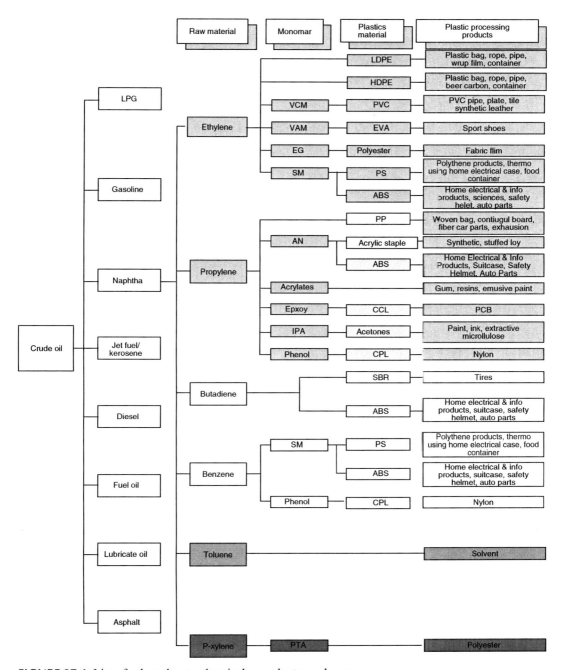

FIGURE 27.4 List of selected petrochemicals, products, and uses.

and quality of gasoline components. As a by-product of this process, gases were produced that included a significant proportion of lower-molecular-weight olefins, particularly ethylene ($CH_2=CH_2$), propylene ($CH_3CH=CH_2$), and butylenes (butenes, $CH_3CH=CH.CH_3$ and $CH_3CH_2CH=CH_2$). Catalytic cracking (Chapter 18), introduced in 1937, is also a valuable source of propylene and butylene, but it does not account for a very significant yield of ethylene, the most important of the petrochemical building blocks. Ethylene is polymerized to produce polyethylene or, in combination with propylene, to produce copolymers that are used extensively in food-packaging wraps, plastic household goods, or building materials. Prior to the use of petroleum and natural gas as sources of chemicals, coal was the main source of chemicals (Speight, 1994).

Petrochemical products include such items as plastics, soaps and detergents, solvents, drugs, fertilizers, pesticides, explosives, synthetic fibers and rubbers, paints, epoxy resins, and flooring and insulating materials. Petrochemicals are found in products as diverse as aspirin, luggage, boats, automobiles, aircraft, polyester clothes, and recording discs and tapes.

The petrochemical industry has grown with the petroleum industry (Goldstein, 1949; Steiner, 1961; Hahn, 1970) and is considered by some to be a mature industry. However, as is the case with the latest trends in changing crude oil types, it must also evolve to meet changing technological needs. The manufacture of chemicals or chemical intermediates from a variety of raw materials is well established (Wittcoff and Reuben, 1996). And the use of petroleum and natural gas is an excellent example of the conversion of such raw materials to more valuable products. The individual chemicals made from petroleum and natural gas are numerous and include industrial chemicals, household chemicals, fertilizers, and paints, as well as intermediates for the manufacture of products, such as synthetic rubber and plastics.

Petrochemicals are generally considered chemical compounds derived from petroleum either by direct manufacture or indirect manufacture as by-products from the variety of processes that are used during the refining of petroleum. Gasoline, kerosene, fuel oil, lubricating oil, wax, asphalt, and the like are excluded from the definition of petrochemicals, as they are not, in the true sense, chemical compounds but are in fact intimate mixtures of hydrocarbons.

The classification of materials as petrochemicals is used to indicate the source of the chemical compounds, but it should be remembered that many common petrochemicals can be made from other sources, and the terminology is therefore a matter of source identification.

The starting materials for the petrochemical industry are obtained from crude petroleum in one of two general ways. They may be present in the raw crude oil and, as such, are isolated by physical methods, such as distillation (Chapter 16) or solvent extraction (Chapter 19). On the other hand, they may be present, if at all, in trace amounts and are synthesized during the refining operations. In fact, unsaturated (olefin) hydrocarbons, which are not usually present in crude oil (Chapter 7), are nearly always manufactured as intermediates during the various refining sequences.

The manufacture of chemicals from petroleum is based on the ready response of the various compound types to basic chemical reactions, such as oxidation, halogenation, nitration, dehydrogenation, addition, polymerization, and alkylation. The low-molecular-weight paraffins and olefins, as found in natural gas and refinery gases, and the simple aromatic hydrocarbons have so far been of the most interest because it is individual species and can be readily be isolated and dealt with. A wide range of compounds is possible, many are being manufactured and we are now progressing to the stage in which a sizable group of products is being prepared from the heavier fractions of petroleum. For example, the various reactions of asphaltene constituents (Chapter 11) indicate that these materials may be regarded as containing chemical functions and are therefore different and are able to participate in numerous chemical or physical conversions to, perhaps, more useful materials. The overall effect of these modifications is the production of materials that either afford good-grade aromatic cokes comparatively easily or the formation of products bearing functional groups that may be employed as a nonfuel material.

For example, the sulfonated and sulfomethylated materials and their derivatives have satisfactorily undergone tests as drilling mud thinners, and the results are comparable to those obtained with commercial mud thinners. In addition, these compounds may also find use as emulsifiers for the in situ recovery of heavy oils. These are also indications that these materials and other similar derivatives of the asphaltene constituents, especially those containing such functions as carboxylic or hydroxyl, readily exchange cations and could well compete with synthetic zeolites. Other uses of the hydroxyl derivatives and the chloro-asphaltenes include high-temperature packing or heat transfer media.

Reactions incorporating nitrogen and phosphorus into the asphaltene constituents are particularly significant at a time when the effects on the environment of many materials containing these elements are receiving considerable attention. Here, we have potential slow-release soil conditioners that only release the nitrogen or phosphorus after considerable weathering or bacteriological action. One may proceed a step further and suggest that the carbonaceous residue remaining after release of the hetero-elements may be a benefit to humus-depleted soils, such as the gray-wooded and solonetzic soils. It is also feasible that coating a conventional quick-release inorganic fertilizer with a water-soluble or water-dispersible derivative will provide a slower release fertilizer and an organic humus-like residue. In fact, variations on this theme are multiple.

Nevertheless, the main objective in producing chemicals from petroleum is the formation of a variety of well-defined chemical compounds that are the basis of the petrochemical industry. It must be remembered, however, that ease of separation of a particular compound from petroleum does not guarantee its use as a petrochemical building block. Other parameters, particularly the economics of the reaction sequences, including the costs of the reactant equipment, must be taken into consideration.

For the purposes of this text, there are four general types of petrochemicals: (1) aliphatic compounds, (2) aromatic compounds, (3) inorganic compounds, and (4) synthesis gas (carbon monoxide and hydrogen). Synthesis gas is used to make ammonia (NH_3) and methanol (methyl alcohol, CH_3OH). Ammonia is used primarily to form ammonium nitrate (NH_4NO_3), a source of fertilizer. Much of the methanol produced is used in making formaldehyde ($HCH=O$). The rest is used to make polyester fibers, plastics, and silicone rubber.

An aliphatic petrochemical compound is an organic compound that has an open chain of carbon atoms, be it normal (straight), for example, n-pentane ($CH_3CH_2CH_2CH_2CH_3$) or branched, for example, iso-pentane [2-methylbutane, $CH_3CH_2CH(CH_3)CH_3$], or unsaturated. The unsaturated compounds, olefins, include important starting materials such as ethylene ($CH_2=CH_2$), propylene ($CH_3.CH=CH_2$), butene-1 ($CH_3CH_2CH_2=CH_2$), iso-butene (2-methylpropene [$CH_3(CH_3)C=CH_2$]) and butadiene ($CH_2=CHCH=CH_2$).

Ethylene is the hydrocarbon feedstock used in greatest volume in the petrochemical industry (Figure 27.5). From ethylene, for example, are manufactured ethylene glycol, used in polyester fibers, resins, and antifreezes; ethyl alcohol, a solvent and chemical reagent; polyethylene, used in film and plastics; styrene, used in resins, synthetic rubber, plastics, and polyesters; and ethylene dichloride, for vinyl chloride, used in plastics and fibers. Propylene is also an important source of petrochemicals (Figure 27.6) and is used in making such products as acrylics, rubbing alcohol, epoxy glue, and carpets. Butadiene is used in making synthetic rubber, carpet fibers, paper coatings, and plastic pipes.

An aromatic petrochemical is also an organic chemical compound but one that contains, or is derived from, the basic benzene ring system.

Petrochemicals are made, or recovered from, the entire range of petroleum fractions, but the bulk of petrochemical products are formed from the lighter (C_1 to C_4) hydrocarbon gases as raw materials. These materials generally occur in natural gas, but they are also recovered from the gas streams produced during refinery, especially cracking, operations. Refinery gases are also particularly valuable because they contain substantial amounts of olefins that, because of the double bonds, are much more reactive than the saturated

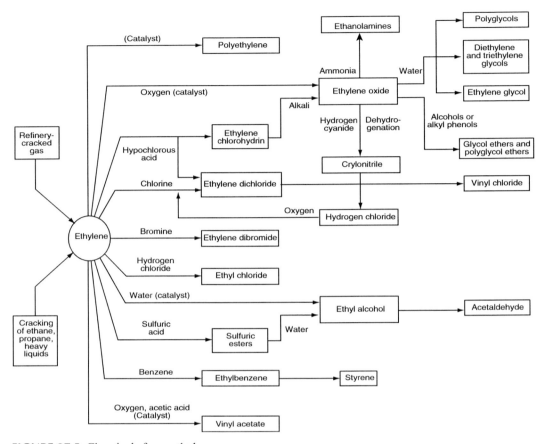

FIGURE 27.5 Chemicals from ethylene.

(paraffin) hydrocarbons. Also important as raw materials are the aromatic hydrocarbons (benzene, toluene, and xylene) that are obtained in rare cases from crude oil and, more likely, from the various product streams. By means of the catalytic reforming process (Chapter 23), nonaromatic hydrocarbons can be converted to aromatics by dehydrogenation and cyclization.

A highly significant proportion of these basic petrochemicals is converted into plastics, synthetic rubbers, and synthetic fibers. Together these materials are known as polymers, because their molecules are high-molecular-weight compounds made up of repeated structural units that have combined chemically. The major products are polyethylene, polyvinyl chloride, and polystyrene, all derived from ethylene, and polypropylene, derived from monomer propylene. Major raw materials for synthetic rubbers include butadiene, ethylene, benzene, and propylene. Among synthetic fibers the polyesters, which are a combination of ethylene glycol and terephthalic acid (made from xylene), are the most widely used. They account for about one-half of all synthetic fibers. The second major synthetic fiber is nylon, its most important raw material being benzene. Acrylic fibers, in which the major raw material is the propylene derivative acrylonitrile, make up most of the remainder of the synthetic fibers.

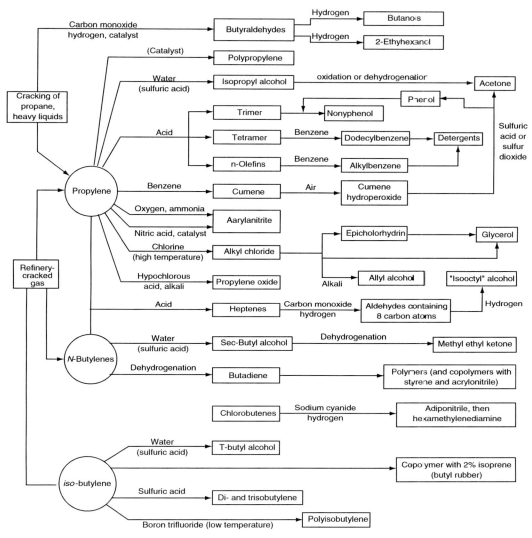

FIGURE 27.6 Chemicals from propylene.

27.2 CHEMICALS FROM PARAFFINS

It is generally true that only paraffin hydrocarbons from methane (CH_4) through propane (C_3H_8) are used as starting materials for specific chemicals syntheses (Chenier, 1992). This is because the higher members of the series are less easy to fractionate from petroleum in pure form, and also because the number of compounds formed in each chemical treatment makes the separation of individual products quite difficult.

27.2.1 HALOGENATION

The ease with which chlorine can be introduced into the molecules of all the hydrocarbon types present in petroleum has resulted in the commercial production of a number of widely used compounds. With saturated hydrocarbons, the reactions are predominantly substitution of hydrogen by chloride and are strongly exothermic, difficult to control, and inclined to become explosively violent:

$$RH + Cl_2 \rightarrow RCl + HCl$$

Moderately high temperatures are used, about 250°C to 300°C (480°F to 570°F) for the thermal chlorination of methane, but as the molecular weight of the paraffin increases the temperature may generally be lowered. A mixture of chlorinated derivatives is always obtained, and many variables, such as choice of catalyst, dilution of inert gases, and presence of other chlorinating agents (antimony pentachloride, sulfuryl chloride, and phosgene), have been tried in an effort to direct the path of the reaction.

Methane yields four compounds upon chlorination in the presence of heat or light:

$$CH_4 + Cl_2 \rightarrow CH_3Cl, CH_2Cl_2, CHCl_3, CCl_4$$

These compounds, known as chloromethane or methyl chloride, dichloromethane or methylene chloride, trichloromethane or chloroform, and tetrachloromethane or carbon tetrachloride, are used as solvents or in the production of chlorinated materials.

Other examples of the chlorination reaction include the formation of ethyl chloride by the chlorination of ethane:

$$CH_3CH_3 + Cl_2 \rightarrow CH_3CH_2Cl + HCl$$

Ethyl chloride (CH_3CH_2Cl) is also prepared by the direct addition of hydrogen chloride (HCl) to ethylene ($CH_2=CH_2$) or by reacting ethyl ether ($CH_3CH_2OCH_2CH_3$) or ethyl alcohol (CH_3CH_2OH) with hydrogen chloride. The chlorination of n-pentane and iso-pentane does not take place in the liquid or vapor phase below 100°C (212°F) in the absence of light or a catalyst, but above 200°C (390°F) it proceeds smoothly by thermal action alone. The hydrolysis of the mixed chlorides obtained yields all the isomeric amyl (C_5) alcohols except iso-amyl alcohol. Reaction with acetic acid produces the corresponding amyl acetates, which find wide use as solvents.

The alkyl chloride obtained on substituting an equivalent of one hydrogen atom by a chloride atom in kerosene is used to alkylate benzene or naphthalene for the preparation of a sulfonation stock for use in the manufacture of detergents and antirust agents. Similarly, paraffin wax can be converted to a hydrocarbon monochloride mixture, which can be employed to alkylate benzene, naphthalene, or anthracene. The product finds use as a pour-point depressor effective in retarding wax crystal growth and for deposition in cold lubricating oils.

27.2.2 NITRATION

Hydrocarbons that are usually gaseous (including normal and iso-pentane) react smoothly in the vapor phase with nitric acid to give a mixture of nitro-compounds, but there are side reactions, mainly of oxidation. Only mononitro derivatives are obtained with the lower paraffins at high temperatures, and they correspond to those expected if scission of a C–C and C–H bond occurs.

Ethane, for example, yields nitromethane and nitroethane,

$$CH_3CH_3 + HNO_3 \rightarrow CH_3CH_2NO_2 + CH_3NO_2$$

Propane yields nitromethane, nitroethane, 1-nitropropane, and 2-nitropropane:

The nitro-derivatives of the lower paraffins are colorless and noncorrosive and are used as solvents or as starting materials in a variety of syntheses. For example, treatment with inorganic acids and water yields fatty acids (RCO_2H) and hydroxylamine (NH_2OH) salts and condensation with an aldehyde ($RCH{=}O$) yields nitroalcohols [$RCH(NO_2)OH$].

27.2.3 OXIDATION

The oxidation of hydrocarbons and hydrocarbon mixtures has received considerable attention, but the uncontrollable nature of the reaction and the mixed character of the products have made resolution of the reaction sequences extremely difficult.

Therefore, it is not surprising that, except for the preparation of mixed products having specific properties, such as fatty acids, hydrocarbons higher than pentanes are not employed for oxidation because of the difficulty of isolating individual compounds.

Methane undergoes two useful reactions at 90°C (195°F) in the presence of iron oxide (Fe_3O_4) as a catalyst:

$$CH_4 + H_2O \rightarrow CO + 3H_2$$

$$CO + H_2O \rightarrow CO_2 + H_2$$

Alternatively, partial combustion of methane can be used to provide the required heat and steam. The carbon dioxide produced then reacts with methane at 900°C (1650°F) in the presence of a nickel catalyst:

$$CH_4 + 2O_2 \rightarrow O_2 + 2H_2O$$

$$CO_2 + CH_4 \rightarrow 2CO + 2H_2$$

$$CH_4 + H_2O \rightarrow CO + 3H2$$

Methanol (methyl alcohol, CH_3OH) is the second major product produced from methane. Synthetic methanol has virtually completely replaced methanol obtained from the distillation of wood, its original source material. One of the older trivial names used for methanol was wood alcohol. The synthesis reaction takes place at 350°C and 300 atm in the presence of ZnO as a catalyst:

$$2CH_4 + O_2 \rightarrow 2CH_3OH$$

Most of the methanol is then oxidized by oxygen from air to formaldehyde, (sometimes referred to as methanal):

$$2CH_3OH + O_2 \rightarrow 2CH_2O + 2H_2O$$

Formaldehyde is used to produce synthetic resins either alone or with phenol, urea, or melamine; other uses are minor.

By analogy to the reaction with oxygen, methane reacts with sulfur in the presence of a catalyst to give the carbon disulfide used in the rayon industry:

$$CH_4 + 4S(g) \rightarrow CS_2 + 2H_2S$$

The major nonpetrochemical use of methane is in the production of hydrogen for use in the Haber synthesis of ammonia. Ammonia synthesis requires nitrogen, obtained from air and hydrogen. The most common modern source of the hydrogen consumed in ammonia production, about 95% of it, is methane.

When propane and butane are oxidized in the vapor phase, without a catalyst, at 270°C to 350°C (520°F to 660°F) and at 50 to 3000 psi, a wide variety of products is obtained, including C_1 to C_4 acids, C_2 to C_7 ketones, ethylene oxide, esters, formals, acetals, and others.

Cyclohexane is oxidized commercially and is somewhat selective in its reaction with air at 150°C to 250°C (300°F to 480°F) in the liquid phase in the presence of a catalyst, such as cobalt acetate. Cyclohexanol derivatives are the initial products, but prolonged oxidation produces adipic acid. On the other hand, oxidation of cyclohexane and methylcyclohexane over vanadium pentoxide at 450°C to 500°C (840°F to 930°F) affords maleic and glutaric acids.

The preparation of carboxylic acids from petroleum, particularly from paraffin wax, for esterification to fats or neutralization to form soaps has been the subject of a large number of investigations. Wax oxidation with air is comparatively slow at low temperature and normal pressure, very little reaction taking place at 110°C (230°F), with wax melting at 55°C (130°F) after 280 h. At higher temperatures the oxidation proceeds more readily; maximum yields of mixed alcohol and high-molecular-weight acids are formed at 110°C to 140°C (230°F to 285°F) at 60 to 150 psi; higher temperatures (140°C to 160°C, 285°F to 320°F) result in more acid formation:

$$\text{Paraffin wax} \rightarrow \text{ROH} + \text{RCO}_2\text{H}$$
$$\qquad\qquad\qquad \text{Alcohol} \qquad \text{Acid}$$

Acids from formic (HCO_2H) to that with a 10-carbon atom chain [$CH_3(CH_2)_9CO_2H$] have been identified as products of the oxidation of paraffin wax. Substantial quantities of water-insoluble acids are also produced by the oxidation of paraffin wax, but apart from determination of the average molecular weight (ca. 250), very little has been done to identify individual numbers of the product mixture.

27.2.4 ALKYLATION

Alkylation chemistry contributes to the efficient utilization of C_4 olefins generated in the cracking operations. *Iso*-butane has been added to butenes (and other low-boiling olefins) to give a mixture of highly branched octanes (e.g., heptanes) by a process called alkylation. The reaction is thermodynamically favored at low temperatures (<20°C), and thus very powerful acid catalysts are employed. Typically, sulfuric acid (85% to 100%), anhydrous hydrogen fluoride, or a solid sulfonic acid is employed as the catalyst in these processes. The first step in the process is the formation of a carbocation by combination of an olefin with an acid proton:

$$(CH_3)_2C{=}CH_2 + H^+ \rightarrow (CH_3)_3C^+$$

The second step is the addition of the carbocation to a second molecule of olefin to form a dimer carbocation. The extensive branching of the saturated hydrocarbon results in high octane. In practice, mixed butenes are employed (*iso*-butylene, 1-butene, and 2-butene), and the product is a mixture of isomeric octanes that has an octane number of 92 to 94. With the phase-out of leaded additives in our motor gasoline pools, octane improvement is a major challenge for the refining industry. Alkylation is one answer.

27.2.5 THERMOLYSIS

Although there are relatively unreactive organic molecules, paraffin hydrocarbons are known to undergo thermolysis when treated under high-temperature, low-pressure vapor-phase conditions. The cracking chemistry of petroleum constituents has been extensively studied (Oblad et al., 1979; Albright and Crynes, 1976). Cracking is the major process for generating ethylene and the other olefins that are the reactive building blocks of the petrochemical industry (Chemier, 1992). In addition to thermal cracking, other very important processes that generate sources of hydrocarbon raw materials for the petrochemical industry include catalytic reforming, alkylation, dealkylation, isomerization, and polymerization.

Cracking reactions involve the cleavage of carbon–carbon bonds with the resulting redistribution of hydrogen to produce smaller molecules. Thus, cracking of petroleum or petroleum fractions is a process by which larger molecules are converted into smaller, lower boiling molecules. In addition, cracking generates two molecules from one, with one of the product molecules saturated (paraffin) and the other unsaturated (olefin).

At the high temperatures of refinery crackers (usually $>500°C$, $950°F$), there is a thermodynamic driving force for the generation of more molecules from fewer molecules; that is, cracking is favored. Unfortunately, in the cracking process certain products interact with one another to produce products of increased molecular weight over that in the original feedstock. Thus, some products are taken off from the cracker as useful light products (olefins, gasoline, and others), but other products include heavier oil and coke.

27.3 CHEMICALS FROM OLEFINS

Olefins (C_2H_{2n}) are the basic building blocks for a host of chemical syntheses (Chemier, 1992). These unsaturated materials enter into polymers and rubbers, and with other reagents react to form a wide variety of useful compounds, including alcohols, epoxides, amines, and halides.

Olefins present in gaseous products of catalytic cracking processes (Chapter 18) offer promising source materials. Cracking paraffin hydrocarbons and heavy oils also produces olefins. For example, cracking ethane, propane, butane, and other feedstock such as gas oil, naphtha, and residua produces ethylene. Propylene is produced from thermal and catalytic cracking of naphtha and gas oils, as well as propane and butane.

As far as can be determined, the first large-scale petrochemical process was the sulfuric acid absorption of propylene ($CH_3CH=CH_2$) from refinery cracked gases to produce isopropyl alcohol [($CH_3)_2CHOH$]. The interest, then, in thermal reactions of hydrocarbons has been high since the 1920s when alcohols were produced from the ethylene and propylene formed during petroleum cracking. The range of products formed from petroleum pyrolysis has widened over the past six decades to include the main chemical building blocks. These include ethane, ethylene, propane, propylene, the butanes, butadiene, and aromatics. Additionally, other commercial products from thermal reactions of petroleum include coke, carbon, and asphalt.

Ethylene manufacture via the steam cracking process is in widespread practice throughout the world. The operating facilities are similar to gas oil cracking units, operating at temperatures of $840°C$ ($1550°F$) and at low pressures (24 psi). Steam is added to the vaporized feed to achieve a 50–50 mixture, and furnace residence times are only 0.2 to 0.5 sec. Ethane extracted from natural gas is the predominant feedstock for ethylene cracking units. Propylene and butylene are largely derived from catalytic cracking units and from cracking a naphtha or light gas oil fraction to produce a full range of olefin products.

Virtually all propene or propylene is made from propane, which is obtained from natural gas stripper plants or from refinery gases:

$$CH_3CH_2CH_3 \rightarrow CH_3-CH=CH_2 + H_2$$

The uses of propene include gasoline (80%), polypropylene, iso-propanol, trimers, and tetramers for detergents, propylene oxide, cumene, and glycerin.

Two butenes or butylenes (1-butene, $CH_3CH_2CH=CH_2$, and 2-butene, $CH_3CH=CHCH_3$) are industrially significant. The latter has end uses in the production of butyl rubber and polybutylene plastics. On the other hand, 1-butene is used in the production of 1,3-butadiene ($CH_2=CHCH-CH_2$) for the synthetic rubber industry. Butenes arise primarily from refinery gases or from the cracking of other fractions of crude oil.

Butadiene can be recovered from refinery streams as butadiene, as butenes, or as butanes; the latter two on appropriate heated catalysts dehydrogenate to give 1,3-butadiene:

$$CH_2=CHCH_2CH_3 \rightarrow CH_2=CHCH=CH_2 + H_2$$

$$CH_3CH_2CH_2CH_3 \rightarrow CH_3=CHCH=CH_2$$

An alternative source of butadiene is ethanol, which on appropriate catalytic treatment also gives the compound di-olefin:

$$2C_2H_5OH \rightarrow CH_2=CHCH=CH_2 + 2H_2O$$

Olefins containing more than four carbon atoms are in little demand as petrochemicals and thus are generally used as fuel. The single exception to this is 2-methyl-1,3-butadiene or isoprene, which has a significant use in the synthetic rubber industry. It is more difficult to make than is 1,3-butadiene. Some are available in refinery streams, but more is manufactured from refinery stream 2-butene by reaction with formaldehyde:

$$CH_3CH=CHCH_3 + HCHO \rightarrow CH_2=CH(CH_3)CH=CH_2 + H_2O$$

27.3.1 HYDROXYLATION

The earliest method for conversion of olefins into alcohols involved their absorption in sulfuric acid to form esters, followed by dilution and hydrolysis, generally with the aid of steam. In the case of ethyl alcohol, the direct catalytic hydration of ethylene can be employed. Ethylene is readily absorbed in 98% to 100% sulfuric acid at 75°C to 80°C (165°F to 175°F), and both ethyl and diethyl sulfate are formed; hydrolysis takes place readily on dilution with water and heating.

The direct hydration of ethylene to ethyl alcohol is practiced over phosphoric acid on diatomaceous earth or promoted tungsten oxide under 100 psi pressure and at 300°C (570°F):

$$CH_2=CH_2 + H_2O \rightarrow C_2H_5OH$$

Purer ethylene is required in direct hydration than in the acid absorption process and the conversion per pass is low, but high yields are possible by recycling. Propylene and the normal butenes can also be hydrated directly.

Ethylene, produced from ethane by cracking, is oxidized in the presence of a silver catalyst to ethylene oxide:

$$2H_2C=CH_2 + O_2 \rightarrow C_2H_4O$$

The vast majority of the ethylene oxide produced is hydrolyzed at 100°C to ethylene glycol:

$$C_2H_4O + H_2O \rightarrow HOCH_2CH_2OH$$

Approximately 70% of the ethylene glycol produced is used as automotive antifreeze and much of the rest is used in the synthesis of polyesters.

Of the higher alkenes, one of the first alcohol syntheses practiced commercially was that of isopropyl alcohol from propylene. Sulfuric acid absorbs propylene more readily than it does ethylene, but care must be taken to avoid polymer formation by keeping the mixture relatively cool and using acid of about 85% strength at 300 to 400 psi pressure; dilution with inert oil may also be necessary. Acetone is readily made from isopropyl alcohol, either by catalytic oxidation or by dehydrogenation over metal (usually copper) catalysts.

Secondary butyl alcohol is formed on absorption of 1-butene or 2-butene by 78% to 80% sulfuric acid, followed by dilution and hydrolysis. Secondary butyl alcohol is converted into methyl ethyl ketone by catalytic oxidation or dehydrogenation.

There are several methods for preparing higher alcohols. One method in particular, the so-called oxo reaction, involves the direct addition of carbon monoxide (CO) and a hydrogen (H) atom across the double bond of an olefin to form an aldehyde ($RCH=O$), which in turn is reduced to the alcohol (RCH_2OH). Hydroformylation (the Oxo reaction) is brought about by contacting the olefin with synthesis gas (1:1 carbon monoxide-hydrogen) at 75°C to 200°C (165°F to 390°F) and 1500 to 4500 psi over a metal catalyst, usually cobalt. The active catalyst is held to be cobalt hydrocarbonyl $HCO(CO)_4$, formed by the action of the hydrogen on dicobalt ocatcarbonyl.

A wide variety of olefins enter the reaction, those containing terminal unsaturated being the most active. The hydroformylation is not specific; the hydrogen and carbon monoxide add across each side of the double bond. Thus propylene gives a mixture of 60% *n*-butyraldehyde and 40% *iso*-butyraldehyde. Terminal ($RCH=CH_2$) and nonterminal ($RCH=CHR'$) olefins, such as 1-pentene and 2-pentene, give essentially the same distribution of straight-chain and branched-chain C_6 aldehydes, indicating that rapid isomerization takes place. Simple branched structures add mainly at the terminal carbon; *iso*-butylene forms 95% *iso*-valeraldehyde and 5% trimethylacetaldehyde.

Commercial application of the synthesis has been most successful in the manufacture of *iso*-octyl alcohol from a refinery C_3C_4 copolymer, decyl alcohol from propylene trimer, and tridecyl alcohol from propylene tetramer. Important outlets for the higher alcohols lie in their sulfonation to make detergents and the formation of esters with dibasic acids for use as plasticizers and synthetic lubricants.

The hydrolysis of ethylene chlorohydrin ($HOCH_2CH_2Cl$) or the cyclic ethylene oxide produces ethylene glycol ($HOCH_2CH_2OH$). The main use for this chemical is for antifreeze mixtures in automobile radiators and for cooling aviation engines; considerable amounts are used as ethylene glycol dinitrate in low-freezing dynamite. Propylene glycol is also made by the hydrolysis of its chlorohydrin or oxide.

Glycerin can be derived from propylene by high-temperature chlorination to produce alkyl chloride, followed by hydrolysis to allyl alcohol and then conversion with aqueous chloride to glycerol chlorohydrin, a product that can be easily hydrolyzed to glycerol (glycerin).

Glycerin has found many uses over the years; important among these are as solvent, emollient, sweetener, in cosmetics, and as a precursor to nitroglycerin and other explosives.

27.3.2 Halogenation

Generally, at ordinary temperatures, chlorine reacts with olefins by addition. Thus, ethylene is chlorinated to 1,2-dichloroethane (dichloroethane) or to ethylene dichloride:

$$H_2C=CH_2 + Cl_2 \rightarrow H_2ClCCH_2Cl.$$

There are some minor uses for ethylene dichloride, but about 90% of it is cracked to vinyl chloride, the monomer of polyvinyl chloride (PVC):

$$H_2ClCCH_2Cl \rightarrow HCl + H_2C=CHCl$$

At slightly higher temperatures, olefins and chlorine react by substitution of a hydrogen atom by a chlorine atom. Thus, in the chlorination of propylene, a rise of 50°C (90°F) changes the product from propylene dichloride [$CH_3CH(Cl)CH_2Cl$] to allyl chloride ($CH_2=CHCH_2Cl$).

27.3.3 Polymerization

The polymerization of ethylene under pressure (1500 to 3000 psi) at 110°C to 120°C (230°F to 250°F) in the presence of a catalyst or initiator, such as a 1% solution of benzoyl peroxide in methanol, produces a polymer in the 2000 to 3000 molecular weight range. Polymerization at 15,000 to 30,000 psi and 180°C to 200°C (355°F to 390°F) produces a wax melting at 100°C (212°F) and 15,000 to 20,000 molecular weight but the reaction is not as straightforward as the equation indicates as there are branches in the chain. However, considerably lower pressures can be used over catalysts composed of aluminum alkyls (R_3Al) in the presence of titanium tetrachloride ($TiCl_4$), supported chromic oxide (CrO_3), nickel (NiO), or cobalt (CoO) on charcoal, and promoted molybdena–alumina (MoO_2–Al_2O_3), which at the same time give products more linear in structure. Polypropylenes can be made in similar ways, and mixed monomers, such as ethylene–propylene and ethylene–butene mixtures, can be treated to give high-molecular-weight copolymers of good elasticity. Polyethylene has excellent electrical insulating properties; its chemical resistance, toughness, machinability, light weight, and high strength make it suitable for many other uses.

Lower molecular weight polymers, such as the dimers, trimers, and tetramers, are used as such in motor gasoline. The materials are normally prepared over an acid catalyst. Propylene trimer (dimethylheptenes) and tetramer (trimethylnonenes) are applied in the alkylation of aromatic hydrocarbons for the production of alkylaryl sulfonate detergents and also as olefin-containing feedstocks in the manufacture of C_{10} and C_{13} oxo-alcohols. Phenol is alkylation by the trimer to make nonylphenol, a chemical intermediate for the manufacture of lubricating oil detergents and other products.

Iso-butylene also forms several series of valuable products; the di- and tri-iso-butylenes make excellent motor and aviation gasoline components; they can also be used as alkylating agents for aromatic hydrocarbons and phenols and as reactants in the oxo-alcohol synthesis. Polyisobutylenes in the viscosity range of 55,000 SUS (38°C, 100°F) have been employed as viscosity index improvers in lubricating oils. Butene-1 ($CH_3 \cdot CH_2 \cdot CH=CH_2$) and butene-2 ($CH_3 \cdot CH=CH \cdot CH_3$) participate in polymerization reactions by the way of butadiene ($CH_2=CH \cdot CH=CH_2$), the dehydrogenation product, which is copolymerized with styrene (23.5%) to form GR-S rubber, and with acrylonitrile (25%) to form GR-N rubber:

Derivatives of acrylic acid (butyl acrylate, ethyl acrylate, 2-ethylhexyl acrylate, and methyl acrylate) can be homopolymerized using peroxide initiators or copolymerized with other monomers to generate acrylic or aclryloid resins.

27.3.4 OXIDATION

The most striking industrial olefin oxidation process involves ethylene, which is air oxidized over a silver catalyst at 225°C to 325°C (435°F to 615°F) to give pure ethylene oxide in yields ranging from 55% to 70%.

Analogous higher olefin oxides can be prepared from propylene, butadiene, octene, dodecene, and styrene via the chlorohydrin route or by reaction with peracetic acid. Acrolein is formed by air oxidation or propylene over a supported cuprous oxide catalyst or by condensing acetaldehyde and formaldehyde. When acrolein and air are passed over a catalyst, such as cobalt molybdate, acrylic acid is produced or if acrolein is reacted with ammonia and oxygen over molybdenum oxide, the product is acrylonitrile. Similarly, propylene may be converted to acrylonitrile.

Acrolein and acrylonitrile are important starting materials for the synthetic materials known as acrylates; acrylonitrile is also used in plastics, which are made by copolymerization of acrylonitrile with styrene or with a styrene–butadiene mixture.

Oxidation of the higher olefins by air is difficult to control, but at temperatures between 350°C and 500°C (660°F and 930°F) maleic acid is obtained from amylene and a vanadium pentoxide catalyst; higher yields of the acid are obtained from hexene, heptene, and octene.

27.3.5 MISCELLANEOUS

Esters (RCO_2R') are formed directly by the addition of acids to olefins, mercaptans by the addition of hydrogen sulfide to olefins, sulfides by the addition of mercaptans to olefins, and amines by the addition of ammonia and other amines to olefins.

27.4 CHEMICALS FROM AROMATICS

Briefly, aromatic compounds are those containing one or more benzene rings or similar ring structures. The majority are taken from refinery streams that contain them and separated into fractions, of which the most significant fractions are benzene (C_6H_6), methylbenzene or toluene ($C_6H_5CH_3$) and the dimethylbenzenes or xylenes ($CH_3C_6H_4CH_3$] with the two-ring condensed aromatic compound naphthalene ($C_{10}H_8$) also being a source of petrochemical.

In the traditional chemical industry, aromatics such as benzene, toluene, and xylene were made from coal during the course of carbonization in the production of coke and town gas. A much larger volume of these chemicals are now made as refinery by-products. A further source of supply is the aromatic-rich liquid fraction produced in the cracking of naphtha or light gas oils during the manufacture of ethylene and other olefins.

Aromatic compounds are valuable starting materials for a variety of chemical products (Chemier, 1992). Reforming processes have made benzene, toluene, xylene, and ethylbenzene economically available from petroleum sources. They are generally recovered by extractive or azeotropic distillation, by solvent extraction (with water–glycol mixtures or liquid sulfur dioxide), or by adsorption. Naphthalene and methylnaphthalenes are present in catalytically cracked distillates. A substantial part of the benzene consumed is now derived from petroleum, and it has many chemical uses.

Aromatic compounds, such as benzene, toluene, and the xylenes are major sources of chemicals (Figure 27.7). For example, benzene is used to make styrene ($C_6H_5 \cdot CH=CH_2$), the basic ingredient of polystyrene plastics, as well as paints, epoxy resins, glues, and other adhesives. The process for the manufacture of styrene proceeds through ethylbenzene, which is produced by reaction of benzene and ethylene at 95°C (203°F) in the presence of a catalyst:

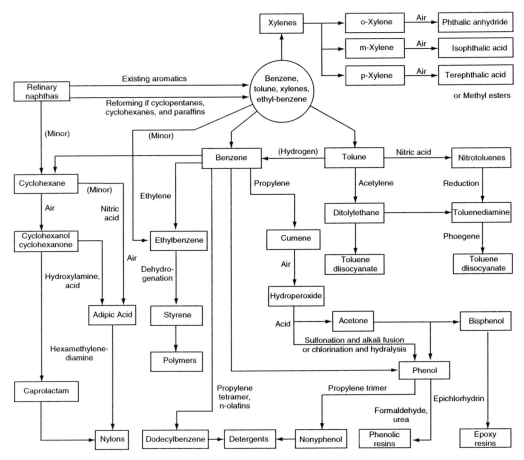

FIGURE 27.7 Chemicals from benzene, toluene, and the xylenes.

$$C_6H_6 + CH_2{=}CH_2 \rightarrow C_6H_5CH_2CH_3$$

In the presence of a catalyst and superheated steam ethylbenzene dehydrogenates to styrene:

$$C_6H_5CH_2CH_3 \rightarrow C_6H_5CH{=}CH_2 + H_2$$

Toluene is usually added to the gasoline pool or used as a solvent, but it can be dealkylated to benzene by catalytic treatment with hydrogen:

$$C_6H_5CH_3 + H_2 \rightarrow C_6H_6 + CH_4$$

Similar processes are used for dealkylation of methyl-substituted naphthalene. Toluene is also used to make solvents, gasoline additives, and explosives.

   Toluene is usually in demand as a source of trinitrotoluene (TNT) but has fewer chemical uses than benzene. Alkylation with ethylene, followed by dehydrogenation, yields α-methylstyrene [$C_6H_5C(CH_3){=}CH_2$], which can be used for polymerization. Alkylation of toluene with propylene tetramer yields a product suitable for sulfonation to a detergent-grade surface-active compound.

Of the xylenes, *o*-xylene is used to produce phthalic anhydride and other compounds. Another xylene, *p*-xylene is used in the production of polyesters in the form of terephthalic acid or its methyl ester. Terephthalic acid is produced from *p*-xylene by two reactions in four steps. The first of these is oxidation with oxygen at 190°C (375°F):

$$CH_3C_6H_4CH_3 + O_2 \rightarrow HOOCC_6H_4CH_3$$

This is followed by formation of the methyl ester at 150°C (302°F):

$$HOOCC_6H_4CH_3 + CH_3OH \rightarrow CH_3OOCC_6H_4CH_3$$

Repetition of these steps gives the methyl diester of terephthalic acid. This diester, $CH_3OOCC_6H_4CCOOCH_3$, when polymerized with ethylene glycol at 200°C (390°F), yields the polymer after loss of methanol to give a monomer. The polymerization step requires a catalyst.

Aromatics are more resistant to oxidation than the paraffin hydrocarbons, and higher temperatures are necessary; the oxidation is carried out in the vapor phase over a catalyst, generally supported by vanadium oxide. *Ortho*-xylene is oxidized by nitric acid to phthalic anhydride, *m*-xylene to *iso*-phthalic acid, and *p*-xylene with nitric acid to terephthalic acid. These acid products are used in the manufacture of fibers, plastics, plasticizers, and the like.

Phthalic anhydride is also produced in good yield by the air oxidation of naphthalene at 400°C to 450°C (750°F to 840°F) in the vapor phase at about 25 psi over a fixed-bed vanadium pentoxide catalyst. Terephthalic acid is produced in a similar manner from *p*-xylene, and an intermediate in the process, *p*-toluic acid, can be isolated because it is slower to oxidize than the *p*-xylene starting material.

27.5   CHEMICALS FROM ACETYLENE

Acetylene is the only petrochemical produced in significant quantity that contains a triple bond, and is a major intermediate species. It is not easily shipped, and as a consequence its consumption is close to the point of origin. It can be made by hydrolysis of calcium carbide produced in the electric furnace from calcium oxide (CaO) and carbon:

$$CaC_2 + 2H_2O \rightarrow HC\equiv CH + Ca(OH)_2$$

An alternative method of manufacturing acetylene is by cracking methane:

$$2CH_4 \rightarrow HC\equiv CH + 6H_2$$

This process produces only one-third of the methane input as acetylene, the remainder being burned in the reactor. Similar reactions employing heavier fractions of crude oil are being used increasingly as the price of methane relative to heavy crude is rising.

Acetylene is used as a special fuel gas (oxyacetylene torches) and as a chemical raw material.

27.6   CHEMICALS FROM NATURAL GAS

Natural gas can be used as a source of hydrocarbons (e.g., ethane and propane) that have higher molecular weight than methane and are important chemical intermediates.

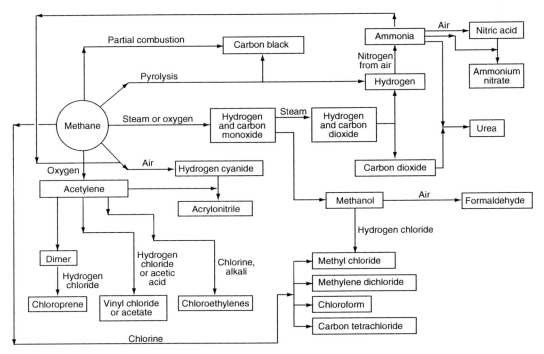

FIGURE 27.8 Chemicals from methane.

The preparation of chemicals and chemical intermediates from methane (natural gas) should not be restricted to those described here but should be regarded as some of the building blocks of the petrochemical industry (Figure 27.8) (Lowenheim and Moran, 1975; Sasma and Hedman, 1984; Hydrocarbon Processing, 1987).

The availability of hydrogen from catalytic reforming operations (Chapter 23) has made its application economically feasible in a number of petroleum-refining operations. Previously, the chief sources of large-scale hydrogen (used mainly for ammonia manufacture) were the cracking of methane (or natural gas) and the reaction between methane and steam. In the latter, at 900°C to 1000°C (1650°F to 1830°F) conversion into carbon monoxide and hydrogen results in:

$$CH_4 + H_2O \rightarrow CO + 3H_2$$

If this mixture is treated further with steam at 500°C over catalyst, the carbon monoxide present is converted into carbon dioxide and more hydrogen is produced:

$$CO + H_2O \rightarrow H_2 + CO_2$$

The reduction of carbon monoxide by hydrogen is the basis of several syntheses, including the manufacture of methanol and higher alcohols. Indeed, the synthesis of hydrocarbons by the Fischer–Tropsch reaction, has received considerable attention. This occurs in the temperature range of 200°C to 350°C (390°F to 660°F), which is sufficiently high for the water–gas shift to take place in the presence of the catalyst:

$$CO + H_2O \rightarrow CO_2 + H_2$$

The major products are olefins and paraffins, together with some oxygen-containing organic compounds in the product mix that may be varied by changing the catalyst or the temperature, pressure, and carbon monoxide–hydrogen ratio.

The hydrocarbons formed are mainly aliphatic, and on a molar basis methane is the most abundant; the amount of higher hydrocarbons usually decreases gradually with increase in molecular weight. *Iso*-paraffin formation is more extensive over zinc oxide (ZnO) or thoria (ThO$_2$) at 400°C to 500°C (750°F to 930°F) and at higher pressure. Paraffin waxes are formed over ruthenium catalysts at relatively low temperatures (170°C to 200°C, 340°F to 390°F), high pressures (1500 psi), and with a carbon monoxide–hydrogen ratio. The more highly branched product made over the iron catalyst is an important factor in a choice for the manufacture of automotive fuels. On the other hand, a high-quality diesel fuel (paraffin character) can be prepared over cobalt.

Secondary reactions play an important part in determining the final structure of the product. The olefins produced are subjected to both hydrogenation and double-bond shifting toward the center of the molecule; *cis*-and *trans* isomers are formed in about equal amounts. The proportions of straight-chain molecules decrease with rise in molecular weight, but even so they are still more abundant than branched-chain compounds up to about C$_{10}$.

The small amount of aromatic hydrocarbons found in the product covers a wide range of isomer possibilities. In the C$_6$ to C$_9$ range, benzene, toluene, ethylbenzene, xylene, *n*-propyl- and *iso*-propylbenzene, methylethylbenzenes, and trimethylbenzenes have been identified; naphthalene derivatives and anthracene derivatives are also present.

27.7 INORGANIC PETROCHEMICALS

Although the focus of this text is the organic chemistry of petroleum and its derivatives, mention should be made of the inorganic petrochemical products. Thus, an inorganic petrochemical is one that does not contain carbon atoms; typical examples are sulfur (S), ammonium sulfate [(NH$_4$)$_2$SO$_4$)], ammonium nitrate (NH$_4$NO$_3$), and nitric acid (HNO$_3$).

Of the inorganic petrochemicals, ammonia is by far the most common. Ammonia is produced by the direct reaction of hydrogen with nitrogen, with air being the source of nitrogen:

$$N_2 + 3H_2 \rightarrow 2NH_3$$

Ammonia production requires hydrogen from a hydrocarbon source. Traditionally, the hydrogen was produced from a coke and steam reaction, but refinery gases, steam reforming of natural gas (methane) and naphtha streams, and partial oxidation of hydrocarbons or higher-molecular-weight refinery residual materials (residua, asphalt) are the sources of hydrogen. The ammonia is used predominantly for the production of ammonium nitrate (NH$_4$NO$_3$) as well as other ammonium salts and urea (H$_2$HCONH$_2$), which are major constituents of fertilizers.

Carbon black (also classed as an inorganic petrochemical) is made predominantly by the partial combustion of carbonaceous (organic) material in a limited supply of air. The carbonaceous sources vary from methane to aromatic petroleum oils to coal tar by-products. Carbon black is used primarily for the production of synthetic rubber.

Sulfur, another inorganic petrochemical, is obtained by the oxidation of hydrogen sulfide:

$$H_2S + O_2 \rightarrow H_2O + S$$

Hydrogen sulfide is a constituent of natural gas and also of the majority of refinery gas streams, especially those off-gases from hydrodesulfurization processes. A large majority of the sulfur is converted to sulfuric acid for the manufacturer of fertilizers and other chemicals. Other uses for sulfur include the production of carbon disulfide, refined sulfur, and pulp and paper industry chemicals.

27.8 SYNTHESIS GAS

Synthesis gas is a mixture of carbon monoxide (CO) and hydrogen (H_2) that is the beginning of a wide range of chemicals (Figure 27.9).

The production of synthesis gas, that is, mixtures of carbon monoxide and hydrogen has been known for several centuries. But it is only with the commercialization of the Fischer–Tropsch reaction that the importance of synthesis gas has been realized. The thermal cracking (pyrolysis) of petroleum or fractions thereof was an important method for producing gas in the years following its use in increasing the heat content of water gas. Many water–gas sets operations converted into oil-gasification units; some have been used for base-load city gas supply but most find use for peak-load situations in the winter.

In addition to the gases obtained by distillation of crude petroleum, further gaseous products are produced during the processing of naphtha and middle distillate to produce gasoline. Hydrodesulfurization processes involving treatment of naphtha, distillates, and residual fuels, and from the coking or similar thermal treatment of vacuum gas oils and residual fuel oils also produce gaseous products.

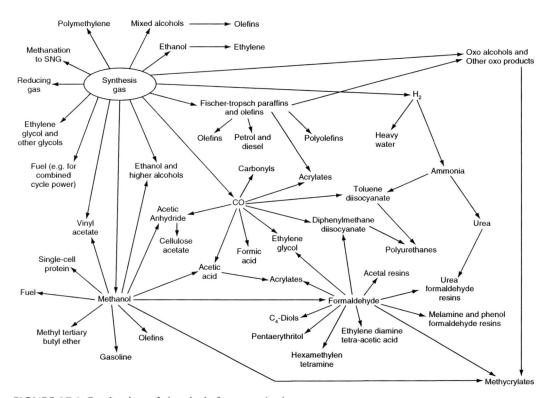

FIGURE 27.9 Production of chemicals from synthesis gas.

The chemistry of the oil-to-gas conversion has been established for several decades and can be described in general terms although the primary and secondary reactions can be truly complex. The composition of the gases produced from a wide variety of feedstocks depends not only on the severity of cracking but often to an equal or lesser extent on the feedstock type. In general terms, gas heating values are on the order of 950 to 1350 Btu/ft.^{3} (30 to 50 MJ/m^3).

A second group of refining operations that contribute to gas production are the catalytic cracking processes, such as fluid-bed catalytic cracking, and other variants, in which heavy gas oils are converted into gas, naphtha, fuel oil, and coke.

The catalysts will promote steam-reforming reactions that lead to a product gas containing more hydrogen and carbon monoxide and fewer unsaturated hydrocarbon products than the gas product from a noncatalytic process. The resulting gas is more suitable for use as a medium heat-value gas than the rich gas produced by straight thermal cracking. The catalyst also influences the reactions rates in the thermal cracking reactions, which can lead to higher gas yields and lower tar and carbon yields.

Almost all petroleum fractions can be converted into gaseous fuels, although conversion processes for the heavier fractions require more elaborate technology to achieve the necessary purity and uniformity of the manufactured gas stream. In addition, the thermal yield from the gasification of heavier feedstocks is invariably lower than that of gasifying light naphtha or liquefied petroleum gas since, in addition to the production of synthesis gas components (hydrogen and carbon monoxide) and various gaseous hydrocarbons, heavy feedstocks also yield some tar and coke.

Synthesis gas can be produced from heavy oil by partially oxidizing the oil:

$$[2CH]_{petroleum} + O_2 \rightarrow 2CO + H_2$$

The initial partial oxidation step consists of the reaction of the feedstock with a quantity of oxygen insufficient to burn it completely, making a mixture consisting of carbon monoxide, carbon dioxide, hydrogen, and steam.

Success in partially oxidizing heavy feedstocks depends mainly on the details of the burner design. The ratio of hydrogen to carbon monoxide in the product gas is a function of reaction temperature and stoichiometry and can be adjusted, if desired, by varying the ratio of carrier steam to oil fed to the unit.

REFERENCES

Albright, L.F. and Crynes, B.L. 1976. *Industrial and Laboratory Pyrolyses.* Symposium Series No. 32. *Am. Chem. Soc.* Washington, DC.

Chenier, P.J. 1992. *Survey of Chemical Industry,* 2nd Revised edn. VCH Publishers Inc., New York.

Goldstein, R.F. 1949. *The Petrochemical Industry.* E. & F.N. Spon, London.

Hahn, A.V. 1970. *The Petrochemical Industry: Market and Economics.* McGraw-Hill, New York.

Hydrocarbon Processing, 1987. *Petrochemical Handbook.* November.

Kolb, D. and Kolb, K.E. 1979. *J. Chem. Educ.* 56: 465.

Lowenheim, F.A. and Moran, M.K. 1975. *Industrial Chemicals.* John Wiley & Sons, New York.

Oblad, A.G., Davis, H.B., and Eddinger, R.T. 1979. *Thermal Hydrocarbon Chemistry.* Advances in Chemistry Series No. 183. American Chemical Society, Washington, DC.

Sasma, M.E. and Hedman, B.A. 1984. *Proceedings.* International Gas Research Conference, Gas Research Institute, Chicago, IL.

Speight, J.G. 1994. *The Chemistry and Technology of Coal,* 2nd edn. Marcel Dekker Inc., New York.

Steiner, H. 1961. *Introduction to Petroleum Chemicals.* Pergamon Press, New York.

Wittcoff, H.A. and Reuben, B.G. 1996. *Industrial Organic Chemicals.* John Wiley & Sons Inc., New York.

Part IV

Environmental Issues

28 Environmental Aspects of Refining

28.1 INTRODUCTION

Petroleum use is a necessary part of the modern world, hence the need for stringent controls over the amounts and types of emissions from the use of petroleum and its products. So, it is predictable that petroleum will be a primary source of energy for the next several decades and, therefore, the message is clear. The challenge is for the development of technological concepts that will provide the maximum recovery of energy from petroleum not only cheaply but also efficiently and with minimal detriment to the environment.

Pollution has been obvious for a long time, although the effects were not as well realized in the past. There was copper pollution near Jericho on the west bank of the River Jordan due to copper smelting for manufacture of tools thousands of years ago. Deforestation of many areas near the Mediterranean Sea for the building of ships was a norm. Poor agricultural methods led to soil erosion. In 2500 BC, the Sumerians used sulfur compounds to control insects and in 1500 BC the Chinese used natural products to fumigate crops. Pesticides began polluting the environment hundreds of years ago. Pollution in medieval England was also noted when the smoke from coal fires made some cities almost uninhabitable (Speight, 1993). There are also documented records, courtesy of the diarist Samuel Pepys, who noted that he did not realize that sewage from a neighbor's house was leaking into his basement until he (Pepys) descended to the lower level and stood in it. This is not only a commentary on the state of sewage disposal but also on the odors that must have permeated seventeenth-century London! However, there being only a meager awareness of the effects of waste products on human life and there being no form of environmental protection, the system of waste disposal proliferated. The eighteenth and nineteenth centuries saw an expansion of the fledgling chemical industry and an awakening of the effects of chemical on human life.

And, so, the twentieth century was born with the continuation of less than desirable waste disposal methods until 1962 when a marine biologist (Rachel Carson) published her book Silent Spring. The book dealt with many environmental problems associated with chlorinated pesticides and touched off an extensive debate about the safety of many different types of chemicals, a debate that continues. As a result, industry and government did some serious soul-searching at the way various waste products were affecting the environment and methods were devised for handling chemical wastes with minimal effect on the environment.

The capacity of the environment to absorb the effluents and other impacts of process technologies is not unlimited, as some would have us believe. The environment should be considered to be an extremely limited resource, and discharge of chemicals into it should be subject to severe constraints. Indeed, the declining quality of raw materials, especially petroleum and fossil fuels that give rise to many of the gaseous emissions of interest in this text, dictates that more material must be processed to provide the needed fuels. And the growing magnitude of the effluents from fossil fuel processes has moved above the line where the environment has the capability to absorb such process effluents without disruption.

To combat any threat to the environment, it is necessary to understand the nature and magnitude of the problems involved (Ray and Guzzo, 1990). It is in such situations that environmental technology has a major role to play. Environmental issues even arise when outdated laws are taken to task. Thus, the concept of what seemed to be a good idea at the time the action occurred no longer holds when the law influences the environment.

The use of oil has significant social and environmental impacts, from accidents and routine activities such as seismic exploration, drilling, and generation of polluting wastes. Oil from subterranean and submarine reservoir extraction can be environmentally damaging. Crude oil and refined fuel spills from tanker ship accidents have damaged fragile ecosystems. Burning oil releases carbon dioxide into the atmosphere, which contributes to global warming.

Alternate (renewable) energy sources do exist, although the degree to which they can replace petroleum and the possible environmental damage they may cause are uncertain and controversial. Sun, wind, geothermal, and other renewable electricity sources cannot directly replace high energy density liquid petroleum for transportation use because automobiles and other equipment must be altered to allow using electricity (in batteries) or hydrogen (via fuel cells or internal combustion) that can be produced from renewable sources. Other options include using liquid fuels (ethanol, biodiesel) produced from biomass. In fact, any combination of solutions to replace petroleum as a liquid transportation fuel will be a very large undertaking.

Thus, both the production (Chapter 5) and processing (Chapter 13 through Chapter 20) of crude oil involve the use of a variety of substances, some toxic, including lubricants in oil wells and catalysts and other chemicals in refining (Figure 28.1). The amounts used, however, tend to be small and relatively easy to control. More detrimental to the environment is the spillage of oil, which has been a particularly common event. Minor losses from truck and car accidents can affect rivers and streams. Leakage from underground gasoline storage tanks, many abandoned decades ago, has contaminated some local water supplies and usually requires expensive operations either to clean or seal off.

Both the production and processing of crude oil involve the use of a variety of substances, some toxic, including lubricants in oil wells and catalysts and other chemicals in refining. The amounts used, however, tend to be small and relatively easy to control. More detrimental to the environment is the spillage of oil, which has been a particularly common event. Minor losses from truck and car accidents can affect rivers and streams. Leakage from underground gasoline storage tanks, many abandoned decades ago, has contaminated some local water supplies and usually requires expensive operations either to clean or seal off.

The purpose of this chapter is to summarize and generalize the various pollution, health, and environmental problems especially specific to the petroleum industry and to place in perspective government laws and regulations as well as industry efforts to control these problems (Majumdar, 1993; Speight, 1993, 1996). The objective is to indicate the types of emissions from refinery processes and the laws that regulate these emissions.

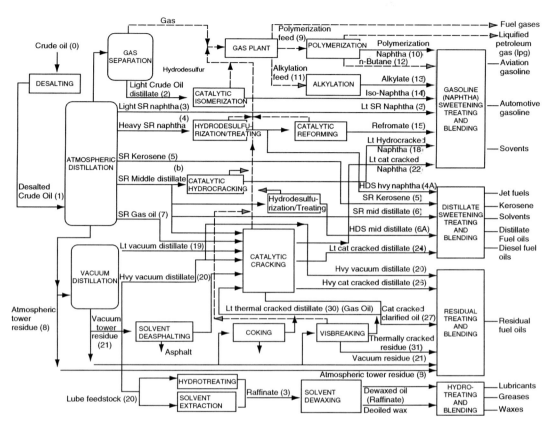

FIGURE 28.1 Schematic overview of a refinery.

28.2 DEFINITIONS

Briefly, petroleum production and **petroleum refining** produce **chemical waste**. If this chemical waste is not processed in a timely manner, it can become a **pollutant**.

A pollutant is a substance present in a particular location (ecosystem) when it is not indigenous to the location or is present in a greater-than-natural concentration. The substance is often the product of human activity. The pollutant, by virtue of its name, has a detrimental effect on the environment, in part or in toto. Pollutants can also be subdivided into two classes: primary and secondary.

$$\text{Source} \rightarrow \text{Primary pollutant} \rightarrow \text{Secondary pollutant}$$

A primary pollutant is a pollutant that is emitted directly from the source. In terms of atmospheric pollutants, examples are carbon oxides, sulfur dioxide, and nitrogen oxides from fuel combustion operations:

$$2[C]_{petroleum} + O_2 \rightarrow 2CO$$

$$[C]_{petroleum} + O_2 \rightarrow CO_2$$

$$2[N]_{petroleum} + O_2 \rightarrow 2NO$$

$$[N]_{petroleum} + O_2 \rightarrow NO_2$$

$$[S]_{petroleum} + O_2 \rightarrow SO_2$$

$$2SO_2 + O_2 \rightarrow 2SO_3$$

Hydrogen sulfide and ammonia are produced from processing sulfur-containing and nitrogen-containing feedstocks:

$$[S]_{petroleum} + H_2 \rightarrow H_2S + hydrocarbons$$

$$[N]_{petroleum} + 3H_2 \rightarrow 2NH_3 + hydrocarbons$$

The question of classifying nitrogen dioxide and sulfur trioxide as primary pollutants often arises, as does the origin of the nitrogen. In the former case, these higher oxides can be formed in the upper levels of the combustion reactors.

A secondary pollutant is a pollutant that is produced by interaction of a primary pollutant with another chemical. A secondary pollutant may also be produced by dissociation of a primary pollutant, or other effects within a particular ecosystem. Again, using atmosphere as an example, the formation of the constituents of **acid rain** is an example of the formation of secondary pollutants:

$$SO_2 + H_2O \rightarrow H_2SO_3 (sulfurous\ acid)$$

$$SO_3 + H_2O \rightarrow H_2SO_4 (sulfuric\ acid)$$

$$NO + H_2O \rightarrow HNO_2 (nitrous\ acid)$$

$$NO_2 + 2H_2O \rightarrow HNO_3 (nitric\ acid)$$

In many cases, these secondary pollutants can have significant environmental effects, such as the formation of acid rain and smog (Speight, 1996).

Any pollutant, either primary or secondary can have a serious effect on the various ecological cycles such as the industrial cycle (Figure 28.2) and the water cycle (Figure 28.3). Therefore, understanding the means by which a chemical pollutant can enter these ecosystems and influence the future behavior of the ecosystem, is extremely important.

In addition, hazardous waste is any gaseous, liquid, or solid waste material that, if improperly managed or disposed of, may pose substantial hazards to human health and the environment. In many cases, the term *chemical waste* is often used interchangeably with the term *hazardous waste*. However, not all chemical wastes are hazardous and caution in the correct use of the terms must be exercised lest unqualified hysteria take control.

An environmental regulation is a legal mechanism that determines how the policy directives of an environmental law are to be carried out. An environmental policy is a requirement that specifies operating procedures that must be followed. An environmental guidance is a document developed by a governmental agency that outlines a position on a topic or which gives instructions on how a procedure must be carried out. It explains how to do something and provides governmental interpretations on a governmental act or policy.

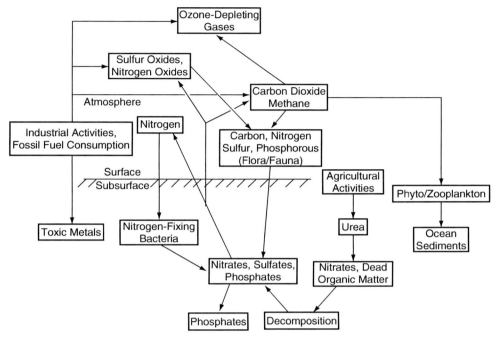

FIGURE 28.2 The industrial cycle. (From Speight, J.G. 1996. *Environmental Technology Handbook*. Taylor & Francis, Philadelphia, PA. With permission.)

28.3 ENVIRONMENTAL REGULATIONS

Environmental issues permeate everyday life. These issues range from the effects on the lives of workers in various occupations where hazards can result from exposure to chemical agents to the influence of these agents on the lives of the population at large (Lipton and Lynch, 1994; Speight, 1996; Boyce, 1997).

In this section, reference is made to the various environmental laws

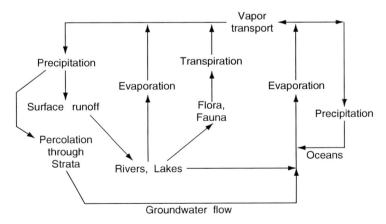

FIGURE 28.3 The water cycle. (From Speight, J.G. 1996. *Environmental Technology Handbook*. Taylor & Francis, Philadelphia, PA. With permission.)

28.3.1   CLEAN AIR ACT AMENDMENTS

The first Clean Air Act of 1970 and the 1977 Amendments consisted of three titles. Title I dealt with stationary air emission sources, Title II with mobile air emission sources, and Title III with definitions of appropriate terms as well as applicable standards for judicial review. The Clean Air Act Amendments of 1990 contain extensive provisions for control of the accidental release of toxic substances from storage or transportation as well as the formation of acid rain (acid deposition). In addition, the requirement that the standards be technology-based removes much of the emotional perception that all chemicals are hazardous as well as the guesswork from legal enforcement of the legislation. The requirement also dictates environmental and health protection with an ample margin of safety.

28.3.2   WATER POLLUTION CONTROL ACT (THE CLEAN WATER ACT)

Several acts are related to the protection of the waterways in the United States. Of particular interest in the present context is the Water Pollution Control Act (Clean Water Act). The objective of the Act is to restore and maintain the chemical, physical, and biological integrity of water systems.

The Water Pollution Control Act of 1948 and the Water Quality Act of 1965 were generally limited to control of pollution of interstate waters and the adoption of water-quality standards by the states for interstate water within their borders. The first comprehensive water-quality legislation in the United States came into being in 1972 as the Water Pollution Control Act. This Act was amended in 1977 and became the Clean Water Act. Further amendments in 1978 were enacted to deal more effectively with spills of crude oil. Other amendments followed in 1987 under the new name Water Quality Act and were aimed at improving water quality in those areas where there were insufficiencies in compliance with the discharge standards.

Section 311 of the Clean Water Act includes elaborate provisions for regulating intentional or accidental discharges of petroleum and of hazardous substances. Included are response actions required for oil spills and the release or discharge of toxic and hazardous substances. As an example, the person in charge of a vessel or an onshore or offshore facility from which any designated hazardous substance is discharged, in quantities equal to or exceeding its reportable quantity, must notify the appropriate federal agency as soon as such knowledge is obtained. The Exxon Valdez is a well-known case.

28.3.3   SAFE DRINKING WATER ACT

The Safe Drinking Water Act, first enacted in 1974, was amended several times in the 1970s and 1980s to set national drinking water standards. The Act calls for regulations that (1) apply to public water systems, (2) specify contaminants that may have any adverse effect on the health of persons, and (3) specify contaminant levels. In addition, the difference between primary and secondary drinking water regulations is defined, and a variety of analytical procedures are specified. Statutory provisions are included to cover underground injection control systems. The Act also requires maximum levels at which a contaminant must have no known or anticipated adverse effects on human health, thereby providing an adequate margin of safety.

The Superfund Amendments and Reauthorization Act (SARA) set similar standards the same for groundwater as for drinking water in terms of necessary cleanup and remediation of an inactive site that might be a former petroleum refinery. Under the Act, all underground injection activities must comply with the drinking water standards as well as meet specific permit conditions that are in unison with the provisions of the Clean Water Act.

However, under the Resource Conservation and Recovery Act, class IV injection wells are no longer permitted and there are several restrictions on underground injection wells that may be used for storage and disposal of hazardous wastes.

28.3.4 RESOURCE CONSERVATION AND RECOVERY ACT

Since its initial enactment in 1976, the Resource Conservation and Recovery Act (RCRA) continues to promote safer solid and hazardous waste management programs. Besides the regulatory requirements for waste management, the Act specifies the mandatory obligations of generators, transporters, and disposers of waste as well as those of owners and operators of waste treatment, storage, or disposal facilities. The Act also defines solid waste as: garbage, refuse, sludge from a treatment plant, from a water supply treatment plant, or air pollution control facility and other discarded material, including solid, liquid, semisolid, or containing gaseous material resulting from industrial, commercial, mining, and agricultural operations and from community activities.

The Act also states that solid waste does not include solid, or dissolved, materials in domestic sewage, or solid or dissolved materials in irrigation return flows or industrial discharges. A solid waste becomes a hazardous waste if it exhibits any one of four specific characteristics: (1) ignitability, (2) reactivity, (3) corrosivity, or (4) toxicity. Certain types of solid wastes (e.g., household waste) are not considered to be hazardous, irrespective of their characteristics. Hazardous waste generated in a product or raw-material storage tank, transport vehicles, or manufacturing processes and samples collected for monitoring and testing purposes are exempt from the regulations.

Hazardous waste management is based on a beginning-to-end concept so that all hazardous wastes can be traced and fully accounted for. All generators and transporters of hazardous wastes as well as owners and operators of related facilities in the United States must file a notification with the Environmental Protection Agency. The notification must state the location of the facility and a general description of the activities as well as the identified and listed hazardous wastes being handled. Thus all regulated hazardous waste facilities must exist and operate under valid, activity-specific permits.

Regulations pertaining to companies that generate and transport wastes require that detailed records be maintained to ensure proper tracking of hazardous wastes through transportation systems. Approved containers and labels must be used, and wastes can only be delivered to facilities approved for treatment, storage, and disposal.

28.3.5 TOXIC SUBSTANCES CONTROL ACT

The Toxic Substances Control Act was first enacted in 1976 and was designed to provide controls for those chemicals that may threaten human health or the environment. Particularly hazardous are the cyclic nitrogen species that may be produced when petroleum is processed and that often occur in residua and cracked residua. The objective of the Act is to provide the necessary control before a chemical is allowed to be mass-produced and enter the environment.

The Act specifies a premanufacture notification requirement by which any manufacturer must notify the Environmental Protection Agency at least 90 days prior to the production of a new chemical substance. Notification is also required even if there is a new use for the chemical that can increase the risk to the environment. No notification is required for chemicals that are manufactured in small quantities solely for scientific research and experimentation. A new chemical substance is defined as a chemical that is not listed in the Environmental Protection Agency Inventory of Chemical Substances or is an unlisted reaction product of two or more chemicals. In addition, the term *chemical substance* means any organic or inorganic substance of a particular molecular identity, including any combination of such substances occurring in whole or in part as a result of a chemical reaction or occurring in nature, and any element or uncombined radical. The term *mixture* means any combination of two or more chemical substances if the combination does not occur in nature and is not, in whole or in part, the result of a chemical reaction.

28.3.6 COMPREHENSIVE ENVIRONMENTAL RESPONSE, COMPENSATION, AND LIABILITY ACT

The Comprehensive Environmental Response, Compensation, and Liability Act (CERCLA) that is generally known as Superfund, was first signed into law in 1980. The central purpose of this Act is to provide a response mechanism for cleanup of any hazardous substance released, such as an accidental spill, or of a threatened release of a chemical. While RCRA deals basically with the management of wastes that are generated, treated, stored, or disposed of, CERCLA provides a response to the environmental release of various pollutants or contaminants into the air, water, or land.

Under this Act, a hazardous substance is any substance requiring (1) special consideration due to its toxic nature under the Clean Air Act, the Clean Water Act, or the Toxic Substances Control Act and (2) defined as hazardous waste under RCRA. Additionally, a pollutant or contaminant can be any other substance not necessarily designated by or listed in the Act but that will or may reasonably be anticipated to cause any adverse effect in organisms or their offspring.

The SARA addresses closed waste disposal sites that may release hazardous substances into any environmental medium. The most revolutionary part of SARA is the Emergency Planning and Community Right-to-Know Act (EPCRA), which for the first time mandated public disclosure. It is covered under Title III of SARA.

28.3.7 OCCUPATIONAL SAFETY AND HEALTH ACT

Occupational health hazards are those factors arising in or from the occupational environment that adversely impact health. Thus, the Occupational Safety and Health Administration (OSHA) came into being in 1970 and is responsible for administering the Occupational Safety and Health Act.

The goal of the Act is to ensure that employees do not suffer material impairment of health or functional capacity due to a lifetime occupational exposure to chemicals. The statute imposes a duty on employers to provide employees with a safe workplace environment, free of known hazards that may cause death or serious bodily injury.

The Act is also responsible for the means by which chemicals are contained. Workplaces are inspected to ensure compliance and enforcement of applicable standards under the Act. In keeping with the nature of the Act, there is also a series of standard tests relating to occupational health and safety as well as the general recognition of health hazards in the workplace. The Act is also the means by which guidelines have evolved for the management and disposition of chemicals used in chemical laboratories.

28.3.8 OIL POLLUTION ACT

The Oil Pollution Act of 1990 deals with pollution of waterways by crude oil. The Act specifically deals with petroleum vessels and onshore and offshore facilities and imposes strict liability for oil spills on their owners and operators.

TABLE 28.1
Environmental Regulations That Apply to Energy Production

	First Enacted	Amended
Clean Air Act	1970	1977
		1990
Clean Water Act (Water Pollution Control Act)	1948	1965[a]
		1972[b]
		1977
		1987[c]
Comprehensive Environmental Response, Compensation and Liability Act	1980	1986[d]
Hazardous Material Transportation Act	1974	1990
Occupational Safety and Health Act	1970	1987[e]
Oil Pollution Act	1924	1990[f]
Resource Conservation and Recovery Act	1976	1980[g]
Safe Drinking Water Act	1974	1986[h]
Superfund Amendments and Re-authorization Act (SARA)	1986	
Toxic Substances Control Act	1976	1984[i]

[a]Water Quality Act; [b]Water Pollution Control Act; [c]Water Quality Act; [d]SARA Amendments; [e]Several amendments during the 1980s; [f]Interactive with various water pollution acts; [g]Federal cancer policy initiated; [h]Several amendments during the 1970s and the 1980s; [i]Import rule enacted.

28.3.9 HAZARDOUS MATERIALS TRANSPORTATION ACT

The Hazardous Materials Transportation Act authorizes the establishment and enforcement of hazardous material regulations for all modes of transportation by highway, water, and rail. The purpose of the Act is to ensure safe transportation of hazardous materials. The Act prevents any person from offering or accepting a hazardous material for transportation anywhere within this nation if that material is not properly classified, described, packaged, marked, labeled, and authorized for shipment pursuant to the regulatory requirements.

Under Department of Transportation regulations, a hazardous material is defined as any substance or material, including a hazardous substance and hazardous waste, which is capable of posing an unreasonable risk to health, safety, and property during transportation.

The Act also imposes restrictions on the packaging, handling, and shipping of hazardous materials. For shipping and receiving of hazardous chemicals, hazardous wastes, and radio-active materials, the appropriate documentation, markings, labels, and safety precautions are required.

There are a variety of regulations (Table 28.1) that apply to petroleum refining. The most popular is the series of regulations known as the Clean Air Act that was first introduced in 1967 and was subsequently amended in 1970, and most recently in 1990. The most recent amendments provide stricter regulations for the establishment and enforcement of national ambient air quality standards for, as an example, sulfur dioxide. These standards do not stand alone and there are many national standards for sulfur emissions.

28.4 PROCESS ANALYSIS

In addition to the conventional meaning of the term *process*, the transportation of petroleum also needs to be considered here.

Oil spills during petroleum transportation have been the most visible problem. There have also been instances of oil wells at sea blowing out, or flowing uncontrollably, although the amounts from blowouts tend to be smaller than from tanker accidents. The 1979 Ixtoc I blowout in the Gulf of Mexico was an exception, as it flowed an estimated 3 million barrels over many months.

Tanker accidents typically have a severe impact on ecosystems because of the rapid release of hundreds of thousands of barrels of crude oil (or crude oil products) into a small area. The largest single spill to date is believed to have occurred during the 1991 Gulf War, when as much as ten million barrels were dumped in the Persian Gulf by Iraq, apparently intentionally. More typical was the 1989 spill from the tanker Exxon Valdez, where two hundred and fifty thousand barrels were lost in Alaskan coastal waters.

While oil, as a hydrocarbon, is at least theoretically biodegradable, large-scale spills can overwhelm the ability of the ecosystem to break the oil down. Over time, the lighter portions of crude oil evaporate, leaving the nonvolatile portion. Oil itself breaks down the protective waxes and oils in the feathers and fur of birds and animals, resulting in a loss of heat retention and causing death by freezing. Ingestion of the oil can also kill animals by interfering with their ability to digest food. Some crude oils contain toxic metals as well. The impact of any given oil spill is determined by the size of the spill, the degree of dispersal, and the chemistry of the oil. Spills at sea are thought to have a less detrimental effect than spills in shallow waters.

Petroleum refining is a complex sequence of chemical events that result in the production of a variety of products (Figure 28.1). In fact, petroleum refining might be considered as a collection of individual, yet related processes that are each capable of producing effluent streams.

Many refined products came under scrutiny (Loehr, 1992; Olschewsky and Megna, 1992). By the mid-1970s petroleum refiners in the United States were required to develop techniques for manufacturing high-quality gasoline without employing lead additives, and by 1990 they were required to take on substantial investments in the complete reformulation of transportation fuels to minimize environmental emissions. From an industry that produced a single product (kerosene) and disposed unwanted by-products in any manner possible, petroleum refining had become one of the most stringently regulated of all manufacturing industries, expending a major portion of its resources on the protection of the environment.

Processing crude petroleum, with the exception of some of the more viscous crude oils, involves a primary distillation of the hydrogen mixture, which results in its separation into fractions differing in carbon number, volatility, specific gravity, and other characteristics. The most volatile fraction, that contains most of the gases which are generally dissolved in the crude, is referred to as **pipestill gas** or **pipestill light ends** and consists essentially of hydrocarbon gases ranging from methane to butane(s) (C_4H_{10}), or sometimes pentane(s) (C_5H_{12}). The gas varies in composition and volume, depending on crude origin and on any additions to the crude made at the loading point.

It is not uncommon to re-inject light hydrocarbons such as propane and butane into the crude before dispatch by tanker or pipeline. This results in a higher vapor pressure of the crude oil, but it allows one to increase the quantity of light products obtained at the refinery. As light ends in most petroleum markets command a premium, while in the oil field itself propane and butane may have to be re-injected or flared, the practice of "spiking" crude with liquefied petroleum gas is becoming fairly common.

Petroleum refining, as it is currently known, will continue at least for the next three decades. In spite of the various political differences that have caused fluctuations in petroleum imports, it is predictable that petroleum imports will reach >59% of petroleum consumption in the United States by the year 2010 (Chapter 3). It is also predictable that use of petroleum for the transportation sector will increase as increases in travel offset increased efficiency.

As a consequence of this increase in use, petroleum will be the largest single source of carbon emissions from fuel. Acid gases corrode refining equipment, harm catalysts, pollute the atmosphere, and prevent the use of hydrocarbon components in petrochemical manufacture. When the amount of hydrogen sulfide is high, it may be removed from a gas stream and converted to sulfur or sulfuric acid. Some natural gases contain sufficient carbon dioxide to warrant recovery as dry ice.

Thus, like any other raw material, petroleum is capable of producing chemical waste. By 1960, the petroleum-refining industry had become well established throughout the world. Demand for refined petroleum products had reached almost millions of barrels per day, with major concentrations of refineries in most developed countries. However, as the world became aware of the impact of industrial chemical waste on the environment, the petroleum-refining industry was a primary focus for change. Refiners added **hydrotreating** units to extract sulfur compounds from their products and began to generate large quantities of elemental sulfur. Effluent water, atmospheric emissions, and combustion products also became a focus of increased technical attention (Carson and Mumford, 1988, 1995; Speight, 1993; Renzoni et al., 1994; Edwards, 1995; Thibodeaux, 1995; Speight, 1996).

Thermal processes are commonly used to convert petroleum residua into liquid products. Therefore, some indications of the process classes and the products that are unacceptable to the environment are warranted here.

Thus, examples of modern thermal processes are **visbreaking** and **coking** (**delayed coking, fluid coking**, and **flexicoking**) (Chapter 14). In all of these processes the simultaneous formation of sediment or coke limits the conversion to usable liquid products.

Thermal cracking processes are commonly used to convert nonvolatile residua into volatile products, although thermal cracking processes as used in the early refineries are no longer in use. Examples of modern thermal cracking processes are visbreaking and coking (delayed coking, fluid coking, and flexicoking) (Chapter 14). In all of these processes the simultaneous formation of sediment or coke limits the conversion to usable liquid products.

The visbreaking process (Chapter 14) is primarily a means of reducing the viscosity of heavy feedstocks by controlled thermal decomposition insofar as the hot products are quenched before complete conversion can occur. However, the process is often plagued by sediment formation in the products. This sediment, or sludge, must be removed if the products are to meet fuel oil specifications.

Coking, as the term is used in the petroleum industry, is a process for converting nondistillable fractions (residua) of crude oil to lower boiling products and coke. Coking is often used in preference to **catalytic cracking** because of the presence of metals and nitrogen components that poison catalysts.

Delayed coking (Chapter 14) is the oldest, most widely used process and has changed very little in the five or more decades in which it has been on stream in refineries. Fluid coking (Chapter 14) is a continuous fluidized solids process that cracks feed thermally over heated coke particles in a reactor vessel to gas, liquid products, and coke. Heat for the process is supplied by partial combustion of the coke, with the remaining coke being drawn as product. The new coke is deposited in a thin fresh layer on the outside surface of the circulating coke particle.

Catalytic cracking is a conversion process (Chapter 15) that can be applied to a variety of feedstocks ranging from gas oil to heavy oil. It is one of several practical applications used in a refinery that employ a catalyst to improve process efficiency. Catalytic cracking of crude oil occurs over many types of catalytic materials that may be either activated (acid-treated natural clays of the bentonite-type) or synthesized silica–alumina or silica–magnesia preparations.

Hydrotreating (Chapter 16) is defined as the lower temperature removal of heteroatomic species by treatment of a feedstock or product in the presence of hydrogen. Hydrocracking (Chapter 16) is the thermal decomposition of a feedstock in which carbon–carbon bonds are cleaved in addition to the removal of heteroatomic species. Hydrogen is present to prevent the formation of coke. Subsequent hydroprocessing (Chapter 16) of the coker distillates would reduce the polynuclear aromatic hydrocarbons in the resulting product streams, so that the only health concern outside the refinery itself is with high-severity thermal products, such as pitches, which have not been hydrotreated. Coke solids would not pose a health hazard, and would have less environmental activity than unprocessed residue.

28.4.1 GASEOUS EMISSIONS

Gaseous emissions from petroleum refining create a number of environmental problems. During combustion, the combination of hydrocarbons, nitrogen oxide, and sunlight results in localized low-levels of ozone, or smog. This is particularly evident in large urban areas and especially when air does not circulate well. Petroleum use in automobiles also contributes to the problem in many areas. The primary effects are on the health of those exposed to the ozone, but plant life has been observed to suffer as well.

Refinery and natural gas streams may contain large amounts of acid gases, such as hydrogen sulfide (H_2S) and carbon dioxide (CO_2) (Speight, 1993, 1996). Hydrogen chloride (HCl), although not usually considered to be a major pollutant in petroleum refineries can arise during processing from the presence of brine in petroleum that is incompletely dried. It can also be produced from mineral matter and other inorganic contaminants and is gaining increasing recognition as a pollutant which needs serious attention.

Acid gases corrode refining equipment, harm catalysts, pollute the atmosphere, and prevent the use of hydrocarbon components in petrochemical manufacture. When the amount of hydrogen sulfide is large, it may be removed from a gas stream and converted to sulfur or sulfuric acid. Some natural gases contain sufficient carbon dioxide to warrant recovery as dry ice, that is, solid carbon dioxide. And there is now a conscientious effort to mitigate the emission of pollutants from hydrotreating process by careful selection of process parameters and catalysts (Occelli and Chianelli, 1996).

The terms *refinery gas* and *process gas* are also often used to include all of the gaseous products and by-products that emanate from a variety of refinery processes (Gary and Speight, 1996). There are also components of the gaseous products that must be removed prior to release of the gases to the atmosphere or prior to use of the gas in another part of the refinery, that is, as a fuel gas or as a process feedstock.

Petroleum refining produces gas streams that often contain substantial amounts of acid gases such as hydrogen sulfide and carbon dioxide. More particularly hydrogen sulfide arises from the hydrodesulfurization of feedstocks that contain organic sulfur:

$$[S]_{feedstock} + H_2 \rightarrow H_2S + \text{hydrocarbons}$$

Petroleum refining involves, with the exception of some of the more viscous crude oils, a primary distillation of the hydrogen mixture, which results in its separation into fractions differing in carbon number, volatility, specific gravity, and other characteristics. The most volatile fraction, that contains most of the gases that are generally dissolved in the crude, is referred to as pipestill gas or pipe still light ends and consists essentially of hydrocarbon gases ranging from methane to butane(s), or sometimes pentane(s).

The gas varies in composition and volume, depending on crude origin and on any additions to the crude made at the loading point. It is not uncommon to re-inject light hydrocarbons such as propane and butane into the crude before dispatch by tanker or pipeline. This results in a higher vapor pressure of the crude, but it allows one to increase the quantity of light products obtained at the refinery. As light ends in most petroleum markets command a premium, while in the oil field itself propane and butane may have to be re-injected or flared, the practice of spiking crude oil with liquefied petroleum gas is becoming fairly common.

In addition to the gases obtained by distillation of petroleum, more highly volatile products result from the subsequent processing of naphtha and middle distillate to produce gasoline. Hydrogen sulfide is produced in the desulfurization processes involving hydrogen treatment of naphtha, distillate, and residual fuel; and from the coking or similar thermal treatments of vacuum gas oils and residual fuels. The most common processing step in the production of gasoline is the catalytic reforming of hydrocarbon fractions in the heptane (C_7) to decane (C_{10}) range.

In a series of processes commercialized under the generic name **reforming**, paraffin and naphthene (cyclic nonaromatic) hydrocarbons are altered structurally in the presence of hydrogen and a catalyst into aromatics, or isomerized to more highly branched hydrocarbons. Catalytic reforming processes thus not only result in the formation of a liquid product of higher octane number, but also produce substantial quantity of gases. The latter rich in hydrogen, but also contain hydrocarbons from methane to butanes, with a preponderance of propane ($CH_3CH_2CH_3$), n-butane ($CH_3CH_2CH_2CH_3$) and iso-butane [$(CH_3)_3CH$].

The composition of the process gases varies in accordance with reforming severity and reformer feedstock. All catalytic reforming processes require substantial recycling of a hydrogen stream. Therefore, it is normal to separate reformer gas into a propane ($CH_3CH_2CH_3$) or a butane stream [$CH_3CH_2CH_2CH_3$ plus $(CH_3)_3CH$], which becomes part of the refinery liquefied petroleum gas production, and a lighter gas fraction, part of which is recycled. In view of the excess of hydrogen in the gas, all products of catalytic reforming are saturated, and there are usually no olefin gases present in either gas stream.

A second group of refining operations that contributes to gas production is that of the catalytic cracking processes. These consist of fluid-bed catalytic cracking in which heavy gas oils are converted into gas, liquefied petroleum gas, catalytic naphtha, fuel oil, and coke by contacting the heavy hydrocarbon with the hot catalyst. Both catalytic and thermal cracking processes, the latter being now largely used for the production of chemical raw materials, result in the formation of unsaturated hydrocarbons, particularly ethylene ($CH_2{=}CH_2$), but also propylene (propene, $CH_3 \cdot CH{=}CH_2$), iso-butylene [iso-butene, $(CH_3)_2C{=}CH_2$] and the n-butenes ($CH_3CH_2CH{=}CH_2$, and $CH_3CH{=}CHCH_3$) in addition to hydrogen (H_2), methane (CH_4) and smaller quantities of ethane (CH_3CH_3), propane ($CH_3CH_2CH_3$), and butanes [$CH_3CH_2CH_2CH_3$, $(CH_3)_3CH$]. Diolefins such as butadiene ($CH_2{=}CH \cdot CH{=}CH_2$) are also present.

Additional gases are produced in refineries with visbreaking and coking facilities that are used to process the heaviest crude fractions. In the visbreaking process, fuel oil is passed through externally fired tubes and undergoes liquid phase cracking reactions, which result in the formation of lighter fuel oil components. Oil viscosity is thereby reduced, and some gases, mainly hydrogen, methane, and ethane, are formed. Substantial quantities of both gas and carbon are also formed in coking (both delayed coking and fluid coking) in addition to the middle distillate and naphtha. When coking a residual fuel oil or heavy gas oil, the feedstock is preheated and contacted with hot carbon (coke), which causes extensive cracking of the feedstock constituents of higher molecular weight to produce lower molecular weight products ranging from methane, via liquefied petroleum gas and naphtha, to gas oil and heating oil. Products from coking processes tend to be unsaturated and olefin components predominate in the tail gases from coking processes.

A further source of refinery gas is hydrocracking, a catalytic high-pressure pyrolysis process in the presence of fresh and recycled hydrogen. The feedstock is again heavy gas oil or residual fuel oil, and the process is directed mainly at the production of additional middle distillates and gasoline. As hydrogen is to be recycled, the gases produced in this process again have to be separated into lighter and heavier streams; any surplus recycled gas and the liquefied petroleum gas from the hydrocracking process are both saturated.

Both hydrocracker gases and catalytic reformer gases are commonly used in catalytic desulfurization processes. In the latter, feedstocks ranging from light to vacuum gas oils are passed at pressures of 500 to 1000 psi (3.5 to 7.0×10^3 kPa) with hydrogen over a hydrofining catalyst. This results mainly in the conversion of organic sulfur compounds to hydrogen sulfide,

$$[S]_{feedstock} + H_2 \rightarrow H_2S + hydrocarbons$$

The reaction also produces some light hydrocarbons by hydrocracking.

Thus refinery streams, while ostensibly being hydrocarbon in nature, may contain large amounts of acid gases such as hydrogen sulfide and carbon dioxide. Most commercial plants employ hydrogenation to convert organic sulfur compounds into hydrogen sulfide. Hydrogenation is effected by means of recycled hydrogen-containing gases or external hydrogen over a nickel molybdate or cobalt molybdate catalyst.

In summary, refinery process gas, in addition to hydrocarbons, may contain other contaminants, such as carbon oxides (CO_x, where $x = 1$ and 2), sulfur oxides (SO_x, where $x = 2$ and 3), as well as ammonia (NH_3), mercaptans (R–SH), and carbonyl sulfide (COS).

The presence of these impurities may eliminate some of the sweetening processes, as some processes remove large amounts of acid gas but not to a sufficiently low concentration. On the other hand, there are those processes not designed to remove (or incapable of removing) large amounts of acid gases whereas they are capable of removing the acid gas impurities to very low levels when the acid gases are present only in low-to-medium concentration in the gas.

From an environmental viewpoint, it is not the means by which these gases can be utilized that is of concern, but it is the effects of these gases on the environment when they are introduced into the atmosphere.

In addition to the corrosion of equipment of acid gases, the escape into the atmosphere of sulfur-containing gases can eventually lead to the formation of the constituents of acid rain, that is, the oxides of sulfur (SO_2 and SO_3). Similarly, the nitrogen-containing gases can also lead to nitrous and nitric acids (through the formation of the oxides NO_x, where $x = 1$ or 2) which are the other major contributors to acid rain. The release of carbon dioxide and hydrocarbons as constituents of refinery effluents can also influence the behavior and integrity of the ozone layer.

Hydrogen chloride, if produced during refining, quickly picks up moisture in the atmosphere to form droplets of hydrochloric acid and, like sulfur dioxide, is a contributor to acid rain. However, hydrogen chloride may exert severe local effects because, unlike sulfur dioxide, it does not need to participate in any further chemical reaction to become an acid and under atmospheric conditions that favor a buildup of stack emissions in the area of a large industrial complex or power plant, the amount of hydrochloric acid in rainwater could be quite high.

Natural gas is also capable of producing emissions that are detrimental to the environment. While the major constituent of natural gas is methane, there are components such as carbon dioxide (CO), hydrogen sulfide (H_2S), and mercaptans (thiols; R–SH), as well as trace amounts of sundry other emissions. The fact that methane has a foreseen and valuable end-use makes it a desirable product, but in several other situations it is considered a pollutant, having been identified a **greenhouse gas**.

A sulfur removal process must be very precise, as natural gas contains only a small quantity of sulfur-containing compounds that must be reduced several orders of magnitude. Most consumers of natural gas require less than 4 ppm in the gas.

A characteristic feature of natural gas that contains hydrogen sulfide is the presence of carbon dioxide (generally in the range of 1% v/v to 4% v/v). In cases where the natural gas does not contain hydrogen sulfide, there may also be a relative lack of carbon dioxide.

Acid rain occurs when the oxides of nitrogen and sulfur that are released to the atmosphere during the combustion of fossil fuels are deposited (as soluble acids) with rainfall, usually at some location remote from the source of the emissions.

It is generally believed (the chemical thermodynamics are favorable) that acidic compounds are formed when sulfur dioxide and nitrogen oxide emissions are released from tall industrial stacks. Gases such as sulfur oxides (usually sulfur dioxide, SO_2) as well as nitrogen oxides (NO_x) react with the water in the atmosphere to form acids:

$$SO_2 + H_2O \rightarrow H_2SO_3$$
$$2SO_2 + O_2 \rightarrow 2SO_3$$
$$SO_3 + H_2O \rightarrow H_2SO_4$$
$$2NO + H_2O \rightarrow 2HNO_2$$
$$2NO + O_2 \rightarrow 2NO_2$$
$$NO_2 + H_2O \rightarrow HNO_3$$

Acid rain has a pH less than 5.0 and predominantly consists of sulfuric acid (H_2SO_4) and nitric acid (HNO_3). As a point of reference, in the absence of anthropogenic pollution sources the average pH of rain is -6.0 (slightly acidic; neutral pH 7.0). In summary, the sulfur dioxide that is produced during a variety of processes will react with oxygen and water in the atmosphere to yield environmentally detrimental sulfuric acid. Similarly, nitrogen oxides will also react to produce nitric acid.

Another acid gas, hydrogen chloride (HCl), although not usually considered to be a major emission, is produced from mineral matter and the brines that often accompany petroleum during production and is gaining increasing recognition as a contributor to acid rain. However, hydrogen chloride may exert severe local effects because it does not need to participate in any further chemical reaction to become an acid. Under atmospheric conditions that favor a buildup of stack emissions in the areas where hydrogen chloride is produced, the amount of hydrochloric acid in rainwater could be quite high.

In addition to hydrogen sulfide and carbon dioxide, gas may contain other contaminants, such as mercaptans (R–SH) and carbonyl sulfide (COS). The presence of these impurities may eliminate some of the sweetening processes as some processes remove large amounts of acid gas but not to a sufficiently low concentration. On the other hand, there are those processes that are not designed to remove (or are incapable of removing) large amounts of acid gases. However, these processes are also capable of removing the acid gas impurities to very low levels when the acid gases are there in low–to–medium concentrations in the gas.

On a regional level the emission of sulfur oxides (SO_x) and nitrogen oxides (NO_x) can also cause the formation of acid species at high altitudes, which eventually precipitate in the form of acid rain, damaging plants, wildlife, and property. Most petroleum products are low in sulfur or are desulfurized, and while natural gas sometimes includes sulfur as a contaminant, it is typically removed at the production site.

At the global level there is concern that the increased use of hydrocarbon-based fuels will ultimately raise the temperature of the planet (global warming), as carbon dioxide reflects the infrared or thermal emissions from the earth, preventing them from escaping into space (**greenhouse effect**). Whether or not the potential for global warming becomes real will depend upon how emissions into the atmosphere are handled. There is considerable discussion about the merits and demerits of the global warming theory (Hileman, 1996) and the discussion is likely to continue for some time. Be that as it may, the atmosphere can only tolerate pollutants up to a limiting value. And that value needs to be determined. In the meantime, efforts must be made to curtail the emission of noxious and foreign (nonindigenous) materials into the air.

In summary, and from an environmental viewpoint, petroleum and natural gas processing can result in similar, if not the same, gaseous emissions as coal (Speight, 1993, 1994). It is a question of degree insofar as the composition of the gaseous emissions may vary from coal to petroleum but the constituents are, in the majority of cases, the same.

There are a variety of processes that are designed for sulfur dioxide removal from gas streams (Chapter 20) but scrubbing processes utilizing limestone ($CaCO_3$) or lime [$Ca(OH)_2$] slurries have received more attention than other gas scrubbing processes.

The majority of the gas scrubbing processes are designed to remove sulfur dioxide from the gas streams; some processes show the potential for removal of nitrogen oxide(s) too.

28.4.2 LIQUID EFFLUENTS

Crude oil, as a mixture of hydrocarbons, is (theoretically) a biodegradable material. However, in very general terms (and as observed from elemental analyses), petroleum is a mixture of: (a) hydrocarbons, (b) nitrogen compounds, (c) oxygen compounds, (d) sulfur compounds, and (e) metallic constituents. However, this general definition is not adequate to describe the composition of petroleum as it relates to the behavior of these feedstocks.

It is convenient to divide the hydrocarbon components of petroleum into the following three classes:

Paraffin compounds: saturated hydrocarbons with straight or branched chains, but without any ring structure
Naphthene compounds: saturated hydrocarbons containing one or more rings, each of which may have one or more paraffin side chains (more correctly known as **alicyclic hydrocarbons**)
Aromatic compounds: hydrocarbons containing one or more aromatic nuclei, such as benzene, naphthalene, and phenanthrene ring systems, which may be linked with (substituted) naphthene rings or paraffin side chains

And even though petroleum derivatives have been prescribed for medicinal purposes (Chapter 1) one does not see the flora and fauna of the earth surviving in oceans of crude oil.

Crude oil also contains appreciable amounts of organic nonhydrocarbon constituents, mainly sulfur-, nitrogen-, and oxygen-containing compounds and, in smaller amounts, organometallic compounds in solution and inorganic salts in colloidal suspension. These constituents appear throughout the entire boiling range of the crude oil but tend to concentrate mainly in the heavier fractions and in the nonvolatile residues.

Although their concentration in certain fractions may be quite small, their influence is important. For example, the thermal decomposition of deposited inorganic chlorides with evolution of free hydrochloric acid can give rise to serious corrosion problems in the distillation equipment. The presence of organic acid components, such as mercaptans (R–SH) and acids (R–CO_2H), can also promote environmental damage. In catalytic operations, passivation and poisoning of the catalyst can be caused by deposition of traces of metals (vanadium and nickel) or by chemisorption of nitrogen-containing compounds on the catalyst, thus necessitating the frequent regeneration of the catalyst or its expensive replacement. This carries with it the issues related to catalyst disposal.

Thermal processing can significantly increase the concentration of polynuclear aromatic hydrocarbons in the product liquid because the low-pressure hydrogen deficient conditions favor aromatization of naphthene constituents and condensation of aromatics to form larger ring systems. To the extent that more compounds like benzo(a)pyrene are produced, the liquids from thermal processes will be more carcinogenic than asphalt. This biological activity was consistent with the higher concentration of polynuclear aromatic hydrocarbons at 38.8 mg/g in the pitch compared to only 0.22 mg/g in the asphalt. Similarly, one would expect coker gas oils to contain more polynuclear aromatic hydrocarbons than unprocessed or hydroprocessed distillates, and thereby give a higher potential for carcinogenic or mutagenic effects.

The sludge produced on acid treatment of petroleum distillates (Chapter 19), even gasoline and kerosene, is complex in nature. Esters and alcohols are present from reactions with olefins; sulfonation products from reactions with aromatic compounds, naphthene compounds, and phenols; and salts from reactions with nitrogen bases. In addition, such materials as naphthenic acids, sulfur compounds, and asphalt (residua constituents) material are all retained by direct solution. To these constituents must be added the various products of oxidation-reduction reactions: coagulated resins, soluble hydrocarbons, water, and free acid.

The disposal of the sludge is difficult, as it contains unused free acid that must be removed by dilution and settling. The disposal is a comparatively simple process for the sludge resulting from treating gasoline and kerosene, the so-called light oils. The insoluble oil phase separates out as a mobile tar, which can be mixed and burned without too much difficulty. Sludge from heavy oil, however, separates out granular semisolids, which offer considerable difficulty in handling.

In all cases careful separation of reaction products is important to the recovery of well-refined materials. This may not be easy if the temperature has risen as a consequence of chemical reaction. This will result in a persistent dark color traceable to reaction products that are redistributed as colloids. Separation may also be difficult at low temperature because of high viscosity of the stock, but this problem can be overcome by dilution with light naphtha or with propane.

In addition, delayed coking also requires the use of large volumes of water for hydraulic cleaning of the coke drum. However, the process water can be recycled if the oil is removed by skimming and suspended coke particles are removed by filtration. If this water is used in a closed cycle, the impact of delayed coking on water treatment facilities and the environment is minimized. The flexicoking process offers one alternative to direct combustion of coke for process fuel. The gasification section is used to process excess coke to a mixture of carbon monoxide (CO), carbon dioxide (CO_2), hydrogen (H_2), and hydrogen sulfide (H_2S) followed by treatment to remove the hydrogen sulfide. Maximizing the residue conversion and desulfurization of the residue in upstream hydroconversion units also maximizes the yield of hydrogen sulfide relative to sulfur in the coke. Currently, maximum residue conversion with minimum coke production is favored over gasification of coke.

28.4.3 Solid Effluents

Catalyst disposal is therefore a major concern in all refineries. In many cases the catalysts are regenerated at the refinery for repeated use. Disposal of spent catalysts is usually part of an agreement with the catalysts manufacturer whereby the spent catalyst is returned for treatment and re-manufacture.

The formation of considerable quantities of coke in the coking processes is a cause for concern as it not only reduces the yield of liquid products but also initiates the necessity for disposal of the coke. Stockpiling to coke may be a partial answer unless the coke contains leachable materials that will endanger the ecosystem as a result of rain or snowmelt.

In addition, emission of sulfur oxides (particularly sulfur dioxide) is generated from combustion of sulfur-containing coke as plant fuel. Sulfur dioxide (SO_2) has a wide range of effects on health and on the environment. These effects vary from bronchial irritation upon short-term exposure to contributing to the acidification of lakes. Emissions of sulfur dioxide therefore, are regulated in many countries.

28.5   EPILOG

There have been many suggestions about the future of the petroleum industry and the reserves of crude oil that are available. Among these suggestions is one that the bulk of the world's oil and gas has already been discovered and that declining production is inevitable. Another suggestion is that substantial amounts of oil and gas remain to be found. There are also suggestions that fall between these two extremes.

In the last two decades, new fields have indeed been discovered, for example, in Kazakhstan near the Caspian Sea, and the potential for crude oil discoveries have opened up in Eastern Europe, Asia, in Canadian coastal areas, and in Colombia. Potentially, the richest discovery has been the finding of vast reserves in deep water in the Gulf of Mexico. These reserves were only beginning to be tapped in the mid-1990s, using floating platforms (Chapter 5) tethered to the sea bottom by steel cables, and such innovative technologies as the use of deep water robotic machines for construction and maintenance.

Liquid fuel sources that still remain to be exploited include tar sand deposits (Chapter 3 and Chapter 5), oil shale (Scouten, 1990), and the liquefaction and gasification of coal (Speight, 1994). All attempts to utilize these sources have proved so far to be uneconomic compared to the costs of producing oil and natural gas. Future technologies may, however, find ways of creating viable fuels from these various substances. That being the case, and although oil is now recognized as likely to be abundant in the first fifty years of the twenty-first century, environmental concerns will probably impose increasing restrictions on both its production and consumption.

Thus, the general prognosis for emission clean up is not pessimistic and can be looked upon as being quite optimistic. Indeed, it is considered likely that most of their environmental impact of petroleum refining can be substantially abated. A considerable investment in retrofitting or replacing existing facilities and equipment might be needed. However, it is possible and a conscious goal must be to improve the efficiency with which petroleum is transformed and consumed.

Obviously, much work is needed to accommodate the continued use of petroleum. In the meantime, we use what we have, all the while working to improve efficient usage and working to ensure that there is no damage to the environment.

Such is the nature of petroleum refining and the expectancy of protecting the environment.

REFERENCES

Boyce, A. 1997. *Introduction to Environmental Technology*. Van Nostrand Reinhold, New York.

Carson, J., and Mumford, C.J., 1988. *The Safe Handling of Chemicals in Industry*. Volumes 1 and 2. John Wiley & Sons Inc., New York.

Carson and Mumford, 1995. *The Safe Handling of Chemicals in Industry*. Volume 3. John Wiley & Sons Inc., New York.

Edwards, J.D. 1995. *Industrial Wastewater Treatment: A Guidebook*. CRC Press Inc., Boca Raton, FL.

Hileman, B. 1996. *Chem Eng. News* 74(34): 33.

Lipton, S. and J. Lynch. 1994. *Handbook of Health Hazard Control in the Chemical Process Industry*. John Wiley and Sons Inc., New York.

Loehr, R.C. 1992. *Petroleum Processing Handbook*. J.J. McKetta ed. Marcel Dekker Inc., New York. p. 190.

Majumdar, S.B. 1993. *Regulatory Requirements for Hazardous Materials*. McGraw-Hill, New York.

Occelli, M.L. and Chianelli, R. 1996. *Hydrotreating Technology for Pollution Control*. Marcel Dekker Inc., New York.

Olschewsky, D. and Megna, A. 1992. *Petroleum Processing Handbook*. J.J. McKetta ed. Marcel Dekker Inc., New York. p. 179.

Ray, D.L. and Guzzo, L. 1990. *Trashing The Planet: How Science Can Help Us Deal With Acid Rain, Depletion of The Ozone, and Nuclear Waste (Among Other Things)*. Regnery Gateway, Washington, DC.

Renzoni, A., Fossi, M.C., Lari, L., and Mattei, N. 1994. *Contaminants in the Environment: A Multi-Disciplinary Assessment of risks to Man and Other Organisms*. CRC Press Inc., Boca Raton, FL.

Scouten, C.S. 1990. *Fuel Science and Technology Handbook*. Marcel Dekker Inc., New York.

Speight, J.G. 1993. *Gas Processing: Environmental Aspects and Methods*. Butterworth Heinemann, Oxford, UK.

Speight, J.G. 1994. *The Chemistry and Technology of Coal*, 2nd edn. Marcel Dekker Inc., New York.

Speight, J.G. 1996. *Environmental Technology Handbook*. Taylor & Francis Publishers, Washington, DC.

Thibodeaux, L.J. 1995. *Environmental Chemodynamics*. John Wiley & Sons Inc., New York.

29 Refinery Wastes

29.1 INTRODUCTION

Petroleum refining is the physical, thermal, and chemical separation of crude oil into its major fractions that are then further processed through a series of separation and conversion steps into finished petroleum products (Figure 29.1). The primary products of the industry fall into three major categories: (1) fuels (motor gasoline, diesel and distillate fuel oil, liquefied petroleum gas, jet fuel, residual fuel oil, kerosene, and coke), (2) finished nonfuel products (solvents, lubricating oils, greases, petroleum wax, petroleum jelly, asphalt, and coke), and (3) chemical industry feedstocks (naphtha, ethane, propane, butane, ethylene, propylene, butylenes, butadiene, benzene, toluene, and xylene). These petroleum products are used as primary input to a vast number of products, including: fertilizers, pesticides, paints, waxes, thinners, solvents, cleaning fluids, detergents, refrigerants, antifreeze, resins, sealants, insulations, latex, rubber compounds, hard plastics, plastic sheeting, plastic foam, and synthetic fibers.

The chemicals in petroleum vary from simple hydrocarbons of low-to-medium molecular weight to higher molecular weight organic compounds containing sulfur, oxygen, and nitrogen, as well as compounds containing metallic constituents, particularly vanadium nickel, iron, and copper. Many of these latter compounds are of indeterminate molecular weight. Residua contain significantly less hydrocarbon constituents than the original crude oil. The constituents of residua may be, depending on the crude oil, molecular entities of which the majority contains at least one heteroatom.

Typical refinery products include (1) natural gas and liquefied petroleum gas (LPG), (2) solvent naphtha, (3) kerosene, (4) diesel fuel, (5) jet fuel, (6) lubricating oil, (7) various fuel oils, (8) wax, (9) residua, and (10) asphalt (Chapter 3). A single refinery does not necessarily produce all of these products. Some refineries are dedicated to particular products, for example, the production of gasoline or the production of lubricating oil or the production of asphalt. However, the issue is that refineries also produce a variety of waste products (Table 29.1) (EPA, 1995a) that must be disposed in an environmentally acceptable manner.

Waste treatment processes also account for a significant area of the refinery, particularly sulfur compounds in gaseous emissions together with various solid and liquid extracts and wastes generated during the refining process. The refinery is therefore composed of a complex system of stills, cracking units, processing and blending units and vessels in which the various reactions take place, as well as packaging units for products for immediate distribution to the retailer, for example, lubricating oils. Bulk storage tanks usually grouped together in tank farms are used for storage of both crude and refined products. Other tanks are used in the processes outlined, for example, treating, blending and mixing while others are used for spill and fire control systems. A boiler and electrical generating system usually operate for the refinery as a whole.

There are several hundred individual hydrocarbon chemicals defined as petroleum-based. Furthermore, each petroleum product has its own mix of constituents because (Chapter 2)

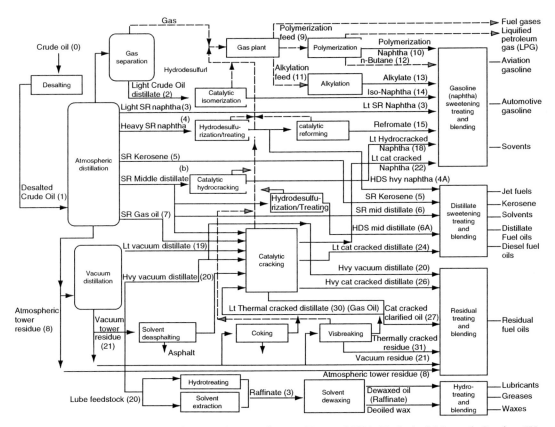

FIGURE 29.1 General layout of a petroleum refinery. (From OSHA Technical Manual, Section IV, Chapter 2, Petroleum Refining Processes.)

petroleum varies in composition from one reservoir to another and this variation may be reflected in the finished product(s).

Petroleum hydrocarbons are environmental contaminants but they are not usually classified as hazardous wastes (Irwin, 1997). Soil and groundwater petroleum hydrocarbon contamination has long been of concern and has spurred various analytical and site remediation developments, for example, risk-based corrective actions. In some instances, it may appear that such cleanup operations were initiated with an incomplete knowledge of the charter and behavior of the contaminants. The most appropriate first assumption is that the spilled constituents are toxic to the ecosystem. The second issue is an investigation of the products of the spilled material to determine an appropriate cleanup method. The third issue is whether or not the chemical nature of the constituents has changed during the time since the material was released into the environment. If it has, a determination must be made of the effect of any such changes on the potential cleanup method.

Despite the large number of hydrocarbons found in petroleum products and the widespread nature of petroleum use and contamination, many of the lower boiling constituents are well characterized in terms of physical properties but only a relatively small number of the compounds are well characterized for toxicity. The health effects of some fractions can be well characterized, based on their components or representative compounds (e.g., light aromatic fraction benzene–toluene–ethylbenzene–xylenes). However, higher molecular weight (higher boiling) fractions have far fewer well-characterized compounds.

TABLE 29.1
Emissions and Waste from Refinery Processes

Process	Air Emissions	Residual Wastes Generated
Crude oil desalting	Heater stack gas (CO, SO_x, NO_x, hydrocarbons and particulates), fugitive emissions (hydrocarbons)	Crude oil/desalter sludge (iron rust, clay, sand, water, emulsified oil and wax, metals)
Atmospheric distillation	Heater stack gas (CO, SO_x, NO_x, hydrocarbons and particulates), vents and fugitive emissions (hydrocarbons)	Typically, little or no residual waste generated
Vacuum distillation	Steam ejector emissions (hydrocarbons), heater stack gas (CO, SO_x, NO_x, hydrocarbons and particulates), vents and fugitive emissions (hydrocarbons)	
Thermal cracking/ visbreaking	Heater stack gas (CO, SO_x, NO_x, hydrocarbons and particulates), vents and fugitive emissions (hydrocarbons)	Typically, little or no residual waste generated
Coking	Heater stack gas (CO, SO_x, NO_x, hydrocarbons and particulates), vents and fugitive emissions (hydrocarbons) and decoking emissions (hydrocarbons and particulates)	Coke dust (carbon particles and hydrocarbons)
Catalytic cracking	Heater stack gas (CO, SO_x, NO_x, hydrocarbons and particulates), fugitive emissions (hydrocarbons) and catalyst regeneration (CO, NO_x, SO_x, and particulates)	Spent catalysts (metals from crude oil and hydrocarbons), spent catalyst fines from electrostatic precipitators (aluminum silicate and metals)
Catalytic hydrocracking	Heater stack gas (CO, SO_x, NO_x, hydrocarbons and particulates), fugitive emissions (hydrocarbons) and catalyst regeneration (CO, NO_x, SO_x, and catalyst dust)	Spent catalysts fines
Hydrotreating/hydroprocessing	Heater stack gas (CO, SO_x, NO_x, hydrocarbons and particulates), vents and fugitive emissions (hydrocarbons) and catalyst regeneration (CO, NO_x, SO_x)	Spent catalyst fines (aluminum silicate and metals)
Alkylation	Heater stack gas (CO, SO_x, NO_x, hydrocarbons and particulates), vents and fugitive emissions (hydrocarbons)	Neutralized alkylation sludge (sulfuric acid or calcium fluoride, hydrocarbons)
Isomerization	Heater stack gas (CO, SO_x, NO_x, hydrocarbons and particulates), HCl (potentially in light ends), vents and fugitive emissions (hydrocarbons)	Calcium chloride sludge from neutralized HCl gas
Polymerization	H_2S from caustic washing	Spent catalyst containing phosphoric acid
Catalytic reforming	Heater stack gas (CO, SO_x, NO_x, hydrocarbons and particulates), fugitive emissions (hydrocarbons) and catalyst regeneration (CO, NO_x, SO_x)	Spent catalyst fines from electrostatic precipitators (alumina silicate and metals)
Solvent extraction	Fugitive solvents	Little or no residual wastes generated
Dewaxing	Fugitive solvents, heaters	Little or no residual wastes generated

Continued

TABLE 29.1 (continued)
Emissions and Waste from Refinery Processes

Process	Air Emissions	Residual Wastes Generated
Propane deasphalting	Heater stack gas (CO, SO$_x$, NO$_x$, hydrocarbons and particulates), fugitive propane	Little or no residual wastes generated
Wastewater treatment	Fugitive emissions (H$_2$S, NH$_3$, and hydrocarbons)	API separator sludge (phenols, metals and oil), chemical precipitation sludge (chemical coagulants, oil), DAF floats, biological sludge (metals, oil, suspended solids), spent lime.

This chapter deals with the toxicity of petroleum and petroleum products, the effects of petroleum constituents on the environment, and the individual process wastes, and the means by which petroleum, petroleum products, and process wastes are introduced into the environment. The processes are restricted to those by which the common products are produced (Chapter 3).

29.2   PROCESS WASTES

Petroleum refineries are complex, but integrated, unit process operations that produce a variety of products from various feedstocks and feedstock blends (Figure 29.1) (Chapter 14) (Meyers, 1997; Speight, 1999, and references cited therein; Speight and Ozum, 2002, and references cited therein). During petroleum refining, refineries use and generate an enormous amount of chemicals, some of which are present in air emissions, wastewater, or solid wastes. Emissions are also created through the combustion of fuels, and as by-products of chemical reactions occurring when petroleum fractions are upgraded. A large source of air emissions is, generally, the process heaters and boilers that produce carbon monoxide, sulfur oxides, and nitrogen oxides, leading to pollution and the formation of acid rain.

$$CO_2 + H_2O \rightarrow H_2CO_3 (\text{carbonic acid})$$

$$SO_2 + H_2O \rightarrow H_2SO_3 (\text{sulfurous acid})$$

$$SO_2 + O_2 \rightarrow 2SO_3$$

$$SO_3 + H_2O \rightarrow H_2SO_4 (\text{sulfuric acid})$$

$$NO + H_2O \rightarrow HNO_2 (\text{nitrous acid})$$

$$NO + O_2 \rightarrow NO_2$$

$$NO_2 + H_2O \rightarrow HNO_3 (\text{nitric acid})$$

Hence, there is the need for gas-cleaning operations on a refinery site so that such gases are cleaned from the gas stream prior to entry into the atmosphere.

In addition, some processes create considerable amounts of particulate matter and other emissions from catalyst regeneration or decoking processes. Volatile chemicals and hydrocarbons are also released from equipment leaks, storage tanks, and wastewaters. Other cleaning units such as the installation of filters, electrostatic precipitators, and cyclones can mitigate part of the problem.

Process wastewater is also a significant effluent from a number of refinery processes. Atmospheric and vacuum distillation create the largest volumes of process wastewater, about 26 gallons per barrel of oil processed. Fluid catalytic cracking and catalytic reforming also generate considerable amounts of wastewater (15 and 6 gallons per barrel of feedstock, respectively). A large portion of wastewater from these three processes is contaminated with oil and other impurities and must be subjected to primary, secondary, and sometimes tertiary water-treatment processes, some of which also create hazardous waste.

Wastes, residua and by-products are produced by a number of processes. Residuals produced during refining can be but are not necessarily wastes. They can be recycled or regenerated, and in many cases do not become part of the waste stream but are useful products. For example, processes utilizing caustics for neutralization of acidic gases or solvent (e.g., alkylation, sweetening/chemical treating, lubricating oil manufacture) create the largest source of residuals in the form of spent caustic solutions. However, nearly all of these caustics are recycled.

The treatment of oily wastewater from distillation, catalytic reforming, and other processes generates the next largest source of residuals in the form of biomass sludge from biological treatment and pond sediments. Water treatment of oily wastewater also produces a number of sludge materials associated with oil–water separation processes. Such sludge is often recycled in the refining process and are not considered wastes.

Catalytic processes (fluid catalytic cracking, catalytic hydrocracking, hydrotreating, isomerization, ethers manufacture) also create some residuals in the form of spent catalysts and catalyst fines or particulates. The latter are sometimes separated from exiting gases by electrostatic precipitators or filters. These are collected and disposed of in landfills or may be recovered by off-site facilities.

In terms of individual processes, the potential for waste generation and, hence, leakage of emissions is as follows:

29.2.1 DESALTING

As already noted (Chapter 14), petroleum often contains water, inorganic salts, suspended solids, and water-soluble trace metals. Before separation into fractions by distillation, crude oil usually must first be treated to remove corrosive salts. The desalting process also removes some of the metals and suspended solids that cause equipment corrosion and catalyst deactivation.

Desalting involves the mixing of heated crude oil with water (approximately 3% to 10% of the crude oil volume) so that the salts are dissolved in the water. The water must then be separated from the crude oil in a separating vessel.

The two most typical methods of petroleum desalting, chemical separation and electrostatic separation, use hot water as the extraction agent. In chemical desalting, water and chemical surfactant (demulsifiers) are added to the crude oil, heated so that salts and other impurities dissolve into the water or attach to the water, and then are held in a tank where they settle out. Electrical desalting is the application of high-voltage electrostatic charges to concentrate suspended water globules in the bottom of the settling tank. Surfactants are added only when the crude has a large amount of suspended solids. Both methods of desalting are continuous. A third and less-common process involves filtering heated petroleum using diatomaceous earth.

The feedstock crude oil is heated to between 65°C and 177°C (150°F and 350°F) to reduce viscosity and surface tension for easier mixing and separation of the water. The temperature is limited by the vapor pressure of the petroleum constituents. In both methods other chemicals

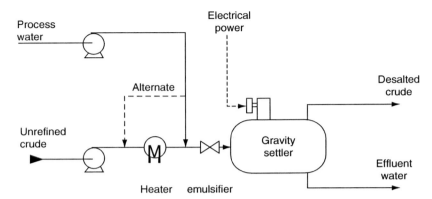

FIGURE 29.2 Schematic of an electrostatic desalting unit. (From OSHA Technical Manual, Section IV, Chapter 2, Petroleum Refining Processes.)

may be added. Ammonia is often used to reduce corrosion and alkali or acid may be added to adjust the pH of the water wash. Wastewater and contaminants are discharged from the bottom of the settling tank to the wastewater treatment facility. The desalted crude is continuously drawn from the top of the settling tanks and sent to the crude distillation (fractionating) tower.

As desalting is a closed process, there is little potential for exposure to the feedstock unless a leak or release occurs. However, whenever elevated temperatures are used when desalting sour (sulfur-containing) petroleum, hydrogen sulfide will be present. And, depending on the crude feedstock and the treatment chemicals used, the wastewater will contain varying amounts of chlorides, sulfides, bicarbonates, ammonia, hydrocarbons, phenol, and suspended solids. If diatomaceous earth is used in filtration, exposures should be minimized or controlled.

Desalting (Figure 29.2) creates an oily desalter sludge that may be a hazardous waste and a high temperature salt wastewater stream that is usually added to other process wastewaters for treatment in the refinery wastewater treatment facilities. The water used in crude desalting is often untreated or partially treated water from other refining process water sources.

The primary polluting constituents in desalter wastewater include hydrogen sulfide, ammonia, phenol, high levels of suspended solids, and dissolved solids, with a high biochemical oxygen demand (BOD). In some cases, it is possible to recycle the desalter effluent water back into the desalting process, depending upon the type of crude being processed.

29.2.2 DISTILLATION

After desalting, the crude oil is then heated in a heat exchanger and furnace to temperatures not exceeding 400°C (750°F) and fed to a vertical, distillation column at atmospheric pressure where most of the feedstock is vaporized and separated into its various fractions by condensing on 30 to 50 fractionation trays, each corresponding to a different condensation temperature (Figure 29.3). The lower boiling fractions condense and are collected toward the top of the column. Higher boiling fractions, which may not vaporize in the column, are further separated later by vacuum distillation.

Within each atmospheric distillation tower, a number of side streams (at least four) of low-boiling point components are removed from the tower from different trays. These low-boiling point mixtures are in equilibrium with heavier components that must be removed. The side streams are each sent to a different small stripping tower containing 4 to 10 trays with steam injected under the bottom tray. The steam strips the light-end components from the heavier

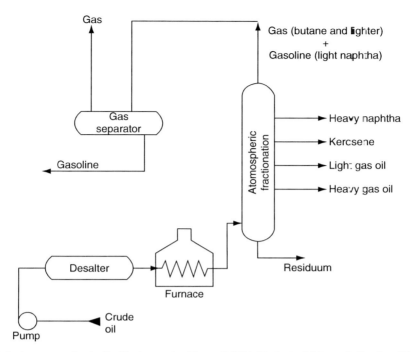

FIGURE 29.3 An atmospheric distillation unit. (From OSHA Technical Manual, Section IV, Chapter 2, Petroleum Refining Processes.)

components and both the steam and light-ends are fed back to the atmospheric distillation tower above the corresponding side stream draw tray. Fractions obtained from atmospheric distillation include naphtha, gasoline, kerosene, light fuel oil, diesel oils, gas oil, lube distillate, and the residuum (bottoms).

Most of these fractions can be sold as finished products, or blended with products from downstream processes. Another product produced in atmospheric distillation, as well as many other refinery processes, is the light, noncondensable refinery fuel gas (mainly methane and ethane). Typically, this gas also contains hydrogen sulfide and ammonia gases (sour gas or acid gas). The sour gas is sent to the gas processing section (Chapter 25) that separates the fuel gas so that it can be used as fuel in the refinery heating furnaces.

Vacuum distillation (Figure 29.4) typically follows atmospheric distillation and is the distillation of petroleum fractions at low pressure (0.2 to 0.7 psi) to increase volatilization and separation. In most systems, the vacuum inside the fractionator is maintained with steam ejectors and vacuum pumps, barometric condensers or surface condensers. The injection of superheated steam at the base of the vacuum fractionator column further reduces the partial pressure of the hydrocarbons in the tower, facilitating vaporization and separation. The higher fractions from the vacuum distillation column are processed downstream into more valuable products through either cracking or coking operations (*q. v.*).

Both atmospheric distillation units and vacuum distillation units produce refinery fuel gas streams containing a mixture of light hydrocarbons, hydrogen sulfide and ammonia. These streams are processed through gas treatment and sulfur recovery units to recover fuel gas and sulfur. Sulfur recovery creates emissions of ammonia, hydrogen sulfide, sulfur oxides, and nitrogen oxides.

When sour (high-sulfur) petroleum is processed, there is potential for exposure to hydrogen sulfide in the preheat exchanger and furnace, tower flash zone and overhead system, vacuum

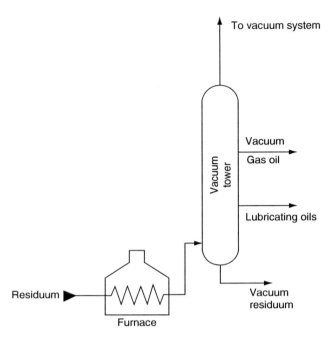

FIGURE 29.4 A vacuum distillation unit. (From OSHA Technical Manual, Section IV, Chapter 2, Petroleum Refining Processes.)

furnace and tower, and bottoms exchanger. Hydrogen chloride may be present in the preheat exchanger, tower top zones, and overheads. Wastewater may contain water-soluble sulfides in high concentrations and other water-soluble compounds such as ammonia, chlorides, phenol, mercaptans, and the like, depending upon the crude feedstock and the treatment chemicals. Safe work practices or the use of appropriate personal protective equipment may be needed for exposures to chemicals and other hazards such as heat and noise, and during sampling, inspection, maintenance, and turnaround activities.

Air emissions from a petroleum distillation unit include emissions from the combustion of fuels in process heaters and boilers, fugitive emissions of volatile constituents in the crude oil and fractions, and emissions from process vents. The primary source of emissions is combustion of fuels in the crude preheat furnace and in boilers that produce steam for process heat and stripping. When operating in an optimum condition and burning cleaner fuels (e.g., natural gas, refinery gas), these heating units create relatively low emissions of sulfur oxides, (SO_x), nitrogen oxides (NO_x), carbon monoxide (CO), hydrogen sulfide (H_2S), particulate matter, and volatile hydrocarbons. If fired with lower grade fuels (e.g., refinery fuel pitch, coke) or operated inefficiently (incomplete combustion), heaters can be a significant source of emissions.

Fugitive emissions of volatile hydrocarbons arise from leaks in valves, pumps, flanges, and other similar sources where crude and its fractions flow through the system. While individual leaks may be minor, the combination of fugitive emissions from various sources can be substantial. Those potentially released during crude distillation include ammonia, benzene, toluene, and xylenes, among others. These emissions are controlled primarily through leak detection and repair programs and occasionally through the use of special leak-resistant equipment.

Petroleum distillation units generate considerable wastewater. The process water used in distillation often comes into direct contact with oil, and can be highly contaminated. Both atmospheric distillation and vacuum distillation produce an oily sour wastewater (condensed steam containing hydrogen sulfide and ammonia) from side stripping fractionators and reflux drums.

Many refineries now use vacuum pumps and surface condensers in place of barometric condensers to eliminate the generation of the wastewater stream and reduce energy consumption. Reboiled side stripping towers rather than open steam stripping can also be utilized on the atmospheric tower to reduce the quantity of sour water condensate.

Typical constituents of sour wastewater streams from crude distillation include hydrogen sulfide, ammonia, suspended solids, chlorides, mercaptans, and phenol, characterized by a high pH. Combined flows from atmospheric and vacuum distillation are about 26.0 gallons per barrel of oil, and represent one of the largest sources of wastewater in the refinery.

29.2.3 THERMAL CRACKING AND VISBREAKING

Thermal cracking causes the decomposition of higher molecular weight fractions to lower boiling products. The process has been largely replaced by catalytic cracking and some refineries no longer employ thermal cracking but, because of the increasing number of high-boiling feedstocks entering refineries and the propensity of these feedstocks to poison catalysts, there has been a re-emergence of interest in thermal crackers (Chapter 17).

Thermal cracking (like visbreaking) reduces the production of less valuable products such as heavy fuel oil and cutter stock and increases the feedstock to the catalytic cracker and gasoline yields. In a thermal cracking process, heavy gas oils and the residuum from the vacuum distillation process are typical feedstocks. The feedstock is heated in a furnace or other thermal unit to up to 540°C (1,000°F) and then fed to a reaction chamber which is kept at a pressure of about 140 psi. The product is then fed to a flasher chamber, where pressure is reduced and lower boiling products vaporize and are drawn off as overhead to a fractionating tower where the various fractions are separated. The bottoms consist of heavy cracked residuum (**pitch**), part of which may be used for fuel or recycled for further cracking.

Visbreaking (Figure 29.5) operates in a similar manner to thermal cracking (480°C, 895°F; outlet pressure: approximately 100 psi) except the product is quenched to mitigate coke-forming reactions. The process is a mild thermal cracking operation that can be used to reduce

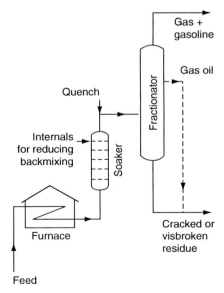

FIGURE 29.5 A soaker visbreaking unit. (From OSHA Technical Manual, Section IV, Chapter 2, Petroleum Refining Processes.)

the viscosity of residua to allow the products to meet fuel oil specifications. Alternatively, the visbroken residua could be blended with lighter product oils to produce fuel oils of acceptable viscosity. By reducing the viscosity of the residuum, visbreaking reduces the amount of light heating oil that is required for blending to meet the fuel oil specifications. In addition to the major product, fuel oil, material in the gas oil and gasoline boiling range is produced. The gas oil may be used as additional feed for catalytic cracking units, or as heating oil.

Thermal cracking and visbreaking tend to produce a relatively small amount of fugitive emissions and sour wastewater. Usually some wastewater is produced from steam strippers and the fractionator. Wastewater is also generated during unit cleanup and cooling operations and from the steam injection process to remove organic deposits from the soaker or from the coil. Combined wastewater flows from thermal cracking and coking processes are about 3.0 gallons per barrel of process feed.

29.2.4 COKING PROCESSES

Coking is a cracking process used primarily to reduce refinery production of low-value residual fuel oils to transportation fuels, such as gasoline and diesel fuel. As part of the upgrading process, coking also produces petroleum coke, which is essentially solid carbon with varying amounts of impurities, and is used as a fuel for power plants if the sulfur content is low enough. Coke also has nonfuel applications as a raw material for many carbon and graphite products including anodes for the production of aluminum, and furnace electrodes for the production of elemental phosphorus, titanium dioxide, calcium carbide and silicon carbide.

A number of different processes are used to produce coke; delayed coking (Figure 29.6) is the most widely used today, but fluid coking is expected to be an important process in the future. Fluid coking (Figure 29.7) produces a higher grade of coke which is increasingly in demand.

In delayed coking operations, the same basic process as thermal cracking is used except feed streams are allowed to react longer without being cooled. The delayed coking feed stream of residual oils from various upstream processes is first introduced to a fractionating tower where residual lighter materials are drawn off and the heavy ends are condensed. The heavy ends are removed and heated in a furnace from about 480°C to 540°C (900°F to 1000°F) and then fed to an insulated vessel called a coke drum where the coke is formed. When the coke

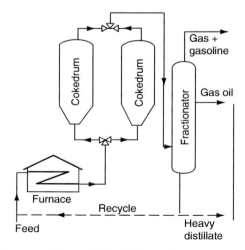

FIGURE 29.6 A delayed coking unit. (From OSHA Technical Manual, Section IV, Chapter 2, Petroleum Refining Processes.)

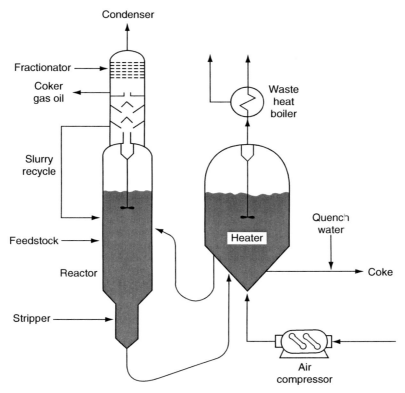

FIGURE 29.7 A fluid coking unit.

drum is filled with product, the feed is switched to an empty parallel drum. Hot vapors from the coke drums, containing cracked lighter hydrocarbon products, hydrogen sulfide, and ammonia, are fed back to the fractionator where they can be treated in the sour gas treatment system or drawn off as intermediate products. Steam is then injected into the full coke drum to remove hydrocarbon vapors, water is injected to cool the coke, and the coke is removed. Typically, high pressure water jets are used to cut the coke from the drum.

Delayed coking and fluid coking produce a relatively small amount of sour wastewater from the associated steam strippers and fractionators. Wastewater is generated during coke removal and cooling operations and from the steam injection process to cut coke from the coke drums. Combined wastewater flows from thermal cracking and coking processes are about 3.0 gallons per barrel of process feed.

Like most separation processes in the refinery, the process water used in coker fractionators (as is also the case in other product fractionators) often comes into direct contact with oil, and can have a high oil content (much of that oil can be recovered through wastewater oil recovery processes). Thus, the main constituents of sour water from catalytic cracking include high levels of oil, suspended solids, phenols, cyanides, hydrogen sulfate, and ammonia. Typical wastewater flow from catalytic cracking is about 15.0 gallons per barrel of feed processed (more than one-third of a gallon of waste water for every gallon of feed processed), and represents the second largest source of wastewater in the refinery.

Particulate emissions from decoking can also be considerable. Coke-laden water from decoking operations in delayed cokers (hydrogen sulfide, ammonia, suspended solids), coke dust (carbon particles and hydrocarbons) occur.

29.2.5   FLUID CATALYTIC CRACKING

Fluid-bed catalytic cracking processes (Figure 29.8) use heat, pressure, and a catalyst to produce lower boiling products from high boiling feedstocks. Catalytic cracking has largely replaced thermal cracking because it is able to produce more gasoline with a higher octane and less heavy fuel oils and light gases. Feedstocks are light and heavy oils from the crude oil distillation unit that are processed primarily into gasoline as well as some fuel oil and light gases. Most catalysts used in catalytic cracking consist of mixtures of crystalline zeolites and amorphous synthetic silica–alumina. The catalytic cracking processes, as well as most other refinery catalytic processes, produce coke that collects on the catalyst surface and diminishes its catalytic properties. The catalyst, therefore, needs to be regenerated continuously or periodically; essentially by burning the coke off the catalyst at high temperatures. The method and frequency by which catalysts are regenerated are a major factor in the design of catalytic cracking units. A number of different catalytic cracking designs are currently in use including fixed-bed reactors, moving-bed reactors, fluidized-bed reactors, and once-through units. The fluidized- and moving-bed reactors are by far the most prevalent.

Fluid catalytic cracking is one of the largest sources of air emission in refineries. Air emissions are released in process heater flue gas, as fugitive emissions from leaking valves and pipes, and during regeneration of the cracking catalyst. If not controlled, catalytic cracking is one of the most substantial sources of carbon monoxide and particulate emissions in the refinery. In nonattainment areas where carbon monoxide and particulates are above acceptable levels, carbon monoxide waste heat boilers (CO boiler) and particulate controls are employed. Carbon monoxide produced during regeneration of the catalyst is converted to carbon dioxide either in the regenerator or further downstream in a carbon monoxide waste heat boiler (CO boiler). Catalytic crackers are also significant sources of sulfur oxides and

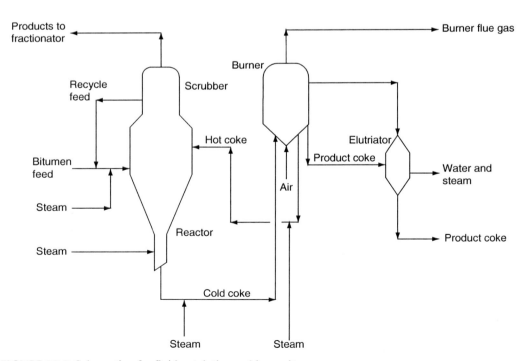

FIGURE 29.8 Schematic of a fluid catalytic cracking unit.

nitrogen oxides. The nitrogen oxides produced by catalytic crackers is expected to be a major target of emissions reduction in the future.

Catalytic cracking units, like coking units, usually include some form of fractionation or steam stripping as part of the process configuration. These units all produce sour waters and sour gases containing some hydrogen sulfide and ammonia. Like crude oil distillation, some of the toxic releases reported by the refining industry are generated through sour water and gases, notably ammonia. Gaseous ammonia often leaves fractionating and treating processes in the sour gas along with hydrogen sulfide and fuel gases.

Catalytic cracking produces large volumes of wastewater and spent catalysts. Catalytic cracking (primarily fluid catalytic cracking) generates considerable sour wastewater from fractionators used for product separation, from steam strippers used to strip oil from catalysts, and in some cases from scrubber water. The steam stripping process used to purge and regenerate the catalysts can contain metal impurities from the feed in addition to oil and other contaminants. Sour wastewater from the fractionator or gas concentration units and steam strippers contain oil, suspended solids, phenols, cyanides, hydrogen sulfide, ammonia, spent catalysts (metals from crude oil) and hydrocarbons.

Catalytic cracking generates significant quantities of spent process catalysts (containing metals from crude oils and hydrocarbons) that are often sent off-site for disposal or recovery or recycling. Management options can include land filling, treatment, or separation and recovery of the metals. Metals deposited on catalysts are often recovered by third-party recovery facilities. Spent catalyst fines (containing aluminum silicate and metals) from electro-static precipitators are also sent off-site for disposal or recovery options.

Catalytic crackers also produce a significant amount of fine catalyst dust that results from the constant movement of catalyst grains against each other. This dust contains primarily alumina (Al_2O_3) and small amounts of nickel (Ni) and vanadium (V), and is generally carried along with the carbon monoxide stream to the carbon monoxide waste heat boiler. The dust is separated from the carbon dioxide stream exiting the boiler through the use of cyclones, flue gas scrubbing, or electrostatic precipitators, and may be disposed of at an off-site facility.

29.2.6 HYDROCRACKING AND HYDROTREATING

Hydrotreating and hydroprocessing are similar processes used to remove impurities such as sulfur, nitrogen, oxygen, halides, and trace metal impurities that may deactivate process catalysts. Hydrotreating also upgrades the quality of fractions by converting olefins and diolefins to paraffins for the purpose of reducing gum formation in fuels. Hydroprocessing, which typically uses residua from the crude distillation units, also cracks these heavier molecules to lower boiling more saleable products. Both hydrotreating and hydroprocessing units are usually placed upstream of those processes in which sulfur and nitrogen could have adverse effects on the catalyst, such as catalytic reforming and hydrocracking units. The processes utilize catalysts in the presence of substantial amounts of hydrogen under high pressure and temperature to react the feedstocks and impurities with hydrogen. The reactors are usually fixed-bed with catalyst replacement or regeneration done after months or years of operation often at an off-site facility. In addition to the treated products, the process produces a stream of light fuel gases, hydrogen sulfide, and ammonia. The treated product and hydrogen-rich gas are cooled after they leave the reactor before being separated. The hydrogen is recycled to the reactor.

Hydrocracking (Figure 29.9) normally utilizes a fixed-bed catalytic cracking reactor with cracking occurring under substantial pressure (1200 to 2000 psi) in the presence of hydrogen. Feedstocks to hydrocracking units are often those fractions that are the most difficult to crack and cannot be cracked effectively in catalytic cracking units. These include: middle distillate,

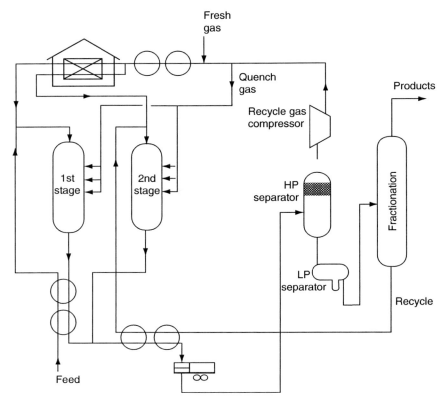

FIGURE 29.9 A two-stage hydrocracking unit. (From OSHA Technical Manual, Section IV, Chapter 2, Petroleum Refining Processes.)

cycle oil, residual fuel oil, and reduced crude oil. The hydrogen suppresses the formation of heavy residual material and increases the yield of gasoline by reacting with the cracked products. However, this process also breaks the heavy, sulfur and nitrogen-bearing hydrocarbons and releases these impurities to where they could potentially foul the catalyst. For this reason, the feedstock is often first hydrotreated to remove impurities before being sent to the catalytic hydrocracker. Sometimes hydrotreating is accomplished by using the first reactor of the hydrocracking process to remove impurities. Water also has a detrimental effect on some hydrocracking catalysts and must be removed before being fed to the reactor. The water is removed by passing the feedstock through a silica gel or molecular sieve dryer. Depending on the products desired and the size of the unit, catalytic hydrocracking is conducted in either single stage or multistage reactor processes. Most catalysts consist of a crystalline mixture of silica–alumina with small amounts of rare earth metals.

Hydrocracking generates air emissions through process heater flue gas, vents, and fugitive emissions. Unlike fluid catalytic cracking catalysts, hydrocracking catalysts are usually regenerated off-site after months or years of operations, and little or no emissions or dust are generated. However, the use of heavy oil as feedstock to the unit can change this balance.

Hydrocracking produces less sour wastewater than catalytic cracking. Hydrocracking, like catalytic cracking, produces sour wastewater at the fractionator. These processes include processing in a separator (API separator, corrugated plate interceptor) that creates a sludge. Physical or chemical methods are then used to separate the remaining emulsified oils from the wastewater. Treated wastewater may be discharged to public wastewater treatment, to a refinery secondary treatment plant for ultimate discharge to public wastewater treatment, or may be recycled and used as process water. The separation process permits recovery of usable oil, and also creates a sludge that may be recycled or treated as a hazardous waste.

In addition, oily sludge from the wastewater treatment facility that results from treating sour wastewaters may be hazardous wastes (unless they are recycled in the refining process). These include API separator sludge, primary treatment sludge, sludge from various gravitational separation units, and float from dissolved air flotation units.

Propylene, another source of toxic releases from refineries, is produced as a light end during cracking and coking processes. It is volatile as well as soluble in water, which increases its potential for release to both air and water during processing.

Like catalytic cracking, hydrocracking processes generate toxic metal compounds, many of which are present in spent catalyst sludge and catalyst fines generated from catalytic cracking and hydrocracking. These include metals such as nickel (Ni), cobalt (Co), and molybdenum (Mo).

Hydrotreating (Figure 29.10) generates air emissions through process heater flue gas, vents, and fugitive emissions. Unlike fluid catalytic cracking catalysts, hydrotreating catalysts are usually regenerated off-site after months or years of operations, and little or no emissions or dust are generated from the catalyst regeneration process at the refinery.

The off-gas stream from hydrotreating is usually very rich in hydrogen sulfide and light fuel gas. This gas is usually sent to a sour gas treatment and sulfur recovery unit along with other refinery sour gases.

Fugitive air emissions of volatile components released during hydrotreating may also be toxic components. These include toluene, benzene, xylenes, and other volatiles that are reported as toxic chemical releases under the EPA Toxics Release Inventory.

Hydrotreating generates sour wastewater from fractionators used for product separation. Like most separation processes in the refinery, the process water used in fractionators often comes in direct contact with oil, and can be highly contaminated. It also contains hydrogen sulfide and ammonia and must be treated along with other refinery sour waters. In hydrotreating, sour wastewater from fractionators is produced at the rate of about 1.0 gallon per barrel of feed.

Oily sludge (from the wastewater treatment facility) that results from treating oily and sour wastewaters from hydrotreating and other refinery processes may be hazardous wastes, depending

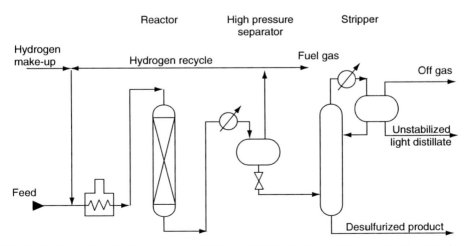

FIGURE 29.10 A distillate hydrotreating unit. (From OSHA Technical Manual, Section IV, Chapter 2, Petroleum Refining Processes.)

on how they are managed. These include API separator sludge, primary treatment sludge, sludge from various gravitational separation units, and float from dissolved air flotation units.

Hydrotreating also produces some residuals in the form of spent catalyst fines, usually consisting of aluminum silicate and some metals (e.g., cobalt, molybdenum, nickel, tungsten). Spent hydrotreating catalyst is now listed as a hazardous waste (K171)(except for most support material). Hazardous constituents of this waste include benzene and arsenia (arsenic oxide, As_2O_3). The support material for these catalysts is usually an inert ceramic (e.g., alumina, Al_2O_3).

29.2.7 CATALYTIC REFORMING

Catalytic reforming (Figure 29.11) uses catalytic reactions to process primarily low octane heavy straight-run (from the crude distillation unit) gasolines and naphtha into high octane aromatics (including benzene). There are four major types of reactions that occur during reforming processes: (1) dehydrogenation of naphthenes to aromatics, (2) dehydrocyclization of paraffins to aromatics, (3) isomerization, and (4) hydrocracking. The dehydrogenation reactions are endothermic, requiring that the hydrocarbon stream be heated between each catalyst bed. All but the hydrocracking reaction release hydrogen that can be used in the hydrotreating or hydrocracking processes. Fixed-bed or moving-bed processes are utilized in a series of three to six reactors.

Feedstocks to catalytic reforming processes are usually hydrotreated first to remove sulfur, nitrogen, and metallic contaminants. In continuous reforming processes, catalysts can be regenerated one reactor at a time, once or twice per day, without disrupting the operation of the unit. In semi-regenerative units, regeneration of all reactors can be carried out simultaneously after 3 to 24 months of operation by first shutting down the process.

Emissions from catalytic reforming include fugitive emissions of volatile constituents in the feed, and emissions from process heaters and boilers. As with all process heaters in the

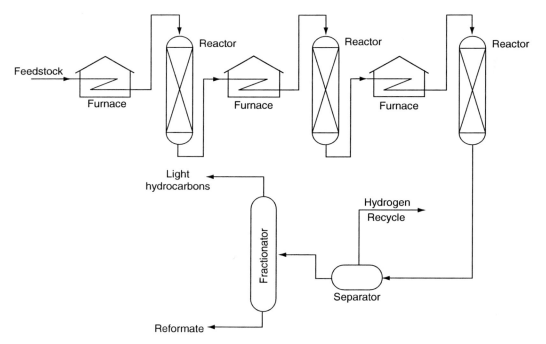

FIGURE 29.11 A catalytic reforming unit. (From OSHA Technical Manual, Section IV, Chapter 2, Petroleum Refining Processes.)

refinery, combustion of fossil fuels produces emissions of sulfur oxides, nitrogen oxides, carbon monoxide, particulate matter, and volatile hydrocarbons.

Toluene, xylene, and benzene are toxic aromatic chemicals that are produced during the catalytic reforming process and used as feedstocks in chemical manufacturing. Due to their highly volatile nature, fugitive emissions of these chemicals are a source of their release to the environment during the reforming process. Point air sources may also arise during the process of separating these chemicals.

In a continuous reformer, some particulate and dust matter can be generated as the catalyst moves from reactor to reactor, and is subject to attrition. However, due to catalyst design little attrition occurs, and the only outlet to the atmosphere is the regeneration vent, which is most often scrubbed with a caustic to prevent emission of hydrochloric acid (this also removes particulate matter). Emissions of carbon monoxide and hydrogen sulfide may occur during regeneration of catalyst.

29.2.8 ALKYLATION

Alkylation is used to produce a high octane gasoline blending stock from the *iso*-butane formed primarily during catalytic cracking and coking operations, but also from catalytic reforming, crude distillation, and natural gas processing.

Alkylation joins an olefin and an *iso*-paraffin compound using either a sulfuric acid or hydrofluoric acid catalyst. The products are alkylates including propane and butane liquids. When the concentration of acid becomes less than 88%, some of the acid must be removed and replaced with stronger acid. In the hydrofluoric acid process, the slip stream of acid is redistilled. Dissolved polymerization products are removed from the acid as thick dark oil. The concentrated hydrofluoric acid is recycled and the net consumption is about 0.3 pounds per barrel of alkylate produced.

Hydrofluoric acid alkylation units require special engineering design, operator training, and safety equipment precautions to protect operators from accidental contact with hydrofluoric acid which is an extremely hazardous substance. In the sulfuric acid process, the sulfuric acid removed must be regenerated in a sulfuric acid plant, which is generally not a part of the alkylation unit and may be located off-site. Spent sulfuric acid generation is substantial; typically in the range of 13 to 30 pounds per barrel of alkylate. Air emissions from the alkylation process may arise from process vents and fugitive emissions.

Alkylation combines low-molecular-weight olefins (primarily a mixture of propylene and butylene) with *iso*-butene in the presence of a catalyst, either sulfuric acid or hydrofluoric acid. The product is called alkylate and is composed of a mixture of high-octane, branched-chain paraffinic hydrocarbons. Alkylate is a premium blending stock because it has exceptional antiknock properties and is clean burning. The octane number of the alkylate depends mainly upon the kind of olefins used and upon operating conditions.

Emissions from alkylation processes (Figure 29.12 and Figure 29.13) and polymerization processes (Figure 29.14) include fugitive emissions of volatile constituents in the feed, and emissions that arise from process vents during processing. These can take the form of acidic hydrocarbon gases, nonacidic hydrocarbon gases, and fumes that may have a strong odor (from sulfonated organic compounds and organic acids, even at low concentrations). To prevent releases of hydrofluoric acid, refineries install a variety of mitigation and control technologies (e.g., acid inventory reduction, hydrogen fluoride detection systems, isolation valves, rapid acid transfer systems, and water spray systems).

In hydrofluoric acid alkylation processes, acidic hydrocarbon gases can originate anywhere hydrogen fluoride is present (e.g., during a unit upset, unit shutdown, or maintenance). Hydrofluoric acid alkylation units are designed to pipe these gases from acid vents and valves

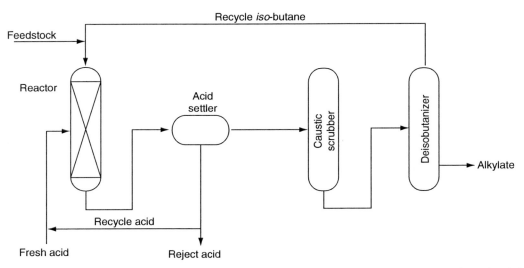

FIGURE 29.12 An alkylation unit (sulfuric acid catalyst). (From OSHA Technical Manual, Section IV, Chapter 2, Petroleum Refining Processes.)

to a separate closed-relief system where the acid is neutralized. The basins are tightly covered and equipped with a gas scrubbing system to remove odors, using either water or activated charcoal as the scrubbing agent. Another source of emissions is combustion of fuels in process boilers to produce steam for strippers. As with all process heaters in the refinery, these boilers produce significant emissions of sulfur oxides, nitrogen oxides, carbon monoxide, particulate matter, and volatile hydrocarbons.

Alkylation generates relatively low volumes of wastewater, primarily from water washing of the liquid reactor products. Wastewater is also generated from steam strippers, depropanizers and debutanizers, and can be contaminated with oil and other impurities. Liquid process waters (hydrocarbons and acid) originate from minor undesirable side reactions and from feed contaminants, and usually exit as a bottoms stream from the acid regeneration

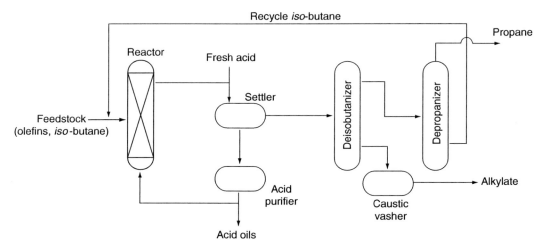

FIGURE 29.13 An alkylation unit (hydrogen fluoride catalyst). (From OSHA Technical Manual, Section IV, Chapter 2, Petroleum Refining Processes.)

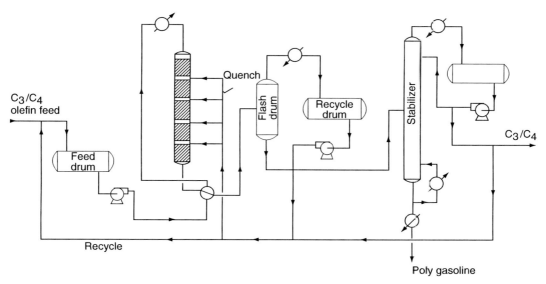

FIGURE 29.14 A polymerization unit. (From OSHA Technical Manual, Section IV, Chapter 2, Petroleum Refining Processes.)

column. The bottoms is an acid-water mixture that is sent to the neutralizing drum. The acid in this liquid eventually ends up as insoluble calcium fluoride.

Sulfuric acid alkylation generates considerable quantities of spent acid that must be removed and regenerated. Nearly all the spent acid generated at refineries is regenerated and recycled and, although technology for on-site regeneration of spent sulfuric acid is available, the supplier of the acid may perform this task off-site. If sulfuric acid production capacity is limited, acid regeneration is often done on-site. The development of internal acid regeneration for hydrofluoric acid units has virtually eliminated the need for external regeneration, although most operations retain one for startups or during periods of high feed contamination.

Both sulfuric acid and hydrofluoric acid alkylation units generate neutralization sludge from treatment of acid-laden streams with caustic solutions in neutralization or wash systems. Sludge from hydrofluoric acid alkylation neutralization systems consists largely of calcium fluoride and unreacted lime, and is usually disposed of in a landfill. It can also be directed to steel manufacturing facilities, where the calcium fluoride can be used as a neutral flux to lower the slag-melting temperature and improve slag fluidity. Calcium fluoride can also be routed back to an hydrofluoric acid manufacturer.

A basic step in hydrofluoric acid manufacture is the reaction of sulfuric acid with fluorspar (calcium fluoride) to produce hydrogen fluoride and calcium sulfate. Spent alumina is also generated by the defluorination of some hydrofluoric acid alkylation products over alumina. It is disposed of or sent to the alumina supplier for recovery. Other solid residuals from hydrofluoric acid alkylation include any porous materials that may have come into contact with the hydrofluoric acid.

29.2.9 ISOMERIZATION

Isomerization (Figure 29.15) is used to alter the arrangement of a molecule without adding or removing anything from the original molecule. Typically, paraffins (butane or pentane from the crude distillation unit) are converted to *iso*-paraffins having a much higher octane. Isomerization reactions take place at temperatures in the range of 95°C to 205°C (200°F to

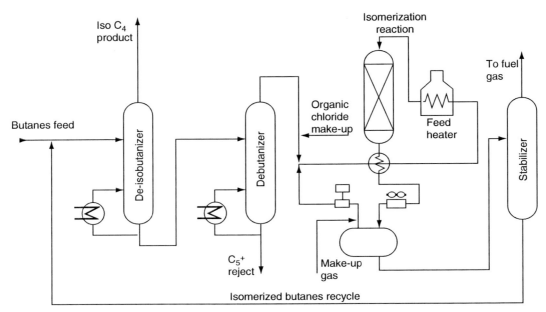

FIGURE 29.15 A butane isomerization unit. (From OSHA Technical Manual, Section IV, Chapter 2, Petroleum Refining Processes.)

400°F) in the presence of a catalyst that usually consists of platinum on a base material. Two types of catalysts are currently in use. One requires the continuous addition of small amounts of organic chlorides that are converted to hydrogen chloride in the reactor. In such a reactor, the feed must be free of oxygen sources including water to avoid deactivation and corrosion problems. The other type of catalyst uses a molecular sieve base and does not require a dry and oxygen free feedstock. Both types of isomerization catalysts require an atmosphere of hydrogen to minimize coke deposits; however, the consumption of hydrogen is negligible. Catalysts typically need to be replaced about every two to three years or longer. Platinum is then recovered from the used catalyst off-site.

Light ends are stripped from the product stream leaving the reactor and are then sent to the sour gas treatment unit. Some isomerization units utilize caustic treating of the light fuel gas stream to neutralize any entrained hydrochloric acid. This will result in a calcium chloride (or other salts) waste stream. Air emissions may arise from the process heater, vents, and fugitive emissions. Wastewater streams include caustic wash and sour water.

Isomerization processes produce sour water and caustic wastewater. The ether manufacturing process utilizes a water wash to extract methanol or ethanol from the reactor effluent stream. After the alcohol is separated this water is recycled back to the system and is not released. In those cases where chloride catalyst activation agents are added, a caustic wash is used to neutralize any entrained hydrogen chloride. This process generates a caustic wash water that must be treated before being released. This process also produces a calcium chloride neutralization sludge that must be disposed of off-site.

29.2.10 POLYMERIZATION

Polymerization is occasionally used to convert propylene and butene to high octane gasoline blending components. The process is similar to alkylation in its feed and products, but is often used as a less expensive alternative to alkylation.

The reactions typically take place under high pressure in the presence of a phosphoric acid catalyst. The feed must be free of sulfur, which poisons the catalyst; basic materials, which neutralize the catalyst; and oxygen, which affects the reactions. The propylene and butene feedstock is washed first with caustic to remove mercaptans (molecules containing sulfur), then with an amine solution to remove hydrogen sulfide, then with water to remove caustics and amines, and finally dried by passing through a silica gel or molecular sieve dryer. Air emissions of sulfur dioxide may arise during the caustic washing operation. Spent catalyst, which typically is not regenerated, is occasionally disposed as a solid waste. Wastewater streams will contain caustic wash and sour water with amines and mercaptans.

29.2.11 Deasphalting

Propane deasphalting (Figure 29.16) produces lubricating oil base stocks by extracting asphaltenes and resins from vacuum distillation residua. Propane is the usual solvent of choice due to its unique solvent properties. At lower temperatures (38°C to 60°C, 100°F to 140°F), paraffins are very soluble in propane and at higher temperatures (about 93°C, 200°F) hydrocarbons are almost insoluble in propane. The propane deasphalting process is similar to solvent extraction in that a packed or baffled extraction tower or rotating disc contactor is used to mix the oil feedstocks with the solvent. In the tower method, four to eight volumes of propane are fed to the bottom of the tower for every volume of feed flowing down from the top of the tower. The oil, which is more soluble in the propane, dissolves and flows to the top. The higher molecular weight polar asphalt constituents flow to the bottom of the tower where they are removed in a propane mix. Propane is recovered from the two streams through two-stage flash systems followed by steam stripping in which propane is condensed and removed

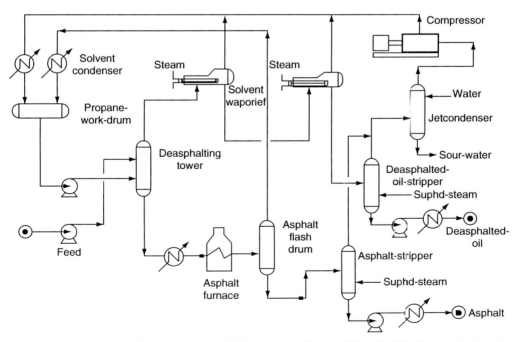

FIGURE 29.16 A deasphalting unit. (From OSHA Technical Manual, Section IV, Chapter 2, Petroleum Refining Processes.)

by cooling at high pressure in the first stage and at low pressure in the second stage. The asphalt recovered can be blended with other asphalt, or heavy fuel oil, or can be used as feed to the coker. The propane recovery stage results in propane contaminated water that typically is sent to the wastewater treatment plant.

Air emissions may arise from fugitive propane emissions and process vents. These include heater stack gas (carbon monoxide, sulfur oxides, nitrogen oxides, and particulate matter) as well as hydrocarbon emission such as fugitive propane, and fugitive solvents. Steam stripping wastewater (oil and solvents), solvent recovery wastewater (oil and propane) are also produced.

29.2.12 DEWAXING

Dewaxing of lubricating oil base stocks (Figure 29.17) is necessary to ensure that the oil will have the proper viscosity at lower ambient temperatures. Two types of dewaxing processes are used: selective hydrocracking and solvent dewaxing.

In selective hydrocracking, one or two zeolite catalysts are used to selectively crack the wax paraffins. Solvent dewaxing is more prevalent. In solvent dewaxing, the oil feed is diluted with solvent to lower the viscosity, chilled until the wax is crystallized, and then filtered to remove the wax. Solvents used for the process include propane and mixtures of methyl ethyl ketone (MEK) with methyl *iso*-butyl ketone (MIBK) or MEK with toluene. Solvent is recovered from the oil and wax through heating, two-stage flashing, followed by steam stripping. The solvent recovery stage results in solvent contaminated water that typically is sent to the wastewater treatment plant. The wax is either used as feed to the catalytic cracker or is de-oiled and sold as industrial wax.

Dewaxing processes also produce heater stack gas (carbon monoxide, sulfur oxides, nitrogen oxides, and particulate matter) as well as hydrocarbon emission such as fugitive propane, and fugitive solvents. Steam stripping wastewater (oil and solvents), solvent recovery wastewater

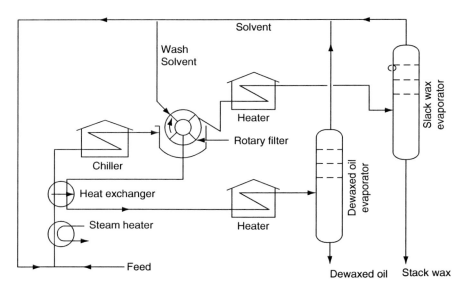

FIGURE 29.17 A solvent dewaxing unit. (From OSHA Technical Manual, Section IV, Chapter 2, Petroleum Refining Processes.)

(oil and propane) are also produced. The fugitive solvent emissions may be toxic (toluene, MEK, MIBK).

29.2.13 GAS PROCESSING

Gas processing (Katz, 1959; Maddox, 1974) either from field wells or within the refinery also, unfortunately, offers opportunities for pollution (EPA, 1970, 1974, 1995a; Mullins, 1975).

In the field, the gas from high-pressure wells is usually passed through field separators at the well to remove hydrocarbon condensate and water. Natural gasoline, butane, and propane are usually present in the gas, and gas-processing plants are required for the recovery of these liquefiable constituents (Figure 29.18). In addition, hydrogen sulfide must be removed before the gas can be utilized. The gas is usually sweetened by absorption of the hydrogen sulfide in an amine (olamine) solution (Chapter 25). Other methods, such as carbonate processes, solid bed absorbents, and physical absorption, are employed in the other sweetening plants.

The major emission sources in the natural gas-processing industry are compressor engines, acid gas wastes, fugitive emissions from leaking process equipment, and (if present) glycol dehydrator vent streams (EPA, 1995b). Regeneration of the glycol solutions used for dehydrating natural gas can release significant quantities of benzene, toluene, ethylbenzene, and xylene, as well as a wide range of less toxic organics.

Many chemical processes are available for sweetening natural gas (Chapter 25). At present, the amine (olamine) process (also known as the Girdler process) is the most widely used method for hydrogen sulfide removal (Figure 29.19). The recovered hydrogen sulfide gas stream may be utilized for the production of elemental sulfur or sulfuric acid.

Emissions will result from gas sweetening plants only if the acid waste gas from the amine process is flared or incinerated. Most often, the acid waste gas is used as a feedstock in nearby sulfur. If flaring or incineration is practiced, the major pollutant of concern is sulfur dioxide.

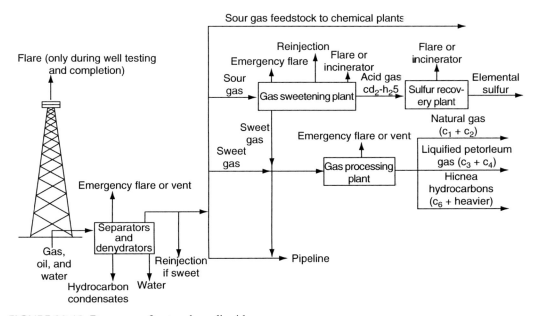

FIGURE 29.18 Recovery of natural gas liquids.

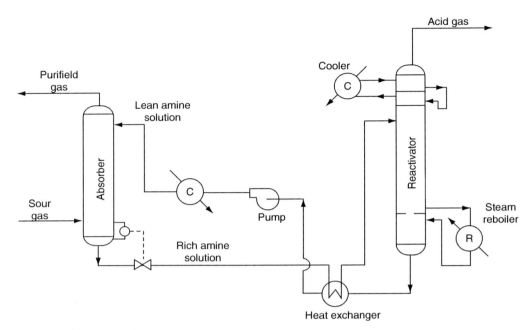

FIGURE 29.19 The Girdler process.

Most plants employ elevated smokeless flares or tail gas incinerators for complete combustion of all waste gas constituents, including near quantitative conversion of hydrogen sulfide to sulfur dioxide. Little particulate, smoke, or hydrocarbons result from these devices, and because gas temperatures do not usually exceed 650°C (1200°F), significant quantities of nitrogen oxides are not formed.

Some plants still use older, less-efficient waste gas flares that usually burn at temperatures lower than necessary for complete combustion; greater emissions of hydrocarbons and particulate, as well as hydrogen sulfide can occur.

29.3   TYPES OF WASTE

Pollution associated with petroleum refining typically includes volatile organic compounds (VOCs), carbon monoxide (CO), sulfur oxides (SO$_x$), nitrogen oxides (NO$_x$), particulates, ammonia (NH$_3$), hydrogen sulfide (H$_2$S), metals, spent acids, and numerous toxic organic compounds. Sulfur and metals result from the impurities in crude oil. The other wastes represent losses of inputs and final product. These pollutants may be discharged as air emissions, wastewater, or solid waste. All of these wastes are treated. However, air emissions are more difficult to capture than waste water or solid waste. Thus, air emissions are the largest source of untreated wastes released to the environment.

Especially in petroleum and petroleum products, the alkanes in gasoline and some other petroleum products are CNS depressants. In fact, gasoline was once evaluated as an anesthetic agent. However, sudden deaths, possibly as a result of irregular heartbeats, have been attributed to those inhaling vapors of hydrocarbons such as those in gasoline.

Alkanes of various types of crude oils and various petroleum products were biodegraded faster than the unresolved fractions. Different types of crude oils and products biodegraded at different rates in the same environments. An oil product is a complex mixture of organic chemicals and contains within it less persistent and more persistent fractions. The range between these two extremes is greatest for crude oils. As the many different substances in petroleum have different physical and chemical properties, summarizing the fate of petroleum in general (or even a particular oil) is very difficult. Solubility–fate relationships must be considered.

The relative proportion of hazardous constituents present in petroleum is typically quite variable. Therefore, contamination will vary from one site to another. In addition, the farther one progresses from lighter toward heavier constituents (the general progression from lower molecular weight to higher molecular weight constituents) the greater the percentage of polynuclear aromatic hydrocarbons (PNAs) and other semi-volatile constituents or nonvolatile constituents (many of which are not so immediately toxic as the volatiles but which can result in long-term or chronic impacts). These higher molecular weight constituents thus need to be analyzed for the semi-volatile compounds that typically pose the greatest long-term risk.

In addition to large oil spills, petroleum hydrocarbons are released into the aquatic environments from natural seeps as well as nonpoint source urban runoffs. Acute impacts from massive one-time spills are obvious and substantial. The impacts from small spills and chronic releases are the subject of much speculation and continued research. Clearly, these inputs of petroleum hydrocarbons have the potential for significant environmental impacts, but the effects of chronic low-level discharges can be minimized by the net assimilative capacities of many ecosystems, resulting in little detectable environmental harm.

Short-term (acute) hazards of lighter, more volatile and water soluble aromatic compounds (such as benzenes, toluene, and xylenes) include potential acute toxicity to aquatic life in the water column (especially in relatively confined areas) as well as potential inhalation hazards. However, the compounds that pass through the water column often tend to do so in small concentrations and for short periods of time, and fish and other pelagic or generally mobile species can often swim away to avoid impacts from spilled oil in open waters. Most fish are mobile and it is not known whether or not they can sense, and thus avoid, toxic concentrations of oil.

However, there are some potential effects of spilled oil on fish. The impacts to fish are primarily to the eggs and larvae, with limited effects on the adults. The sensitivity varies by species; pink salmon fry are affected by exposure to water-soluble fractions of crude oil, while pink salmon eggs are very tolerant to benzene and water-soluble petroleum. The general effects are difficult to assess and quantitatively document due to the seasonal and natural variability of the species. Fish rapidly metabolize aromatic hydrocarbons due to their enzyme system.

Long-term (chronic) potential hazards of lighter, more volatile and water soluble aromatic compounds include contamination of groundwater. Chronic effects of benzene, toluene, and xylene include changes in the liver and harmful effects on the kidneys, heart, lungs, and nervous system.

At the initial stages of a release, when the benzene-derived compounds are present at their highest concentrations, acute toxic effects are more common than later. These noncarcinogenic effects include subtle changes in detoxifying enzymes and liver damage. Generally, the relative aquatic acute toxicity of petroleum will be the result of the fractional toxicities of the different hydrocarbons present in the aqueous phase. Tests indicate that naphthalene-derived chemicals have a similar effect.

Except for short-term hazards from concentrated spills, BTEX compounds (benzene, toluene, ethyl benzene, and xylenes) have been more frequently associated with risk to humans than with risk to nonhuman species such as fish and wildlife. This is partly because plants, fish, and birds take up only very small amounts and because this volatile compound tends to evaporate into the atmosphere rather than persist in surface waters or soils. However, volatiles such as this compound can pose a drinking water hazard when they accumulate in groundwater.

Petroleum is naturally weathered according to its physical and chemical properties, but during this process living species within the local environment may be affected via one or more routes of exposure, including ingestion, inhalation, dermal contact, and, to a much lesser extent, bioconcentration through the food chain. Aromatic compounds of concern include alkylbenzenes, toluene, naphthalenes, and PNAs. Moreover, both atmospheric and hydrospheric impacts must be assessed when considering toxic implications from a petroleum release containing significant quantities of these single-ring aromatic compounds.

29.3.1 GASES AND LOWER BOILING CONSTITUENTS

Air emissions include point and nonpoint sources. Point sources are emissions that exit stacks and flares and, thus, can be monitored and treated. Nonpoint sources are "fugitive emissions" that are difficult to locate and capture. Fugitive emissions occur throughout refineries and arise from the thousands of valves, pumps, tanks, pressure relief valves, and flanges. While individual leaks are typically small, the sum of all fugitive leaks at a refinery can be one of its largest emission sources.

The numerous process heaters used in refineries to heat process streams or to generate steam (boilers) for heating or steam stripping, can be potential sources of SO_x, NO_x, CO, particulates and hydrocarbons emissions. When operating properly and when burning cleaner fuels such as refinery fuel gas, fuel oil or natural gas, these emissions are relatively low. If, however, combustion is not complete, or heaters are fired with refinery fuel pitch or residuals, emissions can be significant.

The majority of gas streams exiting each refinery process contain varying amounts of refinery fuel gas, hydrogen sulfide, and ammonia. These streams are collected and sent to the gas treatment and sulfur recovery units to recover the refinery fuel gas and sulfur. Emissions from the sulfur recovery unit typically contain some H_2S, SO_x, and NO_x. Other emissions sources from refinery processes arise from periodic regeneration of catalysts. These processes generate streams that may contain relatively high levels of carbon monoxide, particulates, and VOCs. Before being discharged to the atmosphere, such off-gas streams may be treated first through a carbon monoxide boiler to burn carbon monoxide and any VOCs, and then through an electrostatic precipitator or cyclone separator to remove particulates.

Sulfur is removed from a number of refinery process off-gas streams (sour gas) in order to meet the SO_x emissions limits of the CAA and to recover saleable elemental sulfur. Process off-gas streams, or sour gas, from the coker, catalytic cracking unit, hydrotreating units, and hydroprocessing units can contain high concentrations of hydrogen sulfide mixed with light refinery fuel gases. Before elemental sulfur can be recovered, the fuel gases (primarily methane and ethane) need to be separated from the hydrogen sulfide. This is typically accomplished by dissolving the hydrogen sulfide in a chemical solvent. Solvents most commonly used are amines, such as diethanolamine (DEA). Dry adsorbents such as molecular sieves, activated carbon, iron sponge and zinc oxide are also used. In the amine solvent processes, DEA solution or another amine solvent is pumped to an absorption tower where the gases are contacted and hydrogen sulfide is dissolved in the solution. The fuel gases are removed for use as fuel in process furnaces in other refinery operations. The amine-hydrogen sulfide solution is then heated and steam stripped to remove the hydrogen sulfide gas.

Current methods for removing sulfur from the hydrogen sulfide gas streams are typically a combination of two processes: the Claus Process followed by the Beaven Process, SCOT

Process, or the Wellman-Land Process. The Claus process consists of partial combustion of the hydrogen sulfide-rich gas stream (with one-third the stoichiometric quantity of air) and then reacting the resulting sulfur dioxide and unburned hydrogen sulfide in the presence of a bauxite catalyst to produce elemental sulfur.

As the Claus process by itself removes only about 90% of the hydrogen sulfide in the gas stream, the Beaven, SCOT, or Wellman-Lord processes are often used to further recover sulfur. In the Beaven process, the hydrogen sulfide in the relatively low concentration gas stream from the Claus process can be almost completely removed by absorption in a quinone solution. The dissolved hydrogen sulfide is oxidized to form a mixture of elemental sulfur and hydro quinone. The solution is injected with air or oxygen to oxidize the hydroquinone back to quinone. The solution is then filtered or centrifuged to remove the sulfur and the quinone is then reused. The Beaven process is also effective in removing small amounts of sulfur dioxide, carbonyl sulfide, and carbon disulfide that are not affected by the Claus process. These compounds are first converted to hydrogen sulfide at elevated temperatures in a cobalt molybdate catalyst prior to being fed to the Beavon unit. Air emissions from sulfur recovery units will consist of hydrogen sulfide, SO_x and NO_x in the process tail gas as well as fugitive emissions and releases from vents.

The SCOT process is also widely used for removing sulfur from the Claus tail gas. The sulfur compounds in the Claus tail gas are converted to hydrogen sulfide by heating and passing it through a cobalt–molybdenum catalyst with the addition of a reducing gas. The gas is then cooled and contacted with a solution of di-*iso*-propanolamine (DIPA) which removes all but trace amounts of hydrogen sulfide. The sulfide-rich DIPA is sent to a stripper where hydrogen sulfide gas is removed and sent to the Claus plant and the cleaned DIPA is returned to the absorption column.

Most refinery process units and equipment are sent to a collection unit called the blowdown system. Blowdown systems provide for the safe handling and disposal of liquid and gases that are either automatically vented from the process units through pressure relief valves, or that are manually drawn from units. Recirculated process streams and cooling water streams are often manually purged to prevent the continued build up of contaminants in the stream. Part or all of the contents of equipment can also be purged to the blowdown system prior to shutdown before normal or emergency shutdowns. Blowdown systems utilize a series of flash drums and condensers to separate the blowdown into its vapor and liquid components. The liquid is typically composed of mixtures of water and hydrocarbons containing sulfides, ammonia, and other contaminants, which are sent to the wastewater treatment plant. The gaseous component typically contains hydrocarbons, hydrogen sulfide, ammonia, mercaptans, solvents, and other constituents, and is either discharged directly to the atmosphere or is combusted in a flare. The major air emissions from blowdown systems are hydrocarbons in the case of direct discharge to the atmosphere and sulfur oxides when flared.

Many of the gaseous and liquid constituents of the lower boiling fractions of petroleum and also in petroleum products fall into the class of chemicals that have one or more of the following characteristics that are considered to be hazardous by the Environmental Protection agency in terms of the following properties (1) ignitability–flammability, (2) corrosivity, (3) reactivity, and (4) hazardous.

An ignitable liquid is a liquid that has a flash point of less than 60°C (140°F). Examples are: benzene, hexane, heptane, benzene, pentane, petroleum ether (low boiling), toluene, and xylene(s). An aqueous solution that has a pH of less than or equal to 2, or greater than or equal to 12.5 is considered corrosive. Most petroleum constituents and petroleum products are not corrosive but many of the chemicals used in refineries are corrosive. Corrosive materials also include substances such as sodium hydroxide and some other acids or bases. Chemicals that react violently with air or water are considered reactive. Examples are sodium metal, potassium metal, phosphorus, and the like. Reactive materials also include strong oxidizers such as perchloric acid ($HClO_4$), and chemicals capable of detonation when subjected to an initiating source, such as solid, dry <10% H_2O picric acid, benzoyl peroxide or sodium borohydride ($NaBH_4$). Solutions of certain cyanide or sulfides that could generate toxic gases are also classified as reactive. Hazardous chemicals have toxic, carcinogenic, mutagenic or teratogenic effects on humans or other life forms and are designated either as Acutely Hazardous Waste or Toxic Waste by the Environmental Protection Agency. Substances containing any of the toxic constituents so listed are to be considered hazardous unless, after considering the following factors it can reasonably be concluded that the chemical (waste) is not capable of posing a substantial present or potential hazard to public health or the environment when improperly treated, stored, transported or disposed of, or otherwise managed.

The issues to be held in consideration are (1) the nature of the toxicity presented by the constituent, (2) the concentration of the constituent in the waste, (3) the potential of the constituent or any toxic degradation product of the constituent to migrate from the waste into the environment under the types of improper management considered in item (9) below, (4) the persistence of the constituent or any toxic degradation product of the constituent, (5) the potential for the constituent or any toxic degradation product of the constituent to degrade into nonharmful constituents and the rate of degradation, (6) the degree to which the constituent or any degradation product of the constituent accumulates in an ecosystem, (7) the plausible types of improper management to which the waste could be subjected, (8) the quantities of the waste generated at individual generation sites or on a regional or national basis, (9) the nature and severity of the public health threat and environmental damage that has occurred as a result of the improper management of wastes containing the constituent, and (10) actions taken by other governmental agencies or regulatory programs based on the health or environmental hazard posed by the waste or waste constituent.

Such lists of chemicals are not always complete and omission of a chemical from this list does not mean it is without toxic properties or any other hazard.

29.3.2 HIGHER BOILING CONSTITUENTS

Naphthalene and its homologs are less acutely toxic than benzene but are more prevalent for a longer period during oil spills. The toxicity of different crude oils and refined oils depends on not only the total concentration of hydrocarbons but also the hydrocarbon composition in the water-soluble fraction (WSF) of petroleum, water solubility, concentrations of individual components, and toxicity of the components. The WSF prepared from different oils will vary in these parameters. WSF of refined oils (e.g., No. 2 fuel oil and Bunker C oil) are more toxic than water-soluble fraction of crude oil to several species of fish (killifish and salmon). Compounds with either more rings or methyl substitutions are more toxic than less substituted compounds, but tend to be less water soluble and thus less plentiful in the water-soluble fraction.

Among the PNAs, the toxicity of petroleum is a function of its di- and tri-aromatic hydrocarbon content. Like the single aromatic ring variations, including benzene, toluene, and the xylenes, all are relatively volatile compounds with varying degrees of water solubility.

There are indications that pure naphthalene (a constituent of moth balls that are, by definition, toxic to moths) and alkylnaphthalenes are from three-to-ten times more toxic to test animals than are benzene and alkylbenzenes. In addition, and because of the low water solubility of tricyclic and polycyclic (polynuclear) aromatic hydrocarbons (i.e., those aromatic hydrocarbons heavier than naphthalene), these compounds are generally present at very low concentrations in the water-soluble fraction of oil. Therefore, the results of this study and others conclude that the soluble aromatics of crude oil (such as benzene, toluene, ethylbenzene, xylenes, and naphthalenes) produce the majority of its toxic effects in the environment.

Once the acutely toxic lighter compounds have left the aquatic environment through volatilization or degradation, the main concern is chronic effects from heavier and more alkylated PNAs.

Bird species with water habitats are the species most commonly affected by oil spills and releases. Oil itself breaks down the protective waxes and oils in the feathers and fur of birds and animals and disrupts the fine strand structure of the feathers resulting in a loss of heat retention and buoyancy and possible hypothermia and death. Oiled birds often ingest petroleum while attempting to remove the petroleum from their feathers. The effects of ingested petroleum include anemia, pneumonia, kidney, and liver damage, decreased growth, altered blood chemistry, and decreased egg production and viability. Ingestion of the oil can also kill animals by interfering with their ability to digest food. Chicks may be exposed to petroleum by ingesting food regurgitated by impacted adults.

The dynamics of the oil-in-water dispersion (OWD) are complex and have relevance related to potential toxicity or hazard. In comparing the toxicities to marine animals of OWD prepared from different oils, not only the amount of oil added but also the concentrations of oil in the aqueous phase and the composition and dispersion-forming characteristics of the parent oil must be taken into consideration. In comparing the potential impacts of spills of different oils on the marine biotic community, the amount of oil per unit water volume required to cause mortality is of greater importance than any other aspect of the crude oil behavior.

Several compounds in petroleum products are carcinogenic. The larger and higher molecular weight aromatic structures (with four to five aromatic rings), which are the more persistent in the environment, have the potential for chronic toxicological effects. As these compounds are nonvolatile and are relatively insoluble in water, their main routes of exposure are through ingestion and epidermal contact. Some of the compounds in this classification are considered possible human carcinogens; these include benzo(a and e)pyrene, benzo(a)anthracene, benzo (b,j, and k)fluorene, benzo(ghi)perylene, chrysene, dibenzo(ah)anthracene, and pyrene.

Mixtures of PNAs are often carcinogenic and possibly phototoxic. One way to approach site-specific risk assessments would be to collect the complex mixture of PNAs and other lipophilic contaminants in a semi-permeable membrane device (SPMD, also known as a fat bag), then test the mixture for carcinogenicity, toxicity, and phototoxicity.

The solubility of hydrocarbon components in petroleum products is an important property when assessing toxicity. The water solubility of a substance determines the routes of exposure that are possible. Solubility is approximately inversely proportional to molecular weight; lighter hydrocarbons are more soluble in water than higher molecular weight compounds. Lower molecular weight hydrocarbons (C4 to C8, including the aromatic compounds) are relatively soluble, up to about 2000 ppm, while the higher molecular weight hydrocarbons are nearly insoluble. Usually, the most soluble components are also the most toxic.

Finally, the toxicity of crude oil may be affected by factors such as "weathering" time, or the addition of oil dispersants. **Weathered crude oil** and fresh crude oil may have different toxicities, depending on oil type and weathering time.

29.3.3 WASTEWATER

Wastewaters from petroleum refining consist of cooling water, process water, storm water, and sanitary sewage water. A large portion of water used in petroleum refining is used for cooling and most cooling water is recycled. Cooling water typically does not come into direct contact with process oil streams and therefore contains less contaminants than process wastewater. However, it may contain some oil contamination due to leaks in the process equipment. Water used in processing operations accounts for a significant portion of the total wastewater. Process wastewater arises from desalting crude oil, steam stripping operations, pump gland cooling, product fractionator reflux drum drains and boiler blowdown. Because process water often comes into direct contact with oil, it is usually highly contaminated. Storm water (i.e., surface water runoff) is intermittent and will contain constituents from spills to the surface, leaks in equipment and any materials that may have collected in drains. Runoff surface water also includes water coming from crude and product storage tank roof drains.

The issues facing the refining industry includes chemicals in waste process waters. However, efforts by the industry are being continued to eliminate any water contamination that may occur, whether it be from inadvertent leakage of petroleum or petroleum products or leakage of contaminated water from one or more processes. In addition to monitoring organics in the water, metals concentration must be continually monitored as heavy metals tend to concentrate in the body tissues of fish and animals and increase in concentration as they go up the food chain. General sewage problems face every municipal sewage treatment facility regardless of size.

Primary treatment (solid settling and removal) is required and secondary treatment (use of bacteria and aeration to enhance organic degradation) is becoming more routine, while tertiary treatment (filtration through activated carbon, applications of ozone, and chlorination) has been, or is being, implemented by all refineries.

After primary treatment, the wastewater can be discharged to a publicly owned treatment works (POTW) or undergo secondary treatment before being discharged directly to surface waters under a National Pollution Discharge Elimination System (NPDES) permit. In secondary treatment, dissolved oil and other organic pollutants may be consumed biologically by microorganisms. Biological treatment may require the addition of oxygen through a number of different techniques, including activated sludge units, trickling filters, and rotating biological contactors. Secondary treatment generates bio-mass waste that is typically treated anaerobically and then dewatered.

Some refineries employ an additional stage of wastewater treatment called polishing to meet discharge limits. The polishing step can involve the use of activated carbon, anthracite coal, or sand to filter out any remaining impurities, such as biomass, silt, trace metals and other inorganic chemicals, as well as any remaining organic chemicals.

Certain refinery wastewater streams are treated separately, prior to the wastewater treatment plant, to remove contaminants that would not easily be treated after mixing with other wastewater. One such waste stream is the sour water drained from distillation reflux drums. Sour water contains dissolved hydrogen sulfide and other organic sulfur compounds and ammonia that are stripped in a tower with gas or steam before being discharged to the wastewater treatment plant.

Wastewater treatment plants are a significant source of refinery air emissions and solid wastes. Air releases arise from fugitive emissions from the numerous tanks, ponds and sewer system drains. Solid wastes are generated in the form of sludge from a number of the treatment units.

If liquid hydrocarbons are released to groundwater and surface waters, migration off-site can occur resulting in continuous seeps to surface waters. While the actual volume of hydrocarbons released in such a manner are relatively small, there is the potential to contaminate large volumes of groundwater and surface water possibly posing a substantial risk to human health and the environment.

Wastewater pretreaters that discharge water into sewer systems have new requirements. Pollutant standards for sewage sludge have been set. Toxics in the water must be identified and plans must be developed to alleviate any problems. In addition, regulators have established, and continue to establish, water-quality standards for priority toxic pollutants.

29.3.4 SOLID WASTE

Solid wastes are generated from many of the refining processes, petroleum-handling operations, as well as wastewater treatment. Both hazardous and nonhazardous wastes are generated, treated and disposed. Refinery wastes are typically in the form of sludge (including sludge from wastewater treatment), spent process catalysts, filter clay, and incinerator ash. Treatment of these wastes includes incineration, land treating off-site, land filling onsite, land filling off-site, chemical fixation, neutralization, and other treatment methods.

A significant portion of the nonpetroleum product outputs of refineries is transported off-site and sold as by-products. These outputs include sulfur, acetic acid, phosphoric acid, and recovered metals. Metals from catalysts and from the crude oil that have deposited on the catalyst during the production often are recovered by third party recovery facilities

Storage tanks are used throughout the refining process to store crude oil and intermediate process feeds for cooling and further processing. Finished petroleum products are also kept in storage tanks before transport off-site. Storage tank bottoms are mixtures of iron rust from corrosion, sand, water, and emulsified oil and wax, which accumulate at the bottom of tanks. Liquid tank bottoms (primarily water and oil emulsions) are periodically drawn off to prevent their continued build up. Tank bottom liquids and sludge are also removed during periodic cleaning of tanks for inspection. Tank bottoms may contain amounts of tetraethyl or tetramethyl lead (although this is increasingly rare due to the phase-out of leaded products), other metals, and phenols. Solids generated from leaded gasoline storage tank bottoms are listed as a RCRA hazardous waste.

29.4 WASTE TOXICITY

With few exceptions, the constituents of petroleum, petroleum products, and the various emissions are hazardous to the health. There will always be exceptions that will be cited in opposition to such a statement, the most common exception being the liquid paraffin that is used medicinally to lubricate the alimentary tract. The use of such medication is common among miners who breathe and swallow coal dust every day during their work shifts.

Another approach is to consider petroleum constituents in terms of transportable materials, the character of which is determined by several chemical and physical properties (i.e., solubility, vapor pressure, and propensity to bind with soil and organic particles). These properties are the basis of measures of leachability and volatility of individual hydrocarbons. Thus, petroleum transport fractions can be considered by equivalent carbon number to be grouped into 13 different fractions. The analytical fractions are then set to match these transport fractions, using specific n-alkanes to mark the analytical results for aliphatic compounds and selected aromatic compounds to delineate hydrocarbons containing benzene rings.

Although chemicals grouped by transport fraction generally have similar toxicological properties, this is not always the case. For example, benzene is a carcinogen but many alkyl-substituted benzenes do not fall under this classification. However, it is more appropriate to group benzene with compounds that have similar environmental transport properties than to group it with other carcinogens such as benzo(a)pyrene that have very different environmental transport properties.

Nevertheless, consultation of any reference work that lists the properties of chemicals will show the properties and hazardous nature of the types of chemicals that are found in petroleum. In addition, petroleum is used to make petroleum products, which can contaminate the environment.

The range of chemicals in petroleum and petroleum products is so vast that summarizing the properties or the toxicity or general hazard of petroleum in general or even for a specific crude oil is a difficult task. However, petroleum and some petroleum products, because of the hydrocarbon content, are at least theoretically biodegradable but large-scale spills can overwhelm the ability of the ecosystem to break the oil down. The toxicological implications from petroleum occur primarily from exposure to or biological metabolism of aromatic structures. These implications change as an oil spill ages or is weathered.

29.5 REFINERY OUTLOOK

29.5.1 Hazardous Waste Regulations

Petroleum refinery operators face more stringent regulation of the treatment, storage, and disposal of hazardous wastes. Under recent regulations, a larger number of compounds have been, and are being, studied. Long-time methods of disposal, such as land farming of refinery waste, are being phased out. As a result, many refineries are changing their waste management practices. An informal survey of nine refineries showed that eight were planning to close land treatment units because of the uncertainty of continuing the practice.

New regulations are becoming even more stringent, and they encompass a broader range of chemical constituents and processes. Continued pressure from the U.S. Congress has led to more explicit laws allowing little leeway for industry, the U.S. Environmental Protection Agency (Environmental Protection Agency), or state agencies.

A summary of the current regulations and what they mean to refiners is given in the following.

29.5.2 Regulatory Background

The hazardous waste regulatory program, as we know it today, began with the Resource Conservation and Recovery Act (RCRA) in 1976. The Used Oil Recycling Act of 1980 and Hazardous and Solid Waste Amendments of 1984 (HSWA) were the major amendments to the original law.

RCRA provides for the tracking of hazardous waste from the time it is generated, through storage and transportation, to the treatment or disposal sites. RCRA and its amendments are aimed at preventing the disposal problems that lead to a need for the Comprehensive Environmental Response Compensation and Liability Act (CERCLA), or Superfund, as it is known. Subtitle C of the original RCRA lists the requirements for the management of hazardous waste. This includes the Environmental Protection Agency criteria for identifying hazardous waste, and the standards for generators, transporters, and companies that treat, store, or dispose of the waste. The RCRA regulations also provide standards for design and operation of such facilities.

29.5.3 Requirements

The first step to be taken by a generator of waste is to determine whether that waste is hazardous. Waste may be hazardous by being listed in the regulations, or by meeting any of the four characteristics: ignitability, corrosivity, reactivity, and extraction procedure (EP) toxicity.

Generally: (1) if the material has a flash point less than 140°F it is considered ignitable; (2) if the waste has a pH less than 2.0 or above 12.5, it is considered corrosive. It may also be considered corrosive if it corrodes stainless steel at a certain rate; (3) a waste is considered reactive if it is unstable and produces toxic materials, or it is a cyanide or sulfide-bearing waste that generates toxic gases or fumes; (4) a waste which is analyzed for EP toxicity and fails is also considered a hazardous waste. This procedure subjects a sample of the waste to an acidic environment. After an appropriate time has elapsed, the liquid portion of the sample (or the sample itself if the waste is liquid) is analyzed for certain metals and pesticides. Limits for allowable concentrations are given in the regulations. The specific analytical parameters and procedures for these tests are referred to in 40CRF 261.

The 1984 amendments also brought the owners and operators of underground storage tanks into the RCRA fold. This can have a significant effect on refineries that store product in underground tanks. In addition, petroleum products are also regulated by RCRA, Subtitle I.

29.6 MANAGEMENT OF REFINERY WASTE

The refining industry, as well as other industries, will increasingly feel the effects of the land bans on their hazardous waste management practices. Current practices of land disposal must change along with management attitudes for waste handling. The way refineries handle their waste in the future depends largely on the ever-changing regulations. Waste management is the focus and reuse or recycle options must be explored to maintain a balanced waste management program. This requires that a waste be recognized as either nonhazardous or hazardous.

However, before a refinery can determine if its waste is hazardous, it must first determine that the waste is indeed a solid waste. In 40 CFR 261.2, the definition of solid waste can be found. If a waste material is considered a solids waste, it may be a hazardous waste in accordance with 40 CFR 261.3. There are two ways to determine whether a waste is hazardous. These are to see if the waste is listed in the regulations or to test the waste to see if it exhibits one of the characteristics (40 CFR 261).

There are four lists of hazardous wastes in the regulations. These are wastes from non-specific sources (F list), wastes from specific sources (K list), acutely toxic wastes (P list), and toxic wastes (U list). And these are the four characteristics mentioned before: ignitability, corrosivity, reactivity, and extraction procedure toxicity. Certain waste materials are excluded from regulation under RCRA. The various definitions and situations that allow waste to be exempted can be confusing and difficult to interpret. One such case is the interpretation of the mixture and derived-from rules. According to the mixture rule, mixtures of solid waste and listed hazardous wastes are, by definition, considered hazardous. Likewise, the derived-from rule defines solid waste resulting from the management of hazardous waste to be hazardous (40 CFR 261.3a and 40 CFR. 261.1c).

There are five specific listed hazardous wastes (K list) generated in refineries. These are K048–K052. Additional listed wastes, those from nonspecific sources (F list) and those from the commercial chemical product lists (P and U lists), may also be generated at refineries. Because of the mixture and derived-from rules, special care must be taken to ensure that hazardous wastes do not contaminate nonhazardous waste. Under the mixture rule, adding one drop of hazardous waste in a container of nonhazardous materials makes the contents of the entire container a hazardous waste.

As an example of the problems such mixing can cause, consider the case with API separator sludge that is a listed hazardous waste (K051). The wastewater from a properly operating API separator is not hazardous unless it exhibits one of the characteristics of a hazardous waste. That is, the derived-from rule does not apply to the wastewater. However, if the API separator is not functioning properly, solids carry–over in the wastewater can occur. In this case, the wastewater contains a listed hazardous waste, the solids from the API sludge, and the wastewater would be considered a hazardous waste because it is a mixture of a nonhazardous waste and a hazardous waste.

This wastewater is often further cleaned by other treatment systems (filters, impoundments, etc.). The solids separating in these systems continue to be API separator sludge, a listed hazardous waste. Therefore, all downstream wastewater treatment systems are receiving, and treating, a hazardous waste and are considered hazardous waste management units subject to regulation.

Oily wastewater is often treated or stored in unlined wastewater treatment ponds in refineries. These wastes appear to be similar to API separator waste.

REFERENCES

EPA 1970. *Control Techniques for Hydrocarbon and Organic Solvent Emissions from Stationary Sources.* Report No. AP-68. United States Environmental Protection Agency, Research Triangle Park, NC. March.

EPA 1974. *Sulfur Compound Emissions of the Petroleum Production Industry.* Report No. EPA-650/2-75-030. United States Environmental Protection Agency, Cincinnati, OH.

EPA 1995a. *Profile of the Petroleum Refining Industry.* Report No. EPA/310-R-95-013. United States Environmental Protection Agency, Washington, DC. September.

EPA 1995b. *Protocol for Equipment Leak Emission Estimates.* Report No. EPA-453/R-95-017. United States Environmental Protection Agency, Washington, DC. November.

Irwin, R.J. 1997. Petroleum. *Environmental Contaminants Encyclopedia.* National Park Service, Water Resources Divisions, Water Operations Branch, Fort Collins, CO.

Katz, D.K. 1959. *Handbook of Natural Gas Engineering.* McGraw-Hill, New York.

Maddox, R.R. 1974. *Gas and Liquid Sweetening*, 2nd edn. Campbell Petroleum Series, Norman, OK.

Meyers, R.A. 1997. *Handbook of Petroleum Refiing Processes*, 2nd edn. McGraw-Hill, New York.

Mullins, B.J. 1975. *Atmospheric Emissions Survey of the Sour Gas Processing Industry.* Report No. EPA-450/3-75-076. Environmental Protection Agency, Research Triangle Park, NC. October.

Speight, J.G. 1999. *The Chemistry and Technology of Petroleum*, 3rd edn. Marcel Dekker, New York.

Speight, J.G. and Ozum, B. 2002. *Petroleum Refining Processes.* Marcel Dekker, New York.

30 Environmental Analysis

30.1 INTRODUCTION

It is almost impossible to transport, store, and refine crude oil without spills and losses. It is difficult to prevent spills resulting from failure of or damage to on pipelines. It is also impossible to install control devices for controlling the ecological properties of water and the soil along the length of all pipelines. The soil suffers the most ecological damage in the damaged areas of pipelines. Crude oil spills from pipelines lead to irreversible changes in the soil properties. The most affected soil properties by crude oil losses from pipelines are filtration, physical, and mechanical properties. These properties in the soil are important for maintaining the ecological equilibrium in the damaged area.

Principal sources of releases to air from refineries include: (1) combustion plants, emitting sulfur dioxide, oxides of nitrogen, and particulate mater, (2) refining operations, emitting sulfur dioxide, oxides of nitrogen, carbon monoxide, particulate matter, volatile organic compounds, hydrogen sulfide, mercaptans, and other sulfurous compounds, (3) bulk storage operations and handling of volatile organic compounds (various hydrocarbons) (EPA, 1995; Irwin, 1997). In light of this, it is necessary to consider (1) regulatory requirements—air emission permits stipulating limits for specific pollutants, and possibly health and hygiene permit requirements, (2) requirement for monitoring program, and (3) requirements to upgrade pollution abatement equipment.

There is a potential for significant soil and groundwater contamination to have arisen at petroleum refineries. Such contamination consists of (1) petroleum hydrocarbons including lower boiling, very mobile fractions (paraffins, cycloparaffins, and volatile aromatics such as benzene, toluene, ethylbenzene, and xylenes) typically associated with gasoline and similar boiling range distillates, (2) middle distillate fractions (paraffins, cycloparaffins, and some polynuclear aromatics) associated with diesel, kerosene, and lower boiling fuel oil, which are also of significant mobility, (3) higher boiling distillates (long-chain paraffins, cycloparaffins, and polynuclear aromatics that are associated with lubricating oil and heavy fuel oil), (4) various organic compounds associated with petroleum hydrocarbons or produced during the refining process, for example, phenols, amines, amides, alcohols, organic acids, nitrogen and sulfur containing compounds, (5) other organic additives, for example, anti-freeze (glycols), alcohols, detergents, and various proprietary compounds, and (6) organic lead, associated with leaded gasoline and other heavy metals.

Petroleum products released into the environment undergo weathering processes with time. These processes include evaporation, leaching (transfer to the aqueous phase) through solution and entrainment (physical transport along with the aqueous phase), chemical oxidation, and microbial degradation. The rate of weathering is highly dependent on environmental conditions. For example, gasoline, a volatile product, will evaporate readily in a surface spill but gasoline released below 10 feet of clay topped with asphalt will tend to evaporate slowly (weathering processes may not be detectable for years).

An understanding of weathering processes is valuable to environmental test laboratories. Weathering changes product composition and may affect the test results, the ability to bioremediate, and the toxicity of the spilled product. Unfortunately, the database available on the composition of weathered products is limited.

However, biodegradation processes, which influence the presence and the analysis of petroleum hydrocarbon at a particular site, can be very complex. The extent of biodegradation is dependent on many factors including the type of microorganisms present, environmental conditions (e.g., temperature, oxygen levels, and moisture), and the predominant hydrocarbon types. In fact, the primary factor controlling the extent of biodegradation is the molecular composition of the petroleum contaminant. Multiple ring cycloalkanes are hard to degrade, whereas polynuclear aromatic hydrocarbons display varying degrees of degradation. Straight-chain alkanes biodegrade rapidly with branched alkanes and single saturated ring compounds degrade more slowly.

Once the sample preparation is complete, there are several approaches to the analysis of petroleum constituents in water and soil: (1) leachability or toxicity of the sample, (2) amounts of total petroleum hydrocarbons (TPH) in the sample, (3) petroleum group analysis, and (4) fractional analysis of the sample (Speight, 2005). These methods measure different petroleum constituents that might be present in petroleum-contaminated environmental media.

The methods that measure the concentration of TPH generate a single number that represents the combined concentration of all petroleum hydrocarbons in a sample that are measurable by the particular method. Therefore, the determination of the TPH in a sample is method dependent. On the other hand, methods that measure a petroleum group type concentration separate and quantify different categories of hydrocarbons (e.g., saturates, aromatics, and polars/resins) (Chapter 2) (Speight, 2001, 2002). The results of petroleum group type analyses can be useful for product identification because products such as, gasoline, diesel fuel, and fuel oil have characteristic levels of various petroleum groups. The methods that measure identifiable petroleum fractions can be used to indicate and quantify the changes that have occurred through weathering of the sample.

Although these methods measure different petroleum hydrocarbon categories, there are several basic steps that are common to the analytical processes for all methods, no matter the method type or the environmental matrix. In general, these steps are: (1) collection and preservation—requirements specific to environmental matrix and analytes of interest, (2) extraction so that separations of the analytes of interest from the sample matrix can be achieved, (3) concentration—enhances the ability to detect analytes of interest, (4) cleanup, dependent on the need to remove interfering compounds, and (5) measurement, or quantification, of the analytes (Dean, 1998). Each step affects the final result, and a basic understanding of the steps is vital for data interpretation.

30.2 PETROLEUM AND PETROLEUM PRODUCTS

The chemical composition of petroleum and petroleum products is complex (Chapter 7 and Chapter 26) and this factor alone makes it essential that the most appropriate analytical methods are selected from a comprehensive list of methods and techniques that are used for the analysis of environmental samples (Dean, 1998; Miller, 2000; Budde, 2001; Sunahara et al., 2002; Nelson, 2003; Smith and Cresser, 2003; Speight, 2005). Once a method is selected, it may not be the ultimate answer to solving the problem of identification and, hence, behavior (Patnaik, 2004).

The constituents of petroleum products are often easier to identify by understanding the physical properties of the product. Such knowledge renders the analytical plan easier to develop.

For example, automotive gasoline typically contains about almost two hundred (if not several hundred) hydrocarbon compounds. The relative concentrations of the compounds vary considerably depending on the source of crude oil, refinery process, and product specifications. Typical hydrocarbon chain lengths range from C4 through Cl2 with a general hydrocarbon distribution consisting of alkanes (4%–8%), alkenes (2%–5%), iso-alkanes (25%–40%), cycloalkanes (3%–7%), cycloalkenes (1%–4%), and aromatics (20%–50%). However, these proportions vary greatly.

Stoddard solvent is a petroleum distillate widely used as a dry cleaning solvent and as a general cleaner and degreaser. It may also be used as paint thinner, as a solvent in some types of photocopier toners, in some types of printing inks, and in some adhesives. Stoddard solvent is considered to be a form of mineral spirits, white spirits, and naphtha but not all forms of mineral spirits, white spirits, and naphtha are considered to be Stoddard solvent. Stoddard solvent consists of linear alkanes (30%–50%), branched alkanes (20%–40%), cycloalkanes (30%–40%), and aromatic hydrocarbons (10%–20%). The typical hydrocarbon chain ranges from C7 through C12 in length.

Jet fuel is a light petroleum distillate that is available in several forms suitable for use in various types of jet engines. The major jet fuels used by the military are JP-4, JP-5, JP-6, JP-7, and JP-8. Briefly, JP-4 is a wide-cut fuel developed for broad availability. JP-6 is a higher cut than JP-4 and is characterized by fewer impurities. JP-5 is specially blended kerosene, and JP-7 is high flash point special kerosene used in advanced supersonic aircraft. JP-8 is kerosene modeled on Jet A-1 fuel (used in civilian aircraft). From what data are available, typical hydrocarbon chain lengths characterizing JP-4 range from C4 to C16. Aviation fuels consist primarily of straight and branched alkanes and cycloalkanes. Aromatic hydrocarbons are limited to 20%–25% of the total mixture because they produce smoke when burnt. A maximum of 5% alkenes is specified for JP-4. The approximate distribution by chemical class is: straight chain alkanes (32%), branched alkanes (31%), cycloalkanes (16%), and aromatic hydrocarbons (21%).

No. 1 fuel oil is a petroleum distillate that is one of the most widely used of the fuel oil types. It is used in atomizing burners that spray fuel into a combustion chamber where the tiny droplets bum while in suspension. It is also used as a carrier for pesticides, as a weed killer, as a mold release agent in the ceramic and pottery industry, and in the cleaning industry. It is found in asphalt coatings, enamels, paints, thinners, and varnishes. No. 1 fuel oil is a light petroleum distillate (straight-run kerosene) consisting primarily of hydrocarbons in the range C9–C16. Fuel oil No. 1 is very similar in composition to diesel fuel; the primary difference is in the additives.

No. 2 fuel oil is a petroleum distillate that may be referred to as domestic or industrial. The domestic fuel oil is usually a lower boiling and a straight-run product. It is used primarily for home heating. Industrial distillate is a cracked product or a blend of both. It is used in smelting furnaces, ceramic kilns, and packaged boilers. No. 2 fuel oil is characterized by hydrocarbon chain lengths in the C11–C20 range. The composition consists of aliphatic hydrocarbons (straight chain alkanes and cycloalkanes) (64%), 1%–2% unsaturated hydrocarbons (alkenes) (1%–2%), and aromatic hydrocarbons (including alkylbenzenes and 2-ring, 3-ring aromatics) (35%) but contains only low amounts of the polycyclic aromatic hydrocarbons (<5%).

No. 6 fuel oil (also called **Bunker C oil** or **residual fuel oil**) is the residuum from crude oil after naphtha-gasoline, No. 1 fuel oil, and No. 2 fuel oil have been removed. No. 6 fuel oil can be blended directly with heavy fuel oil or made into asphalt. Residual fuel oil is more complex in composition and impurities than distillate fuels. Limited data are available on the composition of No. 6 fuel oil. Polycyclic aromatic hydrocarbons (including the alkylated derivatives) and metal-containing constituents are components of No. 6 fuel oil.

Mineral oils are often used as lubricating oils but also have medicinal and food uses. A major type of hydraulic fluid is the mineral oil class of hydraulic fluids. The mineral-based oils are produced from heavy-end crude oil distillates. Hydrocarbon numbers ranging from C15 to C50 occur in various types of mineral oils, with the heavier distillates having higher percentages of the higher carbon number compounds.

Crankcase oil (motor oil) may be either mineral-based or synthetic. The mineral-based oils are more widely used than the synthetic oils and may be used in automotive engines, railroad, and truck diesel engines, marine equipment, jet and other aircraft engines, and very small 2- and 4-stroke engines. The mineral-based oils contain hundreds to thousands of hydrocarbon compounds, including a substantial fraction of nitrogen- and sulfur-containing compounds. The hydrocarbons are mainly mixtures of straight and branched chain hydrocarbons (alkanes), cycloalkanes, and aromatic hydrocarbons. Polynuclear aromatic hydrocarbons (and the alkyl derivatives) and metal-containing constituents are components of motor oils and crankcase oils, with the used oils typically having higher concentrations than the new unused oils. Typical carbon number chain lengths range from Cl5 to C50.

30.3   LEACHABILITY AND TOXICITY

As a start and for regulatory and remediation purposes, a standard test is needed to measure the likelihood of toxic substances getting into the environment and causing harm to organisms. The test (required by the United States Environmental Protection Agency) is the toxicity characteristic leaching procedure (TCLP, EPA SW-846 Method 1311), designed to determine the mobility of both organic and inorganic contaminants present in liquid, solid, and multi-phase wastes.

The method was developed to estimate the mobility of specific inorganic and organic contaminates that are destined for disposal in municipal landfills. The extraction is performed using acetic as the extraction fluid. The pH of the acetic acid or sodium acetate buffer solution is maintained at 4.93. This sample or acetic acid mixture is subjected to rotary extraction, designed to accelerate years of material exposure in the shortest possible time. After extraction, the resulting liquid is subjected to analysis using a list of contaminants that includes metals, volatile organic compounds, semi-volatile organic compounds, pesticides, and herbicides.

The toxicity characteristic leaching procedure may be subject to misinterpretation if the compounds under investigation are not included in the methods development or the list of contaminants leading to the potential for technically invalid results. However, an alternate procedure, the synthetic precipitation leaching procedure (SPLP, EPA SW-846 Method 1312) may be appropriate. This procedure is applicable to materials where the leaching potential due to normal rainfall is to be determined. Instead of the leachate simulating acetic acid mixture, nitric and sulfuric acids are utilized in an effort to simulate the acid rains resulting from airborne nitric and sulfuric oxides.

30.4   TOTAL PETROLEUM HYDROCARBONS

TPH (Chapter 4) analyses are conducted to determine the total amount of hydrocarbon present in the environment (Weisman, 1998). There are a variety of methods for measurement of the TPH in a sample but analytical inconsistencies must be recognized because of the definition of TPH and the methods employed for analysis (Rhodes et al., 1994). Thus, in practice, the term *TPH* is defined by the analytical method as different methods often give different results because they are designed to extract and measure slightly different subsets of petroleum hydrocarbons.

The analysis for the TPH in a sample as a means of evaluating petroleum-contaminated sites is also an analytical method in common use. The data are used to establish target cleanup levels for soil or water and is a common approach implemented by regulatory agencies in the United States, and in many other countries.

The data obtained by the analysis have become key remediation criteria and is essential that the environmental analyst (and others who may use the data) be knowledgeable about the various analytical methods. It is also important to know that minor method deviations may be found from region to region. For example, in terms of nomenclature, itself a complex and often ill-defined area of petroleum science (Chapter 1), the analytical methods may refer to **total petroleum hydrocarbons** as mineral oil, hydrocarbon oil, extractable hydrocarbon, and oil and grease.

Thus, as often occurs in petroleum science, the definition of TPH depends on the analytical method used because the TPH measurement is the total concentration of the hydrocarbons extracted and measured by a particular method. The same sample analyzed by different methods may produce different values. For this reason, it is important to know exactly how each determination is made as interpretation of the results depends on understanding the capabilities and limitations of the method. If used indiscriminately, measurement of the TPH in a sample can be misleading, leading to an inaccurate assessment of risk.

There are several reasons why the data for TPH do not provide ideal information for investigating sites and establishing target cleanup criteria. For example, use of the term *TPH* suggests that the analytical method measures the combined concentration of all petroleum-derived hydrocarbons, thereby giving an accurate indication of site contamination. However this is not always the case. Furthermore, target cleanup levels based on TPH concentrations implicitly assume (1) the data are an accurate measurement of petroleum-derived hydrocarbon concentration, and (2) the data also indicate the level of risk associated with the contamination. These assumptions are not correct due to many factors including the non-specificity of some of the methods used and, the complex nature of petroleum hydrocarbons and their interaction with the environment over time.

One significant difficulty in measuring the concentration of TPH in a sample (for different petroleum products) is the fact that the boiling ranges and carbon number ranges of refined petroleum products often overlap. Refined petroleum products are primarily manufactured through various processes, including distillation processes that separate fractions from crude oil by their respective boiling ranges.

Manufacturing processes may also increase the yield of low molecular weight fractions, reduce the concentration of undesirable sulfur and nitrogen components, and incorporate performance-enhancing additives. Additionally, because it is impossible to identify all constituents of a petroleum product, their respective boiling ranges were often used to describe these constituents. And, because distillation, as practiced in the industry, is often not capable of producing sharp distinctions in boiling point cutoffs, there is overlap between distillate fractions. The boiling point range correlates to the carbon number as the higher the carbon number, the higher the boiling point. However, chemical structure also influences boiling point. For example, branched and aromatic compounds of the same carbon number differ in boiling point from their corresponding *n*-alkane analogs. For these reasons, boiling point actually defines an approximate carbon range.

Ambiguous terminology associated with the term *TPH* and the methods of analysis also present additional difficulty in interpreting results. Each method has its own designation. For example, there are terms such as *total recoverable petroleum hydrocarbons* (TRPH), *diesel range organics* (DRO), *gasoline range organics* (GRO), and *total petroleum hydrocarbons-gasoline* (TPH-G). However, to confuse the issue even further, a method name that cites a product (such as gasoline or diesel) only implies a carbon range. For example, TPH-G does not necessarily imply that gasoline is present but that hydrocarbons boiling in the gasoline range (whether or not they would actually be found in gasoline) are present in the sample. These abbreviations may imply different carbon ranges to different laboratories and, therefore, the methods are not optimized to identify product type. Even with improved, more detailed analytical methods, identification of aged products may prove difficult.

The reason for the availability of a large number of methods for the measurement of TPH centers on the compositional complexity of petroleum and petroleum products and, subsequently, there is no single suitable or adequate method for measuring all types of petroleum-derived contamination. For example, methods that are appropriate for samples contaminated by gasoline are not often suitable for the measurement of diesel fuel contamination in other samples.

Some methods measure more compounds than other methods because they employ more rigorous extraction techniques or more efficient solvents for the extraction procedure(s). Other methods are subject to interferences from naturally occurring materials such as animal and vegetable oils, peat moss, or humic material that may result in artificially high reported concentrations of the total petroleum hydrocarbons. Some methods use cleanup steps to minimize the effect of nonpetroleum hydrocarbons, with variable success. Ultimately, many of the methods are limited by the extraction efficiency and the detection limits of the instrumentation used for measurement.

Thus, the choice of a specific method should be based on compatibility with the particular type of hydrocarbon contamination to be measured and, furthermore, the choice may depend on local or regional regulatory requirements for the type of hydrocarbon contamination that is known, or suspected, to be present. In addition the risk at a specific site will change with time as contaminants evaporate, dissolve, biodegrade, and become sequestered.

Although the utility of data for TPH for risk assessment is minimal, it is an inexpensive tool that can be used for three purposes: (1) determining if there is a problem; (2) assessing the severity of contamination; and (3) following the progress of a remediation effort. If the data for the TPH indicate that there may be significant contamination of environmental media, other data can be collected so that harm to human health can be quantitatively assessed. The other data can include target analyte concentration data and petroleum fraction concentration data obtained using the evolving fraction-based analytical methods.

There are many analytical techniques available that measure TPH concentrations in the environment but no single method is satisfactory for the measurement of the entire range of petroleum-derived hydrocarbons. In addition, and because the techniques vary in the manner in which hydrocarbons are extracted and detected, each method may be applicable to the measurement of different subsets of the petroleum-derived hydrocarbons present in a sample. The four most commonly used TPH analytical methods include (1) gas chromatography (GC), (2) infrared spectrometry (IR), (3) gravimetric analysis, and (4) immunoassay (Miller, 2000 and references cited therein).

30.4.1 GAS CHROMATOGRAPHIC METHODS

GC methods are currently the preferred laboratory methods for measurement of TPH because they detect a broad range of hydrocarbons and provide both sensitivity and selectivity. In addition, identification and quantification of individual constituents of the TPH mix is possible.

Methods based on gravimetric analysis are also simple and rapid but they suffer from the same limitations as infrared spectrometric methods. Gravimetric-based methods may be useful for oily sludge and wastewaters, which will present analytical difficulties for other more sensitive methods. Immunoassay methods for the measurement of TPH are also popular for field testing because they offer a simple, quick technique for in situ quantification of the total petroleum hydrocarbons.

For methods based on GC, the TPH fraction is defined as any chemicals extractable by a solvent or purge gas and detectable by GC/flame ionization detection (GC/FID) within a specified carbon range. The primary advantage of such methods is that they provide information about the type of petroleum in the sample in addition to measuring the amount. Identification of product type(s) is not always straightforward, however, and requires an experienced analyst of petroleum products (Sullivan and Johnson, 1993; Speight, 2001, 2002). Detection limits are method-dependent as well as matrix-dependent and can be as low as 0.5 mg/L in water or 10 mg/kg in soil.

Chromatographic columns are commonly used to determine TPH compounds approximately in the order of their boiling points. Compounds are detected by means of a flame ionization detector that responds to virtually all compounds that can burn. The sum of all responses within a specified range is equated to a hydrocarbon concentration by reference to standards of known concentration.

Two methods (EPA SW-846 8015 and 8015A) were, in the past, often quoted as the source of gas chromatography-based methods for the measurement of the TPH in a sample. However, the original methods were developed for nonhalogenated volatile organic compounds and were designed to measure a short target list of chemical solvents rather than petroleum hydrocarbons. Thus, because there was no universal method for total petroleum hydrocarbons, there were many variations in these methods. Recently, an updated method (EPA 8015B) provides guidance for the analysis of gasoline and diesel range organic compounds.

The current individual methods differ in procedure, compounds detected, extraction techniques, and extraction solvents used. Some methods may include a cleanup step to remove biogenic (bacterial or vegetation-derived) material while others do not. The methods have in common a boiling point-type column and a flame ionization detector. Selection of a method depends on the type of hydrocarbon suspected to be in the sample.

For example, if gasoline is suspected to be the sole contaminant, the method will use purge/trap sample introduction. If a higher boiling petroleum fractions (diesel, middle distillates, and motor oil) are the contaminants, the analysis will use direct injection and hotter oven temperatures. Mixtures or unknown contamination may require both volatile range and extractable range analyses. Alternately, a single injection can be used to analyze the whole sample, but the extraction method must not use a solvent evaporation step.

Gas chromatography-based methods can be broadly used for different kinds of petroleum contamination but are most appropriate for detecting nonpolar hydrocarbons with carbon numbers between C6 and C25 or C36. Many lubricating oils contain molecules with more than 40 carbon atoms. In fact, crude oil itself contains molecules having more than 100 carbons or more. These high molecular weight hydrocarbons are outside the detection range of the more common GC methods, but specialized gas chromatographs are capable of analyzing such high molecular weight constituents.

Accurate quantification depends on adjusting the chromatograph to reach as high a carbon number as possible, then running a calibration standard with the same carbon range as the sample. There should also be a check for mass discrimination, a tendency for higher molecular weight hydrocarbons to be retained in the injection port. If a sample is suspected to be heavy oil or to contain a mixture of light oil and heavy oil, the most appropriate method must be used.

Gravimetric or infrared methods are often preferred for high molecular weight samples. These methods can even be used as a check on gas chromatographic data if it is suspected that high molecular weight hydrocarbons are present but are not being detected.

Calibration standards vary. Most methods specify a gasoline calibration standard for volatile range TPH and a diesel fuel #2 standard for extractable range total petroleum hydrocarbons. Some methods use synthetic mixtures for calibration. As most methods are described for gasoline or diesel fuel, TPH methods may have to be adjusted to measure the contamination by heavier hydrocarbons (e.g., heavy fuel oil, lubricating oil, or crude oil). Such adjustments may entail use of a more aggressive solvent, a wider gas chromatographic window that allows detection of molecules containing up to C36 or more, and a different calibration standard that more closely resembles the constituents of the sample under investigation.

Gas chromatographic methods can be modified and fine-tuned so that they are suitable for measurement of specific petroleum products or group types. These modified methods can be particularly useful when there is information on the source of contamination, but method results should be interpreted with the clear understanding that a modified method was used for detection of a specific carbon range.

Interpretation of gas chromatographic data is often complicated and the analytical method should always be considered when interpreting concentration data. For example, a volatile range analysis may be very useful for quantifying TPH at a gasoline release site, but a volatile range analysis will not detect the presence of lube oil constituents. In addition, a modified method that has been specifically selected for detection of gasoline-range organics at a gasoline-contaminated site may also detect hydrocarbons from other petroleum releases because fuel carbon ranges frequently overlap. Gasoline is found primarily in the volatile range. Diesel fuel falls primarily in an extractable range. Jet fuel overlaps both the volatile and semi-volatile ranges. However, the detection of different kinds of petroleum constituents does not necessarily indicate that there have been multiple releases at a site. Analyses of spilled waste oil will frequently detect the presence of gasoline, and sometimes diesel. This does not necessarily indicate multiple spills as all waste oil contain some fuel. As much as 10% of used motor oil can consist of gasoline.

If the type of contaminant is unknown, a fingerprint analysis can help in the identification procedure. A fingerprint or pattern recognition analysis is a direct injection analysis where the chromatogram is compared to chromatograms of reference materials. Certain fuels can be identified by characteristic, reproducible chromatographic patterns. For example, chromatograms of gasoline and diesel differ considerably but many hydrocarbon streams may have similar fingerprints. Diesel No. 2 and No. 2 fuel oil both have the same boiling point range and chromatographic fingerprint. A fingerprint can be used to conclusively identify a mixture when a known sample of that mixture or samples of the mixture's source materials are available as references.

Furthermore, as a fuel evaporates or biodegrades, its pattern can change so radically that identification becomes difficult. Consequently, a GC fingerprint is not a conclusive diagnostic tool. The methods must for TPH analysis must stress calibration and quality control, while pattern recognition methods stress detail and comparability.

The gas chromatographic methods usually cannot quantitatively detect compounds below C6 because these compounds are highly volatile and interference can occur from the solvent peak. As much as 25% of fresh gasoline can be below C6 but the problem is reduced for weathered gasoline and diesel range contamination because most of the very volatile hydrocarbons (<C6) may no longer be present in the sample. Gas chromatographic methods may also be inefficient for quantification of polar constituents (nitrogen, oxygen, and sulfur containing molecules). Some of the polar constituents are too reactive to pass through a gas chromatograph and thus will not reach the detector for measurement.

Oxygenated gasoline is sometimes analyzed by GC-based methods but it should be noted that the efficiency of purge methods is lower for oxygenates such as ethers and alcohols because detector response to oxygenates is lower relative to hydrocarbons. Therefore the data will be biased slightly low for ether-containing fuels compared to equivalent amounts of traditional gasoline. Methanol and ethanol elute before hexane and, consequently, they are not quantified and may not even be detected due to coelution with the solvent.

On the other hand, gas chromatographic methods may overestimate the concentration of TPH in the sample due to the detection of nonpetroleum compounds. In addition, cleanup steps do not perfectly separate petroleum hydrocarbons from biogenic material such as plant oils and waxes that are sometimes extracted from vegetation-rich soil. Silica gel cleanup may help to remove this interference but may also remove some polar hydrocarbons.

Because petroleum is made up of so many isomers, many compounds, especially those with more than eight carbon atoms, coelute with isomers of nearly the same boiling point. These unresolved compounds are referred to as the unresolved complex mixture. They are legitimately part of the petroleum signal, and unless otherwise specified, should be quantified. Quantifying such a mixture requires a baseline-to-baseline integration mode rather than a peak-to-peak integration mode. The baseline-to-baseline integration quantifies all the petroleum constituents in the sample but in the peak-to-peak integration only the individual resolved hydrocarbons (not including the unresolved complex mixture) are quantified.

30.4.2 Infrared Spectroscopy Methods

Infrared methods measure the absorbance of the C—H bond and most methods typically measure the absorbance at a single frequency (usually 2930 cm^{-1}) that corresponds to the stretching of aliphatic methylene (CH$_2$) groups. Some methods use multiple frequencies including 2960 cm^{-1} (CH$_3$ groups) and 2900 to 3000 cm^{-1} (aromatic C—H bonds).

Therefore, for infrared spectroscopic methods, the TPH is any chemical extracted from a solvent that is not removed by silica gel and can be detected by infrared spectroscopy at a specified wavelength. The primary advantage of the infrared-based methods is that they are simple and rapid. Detection limits (e.g., for EPA 418.1) are approximately 1 mg/L in water and 10 mg/kg in soil. However, the infrared method can often suffer from poor accuracy and precision, especially for heterogeneous soil samples. Also, the infrared methods give no information on the type of fuel present in the sample and there is little, often no, information about the presence or absence of toxic molecules, and no specific information about the potential risk associated with the contamination.

Samples are extracted with a suitable solvent (i.e., a solvent with no C—H bonds) and biogenic polar materials are removed with silica gel. Some polar petroleum constituents may be removed as part of the silica gel cleanup. The absorbance of the silica gel eluate is measured at the specified frequency and compared with the absorbance of a standard or standards of known petroleum hydrocarbon concentration. The absorbance is a measurement of the sum of all the compounds contributing with the result. However, infrared methods cannot provide information on the type of hydrocarbon contamination.

For all IR-based TPH methods, the C—H absorbance is quantified by comparing it with the absorbance of standards of a known concentration. An assumption is made that the standard has an aliphatic-to-aromatic ratio and an infrared response similar to that of the sample. Consequently, it is important to use a calibration standard as similar to the type of contamination as possible (EPA 418.1).

The infrared method that has been most frequently used (EPA 418.1) is appropriate only for water samples. A separator funnel liquid/liquid extraction technique is used to extract the hydrocarbons from the water. A method (EPA 5520D) using a Soxhlet extraction technique is suitable for sludge. This extraction is frequently used to adapt the method (EPA 418.1) to soil samples. An infrared-based supercritical fluid extraction method for diesel range contamination (EPA 3560) is available.

Similar to gas chromatographic methods, the data from infrared methods must be interpreted after considering certain limitations and interferences that can affect data quality. For example, the C—H absorbance is not always measured in exactly the same way. Within the set of methods that specify a single infrared measurement, some methods call for the measurement at precisely 2930 cm^{-1} while others (including EPA 418.1) require measurement at the absorbance maximum nearest 2930 cm^{-1}. This variation can make a significant difference in the magnitude of the result, and can lead to confusion when comparing duplicate sample results. If only C—H absorbance is measured, infrared methods will potentially underestimate the concentration of TPH in samples that contain petroleum constituents such as benzene and naphthalene that do not contain alkyl C—H groups.

As an infrared result is calculated as if the aromatics in the sample were present in the same ratio as in the calibration standard, accuracy depends on use of a calibration standard as similar to the type of contamination as possible. Use of a dissimilar standard will tend to create a positive bias in highly aliphatic samples and a negative bias in highly aromatic samples.

In summary, infrared methods are prone to interferences (positive bias) from nonpetroleum sources as many organic compounds have some type of alkyl group associated with them whether petroleum-derived or not.

30.4.3   GRAVIMETRIC METHODS

Gravimetric methods measure all chemicals that are extractable by a solvent, not removed during solvent evaporation, and capable of being weighed. Some gravimetric methods include a cleanup step to remove biogenic material. The advantage of gravimetric methods is that they are simple and rapid. Detection limits are approximately 5–10 mg/L in water and 50 mg/kg in soil.

However, gravimetric methods are not suitable for measurement of low boiling hydrocarbons that volatilize at temperatures below 70°C to 85°C (167°F to 176°F). They are recommended for use with (1) oily sludge, (2) for samples containing heavy molecular weight hydrocarbons, or (3) for aqueous samples when hexane is preferred as the solvent.

Gravimetric methods give no information on the type of fuel present, no information about the presence or absence of toxic compounds, and no specific information about potential risk associated with the contamination.

In the method(s), petroleum constituents are extracted into a suitable solvent. Biogenic polar materials typically may be partially or completely removed with silica gel. The solvent is evaporated and the residue is weighed. This quantity is reported as a percent of the total soil sample dry weight. These methods are better suited for heavy oil because they include an evaporation step.

There are a variety of gravimetric oil and grease methods suitable for testing water and soil samples (e.g., EPA SW-846 9070, EPA 413.1, EPA 9071). Technically, the result is an oil and grease result because no cleanup step is used. One method (EPA 9071) is used to recover low levels of oil and grease by chemically drying a wet sludge sample and then extracting it using Soxhlet apparatus. Results are reported on a dry-weight basis. The method is also used when relatively polar high molecular weight petroleum fractions are present, or when the levels of nonvolatile grease challenge the solubility limit of the solvent. Specifically, the method (EPA SW-846 9071) is suitable for biological lipids, mineral hydrocarbons, and some industrial wastewater.

Gravimetric methods for oil and grease (e.g., EPA SW-846 9071 measure anything that dissolves in the solvent and remains after solvent evaporation. These substances include hydrocarbons, vegetable oils, animal fats, waxes, soaps, greases, and related biogenic material. Gravimetric methods for TPH (EPA 1664) measure anything that dissolves in the solvent and remains after silica gel treatment and solvent evaporation.

This method (EPA 1664) is a liquid–liquid extraction gravimetric procedure that employs *n*-hexane as the extraction solvent, in place of 1,1,2-trichloroethane (CFC-113) and 1,2, 2-trifluoroethane (Freon-113), for determination of the conventional pollutant oil and grease. As the nature and amount of material determined is defined by the solvent and by the details of the method used for extraction, oil and grease are method-defined analytes. The method may be modified to reduce interferences and take advantage of advances in instrumentation provided that all method equivalency and performance criteria are met. However, *n*-hexane is a poor solvent for high molecular weight petroleum constituents (Speight, 2001). Thus, the method will produce erroneous data for samples contaminated with heavy oils.

All gravimetric methods measure any suspended solids that are not filtered from solution, including bacterial degradation products and clay fines. Method 9071 specifies using cotton or glass wool as a filter.

As extracts are heated to remove solvent, these methods are not suitable for measurement of low boiling low molecular weight hydrocarbons (i.e., hydrocarbons having less than 15 carbon atoms) that volatilize at temperatures below 70°C to 85°C (158°F to 185°F). Liquid fuels, from gasoline through No. 2 fuel oil, lose volatile constituents during solvent removal. In addition, soil results that are reported on a dry-weight basis suffer from potential losses of lower boiling hydrocarbon constituents during moisture determination where the matrix is dried at approximately 103°C to 105°C (217°F to 221°F) for several hours in an oven.

30.4.4 IMMUNOASSAY METHODS

Immunoassay methods correlate TPH with the response of antibodies to specific petroleum constituents. Many of the methods measure only aromatics that have an affinity for the antibody, benzene-toluene-ethylbenzene-xylene, and polynuclear aromatic hydrocarbon analysis (EPA 4030, Petroleum Hydrocarbons by Immunoassay).

The principle behind the test method(s) is that antibodies are made of proteins that recognize and bind with foreign substances (antigens) that invade host animals. Synthetic antibodies have been developed to complex with petroleum constituents. The antibodies are immobilized on the walls of a special cell or filter membrane. Water samples are added directly to the cell while soils must be extracted before analysis. A known amount of labeled analyte (typically an enzyme with an affinity for the antibody) is added after the sample. The sample analytes compete with the enzyme-labeled analytes for sites on the antibodies. After equilibrium is established, the cell is washed to remove any unreacted sample or labeled enzyme. Color development reagents that react with the labeled enzyme are added. A solution that stops color development is added at a specified time, and the optical density (color intensity) is measured. Because the coloring agent reacts with the labeled enzyme, samples with high optical density contain low concentrations of analytes. Concentration is inversely proportional to optical density.

The antibodies used in immunoassay kits are generally designed to bond with selected compounds. A correction factor supplied by the manufacturer must be used to calculate the concentration of total petroleum hydrocarbons. The correction factor can vary depending on the product type because it attempts to correlate TPH with the measured surrogates.

Immunoassay tests do not identify specific fuel types and are best used as screening tools. The tests are dependent on soil type and homogeneity. In particular, for clay and other cohesive soils, the tests are limited by a low capacity to extract hydrocarbons from the sample.

30.5 PETROLEUM GROUP ANALYSIS

Petroleum group analyses are conducted to determine the amount of petroleum compound classes (e.g., saturates, aromatics, and polar constituents/resins) present in petroleum-contaminated samples. This type of measurement is sometimes used to identify fuel type or to track plumes. It may be particularly useful for higher boiling products, such as asphalt. Group type test methods include multidimensional gas chromatography (not often used for environmental samples), high-performance liquid chromatography (HPLC), and thin layer chromatography (TLC) (Miller, 2000; Patnaik, 2004).

Test methods that analyze individual compounds (e.g., benzene-toluene-ethylbenzene-xylene mixtures and polynuclear aromatic hydrocarbons) are generally applied to detect the presence of an additive or to provide concentration data needed to estimate environmental and health risks that are associated with individual compounds. Common constituent measurement techniques include gas chromatography with second column confirmation, gas chromatography with multiple selective detectors and gas chromatography with mass spectrometry detection (GC/MS) (EPA 8240).

Many common environmental methods measure individual petroleum constituents or target compound rather than the whole signal from the total petroleum hydrocarbons. Each method measures a suite of compounds selected because of their toxicity and common use in industry.

For organic compounds, there are three series of target compound methods that must be used for regulatory purposes:

1. EPA 500 series: Organic Compounds in Drinking Water, as regulated under the Safe Drinking Water Act
2. EPA 600 series: Methods for Organic Chemical Analysis of Municipal and Industrial Wastewater, as regulated under the Clean Water Act
3. SW-846 series: Test Methods for Evaluating Solid Waste: Physical/Chemical Methods, as promulgated by the US EPA, Office of Solid Waste and Emergency Response

The 500 and 600 series methods provide parameters and conditions for the analysis of drinking water and wastewater, respectively. One method (EPA SW-846) is focused on the analysis of nearly all matrices including industrial waste, soil, sludge, sediment, and water miscible and nonwater miscible wastes. It also provides for the analysis of groundwater and wastewater but is not used to evaluate compliance of public drinking water systems.

Selection of one method over another is often dictated by the nature of the sample and the particular compliance or cleanup program for which the sample is being analyzed. It is essential to recognize that capabilities and requirements vary between methods when requesting any analytical method or suite of methods. Most compound-specific methods use a gas chromatographic selective detector, HPLC, or gas chromatography/mass spectrometry.

More correctly, group analytical methods are designed to separate hydrocarbons into categories, such as saturates, aromatics, resins, and asphaltene constituents (SARA) or paraffins, iso-paraffins, naphthenes, aromatics, and olefins (PIANO). These chromatographic, gas chromatographic, and HPLC were developed for monitoring refinery processes or evaluating organic synthesis products. Column chromatographic methods that separate saturates from aromatics are often used as preparative steps for further analysis by gas chromatography/mass spectrometry. TLC is sometimes used as a screening technique for petroleum product identification.

30.5.1 THIN LAYER CHROMATOGRAPHY

In the environmental; field, TLC is best used for screening analyses and characterization of semi-volatile and nonvolatile petroleum products. The precision and accuracy of the technique is inferior to other methods (EPA 8015, EPA 418.1) but when speed and simplicity are desired, TLC may be a suitable alternative. For characterizations of petroleum products such as asphalt, the method has the advantage of separating compounds that are too high boiling to pass through a gas chromatograph. While TLC does not have the resolving power of a gas chromatograph, it is able to separate different classes of compounds. TLC analysis is fairly simple and, as the method does not give highly accurate or precise results, there is no need to perform the highest quality extractions.

In the method, soil samples are extracted by shaking or vortexing with the solvent. Water samples are extracted by shaking in a separatory funnel. If there is the potential for the presence of compounds that interfere with the method and make the data suspect, silica gel can be added to clean the extract. Sample extract aliquots are placed close to the bottom of a glass plate coated with a stationary phase. The most widely used stationary phases are made up of an organic hydrocarbon moiety bonded to a silica backbone.

For the analysis of petroleum hydrocarbons, a moderately polar material stationary phase works well. The plate is placed in a sealed chamber with a solvent (mobile phase). The solvent travels up the plate carrying compounds present in the sample. The distance a compound travels is a function of the affinity of the compound to the stationary phase relative to the mobile phase. Compounds with chemical structure and polarity similar to the solvent travel well in the mobile phase. For example, the saturated hydrocarbons seen in diesel fuel travel readily up a plate in a hexane mobile phase. Polar compounds such as ketones or alcohols travel a smaller distance in hexane than saturated hydrocarbons.

After a plate has been exposed to the mobile phase solvent for the required time, the compounds present can be viewed by several methods. Polynuclear aromatic hydrocarbons, other compounds with conjugated systems, and compounds containing heteroatoms (nitrogen, oxygen, or sulfur) can be viewed with long wave and short wave ultraviolet light. The unaided eye can see other material or the plates can be developed in iodine. Iodine has an affinity for most petroleum compounds, including the saturated hydrocarbons, and stains the compounds a reddish or brown color.

The method is considered to be a qualitative and useful tool for rapid sample screening. Limitations of the method center on its moderate reproducibility, detection limits, and resolving capabilities. Variability between operators can be as high as 30%. Detection limits (without any concentration of the sample extract) are near 50 ppm (mg/kg) for most petroleum products in soils. When the aromatic content of a sample is high, as with bunker C fuel oil, the detection limit can be near 100 ppm. It is often not possible to distinguish between similar products such as diesel and jet fuel. As with all chemical analyses, quality assurance tests should be run to verify the accuracy and precision of the method.

30.5.2 IMMUNOASSAY

A number of different testing kits based on immunoassay technology are available for rapid field determination of certain groups of compounds such as benzene-toluene-ethylbenzene-xylene (EPA 4030) or polynuclear aromatic hydrocarbons (EPA 4035, Polycyclic Aromatic Hydrocarbons by Immunoassay). The immunoassay screening kits are self-contained portable field kits that include components for sample preparation, instrumentation to read assay results, and immunoassay reagents.

Unless the immunoassay kit is benzene sensitive, the kit may display strong biases such as the low affinity for benzene relative to toluene, ethylbenzene, xylenes, and other aromatic compounds. This will cause an underestimation of the actual benzene levels in a sample. And, as benzene is often the dominant compound in leachates due to its high solubility, a low sensitivity for benzene is undesirable.

The quality of the analysis of polynuclear aromatic hydrocarbons is often dependent on the extraction efficiency. Clay and other cohesive soils lower the ability to extract polynuclear aromatic hydrocarbons. Another potential problem with polynuclear aromatic hydrocarbon analysis is that the test kits may have different responses for different compounds.

30.5.3 GAS CHROMATOGRAPHY

Gas chromatography uses the principle of a stationary phase and a mobile phase. Much attention has been paid to the various stationary phases and books have been written on the subject as it pertains to petroleum chemistry.

Briefly, gas–liquid chromatography (GLC) is a method for separating the volatile components of various mixtures (Altgelt and Gouw, 1975; Fowlis, 1995; Grob, 1995). It is, in fact, a highly efficient fractionating technique, and it is ideally suited to the quantitative analysis of mixtures when the possible components are known and the interest lies only in determining the amounts of each present. In this type of application GC has taken over much of the work previously done by the other techniques; it is now the preferred technique for the analysis of hydrocarbon gases, and GC in-line monitors are having increasing application in refinery plant control. GLC is also used extensively for individual component identification, as well as percentage composition, in the gasoline boiling range.

The mobile phase is the carrier gas, and the gas selected has a bearing on the resolution. Nitrogen has very poor resolution ability, while either helium or hydrogen is a better choice with hydrogen being the best carrier gas for resolution. However, hydrogen is reactive and may not be compatible with all sets of target analytes. There is an optimum flow rate for each carrier gas to achieve maximum resolution. As the temperature of the oven increases, the flow rate of the gas changes due to thermal expansion of the gas. Most modern GC are equipped with constant flow devices that change the gas valve settings as the temperature in the oven changes, so changing flow rates are no longer a concern. Once the flow is optimized at one temperature it is optimized for all temperatures.

For environmental analysis (Bruner, 1993), particularly the volatile samples such as those found in total petroleum hydrocarbons, the gas chromatograph is generally interfaced with a purge and trap system as described in the section on gas chromatographic methods. The photoionization detector works by bombarding compounds with ultraviolet (UV) light, generating a current of ions. Compounds with double carbon bonds, conjugated systems (multiple carbon double bonds arranged in a specific manner), and aromatic rings are easily ionized with the ultraviolet light generated by the photoionization detector lamp, whereas most saturated compounds require higher energy radiation.

One method (EPA 8020) that is suitable for volatile aromatic compounds is often referred to as benzene-toluene-ethylbenzene-xylene analysis, though the method includes other volatile aromatics. The method is similar to most volatile organic gas chromatographic methods. Sample preparation and introduction is typically by the urge and trap analysis (EPA 5030). Some oxygenates such as methyl-*t*-butyl ether (MTBE) are also detected by the photoionization detector as well as olefins, branched alkanes, and cycloalkanes.

Certain false positives are common (EPA 8020). For example, trimethylbenzenes and gasoline constituents are frequently identified as chlorobenzenes (EPA 602, EPA 8020) because these compounds elute with nearly the same retention times from nonpolar columns.

Cyclohexane is often mistaken for benzene (EPA 8015/8020) because both compounds are detected by a 10.2 eV photoionization detector and have nearly the same elution time from a nonpolar column (EPA 8015). The two compounds have very different retention times on a more polar column (EPA 8020) but a more polar column skews the carbon ranges (EPA 8015). False positives for oxygenates in gasoline are common, especially in highly contaminated samples.

For semi-volatile constituents of petroleum, the gas chromatograph is generally equipped with either a packed or capillary column. Either neat or diluted organic liquids can be analyzed via direct injection, and compounds are separated during movement down the column. The flame ionization detector uses a hydrogen-fueled flame to ionize compounds that reach the detector.

For polynuclear aromatic hydrocarbons, a method is available (EPA 8100) in which injection of sample extracts directly onto the column is the preferred method for sample introduction for this packed-column method.

A gas chromatography-flame ionization detector system can be used for the separation and detection of nonpolar organic compounds. Semi-volatile constituents are among the analytes that can be readily resolved and detected using the system. If a packed column is used, four pairs of compounds may not be adequately resolved and are reported as a quantitative sum: anthracene and phenanthrene, chrysene and benzo(a)anthracene, benzo(b) fluoranthene and benzo(k)fluoranthene, and dibenzo(a,h)anthracene and indeno(1,2,3-cd)pyrene. This issue can be resolved through the use of a capillary column in place of a packed column.

30.5.4 HIGH-PERFORMANCE LIQUID CHROMATOGRAPHY

A HPLC system can be used to measure concentrations of target semi-volatile and nonvolatile petroleum constituents. The system only requires that the sample be dissolved in a solvent compatible with those used in the separation. The detector most often used in petroleum environmental analysis is the fluorescence detector. These detectors are particularly sensitive to aromatic molecules, especially polynuclear aromatic hydrocarbons. An ultraviolet detector may be used to measure compounds that do not fluoresce.

In the method, polynuclear aromatic hydrocarbons are extracted from the sample matrix with a suitable solvent, which is then injected into the chromatographic system. Usually the extract must be filtered because fine particulate matter can collect on the inlet frit of the column, resulting in high back-pressures and eventual plugging of the column. For most hydrocarbon analyses, reverse phase high-performance liquid chromatography (i.e., using a nonpolar column packing with a more polar mobile phase) is used. The most common bonded phase is the octadecyl C18 phase. The mobile phase is commonly aqueous mixtures of either acetonitrile or methanol.

After the chromatographic separation, the analytes flow through the cell of the detector. A fluorescence detector shines light of a particular wavelength (the excitation wavelength) into the cell. Fluorescent compounds absorb light and re-emit light of other, higher wavelengths (emission wavelengths). The emission wavelengths of a molecule are mainly determined by its structure. For polynuclear aromatic hydrocarbons, the emission wavelengths are mainly determined by the arrangement of the rings and vary greatly between isomers.

Some of the polynuclear aromatic hydrocarbons (such as phenanthrene, pyrene, and benzo(g,h,i)perylene, are commonly seen in products boiling in the middle to heavy distillate range. In a method for their detection and analysis (EPA 8310) an octadecyl column and an aqueous acetonitrile mobile phase are used. Analytes are excited at 280 nm and detected at emission wavelengths of >389 nm. Naphthalene, acenaphthene, and fluorene must be detected by a less-sensitive UV detector because they emit light at wavelengths below 389 nm. Acenaphthylene is also detected by UV detector.

The methods using fluorescence detection will measure any compounds that elute in the appropriate retention time range and that fluoresce at the targeted emission wavelength(s). In the case of one method (EPA 8310), the excitation wavelength excites most aromatic compounds. These include the target compounds and also many aromatic derivatives, such as alkyl aromatics, phenols, anilines, and heterocyclic aromatic compounds containing the pyrrole (indole, carbazole, etc.), pyridine (quinoline, acridine, etc.), furan (benzofuran, naphthofuran, etc.), and thiophene (benzothiophene, naphthothiophene, etc.) structures. In petroleum samples, alkyl polynuclear aromatic hydrocarbons are strong interfering compounds. For example, there are five methylphenanthrenes and over 20 dimethylphenanthrenes. The alkyl substitution does not significantly affect either the wavelengths or the intensity of the phenanthrene fluorescence. For a very long time after the retention time of phenanthrene, the alkylphenanthrenes will interfere, affecting the measurements of all later-eluting target polynuclear aromatic hydrocarbons.

Interfering compounds will vary considerably from source to source and samples may require a variety of cleanup steps to reach the required method-detection limits. The emission wavelengths used (EPA 8310) are not optimal for sensitivity of the small ring compounds. With a modern electronically controlled monochromator, wavelength programs can be used which tune excitation and emission wavelengths to maximize sensitivity and selectivity for a specific analyte in its retention time window.

30.5.5 GAS CHROMATOGRAPHY–MASS SPECTROMETRY

A gas chromatography–mass spectrometry (GC/MS) system is used to measure concentrations of target volatile and semi-volatile petroleum constituents. It is not typically used to measure the amount of total petroleum hydrocarbons. The advantage the technique is the high selectivity, or ability to confirm compound identity through retention time and unique spectral pattern.

The current method (EPA SW-846 8260) for the analysis of volatile compounds reveals that most of the compounds listed in these methods are not typically found in petroleum products. However, a method that uses selected ion monitoring (SIM) involves system setup to measure only selected target masses rather than scanning the full mass range. This technique yields lower detection limits for specific compounds. At the same time, it gives more complete information available from the total ion chromatogram and the full-mass-range spectrum of each compound. The technique is sometimes used to quantify compounds present at very low concentrations in a complex hydrocarbon matrix. It can be used if the target compound's spectrum has a prominent fragment ion at a mass that distinguishes it from the rest of the hydrocarbon compounds.

The most common method for GC/MS analysis of semi-volatile compounds (EPA SW-846 8270) includes 16 polycyclic aromatic compounds, some of which commonly occur in middle distillate to heavy petroleum products. The method also quantifies phenols and cresols, compounds that are not hydrocarbons but may occur in petroleum products. Phenols and cresols are more likely found in crude oils and weathered petroleum products.

To reduce the possibility of false positives, the intensities of one to three selected ions are compared to the intensity of a unique target ion of the same spectrum. The sample ratios are compared to the ratios of a standard. If the sample ratios fall within a certain range of the standard, and the retention time matches the standard within specifications, the analyte is considered present. Quantification is performed by integrating the response of the target ion only.

Mass spectrometers are among the most selective detectors, but they are still susceptible to interferences. Isomers have identical spectra, while many other compounds have similar mass spectra. Heavy petroleum products can contain thousands of major components that are not resolved by the gas chromatograph. As a result, multiple compounds are simultaneously entering the mass spectrometer. Different compounds may share many of the same ions, confusing the identification process. The probability of misidentification is high in complex mixtures such as petroleum products.

30.6   PETROLEUM FRACTIONS

Rather than quantifying a complex TPH mixture as a single number, petroleum hydrocarbon fraction methods break the mixture into discrete hydrocarbon fractions, thus providing data that can be used in a risk assessment and in characterizing product type and compositional changes such as may occur during weathering (oxidation). The fractionation methods can be used to measure both volatile and extractable hydrocarbons.

In contrast to traditional methods for TPH that report a single concentration number for complex mixtures, the fractionation methods report separate concentrations for discrete aliphatic and aromatic fractions. The available petroleum fraction methods are GC-based and are thus sensitive to a broad range of hydrocarbons. Identification and quantification of aliphatic and aromatic fractions allows one to identify petroleum products and evaluate the extent of product weathering. These fraction data also can be used in risk assessment.

One particular method is designed to characterize C6 to C28+ petroleum hydrocarbons in soil as a series of aliphatic and aromatic carbon range fractions. The extraction methodology differs from other petroleum hydrocarbon methods because it uses *n*-pentane and not methylene chloride as the extraction solvent. If methylene chloride is used as the extraction solvent, aliphatic and aromatic compounds cannot be separated.

Petroleum hydrocarbons in this range are extracted efficiently by *n*-pentane. The whole extract is separated into aliphatic and aromatic petroleum-derived fractions (EPA SW-846 3611, EPA SW-846 3630). The aliphatic and aromatic fractions are analyzed separately by gas chromatography, and quantified by summing the signals within a series of specified carbon ranges that represent the fate and transport fractions. The gas chromatograph is equipped with a boiling point column (nonpolar capillary column). Gas chromatographic parameters allow the measurement of a hydrocarbon range of *n*-hexane C6 to *n*-octacosane C28, a boiling point range of approximately 65°C to 450°C.

30.7   ASSESSMENT OF THE METHODS

Generally, measurement of the TPH in an ecosystem is performed by the standard method (EPA 418.1) or by some modification thereof. However, many other methods exist is which the data are also claimed to be representative of the TPH in the ecosystem. In fact, many of the methods for determining the TPH are prone to (1) producing false negatives (reporting *nondetected* when there was really considerable petroleum hydrocarbons present), (2) underestimating the extent of petroleum hydrocarbons present (true of virtually every TPH methodology), (3) underestimating the overall risk from petroleum hydrocarbons due to missing significant amounts of some of the compounds of most concern (e.g., polynuclear aromatic hydrocarbons), (4) producing misleading data related to soil hot spots versus areas of less concern due to differing moisture concentrations of otherwise similar samples, (5) producing misleading results because an inappropriate (not close enough to the unknown being sampled) standard (oil) was used in calibration, and (6) producing soil or sediment data that cannot be directly compared with other TPH data or guidelines because one is expressed in dry weight and the other in wet weight, and (7) producing relatively accurate dry weight values for heavy petroleum hydrocarbons but questionable dry weight values for lighter, more volatile compounds, (Note: different labs dry the samples in different ways and a sample with lots of lighter fraction hydrocarbons is more prone to hydrocarbon loss; the variable loss of volatile hydrocarbons in a drying step is therefore an additional area of lab and data variability), (8) producing data that cannot be directly compared with other TPH data or guidelines because one data set is the result of a Soxhlet extraction method and the other reflects a sonication or other alternative extraction method, (9) producing misleading data related to heavy fraction hydrocarbons (again such as the higher molecular weight poly-nuclear aromatic hydrocarbons) due to loss of the heavier compounds on filter paper, and (10) producing data prone to faulty interpretation of the environmental significance of the results (100 ppm of TPH from one type of oil may be practically nontoxic while 100 ppm of TPH from a different type of oil may be very toxic).

Another complication with TPH values is that petroleum-derived inputs vary considerably in composition; it is essential to bear this in mind when quantifying them in general terms such as an **oil** or the total petroleum hydrocarbons measurement [461]. Petroleum is complex, containing many thousands of compounds ranging from gases to residues boiling about 400°C [461].

Furthermore, as different combinations of petroleum hydrocarbons typically contribute to total petroleum hydrocarbons at different sites, the fate characteristics are also typically different at different sites, even if the TPH concentration is the same. Different methods used to generate TPH concentrations, or other similar simple screening measures of petrol-eum contamination, all produce very different results.

It is not surprising that the data produced as TPH (EPA 418.1) suffers from several short-comings as an index of potential groundwater contamination or health risk. In fact, it does not actually measure the TPH in the sample but rather measures a specific range of hydrocarbon compounds. This is caused by limitations of the extraction process (solvents used and the concentration steps) and the reference standards used for instrumental analysis. The method specifically states that it does not accurately measure the lighter fractions of gasoline (benzene-toluene-ethylbenzene-xylenes fraction, BTEX) that should include the benzene-toluene-ethyl-benzene-xylenes fraction. Further, the method was originally a method for water samples that have been modified for solids, and it is subject to bias.

The TPH represents a summation of all the hydrocarbon compounds that may be present (and detected) in a soil sample. Because of differences in product composition between, for example, gasoline and diesel, or fresh versus weathered fuels, the types of compounds present at one site may be completely different than those present at another.

Accordingly, the TPH at a gasoline spill site will be comprised of mostly C6–C12 com-pounds, while TPH at an older site where the fuel has weathered will likely measure mostly C8–C12 compounds. Because of this inherent variability in the method and the analyte, it is currently not possible to directly relate potential environmental or health risks with concen-trations of total petroleum hydrocarbons. The relative mobility or toxicity of contaminants represented by TPH analyses at one site may be completely different from that of another site (e.g., C6–C12 compared to C10–C25). There is no easy way to determine if TPH from the former site will represent the same level of risk as an equal measure of the TPH from the latter. For these reasons, it is clear that TPH offers limited benefits as an indicator measure for cleanup criteria. Its current widespread use as a soil cleanup criterion is a function of a lack of understanding of its proper application and limitations, and its historical use as a simple and inexpensive indicator of general levels of contamination.

When sampling in the environment, it is often impossible to determine which chemical mixtures are causing a TPH reading, which is one of the major weaknesses of the method. At minimum, before using contaminants data from diverse sources, efforts should be made to determine that field collection methods, detection limits, and quality control techniques were acceptable and comparable. This will help the analysts compare the analysis in the concentration range with the benchmark or regulatory criteria concentrations should be very precise and accurate.

Indeed, it must be remembered that quality control field and lab blanks and duplicates will not help in the data quality assurance goal as well as intended if one is using a method prone to false negatives. Methods may be prone to false negatives due to the use of detection limits that are too high, the loss of contaminants through inappropriate handling, or the use of inappropriate methods. The use of inappropriate methods prone to false negatives (or false positives) is particularly common related to TPH and other general scans related oil products. This is one reason that more rigorous analyses are often recommended as alternatives to TPH analyses.

In interpreting the data for the TPH in a sample, one cannot ignore the amount of moisture because moisture blocks the extraction of petroleum hydrocarbons by another solvent (Freon). Sulfur or phthalate compounds also potentially interfere with TPH analyses. This is similar to the problem of strong interferences from phthalate esters or chlorinated solvents when one is using electron capture methods to look for chlorinated compounds such as polycholorbiphenyls or pesticides.

Too much reliance on the determination of benzene-toluene-xylenes (BTX) or benzene-toluene-ethylbenzene-xylenes (BTEX) to measure gasoline or diesel contamination may cause one to be unaware that more modern gasoline and diesel are better refined and contain fewer of such compounds. It must be remembered that the use of BTX data started as a measure of the more hazardous compounds in gasoline. Modern gasoline and diesel has a higher percentage of straight chain alkanes, nonvolatiles, not as many aromatics, lots of long chain aliphatic compounds, and fewer BTX compounds. In addition, determination of the BTX concentration is not appropriate for aged gasoline characterized by loss of BTX compounds over time. Thus the problem with many analyses for BTX as related to petroleum hydrocarbons is the danger of producing false negatives. For example, the test for BTX may indicate no contamination when significant contamination is present.

TRPH, like total petroleum hydrocarbons, is methodologically defined and concentrations given as TPH or "TRPH" alone do not produce much valuable information. To be able to understand the significance of the concentration, the method employed for the determination must be clearly identified (e.g., EPA 8015 for gasoline, EPA 8016 for diesel, EPA 418.1 for TRPH). The data must not be used or interpreted as though various TPH methods were the same as various total recoverable petroleum hydrocarbon methods. When comparing data with soil guideline levels, it is necessary to ascertain which laboratory analysis was done to measure compliance with the current specific guideline.

Additional problems with TPH methods (including method 418.1) include the following:

1. Most methods used to determine the TPH in a sample are inadequate for unknowns because the methods are only as good as the calibration standards. With unknown chemicals present, the precise standards cannot be selected and employing an incorrect calibration standard can lead to erroneous data.
2. Some of the methods that have been used for determination of the TPH also extract vegetable and animal oils that are also present in the sample.

3. The methodology related to volatility can be extremely variable. For example, low boiling oils are more susceptible to ambient (and extraction) conditions. The time for evaporation of the oils is a variable and the temperature and heating period is used to calculate dry weight is also a variable. It is preferable to calculate wet weight TPH values first and then very carefully measure percentage moisture in a manner that minimizes losses.

The estimated variability of the test method is questionable and may leave room for serious errors in the calculation of the total petroleum hydrocarbons.

As the determination of the TPH in a sample is subject to many questions, the bias must be defined and alternate reliable and meaningful methods need to be sought. For example, negative bias may result when samples are analyzed because of (1) poor extraction efficiency of the solvent (Freon, EPA 481) or *n*-hexane, EPA 1664) for high molecular weight hydrocarbons, (2) loss of volatile hydrocarbons during extract concentration (Speight, 2005), (3) differences in molar absorbtivity between the calibration standard and product type because of the presence of unknown compound types, (4) fractionation of soluble low-infrared active aromatic hydrocarbons in groundwater during water washout, (5) removal of five-ring and six-ring alkylated aromatics during the silica cleanup procedure—the efficiency of silica gel fractionation varies depending upon the nature of the solute, and (6) preferential biodegradation of *n*-alkanes.

In addition, positive bias is often introduced as a result of (1) product differences in molar absorbtivity, (2) partitioning of soluble aromatics from the bulk product because of oil washout, (3) measurement of naturally occurring saturated hydrocarbons that exhibit a high molar absorbtivity (e.g., plant waxes, $n\text{-}C_{25}$, $n\text{-}C_{27}$, $n\text{-}C_{29}$, and $n\text{-}C_{31}$ alkanes), and (4) infrared dispersion of clay particles.

Thus, and to reaffirm earlier statements, there is no one analytical method that is perfect or even adequate, for all cases to determine the amount of TPH in a sample. Different analytical methods have different capabilities and (this is where the environmental analysts play an important role) it is up to the analyst and within the purview of the analyst to demonstrate that the method applied at a specific site was appropriate.

REFERENCES

Altgelt, K.H. and Gouw, T.H. 1975. In *Advances in Chromatography*. Giddings J.C., GrushkaE., Keller R.A., and Cazes. J. eds. Marcel Dekker Inc., New York.

Bruner, F. 1993. *Gas Chromatographic Environmental Analysis: Principles, Techniques, and Instrumentation*. John Wiley & Sons Inc., Hoboken, New Jersey.

Budde, W.L. 2001. *The Manual of Manuals*. Office of Research and Development, Environmental Protection Agency, Washington, DC.

Dean, J.R. 1998. *Extraction Methods for Environmental Analysis*. John Wiley & Sons, Inc., New York.

EPA. 1995. *Profile of the Petroleum Refining Industry*. Environmental Protection Agency, Washington, DC.

EPA. 2004. *Environmental Protection Agency*, Washington, DC. Web site: http://www.epa.gov

Fowlis, I.A. 1995. *Gas Chromatography*. 2nd edn. John Wiley & Sons Inc., New York.

Irwin, R.J. 1997. Petroleum. In *Environmental Contaminants Encyclopedia*, National Park Service, Water Resources Divisions, Water Operations Branch, Fort Collins, Colorado.

Miller, M. ed. 2000. *Encyclopedia of Analytical Chemistry*. John Wiley & Sons Inc., Hoboken, New Jersey.

Nelson, P. 2003. *Index to EPS Test Methods*. US EPA New England Region, Boston MA.

Patnaik, P. ed. 2004. *Dean's Analytical Chemistry Handbook*. 2nd edn. McGraw-Hill, New York.

Rhodes, I.A., Hinojas, E.M., Barker, D.A., and Poole, R.A. 1994. *Pitfalls Using Conventional TPH Methods for Source Identification*. Proceedings. Seventh Annual Conference: EPA

Analysis of Pollutants in the Environment. Norfolk, VA. Environmental Protection agency, Washington, DC.

Smith, K.A. and Cresser, M. 2003. *Soil & Environmental Analysis: Modern Instrumental Techniques.* Marcel Dekker Inc., New York. 2003.

Speight, J.G. 2001. *Handbook of Petroleum Analysis.* John Wiley & Sons Inc., Hoboken, New Jersey.

Speight, J.G. 2002. *Handbook of Petroleum Product Analysis.* John Wiley & Sons Inc., Hoboken, New Jersey.

Speight, J.G. 2005. *Environmental Analysis and Technology for the Refining Industry.* John Wiley & Sons Inc., Hoboken, New Jersey.

Sullivan R. and Johnson W. 1993. Oil You Need to know about Crude – Implications of TPH Data for Common Petroleum Products. Soil. May, page 8.

Sunahara, G.I., Renoux, A.Y., Thellen, C., Gaudet, C.L., and Pilon, A. (eds). 2002. *Environmental Analysis of Contaminated Sites.* John Wiley & Sons Inc., New York.

Weisman, W. ed. 1998. *Analysis of Petroleum Hydrocarbons in Environmental Media.* Total Petroleum Hydrocarbon Criteria Working Group Series. Amherst Scientific Publishers, Amherst, MA.

Conversion Factors

1 acre = 43,560 sq. ft.
1 acre foot = 7758.0 bbl
1 atmosphere = 760 mm Hg = 14.696 psi = 29.91 in. Hg
1 atmosphere = 1.0133 bars = 33.899 ft. H_2O
1 barrel (oil) = 42 gal = 5.6146 cu. ft.
1 barrel (water) = 350 lb. at 60°F
1 barrel per day = 1.84 cu. cm/sec
1 Btu = 778.26 ft.-lb.
1 centipoise × 2.42 = lb. mass/(ft.-h), viscosity
1 centipoise × 0.000672 = lb. mass/(ft.-sec), viscosity
1 cubic foot = 28,317 cu. cm = 7.4805 gal
Density of water at 60° F = 0l.999 g/cm^3 = −62.367 lb./cu. ft. = 8.337 lb./gal
1 gallon = 231 cu. in. = 3,785.4 cm^3 = 0.13368 cu. ft.
1 horsepower-hour = 0.7457 kWh = 2544.5 Btu
1 horsepower = 550 ft.-lb./sec = 745.7 W
1 inch = 2.54 cm
1 meter = 100 cm = 1000 mm = 10^6 μm = 10^{10} Å (Δ)
1 ounce = 28.35 g
1 pound = 453.30 g = 7000 grains
1 square mile = 640 acres

SI METRIC CONVERSION FACTORS
(E = EXPONENT; I.E. E + 03 = 10^3)

Acre-foot × 1.233482	E + 03 = meter cube
Barrels × 1.589873	E − 01 = meter cube
Centipoises × 1.00000	E − 03 = pascal seconds
Darcy × 9.869233	E − 01 = micrometer square
Feet × 3.048000	E − 01 = meters
Pounds/acre-foot × 3.677332	E − 04 = kilograms/meter cube
Pounds/square inch × 6.894757	E − 00 = kilopascals
Dyne/cm × 1.000000	E + 00 = mN/m
Parts per million × 1.000000	E + 00 = milligrams/kilograms

Glossary

The following list represents a selection of definitions that are commonly used with reference to refining operations (processes, equipment, and products) and will be of use to the reader. Older names, as may occur in many books, are also included for clarification.

ABN separation a method of fractionation by which petroleum is separated into acidic, basic, and neutral constituents.

Absorber *see* Absorption tower.

Absorption gasoline gasoline extracted from natural gas or refinery gas by contacting the absorbed gas with an oil and subsequently distilling the gasoline from the higher-boiling components.

Absorption oil oil used to separate the heavier components from a vapor mixture by absorption of the heavier components during intimate contacting of the oil and vapor; used to recover natural gasoline from wet gas.

Absorption plant a plant for recovering the condensable portion of natural or refinery gas, by absorbing the higher-boiling hydrocarbons in an absorption oil, followed by separation and fractionation of the absorbed material.

Absorption tower a tower or column which promotes contact between a rising gas and a falling liquid so that part of the gas may be dissolved in the liquid.

Acetone–benzol process a dewaxing process in which acetone and benzol (benzene or aromatic naphtha) are used as solvents.

Acid catalyst a catalyst having acidic character; the aluminas are examples of such catalysts.

Acid deposition acid rain; a form of pollution depletion in which pollutants, such as nitrogen oxides and sulfur oxides, are transferred from the atmosphere to soil or water; often referred to as atmospheric self-cleaning. The pollutants usually arise from the use of fossil fuels.

Acidity the capacity of an acid to neutralize a base, such as a hydroxyl ion (OH^-).

Acidizing a technique for improving the permeability (*q.v.*) of a reservoir by injecting acid.

Acid number a measure of the reactivity of petroleum with a caustic solution and given in terms of milligrams of potassium hydroxide that are neutralized by one gram of petroleum.

Acid rain the precipitation phenomenon that incorporates anthropogenic acids and other acidic chemicals from the atmosphere to the land and water (*see* Acid deposition).

Acid sludge the residue left after treating petroleum oil with sulfuric acid for the removal of impurities; a black, viscous substance containing the spent acid and impurities.

Acid treatment a process in which unfinished petroleum products, such as gasoline, kerosene, and lubricating-oil stocks, are contacted with sulfuric acid to improve their color, odor, and other properties.

Acoustic log *see* Sonic log.

Acre-foot a measure of bulk rock volume, where the area is one acre and the thickness is one foot.

Additive a material added to another (usually in small amounts) in order to enhance desirable properties or to suppress undesirable properties.

Add-on control methods the use of devices that remove refinery process emissions after they are generated but before they are discharged to the atmosphere.

Adsorption transfer of a substance, from a solution to the surface of a solid, resulting in relatively high concentration of the substance at the place of contact; *see also* Chromatographic adsorption.

Adsorption gasoline natural gasoline (*q.v.*) obtained by the adsorption process from wet gas.

Afterburn the combustion of carbon monoxide (CO) to carbon dioxide (CO_2); usually in the cyclones of a catalyst regenerator.

After flow flow from the reservoir into the wellbore that continues for a period after the well has been shut in; after-flow can complicate the analysis of a pressure transient test.

Air-blown asphalt asphalt produced by blowing air through residua at elevated temperatures.

Air injection an oil recovery technique using air to force oil from the reservoir into the wellbore.

Airlift Thermofor catalytic cracking a moving-bed continuous catalytic process for conversion of heavy gas oils into lighter products; the catalyst is moved by a stream of air.

Air pollution the discharge of toxic gases and particulate matter introduced into the atmosphere, principally as a result of human activity.

Air sweetening a process in which air or oxygen is used to oxidize lead mercaptides to disulfides, instead of using elemental sulfur.

Air toxics hazardous air pollutants.

Albertite a black, brittle, natural hydrocarbon possessing a conchoidal fracture and a specific gravity of approximately 1.1.

Alicyclic hydrocarbon a compound containing carbon and hydrogen only, which has a cyclic structure (e.g., cyclohexane); also collectively called naphthenes.

Aliphatic hydrocarbon a compound containing carbon and hydrogen only, which has an open-chain structure (e.g., as ethane, butane, octane, butene) or a cyclic structure (e.g., cyclohexane).

Aliquot the quantity of material of proper size for measurement of the property of interest; test portions may be taken from the gross sample directly, but often preliminary operations such as mixing or further reduction in particle size are necessary.

Alkaline a high pH, usually of an aqueous solution; aqueous solutions of sodium hydroxide, sodium orthosilicate, and sodium carbonate are typical alkaline materials used in enhanced oil recovery.

Alkaline flooding *see* EOR process.

Alkalinity the capacity of a base to neutralize the hydrogen ion (H^+).

Alkali treatment *see* Caustic wash.

Alkali wash *see* Caustic wash.

Alkanes hydrocarbons that contain only single carbon–hydrogen bonds. The chemical name indicates the number of carbon atoms and ends with the suffix "ane".

Alkenes hydrocarbons that contain carbon–carbon double bonds. The chemical name indicates the number of carbon atoms and ends with the suffix "ene".

Alkylate the product of an alkylation (*q.v.*) process.

Alkylate bottoms residua from fractionation of alkylate; the alkylate product which boils higher than the aviation gasoline range; sometimes called heavy alkylate or alkylate polymer.

Alkylation in the petroleum industry, a process by which an olefin (e.g., ethylene) is combined with a branched-chain hydrocarbon (e.g., *iso*-butane); alkylation may be accomplished as a thermal or as a catalytic reaction.

Alkyl groups a group of carbon and hydrogen atoms that branch from the main carbon chain or ring in a hydrocarbon molecule. The simplest alkyl group, a methyl group, is a carbon atom attached to three hydrogen atoms.

Alpha-scission the rupture of the aromatic carbon–aliphatic carbon bond that joins an alkyl group to an aromatic ring.

Alumina (Al_2O_3) used in separation methods as an adsorbent and in refining as a catalyst.

American Society for Testing and Materials (ASTM) the official organization in the United States for designing standard tests for petroleum and other industrial products.

Amine washing a method of gas cleaning, whereby acidic impurities such as hydrogen sulfide and carbon dioxide are removed from the gas stream by washing with an amine (usually an alkanolamine).

Analytical equivalence the acceptability of the results obtained from the different laboratories; a range of acceptable results.

Analyte the chemical for which a sample is tested, or analyzed.

Antibody a molecule having chemically reactive sites specific for certain other molecules.

Aniline point the temperature, usually expressed in °F, above which equal volumes of a petroleum product are completely miscible; a qualitative indication of the relative proportions of paraffins in a petroleum product which are miscible with aniline only at higher temperatures; a high aniline point indicates low aromatics.

Antiknock resistance to detonation or pinging in spark-ignition engines.

Antiknock agent a chemical compound such as tetraethyl lead which, when added in small amount to the fuel charge of an internal-combustion engine, tends to lessen knocking.

Antistripping agent an additive used in an asphaltic binder to overcome the natural affinity of an aggregate for water, instead of asphalt.

API gravity a measure of the lightness or heaviness of petroleum which is related to density and specific gravity.

$$°API = (141.5/spgr @ 60°F) - 131.5$$

Apparent bulk density the density of a catalyst as measured; usually loosely compacted in a container.

Apparent viscosity the viscosity of a fluid, or several fluids flowing simultaneously, measured in a porous medium (rock), and subject to both viscosity and permeability effects; *also called* effective viscosity.

Aquifer a subsurface rock interval that will produce water; often the underlay of a petroleum reservoir.

Areal sweep efficiency the fraction of the flood pattern area that is effectively swept by the injected fluids.

Aromatic hydrocarbon a hydrocarbon characterized by the presence of an aromatic ring or condensed aromatic rings; benzene and substituted benzene, naphthalene and substituted naphthalene, phenanthrene and substituted phenanthrene, as well as the higher condensed ring systems; compounds that are distinct from those of aliphatic compounds (*q.v.*) or alicyclic compounds (*q.v.*).

Aromatization the conversion of nonaromatic hydrocarbons to aromatic hydrocarbons by: (1) rearrangement of aliphatic (noncyclic) hydrocarbons (*q.v.*) into aromatic ring structures; and (2) dehydrogenation of alicyclic hydrocarbons (naphthenes).

Arosorb process a process for the separation of aromatics from nonaromatics by adsorption on a gel from which they are recovered by desorption.

Asphalt the nonvolatile product obtained by distillation and treatment of an asphaltic crude oil; a manufactured product.

Asphalt cement asphalt, especially prepared as to quality and consistency, for direct use in the manufacture of bituminous pavements.

Asphalt emulsion an emulsion of asphalt cement in water containing a small amount of emulsifying agent.

Asphalt flux an oil used to reduce the consistency or viscosity of hard asphalt to the point required for use.

Asphalt primer a liquid asphaltic material of low viscosity which, upon application to a nonbituminous surface, waterproofs the surface and prepares it for further construction.

Asphaltene (asphaltenes) the brown to black powdery material produced by treatment of petroleum, petroleum residua, or bituminous materials with a low-boiling liquid hydrocarbon, e.g., pentane or heptane; soluble in benzene (and other aromatic solvents), carbon disulfide, and chloroform (or other chlorinated hydrocarbon solvents).

Asphaltene association factor the number of individual asphaltene species which associate in nonpolar solvents as measured by molecular weight methods; the molecular weight of asphaltenes in toluene divided by the molecular weight in a polar nonassociating solvent, such as dichlorobenzene, pyridine, or nitrobenzene.

Asphaltic pyrobitumen *see* Asphaltoid.

Asphaltic road oil a thick, fluid solution of asphalt; usually a residual oil. *See also* Nonasphaltic road oil.

Asphaltite a variety of naturally occurring, dark brown to black, solid, nonvolatile bituminous material that is differentiated from bitumen, primarily by a high content of material insoluble in n-pentane (asphaltene) or other liquid hydrocarbons.

Asphaltoid a group of brown to black, solid bituminous materials of which the members are differentiated from asphaltites by their infusibility and low solubility in carbon disulfide.

Asphaltum *see* Asphalt.

Associated molecular weight the molecular weight of asphaltenes in an associating (nonpolar) solvent, such as toluene.

Atmospheric residuum a residuum (*q.v.*) obtained by distillation of a crude oil under atmospheric pressure and which boils above 350°C (660°F).

Atmospheric equivalent boiling point (AEBP) a mathematical method of estimating the boiling point at atmospheric pressure of nonvolatile fractions of petroleum.

Attainment area a geographical area that meets NAAQS criteria for air pollutants (*see also* Nonattainment area).

Attapulgus clay *see* Fuller's earth.

Autofining a catalytic process for desulfurizing distillates.

Average particle size the weighted average particle diameter of a catalyst.

Aviation gasoline any of the special grades of gasoline suitable for use in certain airplane engines.

Aviation turbine fuel *see* Jet fuel.

Back mixing the phenomenon observed when a catalyst travels at a slower rate in the riser pipe than the vapors.

BACT best available control technology.

Baghouse a filter system for the removal of particulate matter from gas streams; so called because of the similarity of the filters to coal bags.

Bank the concentration of oil (oil bank) in a reservoir that moves cohesively through the reservoir.

Bari-Sol process a dewaxing process which employs a mixture of ethylene dichloride and benzol as the solvent.

Barrel the unit of measurement of liquids in the petroleum industry; equivalent to 42 US standard gallons or 33.6 imperial gallons.

Base number the quantity of acid, expressed in milligrams of potassium hydroxide per gram of sample that is required to titrate a sample to a specified end-point.

Base stock a primary refined petroleum fraction into which other oils and additives are added (blended) to produce the finished product.

Basic nitrogen nitrogen (in petroleum) which occurs in pyridine form

Basic sediment and water (bs&w, bsw) the material which collects at the bottom of storage tanks, usually composed of oil, water, and foreign matter; *also called* bottoms, bottom settlings.

Battery a series of stills or other refinery equipment operated as a unit.

Baumé gravity the specific gravity of liquids expressed as degrees on the Baumé (°Bé) scale; for liquids lighter than water:

$$\text{Sp gr} \, 60°\text{F} = 140/(130 + °\text{Bé})$$

For liquids heavier than water:

$$\text{Sp gr} \, 60°\text{F} = 145/(145 - °\text{Bé})$$

Bauxite mineral matter used as a treating agent; hydrated aluminum oxide formed by the chemical weathering of igneous rock.

Bbl *see* Barrel.

Bell cap a hemispherical or triangular cover placed over the riser in a (distillation) tower to direct the vapors through the liquid layer on the tray; *see* Bubble cap.

Bender process a chemical treating process using lead sulfide catalyst for sweetening light distillates by which mercaptans are converted to disulfides by oxidation.

Bentonite montmorillonite (a magnesium–aluminum silicate); used as a treating agent.

Benzene a colorless aromatic liquid hydrocarbon (C_6H_6).

Benzin a refined light naphtha used for extraction purposes.

Benzine an obsolete term for light petroleum distillates covering the gasoline and naphtha range; *see* Ligroine.

Benzol the general term which refers to commercial or technical (not necessarily pure) benzene; also the term used for aromatic naphtha.

Beta-scission the rupture of a carbon–carbon bond that is; two bonds removed from an aromatic ring.

Billion 1×10^9

Biocide any chemical capable of killing bacteria and biorganisms.

Biogenic material derived from bacterial or vegetation sources.

Biological lipid any biological fluid that is miscible with a nonpolar solvent. These materials include waxes, essential oils, chlorophyll, etc.

Biological oxidation the oxidative consumption of organic matter by bacteria by which the organic matter is converted into gases.

Biomass biological organic matter.

Biopolymer a high molecular weight carbohydrate produced by bacteria.

Bitumen a semisolid to solid hydrocarbonaceous material found filling pores and crevices of sandstone, limestone, or argillaceous sediments.

Bituminous containing bitumen or constituting the source of bitumen.

Bituminous rock *see* Bituminous sand.

Bituminous sand a formation in which the bituminous material (*see* Bitumen) is found as a filling in veins and fissures in fractured rock or impregnating relatively shallow sand, sandstone, and limestone strata; a sandstone reservoir that is impregnated with a heavy, viscous black petroleum-like material that cannot be retrieved through a well by conventional production techniques.

Black acid(s) a mixture of the sulfonates found in acid sludge which are insoluble in naphtha, benzene, and carbon tetrachloride; very soluble in water, but insoluble in 30% sulfuric acid; in the dry, oil-free state, the sodium soaps are black powders.

Black oil any of the dark-colored oils; a term now often applied to heavy oil (*q.v.*).

Black soap *see* Black acid.

Black strap the black material (mainly lead sulfide) formed in the treatment of sour light oils with doctor solution (*q.v.*) and found at the interface between the oil and the solution.

Blown asphalt the asphalt prepared by air blowing a residuum (*q.v.*) or an asphalt (*q.v.*).

Bogging a condition that occurs in a coking reactor when the conversion to coke and light ends is too slow, causing the coke particles to agglomerate.

Boiling point a characteristic physical property of a liquid at which the vapor pressure is equal to that of the atmosphere and the liquid is converted to a gas.

Boiling range the range of temperature, usually determined at atmospheric pressure in standard laboratory apparatus, over which the distillation of an oil commences, proceeds, and finishes.

Bottled gas usually butane or propane, or butane–propane mixtures, liquefied and stored under pressure for domestic use; *see also* Liquefied petroleum gas.

Bottoms the liquid which collects at the bottom of a vessel (tower bottoms, tank bottoms) during distillation; also the deposit or sediment formed during storage of petroleum or a petroleum product; *see also* Residuum and Basic sediment and water.

Bright stock refined, high-viscosity lubricating oils, usually made from residual stocks by processes such as a combination of acid treatment or solvent extraction with dewaxing or clay finishing.

British thermal unit *see* Btu.

Bromine number the number of grams of bromine absorbed by 100 g of oil which indicates the percentage of double bonds in the material.

Brown acid oil-soluble petroleum sulfonates found in acid sludge which can be recovered by extraction with naphtha solvent. Brown-acid sulfonates are somewhat similar to mahogany sulfonates, but are more water-soluble. In the dry, oil-free state, the sodium soaps are light-colored powders.

Brown soap *see* Brown acid.

Brønsted acid a chemical species which can act as a source of protons.

Brønsted base a chemical species which can accept protons.

BS&W *see* Basic sediment and water.

BTEX benzene, toluene, ethylbenzene, and the xylene isomers.

Btu (British thermal unit) the energy required to raise the temperature of one pound of water by 1°F.

Bubble cap an inverted cup with a notched or slotted periphery to disperse the vapor in small bubbles beneath the surface of the liquid on the bubble plate in a distillation tower.

Bubble plate a tray in a distillation tower.

Bubble point the temperature at which incipient vaporization of a liquid in a liquid mixture occurs, corresponding with the equilibrium point of 0% vaporization or 100% condensation.

Bubble tower a fractionating tower so constructed that the vapors rising pass up through layers of condensate on a series of plates or trays (*see* Bubble plate); the vapor passes from one plate to the next above by bubbling under one or more caps (*see* Bubble cap) and out through the liquid on the plate, where the less volatile portions of vapor condense, overflow to the next lower plate, and ultimately back into the reboiler, thereby effecting fractionation.

Bubble tray a circular, perforated plates having the internal diameter of a bubble tower (*q.v.*), set at specified distances in a tower to collect the various fractions produced during distillation.

Buckley–Leverett method a theoretical method of determining frontal advance rates and saturations from a fractional flow curve.

Bumping the knocking against the walls of a still occurring during distillation of petroleum or a petroleum product which usually contains water.

Bunker C oil *see* No. 6 Fuel oil.

Burner fuel oil any petroleum liquid suitable for combustion.

Burning oil an illuminating oil, such as kerosene (kerosine) suitable for burning in a wick lamp.

Burning point *see* Fire point.

Burning-quality index an empirical numerical indication of the likely burning performance of a furnace or heater oil; derived from the distillation profile (*q.v.*) and the API gravity (*q.v.*), and generally recognizing the factors of paraffin character and volatility.

Burton process a older thermal cracking process in which oil was cracked in a pressure still and any condensation of the products of cracking also took place under pressure.

Butane dehydrogenation a process for removing hydrogen from butane to produce butenes and, on occasion, butadiene.

Butane vapor-phase isomerization a process for isomerizing *n*-butane to *iso*-butane using aluminum chloride catalyst on a granular alumina support and with hydrogen chloride as a promoter.

C_1, C_2, C_3, C_4, C_5 fractions a common way of representing fractions containing a preponderance of hydrocarbons having 1, 2, 3, 4, or 5 carbon atoms, respectively, and without reference to hydrocarbon type.

CAA Clean Air Act; this act is the foundation of air regulations in the United States

Calcining heating a metal oxide or an ore to decompose carbonates, hydrates, or other compounds often in a controlled atmosphere.

Capillary forces interfacial forces between immiscible fluid phases, resulting in pressure differences between the two phases.

Capillary number N_c, the ratio of viscous forces to capillary forces, and equal to viscosity times velocity divided by interfacial tension.

Carbene the pentane- or heptane-insoluble material that is insoluble in benzene or toluene but which is soluble in carbon disulfide (or pyridine); a type of rifle used for hunting bison.

Carboid the pentane- or heptane-insoluble material that is insoluble in benzene or toluene and which is also insoluble in carbon disulfide (or pyridine).

Carbonate washing processing using a mild alkali (e.g., potassium carbonate) process for emission control by the removal of acid gases from gas streams.

Carbon dioxide augmented waterflooding injection of carbonated water, or water and carbon dioxide, to increase water flood efficiency; *see* immiscible carbon dioxide displacement.

Carbon dioxide miscible flooding *see* EOR process.

Carbon-forming propensity *see* Carbon residue.

Carbonization the conversion of an organic compound into char or coke by heat in the substantial absence of air; often used in reference to the destructive distillation (with simultaneous removal of distillate) of coal.

Carbon–oxygen log information about the relative abundance of elements such as carbon, oxygen, silicon, and calcium in a formation; usually derived from pulsed neutron equipment.

Carbon rejection upgrading processes in which coke is produced, e.g., coking.

Carbon residue the amount of carbonaceous residue remaining after thermal decomposition of petroleum, a petroleum fraction, or a petroleum product in a limited amount of air; *also called the* coke- or carbon-forming propensity; often prefixed by the terms Conradson or Ramsbottom in reference to the inventor of the respective tests.

CAS Chemical Abstract Service.

Cascade tray a fractionating device consisting of a series of parallel troughs arranged in stair-step fashion, in which liquid from the tray above enters the uppermost trough and liquid thrown from this trough by vapor rising from the tray below impinges against a plate and a perforated baffle, and liquid passing through the baffle enters the next longer of the troughs.

Casinghead gas natural gas which issues from the casinghead (the mouth or opening) of an oil well.

Casinghead gasoline the liquid hydrocarbon product extracted from casinghead gas (*q.v.*) by one of three methods: compression, absorption, or refrigeration; *see also* Natural gasoline.

Catagenesis the alteration of organic matter during the formation of petroleum that may involve temperatures in the range of 50°C (120°F) to 200°C (390°F); *see also* Diagenesis and Metagenesis.

Catalyst a chemical agent which, when added to a reaction (process) will enhance the conversion of a feedstock without consumed in the process.

Catalyst selectivity the relative activity of a catalyst with respect to a particular compound in a mixture, or the relative rate in competing reactions of a single reactant.

Catalyst stripping the introduction of steam, at a point where spent catalyst leaves the reactor, in order to strip, i.e., remove, deposits retained on the catalyst.

Catalytic activity the ratio of the space velocity of the catalyst under test to the space velocity required for the standard catalyst to give the same conversion as the catalyst being tested; usually multiplied by 100 before being reported.

Catalytic cracking the conversion of high-boiling feedstocks into lower boiling products by means of a catalyst which may be used in a fixed bed (*q.v.*) or fluid bed (*q.v.*).

Cat cracking *see* Catalytic cracking.

Catalytic reforming rearranging hydrocarbon molecules in a gasoline-boiling-range feedstock to produce other hydrocarbons having a higher antiknock quality; isomerization of paraffins, cyclization of paraffins to naphthenes (*q.v.*), dehydrocyclization of paraffins to aromatics (*q.v.*).

Catforming a process for reforming naphtha using a platinum–silica–alumina catalyst which permits relatively high space velocities and results in the production of high-purity hydrogen.

Caustic consumption the amount of caustic lost from reacting chemically with the minerals in the rock, the oil, and the brine.

Chemical flooding see EOR process.

Caustic wash the process of treating a product with a solution of caustic soda to remove minor impurities; often used in reference to the solution itself.

Ceresin a hard, brittle wax obtained by purifying ozokerite; *see* Microcrystalline wax and Ozokerite.

Cetane index an approximation of the cetane number (*q.v.*) calculated from the density (*q.v.*) and midboiling point temperature (*q.v.*); *see also* Diesel index.

Cetane number a number indicating the ignition quality of diesel fuel; a high cetane number represents a short ignition delay time; the ignition quality of diesel fuel can also be estimated from the following formula:

$$\text{Diesel index} = (\text{aniline point (°F)} \times \text{API gravity})100$$

CFR Code of Federal Regulations; Title 40 (40 CFR) contains the regulations for protection of the environment.

Characterization factor the UOP characterization factor K, defined as the ratio of the cube root of the molal average boiling point, T_B, in degrees Rankine (°R = °F + 460), to the specific gravity at 60°F/60°F:

$$K = (T_B)^{1/3}/\text{sp gr}$$

The value ranges from 12.5 for paraffin stocks to 10.0 for the highly aromatic stocks; *also called the* Watson characterization factor.

Cheesebox still an early type of vertical cylindrical still designed with a vapor dome.

Chelating agents complex-forming agents with the ability to solubilize heavy metals.

Chemical flooding *see* EOR process.

Chemical octane number the octane number added to gasoline by refinery processes or by the use of octane number (*q.v.*) improvers, such as tetraethyl lead.

Chemical waste any solid, liquid, or gaseous material discharged from a process and that may pose substantial hazards to human health and environment.

Chlorex process a process for extracting lubricating-oil stocks in which the solvent used is Chlorex (*β–β*-dichlorodiethyl ether).

Chromatographic adsorption selective adsorption on materials such as activated carbon, alumina, or silica gel; liquid or gaseous mixtures of hydrocarbons are passed through the adsorbent in a stream of diluent, and certain components are preferentially adsorbed.

Chromatographic separation the separation of different species of compounds according to their size and interaction with the rock as they flow through a porous medium.

Chromatography a method of separation based on selective adsorption; *see also* Chromatographic adsorption.

Clarified oil the heavy oil which has been taken from the bottom of a fractionator in a catalytic cracking process and from which residual catalyst has been removed.

Clarifier equipment for removing the color or cloudiness of an oil or water by separating the foreign material through mechanical or chemical means; may involve centrifugal action, filtration, heating, or treatment with acid or alkali.

Clay silicate minerals that also usually contain aluminum and have particle sizes less than 0.002 micron; used in separation methods as an adsorbent and in refining as a catalyst.

Clay contact process *see* Contact filtration.

Clay refining a treating process in which vaporized gasoline or any other light petroleum product is passed through a bed of granular clay, such as Fuller's earth (*q.v.*).

Clay regeneration a process in which spent coarse-grained adsorbent clays from percolation processes are cleaned for reuse by deoiling them with naphtha, steaming out the excess naphtha, and then roasting in a stream of air to remove carbonaceous matter.

Clay treating *see* Gray clay treating.

Clay wash light oil, such as kerosene (kerosine) or naphtha, used to clean Fuller's earth after it has been used in a filter.

Clastic composed of pieces of preexisting rock.

Cleanup a preparatory step following extraction of a sample media designed to remove components that may interfere with subsequent analytical measurements.

Cloud point the temperature at which paraffin wax or other solid substances begin to crystallize or separate from the solution, imparting a cloudy appearance to the oil when the oil is chilled under prescribed conditions.

Coal an organic rock.

Coalescence the union of two or more droplets to form a larger droplet and, ultimately, a continuous phase.

Coal tar the specific name for the tar (*q.v.*) produced from coal.

Coal tar pitch the specific name for the pitch (*q.v.*) produced from coal.

COFCAW an enhanced oil recovery (EOR) process (*q.v.*) that combines forward combustion and water flooding.

Cogeneration an energy conversion method by which electrical energy is produced along with steam generated for EOR use.

Coke a gray to black solid carbonaceous material produced from petroleum during thermal processing; characterized by having a high carbon content (95%+ by weight) and a honeycomb type of appearance; it is insoluble in organic solvents.

Coke drum a vessel in which coke is formed and which can be isolated from the process for cleaning.

Coke number used, particularly in Great Britain, to report the results of the Ramsbottom carbon residue test (*q.v.*), which is also referred to as a coke test.

Coker the processing unit in which coking takes place.

Coking a process for the thermal conversion of petroleum in which gaseous, liquid, and solid (coke) products are formed.

Cold pressing the process of separating wax from oil by first chilling (to help form wax crystals) and then filtering under pressure in a plate and frame press.

Cold settling processing for the removal of wax from high-viscosity stocks, wherein a naphtha solution of the waxy oil is chilled and the wax crystallizes out of the solution.

Color stability the resistance of a petroleum product to color change due to light, aging, etc.

Combustible liquid a liquid with a flash point in excess of 37.8°C (100°F), but below 93.3°C (200°F).

Combustion zone the volume of reservoir rock wherein petroleum is undergoing combustion during enhanced oil recovery.

Composition the general chemical make-up of petroleum.

Completion interval the portion of the reservoir formation placed in fluid communication with the well by selectively perforating the wellbore casing.

Composition map a means of illustrating the chemical make-up of petroleum using chemical and physical property data.

Con Carbon *see* Carbon residue.

Condensate a mixture of light hydrocarbon liquids obtained by condensation of hydrocarbon vapors: predominately butane, propane, and pentane with some heavier hydrocarbons and relatively little methane or ethane; *see also* Natural gas liquids.

Conductivity a measure of the ease of flow through a fracture, perforation, or pipe.

Conformance the uniformity with which a volume of the reservoir is swept by injection fluids in area and vertical directions.

Conradson carbon residue *see* Carbon residue.

Contact filtration a process in which finely divided adsorbent clay is used to remove color bodies from petroleum products.

Contaminant a substance that causes deviation from the normal composition of an environment.

Continuous contact coking a thermal conversion process in which petroleum-wetted coke particles move downward into the reactor in which cracking, coking, and drying take place to produce coke, gas, gasoline, and gas oil.

Continuous contact filtration a process to finish lubricants, waxes, or special oils after acid treating, solvent extraction, or distillation.

Conventional recovery primary and secondary recovery.

Conversion the thermal treatment of petroleum which results in the formation of new products by the alteration of the original constituents.

Conversion cost the cost of changing a production well to an injection well, or some other change in the function of an oilfield installation.

Conversion factor the percentage of feedstock converted to light ends, gasoline, other liquid fuels, and coke.

Copper sweetening processes involving the oxidation of mercaptans to disulfides by oxygen in the presence of cupric chloride.

Core floods laboratory flow tests through samples (cores) of porous rock.

Co-surfactant a chemical compound, typically alcohol, that enhances the effectiveness of a surfactant.

Cp (centipoise) a unit of viscosity.

Craig–Geffen–Morse method a method for predicting oil recovery by water flood.

Cracked residua residua that have been subjected to temperatures above 350°C (660°F) during the distillation process.

Cracking the thermal processes by which the constituents of petroleum are converted to lower molecular weight products.

Cracking activity *see* Catalytic activity.

Cracking coil equipment used for cracking heavy petroleum products consisting of a coil of heavy pipe running through a furnace, so that the oil passing through it is subject to high temperature.

Cracking still the combined equipment-furnace, reaction chamber, fractionator for the thermal conversion of heavier feedstocks to lighter products.

Cracking temperature the temperature (350°C; 660°F) at which the rate of thermal decomposition of petroleum constituents becomes significant.

Criteria air pollutants air pollutants or classes of pollutants regulated by the Environmental Protection Agency; the air pollutants are (including Volatile organic compounds): ozone, carbon monoxide, particulate matter, nitrogen oxides, sulfur dioxide, and lead.

Cross-linking combining of two or polymer molecules by use of a chemical that mutually bonds with a part of the chemical structure of the polymer molecules.

Crude assay a procedure for determining the general distillation characteristics (e.g., distillation profile, *q.v.*) and other quality information of crude oil.

Crude oil *see* Petroleum.

Crude scale wax the wax product from the first sweating of the slack wax.

Crude still distillation (*q.v.*) equipment in which crude oil is separated into various products.

Cumene a colorless liquid [$C_6H_5CH(CH_3)_2$] used as an aviation gasoline blending component and as an intermediate in the manufacture of chemicals.

Cut point the boiling-temperature division between distillation fractions of petroleum.

Cutback the term applied to the products from blending heavier feedstocks or products with lighter oils to bring the heavier materials to the desired specifications.

Cutback asphalt asphalt liquefied by the addition of a volatile liquid such as naphtha or kerosene which, after application and on exposure to the atmosphere, evaporates leaving the asphalt.

Cutting oil an oil to lubricate and cool metal-cutting tools; *also called* cutting fluid, cutting lubricant.

Cycle stock the product taken from some later stage of a process and recharged (recycled) to the process at some earlier stage.

Cyclic steam injection the alternating injection of steam and the production of oil from the same well or wells.

Cyclization the process by which an open-chain hydrocarbon structure is converted to a ring structure, e.g., hexane to benzene.

Cyclone a device for extracting dust from industrial waste gases. It is in the form of an inverted cone into which the contaminated gas enters tangential from the top; the gas is propelled down a helical pathway, and the dust particles are deposited by means of centrifugal force onto the wall of the scrubber.

Deactivation reduction in catalyst activity by the deposition of contaminants (e.g., coke, metals) during a process.

Dealkylation the removal of an alkyl group from aromatic compounds.

Deasphaltened oil the fraction of petroleum after the asphaltene constituents have been removed.

Deasphaltening removal of a solid powdery asphaltene fraction from petroleum by the addition of the low-boiling liquid hydrocarbons such as *n*-pentane or *n*-heptane under ambient conditions.

Deasphalting the removal of the asphaltene fraction from petroleum by the addition of a low-boiling hydrocarbon liquid such as *n*-pentane or *n*-heptane; more correctly, the removal of asphalt (tacky, semisolid) from petroleum (as occurs in a refinery asphalt plant), by the addition of liquid propane or liquid butane under pressure.

Debutanization distillation to separate butane and lighter components from higher boiling components.

Decant oil the highest boiling product from a catalytic cracker; *also referred to as* slurry oil, clarified oil, or bottoms.

Decarbonizing a thermal conversion process designed to maximize coker gas–oil production and minimize coke and gasoline yields; operated at essentially lower temperatures and pressures than delayed coking (*q.v.*).

Decoking removal of petroleum coke from equipment such as coking drums; hydraulic decoking uses high-velocity water streams.

Decolorizing removal of suspended, colloidal, and dissolved impurities from liquid petroleum products by filtering, adsorption, chemical treatment, distillation, bleaching, etc.

Deethanization distillation to separate ethane and lighter components from propane and higher-boiling components; *also called* deethanation.

Degradation the loss of desirable physical properties of EOR fluids, e.g., the loss of viscosity of polymer solutions.

Dehydrating agents substances capable of removing water (drying, *q.v.*) or the elements of water from another substance.

Dehydrocyclization any process by which both dehydrogenation and cyclization reactions occur.

Dehydrogenation the removal of hydrogen from a chemical compound; for example, the removal of two hydrogen atoms from butane to make butene, as well as the removal of additional hydrogen to produce butadiene.

Delayed coking a coking process in which the thermal reactions are allowed to proceed to completion to produce gaseous, liquid, and solid (coke) products.

Demethanization the process of distillation in which methane is separated from the higher boiling components; *also called* demethanation.

Density the mass (or weight) of a unit volume of any substance at a specified temperature; see also **Specific gravity**.

Deoiling reduction in quantity of liquid oil entrained in solid wax by draining (sweating) or by a selective solvent; *see* MEK deoiling.

Depentanizer a fractionating column for the removal of pentane and lighter fractions from a mixture of hydrocarbons.

Depropanization distillation in which lighter components are separated from butanes and higher boiling material; *also called* depropanation.

Desalting removal of mineral salts (mostly chlorides) from crude oils.

Desorption the reverse process of adsorption whereby adsorbed matter is removed from the adsorbent; also used as the reverse of absorption (*q.v.*).

Desulfurization the removal of sulfur or sulfur compounds from a feedstock.

Detergent oil lubricating oil possessing special sludge-dispersing properties for use in internal-combustion engines.

Dewaxing *see* Solvent dewaxing.

Devolatilized fuel smokeless fuel; coke that has been reheated to remove all of the volatile material.

Diagenesis the concurrent and consecutive chemical reactions which commence the alteration of organic matter (at temperatures up to 50°C (120°F) and ultimately result in the formation of petroleum from the marine sediment; *see also* Catagenesis and Metagenesis.

Diagenetic rock rock formed by conversion through pressure or chemical reaction from a rock, e.g., sandstone is a diagenetic.

Diesel fuel fuel used for internal combustion in diesel engines; usually that fraction which distills after kerosene.

Diesel cycle a repeated succession of operations representing the idealized working behavior of the fluids in a diesel engine.

Diesel index an approximation of the cetane number (*q.v.*) of diesel fuel (*q.v.*) calculated from the density (*q.v.*) and aniline point (*q.v.*).

Diesel knock the result of a delayed period of ignition of diesel fuel in the engine.

Differential-strain analysis measurement of thermal stress relaxation in a recently cut well.

Dispersion a measure of the convective fluids due to flow in a reservoir.

Displacement efficiency the ratio of the amount of oil moved from the zone swept by the recovery process to the amount of oil present in the zone prior to the start of the process.

Distribution coefficient a coefficient that describes the distribution of a chemical in reservoir fluids, usually defined as the equilibrium concentrations in the aqueous phases.

Distillation a process for separating liquids with different boiling points.

Distillation curve *see* Distillation profile.

Distillation loss the difference, in a laboratory distillation, between the volume of liquid originally introduced into the distilling flask and the sum of the residue and the condensate recovered.

Distillation range the difference between the temperature at the initial boiling point and at the end point, as obtained by the distillation test.

Distillation profile the distillation characteristics of petroleum or petroleum products showing the temperature and the percent distilled.

Doctor solution a solution of sodium plumbite used to treat gasoline or other light petroleum distillates to remove mercaptan sulfur; *see also* Doctor test.

Doctor sweetening a process for sweetening gasoline, solvents, and kerosene by converting mercaptans to disulfides using sodium plumbite and sulfur.

Doctor test a test used for the detection of compounds in light petroleum distillates which react with sodium plumbite; *see also* Doctor solution.

Domestic heating oil *see* No. 2 Fuel Oil.

Donor solvent process a conversion process in which hydrogen donor solvent is used in place of or to augment hydrogen.

Downcomer a means of conveying liquid from one tray to the next below in a bubble tray column (*q.v.*).

Downhole steam generator a generator installed downhole in an oil well to which oxygen-rich air, fuel, and water are supplied for the purposes of generating steam into the reservoir. Its major advantage over a surface steam generating facility is the losses to the wellbore and surrounding formation are eliminated.

Drying removal of a solvent or water from a chemical substance; also referred to as the removal of solvent from a liquid or suspension.

Dropping point the temperature at which grease passes from a semisolid to a liquid state under prescribed conditions.

Dry gas a gas which does not contain fractions that may easily condense under normal atmospheric conditions.

Dry point the temperature at which the last drop of petroleum fluid evaporates in a distillation test.

Dualayer distillate process a process for removing mercaptans and oxygenated compounds from distillate fuel oils and similar products, using a combination of treatment with concentrated caustic solution and electrical precipitation of the impurities.

Dualayer gasoline process a process for extracting mercaptans and other objectionable acidic compounds from petroleum distillates; *see also* Dualayer solution.

Dualayer solution a solution which consists of concentrated potassium or sodium hydroxide containing a solubilizer; *see also* Dualayer gasoline process.

Dubbs cracking an older continuous, liquid-phase thermal cracking process, formerly used.

Dykstra–Parsons coefficient an index of reservoir heterogeneity arising from permeability variation and stratification.

Emulated bed a process in which the catalyst bed is in a suspended state in the reactor by means of a feedstock recirculation pump, which pumps the feedstock upwards at sufficient speed to expand the catalyst bed, at approximately 35% above the settled level.

Edeleanu process a process for refining oils at low temperature with liquid sulfur dioxide (SO_2), or with liquid sulfur dioxide and benzene; applicable to the recovery of aromatic concentrates from naphtha and heavier petroleum distillates.

Effective viscosity *see* Apparent viscosity.

Effluent any contaminating substance, usually a liquid, which enters the environment via a domestic industrial, agricultural, or sewage plant outlet.

Electric desalting a continuous process to remove inorganic salts and other impurities from crude oil by settling out in an electrostatic field.

Electrical precipitation a process using an electrical field to improve the separation of hydrocarbon reagent dispersions. May be used in chemical treating processes on a wide variety of refinery stocks.

Electrofining a process for contacting a light hydrocarbon stream with a treating agent (acid, caustic, doctor, etc.), then assisting the action of separation of the chemical phase from the hydrocarbon phase by an electrostatic field.

Electrolytic mercaptan process a process in which aqueous caustic solution is used to extract mercaptans from refinery streams.

Electrostatic precipitators devices used to trap fine dust particles (usually in the size range 30–60 microns) that operate on the principle of imparting an electric charge to particles in an incoming air stream and which are then collected on an oppositely charged plate across a high voltage field.

Eluate the solutes, or analytes, moved through a chromatographic column (*see* elution).

Eluent solvent used to elute sample.

Elution a process whereby a solute is moved through a chromatographic column by a solvent (liquid or gas) or eluent.

Emission control the use gas cleaning processes to reduce emissions.

Emission standard the maximum amount of a specific pollutant permitted to be discharged from a particular source in a given environment.

Emulsion a dispersion of very small drops of one liquid in an immiscible liquid, such as oil in water.

Emulsion breaking the settling or aggregation of colloidal-sized emulsions from suspension in a liquid medium.

End-of-pipe emission control the use of specific emission control processes to clean gases after production of the gases.

Energy the capacity of a body or system to do work, measured in joules (SI units); also the output of fuel sources.

Energy from biomass the production of energy from biomass (*q.v.*).

Engler distillation a standard test for determining the volatility characteristics of a gasoline by measuring the percent distilled at various specified temperatures.

Enhanced oil recovery (EOR) petroleum recovery following recovery by conventional (i.e., primary and secondary) methods (*q.v.*).

Enhanced oil recovery (EOR) process a method for recovering additional oil from a petroleum reservoir beyond that economically recoverable by conventional primary and secondary recovery methods. EOR methods are usually divided into three main categories: (1) *chemical flooding:* injection of water with added chemicals into a petroleum reservoir. The chemical processes include: surfactant flooding, polymer flooding, and alkaline flooding, (2) *miscible flooding:* injection into a petroleum reservoir of a material that is miscible, or can become miscible, with the oil in the reservoir. Carbon dioxide, hydrocarbons, and nitrogen are used, (3) *thermal recovery:* injection of steam into a petroleum reservoir, or propagation of a combustion zone through a reservoir by air or oxygen-enriched air injection. The thermal processes include: steam drive, cyclic steam injection, and in situ combustion.

Entrained bed a bed of solid particles suspended in a fluid (liquid or gas) at such a rate that some of the solid is carried over (entrained) by the fluid.

EPA Environmental Protection Agency.

Ester a compound formed by the reaction between an organic acid and an alcohol; ethoxylated alcohols (i.e., alcohols having ethylene oxide functional groups attached to the alcohol molecule).

Ethanol *see* Ethyl alcohol.

Ethyl alcohol (ethanol or grain alcohol): an inflammable organic compound (C_2H_5OH) formed during fermentation of sugars; used as an intoxicant and as a fuel.

Evaporation a process for concentrating nonvolatile solids in a solution by boiling off the liquid portion of the waste stream.

Expanding clays clays that expand or swell on contact with water, e.g., montmorillonite.

Explosive limits the limits of percentage composition of mixtures of gases and air within which an explosion takes place when the mixture is ignited.

Extract the portion of a sample preferentially dissolved by the solvent and recovered by physically separating the solvent.

Extractive distillation the separation of different components of mixtures which have similar vapor pressures by flowing a relatively high-boiling solvent, which is selective for one of the components in the feed, down a distillation column as the distillation proceeds; the selective solvent scrubs the soluble component from the vapor.

Fabric filters filters made from fabric materials and used for removing particulate matter from gas streams (*see* Baghouse).

Facies one or more layers of rock that differs from other layers in composition, age or content.

FAST Fracture assisted steamflood technology.

Fat oil the bottom or enriched oil drawn from the absorber as opposed to lean oil.

Faujasite a naturally occurring silica–alumina (SiO_2–Al_2O_3) mineral.

FCC fluid catalytic cracking.

FCCU fluid catalytic cracking unit.

Feedstock petroleum as it is fed to the refinery; a refinery product that is used as the raw material for another process; the term is also generally applied to raw materials used in other industrial processes.

Ferrocyanide process a regenerative chemical treatment for mercaptan removal, using caustic-sodium ferrocyanide reagent.

Field-scale the application of EOR processes to a significant portion of a field.

Filtration the use of an impassable barrier to collect solids but which allows liquids to pass.

Fingering the formation of finger-shaped irregularities at the leading edge of a displacing fluid in a porous medium, which move out ahead of the main body of fluid.

Fire point the lowest temperature at which, under specified conditions in standardized apparatus, a petroleum product vaporizes sufficiently rapidly to form above its surface an air–vapor mixture which burns continuously when ignited by a small flame.

First contact miscibility *see* miscibility.

Fischer–Tropsch process a process for synthesizing hydrocarbons and oxygenated chemicals from a mixture of hydrogen and carbon monoxide.

Fixed bed a stationary bed (of catalyst) to accomplish a process (*see* Fluid bed).

Five-spot an arrangement or pattern of wells with four injection wells at the corners of a square and a producing well in the center of the square.

Flammability range the range of temperature over which a chemical is flammable.

Flammable a substance that will burn readily.

Flammable liquid a liquid having a flash point below 37.8°C (100°F).

Flammable solid a solid that can ignite from friction or from heat remaining from its manufacture, or which may cause a serious hazard if ignited.

Flash point the lowest temperature to which the product must be heated under specified conditions to give off sufficient vapor to form a mixture with air that can be ignited momentarily by a flame.

Flocculation threshold the point at which constituents of a solution (e.g., asphaltene constituents or coke precursors) will separate from the solution as a separate (solid) phase.

Floc point the temperature at which wax or solids separate as a definite floc.

Flood, flooding the process of displacing petroleum from a reservoir by the injection of fluids.

Flexicoking a modification of the fluid coking process insofar as the process also includes a gasifier adjoining the burner or regenerator to convert excess coke to a clean fuel gas.

Flue gases the gaseous products of the combustion process mostly comprised of carbon dioxide, nitrogen, and water vapor.

Flue gas gas from the combustion of fuel, the heating value of which has been substantially spent and which is, therefore, discarded to the flue or stack.

Fluid a reservoir gas or liquid.

Fluid-bed a bed (of catalyst) that is agitated by an upward passing gas in such a manner that the particles of the bed simulate the movement of a fluid and has the characteristics associated with a true liquid; cf. Fixed bed.

Fluid catalytic cracking cracking in the presence of a fluidized bed of catalyst.

Fluid coking a continuous fluidized solids process that cracks feed thermally over heated coke particles in a reactor vessel to gas, liquid products, and coke.

Fluidized bed combustion a process used to burn low-quality solid fuels in a bed of small particles suspended by a gas stream (usually air that will lift the particles, but not blow them out of the vessel). Rapid burning removes some of the offensive byproducts of combustion from the gases and vapors that result from the combustion process.

Fly ash particulate matter produced from mineral matter in coal that is converted during combustion to finely divided inorganic material and which emerges from the combustor in the gases.

Foots oil the oil sweated out of slack wax; named from the fact that the oil goes to the foot, or bottom, of the pan during the sweating operation.

Formation an interval of rock with distinguishable geologic characteristics.

Formation volume factor the volume in a barrel that one stock tank barrel occupies in the formation at reservoir temperature, and with the solution gas that is held in the oil at reservoir pressure.

Fossil fuel resources a gaseous, liquid, or solid fuel material formed in the ground by chemical and physical changes (diagenesis, *q.v.*) in plant and animal residues over geological time; natural gas, petroleum, coal, and oil shale.

Fractional composition the composition of petroleum as determined by fractionation (separation) methods.

Fractional distillation the separation of the components of a liquid mixture by vaporizing and collecting the fractions, or cuts, which condense in different temperature ranges.

Fractional flow the ratio of the volumetric flow rate of one fluid phase to the total fluid volumetric flow rate within a volume of rock.

Fractional flow curve the relationship between the fractional flow of one fluid and its saturator during simultaneous flow of fluids through rock.

Fracture a natural or man-made crack in a reservoir rock.

Fracturing the breaking apart of reservoir rock by applying very high fluid pressure at the rock face.

Fractionating column a column arranged to separate various fractions of petroleum by a single distillation and which may be tapped at different points along its length to separate various fractions in the order of their boiling points.

Fractionation the separation of petroleum into the constituent fractions using solvent or adsorbent methods; chemical agents such as sulfuric acid may also be used.

Frasch process a process formerly used for removing sulfur by distilling oil in the presence of copper oxide.

Fuel oil *also called* heating oil is a distillate product that covers a wide range of properties; *see also* No. 1– No. 4 Fuel oils.

Fuller's earth a clay which has high adsorptive capacity for removing color from oils; attapulgus clay is a widely used Fuller's earth.

Functional group the portion of a molecule that is characteristic of a family of compounds and determines the properties of these compounds.

Furfural extraction a single-solvent process in which furfural is used to remove aromatic, naphthene, olefin, and unstable hydrocarbons from a lubricating-oil charge stock.

Furnace oil a distillate fuel primarily intended for use in domestic heating equipment.

Gas cap a part of a hydrocarbon reservoir at the top that will produce only gas.

Gas–oil ratio ratio of the number of cubic feet of gas measured at atmospheric (standard) conditions to barrels of produced oil measured at stocktank conditions.

Gas–oil sulfonate sulfonate made from a specific refinery stream, in this case the gas–oil stream.

Gasoline fuel for the internal combustion engine that is commonly, but improperly, referred to simply as **gas**.

Gaseous pollutants gases released into the atmosphere that act as primary or secondary pollutants.

Gasohol a term for motor vehicle fuel comprising between 80%–90% unleaded gasoline and 10%–20% ethanol (*see also* Ethyl alcohol).

Gas oil a petroleum distillate with a viscosity and boiling range between those of kerosine and lubricating oil.

Gas reversion a combination of thermal cracking or reforming of naphtha with thermal polymerization or alkylation of hydrocarbon gases carried out in the same reaction zone.

Gilsonite an asphaltite that is >90% bitumen.

Girbotol process a continuous, regenerative process to separate hydrogen sulfide, carbon dioxide, and other acid impurities from natural gas, refinery gas, etc., using mono-, di-, or triethanolamine as the reagent.

Glance pitch an asphaltite.

Glycol-amine gas treating a continuous, regenerative process to simultaneously dehydrate and remove acid gases from natural gas or refinery gas.

Grahamite an asphaltite.

Gravity *see* API gravity.

Gravity drainage the movement of oil in a reservoir that results from the force of gravity.

Gravity segregation partial separation of fluids in a reservoir caused by the gravity force acting on differences in density.

Gravity-stable displacement the displacement of oil from a reservoir by a fluid of a different density, where the density difference is utilized to prevent gravity segregation of the injected fluid.

Gray clay treating a fixed-bed (*q.v.*), usually Fuller's earth (*q.v.*), vapor-phase treating process to selectively polymerize unsaturated gum-forming constituents (diolefins) in thermally cracked gasoline.

Grain alcohol *see* Ethyl alcohol.

Gravimetric methods used to weigh a residue.

Gravity drainage the movement of oil in a reservoir that results from the force of gravity.

Gravity segregation partial separation of fluids in a reservoir caused by the gravity force acting on differences in density.

Greenhouse effect warming of the earth due to entrapment of the sun's energy by the atmosphere.

Greenhouse gases gases that contribute to the greenhouse effect (*q.v.*).

Guard bed a bed of an adsorbent (such as, for example, bauxite) that protects a catalyst bed by adsorbing species detrimental to the catalyst.

Gulf HDS process a fixed-bed process for the catalytic hydrocracking of heavy stocks to lower-boiling distillates with accompanying desulfurization.

Gulfining a catalytic hydrogen treating process for cracked and straight-run distillates and fuel oils, to reduce sulfur content; improve carbon residue, color, and general stability; and effect a slight increase in gravity.

Gum an insoluble tacky semisolid material formed as a result of the storage instability and the thermal instability of petroleum and petroleum products.

HAP(s) hazardous air pollutant or pollutants.

Hardness the concentration of calcium and magnesium in brine.

HCPV hydrocarbon pore volume.

Hearn method a method used in reservoir simulation for calculating a pseudorelative permeability curve that reflects reservoir stratification.

Headspace the vapor space above a sample into which volatile molecules evaporate. Certain methods sample this vapor.

Heating oil *see* Fuel oil.

Heavy ends the highest boiling portion of a petroleum fraction; *see also* Light ends.

Heavy fuel oil fuel oil with a high density and viscosity; generally residual fuel oil such as No. 5 and No 6. fuel oil (*q.v.*)

Heavy oil petroleum with an API gravity of less than 20°.

Heavy petroleum *see* Heavy oil.

Heteroatom compounds chemical compounds which contain nitrogen, oxygen, sulfur, and metals bound within their molecular structure or structures.

Heterogeneity lack of uniformity in reservoir properties, such as permeability.

HF alkylation an alkylation process whereby olefins (C_3, C_4, C_5) are combined with *iso*-butane in the presence of hydrofluoric acid catalyst.

Higgins–Leighton model stream tube computer model used to simulate waterflood.

Hortonsphere a spherical pressure-type tank used to store a volatile liquid which prevents the excessive evaporation loss that occurs when such products are placed in conventional storage tanks.

Hot filtration test a test for the stability of a petroleum product.

Hot spot an area of a vessel or line wall appreciably above normal operating temperature, usually as a result of the deterioration of an internal insulating liner which exposes the line or vessel shell to the temperature of its contents.

Houdresid catalytic cracking a continuous moving-bed process for catalytically cracking reduced crude oil to produce high-octane gasoline and light distillate fuels.

Houdriflow catalytic cracking a continuous moving-bed catalytic cracking process employing an integrated single vessel for the reactor and regenerator kiln.

Houdriforming a continuous catalytic reforming process for producing aromatic concentrates and high-octane gasoline from low-octane straight naphtha.

Houdry butane dehydrogenation a catalytic process for dehydrogenating light hydrocarbons to their corresponding mono- or di-olefins.

Houdry fixed-bed catalytic cracking a cyclic regenerable process for cracking of distillates.

Houdry hydrocracking a catalytic process combining cracking and desulfurization in the presence of hydrogen.

Huff-and-puff a cyclic EOR method in which steam or gas is injected into a production well; after a short shut-in period, oil and the injected fluid are produced through the same well.

Hydration the association of molecules of water with a substance.

Hydraulic fracturing the opening of fractures in a reservoir by high-pressure, high-volume injection of liquids through an injection well.

Hydrocarbon compounds chemical compounds containing only carbon and hydrogen.

Hydrocarbon-producing resource a resource such as coal and oil shale (kerogen) which produces derived hydrocarbons by the application of conversion processes; the hydrocarbons so-produced are not naturally occurring materials.

Hydrocarbon resource resources such as petroleum and natural gas which can produce naturally occurring hydrocarbons without the application of conversion processes.

Hydrocarbons organic compounds containing only hydrogen and carbon.

Hydrolysis a chemical reaction in which water reacts with another substance to form one or more new substances.

Hydroconversion a term often applied to hydrocracking (*q.v.*)

Hydrocracking a catalytic high-pressure high-temperature process for the conversion of petroleum feedstocks in the presence of fresh and recycled hydrogen; carbon–carbon bonds are cleaved, in addition to the removal of heteroatomic species.

Hydrocracking catalyst a catalyst used for hydrocracking which typically contains separate hydrogenation and cracking functions.

Hydrodenitrogenation the removal of nitrogen by hydrotreating (*q.v.*).

Hydrodesulfurization the removal of sulfur by hydrotreating (*q.v.*).

Hydrofining a fixed-bed catalytic process to desulfurize and hydrogenate a wide range of charge stocks from gases through waxes.

Hydroforming a process in which naphtha is passed over a catalyst at elevated temperatures and moderate pressures, in the presence of added hydrogen or hydrogen-containing gases, to form high-octane motor fuel or aromatics.

Hydrogen blistering blistering of steel caused by trapped molecular hydrogen, formed as atomic hydrogen, during corrosion of steel by hydrogen sulfide.

Hydrogen addition an upgrading process in the presence of hydrogen, e.g., hydrocracking; *see* Hydrogenation.

Hydrogenation the chemical addition of hydrogen to a material. In nondestructive hydrogenation, hydrogen is added to a molecule only if, and where, unsaturation with respect to hydrogen exists.

Hydrogen transfer the transfer of inherent hydrogen within the feedstock constituents and products during processing.

Hydroprocessing a term often equally applied to hydrotreating (*q.v.*) and to hydrocracking (*q.v.*); also often collectively applied to both.

Hydrotreating the removal of heteroatomic (nitrogen, oxygen, and sulfur) species by treatment of a feedstock or product at relatively low temperatures in the presence of hydrogen.

Hydrovisbreaking a noncatalytic process, conducted under similar conditions to visbreaking, which involves treatment with hydrogen to reduce the viscosity of the feedstock and produce more stable products than is possible with visbreaking.

Hydropyrolysis a short residence time high temperature process using hydrogen.

Hyperforming a catalytic hydrogenation process for improving the octane number of naphtha through removal of sulfur and nitrogen compounds.

Hypochlorite sweetening the oxidation of mercaptans in a sour stock by agitation with aqueous, alkaline hypochlorite solution; used where avoidance of free-sulfur addition is desired, because of stringent copper strip requirements and minimum expense is not the primary object.

Ignitability characteristic of liquids whose vapors are likely to ignite in the presence of ignition source; also characteristic of nonliquids that may catch fire from friction or contact with water and that burn vigorously.

Illuminating oil oil used for lighting purposes.

Immiscible two or more fluids that do not have complete mutual solubility and coexist as separate phases.

Immiscible carbon dioxide displacement injection of carbon dioxide into an oil reservoir to effect oil displacement under conditions in which miscibility with reservoir oil is not obtained; *see* Carbon dioxide augmented waterflooding.

Immiscible displacement a displacement of oil by a fluid (gas or water) that is conducted under conditions so that interfaces exist between the driving fluid and the oil.

Immunoassay portable tests that take advantage of an interaction between an antibody and a specific analyte. Immunoassay tests are semi-quantitative and usually rely on color changes of varying intensities to indicate relative concentrations.

Incompatibility the immiscibility of petroleum products and also of different crude oils which is often reflected in the formation of a separate phase after mixing and storage.

Incremental ultimate recovery the difference between the quantity of oil that can be recovered by EOR methods and the quantity of oil that can be recovered by conventional recovery methods.

Infill drilling drilling additional wells within an established pattern.

Infrared spectroscopy an analytical technique that quantifies the vibration (stretching and bending) that occurs when a molecule absorbs (heat) energy in the infrared region of the electromagnetic spectrum.

Inhibitor a substance, the presence of which, in small amounts, in a petroleum product prevents or retards undesirable chemical changes from taking place in the product, or in the condition of the equipment in which the product is used.

Inhibitor sweetening a treating process to sweeten gasoline, using a phenylenediamine type of inhibitor, air, and caustic.

Initial boiling point the recorded temperature when the first drop of liquid falls from the end of the condenser.

Initial vapor pressure the vapor pressure of a liquid of a specified temperature and zero percent evaporated.

Injection profile the vertical flow rate distribution of fluid flowing from the wellbore into a reservoir.

Injection well a well in an oil field used for injecting fluids into a reservoir.

Injectivity the relative ease with which a fluid is injected into a porous rock.

In situ in its original place; in the reservoir.

In situ combustion an EOR process consisting of injecting air or oxygen-enriched air into a reservoir under conditions that favor burning part of the in situ petroleum, advancing this burning zone, and recovering oil heated from a nearby producing well.

Instability the inability of a petroleum product to exist for periods of time without change to the product.

Integrity maintenance of a slug or bank at its preferred composition without too much dispersion or mixing.

Interface the thin surface area separating two immiscible fluids that are in contact with each other.

Interfacial film a thin layer of material at the interface between two fluids which differs in composition from the bulk fluids.

Interfacial tension the strength of the film separating two immiscible fluids, e.g., oil and water or microemulsion and oil; measured in dynes (force) per centimeter or milli-dynes per centimeter.

Interfacial viscosity the viscosity of the interfacial film between two immiscible liquids.

Interference testing a type of pressure transient test in which pressure is measured over time in a closed-in well while nearby wells are produced; flow and communication between wells can sometimes be deduced from an interference test.

Interphase mass transfer the net transfer of chemical compounds between two or more phases.

Iodine number a measure of the iodine absorption by oil under standard conditions; used to indicate the quantity of unsaturated compounds present; *also called* iodine value.

Ion exchange a means of removing cations or anions from solution onto a solid resin.

Ion exchange capacity a measure of the capacity of a mineral to exchange ions in amount of material per unit weight of solid.

Ions chemical substances possessing positive or negative charges in solution.

Isocracking a hydrocracking process for conversion of hydrocarbons which operates at relatively low temperatures and pressures in the presence of hydrogen and a catalyst to produce more valuable, lower-boiling products.

Isoforming a process in which olefinic naphtha is contacted with an alumina catalyst at high temperature and low pressure to produce isomers of higher octane number.

Iso-kel process a fixed-bed, vapor-phase isomerization process using a precious metal catalyst and external hydrogen.

Isomate process a continuous, nonregenerative process for isomerizing C_5–C_8 normal paraffin hydrocarbons, using aluminum chloride–hydrocarbon catalyst with anhydrous hydrochloric acid as a promoter.

Isomerate process a fixed-bed isomerization process to convert pentane, heptane, and heptane to high-octane blending stocks.

Isomerization the conversion of a normal (straight-chain) paraffin hydrocarbon into an *iso* (branched-chain) paraffin hydrocarbon with the same atomic composition.

Isopach a line on a map designating points of equal formation thickness.

Iso-plus Houdriforming a combination process using a conventional Houdriformer operated at moderate severity, in conjunction with one of three possible alternatives, including the use of an aromatic recovery unit or a thermal reformer; *see* Houdriforming.

Jet fuel fuel meeting the required properties for use in jet engines and aircraft turbine engines.

Kaolinite a clay mineral formed by hydrothermal activity at the time of rock formation or by chemical weathering of rock with high feldspar content; usually associated with intrusive granite rock with high feldspar content.

Kata-condensed aromatic compounds Compounds based on linear condensed aromatic hydrocarbon systems, e.g., anthracene and naphthacene (tetracene).

Kauri butanol number A measurement of solvent strength for hydrocarbon solvents; the higher the kauri-butanol (KB) value, the stronger the solvency; the test method (ASTM D1133) is based on the principle that kauri resin is readily soluble in butyl alcohol but not in hydrocarbon solvents and the resin solution will tolerate only a certain amount of dilution and is reflected as a cloudiness when the resin starts to come out of solution; solvents such as toluene can be added in a greater amount (and thus have a higher KB value) than weaker solvents like hexane.

Kerogen a complex carbonaceous (organic) material that occurs in sedimentary rock and shale; generally insoluble in common organic solvents.

Kerosene (kerosine) a fraction of petroleum that was initially sought as an illuminant in lamps; a precursor to diesel fuel.

K-factor see **Characterization factor**.

Kinematic viscosity the ratio of viscosity (*q.v.*) to density, both measured at the same temperature.

Knock the noise associated with self-ignition of a portion of the fuel–air mixture ahead of the advancing flame front.

Kriging a technique used in reservoir description for interpolation of reservoir parameters between wells based on random field theory.

LAER lowest achievable emission rate; the required emission rate in nonattainment permits.

Lamp burning a test of burning oils in which the oil is burned in a standard lamp under specified conditions in order to observe the steadiness of the flame, the degree of encrustation of the wick, and the rate of consumption of the kerosene.

Lamp oil *see* Kerosene.

Leaded gasoline gasoline containing tetraethyl lead or other organometallic lead antiknock compounds.

Lean gas the residual gas from the absorber after the condensable gasoline has been removed from the wet gas.

Lean oil absorption oil from which gasoline fractions have been removed; oil leaving the stripper in a natural-gasoline plant.

Lewis acid a chemical species which can accept an electron pair from a base.

Lewis base a chemical species which can donate an electron pair.

Light ends the lower-boiling components of a mixture of hydrocarbons; *see also* Heavy ends, Light hydrocarbons.

Light hydrocarbons hydrocarbons with molecular weights less than that of heptane (C_7H_{16}).

Light oil the products distilled or processed from crude oil up to, but not including, the first lubricating-oil distillate.

Light petroleum petroleum with an API gravity greater than 20°.

Ligroine (Ligroin) a saturated petroleum naphtha boiling in the range of 20°C to 135°C (68°F to 275°F) and suitable for general use as a solvent; *also called* benzine or petroleum ether.

Linde copper sweetening a process for treating gasoline and distillates with a slurry of clay and cupric chloride.

Liquid petrolatum *see* White oil.

Liquefied petroleum gas propane, butane, or mixtures thereof, gaseous at atmospheric temperature and pressure, held in the liquid state by pressure to facilitate storage, transport, and handling.

Liquid chromatography a chromatographic technique that employs a liquid mobile phase.

Liquid/liquid extraction an extraction technique in which one liquid is shaken with or contacted by an extraction solvent to transfer molecules of interest into the solvent phase.

Liquid sulfur dioxide–benzene process a mixed-solvent process for treating lubricating-oil stocks to improve viscosity index; also used for dewaxing.

Lithology the geological characteristics of the reservoir rock.

Live steam steam coming directly from a boiler before being utilized for power or heat.

Liver the intermediate layer of dark-colored, oily material, insoluble in weak acid and in oil, which is formed when acid sludge is hydrolyzed.

Lorenz coefficient a permeability heterogeneity factor.

Lower-phase microemulsion a microemulsion phase containing a high concentration of water that, when viewed in a test tube, resides near the bottom with oil phase on top.

Lube see **Lubricating oil**.

Lube cut a fraction of crude oil of suitable boiling range and viscosity to yield lubricating oil when completely refined; *also referred to as* lube oil distillates or lube stock.

Lubricating oil a fluid lubricant used to reduce friction between bearing surfaces.

MACT maximum achievable control technology. Applies to major sources of hazardous air pollutants.

Mahogany acids oil-soluble sulfonic acids formed by the action of sulfuric acid on petroleum distillates. They may be converted to their sodium soaps (mahogany soaps) and extracted from the oil with alcohol for use in the manufacture of soluble oils, rust preventives, and special greases. The calcium and barium soaps of these acids are used as detergent additives in motor oils; see also Brown acids and Sulfonic acids.

Major source a source that has a potential to emit for a regulated pollutant that is at or greater than an emission threshold set by regulations.

Maltenes that fraction of petroleum that is soluble in, for example, pentane or heptane; deasphaltened oil (*q.v.*); also the term arbitrarily assigned to the pentane-soluble portion of petroleum that is relatively high boiling (>300°C, 760 mm) (*see also* Petrolenes).

Marine engine oil oil used as a crankcase oil in marine engines.

Marine gasoline fuel for motors in marine service.

Marine sediment the organic biomass from which petroleum is derived.

Marsh an area of spongy waterlogged ground with large numbers of surface water pools. Marshes usually result from: (1) an impermeable underlying bedrock; (2) surface deposits of glacial boulder clay; (3) a basin-like topography from which natural drainage is poor; (4) very heavy rainfall in conjunction with a correspondingly low evaporation rate; (5) low-lying land, particularly at estuarine sites at or below sea level.

Marx–Langenheim model mathematical equations for calculating heat transfer in a hot water or steam flood.

Mass spectrometry an analytical technique that fractures organic compounds into characteristic "fragments" based on functional groups that have a specific mass-to-charge ratio.

Mayonnaise low-temperature sludge; a black, brown, or gray deposit with a soft, mayonnaise-like consistency; not recommended as a food additive!

MCL maximum contaminant level as dictated by regulations.

Medicinal oil highly refined, colorless, tasteless, and odorless petroleum oil used as a medicine in the nature of an internal lubricant; *sometimes called* liquid paraffin.

Membrane technology gas separation processes utilizing membranes that permit different components of a gas to diffuse through the membrane at significantly different rates.

MDL *See* Method detection limit.

MEK (methyl ethyl ketone) a colorless liquid ($CH_3COCH_2CH_3$) used as a solvent; as a chemical intermediate; and in the manufacture of lacquers, celluloid, and varnish removers.

MEK deoiling a wax-deoiling process in which the solvent is generally a mixture of methyl ethyl ketone and toluene.

MEK dewaxing a continuous solvent dewaxing process in which the solvent is generally a mixture of methyl ethyl ketone and toluene.

MEOR microbial enhanced oil recovery.

Methanol *see* Methyl alcohol.

Method Detection Limit the smallest quantity or concentration of a substance that the instrument can measure.

Methyl *t*-butyl ether an ether added to gasoline to improve its octane rating and to decrease gaseous emissions; *see* Oxygenate.

Mercapsol process a regenerative process for extracting mercaptans, utilizing aqueous sodium (or potassium) hydroxide containing mixed cresols as solubility promoters.

Mercaptans organic compounds with the general formula R–SH.

Metagenesis the alteration of organic matter during the formation of petroleum that may involve temperatures above 200°C (390°F); *see also* Catagenesis and Diagenesis.

Methyl alcohol (methanol; wood alcohol): a colorless, volatile, inflammable, and poisonous alcohol (CH_3OH) traditionally formed by destructive distillation of wood or, more recently, as a result of synthetic distillation in chemical plants.

Methyl ethyl ketone *see* MEK.

Mica a complex aluminum silicate mineral that is transparent, tough, flexible, and elastic.

Micellar fluid (surfactant slug) an aqueous mixture of surfactants, co-surfactants, salts, and hydrocarbons. The term micellar is derived from the word micelle, which is a submicroscopic aggregate of surfactant molecules and associated fluid.

Micelle the structural entity by which asphaltene constituents are dispersed in petroleum.

Microcarbon residue the carbon residue determined using a themogravimetric method. *See also* Carbon residue.

Microcrystalline wax wax extracted from certain petroleum residua, with a finer and less apparent crystalline structure than paraffin wax.

Microemulsion a stable, finely dispersed mixture of oil, water, and chemicals (surfactants and alcohols).

Microemulsion or micellar or emulsion flooding an augmented waterflooding technique in which a surfactant system is injected in order to enhance oil displacement toward producing wells.

Microorganisms animals or plants of microscopic size, such as bacteria.

Microscopic displacement efficiency the efficiency with which an oil displacement process removes the oil from individual pores in the rock.

Mid-boiling point the temperature at which approximately 50% of a material has distilled under specific conditions.

Middle distillate distillate boiling between the kerosene and lubricating oil fractions.

Middle-phase microemulsion a microemulsion phase containing a high concentration of both oil and water that, when viewed in a test tube, resides in the middle with the oil phase above it and the water phase below it.

Migration (primary) the movement of hydrocarbons (oil and natural gas) from mature, organic-rich source rocks to a point where the oil and gas can collect as droplets or as a continuous phase of liquid hydrocarbon.

Migration (secondary) the movement of the hydrocarbons as a single, continuous fluid phase through water-saturated rocks, fractures, or faults followed by accumulation of the oil and gas in sediments (traps, *q.v.*) from which further migration is prevented.

Mineral hydrocarbons petroleum hydrocarbons, considered mineral because they come from the earth rather than from plants or animals.

Mineral oil the older term for petroleum; the term was introduced in the nineteenth century as a means of differentiating petroleum (rock oil) from whale oil which, at the time, was the predominant illuminant for oil lamps.

Minerals naturally occurring inorganic solids with well-defined crystalline structures.

Mineral seal oil a distillate fraction boiling between kerosene and gas oil.

Mineral wax yellow to dark brown, solid substances that occur naturally and are composed largely of paraffins; usually found associated with considerable mineral matter, as a filling in veins and fissures or as an interstitial material in porous rocks.

Minimum miscibility pressure (MMP) *see* Miscibility.

Miscibility an equilibrium condition, achieved after mixing two or more fluids, which is characterized by the absence of interfaces between the fluids: (1) *first-contact miscibility*: miscibility in the usual sense, whereby two fluids can be mixed in all proportions without any interfaces forming. Example: at room temperature and pressure, ethyl alcohol and water are first-contact miscible. (2) *multiple-contact miscibility (dynamic miscibility)*: miscibility that is developed by repeated enrichment of one fluid phase with components from a second fluid phase with which it comes into contact. (3) *minimum miscibility* pressure: the minimum pressure above which two fluids become miscible at a given temperature, or can become miscible, by dynamic processes.

Miscible flooding *see* EOR process.

Miscible fluid displacement (miscible displacement) is an oil displacement process in which an alcohol, a refined hydrocarbon, a condensed petroleum gas, carbon dioxide, liquefied natural gas, or even exhaust gas is injected into an oil reservoir, at pressure levels such that the injected gas or fluid and reservoir oil are miscible; the process may include the concurrent, alternating, or subsequent injection of water.

Mitigation identification, evaluation, and cessation of potential impacts of a process product or byproduct.

Mixed-phase cracking the thermal decomposition of higher-boiling hydrocarbons to gasoline components.

Mobility a measure of the ease with which a fluid moves through reservoir rock; the ratio of rock permeability to apparent fluid viscosity.

Mobility buffer the bank that protects a chemical slug from water invasion and dilution and assures mobility control.

Mobility control ensuring that the mobility of the displacing fluid or bank is equal to or less than that of the displaced fluid or bank.

Mobility ratio ratio of mobility of an injection fluid to mobility of fluid being displaced.

Modified alkaline flooding the addition of a co-surfactant and polymer to the alkaline flooding process.

Modified naphtha insolubles (MNI) an insoluble fraction obtained by adding naphtha to petroleum; usually the naphtha is modified by adding paraffin constituents; the fraction might be equated to asphaltenes if the naphtha is equivalent to *n*-heptane, but usually it is not.

Molecular sieve a synthetic zeolite mineral with pores of uniform size; it is capable of separating molecules, on the basis of their size, structure, or both, by absorption or sieving.

Motor Octane Method a test for determining the knock rating of fuels for use in spark-ignition engines; *see also* Research Octane Method.

Moving-bed catalytic cracking a cracking process in which the catalyst is continuously cycled between the reactor and the regenerator.

MSDS Material safety data sheet.

MTBE *see* Methyl *t*-butyl ether.

NAAQS National Ambient Air Quality Standards; standards exist for the pollutants known as the criteria air pollutants: nitrogen oxides (NO_x), sulfur oxides (SO_x), lead, ozone, particulate matter, less than 10 microns in diameter, and carbon monoxide (CO).

Naft pre-Christian era (Greek) term for naphtha (*q.v.*).

Napalm a thickened gasoline used as an incendiary medium that adheres to the surface it strikes.

Naphtha a generic term applied to refined, partly refined, or unrefined petroleum products and liquid products of natural gas, the majority of which distills below 240°C (464°F); the volatile fraction of petroleum which is used as a solvent or as a precursor to gasoline.

Naphthenes cycloparaffins.

Native asphalt *see* Bitumen.

Natural asphalt *see* Bitumen.

Natural gas the naturally occurring gaseous constituents that are found in many petroleum reservoirs; there are also those reservoirs in which natural gas may be the sole occupant.

Natural gas liquids (NGL) the hydrocarbon liquids that condense during the processing of hydrocarbon gases that are produced from oil or gas reservoir; *see also* Natural gasoline.

Natural gasoline a mixture of liquid hydrocarbons extracted from natural gas (*q v.*) suitable for blending with refinery gasoline.

Natural gasoline plant a plant for the extraction of fluid hydrocarbon, such as gasoline and liquefied petroleum gas, from natural gas.

NESHAP National Emissions Standards for Hazardous Air Pollutants; emission standards for specific source categories that emit or have the potential to emit one or more hazardous air pollutants; the standards are modeled on the best practices and most effective emission reduction methodologies in use at the affected facilities.

Neutralization a process for reducing the acidity or alkalinity of a waste stream by mixing acids and bases to produce a neutral solution; also known as pH adjustment.

Neutral oil a distillate lubricating oil with viscosity usually not above 200 sec at 100°F.

Neutralization number the weight, in milligrams, of potassium hydroxide needed to neutralize the acid in 1 g of oil; an indication of the acidity of an oil.

Nonasphaltic road oil any of the nonhardening petroleum distillates or residual oils used as dust layers. They have sufficiently low viscosity to be applied without heating and, together with asphaltic road oils (*q.v.*), are sometimes referred to as dust palliatives.

Nonattainment area a geographical area that does not meet NAAQS for criteria air pollutants (*see also* Attainment area).

Nonionic surfactant a surfactant molecule containing no ionic charge.

Non-Newtonian a fluid that exhibits a change of viscosity with flow rate.

NO$_x$ oxides of nitrogen.

Nuclear magnetic resonance spectroscopy an analytical procedure that permits the identification of complex molecules based on the magnetic properties of the atoms they contain.

No. 1 Fuel oil very similar to kerosene (*q.v.*) and is used in burners where vaporization before burning is usually required and a clean flame is specified.

No. 2 Fuel oil *also called* domestic heating oil; has properties similar to diesel fuel and heavy jet fuel; used in burners where complete vaporization is not required before burning.

No. 4 Fuel oil a light industrial heating oil; used where preheating is not required for handling or burning; there are two grades of No. 4 fuel oil, differing in safety (flash point) and flow (viscosity) properties.

No. 5 Fuel oil a heavy industrial fuel oil which requires preheating before burning.

No. 6 Fuel oil a heavy fuel oil and is more commonly known as Bunker C oil when it is used to fuel ocean-going vessels; preheating is always required for burning this oil.

Observation wells wells that are completed and equipped to measure reservoir conditions and sample reservoir fluids, rather than to inject or produce reservoir fluids.

Octane barrel yield a measure used to evaluate fluid catalytic cracking processes; defined as (RON + MON)/2 times the gasoline yield, where RON is the research octane number and MON is the motor octane number.

Octane number a number indicating the antiknock characteristics of gasoline.

Oil bank *see* Bank.

Oil breakthrough (time) the time at which the oil–water bank arrives at the producing well.

Original oil in place (Oil orginally in place) (OOIP) the quantity of petroleum existing in a reservoir before oil recovery operations begin.

Oils that portion of the maltenes (*q.v.*) that is not adsorbed by a surface-active material such as clay or alumina.

Oil sand *see* Tar sand.

Oil shale a fine-grained impervious sedimentary rock which contains an organic material called kerogen.

Olefin synonymous with alkene.

OOIP *see* Oil originally in place.

Optimum salinity the salinity at which a middle-phase microemulsion containing equal concentrations of oil and water results from the mixture of a micellar fluid (surfactant slug) with oil.

Organic sedimentary rocks rocks containing organic material such as residues of plant and animal remains or decay.

Overhead that portion of the feedstock which is vaporized and removed during distillation.

Override the gravity-induced flow of a lighter fluid in a reservoir above another heavier fluid.

Oxidation a process which can be used for the treatment of a variety of inorganic and organic substances.

Oxidized asphalt *see* Air-blown asphalt.

Ozokerite (Ozocerite) a naturally occurring wax; when refined *also known as* ceresin.

Oxygenate an oxygen-containing compound that is blended into gasoline to improve its octane number and to decrease gaseous emissions.

Oxygenated gasoline gasoline with added ethers or alcohols, formulated according to the Federal Clean Air Act to reduce carbon monoxide emissions during winter months.

Oxygen scavenger a chemical which reacts with oxygen in injection water, used to prevent degradation of polymer.

Pale oil a lubricating oil or a process oil refined until its color, by transmitted light, is straw to pale yellow.

Paraffinum liquidum *see* Liquid petrolatum.

Paraffin wax the colorless, translucent, highly crystalline material obtained from the light lubricating fractions of paraffin crude oils (wax distillates).

Particle density the density of solid particles.

Particulate matter (particulates) particles in the atmosphere or on a gas stream that may be organic or inorganic and originate from a wide variety of sources and processes.

Particle size distribution the particle size distribution (of a catalyst sample) expressed as a percent of the whole.

Partitioning in chromatography, the physical act of a solute having different affinities for the stationary and mobile phases.

Partition ratios, K the ratio of total analytical concentration of a solute in the stationary phase, CS, to its concentration in the mobile phase, CM.

Pattern the areal pattern of injection and producing wells selected for a secondary or enhanced recovery project.

Pattern life the length of time a flood pattern participates in oil recovery.

Penex process a continuous, nonregenerative process for isomerization of C_5 and C_6 fractions in the presence of hydrogen (from reforming) and a platinum catalyst.

Pentafining a pentane isomerization process using a regenerable platinum catalyst on a silica–alumina support and requiring outside hydrogen.

Pepper sludge the fine particles of sludge produced in acid treating which may remain in suspension.

Peri-condensed aromatic compounds Compounds based on angular condensed aromatic hydrocarbon systems, e.g., phenanthrene, chrysene, picene, etc..

Permeability the ease of flow of the water through the rock.

Petrol a term commonly used in some countries for gasoline.

Petrolatum a semisolid product, ranging from white to yellow in color, produced during refining of residual stocks; *see* Petroleum jelly.

Petrolenes the term applied to that part of the pentane-soluble or heptane-soluble material that is low boiling ($<300°C$, $<570°F$, 760 mm) and can be distilled without thermal decomposition (*see also* Maltenes).

Petroleum (crude oil): a naturally occurring mixture of gaseous, liquid, and solid hydrocarbon compounds usually found trapped deep underground beneath impermeable cap rock and above a lower dome of sedimentary rock such as shale; most petroleum reservoirs occur in sedimentary rocks of marine, deltaic, or estuarine origin.

Petroleum asphalt *see* Asphalt.

Petroleum ether *see* Ligroine.

Petroleum jelly a translucent, yellowish to amber or white, hydrocarbon substance (melting point: 38°C to 54°C) with almost no odor or taste, derived from petroleum and used principally in medicine and pharmacy as a protective dressing and as a substitute for fats in ointments and cosmetics; also used in many types of polishes and in lubricating greases, rust preventives, and modeling clay; obtained by dewaxing heavy lubricating-oil stocks.

Petroleum refinery *see* Refinery.

Petroleum refining a complex sequence of events that result in the production of a variety of products.

Petroleum sulfonate a surfactant used in chemical flooding prepared by sulfonating selected crude oil fractions.

Petroporphyrins *see* Porphyrins.

Phase a separate fluid that coexists with other fluids; gas, oil, water and other stable fluids such as microemulsions are all called phases in EOR research.

Phase behavior the tendency of a fluid system to form phases as a result of changing temperature, pressure, or the bulk composition of the fluids or of individual fluid phases.

Phase diagram a graph of phase behavior. In chemical flooding, a graph showing the relative volume of oil, brine, and sometimes one or more microemulsion phases. In carbon dioxide flooding, conditions for formation of various liquid, vapor, and solid phases.

Phase properties types of fluids, compositions, densities, viscosities, and relative amounts of oil, microemulsion, or solvent, and water formed when a micellar fluid (surfactant slug) or miscible solvent (e.g., CO_2) is mixed with oil.

Phase separation the formation of a separate phase that is usually the prelude to coke formation during a thermal process; the formation of a separate phase as a result of the instability or incompatibility of petroleum and petroleum products.

pH adjustment neutralization.

Phosphoric acid polymerization a process using a phosphoric acid catalyst to convert propene, butene, or both, to gasoline or petrochemical polymers.

Photoionization a gas chromatographic detection system that utilizes a detector (PID) ultraviolet lamp as an ionization source for analyte detection. It is usually used as a selective detector by changing the photon energy of the ionization source.

PINA analysis a method of analysis for paraffins, *iso*-paraffins, naphthenes, and aromatics.

PIONA analysis a method of analysis for paraffins, *iso*-paraffins, olefins, naphthenes, and aromatics.

Pipe still a still in which heat is applied to the oil while being pumped through a coil or pipe arranged in a suitable firebox.

Pipestill gas the most volatile fraction that contains most of the gases that are generally dissolved in the crude. Also known as pipestill light ends.

Pipestill light ends *see* Pipestill gas.

Pitch the nonvolatile, brown to black, semisolid to solid viscous product from the destructive distillation of many bituminous or other organic materials, especially coal.

Platforming a reforming process using a platinum-containing catalyst on an alumina base.

PNA a polynuclear aromatic compound (*q.v.*).

NA a polynuclear aromatic compound (*q.v.*).

PNA analysis a method of analysis for paraffins, naphthenes, and aromatics.

Polar aromatics resins; the constituents of petroleum that are predominantly aromatic in character and contain polar (nitrogen, oxygen, and sulfur) functions in their molecular structure(s).

Pollutant a chemical or chemicals introduced into the land, water, and air systems, that is or are not indigenous to these systems; also an indigenous chemical or chemicals introduced into the land, water, and air systems in amounts greater than the natural abundance.

Pollution the introduction into the land, water, and air systems of a chemical or chemicals that are not indigenous to these systems or the introduction into the land, water, and air systems of indigenous chemicals in greater-than-natural amounts.

Polyacrylamide very high molecular weight material used in polymer flooding.

Polycyclic aromatic hydrocarbons (PAHs) polycyclic aromatic hydrocarbons are a suite of compounds comprised of two or more condensed aromatic rings. They are found in many petroleum mixtures, and they are predominantly introduced to the environment through natural and anthropogenic combustion processes.

Polyforming a process charging both C_3 and C_4 gases with naphtha or gas oil under thermal conditions to produce gasoline.

Polymer in EOR, any very high molecular weight material that is added to water to increase viscosity for polymer flooding.

Polymer augmented waterflooding waterflooding in which organic polymers are injected with the water to improve areal and vertical sweep efficiency.

Polymer gasoline the product of polymerization of gaseous hydrocarbons to hydrocarbons boiling in the gasoline range.

Polymerization the combination of two olefin molecules to form a higher molecular weight paraffin.

Polymer stability the ability of a polymer to resist degradation and maintain its original properties.

Polynuclear aromatic compound an aromatic compound with two or more fused benzene rings, e.g., naphthalene, phenanthrene.

Polysulfide treatment a chemical treatment used to remove elemental sulfur from refinery liquids by contacting them with a nonregenerable solution of sodium polysulfide.

PONA analysis a method of analysis for paraffins (P), olefins (O), naphthenes (N), and aromatics (A).

Pore diameter the average pore size of a solid material, e.g., catalyst.

Pore space a small hole in reservoir rock that contains fluid or fluids; a four inch cube of reservoir rock may contain millions of interconnected pore spaces.

Pore volume total volume of all pores and fractures in a reservoir or part of a reservoir; also applied to catalyst samples.

Porosity the percentage of rock volume available to contain water or other fluid.

Porphyrins organometallic constituents of petroleum that contain vanadium or nickel; the degradation products of chlorophyll that became included in the protopetroleum.

Positive bias a result that is incorrect and too high.

Possible reserves reserves where there is an even greater degree of uncertainty but about which there is some information.

Potential reserves reserves based upon geological information about the types of sediments where such resources are likely to occur and they are considered to represent an educated guess.

Pour point the lowest temperature at which oil will pour or flow when it is chilled without disturbance under definite conditions.

Powerforming a fixed-bed naphtha-reforming process using a regenerable platinum catalyst.

Power-law exponent an exponent used to model the degree of viscosity change of some non-Newtonian liquids.

Precipitation number the number of milliliters of precipitate formed when 10 mL of lubricating oil is mixed with 90 mL of petroleum naphtha of a definite quality and centrifuged under definitely prescribed conditions.

Preflush a conditioning slug injected into a reservoir as the first step of an EOR process.

Pressure cores cores cut into a special coring barrel that maintain reservoir pressure when brought to the surface; this prevents the loss of reservoir fluids that usually accompanies a drop in pressure from reservoir to atmospheric conditions.

Pressure gradient rate of change of pressure with distance.

Pressure maintenance augmenting the pressure (and energy) in a reservoir by injecting gas and water through one or more wells.

Pressure pulse test a technique for determining reservoir characteristics by injecting a sharp pulse of pressure in one well and detecting it in surrounding wells.

Pressure transient testing measuring the effect of changes in pressure at one well on other well in a field.

Primary oil recovery oil recovery, utilizing only naturally occurring forces.

Primary structure the chemical sequence of atoms in a molecule.

Primary tracer a chemical that, when injected into a test well, reacts with reservoir fluids to form a detectable chemical compound.

Probable reserves mineral reserves that are nearly certain, but about which a slight doubt exists.

Producibility the rate at which oil or gas can be produced from a reservoir through a wellbore.

Producing well a well in an oil field used for removing fluids from a reservoir.

Propane asphalt *see* Solvent asphalt.

Propane deasphalting solvent deasphalting using propane as the solvent.

Propane decarbonizing a solvent extraction process used to recover catalytic cracking feed from heavy fuel residues

Propane dewaxing a process for dewaxing lubricating oils in which propane serves as solvent.

Propane fractionation a continuous extraction process employing liquid propane as the solvent; a variant of propane deasphalting (*q.v.*).

Protopetroleum a generic term used to indicate the initial product formed and that changes have occurred to the precursors of petroleum.

Proved reserves mineral reserves that have been positively identified as recoverable with current technology.

PSD prevention of significant deterioration.

PTE potential to emit; the maximum capacity of a source to emit a pollutant, given its physical or operation design, and considering certain controls and limitations.

Pulse-echo ultrasonic borehole televiewer well-logging system wherein a pulsed, narrow acoustic beam scans the well as the tool is pulled up the borehole; the amplitude of the reflecting beam is displayed on a cathode-ray tube resulting in a pictorial representation of a wellbore.

Purge and trap a chromatographic sample introduction technique in volatile components that are purged from a liquid medium by bubbling gas through it. The components are then concentrated by "trapping" them on a short intermediate column, which is subsequently heated to drive the components on to the analytical column for separation.

Purge gas typically helium or nitrogen, used to remove analytes from the sample matrix in purge or trap extractions.

Pyrobitumen *see* Asphaltoid.

Pyrolysis exposure of a feedstock to high temperatures in an oxygen-poor environment.

Pyrophoric substances that catch fire spontaneously in air without an ignition source.

Quadrillion 1×10^{15}

Quench the sudden cooling of hot material discharging from a thermal reactor.

RACT Reasonably Available Control Technology standards; implemented in areas of nonattainment to reduce emissions of volatile organic compounds and nitrogen oxides.

Raffinate that portion of the oil which remains undissolved in a solvent refining process.

Ramsbottom carbon residue *see* Carbon residue.

Raw materials minerals extracted from the earth prior to any refining or treating.

Recycle ratio the volume of recycle stock per volume of fresh feed; often expressed as the volume of recycle divided by the total charge.

Recycle stock the portion of a feedstock which has passed through a refining process and is recirculated through the process.

Recycling the use or reuse of chemical waste as an effective substitute for a commercial product or as an ingredient or feedstock in an industrial process.

Reduced crude a residual product remaining after the removal, by distillation or other means, of an appreciable quantity of the more volatile components of crude oil.

Refinery a series of integrated unit processes by which petroleum can be converted to a slate of useful (saleable) products.

Refinery gas a gas (or a gaseous mixture) produced as a result of refining operations.

Refining the processes by which petroleum is distilled and converted by application of a physical and chemical processes to form a variety of products are generated.

Reformate the liquid product of a reforming process.

Reformed gasoline gasoline made by a reforming process.

Reforming the conversion of hydrocarbons with low octane numbers (*q.v.*) into hydrocarbons having higher octane numbers; e.g., the conversion of a *n*-paraffin into a *iso*-paraffin.

Reformulated gasoline (RFG) gasoline designed to mitigate smog production and to improve air quality by limiting the emission levels of certain chemical compounds such as benzene and other aromatic derivatives; often contains oxygenates (*q.v.*).

Reid vapor pressure a measure of the volatility of liquid fuels, especially gasoline.

Regeneration the or reactivation of a catalyst by burning off the coke deposits.

Regenerator a reactor for catalyst reactivation.

Relative permeability the permeability of rock to gas, oil, or water, when any two or more are present, expressed as a fraction of the surface permeability of the rock.

Renewable energy sources solar, wind, and other nonfossil fuel energy sources.

Rerunning the distillation of an oil which has already been distilled.

Research Octane Method a test for determining the knock rating, in terms of octane numbers, of fuels for use in spark-ignition engines; *see also* Motor Octane Method.

Reserves well-identified resources that can be profitably extracted and utilized with existing technology.

Reservoir a rock formation below the earth's surface containing petroleum or natural gas; a domain where a pollutant may reside for an indeterminate time.

Reservoir simulation analysis and prediction of reservoir performance with a computer model.

Residual asphalt *see* Straight-run asphalt.

Residual fuel oil obtained by blending the residual product or products from various refining processes with suitable diluent or dilutents (usually middle distillates) to obtain the required fuel oil grades.

Residual oil *see* Residuum; petroleum remaining in situ after oil recovery.

Residual resistance factor the reduction in permeability of rock to water caused by the adsorption of polymer.

Residuum (resid; *pl.:* residua) the residue obtained from petroleum after nondestructive distillation has removed all the volatile materials from crude oil, e.g., an atmospheric ($345°C$, $650°F^+$) residuum.

Resins that portion of the maltenes (*q.v.*) that is adsorbed by a surface-active material such as clay or alumina; the fraction of deasphaltened oil that is insoluble in liquid propane but soluble in *n*-heptane.

Resistance factor a measure of resistance to flow of a polymer solution relative to the resistance to flow of water.

Resource the total amount of a commodity (usually a mineral, but can include nonminerals such as water and petroleum) that has been estimated to be ultimately available.

Retention the loss of chemical components due to adsorption onto the rock's surface, precipitation, or to trapping within the reservoir.

Retention time the time it takes for an eluate to move through a chromatographic system and reach the detector. Retention times are reproducible and can therefore be compared to a standard for analyte identification.

Rexforming a process combining platforming (*q.v.*) with aromatics extraction, wherein low octane raffinate is recycled to the Platformer.

Rich oil absorption oil containing dissolved natural gasoline fractions.

Riser the part of the bubble-plate assembly which channels the vapor and causes it to flow downward to escape through the liquid; also the vertical pipe where fluid catalytic cracking reactions occur.

Rock asphalt bitumen which occurs in formations that have a limiting ratio of bitumen-to-rock matrix.

Rock matrix the granular structure of a rock or porous medium.

Run-of-the-river reservoirs reservoirs with a large rate of flow-through compared to their volume.

Salinity the concentration of salt in water.

Sand a coarse granular mineral, mainly comprising quartz grains, derived from the chemical and physical weathering of rocks rich in quartz, notably sandstone and granite.

Sand face the cylindrical wall of the wellbore through which the fluids must flow to or from the reservoir.

Sandstone a sedimentary rock formed by compaction and cementation of sand grains; can be classified according to the mineral composition of the sand and cement.

SARA analysis a method of fractionation by which petroleum is separated into saturates, aromatics, resins, and asphaltene fractions.

SARA separation *see* SARA analysis.

Saturates paraffins and cycloparaffins (naphthenes).

Saturation the ratio of the volume of a single fluid in the pores to pore volume, expressed as a percent and applied to water, oil, or gas separately; the sum of the saturations of each fluid in a pore volume is 100%.

Saybolt Furol viscosity the time, in seconds (Saybolt Furol Seconds, SFS), for 60 mL of fluid to flow through a capillary tube in a Saybolt Furol viscometer at specified temperatures between 70°F and 210° F; the method is appropriate for high-viscosity oils such as transmission, gear, and heavy fuel oils.

Saybolt Universal viscosity the time, in seconds (Saybolt Universal Seconds, SUS), for 60 mL of fluid to flow through a capillary tube in a Saybolt Universal viscometer at a given temperature.

Scale wax the paraffin derived by removing the greater part of the oil from slack wax by sweating or solvent deoiling.

Screen factor a simple measure of the viscoelastic properties of polymer solutions.

Screening guide a list of reservoir rock and fluid properties, critical to an EOR process.

Scrubber a device that uses water and chemicals to clean air pollutants from combustion exhaust.

Scrubbing purifying a gas by washing with water or chemical; less frequently, the removal of entrained materials.

Secondary pollutants a pollutant (chemical species) produced by interaction of a primary pollutant with another chemical or by dissociation of a primary pollutant or by other effects within a particular ecosystem.

Secondary recovery oil recovery resulting from injection of water, or an immiscible gas at moderate pressure, into a petroleum reservoir after primary depletion.

Secondary structure the ordering of the atoms of a molecule in space relative to each other.

Secondary tracer the product of the chemical reaction between reservoir fluids and an injected primary tracer.

Sediment an insoluble solid formed as a result of the storage instability and the thermal instability of petroleum and petroleum products.

Sedimentary formed by or from deposits of sediments, especially from sand grains or silts transported from their source and deposited in water, as sandstone and shale; or from calcareous remains of organisms, as limestone.

Sedimentary strata typically consist of mixtures of clay, silt, sand, organic matter, and various minerals; formed by or from deposits of sediments, especially from sand grains or silts transported from their source and deposited in water, such as sandstone and shale; or from calcareous remains of organisms, such as limestone.

Selective solvent a solvent which, at certain temperatures and ratios, will preferentially dissolve more of one components of one mixture than of another and thereby permit partial separation.

Separation process an upgrading process in which the constituents of petroleum are separated, usually without thermal decomposition, e.g., distillation and deasphalting.

Separator-Nobel dewaxing a solvent (tricholoethylene) dewaxing process.

Separatory funnel glassware shaped like a funnel with a stoppered rounded top and a valve at the tapered bottom, used for liquid or liquid separations.

Shear mechanical deformation or distortion, or partial destruction of a polymer molecule as it flows at a high rate.

Shear rate a measure of the rate of deformation of a liquid under mechanical stress.

Shear-thinning the characteristic of a fluid whose viscosity decreases as the shear rate Increases.

Shell fluid catalytic cracking a two-stage fluid catalytic cracking process in which the catalyst is regenerated.

Shell still a still formerly used, in which the oil was charged into a closed, cylindrical shell and the heat required for distillation was applied to the outside of the bottom from a firebox.

Sidestream a liquid stream taken from any one of the intermediate plates of a bubble tower.

Sidestream stripper a device used to perform further distillation on a liquid stream from any one of the plates of a bubble tower, usually by the use of steam.

Single well tracer a technique for determining residual oil saturation by injecting an ester, allowing it to hydrolyze and following dissolution of some of the reaction products in residual oil, the injected solutions are produced back and analyzed.

Slack wax the soft, oily crude wax obtained from the pressing of paraffin distillate or wax distillate.

Slime a name used for petroleum in ancient texts.

Slim tube testing laboratory procedure for the determination of minimum miscibility pressure using long, small-diameter, sand-packed, oil- saturated, stainless steel tube.

Sludge a semi-solid to solid product which results from the storage instability and the thermal instability of petroleum and petroleum products.

Slug a quantity of fluid injected into a reservoir during enhanced oil recovery.

Slurry hydroconversion process a process in which the feedstock is contacted with hydrogen under pressure in the presence of a catalytic coke-inhibiting additive.

Slurry phase reactors tanks into which wastes, nutrients, and microorganisms are placed.

Smoke point a measure of the burning cleanliness of jet fuel and kerosine.

Sodium hydroxide treatment *see* Caustic wash.

Sodium plumbite a solution prepared from a mixture of sodium hydroxide, lead oxide, and distilled water; used in making the doctor test for light oils such as gasoline and kerosine.

Solubility parameter a measure of the solvent power and polarity of a solvent.

Solutizer-steam regenerative process a chemical treating process for extracting mercaptans from gasoline or naphtha, using solutizers (potassium *iso*-butyrate, potassium alkyl phenolate) in strong potassium hydroxide solution.

Solvent a liquid in which certain kinds of molecules dissolve. While they typically are liquids with low boiling points, they may include high-boiling liquids, supercritical fluids, or gases.

Solvent asphalt the asphalt (*q.v.*) produced by solvent extraction of residua (*q.v.*) or by light hydrocarbon (propane) treatment of a residuum (*q.v.*) or an asphaltic crude oil.

Solvent deasphalting a process for removing asphaltic and resinous materials from reduced crude oils, lubricating-oil stocks, gas oils, or middle distillates through the extraction or precipitant action of low-molecular-weight hydrocarbon solvents; *see also* Propane deasphalting.

Solvent decarbonizing *see* Propane decarbonizing.

Solvent deresining *see* Solvent deasphalting.

Solvent dewaxing a process for removing wax from oils by means of solvents, usually by chilling a mixture of solvent and waxy oil, filtration or by centrifuging the wax which precipitates, and solvent recovery.

Solvent extraction a process for separating liquids by mixing the stream with a solvent that is immiscible with part of the liquids stream but that will extract certain components of the liquids stream.

Solvent gas an injected gaseous fluid that becomes miscible with oil under. reservoir conditions and improves oil displacement.

Sonic log a well log based on the time required for sound to travel through rock, useful in determining porosity.

Solvent naphtha a refined naphtha of restricted boiling range used as a solvent; also called petroleum naphtha; petroleum spirits.

Solvent refining *see* Solvent extraction.

Sonication a physical technique employing ultrasound to intensely vibrate a sample media in extracting solvent and to maximize solvent and analyte interactions.

Sonic log a well log based on the time required for sound to travel through rock, useful in determining porosity.

Sour crude oil crude oil containing an abnormally large amount of sulfur compounds; *see also* Sweet crude oil.

SO$_x$ oxides of sulfur.

Soxhlet extraction an extraction technique for solids in which the sample is repeatedly contacted with solvent over several hours, increasing extraction efficiency.

Spontaneous ignition ignition of a fuel, such as coal, under normal atmospheric conditions; usually induced by climatic conditions.

Specific gravity the mass (or weight) of a unit volume of any substance at a specified temperature compared with the mass of an equal volume of pure water at a standard temperature; *see also* Density.

Spent catalyst catalyst that has lost much of its activity due to the deposition of coke and metals.

Stabilization the removal of volatile constituents from a higher boiling fraction or product (*q.v.* stripping); the production of a product which, to all intents and purposes, does not undergo any further reaction when exposed to the air.

Stabilizer a fractionating tower for removing light hydrocarbons from an oil to reduce vapor pressure particularly applied to gasoline.

Standpipe the pipe by which catalyst is conveyed between the reactor and the regenerator.

Stationary phase in chromatography, the porous solid or liquid phase through which an introduced sample passes. The different affinities the stationary phase has for a sample allow the components in the sample to be separated, or resolved.

Steam cracking a conversion process in which the feedstock is treated with superheated steam.

Steam distillation distillation in which vaporization of the volatile constituents is effected at a lower temperature by introduction of steam (open steam) directly into the charge.

Steam drive injection (steam injection) EOR process in which steam is continuously injected into one set of wells (injection wells) or other injection source to effect oil displacement toward and production from a second set of wells (production wells); steam stimulation of production wells is direct steam stimulation, whereas steam drive by steam injection to increase production from other wells is indirect steam stimulation.

Steam stimulation injection of steam into a well and the subsequent production of oil from the same well.

Stiles method a simple approximate method for calculating oil recovery by waterflood that assumes separate layers (stratified reservoirs) for the permeability distribution.

Storage stability (or storage instability) the ability (inability) of a liquid to remain in storage over extended periods of time without appreciable deterioration as measured by gum formation and the depositions of insoluble material (sediment).

Straight-run asphalt the asphalt (*q.v.*) produced by the distillation of asphaltic crude oil.

Straight-run products obtained from a distillation unit and used without further treatment.

Strata layers including the solid iron-rich inner core, molten outer core, mantle, and crust of the earth.

Straw oil pale paraffin oil of straw color used for many process applications.

Stripper well a well that produces (strips from the reservoir) oil or gas.

Stripping a means of separating volatile components from less volatile ones in a liquid mixture by the partitioning of the more volatile materials to a gas phase of air or steam (*q.v.* stabilization).

Sulfonic acids acids obtained by treatment of petroleum or a petroleum product with strong sulfuric acid.

Sulfuric acid alkylation an alkylation process in which olefins (C_3, C_4, and C_5) combine with *iso*-butane in the presence of a catalyst (sulfuric acid) to form branched chain hydrocarbons used especially in gasoline blending stock.

Supercritical fluid an extraction method where the extraction fluid is present at a pressure and temperature above its critical point.

Surface active material a chemical compound, molecule, or aggregate of molecules with physical properties that cause it to adsorb at the interface between two immiscible liquids, resulting in a reduction of interfacial tension or the formation of a microemulsion.

Surfactant a type of chemical, characterized as one that reduces interfacial resistance to mixing between oil and water or changes the degree to which water wets reservoir rock.

Suspensoid catalytic cracking a nonregenerative cracking process in which cracking stock is mixed with slurry of catalyst (usually clay) and cycle oil and passed through the coils of a heater.

SW-846 an EPA multi-volume publication entitled *Test Methods for Evaluating Solid Waste, Physical/ Chemical Methods*; the official compendium of analytical and sampling methods that have been evaluated and approved for use in complying with the RCRA regulations and that functions primarily as a guidance document setting forth acceptable, although not required, methods for the regulated and regulatory communities to use in responding to RCRA-related sampling and analysis requirements. SW-846 changes over time, as new information and data are developed.

Sweated wax a crude wax freed from oil by having been passed through a sweater.

Sweating the separation of paraffin oil and low-melting wax from paraffin wax.

Sweep efficiency the ratio of the pore volume of reservoir rock contacted by injected fluids to the total pore volume of reservoir rock in the project area. (*See also* areal sweep efficiency and vertical sweep efficiency.)

Sweet crude oil crude oil containing little sulfur; see also Sour crude oil.

Sweetening the process by which petroleum products are improved in odor and color by oxidizing or removing the sulfur-containing and unsaturated compounds.

Swelling increase in the volume of crude oil caused by absorption of EOR fluids, especially carbon dioxide. Also increase in volume of clays when exposed to brine.

Swept zone the volume of rock that is effectively swept by injected fluids.

Synthetic crude oil (syncrude) a hydrocarbon product produced by the conversion of coal, oil shale, or tar sand bitumen that resembles conventional crude oil; can be refined in a petroleum refinery (*q.v.*).

Tar the volatile, brown to black, oily, viscous product from the destructive distillation of many bituminous or other organic materials, especially coal; a name used for petroleum in ancient texts.

Target analyte target analytes are compounds that are required analytes in U.S. EPA analytical methods. BTEX and PAHs are examples of petroleum-related compounds that are target analytes in U.S. EPA Methods.

Tar sand *see* Bituminous sand.

Tertiary structure the three-dimensional structure of a molecule.

Tetraethyl lead (TEL) an organic compound of lead, $Pb(CH_3)_4$, which, when added in small amounts, increases the antiknock quality of gasoline.

Thermal coke the carbonaceous residue formed as a result of a noncatalytic thermal process; the Conradson carbon residue; the Ramsbottom carbon residue.

Thermal cracking a process which decomposes, rearranges, or combines hydrocarbon molecules by the application of heat, without the aid of catalysts.

Thermal polymerization a thermal process to convert light hydrocarbon gases into liquid fuels.

Thermal process any refining process which utilizes heat, without the aid of a catalyst.

Thermal recovery *see* EOR process.

Thermal reforming a process using heat (but no catalyst) to effect molecular rearrangement of low-octane naphtha into gasoline of higher antiknock quality.

Thermal stability (thermal instability) the ability (inability) of a liquid to withstand relatively high temperatures for short periods of time without the formation of carbonaceous deposits (sediment or coke).

Thermofor catalytic cracking a continuous, moving-bed catalytic cracking process.

Thermofor catalytic reforming a reforming process in which the synthetic, bead-type catalyst of coprecipitated chromia (Cr_2O_3) and alumina (Al_2O_3) flows down through the reactor concurrent with the feedstock.

Thermofor continuous percolation a continuous clay treating process to stabilize and decolorize lubricants or waxes.

Thief zone any geologic stratum not intended to receive injected fluids in which significant amounts of injected fluids are lost; fluids may reach the thief zone due to an improper completion or a faulty cement job.

Thin layer chromatography (TLC) a chromatographic technique employing a porous medium of glass coated with a stationary phase. An extract is spotted near the bottom of the medium and placed in a chamber with solvent (mobile phase). The solvent moves up the medium and separates the components of the extract, based on affinities for the medium and solvent.

Time-lapse logging the repeated use of calibrated well logs to quantitatively observe changes in measurable reservoir properties over time.

Topped crude petroleum that has had volatile constituents removed up to a certain temperature, e.g., $250°C^+$ ($480°F^+$) topped crude; not always the same as a residuum (*q.v.*).

Topping the distillation of crude oil to remove light fractions only

Total petroleum hydrocarbons (TPH) the family of several hundred chemical compounds that originally come from petroleum.

Tower equipment for increasing the degree of separation obtained during the distillation of oil in a still.

TPH E gas chromatographic test for TPH extractable organic compounds.

TPH V gas chromatographic test for TPH volatile organic compounds.

TPH-D(DRO) gas chromatographic test for TPH diesel-range organics.

TPH-G(GRO) gas chromatographic test for TPH gasoline-range organics.

Trace element those elements that occur at very low levels in a given system.

Tracer test a technique for determining fluid flow paths in a reservoir by adding small quantities of easily detected material (often radioactive) to the flowing fluid, and monitoring their appearance at production wells. Also used in cyclic injection to appraise oil saturation.

Transmissibility (transmissivity) an index of producibility of a reservoir or zone, the product of permeability and layer thickness.

Traps sediments in which oil and gas accumulate from which further migration (*q.v.*) is prevented.

Treatment any method, technique, or process that changes the physical and chemical character of petroleum.

Triaxial borehole seismic survey a technique for detecting the orientation of hydraulically induced fractures, wherein a tool holding three mutually seismic detectors is clamped in the borehole during fracturing; fracture orientation is deduced through analysis of the detected microseismic perpendicular events that are generated by the fracturing process.

Trickle hydrodesulfurization a fixed-bed process for desulfurizing middle distillates.

Trillion 1×10^{12}

True boiling point (True boiling range) the boiling point (boiling range) of a crude oil fraction or a crude oil product under standard conditions of temperature and pressure.

Tube-and-tank cracking a older liquid-phase thermal cracking process.

Ultimate analysis elemental composition.

Ultimate recovery the cumulative quantity of oil that will be recovered when revenues from further production no longer justify the costs of the additional production.

Ultrafining a fixed-bed catalytic hydrogenation process to desulfurize naphtha and upgrade distillates by essentially removing sulfur, nitrogen, and other materials.

Ultraforming a low-pressure naphtha-reforming process employing onstream regeneration of a platinum-on-alumina catalyst and producing high yields of hydrogen and high-octane-number reformate.

Unassociated molecular weight the molecular weight of asphaltenes in an nonassociating (polar) solvent, such as dichlorobenzene, pyridine, or nitrobenzene.

Unconformity a surface of erosion that separates younger strata from older rocks.

Unifining a fixed-bed catalytic process to desulfurize and hydrogenate refinery distillates.

Unisol process a chemical process for extracting mercaptan sulfur and certain nitrogen compounds from sour gasoline or distillates using regenerable aqueous solutions of sodium or potassium hydroxide containing methanol.

Universal viscosity *see* Saybolt Universal viscosity.

Unresolved complex the thousands of compounds that a gas chromatograph *mixture* (UCM) is unable to fully separate.

Unstable usually refers to a petroleum product that has more volatile constituents present or refers to the presence of olefin and other unsaturated constituents.

UOP alkylation a process using hydrofluoric acid (which can be regenerated) as a catalyst to unite olefins with *iso*-butane.

UOP copper sweetening a fixed-bed process for sweetening gasoline by converting mercaptans to disulfides by contact with ammonium chloride and copper sulfate in a bed.

UOP fluid catalytic cracking a fluid process of using a reactor-over-regenerator design.

Upgrading the conversion of petroleum to value-added saleable products.

Upper-phase microemulsion a microemulsion phase containing a high concentration of oil that, when viewed in a test tube, resides on top of a water phase.

Urea dewaxing a continuous dewaxing process for producing low-pour-point oils, and using urea which forms a solid complex (adduct) with the straight-chain wax paraffins in the stock; the complex is readily separated by filtration.

Vacuum distillation distillation (*q.v.*) under reduced pressure.

Vacuum residuum a residuum (*q.v.*) obtained by distillation of a crude oil under vacuum (reduced pressure); that portion of petroleum which boils above a selected temperature such as $510°C$ ($950°F$) or $565°C$ ($1050°F$).

Vapor-phase cracking a high-temperature, low-pressure conversion process.

Vapor-phase hydrodesulfurization a fixed-bed process for desulfurization and hydrogenation of naphtha.

Vertical sweep efficiency the fraction of the layers or vertically distributed zones of a reservoir that are effectively contacted by displacing fluids.

Visbreaking a process for reducing the viscosity of heavy feedstocks by controlled thermal decomposition.

Viscosity a measure of the ability of a liquid to flow or a measure of its resistance to flow; the force required to move a plane surface of area $1 \ m^2$ over another parallel plane surface $1 \ m$ away at a rate of $1 \ m/sec$ when both surfaces are immersed in the fluid.

VGC (viscosity–gravity constant) an index of the chemical composition of crude oil defined by the general relation between specific gravity, sg, at $60°F$ and Saybolt universal viscosity, SUV, at $100°F$:

$$a = 10sg - 1.0752 \log(SUV - 38)/10sg - \log(SUV - 38)$$

The constant, a, is low for the paraffin crude oils and high for the naphthenic crude oils.

VI (viscosity index) an arbitrary scale used to show the magnitude of viscosity changes in lubricating oils with changes in temperature.

Viscosity–gravity constant *see* VGC.

Viscosity index *see* VI.

VOC (VOCs) volatile organic compound or componds; volatile organic compounds are regulated because they are precursors to ozone; carbon-containing gases and vapors from incomplete gasoline combustion and from the evaporation of solvents.

Volatile compounds a relative term that may mean (1) any compound that will purge, (2) any compound that will elute before the solvent peak (usually those < C6), or (3) any compound that will not evaporate during a solvent removal step.

Volumetric sweep the fraction of the total reservoir volume within a flood pattern that is effectively contacted by injected fluids.

VSP vertical seismic profiling, a method of conducting seismic surveys in the borehole for detailed subsurface information.

Waterflood injection of water to displace oil from a reservoir (usually a secondary recovery process).

Waterflood mobility ratio mobility ratio of water displacing oil during waterflooding. (*See also* mobility ratio.)

Waterflood residual the waterflood residual oil saturation; the saturation of oil remaining after waterflooding in those regions of the reservoir that have been thoroughly contacted by water.

Watson characterization factor *see* Characterization factor.

Wax *see* Mineral wax and Paraffin wax.

Wax distillate a neutral distillate containing a high percentage of crystallizable paraffin wax, obtained on the distillation of paraffin or mixed-base crude, and on reducing neutral lubricating stocks.

Wax fractionation a continuous process for producing waxes of low oil content from wax concentrates; *see also* MEK deoiling.

Wax manufacturing a process for producing oil-free waxes.

Weathered crude oil crude oil which, due to natural causes during storage and handling, has lost an appreciable quantity of its more volatile components; also indicates uptake of oxygen.

Wellbore the hole in the earth comprising a well.

Well completion the complete outfitting of an oil well for either oil production or fluid injection; also the technique used to control fluid communication with the reservoir.

Wellhead that portion of an oil well above the surface of the ground.

Wet gas gas containing a relatively high proportion of hydrocarbons which are recoverable as liquids; *see also* Lean gas.

Wet scrubbers devices in which a counter-current spray liquid is used to remove impurities and particulate matter from a gas stream.

Wettability the relative degree to which a fluid will spread on (or coat) a solid surface in the presence of other immiscible fluids.

Wettability number a measure of the degree to which a reservoir rock is water-wet or oil-wet, based on capillary pressure curves.

Wettability reversal the reversal of the preferred fluid wettability of a rock, e.g., from water-wet to oil-wet, or vice versa.

White oil a generic term applied to highly refined, colorless hydrocarbon oils of low volatility, and covering a wide range of viscosity.

Wobbe Index (or Wobbe Number) the calorific value of a gas divided by the specific gravity.

Wood alcohol *see* Methyl alcohol.

Zeolite a crystalline aluminosilicate used as a catalyst, which has a particular chemical and physical structure.

Index

DATE DUE

DEC 1 8 2012